ORGANIC SYNTHESES

ORGANIC SYNTHESES

Collective Volumes I-VIII

CUMULATIVE INDICES

Edited by

Jeremiah P. Freeman

JOHN WILEY & SONS, INC.

NEW YORK / CHICHESTER / BRISBANE / TORONTO / SINGAPORE

ORGANIC SYNTHESES

VOLUME	EDITOR-IN-CHIEF	PAGES
I*	† ROGER ADAMS	84
II*	† JAMES BRYANT CONANT	100
III*	† HANS THACHER CLARKE	105
IV*	† OLIVER KAMM	89
V*	† CARL SHIPP MARVEL	110
VI*	† HENRY GILMAN	120
VII*	† FRANK C. WHITMORE	105
VIII*	† ROGER ADAMS	139
IX*	† JAMES BRYANT CONANT	108
Collective Vol. 1	A revised edition of Annual Volumes I–IX † HENRY GILMAN, *Editor-in-Chief* 2nd Edition revised by † A. H. BLATT	580
X*	† HANS THACHER CLARKE	119
XI*	† CARL SHIPP MARVEL	106
XII*	† FRANK C. WHITMORE	96
XIII*	† WALLACE H. CAROTHERS	119
XIV*	† WILLIAM W. HARTMAN	100
XV*	† CARL R. NOLLER	104
XVI*	† JOHN R. JOHNSON	104
XVII*	† L. F. FIESER	112
XVIII*	† REYNOLD C. FUSON	103
XIX*	† JOHN R. JOHNSON	105
Collective Vol. II	A revised edition of Volumes X–XIX † A. H. BLATT, *Editor-in-Chief*	654
20*	† CHARLES F. H. ALLEN	113
21*	† NATHAN L. DRAKE	120
22*	† LEE IRVIN SMITH	114
23*	† LEE IRVIN SMITH	124
24*	† NATHAN L. DRAKE	119
25*	† WERNER E. BACHMANN	120
26*	† HOMER ADKINS	124
27*	† R. L. SHRINER	121
28*	† H. R. SNYDER	121
29*	† CLIFF S. HAMILTON	119
Collective Vol. III	A revised edition of Annual Volumes 20–29 † E. C. HORNING, *Editor-in-Chief*	890
30*	† ARTHUR C. COPE	115
31*	† R. S. SCHREIBER	122
32*	RICHARD T. ARNOLD	119
33*	CHARLES C. PRICE	115
34*	WILLIAM S. JOHNSON	121

Out of print.

† *Deceased.*

VOLUME	EDITOR-IN-CHIEF	PAGES
35*	†T. L. Cairns	122
36*	N. J. Leonard	120
37*	James Cason	109
38*	†John C. Sheehan	120
39*	†Max Tishler	114
Collective Vol. IV	A revised edition of Annual Volumes 30–39	1036
	Norman Rabjohn, *Editor-in-Chief*	
40*	†Melvin S. Newman	114
41*	John D. Roberts	118
42*	Virgil Boekelheide	118
43*	B. C. McKusick	124
44*	†William E. Parham	131
45*	William G. Dauben	118
46*	E. J. Corey	146
47*	William D. Emmons	140
48*	†Peter Yates	164
49*	Kenneth B. Wiberg	124
Collective Vol. V	A revised edition of Annual Volumes 40–49	1234
	Henry E. Baumgarten, *Editor-in-Chief*	
Cumulative Indices to Collective Volumes, I, II, III, IV, V		
	†Ralph L. and †Rachel H. Shriner, *Editors*	
50*	Ronald Breslow	136
51*	Richard E. Benson	209
52*	Herbert O. House	192
53*	Arnold Brossi	193
54*	Robert E. Ireland	155
55*	Satoru Masamune	150
56*	George H. Büchi	144
57*	Carl R. Johnson	135
58*	†William A. Sheppard	216
59*	Robert M. Coates	267
Collective Vol. VI	A revised edition of Annual Volumes 50–59	1208
	Wayland E. Noland, *Editor-in-Chief*	
60*	Orville L. Chapman	140
61*	†Robert V. Stevens	165
62	Martin F. Semmelhack	269
63	Gabriel Saucy	291
64	Andrew S. Kende	308

Out of print.
† *Deceased.*

VOLUME	EDITOR-IN-CHIEF	PAGES
Collective Vol. VII	A revised edition of Annual Volumes 60–64 JEREMIAH P. FREEMAN, *Editor-in-Chief*	602
65	EDWIN VEDEJS	278
66	CLAYTON H. HEATHCOCK	265
67	BRUCE E. SMART	289
68	JAMES D. WHITE	318
69	LEO A. PAQUETTE	328
Reaction Guide to Collective Volumes I–VII and Annual Volumes 65–68 DENNIS C. LIOTTA AND MARK VOLMER, *Editors*		854
Collective Vol. VIII	A revised edition of Annual Volumes 65–69 JEREMIAH P. FREEMAN, *Editor-in-Chief*	696
70	ALBERT I. MEYERS	305
71	LARRY E. OVERMAN	285
72	DAVID L. COFFEN	333

Collective Volumes, Collective Indices, Annual Volumes 70–72, and Reaction Guide are available from John Wiley & Sons, Inc.

Out of print.
† *Deceased.*

PREFACE

This volume constitutes an update and some revisions to the previous classified indices of the first five *Collective Volumes* of *Organic Syntheses*, edited by Ralph and Rachel Shriner and published in 1976, updated to include all eight presently existing Collective Volumes. The Preface of the earlier Volume is included following the present preface.

In this volume the basic organization has been retained but some significant changes or additions have been made. In place of the Cumulative Contents Index, which is based on the title names of the preparations, that index in this volume contains all compounds for which complete directions are provided. Thus all the compounds prepared in a multistep procedure will be found in this index along with the final product. All the compounds in this index are also found in the Cumulative Contents Index, which is arranged alphabetically by the current Chemical Abstracts (CA) Index name. (Since names were changed between the eighth and ninth CA indices, in all cases we have tried to include the latest version.) In addition, the name is followed by the term "*Hazard*" if the compound appears in the latest listing of Occupational Safety and Health Administration (OSHA) Toxic Substances (1988).

In some procedures a starting material's preparation has been described in abbreviated form in a Note. Such preparations will be found in the Solvents and Reagents Index together with procedures describing catalyst preparations or reagents (e.g., diazomethane), which have been prepared for use in a subsequent procedure. (The placement of a procedure in this index rather than the Preparations Index was somewhat arbitrary and both should be consulted.) This index also contains information about the purification of solvents and reagents, or starting materials used in procedures, in addition to assay methods for many of these reagents.

Another change is the inclusion in this volume of a Hazards and Waste Disposal Index, which directs the reader to *Warnings* concerning hazards involved in certain procedures, which are due to the explosive, toxic, or other properties of some compounds. Some of these warnings were published subsequent to the publication of the procedure; some reflect experience of users, and many reflect information about compounds only available in recent years, such as carcinogenicity. As much as possible such hazards, which are associated with procedures published in the early volumes, have been updated and can be found in this index.

All users of *Organic Syntheses* should give special attention to the numerous precautions, warnings, and hazards cited in the procedures and notes. Special warning notices have been added in each volume (annual and collective) on the basis of information supplied by chemists using the procedures. *Organic Syntheses* is written for competently trained chemists, but the occasional appearance of hazard notices suggests that all chemicals should be treated with due caution and all procedures should be carried out with all modern safety practices. Neither *Organic Syntheses,*

Inc. nor the publishers assume any liability with respect to the use of the pre-parations. The procedures have been carried out by the submitting authors many times and checked by a senior editor and his/her co-workers in order to assure reproducibility.

Until recently very little information was provided in the procedures about the disposal of waste from the preparations or for the trapping of toxic gases except to urge the use of efficient fume hoods. To the extent possible such information is included in this index as well as references to other appropriate current volumes treating this issue.

In recent volumes more examples have been included in tabular form so an index of these tables has been assembled.

Additional information about the individual indices may be found in forwards to each of those indices.

Finally, the Shriners' outline of the origin and development of these volumes has been retained. It has been updated to reflect recent developments and may be found immediately following the original Preface.

Compiling an index is not a one-person job and several people who contributed to this project must be acknowledged. The project was immensely aided by Mrs. Rita Egendoerfer, who scanned the previous cumulative indices into computer form. Dr. Theo Greene, the Assistant Editor of *Organic Syntheses*, Allison Barbeau, Emily Lehrman, and Harvey Leo, undergraduates at Notre Dame, did much of the CAS on-line work to obtain CAS names, registry numbers, and molecular formulas. Finally, Mrs. Myra Martin cleaned up the rough copy and prepared the final copy for publication. Her exceptional thoroughness and attention to detail have made this work as error-free as we can make it.

JEREMIAH P. FREEMAN

Notre Dame, Indiana
May, 1994

PREFACE
Collective Volumes I–V

This volume constitutes a single reference source to the classified indices in all five *Collective Volumes* of *Organic Syntheses*. Thus, users of *Organic Syntheses* may now consult this one set of *Cumulative Indices* instead of those in five separate volumes. In addition to the time-saving factor, it is hoped that the usefulness of the *Collective Volumes* will be enhanced.

Annual Volumes of *Organic Syntheses* have been published since 1921. As these volumes were published and distributed, chemists from all parts of the world reported improvements and helpful modifications and precautions to the secretaries of the Editorial Boards. These comments were carefully collected and the editor of each Collective Volume incorporated them into the *Collective Volumes* at ten-year intervals. Additional references to methods of preparation were also inserted. Thus the *Collective Volumes* are more than mere reprints of the annual volumes. The publication of the *Collective Volumes* extended over a 41-year span:

Collective Volume	Editor-in-Chief	Revision of Annual Volumes	Pages
I (1932)	Henry Gilman	1–9	564
I (1941, Revised)	Henry Gilman and A. Harold Blatt	1–9	580
II (1943)	A. Harold Blatt	10–19	654
III (1955)	Evan C. Horning	20–29	890
IV (1963)	Norman Rabjohn	30–39	1036
V (1973)	Henry E. Baumgarten	40–49	1234

The preface to each collective volume, written by the Editor-in-Chief of that volume, contains valuable information and should be consulted. Naturally, over the years there have been changes in editorial policy, selection of compounds for checking, illustrations of useful reactions, purity of products, and changes in nomenclature. In the early *Annual Volumes*, common names were used for the products of the syntheses. In later volumes, some titles were based on the international systems of nomenclature, such as, the Geneva System (1892), International Union of Chemistry (IUC) rules (1940), and names adopted by the International Union of Pure and Applied Chemistry (IUPAC) (1949). After about 1940 some titles of the preparations were those in use in indices of *Chemical Abstracts* and these names were also added beneath the common title name. The *Collective Volumes* retain the titles from the *Annual Volumes* plus the CA index name, but no uniform system of nomenclature is used in all five volumes.

To help solve nomenclature problems, to facilitate locating compounds in the *Organic Syntheses Collective Volumes*, and to assist in later literature surveys, Index No. 1, Cumulative Contents Index, has an alphabetical index of the preparations according to the main title names followed by the latest Chemical Abstracts Index name. Index No. 2 has the compounds alphabetized by the CA Index Name followed by the common name. The CA Registry Numbers are given in both indices. All citations are to *Collective Volume* numbers and pages in that *Collective Volume*.

The sequence of the classified indices in this volume corresponds to frequency of usage. Thus, specific compounds (Indices Nos. 1 and 2) are the most sought after, followed by Type of Reaction (No. 3) and Type of Compound (No. 4). Users of the Cumulative Index should note that each of the indices has an introductory paragraph describing the material in that particular index. The coordination and consolidation of the indices has posed some problems. Most of the preparations in the first 20 Annual Volumes were single-step procedures to produce a specific compound. Later, multistep syntheses were selected that involved one, two, or more intermediates. Beginning in 1961 with *Annual Volume* 41, sections on the merits of the preparation were introduced. Also, emphasis was placed on unique model procedures illustrating important types of reactions rather than a specific compound. The accompanying discussion sections are difficult to index, but by consulting the Name Index, the Reaction Type Index or the Type of Compound Index, information may be found concerning the scope of the reactions. Duplicate entries have been consolidated and minor reagents used in the work-up of the products are not indexed. The purification of solvents and special reagents, not in the original *Collective Volumes I* and *II,* have been indexed, combined with *III,IV*, and *V*, and inserted in Index No. 6. Author Indices for *Collective Volumes I, II*, and *III* have been compiled and integrated with those of *Collective Volumes IV* and *V* to give a complete Index (No. 8).

Especial attention of all users of *Organic Syntheses* is called to the numerous precautions, warnings, and hazards cited in the procedures and notes. Special warning notices have been added in each volume (annual and collective) on the basis of information supplied by chemists using the procedures. *Organic Syntheses* is written for competently trained chemists, but the occasional appearance of hazard notices suggests that all procedures and chemicals should be treated with due caution and carried out with all modern safety practices. Neither the Editorial Board nor the publishers assume any liability with respect to the use of the preparations. The procedures have been carried out by the submitting authors many times and checked by a senior editor and his co-workers in order to assure reproducibility.

For information concerning hazards of individual compounds chemists should consult, *The Toxic Substances List 1974 Edition*, published by the National Institute for Occupational Safety and Health (USPHS), Rockville, Md. 20852. The first part of this volume has explanations for the selection of toxicity basis and reprints of *Rules and Regulations* from the Federal Register (1972–1974). Then follow 817 pages with 13,000 names of inorganic and organic chemicals and 29,000 synonyms and codes. Arrangement is alphabetical with TSL Compound Number. Toxicity data for various species of animals, routes of administration and information concerning possible hazardous compounds are given. The *"Toxic Substance Prime Name"* is the Chemical Abstracts Index Name with the Chemical Abstracts Service (CAS) Registry Numbers (See p. 2) and the Wiswesser Line Notation (WLN); in the book, a short

but comprehensive description of WLN is given and an alphanumeric list of WLN names with citation of 6000 compound numbers in TSL. This volume is updated annually.

Since the editing of a *Collective Volume* takes 3 or 4 years of the "spare time" of the editor-in-chief, there is a time lag in publication. To aid users, Appendix A lists the alphabetized contents indices of the most recent *Annual Volumes*, 50 through 54 (1970–1974). Appendix B outlines the development and operations of *Organic Syntheses, Inc.* Special attention is called to the article in Annual Volume 50, "Fifty Years of Organic Syntheses" by Roger Adams (1889–1971), one of the founders of *Organic Syntheses* and its leader during the years 1920–1971.

The editors of the current *Annual Volumes* of *Organic Syntheses* invite the submission of interesting syntheses of compounds of research utility and also procedures illustrating new unique reactions. Suggestions for time-saving modifications and real improvements in published procedures are desired for inclusion in the next collective volume. All correspondence should be sent to the current secretary of the Board of Editors, Wayland E. Noland, School of Chemistry, University of Minnesota, Minneapolis, Minn. 55455. Each recent *Annual Volume* has information concerning the "Submission of Preparations", and the secretary will provide a style guide for preparing the written procedures. Authors should consult these *Cumulative Indices* to avoid duplicating preparations of compounds or types of reactions already published.

This volume is to recognize and express appreciation to the organic chemists who have edited the *Collective Volumes of Organic Syntheses* and served as secretaries to the Board of Editors. Their altruistic labors have benefitted thousands of chemists.

Henry Gilman, of Iowa State University, edited *Annual Volume* 6 in 1926 and *Collective Volume I* in 1932.

Charles F. H. Allen, of Eastman Kodak Co., served as the first Secretary to the Board of Editors, 1927–1937, and edited *Annual Volume* 20 in 1940.

A. Harold Blatt, of Queens College was the second Secretary to the Board of Editors, 1938–1943. He collaborated with Henry Gilman to edit a revised edition of *Collective Volume I* in 1941 and then edited *Collective Volume II* in 1943.

Evan C. Horning, now at Baylor College of Medicine, was Secretary to the Editorial Board, 1944–1949, and edited *Collective Volume III* in 1953.

Norman Rabjohn, at the University of Missouri, was Secretary to the Editorial Board, 1950–1958, and edited *Collective Volume IV* in 1963.

Henry Baumgarten, at the University of Nebraska, was Secretary to the Editorial Board, 1959–1968, and edited *Collective Volume V* in 1973.

All these received great help from the editors of *Annual Volumes* 1–49; their names are listed on the inside of the front cover and they are members of the Advisory Board shown on the title page.

RALPH L. SHRINER
RACHEL H. SHRINER

Dallas, Texas
June, 1975

HISTORY OF ORGANIC SYNTHESES

Origin, Development, Organization, Operations

Prior to 1914, the industrial production of organic chemicals in the United States was very limited both in the number of compounds and quantities. Petroleum refining was primarily by distillation; there were no cracking processes and no petrochemical plants. Replacement of beehive coke ovens by byproduct coking ovens to recover aromatic chemicals had just started. Most organic compounds were imported from Europe; research chemicals for use in universities and industrial laboratories were imported from Germany (Kahlbaum's Chemicals), Great Britain (Boots Ltd.), and France. There were only a few small scientific supply houses that distributed small amounts of imported chemicals. Indeed, organic research in universities and industry was limited to a few schools and very few companies (1). In 1914, the outbreak of the war in Europe led to embargoes, blockades, and destruction of shipping, which meant that chemical supplies in the United States were quickly exhausted. The escalation of World War I (2), with United States involvement in 1917, demanded immediate production of tremendous amounts of food, grains, meat, oils, coke, iron, steel, nonferrous metals, ships, trucks, guns, tanks, airplanes, gasoline, kerosene, lubricating oils, war gases, phenol, toluene, glycerol and nitric acid, protective agents, dyes, and drugs. Since all the industrial plants and laboratories were in use, the chemistry staff at the universities began to increase their "student preps" to make chemicals needed for research. Clarence G. Derick of the Chemistry Department at the University of Illinois in Urbana, actually initiated "Summer Preps" with about five students in 1914 before the war started. In the summer of 1915, Ernest H. Volwiler, a graduate student, joined Derick's prep group and was placed in charge during 1916 and 1917. Oliver Kamm, a member of the teaching staff after 1915, also helped in the prep work.

Carl S. Marvel, who started graduate study in 1915, began making compounds in June of 1916 and worked full time until August 1919. He was a most skillful operator and "speedily" (3) built up a reputation for modifying poor procedures so that they would work. Roger Adams joined the chemistry staff in 1916, and enthusiastically took up the idea of synthesizing research chemicals in larger quantities: one-half to several kilos. The compounds made during 1917–1918 were those needed in the World War I effort. Dr. William A. Noyes, Head of the Chemistry Department of the University of Illinois, persuaded the Illinois administration to provide a revolving "Organic Chemical Manufactures" fund, which was used to purchase chemicals and to pay the summer preps chemists. These graduate students, numbering from 10

to 12, worked full time, 8–10 h/day for the 8-week summer session. Their pay started at 25¢/h in 1915 and gradually rose over the years, but the students received one unit of graduate credit for their work. Adams and Marvel put the operation on a sound cost basis by requiring all students making preps to keep careful notebook records of the cost of chemicals, apparatus, and the time needed for each preparation. The compounds made were then sold to anyone who needed them and the money returned to the fund. In 1917 Roger Adams (4) published a list of 43 organic chemicals available for purchase, and in 1918 a note listing 59 compounds as available at once, 37 to be made, and 29 more that would probably be available by the end of the summer. When the importation of dyes for sensitizing photographic film stopped in 1914, Hans T. Clarke, who had just joined the research division of the Eastman Kodak Co., was called on to synthesize the dyes. The lack of organic raw materials for this project and others led Clarke and C. E. K. Mees to recommend to George Eastman the formation of an Eastman Organic Chemicals Division. It would assist research chemists by repackaging commercial chemicals in small lots, purifying industrial chemicals, and synthesizing any needed but nonavailable chemicals. Clarke visited Adams and Marvel at the University of Illinois and spent several weeks observing how "Summer Preps" was operated. The Eastman Organic Chemicals Division began operations at the end of the 1918 and contributed greatly to the advancement of organic chemical research. Its synthesis group worked out many good procedures and designed unique laboratory apparatus and techniques. After Clarke left in 1928, William W. Hartman took charge.

 The production and distribution of Pyrex laboratory glassware by the Corning Glass Works, Corning, New York in 1915 was a very important factor in the preps work. Pyrex™ labware was far superior to the old lime-soda glass against breakage by mechanical or thermal shock, and resistance to reagents. It surpassed even the Jena glass that had been imported from Germany prior to 1914. Pyrex™ round-bottomed reaction flasks became available in large 5, 12, and 22 L and smaller sizes. Glass blowing with Pyrex was easily mastered; hence, special distilling flasks, fractionating columns, the now familiar three-necked flasks for use with a mechanical stirrer, and the reflux condenser and dropping funnel were made and used as standard items in the prep labs. Also in 1914, when shipments of laboratory porcelain ware from Germany ceased, the Coors Porcelain Company of Golden, Colorado, converted their ovenware and pottery plant to chemical porcelainware. High-quality Coors U.S.A.™ glazed laboratory porcelain evaporating dishes, Buchner funnels, casseroles, mortars and pestles became available. The armistice of November 11, 1918 ended the war but did not end the shortage of research chemicals. Hence, the synthesis of special research compounds, not available commercially, was continued during the summers under the direction of Carl S. Marvel who became a member of the Organic faculty at Illinois after completing his graduate study. The expanding organic and biochemical research divisions of universities and commercial concerns requested the compounds to be made in the "preps" lab. About 1940, Harold R. Snyder took over operations from Carl S. Marvel and carried the synthetic work through the difficult World War II years (1941–1946). The prep group made unclassified starting compounds and intermediates needed by any of the various war-time agencies. Leonard E. Miller of the Organic Chemistry Department at Illinois directed the Summer Prep work during 1948–1950. After 1950 the

program was discontinued because by that time many organic and biochemical supply companies had been established for the synthesis of specialty chemicals. The Summer Prep operation had provided a superior education for over 500 graduate students (and some seniors) for 36 years. Other universities also incorporated advanced organic preparations in their graduate programs. These well-trained chemists contributed to the pool of expert synthetic organic chemists for the organic chemical industries that had established real research laboratories from about 1922 onward.

The foregoing account is incomplete, however. What were the sources of the procedures, operating directions, techniques for carrying out reactions, isolating, and purifying the products? Most of the compounds made were not new; they had been described in the various journals, both American and European; some were described in patents. Beilsein's "Handbuch" gave only a sentence or two summarizing the method. Houben-Weyl's "Methoden" were likewise limited. The previous literature procedures were so incomplete that frequently a synthesis, using what seemed to be a simple reaction, became a research problem of weeks or months. Four laboratory manuals available in 1915–1916 that proved helpful were Ludwig Gatterman's *Die Praxis der organischen Chemie*, 1st ed., 1894, later revised by H. Wieland (21st to 24th eds.). L. Vanino's *Handbuch der preparativen Chemie*, Part II, summarized the literature preparations of several hundred organic compounds. E. Fischer's *Anleitung zur Darstellung Organische Preparative* (1908) was useful as was J. B. Cohen's *Practical Organic Chemistry*, 2nd ed. (1908). These manuals, designed for the first course in organic chemistry, were very useful but limited in scope. It was common experience that many procedures in the chemical literature could not be duplicated; indeed, certain procedures were hazardous. Hence, from the very beginning of Summer Preps in 1914, and continuing through all the years, each student had to write out in detail the procedures they used, add precautionary notes, and references to the literature. The procedures were carefully filed and used in succeeding years; each prep person added their observations plus data on yield and purity. The first batch of directions culminated in the publication of four pamphlets; *Organic Chemical Reagents*, by Roger Adams, O. Kamm, and C. S. Marvel. These were bulletins published by the University of Illinois Press, Urbana, Illinois, from 1919 to 1922, containing directions for preparing a total of 111 compounds. Although not advertised, these bulletins were quickly sold out, as their availability became known at meetings of the Organic Division of the ACS and from citations in articles published in the journals.

The success of these little booklets, and the accumulation of several hundred additional good directions for the syntheses of organic compounds, led Roger Adams (6) to consider the publication of an annual volume of satisfactory methods. He discussed this project with James B. Conant of Harvard, Hans T. Clarke of Eastman Kodak, and Oliver Kamm of Parke Davis. The unique feature was the preparation of sets of directions which, if carefully followed, could be duplicated by an advanced student (senior or graduate). Moreover, before publication, each preparation must be checked in the laboratory of an editor and always in a laboratory other than that of the submitter. In addition, the fact that this original group represented both industrial and university laboratories constituted excellent support for the project. The first annual volume of *Organic Syntheses* was published in 1921. The

procedures were collected, checked, and edited by the first Board of Editors; Roger Adams (University of Illinois), James B. Conant (Harvard), Hans T. Clarke (Eastman Kodak Co.), and Oliver Kamm (Parke Davis). Publication was made possible through the friendship of Mr. Edward P. Hamilton of John Wiley & Sons, Inc. This was a most unusual publication venture for those times; there was no assurance that the publisher could recover the costs of the printing, binding, and distribution of this slender little "pamphlet" of 84 pages.

Each of the first four members of the Editorial Board acted as Editor-in-Chief of one or two volumes. Then the Editorial Board was expanded during the next 10 years to include Carl S. Marvel (University of Illinois), Frank C. Whitmore (Northwestern University, Pennsylvania State University), Henry Gilman (Iowa State University), and Carl R. Noller (Stanford). In 1929, C. F. H. Allen was appointed Secretary to the Board when the number of chemists contributing preparations rose from 8 to 24, thereby causing a great increase in correspondence and record keeping. Each of the new editors took turns in preparing volumes. The policy of changing membership on the editorial board by selecting additional organic chemists to serve on the active board and moving those who had already served a term and edited one or more volumes to an Advisory Board of Editors was adopted. A new secretary to the Board of Editors was appointed every 10 years. Beginning with *Collective Volume 11* the retiring secretary became the Editor-in-Chief of the collective volume for the years in which he served. Thus, this project involved many different university and industrial research chemists so as to make it representative of as many institutions as possible. These policies continue today (6). In addition to the first 8 editors mentioned above, there are 51 other organic chemists who have served on the Boards of Editors. Their names are listed on the title page of this volume. They are an enthusiastic group of chemists working with their students in universities and coworkers in industry, dovetailing their regularly assigned work with writing up procedures, and editing and checking them in their "spare time." None of the contributors of procedures, editors, or checkers received any pay or any of the royalties from the sale of the volumes. The starting chemicals needed for checking procedures were contributed by the chemistry departments of the universities or the research departments of industrial companies, and the products of the syntheses then were added to the research stocks of the contributors or editors. The products were always more valuable than the crude commercial starting materials so this was an economical way of getting valuable intermediates for research.

From 1921 to 1939 the *Organic Syntheses* Editorial Boards operated in a very informal fashion. However, changes in the income tax laws led to the formal incorporation of *Organic Syntheses* as a "Membership Corporation" under the laws of the State of New York on December 11, 1939. The certificate specified: The purposes for which the corporation is to be formed are the following: To collaborate in the writing, editing, and causing to be published from time to time of books and articles dealing with the methods of preparation of organic chemicals and other subject matter connected with organic chemistry; the royalties or other proceeds received from them to be placed in a fund, the principal and income thereof to be used exclusively (apart from bona fide expenses of operation of the corporation) for the establishment of fellowships, scholarships and other benefits for students in organic chemistry in various colleges and universities; to acquire property, both real

and personal, for the conduct of its corporate purposes. The corporation is to be
organized and operated exclusively for strictly scientific, educational and charitable purposes, and not for pecuniary profit, and no part of its net earnings will inure
to the benefit of any member, director, or officer other than as reasonable compen-
sation for services in effecting one or more of such purposes, or to any other
individual except as a proper beneficiary of its strictly charitable purposes, and no
part of its activities will be the carrying on of propaganda or otherwise attempting to
influence legislation. The First Board of Directors consisted of Roger Adams
(University of Illinois), President; William W. Hartman (Eastman Kodak Co.),
Treasurer; A. Harold Blatt (Queens College), Secretary to the Editorial Board; Louis
Fieser (Harvard), and John R. Johnson (Cornell). Royalties from the sale of the
Annual Volumes and *Collective Volumes* were paid to the *Organic Syntheses* treasurer
and used to pay postage and typing expenses in collecting preparations and editing
the volumes. Periodically any balance in the fund was invested in stocks in the grow-
ing chemical industries. A set of By-Laws of the Corporation was adopted and filed
with the State of New York. They were amended from time to time as conditions
changed, but always conformed to the above cited nonprofit purposes. The Board of
Directors for 1992–1994 consists of the following officers and members:

Robert M. Coates (Illinois), William G. Dauben (California, Berkeley),
William D. Emmons (Vice-president, Philadelphia), Jeremiah P. Freeman
(Secretary, Notre Dame), Clayton H. Heathcock (California, Berkeley), Carl
R. Johnson (Treasurer, Wayne State), Andrew S. Kende (President, Roches-
ter), Nelson J. Leonard (Pasadena), Blaine C. McKusick (Wilmington),
Wayland E. Noland (Minnesota), John D. Roberts (Cal Tech).

The corporation membership is composed of all the past and present editors.
The Board of Directors has responsibility for

1. Supervising all operations of the corporation so that they are in conformity
   with the Certificate of Incorporation and the By-Laws.
2. Authorizing those expenditures from the Treasurer's funds that are essential
   to its scientific, educational, and charitable purposes.
3. Conforming to Acts of Congress concerning nonprofit corporations.

For efficient operation the Board has delegated certain duties to its officers, com-
mittees, current active Editorial Board, Editors of Annual Volumes, Editors of
Collective Volumes and Editors of Cumulative Indices. Expenditures from its funds
(royalties plus investment income) have been used for

1. Expenses for the current Annual Volumes of *Organic Syntheses*.
2. Secretarial help for the Editors-in-Chief of the *Collective Volumes* and *Cumulative Index*.
3. A biennial award to an outstanding organic chemist who presents a scientific
   educational lecture at the biennial Symposium of the Division of Organic
   Chemistry of the American Chemical Society (a nonprofit corporation char-

tered by an Act of Congress.) This award, known as the Roger Adams Award, amounts to $25,000, a fourth of which being contributed by Organic Reactions, Inc., a nonprofit corporation, also founded by Roger Adams.

4. Subsidies to enable undergraduate or graduate students majoring in any field of chemistry and to postdoctoral fellows and research associates in chemistry at schools in the United States and Canada to purchase volumes of *Organic Syntheses* at one-half the list price (7).

5. Establishment of a lectureship program at the home institution (or one designated) of members of the active Board of Editors.

6. Special awards in recognition of outstanding contributions for the advancement of the purposes of Organic Syntheses, Inc.

7. Corporation expenses for accounting and legal assistance to the Treasurer of the Corporation.

To complete the story of 72 years work of *Organic Syntheses*, the prefaces to the *Annual Volumes*, the *Collective Volumes*, and this volume and its dedication page should be read.

(Reprinted from *Cumulative Indices, Organic Syntheses, Collective Volumes I–V.*)

R. L. SHRINER, RACHEL H. SHRINER

Dallas, Texas
May 1975

REFERENCES

1. Fisher, Harry L. "Organic Chemistry, 1876–1951," in "Chemistry: Key to Better Living," *Diamond Jubilee Volume*, pp. 52–57 (1951) American Chemical Society, Washington, DC.

2. Browne, Charles Albert, and Weeks, Mary Elvira, "The American Chemical Society and the First World War," in *A History of the American Chemical Society, Seventy-five Eventful Years*, Chap. IX, pp. 108–126 (1951).

3. Carl S. Marvel was known to all organic chemists as "Speed" Marvel. There are many legends as to origin of this nickname; lecturing, eating, driving a Marmon car, hunting, trapshooting, birding, and the present text.

4. Adams, Roger, *J. Ind. Eng. Chem. 9*, 685 (1917).

5. Adams, Roger, *J. Am. Chem. Soc.*, 40, 869 (1918).

6. Adams, Roger, "Fifty years of Organic Syntheses," *Org. Syn., 50*, (1970).

(*Supplement*)

Since 1975 some significant changes have been made in the operation of *Organic Syntheses*. Junior checkers (students associated with members of the Board of Editors) now receives an honorarium for their efforts. This change recognizes the more complex and sophisticated procedures that now appear in these volumes.

Also, because of the greatly increased cost of chemicals involved in the checking process, checking editors are now reimbursed for their costs; it is no longer reasonable to expect their own departments to absorb these expenses.

Beginning with Volume 62 (1984), each Annual Volume has appeared in two formats: the traditional bound volume complete with index, and a paper-back version without indices, which is furnished free-of-charge to the Division of Organic Chemistry of the ACS and several foreign societies for distribution to its members. In addition the publication of Collective Volumes has now been changed to a 5-year rather than a 10-year cycle to reflect the accelerated pace of work in the field.

JEREMIAH P. FREEMAN

Notre Dame, Indiana
May, 1994

CONTENTS

CUMULATIVE CONTENTS INDEX 1
(*According to Names of Compounds Prepared*)

CUMULATIVE CONTENTS INDEX 141
(*According to Chemical Abstracts Index Names*)

SOLVENTS AND REAGENTS INDEX 279
(*Purification and/or Preparation*)

TABLES INDEX 299

TYPE OF REACTION INDEX 307

TYPE OF COMPOUND INDEX 385

FORMULA INDEX 491

APPARATUS INDEX 571

HAZARD AND WASTE DISPOSAL INDEX 577

CUMULATIVE AUTHOR INDEX 587

ORGANIC SYNTHESES

CUMULATIVE CONTENTS INDEX

According to Names of Compounds Prepared

All compounds for which complete directions and characterization are found in the Procedure section in Volumes 1–69 are included in this index. Thus intermediates and analogs as well as title compounds will be found here. Each name is printed as found in the procedures in boldfaced capital letters. These names were those that were in common use at the time of checking and editing, but may not be those most familiar to the current user. No attempt has been made to include synonyms. Rather beneath each such name appears the Chemical Abstracts Service (CAS) Index name and its registry number. Both the names and registry numbers are those currently found in the CAS computer access. (Because of changes in CAS nomenclature between the Eighth and Ninth Collective Indices the names may not correspond to those found in the printed Registry Handbooks.)

If a compound in this index is also listed in the *Toxic Substance List 1987 edition*, it is so indicated by the addition of the word *Hazard* after the name. This listing is only to alert the user to a potential problem in the handling and disposal of such compounds; detailed information about such matters will be found in the appropriate OSHA bulletins.

A

ABIETIC ACID, IV, 1
    1-Phenanthrenecarboxylic acid, 1,2,3,4,4a,4b,5,6,10,10a-decahydro-1,4a-dimethyl-7 methylethyl)-, [1R(1α,4aβ,4bα,10aα)]- [514-10-3]
ACENAPHTHENEQUINONE, III, 1
    1,2-Acenaphthalenedione [82-86-0]
ACENAPHTHENOL-7, III, 3
    1-Acenaphthylenol,1,2-dihydro- [6306-07-6]
ACENAPHTHENOL ACETATE, III, 3
    1-Acenaphthylenol,1,2-dihydro-, acetate [14966-36-0]
β-(3-ACENAPHTHOYL)PROPIONIC ACID, III, 6
    5-Acenaphthylenebutanoic acid, 1,2-dihydro-γ-oxo- [16294-60-3]
ACETAL, I, 1; *Hazard*
    Ethane, 1,1-diethoxy [105-57-7]
ACETAMIDE, I, 3
    Acetamide [60-35-5]

ACETAMIDINE HYDROCHLORIDE, I, 5
Ethanimidamide, monohydrochloride [124-42-5]
ACETAMIDOACETONE, V, 27
Acetamide, *N*-(2-oxopropyl)- [7737-16-8]
3-ACETAMIDO-2-BUTANONE, IV, 5
Acetamide, *N*-(1- methyl-2-oxopropyl)- [6628-81-5]
2-(2-ACETAMIDOETHYL)-4,5-DIMETHOXYACETOPHENONE, VI, 1
Acetamide, *N*-[2-(2-acetyl-4,5-dimethoxyphenyl)ethyl] [57621-03-1]
S-ACETAMIDOMETHYL-L-CYSTEINE HYDROCHLORIDE, VI, 5
L-Cysteine, *S*-[(acetylamino)methyl]-, monohydrochloride [28798-28-9]
**2-ACETAMIDO-3,4,6-TRI-*O*-ACETYL-2-DEOXY-α-D-GLUCOPYRANOSYL
   CHLORIDE, V,** 1
α-D-Glucopyranosyl chloride, 2-(acetylamino)-2-deoxy-3,4,6-triacetate
   [3068-34-6]
***p*-ACETAMINOBENZENESULFINIC ACID, I,** 7
Benzenesulfinic acid, 4-(acetylamino)- [710-24-7]
***p*-ACETAMINOBENZENESULFONYL CHLORIDE, I,** 8
Benzenesulfonyl chloride, 4-(acetylamino)- [121-60-8]
α-ACETAMINOCINNAMIC ACID, II, 1
2-Propenoic acid, 2-(acetylamino)-3-phenyl- [5469-45-4]
ACETIC FORMIC ANHYDRIDE, VI, 8
Acetic acid, anhydride with formic acid [2258-42-6]
ACETOACETANILIDE, III, 10; *Hazard*
Butanamide, 3-oxo-*N*-phenyl- [102-01-2]
ACETOBROMOGLUCOSE, III, 11
α-D-Glucopyranosyl bromide, tetraacetate [572-09-8]
ACETO *p*-CYMENE, II, 3
Ethanone, 1-[2-methyl-5-(1-methylethyl)phenyl]- [1202-08-0]
ACETOL, II, 5
2-Propanone, 1-hydroxy- [116-09-6]
"ACETONE-ANIL"(2,2,4-TRIMETHYL-1,2-DIHYDROQUINOLINE), III, 329
Quinoline, 1,2-dihydro-2,2,4-trimethyl- [147-47-7]
ACETONE AZINE, VI, 10
2-Propanone, (1-methylethylidene)-, hydrazone [627-70-3]
ACETONE CARBOXYMETHOXIME, III, 172
Acetic acid, [[(1-methylethylidene)amino]oxy]-[5382-89-8]
ACETONE CYANOHYDRIN, II, 7, 29; **III,** 324; *Hazard*
Propanenitrile, 2-hydroxy-2-methyl- [75-86-5]
ACETONE CYANOHYDRIN NITRATE, V, 839
Propanenitrile, 2-methyl-2-nitrooxy- [40561-27-1]
ACETONE DIBUTYL ACETAL, V, 5
Butane, 1,1'-[(1-methylethylidene)bis(oxy)]bis- [141-72-0]
ACETONEDICARBOXYLIC ACID, I, 10
Pentanedioic acid, 3-oxo- [542-05-2]
ACETONE HYDRAZONE, VI, 10
2-Propanone, hydrazone [5281-20-9]

ACETONE TRIMETHYLSILYL ENOL ETHER, VIII, 1
  Silane, trimethyl[(1-methylethenyl)oxy]- [1833-53-0]
ACETONYLACETONE, II, 219
  2,5-Hexanedione [110-13-4]
ACETOPHENONE *N,N*-DIMETHYLHYDRAZONE, VI, 12
  Ethanone, 1-phenyl-, dimethylhydrazone [13466-32-5]
ACETOPHENONE HYDRAZONE, VI, 12
  Ethanone, 1-phenyl-, hydrazone [13466-30-3]
ACETOPHENONE TRIMETHYLSILYL ENOL ETHER, VIII, 324
  Silane, trimethyl[(1-phenylethenyl)oxy]- [13735-81-4]
2-ACETOTHIENONE, II, 8; **III,** 14
  Ethanone, 1-(2-thienyl)- [88-15-3]
ACETOXIME, I, 318
  2-Propanone, oxime [127-06-0]
4-ACETOXYAZETIDIN-2-ONE, VIII, 3
  2-Azetidinone, 4-(acetyloxy)- [28562-53-0]
1-ACETOXY-4-BENZYLAMINO-2-BUTENE, VIII, 9
  2-Buten-1-ol, 4-[(phenylmethyl)amino]-, acetate (ester), (*E*)- [130892-14-7]
(*S*)-(–)-2-ACETOXY-1-BROMOPROPANE, VII, 356
  2-Propanol, 1-bromo-, acetate, (*S*)- [39968-99-5]
1-ACETOXY-4-CHLORO-2-BUTENE, VIII, 9
  2-Buten-1-ol, 4-chloro-, acetate, (*E*)- [34414-28-3]
4-ACETOXY-3-CHLORO-1-BUTENE, VIII, 9
  3-Buten-1-ol, 2-chloro-, acetate [96039-67-7]
***cis*-1-ACETOXY-4-CHLORO-2-CYCLOHEXENE, VIII,** 6
  2-Cyclohexen-1-ol, 4-chloro-, acetate, *cis*- [82736-39-8]
3β-ACETOXY-5α-CYANOCHOLESTAN-7-ONE, VI, 14
  Cholestane-5-carbonitrile, 3-(acetyloxy)-7-oxo-, (3β,5α)- [2827-02-3]
4-ACETOXY-2-CYCLOPENTEN-1-ONE, VIII, 15
  2-Cyclopenten-1-one, 4-(acetyloxy)- [768-48-9]
**1-ACETOXY-4-(DICARBOMETHOXYMETHYL)-2-CYCLOHEXENE, *cis*-,
  VIII,** 5
  Propanedioic acid, [4-(acetyloxy)-2-cyclohexen-1-yl]-, dimethyl ester, *cis*-
    [82736-52-5]
**1-ACETOXY-4-(DICARBOMETHOXYMETHYL)-2-CYCLOHEXENE, *trans*-,
  VIII,** 5
  Propanedioic acid, [4-(acetyloxy)-2-cyclohexen-1-yl]-, dimethyl ester, *trans*-
    [82736-53-6]
1-ACETOXY-4-DIETHYLAMINO-2-BUTENE, VIII, 9
  2-Buten-1-ol, 4-(diethylamino)-, acetate (ester) [82736-47-8]
3β-ACETOXYETIENIC ACID, V, 8
  Androst-5-ene-17-carboxylic acid, 3-(acetyloxy)-, (3β,17β)- [7150-18-7]
3-ACETOXY-5-HYDROXYCYCLOPENT-1-ENE, *cis*-, VIII, 134
  4-Cyclopenten-1,3-diol, monoacetate, *cis*- [60410-18-6]
3β-ACETOXY-20β-HYDROXY-5-PREGNENE, V, 692
  Pregn-5-ene-3,20-diol, 3-acetate, (3β,20*S*) [53603-96-6]; (3β,20*R*) [14553-79-8]

(1R,2r,3S)-3-ACETOXY-2-NITRO-1-CYCLOHEXYL ACETATE, VIII, 332
  (1R,2r,3S)-1,3-Cyclohexanediol, 2-nitro-, diacetate (ester), (1α,2β,3α)- [51269-14-8]
2-ACETOXY-trans-1-PHENYLPROPENE, VI, 692
  1-Propen-2-ol, 1-phenyl, acetate, (E)- [19980-44-0]
7-ACETOXY-4,4,6,7-TETRAMETHYLBICYCLO[4.2.0]OCTAN-2-ONE, VI, 1024
  Bicyclo[4.2.0]octan-2-one, 7-(acetyloxy)-4,4,6,7-tetramethyl- [66016-89-5]
ACETYLACETONE, III, 16; *Hazard*
  2,4-Pentanedione [123-54-6]
N-(p-ACETYLAMINOPHENYL)RHODANINE, IV, 6
  Acetamide, N-[4-(4-oxo-2-thioxo-3-thiazolidinyl)phenyl]- [53663-36-8]
9-ACETYLANTHRACENE, IV, 8
  Ethanone, 1-(9-anthracenyl)- [784-04-3]
ACETYLBENZOYL, III, 20
  1,2-Propanedione, 1-phenyl- [579-07-7]
p-ACETYL-α-BROMOHYDROCINNAMIC ACID, VI, 21
  Benzenepropanoic acid, 4-acetyl-α-bromo- [18910-19-5]
3-ACETYL-4- and 5-CHLORO-2-OXAZOLIDINONE, VII, 5
  2-Oxazolidinone, 3-acetyl-5-chloro- [60759-48-0]
  2-Oxazolidinone, 3-acetyl-4-chloro- [74348-55-3]
α-ACETYL-δ-CHLORO-γ-VALEROLACTONE, IV, 10
  2(3H)-Furanone, 3-acetyl-5-(chloromethyl)dihydro- [3154-75-4]
1-ACETYLCYCLOHEXANOL, IV, 13
  Ethanone, 1-(1-hydroxycyclohexyl)- [1123-27-9]
1-ACETYLCYCLOHEXENE, III, 22
  Ethanone, 1-(1-cyclohexen-1-yl)- [932-66-1]
2-ACETYL-2-CYCLOHEXEN-1-ONE, VI, 23
  2-Cyclohexen-1-one, 2-acetyl- [52784-38-0]
2-ACETYL-1,3-CYCLOPENTANEDIONE, VI, 28
  1,3-Cyclopentanedione, 2-acetyl- [3859-39-0]
**2-ACETYL-6,7-DIMETHOXY-1-METHYLENE-1,2,3,4-TETRAHYDROISO-
  QUINOLINE, VI**, 1
  Isoquinoline, 2-acetyl-1,2,3,4-tetrahydro-6,7-dimethoxy-1-methylene-
    [57621-04-2]
3-ACETYL-2,4-DIMETHYLFURAN, VI, 31
  Ethanone, 1-(2,4-dimethyl-3-furanyl)- [32933-07-6]
**4-ACETYL-7,8-DIMETHYL-2-OXA-4-AZABICYCLO[4.3.0]NON-7-EN-3-ONE,
  VII**, 5
  2(3H)-Benzoxazolone, 3-acetyl-3a,4,7,7a-tetrahydro-5,6-dimethyl- [65948-43-8]
ACETYLENEDICARBOXYLIC ACID, II, 10
  2-Butynedioic acid [142-45-0]
2-ACETYLFLUORENE, III, 23
  Ethanone, 1-(9H-fluoren-2-yl)- [781-73-7]
ACETYLGLYCINE, II, 11
  Glycine, N-acetyl- [543-24-8]
3-ACETYL-4-HYDROXY-5,5-DIMETHYLFURAN-2(5H)-ONE, VIII, 71
  2(5H)-Furanone, 3-acetyl-4-hydroxy-5,5-dimethyl- [22609-93-4]

N-ACETYLISATIN, III, 456
  1*H*-Indole-2,3-dione, 1-acetyl- [574-17-4]
ACETYLMANDELIC ACID, I, 12
  Benzeneacetic acid, α-(acetyloxy)- [5438-68-6]
ACETYLMANDELYL CHLORIDE, I, 12
  Benzeneacetyl chloride, α-(acetyloxy)- [1638-63-7]
2-ACETYL-6-METHOXYNAPHTHALENE, VI, 34
  Ethanone, 1-(6-methoxy-2-naphthalenyl)- [3900-45-6]
ACETYL METHYLUREA, II, 462
  Acetamide, *N*-[(methylamino)carbonyl]-[623-59-6]
3-ACETYL-2-OXAZOLIDINONE, VII, 5
  2-Oxazolidinone, 3-acetyl- [1432-43-5]
3-ACETYL-2(3*H*)-OXAZOLONE, VII,.4
  2(3*H*)-Oxazolone, 3-acetyl- [60759-49-1]
3-ACETYLOXINDOLE, V, 12
  2*H*-Indol-2-one, 3-acetyl-1,3-dihydro- [17266-70-5]
2-(1-ACETYL-2-OXOPROPYL)BENZOIC ACID, VI, 36
  Benzoic acid, 2-(1-acetyl-2-oxopropyl)- [52962-26-2]
5-ACETYL-1,2,3,4,5-PENTAMETHYL-2,4-CYCLOPENTADIENE, VI, 39
  Ethanone, 1-(1,2,3,4,5-pentamethyl-2,4-cyclopentadien-1-yl)-[15971-76-3]
9-ACETYLPHENANTHRENE, III, 26
  Ethanone, 1-(9-phenanthrenyl)- [2039-77-2]
N-ACETYLPHENYLALANINE, II, 493
  Phenylalanine, *N*-acetyl- [2018-61-3]
2-*p*-ACETYLPHENYLHYDROQUINONE and DIACETATE, IV, 15
  Ethanone, 1-(2',5'-dihydroxy[1,1'-biphenyl]-4-yl)- [3948-13-8]
N-ACETYL-*N*-PHENYLHYDROXYLAMINE, VIII, 16
  Acetamide, *N*-hydroxy-*N*-phenyl- [1795-83-1]
4-ACETYLPYRIDINE OXIME, VII, 149
  Ethanone, 1-(4-pyridinyl)- oxime [1194-99-6]
4-ACETYLPYRIDINE OXIME TOSYLATE, VII, 149
  Ethanone, 1-(4-pyridinyl)-*O*-[(4-methylphenyl)sulfonyl]-, oxime [74209-52-2]
δ-ACETYLVALERIC ACID, IV, 19
  Heptanoic acid, 6-oxo-[3128-07-2]
ACETYLTRIMETHYLSILANE, VIII, 19
  Silane, acetyltrimethyl-[13411-48-8]
ACID AMMONIUM *o*-SULFOBENZOATE, I, 14
  Benzoic acid, 2-sulfo-, monoammonium salt [6939-89-5]
ACONITIC ACID, II, 12
  1-Propene-1,2,3-tricarboxylic acid [499-12-7]
ACRIDONE, II, 15
  9(10*H*)-Acridinone [578-95-0]
ACROLEIN, I, 15; *Hazard*
  2-Propenal [107-02-8]
ACROLEIN ACETAL, II, 17; IV, 21
  1-Propene, 3,3-diethoxy- [3054-95-3]

ACRYLIC ACID, III, 30; *Hazard*
  2-Propenoic acid [79-10-7]
ADAMANTANE, V, 16
  Tricyclo[3.3.1.1^{3,7}]decane [281-23-2]
2-ADAMANTANECARBONITRILE, VI, 41
  Tricyclo[3.3.1.1^{3,7}]decane-2-carbonitrile [35856-00-9]
1-ADAMANTANECARBOXYLIC ACID, V, 20
  Tricyclo[3.3.1.1^{3,7}]decane-1-carboxylic acid [828-51-3]
1-ADAMANTANOL, VI, 958
  Tricyclo[3.3.1.1^{3,7}]decan-1-ol [768-95-6]
ADAMANTANONE, VI, 48
  Tricyclo[3.3.1.1^{3,7}]decanone [700-58-3]
ADAMANTYLIDENEADAMANTANE, VII, 1
  Tricyclo[3.3.1.1^{3,7}]decane, tricyclo[3.3.1.1^{3,7}]decylidene- [30541-56-1]
ADIPIC ACID, 1, 18; *Hazard*
  Hexanedioic acid [124-04-9]
β-ALANINE, II, 19; **III**, 34
  β-Alanine [107-95-9]
DL-ALANINE, I, 21
  DL-Alanine [302-72-7]
ALLANTOIN, II, 21
  Urea, (2,5-dioxo-4-imidazolidinyl)- [97-59-6]
ALLENE, V, 22
  1,2-Propadiene [463-49-0]
ALLOXAN MONOHYDRATE, III, 37; **IV**, 23
  2,4,6(1*H*,3*H*,5*H*)-Pyrimidinetrione, 5,5-dihydroxy- [3237-50-1]
ALLOXANTIN DIHYDRATE, III, 42; **IV**, 25
  [5,5'-Bipyrimidine]-2,2',4,4',6,6'(1*H*,1'*H*,3*H*,3'*H*,5*H*,5'*H*)-hexone, 5,5'-dihydroxy,
    dihydrate [6011-27-4]
ALLYL ALCOHOL, I, 42; *Hazard*
  2-Propen-1-ol [107-18-6]
ALLYLAMINE, II, 24; *Hazard*
  2-Propen-1-amine [107-11-9]
ALLYL BROMIDE, I, 27; *Hazard*
  1-Propene, 3-bromo- [106-95-6]
α-ALLYL-β-BROMOETHYL ETHYL ETHER, IV, 750
  1-Pentene, 5-bromo-4-ethoxy- [22089-55-0]
ALLYL CYANIDE, I, 46; **III**, 852
  3-Butenenitrile [109-75-1]
2-ALLYLCYCLOHEXANONE, III, 44; **V**, 25
  Cyclohexanone, 2-(2-propenyl)- [94-66-6]
ALLYL LACTATE, III, 46
  Propanoic acid, 2-hydroxy-, 2-propenyl ester [5349-55-3]
2-ALLYL-3-METHYLCYCLOHEXANONE, VI, 51
  Cyclohexanone, 3-methyl-2-(2-propenyl)- [56620-95-2]

ALLYLTRIBUTYLTIN, VIII, 23
    Stannane, tributyl-2-propenyl- [24850-33-7]
AMINOACETAL, III, 50
    Ethanamine, 2,2-diethoxy- [645-36-3]
AMINOACETONE SEMICARBAZONE HYDROCHLORIDE, V, 27
    Hydrazinecarboxamide, 2-(2-amino-1-methylethylidene)-, monohydrochloride
    [10469-70-2]
AMINOACETONITRILE HYDROGEN SULFATE, I, 298
    Acetonitrile, amino-, sulfate [151-63-3]
9-AMINOACRIDINE, III, 53; *Potential poison*
    9-Acridinamine [90-45-9]
2-AMINO-4-ANILINO-6-(CHLOROMETHYL)-*s*-TRIAZINE, IV, 29
    1,3,5-Triazine-2,4-diamine, 6-(chloromethyl-*N*-phenyl) [30355-60-3]
***o*-AMINOBENZALDEHYDE, III,** 56
    Benzaldehyde, 2-amino [529-23-7]
***p*-AMINOBENZALDEHYDE, IV,** 31
    Benzaldehyde, 4-amino- [556-18-3]
***m*-AMINOBENZALDEHYDE DIMETHYLACETAL, III,** 59
    Benzenamine, 3-(dimethoxymethyl)- [53663-37-9]
2-AMINOBENZOPHENONE, IV, 34
    Methanone, (2-aminophenyl)phenyl- [2835-77-0]
***o*-AMINOBENZYL ALCOHOL, III,** 60
    Benzenemethanol, 2-amino- [5344-90-1]
***o*-AMINOBIPHENYL, V,** 830
    [1,1'-Biphenyl]-2-amine [90-41-5]
2-AMINO-5-BROMO-3-NITROPYRIDINE, V, 347
    Pyridine, 2-amino-5-bromo-3-nitro- [6945-68-2]
2-AMINO-5-BROMOPYRIDINE, V, 346
    2-Pyridinamine, 5-bromo- [1072-97-5]
γ-AMINOBUTYRIC ACID, II, 25
    Butanoic acid, 4-amino- [56-12-2]
α-AMINOCAPROIC ACID, I, 48
    L-Norleucine [5157-09-5]
ε-AMINOCAPROIC ACID, II, 28; **IV,** 39
    Hexanoic acid, 6-amino- [60-32-2]
4-AMINOCYCLOHEXANECARBOXYLIC ACID, cis–trans **MIXTURE, V,** 670
    Cyclohexanecarboxylic acid, 4-amino- [1776-53-0]
1-(AMINOMETHYL)CYCLOHEXANOL, IV, 224
    Cyclohexanol, 1-(aminomethyl)- [4000-72-0]
2-AMINO-*p*-CYMENE, III, 63
    Benzenamine, 2-methyl-5-(1-methylethyl)- [2051-53-8]
α-AMINODIETHYLACETIC ACID, III, 66
    Butanoic acid, 2-amino-2-ethyl- [2566-29-2]
5-AMINO-2,3-DIHYDRO-1,4-PHTHALAZINEDIONE, III, 69
    1,4-Phthalazinedione, 5-amino-2,3-dihydro- [521-31-3]

6-AMINO-3,4-DIMETHYL-*cis*-3-CYCLOHEXEN-1-OL, VII, 5
3-Cyclohexen-1-ol, 6-amino-3,4-dimethyl-, *cis*- [65948-45-0]
4-AMINO-2,6-DIMETHYLPYRIMIDINE, III, 71
4-Pyrimidinamine, 2,6-dimethyl- [461-98-3]
***trans*-1-AMINO-2,3-DIPHENYLAZIRIDINE, VI,** 56, 679
1-Aziridinamine, 2,3-diphenyl-, *trans*- [28161-60-6]
2-AMINOFLUORENE, II, 447; **V,** 30
9*H*-Fluoren-2-amine [153-78-6]
AMINOGUANIDINE BICARBONATE, III, 73
Carbonic acid, compound, with hydrazinecarboximidamide(1:1) [2582-30-1]
α-AMINOISOBUTYRIC ACID, III, 29
Alanine, 2-methyl- [62-57-7]
AMINOMALONONITRILE *p*-TOLUENESULFONATE, V, 32
Propanedinitrile, amino-, mono(4-methylbenzenesulfonate) [5098-14-6]
1-AMINO-2-METHOXYMETHYLPYRROLIDINE, (*R*)-(+)- (RAMP), VIII, 26
1-Pyrrolidinamine, 2-(methoxymethyl)-, (*R*)-(+)- [72748-99-3]
1-AMINO-2-METHOXYMETHYLPYRROLIDINE, (*S*)-(−)- (SAMP), VIII, 26, 403
1-Pyrrolidinamine, 2-(methoxymethyl)-, (*S*)-(−)- [59983-30-0]
2-AMINO-6-METHYLBENZOTHIAZOLE, III, 76
2-Benzothiazolamine, 6-methyl- [2536-91-6]
1-AMINO-1-METHYLCYCLOHEXANE, V, 35
Cyclohexanamine, 1-methyl- [6526-78-9]
1-(AMINOMETHYL)CYCLOHEXANOL, IV, 224
Cyclohexanol, 1-(aminomethyl)- [4000-72-0]
2-AMINO-4-METHYLTHIAZOLE, II, 31
2-Thiazolamine, 4-methyl- [1603-91-4]
3-AMINO-2-NAPHTHOIC ACID, III, 7
2-Naphthalenecarboxylic acid, 3-amino- [5959-52-4]
1,2-AMINONAPHTHOL HYDROCHLORIDE, II, 33
2-Naphthalenol, 1-amino-, hydrochloride [1198-27-2]
1,4-AMINONAPHTHOL HYDROCHLORIDE, I, 49; **II,** 39
1-Naphthalenol, 4-amino-, hydrochloride [5959-56-8]
1-AMINO-2-NAPHTHOL-4-SULFONIC ACID, II, 42; **III,** 635
1-Naphthalenesulfonic acid, 4-amino-3-hydroxy- [116-63-2]
***o*-AMINO-*p*'-NITROBIPHENYL, V,** 830; *Potential poison*
[1,1'-Biphenyl]-2-amine, 4'-nitro- [6272-52-2]
2-AMINO-4-NITROPHENOL, III, 82
Phenol, 2-amino-4-nitro- [99-57-0]
2-AMINO-3-NITROTOLUENE, IV, 42
Benzenamine, 2-methyl-6-nitro- [570-24-1]
2-AMINO-5-NITROTOLUENE, IV, 42
Benzenamine, 2-methyl-4-nitro- [99-52-5]
DL-α-AMINOPHENYLACETIC ACID, III, 84
Benzeneacetic acid, α-amino-(±)- [2835-06-5]
***p*-AMINOPHENYLACETIC ACID, I,** 52
Benzeneacetic acid, 4-amino- [1197-55-3]

p-AMINOPHENYL DISULFIDE, III, 86
Benzenamine, 4,4'-dithiobis- [722-27-0]
DL-α-AMINO-α-PHENYLPROPIONIC ACID, III, 88
Benzeneacetic acid, α-amino-α-methyl-(±)- [6945-32-0]
DL-β-AMINO-β-PHENYLPROPIONIC ACID, III, 91
Benzenepropanoic acid, β–amino-, (±)- [3646-50-2]
1-AMINO-2-PHENYLAZIRIDINE, VI, 56
1-Aziridinamine, 2-phenyl-, (±)- [19615-20-4]
1-AMINO-2-PHENYLAZIRIDINIUM ACETATE, VI, 56
1-Aziridinamine, 2-phenyl-, (±)-, monoacetate [37079-43-9]
β-AMINOPROPIONITRILE, III, 93
Propanenitrile, 3-amino- [151-18-8]
3(5)-AMINOPYRAZOLE, V, 39
1*H*-Pyrazol-3-amine [1820-80-0]
3-AMINO-3-PYRAZOLINE SULFATE, V, 39
3-Pyrazolamine, 4,5-dihydro-, sulfate [29574-26-3]
3-AMINOPYRIDINE, IV, 45
3-Pyridinamine [462-08-8]
1-AMINOPYRIDINIUM IODIDE, V, 43
Pyridinium, 1-amino-, iodide [6295-87-0]
4-AMINO-1,2,5,6-TETRAHYDRO-1-PHENYLPHOSPHORIN-3-CARBONI-
TRILE, VI, 932
3-Phosphorincarbonitrile, 4-amino-1,2,5,6-tetrahydro-1-phenyl- [84819-76-1]
p-AMINOTETRAPHENYLMETHANE, IV, 47
Benzenamine, 4-(triphenylmethyl)- [22948-06-7]
7-AMINOTHEOPHYLLINE, VII, 8
1*H*-Purine-2,6-dione, 7-amino-3,7-dihydro-1,3-dimethyl- [81281-58-5]
3-AMINO-1*H*-1,2,4-TRIAZOLE, III, 95; *Hazard*
1*H*-1,2,4-Triazol-3-amine [61-82-5]
4-AMINO-4*H*-1,2,4-TRIAZOLE, III, 96
4*H*-1,2,4-Triazol-4-amine [584-13-4]
3-AMINO-2,4,6-TRIBROMOBENZOIC ACID, IV, 947
Benzoic acid, 3-amino-2,4,6-tribromo- [6628-84-8]
4-AMINOVERATROLE, II, 44
Benzenamine, 3,4-dimethoxy- [6315-89-5]
AMMONIUM SALT of AURIN TRICARBOXYLIC ACID, I, 54
Benzoic acid, 5-[(3-carboxy-4-hydroxyphenyl)(3-carboxy-4-oxo-2,5-cyclohexadien-
1-ylidene)methyl]-2-hydroxy-, triammonium salt [569-58-4]
AMYLBENZENE, II, 47
Benzene, pentyl [538-68-1]
AMYL BORATE, II, 107
Boric acid, tripentyl ester [621-78-3]
ANDROSTAN-17-OL, (5α,17β)-, VI, 62
5α-Androstan-17β-ol [1225-43-0]
ANHYDRO-2-HYDROXYMERCURIBENZOIC ACID, I, 57; *Potential poison*
Mercury, [2-hydroxybenzoato(2–), O^1, O^2] [5970-32-1]

ANHYDRO-2-HYDROXYMERCURI-3-NITROBENZOIC ACID, I, 56; *Potential poison*
  Mercury, [(3-nitrobenzoato-)-C2, 01] [53663-14-2]
o-**ANISALDEHYDE, V,** 46; **VI,** 64
  Benzaldehyde, 2-methoxy- [135-02-4]
ANISOLE, I, 58; *Hazard*
  Benzene, methoxy- [100-66-3]
[18]ANNULENE, VI, 68
  1,3,5,7,9,11,13,15,17-Cyclooctadecanonaene [2040-73-5]
9-ANTHRALDEHYDE, III, 98
  9-Anthracenecarboxaldehyde [642-31-9]
ANTHRAQUINONE, II, 554
  9,10-Anthracenedione [84-65-1]
ANTHRONE, I, 60
  9(10*H*)-Anthracenone [90-44-8]
D-ARABINOSE, III, 101
  D-Arabinose [10323-20-3]
L-ARABINOSE, I, 67
  L-Arabinose [5328-37-0]
L-ARGININE HYDROCHLORIDE, II, 49
  L-Arginine, monohydrochloride [1119-34-2]
ARSANILIC ACID, I, 70; *Potential poison*
  Arsonic acid, (4-aminophenyl)- [98-50-0]
ARSENOACETIC ACID, I, 73; *Potential poison*
  Acetic acid, 2,2'-(1,2-diarsenediyl)bis- [544-27-4]
ARSONOACETIC ACID, I, 73; *Potential poison*
  Acetic acid, arsono- [107-38-0]
p-**ARSONOPHENOXYACETIC ACID, I,** 75; *Potential poison*
  Acetic acid, (4-arsonophenoxy)- [53663-15-3]
DL-ASPARTIC ACID, IV, 55
  DL-Aspartic acid [617-45-8]
ATROLACTIC ACID, IV, 58
  Benzeneacetic acid, α-hydroxy-α-methyl- [515-30-0]
ATROPALDEHYDE, VII, 13
  Benzeneacetaldehyde, α-methylene- [4432-63-7]
ATROPALDEHYDE DIETHYL ACETAL, VII, 13
  Benzene, [1-(diethoxymethyl)ethenyl]- [80234-04-4]
AZELAIC ACID, II, 53
  Nonanedioic acid [123-99-9]
AZELANITRILE, IV, 62
  Nonanedinitrile [1675-69-0]
AZETIDINE, VI, 75
  Azetidine [503-29-7]
4-AZIDO-3-CHLORO-5-ISOPROPOXY-2(5*H*)-FURANONE, (in solution), VIII, 116
  2(5*H*)-Furanone, 4-azido-3-chloro-5-(1-methylethoxy)- [126773-43-1]

(1-AZIDO-3,3-DIMETHOXY-1-PROPENYL)BENZENE, VI, 893
    Benzene, (1-azido-3,3-dimethoxy-1-propenyl)- [56900-67-5]
(1-AZIDO-2-IODO-3,3-DIMETHOXYPROPYL)BENZENE, VI, 893
    Benzene, (1-azido-2-iodo-3,3-dimethoxypropyl)- [56900-66-4]
o-AZIDO-*p*'-NITROBIPHENYL, **V**, 830
    1,1'-Biphenyl, 2-azido-4'-nitro- [14191-25-4]
2-AZIDO-2-PHENYLADAMANTANE, VII, 433
    Tricyclo[3.3.1.1^{3,7}]decane, 2-azido-2-phenyl- [65218-96-4]
AZLACTONE of α-ACETAMINOCINNAMIC ACID, II, 1
    5(4*H*)-Oxazolone, 4-(phenylmethylene)-2-methyl- [881-90-3]
AZLACTONE of α-BENZOYLAMINOCINNAMIC ACID, II, 490
    5(4*H*)-Oxazolone, 4-(phenylmethylene)-2-phenyl- [842-74-0]
**AZLACTONE of α-BENZOYLAMINO-β-(3,4-DIMETHOXYPHENYL)ACRYLIC
    ACID, II**, 55
    5(4*H*)-Oxazolone, 4-[(3,4-dimethoxyphenyl)methylene]-2-phenyl-(*E*)-
        [25349-38-6]; (*Z*)- [25349-37-5]
AZOBENZENE, III, 103
    Diazene, diphenyl- [103-33-3]
1,1'-AZO-bis-1-CYCLOHEXANENITRILE, IV, 66
    Cyclohexanecarbonitrile,1,1'-azobis- [2094-98-6]
AZOETHANE, VI, 78
    Hydrazono, diethyl [38534-43-9]
AZOXYBENZENE, II, 57
    Diazene, diphenyl-, 1-oxide [495-48-7]
AZULENE, VII, 15
    Azulene [275-51-4]

B

BARBITURIC ACID, II, 60
    2,4,6(1*H*,3*H*,5*H*)-Pyrimidinetrione [67-52-7]
BENZALACETONE, I, 77
    3-Buten-2-one, 4-phenyl- [122-57-6]
BENZALACETONE DIBROMIDE, III, 105
    2-Butanone, 3,4-dibromo-4-phenyl- [6310-44-7]
BENZALACETOPHENONE, I, 78
    2-Propen-1-one, 1,3-diphenyl- [94-41-7]
BENZALACETOPHENONE DIBROMIDE, I, 205
    1-Propanone, 2,3-dibromo-1,3-diphenyl- [611-91-6]
1-[α-(BENZALAMINO)BENZYL]-2-NAPHTHOL, I, 381
    2-Naphthalenol, 1-[α-(benzylideneamino)benzyl]-, (*E*)- [24609-80-1]
BENZALANILINE, I, 80
    Benzenamine, *N*-(phenylmethylene)- [538-51-2]
BENZALAZINE, II, 395
    Benzaldehyde, (phenylmethylene)hydrazone, (*E,E*)- [28867-76-7]

BENZALBARBITURIC ACID, III, 39
  2,4,6(1*H*,3*H*,5*H*)-Pyrimidinetrione, 5-(phenylmethylene)- [27402-47-7]
BENZALDEHYDE TOSYLHYDRAZONE, VII, 438
  Benzenesulfonic acid, 4-methyl-, hydrazide, phenylmethylene- [1666-17-7]
BENZALPHTHALIDE, II, 61
  1(3*H*)-Isobenzofuranone, 3-(phenylmethylene)- [575-61-1]
BENZALPINACOLONE, I, 81
  1-Penten-3-one, 4,4-dimethyl-1-phenyl- [538-44-3]
BENZANILIDE, I, 82
  Benzamide, *N*-phenyl- [93-98-1]
BENZ[*a*]ANTHRACENE, VII, 18
  Benz[*a*]anthracene [56-55-3]
BENZANTHRONE, II, 62
  7*H*-Benz[*de*]anthracen-7-one [82-05-3]
BENZENE, V, 998; *Potential poison*
  Benzene [71-43-2]
BENZENEBORONIC ANHYDRIDE, IV, 68
  Boroxin, triphenyl- [3262-89-3]
BENZENEDIAZONIUM-2-CARBOXYLATE, V, 54
  Benzenediazonium, 2-carboxy-, hydroxide, inner salt [1608-42-0]
BENZENESELENENYL CHLORIDE, VI, 533
  Benzeneselenenyl chloride [5707-04-0]
BENZENESULFONYL CHLORIDE, I, 84
  Benzenesulfonyl chloride [98-09-9]
BENZHYDRYL β-CHLOROETHYL ETHER, IV, 72
  Benzene, 1,1'-[(2-chloroethoxy)methylene]bis- [32669-06-0]
BENZIL, I, 87
  Ethanedione, diphenyl- [134-81-6]
BENZILIC ACID, I, 89
  Benzeneacetic acid, α-hydroxy-α-phenyl- [76-93-7]
BENZIMIDAZOLE, II, 65
  1*H*-Benzimidazole [51-17-2]
BENZOBARRELENE, VI, 82
  1,4-Ethenonaphthalene, 1,4-dihydro- [7322-47-6]
BENZOCYCLOPROPENE, VI, 87
  Bicyclo[4.1.0]hepta-1,3,5-triene [4646-69-9]
1,2-BENZO-3,4-DIHYDROCARBAZOLE, IV, 885
  5*H*-Benzo[*a*]carbazole, 6,11-dihydro- [21064-49-3]
BENZOFURAZAN OXIDE, IV, 74
  Benzofurazan, 1-oxide [480-96-6]
BENZOGUANAMINE, IV, 78
  1,3,5-Triazine-2,4-diamine, 6-phenyl- [91-76-9]
BENZOHYDROL, I, 90
  Benzenemethanol, α-phenyl- [91-01-0]
BENZOHYDROXAMIC ACID, II, 67
  Benzamide, *N*-hydroxy- [495-18-1]

BENZOHYDROXAMOYL CHLORIDE, V, 504
Benzenecarboximidoyl chloride, *N*-hydroxy- [698-16-8]
BENZOIC ANHYDRIDE, I, 91
Benzoic acid, anhydride [93-97-0]
BENZOIN, I, 94
Ethanone, 2-hydroxy-1,2-diphenyl- [119-53-9]
BENZOIN ACETATE, II, 69
Ethanone, 2-(acetyloxy)-1,2-diphenyl- [574-06-1]
BENZOPHENONE, I, 95
Methanone, diphenyl- [119-61-9]
BENZOPHENONE CYANOHYDRIN, VII, 20
Benzeneacetonitrile, α-hydroxy-α-phenyl- [4746-48-9]
BENZOPHENONE OXIME, II, 70
Methanone, diphenyl-, oxime [574-66-3]
BENZOPINACOL, II, 71
1,2-Ethandiol, 1,1,2,2-tetraphenyl- [464-72-2]
β-BENZOPINACOLONE, II, 73
Ethanone, tetraphenyl- [466-37-5]
BENZOQUINONE *O*-TRIMETHYLSILYL CYANOHYDRIN, VII, 517
2,5-Cyclohexadiene-1-carbonitrile, 4-oxo-1-[(trimethylsilyl)oxy]- [40861-57-2]
1,2,3-BENZOTHIADIAZOLE 1,1-DIOXIDE, V, 60
1,2,3-Benzothiadiazole, 1,1-dioxide [37150-27-9]
1,2,3-BENZOTRIAZOLE, III, 106; *Hazard*
1*H*-Benzotriazole [95-14-7]
3,1-BENZOXAZEPINE, VII, 23
3,1-Benzoxazepine [15123-59-8]
BENZOYLACETANILIDE, III, 108; **IV**, 80
Benzenepropanamide, β-oxo-*N*-phenyl- [959-66-0]
β-BENZOYLACRYLIC ACID, III, 109
2-Butenoic acid, 4-oxo-4-phenyl- [583-06-2]
ε-BENZOYLAMINO-α-BROMOCAPROIC ACID, II, 74
Hexanoic acid, 6-(benzoylamino)-2-bromo- [1700-05-6]
ε-BENZOYLAMINOCAPROIC ACID, II, 76
Hexanoic acid, 6-(benzoylamino)- [956-09-2]
ε-BENZOYLAMINO-α-CHLOROCAPROIC ACID, VI, 90
Hexanoic acid, 6-benzoylamino-2-chloro- [5107-15-3]
α-BENZOYLAMINOCINNAMIC ACID, II, 491
2-Propenoic acid, 2-benzoylamino-3-phenyl- [1155-48-2]
γ-BENZOYLBUTYRIC ACID, II, 82
Benzenepentanoic acid, δ-oxo- [1501-05-9]
BENZOYLCHOLINE CHLORIDE, IV, 84
Ethanaminium, 2-(benzoyloxy)-*N,N,N*-trimethyl-, chloride [2964-09-2]
BENZOYLCHOLINE IODIDE, IV, 84
Ethanaminium, 2-(benzoyloxy)-*N,N,N*-trimethyl-, iodide [17518-43-3]
BENZOYL CYANIDE, III, 112
Benzeneacetonitrile, α-oxo- [613-90-1]

2-BENZOYL-1,2-DIHYDROISOQUINALDONITRILE, VI, 115 or **1-CYANO-2-BENZOYL-1,2-DIHYDROISOQUINOLINE, IV**, 641
1-Isoquinolinecarbonitrile, 2-benzoyl-1,2-dihydro- [844-25-7]
1-BENZOYL-2-(1,3-DIBENZOYL-4-IMIDAZOLIN-2-YL)IMIDAZOLE, VII, 287
2,2'-Bi-1H-imidazole, 1,1',3-tribenzoyl-2,3-dihydro- [62457-77-6]
BENZOYL DISULFIDE, III, 116
Disulfide, dibenzoyl [644-32-6]
BENZOYLENE UREA, II, 79
2,4(1H,3H)-Quinazolinedione [86-96-4]
BENZOYL FLUORIDE, V, 66
Benzoyl fluoride [455-32-3]
BENZOYLFORMIC ACID, I, 244; **III**, 114
Benzeneacetic acid, α-oxo- [611-73-4]
3-BENZOYLINDOLE, VI, 109
Methanone, 1H-indol-3-ylphenyl- [15224-25-6]
DL-ϵ-**BENZOYLLYSINE, II**, 374
Lysine, N^6-benzoyl-, DL- [5107-18-6]
o-**BENZOYLOXYACETOPHENONE, IV**, 478
Ethanone, 1-[2-(benzoyloxy)phenyl]- [4010-33-7]
3-BENZOYLOXYCYCLOHEXENE, V, 70
2-Cyclohexen-1-ol, benzoate [3352-93-0]
BENZOYLPIPERIDINE, I, 99
Piperidine, 1-benzoyl- [776-75-0]
1-(2-BENZOYLPROPANOYL)PIPERIDINE, VIII, 326
Piperidine, 1-(2-methyl-1,3-dioxo-3-phenylpropyl)-, (±)- [99114-34-8]
β-**BENZOYLPROPIONIC ACID, II**, 81
Benzenebutanoic acid, γ-oxo- [2051-95-8]
3-BENZOYLPYRIDINE, IV, 88
Methanone, phenyl-3-pyridinyl- [5424-19-1]
4-BENZOYLPYRIDINE, IV, 88
Methanone, phenyl-4-pyridinyl- [14548-46-0]
BENZYLACETOPHENONE, I, 101
1-Propanone, 1,3-diphenyl- [1083-30-3]
N-**BENZYLACRYLAMIDE, V**, 73
2-Propenamide, N-(phenylmethyl)- [13304-62-6]
2-BENZYLAMINOPYRIDINE, IV, 91
2-Pyridinamine, N-(phenylmethyl)- [6935-27-9]
5-BENZYLAMINOTETRAZOLE, VII, 27
1H-Tetrazol-5-amine, N-(phenylmethyl)- [14832-58-7]
BENZYLANILINE, I, 102
Benzenemethanamine, N-phenyl- [103-32-2]
N-**BENZYL-2-AZANORBORNENE, VIII**, 31
2-Azabicyclo[2.2.1]hept-5-ene, 2-(phenylmethyl)- [112375-05-0]
BENZYL BENZOATE, I, 104
Benzoic acid, phenylmethyl ester [120-51-4]

1-BENZYL-2-BENZOYL-1,2-DIHYDROISOQUINALDONITRILE, VI, 118
  1-Isoquinolinecarbonitrile, 2-benzoyl-1,2-dihydro-1-(phenylmethyl)- [16576-37-2]
BENZYL *trans*-1,3-BUTADIENE-1-CARBAMATE, VI, 95
  Carbamic acid, 1,3-butadienyl-, phenylmethyl ester, (*E*)- [71616-72-3]
BENZYL CARBAMATE, III, 167
  Carbamic acid, phenylmethyl ester [621-84-1]
2-BENZYL-2-CARBOMETHOXYCYCLOPENTANONE, V, 75
  Cyclopentanecarboxylic acid, 2-oxo-1-(phenylmethyl)-, methyl ester [10386-81-9]
BENZYL CHLOROMETHYL ETHER, VI, 101
  Benzene, (chloromethoxy)methyl- [3587-60-8]
BENZYL CHLOROMETHYL KETONE, III, 119
  2-Propanone, 1-chloro-3-phenyl- [937-38-2]
BENZYL CYANIDE, I, 107; *Hazard*
  Benzeneacetonitrile [140-29-4]
2-BENZYLCYCLOPENTANONE, V, 76
  Cyclopentanone, 2-(phenylmethyl)- [2867-63-2]
BENZYL 2,3-EPOXYPROPYL ETHER, VIII, 33
  Oxirane, [(phenylmethoxy)methyl]- [2930-05-4]
7-BENZYLIDENEAMINOTHEOPHYLLINE, VII, 9
  1*H*-Purine-2,6-dione, 3,7-dihydro-1,3-dimethyl-7-[(phenylmethylene)amino]-
    [81281-59-6]
***N*-BENZYLIDENEBENZENESULFONAMIDE, VIII**, 546
  Benzenesulfonamide, *N*-(phenylmethylene)- [13909-34-7]
***N*-BENZYLIDENEETHYLAMINE, V**, 758
  Ethanamine, *N*-(phenylmethylene)- [6852-54-6]
***N*-BENZYLIDENEMETHYLAMINE, IV**, 605
  Methanamine, *N*-(phenylmethylene)- [622-29-7]
α-BENZYLIDENE-γ-PHENYL-Δ$^{\beta,\gamma}$-BUTENOLIDE, V, 80
  2(3*H*)-Furanone, 5-phenyl-3-(phenylmethylene)- [4361-96-0]
1-BENZYLINDOLE, VI, 104, 106
  1*H*-Indole, 1-(phenylmethyl)- [3377-71-7]
3-BENZYLINDOLE, VI, 109
  1*H*-Indole, 3-(phenylmethyl)- [16886-10-5]
BENZYL ISOCYANIDE, VII, 27
  Benzene, (isocyanomethyl)- [10340-91-7]
1-BENZYLISOQUINOLINE, VI, 115
  Isoquinoline, 1-(phenylmethyl)- [6907-59-1]
2-BENZYL-2-METHYLCYCLOHEXANONE, VI, 121
  Cyclohexanone, 2-methyl-2-(phenylmethyl)- [1206-21-9]
2-BENZYL-6-METHYLCYCLOHEXANONE, VI, 121
  Cyclohexanone, 2-methyl-6-(phenylmethyl)- [24785-76-0]
3-BENZYL-3-METHYLPENTANENITRILE, IV, 95
3-BENZYL-3-METHYLPENTANOIC ACID, IV, 93
  Benzenebutanoic acid, β-ethyl-β-methyl- [53663-16-4]
4-BENZYLOXY-2-BUTANONE, VII, 386
  2-Butanone, 4-phenylmethoxy- [6278-91-7]

BENZYLOXYCARBONYL-L-ALANYL-L-CYSTEINE METHYL ESTER, VII, 30
  L-Cysteine, *N*-[*N*-[(phenylmethoxy)carbonyl]-L-alanyl-, methyl ester [34804-78-3]
**BENZYLOXYCARBONYL-L-ASPARTYL-(*tert*-BUTYL ESTER)-L-PHENYL-
ALANYL-L-VALINE METHYL ESTER, VII**, 30
  L-Valine, *N*-[*N*-[*N*-[(phenylmethoxy)carbonyl]-L-aspartyl]-L-phenylalanyl]-, 4- (1,1-
  dimethylethyl) 1-methyl ester [57850-41-6]
5-BENZYLOXY-3-HYDROXY-3-METHYLPENTANOIC-2-^{13}C ACID, VII, 386
  Pentanoic-2-^{13}C acid, 3-hydroxy-3-methyl-5-(phenylmethoxy)- [57830-65-6]
4-BENZYLOXYINDOLE, VII, 34
  1*H*-Indole, 4-(phenylmethoxy)-[20289-26-3]
(*E*)-6-BENZYLOXY-2-NITRO-β-PYRROLIDINOSTYRENE, VII, 34
  Pyrrolidine, 1-[2-[2-nitro-6-(phenylmethoxy)phenyl]ethenyl]-, (*E*)- [99474-12-1]
6-BENZYLOXY-2-NITROTOLUENE, VII, 34
  Toluene, 6-benzyloxy-2-nitro- [20876-37-3]
1-BENZYLOXY-4-PENTEN-2-OL, VIII, 33
  4-Penten-2-ol, 1-(phenylmethoxy)- [58931-16-1]
***cis*-2-BENZYL-3-PHENYLAZIRIDINE, V**, 83
  Aziridine, 2-phenyl-3-(phenylmethyl)-, *cis*- [1605-08-9]
BENZYLPHTHALIMIDE, II, 83
  1*H*-Isoindole-1,3(2*H*)-dione, 2-(phenylmethyl)- [2142-01-0]
1-BENZYLPIPERAZINE, V, 88
  Piperazine, 1-(phenylmethyl)- [2759-28-6]
BENZYL SULFIDE, VI, 130
  Benzene, 1,1'-[thiobis(methylene)]bis- [538-74-9]
7-BENZYLTHIOMENTHONE, VIII, 304
  Cyclohexanone, 5-methyl-2-[1-methyl-1-(phenylmethylthio)ethyl]-, (2*R-trans*)-
  [79563-58-9]; (2*S-cis*)- [79618-04-5]
BENZYLTRIMETHYLAMMONIUM ETHOXIDE, IV, 98
  Benzenemethanaminium, *N,N,N*-trimethyl-, ethoxide [27292-06-4]
***N*-BENZYL-*N*-[(TRIMETHYLSILYL)METHYL]AMINE, VIII**, 231
  Benzenemethanamine, *N*-[(trimethylsilyl)methyl]- [53215-95-5]
BETAINE HYDRAZIDE HYDROCHLORIDE, II, 85
  Ethanaminium, 2-hydrazino-*N,N,N*-trimethyl-2-oxo-, chloride [123-46-6]
BIACETYL MONOXIME, II, 205
  2,3-Butanedione, monooxime [57-71-6]
BIALLYL, III, 121
  1,5-Hexadiene [592-42-7]
BICYCLO[1.1.0]BUTANE, VI, 133
  Bicyclo[1.1.0]butane [157-33-5]
BICYCLO[2.2.1]HEPTEN-7-ONE, V, 91
  Bicyclo[2.2.1]hept-2-en-7-one [694-71-3]
BICYCLO[3.3.1]NONAN-9-ONE, VI, 137
  Bicyclo[3.3.1]nonan-9-one [17931-55-4]
BICYCLO[4.3.0]NON-1-EN-4-ONE, VIII, 38
  5*H*-Indenone, 1,2,3,3a,4,6-hexahydro- [131712-16-8]

exo-cis-**BICYCLO[3.3.0]OCTANE-2-CARBOXYLIC ACID, V**, 93
  1-Pentalenecarboxylic acid, octahydro-, (1α,3aα,6aα)- [18209-43-3]
BICYCLO[3.3.0]OCTANE-2,6-DIOL, VIII, 43
  1,4-Pentalenediol, octahydro- [17572-86-0]
BICYCLO[3.3.0]OCTANE-2,6-DIONE, VIII, 43
  1,4-Pentalenedione, hexahydro- [77483-80-8]
BICYCLO[3.3.0]OCTANE-3,7-DIONE, *cis*-, VII, 50
  2,5-(1*H*,3*H*)-Pentalenedione, tetrahydro-, *cis*- [51716-63-3]
BICYCLO[3.2.1]OCTAN-3-ONE, VI, 142
  Bicyclo[3.2.1]octan-3-one [14252-05-2]
BICYCLO[2.1.0]PENTANE, V, 96
  Bicyclo[2.1.0]pentane [185-94-4]
BICYCLO[2.1.0]PENT-2-ENE, VI, 145
  Bicyclo[2.1.0]pent-2-ene [5164-35-2]
1,1'-BINAPHTHALENE-2,2'-DIOL, VIII, 46, 50, 57
  [1,1'-Binaphthalene]-2,2'-diol,(±)- [602-09-5]; (*R*)-(+)- [18531-94-7]; (*S*)-(−)-
  [18531-99-2]
1,1'-BINAPHTHYL-2,2'-DIYL HYDROGEN PHOSPHATE, (*R*)-(−)- and (*S*)-(+)-,
  VIII, 46, 50
  (*R*)-(−) and (*S*)-(+)-Dinaphtho[2,1-*d*:1'2'-*f*][1,3,2]dioxaphosphepin, 4-hydroxy-4-
  oxide; (*R*)- [39648-67-4] and (*S*)- [35193-64-7]
BIPHENYL, VI, 150, 490; *Hazard*
  1,1'-Biphenyl [92-52-4]
BIPHENYLENE, V, 54
  Biphenylene [259-79-0]
2,2'-BIPYRIDINE, V, 102
  2,2'-Bipyridine [366-18-7]
4,4'-BIS(ACETAMIDO)AZOBENZENE, V, 341
  Acetanilide, 4',4'''-azo-bis- [15446-39-6]
1,1-BIS(BROMOMETHYL)CYCLOPROPANE, VI, 153
  Cyclopropane, 1,1-bis(bromomethyl)- [29086-41-7]
1,2-BIS(BUTYLTHIO)BENZENE, V, 107
  Benzene, 1,2-bis(butylthio)- [53663-38-0]
BIS(CHLOROMETHYL) ETHER, IV, 101; (*Hazard Note*), **V**, 218
  Methane, oxybis[chloro]- [542-88-1]
BIS(β-CYANOETHYL)AMINE, III, 93
  Propanenitrile, 3,3'-iminobis- [111-94-4]
BIS(2-CYANOETHYL)PHENYLPHOSPHINE, VI, 932
  Propanenitrile, 3,3'-(phenylphosphinidene)bis- [15909-92-9]
BIS-3,4-DICHLOROBENZOYL PEROXIDE, V, 52
4,4'-BIS(DIMETHYLAMINO)BENZIL, V, 111
  Ethanedione, bis[4-(dimethylamino)phenyl]- [17078-27-2]
1,4-BIS(DIMETHYLAMINO)-2,3-DIMETHOXYBUTANE, (*S,S*)-(+)-, [DDB],
  VII, 41
  1,4-Butanediamine, 2 ,3-dimethoxy-*N,N,N',N'*-tetramethyl-, (*S,S*)-(+)- [26549-21-3]

BIS(DIMETHYLAMINO)METHANE, V, 435; **VI**, 474
Methanediamine, *N,N,N',N'*-tetramethyl- [51-80-9]
BIS(3-DIMETHYLAMINOPROPYL)PHENYLPHOSPHINE, VI, 776
1-Propanamine, 3,3'-(phenylphosphinidene)bis[*N,N*-dimethyl]- [32357-32-7]
***N*-[2,4-BIS(1,3-DIPHENYLIMIDAZOLIDIN-2-YL)-5-
HYDROXYPHENYL]ACETAMIDE, VII**, 162
Acetamide, *N*-[2,4-bis(1,3-diphenyl-2-imidazolidinyl)-5-hydroxyphenyl]- [67149-
22-8], **VII**, 162
BIS(1,3-DIPHENYLIMIDAZOLIDINYLIDENE-2), V, 115
Imidazolidine, 2-(1,3-diphenyl-2-imidazolidinylidene)-1,3-diphenyl- [2179-89-7]
2,2'-BIS(DIPHENYLPHOSPHINO)-1,1'-BINAPHTHYL (BINAP), (*R*)-(+)- and (*S*)-
(−)-, **VIII**, 57, 183
Phosphine, [1,1'-binaphthalene]-2,2'-diylbis-[diphenyl-, (*R*)- and (*S*)-]; (*R*)- [76189-
55-4], (*S*)- [76189-56-5]
(±)-2,2'-BIS(DIPHENYLPHOSPHINYL)-1,1'-BINAPHTHYL [(±)BINAPO],
VIII, 58
Phosphine oxide, [1,1'-binaphthalene]-2,2'-diylbis-[diphenyl- [130164-89-5]
1,2-BIS(2-IODOETHOXY)ETHANE, VIII, 153
Ethane, 1,2-bis(2-iodoethoxy)- [36839-55-1]
***N,N'*-BIS(METHOXYCARBONYL)SULFUR DIIMIDE, VIII**, 427
Sulfur diimide, dicarboxy-, dimethyl ester [16762-82-6]
3-BIS(METHYLTHIO)-1-HEXEN-4-OL, VI, 683
5-Hexen-3-ol, 4,6-bis(methylthio)- [53107-07-6]
1,3-BIS(METHYLTHIO)-2-METHOXYPROPANE, VI, 683
Propane, 2-methoxy-1,3-bis(methylthio)- [31805-84-2]
1,3-BIS(METHYLTHIO)-2-PROPANOL, VI, 683
2-Propanol, 1,3-bis(methylthio)- [31805-83-1]
3,3-BIS(METHYLTHIO)-1-(2-PYRIDINYL)-2-PROPEN-1-ONE, VII, 476
2-Propen-1-one, 3,3-bis(methylthio)-1-(2-pyridinyl)- [78570-34-0]
1,4-BIS(4-PHENYLBUTADIENYL)BENZENE, V, 985
Benzene, 1,4-bis(4-phenyl-1,3-butadienyl)- [10162-88-6]
BIS(PHENYLTHIO)METHANE, VI, 737
Benzene, 1,1'-[methylenebis(thio)]bis- [3561-67-9]
4,4-BIS(PHENYLTHIO)-3-METHOXY-1-BUTENE, VI, 737
Benzene, 1,1'-[(2-methoxy-3-butenylidene)bis(thio)]bis- [60466-65-1]
(*E*)-1,2-BIS(TRIBUTYLSTANNYL)ETHYLENE, VIII, 268
Stannane, 1,2-ethenediylbis[dibutyl-, (*E*)- [14275-61-7]
BIS(2,2,2-TRICHLOROETHYL) AZODICARBOXYLATE, VII, 56
Diazenedicarboxylic acid, bis (2,2,2-trichloroethyl) ester [38857-88-4]
BIS(2,2,2-TRICHLOROETHYL) HYDRAZODICARBOXYLATE, VII, 56
1,2-Hydrazinedicarboxylic acid, bis- (2,2,2-trichloroethyl) ester [38858-02-5]
BIS(TRIFLUOROMETHYL)DIAZOMETHANE, VI, 161
Propane, 2-diazo-1,1,1,3,3,3-hexafluoro- [684-23-1]
**BIS[2,2,2-TRIFLUORO-1-PHENYL-1-(TRIFLUOROMETHYL)ETHOXY]
DIPHENYLSULFURANE, VI**, 163
Sulfur, bis[α,α-bis(trifluoromethyl)benzenemethanolato]diphenyl- [32133-82-7]

1,4-BIS(TRIMETHYLSILYL)BUTA-1,3-DIYNE, VIII, 63
    Silane, 1,3-butadiyne-1,4-diylbis[trimethyl]- [4526-07-2]
1,2-BIS(TRIMETHYLSILYLOXY)CYCLOBUTENE, VI, 167
    Silane, (1-cyclobuten-1,2-ylendioxy)bis[trimethyl]- [17082-61-0]
2-BORNENE, VI, 172
    Bicyclo[2.2.1]hept-2-ene, 1,7,7-trimethyl- [464-17-5]
BROMAL, II, 87
    Acetaldehyde, tribromo- [115-17-3]
BROMOACETAL, III, 123
    Ethane, 2-bromo-1,1-diethoxy- [2032-35-1]
***N*-BROMOACETAMIDE, IV**, 104
    Acetamide, *N*-bromo- [79-15-2]
3-BROMO-4-ACETAMINOTOLUENE, I, 111
    Acetamide, *N*-(2-bromo-4-methylphenyl)- [614-83-5]
BROMOACETIC ACID, III, 381; *Hazard*
    Acetic acid, bromo- [79-08-3]
BROMOACETONE, II, 88
    2-Propanone, 1-bromo- [598-31-2]
3-BROMOACETOPHENONE, V, 117
    Ethanone,1-(3-bromophenyl)- [2142-63-4]
***p*-BROMOACETOPHENONE, I**, 109
    Ethanone, 1-(4-bromophenyl)- [99-90-1]
2-BROMOACETYL-6-METHOXYNAPHTHALENE, VI, 175
    Ethanone, 2-bromo-1-(6-methoxy-2-naphthalenyl)- [10262-65-4]
2-BROMOALLYLAMINE, V,121
    2-Propen-1-amine, 2-bromo- [6943-51-7]
***N*-(2-BROMOALLYL)ETHYLAMINE, V**, 124
    2-Propen-1-amine, 2-bromo-*N*-ethyl- [871-23-8]
3-BROMO-4-AMINOTOLUENE, I, 111
    Benzenamine, 2-bromo-4-methyl- [583-68-6]
α-BROMOBENZALACETONE, III, 125
    3-Buten-2-one, 3-bromo-4-phenyl- [31207-17-7]
α-BROMOBENZALACETOPHENONE, III, 126
    2-Propen-1-one, 2-bromo-1,3-diphenyl- [6935-75-7]
***m*-BROMOBENZALDEHYDE, II**, 132
    Benzaldehyde, 3-bromo- [3132-99-8]
***p*-BROMOBENZALDEHYDE, II**, 89, 442
    Benzaldehyde, 4-bromo- [1122-91-4]
***p*-BROMOBIPHENYL, I**, 113; *Hazard*
    1,1' -Biphenyl, 4-bromo- [92-66-0]
α-BROMO-γ-BUTYROLACTONE, V, 255
    2(3*H*)-Furanone, 3-bromodihydro- [5061-21-2]
α-BROMOCAPROIC ACID(2-BROMOHEXANOIC ACID), I, 115; **II**, 95; **VI**, 192
    Hexanoic acid, 2-bromo- [616-05-7]
1-BROMO-3-CHLOROCYCLOBUTANE, VI, 179, 272
    Cyclobutane, 1-bromo-3-chloro- [4935-03-9]

1-BROMO-3-CHLORO-2,2-DIMETHOXYPROPANE, VI, 361
  Propane, 1-bromo-3-chloro-2,2-dimethoxy- [22089-54-9]
α-BROMOCINNAMIC ALDEHYDE, III, 731
  2-Propenal, 2-bromo-3-phenyl- [5443-49-2]
α-BROMOCINNAMIC ALDEHYDE ACETAL, III, 732
2-BROMO-2-CYCLOPENTEN-1-ONE, VII, 271
  2-Cyclopenten-1-one, 2-bromo- [10481-34-2]
2-BROMO-2-CYCLOPENTENONE ETHYLENE KETAL, VII, 271, 495
  1,4-Dioxaspiro[4.4]non-6-ene, 6-bromo- [68241-78-1]
BROMOCYCLOPROPANE, V, 126
  Cyclopropane, bromo- [4333-56-6]
6-BROMO-3,4-DIMETHOXYBENZALDEHYDE CYCLOHEXYLIMINE,
    VIII, 586
  Cyclohexanamine, N-[(2-bromo-4,5-dimethoxyphenyl)methylene]- [73252-55-8]
4-BROMO-N,N-DIMETHYL-3-(TRIFLUOROMETHYL)ANILINE, VI, 181
  Benzenamine, 4-bromo-N,N-dimethyl-3-(trifluoromethyl)- [51332-24-2]
erythro-**2-BROMO-1,2-DIPHENYLETHANOL, VI**, 184
  Benzenethanol, β-bromo-α-phenyl-, (R*, S*)- [10368-43-1]
2-BROMO-3,3-DIPHENYL-2-PROPEN-1-YL ACETATE, VI, 187
  2-Propen-1-ol, 2-bromo-3,3-diphenyl-, acetate [14310-15-7]
p-**BROMODIPHENYLMETHANE, V**, 130
  Benzene, 1-bromo-4-(phenylmethyl)- [2116-36-1]
2-BROMOETHANOL, I, 117
  Ethanol, 2-bromo- [540-51-2]
(1-BROMOETHENYL)TRIMETHYLSILANE, VI, 1033
  Silane, (1-bromoethenyl)trimethyl [13683-41-5]
1-BROMO-1-ETHOXYCYCLOPROPANE, VIII, 556
  Cyclopropane, 1-bromo-1-ethoxy- [95631-62-2]
β-BROMOETHYLAMINE HYDROBROMIDE, II, 91
  Ethanamine, 2-bromo-, hydrobromide [2576-47-8]
2-(2-BROMOETHYL)-1,3-DIOXANE, VII, 59
  1,3-Dioxane, 2-(2-bromoethyl)- [33884-43-4]
β-BROMOETHYLPHTHALIMIDE, I, 119; **IV**, 106
  1H-Isoindole-1,3(2H)-dione, 2-(2-bromoethyl)- [574-98-1]
1-BROMO-2-FLUOROBENZENE, V, 133
  Benzene, 1-bromo-2-fluoro- [1072-85-1]
1-BROMO-2-FLUOROHEPTANE, V, 136
  Heptane, 1-bromo-2-fluoro- [1786-32-9]
α-BROMOHEPTALDEHYDE, III, 127
  Heptanal, 2-bromo- [16486-84-3]
α-BROMOHEPTALDEHYDE DIMETHYL ACETAL, III, 128
  Heptane, 2-bromo-1,1-dimethoxy- [18207-18-6]
4-BROMO-2-HEPTENE, IV, 108
  2-Heptene, 4-bromo- [22118-57-6]
2-BROMOHEXANOIC ACID (see α-bromocaproic acid)
  Hexanoic acid, 2-bromo- [616-05-7]

2-BROMOHEXANOYL CHLORIDE, VI, 190
Hexanoyl chloride, 2-bromo- [42768-46-7]
10-BROMO-11-HYDROXY-10,11-DIHYDROFARNESYL ACETATE, VI, 560
2,6-Dodecadiene-1,11-diol, 10-bromo-3,7,11-trimethyl-, 1-acetate, *(E,E)*-
[54795-59-4]
3-BROMO-4-HYDROXYTOLUENE, III, 130
Phenol, 2-bromo-4-methyl- [6627-55-0]
α-BROMOISOBUTYRYL BROMIDE, IV, 348
Propanoyl bromide, 2-bromo-2-methyl- [20769-85-1]
α-BROMOISOCAPROIC ACID, II, 95; **III,** 523
Pentanoic acid, 2-bromo-4-methyl- [49628-52-6]
α-BROMOISOVALERIC ACID, II, 93; **III,** 848
Butanoic acid, 2-bromo-3-methyl- [565-74-2]
***p*-BROMOMANDELIC ACID, IV,** 110
Benzeneacetic acid, 4-bromo-α-hydroxy- [6940-50-7]
BROMOMESITYLENE, II, 95
Benzene, 2-bromo-1,3,5-trimethyl- [576-83-0]
BROMOMETHANESULFONYL BROMIDE, VIII, 212
Methanesulfonyl bromide, bromo- [54730-18-6]
α-BROMO-β-METHOXYBUTYRIC ACID, III, 813
Butanoic acid, 2-bromo-3-methoxy-, *erythro*- [26839-93-0]; *threo*- [26839-91-8]
α-BROMO-β-METHOXYPROPIONIC ACID, III, 775
Propanoic acid, 2-bromo-3-methoxy- [65090-78-0]
α-(BROMOMETHYL)ACRYLIC ACID, VII, 210
2-Propenoic acid, 2-(bromomethyl)- [72707-66-5]
2-BROMO-4-METHYLBENZALDEHYDE, V, 139
Benzaldehyde, 2-bromo-4-methyl- [824-54-4]
2-BROMO-3-METHYLBENZOIC ACID, IV, 114
Benzoic acid, 2-bromo-3-methyl- [53663-39-1]
**1-BROMO-1-METHYL-2-(BROMOMETHYLSULFONYL)CYCLOHEXANE,
VIII,** 212
Cyclohexane, 1-bromo-[(bromomethyl)sulfonyl]-1-methyl- [120696-44-8]
1-BROMO-3-METHYL-2-BUTANONE, VI, 193
2-Butanone, 1-bromo-3-methyl [19967-55-6]
2-(BROMOMETHYL)-2-(CHLOROMETHYL)-1,3-DIOXANE, VIII, 173
1,3-Dioxane, 2-(bromomethyl)-2-(chloromethyl)- [60935-30-0)]
***N*-BROMOMETHYLPHTHALIMIDE, VIII,** 536
1*H*-Isoindole-1,3-(2*H*)-dione, 2-(bromomethyl)- [5332-26-3]
α-BROMO-β-METHYLVALERIC ACID, II, 95; **III,** 496
Pentanoic acid, 2-bromo-3-methyl- [42880-22-8]
α-BROMONAPHTHALENE, I, 121
Naphthalene, 1-bromo- [90-11-9]
2-BROMONAPHTHALENE, V, 142
Naphthalene, 2-bromo- [580-13-2]
6-BROMO-2-NAPHTHOL, III, 132
2-Naphthalenol, 6-bromo- [15231-91-1]

m-BROMONITROBENZENE, I, 123
  Benzene, 1-bromo-3-nitro- [585-79-5]
2-BROMO-3-NITROBENZOIC ACID, I, 125
  Benzoic acid, 2-bromo-3-nitro- [573-54-6]
2-BROMO-4-NITROTOLUENE, IV, 114
  Benzene, 2-bromo-1-methyl-4-nitro- [7745-93-9]
p-BROMOPHENACYL BROMIDE, I, 127
  Ethanone, 2-bromo-1-(4-bromophenyl)- [99-73-0]
9-BROMOPHENANTHRENE, III, 134
  Phenanthrene, 9-bromo- [573-17-1]
o-BROMOPHENOL, II, 97
  Phenol, 2-bromo- [95-56-7]
p-BROMOPHENOL, I, 128
  Phenol, 4-bromo- [106-41-2]
α-BROMOPHENYLACETIC ACID, VI, 403
  Benzeneacetic acid, α-bromo- [4870-65-9]
α-BROMO-α-PHENYLACETONE (in solution), III, 343
  2-Propanone, 1-bromo-1-phenyl- [23022-83-5]
α-BROMO-α-PHENYLACETONITRILE (in solution), III, 347
  Benzeneacetonitrile, α-bromo- [5798-79-8]
1-*p*-BROMOPHENYL-1-PHENYL-2,2,2-TRICHLOROETHANE, V, 131
  Benzene, 1-bromo-4-(2,2,2-trichloro-1-phenylethyl)- [39211-93-3]
α-BROMO-β-PHENYLPROPIONIC ACID, III, 705
  Benzenepropanoic acid, α-bromo- [42990-49-8]
p-BROMOPHENYLUREA, IV, 49
  Urea, (4-bromophenyl)- [1967-25-5]
(2)3-BROMOPHTHALIDE, III, 705; V, 145
  1(3*H*)-Isobenzofuranone, 3-bromo- [6940-49-4]
β-BROMOPROPIONIC ACID, I, 131
  Propanoic acid, 3-bromo- [590-92-1]
3-BROMOPYRENE, V, 147
  Pyrene, 1-bromo- [1714-29-0]
2-BROMOPYRIDINE, III, 136
  Pyridine, 2-bromo- [109-04-6]
4-BROMORESORCINOL, II, 100
  1,3-Benzenediol, 4-bromo- [6626-15-9]
3-BROMOTHIOPHENE, V, 149
  Thiophene, 3-bromo- [872-31-1]
o-BROMOTOLUENE, I, 135
  Benzene, 1-bromo-2-methyl- [95-46-5]
m-BROMOTOLUENE, I, 133
  Benzene, 1-bromo-3-methyl- [591-17-3]
p-BROMOTOLUENE, I, 136
  Benzene, 1-bromo-4-methyl- [106-38-7]
4-BROMO-*o*-XYLENE, III, 138
  Benzene, 4-bromo-1,2-dimethyl- [583-71-1]

1,3-BUTADIENE, II, 102; *Hazard*
  1,3-Butadiene [106-99-0]
***trans*, *trans*-1,3-BUTADIENE-1,4-DIYL DIACETATE, VI**, 196
  1,3-Butadiene-1,4-diol, diacetate, *E,E*- [15910-11-9]
***erythro*-2,3-BUTANEDIOL MONOMESYLATE, VI**, 312
  2,3-Butanediol, monomethanesulfonate, (*R**-,*S**-) [35405-98-2]
BUTEN-3-YNYL-2,6,6-TRIMETHYL-1-CYCLOHEXENE, (*E*)-, VII, 63
  Cyclohexene, 2-(1-buten-3-ynyl)-1,3,3-trimethyl-, (*E*)- [73395-75-2]
17β-*tert*-BUTOXY-5α-ANDROSTAN-3-ONE, VII, 66
  Androstan-3-one, 17-(1,1-dimethylethoxy)-, (5α,17β)- [87004-41-9]
17β-*tert*-BUTOXY-5α-ANDROST-2-ENE, VII, 66
  Androst-2-ene, 17-(1,1-dimethylethoxy)-, (5α,17β)- [87004-43-1]
***N*-*tert*-BUTOXYCARBONYL-L-LEUCINAL, VIII**, 68
  Carbamic acid, (1-formyl-3-methylbutyl)-, 1,1-dimethylethyl ester, (*S*)- [58521-45-2]
***tert*-BUTOXYCARBONYL-L-LEUCINE-*N*-METHYL-*O*-METHYLCARBOX-**
  AMIDE, VIII, 68
  Carbamic acid, [1-[(methoxymethylamino)carbonyl]-3-methylbutyl]-, 1,1-
  dimethylethyl ester, (*S*)- [87694-50-6]
2-*tert*-BUTOXYCARBONYLOXYIMINO-2-PHENYLACETONITRILE, VI, 199,
  718
  Benzeneacetonitrile, α-[[[(1,1-dimethylethoxy)carbonyl]carbonyl]oxy]imino]-
  [58632-95-4]
***N*-*tert*-BUTOXYCARBONYL-L-PHENYLALANINE, VII**, 70, 75
  Phenylalanine, *N*-[(1,1-dimethylethoxy)carbonyl]-,L- [13734-34-4]
***N*-*tert*-BUTOXYCARBONYL-L-PROLINE, VI**, 203
  1,2-Pyrrolidinedicarboxylic acid, 1-(1,1-dimethylethyl) ester, (*S*)- [15761-39-4]
***o*-BUTOXYNITROBENZENE, III**, 140
  Benzene, 1-butoxy-2-nitro- [7252-51-9]
7-*tert*-BUTOXYNORBORNADIENE, V, 151
  Bicyclo[2.2.1]hepta-2,5-diene, 7-(1,1-dimethylethoxy)- [877-06-5]
2-*tert*-BUTOXYTHIOPHENE, V, 642
  Thiophene, 2-(1,1-dimethylethoxy)- [23290-55-3]
***tert*-BUTYL ACETATE, III**, 141; **IV**, 263; *Hazard*
  Acetic acid, 1,1-dimethylethyl ester [540-88-5]
***tert*-BUTYL ACETOACETATE, V**, 155
  Butanoic acid, 3-oxo-, 1,1-dimethylethyl ester [1694-31-1]
***S*-*tert*-BUTYL ACETOTHIOACETATE, VIII**, 71
  Butanethioic acid, 3-oxo-, *S*-(1,1-dimethylethyl) ester [15925-47-0]
BUTYLACETYLENE, IV, 117
  1-Hexyne [693-02-7]
BUTYL ACRYLATE, III, 146; *Hazard*
  2-Propenoic acid, butyl ester [141-32-2]
***tert*-BUTYLAMINE, III**, 148
  2-Propanamine, 2-methyl- [75-64-9]
***tert*-BUTYLAMINE HYDROCHLORIDE, III**, 153
  2-Propanamine, 2-methyl-, hydrochloride [10017-37-5]

tert-BUTYL AZIDOACETATE, V, 586
  Acetic acid, azido-, 1,1-dimethylethyl ester [6367-36-8]
tert-BUTYL AZIDOFORMATE, V, 157; VI, 207; *Hazard*
  Carbonazidic acid, 1,1-dimethylethyl ester [1070-19-5]
tert-BUTYL AZODIFORMATE, V, 160
  Diazenedicarboxylic acid, bis(1,1-dimethylethyl) ester [870-50-8]
N-tert-BUTYLBENZALDIMINE, V, 191
  2-Propanamine, 2-methyl-*N*-(phenylmethylene)- [6852-58-0]
BUTYLBENZENE, III, 157; *Hazard*
  Benzene, butyl- [104-51-8]
BUTYL BORATE, II, 106; *Hazard*
  Boric acid(H$_3$BO$_3$), tributyl ester [688-74-4]
2-BUTYLBORNENE, VII, 77
  Bicyclo[2.2.1]hept-2-ene, 2-butyl-1,7,7-trimethyl- [87068-33-5]
BUTYL BROMIDE, I, 28, 37; *Hazard*
  Butane, 1-bromo- [109-65-9]
sec-BUTYL BROMIDE, I, 38; II, 359
  Butane, 2-bromo- [78-76-2]
tert-BUTYL BROMIDE, I, 38
  Propane, 2-bromo-2-methyl- [507-19-7]
BUTYL BUTYRATE, I, 138; *Hazard*
  Butanoic acid, butyl ester [109-21-7]
p-tert-BUTYLCALIX[4]ARENE, VIII, 75
  Pentacyclo[19.3.1.1^{3,7},1^{9,13},1^{15,19}]octacosa-
        1(25),3,5,7(28),9,11,13(27),15,17,19(26),
           21,23-dodecaene-25,26,27,28-tetraol, 5,11,17,23-tetrakis
              (1,1-dimethylethyl)- [60705-62-6]
p-tert-BUTYLCALIX[6]ARENE, VIII, 77
  Heptacyclo[31.3.1.1^{3,7}.1^{9,13}.1^{15,19}.1^{21,25}.1^{27,31}]-octatetraconta-
        1(37),3,5,7(42),9,11,13(41),15,17,19,(40),21,23,25(39),
           27,29,31(38),33,35-octadecaene-37,38,39,40,41,42-hexol,
              5,11,17,23,29,35-hexakis(1,1-dimethylethyl)- [78092-53-2]
p-tert-BUTYLCALIX[8]ARENE VIII, 80
  Nonacyclo[43.3.1.1^{3,7}.1^{9,13}.1^{15,19}.1^{21,25}.1^{27,31}.1^{33,37}.1^{39,43}]hexapentaconta-
        1(49),3,5,7(56),9,11,13(55),15,17,19(54),21,23,24(53),
           27,29,31(52),33,35,37(51),39,41,43(50),45,47-tetracosaene-
              49,50,51,52,53,54,55,56-octol,
                 5,11,17,23,29,35,41,47-
                    octakis(1,1-dimethyethyl)- [68971-82-4]
sec-BUTYL-α-CAPROYLPROPIONATE, IV, 120
  Octanoic acid, 2-methyl-3-oxo-, 1-methylpropyl ester [53663-40-4]
BUTYL CARBAMATE, I, 140
  Carbamic acid, butyl ester [592-35-8]
tert-BUTYL CARBAMATE, V, 162
  Carbamic acid, 1,1-dimethylethyl ester [4248-19-5]

tert-**BUTYL CARBAZATE, V,** 166
  Hydrazinecarboxylic acid, 1,1-dimethylethyl ester [870-46-2]
tert-**BUTYLCARBONIC DIETHYLPHOSPHORIC ANHYDRIDE, VI,** 207
  Carbonic acid, monoanhydride with diethyl phosphate, *tert*-butyl ester
    [14618-58-7]
BUTYL CHLORIDE, I, 142; *Hazard*
  Butane, 1-chloro- [109-69-3]
tert-**BUTYL CHLORIDE, I,** 144; *Hazard*
  Propane, 2-chloro-2-methyl- [507-20-0]
tert-**BUTYL CHLOROACETATE, III,** 144
  Acetic acid, chloro-, 1,1-dimethylethyl ester [107-59-5]
tert-**BUTYL CINNAMATE, III,** 144
  2-Propenoic acid, 3-phenyl-, 1,1-dimethylethyl ester [14990-09-1]
sec-**BUTYL CROTONATE, V,** 762
  2-Butenoic acid, 1-methylpropyl ester, (*E*)- [10371-45-6]
tert-**BUTYL CYANOACETATE, V,** 171
  Acetic acid, cyano-, 1,1-dimethylethyl ester [1116-98-9]
tert-**BUTYLCYANOKETENE, VI,** 210
  Butanenitrile, 2-carbonyl-3,3-dimethyl- [29342-22-1]
3-BUTYLCYCLOBUTENONE, VIII, 82
  2-Cyclobuten-1-one, 3-butyl- [38425-48-8]
S-tert-**BUTYL CYCLOHEXANECARBOTHIOATE, VII,** 81
  Cyclohexanecarbothioic acid, *S*-(1,1-dimethylethyl) ester [54829-37-7]
tert-**BUTYL CYCLOHEXANECARBOXYLATE, VII,** 87
  Cyclohexanecarboxylic acid, 1,1-dimethylethyl ester [16537-05-6]
cis-**4-***tert*-**BUTYLCYCLOHEXANOL, VI,** 215
  Cyclohexanol, 4-(1,1-dimethylethyl)-, *cis*- [937-05-3]
trans-**4-***tert*-**BUTYLCYCLOHEXANOL, V,** 175
  Cyclohexanol, 4-(1,1-dimethylethyl)-, *trans*- [21862-63-5]
4-*tert*-**BUTYLCYCLOHEXANONE, VI,** 218, 220
  Cyclohexanone, 4-(1,1-dimethylethyl)- [98-53-3]
4-*tert*-**BUTYLCYCLOHEXEN-1-YL TRIFLUOROMETHANESULFONATE,**
    VIII, 97
  Methanesulfonic acid, trifluoro, 4-(1,1-dimethylethyl)-1-cyclohexen-1-yl ester
    [77412-96-5]
1-(4-*tert*-**BUTYLCYCLOHEXEN-1-YL)-2-PROPEN-1-ONE, VIII,** 97
  2-Propen-1-one, 1-[4-(1,1-dimethylethyl)-1-cyclohexen-1-yl]- [92622-56-5]
S-tert-**BUTYL-L-CYSTEINE *tert*-BUTYL ESTER, VI,** 255
  L-Cysteine, *S*-(1,1-dimethylethyl)-, 1,1-dimethylethyl ester [45157-84-4]
2-*tert*-**BUTYL-1,3-DIAMINOPROPANE, VI,** 223
  1,3- Propanediamine, 2-(1,1-dimethylethyl)- [56041-82-8]
tert-**BUTYL DIAZOACETATE, V,** 179
  Acetic acid, diazo-, 1,1-dimethylethyl ester [35059-50-8]
tert-**BUTYL α-DIAZOACETOACETATE, V,** 180
  Butanoic acid, 2-diazo-3-oxo-, 1,1-dimethylethyl ester [13298-76-5]

3-BUTYL-4,4-DICHLOROCYCLOBUTENONE, VIII, 82
    2-Cyclobuten-1-one, 3-butyl-4,4-dichloro- [72284-70-9]
***S-tert*-BUTYL ESTER OF CHOLIC ACID, VII**, 81
    Cholane-24-thioic acid, 3,7,12-trihydroxy-, *S*-(1,1-dimethylethyl) ester
        (3α,5β,7α,12α)- [58587-05-6]
***tert*-BUTYL *N*-(1-ETHOXYCYCLOPROPYL)CARBAMATE, VI**, 226
    Carbamic acid, (1-ethoxycyclopropyl)-, 1,1-dimethylethyl ester [28750-48-3]
***tert*-BUTYL ETHYL FUMARATE, VII**, 93
    2-Butenedioic acid, (*E*)-, ethyl (1,1-dimethylethyl) ester [100922-16-5]
BUTYL GLYOXYLATE, IV, 124
    Acetic acid, oxo-, butyl ester [6295-06-3]
***N-tert*-BUTYLHYDROXYLAMINE, VI**, 804
    2-Propanamine, *N*-hydroxy-2-methyl- [16649-50-6]
8-BUTYL-2-HYDROXYTRICYCLO[7.3.1.0^{2,7}]TRIDECAN-13-ONE, VIII, 87
    5,9-Methanobenzocycloocten-11-one, 10-butyldodecahydro-4a-hydroxy- [24133-22-0]
***tert*-BUTYL HYPOCHLORITE, IV**, 125; **V**, 184; *Warning*, **V**, 183
    Hypochlorous acid, 1,1-dimethylethyl ester [507-40-4]
***tert*-BUTYL ISOBUTYRATE, III**, 143
    Propanoic acid, 2-methyl-, 1,1-dimethylethyl ester [16889-72-8]
***tert*-BUTYL ISOCYANIDE, VI**, 232
    Propane, 2-isocyano-2-methyl- [7188-38-7]
***sec*-BUTYL ISOPROPYL DISULFIDE, VI**, 235
    Disulfide, 1-methylethyl 1-methylpropyl [67421-86-7]
***tert*-BUTYL ISOVALERATE, III**, 144
    Butanoic acid, 3-methyl-, 1,1-dimethylethyl ester [16792-03-3]
***tert*-BUTYLMALONONITRILE, VI**, 223
    Propanedinitrile, 1,1-dimethylethyl- [4210-60-0]
2-BUTYL-2-METHYLCYCLOHEXANONE, V, 187
    Cyclohexanone, 2-butyl-2-methyl- [1197-78-0]
***tert*-BUTYL 2-METHYLENEDODECANOATE, IV**, 616, 618
***sec*-BUTYL 3-METHYLHEPTANOATE, V**, 763
    Heptanoic acid, 3-methyl-, 1-methylpropyl ester [16253-72-8]
3-BUTYL-2-METHYL-1-HEPTEN-3-OL, VI, 240
    5-Nonanol, 5-(2-propenyl)- [76071-61-9]
1-BUTYL-10-METHYL-Δ$^{1(9)}$-2-OCTALONE, VI, 242
    2(3*H*)-Naphthalenone, 1-butyl-4,4a,5,6,7,8-hexahydro-4a-methyl- [66252-93-5]
***tert*-BUTYL *S*-METHYLTHIOLCARBONATE, V**, 168
    Carbonic acid, thio-, *O-tert*-butyl *S*-methyl ester [29518-83-0]
BUTYL NITRITE, II, 108; *Hazard*
    Nitrous acid, butyl ester [544-16-1]
9-BUTYL-1,2,3,4,5,6,7,8-OCTAHYDROACRIDINE, VIII, 87
    Acridine, 9-butyl-1,2,3,4,5,6,7,8-octahydro- [99922-90-4]
9-BUTYL-1,2,3,4,5,6,7,8-OCTAHYDROACRIDINE *N*-OXIDE, VIII, 87
    Acridine, 9-butyl-1,2,3,4,5,6,7,8-octahydro-, 10-oxide [136528-61-5]
9-BUTYL-1,2,3,4,5,6,7,8-OCTAHYDROACRIDIN-4-OL, VIII, 87
    4-Acridinol, 9-butyl-1,2,3,4,5,6,7,8-octahydro- [99922-91-5]

tert-**BUTYL**-*tert*-**OCTYLAMINE, VIII**, 93
  2-Pentanamine, *N*-(1,1-dimethylethyl)-2,4,4-trimethyl- [90545-94-1]
N-*tert*-**BUTYL**-*N*-*tert*-**OCTYL**-*O*-*tert*-**BUTYLHYDROXYLAMINE, VIII**, 93
  2-Pentanamine, *N*-(1,1-dimethylethoxy)-*N*-(1,1-dimethylethyl)- 2,4,4-trimethyl-
    [90545-93-0]
3-BUTYL-2,4-PENTANEDIONE, VI, 245
  2,4-Pentanedione, 3-butyl- [1540-36-9]
tert-**BUTYL PHENYLCARBONATE, V**, 166
  Carbonic acid, 1,1-dimethylethyl phenyl ester [6627-89-0]
tert-**BUTYL PHENYL KETONE, VI**, 248
  1-Propanone, 2,2-dimethyl-1-phenyl- [938-16-9]
2-*tert*-**BUTYL-3-PHENYLOXAZIRANE, V**, 191
  Oxaziridine, 2-(1,1-dimethylethyl)-3-phenyl- [7731-34-2]
tert-**BUTYLPHTHALIMIDE, III**, 152
  1*H*-Isoindole-1,3(2*H*)-dione, 2-(1,1-dimethylethyl)- [2141-99-3]
BUTYL PHOSPHATE, II, 109
  Phosphoric acid, tributyl ester [126-73-8]
tert-**BUTYL PROPIONATE, III**, 143
  Propanoic acid, 1,1-dimethylethyl ester [51233-80-8]
1-BUTYLPYRROLIDINE, III, 159
  Pyrrolidine, 1-butyl- [767-10-2]
O-*tert*-**BUTYL**-L-**SERYL**-*S*-*tert*-**BUTYL**-L-**CYSTEINE** *tert*-**BUTYL ESTER**,
    VI, 252
  L-Cysteine, *S*-(1,1-dimethylethyl)-*N*-[*O*-(1,1-dimethylethyl)-L-seryl]-, 1,1-
    dimethylethyl ester [67942-95-4]
BUTYL SULFATE, II, 111
  Sulfuric acid, dibutyl ester [625-22-9]
BUTYL SULFITE, II, 112
  Sulfurous acid, dibutyl ester [626-85-7]
tert-**BUTYL** *p*-**TOLUATE, VI**, 259
  Benzoic acid, 4-methyl-, 1,1-dimethylethyl ester [13756-42-8]
BUTYL *p*-**TOLUENESULFONATE, I**, 145
  Benzenesulfonic acid, 4-methyl-, butyl ester [778-28-9]
S-*tert*-**BUTYL 3a,7a,12a-TRIHYDROXY-5β-CHOLANE-24-THIOATE, VII**, 81
  Cholane-24-thioic acid, 3,7,12-trihydroxy-, *S*-(1,1-dimethylethyl) ester,
    (3α,5β,7α,12α)- [58587-05-6]
tert-**BUTYLUREA, III**, 256
  Urea, (1,1-dimethylethyl)- [1118-12-3]
4-*tert*-**BUTYL-1-VINYLCYCLOHEXENE, VIII**, 97
  Cyclohexene, 4-(1,1-dimethylethyl)-1-ethenyl- [33800-81-6]
2-BUTYN-1-OL, IV, 128
  2-Butyn-1-ol [764-01-2]
3-BUTYN-1-YL TRIFLUOROMETHANESULFONATE, VI, 324
  Methanesulfonic acid, trifluoro-, 3-butynyl ester [32264-79-2]
BUTYRCHLORAL, IV, 130
  Butanal, 2,2,3-trichloro- [76-36-8]

BUTYROIN, II, 114; **VII**, 95
   4-Octanone, 5-hydroxy- [496-77-5]
γ-BUTYROLACTONE-γ-CARBOXYLIC ACID, (S)-(+)-, **VII**, 99
   2-Furancarboxylic acid, tetrahydro-5-oxo-, (S)- [21461-84-7]
3-BUTYROYL-1-METHYLPYRROLE, VII, 102
   1-Butanone, 1-(1-methyl-1H-pyrrol-3-yl)- [62128-46-5]

C

(+)-(1S)-10-CAMPHORSULFONAMIDE, VIII, 104
   Bicyclo[2.2.1]heptane-1-methanesulfonamide, 7,7-dimethyl-2-oxo-, (1S)-
      [60933-63-3]
DL-10-CAMPHORSULFONIC ACID (REYCHLER'S ACID), V, 194
   Bicyclo[2.2.1]heptane-1-methanesulfonic acid, 7,7-dimethyl-2-oxo-, (±)- [5872-08-2]
DL-10-CAMPHORSULFONYL CHLORIDE, V, 196, 878
   Bicyclo[2.2.1]heptane-1-methanesulfonyl chloride, 7,7-dimethyl-2-oxo-, (±)-
      [6994-93-0]
(−)-(CAMPHORSULFONYL)IMINE, VIII, 104, 110
   3H-3a,6-Methano-2,1-benzisothiazole, 4,5,6,7-tetrahydro-8,8-dimethyl- 2,2-dioxide,
      (3aS)- [60886-80-8]
(+)-(2R,8aS)-10-(CAMPHORYLSULFONYL)OXAZIRIDINE, VIII, 104
   4H-4a,7-Methanooxazirino-[3,2-i][2,1]benzisothiazole, tetrahydro-9,9-dimethyl-,
      3,3-dioxide, [4aS-(4aα,7α,8aR*)]- [104322-63-6]
(−)-D-2,10-CAMPHORSULTAM, VIII, 110
   3H-3a,6-Methano-2,1-benzisothiazole, 8,8-dimethyl-, 2,2-dioxide, 3aS-
      (3aα,6α,7aβ)]- [94594-90-8]
D-CAMPHOR 2,4,6-TRIISOPROPYLBENZENESULFONYLHYDRAZONE,
      VII, 77
   Benzenesulfonic acid, 2,4,6-tris(1-methylethyl)- (1,7,7-trimethylbicyclo[2.2.1]hept-
      2-ylidene)hydrazide, 1R- [87068-34-6]
CAPROIC ANHYDRIDE, III, 164
   Hexanoic acid, anhydride [2051-49-2]
(S)-1-CARBAMOYL-2-METHOXYMETHYLPYRROLIDINE, VIII, 26
   1-Pyrrolidinecarboxamide, 2-(methoxymethyl)-, (S)- [95312-82-6]
2-CARBETHOXYCYCLOOCTANONE, V, 198
   Cyclooctanecarboxylic acid, 2-oxo-, ethyl ester [4017-56-5]
β-CARBETHOXY-γ,γ-DIPHENYLVINYLACETIC ACID, IV, 132
   Butanedioic acid, (diphenylmethylene)-, ethyl ester [5438-22-2]
1-CARBETHOXYMETHYL-4-CARBETHOXYPIPERIDINE, V, 989
   1-Piperidineacetic acid, 4-carboxy-, diethyl ester [1838-39-7]
CARBOBENZOXYGLYCINE, III, 168
   Glycine, N-(phenylmethoxy)carbonyl- [1138-80-3]
N-CARBOBENZYLOXY-L-ASPARAGINYL-L-LEUCINE METHYL ESTER,
      VI, 263
   L-Leucine, N-[N²-[(phenylmethoxy)carbonyl]-L-asparaginyl]-, methyl ester
      [14317-83-0]

N-CARBOBENZYLOXY-3-HYDROXY-L-PROLYLGLYCYLGLYCINE ETHYL ESTER, VI, 263
   Glycine, N-[N-[3-hydroxy-1-[(phenylmethoxy)carbonyl]-L-prolyl]glycyl]-, ethyl ester [57621-06-4]
3-CARBOETHOXYCOUMARIN, III, 165
   2H-1-Benzopyran-3-carboxylic acid, 2-oxo-, ethyl ester [1846-76-0]
2-CARBOETHOXYCYCLOPENTANONE, II, 116
   Cyclopentanecarboxylic acid, 2-oxo-, ethyl ester [611-10-9]
β-CARBOETHOXY-γ,γ-DIPHENYLVINYLACETIC ACID, IV, 132
   Butanedioic acid, (diphenylmethylene)- ethyl ester [5438-22-2]
1-CARBOMETHOXY-1-METHYLETHYL 3-OXOBUTANOATE, VIII, 71
   Butanoic acid, 3-oxo-, 2-methoxy-1,1-dimethyl-2-oxoethyl ester [110451-07-5]
β-CARBOMETHOXYPROPIONYL CHLORIDE, III, 169
   Butanoic acid, 4-chloro-4-oxo-, methyl ester [1490-25-1]
N-CARBOMETHOXYPYRROLIDINE, VII, 307
   1-Pyrrolidinecarboxylic acid, methyl ester [56475-80-0]
2-CARBOMETHOXY-3-VINYLCYCLOPENTANONE, VIII, 112
   Cyclopentanecarboxylic acid, 2-ethenyl-5-oxo-, methyl ester [75351-19-8]
CARBONYL CYANIDE, VI, 268
   Propanedinitrile, oxo- [1115-12-4]
1,1'-CARBONYLDIIMIDAZOLE, V, 201
   1H-Imidazole, 1,1'-carbonylbis- [530-62-1]
o-CARBOXYCINNAMIC ACID, IV, 136
   2-Propenoic acid, 3-(2-carboxyphenyl)- [612-40-8]
3-CARBOXY-5-DODECEN-2-ONE, (E)-, VIII, 235
   4-Undecenoic acid, 2-acetyl- [133538-61-1]
CARBOXYMETHOXYLAMINE HEMIHYDROCHLORIDE, III, 172
   Acetic acid, (aminooxy)-, hydrochloride (2:1) [2921-14-4]
o-CARBOXYPHENYLACETONITRILE, III, 174
   Benzoic acid, 2-(cyanomethyl)- [6627-91-4]
β-(o-CARBOXYPHENYL) PROPIONIC ACID, IV, 136
   Benzenepropanoic acid, 2-carboxy- [776-79-4]
(E)-(CARBOXYVINYL)TRIMETHYLAMMONIUM BETAINE, VIII, 536
   Ethenaminium, 2-carboxy-N,N,N-trimethyl-, hydroxide, inner salt, (E)- [54299-83-1]
CASEIN, II, 120
   Casein [9000-71-9]
CATECHOL, I, 149; *Hazard*
   1,2-Benzenediol [120-80-9]
CELLOBIOSE, II, 122
   D-Glucose, 4-O-α-D-glucopyranosyl- [528-50-7]
α-CELLOBIOSE OCTAACETATE, II, 124
   α-D-Glucopyranose, 4-O-(2,3,4,6-tetra-O-acetyl-β-D-glucopyranosyl)-, tetraacetate [5346-90-7]
CETYLMALONIC ESTER, IV, 141
   Propanedioic acid, hexadecyl-, diethyl ester [41433-81-2]

CHELIDONIC ACID, II, 126
   4H-Pyran-2,6-dicarboxylic acid, 4-oxo- [99-32-1]
CHLOROACETAMIDE, I, 153
   Acetamide, 2-chloro- [79-07-2]
CHLOROACETONITRILE, IV, 144
   Acetonitrile, chloro- [107-14-2]
***p*-CHLOROACETYLACETANILIDE, III,** 183
   Acetamide, *N*-[4-(chloroacetyl)phenyl]- [140-49-8]
α-CHLOROACETYL ISOCYANATE, V, 204
   Acetyl isocyanate, chloro- [4461-30-7]
9-CHLOROACRIDINE, III, 53
   Acridine, 9-chloro- [1207-69-8]
3-(*o*-CHLOROANILINO)PROPIONITRILE, IV, 146
   Propanenitrile, 3-[(2-chlorophenyl)amino]- [94-89-3]
9-CHLOROANTHRACENE, V, 206
   Anthracene, 9-chloro- [716-53-0]
α-CHLOROANTHRAQUINONE, II, 128
   9,10-Anthracenedione, 1-chloro- [82-44-0]
***m*-CHLOROBENZALDEHYDE, II,** 130
   Benzaldehyde, 3-chloro- [587-04-2]
***p*-CHLOROBENZALDEHYDE, II,** 133
   Benzaldehyde, 4-chloro- [104-88-1]
***o*-CHLOROBENZOIC ACID, II,** 135
   Benzoic acid, 2-chloro- [118-91-2]
***p*-CHLOROBENZOIC ANHYDRIDE, III,** 29
   Benzoic acid, 4-chloro-, anhydride [790-41-0]
CHLORO-*p*-BENZOQUINONE, IV, 148
   2,5-Cyclohexadiene-1,4-dione, 2-chloro- [695-99-8]
***o*-CHLOROBENZOYL CHLORIDE, I,** 155
   Benzoyl chloride, 2-chloro- [609-65-4]
(2-CHLOROBENZOYL)FERROCENE, VI, 625
   Ferrocene, (2-chlorobenzoyl)- [49547-67-3]
***N*-CHLOROBETAINYL CHLORIDE, IV,** 154
   Ethanaminium, 2-chloro-*N,N,N*-trimethyl-2-oxo-, chloride [53684-57-4]
3-CHLOROBICYCLO[3.2.1]OCT-2-ENE, VI, 142
   Bicyclo[3.2.1]oct-2-ene, 3-chloro- [35242-17-2]
***o*-CHLOROBROMOBENZENE, III,** 185
   Benzene, 1-bromo-2-chloro- [694-80-4]
3-CHLORO-2-BUTEN-1-OL, IV, 128
   2-Buten-1-ol, 3-chloro- [40605-42-3]
4-CHLOROBUTYL BENZOATE, III, 187
   1-Butanol, 4-chloro-, benzoate [946-02-1]
γ-CHLOROBUTYRONITRILE, I, 156
   Butanenitrile, 4-chloro- [628-20-6]
α-CHLOROCROTONALDEHYDE, IV, 131
   2-Butenal, 2-chloro- [53175-28-3]

7-CHLORO-7-CYANOBICYCLO[4.2.0]OCTAN-8-ONE, VIII, 116
Bicyclo[4.2.0]octane-7-carbonitrile, 7-chloro-8-oxo-, (1α,6α,7β)- [89937-15-5]
3-CHLOROCYCLOBUTANECARBOXYLIC ACID, VI, 271
Cyclobutanecarboxylic acid, 3-chloro- [35207-71-7]
2-CHLOROCYCLOHEXANOL, I, 158
Cyclohexanol, 2-chloro- [1561-86-0]
2-CHLOROCYCLOHEXANONE, III, 188
Cyclohexanone, 2-chloro- [822-87-7]
***N*-CHLOROCYCLOHEXYLIDENEIMINE, V**, 208
Cyclohexanimine, *N*-chloro- [6681-70-5]
2-CHLOROCYCLOOCTANONE OXIME, V, 266
Cyclooctanone, 2-chloro- oxime [10499-33-9]
***trans*-2-CHLOROCYCLOPENTANOL, IV**, 157
Cyclopentanol, 2-chloro-, *trans*- [20377-80-4]
3-CHLOROCYCLOPENTENE, IV, 238
Cyclopentene, 3-chloro- [96-40-2]
**2-CHLORO-2-CYCLOPROPYLETHENE-1,1-DICARBOXYLIC ACID,
MONOETHYL ESTER, VIII**, 247
Propanedioic acid, (chlorocyclopropylmethylene)-, monoethyl ester [123844-19-9]
3-CHLORO-2,5-DL-*tert*-BUTYL-1,4-BENZOQUINONE, VI, 210
2,5-Cyclohexadiene-1,4-dione, 3-chloro-2,5-bis(1,1-dimethylethyl)- [33611-70-0]
1-CHLORO-1,2-DICYANOCYCLOBUTANE, VI, 427
Cyclobutanedicarbonitrile, 1-chloro- [3716-98-1]
CHLORODIISOPROPYLPHOSPHINE, V, 211
Phosphinous chloride, bis(1-methylethyl)- [40244-90-4]
1-CHLORO-2,6-DINITROBENZENE, IV, 160
Benzene, 2-chloro-1,3-dinitro- [606-21-3]
1-CHLORO-1,2-DIPHENYL-3-METHYLCYCLOPROPANE, VII, 203
Benzene, 1,1'-(1-chloro-3-methylcyclopropane-1,2-diyl)bis- (1α,2α,3β)-
[80361-87-3]; (1α,2β,3α)- [80408-28-2]
2-CHLOROETHYL BENZOATE, IV, 84
Ethanol, 2-chloro-, benzoate [939-55-9]
β-CHLOROETHYL METHYL SULFIDE, II, 136
Ethane, 1-chloro-2-(methylthio)- [542-81-4]
2-CHLORO-1-FORMYL-1-CYCLOHEXENE, V, 215
1-Cyclohexene-1-carboxaldehyde, 2-chloro- [1680-73-5]
(Z)-4-CHLORO-4-HEXENYL TRIFLUOROACETATE, VI, 273
Acetic acid, trifluoro-, (Z)-4-chloro-4-hexenyl ester [28077-77-2]
6-CHLORO-2-HEXYNE, VI, 273
2-Hexyne, 6-chloro- [28077-73-8]
α-CHLOROHYDRINDENE, II, 336
1*H*-Indene, 1-chloro-2,3-dihydro- [35275-62-8]
***o*-CHLOROHYDROCINNAMONITRILE, V**, 264
Benzenepropanenitrile, 2-chloro- [7315-17-5]
7-CHLORO-4-HYDROXY-3-QUINOLINECARBOXYLIC ACID, III, 273
3-Quinolinecarboxylic acid, 7-chloro-4-hydroxy- [86-47-5]

3-CHLOROINDAZOLE, III, 476
  1*H*-Indazole, 3-chloro- [29110-74-5]
α²-CHLOROISODURENE, III, 557
  Benzene, 2-(chloromethyl)-1,3,5-trimethyl- [1585-16-6]
CHLOROISONITROSOACETOPHENONE, III, 191
  Benzeneethanimidoyl chloride, *N*-hydroxy-α-oxo- [4937-87-5]
2-CHLOROLEPIDINE, III, 194
  Quinoline, 2-chloro-4-methyl- [634-47-9]
p-**CHLOROMERCURIBENZOIC ACID, I,** 159; *Potential poison*
  Mercurate (1–), (4-carboxylatophenyl)chloro-, hydrogen [90-03-9]
o-**CHLOROMERCURIPHENOL, I,** 161; *Potential poison*
  Mercury, chloro(2-hydroxyphenyl)- [59-85-8]
2-CHLORO-2-METHYLBUTANE, I, 144; **VII,** 425
  Butane, 2-chloro-2-methyl- [594-36-5]
2-CHLORO-2-METHYLCYCLOHEXANONE, IV, 162
  Cyclohexanone, 2-chloro-2-methyl- [10409-46-8]
6-CHLOROMETHYL-2,2-DIMETHYL-1,3-DIOXEN-4-ONE, VIII, 192
  4*H*-1,3-Dioxin-4-one, 6-(chloromethyl)-2,2-dimethyl- [81956-31-2]
4-CHLOROMETHYL-3-ETHYL-5-METHYLISOXAZOLE, VI, 781
  Isoxazole, 4-chloromethyl-3-ethyl-5-methyl- [40500-39-8]
1-CHLOROMETHYLNAPHTHALENE, III, 195
  Naphthalene, 1-(chloromethyl)- [86-52-2]
CHLOROMETHYLPHOSPHONOTHIOIC DICHLORIDE, V, 218
  Phosphonothioic dichloride, (chloromethyl)- [1983-27-3]
3-CHLORO-2-METHYLTETRAHYDROPYRAN, VI, 675
  2*H*-Pyran, 3-chlorotetrahydro-2-methyl-, *cis*- [53107-04-3] and *trans*- [53107-05-4]
2-CHLOROMETHYLTHIOPHENE, III, 197
  Thiophene, 2-(chloromethyl)- [765-50-4]
N-**CHLOROMORPHOLINE, VIII,** 167
  Morpholine, 4-chloro- [23328-69-0]
2-CHLORONICOTINONITRILE, IV, 166
  3-Pyridinecarbonitrile, 2-chloro- [6602-54-6]
m-**CHLORONITROBENZENE, I,** 162
  Benzene, 1-chloro-3-nitro- [121-73-3]
1-CHLORONORBORNANE, VI, 845
  Bicyclo[2.2.1]heptane, 1-chloro- [765-67-3]
5-CHLORO-2-PENTANONE, IV, 597
  2-Pentanone, 5-chloro- [5891-21-4]
m-**CHLOROPERBENZOIC ACID, VI,** 276
  Benzocarboperoxoic acid, 3-chloro- [937-14-4]
p-**CHLOROPHENOXYMETHYL CHLORIDE, V,** 221
  Benzene, 1-chloro-4-(chloromethoxy)- [21151-56-4]
α-**CHLOROPHENYLACETIC ACID, IV,** 169
  Benzeneacetic acid, α-chloro- [4755-72-0]
4-(4-CHLOROPHENYL)BUTAN-2-ONE, VII, 105
  2-Butanone, 4-(4-chlorophenyl)- [3506-75-0]

o-**CHLOROPHENYLCYANAMIDE, IV**, 172
Cyanamide, (2-chlorophenyl)- [45765-25-1]
CHLOROPHENYLDIAZIRINE (in solution), VI, 276
3*H*-Diazirine, 3-chloro-3-phenyl- [4460-46-2]
p-**CHLOROPHENYL ISOTHIOCYANATE, I**, 165; **V**, 223
Benzene, 1-chloro-4-isothiocyanato- [2131-55-7]
3-(4-CHLOROPHENYL)-5-(4-METHOXYPHENYL)ISOXAZOLE, VI, 278
Isoxazole, 3-(4-chlorophenyl)-5-(4-methoxyphenyl)- [24097-19-6]
m-**CHLOROPHENYLMETHYLCARBINOL, III**, 200
Benzenemethanol, 3-chloro-α-methyl [6939-95-3]
α-**(4-CHLOROPHENYL)-γ-PHENYLACETOACETONITRILE, IV**, 174
Benzenebutanenitrile, α-(4-chlorophenyl)-β-oxo- [35741-47-0]
1-(*p*-CHLOROPHENYL)-3-PHENYL-2-PROPANONE, IV, 176
2-Propanone, 1-(4-chlorophenyl)-3-phenyl- [35730-03-1]
p-**CHLOROPHENYL SALICYLATE, IV**, 178
Benzoic acid, 2-hydroxy-, 4-chlorophenyl ester [2944-58-3]
o-**CHLOROPHENYLTHIOUREA, IV**, 180
Thiourea, (2-chlorophenyl)- [5344-82-1]
N-**CHLOROPIPERIDINE, VI**, 968
Piperidine, 1-chloro- [2156-71-0]
(*S*)-2-CHLOROPROPANOIC ACID, VIII, 119, 434
Propanoic acid, 2-chloro-, (*S*)- [29617-66-1]
(*S*)-2-CHLOROPROPAN-1-OL, VIII, 434
1-Propanol, 2-chloro-, (*S*)-(+)- [19210-21-0]
β-**CHLOROPROPIONALDEHYDE ACETAL, II**, 137
Propane, 3-chloro-1,1-diethoxy- [35573-93-4]
β-**CHLOROPROPIONIC ACID, I**, 166
Propanoic acid, 3-chloro- [107-94-8]
γ-**CHLOROPROPYL ACETATE, III**, 203
1-Propanol, 3-chloro-, acetate [628-09-1]
3-CHLOROPROPYLDIPHENYLSULFONIUM TETRAFLUOROBORATE, VI, 364
Sulfonium, 3-chloropropyldiphenyl-, tetrafluoroborate (1−) [33462-80-5]
2-CHLOROPYRIMIDINE, IV, 182
Pyrimidine, 2-chloro- [1722-12-9]
CHLOROPYRUVIC ACID, V, 636
Propanoic acid, 3-chloro-2-oxo- [3681-17-2]
m-**CHLOROSTYRENE, III**, 204
Benzene, 1-chloro-3-ethenyl- [2039-85-2]
3-CHLOROTHIETANE 1,1-DIOXIDE, VII, 491
Thietane, 3-chloro-, 1,1-dioxide [15953-83-0]
3-CHLOROTHIETE 1,1-DIOXIDE, VII, 491
2*H*-Thiete, 3-chloro-, 1,1-dioxide [90344-86-8]
2-CHLOROTHIIRANE 1,1-DIOXIDE, V, 231
Thiirane, chloro-, 1,1-dioxide [10038-13-8]
o-**CHLOROTOLUENE, I**, 170; *Hazard*
Benzene, 1-chloro-2-methyl- [95-49-8]

p-CHLOROTOLUENE, I, 170; *Hazard*
  Benzene, 1-chloro-4-methyl- [106-43-4]
1-CHLORO-1-(TRICHLOROETHENYL)CYCLOPROPANE, VIII, 124, 373
  Cyclopropane, 1-chloro-1-(trichloroethenyl)- [82979-27-9]
1-CHLORO-1,4,4-TRIFLUOROBUTADIENE, V, 235
  1,3-Butadiene, 1-chloro-1,4,4-trifluoro- [764-14-7]
1-CHLORO-2,3,3-TRIFLUOROCYCLOBUTENE, V, 394
  Cyclobutene, 1-chloro-2,3,3-trifluoro- [694-62-2]
2-CHLORO-1,1,2-TRIFLUOROETHYL ETHYL ETHER, IV, 184
  Ethane, 2-chloro-1-ethoxy-1,1,2-trifluoro- [310-71-4]
3-CHLORO-2,2,3-TRIFLUOROPROPIONIC ACID, V, 239
  Propanoic acid, 3-chloro-2,2,3-trifluoro- [425-97-8]
1-CHLORO-*N,N*,2-TRIMETHYLPROPENYLAMINE, VI, 282; VIII, 441
  1-Propen-1-amine, 1-chloro-*N,N*,2-trimethyl- [26189-59-3]
1-CHLORO-*N,N*,2-TRIMETHYLPROPYLIDENIMINIUM CHLORIDE,
    VI, 282
  Methanaminium, *N*-(1-chloro-2-methylpropylidene)-*N*-methyl- chloride
    [52851-35-1]
β-CHLOROVINYL ISOAMYL KETONE, IV, 186
  1-Hepten-3-one, 1-chloro-6-methyl- [18378-90-0]
CHOLAN-24-AL, V, 242
  Cholan-24-al [26606-02-0]
CHOLESTA-3,5-DIENE, VIII, 126
  Cholesta-3,5-diene [747-90-0]
CHOLESTA-3,5-DIEN-3-YL TRIFLUOROMETHANESULFONATE, VIII, 126
  Cholesta-3,5-dien-3-ol, trifluoromethanesulfonate [95667-40-6]
CHOLESTANE, VI, 289
  Cholestane, (5α)- [481-21-0]
CHOLESTANONE, II, 139
  Cholestan-3-one, (5α)- [566-88-1]
CHOLESTANYL METHYL ETHER, V, 245
  Cholestane, 3-methoxy-, (3β,5α)- [1981-90-4]
Δ⁴-CHOLESTEN-3,6-DIONE, IV, 189
  Cholest-4-ene-3,6-dione [984-84-9]
5β-CHOLEST-3-ENE, VI, 293
  Cholest-3-ene, 5β- [13901-20-7]
5β-CHOLEST-3-ENE-5-ACETALDEHYDE, VI, 298
  Cholest-3-ene-5-acetaldehyde, 5β- [56101-55-4]
Δ⁴-CHOLESTEN-3-ONE, III, 207; IV, 192; 195
  Cholest-4-en-3-one [601-57-0]
Δ⁵-CHOLESTEN-3-ONE, IV, 195
  Cholest-5-en-3-one [601-54-7]
CHOLEST-4-EN-3-ONE *p*-TOLUENESULFONYLHYDRAZONE, VI, 293
  Cholest-4-en-3-one, *p*-toluenesulfonylhydrazone [21301-41-7]
3β-CHOLEST-4-ENYL VINYL ETHER, VI, 298
  Cholest-4-ene, 3-(ethenyloxy)-, (3β)- [56101-54-3]

CHOLESTEROL, IV, 195
   Cholest-5-en-3-ol, (3β)- [57-88-5]
CHOLESTEROL DIBROMIDE, IV, 195
   5α-Cholestan-3β-ol, 5,6β-dibromo- [1857-80-3]
CINNAMONITRILE, VI, 304; **VII**, 108
   2-Propenenitrile, 3-phenyl- [4360-47-8]
CINNAMYL BROMIDE, V, 249
   Benzene, (3-bromo-1-propenyl)- [4392-24-9]
CITRACONIC ACID, II, 140
   2-Butenedioic acid, 2-methyl-, (Z)- [498-23-7]
CITRACONIC ANHYDRIDE, II, 140
   2,5-Furandione, 3-methyl- [616-02-4]
CITRONELLAL, (R)-(+)-, VIII, 183
   6-Octenal, 3,7-dimethyl, (R)-(+)- [2385-77-5]
COUMALIC ACID, IV, 201
   2H-Pyran-5-carboxylic acid, 2-oxo- [500-05-0]
COUMARILIC ACID, III, 209
   2-Benzofurancarboxylic acid [496-41-3]
COUMARIN DIBROMIDE, III, 209
   2H-1-Benzopyran-2-one, 3,4-dibromo-3,4-dihydro- [55077-11-7]
COUMARONE, V, 251
   Benzofuran [271-89-6]
COUPLING of o-TOLUIDINE and CHICAGO ACID, II, 145
   1,3-Naphthalenedisulfonic acid, 6,6'-[(3,3'-dimethyl[1,1'-biphenyl]-
      4,4'diyl)bis(azo)]bis[4-amino-5-hydroxy-, tetrasodium salt [314-13-6]
CREATININE, I, 172
   4H-Imidazol-4-one, 2-amino-1,5-dihydro-1-methyl- [60-27-5]
CREOSOL, IV, 203
   Phenol, 2-methoxy-4-methyl- [93-51-6]
p-CRESOL, I, 175
   Phenol, 4-methyl- [106-44-5]
γ-CROTONOLACTONE, V, 255; **VIII**, 397
   2(5H)-Furanone [497-23-4]
CROTYL DIAZOACETATE, V, 258
   Acetic acid, diazo-, 2-butenyl ester, (E)- [14746-03-3]
18-CROWN-6, VI, 301
   1,4,7,10,13,16-Hexaoxacyclooctadecane [17455-13-9]
CUPFERRON, I, 177
   Benzenamine, N-hydroxy-N-nitroso-, ammonium salt [135-20-6]
CYANOACETAMIDE, I, 179
   Acetamide, 2-cyano- [107-91-5]
1-CYANOBENZOCYCLOBUTENE, V, 263
   Bicyclo[4.2.0]octa-1,3,5-triene-7-carbonitrile [6809-91-2]
1-CYANO-2-BENZOYL-1,2-DIHYDROISOQUINOLINE, IV, 641 **or 2-BENZOYL-**
      1,2-DIHYDROISOQUINALDONITRILE, VI, 115
   1-Isoquinolinecarbonitrile, 2-benzoyl-1,2-dihydro- [844-25-7]

2-(1-CYANOCYCLOHEXYL)DIAZENECARBOXYLIC ACID, METHYL ESTER,
  VI, 334
  Diazenecarboxylic acid, (1-cyanocyclohexyl)-, methyl ester [33670-04-1]
N-2-CYANOETHYLANILINE, IV, 205
  Propanenitrile, 3-(phenylamino)- [1075-76-9]
CYANOGEN BROMIDE, II, 150
  Cyanogen bromide [(CN)Br] [506-68-3]
CYANOGEN IODIDE, IV, 207
  Iodine cyanide, [I(CN)] [506-78-5]
7-CYANOHEPTANAL, V, 266
  Octanenitrile, 8-oxo- [13050-09-4]
1-CYANO-6-METHOXY-3,4-DIHYDRONAPHTHALENE, VI, 307
  Naphthalenecarbonitrile, 3,4-dihydro-6-methoxy- [6398-50-1]
2-CYANO-6-METHYLPYRIDINE, V, 269
  2-Pyridinecarbonitrile, 6-methyl- [1620-75-3]
3-CYANO-6-METHYL-2(1)-PYRIDONE, IV, 210
  3-Pyridinecarbonitrile, 1,2-dihydro-6-methyl-2-oxo- [4241-27-4]
ω-CYANOPELARGONIC ACID, III, 768
  Nonanoic acid, 9-cyano- [5810-19-5]
9-CYANOPHENANTHRENE, III, 212
  9-Phenanthrenecarbonitrile [2510-55-6]
α-CYANO-β-PHENYLACRYLIC ACID, I, 181
  2-Propenoic acid, 2-cyano-3-phenyl- [1011-92-3]
1-CYANO-3-PHENYLUREA, IV, 213
  Urea, N-cyano-N'-phenyl- [41834-91-7]
CYCLOBUTADIENEIRON TRICARBONYL, VI, 310
  Iron, tricarbonyl(η^4-1,3-cyclobutadiene)- [12078-17-0]
CYCLOBUTANECARBOXALDEHYDE, VI, 312
  Cyclobutanecarboxaldehyde [2987-17-9]
CYCLOBUTANECARBOXAMIDE, VIII, 132
  Cyclobutanecarboxamide [1503-98-6]
CYCLOBUTANECARBOXYLIC ACID, III, 213
  Cyclobutanecarboxylic acid [3721-95-7]
1,1-CYCLOBUTANEDICARBOXYLIC ACID, III, 213
  1,1-Cyclobutanedicarboxylic acid [5445-51-2]
1,2-CYCLOBUTANEDIOL, VII, 129
  1,2-Cyclobutanediol, cis- [35358-33-9], and trans- [35358-34-0]
1,2-CYCLOBUTANEDIONE, VII, 112
  1,2-Cyclobutanedione [33689-28-0]
CYCLOBUTANOL, VII, 114, 117
  Cyclobutanol [2919-23-5]
CYCLOBUTANONE, VI, 316, 320, 324; **VII**, 114
  Cyclobutanone [1191-95-3]
CYCLOBUTENE, VII, 117
  Cyclobutene [822-35-5]

CYCLOBUTYLAMINE, V, 273
  Cyclobutanamine [2516-34-9]
CYCLOBUTYLAMINE HYDROCHLORIDE, VIII, 132
  Cyclobutanamine hydrochloride [6291-01-6]
2-CYCLOBUTYL-*cis*-4-*trans*-5-DIMETHYL-1,3-DIOXOLANE, VI, 312
  1,3-Dioxolane, 2-cyclobutyl-4,5-dimethyl-, (2α,4α,5β)- [36416-29-2]
CYCLOBUTYL TOSYLATE, VII, 117
  Cyclobutanol, 4-methylbenzenesulfonate [10437-85-1]
1,2-CYCLODECANEDIOL, IV, 216
  1,2-Cyclodecanediol [21014-77-7]
CYCLODECANONE, IV, 218; **V,** 277
  Cyclodecanone [1502-06-3]
***cis*-CYCLODODECENE, V,** 281
  Cyclododecene, (Z)- [1129-89-1]
CYCLODODECENE, (E)-, VII, 121
  Cyclododecene, (E)- [1486-75-5]
CYCLODODECYL MERCAPTAN, VII, 124
  Cyclododecanethiol [7447-11-2]
CYCLOHEPTANONE, IV, 221
  Cycloheptanone [502-42-1]
2-CYCLOHEPTEN-1-ONE, VI, 327
  2-Cyclohepten-1-one [1121-66-0]
2-CYCLOHEPTEN-1-YL ACETATE, VIII, 137
  2-Cyclohepten-1-ol, acetate [826-13-1]
1,3-CYCLOHEXADIENE, V, 285
  1,3-Cyclohexadiene [592-57-4]
CYCLOHEXANECARBONITRILE, VI, 305, 334
  Cyclohexanecarbonitrile [766-05-2]
CYCLOHEXANECARBOXALDEHYDE, VI, 338
  Cyclohexanecarboxaldehyde [2043-61-0]
CYCLOHEXANECARBOXYLIC ACID CHLORIDE, VII, 81
  Cyclohexanecarbonyl chloride [2719-27-9]
***cis*-1,2-CYCLOHEXANEDIMETHANOL, VI,** 482
  cis-1,2-Cyclohexanedimethanol [5059-76-7]
**1,2-CYCLOHEXANEDIMETHANOL DIMETHANESULFONATE, *cis*-,
    VI,** 482
  1,2-Cyclohexanedimethanol, dimethanesulfonate, *cis*- [66347-68-0]
***cis*-1,2-CYCLOHEXANEDIOL, VI,** 342, 348
  cis-1,2-Cyclohexanedlol [1792-81-0]
***trans*-1,2-CYCLOHEXANEDIOL, III,** 217; **VI,** 348
  1,2-Cyclohexanediol, *trans*- [1460-57-7]
1,2-CYCLOHEXANEDIONE, IV, 229
  1,2-Cyclohexanedione [765-87-7]
1,4-CYCLOHEXANEDIONE, V, 288
  1,4-Cyclohexanedione [637-88-7]

1,2-CYCLOHEXANEDIONE DIOXIME, IV, 229
   1,2-Cyclohexanedione, dioxime [492-99-9]
CYCLOHEXANOL, VI, 353; *Hazard*
   Cyclohexanol [108-93-0]
CYCLOHEXANONE DIALLYL ACETAL, V, 292
   Cyclohexane, 1,1-bis(2-propenyloxy)- [53608-84-7]
CYCLOHEXANONE DIETHYL KETAL, VIII, 579
   Cyclohexane, 1,1-diethoxy- [1670-47-9]
CYCLOHEXENE, I, 183; **II**, 152; *Hazard*
   Cyclohexene [110-83-8]
CYCLOHEXENE OXIDE, I, 185
   7-Oxabicyclo[4.1.0]heptane [286-20-4]
CYCLOHEXENE SULFIDE, IV, 232
   7-Thiabicyclo[4.1.0]heptane [286-28-2]
2-CYCLOHEXENONE, V, 294
   2-Cyclohexen-1-one [930-68-7]
1-CYCLOHEXENYLACETONITRILE, IV, 234
   1-Cyclohexene-1-acetonitrile [6975-71-9]
CYCLOHEXYLBENZENE, II, 151
   Benzene, cyclohexyl- [827-52-1]
3-CYCLOHEXYL-2-BROMOPROPENE, I, 186
   Cyclohexane, (2-bromo-2-propenyl)- [53608-85-8]
CYCLOHEXYLCARBINOL, I, 188
   Cyclohexanemethanol [100-49-2]
CYCLOHEXYLIDENEACETALDEHYDE, VI, 358
   Acetaldehyde, cyclohexylidene-[1713-63-9]
CYCLOHEXYLIDENEACETONITRILE, VII, 108
   Acetonitrile, cyclohexylidene-[4435-18-1]
CYCLOHEXYLIDENECYANOACETIC ACID, IV, 234
   Acetic acid, cyanocyclohexylidene- [37107-50-9]
CYCLOHEXYLIDENECYCLOHEXANE, V, 297
   Cyclohexane, cyclohexylidene- [4233-18-5]
CYCLOHEXYL ISOCYANIDE, V, 300
   Cyclohexane, isocyano- [931-53-3]
CYCLOHEXYL METHYL ETHER, VI, 355
   Cyclohexane, methoxy- [931-56-6]
CYCLOHEXYL METHYL KETONE, V, 775
   Ethanone, 1-cyclohexyl-[823-76-7]
2-CYCLOHEXYLOXYETHANOL, V, 303
   Ethanol, 2-(cyclohexyloxy)- [1817-88-5]
α-CYCLOHEXYLPHENYLACETONITRILE, III, 219
   Benzeneacetonitrile, α-cyclohexyl- [3893-23-0]
3-CYCLOHEXYLPROPYNE, I, 191
   Cyclohexane, 2-propynyl- [17715-00-3]
CYCLOHEXYLUREA, V, 801
   Urea, cyclohexyl- [698-90-8]

1,2-CYCLONONADIENE, V, 306
  1,2-Cyclononadiene [1123-11-1]
CYCLOOCTANONE, V, 310
  Cyclooctanone [502-49-8]
***trans*-CYCLOOCTENE, V**, 315
  Cyclooctene, (*E*)- [931-89-5]
CYCLOOCTENE OXIDE, *cis*-, VII, 126
  9-Oxabicyclo[6.1.0]nonane, *cis*- [4925-71-7]
1-(CYCLOOCTYLIDENEMETHYL)PIPERIDINE, VII, 473
  Piperidine, 1-(cyclooctylidenemethyl)- [65226-88-2]
CYCLOPENTANECARBOXALDEHYDE, V, 320
  Cyclopentanecarboxaldehyde [872-53-7]
CYCLOPENTANONE, I, 192
  Cyclopentanone [120-92-3]
2-CYCLOPENTENE-1,4-DIONE, V, 324
  4-Cyclopentene-1,3-dione [930-60-9]
2-CYCLOPENTENONE, V, 326
  2-Cyclopenten-1-one [930-30-3]
CYCLOPROPANECARBOXALDEHYDE, VII, 129
  Cyclopropanecarboxaldehyde [1489-69-6]
CYCLOPROPANECARBOXYLIC ACID, III, 221
  Cyclopropanecarboxylic acid [1759-53-1]
CYCLOPROPANE-1,1-DICARBOXYLIC ACID, VII, 411
  1,1-Cyclopropanedicarboxylic acid [598-10-7]
CYCLOPROPANE-1,2-DICARBOXYLIC ACID, (+)-(1*S*,2*S*)-, VIII, 141
  1,2-Cyclopropanedicarboxylic acid, (1*S-trans*)- [14590-54-6]
CYCLOPROPANONE ETHYL HEMIACETAL, VII, 131
  Cyclopropanol, 1-ethoxy- [13837-45-1]
CYCLOPROPENONE, VI, 361
  2-Cyclopropen-1-one [2961-80-0]
CYCLOPROPENONE 1,3-PROPANEDIOL KETAL, VIII, 173
  4,8-Dioxaspiro[2.5]oct-1-ene [60935-21-9]
CYCLOPROPYLBENZENE, V, 328; **PHENYLCYCLOPROPANE, V**, 929
  Benzene, cyclopropyl- [873-49-4]
CYCLOPROPYL CYANIDE: III, 223
  Cyclopropanecarbonitrile [5500-21-0]
CYCLOPROPYLDIPHENYLSULFONIUM TETRAFLUOROBORATE, VI, 364
  Sulfonium, cyclopropyldiphenyl-, tetrafluoroborate (1–) [33462-81-6]
CYCLOUNDECANECARBOXYLIC ACID, VII, 135
  Cycloundecanecarboxylic acid [4373-07-3]
CYCLOUNDECANONE, VI, 368
  Cycloundecanone [878-13-7]
***m*-CYMENE, V**, 332; *Hazard*
  Benzene, 1-methyl-3-(1-methylethyl)- [535-77-3]
CYSTEIC ACID MONOHYDRATE, III, 226
  L-Alanine, 3-sulfo-, hydrate [53643-49-5]

L-CYSTINE, I, 194
    L-Cystine [56-89-3]

D

1-DECALOL, VI, 371
    1-Naphthalenol, decahydro- [529-32-8]
DECAMETHYLENE BROMIDE, III, 227
    Decane, 1,10-dibromo- [4101-68-2]
DECAMETHYLENEDIAMINE, III, 229
    1,10-Decanediamine [646-25-3]
DECAMETHYLENE GLYCOL, II, 154
    1,10-Decanediol [112-47-0]
DECANAL, VI, 373
    Decanal [112-31-2]
DECANE, VI, 376
    Decane [124-18-5]
2-DECANONE, VII, 137
    2-Decanone [693-54-9]
(*E*)-1-DECENYLDIISOBUTYLALANE, VIII, 295
    Aluminum, 1-decenylbis(2-methylpropyl)-, (*E*)- [107441-86-1]
9-DECYN-1-OL, VIII, 146
    9-Decyn-1-ol [17643-36-6]
DEHYDROACETIC ACID, III, 231
    2*H*-Pyran-2,4(3*H*)-dione, 3-acetyl-6-methyl- [520-45-6]
DEOXYANISOIN, V, 339
    Ethanone, 1,2-bis(4-methoxyphenyl)- [120-44-5]
3-DEOXY-1,2:5,6-DI-*O*-ISOPROPYLIDENE-α-D-ribo-HEXOFURANOSE, VII, 139
    α-D-ribo-Hexofuranose, 3-deoxy-1,2:5,6-bis-*O*-(1-methylethylidene)- [4613-62-1]
**1-DEOXY-2,3,4,6-TETRA-*O*-ACETYL-1-(2-CYANOETHYL)-α-D-
        GLUCOPYRANOSE, VIII,** 148
    D-Glycero-D-ido-nonononitrile, 4,8-anhydro-2,3-dideoxy-, 5,6,7,9-tetraacetate
        [86563-27-1]
DESOXYBENZOIN, II, 156
    Ethanone, 1,2-diphenyl- [451-40-1]
nor-DESOXYCHOLIC ACID, III, 234
    24-Norcholan-23-oic acid, 3,12-dihydroxy-, (3α,5β,12α)- [53608-86-9]
DESYL CHLORIDE, II, 159
    Ethanone, 2-chloro-1,2-diphenyl- [447-31-4]
DIACETONAMINE HYDROGEN OXALATE, I, 196
    2-Pentanone, 4-amino-4-methyl-, ethanedioate (2:1) [53608-87-0]
DIACETONE ALCOHOL, I, 199; *Hazard*
    2-Pentanone, 4-hydroxy-4-methyl- [123-42-2]
***trans*-7,8-DIACETOXYBICYCLO[4.2.0]OCTA-2,4-DIENE, VI,** 196
    Bicyclo[4.2.0]octa-2,4-diene-7,8-diol, diacetate, (1α,6α,7α,8β)- [42301-50-8]
2,6-DIACETOXYBICYCLO[3.3.0]OCTANE, VIII, 43
    1,4-Pentalenediol, octahydro-, diacetate [17572-85-9]

3,12-DIACETOXY-nor-CHOLANIC ACID, III, 234
24-Norcholan-23-oic acid, 3,12-bis(acetyloxy)- (3α,5β,12α)- [63714-58-9]
3,12-DIACETOXY-bisnor-CHOLANYLDIPHENYLETHYLENE, III, 237
Chol-23-ene-3,12-diol, 24,24-diphenyl-, diacetate, (3α,5β,12α)- [53608-88-1]
DIACETYL D-TARTARIC ANHYDRIDE, IV, 242
2,5-Furandione, 3,4-bis(acetyloxy)dihydro-, (3*R-trans*)- [19523-83-2]
DIALLYLAMINE, I, 201
2-Propen-1-amine, *N*-2-propenyl- [124-02-7]
DIALLYLCYANAMIDE, I, 203
Cyanamide, di-2-propenyl- [538-08-9]
DIALURIC ACID MONOHYDRATE, IV, 29
2,4,6-(1*H*,3*H*,5*H*)-Pyrimidinetrione, 5-hydroxy- [444-15-5]
DIAMANTANE, VI, 378
3,5,1,7-[1,2,3,4]-Butanetetraylnaphthalene, decahydro- [2292-79-7]
4,4'-DIAMINOAZOBENZENE, V, 341
Benzenamine, 4,4'-azobis- [538-41-0]
DIAMINOBIURET, III, 404
Imidodicarbonic dihydrazide [4375-11-5]
2,3-DIAMINO-5-BROMOPYRIDINE, V, 346
2,3-Pyridinediamine, 5-bromo- [38875-53-5]
2,5-DIAMINO-3,4-DICYANOTHIOPHENE, IV, 243
3,4-Thiophenedicarbonitrile, 2,5-diamino- [17989-89-8]
1,5-DIAMINO-2,4-DINITROBENZENE, V, 1068
1,3-Benzenediamine, 4,6-dinitro- [4987-96-6]
4,4'-DIAMINODIPHENYLSULFONE, III, 239
Benzenamine, 4,4'-sulfonylbis- [80-08-0]
2,4-DIAMINO-6-HYDROXYPYRIMIDINE, IV, 245
4(1*H*)-Pyrimidinone, 2,6-diamino- [56-06-4]
DIAMINOMALEONITRILE(HYDROGEN CYANIDE TETRAMER), V, 344
2-Butenedinitrile, 2,3-diamino, (Z)- [1187-42-4]
1,2-DIAMINO-4-NITROBENZENE, III, 242
1,2-Benzenediamine, 4-nitro- [99-56-9]
2,3-DIAMINOPYRIDINE, V, 346
2,3-Pyridinediamine [452-58-4]
2,4-DIAMINOTOLUENE, II, 160
1,3-Benzenediamine, 4-methyl- [95-80-7]
DIAMINOURACIL HYDROCHLORIDE, IV, 247
2,4(1*H*,3*H*)-Pyrimidinedione, 4,5-diamino-, monohydrochloride [53608-89-2]
4,13-DIAZA-18-CROWN-6, VIII, 152
1,4,10,13-Tetraoxa-7,16-diazacyclooctadecane [23978-55-4]
1,10-DIAZACYCLOOCTADECANE, VI, 382
1,10-Diazacyclooctadecane [296-30-0]
3,6-DIAZIDO-2,5-DI-*tert*-BUTYL-1,4-BENZOQUINONE, VI, 211
2,5-Cyclohexadiene-1,4-dione, 2,5-diazido-3,6-bis(1,1-dimethylethyl)- [29342-21-0]
DIAZOACETOPHENONE, VI, 386
Ethanone, 2-diazo-1-phenyl- [3282-32-4]

1-(DIAZOACETYL)NAPHTHALENE, VI, 613
Ethanone, 2-diazo-1-naphthalenyl- [4372-76-3]
DIAZOAMINOBENZENE, II, 163
1-Triazene, 1,3-diphenyl- [136-35-6]
16-DIAZOANDROST-5-EN-3β-OL-17-ONE, VI, 840
Androst-5-en-17-one, 16-diazo-3-hydroxy-, (3β)- [26003-42-9]
2-DIAZOCYCLOHEXANONE, VI, 389
Cyclohexanone, 2-diazo- [3242-56-6]
1-DIAZO-4-PHENYL-2-BUTANONE, VIII, 196
2-Butanone, 1-diazo-4-phenyl- [10290-42-3]
2-DIAZOPROPANE, VI, 392
Propane, 2-diazo- [2684-60-8]
DIBENZALACETONE, II, 167
1,4-Pentadien-3-one, 1,5-diphenyl- [538-58-9]
DIBENZO-18-CROWN-6 POLYETHER, VI, 395
Dibenzo[b,k][1,4,7,10,13,16]hexaoxacyclooctadecin, 6,7,9,10,17,18,20,21-
octahydro- [14187-32-7]
1,4-DIBENZOYLBUTANE, II, 169
1,6-Hexanedione, 1,6-diphenyl- [3375-38-0]
***trans*-1,2-DIBENZOYLCYCLOPROPANE, VI**, 401
Methanone, 1,2-cyclopropanediylbis[phenyl-], *trans*- [38400-84-9]
DIBENZOYLDIBROMOMETHANE, II, 244
1,3-Propanedione, 2,2-dibromo-1,3-diphenyl- [16619-55-9]
***sym*-DIBENZOYLDIMETHYLHYDRAZINE, II**, 209
Benzoic acid, hydrazide, 2-benzoyl-1,2-dimethyl- [1226-43-3]
***trans*-DIBENZOYLETHYLENE, III**, 248
2-Butene-1,4-dione, 1,4-diphenyl-, (*E*)- [959-28-4]
***sym*-DIBENZOYLHYDRAZINE, II**, 208
Benzoic acid, 2-benzoylhydrazide [787-84-8]
**2-(1,3-DIBENZOYLIMIDAZOLIDIN-2-YL)IMIDAZOLE HYDROCHLORIDE,
VII**, 287
Imidazolidine, 1,3-dibenzoyl-2(1*H*-imidazol-2-yl)-, monohydrochloride [65276-01-9]
**2-(1,3-DIBENZOYLIMIDAZOLIN-2-YL)IMIDAZOLE HYDROCHLORIDE,
VII**, 287
2,2'-Bi-1*H*-imidazole, 1,3-dibenzoyl-2,3-dihydro-, monohydrochloride [65276-00-8]
DIBENZOYLMETHANE, I, 205; **III**, 251
1,3-Propanedione, 1,3-diphenyl- [120-46-7]
***N,N'*-DIBENZYL-4,13-DIAZA-18-CROWN-6, VIII**, 152
1,4,10,13-Tetraoxa-7,16-diazacyclooctadecane, 7,16-bis(phenylmethyl)-
[69703-25-9]
1,4-DI-*O*-BENZYL-2,3-DI-*O*-ISOPROPYLIDENE-L-THREITOL, VIII, 155
1,3-Dioxolane, 2,2-dimethyl-4,5-bis[(phenylmethoxy)methyl]-, (4*R-trans*)-
[91604-40-9]; (4*S-trans*)- [68394-39-8]
1,10-DIBENZYL-4,7-DIOXA-1,10-DIAZADECANE, VIII, 152
Benzenemethanamine, *N,N'*-[1,2-ethanediylbis(oxy-2,1-ethanediyl)]bis-
[66582-26-1]

1,4-DI-*O*-BENZYL-L-THREITOL, VIII, 155
    2,3-Butanediol, 1,4-bis(phenylmethoxy)-, [*S*(*R**,*R**)]- [17401-06-8]
DIBROMOACETONITRILE, IV, 254
    Acetonitrile, dibromo- [3252-43-5]
2,2-DIBROMOACETYL-6-METHOXYNAPHTHALENE, VI, 175
    Ethanone, 2,2-dibromo-1-(6-methoxy-2-naphthalenyl)- [52997-56-5]
2,6-DIBROMOANILINE, III, 262
    Benzenamine, 2,6-dibromo- [608-30-0]
9,10-DIBROMOANTHRACENE, I, 207
    Anthracene, 9,10-dibromo- [523-27-3]
4,4-DIBROMOBIBENZYL, IV, 257
    Benzene, 1,1'-(1,2-ethanediyl)bis[4-bromo- [19829-56-2]
9,9-DIBROMOBICYCLO[6.1.0]NONANE, V, 306
    Bicyclo[6.1.0]nonane, 9,9-dibromo- [1196-95-8]
2,2'-DIBROMO-1,1'-BINAPHTHYL, VIII, 57
    1,1'-Binaphthalene, 2,2'-dibromo- [74866-28-7]
4,4'-DIBROMOBIPHENYL, IV, 256
    1,1'-Biphenyl,4,4'-dibromo- [92-86-4]
1,3-DIBROMO-2-BUTANONE, VI, 711
    2-Butanone, 1,3-dibromo- [815-51-0]
5α,6β-DIBROMOCHOLESTAN-3-ONE, IV, 197
    Cholestan-3-one, 5,6-dibromo-, (5α,6β)- [2515-09-5]
1,2-DIBROMOCYCLOBUTANE, VII, 119
    Cyclobutane, 1,2-dibromo- [89033-70-5]
2,12-DIBROMOCYCLODODECANONE, VI, 368
    Cyclododecanone, 2,12-dibromo- [24459-40-3]
1,2-DIBROMOCYCLOHEXANE, II, 171
    Cyclohexane, 1,2-dibromo- [5401-62-7]
2-(2,2-DIBROMOCYCLOPROPYL)-6,6-DIMETHYLBICYCLO[2.2.1]HEPT-2-
    ENE, VIII, 223
    Bicyclo[3.1.1]hept-2-ene, 2-(2,2-dibromocyclopropyl)-6,6-dimethyl-,
        [1α, 2(*S**), 5α]- [108404-80-4]; [1*R*-[1α, 2(*R**), 5α]]- [108450-07-3]
α,α'-DIBROMODIBENZYL KETONE, V, 514
    2-Propanone, 1,3-dibromo-1,3-diphenyl- [958-79-2]
α,α'-DIBROMODIBENZYL SULFONE, VI, 403
    Benzene, 1,1'-[sulfonylbis(bromomethylene)]bis-, (*R**,*R**)- and (*R**,*S**)-
        [21966-50-7]
7,7-DIBROMO-1,6-DIMETHYLBICYCLO[4.1.0]HEPT-3-ENE, VII, 200
    Bicyclo[4.1.0]hept-3-ene, 7,7-dibromo-1,6-dimethyl- [38749-438]
4,5-DIBROMO-1,2-DIMETHYL-1,2-EPOXYCYCLOHEX-4-ENE, V, 468
    7-Oxabicyclo[4.1.0]heptane, 3,4-dibromo-1,6-dimethyl-, 1α,3α,4β,6α-, (±)-
        [137053-60-2]
α,α'-DIBROMODINEOPENTYL KETONE, VI, 991
    4-Heptanone, 3,5-dibromo-2,2,6,6-tetramethyl- [23438-05-3]
1,1-DIBROMO-2,2-DIPHENYLCYCLOPROPANE, VI, 187
    Benzene, 1,1'-(2,2-dibromocyclopropylidene)bis [17343-74-7]

α,β-**DIBROMOETHYL ETHYL ETHER, IV,** 749
   Ethane, 1,2-dibromo-1-ethoxy- [2983-26-8]
β,β'-**DIBROMOISOBUTYRIC ACID, VII,** 319,
   Propanoic acid, 3-bromo-2-(bromomethyl)- [41459-42-1]
2,6-DIBROMO-4-NITROPHENOL, II, 173
   Phenol, 2,6-dibromo-4-nitro- [99-28-5]
2,4-DIBROMO-3-PENTANONE, VI, 512, 520
   3-Pentanone, 2,4-dibromo- [815-60-1]
1,3-DIBROMO-1-PHENYLPROPANE, V, 328
   Benzene, (1,3-dibromopropyl)- [17714-42-0]
erythro-α,β-**DIBROMO-β-PHENYLPROPIONIC ACID, IV,** 961; **VII,** 173
   Benzenepropanoic acid, α,β-dibromo-, (*R**,*S**)- [31357-31-0]
2,3-DIBROMOPROPENE, I, 209
   1-Propene, 2,3-dibromo- [513-31-5]
2,3-DIBROMOPROPIONALDEHYDE DIETHYL ACETAL, VI, 954
   Propane, 2,3-dibromo-1,1-diethoxy- [10160-86-8]
2,6-DIBROMOQUINONE-4-CHLOROIMIDE, II, 175
   2,5-Cyclohexadien-1-one, 2,6-dibromo-4-(chloroimino)- [537-45-1]
α,β-**DIBROMOSUCCINIC ACID, II,** 177
   Butanedioic acid, 2,3-dibromo- [526-78-3]
3,5-DIBROMOSULFANILAMIDE, III, 263
   Benzenesulfonamide, 4-amino-3,5-dibromo- [39150-45-3]
(*E*)-**1,2-DI-*tert*-BUTOXY-1-CHLOROETHENE, VIII,** 161
   Propane, 2,2'-[(1-chloro-1,2-ethenediyl)bis(oxy)]bis[2-methyl-, (*E*)- [70525-93-8]
1,2-DI-*tert*-BUTOXY-1,2-DICHLOROETHANE, VIII, 161
   Propane, 2,2'-[(1,2-dichloro-1,2-ethanediyl)]bis(oxy)]bis[2-methyl-, (*R**,*R**)-(±)-
      [68470-80-4]
   Propane, 2,2'-[(1,2-dichloro-1,2-ethanediyl)bis(oxy)]bis[2-methyl-, (*R**,*S**)-
      [68470-81-5]
2,3-DI-*tert*-BUTOXY-1,4-DIOXANE, VIII, 161
   1,4-Dioxane, 2,3-bis(1,1-dimethylethoxy)-, *cis*- [68470-78-0]
   1,4-Dioxane, 2,3-bis(1,1-dimethylethoxy)-, *trans*- [68470-79-1]
DI-*tert*-BUTOXYETHYNE, VIII, 161
   Propane, 2,2'-[1,2-ethynediylbis(oxy)]bis[2-methyl]- [66478-63-5]
β-**DIBUTYLAMINOETHYLAMINE, III,** 254
   1,2-Ethanediamine, *N,N*-dibutyl- [3529-09-7]
γ-**DIBUTYLAMINOPROPYLAMINE, III,** 256
   1,3-Propanediamine, *N,N*-dibutyl- [102-83-0]
1,2-DIBUTYLBENZENE, VI, 407
   Benzene, 1,2-dibutyl- [17171-73-2]
2,6-DI-*tert*-BUTYL-1,4-BENZOQUINONE, VI, 412
   2,5-Cyclohexadiene-1,4-dione, 2,6-bis(1,1-dimethylethyl)- [719-22-2]
DIBUTYLCARBINOL, II, 179
   5-Nonanol [623-93-8]
DI-*tert*-BUTYLCYCLOPROPRENONE, VI, 991
   Cyclopropenone, 2,3-bis(1,1-dimethylethyl)- [19985-79-6]

DI-*tert*-BUTYL DIAZOMALONATE, VI, 414
Propanedioic acid, diazo-, bis(1,1-dimethylethyl) ester [35207-75-1]
DI-*tert*-BUTYL DICARBONATE, VI, 418
Dicarbonic acid, bis(1,1-dimethylethyl) ester [24424-99-5]
2,5-DI-*tert*-BUTYL-3,6-DICHLORO-1,4-BENZOQUINONE, VI, 210
2,5-Cyclohexadiene-1,4-dione, 2,5-dichloro-3,6-bis(1,1-dimethylethyl)- [33611-73-3]
2,5-DI-*tert*-BUTYL-5,6-DICHLORO-2-CYCLOHEXENE-1,4-DIONE, VI, 210
2-Cyclohexene-1,4-dione, 5,6-dichloro-2,5-bis(1,1-dimethylethyl)- [33611-72-2]
DIBUTYLDIVINYLTIN, IV, 258
Stannane, dibutyldiethenyl- [7330-43-0]
DI-*tert*-BUTYL MALONATE, IV, 261
Propanedioic acid, bis(1,1-dimethylethyl) ester [541-16-2]
DI-*tert*-BUTYL METHYLENEMALONATE, VII, 142
Propanedioic acid, methylene-, bis(1,1-dimethylethyl) ester [86633-09-2]
2,6-DI-*tert*-BUTYL-4-METHYLPYRIDINE, VII, 144
Pyridine, 2,6-bis(1,1-dimethylethyl)-4-methyl- [38222-83-2]
2,6-DI-*tert*-BUTYL-4-METHYLPYRYLIUM
TRIFLUOROMETHANESULFONATE, VII, 144
Pyrylium, 2,6-bis(1,1-dimethylethyl)-4-methyl-, salt with trifluoromethanesulfonic
acid [59643-43-5]
DI-*tert*-BUTYL NITROXIDE, V, 355
Nitroxide, bis(1,1-dimethylethyl)- [2406-25-9]
DI-*tert*-BUTYL TRICARBONATE, VI, 418
Tricarbonic acid, bis(1,1-dimethylethyl)ester [24424-95-1]
2,5-DI-*tert*-BUTYL-3,5,6-TRICHLORO-2-CYCLOHEXENE-1,4-DIONE,
VI, 210
2-Cyclohexene-1,4-dione, 3,5,6-trichloro-2,5-bis(1,1-dimethylethyl)-, *cis*
[117257-59-7]; *trans* [117257-58-6]
3,5-DICARBETHOXY-2,6-DIMETHYLPYRIDINE, II, 215
3,5-Pyridinedicarboxylic acid, 2,6-dimethyl-, diethyl ester [1149-24-2]
DI-β-CARBOETHOXYETHYLMETHYLAMINE, III, 258
β-Alanine, *N*-(3-ethoxy-3-oxopropyl)-*N*-methyl-, ethyl ester [6315-60-2]
α,α-DICHLOROACETAMIDE, III, 260
Acetamide, 2,2-dichloro- [683-72-7]
DICHLOROACETIC ACID, II, 181
Acetic acid, dichloro- [79-43-6]
α,γ-DICHLOROACETONE, I, 211
2-Propanone, 1,3-dichloro- [534-07-6]
α,α-DICHLOROACETOPHENONE, III, 538
Ethanone, 2,2-dichloro-1-phenyl- [2648-61-5]
3,5-DICHLORO-2-AMINOBENZOIC ACID, IV, 872
Benzoic acid, 2-amino-3,5-dichloro- [2789-92-6]
2,6-DICHLOROANILINE, III, 262
Benzenamine, 2,6-dichloro- [608-31-1]
7,7-DICHLOROBICYCLO[4.1.0]HEPT-3-ENE, VI, 87
Bicyclo[4.1.0]hept-3-ene, 7,7-dichloro- [16554-84-0]

7,7-DICHLOROBICYCLO[3.2.0]HEPT-2-EN-6-ONE, VI, 1037
  Bicyclo[3.2.0]hept-2-en-6-one, 7,7-dichloro- [5307-99-3]
***exo*-3,4-DICHLOROBICYCLO[3.2.1]OCT-2-ENE, VI,** 142
  Bicyclo[3.2.1]oct-2-ene, 3,4-dichloro-, *exo*- [2394-47-0]
3,4-DICHLOROBIPHENYL, V, 51
  1,1'-Biphenyl, 3,4-dichloro [2974-92-7]
***cis*-3,4-DICHLOROCYCLOBUTENE, VI,** 422
  Cyclobutene, 3,4-dichloro-, *cis*- [2957-95-1]
***cis*-1,2-DICHLOROCYCLOHEXANE, VI,** 424
  Cyclohexane, 1,2-dichloro-, *cis*- [10498-35-8]
4,4'-DICHLORODIBUTYL ETHER, IV, 266
  Butane, 1,1'-oxybis[4-chloro]- [6334-96-9]
1,1-DICHLORO-2,2-DIFLUOROETHYLENE, IV, 268
  Ethene, 1,1-dichloro-2,2-difluoro- [79-35-6]
β,β-DICHLORO-*p*-DIMETHYLAMINOSTYRENE, V, 361
  Benzenamine, 4-(2,2-dichloroethenyl)-*N,N*-dimethyl- [6798-58-9]
1,5-DICHLORO-2,4-DINITROBENZENE, V, 1067
  Benzene, 1,5-dichloro-2,4-dinitro- [3698-83-7]
***trans*-2,3-DICHLORO-1,4-DIOXANE, VIII,** 161
  1,4-Dioxane, 2,3-dichloro-, *trans*- [3883-43-0]
2,2'-DICHLORO-α,α'-EPOXYBIBENZYL, V, 358
  Oxirane, 2,3-bis(2-chlorophenyl)- [53608-92-7]
2,2-DICHLOROETHANOL, IV, 271
  Ethanol, 2,2-dichloro- [598-38-9]
1,7-DICHLORO-4-HEPTANONE, IV, 279
  4-Heptanone, 1,7-dichloro- [40624-07-5]
3,4-DICHLORO-5-ISOPROPOXY-2(5*H*)-FURANONE, VIII, 116
  2(5*H*)-Furanone, 3,4-dichloro-5-(1-methylethoxy)- [29814-12-8]
2,4-DICHLOROMETHOXYBENZENE, VIII, 167
  Benzene, 2,4-dichloro-1-methoxy- [553-82-2]
7,7-DICHLORO-1-METHYLBICYCLO[3.2.0]HEPTAN-6-ONE, VIII, 377
  Bicyclo[3.2.0]heptan-6-one, 7,7-dichloro-1-methyl- [51284-43-6]
DICHLOROMETHYLENETRIPHENYLPHOSPHORANE, V, 361
  Phosphorane, (dichloromethylene)triphenyl- [6779-08-4]
DICHLOROMETHYL METHYL ETHER, V, 365
  Methane, dichloromethoxy- [4885-02-3]
2,6-DICHLORONITROBENZENE, V, 367
  Benzene, 1,3-dichloro-2-nitro- [601-88-7]
2,2-DICHLORONORBORNANE, VI, 845
  Bicyclo[2.2.1]heptane, 2,2-dichloro- [19916-65-5]
2,6-DICHLOROPHENOL, III, 267; *Hazard*
  Phenol, 2,6-dichloro- [87-65-0]
DI-(*p*-CHLOROPHENYL)ACETIC ACID, III, 270; *Correction*, **V,** 370
  Benzeneacetic acid, 4-chloro-α-(4-chlorophenyl)- [83-05-6]
1,1-DICHLORO-2-PHENYLCYCLOPROPANE, VII, 12
  Benzene, (2,2-dichlorocyclopropyl)- [2415-80-7]

1,1-DI-(*p*-CHLOROPHENYL)-2,2-DICHLOROETHYLENE, III, 270
   Benzene, 1,1'-(dichloroethenylidene)bis[4-chloro]- [72-55-9]
4,7-DICHLOROQUINOLINE, III, 272
   Quinoline, 4,7-dichloro- [86-98-6]
3,5-DICHLOROSULFANILAMIDE, III, 262
   Benzenesulfonamide, 4-amino-3,5-dichloro- [22134-75-4]
2,3-DICHLOROTETRAHYDROPYRAN, VI, 675
   2*H*-Pyran, 2,3-dichlorotetrahydro- [5631-95-8]
3,4-DICHLORO-1,2,3,4-TETRAMETHYLCYCLOBUTENE, V, 370
   Cyclobutene, 3,4-dichloro-1,2,3,4-tetramethyl- [1194-30-5]
3,3-DICHLOROTHIETANE 1,1-DIOXIDE, VII, 491
   Thietane, 3,3-dichloro-, 1,1-dioxide [90344-85-7]
11,11-DICHLOROTRICYCLO[4.4.1.0^{1,6}]UNDECA-3,8-DIENE, VI, 731
   Tricyclo[4.4.1.0^{1,6}]undeca-3,8-diene, 11,11-dichloro- [39623-22-8]
6-[(*E*)-1,2-DICHLOROVINYL]-3-ETHOXY-6-METHYL-2-CYCLOHEXEN-1-
       ONE, VII, 241
   2-Cyclohexen-1-one, 6-(1,2-dichloroethenyl)-3-ethoxy-6-methyl- [73843-25-1]
4-[(*E*)-1,2-DICHLOROVINYL]-4-METHYL-2-CYCLOHEXEN-1-ONE, VII, 241
   2-Cyclohexen-1-one, 4-(1,2-dichloroethenyl)-4-methyl- [73843-27-3]
DICINNAMALACETONE, VII, 60
   Nona-1,3,6,8-tetraen-5-one, 1,9-diphenyl- [622-21-9]
1,1'-DICYANO-1,1'-BICYCLOHEXYL, IV, 273
   [1,1'-Bicyclohexyl]-1,1'-dicarbonitrile [18341-40-7]
1,2-DICYANOCYCLOBUTENE, VI, 427
   1-Cyclobutene-1,2-dicarbonitrile [3716-97-0]
1,2-DI-1-(1-CYANO)CYCLOHEXYLHYDRAZINE, IV, 274
   Cyclohexanecarbonitrile, 1,1'-hydrazobis- [17643-01-5]
α,α'-DICYANO-β-ETHYL-β-METHYLGLUTARIMIDE, IV, 441
   3,5-Piperidinedicarbonitrile, 4-ethyl-4-methyl-2,6-dioxo- [1135-62-2]
α,α'-DICYANO-β-METHYLGLUTARAMIDE, III, 591
   Glutaramide, 2,4-dicyano-3-methyl- [5447-66-5]
DICYANOKETENE ETHYLENE ACETAL, IV, 276
   Propanedinitrile, 1,3-dioxolan-2-ylidene- [5694-65-5]
5,5-DICYANO-4-PHENYL-2-CYCLOPENTEN-1-ONE 1,3-PROPANEDIOL
       KETAL, VIII, 173
   6,10-Dioxaspiro[4,5]dec-3-ene-1,1-dicarbonitrile, 2-phenyl- [88442-12-0]
2,3-DICYANO-1,4,4a,9a-TETRAHYDROFLUORENE, VI, 427
   1*H*-Fluorene-2,3-dicarbonitrile, 4,4a,9,9a-tetrahydro- [52477-65-3]
DICYCLOHEXYL-18-CROWN-6 POLYETHER, VI, 395
   Dibenzo[*b,k*][1,4,7,10,13,16]hexaoxacyclooctadecin, eicosahydro- [16069-36-6]
DICYCLOPROPYL KETONE, IV, 278
   Methanone, dicyclopropyl- [1121-37-5]
DIDEUTERIODIAZOMETHANE, VI, 432
   Methane-d_2, diazo- [14621-84-2]
5,6-DIETHOXYBENZOFURAN-4,7-DIONE, VIII, 179
   4,7-Benzofurandione, 5,6-diethoxy- [138225-13-5]

cis and *trans*-2,6-DIETHOXY-1,4-OXATHIANE 4,4-DIOXIDE, VI, 976
    1,4-Oxathiane, 2,6-diethoxy-, 4,4-dioxide, *cis*- [40263-59-0]
2,2-DIETHOXY-2-(4-PYRIDYL)ETHYLAMINE, VII, 149
    4-Pyridineethanamine, β,β-diethoxy- [74209-44-2]
DIETHYL ACETAMIDOMALONATE, V, 373
    Propanedioic acid, (acetylamino)-, diethyl ester [1068-90-2]
DIETHYL ACETONEDICARBOXYLATE, I, 237
    Pentanedioic acid, 3-oxo-, diethyl ester [105-50-0]
DIETHYL ACETOSUCCINATE, II, 262
    Propanedioic acid, acetyl-, diethyl ester [570-08-1]
DIETHYL ACETYLENEDICARBOXYLATE, IV, 330
    2-Butynedioic acid, diethyl ester [762-21-0]
DIETHYL α-ACETYL-β-KETOPIMELATE, V, 384
    Heptanedioic acid, 2-acetyl-3-oxo-, diethyl ester [61983-62-8]
DIETHYL ADIPATE, II, 264
    Hexanedioic acid, diethyl ester [141-28-6]
DIETHYL (2S,3R)-(+)-3-ALLYL-2-HYDROXYSUCCINATE, VII, 153
    Butanedioic acid, 2-hydroxy-3-(2-propenyl)-, (S-(R*,S*))-, diethyl ester [73837-97-5]
DIETHYLALUMINUM CYANIDE, VI, 436
    Aluminum, (cyano-C)diethyl- [5804-85-3]
DIETHYLAMINOACETONITRILE, III, 275
    Acetonitrile, (diethylamino)- [3010-02-4]
1-DIETHYLAMINO-3-BUTANONE, IV, 281
    2-Butanone, 4-(diethylamino)- [3299-38-5]
β-DIETHYLAMINOETHYL ALCOHOL, II, 183
    Ethanol, 2-(diethylamino)- [100-37-8]
DIETHYL AMINOMALONATE and HYDROCHLORIDE, V, 376
    Propanedioic acid, amino-, diethyl ester [6829-40-9]
    Propanedioic acid, amino-, diethyl ester, hydrochloride [13433-00-6]
DIETHYL AMINOMETHYLPHOSPHONATE, VIII, 451
    Phosphonic acid, (aminomethyl)-, diethyl ester [50917-72-1]
DIETHYLAMINOSULFUR TRIFLUORIDE, VI, 440
    Sulfur, (diethylaminato)trifluoro- [38078-09-0]
DIETHYL AZODICARBOXYLATE, III, 375; IV, 411; *Warning*, IV, 412; V, 544
    Diazenedicarboxylic acid, diethyl ester [1972-28-7]
DIETHYL BENZALMALONATE, III, 377
    Propanedioic acid, (phenylmethylene)-, diethyl ester [5292-53-5]
N,N'-DIETHYLBENZIDINE, IV, 283
    [1,1'-Biphenyl]-4,4'-diamine, *N,N'*-diethyl- [6290-86-4]
DIETHYL [O-BENZOYL]ETHYLTARTRONATE, V, 379
    Propanedioic acid, (benzoyloxy)ethyl-, diethyl ester [6259-78-5]
DIETHYL BENZOYLMALONATE, IV, 285
    Propanedioic acid, benzoyl-, diethyl ester [1087-97-4]
DIETHYL *N*-BENZYLIDENAMINOMETHYLPHOSPHONATE, VIII, 451
    Phosphonic acid, [[(phenylmethylene)amino]methyl]-, diethyl ester [50917-73-2]

DIETHYL BENZYLMALONATE, III, 705
Propanedioic acid, (phenylmethyl)-, diethyl ester [607-81-8]
DIETHYL BROMOMALONATE, I, 245
Propanedioic acid, bromo-, diethyl ester [685-87-0]
DIETHYL BUTYLMALONATE, I, 250
Propanedioic acid, butyl-, diethyl ester [133-08-4]
DIETHYL *sec*-BUTYLMALONATE, III, 495
Propanedioic acid, (1-methylpropyl)-, diethyl ester [83-27-2]
DIETHYL *tert*-BUTYLMALONATE, VI, 442
Propanedioic acid, (1,1-dimethylethyl)-, diethyl ester [759-24-0]
DIETHYL BIS(HYDROXYMETHYL)MALONATE, V, 381
Propanedioic acid, bis(hydroxymethyl)-, diethyl ester [20605-01-0]
DIETHYL α-CARBETHOXY-β-(*m*-CHLOROANILINO)ACRYLATE, III, 272
Propanedioic acid, [[(3-chlorophenyl)amino]methylene], diethyl ester [3412-99-5]
DIETHYL 2-CHLORO-2-CYCLOPROPYLETHENE-1,1-DICARBOXYLATE, VIII, 247
Propanedioic acid, (chlorocyclopropylmethylene)-, diethyl ester [123844-18-8]
N,N-DIETHYL-(*E*)-CITRONELLALENAMINE, (*R*)-(−)-, VIII, 183
1,6-Octadien-1-amine, N,N-diethyl-3,7-dimethyl- [*R*-(*E*)]- [67392-56-7]
DIETHYL 1,1-CYCLOBUTANEDICARBOXYLATE, IV, 288
1,1-Cyclobutanedicarboxylic acid, diethyl ester [3779-29-1]
DIETHYL 2-(CYCLOHEXYLAMINO)VINYLPHOSPHONATE, VI, 448
Phosphonic acid, [2-(cyclohexylamino)ethenyl]-, diethyl ester [20061-84-1]
DIETHYL Δ²-CYCLOPENTENYLMALONATE, IV, 291
Propanedioic acid, 2-cyclopenten-1-yl-, diethyl ester [53608-93-8]
DIETHYL CYCLOPROPYLCARBONYLMALONATE, VIII, 248
Propanedioic acid, (cyclopropylcarbonyl)-, diethyl ester [7394-16-3]
DIETHYL 2,3-DIAZABICYCLO[2.2.1]HEPTANE-2,3-DICARBOXYLATE, V, 97
2,3-Diazabicyclo[2.2.1]heptane-2,3-dicarboxylic acid, diethyl ester [18860-71-4]
DIETHYL 2,3-DIAZABICYCLO[2.2.1]HEPT-5-ENE-2,3-DICARBOXYLATE, V, 96
2,3-Diazabicyclo[2.2.1]hept-5-ene-2,3-dicarboxylic acid, diethyl ester [14011-60-0]
DIETHYL α,δ-DIBROMOADIPATE, III, 623
Hexanedioic acid, 2,5-dibromo-, diethyl ester [869-10-3]
DIETHYL 2,2-DIETHOXYETHYLPHOSPHONATE, VI, 448
Phosphonic acid, 2,2-diethoxyethyl-, diethyl ester [7598-61-0]
DIETHYL 2,5-DIMETHYLPHENYLHYDROXYMALONATE, III, 326
Propanedioic acid, (2,5-dimethylphenyl)hydroxy-, diethyl ester [83026-12-4]
DIETHYL DIOXOSUCCINATE, VIII, 597
Butanedioic acid, dioxo-, diethyl ester [59743-08-7]
DIETHYL ENANTHYLSUCCINATE, IV, 430
Butanedioic acid, (1-oxoheptyl)-, diethyl ester [41117-78-6]
DIETHYL ETHOXALYLPROPIONATE, II, 272
Pentanedioic acid, 2-oxo-, diethyl ester [5965-53-7]
DIETHYL ETHOXYMETHYLENEMALONATE, III, 395
Propanedioic acid, (ethoxymethylene)-, diethyl ester [87-13-8]

DIETHYL ETHYLIDENEMALONATE, IV, 293
 Propanedioic acid, ethylidene-, diethyl ester [1462-12-0]
DIETHYL FORMYLMETHYLPHOSPHONATE, VI, 448
 Phosphonic acid, (2-oxoethyl)-, diethyl ester [1606-75-3]
N,N-**DIETHYLGERANYLAMINE, VIII**, 183, 188
 2,6-Octadien-1-amine, *N,N*-diethyl-3,7-dimethyl-, (*E*)- [40267-53-6]
DIETHYL *cis*-HEXAHYDROPHTHALATE, IV, 304
 1,2-Cyclohexanedicarboxylic acid, diethyl ester, *cis*- [17351-07-4]
DIETHYL 1,16-HEXADECANEDICARBOXYLATE, III, 401
 Octadecanedioic acid, diethyl ester [1472-90-8]
DIETHYL HYDRAZODICARBOXYLATE, III, 375; **IV**, 411
 1,2-Hydrazinedicarboxylic acid, diethyl ester [4114-28-7]
DIETHYL HYDROXYMETHYLPHOSPHONATE, VII, 160
 Phosphonic acid, (hydroxymethyl)-, diethyl ester [3084-40-0]
DIETHYL ISOPROPYLIDENEMALONATE, VI, 442
 Propanedioic acid, (1-methylethylidene)-, diethyl ester [6802-75-1]
DIETHYL β-KETOPIMELATE, V, 384
 Heptanedioic acid, 3-oxo, diethyl ester [40420-22-2]
DIETHYL MERCAPTOACETAL, IV, 295
 Ethanethiol, 2,2-diethoxy- [53608-94-9]
DIETHYL 5-METHYLCOPROST-3-EN-3-YL PHOSPHATE, VI, 762
 Cholest-3-en-3-ol, 5-methyl-, diethyl hydrogen phosphate [23931-37-5]
DIETHYL METHYLENEMALONATE, IV, 298
 Propanedioic acid, methylene-, diethyl ester [3377-20-6]
DIETHYL METHYLMALONATE, II, 279
 Propanedioic acid, methyl-, diethyl ester [609-08-5]
DIETHYL 1-METHYLTHIOL-3-PHTHALIMIDOPROPANE-3,3-
 DICARBOXYLATE, II, 384
N,N-**DIETHYLNERYLAMINE, VIII**, 183, 190
 2,6-Octadien-1-amine, *N,N*-diethyl-3,7-dimethyl-, (*Z*)- [40137-00-6]
DIETHYL OXALATE, I, 261
 Ethanedioic acid, diethyl ester [95-92-1]
DIETHYL OXIMINOMALONATE, V, 373
 Propanedioic acid, (hydroxyimino)-, diethyl ester [6829-41-0]
DIETHYL OXOMALONATE, I, 266
 Propanedioic acid, oxo-, diethyl ester [609-09-6]
DIETHYL γ-OXOPIMELATE, IV, 302
 Heptanedioic acid, 4-oxo-, diethyl ester [6317-49-3]
DIETHYL PHENYLMALONATE, II, 288
 Propanedioic acid, phenyl-, diethyl ester [83-13-6]
DIETHYL PHENYLPHOSPHONATE, VI, 451
 Phosphonic acid, phenyl-, diethyl ester [1754-49-0]
6-DIETHYLPHOSPHONOMETHYL-2,2-DIMETHYL-1,3-DIOXEN-4-ONE, VIII, 192
 Phosphonic acid, [(2,2-dimethyl-4-oxo-4*H*-1,3-dioxin-6-yl)methyl]-, diethyl ester
  [81956-28-7]
DIETHYL PHTHALIMIDOMALONATE, I, 271
 Propanedioic acid, (1,3-dihydro-1,3-dioxo-2*H*-isoindol-2-yl)-, diethyl ester [5680-61-5]

DIETHYL PHTHALIMIDOMETHYLPHOSPHONATE, VIII, 451
Phosphonic acid, [(1,3-dihydro-1,3-dioxo-2*H*-isoindol-2-yl)-methyl]-, diethyl ester [33512-26-4]
DIETHYL PIMELATE, II, 536
Heptanedioic acid, diethyl ester [2050-20-6]
DIETHYL PROPIONYLSUCCINATE, VI, 615
Butanedioic acid, 1-oxopropyl-, diethyl ester [4117-76-4]
DIETHYL SEBACATE, II, 277
Decanedioic acid, diethyl ester [110-40-7]
N,N-**DIETHYLSELENOUREA, IV**, 360
Selenourea, *N,N*-diethyl- [15909-81-6]
DIETHYL SUCCINATE, V, 993
Butanedioic acid, diethyl ester [123-25-1]
N,N'-**DIETHYLSULFAMIDE, VI**, 78
Sulfamide, *N,N'*-diethyl- [6104-21-8]
DIETHYL (2R,3R)-(+)-TARTRATE, VII, 41, 461
Tartaric acid, (*R,R*)-(+)-, diethyl ester [608-84-4]
DIETHYL *cis*-Δ⁴-TETRAHYDROPHTHALATE, IV, 304
4-Cyclohexene-1,2-dicarboxylic acid, diethyl ester, *cis*- [4841-85-4]
DIETHYL *trans*-Δ⁴-TETRAHYDROPHTHALATE, VI, 454
4-Cyclohexene-1,2-dicarboxylic acid, diethyl ester, *trans*- [5048-50-0]
DIETHYL [(2-TETRAHYDROPYRANYLOXY)METHYL]PHOSPHONATE, VII, 160
Phosphonic acid, [(tetrahydro-2*H*-pyran-2-yl)oxy]methyl]-, diethyl ester [71885-51-3]
DIETHYLTHIOCARBAMYL CHLORIDE, IV, 307
Carbamothioic chloride, diethyl- [88-11-9]
N,N-**DIETHYL-1,2,2-TRICHLOROVINYLAMINE, V**, 387
Ethenamine, 1,2,2-trichloro-*N,N*-diethyl- [686-10-2]
DIETHYL *trans*-4-TRIMETHYLSILYLOXY-4-CYCLOHEXENE-1,2-DICARBOXYLATE, VI, 445
4-Cyclohexene-1,2-dicarboxylic acid, 4-[(trimethylsilyl)oxy]-,diethyl ester, *trans*- [61692-31-7]
DIETHYLZINC, II, 184
Zinc, diethyl- [557-20-0]
4,4'-DIFLUOROBIPHENYL, II, 188
1,1'-Biphenyl, 4,4'-difluoro- [398-23-2]
β,β-**DIFLUOROSTYRENE, V**, 390
Benzene, (2,2-difluoroethenyl)- [405-42-5]
2,2-DIFLUOROSUCCINIC ACID, V, 393
Butanedioic acid, 2,2-difluoro- [665-31-6]
α,α-**DIFLUOROTOLUENE, V**, 396
Benzene, (difluoromethyl)- [455-31-2]
N-(**2,4-DIFORMYL-5-HYDROXYPHENYL)ACETAMIDE, VII**, 162
Acetamide, *N*-(2,4-diformyl-5-hydroxyphenyl)- [67149-23-9]
9,10-DIHYDROANTHRACENE, V, 398
Anthracene, 9,10-dihydro- [613-31-0]

3,4-DIHYDRO-1(2H)-AZULENONE, VIII, 196
  1(2H)-Azulenone, 3,4-dihydro- [52487-41-9]
1,4-DIHYDROBENZOIC ACID, V, 400
  2,5-Cyclohexadiene-1-carboxylic acid [4794-04-1]
DIHYDROCARVONE, VI, 459
  2-Cyclohexen-1-one, 2-methyl-5-(1-methylethyl)- [43205-82-9]
DIHYDROCHOLESTEROL, II, 191
  Cholestan-3-ol, (3β,5α)- [80-97-7]
1,4-DIHYDRO-3,5-DICARBETHOXY-2,6-DIMETHYLPYRIDINE, II, 214
  3,5-Pyridinedicarboxylic acid, 1,4-dihydro-2,6-dimethyl-, diethyl ester [1149-23-1]
2,5-DIHYDRO-2,5-DIMETHOXYFURAN, V, 403
  Furan, 2,5-dihydro-2,5-dimethoxy- [332-77-4]
9,10-DIHYDROFULVALENE, VIII, 298
  Bi-2,4-cyclopentadien-1-yl [21423-86-9]
4,5-DIHYDRO-5-IODOMETHYL-4-PHENYL-2(3H)-FURANONE, *cis-* and *trans-*
  VII, 164
  2(3H)-Furanone, dihydro-5-(iodomethyl)-4-phenyl-, *cis-* [67279-70-3] and *trans-*
  [67279-69-0]
1,3-DIHYDROISOINDOLE, V, 406
  1H-Isoindole, 2,3-dihydro- [496-12-8]
DIHYDROJASMONE, VIII, 620
  2-Cyclopenten-1-one, 3-methyl-2-pentyl- [1128-08-1]
3,4-DIHYDRO-2-METHOXY-4-METHYL-2H-PYRAN, IV, 311
  2H-Pyran, 3,4-dihydro-2-methoxy-4-methyl- [53608-95-0]
3,4-DIHYDROMETHYLENE-2H-1-BENZOPYRAN, VIII, 512
  2H-1-Benzopyran, 3,4-dihydro-2-methylene- [74104-13-5]
3,4-DIHYDRO-1,2-NAPHTHALIC ANHYDRIDE, II, 194
  Naphtho[1,2-c]furan-1,3-dione, 4,5-dihydro- [37845-14-0]
9,10-DIHYDROPHENANTHRENE, IV, 313
  Phenanthrene, 9,10-dihydro- [776-35-2]
***trans*-9,10-DIHYDRO-9,10-PHENANTHRENEDIOL, VI**, 887
  9,10-Phenanthrenediol, 9,10-dihydro-, *trans-* [25061-61-4]
4,5-DIHYDRO-2-[(1-PHENYLMETHYL-2-PROPENYL)THIO]THIAZOLE, VI, 705
  Thiazole, 4,5-dihydro-2-[[1-(phenylmethyl)-2-propenyl]thio]- [52534-82-4]
***trans*-1,2-DIHYDROPHTHALIC ACID, VI**, 461
  3,5-Cyclohexadiene-1,2-dicarboxylic acid, *trans-* [5675-13-8]
4,5-DIHYDRO-2-(2-PROPENYLTHIO)THIAZOLE, VI, 705
  Thiazole, 4,5-dihydro-2-(2-propenylthio)- [3571-74-2]
2,3-DIHYDROPYRAN, III, 276
  2H-Pyran, 3,4-dihydro- [110-87-2]
5,6-DIHYDRO-2H-PYRAN-2-ONE, VI, 462
  2H-Pyran-2-one, 5,6-dihydro- [3393-45-1]
DIHYDRORESORCINOL, III, 278
  1,3-Cyclohexanedione [504-02-9]
1,3-DIHYDRO-3,5,7-TRIMETHYL-2H-AZEPIN-2-ONE, V, 408
  2H-Azepin-2-one, 1,3-dihydro-3,5,7-trimethyl- [936-85-6]

2,5-DIHYDROXYACETOPHENONE, III, 280
Ethanone, 1-(2,5-dihydroxyphenyl)- [490-78-8]
2,6-DIHYDROXYACETOPHENONE, III, 281
Ethanone, 1-(2,6-dihydroxyphenyl)- [699-83-2]
2,5-DIHYDROXY p-BENZENEDIACETIC ACID, III, 286
1,4-Benzenediacetic acid, 2,5-dihydroxy- [5488-16-4]
3,5-DIHYDROXYBENZOIC ACID, III, 288
Benzoic acid, 3,5-dihydroxy- [99-10-5]
3,3'-DIHYDROXYBIPHENYL, V, 412
[1,1'-Biphenyl]-3,3'-diol [612-76-0]
2,4-DIHYDROXY-5-BROMOBENZOIC ACID, II, 100
Benzoic acid, 5-bromo-2,4-dihydroxy- [7355-22-8]
DIHYDROXYCYCLOPENTENE, V, 414
3-Cyclopentene-1,2-diol, *cis*- [694-29-1]
3,5-DIHYDROXYCYCLOPENTENE, V, 414
4-Cyclopentene-1,3-diol [4157-01-1]
1,3-DIHYDROXY-2-NAPHTHOIC ACID, III, 638
2-Naphthalenecarboxylic acid, 1,3-dihydroxy- [3147-58-8]
9,10-DIHYDROXYSTEARIC ACID, IV, 317
Octadecanoic acid, 9,10-dihydroxy- [120-87-6]
1,4-DIIODOBUTANE, IV, 321
Butane, 1,4-diiodo- [628-21-7]
1,6-DIIODOHEXANE, IV, 323
Hexane, 1,6-diiodo- [629-09-4]
2,6-DIIODO-p-NITROANILINE, II, 196
Benzenamine, 2,6-diiodo-4-nitro- [5398-27-6]
DIISOPROPYL ETHYLPHOSPHONATE, IV, 326
Phosphonic acid, ethyl-, bis(1-methylethyl) ester [1067-69-2]
1,2:5,6-DI-(O-ISOPROPYLIDENE-3-O-(S-METHYL DITHIOCARBONATE)-
α-D-GLUCOFURANOSE, VII, 139
Glucofuranose, 1,2:5,6-di-O-isopropylidene-, S-methyldithiocarbonate, α-D-
[16667-96-2]
2,3-DI-O-ISOPROPYLIDENE-L-THREITOL, VIII, 155
1,3-Dioxolane-4,5-dimethanol, 2,2-dimethyl-, (4S-trans)- [50622-09-8]
DIISOPROPYL (2S,3S)-2,3-O-ISOPROPYLIDENETARTRATE, VIII, 201
1,3-Dioxolane-4,5-dicarboxylic acid, 2,2-dimethyl-, bis(1-methylethyl) ester,
(4R-trans)- [81327-47-1]
DIISOPROPYL METHYLPHOSPHONATE, IV, 325
Phosphonic acid, methyl, bis(1-methylethyl) ester [1445-75-6]
DIISOVALERYLMETHANE, III, 291
4,6-Nonanedione, 2,8-dimethyl- [7307-08-6]
7,16-DIKETODOCOSANEDIOIC ACID and DISODIUM SALT, V, 534, 536
Docosanedioic acid, 7,16-dioxo-, disodium salt [134507-60-1]
(−)-DIMENTHYL (1S,2S)-CYCLOPROPANE-1,2-DICARBOXYLATE, VIII, 142
1,2-Cyclopropanedicarboxylic acid, bis[5-methyl-2-(1-methylethyl)cyclohexyl]
ester [1S-[1α(1S*,2S*,5R*)], 2β,5α]]- [96149-01-8]

(–)-DIMENTHYL SUCCINATE, VIII, 141
   Butanedioic acid, bis[5-methyl-2-(1-methylethyl)cyclohexyl] ester
   [1R-[1α(1R*,2S*,5R*)], 2β,5α]]- [34212-59-4]
1,2-DIMERCAPTOBENZENE, V, 419
   1,2-Benzenedithiol [17534-15-5]
DIMESITYLMETHANE, V, 422
   Benzene, 1,1'-methylenebis[2,4,6-trimethyl- [733-07-3]
2,4-DIMETHOXYBENZONITRILE, VI, 465
   Benzonitrile, 2,4-dimethoxy- [4107-65-7]
2,6-DIMETHOXYBENZONITRILE, III, 293
   Benzonitrile, 2,6-dimethoxy- [16932-49-3]
7,7-DIMETHOXYBICYCLO[2.2.1]HEPTENE, V, 424
   Bicyclo[2.2.1]hept-2-ene, 7,7-dimethoxy- [875-04-7]
3,3'-DIMETHOXYBIPHENYL, III, 295
   1,1'-Biphenyl, 3,3'-dimethoxy- [6161-50-8]
4,4'-DIMETHOXY-1,1'-BIPHENYL, VI, 468, 490
   1,1'-Biphenyl, 4,4'-dimethoxy- [2132-80-1]
2,3-DIMETHOXYCINNAMIC ACID, IV, 327
   2-Propenoic acid, 3-(2,3-dimethoxyphenyl)- [7461-60-1]
3,3-DIMETHOXYCYCLOPROPENE, VI, 361
   Cyclopropene, 3,3-dimethoxy- [23529-83-1]
3,8-DIMETHOXY-4,5,6,7-DIBENZO-1,2-DIOXACYCLOOCTANE, V, 493
   Dibenzo[d,f][1,2]-dioxocin, 5,8-dihydro-5,8-dimethoxy- [6623-54-7]
6,7-DIMETHOXY-3,4-DIHYDRO-2-NAPHTHOIC ACID, III, 300
   2-Naphthalenecarboxylic acid, 3,4-dihydro-6,7-dimethoxy- [53684-50-7]
6,6-DIMETHOXYHEXANAL, VII, 168
   Hexanal, 6,6-dimethoxy- [55489-11-7]
6,7-DIMETHOXY-3-ISOCHROMANONE, VI, 471
   3H-2-Benzopyran-3-one, 1,4-dihydro-6,7-dimethoxy- [16135-41-4]
6,7-DIMETHOXY-1-METHYL-3,4-DIHYDROISOQUINOLINE, VI, 1
   Isoquinoline, 3,4-dihydro-6,7-dimethoxy-1-methyl- [4721-98-6]
2-(DIMETHOXYMETHYL)-3-PHENYL-2H-AZIRINE, VI, 893
   2H-Azirine, 2-(dimethoxymethyl)-3-phenyl- [56900-68-6]
**(S)-6,7-DIMETHOXY-1-METHYL-1,2,3,4-TETRAHYDROISOQUINOLINE
   (SALSOLIDINE), VIII**, 573
   Isoquinoline, 1,2,3,4-tetrahydro-6,7-dimethoxy-1-methyl-, (S)- [493-48-1]
N-[2-(3,4-DIMETHOXYPHENYL)ETHYL]ACETAMIDE, VI, 1
   Acetamide, N-[2-(3,4-dimethoxyphenyl)ethyl]- [6275-29-2]
3,4-DIMETHOXYPHENYLPYRUVIC ACID, II, 333, 335
   Benzenepropanoic acid, 3,4-dimethoxy-α-oxo- [2460-33-5]
trans-4,4'-DIMETHOXYSTILBENE, V, 428
   Benzene, 1,1'-(1,2-ethenediyl)bis[4-methoxy-, (E)- [15638-14-9]
7,7-DIMETHOXY-1,2,3,4-TETRACHLOROBICYCLO[2.2.1]HEPTENE, V, 425
   Bicyclo[2.2.1]heptene, 1,2,3,4-tetrachloro-7,7-dimethoxy- [19448-78-3]
5,5-DIMETHOXY-1,2,3,4-TETRACHLOROCYCLOPENTADIENE, V, 424
   1,3-Cyclopentadiene, 1,2,3,4-tetrachloro-5,5-dimethoxy- [2207-27-4]

6,7-DIMETHOXY-1,2,3,4-TETRAHYDRO-2-[(1-*tert*-BUTOXY-3-METHYL)-2-BUTYLIMINOMETHYL]ISOQUINOLINE, VIII, 573

  Isoquinoline 2-[[[1-[{1,1-dimethylethoxy)methyl]-2-methylpropyl]imino]methyl]-1,2,3,4-tetrahydro-6,7-dimethoxy-, (*S*)- [90482-03-4]

(*R,R*)-(+)-2,3-DIMETHOXY-*N,N,N',N'*-TETRAMETHYLSUCCINIC ACID DIAMIDE, VII, 41

  Butanediamide, 2,3-dimethoxy-*N,N,N',N'*-tetramethyl-, (*R,R*)- [26549-29-1]

2,4-DIMETHYL-3-ACETYL-5-CARBETHOXYPYRROLE, III, 513

  1*H*-Pyrrole-2-carboxylic acid, 4-acetyl-3,5-dimethyl-, ethyl ester [2386-26-7]

DIMETHYL ACETYLENEDICARBOXYLATE, IV, 329

  2-Butynedioic acid, dimethyl ester [762-42-5]

β,β-DIMETHYLACRYLIC ACID, III, 302

  2-Butenoic acid, 3-methyl- [541-47-9]

[3-(DIMETHYLAMINO)-2-AZAPROP-2-EN-1-YLIDENE] DIMETHYLAMMONIUM CHLORIDE, VII, 197

  Methanaminium, [[[(dimethylamino)methylene]amino]methylene] *N*-methyl-, chloride [20353-93-9]

p-DIMETHYLAMINOBENZALDEHYDE, I, 214; **IV**, 331

  Benzaldehyde, 4-(dimethylamino)- [100-10-7]

p-DIMETHYLAMINOBENZOPHENONE, I, 217

  Methanone, [4-(dimethylamino)phenyl]phenyl- [530-44-9]

β-DIMETHYLAMINOETHYL CHLORIDE HYDROCHLORIDE, IV, 333

  Ethanamine, 2-chloro-*N,N*-dimethyl-, hydrochloride [4584-46-7]

6-(DIMETHYLAMINO)FULVENE, V, 431

  Methanamine, 1-(2,4-cyclopentadien-1-ylidene)-*N,N*-dimethyl- [696-68-4]

2-DIMETHYLAMINO-2'-METHOXYBENZHYDROL, V, 46

N,N-DIMETHYLAMINOMETHYLFERROCENE METHIODIDE, V, 434

  Methanaminium,1-ferrocenyl-*N,N,N*-trimethyl-, iodide [12086-40-7]

1-(DIMETHYLAMINO)-4-METHYL-3-PENTANONE, VI, 474

  3-Pentanone, 1-(dimethylamino)-4-methyl- [5782-64-9]

(2-DIMETHYLAMINO-5-METHYLPHENYL)DIPHENYLCARBINOL, VI, 478

  Benzenemethanol, 2-(dimethylamino)-5-methyl-α,α-diphenyl- [23667-05-2]

[5-(DIMETHYLAMINO)-2,4-PENTADIENYLIDENE]DIMETHYLAMMONIUM CHLORIDE (in solution), VII, 15

  Methanaminium, *N*-5-[(dimethylamino)-2,4-pentadienylidene]-*N*-methyl-, chloride, (*E,E*)- [70669-79-3]

α-*N,N*-DIMETHYLAMINOPHENYLACETONITRILE, V, 437

  Benzeneacetonitrile, α-(dimethylamino)- [827-36-1]

N-(p-DIMETHYLAMINOPHENYL)-α-(o-NITROPHENYL)NITRONE, V, 826

  Nitrone, *N*-(p-dimethylaminophenyl)-α-(o-nitrophenyl)- [13664-79-4]

β-[2-(*N,N*-DIMETHYLAMINO)PHENYL]STYRENE, (Z)-, VII, 172

  Benzenamine, *N,N*-dimethyl-2-(2-phenylethenyl)-, (Z)- [70197-43-2]

β-DIMETHYLAMINOPROPIOPHENONE HYDROCHLORIDE, III, 305

  1-Propanone, 3-(dimethylamino)-1-phenyl-, hydrochloride [879-72-1]

2-(DIMETHYLAMINO)PYRIMIDINE, IV, 336

  2-Pyrimidinamine, *N,N*-dimethyl- [5621-02-3]

3,4-DIMETHYLANILINE, III, 307
  Benzenamine, 3,4-dimethyl- [95-64-7]
2,3-DIMETHYLANTHRAQUINONE, III, 310
  9,10-Anthracenedione, 2,3-dimethyl- [6531-35-7]
DIMETHYL AZODICARBOXYLATE, IV, 411; *Warning*, **IV**, 412
  Diazenedicarboxylic acid, dimethyl ester [2446-84-6]
4,5-DIMETHYL-*o*-BENZOQUINONE, VI, 480
  3,5-Cyclohexadiene-1,2-dione, 4,5-dimethyl- [4370-50-7]
3,3-DIMETHYL-*cis*-BICYCLO[3.2.0]HEPTAN-2-OL, VII, 177
  Bicyclo[3.2.0]heptan-2-ol, 3,3-dimethyl- [71221-67-5]
3,3-DIMETHYL-*cis*-BICYCLO[3.2.0]HEPTAN-2-ONE, VII, 177
  Bicyclo[3.2.0]heptan-2-one, 3,3-dimethyl- [71221-70-0]
***cis*-1,5-DIMETHYLBICYCLO[3.3.0]OCTANE-3,7-DIONE, VII**, 50
  2,5-(1*H*,3*H*)-Pentalenedione, tetrahydro-3a,6a-dimethyl-, *cis*- [21170-10-5]
***cis*-7,8-DIMETHYLBICYCLO[4.2.0]OCT-7-ENE, VI**, 482
  Bicyclo[4.2.0]oct-7-ene, 7,8-dimethyl-, *cis*- [53225-88-0]
3,3'-DIMETHYLBIPHENYL, III, 295
  1,1'-Biphenyl, 3,3'-dimethyl- [612-75-9]
4,4'-DIMETHYL-1,1'-BIPHENYL, VI, 488
  1,1'-Biphenyl, 4,4'-dimethyl- [613-33-2]
2,3-DIMETHYL-1,3-BUTADIENE, III, 312
  1,3-Butadiene, 2,3-dimethyl- [513-81-5]
N,N-DIMETHYL-N'-(1-*tert*-BUTOXY-3-METHYL-2-BUTYL)FORMAMIDINE,
    (S)-, VIII, 204, 573
  Methanimidamide, N'-[1-[(1,1-dimethylethoxy)methyl]-2-methylpropyl]- N,N-
    dimethyl-, (S) [90482-06-7]
3,5-DIMETHYL-4-CARBETHOXY-2-CYCLOHEXEN-1-ONE, III, 317
  2-Cyclohexene-1-carboxylic acid, 2,6-dimethyl-4-oxo-, ethyl ester [6102-15-4]
2,4-DIMETHYL-5-CARBETHOXYPYRROLE, II, 198
  1*H*-Pyrrole-2-carboxylic acid, 3,5-dimethyl-, ethyl ester [2199-44-2]
N,N-DIMETHYL(CHLOROMETHYLENE)AMMONIUM CHLORIDE, VIII, 498
  Methanaminium, N-(chloromethylene)-N-methyl-, chloride [3724-43-4]
DIMETHYL 3-CHLORO-2-PENTENEDIOATE, VI, 505
  2-Pentenedioic acid, 3-chloro-, dimethyl ester, (E)- [66016-87-3]; (Z)- [66016-88-4]
N,N-DIMETHYL-5β-CHOLEST-3-ENE-5-ACETAMIDE, VI, 491
  Cholest-3-ene-5-acetamide, N,N-dimethyl-, (5β)- [56255-03-9]
4,6-DIMETHYLCOUMALIN, IV, 337
  2*H*-Pyran-2-one, 4,6-dimethyl- [675-09-2]
1,2-DIMETHYL-1,4-CYCLOHEXADIENE, V, 467
  1,4-Cyclohexadiene, 1,2-dimethyl- [17351-28-9]
N,N-DIMETHYLCYCLOHEXANECARBOXAMIDE, IV, 339; **VI**, 492
  Cyclohexanecarboxamide, N,N-dimethyl- [17566-51-7]
2,2-DIMETHYL-1,3-CYCLOHEXANEDIONE, VIII, 312
  1,3-Cyclohexanedione, 2,2-dimethyl- [562-13-0]
5,5-DIMETHYL-1,3-CYCLOHEXANEDIONE, II, 200
  1,3-Cyclohexanedione, 5,5-dimethyl- [126-81-8]

DIMETHYL CYCLOHEXANONE-2,6-DICARBOXYLATE, V, 439
   1,3-Cyclohexanedicarboxylic acid, 2-oxo-, dimethyl ester [25928-05-6]
3,5-DIMETHYL-2-CYCLOHEXEN-1-ONE, III, 317
   2-Cyclohexen-1-one, 3,5-dimethyl- [1123-09-7]
4,4-DIMETHYL-2-CYCLOHEXEN-1-ONE, VI, 496
   2-Cyclohexen-1-one, 4,4-dimethyl- [1073-13-8]
N,N-**DIMETHYLCYCLOHEXYLAMINE, VI**, 499
   Cyclohexanamine, *N,N*-dimethyl- [98-94-2]
N,N-**DIMETHYLCYCLOHEXYLMETHYLAMINE, IV**, 339; *Hazard*
   Cyclohexanemethanamine, *N,N*-dimethyl- [16607-80-0]
4,4-DIMETHYL-2-CYCLOPENTEN-1-ONE, VIII, 208
   2-Cyclopenten-1-one, 4,4-dimethyl- [22748-16-9]
DIMETHYL DECANEDIOATE, VII, 177
   Decanedioic acid, dimethyl ester [106-79-6]
2,4-DIMETHYL-3,5-DICARBETHOXYPYRROLE, II, 202
   1*H*-Pyrrole-2,4-dicarboxylic acid, 3,5-dimethyl-, diethyl ester [2436-79-5]
2,7-DIMETHYL-2,7-DINITROOCTANE, V, 445
   Octane, 2,7-dimethyl-2,7-dinitro- [53684-51-8]
6,6-DIMETHYL-5,7-DIOXASPIRO[2.5]OCTANE-4,8-DIONE, VII, 411
   5,7-Dioxaspiro[2.5]octane-4,8-dione, 6,6-dimethyl- [5617-70-9]
3,3-DIMETHYL-1,5-DIPHENYLPENTANE-1,5-DIONE, VIII, 210
   1,5-Pentanedione, 3,3-dimethyl-1,5-diphenyl- [42052-44-8]
2,6-DIMETHYL-3,5-DIPHENYL-2*H*-PYRAN-4-ONE, V, 450
   4*H*-Pyran-4-one, 2,6-dimethyl-3,5-diphenyl- [33731-54-3]
(*E*)-7,11-DIMETHYL-6,10-DODECADIEN-2-YN-1-OL, VIII, 226
   6,10-Dodecadien-2-yn-1-ol, 7,11-dimethyl-, (*E*)- [16933-56-5]
N,N-**DIMETHYLDODECYLAMINE OXIDE, VI**, 501
   1-Dodecanamine, *N,N*-dimethyl-, *N*-oxide [1643-20-5]
1,2-DIMETHYLENECYCLOHEXANE, VIII, 212
   Cyclohexane, 1,2-bis(methylene)- [2819-48-9]
1,2-DIMETHYL-1,2-EPOXYCYCLOHEX-4-ENE, V, 468
   7-Oxabicyclo[4.1.0]hept-2-ene, 1,6-dimethyl- [57338-10-0]
2,2-DIMETHYLETHYLENIMINE, III, 148
   Aziridine, 2,2-dimethyl- [2658-24-4]
DIMETHYLETHYNYLCARBINOL, III, 320
   3-Butyn-2-ol, 2-methyl- [115-19-5]
DIMETHYLFURAZAN, IV, 342
   Furazan, dimethyl- [4975-21-7]
β,β-**DIMETHYLGLUTARIC ACID, IV**, 345
   Pentanedioic acid, 3,3-dimethyl- [4839-46-7]
DIMETHYLGLYOXIME, II, 204
   2,3-Butanedione, dioxime [95-45-4]
4,4-DIMETHYL-1,6-HEPTADIEN-3-OL, VII, 177
   1,6-Heptadien-3-ol, 4,4-dimethyl- [58144-16-4]
4,6-DIMETHYL-1-HEPTEN-4-OL, V, 452
   1-Hepten-4-ol, 4,6-dimethyl- [32189-75-6]

DIMETHYL 3,3a,3b,4,6a,7a-HEXAHYDRO-3,4,7-METHENO-7*H*-CYCLOPENTA[*a*]PENTALENE-7,8-DICARBOXYLATE, VIII, 298
  3,4,7-Metheno-7*H*-cyclopenta[*a*]pentalene-7,8-dicarboxylic acid,
  3,3a,3b,4,6a,7a-hexahydro-, dimethyl ester [53282-97-6]
DIMETHYL (*E*)-2-HEXENEDIOATE, VIII, 112
  2-Hexenedioic acid, dimethyl ester, (*E*)- [113327-79-0]
5,5-DIMETHYLHYDANTOIN, III, 323
  2,4-lmidazolidinedione, 5,5-dimethyl- [77-71-4]
***unsym*-DIMETHYLHYDRAZINE, II**, 213; *Hazard*
  Hydrazine, 1,1-dimethyl- [57-14-7]
***sym*-DIMETHYLHYDRAZINE DIHYDROCHLORIDE, II**, 208
  Hydrazine, 1,2-dimethyl-, dihydrochloride [306-37-6]
***unsym*-DIMETHYLHYDRAZINE HYDROCHLORIDE, II**, 211
  Hydrazine, 1,1-dimethyl-, monohydrochloride [593-82-8]
***N,N*-DIMETHYLHYDROXYLAMINE HYDROCHLORIDE, IV**, 612
  Methanamine, *N*-hydroxy-*N*-methyl-, hydrochloride [16645-06-0]
2,4-DIMETHYL-3-HYDROXYPENTANOIC ACID, (2*SR*,3*RS*)-, VII, 185
  Pentanoic acid, 3-hydroxy-2,4-dimethyl-, (*R**,*S**)-(±)- [64869-26-7]
2,4-DIMETHYL-3-HYDROXYPENTANOIC ACID, (2*SR*,3*SR*)- VII, 190
  Pentanoic acid, 3-hydroxy-2,4-dimethyl-, (*R**,*R**)- [73198-99-9]
1,3-DIMETHYLIMIDAZOLE-2-THIONE, VII, 195
  2*H*-Imidazole-2-thione, 1,3-dihydro-1,3-dimethyl- [6596-81-2]
1,3-DIMETHYLIMIDAZOLIUM IODIDE, VII, 195
  1*H*-Imidazolium, 1,3-dimethyl-, iodide [4333-62-4]
DIMETHYL 2,3-*O*-ISOPROPYLIDENE-L-TARTRATE, VIII, 155
  1,3-Dioxolane-4,5-dicarboxylic acid, 2,2-dimethyl-, dimethyl ester, (4*R-trans*)-
    [37031-29-1] or (4*S-trans*)-[37031-30-4]
DIMETHYLKETENE, IV, 348
  1-Propen-1-one, 2-methyl- [598-26-5]
DIMETHYLKETENE β-LACTONE DIMER, V, 456
  2-Oxetanone, 3,3-dimethyl-4-(1-methylethylidene)- [3173-79-3]
2,5-DIMETHYLMANDELIC ACID, III, 326
  Benzeneacetic acid, α-hydroxy-2,5-dimethyl- [576-40-5]
1,3-DIMETHYL-3-METHOXY-4-PHENYLAZETIDINONE, VIII, 216
  2-Azetidinone, 3-methoxy-1,3-dimethyl-4-phenyl- [82918-98-7]
**DIMETHYL 3-METHYLENECYCLOBUTANE-1,2-DICARBOXYLATE,
    V**, 459
  1,2-Cyclobutanedicarboxylic acid, 3-methylene-, dimethyl ester [53684-52-9]
DIMETHYL NITROSUCCINATE, VI, 503
  Butanedioic acid, 2-nitro-, dimethyl ester [28081-31-4]
(*E*)-4,8-DIMETHYL-1,3,7-NONATRIENE, VII, 259
  1,3,7-Nonatriene, 4,8-dimethyl-, (*E*)- [19945-61-0]
DIMETHYL OCTADECANEDIOATE, V, 463
  Octadecanedioic acid, dimethyl ester [1472-93-1]
3,7-DIMETHYL-1,6-OCTADIEN-3-AMINE, VII, 507
  1,6-Octadien-3-amine, 3,7-dimethyl- [59875-02-4]

2α,4α-DIMETHYL-8-OXABICYCLO[3.2.1]OCT-6-EN-3-ONE, VI, 512
    8-Oxabicyclo[3.2.1]oct-6-en-3-one, 2,4-dimethyl- (endo, endo) [37081-58-6]
2,7-DIMETHYLOXEPIN, V, 467
    Oxepin, 2,7-dimethyl- [1487-99-6]
1,3-DIMETHYL-5-OXOBICYCLO[2.2.2]OCTANE-2-CARBOXYLIC ACID,
    VIII, 219
    1,3-Dimethyl-5-oxobicyclo[2.2.2]octane-2-carboxylic acid, 1α,2β,3α,4α-
        [121829-82-1]
1,3-DIMETHYL-5-OXOBICYCLO[2.2.2]OCTANE-2-CARBOXYLIC ACID,
    METHYL ESTER, VIII, 219
    1,3-Dimethyl-5-oxobicyclo[2.2.2]octane-2-carboxylic acid, methyl ester,
        1α,2β,3α,4α- [121917-73-5]
2,2-DIMETHYL-5-OXOHEXANAL, VI, 497
    Hexanal, 2,2-dimethyl-5-oxo- [13544-11-1]
2,2-DIMETHYL-4-OXOPENTANAL, VIII, 208
    Pentanal, 2,2-dimethyl-4-oxo- [61031-76-3]
DIMETHYL 2,3-PENTADIENEDIOATE, VI, 505
    2,3-Pentadienoic acid, dimethyl ester [1712-36-3]
2,2-DIMETHYL-4-PENTENAL, VII, 177
    4-Pentenal, 2,2-dimethyl- [5497-67-6]
5,5-DIMETHYL-2-PENTYLTETRAHYDROFURAN, IV, 350
    Furan, tetrahydro-2,2-dimethyl-5-pentyl- [53684-53-0]
α,α-DIMETHYL-β-PHENETHYLAMINE, V, 471
    Benzeneethanamine, α,α-dimethyl- [122-09-8]
2,2-DIMETHYL-4-PHENYLBUTYRIC ACID, VI, 517
    Benzenebutanoic acid, α,α-dimethyl- [4374-44-1]
2,5-DIMETHYL-3-PHENYL-2-CYCLOPENTEN-1-ONE, VI, 520
    2-Cyclopenten-1-one, 2,5-dimethyl-3-phenyl- [36461-43-5]
2',6'-DIMETHYLPHENYL (2SR,3SR)-2,4-DIMETHYL-
    3-HYDROXYPENTANOATE, VII, 190
    Pentanoic acid, 3-hydroxy-2,4-dimethyl-, (R*,R*)-, 2,6-dimethylphenyl ester
        [73198-92-2]
2',6'-DIMETHYLPHENYL PROPANOATE, VII, 190
    Propanoic acid, 2,6-dimethylphenyl ester [51233-80-8]
2,2-DIMETHYL-3-PHENYLPROPIONALDEHYDE, VI, 526
    Benzenepropanal, α,α-dimethyl- [1009-62-7]
2,6-DIMETHYLPHENYLTHIOUREA, V, 802
    Thiourea, (2,6-dimethylphenyl)- [6396-76-5]
DIMETHYL-2-PROPYNYLSULFONIUM BROMIDE, VI, 31
    Sulfonium, dimethyl-2-propynyl-, bromide [23451-62-9]
3,5-DIMETHYLPYRAZOLE, IV, 351
    1H-Pyrazole, 3,5-dimethyl- [67-51-6]
2,6-DIMETHYLPYRIDINE, II, 214
    Pyridine, 2,6-dimethyl- [108-48-5]
2,4-DIMETHYLPYRROLE, II, 217
    1H-Pyrrole, 2,4-dimethyl- [625-82-1]

2,5-DIMETHYLPYRROLE, II, 219
1*H*-Pyrrole, 2,5-dimethyl- [625-84-3]
2,2-DIMETHYLPYRROLIDINE, IV, 354
Pyrrolidine, 2,2-dimethyl- [35018-15-6]
1,5-DIMETHYL-2-PYRROLIDONE, III, 328
2-Pyrrolidinone, 1,5-dimethyl- [5075-92-3]
5,5-DIMETHYL-2-PYRROLIDONE, IV, 357
2-Pyrrolidinone, 5,5-dimethyl- [5165-28-6]
2,4-DIMETHYLQUINOLINE, III, 329
Quinoline, 2,4-dimethyl- [1198-37-4]
***N,N*-DIMETHYLSELENOUREA, IV**, 359
Selenourea, *N,N*-dimethyl- [5117-16-8]
2,3-DIMETHYL-1,4,5,6-TETRAHYDROANTHRAQUINONE, III, 310
9,10-Anthracenedione, 1,4,4a,9a-tetrahydro-2,3-dimethyl-, *cis*- [55538-11-9]
DIMETHYL *cis*-Δ⁴-TETRAHYDROPHTHALATE, IV, 304
4-Cyclohexene-1,2-dicarboxylic acid, dimethyl ester, *cis*- [4841-84-3]
7,9-DIMETHYL-*cis*-8-THIABICYCLO[4.3.0]NONANE 8,8-DIOXIDE, VI, 482
Benzo[*c*]thiophene 2,2-dioxide, octahydro-1,3-dimethyl-, *cis*- [60090-27-9]
2,4-DIMETHYLTHIAZOLE, III, 332
Thiazole, 2,4-dimethyl- [541-58-2]
***N,N*-DIMETHYL-*N'*-*p*-TOLYLFORMAMIDINE, VII**, 197
Methanimidamide, *N,N*-dimethyl-*N'*-(4-methylphenyl)-[7549-96-4]
3,7-DIMETHYL-3-TRICHLOROACETAMIDO-1,6-OCTADIENE, VI, 507
Acetamide, 2,2,2-trichloro-*N*-(1-ethenyl-1,5-dimethyl-4-hexenyl)- [51479-78-8]
(1*R*)-9,9-DIMETHYLTRICYCLO[6.1.1.0²ʻ⁶]DECA-2,5-DIENE, VIII, 223
4,6-Methano-2*H*-indene, 4,5,6,7-tetrahydro-5,5-dimethyl-, (4*R*)- [108404-79-1]
1,6-DIMETHYLTRICYCLO[4.1.0.0²ʼ⁷]HEPT-3-ENE, VII, 200
Tricyclo[4.1.0.0²ʼ⁷]hept-3-ene, 1,6-dimethyl- [61772-32-5]
(*E*)-6,10-DIMETHYL-5,9-UNDECADIEN-1-YNE, VIII, 226
5,9-Undecadien-1-yne, 6,10-dimethyl-, (*E*)- [22850-55-1]
***asym*-DIMETHYLUREA, IV**, 361
Urea, *N,N*-dimethyl- [598-94-7]
DINEOPENTYL KETONE, VI, 991
4-Heptanone, 2,2,6,6-tetramethyl- [4436-99-1]
2,4-DINITROANILINE, II, 221
Benzenamine, 2,4-dinitro- [97-02-9]
2,6-DINITROANILINE, IV, 364
Benzenamine, 2,6-dinitro- [606-22-4]
3,5-DINITROANISOLE, I, 219
Benzene, 1-methoxy-3,5-dinitro- [5327-44-6]
2,4-DINITROBENZALDEHYDE, II, 223
Benzaldehyde, 2,4-dinitro- [528-75-6]
3,5-DINITROBENZALDEHYDE, VI, 529
Benzaldehyde, 3,5-dinitro- [14193-18-1]
***o*-DINITROBENZENE, II**, 226
Benzene, 1,2-dinitro- [528-29-0]

p-DINITROBENZENE, II, 225
 Benzene, 1,4-dinitro- [100-25-4]
2,4-DINITROBENZENESULFENYL CHLORIDE, V, 474
 Benzenesulfenyl chloride, 2,4-dinitro- [528-76-7]
2,5-DINITROBENZOIC ACID, III, 334
 Benzoic acid, 2,5-dinitro- [610-28-6]
3,5-DINITROBENZOIC ACID, III, 337
 Benzoic acid, 3,5-dinitro- [99-34-3]
p,p'-DINITROBIBENZYL, IV, 367
 Benzene, 1,1'-(1,2-ethanediyl)bis[4-nitro- [736-30-1]
2,2'-DINITROBIPHENYL, III, 339
 1,1-Biphenyl, 2,2'-dinitro- [2436-96-6]
1,4-DINITROBUTANE, IV, 368
 Butane, 1,4-dinitro- [4286-49-1]
2,3-DINITRO-2-BUTENE, IV, 374
 2-Butene, 2,3-dinitro-, (*E*)- [24335-43-1]; (*Z*)- [24335-44-2]
DINITRODURENE, II, 254
 Benzene, 1,2,4,5-tetramethyl-3,6-dinitro- [5465-13-4]
3,4-DINITRO-3-HEXENE, IV, 372
 3-Hexene, 3,4-dinitro- [53684-54-1]
2,4-DINITROIODOBENZENE, V, 478
 Benzene, 1-iodo-2,4-dinitro- [709-49-9]
1,4-DINITRONAPHTHALENE, III, 341
 Naphthalene, 1,4-dinitro- [6921-26-2]
1,5-DINITROPENTANE, IV, 370
 Pentane, 1,5-dinitro- [6848-84-6]
2,4-DINITROPHENYL BENZYL SULFIDE, V, 474
 Sulfide, benzyl 2,4-dinitrophenyl- [7343-61-5]
DI-*o*-NITROPHENYL DISULFIDE, I, 220
 Disulfide, bis(2-nitrophenyl)-[1155-00-6]
2,4-DINITROPHENYLHYDRAZINE, II, 228
 Hydrazine, (2,4-dinitrophenyl)- [119-26-6]
1,3-DINITROPROPANE, IV, 370
 Propane, 1,3-dinitro- [6125-21-9]
3,5-DINITRO-*o*-TOLUNITRILE, V, 480
 Benzonitrile, 2-methyl-3,5-dinitro- [948-31-2]
1,4-DIOXASPIRO[4.5]DECANE, V, 303
 1,4-Dioxaspiro[4.5]decane [177-10-6]
1,6-DIOXO-8a-METHYL-1,2,3,4,6,7,8,8a-OCTAHYDRONAPHTHALENE, V, 486
 1,6(2*H*,7*H*)-Naphthalenedione, 3,4,8,8a-tetrahydro-8a-methyl- [20007-72-1]
2,6-DIOXO-1-PHENYL-4-BENZYL-1,4-DIAZABICYCLO[3.3.0]OCTANE,
  VIII, 231
 Pyrrolo[3,4-*c*]pyrrole-1,3(2*H*,3a*H*)-dione, tetrahydro-2-phenyl-5-(phenylmethyl)-,
  cis- [87813-00-1]
DIPHENALDEHYDE, V, 489
 [1,1'-Biphenyl]-2,2'-dicarboxaldehyde [1210-05-5]

DIPHENALDEHYDIC ACID, V, 493
  [1,1'-Biphenyl]-2-carboxylic acid, 2'-formyl- [6720-26-9]
DIPHENIC ACID, I, 222
  [1,1'-Biphenyl]-2,2'-dicarboxylic acid [482-05-3]
DIPHENYLACETALDEHYDE, IV, 375
  Benzeneacetaldehyde, α-phenyl- [947-91-1]
DIPHENYLACETIC ACID, I, 224
  Benzeneacetic acid, α-phenyl- [117-34-0]
DIPHENYLACETONE, III, 343
  2-Propanone, 1,1-diphenyl- [781-35-1]
DIPHENYLACETONITRILE, III, 347
  Benzeneacetonitrile, α-phenyl- [86-29-3]
DIPHENYLACETYL CHLORIDE VI, 549
  Benzeneacetyl chloride, α-phenyl- [1871-76-7]
DIPHENYLACETYLENE, III, 350; **IV,** 377; **V,** 606
  Benzene, 1,1'-(1,2-ethynediyl)bis- [501-65-5]
1,4-DIPHENYL-5-AMINO-1,2,3-TRIAZOLE, IV, 380
  1*H*-1,2,3-Triazol-5-amine, 1,4-diphenyl- [29704-63-0]
N,N'-**DIPHENYLBENZAMIDINE, IV,** 383
  Benzenecarboximidamide, *N,N'*-diphenyl- [2556-46-9]
DIPHENYL-*p*-BROMOPHENYLPHOSPHINE, V, 496
  Phosphine, (4-bromophenyl)diphenyl- [734-59-8]
1,4-DIPHENYL-1,3-BUTADIENE, II, 229; **V,** 499
  Benzene, 1,1'-(1,3-butadiene-1,4-diyl)bis- [886-65-7]
2,3-DIPHENYL-1,3-BUTADIENE, VI, 531
  1,3-Butadiene, 2,3-diphenyl- [2548-47-2]
DIPHENYLCARBODIIMIDE (METHODS I and II), V, 501, 504
  Benzenamine, *N,N'*-methanetetraylbis- [622-16-2]
α,β-**DIPHENYLCINNAMONITRILE, IV,** 387
  Benzeneacetonitrile, α-(diphenylmethylene)- [6304-33-2]
1,1-DIPHENYLCYCLOPROPANE, V, 509
  Benzene, 1,1'-cyclopropylidenebis- [3282-18-6]
DIPHENYLCYCLOPROPENONE, V, 514
  2-Cyclopropen-1-one, 2,3-diphenyl- [886-38-4]
DIPHENYL (CYCLOUNDECYL-1-PYRROLIDINYLMETHYLENE)-
  PHOSPHORAMIDATE, VII, 135
  Phosphoramidic acid, (cycloundecyl-1-pyrrolidinylmethylene)-, diphenyl ester
  [62914-02-7]
DIPHENYLDIACETYLENE, V, 517
  Benzene, 1,1'-(1,3-butadiyne-1,4-diyl)bis- [886-66-8]
DIPHENYLDIAZOMETHANE, III, 351
  Benzene, 1,1'-(diazomethylene)bis- [883-40-9]
DIPHENYL DISELENIDE, VI, 533
  Diselenide, diphenyl- [1666-13-3]
1,1-DIPHENYLETHANE, VI, 537
  Benzene, 1,1'-(1,1-ethanediyl)bis- [612-00-0]

1,1-DIPHENYLETHYLENE, I, 226
  Benzene, 1,1'-ethenylidenebis- [530-48-3]
4,5-DIPHENYLGLYOXALONE, II, 231
  2*H*-lmidazol-2-one, 1,5-dihydro-4,5-diphenyl- [53684-56-3]
2-(DIPHENYLHYDROXYMETHYL)PYRROLIDINE, VI, 542
  2-Pyrrolidinemethanol, α,α-diphenyl-, ($\pm$)- [22348-32-9]
2,3-DIPHENYLINDONE (2,3-DIPHENYL-1-INDENONE), III, 353
  1*H*-Inden-1-one, 2,3-diphenyl- [1801-42-9]
DIPHENYLIODONIUM IODIDE, III, 355
  Iodonium, diphenyl, iodide [2217-79-0]
DIPHENYLKETENE, III, 356; **VI,** 549
  Ethenone, diphenyl [525-06-4]
DIPHENYL KETIMINE, V, 520
  Benzenemethanimine, α-phenyl- [1013-88-3]
DIPHENYLMERCURY, I, 228
  Mercury, diphenyl- [587-85-9]
DIPHENYLMETHANE, II, 232
  Benzene, 1,1'-methylenebis- [101-81-5]
DIPHENYLMETHANE IMINE HYDROCHLORIDE, II, 234
  Benzenemethanimine, α-phenyl-, hydrochloride [5319-67-5]
1,4-DIPHENYL-1-METHYL-3-AZA-1,4-BUTADIENE, VIII, 451, 455
  1-Propen-1-amine, 2-phenyl-*N*-(phenylmethylene)- [64244-34-4]
1,2-DIPHENYL-3-METHYLCYCLOPROPENE, VII, 203
  Benzene, 1,1'-(3-methyl-1-cyclopropene-1,2-diyl)bis- [51425-87-7]
DIPHENYLMETHYL VINYL ETHER, VI, 552
  Benzene, 1,1'-[(ethenyloxy)methylene]bis- [23084-88-0]
N,α-**DIPHENYLNITRONE, V,** 1124
  Benzenamine, *N*-(phenylmethylene)-, *N*-oxide [1137-96-8]
1,1-DIPHENYLPENTANE, V, 523
  Benzene, 1,1'-pentylidenebis- [1726-12-1]
O-**DIPHENYLPHOSPHINYLHYDROXYLAMINE, VII,** 8
  Hydroxylamine, *O*-(diphenylphosphinyl)- [72804-96-7]
DIPHENYL PHOSPHORAZIDATE, VII, 135, 206
  Phosphorazidic acid, diphenyl ester [26386-88-9]
trans-**2,3-DIPHENYL-1-PHTHALIMIDOAZIRIDINE , VI,** 56
  1*H*-Isoindole-1,3(2*H*)-dione, *trans*-($\pm$)-2-(2,3-diphenyl-1-aziridinyl-
    [37079-32-6]
α,β-**DIPHENYLPROPIONIC ACID, V,** 526
  Benzenepropanoic acid, α-phenyl- [3333-15-1]
DIPHENYLPROPIOPHENONE, II, 236
  1-Propanone, 1,3,3-triphenyl- [606-86-0]
2,4-DIPHENYLPYRROLE, III, 358
  1*H*-Pyrrole, 2,4-diphenyl- [3274-56-4]
2,4-DIPHENYL-2-PYRROLINE, III, 358
DIPHENYL SELENIDE, II, 238
  Benzene, 1,1'-selenobis- [1132-39-4]

DIPHENYLSELENIUM DICHLORIDE, II, 240
   Selenium, dichlorodiphenyl- [2217-81-4]
DIPHENYL SUCCINATE, IV, 390
   Butanedioic acid, diphenyl ester [621-14-7]
2,3-DIPHENYLSUCCINONITRILE, IV, 392
   Butanedinitrile, 2,3-diphenyl- [5424-86-2]
DIPHENYL SULFIDE, II, 242
   Benzene, 1,1'-thiobis- [139-66-2]
DIPHENYLTHIOCARBAZIDE, III, 360
   Carbohydrazide, 1,5-diphenyl-3-thio- [622-03-7]
α,α'-DIPHENYLTHIODIGLYCOLIC ACID, VI, 403
   Benzeneacetic acid, α,α'-thiobis-, (R^*,R^*)-(±)-[3442-23-7]; (R^*,S^*)-[2845-49-0]
DIPHENYL TRIKETONE, II, 244
   Propanetrione, diphenyl- [643-75-4]
sym-**DIPHENYLUREA, I**, 453
   Urea, *N,N'*-diphenyl- [102-07-8]
2,3-DIPHENYLVINYLENE SULFONE, VI, 555
   Thiirene, diphenyl-, 1,1-dioxide [5162-99-2]
DISPIRO[5.1.5.1]TETRADECANE-7,14-DIONE, V, 297
   Dispiro[5.1.5.1]tetradecane-7,14-dione [950-21-0]
1,3-DITHIANE, VI, 556
   1,3-Dithiane [505-23-7]
p-**DITHIANE, IV**, 396
   1,4-Dithiane [505-29-3]
3,7-DITHIANONANE-1,9-DIOL, VIII, 592
   Ethanol, 2,2'-[1,3-propanediylbis(thio)]bis- [16260-48-3]
3,7-DITHIANONANE-1,9-DITHIOL, VIII, 592
   Ethanethiol, 2,2'-[1,3-propanediylbis(thio)]bis- [25676-62-4]
1-(1,3-DITHIAN-2-YL)-2-CYCLOHEXEN-1-OL, VIII, 309
   2-Cyclohexen-1-ol, 1-(1,3-dithian-2-yl)- [53178-46-4]
1,4-DITHIASPIRO[4.11]HEXADECANE, VII, 124
   1,4-Dithiaspiro[4.11]hexadecane [16775-67-0]
5,9-DITHIASPIRO[3.5]NONANE, VI, 316
   5,9-Dithiaspiro[3.5]nonane [15077-16-4]
2,2'-DITHIENYL SULFIDE, VI, 558
   Thiophene, 2,2'-thiobis- [3988-99-6]
2,4-DITHIOBIURET, IV, 504; *Hazard*
   Dithioimidodicarbonic diamide [541-53-7]
DITHIZONE, III, 360
   Diazenecarbothioic acid, phenyl-, 2-phenylhydrazide [60-10-6]
1,1-Di-*p*-TOLYLETHANE, I, 229
   Benzene, 1,1'-ethylidenebis[4-methyl-] [530-45-0]
DI-*p*-TOLYLMERCURY, I, 231
   Mercury, bis(4-methylphenyl)- [537-64-4]
trans-**1,2-DIVINYLCYCLOBUTANE, V**, 528
   Cyclobutane, 1,2-diethenyl-, *trans*- [6553-48-6]

cis-1,2-DIVINYLCYCLOBUTANE, V, 528
  Cyclobutane, 1,2-diethenyl-, *cis*- [16177-46-1]
DOCOSANEDIOIC ACID, V, 533
  Docosanedioic acid [505-56-6]
DODECANE, VI, 376
  Dodecane [112-40-3]
trans-2-DODECENOIC ACID, IV, 398
  2-Dodecenoic acid, (*E*)- [32466-54-9]
5-DODECEN-2-ONE, (*E*)-, VIII, 235
  5-Dodecen-2-one, (*E*)- [81953-05-1]
DODECYL BROMIDE, I, 29; II, 246; III, 531
  Dodecane, 1-bromo- [143-15-7]
DODECYL MERCAPTAN, III, 363
  1-Dodecanethiol [112-55-0]
DURENE, II, 248; VI, 703
  Benzene, 1,2,4,5-tetramethyl- [95-93-2]
DUROQUINONE, II, 254
  2,5-Cyclohexadiene-1,4-dione, 2,3,5,6-tetramethyl- [527-17-3]
DYPNONE, III, 367
  2-Buten-1-one, 1,3-diphenyl- [495-45-4]

E

EPIBROMOHYDRIN, II, 256
  Oxirane, (bromomethyl)- [3132-64-7]
EPICHLOROHYDRIN, I, 233; II, 256
  Oxirane, (chloromethyl)- [106-89-8]
2,3-EPOXYCYCLOHEXANONE, VI, 679
  7-Oxabicyclo[4.1.0]heptan-2-one [6705-49-3]
10,11-EPOXYFARNESYL ACETATE, VI, 560
  2,6-Nonadien-1-ol, 9-(3,3-dimethyloxiranyl)-3,7-dimethyl-, acetate, (*E,E*)- [50502-44-8]
ERUCIC ACID, II, 258
  13-Docosenoic acid, (*Z*)- [112-86-7]
D-ERYTHRONOLACTONE, VII, 297
  2(3*H*)-Furanone, dihydro-3,4-dihydroxy-, [*R*-(*R**,*S**)]- [21730-93-8]
ETHANEDITHIOL, IV, 401
  1,2-Ethanedithiol [540-63-6]
(1-ETHENYL)HEPTANYL 3-KETOBUTANOATE, VIII, 235
  Butanoic acid, 3-oxo-, 1-ethenylheptyl ester [133538-60-0]
ETHOXYACETIC ACID, II, 260
  Acetic acid, ethoxy- [627-03-2]
ETHOXYACETYLENE, IV, 404
  Ethyne, ethoxy- [927-80-0]
1-ETHOXY-1-BUTYNE, VI, 564
  1-Butyne, 1-ethoxy- [14272-91-4]
3-ETHOXY-2-CYCLOHEXENONE, V, 539
  2-Cyclohexen-1-one, 3-ethoxy- [5323-87-5]

2-(1-ETHOXYCYCLOHEXYL)-2-(TRIMETHYLSILILOXY)CYCLOBUTANONE, VIII, 579

   Cyclobutanone, 2-(1-ethoxycyclohexyl)-2-[(trimethylsilyl)oxy]- [69152-09-6]

2-[(1'-ETHOXY)-1-ETHOXY]BUTANENITRILE, VII, 381

   Butanenitrile, 2-(1-ethoxy-ethoxy)-, (*R*,R**)- [72658-42-5] or (*R*,S**)- [72658-43-6]

β-ETHOXYETHYL BROMIDE, III, 370

   Ethane, 1-bromo-2-ethoxy- [592-55-2]

4-ETHOXY-3-HYDROXYBENZALDEHYDE, VI, 567

   Benzaldehyde, 4-ethoxy-3-hydroxy- [2539-53-9]

(*E*)-1-ETHOXY-1-(1-HYDROXY-2-BUTENYL)CYCLOPROPANE, VIII, 556

   Cyclopropanemethanol, 1-ethoxy-α-1-propenyl-, (*E*)- [130719-17-4]

4-ETHOXY-3-METHOXYBENZALDEHYDE ETHYLENE ACETAL, VI, 567

   1,3-Dioxolane, 2-(4-ethoxy-3-methoxyphenyl)- [52987-93-6]

3-ETHOXY-6-METHYL-2-CYCLOHEXEN-1-ONE, VII, 208, 241

   2-Cyclohexen-1-one, 3-ethoxy-6-methyl- [62952-33-4]

2-ETHOXY-1-NAPHTHALDEHYDE, III, 98

   1-Naphthalenecarboxaldehyde, 2-ethoxy- [19523-57-0]

***p*-ETHOXYPHENYLUREA, IV**, 52

   Urea, (4-ethoxyphenyl)- [150-69-6]

β-ETHOXYPROPIONALDEHYDE ACETAL, III, 371

   Propane, 1,1,3-triethoxy- [7789-92-6]

β-ETHOXYPROPIONITRILE, III, 372

   Propanenitrile, 3-ethoxy- [2141-62-0]

2-ETHOXYPYRROLIN-5-ONE, VI, 226

   2*H*-Pyrrol-2-one, 5-ethoxy-3,4-dihydro- [29473-56-1]

***cis*-1-ETHOXY-2-*p*-TOLYLCYCLOPROPANE, VI**, 571

   Benzene, 1-(2-ethoxycyclopropyl)-4-methyl-, *cis*- [40237-67-0]

***trans*-1-ETHOXY-2-*p*-TOLYLCYCLOPROPANE, VI**, 571

   Benzene, 1-(2-ethoxycyclopropyl)-4-methyl-, *trans*- [40489-59-6]

1-ETHOXY-1-TRIMETHYLSILOXYCYCLOPROPANE, VII, 131

   Silane, [(1-ethoxycyclopropyl)oxy]trimethyl- [27374-25-0]

ETHYL ACETOACETATE, I, 235

   Butanoic acid, 3-oxo-, ethyl ester [141-97-9]

ETHYL ACETOGLUTARATE, II, 263

   Pentanedioic acid, 2-acetyl-, ethyl ester [1501-06-0]

ETHYL ACETOPYRUVATE, I, 238

   Pentanoic acid, 2,4-dioxo-, ethyl ester [615-79-2]

ETHYL 4-ACETOXYBENZOATE, VI, 576

   Benzoic acid, 4-(acetyloxy)-, ethyl ester [13031-45-3]

ETHYL α-ACETYL-β-(2,3-DIMETHOXYPHENYL)ACRYLATE, IV, 408

   2-Propenoic acid, 2-acetyl-3-(2,3-dimethoxyphenyl)-, ethyl ester [19411-81-5]

ETHYL α-ACETYL-β-(2,3-DIMETHOXYPHENYL)PROPIONATE, IV, 408

   Benzenepropanoic acid, α-acetyl-2,3-dimethoxy-, ethyl ester [53608-80-3]

***N*-ETHYLALLENIMINE, V**, 541

   Aziridine, 1-ethyl-2-methylene- [872-39-9]

4-ETHYL-4-ALLYL-2-CYCLOHEXEN-1-ONE, (*R*)-, VIII, 241

   2-Cyclohexen-1-one, 4-ethyl-4-(2-propenyl)-, (*R*)- [122444-62-6]

ETHYL *p*-AMINOBENZOATE, I, 240; **II**, 300
Benzoic acid, 4-amino-, ethyl ester [94-09-7]
ETHYL 4-AMINO-3-METHYLBENZOATE, VI, 581
Benzoic acid, 4-amino-3-methyl-, ethyl ester [40800-65-5]
ETHYL 4-AMINO-3-(METHYLTHIOMETHYL)BENZOATE, VI, 581
Benzoic acid, 4-amino-3-[(methylthio)methyl]- [50461-34-2]
ETHYL β-ANILINOCROTONATE, III, 374
2-Butenoic acid, 3-(phenylamino)-, ethyl ester [31407-07-5]
ETHYL 3-(1-AZETIDINYL)PROPIONATE, VI, 75
1-Azetidinepropanoic acid, ethyl ester [7730-42-9]
ETHYL BENZOYLACETATE, II, 266; **III**, 379; **IV**, 415
Benzenepropanoic acid, β-oxo, ethyl ester [94-02-0]
ETHYL BENZOYLACETOACETATE, II, 266
Benzenepropanoic acid, α-acetyl-β-oxo, ethyl ester [569-37-9]
ETHYL BENZOYLDIMETHYLACETATE, II, 268
Benzenepropanoic acid, α,α-dimethyl-β oxo-, ethyl ester [25491-42-3]
ETHYL BENZOYLFORMATE, I, 241
Benzeneacetic acid, α-oxo-, ethyl ester [1603-79-8]
ETHYL *N*-BENZYLCARBAMATE, IV, 780
Carbamic acid, (phenylmethyl)-, ethyl ester [2621-78-5]
ETHYL 3-BENZYL-2-CYANO-3-METHYLPENTANOATE, IV, 94
ETHYL BROMIDE, I, 29, 36
Ethane, bromo- [74-96-4]
ETHYL BROMOACETATE, III, 381; *Hazard*
Acetic acid, bromo-, ethyl ester [105-36-2]
ETHYL γ-BROMOBUTYRATE, V, 545
Butanoic acid, 4-bromo-, ethyl ester [2969-81-5]
ETHYL α-(BROMOMETHYL)ACRYLATE, VII, 210; **VIII**, 266
2-Propenoic acid, 2-(bromomethyl)-, ethyl ester [17435-72-2]
ETHYL β-BROMOPROPIONATE, I, 246
Propanoic acid, 3-bromo-, ethyl ester [539-74-2]
ETHYL BUTYLACETOACETATE, I, 248
Hexanoic acid, 2-acetyl-, ethyl ester [1540-29-0]
ETHYL BUTYLCYANOACETATE, III, 385
Hexanoic acid, 2-cyano-, ethyl ester [7391-39-1]
ETHYL *sec*-BUTYLIDENECYANOACETATE, IV, 94
2-Pentenoic acid, 2-cyano-4-methyl-, ethyl ester [868-47-3]
ETHYL *tert*-BUTYL MALONATE, IV, 417
Propanedioic acid, (1,1-dimethyl) ethyl ester [32864-38-3]
ETHYL 2-BUTYRYLACETATE, VII, 213
Pentanoic acid, 4-methyl-3-oxo-, ethyl ester [7152-15-0]
ETHYL α-CARBETHOXY-β-(*m*-CHLOROANILINO)ACRYLATE, III, 272
Propanedioic acid, [[(3-chlorophenyl)amino]methylene], diethyl ester [3412-99-5]
ETHYL CHLOROACETATE, II, 263; *Hazard*
Acetic acid, chloro-, ethyl ester [105-39-5]

N-ETHYL-*p*-CHLOROANILINE, IV, 420
   Benzenamine, 4-chloro-*N*-ethyl- [13519-75-0]
ETHYL CHLOROFLUOROACETATE, IV, 423
   Acetic acid, chlorofluoro-, ethyl ester [401-56-9]
N-ETHYL-*p*-CHLOROFORMANILIDE, IV, 420
   Formamide, *N*-(4-chlorophenyl)-*N*-ethyl- [26772-93-0]
ETHYL α-CHLOROPHENYLACETATE, IV, 169
   Benzeneacetic acid, α-chloro-, ethyl ester [4773-33-5]
ETHYL 5β-CHOLEST-3-ENE-5-ACETATE, VI, 584
   Cholest-3-ene-5-acetic acid, ethyl ester, (5β)- [56101-56-5]
2-ETHYLCHROMONE, III, 387
   4*H*-1-Benzopyran-4-one, 2-ethyl- [14736-30-2]
ETHYL CINNAMATE, I, 252
   2-Propenoic acid, 3-phenyl-, ethyl ester [103-36-6]
ETHYL CYANOACETATE, I, 254
   Acetic acid, cyano-, ethyl ester [105-56-6]
ETHYL 1-CYANO-2-METHYLCYCLOHEXANECARBOXYLATE, VI, 586
   Cyclohexanecarboxylic acid, 1-cyano-2-methyl-, ethyl ester [5231-79-8]
ETHYL (*E*)-2-CYANO-6-OCTENOATE, VI, 586
   6-Octenoic acid, 2-cyano-, ethyl ester, (*E*)- [25143-86-6]
ETHYL CYANO(PENTAFLUOROPHENYL)ACETATE, VI, 873
   Benzeneacetic acid, α-cyano-2,3,4,5,6-pentafluoro-, ethyl ester [2340-87-6]
ETHYL α-CYANO-β-PHENYLACRYLATE, I, 451
   2-Propenoic acid, 2-cyano-3-phenyl-, ethyl ester [2025-40-3]
ETHYL CYCLOHEXYLIDENEACETATE, V, 547
   Acetic acid, cyclohexylidene-, ethyl ester [1552-92-7]
ETHYL 4-CYCLOHEXYLIDENE-4-(TRIMETHYLSILOXY)BUTANOATE,
   VIII, 578
   Butanoic acid, 4-cyclohexylidene-4-(trimethylsilyl)oxy-, ethyl ester [65213-35-6]
ETHYL 4-CYCLOHEXYL-4-OXOBUTANOATE, VIII, 578
   Cyclohexanebutanoic acid, γ-oxo-, ethyl ester [54966-52-8]
ETHYL CYCLOPROPYLPROPIOLATE, VIII, 247
   2-Propynoic acid, 3-cyclopropyl-, ethyl ester [123844-20-2]
ETHYL (*E,Z*)-2,4-DECADIENOATE, VIII, 251
   2,4-Decadienoic acid, ethyl ester [3025-30-7]
ETHYL 3,4-DECADIENOATE, VIII, 251
   3,4-Decadienoic acid, ethyl ester [36186-28-4]
ETHYL DIACETYLACETATE, III, 390
   Butanoic acid, 2-acetyl-3-oxo-, ethyl ester [603-69-0]
ETHYL DIAZOACETATE, III, 392; IV, 424
   Acetic acid, diazo-, ethyl ester [623-73-4]
ETHYL α,β-DIBROMO-β-PHENYLPROPIONATE, II, 270
   Benzenepropanoic acid, α,β-dibromo-, ethyl ester [5464-70-0]
ETHYL 3,5-DICHLORO-4-HYDROXYBENZOATE, III, 267
   Benzoic acid, 3,5-dichloro-4-hydroxy-, ethyl ester [17302-82-8]

ETHYL DIETHOXYACETATE, IV, 427
Acetic acid, diethoxy, ethyl ester [6065-82-3]
ETHYL 3,3-DIETHOXYPROPANOATE, VIII, 254
Propanoic acid, 3,3-diethoxy-, ethyl ester [10601-80-6]
ETHYL 1,3-DIHYDROXY-2-NAPHTHOATE, III, 638
2-Naphthalenecarboxylic acid, 1,3-dihydroxy-, ethyl ester [6843-89-6]
ETHYL 6,7-DIMETHOXY-3-METHYLINDENE-2-CARBOXYLATE, V, 550
1*H*-Indene-2-carboxylic acid, 6,7-dimethoxy-3-methyl-, ethyl ester [5415-54-3]
ETHYL *p*-DIMETHYLAMINOPHENYLACETATE, V, 552
Benzeneacetic acid, 4-(dimethylamino)-, ethyl ester [17078-29-4]
1-ETHYL-3-(3-DIMETHYLAMINO)PROPYLCARBODIIMIDE, V, 555
1,3-Propanediamine, *N*'-(ethylcarbonimidoyl)-*N,N*-dimethyl- [1892-57-5]
1-ETHYL-3-(3-DIMETHYLAMINO)PROPYLCARBODIIMIDE
   HYDROCHLORIDE, V, 555
1,3-Propanediamine, *N*'-(ethylcarbonimidoyl)-*N,N*-dimethyl-, monohydrochloride
   [25952-53-8]
1-ETHYL-3-(3-DIMETHYLAMINO)PROPYLCARBODIIMIDE METHIODIDE,
   V, 555
1-Propanaminium, 3-[(ethylcarbonimidoyl)amino]-*N,N,N*-trimethyl-, iodide
   [22572-40-3]
ETHYL 2,4-DIPHENYLBUTANOATE, V, 559
Benzenebutanoic acid, α-phenyl-, ethyl ester [53608-81-4]
ETHYL 2,3-DIPHENYL-5-METHYLISOXAZOLIDINE-4-CARBOXYLATE, V, 1126
Isoxazolidine-4-carboxylic acid, 5-methyl-2,3-diphenyl-, ethyl ester [19744-10-6]
ETHYL 2-(DIPHENYLSILYLMETHYL)DECANOATE, VIII, 474
Decanoic acid, 2-(methyldiphenylsilyl)-, ethyl ester [89638-16-4]
ETHYLENE CYANOHYDRIN, I, 256; *Hazard*
Propanenitrile, 3-hydroxy-[109-78-4]
2,2-(ETHYLENEDITHIO)CYCLOHEXANONE, VI, 590
1,4-Dithiaspiro[4.5]decan-6-one [27694-08-2]
ETHYENE DITHIOTOSYLATE, VI, 1016
Benzenesulfonothioic acid, 4-methyl-, *S,S*'-1,2-ethanediyl ester [2225-23-2]
ETHYLENE SULFIDE, V, 562
Thiirane [420-12-2]
ETHYLENE THIOUREA, III, 394
2-lmidazolidinethione [96-45-7]
ETHYLENIMINE, IV, 433; *Hazard*
Aziridine [151-56-4]
ETHYL ETHOXYACETATE, II, 260
Acetic acid, ethoxy-, ethyl ester [817-95-8]
ETHYL ETHOXALYLPROPIONATE, II, 272
Pentanedioic acid, 2-oxo-, diethyl ester [5965-53-7]
ETHYL 4-ETHYL-2-METHYL-3-HYDROXYOCTANOATE, IV, 444
ETHYL 3-ETHYL-5-METHYL-4-ISOXAZOLECARBOXYLATE,VI, 592
4-Isoxazolecarboxylic acid, 3-ethyl-5-methyl-, ethyl ester [53064-41-8]

ETHYL 4-ETHYL-2-METHYL-2-OCTENOATE and ETHYL 4-ETHYL-2-METHYL-3-OCTENOATE, IV, 444
ETHYL (1-ETHYLPROPENYL)METHYLCYANOACETATE, III, 397
   3-Pentenoic acid, 2-cyano-3-ethyl-2-methyl-, ethyl ester [53608-83-6]
ETHYL (1-ETHYLPROPYLIDENE)CYANOACETATE, III, 399
   2-Pentenoic acid, 2-cyano-3-ethyl-, ethyl ester [868-04-2]
ETHYL 2-FLUOROHEXANOATE, (*RS*)-, (*R*)-, and (*S*)-, VIII, 258
   Hexanoic acid, 2-fluoro-, ethyl ester, (*RS*)-, (*R*)-, and (*S*)-; (*RS*)- [17841-31-5];
      (*R*)- [124439-29-8]; (*S*)- [124439-31-2]
ETHYL α-(HEXAHYDROAZEPINYLIDENE-2)ACETATE, VIII, 263
   Acetic acid, (hexahydro-2*H*-azepin-2-ylidene)-, ethyl ester, (*Z*)- [70912-51-5]
2-ETHYLHEXANONITRILE, IV, 436
   Hexanenitrile, 2-ethyl- [4528-39-6]
3-ETHYL-1-HEXYNE, VI, 595
   1-Hexyne, 3-ethyl- [76347-58-5]
ETHYL HYDRAZINECARBOXYLATE, III, 404
   Hydrazinecarboxylic acid, ethyl ester [4114-31-2]
ETHYL HYDROGEN SEBACATE, II, 276
   Decanedioic acid, monoethyl ester [693-55-0]
ETHYL (*S*)-(+)-3-HYDROXYBUTANOATE, VII, 215
   Butanoic acid, 3-hydroxy-, (*S*)-, ethyl ester [56816-01-4]
ETHYL 4-HYDROXYCROTONATE, VII, 221
   2-Butenoic acid, 4-hydroxy-, (*E*)-, ethyl ester [10080-68-9]
ETHYL 1-HYDROXYCYCLOHEXYLACETATE, VI, 598
   Cyclohexaneacetic acid, 1-hydroxy-, ethyl ester [1127-01-1]
ETHYL *threo*-[1-(2-HYDROXY-1,2-DIPHENYL)ETHYL]CARBAMATE, VII, 223
   Carbamic acid, (2-hydroxy-1,2-diphenylethyl)-, (*R**,*R**)-, ethyl ester [73197-89-4]
ETHYL β-HYDROXY-β,β–DIPHENYLPROPIONATE, V, 564
   Benzenepropanoic acid, β-hydroxy-β-phenyl-, ethyl ester [894-18-8]
ETHYL α-(HYDROXYMETHYL)ACRYLATE, VIII, 265
   Propenoic acid, 2-(hydroxymethyl)-, ethyl ester [10029-04-6]
3-ETHYL-4-HYDROXYMETHYL-5-METHYLISOXAZOLE, VI, 781
   4-Isoxazolemethanol, 3-ethyl-5-methyl- [53064-42-9]
***N*-ETHYLIDENECYCLOHEXYLAMINE, VI**, 901
   Cyclohexanamine, *N*-ethylidene- [1193-93-7]
ETHYL INDOLE-2-CARBOXYLATE, V, 567
   1*H*-Indole-2-carboxylic acid, ethyl ester [3770-50-1]
ETHYL 4-IODOBUTYRATE, VIII, 274
   Butanoic acid, 4-iodo-, ethyl ester [7425-53-8)
ETHYL ISOCROTONATE, VII, 227
   2-Butenoic acid, (*Z*)-, ethyl ester [6776-19-8]
ETHYL ISOCYANIDE, IV, 438
   Ethane, isocyano- [624-79-3]
ETHYL ISOCYANOACETATE, VI, 620
   Acetic acid, isocyano-, ethyl ester [2999-46-4]

ETHYL ISODEHYDROACETATE, IV, 549
2*H*-Pyran-5-carboxylic acid, 4,6-dimethyl-2-oxo-, ethyl ester [3385-34-0]
ETHYL α-ISOPROPYLACETOACETATE, III, 405
Butanoic acid, 2-acetyl-3-methyl-, ethyl ester [1522-46-9]
ETHYL α-(ISOPROPYLIDENEAMINOOXY)PROPIONATE, V, 1031
Propanoic acid, 2-[[(1-methylethylidene)amino]oxy]-, ethyl ester [54716-29-9]
ETHYL LINOLEATE, III, 526
9,12-Octadecadienoic acid, ethyl ester, (*Z,Z*)- [544-35-4]
ETHYL LINOLENATE, III, 532
9,12,15-Octadecatrienoic acid, ethyl ester, (*Z,Z,Z*)- [1191-41-9]
ETHYL MANDELATE, IV, 169
Benzeneacetic acid, α-hydroxy-, ethyl ester [774-40-3]
**ETHYL 1-(*p*-METHOXYPHENYL)-2-OXOCYCLOHEXANECARBOXYLATE,
    VII,** 229
Cyclohexanecarboxylic acid, 1-(4-methoxyphenyl)-2-oxo-, ethyl ester [95793-86-5]
ETHYL *N*-METHYLCARBAMATE, II, 278
Carbamic acid, methyl-, ethyl ester [105-40-8]
ETHYL 3-METHYLCOUMARILATE, IV, 590
Benzofurancarboxylic acid, 3-methyl-, ethyl ester [22367-82-4]
β-ETHYL-β-METHYLGLUTARIC ACID, IV, 441
Pentanedioic acid, 3-ethyl-3-methyl- [5345-01-7]
ETHYL 2-METHYLINDOLE-5-CARBOXYLATE, VI, 601
1*H*-Indole-5-carboxylic acid, 2-methyl-, ethyl ester [53600-12-7]
2-(3-ETHYL-5-METHYL-4-ISOXAZOLYLMETHYL)CYCLOHEXANONE, VI,
    781
Cyclohexanone, 2-[(3-ethyl-5-methyl-4-isoxazolyl)methyl)]- [53984-03-5]
ETHYL 2-METHYL-3-METHYLTHIOINDOLE-5-CARBOXYLATE, VI, 601
1*H*-Indole-5-carboxylic acid, 2-methyl-3-(methylthio)-, ethyl ester
ETHYL *N*-(2-METHYL-3-NITROPHENYL)FORMIMIDATE, VIII, 493
Methanimidic acid, *N*-(2-methyl-3-nitrophenyl)- [115118-93-9]
ETHYL 4-METHYL-(*E*)-4,8-NONADIENOATE, VI, 606
4,8-Nonadienoic acid, 4-methyl-, ethyl ester, (*E*)- [53359-96-9]
4-ETHYL-2-METHYL-2-OCTENOIC ACID, IV, 444
2-Octenoic acid, 4-ethyl-2-methyl- [6975-97-9]
5-ETHYL-2-METHYLPYRIDINE, IV, 451
Pyridine, 5-ethyl-2-methyl- [104-90-5]
ETHYL 6-METHYLPYRIDINE-2-ACETATE, VI, 611
2-Pyridineacetic acid, 6-methyl-, ethyl ester [5552-83-0]
α-ETHYL-α-METHYLSUCCINIC ACID, V, 572
Butanedioic acid, 2-ethyl-2-methyl- [631-31-2]
ETHYL α-NAPHTHOATE, II, 282
1-Naphthalenecarboxylic acid, ethyl ester [3007-97-4]
ETHYL 1-NAPHTHYLACETATE, VI, 613
1-Naphthaleneacetic acid, ethyl ester [2122-70-5]
ETHYL α-NITROBUTYRATE, IV, 454
Butanoic acid, 2-nitro-, ethyl ester [2531-81-9]

ETHYL (*E*)-4-(4-NITROPHENYL)-4-OXO-2-BUTENOATE, VIII, 268
2-Butenoic acid, 4-(4-nitrophenyl)-4-oxo-, ethyl ester [131504-53-5]
ETHYL *o*-NITROPHENYLPYRUVATE POTASSIUM SALT, V, 567
Propanoic acid, 2-nitro-α-oxo- [5461-32-5]
ETHYL *N*-NITROSO-*N*-BENZYLCARBAMATE, IV, 780
Carbamic acid, nitroso(phenylmethyl)-, ethyl ester [6558-76-5]
**ETHYL *N*-NITROSO-*N*-(*p*-TOLYLSULFONYLMETHYL)CARBAMATE,
    VI**, 981
Carbamic acid, [[(4-methylphenyl)sulfonyl]methyl]nitroso-, ethyl ester
    [2951-53-3]
ETHYL ORTHOCARBONATE, IV, 457
Ethane, 1,1',1'',1'''-[methanetetrayltetrakis(oxy)]tetrakis- [78-09-1]
ETHYL ORTHOFORMATE, I, 258
Ethane, 1,1',1''-[methylidynetris(oxy)]tris- [122-51-0]
ETHYL OXALAMIDRAZONATE, VIII, 597
Acetic acid, hydrazinoimino-, ethyl ester [53085-26-0]
ETHYL OXALYLSUCCINATE, III, 510; **V**, 687 (see **TRIETHYL
    OXALYLSUCCINATE**)
1,2,3-Propanetricarboxylic acid, 1-oxo-, triethyl ester [42126-21-6]
ETHYL 4-OXOHEXANOATE, VI, 615
Hexanoic acid, 4-oxo-, ethyl ester [3249-33-0]
ETHYL 5-OXO-6-METHYL-6-HEPTENOATE, VIII, 274
6-Heptenoic acid, 6-methyl-5-oxo-, ethyl ester [130892-17-0]
ETHYL 2,3-PENTADIENOATE, VII, 232
2,3-Pentadienoic acid, ethyl ester [74268-51-2]
ETHYL β,β-PENTAMETHYLENEGLYCIDATE, IV, 459
1-Oxaspiro[2.5]octane-2-carboxylic acid, ethyl ester [6975-17-3]
ETHYL PHENYLACETATE, I, 270
Benzeneacetic acid, ethyl ester [101-97-3]
ETHYL α-PHENYLACETOACETATE, II, 284
Benzeneacetic acid, α-acetyl-, ethyl ester [5413-05-8]
ETHYL γ-PHENYLBUTYRATE, II, 196
Benzenebutanoic acid, ethyl ester [10031-93-3]
ETHYL PHENYLCYANOACETATE, IV, 461
Benzeneacetic acid, α-cyano-, ethyl ester [4553-07-5]
ETHYL β-PHENYL-β-CYANOPROPIONATE, IV, 804
Benzenepropanoic acid, β-cyano-, ethyl ester [14025-83-3]
ETHYL PHENYLCYANOPYRUVATE, II, 287
Benzenepropanoic acid, β-cyano-α-oxo-, ethyl ester [6362-63-6]
ETHYL(1-PHENYLETHYLIDENE)CYANOACETATE, IV, 463
2-Butenoic acid, 2-cyano-3-phenyl-, ethyl ester [18300-89-5]
ETHYL *N*-PHENYLFORMIMIDATE, IV, 464
Methanimidic acid, *N*-phenyl-, ethyl ester [6780-49-0]
ETHYL β-PHENYL-β-HYDROXYPROPIONATE, III, 408
Benzenepropanoic acid, β-hydroxy-, ethyl ester [5764-85-2]

N-ETHYLPIPERIDINE, V, 575
    Piperidine, 1-ethyl- [766-09-6]
4-ETHYLPYRIDINE, III, 410
    Pyridine, 4-ethyl- [536-75-4]
ETHYL 2-PYRIDYLACETATE, III, 413
    2-Pyridineacetic acid, ethyl ester [2739-98-2]
ETHYL PYRROLE-2-CARBOXYLATE, VI, 618
    1*H*-Pyrrole-2-carboxylic acid, ethyl ester [2199-43-1]
ETHYL β-PYRROLIDINOCROTONATE, VI, 592
    2-Butenoic acid, 3-(1-pyrrolidinyl)-, ethyl ester [54716-02-8]
ETHYL α-(1-PYRROLIDYL)PROPIONATE, IV, 466
    1-Pyrrolidineacetic acid, α-methyl-, ethyl ester [26846-86-6]
ETHYL PYRUVATE, IV, 467
    Propanoic acid, 2-oxo-, ethyl ester [617-35-6]
ETHYL TETROLATE, VII, 226
    2-Butynoic acid, ethyl ester [4341-76-8]
ETHYL THIAZOLE-4-CARBOXYLATE, VI, 620
    4-Thiazolecarboxylic acid, ethyl ester [14527-43-6]
ETHYL THIOAMIDOOXALATE, VIII, 597
    Acetic acid, aminothioxo-, ethyl ester [16982-21-1]
O-ETHYL THIOFORMATE, VI, 620,
    Methanethioic acid, *O*-ethyl ester [29392-46-9]
N-ETHYL-*m*-TOLUIDINE, II, 290
    Benzenamine, *N*-ethyl-3-methyl- [102-27-2]
ETHYL *N*-(*p*-TOLYLSULFONYLMETHYL)CARBAMATE, VI, 981
    Carbamic acid, [[(4-methylphenyl)sulfonyl]methyl]-, ethyl ester [2850-26-2]
ETHYL (*E*)-3-(TRIBUTYLSTANNYL)PROPENOATE, VIII, 268
    2-Propenoic acid, 3-(tributylstannyl)-, ethyl ester, (*E*)- [106335-84-6]
ETHYL TRIDECYLATE, II, 292
    Tridecanoic acid, ethyl ester [28267-29-0]
ETHYL TRIMETHYLSILYLACETATE, VII, 512
    Acetic acid, (trimethylsilyl)-, ethyl ester [4071-88-9]
ETHYL 3-[3-(TRIMETHYLSILOXY)CYCLOHEX-2-ENYL]PROPIONATE,
    VIII, 277
    2-Cyclohexene-1-propanoic acid, 3-[(trimethylsilyl)oxy]-, ethyl ester [90147-64-1]
5-ETHYL-1,4-UNDECADIENE, (*E*)-, VII, 236
    1,4-Undecadiene, 5-ethyl-, (*E*)- [107445-19-2]
1-ETHYNYLCYCLOHEXANOL, III, 416
    Cyclohexanol, 1-ethynyl- [78-27-3]
1,1'-ETHYNYLENE-BIS-CYCLOHEXANOL, IV, 471
    Cyclohexanol, 1,1'-(1,2-ethynediyl)bis- [78-54-6]
8-ETHYNYL-8-METHYL-1,4-DIOXASPIRO[4.5]DEC-6-ENE, VII, 241
    1,4-Dioxaspiro[4.5]dec-6-ene, 8-ethynyl-8-methyl- [73843-26-2]
ETHYNYL *p*-TOLYL SULFONE, VIII, 281
    Benzene, 1-(ethynylsulfonyl)-4-methyl- [13894-21-8]

o-EUGENOL, III, 418
    Phenol, 2-methoxy-6-(2-propenyl)- [579-60-2]

F

α-FARNESENE, VII, 245
    1,3,6,10-Dodecatetraene, 3,7,11-trimethyl- [502-61-4]
FARNESYL ACETATE, VI, 560
    2,6,10-Dodecatrien-1-ol, 3,7,11-trimethyl-, acetate, (*E,E*)- [4121-17-0]
FERROCENE, IV, 473
    Ferrocene [102-54-5]
FERROCENECARBOXYLIC ACID, VI, 625
    Ferrocene, carboxy- [1271-42-7]
FERROCENYLACETONITRILE, V, 578
    Ferrocene, (cyanomethyl)- [1316-91-2]
FLAVONE, IV, 478
    4*H*-1-Benzopyran-4-one, 2-phenyl- [525-82-6]
9-FLUORENECARBOXYLIC ACID, IV, 482
    9*H*-Fluorene-9-carboxylic acid [1989-33-9]
FLUORENONE-2-CARBOXYLIC ACID, III, 420
    9*H*-Fluorene-2-carboxylic acid, 9-oxo- [784-50-9]
1-FLUOROADAMANTANE, VI, 628
    Tricyclo[3.3.1.1^{3,7}]decane, 1-fluoro- [768-92-3]
FLUOROBENZENE, II, 295
    Benzene, fluoro- [462-06-6]
p-FLUOROBENZOIC ACID, II, 299
    Benzoic acid, 4-fluoro- [456-22-4]
2-FLUOROHEPTANOIC ACID, V, 580
    Heptanoic acid, 2-fluoro- [1578-58-1]
2-FLUOROHEPTYL ACETATE, V, 580
    1-Heptanol, 2-fluoro-, acetate [1786-44-3]
16α-FLUORO-3-METHOXY-1,3,5(10)-ESTRATRIEN-17-ONE, VIII, 288
    1,3,5(10)-Estratrien-17-one, 16α-fluoro-3-methoxy- [2383-28-0]
N-FLUOROPYRIDINIUM TRIFLATE, VIII, 287
    Pyridinium, 1-fluoro-, salt with trifluoromethanesulfonic acid (1:1) [107263-95-6]
FORMAMIDINE ACETATE, V, 582
    Methanimidamide, monoacetate [3473-63-0]
4-FORMYLBENZENESULFONAMIDE, VI, 631
    Benzenesulfonamide, 4-formyl- [3240-35-5]
(*S*)-*N*-FORMYL-*O-tert*-BUTYL-L-VALINOL, VIII, 204
    Formamide, *N*-[1-[(1,1-dimethylethoxy)-methyl]-2-methylpropyl]-, (*S*)-
        [90482-04-5]
N-FORMYL-α,α-DIMETHYL-β-PHENETHYLAMINE, V, 471
    Formamide, *N*-(2,2-dimethyl-1-phenylethyl)- [52117-13-2]
N-FORMYLGLYCINE ETHYL ESTER, VI, 620
    Glycine, *N*-formyl-, ethyl ester [3154-51-6]

(S)-(−)-1-FORMYL-2-HYDROXYMETHYLPYRROLIDINE, VIII, 26
1-Pyrrolidinecarboxaldehyde, 2-(hydroxymethyl)-, (S)-(−)- [55456-46-7]
(S)-(−)-1-FORMYL-2-METHOXYMETHYLPYRROLIDINE, VIII, 26
1-Pyrrolidinecarboxaldehyde, 2-(methoxymethyl)-, (S)-(−)- [63126-45-4]
FORMYL-DL-O-METHYLTHREONINE, III, 814
5-FORMYL-4-PHENANTHROIC ACID, IV, 484
4-Phenanthrenecarboxylic acid, 5-formyl- [5684-15-1]
o-FORMYLPHENOXYACETIC ACID, V, 251
Acetic acid, 2-formylphenoxy- [6280-80-4]
1-FORMYL-3-THIOSEMICARBAZIDE, V, 1070
Semicarbazide, 1-formyl-3-thio- [2302-84-3]
(S)-N-FORMYL-L-VALINOL, VIII, 204
Formamide, 1-hydroxymethyl-2-methylpropyl-, (S)- [89876-66-4]
FUMARAMIDE, IV, 486
2-Butenediamide, (E)- [627-64-5]
FUMARIC ACID, II, 302
2-Butenedioic acid, (E)- [110-17-8]
FUMARONITRILE, IV, 486
2-Butenedinitrile, (E)- [764-42-1]
FUMARYL CHLORIDE, III, 422
2-Butenedioyl dichloride, (E)- [627-63-4]
FURAN, I, 274
Furan [110-00-9]
2-FURANCARBOXYLIC ACID, I, 276; **2-FUROIC ACID, IV,** 493
2-Furancarboxylic acid [88-14-2]
2(5H)-FURANONE, VIII, 397; **γ-CROTONOLACTONE, V,** 255
2(5H)-Furanone [497-23-4]
FURFURAL, I, 280
2-Furancarboxaldehyde [98-01-1]
2-FURFURALACETONE, I, 283
3-Buten-2-one, 4-(2-furanyl)- [623-15-4]
FURFURAL DIACETATE, IV, 489
Methanediol, 2-furanyl, diacetate [613-75-2]
2-FURFURYL MERCAPTAN, IV, 491
2-Furanmethanethiol [98-02-2]
FURYLACRYLIC ACID, III, 425
2-Propenoic acid, 3-(2-furanyl)- [539-47-9]
3-(2-FURYL)ACRYLONITRILE, V, 585
2-Propenenitrile, 3-(2-furanyl)- [7187-01-1]
2-FURYLCARBINOL, I, 276
2-Furanmethanol [98-00-0]
4-(2-FURYL)-4-HYDROXY-2,3-DIETHOXYCYCLOBUTENONE, VIII, 179
2-Cyclobuten-1-one, 2,3-diethoxy-4-(2-furanyl)-4-hydroxy- [138225-12-4]
2-FURYLMETHYL ACETATE, I, 285
2-Furanmethanol, acetate [623-17-6]

FURYL PHOSPHORODICHLORIDATE, VIII, 396
  Phosphorodichloridic acid, 2-furanyl ester [105262-70-2]
FURYL *N,N,N',N'*-TETRAMETHYLDIAMIDOPHOSPHATE, VIII, 396
  Phosphorodiamidic acid, tetramethyl-, 2-furanyl ester [105262-58-6]

G

GALLACETOPHENONE, II, 304
  Ethanone,1-(2,3,4-trihydroxyphenyl)- [528-21-2]
β-GENTIOBIOSE OCTAACETATE, III, 428
  β-D-Glucopyranose, 6-*O*-(2,3,4,6-tetra-*O*-acetyl-β–D-glucopyranosyl)-, tetraacetate
  [4613-78-9]
GERANIAL, VII, 258
  2,6-Octadienal, 3,7-dimethyl- [5392-40-5]
GERANIOL TRICHLOROACETIMIDATE, VI, 508
  Ethanimidic acid, 2,2,2-trichloro-, 3,7-dimethyl-2,6-octadienyl ester, (*E*)- [51479-75-5]
GERANYL CHLORIDE, VI, 634, 638; **VIII**, 226, 616
  2,6-Octadiene, 1-chloro-3,7-dimethyl-, (*E*)- [5389-87-7]
D-GLUCOSAMINE HYDROCHLORIDE, III, 430
  D-Glucose, 2-amino-2-deoxy-, hydrochloride [66-84-2]
β–D-GLUCOSE-1,2,3,4-TETRAACETATE, III, 432
  β-D-Glucopyranose, 1,2,3,4-tetraacetate [13100-46-4]
β–D-GLUCOSE-2,3,4,6-TETRAACETATE, III, 434
  β–D-Glucopyranose, 2,3,4,6-tetraacetate [3947-62-4]
GLUTACONALDEHYDE SODIUM SALT, VI, 640
  2-Pentenedial, ion(1⁻), sodium [24290-36-6]
D-GLUTAMIC ACID, I, 286
  D-Glutamic acid [6893-26-1]
GLUTARIC ACID, I, 289; **IV**, 496
  Pentanedioic acid [110-94-1]
GLUTARIMIDE, IV, 496
  2,6-Piperidinedione [1121-89-7]
DL-GLYCERALDEHYDE, II, 305
  Propanal, 2,3-dihydroxy-, (±)- [56-82-6]
DL-GLYCERALDEHYDE DIETHYL ACETAL, II, 307
  1,2-Propanediol, 3,3-diethoxy-, (±)- [10487-05-5]
GLYCEROL α,γ-DIBROMOHYDRIN, II, 308
  2-Propanol, 1,3-dibromo- [96-21-9]
GLYCEROL α,γ-DICHLOROHYDRIN, I, 292
  2-Propanol, 1,3-dichloro- [96-23-1]
GLYCEROL α-MONOCHLOROHYDRIN, I, 294
  1,2-Propanediol, 3-chloro- [96-24-2]
α-GLYCERYL PHENYL ETHER, I, 296
  1,2-Propanediol, 3-phenoxy- [538-43-2]
GLYCINE, I, 298
  Glycine [56-40-6]
GLYCINE *tert*-BUTYL ESTER, V, 586
  Glycine, 1,1-dimethylethyl ester [6456-74-2]

GLYCINE ETHYL ESTER HYDROCHLORIDE, II, 310
    Glycine, ethyl ester, hydrochloride [623-33-6]
GLYCOLONITRILE, III, 436
    Acetonitrile, hydroxy- [107-16-4]
GLYOXAL BISULFITE, III, 438
    1,2-Ethanedisulfonic acid, 1,2-dihydroxy- [18381-20-9]
GLYOXYLIC ACID (and ACID CHLORIDE) *p*-TOLUENESULFONYLHYDRA-
    ZONE, V, 258
    Glyoxylic acid *p*-toluenesulfonylhydrazone [14661-68-8]
    Glyoxylic acid chloride 2-(*p*-toluenesulfonylhydrazone) [14661-69-9]
GUANIDOACETIC ACID, III, 440
    Glycine, *N*-(aminoiminomethyl)- [352-97-6]
GUANYLTHIOUREA, IV, 502
    Thiourea, (aminoiminomethyl)- [2114-02-5]
GUAIACOL ALLYL ETHER, III, 418
    Benzene, 1-allyloxy-2-methoxy- [4125-43-3]
D-GULONIC-γ-LACTONE, IV, 506
    D-Gulonic acid, γ-lactone [6322-07-2]

H

HEMIMELLITENE, IV, 508
    Benzene, 1,2,3-trimethyl- [526-73-8]
HEMIN, III, 442
    Ferrate (2⁻), chloro[7,12-diethenyl-3,8,13,17-tetramethyl-21H,23H-porphine-
    2,18-dipropanoato(4–)-N^{21},N^{22},N^{23},N^{24}-, (SP-5-13)- [16009-13-5]
HENDECANEDIOIC ACID, IV, 510
    Undecanedioic acid [1852-04-6]
***unsym*-HEPTACHLOROPROPANE, II**, 312
    Propane, 1,1,1,2,2,3,3-heptachloro- [594-89-8]
HEPTALDEHYDE ENOL ACETATE, III, 127
    1-Hepten-1-ol, acetate (*E*)- [17574-85-5]; (*Z*)- [17574-86-6]
HEPTALDOXIME, II, 313
    Heptanal, oxime [629-31-2]
HEPTAMIDE, IV, 513
    Heptanamide [628-62-6]
HEPTANAL, VI, 644, 650
    Heptanal [111-71-7]
2,5-HEPTANEDIONE, VI, 648
    2,5-Heptanedione [1703-51-1]
HEPTANOIC ACID, II, 315
    Heptanoic acid [111-14-8]
2-HEPTANOL, II, 317
    2-Heptanol [543-49-7]
4-HEPTANONE, V, 589; *Hazard*
    4-Heptanone [123-19-3]
HEPTOIC ANHYDRIDE, III, 28
    Heptanoic acid, anhydride [626-27-7]

HEPTYL ALCOHOL, I, 304
   1-Heptanol [111-70-6]
HEPTYLAMINE, II, 318; *Hazard*
   1-Heptanamine [111-68-2]
3-HEPTYL-5-CYANOCYTOSINE, IV, 515
   5-Pyrimidinecarbonitrile, 1-heptyl-1,2,3,6-tetrahydro-6-imino-2-oxo- [53608-90-5]
2-HEPTYL-2-CYCLOHEXENONE, VII, 249
   2-Cyclohexen-1-one, 2-heptyl- [87588-68-9]
o-**HEPTYLPHENOL, III**, 444
   Phenol, 2-heptyl- [5284-22-0]
N-**HEPTYLUREA, IV**, 515
   Urea, heptyl- [42955-46-4]
3-HEPTYLUREIDOMETHYLENEMALONONITRILE, IV, 515
1,4,7,10,13,16-HEXAAZACYCLOOCTADECANE, VI, 652
   1,4,7,10,13,16-Hexaazacyclooctadecane [296-35-5]
9,10,12,13,15,16-HEXABROMOSTEARIC ACID, III, 531
   Octadecanoic acid, 9,10,12,13,15,16-hexabromo- [4167-08-2]
**HEXACARBONYL (1-METHYL-2-PROPYNYLIUM)DICOBALT
   TETRAFLUOROBORATE, VIII**, 460
   Cobalt(1+), hexacarbonyl[μ-[(2,3-η:2,3-η)-1-methyl-2-propynylium]]di-, (Co-Co),
   tetrafluoroborate(1⁻) [62866-98-2]
(5Z,7E)-5,7-HEXADECADIENE, VIII, 295
   5,7-Hexadecadiene, (5Z,7E)- [86883-79-6]
HEXADECANE, II, 320
   Hexadecane [544-76-3]
HEXADECYL IODIDE, II, 322
   Hexadecane, 1-iodo- [544-77-4]
HEXADEHYDRO[18]ANNULENE, VI, 68
   1,3,7,9,13,15-Cyclooctadecahexayne [7092-39-9]
(Z,Z)-2,4-HEXADIENEDINITRILE, VI, 306, 662
   (Z,Z)-2,4-Hexadienedinitrile [1557-59-1]
1,5-HEXADIEN-3-OL, V, 608
   1,5-Hexadien-3-ol [924-41-4]
HEXAFLUOROACETONE, VII, 251; *Hazard*
   2-Propanone, 1,1,1,3,3,3-hexafluoro- [684-16-2]
HEXAFLUOROACETONE HYDRAZONE, VI, 161
   2-Propanone, 1,1,1,3,3,3-hexafluoro-, hydrazone [1588-41-6]
HEXAFLUOROACETONE IMINE, VI, 664
   2-Propanimine, 1,1,1,3,3,3-hexafluoro- [1645-75-6]
HEXAHYDRO-2-(1*H*)-AZOCINONE, VII, 254
   2(1*H*)-Azocinone, hexahydro- [673-66-5]
cis-**4,4a,5,6,7,8-HEXAHYDRO-4a,5-DIMETHYL-2(3*H*)-NAPHTHALENONE,
   VI**, 666
   2(3*H*)-Naphthalenone, 4,4a,5,6,7,8-hexahydro-4a,5-dimethyl-, *cis*-, (±)-
   [20536-80-5]

HEXAHYDRO-6-ETHYL-3-(HYDROXYMETHYL)-6-ALLYL-2-
  PHENYL[2S,3S,8aR]- 5-OXO-5H-OXAZOLO[3,2-a]PYRIDINE, VIII, 241
  5H-Oxazolo[3,2-a]pyridin-5-one, 6-ethylhexahydro-3-(hydroxymethyl)- 8a-methyl-
  2-phenyl-6-(2-propenyl)-, [2S-(2α,3β,6α, 8aβ)] [122444-61-5];
  [2S-(2α,3β,6β, 8aβ)] [122518-73-4]
HEXAHYDROGALLIC ACID, V, 591
  Cyclohexanecarboxylic acid, 3,4,5-trihydroxy-, (1α,3α,4α,5α)- [53796-39-7]
HEXAHYDROGALLIC ACID TRIACETATE, V, 591
  Cyclohexanecarboxylic acid, 3,4,5-tris(acetyloxy)-, (1α,3α,4α,5α)- [53796-40-0]
(3aS,7aS)-(+)-2,3,3a,4,7,7a-HEXAHYDRO-3a-HYDROXY-7a-METHYL-1H-
  INDENE- 1,5(6H)-DIONE, VII, 363
  1H-Indene-1,5(4H)-dione, hexahydro-3a-hydroxy-7a-methyl-, (3aS-cis)-
  [33879-04-8]
HEXAHYDRO-3-(HYDROXYMETHYL)-8a-METHYL-2-PHENYL[2S,3S,8aR]-5-
  OXO- 5H-OXAZOLO-[3,2-a]PYRIDINE, VIII, 241
  5H-Oxazolo[3,2-a]pyridin-5-one, hexahydro-3-hydroxymethyl)-8a-methyl-
  2-phenyl-, [2S-(2α,3β,8aβ)]- [116950-01-7]
3,3a,3b,4,6a,7a-HEXAHYDRO-3,4,7-METHENO-7H-CYCLOPENTA[a]PENTAL-
  ENE- 7,8-DICARBOXYLIC ACID, VIII, 298
  3,4,7-Metheno-7H-cyclopenta[a]pentalene-7,8-dicarboxylic acid, 3,3a,3b,4,6a,7a-
  hexahydro- [61206-25-5]
HEXAHYDRO-1,3,3,6-TETRAMETHYL-2,1-BENZISOXAZOLINE, VI, 670
  2,1-Benzisoxazoline, hexahydro-1,3,3,6-tetramethyl-, stereoisomer [6603-39-0]
HEXAHYDRO-4,4,7-TRIMETHYL-4H-1,3-BENZOXATHIIN, VIII, 304
  4H-1,3-Benzoxathiin, hexahydro-4,4,7-trimethyl- [59324-06-0]
HEXAHYDRO-1,3,5-TRIPROPIONYL-s-TRIAZINE, IV, 518
  1,3,5-Triazine, hexahydro-1,3,5-tris(1-oxopropyl)- [30805-19-7]
HEXAHYDROXYBENZENE, V, 595
  Benzenehexol [608-80-0]
1,4,7,10,13,16-HEXAKIS(p-TOLYLSULFONYL)-1,4,7,10,13,16-
  HEXAAZACYCLOOCTADECANE, VI, 652
  1,4,7,10,13,16-Hexaazacyclooctadecane, 1,4,7,10,13,16-
  hexakis[(methylphenyl)sulfonyl]- [52601-75-9]
HEXALDEHYDE, II, 323
  Hexanal [66-25-1]
HEXAMETHYLBENZENE, II, 250; **IV**, 520
  Benzene, hexamethyl- [87-85-4]
2,3,4,5,6,6-HEXAMETHYL-2,4-CYCLOHEXADIEN-1-ONE, V, 598
  2,4-Cyclohexadien-1-one, 2,3,4,5,6,6-hexamethyl- [3854-96-4]
HEXAMETHYL DEWAR BENZENE, VII, 256
  Bicyclo[2.2.0]hexa-2,5-diene, 1,2,3,4,5,6-hexamethyl- [7641-77-2]
HEXAMETHYLENE CHLOROHYDRIN, III, 446
  1-Hexanol, 6-chloro- [2009-83-8]
HEXAMETHYLENE DIISOCYANATE, IV, 521
  Hexane, 1,6-diisocyanato- [822-06-0]

HEXAMETHYLENE GLYCOL, II, 325
   1,6-Hexanediol [629-11-8]
HEXAMETHYLPHOSPHOROUS TRIAMIDE, V, 602
   Phosphorous triamide, hexamethyl-[1608-26-0]
1-HEXANOL (HEXYL ALCOHOL), I, 306; **VI**, 919
   1-Hexanol [111-27-3]
HEXAPHENYLBENZENE, V, 604
   1,1':2',1''-Terphenyl, 3',4',5',6'-tetraphenyl- [992-04-1]
1,2,4,5-HEXATETRAENE (in solution), VII, 485
   1,2,4,5-Hexatetraene [29776-96-3]
1,3,5-HEXATRIENE, V, 608
   1,3,5-Hexatriene [2235-12-3]
(E)-4-HEXEN-1-OL, VI, 675
   (E)-4-Hexen-1-ol [6126-50-7]
(E)-1-HEXENYL-1,3,2-BENZODIOXABOROLE, VIII, 532
   1,3,2-Benzodioxaborole, 2-(1-hexenyl)-, (E)- [37490-22-5]
3-HEXYLCYCLOBUTANONE, VIII, 306
   Cyclobutanone, 3-hexyl- [138173-74-7]
HEXYL FLUORIDE, IV, 525
   Hexane, 1-fluoro- [373-14-8]
5-HEXYNAL, VI, 679
   5-Hexynal [29329-03-1]
HIPPURIC ACID, II, 328
   Glycine, N-benzoyl- [495-69-2]
L-HISTIDINE MONOHYDROCHLORIDE, II, 330
   L-Histidine, monohydrochloride [645-35-2]
HOMOGERANIOL, VII, 258
   3,7-Nonadien-1-ol, 4,8-dimethyl-, (E)- [459-88-1]
HOMOPHTHALIC ACID, III, 449; **V**, 612
   Benzeneacetic acid, 2-carboxy- [89-51-0]
HOMOPHTHALIC ACID ANHYDRIDE, III, 449
   1H-2-Benzopyran-1,3(4H)-dione [703-59-3]
HOMOVERATRIC ACID, II, 333
   Benzeneacetic acid, 3,4-dimethoxy- [93-40-3]
o-HYDRAZINOBENZOIC ACID HYDROCHLORIDE, III, 475
   Benzoic acid, 2-hydrazino-, hydrochloride [33906-30-8]
α-HYDRINDONE, II, 336
   1H-Inden-1-one, 2,3-dihydro- [83-33-0]
HYDROCINNAMIC ACID, I, 311
   Benzenepropanoic acid [501-52-0]
HYDROQUINONE DIACETATE, III, 452
   1,4-Benzenediol, diacetate [1205-91-0]
α-HYDROXYACETOPHENONE, VII, 263
   Ethanone, 2-hydroxy-1-phenyl- [582-24-1]
α-HYDROXYACETOPHENONE DIMETHYL ACETAL, VII, 263
   Ethanone, 2-hydroxy-1-phenyl-, dimethylacetal [28203-05-6]

m-HYDROXYBENZALDEHYDE, III, 453, 564
    Benzaldehyde, 3-hydroxy- [100-83-4]
5-(p-HYDROXYBENZAL)HYDANTOIN, V, 627
    2,4-Imidazolidinedione, 5-[(4-hydroxyphenyl)methylene]- [80171-33-1]
p-HYDROXYBENZOIC ACID, II, 341
    Benzoic acid, 4-hydroxy- [99-96-7]
2-HYDROXYBUTANENITRILE, VII, 381
    Butanenitrile, 2-hydroxy- [4476-02-2]
4-HYDROXY-1-BUTANESULFONIC ACID SULTONE, IV, 529
    1,2-Oxathiane, 2,2-dioxide [1633-83-6]
3-(1-HYDROXYBUTYL)-1-METHYLPYRROLE, VII, 102
    1H-Pyrrole-3-methanol, 1-methyl-α-propyl- [70702-66-8]
2-HYDROXYCINCHONINIC ACID, III, 456
    4-Quinolinecarboxylic acid, 1,2-dihydro-2-oxo- [15733-89-8]
3-HYDROXYCINCHONINIC ACID, V, 636
    4-Quinolinecarboxylic acid, 3-hydroxy- [118-13-8]
2-HYDROXYCYCLOBUTANONE, VI, 167
    Cyclobutanone, 2-hydroxy- [17082-63-2]
3-HYDROXY-1-CYCLOHEXENE-1-CARBOXALDEHYDE, VIII, 309
    1-Cyclohexene-1-carboxaldehyde, 3-hydroxy- [67252-14-6]
o-HYDROXYDIBENZOYLMETHANE, IV, 479
    1,3-Propanedione, 1-(2-hydroxyphenyl)-3-phenyl- [1469-94-9]
2-HYDROXY-3,5-DIIODOBENZOIC ACID, II, 343
    Benzoic acid, 2-hydroxy-3,5-diiodo- [133-91-5]
(S)-(+)-3-HYDROXY-2,2-DIMETHYLCYCLOHEXANONE, VIII, 312
    Cyclohexanone, 3-hydroxy-2,2-dimethyl-, (S)- [87655-21-8]
N-(3-HYDROXY-3,3-DIPHENYLPROPYLIDENE)CYCLOHEXYLAMINE, VI, 901
    Benzenemethanol, α-[2-cyclohexylimino)ethyl]-α-phenyl- [1235-46-7]
β-(2-HYDROXYETHYLMERCAPTO)PROPIONITRILE, III, 458
    Propanenitrile, 3-[(2-hydroxyethyl)thio]- [15771-37-6]
β-HYDROXYETHYL METHYL SULFIDE, II, 345
    Ethanol, 2-(methylthio)- [5271-38-5]
3-HYDROXYGLUTARONITRILE, V, 614
    Pentanedinitrile, 3-hydroxy- [13880-89-2]
trans-4-HYDROXY-2-HEXENAL, VI, 683
    2-Hexenal, 4-hydroxy-, (E)- [17427-21-3]
HYDROXYHYDROQUINONE TRIACETATE, I, 317
    1,2,4-Benzenetriol, triacetate [613-03-6]
2-HYDROXYIMINO-2-PHENYLACETONITRILE, VI, 199
    Benzeneacetonitrile, α-(hydroxyimino)- [825-52-5]
2-HYDROXYISOPHTHALIC ACID, V, 617
    1,3-Benzenedicarboxylic acid, 2-hydroxy- [606-19-9]
N-(HYDROXYMETHYL)ACETAMIDE, VI, 5
    Acetamide, N-(hydroxymethyl) [625-51-4]
2-(HYDROXYMETHYL)ALLYLTRIMETHYLSILANE, VII, 266
    2-Propen-1-ol, 2-[(trimethylsilyl)methyl]- [81302-80-9]

1-(HYDROXYMETHYL)CYCLOHEXANOL, VIII, 315
    Cyclohexanemethanol, 1-hydroxy- [15753-47-6]
2-HYDROXYMETHYL-2-CYCLOPENTEN-1-ONE, VII, 271
    2-Cyclopentenone, 2-hydroxymethyl- [68882-71-3]
2-HYDROXYMETHYL-2-CYCLOPENTEN-1-ONE ETHYLENE KETAL, VII, 274
    1,4-Dioxaspiro[4.4]non-6-ene-6-methanol [80963-19-5]
2-HYDROXYMETHYLENECYCLOHEXANONE, IV, 536, 537
    Cyclohexanone, 2-hydroxymethylene- [823-45-0]
HYDROXYMETHYLFERROCENE, V, 621
    Ferrocene, (hydroxymethyl)- [1273-86-5]
4(5)-HYDROXYMETHYLIMIDAZOLE HYDROCHLORIDE, III, 460
    1H-Imidazole-4-methanol, monohydrochloride [32673-41-9]
2-HYDROXY-3-METHYLISOCARBOSTYRIL, V, 623
    1(2H)-Isoquinolinone, 2-hydroxy-3-methyl- [7114-79-6]
**1-HYDROXYMETHYL-4-(1-METHYLCYCLOPROPYL)-1-CYCLOHEXENE,
VIII,** 321
    1-Cyclohexene-1-methanol, 4-(1-methylcyclopropyl)- [98678-72-9]
2-HYDROXY-2-METHYLPENTAN-3-ONE, VII, 381
    3-Pentanone, 2-hydroxy-2-methyl- [2834-17-5]
3-HYDROXY-3-METHYL-1-PHENYL-1-BUTANONE, VIII, 210, 323
    1-Butanone, 3-hydroxy-3-methyl-1-phenyl- [43108-74-3]
1-(3-HYDROXY-2-METHYL-3-PHENYLPROPANOYL)PIPERIDINE, *erythro-*,
VIII, 326
    Piperidine, 1-(3-hydroxy-2-methyl-1-oxo-3-phenylpropyl)-, (R^*,R^*)-(±)-
    [99114-36-0]
1-(3-HYDROXY-2-METHYL-3-PHENYLPROPANOYL)PIPERIDINE, *threo-* **VIII,**
326
    Piperidine, 1-(3-hydroxy-2-methyl-1-oxo-3-phenylpropyl)-, (R^*,S^*)-(±)-
    [116596-04-4]
3-HYDROXY-2-METHYLPROPIOPHENONE, V, 624
    1-Propanone, 1-(3-hydroxy-2-methylphenyl)- [3338-15-6]
(S)-(+)-2-HYDROXYMETHYLPYRROLIDINE, VIII, 26
    2-Pyrrolidinemethanol, (S)-(+)- [23356-96-9]
2-HYDROXY-1-NAPHTHALDEHYDE, III, 463
    1-Naphthalenecarboxaldehyde, 2-hydroxy- [708-06-5]
2-HYDROXY-1,4-NAPHTHOQUINONE, III, 465
    1,4-Naphthalenedione, 2-hydroxy- [83-72-7]
6-HYDROXYNICOTINIC ACID, IV, 532
    3-Pyridinecarboxylic acid, 1,6-dihydro-6-oxo- [5006-66-6]
2-HYDROXY-5-NITROBENZYL CHLORIDE, III, 468
    Phenol, 2-(chloromethyl)-4-nitro- [2973-19-5]
3-HYDROXY-2-NITROCYCLOHEXYL ACETATE, (1S,2S,3R)-, VIII, 332
    1,3-Cyclohexanediol, 2-nitro-, 1-acetate, [1S-(1α,2β,3α)]- [108186-61-4]
4-HYDROXYNONA-1,2-DIENE, VII, 276
    1,2-Nonadien-4-ol [73229-28-4]

3-HYDROXY-1-NONENE, VIII, 237
1-Nonen-3-ol [21964-44-3]
17β-HYDROXY-5-OXO-3,5-*seco*-4-NORANDROSTANE-3-CARBOXYLIC ACID, VI, 690
1*H*-Benz[*e*]indene-6-propanoic acid, dodecahydro-3-hydroxy-3a,6-dimethyl-7-oxo-, [3*S*-(3α,3aα,5aβ,6β,9aα,9bβ)]- [1759-35-9]
5-HYDROXYPENTANAL, III, 470
Pentanal, 5-hydroxy- [4221-03-8]
α-HYDROXYPHENAZINE, III, 754
1-Phenazinol [528-71-2]
***threo*-4-HYDROXY-3-PHENYL-2-HEPTANONE, VI**, 692
2-Heptanone, 4-hydroxy-3-phenyl-, (*R*,R**)- [42052-62-0]
(2*R**,3*R**)-3-HYDROXY-3-PHENYL-2-METHYLPROPANOIC ACID, VIII, 339
Benzenepropanoic acid, β-hydroxy-α-methyl- (2*R**,3*S**)- [14366-87-1]
***p*-HYDROXYPHENYLPYRUVIC ACID, V**, 627
Benzenepropanoic acid, 4-hydroxy-α-oxo- [156-39-8]
(1-HYDROXY-2-PROPENYL)TRIMETHYLSILANE, VIII, 501
2-Propen-1-ol, 1-(trimethylsilyl)- [95061-68-0]
β-HYDROXYPROPIONIC ACID, I, 321
Propanoic acid, 3-hydroxy- [503-66-2]
3-HYDROXYPYRENE, V, 632
1-Pyrenol [5315-79-7]
3-HYDROXYQUINOLINE, V, 635
3-Quinolinol [580-18-7]
3-HYDROXYTETRAHYDROFURAN, IV, 534
3-Furanol, tetrahydro- [453-20-3]
7-HYDROXY-4,4,6,7-TETRAMETHYLBICYCLO[4.2.0]OCTAN-2-ONE, VI, 1024
Bicyclo[4.2.0]octan-2-one, 7-hydroxy-4,4,6,7-tetramethyl- [61879-76-3]
2-HYDROXYTHIOPHENE, V, 642
Thiophen-2-ol [17236-58-7]
3-HYDROXY-1,7,7-TRIMETHYLBICYCLO[2.2.1]HEPTAN-2-ONE (endo isomer), VII, 277
Bicyclo[2.2.1]heptan-2-one, 3-hydroxy-1,7,7-trimethyl-, *endo*- [21488-68-6]
6-HYDROXY-3,5,5-TRIMETHYL-2-CYCLOHEXEN-1-ONE, VII, 282
2-Cyclohexen-1-one, 6-hydroxy-3,5,5-trimethyl- [61592-66-3]
3-HYDROXY-2,2,3-TRIMETHYLOCTAN-4-ONE, VIII, 343
4-Octanone, 3-hydroxy-2,2,3,-trimethyl- [85083-71-2]
5-HYDROXY-2,4,6-TRIMETHYL-2-TRIMETHYLSILOXYHEPTAN-3-ONE, VII, 185
3-Heptanone, 5-hydroxy-2,4,6-trimethyl-2-[(trimethylsilyloxy)]-, (*R*,S**)-(+)- [64869-24-5]
11-HYDROXYUNDECANOIC LACTONE, VI, 698
Oxacyclododecan-2-one [1725-03-7]
HYDROXYUREA, V, 645
Urea, hydroxy- [127-07-1]

I

IMIDAZOLE, III, 471
   1*H*-Imidazole [288-32-4]
IMIDAZOLE-2-CARBOXALDEHYDE, VII, 287
   1*H*-Imidazole-2-carboxaldehyde [10111-08-7]
IMIDAZOLE-4,5-DICARBOXYLIC ACID, III, 471
   1*H*-Imidazole-4,5-dicarboxylic acid [570-22-9]
3-IMINO-1-(*p*-TOLYLSULFONYL)PYRAZOLIDINE, V, 40
   3-Pyrazoline, 3-amino-1-(*p*-tolylsulfonyl)- [1018-36-6]
2-INDANONE, V, 647
   2*H*-Inden-2-one, 1,3-dihydro- [615-13-4]
INDAZOLE, III, 475; **IV**, 536; **V**, 650
   1*H*-lndazole [271-44-3]
INDAZOLONE, III, 476
   3*H*-3-Indazolone [5686-93-1]
INDOLE, III, 479
   1*H*-Indole [120-72-9]
INDOLE-3-ACETIC ACID, V, 654
   1*H*-Indole-3-acetic acid [87-51-4]
INDOLE-3-ALDEHYDE, IV, 539
   1*H*-Indole-3-carboxaldehyde [487-89-8]
INDOLE-3-CARBONITRILE, V, 656
   1*H*-Indole-3-carbonitrile [5457-28-3]
***p*-IODOANILINE, II**, 347
   Benzenamine, 4-iodo- [540-37-4]
5-IODOANTHRANILIC ACID, II, 349
   Benzoic acid, 2-amino-5-iodo- [5326-47-6]
IODOBENZENE, I, 323; **II**, 351
   Benzene, iodo- [591-50-4]
IODOBENZENE DICHLORIDE, III, 482
   Iodine, (dichlorophenyl)- [932-72-9]
***m*-IODOBENZOIC ACID, II**, 353
   Benzoic acid, 3-iodo- [618-51-9]
***p*-IODOBENZOIC ACID, I**, 325
   Benzoic acid, 4-iodo- [619-58-9]
***p*-IODOBENZYL ALCOHOL, III**, 652
   Benzenemethanol, 4-iodo- [18282-51-4]
***o*-IODOBROMOBENZENE, V**, 1120
   Benzene, 1-bromo-2-iodo- [583-55-1]
IODOCYCLOHEXANE, IV, 324, 543; **VI**, 830
   Cyclohexane, iodo- [626-62-0]
6-IODO-3,4-DIMETHOXYBENZALDEHYDE CYCLOHEXYLIMINE,
   VIII, 586
   Cyclohexanamine, *N*-[(2-iodo-4,5-dimethoxyphenyl)methylene]- [61599-78-8]
IODODURENE, VI, 700
   Benzene, 3-iodo-1,2,4,5-tetramethyl- [2100-25-6]

2-IODOETHYL BENZOATE, IV, 84
Ethanol, 2-iodo-, benzoate [39252-69-2]
1-IODOHEXENE, (Z)-, VII, 290
1-Hexene, 1-iodo-, (Z)- [16538-47-9]
***endo*-7-IODOMETHYLBICYCLO[3.3.1]NONAN-3-ONE, VI**, 958
Bicyclo[3.3.1]nonan-9-one, 7-(iodomethyl)-, *endo*- [29817-49-0]
***o*-IODOPHENOL, I**, 326
Phenol, 2-iodo- [533-58-4]
***p*-IODOPHENOL, II**, 355
Phenol, 4-iodo- [540-38-5]
(*E*)-1-IODO-4-PHENYL-2-BUTENE, VI, 704
Benzene, (4-iodo-2-butenyl)-, (*E*)- [52534-83-5]
IODOSOBENZENE, III, 483; **V**, 658
Benzene, iodosyl- [536-80-1]
IODOSOBENZENE DIACETATE, V, 660
Iodine, bis(acetato-*O*)phenyl- [3240-34-4]
***p*-IODOBENZYL ALCOHOL, III**, 652
Benzenemethanol, 4-iodo- [18282-51-4]
2-IODOTHIOPHENE, II, 357; **IV**, 545
Thiophene, 2-iodo- [3437-95-4]
IODOTRIMETHYLSILANE, VI, 353
Silane, iodotrimethyl- [16029-98-4]
1-IODO-3-TRIMETHYLSILYLPROPANE, VIII, 486
Silane, (3-iodopropyl)trimethyl- [18135-48-3]
4-IODOVERATROLE, IV, 547
Benzene, 4-iodo-1,2-dimethoxy- [5460-32-2]
IODOXYBENZENE, III, 485; **V**, 665
Benzene, iodyl- [696-33-3]
2-IODO-*p*-XYLENE, VI, 709
Benzene, 2-iodo-1,4-dimethyl- [1122-42-5]
ISATIN, I, 327; *Warning*
1*H*-Indole-2,3-dione [91-56-5]
ISATOIC ANHYDRIDE, III, 488
2*H*-3,1-Benzoxazine-2,4(1*H*)-dione [118-48-9]
ISOAMYL BROMIDE, I, 27
Butane, 1-bromo-3-methyl- [107-82-4]
ISOBUTYL BROMIDE, II, 358
Propane, 1-bromo-2-methyl- [78-77-3]
ISOBUTYRAMIDE, III, 490
Propanamide, 2-methyl- [563-83-7]
ISOBUTYRONITRILE, III, 493; *Hazard*
Propanenitrile, 2-methyl- [78-82-0]
ISOCROTONIC ACID, VI, 711
2-Butenoic acid, (Z)- [503-64-0]
3-ISOCYANATOPROPANOYL CHLORIDE, VI, 715
Propanoyl chloride, 3-isocyanato- [3729-19-9]

2-ISOCYANOCYCLOHEXANOL, *trans-,* **VII,** 294
  Cyclohexanol, 2-isocyano-, *trans-* [83152-97-0]
[(*trans*-2-ISOCYANOCYCLOHEXYL)OXY]TRIMETHYLSILANE, VII, 294
  Silane, [(2-isocyanocyclohexyl)oxy]trimethyl-, *trans-* [83152-87-8]
ISODEHYDROACETIC ACID, IV, 549
  2*H*-Pyran-3-carboxylic acid, 4,6-dimethyl-2-oxo- [33953-26-3]
ISODURENE, II, 360
  Benzene, 1,2,3,5-tetramethyl- [527-53-7]
DL-ISOLEUCINE, III, 495
  DL-Isoleucine [443-79-8]
(+)-ISOMENTHONE, VIII, 386
  Cyclohexanone, 5-methyl-2-(1-methylethyl)-, (2*R-cis*)- [1196-31-2]
ISONITROSOACETANILIDE, I, 327
  Glyoxylanilide, 2-oxime [1769-41-1]
ISONITROSOPROPIOPHENONONE, II, 363
  1,2-Propanedione, 1-phenyl-, 2-oxime [119-51-7]
ISOPHORONE OXIDE, IV, 552
  7-Oxabicyclo[4.1.0]heptan-2-one, 4,4,6-trimethyl- [10276-21-8]
ISOPHTHALALDEHYDE, V, 668
  1,3-Benzenedicarboxaldehyde [626-19-7]
(−)-ISOPINOCAMPHEOL, VI, 719
  Bicyclo[3.1.1]heptan-3-ol, 2,6,6-trimethyl-, [1*R*-(1α,2β,3α,5α)]- [1196-00-5]
ISOPRENE CYCLIC SULFONE, III, 499
  Thiophene, 2,5-dihydro-3-methyl-, 1,1-dioxide [1193-10-8]
cis-4-*exo*-**ISOPROPENYL-1,9-DIMETHYL-8-(TRIMETHYLSILYL)-**
  BICYCLO[4.3.0]NON-8-EN-2-ONE, VIII, 347
  4*H*-Inden-4-one, 1,3a,5,6,7,7a,-hexahydro-3,3a-dimethyl-6-(1-methylethenyl)-
    2-trimethylsilyl-, (3aα,6α,7aα)- [77494-23-6]
(ISOPROPOXYDIMETHYLSILYL)METHYL CHLORIDE, VIII, 317
  Silane, (chloromethyl)isopropoxydimethyl- [18171-11-4]
[(ISOPROPOXYDIMETHYLSILYL)METHYL]CYCLOHEXANOL, VIII, 315
  Cyclohexanol, 1-[[dimethyl(1-methoxyethoxy)silyl]methyl]- [138080-23-6]
(ISOPROPOXYDIMETHYLSILYL)METHYLMAGNESIUM CHLORIDE, VIII, 315
  Magnesium, chloro[[dimethyl(1-methylethoxy)silyl]methyl]- [122588-50-5]
2-ISOPROPYLAMINOETHANOL, III, 501
  Ethanol, 2-[(1-methylethyl)amino]- [109-56-8]
ISOPROPYL BROMIDE, I, 37
  Propane, 2-bromo- [75-26-3]
ISOPROPYLIDENACETOPHENONE, VIII, 210
  2-Buten-1-one, 3-methyl-1-phenyl- [5650-07-7]
(±)- and (+)- and (−)-α-(ISOPROPYLIDENEAMINOOXY)PROPIONIC ACID, V,
  1032
  Propanoic acid, 2-[[(1-methylethylidene)amino]oxy]- (±)- [2009-90-7]; (*S*)-
    [2009-89-4]
2,3-*O*-ISOPROPYLIDENE-D-ERYTHRONOLACTONE, VII, 297
  Furo[3,4-*d*]-1,3-dioxol-4(3a*H*)-one, dihydro-2,2-dimethyl-, (3a*R-cis*)- [25581-41-3]

DL-ISOPROPYLIDENEGLYCEROL, III, 502
    1,3-Dioxolane-4-methanol, 2,2-dimethyl-, (±)- [22323-83-7]
ISOPROPYLIDENE α-(HEXAHYDROAZEPINYLIDENE-2)MALONATE, VIII, 263
    1,3-Dioxane-4,6-dione, 5-(hexahydro-2*H*-azepin-2-ylidene)-2,2-dimethyl-
    [70912-54-8]
ISOPROPYL LACTATE, II, 365
    Propanoic acid, 2-hydroxy, 1-methylethyl ester [617-51-6]
ISOPROPYL THIOCYANATE, II, 366
    Thiocyanic acid, 1-methylethyl ester [625-59-2]
3-ISOQUINUCLIDONE, V, 670
    2-Azabicyclo[2.2.2]octan-3-one [3306-69-2]
o-**ISOTHIOCYANATO-(*E*)-CINNAMALDEHYDE, VII,** 302
    2-Propenal, 3-(2-isothiocyanatophenyl)-, (*E*)- [19908-01-1]
β–**ISOVALEROLACTAM, V,** 673
    2-Azetidinone, 4,4-dimethyl- [4879-95-2]
β–**ISOVALEROLACTAM-*N*-SULFONYL CHLORIDE, V,** 673
    Azetidinesulfonyl chloride, 2,2-dimethyl-4-oxo- [17174-96-8]
ITACONIC ACID, II 368
    Butanedioic acid, methylene- [97-65-4]
ITACONIC ANHYDRIDE, II 368
    2,5-Furandione, dihydro-3-methylene- [2170-03-8]
ITACONYL CHLORIDE, IV, 554
    Butanedioyl dichloride, methylene- [1931-60-8]

J

JULOLIDINE, III, 504
    1*H*,5*H*-Benzo[*ij*]quinolizine, 2,3,6,7-tetrahydro- [479-59-4]

K

KETENE, I, 330; **V,** 679
    Ethenone [463-51-4]
KETENE DIETHYLACETAL, III, 506
    Ethene, 1,1-diethoxy- [2678-54-8]
KETENE DIMER, III, 508
    2-Oxetanone, 4-methylene- [674-82-8]
KETENE DI-(2-METHOXYETHYL)ACETAL, V, 684
    2,5,7,10-Tetraoxaundecane, 6-methylene- [5130-02-9]
α-**KETOGLUTARIC ACID, III,** 510; **V,** 687
    Pentanedioic acid, 2-oxo- [328-50-7]
6-KETOHENDECANEDIOIC ACID, IV, 555
    Undecanedioic acid, 6-oxo- [3242-53-3]
2-KETOHEXAMETHYLENIMINE, II, 371
    2*H*-Azepin-2-one, hexahydro- [105-60-2]
β-**KETOISOOCTALDEHYDE DIMETHYL ACETAL, IV,** 558
    2-Hexanone, 1,1-dimethoxy-5-methyl- [53684-58-5]
KETOPANTOYL LACTONE, VII, 417
    2,3-Furandione, dihydro-4,4-dimethyl- [13031-04-4]

DL-**KETOPINIC ACID, V,** 689
    Bicyclo[2.2.1]heptane-1-carboxylic acid, 7,7-dimethyl-2-oxo- [464-78-8]
KRYPTOPYRROLE, III, 513
    1H-Pyrrole, 3-ethyl-2,4-dimethyl- [517-22-6]

L

LACTAMIDE, III, 516
    Propanamide, 2-hydroxy- [2043-43-8]
18,20-LACTONE OF 3β-ACETOXY-20β-HYDROXY-5-PREGNEN-18-OIC ACID,
    V, 692
    Pregn-5-en-18-oic acid, 3-(acetyloxy)-20-hydroxy-, γ-lactone, (3β,20R)-
        [3020-10-8]
LAURONE, IV, 560
    12-Tricosanone [540-09-0]
LAURYL ALCOHOL, II, 372
    1-Dodecanol [112-53-8]
LAURYL BROMIDE, (see DODECYL BROMIDE), I, 29; **II,** 246; **III,** 364
    Dodecane, 1-bromo- [143-15-7]
LAURYLMETHYLAMINE, IV, 564
    1-Dodecanamine, N-methyl- [7311-30-0]
LEPIDINE, III, 519
    Quinoline, 4-methyl- [491-35-0]
DL-**LEUCINE, III,** 523
    DL-Leucine [328-39-2]
LEVOPIMARIC ACID, V, 699
    1-Phenanthrenecarboxylic acid, 1,2,3,4,4a,4b,5,9,10,10a-decahydro- 1,4a-dimethyl-
        7-(1-methylethyl)-, [1R-(1α,4aβ,4bα,10aα)]- [79-54-9]
LEVULINIC ACID, I, 335
    Pentanoic acid, 4-oxo- [123-76-2]
LINOLEIC ACID, III, 526
    9,12-Octadecadienoic acid, (Z,Z)- [60-33-3]
LINOLENIC ACID, III, 531
    9,12,15-Octadecatrienoic acid, (Z,Z,Z)- [463-40-1]
DL-**LYSINE MONOHYDROCHLORIDE, II,** 374
    DL-Lysine, monohydrochloride [70-53-1]
DL-**LYSINE DIHYDROCHLORIDE, II,** 374
    DL-Lysine, dihydrochloride [617-68-5]

M

MALEANILIC ACID, V, 944
    2-Butenoic acid, 4-oxo-4-(phenylamino)- [37902-58-2]
MALONIC ACID, II, 376
    Propanedioic acid [141-82-2]
MALONONITRILE, II, 379; **III,** 535
    Propanedinitrile [109-77-3]
MALONYL DICHLORIDE, IV, 263
    Propanedioyl dichloride [1663-67-8]

MANDELAMIDE, III, 536
  Benzeneacetamide, α-hydroxy- [4410-31-5]
MANDELIC ACID, I, 336; **III** 538
  Benzeneacetic acid, α-hydroxy- [90-64-21]
D-MANNOSE, III, 541
  D-Mannose [3458-28-4]
***cis-p*-MENTH-8-EN-1-YL METHYL ETHER, VII,** 304
  Cyclohexane, 1-methoxy-1-methyl-4-(1-methylethenyl)-, *cis*- [24655-72-9]
***trans-p*-MENTH-8-EN-1-YL METHYL ETHER, VII,** 304
  Cyclohexane, 1-methoxy-1-methyl-4-(1-methylethenyl)-, *trans*- [24655-71-8]
(–)-MENTHONE, I, 340
  Cyclohexanone, 5-methyl-2-(1-methylethyl)-, (2*S-trans*)- [14073-97-3]
(–)-MENTHOXYACETIC ACID, III, 544
  Acetic acid, [[5-methyl-2-(1-methylethyl)cyclohexyl]oxy]-, [1*R*-(1α,2β,5α)]-
  [40248-63-3]
(–)-MENTHOXYACETYL CHLORIDE, III, 547
  Acetyl chloride, [[5-methyl-2-(1-methylethyl)cyclohexyl]oxy]-, [1*R*-(1α,2β,5α)]-
  [15356-62-4]
(–)-MENTHYL CINNAMATE, VIII, 350
  2-Propenoic acid, 3-phenyl-, 5-methyl-2-(1-methylethyl)cyclohexyl ester,
  [1*R*-(1α,2β,5α)]- [16205-99-5]
(–)-MENTHYL NICOTINATE, VIII, 350
  3-Pyridinecarboxylic acid, 5-methyl-2-(1-methylethyl)cyclohexyl ester, [1*R*-
  (1α,2β,5α)]- [133005-61-5]
(*S*)-(–)-MENTHYL *p*-TOLUENESULFINATE, VII, 495
  Menthol, (–)-, (*S*)-*p*-toluenesulfinate, (–)- [1517-82-4]
2-MERCAPTO-4-AMINO-5-CARBETHOXYPYRIMIDINE, IV, 566
  5-Pyrimidinecarboxylic acid, 4-amino-1,2-dihydro-2-thioxo-, ethyl ester [774-07-2]
2-MERCAPTOBENZIMIDAZOLE, IV, 569
  2*H*-Benzimidazole-2-thione, 1,3-dihydro- [583-39-1]
2-MERCAPTO-4-HYDROXY-5-CYANOPYRIMIDINE, IV, 566
  5-Pyrimidinecarbonitrile, 4-hydroxy-2-mercapto- [23945-49-5]
2-(1-MERCAPTO-1-METHYLETHYL)-5-METHYLCYCLOHEXANOL,
    VIII, 304
  Cyclohexanol, 2-(1-mercapto-1-methylethyl)-5-methyl-, [1*R*-(1a,2a,5a)]-
  [79563-68-1]; [1*R*-(1a,2b,5a)]-[79563-59-0]; [1*S*-(1a,2a,5b)]- [79563-67-0]
2-MERCAPTOPYRIMIDINE and HYDROCHLORIDE, V, 703
  2(1*H*)-Pyrimidinethione [1450-85-7]
MERCURY DI-β-NAPHTHYL, II, 381; *Hazard*
  Mercury, di-2-naphthalenyl- [19510-26-0]
MESACONIC ACID, II, 382
  2-Butenedioic acid, 2-methyl-, (*E*)- [498-24-8]
MESITALDEHYDE, III, 549
  Benzaldehyde, 2,4,6-trimethyl- [487-68-3]
MESITOIC ACID, III, 553, 555; **V,** 706
  Benzoic acid, 2,4,6-trimethyl- [480-63-7]

MESITOYL CHLORIDE, III, 555
   Benzoyl chloride, 2,4,6-trimethyl- [938-18-1]
MESITYLACETIC ACID, III, 557
   Benzeneacetic acid, 2,4,6-trimethyl- [4408-60-0]
MESITYLACETONITRILE, III, 557
   Benzeneacetonitrile, 2,4,6-trimethyl- [34688-71-6]
MESITYLENE, I, 341
   Benzene, 1,3,5-trimethyl- [108-67-8]
MESITYL OXIDE, I, 345
   3-Penten-2-one, 4-methyl- [141-79-7]
METHACRYLAMIDE, III, 560
   2-Propenamide, 2-methyl- [79-39-0]
METHALLYLBENZENE, VI, 722
   Benzene, (2-methyl-2-propenyl)- [3290-53-7]
METHANESULFINYL CHLORIDE, V, 709
   Methanesulfinyl chloride [676-85-7]
METHANESULFONYL CHLORIDE, IV, 571
   Methanesulfonyl chloride [124-63-0]
METHANESULFONYL CYANIDE, VI, 727
   Methanesulfonyl cyanide [24225-08-9]
erythro-**2-METHANESULFONYLOXY-2-BUTYL CYCLOBUTANECARBOXY-
   LATE, VI**, 312
   Cyclobutanecarboxylic acid, 1-methyl-2-[(methylsulfonyl)oxy]propyl ester,
      (*R**,*S**)- [35358-24-8]
1,6-METHANO[10]ANNULENE, VI, 731
   Bicyclo[4.4.1]undeca-1,3,5,7,9-pentaene [2443-46-1]
DL-METHIONINE, II, 384
   DL-Methionine [59-51-8]
L-METHIONINE, VI, 253
   L-Methionine [63-68-3]
METHOXYACETONITRILE, II, 387
   Acetonitrile, methoxy- [1738-36-9]
α-METHOXYACETOPHENONE, III, 562
   Ethanone, 2-methoxy-1-phenyl- [4079-52-1]
METHOXYACETYLENE, IV, 406
   Ethyne, methoxy- [6443-91-0]
m-**METHOXYBENZALDEHYDE, III**, 564
   Benzaldehyde, 3-methoxy- [591-31-1]
2-METHOXY-*N*-CARBOMETHOXYPYRROLIDINE, VII, 307
   1-Pyrrolidinecarboxylic acid, 2-methoxy-, methyl ester [56475-88-8]
1-(2-METHOXYCARBONYLPHENYL)PYRROLE, V, 716
   Benzoic acid, 2-(1*H*-pyrrol-1-yl)-, methyl ester [10333-67-2]
**4-METHOXYCARBONYL-1,1,6-TRIMETHYL-1,4,4a,5,6,7,8,8a-OCTAHYDRO-
   2,3-BENZOPYRONE, (4*RS*,4a*RS*,6*RS*,8a*RS*), VIII**, 353
   1*H*-2-Benzopyran-4-carboxylic acid, octahydro-1,1,6-trimethyl-3-oxo-, methyl
      ester, (4α,4aα,6β,8aβ)-, (±)- [138233-55-3]

4-METHOXYCARBONYL-1,1,6-TRIMETHYL-1,4,4a,5,6,7,8,8a-OCTAHYDRO-2,3- BENZOPYRONE, (4S,4aS,6S,8aS), **VIII**, 353

1H-2-Benzopyran-4-carboxylic acid, octahydro-1,1,6-trimethyl-3-oxo-, methyl ester, [4R-(4α,4aα,6β,8aβ)]- [121114-94-1]

4-METHOXYCARBONYL-1,1,6-TRIMETHYL-1,4,4a,5,6,7,8,8a-OCTAHYDRO-2,3- BENZOPYRONE, (4R,4aR,6R,8aR), **VIII**, 353

1H-2-Benzopyran-4-carboxylic acid, octahydro-1,1,6-trimethyl-3-oxo-, methyl ester, [4S-(4α,4aα,6β,8aβ)]- [138147-65-6)

1-METHOXY-1,4-CYCLOHEXADIENE, **VI**, 996

1,4-Cyclohexadiene, 1-methoxy- [2886-59-1]

5β-METHOXYCYCLOHEXAN-1-ONE-3β,4β-DICARBOXYLIC ACID ANHYDRIDE, **VII**, 312

1,3,5(4H)-Isobenzofurantrione, tetrahydro-7-methoxy-, (3aα,7α,7aα)- [87334-37-0]

2-METHOXYCYCLOOCTANONE OXIME, **V**, 267

Cyclooctanone, 2-methoxy-, oxime [10499-36-2]

2-METHOXYDIPHENYL ETHER, **III**, 566

Benzene, 1-methoxy-2-phenoxy- [1695-04-1]

5-(3-METHOXY-4-HYDROXYBENZAL)CREATININE, **III**, 587

4-Imidazolidinone, 2-imino-1-methyl-5-vanillylidene- [29974-40-1]

5-(3-METHOXY-4-HYDROXYBENZYL)CREATININE, **III**, 587

6-METHOXY-7-METHOXYCARBONYL-1,2,3,4-TETRAHYDRONAPHTHALENE, **VIII**, 444

2-Naphthalenecarboxylic acid, 5,6,7,8-tetrahydro-3-methoxy-, methyl ester [78112-34-2]

1-METHOXY-2-METHYLPYRIDINIUM METHYL SULFATE, **V**, 270

Pyridinium, 1-methoxy-2-methyl-, methyl sulfate [55369-05-6]

(S)-(+)-2-METHOXYMETHYLPYRROLIDINE, **VIII**, 26

Pyrrolidine, 2-(methoxymethyl)-, (S)-(+)- [63126-47-6]

6-METHOXY-2-NAPHTHOL, **V**, 918

2-Naphthalenol, 6-methoxy- [5111-66-0]

2-METHOXY-1,4-NAPHTHOQUINONE, **III**, 465

1,4-Naphthalenedione, 2-methoxy- [2348-82-5]

6-METHOXY-8-NITROQUINOLINE, **III**, 568

Quinoline, 6-methoxy-8-nitro- [85-81-4]

4-METHOXY-3-PENTEN-2-ONE, **VIII**, 357

3-Penten-2-one, 4-methoxy- [2845-83-2]

α-METHOXYPHENAZINE, **III**, 753

Phenazine, 1-methoxy- [2876-17-7]

o-**METHOXYPHENYLACETONE**, **IV**, 573

2-Propanone, 1-(2-methoxyphenyl)- [5211-62-1]

p-**METHOXYPHENYLACETONITRILE**, **IV**, 576

Benzeneacetonitrile, 4-methoxy- [104-47-2]

p-**METHOXYPHENYLLEAD TRIACETATE**, **VII**, 229

Plumbane, tris(acetyloxy)(4-methoxyphenyl)- [18649-43-9]

1-(*o*-METHOXYPHENYL)-2-NITRO-1-PROPENE, **IV**, 573

Benzene, 1-methoxy-2-(2-nitro-1-propenyl)- [6306-34-9]

1-(p-METHOXYPHENYL)-5-PHENYL-1,3,5-PENTANETRIONE, V, 718
  1,3,5-Pentanetrione, 1-(4-methoxyphenyl)-5-phenyl- [1678-17-7]
2-(p-METHOXYPHENYL)-6-PHENYL-4-PYRONE, V, 721
  4H-1-Pyran-4-one, 2-(4-methoxyphenyl)-6-phenyl- [14116-43-9]
1-(4-METHOXYPHENYL)-1,2,5,6-TETRAHYDROPYRIDINE, VIII, 358
  Pyridine, 1,2,3,6-tetrahydro-1-(4-methoxyphenyl)- [133157-31-0]
(Z)-2-METHOXY-1-PHENYLTHIO-1,3-BUTADIENE, VI, 737
  Benzene, [(2-methoxy-1,3-butadienyl)thio]-, (Z)- [60466-66-2]
**N-(4-METHOXYPHENYL)-(Z)-4-(TRIMETHYLSILYL)-2-BUTENAMINE,
  VIII**, 358
  Benzenamine, 4-methoxy-N-[4-(trimethylsilyl)-3-butenyl]-, (Z)- [133157-30-9]
6-METHOXY-β-TETRALONE, VI, 744
  2(1H)-Naphthalenone, 3,4-dihydro-6-methoxy- [2472-22-2]
1-METHOXY-3-TRIMETHYLSILOXY-1,3-BUTADIENE, trans-, VII, 312
  Silane, [(3-methoxy-1-methylene-2-propenyl)oxy]trimethyl- [59414-23-1]
3-METHOXY-17-TRIMETHYLSILOXY-1,3,5(10)-ESTRATETRAENE, VIII, 286
  Silane, [(3-methoxyestra-1,3,5(10),16-tetraen-17-yl)oxy]trimethyl- [115419-13-1]
1-METHOXYVINYLLITHIUM, VIII, 19
  Lithium, 1-methoxyethenyl- [42722-80-5]
1-(METHOXYVINYL)TRIMETHYLSILANE, VIII, 19
  Silane, (1-methoxyethenyl)trimethyl- [79678-01-6]
METHYL 4-ACETOXYBENZOATE, VI, 576
  Benzoic acid, 4-(acetyloxy)-, methyl ester [24262-66-6]
4-METHYL-7-ACETOXYCOUMARIN, III, 283
  2H-1-Benzopyran-2-one, 7-acetyloxy-4-methyl- [2747-05-9]
METHYL p-ACETYLBENZOATE, IV, 579
  Benzoic acid, 4-acetyl-, methyl ester [3609-53-8]
METHYLAMINE HYDROCHLORIDE, I, 347
  Methanamine, hydrochloride [593-51-1]
1-METHYLAMINOANTHRAQUINONE, III, 573
  9,10-Anthracenedione, 1-(methylamino)- [82-38-2]
1-METHYLAMINO-4-BROMOANTHRAQUINONE, III, 575
  9,10-Anthracenedione, 1-bromo-4-(methylamino)- [128-93-8]
METHYL AMYL KETONE, I, 351
  2-Heptanone [110-43-0]
β-METHYLANTHRAQUINONE, I, 353
  9,10-Anthracenedione, 2-methyl- [84-54-8]
METHYL BENZENESULFINATE, V, 723
  Benzenesulfinic acid, methyl ester [670-98-4]
**METHYL 4-O-BENZOYL-6-BROMO-6-DEOXY-α-D-GLUCOPYRANOSIDE,
  VIII**, 363
  α-D-Glucopyranoside, methyl 6-bromo-6-deoxy, 4-benzoate [10368-81-7]
o-METHYLBENZYL ACETATE, IV, 582
  Benzenemethanol, 2-methyl-, acetate [17373-93-2]
o-METHYLBENZYL ALCOHOL, IV, 582
  Benzenemethanol, 2-methyl- [89-95-2]

2-METHYLBENZYLDIMETHYLAMINE, IV, 585
Benzenemethanamine, *N,N*-2-trimethyl- [4525-48-8]
METHYL 4,6-*O*-BENZYLIDENE-α-D-GLUCOPYRANOSIDE, VIII, 363
α-D-Glucopyranoside, methyl 4,6-*O*-(phenylmethylene)- [3162-96-7]
***N*-(α-METHYLBENZYLIDENE)METHYLAMINE, VI**, 818
Methanamine, *N*-(1-phenylethylidene)- [6907-71-7]
METHYL BENZYL KETONE, II, 389
2-Propanone, 1-phenyl- [103-79-7]
4-METHYLBICYCLO[2.2.2]OCTANE-2,6-DIONE, VIII, 468
Bicyclo[2.2.2]octane-2,6-dione, 4-methyl- [119986-98-0]
6-METHYLBICYCLO[4.2.0]OCTAN-2-ONE, VII, 315
Bicyclo[4.2.0]octan-2-one, 6-methyl- [13404-66-5]
(*R*)-(−)-METHYL 1,1'-BINAPHTHYL-2,2'-DIYL PHOSPHATE, VIII, 46
Dinaphtho[2,1-*d*:1'2'-*f*]dioxaphosphepin, 4-methoxy-, 4-oxide, (*R*)- [86334-02-3]
2-METHYLBIPHENYL, VI, 747
1,1'-Biphenyl, 2-methyl- [643-58-3]
10-METHYL-10,9-BORAZAROPHENANTHRENE, V, 727
Dibenz[*c,e*][1,2]azaborine, 5,6-dihydro-6-methyl- [15813-13-5]
METHYL 4-BROMO-1-BUTANIMIDATE HYDROCHLORIDE, VIII, 415
Butanimidic acid, 4-bromo-, methyl ester, hydrochloride [21367-90-8]
***N*-METHYL-α-BROMOBUTYRANILIDE, IV**, 620
METHYL α-BROMO-β-METHOXYPROPIONATE, III, 774
Propanoic acid, 2-bromo-3-methoxy-, methyl ester [27704-96-7]
METHYL α-(BROMOMETHYL)ACRYLATE, VII, 319
2-Propenoic acid, 2-(bromomethyl)-, methyl ester [4224-69-5]
METHYL β-BROMOPROPIONATE, III, 576
Propanoic acid, 3-bromo-, methyl ester [3395-91-3]
METHYL 5-BROMOVALERATE, III, 578
Pentanoic acid, 5-bromo-, methyl ester [5454-83-1]
METHYL BUTADIENOATE, V, 734
2,3-Butadienoic acid, methyl ester [18913-35-4]
(*E*)-2-(METHYL-1,3-BUTADIENYL)DIMETHYLALANE, VII, 245
Aluminum, dimethyl-(2-methyl-1,3-butadienyl)- [96160-49-5]
(*S*)-2-METHYLBUTANAL, VIII, 367
Butanal, 2-methyl-, (*S*)- [1730-97-8]
2-METHYLBUTANAL-1-*d*, VI, 751
Butanal-1-*d*, 2-methyl- [25132-57-4]
**(3-METHYL-2-BUTENYL)PROPANEDIOIC ACID, DIMETHYL ESTER,
VIII**, 381
Propanedioic acid, (3-methyl-2-butenyl)-, dimethyl ester [43219-18-7]
**(3-METHYL-2-BUTENYL)(2-PROPYNYL)PROPANEDIOIC ACID, DIMETHYL
ESTER, VIII**, 381
Propanedioic acid, (3-methyl-2-butenyl)-2-propynyl-, dimethyl ester
[107473-14-3]
3-METHYL-2-BUTEN-2-YL TRIFLATE, VI, 757
Methanesulfonic acid, trifluoro-, 1,2-dimethyl-1-propenyl ester [28143-80-8]

N-(2-METHYLBUTYLIDENE-1-*d*)-1,1,3,3-TETRAMETHYLBUTYLAMINE, VI, 751
   2-Pentanamine, 2,4,4-trimethyl-*N*-(2-methylethylidene-1-*d*)- [34668-70-7]
N-METHYLBUTYLAMINE, V, 736
   1-Butanamine, *N*-methyl- [110-68-9]
O-METHYLCAPROLACTIM, IV, 588
   2*H*-Azepine, 3,4,5,6-tetrahydro-7-methoxy- [2525-16-8]
4-METHYLCARBOSTYRIL, III, 580
   2(1*H*)-Quinolinone, 4-methyl- [607-66-9]
METHYL (CARBOXYSULFAMOYL)TRIETHYLAMMONIUM HYDROXIDE, VI, 788
   Ethanaminium, *N*,*N*-diethyl-*N*-[[methoxycarbonyl)amino]sulfonyl]-, inner salt [29684-56-8]
METHYL 4-CHLORO-2-BUTYNOATE, VIII, 371
   2-Butynoic acid, 4-chloro-, methyl ester [41658-12-2]
METHYL 2-CHLORO-2-CYCLOPROPYLIDENACETATE, VIII, 373, 374
   Acetic acid, chlorocyclopropylidene-, methyl ester [82979-45-1]
METHYL (CHLOROSULFONYL)CARBAMATE, VI, 788
   Carbamic acid, (chlorosulfonyl)-, methyl ester [36914-92-8]
5-METHYLCOPROST-3-ENE, VI, 762
   Cholest-3-ene, 5-methyl-, (5β)- [23931-38-6]
METHYL COUMALATE, IV, 532
   2*H*-Pyran-5-carboxylic acid, 2-oxo-, methyl ester [6018-41-3]
3-METHYLCOUMARILIC ACID, IV, 591
   Benzofurancarboxylic acid, 3-methyl- [24673-56-1]
4-METHYLCOUMARIN, III, 581
   2*H*-1-Benzopyran-2-one, 4-methyl- [607-71-6]
3-METHYLCOUMARONE, IV, 590
   Benzofuran, 3-methyl- [21535-97-7]
METHYL 2-(1-CYANOCYCLOHEXYL)DIAZENECARBOXYLATE, VI, 334
   Diazenecarboxylic acid, 2-(1-cyanocyclohexyl)-, methyl ester [33670-04-1]
METHYL 2-(1-CYANOCYCLOHEXYL)HYDRAZINECARBOXYLATE, VI, 334
   Hydrazinecarboxylic acid, 2-(1-cyanocyclohexyl)-, methyl ester [61827-29-0]
METHYL ω-CYANOPELARGONATE, III, 584
   Nonanoic acid, 9-cyano-, methyl ester [53663-26-6]
1-METHYLCYCLOHEXANECARBOXYLIC ACID, V, 739
   Cyclohexanecarboxylic acid, 1-methyl- [1123-25-7]
2-METHYL-1,3-CYCLOHEXANEDIONE, V, 743
   1,3-Cyclohexanedione, 2-methyl- [1193-55-1]
1-METHYLCYCLOHEXANOL, VI, 766
   Cyclohexanol, 1-methyl- [590-67-0]
3-METHYLCYCLOHEXANONE-3-ACETIC ACID, VIII, 467
   Cyclohexaneacetic acid, 1-methyl-3-oxo-, (±)- [119986-97-9]
3-METHYLCYCLOHEXENE, VI, 174, 769
   Cyclohexene, 3-methyl [591-48-0]
3-METHYL-2-CYCLOHEXEN-1-OL, VI, 769
   2-Cyclohexen-1-ol, 3-methyl- [21378-21-2]

trans-2-METHYL-1-CYCLOHEXEN-1-YL ACETATE, **VI**, 121
  1-Cyclohexen-1-ol, 2-methyl-, acetate [75411-49-3]
2-METHYL-2-CYCLOHEXENONE, IV, 162
  2-Cyclohexen-1-one, 2-methyl- [1121-18-2]
METHYL 2,4-CYCLOPENTADIENE-1-ACETATE (in solution), VII, 339
  2,4-Cyclopentadiene-1-acetic acid, methyl ester [37455-98-4]
METHYL CYCLOPENTANECARBOXYLATE, IV, 594
  Cyclopentanecarboxylic acid, methyl ester [4630-80-2]
cis-1-METHYLCYCLOPENTANE-1,2-DICARBOXYLIC ACID, **VIII**, 377
  1,2-Cyclopentanedicarboxylic acid, 1-methyl-, *cis*-(±)- [70433-31-7]
2-METHYLCYCLOPENTANE-1,3-DIONE, V, 747; **VI**, 30, 774
  1,3-Cyclopentanedione, 2-methyl- [765-69-5]
2-METHYLCYCLOPENTANE-1,3,5-TRIONE HYDRATE, V, 748
  Cyclopentane-1,2,4-trione, 3-methyl- [4505-54-8]
METHYL CYCLOPROPYL KETONE, IV, 597
  Ethanone, 1-cyclopropyl- [765-43-5]
METHYL 1-CYCLOUNDECENECARBOXYLATE, VI, 368
  1-Cycloundecene-1-carboxylic acid, methyl ester [3667-72-9]
2-METHYL-2,5-DECANEDIOL, IV, 601
  2,5-Decanediol, 2-methyl- [53731-34-3]
METHYL DESOXYCHOLATE, III, 237
  Cholan-24-oic acid, 3a,12a-dihydroxy-, methyl ester [20231-66-7]
METHYL β,β-DIBROMOISOBUTYRATE, VII, 319
  Propanoic acid, 3-bromo-2-(bromomethyl)-, methyl ester [22262-60-8]
METHYL *N,N*-DICHLOROCARBAMATE, VIII, 427
  Carbamic acid, dichloro-, methyl ester [16487-46-0]
METHYL 3,3-DIETHOXYPROPANOATE, VIII, 255
  Propanoic acid, 3,3-diethoxy-, methyl ester [127581-78-6]
**METHYL 2-(DIETHYLPHOSPHORYLOXY)-1-CYCLOHEXENE-
  1-CARBOXYLATE, VII**, 351
  1-Cyclohexene-1-carboxylic acid, 2-[(diethoxyphosphinyl)oxy]-, methyl ester
    [71712-64-6]
METHYL DIFORMYLACETATE, VII, 323
  2-Propenoic acid, 2-formyl-3-hydroxy-, methyl ester [39947-70-1]
N-METHYL-3,4-DIHYDROXYPHENYLALANINE, **III**, 586
  DL-Tyrosine, 3-hydroxy-*N*-methyl- [53663-27-7]
N-METHYL-2,3-DIMETHOXYBENZYLAMINE, **IV**, 603
  Benzenemethanamine, 2,3-dimethoxy-*N*-methyl- [53663-28-8]
METHYL 6,6-DIMETHOXYHEXANOATE, VII, 168
  Hexanoic acid, 6,6-dimethoxy-, methyl ester [25176-55-0]
METHYL 5,5-DIMETHOXY-3-METHYL-2,3-EPOXYPENTANOATE, IV, 649
**METHYL 1,3-DIMETHYL-5-OXOBICYCLO[2.2.2]OCTANE-2-CARBOXYLATE,
  VIII**, 219
  Bicyclo[2.2.2]octane-2-carboxylic acid, 1,3-dimethyl-5-oxo-, methyl ester,
    1α,2β,3α,4α- [121917-73-5]
METHYL 3,5-DINITROBENZOATE, V, 798
  Benzoic acid, 3,5-dinitro-, methyl ester [2702-58-1]

N-METHYL-1,2-DIPHENYLETHYLAMINE, IV, 605
    Benzeneethanamine, *N*-methyl-α-phenyl- [53663-25-5]
N-METHYL-1,2-DIPHENYLETHYLAMINE HYDROCHLORIDE, IV, 605
    Benzeneethanamine, *N*-methyl-α-phenyl-, hydrochloride [7400-77-3]
2-METHYLDODECANOIC ACID, IV, 618
    Dodecanoic acid, 2-methyl- [2874-74-0]
trans-2-METHYL-2-DODECENOIC ACID, IV, 608
    2-Dodecenoic acid, 2-methyl-, (*E*)- [53663-29-9]
METHYLENAMINOACETONITRILE TRIMER, I, 355
    Acetonitrile, (methyleneamino)- [109-82-0]
METHYLENE BROMIDE, I, 357
    Methane, dibromo- [74-95-3]
3-METHYLENECYCLOBUTANE-1,2-DICARBOXYLIC ANHYDRIDE, V, 459
METHYLENECYCLOHEXANE, IV, 612; V, 751
    Cyclohexane, methylene- [1192-37-6]
METHYLENECYCLOHEXANE OXIDE, V, 755
    1-Oxaspiro[2.5]octane [185-70-6]
METHYLENECYCLOPROPANE, VI, 320
    Methylene, cyclopropyl- [19527-12-9]
4,5-METHYLENEDIOXYBENZOCYCLOBUTENE, VII, 326
    Cyclobuta-[*f*]-1,3-benzodioxole, 5,6-dihydro- [61099-23-8]
2-METHYLENEDODECANOIC ACID, IV, 616
    Dodecanoic acid, 2-methylene- [52756-21-5]
METHYLENE IODIDE, I, 358
    Methane, diiodo- [75-11-6]
3-METHYLENE-4-ISOPROPYL-1,1-CYCLOPENTANEDICARBOXYLIC ACID,
    DIMETHYL ESTER, VIII, 381
    1,1-Cyclopentanedicarboxylic acid, 3-methylene-4-(1-methylethyl)-, dimethyl ester
    [107473-16-5]
3-METHYLENE-*cis*-*p*-MENTHANE, (+)-, VIII, 386
    Cyclohexane, 4-methyl-2-methylene-1-(1-methylethyl)-, (1*R-cis*)- [122331-74-2]
2-METHYLENE-1-OXO-1,2,3,4-TETRAHYDRONAPHTHALENE, VII, 332
    1(2*H*)-Naphthalenone, 3,4-dihydro-2-methylene- [13203-73-1]
METHYL (3*RS*,4*RS*)-4,5-EPOXY-3-PHENYLPENTANOATE, VII, 164
    Oxiranepropanoic acid, β-phenyl-, methyl ester, (±)- [107445-25-0]
4-METHYLESCULETIN, I, 360
    2*H*-1-Benzopyran-2-one, 6,7-dihydroxy-4-methyl- [529-84-0]
2-METHYL-4-ETHOXALYLCYCLOPENTANE-1,3,5-TRIONE, V, 747
    Cyclopentaneglyoxylic acid, 3-methyl-2,4,5-trioxo-, ethyl ester [781-38-4]
DL-METHYLETHYLACETIC ACID, I, 361
    Butanoic acid, 2-methyl-, (±)- [600-07-7]
N-METHYLETHYLAMINE, V, 758
    Ethanamine, *N*-methyl- [624-78-2]
1-METHYL-3-ETHYLOXINDOLE, IV, 620
    2*H*-Indol-2-one, 3-ethyl-1,3-dihydro-1-methyl- [2525-35-1]
1-METHYL-2-ETHYNYL-*endo*-3,3-DIMETHYL-2-NORBORNANOL, VIII, 391
    Bicyclo[2.2.1]heptan-2-ol, 2-ethynyl-1,3,3-trimethyl-, (1*R-endo*)- [131062-94-7]

9-METHYLFLUORENE, IV, 623
9*H*-Fluorene, 9-methyl- [2523-37-7]
***N*-METHYLFORMANILIDE, III**, 590
Formamide, *N*-methyl-*N*-phenyl- [93-61-8]
3-METHYLFURAN, IV, 628
Furan, 3-methyl- [930-27-8]
3-METHYL-2(5*H*)-FURANONE, VIII, 396
2(5*H*)-Furanone, 3-methyl- [22122-36-7]
5-METHYLFURFURAL, II, 393
2-Furancarboxaldehyde, 5-methyl- [620-02-0]
5-METHYLFURFURYLDIMETHYLAMINE, IV, 626
2-Furanmethanamine, *N,N*,5-trimethyl- [14496-35-6]
3-METHYL-2-FUROIC ACID, IV, 628
2-Furancarboxylic acid, 3-methyl- [4412-96-8]
α–METHYL D-GLUCOSIDE, I, 364
α-D-Glucopyranoside, methyl [97-30-3]
β-METHYLGLUTARALDEHYDE, IV, 661
Pentanedial, 3-methyl- [6280-15-5]
β–METHYLGLUTARIC ACID, III, 591; **IV**, 630
Pentanedioic acid, 3-methyl- [626-51-7]
β–METHYLGLUTARIC ANHYDRIDE, IV, 630
2*H*-Pyran-2,6(3*H*)-dione, dihydro-4-methyl- [4166-53-4]
METHYLGLYOXAL ω-PHENYLHYDRAZONE, IV, 633
Propanal, 2-oxo-, 1-(phenylhydrazone) [5391-74-2]
2-METHYL-1,6-HEPTADIEN-3-OL, VI, 606
1,6-Heptadien-3-ol, 2-methyl- [53268-46-5]
3-METHYL-2,4-HEPTANEDIONE, VI, 776
2,4-Heptanedione, 3-methyl- [13152-54-0]
3-METHYLHEPTANOIC ACID, V, 762
Heptanoic acid, 3-methyl- [53663-30-2]
4-METHYL-3-HEPTANONE, (*S*)-(+)-, VIII, 403
3-Heptanone, 4-methyl-, (*S*)- [51532-30-0]
4-METHYL-3-HEPTANONE SAMP-HYDRAZONE, (*S*)-(+)-, VIII, 403
1-Pyrrolidinamine, *N*-(1-ethyl-2-methylpentylidene)-2-methoxymethyl)-,
[*S*-[*R**,*R**-(*Z*)]- [69943-24-4]
1-METHYL-4,4a,5,6,7,8-HEXAHYDRONAPHTHALEN-2(3*H*)-ONE, VI, 781
2(3*H*)-Naphthalenone, 4,4a,5,6,7,8-hexahydro-1-methyl- [5164-37-4]
2-METHYL-1-HEXANOL, VI, 786
1-Hexanol, 2-methyl- [624-22-6]
5-METHYL-5-HEXEN-2-ONE, V, 767
5-Hexen-2-one, 5-methyl- [3240-09-3]
METHYL *N*-HEXYLCARBAMATE, VI, 788
Carbamic acid, hexyl-, methyl ester [22139-32-8]
METHYL HEXYL CARBINOL, I, 366, 418
2-Octanol [123-96-6]
METHYL 2-HEXYNOATE, VI, 791
2-Hexynoic acid, methyl ester [18937-79-6]

METHYL HOMOVERATRATE, II, 333
  Benzeneacetic acid, 3,4-dimethoxy-, methyl ester [15964-79-1]
METHYLHYDRAZINE SULFATE, II, 395
  Hydrazine, methyl-, sulfate (1:1) [302-15-8]
METHYL HYDROGEN HENDECANEDIOATE, IV, 635
  Undecanedioic acid, monomethyl ester [3927-60-4]
METHYL HYDROGEN SUCCINATE, III, 169
  Butanedioic acid, monomethyl ester [3878-55-5]
4-METHYL-7-HYDROXY-8-ACETYLCOUMARIN, III, 283
  2*H*-1-Benzopyran-2-one, 8-acetyl-7-hydroxy-4-methyl- [2555-29-5]
METHYL 4-HYDROXY-2-BUTYNOATE, VII, 334
  2-Butynoic acid, 4-hydroxy-, methyl ester [31555-05-2]
4-METHYL-7-HYDROXYCOUMARIN, III, 282
  2*H*-1-Benzopyran-2-one, 7-hydroxy-4-methyl- [90-33-5]
METHYL (1*R*,5*R*)-5-HYDROXY-2-CYCLOPENTENE-1-ACETATE, VII, 339
  2-Cyclopentene-1-acetic acid, 5-hydroxy-, methyl ester, (1*R-trans*)- [49825-99-2]
METHYL 7-HYDROXYHEPT-5-YNOATE, VIII, 415
  5-Heptynoic acid, 7-hydroxy-, methyl ester [50781-91-4]
METHYL 3-HYDROXY-2-METHYLENEPENTANOATE, VIII, 420
  Pentanoic acid, 3-hydroxy-2-methylene-, methyl ester [18052-21-6]
METHYL *anti*-3-HYDROXY-2-METHYLPENTANOATE, VIII, 420
  Pentanoic acid, 3-hydroxy-2-methyl-, methyl ester, (*R**,*R**)-(±)- [100992-75-4]
4-METHYL-6-HYDROXYPYRIMIDINE, IV, 638
  4(1*H*)-Pyrimidinone, 6-methyl- [3524-87-6]
2-METHYL-4-HYDROXYQUINOLINE, III, 593
  4-Quinolinol, 2-methyl- [607-67-0]
METHYLIMINODIACETIC ACID, II, 397
  Glycine, *N*-(carboxymethyl)-*N*-methyl- [4408-64-4]
1-METHYL-2-IMINO-β-NAPHTHOTHIAZOLINE, III, 595
  Naphtho[1,2-*d*]thiazol-2(1*H*)-imine, 1-methyl- [53663-31-3]
2-METHYL-1-INDANONE, V, 624
  1*H*-Inden-1-one, 2,3-dihydro-2-methyl- [17496-14-9]
1-METHYLINDOLE, V, 769
  1*H*-Indole, 1-methyl- [603-76-9]
2-METHYLINDOLE, III, 597
  1*H*-Indole, 2-methyl- [95-20-5]
METHYL IODIDE, II, 399
  Methane, iodo- [74-88-4]
METHYL 2-IODO-3-NITROPROPIONATE, VI, 799
METHYL (*trans*-2-IODO-1-TETRALIN)CARBAMATE, VI, 795
  Carbamic acid, (1,2,3,4-tetrahydro-2-iodo-1-naphthalenyl)-, methyl ester, *trans*-
    [1210-13-5]
METHYL ISOCYANIDE, V, 772
  Methane, isocyano- [593-75-9]
METHYL ISOPROPYL CARBINOL, II, 406
  2-Butanol, 3-methyl- [598-75-4]

METHYL ISOPROPYL KETONE, II, 408
  2-Butanone, 3-methyl- [563-80-4]
1-METHYLISOQUINOLINE, IV, 641
  Isoquinoline, 1-methyl- [1721-93-3]
METHYL ISOTHIOCYANATE, III, 599
  Methane, isothiocyanato- [556-61-6]
***S*-METHYL ISOTHIOUREA SULFATE, II**, 411
  Carbamimidothioic acid, methyl ester, sulfate (2:1) [867-44-7]
METHYLISOUREA HYDROCHLORIDE, IV, 645
  Carbamimidic acid, methyl ester, monohydrochloride [5329-33-9]
METHYL 4-KETO-7-METHYLOCTANOATE, III, 601
  Octanoic acid, 7-methyl-4-oxo-, methyl ester [53663-32-4]
METHYLLITHIUM, LOW-HALIDE, VII, 346
  Lithium, methyl [917-54-4]
α–METHYL MANNOSIDE, I, 371
  α-D-Mannoside, methyl [27939-30-6]
2-METHYLMERCAPTO-*N*-METHYL-Δ²-PYRROLINE, V, 780
  1*H*-Pyrrole, 4,5-dihydro-1-methyl-2-(methylthio)- [25355-52-6]
[(METHYL)(METHOXY)CARBENE]PENTACARBONYL CHROMIUM(0), VIII, 216
  Chromium, pentacarbonyl(1-methoxyethylidene)-, (OC-6-21)- [20540-69-6]
N-METHYL-(3-METHOXY-4-HYDROXY)PHENYLALANINE, III, 588
METHYL N-(2-METHYL-2-BUTENYL)CARBAMATE, VIII, 427
  Carbamic acid, (2-methyl-2-butenyl)-, methyl ester [86766-65-6]
METHYL 2-METHYL-1-CYCLOHEXENE-1-CARBOXYLATE, VII, 351
  1-Cyclohexene-1-carboxylic acid, 2-methyl-, methyl ester [25662-38-8]
METHYL 3-METHYL-2-FUROATE, IV, 649
  2-Furancarboxylic acid, 3-methyl-, methyl ester [6141-57-7]
METHYL γ-METHYL-γ-NITROVALERATE, IV, 652
  Pentanoic acid, 4-methyl-4-nitro-, methyl ester [16507-02-1]
(1*RS*,2*SR*,5*R*)-5-METHYL-2-(1-METHYL-1-PHENYLETHYL)CYCLOHEXANOL
  (see 8-PHENYLMENTHOL), VIII, 524
(2*RS*,5*R*)-5-METHYL-2-(1-METHYL-1-PHENYLETHYL)CYCLOHEXANONE,
  VIII, 523
  Cyclohexanone, 5-methyl-2-(1-methyl-1-phenylethyl)-, (2*S-cis*)- [65337-06-6];
    (2*R-trans*)- [57707-92-3]
(1*R*,2*S*,5*R*)-5-METHYL-2-[(1-METHYL-1-PHENYLETHYL)]CYCLOHEXYL
  CHLOROACETATE, VIII, 522
  Acetic acid, chloro-, 5-methyl-2-(1-methyl-1-phenylethyl)cyclohexyl ester,
    [1*R*-(1α,2β,5α)]- [71804-27-8]
5-METHYL-2-[1-METHYL-1-(PHENYLMETHYLTHIO)ETHYL]-
  CYCLOHEXANONE, *cis*- and *trans*-(7-BENZYLTHIOMENTHONE),
  VIII, 304
  Cyclohexanone, 5-methyl-2-[1-methyl-1-(phenylmethylthio)ethyl]-, (2*R-trans*)-
    [79563-58-9]; (2*S-cis*)- [79618-04-5]
METHYL 2-METHYL-5-TRIMETHYLSILYL-2-VINYLPENTANOATE, VIII, 486
  Pentanoic acid, 2-ethenyl-2-methyl-5-trimethylsilyl-, methyl ester [88729-80-0]

N-METHYLMORPHOLINE N-OXIDE, VI, 342
  Morpholine, 4-methyl-, 4-oxide- [7529-22-8]
METHYL MYRISTATE, III, 605
  Tetradecanoic acid, methyl ester [124-10-7]
N-METHYL-1-NAPHTHYLCYANAMIDE, III, 608
  Cyanamide, methyl-1-naphthalenyl- [53663-33-5]
1-METHYL-1-(1-NAPHTHYL)-2-THIOUREA, III, 609
  Thiourea, N-methyl-N-1-naphthalenyl- [53663-34-6]
METHYL NITRATE, II, 412
  Nitric acid, methyl ester [598-58-3]
METHYL NITROACETATE, VI, 797
  Acetic acid, nitro-, methyl ester [2483-57-0]
METHYL (E)-3-NITROACRYLATE, VI, 799
  2-Propenoic acid, 3-nitro-, methyl ester, (E)- [52745-92-3]
METHYL m-NITROBENZOATE, I, 372
  Benzoic acid, 3-nitro-, methyl ester [618-95-1]
2-METHYL-4'-NITROBIPHENYL, VIII, 430
  1,1'-Biphenyl, 2-methyl-4'-nitro- [33350-73-1]
2-METHYL-2-NITROPROPANE, VI, 803
  Propane, 2-methyl-2-nitro- [594-70-7]
3-METHYL-4-NITROPYRIDINE-1-OXIDE, IV, 654
  Pyridine, 3-methyl-4-nitro, 1-oxide [1074-98-2]
2-METHYL-2-NITROSOPROPANE and DIMER, VI, 803
  Propane, 2-methyl-2-nitroso- [917-95-3]
4-METHYL-4-NITROVALERIC ACID, V, 445
  Valeric acid, 4-methyl-4-nitro- [32827-16-0]
**10-METHYL-$\Delta^{1(9)}$-2-OCTALONE N,N-DIMETHYLHYDRAZONE, (E)-,
    VI**, 242
  2(3H)-Naphthalenone, 4,4a,5,6,7,8-hexahydro-4a-methyl-,
    dimethylhydrazone, (E)- [66252-92-4]
(Z)-3-METHYL-2-OCTEN-1-OL, VIII, 509
  2-Octen-1-ol, 3-methyl-, (Z)- [30804-78-5]
METHYL OXALATE, II, 414
  Ethanedioic acid, dimethyl ester [553-90-2]
3-METHYLOXINDOLE, IV, 657
  2H-Indol-2-one, 1,3-dihydro-3-methyl- [1504-06-9]
(R)-METHYLOXIRANE, VIII, 434
  Oxirane, methyl- (R)- [15448-47-2]
(S)-(–)-METHYLOXIRANE, VII, 356
  Oxirane, methyl-, (S)- [16088-62-3]
2-METHYL-2-(3-OXOBUTYL)-1,3-CYCLOHEXANEDIONE, VII, 368
  1,3-Cyclohexanedione, 2-methyl-2-(3-oxobutyl)- [5073-65-4]
2-METHYL-2-(3-OXOBUTYL)-1,3-CYCLOPENTANEDIONE, VII, 363
  1,3-Cyclopentanedione, 2-methyl-2-(3-oxobutyl)- [25112-78-1]
METHYL 2-OXOCYCLOHEXANECARBOXYLATE, VII, 351
  Cyclohexanecarboxylic acid, 2-oxo-, methyl ester [41302-34-5]

METHYL 6-OXODECANOATE, VIII, 441
  Decanoic acid, 6-oxo-, methyl ester [61820-00-6]
METHYL 7-OXOHEPTANOATE, VI, 807
  Heptanoic acid, 7-oxo-, methyl ester [35376-00-2]
METHYL 6-OXOHEXANOATE, VII, 168
  Hexanoic acid, 6-oxo-, methyl ester [6654-36-0]
METHYL 7-OXOOCTANOATE, VI, 807
  Octanoic acid, 7-oxo-, methyl ester [16493-42-8]
**METHYL 2-OXO-5,6,7,8-TETRAHYDRO-2*H*-1-BENZOPYRAN-3-
    CARBOXYLATE, VIII,** 444
  2*H*-1-Benzopyran-3-carboxylic acid, 5,6,7,8-tetrahydro-2-oxo-, methyl ester
    [85531-80-2]
METHYL PALMITATE, III, 605
  Hexadecanoic acid, methyl ester [112-39-0]
3-METHYL-1,5-PENTANEDIOL, IV, 660
  1,5-Pentanediol, 3-methyl- [4457-71-0]
3-METHYLPENTANE-2,4-DIONE, V, 785
  2,4-Pentanedione, 3-methyl- [815-57-6]
3-METHYLPENTANOIC ACID, II, 416
  Pentanoic acid, 3-methyl- [105-43-1]
**3-METHYL-2-PENTYL-2-CYCLOPENTEN-1-ONE(DIHYDROJASMONE),
    VIII,** 620
  2-Cyclopenten-1-one, 3-methyl-2-pentyl- [1128-08-1]
METHYL PHENYLACETYLACETATE, VII, 359
  Benzenebutanoic acid, β-oxo-, methyl ester [37779-49-0]
β-METHYL-β-PHENYL-α,α'-DICYANOGLUTARIMIDE, IV, 662
  3,5-Piperidinedicarbonitrile, 4-methyl-2,6-dioxo-4-phenyl- [6936-95-4]
β-METHYL-β-PHENYLGLUTARIC ACID, IV, 664
  Pentanedioic acid, 3-methyl-3-phenyl- [4160-92-3]
α-METHYL-α-PHENYLHYDRAZINE, II, 418
  Hydrazine, 1-methyl-1-phenyl- [618-40-6]
1-METHYL-3-PHENYLINDANE, IV, 665
  1*H*-Indene, 2,3-dihydro-1-methyl-3-phenyl- [6416-39-3]
2-METHYL-2-PHENYL-4-PENTENAL, VIII, 451
  Benzeneacetaldehyde, α-methyl-α-2-propenyl- [24401-39-6]
3-METHYL-1-PHENYLPHOSPHOLENE 1,1-DICHLORIDE, V, 787
  1*H*-Phosphole, 1,1-dichloro-1,1,2,3-tetrahydro-4-methyl-1-phenyl- [17154-12-0]
3-METHYL-1-PHENYLPHOSPHOLENE OXIDE, V, 787
  1*H*-Phosphole, 2,3-dihydro-4-methyl-1-phenyl-, 1-oxide [707-61-9]
**2-METHYL-3-PHENYLPROPANAL (2-METHYL- 3-PHENYLPROPIONALDE-
    HYDE), VI,** 815; **VII,** 361
  Benzenepropanal, α-methyl- [5445-77-2]
METHYL (Z)-3-(PHENYLSULFONYL)PROP-2-ENOATE, VIII, 458
  2-Propenoic acid, 3-(phenylsulfonyl)-, methyl ester, (Z)- [91077-67-7]
METHYL PHENYL SULFOXIDE, V, 791
  Benzene, (methylsulfinyl)- [1193-82-4]

N-METHYL-2-PHENYL-Δ²-TETRAHYDROPYRIDINE, VI, 818
Pyridine, 1,2,3,4-tetrahydro-1-methyl-6-phenyl- [30609-85-9]
1-(2-METHYLPROPENYL)PYRROLIDINE, VI, 496
Pyrrolidine, 1-(2-methyl-1-propenyl)- [2403-57-8]
N-(2-METHYLPROPYLIDENE)-*tert*-BUTYLAMINE, VI, 526
2-Propanamine, 2-methyl-*N*-(2-methylpropylidene)- [6852-60-4]
2-(1-METHYL-2-PROPYNYL)CYCLOHEXANONE, VIII, 460
Cyclohexanone, 2-(1-methyl-2-propynyl)-, (*R**,*R**)-(±)- [130719-23-1] (*R**,*S**)-(±)-
[130719-24-3]
1-METHYL-2-PYRIDONE, II, 419
2(1*H*)-Pyridinone, 1-methyl- [694-85-9]
4-METHYLPYRIMIDINE, V, 794
Pyrimidine, 4-methyl- [3438-46-8]
METHYL PYRUVATE, III, 610
Propanoic acid, 2-oxo-, methyl ester [600-22-6]
METHYL RED, I, 374
Benzoic acid, 2-[[4-(dimethylamino)phenyl]azo]- [493-52-7]
METHYL SEBACAMATE, III, 613
Decanoic acid, 10-amino-10-oxo-, methyl ester [53663-35-7]
trans-METHYLSTYRYLCARBINOL, IV, 773
3-Buten-2-ol, 4-phenyl- [17488-65-2]
METHYLSUCCINIC ACID, III, 615
Butanedioic acid, methyl- [498-21-5]
2-(METHYLSULFINYL)ACETOPHENONE, V, 939
Ethanone, 2'-methylsulfinyl-1-phenyl- [26524-93-6]
(+)-(7a*S*)-7a-METHYL-2,3,7,7a-TETRAHYDRO-1*H*-INDENE-1,5(6*H*)-DIONE,
VII, 363
1*H*-Indene-1,5(6*H*)-dione, 7a-methyl-2,3,7,7a-tetrahydro-, (*S*)- [17553-86-5]
8a-METHYL-3,4,8,8a-TETRAHYDRO-1,6(2*H*,7*H*)-NAPHTHALENEDIONE, (*S*)-
[33878-99-8], VII, 368
1,6(2*H*,7*H*)-Naphthalenedione, 3,4,8,8a-tetrahydro-8a-methyl- [20007-72-1]
METHYL 2-THIENYL SULFIDE, IV, 667
Thiophene, 2-(methylthio)- [5780-36-9]
METHYL β-THIODIPROPIONATE, IV, 669
Propanoic acid, 3,3'-thiobis-, dimethyl ester [4131-74-2]
1-METHYLTHIOL-3-PHTHALIMIDOPROPANE-3,3-DICARBOXYLIC ACID,
II, 385
3-METHYLTHIOPHENE, IV, 671
Thiophene, 3-methyl- [616-44-4]
2-METHYLTHIO-2-PHENYL-1,3-DITHIANE, VI, 109
1,3-Dithiane, 2-methylthio-2-phenyl- [34858-82-7]
N-METHYLTHIOPYRROLIDONE, V, 780; VII, 372
2-Pyrrolidinethione, 1-methyl- [10441-57-3]
4'-(METHYLTHIO)-2,2':6',2''-TERPYRIDINE, VII, 476
2,2':6',2''-Terpyridine, 4'-(methylthio)- [78570-35-1]

METHYLTHIOUREA, III, 617
    Thiourea, methyl- [598-52-7]
2-METHYL-3-(*p*-TOLUENESULFONAMIDO)-2-PENTANOL, VII, 375
    Benzenesulfonamide, *N*-(1-ethyl-2-hydroxy-2-methylpropyl)-4-methyl-
    [87291-33-6]
METHYL *p*-TOLUENESULFONATE, I, 146
    Benzenesulfonic acid, 4-methyl-, methyl ester [80-48-8]
METHYL *p*-TOLYL SULFONE, IV, 674
    Benzene, 1-methyl-4-(methylsulfonyl)- [3185-99-7]
(*S*)-(–)-METHYL *p*-TOLYL SULFOXIDE, VIII, 464
    Benzene, 1-methyl-4-(methylsulfinyl)-, (*S*)- [5056-07-5]
1-METHYL-3-*p*-TOLYLTRIAZENE, V, 797
    1-Triazene, 1-methyl-3-(4-methylphenyl)- [21124-13-0]
4-METHYLTRICYCLO[2.2.2.0³,⁵]OCTANE-2,6-DIONE, VIII, 467
    Tricyclo[3.2.1.0²,⁷]-octane-6,8-dione, 2-methyl- [119986-99-1]
2-METHYL-2-TRIMETHYLSILOXYPENTAN-3-ONE, VII, 185, 381
    3-Pentanone, 2-methyl-2-[(trimethylsilyl)oxy]- [72507-50-7]
1-METHYL-1-(TRIMETHYLSILYL)ALLENE, VIII, 471
    Silane, trimethyl(1-methyl-1,2-propadienyl)- [74542-82-8]
2-METHYL-5-TRIMETHYLSILYL-2-VINYLPENTANOIC ACID, VIII, 488
    Pentanoic acid, 2-ethenyl-2-methyl-5-trimethylsilyl- [88729-72-0]
METHYLTRIISOPROPOXYTITANIUM, VIII, 495
    Titanium, methyltris(2-propanolato)-, (T-4)- [18006-13-8]
2-METHYL-2-UNDECENE, VIII, 474
    2-Undecene, 2-methyl- [56888-88-1]
6-METHYLURACIL, II, 422
    2,4(1*H*,3*H*)-Pyrimidinedione, 6-methyl- [626-48-2]
β-METHYL-δ-VALEROLACTONE, IV, 677
    2*H*-Pyran-2-one, tetrahydro-4-methyl- [1121-84-2]
***trans*-3-METHYL-2-VINYLCYCLOHEXANONE, VIII,** 479
    Cyclohexanone, 2-ethenyl-3-methyl-, *trans*- [110222-94-1]
2-METHYL-2-VINYLCYCLOPENTANONE, VIII, 486
    Cyclopentanone, 2-ethenyl-2-methyl- [88729-76-4]
MEVALONOLACTONE-2-¹³C, (*R*,*S*)-, VII, 386
    2*H*-Pyran-2-one-3-¹³C, tetrahydro-4-hydroxy-4-methyl- [53771-22-5]
MONOBENZALPENTAERYTHRITOL, IV, 679
    1,3-Dioxane-5,5-dimethanol, 2-phenyl- [2425-41-4]
MONOBROMOPENTAERYTHRITOL, IV, 681
    1,3-Propanediol, 2-(bromomethyl)-2-(hydroxymethyl)- [19184-65-7]
MONOCHLOROMETHYL ETHER, I, 377
    Methane, chloromethoxy- [107-30-2]
***cis*,*cis*-MONOMETHYL MUCONATE, VIII,** 490
    2,4-Hexadienoic acid, monomethyl ester, (*Z*,*Z*)- [61186-96-7]
MONOPERPHTHALIC ACID, III, 619; **V,** 805
    Benzenecarboperoxoic acid, 2-carboxy- [2311-91-3]

MONOVINYLACETYLENE, IV, 683
1-Buten-3-yne [689-97-4]
1-MORPHOLINO-1-CYCLOHEXENE, V, 808
Morpholine, 4-(1-cyclohexen-1-yl)- [670-80-4]
α-**MORPHOLINOSTYRENE, VI,** 520
Morpholine, 4-(1-phenylethenyl)- [7196-01-2]
MUCOBROMIC ACID, III, 621; **IV,** 688
2-Butenoic acid, 2,3-dibromo-4-oxo-, (Z)- [488-11-9]
MUCONIC ACID, III, 623
2,4-Hexadienedioic acid [505-70-4]
MYRISTIC ACID, I, 379; **III,** 605
Tetradecanoic acid [544-63-8]

N

1-NAPHTHALDEHYDE, IV, 690
1-Naphthalenecarboxaldehyde [66-77-3]
β-**NAPHTHALDEHYDE, III,** 626
2-Naphthalenecarboxaldehyde [66-99-9]
NAPHTHALENE, VI, 152, 821; *Hazard*
Naphthalene [91-20-3]
2,3-NAPHTHALENEDICARBOXYLIC ACID, V, 810
2,3-Naphthalenedicarboxylic acid [2169-87-1]
2,6-NAPHTHALENEDICARBOXYLIC ACID, V, 813
2,6-Naphthalenedicarboxylic acid [1141-38-4]
NAPHTHALENE-1,5-DISULFONYL CHLORIDE, IV, 693
1,5-Naphthalenedisulfonyl dichloride [1928-01-4]
1,5-NAPHTHALENEDITHIOL, IV, 695
1,5-Naphthalenedithiol [5325-88-2]
2-NAPHTHALENETHIOL, VI, 824
2-Naphthalenethiol [91-60-1]
1,2-NAPHTHALIC ANHYDRIDE, II, 423
Naphtho[1,2-*c*]furan-1,3-dione [5343-99-7]
α–**NAPHTHOIC ACID, II,** 425
1-Naphthalenecarboxylic acid [86-55-5]
β–**NAPHTHOIC ACID, II,** 428
2-Naphthalenecarboxylic acid [93-09-4]
β–**NAPHTHOL PHENYLAMINOMETHANE, 1,** 381
2-Naphthalenol, 1-(aminophenylmethyl)- [481-82-3]
α-**NAPHTHONITRILE, III,** 631
1-Naphthalenecarbonitrile [86-53-3]
1,2-NAPHTHOQUINONE, II, 430
1,2-Naphthalenedione [524-42-5]
1,4-NAPHTHOQUINONE, I, 383; **IV,** 698
1,4-Naphthalenedione [130-15-4]
1,2-NAPHTHOQUINONE-4-SULFONATE, AMMONIUM SALT, III, 633
1-Naphthalenesulfonic acid, 3,4-dihydro-3,4-dioxo-, ammonium salt [53684-60-9]

1,2-NAPHTHOQUINONE-4-SULFONATE POTASSIUM SALT, III, 633
1-Naphthalenesulfonic acid, 3,4-dihydro-3,4-dioxo, potassium salt [5908-27-0]
NAPHTHORESORCINOL, III, 637
1,3-Naphthalenediol [132-86-5]
***O*-2-NAPHTHYL DIMETHYLTHIOCARBAMATE, VI,** 824
Carbamothioic acid, dimethyl-, *O*-2-naphthalenyl ester [2951-24-8]
(*S*)-(−)-α-(1-NAPHTHYL)ETHYLAMINE, VI, 826, 828
1-Naphthalenethanamine, α-methyl-, (*S*)- [10420-89-0]
α-NAPHTHYL ISOTHIOCYANATE, IV, 700
Naphthalene, 1-isothiocyanato- [551-06-4]
β–NAPHTHYLMERCURIC CHLORIDE, II, 432; *Hazard*
Mercury, chloro-2-naphthalenyl- [39966-41-1]
***N*-β-NAPHTHYLPIPERIDINE, V,** 816
Piperidine, 1-(2-naphthalenyl)- [5465-85-0]
NEOPENTYL ALCOHOL, V, 818
1-Propanol, 2,2-dimethyl- [75-84-3]
NEOPENTYL IODIDE, VI, 830
Propane, 1-iodo-2,2-dimethyl- [15501-33-4]
NEOPENTYL PHENYL SULFIDE, VI, 833
Benzene, [(2,2-dimethylpropyl)thio]- [7210-80-2]
NEOPHYL CHLORIDE, IV, 702
Benzene, (2-chloro-1,1-dimethylethyl)- [515-40-2]
NICOTINAMIDE-1-OXIDE, IV, 704
3-Pyridinecarboxamide, 1-oxide [1986-81-8]
NICOTINIC ACID, I, 385
3-Pyridinecarboxylic acid [59-67-6]
NICOTINIC ANHYDRIDE, V, 822
3-Pyridinecarboxylic acid, anhydride [16837-38-0]
NICOTINONITRILE, IV, 706
3-Pyridinecarbonitrile [100-54-9]
***m*-NITROACETOPHENONE, II,** 434
Ethanone, 1-(3-nitrophenyl)- [121-89-1]
***o*-NITROACETOPHENONE, IV,** 708
Ethanone, 1-(2-nitrophenyl)- [577-59-3]
***p*-NITRO-*p*'-ACETYLAMINODIPHENYLSULFONE, III,** 239
Acetamide, *N*-[4-[(4-nitrophenyl)sulfonyl]phenyl]- [1775-37-7]
1-NITRO-2-ACETYLAMINONAPHTHALENE, II, 438
Acetamide, *N*-(1-nitro-2-naphthalenyl)- [5419-82-9]
***o*-NITROANILINE, I,** 388
Benzenamine, 2-nitro- [88-74-4]
9-NITROANTHRACENE, IV, 711
Anthracene, 9-nitro- [602-60-8]
NITROANTHRONE, I, 390
9(10*H*)-Anthracenone, 10-nitro- [6313-44-6]
NITROBARBITURIC ACID, II, 440
2,4,6(1*H*,3*H*,5*H*)-Pyrimidinetrione, 5-nitro- [480-68-2]

o-**NITROBENZALANILINE, V,** 941
  Benzenamine, *N*-[(2-nitrophenyl)methylene]- [17064-77-6]
o-**NITROBENZALDEHYDE, III,** 641; **V,** 825
  Benzaldehyde, 2-nitro- [552-89-6]
m-**NITROBENZALDEHYDE, III,** 644; *Warning,* **III,** 645
  Benzaldehyde, 3-nitro- [99-61-6]
p-**NITROBENZALDEHYDE, II,** 441
  Benzaldehyde, 4-nitro- [555-16-8]
m-**NITROBENZALDEHYDE DIMETHYLACETAL, III,** 644
  Benzene, 1-(dimethoxymethyl)-3-nitro- [3395-79-7]
o-**NITROBENZALDIACETATE, III,** 641; **IV,** 713
  Methanediol, (2-nitrophenyl)-, diacetate [6345-63-7]
p-**NITROBENZALDIACETATE, II,** 441; **IV,** 713
  Methanediol, (4-nitrophenyl)-, diacetate (ester) [2929-91-1]
m-**NITROBENZAZIDE, IV,** 715
  Benzoyl azide, 3-nitro- [3532-31-8]
2-**NITROBENZENESULFINIC ACID, V,** 60
  Benzenesulfinic acid, 2-nitro- [13165-79-2]
o-**NITROBENZENESULFONYL CHLORIDE, II,** 471
  Benzenesulfonyl chloride, 2-nitro- [1694-92-4]
m-**NITROBENZOIC ACID, I,** 391
  Benzoic acid, 3-nitro- [121-92-6]
p-**NITROBENZOIC ACID, I,** 392
  Benzoic acid, 4-nitro- [62-23-7]
p-**NITROBENZONITRILE, III,** 646
  Benzonitrile, 4-nitro- [619-72-7]
m-**NITROBENZOYL CHLORIDE, IV,** 715
  Benzoyl chloride, 3-nitro- [121-90-4]
p-**NITROBENZOYL CHLORIDE, I,** 394
  Benzoyl chloride, 4-nitro- [122-04-3]
p-**NITROBENZOYL PEROXIDE, III,** 649
  Peroxide, bis(4-nitrobenzoyl)- [1712-84-1]
p-**NITROBENZYL ACETATE, III,** 650
  Benzenemethanol, 4-nitro-, acetate (ester) [619-90-9]
p-**NITROBENZYL ALCOHOL, III,** 652
  Benzenemethanol, 4-nitro- [619-73-8]
p-**NITROBENZYL BROMIDE, II,** 443
  Benzene, 1-(bromomethyl)-4-nitro- [100-11-8]
p-**NITROBENZYL CYANIDE, I,** 396
  Benzeneacetonitrile, 4-nitro- [555-21-5]
p-**NITROBENZYL FLUORIDE, VI,** 835
  Benzene, 1-(fluoromethyl)-4-nitro- [500-11-8]
o-**NITROBENZYLPYRIDINIUM BROMIDE, V,** 825
  Pyridinium, 1-(2-nitrobenzyl)-, bromide [13664-80-7]
m-**NITROBIPHENYL, IV,** 718
  1,1'-Biphenyl, 3-nitro- [2113-58-8]

2-NITROCARBAZOLE, V, 829
  9*H*-Carbazole, 2-nitro- [14191-22-1]
o-**NITROCINNAMALDEHYDE, IV,** 722
  2-Propenal, 3-(2-nitrophenyl)- [1466-88-2]
m-**NITROCINNAMIC ACID, I,** 398; **IV,** 731
  2-Propenoic acid, 3-(3-nitrophenyl)- [555-68-0]
(1R,2r,3S)-2-NITROCYCLOHEXANE-1,3-DIOL, VIII, 332
  1,3-Cyclohexanediol, 2-nitro-, (1α,2β,3α)- [38150-01-5]
1-NITROCYCLOOCTENE, VI, 837
  Cyclooctene, 1-nitro- [1782-03-2]
2-NITRO-*p*-CYMENE, III, 653
  Benzene, 1-methyl-4-(1-methylethyl)-2-nitro- [943-15-7]
5-NITRO-2,3-DIHYDRO-1,4-PHTHALAZINEDIONE, III, 656
  1,4-Phthalazinedione, 2,3-dihydro-5-nitro- [3682-15-3]
m-**NITRODIMETHYLANILINE, III,** 658
  Benzenamine, *N,N*-dimethyl-3-nitro- [619-31-8]
o-**NITRODIPHENYL ETHER, II,** 446
  Benzene, 1-nitro-2-phenoxy- [2216-12-8]
p-**NITRODIPHENYL ETHER, II,** 445
  Benzene, 1-nitro-4-phenoxy- [620-88-2]
2-NITROETHANOL, V, 833
  Ethanol, 2-nitro- [625-48-9]
m-**NITROETHYLBENZENE, VII,** 393
  Benzene, 1-ethyl-3-nitro- [7369-50-8]
2-NITROFLUORENE, II, 447
  9*H*-Fluorene, 2-nitro- [607-57-8]
NITROGUANIDINE, I, 399; *Hazard*
  Guanidine, nitro- [556-88-7]
5-NITROHEPTAN-2-ONE, VI, 648
  2-Heptanone, 5-nitro- [42397-25-1]
5-NITROINDAZOLE, III, 660
  1*H*-Indazole, 5-nitro- [5401-94-5]
4-NITROINDOLE, VIII, 493
  1*H*-Indole, 4-nitro- [4769-97-5]
NITROMESITYLENE, II, 449
  Benzene, 1,3,5-trimethyl-2-nitro- [603-71-4]
NITROMETHANE, I, 401
  Methane, nitro- [75-52-5]
2-NITRO-4-METHOXYACETANILIDE, III, 661
  Acetamide, (4-methoxy-2-nitrophenyl)- [119-81-3]
2-NITRO-4-METHOXYANILINE, III, 661
  Benzenamine, 4-methoxy-2-nitro- [96-96-8]
2-NITRO-6-METHOXYBENZONITRILE, III, 293
  Benzonitrile, 2-methoxy-6-nitro- [38469-85-1]
1-(NITROMETHYL)CYCLOHEXANOL, IV, 221, 224
  Cyclohexanol, 1-nitromethyl- [3164-73-6]

N-NITROMORPHOLINE, V, 839
   Morpholine, 4-nitro- [4164-32-3]
1-NITRO-2-NAPHTHOL, II, 451
   2-Naphthalenol, 1-nitro- [550-60-7]
4-NITRO-1-NAPHTHYLAMINE, III, 664
   1-Naphthalenamine, 4-nitro- [776-34-1]
1-NITROOCTANE, IV, 724
   Octane, 1-nitro- [629-37-8]
m-NITROPHENOL, I, 404
   Phenol, 3-nitro- [554-84-7]
p-NITROPHENYLACETIC ACID, I, 406
   Benzeneacetic acid, 4-nitro- [104-03-0]
p-NITROPHENYLARSONIC ACID, III, 665
   Arsonic acid, (4-nitrophenyl)- [98-72-6]
o-NITROPHENYL AZIDE, IV, 75
   Benzene, 1-azido-2-nitro- [1516-58-1]
1-(*p*-NITROPHENYL)-1,3-BUTADIENE, IV, 727
   Benzene, 1-(1,3-butadienyl)-4-nitro- [20264-89-5]
1-(*p*-NITROPHENYL)-4-CHLORO-2-BUTENE, IV, 727
trans-o-NITRO-α-PHENYLCINNAMIC ACID, IV, 730
   Benzeneacetic acid, α-[(2-nitrophenyl)methylene]-, (*E*)- [19319-35-8]
1-(*m*-NITROPHENYL)-3,3-DIMETHYLTRIAZENE, IV, 718
   1-Triazene, 3,3-dimethyl-1-(3-nitrophenyl)- [20241-06-9]
m-NITROPHENYL DISULFIDE, V, 843
   Disulfide, bis(3-nitrophenyl)-[537-91-7]
3'-NITRO-1-PHENYLETHANOL, VIII, 495
   Benzenemethanol, α-methyl-3-nitro- [5400-78-2]
p-NITROPHENYL ISOCYANATE, II, 453
   Benzene, 1-isocyanato-4-nitro- [100-28-7]
p-NITROPHENYL SULFIDE, III, 667
   Benzene, 1,1'-thiobis[4-nitro- [1223-31-0]
o-NITROPHENYLSULFUR CHLORIDE, II, 455
   Benzenesulfenyl chloride, 2-nitro- [7669-54-7]
3-NITROPHTHALIC ACID, I, 408
   1,2-Benzenedicarboxylic acid, 3-nitro- [603-11-2]
4-NITROPHTHALIC ACID, II, 457
   1,2-Benzenedicarboxylic acid, 4-nitro- [610-27-5]
3-NITROPHTHALIC ANHYDRIDE, I, 410
   1,3-Isobenzofurandione, 4-nitro- [641-70-3]
4-NITROPHTHALIMIDE, II, 459
   1*H*-Isoindole-1,3(2*H*)-dione, 5-nitro- [89-40-7]
2-NITROPROPENE, VII, 396, 414
   1-Propene, 2-nitro- [4749-28-4]
NITROSOBENZENE, III, 668
   Benzene, nitroso- [586-96-9]

p-**NITROSODIETHYLANILINE and HYDROCHLORIDE, II**, 224; *Hazard*
    Benzenamine, *N,N*-diethyl-4-nitroso- [120-22-9]
    Benzenamine, *N,N*-diethyl-4-nitroso-, hydrochloride [58066-98-1]
N-**NITROSO-β-METHYLAMINOISOBUTYL METHYL KETONE, III**, 244;
    Hazard
    2-Pentanamine, 4-methyl-4-(methylnitrosoamino)- [16339-21-2]
NITROSODIMETHYLAMINE, II, 211; *Hazard*
    Methanamine, *N*-methyl-*N*-nitroso- [62-75-9]
p-**NITROSODIMETHYLANILINE HYDROCHLORIDE, I**, 410; **II**, 223
    Benzenamine, *N,N*-dimethyl-4-nitroso-, hydrochloride [42344-05-8]
N-**NITROSOMETHYLANILINE, II**, 460; *Hazard*
    Benzenamine, *N*-methyl-*N*-nitroso- [614-00-6]
NITROSOMETHYLUREA, II, 461; *Hazard; Warning*, **II**, 165, 461
    Urea, *N*-methyl-*N*-nitroso- [684-93-5]
NITROSOMETHYLURETHANE, II, 464; *Hazard; Warning*, **V**, 842
    Carbamic acid, methylnitroso-, ethyl ester [615-53-2]
NITROSO-β-NAPHTHOL, I, 411
    2-Naphthalenol, 1-nitroso- [131-91-9]
2-NITROSO-5-NITROTOLUENE, III, 334
    Benzene, 2-methyl-4-nitro-1-nitroso- [57610-10-3]
NITROSO-tert-OCTANE, VIII, 93
    Pentane, 2,2,4-trimethyl-4-nitroso- [31044-98-1]
N-**NITROSO-*N*-(2-PHENYLETHYL)BENZAMIDE, V**, 336
N-**NITROSO-*N*-PHENYLGLYCINE, V**, 962; *Hazard*
    Glycine, *N*-nitroso-*N*-phenyl- [6415-68-5]
NITROSOTHYMOL, I, 511
    Phenol, 5-methyl-2-(1-methylethyl)-4-nitroso- [2364-54-7]
β-**NITROSTYRENE, I**, 413
    Benzene, (2-nitroethenyl)- [102-96-5]
m-**NITROSTYRENE, IV**, 731
    Benzene, 1-ethenyl-3-nitro- [586-39-0]
2-NITROTHIOPHENE, II, 466
    Thiophene, 2-nitro- [609-40-5]
m-**NITROTOLUENE, I**, 415
    Benzene, 1-methyl-3-nitro- [99-08-1]
4-NITRO-2,2,4-TRIMETHYLPENTANE, V, 845
    Pentane, 2,2,4-trimethyl-4-nitro- [5342-78-9]
NITROUREA, I, 417; *Hazard*
    Urea, nitro- [556-89-8]
6-NITROVERATRALDEHYDE, IV, 735
    Benzaldehyde, 4,5-dimethoxy-2-nitro- [20357-25-9]
NONADECANOIC ACID, VII, 397
    Nonadecanoic acid [646-30-0]
2,4-NONANEDIONE, V, 848
    2,4-Nonanedione [6175-23-1]

(1R)-NOPADIENE, VIII, 223
  Bicyclo[3.1.1]hept-2-ene, 2-ethenyl-6,6-dimethyl- (1R)- [30293-06-2]
(1R)-NOPYL 4-METHYLBENZENESULFONATE, VIII, 223
  Bicyclo[3.1.1]hept-2-ene-2-ethanol, 6,6-dimethyl-, 4-methylbenzenesulfonate, (1R)-
  [81600-63-7]
D-**NORANDROST-5-EN-3β-OL-16α-CARBOXYLIC ACID, VI,** 840
  D-Norandrost-5-ene-16-carboxylic acid, 3-hydroxy-, (3β,16α)- [40013-51-2]
D-**NORANDROST-5-EN-3β-OL-16β-CARBOXYLIC ACID, VI,** 840
  D-Norandrost-5-ene-16-carboxylic acid, 3-hydroxy-, [3β,13α,16β]- [50764-17-5]
1-NORBORNANECARBOXYLIC ACID, VI, 845
  Bicyclo[2.2.1]heptane-1-carboxylic acid [18720-30-4]
2-NORBORNANONE, V, 852
  Bicyclo[2.2.1]heptan-2-one [497-38-1]
NORBORNYLENE, IV, 738
  Bicyclo[2.2.1]hept-2-ene [498-66-8]
2-exo-NORBORNYL FORMATE, V, 852
  Bicyclo[2.2.1]heptan-2-ol, formate, exo- [41498-71-9]
NORCARANE, V, 855
  Bicyclo[4.1.0]heptane [286-08-8]
exo/endo-7-NORCARANOL, V, 859
  Bicyclo[4.1.0]heptan-7-ol, (1α,6α,7α)- [13830-44-9]
  Bicyclo[4.1.0]heptan-7-ol, (1α,6α,7β)- [931-31-7]
NORTRICYCLANOL, V, 863
  Tricyclo[2.2.1.0^{2,6}]heptan-3-ol [695-04-5]
NORTRICYCLANONE, V, 866
  Tricyclo[2.2.1.0^{2,6}]heptanone [695-05-6]
NORTRICYCLYL ACETATE, V, 863
  Tricyclo[2.2.1.0^{2,6}]heptan-3-ol, acetate [6555-48-2]

O

**6aR-(6aα,9a,10aβ)]-OCTAHYDRO-3,3,6,6,9-PENTAMETHYL-1H,6H-[1,3]-DIOX-
  INO[4,5-c']BENZOPYRAN-1-ONE, VIII,** 353
  1H,6H-[1,3]-Dioxino[4,5-c][2]benzopyran-1-one, octahydro-3,3,6,6,9-pentamethyl-,
  [6aR-(6aα,9α,10aβ)]- [78394-10-2]
Δ^{9,10}-OCTALIN, VI, 852
  Naphthalene, 1,2,3,4,5,6,7,8-octahydro- [493-03-8]
2-OCTALONE, V, 872
  2(3H)-Naphthalenone, 4,4a,5,6,7,8-hexahydro- [1196-55-0]
OCTANAL, V, 872
  Octanal [124-13-0]
1-OCTANOL, VI, 919
  1-Octanol [111-87-5]
DL-, D- **and** L-**OCTANOL-2, I,** 366, 418
  2-Octanol, (S)- [6169-06-8]
  2-Octanol, (R)- [5978-70-1]
  2-Octanol, (RS)- [123-96-6]

2,2,7,7,12,12,17,17-OCTAMETHYL-21,22,23,24-
TETRAOXAPERHYDROQUATERENE, VI, 856
21,22,23,24-Tetraoxapentacyclo[16.2.1.1^{3,6}.1^{8,11}.1^{13,16}]tetracosa-3,5,8,10,13,15,18,20-
octaene [22900-44-3]

(Z)-2-OCTEN-1-OL, VIII, 508
2-Octen-1-ol, (Z)-[26001-58-1]

OCTYL BROMIDE, I, 30
Octane, 1-bromo- [111-83-1]

γ-(OCTYL)-γ-BUTYROLACTONE, VII, 400
2(3H)-Furanone, dihydro-5-octyl- [2305-05-7]

sec-**OCTYL HYDROGEN PHTHALATE, I**, 418
1,2-Benzenedicarboxylic acid, monooctyl ester [5393-19-1]

tert-**OCTYL HYDROPEROXIDE, V**, 818
Hydroperoxide, 1,1,3,3-tetramethylbutyl- [5809-08-5]

(R)-(+)-1-OCTYN-3-OL, VII, 402
1-Octyn-3-ol, (R)- [32556-70-0]

OLEOYL CHLORIDE, IV, 739
9-Octadecenoyl chloride, (Z)- [112-77-6]

OLEYL ALCOHOL, II, 468; **III**, 671
9-Octadecen-1-ol, (Z)- [143-28 2]

ORCINOL DIMETHYL ETHER, VI, 859
Benzene, 1,3-dimethoxy-5-methyl- [4179-19-5]

ORCINOL MONOMETHYL ETHER, VI, 859
Phenol, 3-methoxy-5-methyl- [3209-13-0]

ORTHANILIC ACID, II, 471
Benzenesulfonic acid, 2-amino- [88-21-1]

8-OXABICYCLO[4.3.0]NONAN-7-ONE, (1R,6S)-(+)-, VII, 406
1(3H)-Isobenzofuranone, hexahydro-, (3aS-cis)- [65376-02-5]

2-OXA-7,7-DICHLORONORCARANE, V, 874
2-Oxabicyclo[4.1.0]heptane, 7,7-dichloro- [7556-13-0]

OXALIC ACID (ANHYDROUS), I, 421
Ethanedioic acid [144-62-7]

OXASPIROPENTANE, VI, 320
Oxaspiro[2.2]pentane [157-41-5]

11-OXATRICYCLO[4.4.1.0^{1,6}]UNDECA-3,8-DIENE, VI, 862
4a,8a-Epoxynaphthalene, 1,4,5,8-tetrahydro- [16573-72-1]

2-OXAZOLIDINONE, VII, 4
2-Oxazolidinone [497-25-6]

1,6-OXIDO[10]ANNULENE, VI, 736, 862
11-Oxabicyclo[4.4.1]undeca-1,3,5,7,9-pentaene [4759-11-9]

16-OXIMINOANDROST-5-EN-3β-OL-17-ONE, VI, 840
Androst-5-ene-16,17-dione, 3-hydroxy-, 16-oxime, (3β)- [21242-37-5]

OXIMINOMALONONITRILE, V, 32
Propanedinitrile, (hydroxyimino)- [36568-05-5]

S-**(2-OXOBUT-3-YL) BUTANETHIOATE, VI**, 776
Butanethioic acid, S-(1-methyl-2-oxopropyl) ester [32425-98-2]

6-OXODECANAL, VIII, 498
Decanal, 6-oxo- [63049-53-6]
6-OXODECANOIC ACID, VIII, 499
Decanoic acid, 6-oxo- [4144-60-9]
D-2-OXO-7,7-DIMETHYL-1-VINYLBICYCLO[2.2.1]HEPTANE, V, 877
Bicyclo[2.2.1]heptan-2-one, 1-ethenyl-7,7-dimethyl-, (+)- [53585-70-9]
OXODIPEROXYMOLYBDENUM(PYRIDINE) (HEXAMETHYLPHOSPHORIC TRIAMIDE), VII, 277
Molybdenum, (hexamethylphosphoric triamide-*O*)oxodiperoxy(pyridine)-
[23319-63-3]
(4S)-3-[(1-OXO-*syn*-2-METHYL-3-HYDROXY)-3-PHENYLPROPYL]-4-(PHENYLMETHYL)-2-OXAZOLIDINONE, VIII, 339
2-Oxazolidinone, 3-[3-hydroxy-2-methyl-1-oxo-3-phenylpropyl]-
4-(phenylmethyl)-, [4S-[3(2R*, 3R*), 4R*]]- [133467-37-5]
2-OXO-1-PHENYL-3-PYRROLIDINECARBOXYLIC ACID, VII, 411
3-Pyrrolidinecarboxylic acid, 2-oxo-1-phenyl- [56137-52-1]
4-OXO-1-(PHENYLSULFONYL)-*cis*-BICYCLO[4.3.0]NON-2-ENE, VIII, 39
5*H*-Inden-5-one, 1,2,3,3a,4,7a-hexahydro-7a-(phenylsulfonyl)-, *cis*- [131712-15-7]
(1-OXOPROPENYL)TRIMETHYLSILANE, VIII, 501
Silane, trimethyl(1-oxo-2-propenyl)- [51023-60-0]
2-(2-OXOPROPYL)CYCLOHEXANONE, VII, 414
Cyclohexanone, 2-(2-oxopropyl)- [6126-53-0]
(S)-3-(1-OXOPROPYL)-4-(PHENYLMETHYL)-2-OXAZOLIDINONE, VIII, 339
2-Oxazolidinone, 3-(1-oxopropyl)-4-(phenylmethyl)-, (S)- [101711-78-8]
4-OXO-4-(3-PYRIDYL)BUTYRONITRILE, VI, 866
3-Pyridinebutanenitrile, γ-oxo- [36740-10-0]
OZONE, III, 673
Ozone [10028-15-6]

P

PALMITIC ACID, III, 605
Hexadecanoic acid [57-10-3]
PANTOYL LACTONE, D-(–)-, VII, 417
2(3*H*)-Furanone, dihydro-3-hydroxy-4,4-dimethyl- [599-04-2]
PARABANIC ACID, IV, 744
Imidazolidinetrione [120-89-8]
[2.2]PARACYCLOPHANE, V, 883
Tricyclo[8.2.2.2⁴ᐧ⁷] hexadeca-4,6,10,12,13,15-hexaene [1633-22-3]
PELARGONIC ACID, II, 474
Nonanoic acid [112-05-0]
2,3,4,5,6-PENTA-*O*-ACETYL-D-GLUCONIC ACID, V, 887
D-Gluconic acid, pentaacetate [17430-71-6]
PENTAACETYL D-GLUCONONITRILE, III, 690
D-Gluconnitrile, 2,3,4,5,6-pentaacetate [6272-51-1]
2,3,4,5,6-PENTA-*O*-ACETYL-D-GLUCONYL CHLORIDE, V, 887
D-Gluconyl chloride, pentaacetate [53555-69-4]

PENTACHLORBENZOIC ACID, V, 890
Benzoic acid, pentachloro- [1012-84-6]
1,2,3,4,5-PENTACHLORO-5-ETHYLCYCLOPENTADIENE, V, 893
1,3-Cyclopentadiene, 1,2,3,4,5-pentachloro-5-ethyl- [16177-48-3]
PENTADECANAL, VI, 869
Pentadecanal [2765-11-9]
PENTADECANAL DIMETHYL ACETAL, VI, 869
Pentadecane, 1,1-dimethoxy- [52517-73-4]
1,4-PENTADIENE, IV, 746
1,4-Pentadiene [591-93-5]
trans-**2,4-PENTADIENOIC ACID, VI**, 95
2,4-Pentadienoic acid, (*E*)- [21651-12-7]
PENTAERYTHRITOL, I, 425
1,3-Propanediol, 2,2-bis(hydroxymethyl)- [115-77-5]
PENTAERYTHRITYL BROMIDE (or TETRABROMIDE), II, 476; **IV**, 753
Propane, 1,3-dibromo-2,2-bis(bromomethyl)- [3229-00-3]
PENTAERYTHRITYL IODIDE, II, 476
Propane, 1,3-diiodo-2,2-bis(iodomethyl)- [1522-88-9]
(PENTAFLUOROPHENYL)ACETONITRILE, VI, 873
Benzeneacetonitrile, 2,3,4,5,6-pentafluoro- [653-30-5]
PENTAFLUOROPHENYLCOPPER TETRAMER, VI, 875
Copper, tetrakis(pentafluorophenyl)tetra- [34077-61-7]
PENTAMETHYLBENZENE, II, 250
Benzene, pentamethyl- [700-12-9]
1,2,3,4,5-PENTAMETHYLCYCLOPENTADIENE, VII, 505
1,3-Cyclopentadiene, 1,2,3,4,5-pentamethyl- [4045-44-7]
PENTAMETHYLENE BROMIDE, I, 428; **III**, 692
Pentane, 1,5-dibromo- [111-24-0]
3,3-PENTAMETHYLENEDIAZIRIDINE, V, 897
1,2-Diazaspiro[2.5]oct-1-ane [185-79-5]
3,3-PENTAMETHYLENEDIAZIRINE, V, 897
1,2-Diazaspiro[2.5]oct-1-ene [930-82-5]
PENTANE, II, 478; *Hazard*
Pentane [109-66-0]
1,5-PENTANEDIOL, III, 693
1,5-Pentanediol [111-29-5]
3-PENTANONE SAMP-HYDRAZONE, VIII, 403
1-Pyrrolidinamine, *N*-(1-ethylpropylidene)-2-(methoxymethyl)-, (*S*)- [59983-36-7]
2-PENTENE, I, 430
2-Pentene [109-68-2]
3-PENTEN-2-OL, III, 696
3-Penten-2-ol [1569-50-2]
4-PENTEN-1-OL, III, 698
4-Penten-1-ol [821-09-0]
trans-**3-PENTEN-2-ONE, VI**, 883
3-Penten-2-one, (*E*)- [3102-33-8]

4-PENTYLBENZOYL CHLORIDE, VII, 420
  Benzoyl chloride, 4-pentyl- [49763-65-7]
2-*tert*-PENTYLCYCLOPENTANONE, VII, 424
  Cyclopentanone, 2-(1,1-dimethylpropyl)- [25184-25-2]
(*E*)-3-PENTYL-2-NONENE-1,4-DIOL, VIII, 507
  2-Nonene-1,4-diol, 3-pentyl- [138149-15-2]
4-PENTYN-1-OL, IV, 755
  4-Pentyn-1-ol [5390-04-5]
PERBENZOIC ACID or PEROXYBENZOIC ACID, I, 431; **V**, 900, 904
  Benzenecarboperoxoic acid [93-59-4]
PERCHLOROFULVALENE, V, 901
  1,3-Cyclopentadiene, 1,2,3,4-tetrachloro-5-(2,3,4,5-tetrachloro-2,4-cyclopentadien-
    1-ylidene)- [6298-65-3]
PERHYDRO-9*b*-BORAPHENALENE, VII, 427
  9b-Boraphenalene, dodecahydro- [16664-33-8]; *cis,cis*- (3aα,6aα,9aα)-,
    [1130-59-2]; *cis,trans*- (3aα,6aα,9aβ)- [2938-53-6]
PERHYDRO-9*b*-PHENALENOL, VII, 427
  9b*H*-Phenalen-9b-ol, dodecahydro-, 3aα,6aα,9aβ,9bα- [16664-34-9]
PHENACYLAMINE HYDROCHLORIDE, V, 909
  Ethanone, 2-amino-1-phenyl-, hydrochloride [5468-37-1]
PHENACYL BROMIDE, II, 480; *Hazard*
  Ethanone, 2-bromo-1-phenyl- [70-11-1]
PHENANTHRENE, IV, 413; *Hazard*
  Phenanthrene [85-01-8]
PHENANTHRENE-9-ALDEHYDE, III, 701
  9-Phenanthrenecarboxaldehyde [4707-71-5]
PHENANTHRENE-9,10-OXIDE, VI, 887
  Phenanthro[9,10-*b*]oxirene, 1a,9b-dihydro- [585-08-0]
PHENANTHRENEQUINONE, IV, 757
  9,10-Phenanthrenedione [84-11-7]
2-PHENANTHRENESULFONIC ACID, II, 482
  2-Phenanthrenesulfonic acid [41105-40-2]
3-PHENANTHRENESULFONIC ACID, II, 482
  3-Phenanthrenesulfonic acid [2039-95-4]
PHENOXTHIN, II, 485
  Phenoxathiin [262-20-4]
2-PHENOXYMETHYL-1,4-BENZOQUINONE, VI, 890
  2,5-Cyclohexadiene-1,4-dione, 2-(phenoxymethyl)- [7714-50-3]
1-PHENOXY-1-PHENYLETHENE, VIII, 512
  Benzene, (1-phenoxyethenyl)- [19928-57-5]
PHENOXYETHYL BROMIDE, I, 436
  Benzene, (2-bromoethoxy)- [589-10-6]
γ-PHENOXYPROPYL BROMIDE, I, 435
  Benzene, (3-bromopropoxy)- [588-63-6]
PHENYLACETAMIDE, IV, 760
  Benzeneacetamide [103-81-1]

PHENYLACETIC ACID, I, 436; **IV**, 761
  Benzeneacetic acid [103-82-2]
α-**PHENYLACETOACETONITRILE, II**, 487
  Benzeneacetonitrile, α-acetyl- [4468-48-8]
PHENYLACETYLENE, I, 438; **IV**, 763
  Benzene, ethynyl- [536-74-3]
PHENYLACETYL ISOTHIOCYANATE, V, 1051
  Benzeneacetyl isothiocyanate [29313-32-4]
2-PHENYL-2-ADAMANTANAMINE HYDROCHLORIDE, VII, 433
  Tricyclo[3.3.1.1^{3,7}]decan-2-amine, 2-phenyl-, hydrochloride [73032-81-2]
2-PHENYL-2-ADAMANTANOL, VII, 433
  Tricyclo[3.3.1.1^{3,7}]decan-2-ol, 2-phenyl- [29480-18-0]
DL-β-**PHENYLALANINE, II**, 489; **III**, 705
  DL-Phenylalanine [150-30-1]
(*S*)-(+)-**PHENYLALANOL, VIII**, 528
  Benzenepropanol, β-amino- (*S*)- [3182-95-4]
γ-**PHENYLALLYLSUCCINIC ACID, IV**, 766
  Butanedioic acid, (3-phenyl-2-propenyl)- [5671-91-0]
γ-**PHENYLALLYLSUCCINIC ANHYDRIDE, IV**, 766
  2,5-Furandione, dihydro-3-(3-phenyl-2-propenyl)- [7508-06-7]
1-PHENYL-3-AMINO-5-PYRAZOLONE, III, 708
  3*H*-Pyrazol-3-one, 5-amino-2,4-dihydro-2-phenyl- [4149-06-8]
4-PHENYL-5-ANILINO-1,2,3-TRIAZOLE, IV, 380
  1*H*-1,2,3-Triazol-4-amine, *N*-5-diphenyl- [53684-55-2]
N-**PHENYLANTHRANILIC ACID, II**, 15
  Benzoic acid, 2-phenylamino- [91-40-7]
PHENYLARSONIC ACID, II, 494; *Hazard*
  Arsonic acid, phenyl- [98-05-5]
PHENYL AZIDE, III, 710
  Benzene, azido- [622-37-7]
3-PHENYL-2*H*-AZIRINE-2-CARBOXALDEHYDE, VI, 893
  2*H*-Azirine-2-carboxaldehyde, 3-phenyl- [42970-55-8]
p-**PHENYLAZOBENZOIC ACID, III**, 711
  Benzoic acid, 4-(phenylazo)- [1562-93-2]
p-**PHENYLAZOBENZOYL CHLORIDE, III**, 712
  Benzoyl chloride, 4-(phenylazo)- [104-24-5]
N-**PHENYLBENZAMIDINE, IV**, 769
  Benzenecarboximidamide, *N*-phenyl- [1527-91-9]
PHENYL BENZENESELENOSULFONATE, VIII, 543
  Benzenesulfonoselenoic acid, Se-phenyl ester [60805-71-2]
PHENYLBENZOYLDIAZOMETHANE, II, 496
  Ethanone, diazodiphenyl- [3469-17-8]
α-**PHENYL-β-BENZOYLPROPIONITRILE, II**, 498
  Benzenebutanenitrile, γ-oxo-α-phenyl- [6268-00-4]
PHENYLBROMOETHYNE, V, 921
  Benzene, (bromoethynyl)- [932-87-6]

trans-1-PHENYL-1,3-BUTADIENE, IV, 771
Benzene, 1,3-butadienyl-, (*E*)- [16939-57-4]
PHENYL *tert*-BUTYL ETHER (METHOD I), V, 924; (METHOD II), V, 926
Benzene, (1,1-dimethylethoxy)- [6669-13-2]
γ-PHENYLBUTYRIC ACID, II, 499
Benzenebutanoic acid [1821-12-1]
2-PHENYLBUTYRONITRILE, VI, 897
Benzeneacetonitrile, α-ethyl- [769-68-6]
α-PHENYL-α-CARBETHOXYGLUTARONITRILE, IV, 776
Benzeneacetic acid, α-cyano-α-(2-cyanoethyl)-, ethyl ester [53555-70-7]
4-PHENYL-1-CARBETHOXYSEMICARBAZIDE, VI, 936
Hydrazinecarboxylic acid, 2-[(phenylamino)carbonyl]-, ethyl ester [537-47-3]
β-PHENYLCINNAMALDEHYDE, VI, 901
2-Propenal, 3,3-diphenyl- [1210-39-5]
PHENYL CINNAMATE, III, 714
2-Propenoic acid, 3-phenyl-, phenyl ester [2757-04-2]
α-PHENYLCINNAMIC ACID, IV, 777
Benzeneacetic acid, α-(phenylmethylene)- [3368-16-9]
β-PHENYLCINNAMIC ACID, V, 509
2-Propenoic acid, 3,3-diphenyl- [606-84-8]
α-PHENYLCINNAMONITRILE, III, 715
Benzeneacetonitrile, α-(phenylmethylene)- [2510-95-4]
PHENYL CYANATE, VII, 435
Cyanic acid, phenyl ester [1122-85-6]
2-PHENYLCYCLOHEPTANONE, IV, 780
Cycloheptanone, 2-phenyl- [14996-78-2]
PHENYLCYCLOHEXANOL, (+)-(1*S*,2*R*)-*trans*-2-, VIII, 516
Cyclohexanol, 2-phenyl- (1*S-trans*)- [34281-92-0]
PHENYLCYCLOHEXANOL, (−)-(1*R*,2*S*)-*trans*-2-, VIII, 516
Cyclohexanol, 2-phenyl-, (1*R-trans*)- [98919-68-7]
trans-2-PHENYLCYCLOHEXYL CHLOROACETATE (±), (+)-, (−), VIII, 516
Acetic acid, chloro-, 2-phenylcyclohexyl ester, *trans*-(±)- [121906-81-8]
1-PHENYLCYCLOPENTANECARBOXALDEHYDE, VI, 905
Cyclopentanecarboxaldehyde, 1-phenyl- [21573-69-3]
1-PHENYLCYCLOPENTYLAMINE, VI, 910
Cyclopentanamine, 1-phenyl- [17380-74-4]
PHENYLCYCLOPROPANE or CYCLOPROPYLBENZENE, V, 328, 929
Benzene, cyclopropyl- [873-49-4]
cis-2-PHENYLCYCLOPROPANECARBOXYLIC ACID, VI, 913
Cyclopropanecarboxylic acid, 2-phenyl-, *cis*- [939-89-9]
trans-2-PHENYLCYCLOPROPANECARBOXYLIC ACID, VI, 915
Cyclopropanecarboxylic acid, 2-phenyl-, *trans*- [939-90-2]
1-PHENYLDIALIN (1-PHENYL-3,4-DIHYDRONAPHTHALENE), III, 729
Naphthalene, 1,2-dihydro-4-phenyl- [7469-40-1]
PHENYLDIAZOMETHANE, VII, 438
Benzene, diazomethyl [766-91-6]

PHENYLDICHLOROPHOSPHINE, IV, 784
Phosphonous dichloride, phenyl- [644-97-3]
1-PHENYL-3,4-DIHYDRO-6,7-METHYLENEDIOXYISOQUINOLINE, VII, 326
1,3-Dioxolo[4,5-g]isoquinoline, 7,8-dihydro-5-phenyl- [55507-10-3]
4-PHENYL-m-DIOXANE, IV, 786
1,3-Dioxane, 4-phenyl- [772-00-9]
2-PHENYL-1,3-DITHIANE, VI, 109
1,3-Dithiane, 2-phenyl- [5425-44-5]
3-(2-PHENYL-1,3-DITHIAN-2-YL)-1H-INDOLE, VI, 109
1H-Indole, 3-(2-phenyl-1,3-dithian-2-yl)- [57621-00-8]
o-PHENYLENE CARBONATE, IV, 788
1,3-Benzodioxol-2-one [2171-74-6]
o-PHENYLENEDIAMINE, II, 501
1,2-Benzenediamine [95-54-5]
α-PHENYLETHYLAMINE, II, 503; **III**, 717
Benzenemethanamine, α-methyl- [98-84-0]
d- and l-α-PHENYLETHYLAMINE, II, 506; **V**, 932
Benzenemethanamine, α-methyl-, (R)- [3886-69-9]; (S)- [2627-86-3]
β-PHENYLETHYLAMINE, III, 720
Benzeneethanamine [64-04-0]
N-(2-PHENYLETHYL)BENZAMIDE, V, 336
Benzamide, N-(2-phenylethyl)- [3278-14-6]
2-PHENYLETHYL BENZOATE, V, 336
Benzoic acid, 2-phenylethyl ester [94-47-3]
β-PHENYLETHYLDIMETHYLAMINE, III, 723
Benzeneethanamine, N,N-dimethyl- [1126-71-2]
PHENYLETHYLENE, (STYRENE), I, 440; *Harard*
Benzene, ethenyl- [100-42-5]
2-PHENYLFURO[3,2-b]PYRIDINE, VI, 916
Furo[3,2-b]pyridine, 2-phenyl- [18068-82-1]
α-PHENYLGLUTARIC ANHYDRIDE, IV, 790
2H-Pyran-2,6(3H)-dione, dihydro-3-phenyl- [2959-96-8]
PHENYLGLYOXAL, II, 509; **V**, 937
Benzeneacetaldehyde, α-oxo- [1074-12-0]
PHENYLGLYOXAL HEMIMERCAPTAL, V, 937
Ethanone, 2-hydroxy-2-(methylthio)-1-phenyl- [13603-49-1]
4-PHENYL-6-HEPTEN-2-ONE, VII, 443
6-Hepten-2-one, 4-phenyl- [69492-29-1]
PHENYL HEPTYL KETONE, VI, 919
1-Octanone, 1-phenyl- [1674-37-9]
PHENYLHYDRAZINE, I, 422; *Hazard*
Hydrazine, phenyl- [100-63-0]
β-PHENYLHYDROXYLAMINE, I, 445; **VIII**, 16
Benzenamine, N-hydroxy- [100-65-2]
2-PHENYLINDAZOLE, V, 941
2H-Indazole, 2-phenyl- [3682-71-1]

2-PHENYLINDOLE, III, 725
   1*H*-Indole, 2-phenyl- [948-65-2]
PHENYL ISOTHIOCYANATE, I, 447
   Benzene, isothiocyanato- [103-72-0]
***N*-PHENYLMALEIMIDE, V**, 944
   1*H*-Pyrrole-2,5-dione, 1-phenyl- [941-69-5]
8-PHENYLMENTHOL, (–)-, VIII, 522
   Cyclohexanol, 5-methyl-2-(1-methyl-1-phenylethyl)- [1*R*-(1α,2β,5α)]- [65253-04-5]
PHENYLMETHYLGLYCIDIC ESTER, III, 727
   Oxiranecarboxylic acid, 3-methyl-3-phenyl-, ethyl ester [77-83-8]
(*S*)-4-(PHENYLMETHYL)-2-OXAZOLIDINONE, VIII, 339, 528
   2-Oxazolidinone, 4-(phenylmethyl)-, (*S*)- [90719-32-7]
1-PHENYLNAPHTHALENE, III, 729
   Naphthalene, 1-phenyl- [605-02-7]
PHENYLNITROMETHANE, II, 512
   Benzene, (nitromethyl)- [622-42-4]
(1Z,3E)-1-PHENYL-1,3-OCTADIENE, VIII, 532
   Benzene, 1,3-octadienyl-, (Z,E)- [39491-66-2]
2-PHENYL-5-OXAZOLONE, V, 946
   5(4*H*)-Oxazolone, 2-phenyl- [1199-01-5]
1-PHENYL-1,3-PENTADIYNE, VI, 925
   Benzene, 1,3-pentadiynyl- [4009-22-7]
1-PHENYL-1,4-PENTADIYNE, VI, 925
   Benzene, 1,4-pentadiynyl- [6088-96-6]
1-PHENYL-2,4-PENTANEDIONE, VI, 928
   2,4-Pentanedione, 1-phenyl- [3318-61-4]
(*R*)-1-PHENYL-1-PENTANOL, VII, 447
   Benzenemethanol, α-butyl-, (*R*)- [19641-53-3]
3-PHENYL-4-PENTENAL, VIII, 536
   Benzenepropanal, β-ethenyl- [939-21-9]
3-PHENYL-4-PENTENOIC ACID, VII, 164
   Benzenepropanoic acid, β-ethenyl- [5703-57-1]
1-PHENYL-1-PENTEN-4-YN-3-OL, IV, 792
   1-Penten-4-yn-3-ol, 1-phenyl [14604-31-0]
2-PHENYLPERFLUOROPROPENE, V, 949
   Benzene, [2,2-difluoro-1-(trifluoromethyl)ethenyl]- [1979-51-7]
9-PHENYLPHENANTHRENE, V, 952
   Phenanthrene, 9-phenyl- [844-20-2]
1-PHENYL-4-PHOSPHORINANONE, VI, 932
   4-Phosphorinanone, 1-phenyl- [23855-87-0]
1-PHENYLPIPERIDINE, IV, 795
   Piperidine, 1-phenyl- [4096-20-2]
3-PHENYL-1-PROPANOL, IV, 798
   Benzenepropanol [122-97-4]
PHENYLPROPARGYL ALDEHYDE, III, 731
   2-Propynal, 3-phenyl- [2579-22-8]

PHENYLPROPARGYLALDEHYDE DIETHYL ACETAL, IV, 801
  Benzene, (3,3-diethoxy-1-propynyl)- [6142-95-6]
(*E*)-3-[(*E*)-3-PHENYL-2-PROPENOXY]ACRYLIC ACID, VIII, 536
  2-Propenoic acid, 3-[(3-phenyl-2-propenyl)oxy]-, (*E,E*)- [88083-18-5]
PHENYLPROPIOLIC ACID, II, 515
  2-Propynoic acid, 3-phenyl- [637-44-5]
α-PHENYLPROPIONALDEHYDE, III, 733
  Benzeneacetaldehyde, α-methyl- [93-53-8]
3-PHENYLPROPIONALDEHYDE, VII, 451
  Benzenepropanal [104-53-0]
2-PHENYL-3-PROPYLISOXAZOLIDINE-4,5-*cis*-DICARBOXYLIC ACID
N-PHENYLIMIDE, V, 957
  2*H*-Pyrrolo[3,4-*d*]isoxazole-4,6(3*H*,5*H*)-dione, dihydro-2,5-diphenyl-3-propyl-
  [53555-71-8]
2-PHENYLPYRIDINE, II, 517
  Pyridine, 2-phenyl- [1008-89-5]
PHENYLPYRUVIC ACID, II, 519
  Benzenepropanoic acid, α-oxo- [156-06-9]
4-PHENYLSEMICARBAZIDE, I, 450
  Hydrazinecarboxamide, N-phenyl- [537-47-3]
PHENYLSUCCINIC ACID, I, 451; **IV,** 804
  Butanedioic acid, phenyl- [635-51-8]
***trans*-2-(PHENYLSULFONYL)-4-CHLOROMERCURI)CYCLOHEXENE, VIII,**
  540; *Hazard*
  Mercury, chloro[2-(phenylsulfonyl)-3-cyclohexen-1-yl]-, *trans*- [102815-53-2]
2-(PHENYLSULFONYL)-1,3-CYCLOHEXADIENE, VIII, 540
  Benzene, (1,5-cyclohexadien-1-ylsulfonyl)- [102860-22-0]
1-(PHENYLSULFONYL)CYCLOPENTENE, VIII, 38, 543
  Benzene, (1-cyclopenten-1-ylsulfonyl)- [64740-90-5]
(±)-*trans*-2-(PHENYLSULFONYL)-3-PHENYLOXAZIRIDINE, VIII, 546
  Oxaziridine, 3-phenyl-2-(phenylsulfonyl)- [63160-13-4]
PHENYLSULFUR TRIFLUORIDE, V, 959
  Sulfur, trifluorophenyl- [672-36-6]
3-PHENYLSYDNONE, V, 962
  Sydnone, 3-phenyl- [120-06-9]
4-(1-PHENYL-5-TETRAZOLYLOXY)BIPHENYL, VI, 150
  1*H*-Tetrazole, 5-[(1,1'-biphenyl)-4-yloxy]-1-phenyl- [17743-27-0]
PHENYL THIENYL KETONE, II, 520
  Methanone, phenyl-2-thienyl- [135-00-2]
1-PHENYL-2-THIOBIURET, V, 966
  Thioimidodicarbonic diamide[(H_2C(O)NHC(S)(NH$_2$)], N'-phenyl- [53555-72-9]
1-PHENYLTHIO-2-BROMOETHANE (in solution), VII, 453
  Benzene, [(2-bromoethyl)thio]- [4837-01-8]
1-PHENYL-2-THIO-4-METHYLISOBIURET, V, 966
(PHENYLTHIO)NITROMETHANE, VIII, 550
  Benzene, [(nitromethyl)thio]- [60595-16-6]

α-PHENYLTHIOUREA, III, 735; *Hazard*
    Thiourea, phenyl- [103-85-5]
4-PHENYL-1,2,4-TRIAZOLINE-3,5-DIONE, VI, 936
    3*H*-1,2,4-Triazole-3,5(4*H*)-dione, 4-phenyl- [4233-33-4]
1-PHENYL-2,2,2-TRICHLOROETHANOL, V, 130
    Benzenemethanol, α-trichloromethyl- [2000-43-3]
PHENYL(TRICHLOROMETHYL)MERCURY, V, 969; *Hazard*
    Mercury, phenyl(trichloromethyl)- [3294-57-3]
PHENYLTRIMETHYLAMMONIUM TRIBROMIDE, VI, 175
    Benzenaminium, *N*,*N*,*N*-trimethyl-, (tribromide) [4207-56-1]
PHENYLUREA, I, 453
    Urea, phenyl- [64-10-8]
2-PHENYL-2-VINYLBUTYRONITRILE, VI, 940
    Benzeneacetonitrile, α-ethenyl-α-ethyl- [13312-96-4]
PHENYL VINYL SULFIDE, VII, 453
    Benzene, (ethenylthio) [1822-73-7]
PHENYL VINYL SULFONE, VII, 453
    Benzene, (ethenylsulfonyl) [5535-48-8]
PHENYL VINYL SULFOXIDE, VII, 453
    Benzene, (ethenylsulfinyl) [20451-53-0]
PHLOROACETOPHENONE, II, 522
    Ethanone, 1-(2,4,6-trihydroxyphenyl)- [480-66-0]
PHLOROGLUCINOL, I, 455
    1,3,5-Benzenetriol [108-73-6]
o-PHTHALALDEHYDE, IV, 807
    1,2-Benzenedicarboxaldehyde [643-79-8]
PHTHALALDEHYDIC ACID, II, 523; III, 737
    Benzoic acid, 2-formyl- [119-67-5]
PHTHALIDE, II, 526
    1(3*H*)-lsobenzofuranone [87-41-2]
PHTHALIMIDE, I, 457
    1*H*-lsoindole-1,3(2*H*)-dione [85-41-6]
α-PHTHALIMIDO-*o*-TOLUIC ACID, IV, 810
    Benzoic acid, 2-[(1,3-dihydro-1,3-dioxo-2*H*-isoindol-2-yl)methyl]- [53663-18-6]
sym-o-PHTHALYL CHLORIDE, II, 528
    1,2-Benzenedicarbonyl dichloride [88-95-9]
unsym-o-PHTHALYL CHLORIDE, II, 528
    1(3*H*)-Isobenzofuranone, 3,3-dichloro- [601-70-7]
N-PHTHALYL-L-β-PHENYLALANINE, V, 973
    2*H*-Isoindole-2-acetic acid, 1,3-dihydro-1,3-dioxo-α-(phenylmethyl)-, (*S*)- [5123-55-7]
PICOLINIC ACID HYDROCHLORIDE, III, 740
    2-Pyridinecarboxylic acid, hydrochloride [636-80-6]
PIMELIC ACID, II, 531
    Heptanedioic acid [111-16-0]
PINACOL HYDRATE, I, 459
    2,3-Butanediol, 2,3-dimethyl-, hexahydrate [6091-58-3]

PINACOLONE, I, 462
2-Butanone, 3,3-dimethyl- [75-97-8]
3-PINANAMINE, VI, 943
Bicyclo[3.1.1]heptan-3-amine, 2,2,6-trimethyl- [17371-27-6]
B-3-PINANYL-9-BORABICYCLO[3.3.1]NONANE, VII, 402
9-Borabicyclo[3.3.1]nonane, 9-(2,6,6-trimethylbicyclo[3.1.1]hept-3-yl)-
[73624-47-2]
PINENE, (–)-α-, VIII, 553
Bicyclo[3.1.1]hept-2-ene, 2,6,6-trimethyl-, (1S)- [7785-26-4]
trans-**PINOCARVEOL, VI,** 946, 948
Bicyclo[3.1.1]heptan-3-ol, 6,6-dimethyl-2-methylene-, (1α,3β,5α)- [6712-79-4]
PIPERONYLIC ACID, II, 538
1,3-Benzodioxole-5-carboxylic acid [94-53-1]
POLYMERIC BENZYLAMINE, VI, 951
POLYMERIC CARBODIIMIDE, VI, 951
POLYMERIC UREA, VI, 951
POTASSIUM ANTHRAQUINONE-α-SULFONATE, II, 539
1-Anthracenesulfonic acid, 9,10-dihydro-9,10-dioxo-, potassium salt [30845-78-4]
(S)-(+)-PROPANE-1,2-DIOL, VII, 356
1,2-Propanediol, (S)- [4254-15-3]
2-(1-PROPENYL)CYCLOBUTANONE, (E)-, VIII, 556
Cyclobutanone, 2-(1-propenyl)-, (E)- [63049-06-9]
2-(E)-PROPENYLCYCLOHEXANOL, VII, 456
Cyclohexanol, 2-(1-propenyl)-, [1α,2α(E)]- [76123-38-1]
trans-**2-(2-PROPENYL)CYCLOPENTANOL, VII,** 501
Cyclopentanol, 2-(2-propenyl)-, *trans*- [74743-89-8]
cis-**2-(2-PROPENYL)CYCLOPENTYLAMINE, VII,** 501
Cyclopentanamine, 2-(2-propenyl)-, *cis*- [81097-02-1]
PROPIOLALDEHYDE, IV, 813
2-Propynal [624-67-9]
PROPIOLALDEHYDE DIETHYL ACETAL, VI, 954
1-Propyne, 3,3-diethoxy- [10160-87-9]
PROPIONALDEHYDE, II, 541
Propanal [123-38-6]
8-PROPIONYL-(E)-5-NONENOLIDE, VIII, 562
2H-Oxecin-2-one, 3,4,5,8,9,10-hexahydro-9-(1-oxopropyl)-, (E)- [114633-68-0]
β-PROPIONYLPHENYLHYDRAZINE, IV, 657
Propanoic acid, 2-phenylhydrazide [20730-02-3]
o-**PROPIOPHENOL, II,** 543
1-Propanone, 1-(2-hydroxyphenyl)- [610-99-1]
p-**PROPIOPHENOL, II,** 543
1-Propanone, 1-(4-hydroxyphenyl)- [70-70-2]
2-PROPYL-1-AZACYCLOHEPTANE, VIII, 568
1H-Azepine, hexahydro-2-propyl-, (±)- [85028-29-1]
PROPYLBENZENE, I, 471
Benzene, propyl- [103-65-1]

PROPYL BROMIDE, I, 37; **II**, 359
  Propane, 1-bromo- [106-94-5]
γ-PROPYLBUTYROLACTONE, III, 742
  2(3*H*)-Furanone, dihydro-5-propyl- [105-21-5]
L-PROPYLENE GLYCOL, II, 545
  1,2-Propanediol, (*R*)- [4254-14-2]
(2*S*,3*S*)-3-PROPYLOXIRANEMETHANOL, VII, 461
  Oxiranemethanol, 3-propyl-, (2*S*)-*trans*- [89321-71-1]
3-(1-PROPYL)-2-PYRAZOLIN-5-ONE, VI, 791
  2-Pyrazolin-5-one, 3-propyl- [29211-70-9]
PROPYL SULFIDE, II, 547
  Propane, 1,1-thiobis- [111-47-7]
4-PROTOADAMANTANONE, VI, 958
  2,5-Methano-1*H*-inden-7(4*H*)-one, hexahydro- [27567-85-7]
PROTOCATECHUALDEHYDE, II, 549
  Benzaldehyde, 3,4-dihydroxy- [139-85-5]
PROTOCATECHUIC ACID, III, 745
  Benzoic acid, 3,4-dihydroxy- [99-50-3]
PSEUDOIONONE, III, 747
  3,5,9-Undecatrien-2-one, 6,10-dimethyl- [141-10-6]
PSEUDOPELLETIERINE, IV, 816
  9-Azabicyclo[3.3.1]nonan-3-one, 9-methyl- [552-70-5]
PSEUDOTHIOHYDANTOIN, III, 751
  4(5*H*)-Thiazolone, 2-amino- [556-90-1]
PUTRESCINE DIHYDROCHLORIDE, IV, 819
  1,4-Butanediamine, dihydrochloride [333-93-7]
PYOCYANINE, III, 753
  1(5*H*)-Phenazinone, 5-methyl- [85-66-5]
2*H*-PYRAN-2-ONE, VI, 462
  2*H*-Pyran-2-one [504-31-4]
2,3-PYRAZINEDICARBOXYLIC ACID, IV, 824
  2,3-Pyrazinedicarboxylic acid [89-01-0]
PYRIDINE-*N*-OXIDE and HYDROCHLORIDE, IV, 828
  Pyridine, 1-oxide [694-59-7]
4-PYRIDINESULFONIC ACID, V, 977
  4-Pyridinesulfonic acid [5402-20-0]
1-(α-PYRIDYL)-2-PROPANOL, III, 757
  2-Pyridineethanol, α-methyl- [5307-19-7]
PYROGALLOL 1-MONOMETHYL ETHER, III, 759
  1,2-Benzenediol, 3-methoxy- [934-00-9]
PYROMELLITIC ACID, II, 551
  1,2,4,5-Benzenetetracarboxylic acid [89-05-4]
α-PYRONE, V, 982
  2*H*-Pyran-2-one [504-31-4]
PYRROLE, I, 473
  1*H*-Pyrrole [109-97-7]

2-PYRROLEALDEHYDE, IV, 831
1*H*-Pyrrole-2-carboxaldehyde [1003-29-8]
1-(*N*-PYRROLIDINO)-1-CYCLODODECENE, VII, 135
Pyrrolidine, 1-(1-cyclododecen-1-yl)- [25769-05-5]
1-PYRROLIDINOCYCLOHEXENE, VI, 1014
Pyrrolidine, 1-(1-cyclohexen-1-yl)- [1125-99-1]
2-(1-PYRROLIDYL)PROPANOL, IV, 834
1-Pyrrolidineethanol, β-methyl- [53663-19-7]
PYRROL-2-YL TRICHLOROMETHYL KETONE, VI, 618
Ethanone, 2,2,2-trichloro-1-(1*H*-pyrrol-2-yl)- [35302-72-8]
PYRUVIC ACID, I, 475
Propanoic acid, 2-oxo- [127-17-3]
PYRUVOYL CHLORIDE, VII, 467
Propanoyl chloride, 2-oxo- [5704-66-5]

Q
QUADRICYCLANE, VI, 962
Tetracyclo[3.2.0.0^{2,7}.0^{4,6}]heptane [278-06-8]
QUINACETOPHENONE DIMETHYL ETHER, IV, 836
Ethanone, 2,5-dimethoxyphenyl- [1201-38-3]
QUINACETOPHENONE MONOMETHYL ETHER, IV, 837
Ethanone, 1-(2-hydroxy-5-methoxyphenyl)- [705-15-7]
QUINIZARIN, I, 476
9,10-Anthracenedione, 1,4-dihydroxy- [81-64-1]
QUINOLINE, I, 478
Quinoline [91-22-5]
QUINONE, I, 482; **II**, 553; *Hazard*
2,5-Cyclohexadiene-1,4-dione [106-51-4]
QUINOXALINE, IV, 824
Quinoxaline [91-19-0]
***p*-QUINQUEPHENYL, V**, 985
1,1':4',1":4",1"':4"',1""-Quinquephenyl [3073-05-0]
3-QUINUCLIDONE HYDROCHLORIDE, V, 989
1-Azabicyclo[2.2.2]octan-3-one, hydrochloride [1193-65-3]

R
REINECKE SALT, II, 555
Chromate(1⁻), diamminetetrakis(thiocyanato-*N*)-, ammonium, (OC-6 11)- [13573-16-5]
RESACETOPHENONE, III, 761
Ethanone, 1-(2,4-dihydroxyphenyl)- [89-84-9]
β-RESORCYLIC ACID, II, 557
Benzoic acid, 2,4-dihydroxy- [89-86-1]
RHODANINE, III, 763
4-Thiazolidinone, 2-thioxo- [141-84-4]
RICINELAIDIC ACID, VII, 470
9-Octadecenoic acid, 12-hydroxy- [*R*-(*E*)]- [540-12-5]

RICINELAIDIC ACID LACTONE, VII, 470
  Oxatridec-10-en-2-one, (*E*)- [79894-06-7]
RICINELAIDIC ACID *S*-(2-PYRIDYL)CARBOTHIOATE, VII, 470
  9-Octadecenethioic acid, 12-hydroxy-, *S*-2-pyridyl ester, [*R*-(*E*)]- [100819-69-0]
RUTHENOCENE, V, 1001
  Ruthenocene [1287-13-4]

S

SALICYL-*o*-TOLUIDE, III, 765
  Benzamide, 2-hydroxy-*N*-(2-methylphenyl)- [7133-56-4]
SALSOLIDINE, (–)-, VIII, 573
  Isoquinoline, 1,2,3,4-tetrahydro-6,7-dimethoxy-1-methyl-, (*S*)- [493-48-1]
SEBACIL, IV, 838
  1,2-Cyclodecanedione [96-01-5]
SEBACOIN, IV, 840
  Cyclodecanone, 2-hydroxy- [96-00-4]
SEBACONITRILE, III, 768
  Decanedinitrile [1871-96-1]
SEBACOYLDICYCLOHEXANONE, V, 533
  1,10-Decanedione, 1,10-bis(2-oxocyclohexyl)- [17343-93-0]
SELENOPHENOL, III, 771
  Benzeneselenol [645-96-5]
DL-SERINE, III, 774
  DL-Serine [302-84-1]
SODIUM 2-AMINOBENZENESULFINATE, V, 61
  Benzenesulfinic acid, 2-amino-, monosodium salt [50827-53-7]
SODIUM *p*-ARSONO-*N*-PHENYLGLYCINAMIDE, I, 488; *Hazard*
  Arsonic acid, [4-[(2-amino-2-oxoethyl)amino]phenyl]-, disodium salt [834-03-7]
SODIUM 2-BROMOETHANESULFONATE, II, 558
  Ethanesulfonic acid, 2-bromo-, sodium salt [4263-52-9]
SODIUM DICARBONYL(CYCLOPENTADIENYL)FERRATE, VIII, 479
  Ferrate(1–), dicarbonyl(η^5-2,4-cyclopentadien-1-yl)-, sodium [12152-20-4]
SODIUM *p*-HYDROXYPHENYLARSONATE, I, 490; *Hazard*
  Arsonic acid, (4-hydroxyphenyl)-, sodium salt [53663-20-0]
SODIUM NITROMALONALDEHYDE MONOHYDRATE, IV, 844; *Warning*, **V**, 1004
  Propanedial, nitro-, ion(1–), sodium, monohydrate [53821-72-0]
SODIUM β-STYRENESULFONATE, IV, 846
  Ethenesulfonic acid, 2-phenyl-, sodium salt [2039-44-3]
SODIUM-*p*-TOLUENESULFINATE, I, 492
  Benzenesulfinic acid, 4-methyl-, sodium salt [824-79-3]
SORBIC ACID, III, 783
  2,4-Hexadienoic acid, (*E,E*)- [110-44-1]
SPIRO[4.5]DECANE-1,4-DIONE, VIII, 578
  Spiro[4.5]decane-1,4-dione [39984-92-4]
SPIRO[5,7]TRIDECA-1,4-DIEN-3-ONE, VII, 473
  Spiro[5,7]trideca-1,4-dien-3-one [41138-71-0]

SPIRO[5.7]TRIDEC-1-EN-3-ONE, VII, 473
Spiro[5.7]tridec-1-en-3-one [60033-39-8]
STEAROLIC ACID, III, 785; **IV**, 851
9-Octadecynoic acid [506-24-1]
STEARONE, IV, 854
18-Pentatriacontanone [504-53-0]
cis-**STILBENE, IV**, 857
Benzene, 1,1'-(1,2-ethenediyl)bis-, (*Z*)- [645-49-8]
trans-**STILBENE, III**, 786
Benzene, 1,1'-(1,2-ethenediyl)bis-, (*E*)- [103-30-0]
STILBENE DIBROMIDE, III, 350
Benzene, 1,1'-(1,2-dibromoethylidene)bis- [40957-21-9]
trans-**STILBENE OXIDE, IV**, 860
Oxirane, 2,3-diphenyl-, *trans*- [1439-07-2]
STYRENE GLYCOL DIMESYLATE, VI, 56
1,2-Ethanediol, 1-phenyl-, dimethanesulfonate [32837-95-9]
STYRENE OXIDE, I, 494
Oxirane, phenyl- [96-09-3]
β-**STYRENESULFONYL CHLORIDE, IV**, 846
Ethenesulfonyl chloride, 2-phenyl- [4091-26-3]
STYRYLPHOSPHONIC DICHLORIDE, V, 1005
Phosphonic dichloride, (2-phenylethenyl)- [4708-07-0]
SUCCINIC ANHYDRIDE, II, 560
2,5-Furandione, dihydro- [108-30-5]
SUCCINIMIDE, II, 562
2,5-Pyrrolidinedione [123-56-8]
N-**SULFINYLANILINE, V**, 504
Benzenamine, *N*-sulfinyl- [1122-83-4]
o-**SULFOBENZOIC ANHYDRIDE, I**, 495
3*H*-2,1-Benzoxathiol-3-one, 1,1-dioxide [81-08-3]
α-**SULFOPALMITIC ACID, IV**, 862
Hexadecanoic acid, 2-sulfo- [1782-10-1]
SYRINGIC ALDEHYDE, IV, 866
Benzaldehyde, 4-hydroxy-3,5-dimethoxy- [134-96-3]

T

DL-**TARTARIC ACID, I**, 497
Butanedioic acid, 2,3-dihydroxy, (*R**,*R**)-(±)- [133-37-9]
TAURINE, II, 563
Ethanesulfonic acid, 2-amino- [107-35-7]
TEREPHTHALALDEHYDE, III, 788
1,4-Benzenedicarboxaldehyde [623-27-8]
TEREPHTHALIC ACID, III, 791
1,4-Benzenedicarboxylic acid [100-21-0]
2,2':6',2''-TERPYRIDINE, VII, 476
2,2':6',2''-Terpyridine [1148-79-4]

1,3,4,6-TETRA-*O*-ACETYL-2-DEOXY-α-D-GLUCOPYRANOSE, VIII, 583
α-D-*arabino*-Hexopyranose, 2-deoxy-, tetraacetate [16750-06-4]
TETRAACETYLETHANE, IV, 869
2,5-Hexanedione, 3,4-diacetyl- [5027-32-7]
2,3,4,6-TETRA-*O*-ACETYL-D-GLUCONIC ACID MONOHYDRATE, V, 887
D-Gluconic acid, 2,3,4,6-tetraacetate [61259-49-2]
2,4,4,6-TETRABROMO-2,5-CYCLOHEXADIEN-1-ONE, VI, 181
2,5-Cyclohexadien-1-one, 2,4,4,6-tetrabromo- [20244-61-5]
3,4,8,9-TETRABROMO-11-OXATRICYCLO[4.4.1.0^{1,6}]UNDECANE, VI, 862
14a,8a-Epoxynaphthalene, 2,3,6,7-tetrabromooctahydro- [16573-82-3]
TETRABROMOSTEARIC ACID, III, 526
Octadecanoic acid, 9,10,11,12-tetrabromo- [18464-04-5]
α,α,α',α'-TETRABROMO-*o*-XYLENE, IV, 807
Benzene, 1,2-bis(dibromomethyl) [13209-15-9]
α,α,α',α'-TETRABROMO-*p*-XYLENE, III, 788
Benzene, 1,4-bis(dibromomethyl) [1592-31-0]
TETRACHLOROBENZOBARRELENE, VI, 82
1,4-Ethenonaphthalene, 5,6,7,8-tetrachloro-1,4-dihydro- [13454-02-9]
DL-4,4',6,6'-TETRACHLORODIPHENIC ACID, IV, 872
[1,1'-Biphenyl]-2,2'-dicarboxylic acid, 4,4',6,6'-tetrachloro-, (±)- [53663-22-2]
TETRACYANOETHYLENE, IV, 877; **V**, 1008, 1014
Ethenetetracarbonitrile [670-54-2]
TETRACYANOETHYLENE OXIDE, V, 1007
Oxiranetetracarbonitrile [3189-43-3]
2-TETRADECYL-*sym*-TRITHIANE, VI, 869
1,3,5-Trithiane, 2-tetradecyl- [24644-07-3]
1,1,1',1'-TETRAETHOXYETHYL POLYSULFIDE, IV, 295
TETRAETHYL 1,2,3,4-BUTANETETRACARBOXYLATE, VII, 479
1,2,3,4-Butanetetracarboxylic acid, tetraethyl ester [4373-15-3]
TETRAETHYL ETHYLENETETRACARBOXYLATE, II, 273
Ethenetetracarboxylic acid, tetraethyl ester [6174-95-4]
TETRAETHYL PROPANE-1,1,2,3-TETRACARBOXYLATE, I, 272
1,1,2,3-Propanetetracarboxylic acid, tetraethyl ester [635-03-0]
TETRAETHYL PROPANE-1,1,3,3-TETRACARBOXYLATE, I, 290
1,1,3,3-Propanetetracarboxylic acid, tetraethyl ester [2121-66-6]
TETRAETHYLTIN, IV, 881
Stannane, tetraethyl- [597-64-8]
TETRAHYDRO-BINOR-S, VI, 379
1,2,3,4-TETRAHYDROCARBAZOLE, IV, 884
1*H*-Carbazole, 2,3,4,9-tetrahydro- [942-01-8]
1,2,3,4-TETRAHYDRO-β-CARBOLINE, VI, 965
1*H*-Pyrido[3,4-*b*]indole, 2,3,4,9-tetrahydro- [16502-01-5]
***endo*-TETRAHYDRODICYCLOPENTADIENE, V**, 16
4,7-Methano-1*H*-indene, octahydro-, 3aα,4α,7α,7aα- [2825-83-4]
TETRAHYDROFURAN, II, 566; *Hazard*; *Warning* on PURIFICATION, **V**, 976
Furan, tetrahydro- [109-99-9]

TETRAHYDROFURFURYL BROMIDE, III, 793
Furan, 2-(bromomethyl)tetrahydro- [1192-30-9]
TETRAHYDROFURFURYL CHLORIDE, III, 698
Furan, 2-(chloromethyl)tetrahydro- [3003-84-7]
β-(TETRAHYDROFURYL)PROPIONIC ACID, III, 742
2-Furanpropanoic acid, tetrahydro- [935-12-6]
4,5,6,7-TETRAHYDROINDAZOLE, IV, 536
1*H*-Indazole, 4,5,6,7-tetrahydro- [2305-79-5]
1,4,5,8-TETRAHYDRONAPHTHALENE, VI, 731
Naphthalene, 1,4,5,8-tetrahydro- [493-04-9]
1,2,3,4-TETRAHYDRONAPHTHALENE(1,2)IMINE, VI, 967
1*H*-Naphth[1,2-*b*]azirine, 1a,2,3,7b-tetrahydro- [1196-87-8]
ar-TETRAHYDRO-α-NAPHTHOL, IV, 887
1-Naphthalenol, 5,6,7,8-tetrahydro- [529-35-1]
ac-TETRAHYDRO-β-NAPHTHYLAMINE, I, 499
2-Naphthalenamine, 1,2,3,4-tetrahydro- [2954-50-9]
***cis*-Δ⁴-TETRAHYDROPHTHALIC ANHYDRIDE, IV**, 890
1,3-Isobenzofurandione, 3a,4,7,7a-tetrahydro-, *cis*- [935-79-5]
TETRAHYDROPYRAN, III, 794
2*H*-Pyran, tetrahydro- [142-68-7]
TETRAHYDRO-2-(2-PROPYNYLOXY)-2*H*-PYRAN, VII, 334
2*H*-Pyran, tetrahydro-2-(2-propynyloxy)- [6089-04-9]
2,3,4,5-TETRAHYDROPYRIDINE TRIMER, VI, 968
Pyridine, 2,3,4,5-tetrahydro-, trimer [27879-53-4]
TETRAHYDROTHIOPHENE, IV, 892
Thiophene, tetrahydro- [110-01-0]
TETRAHYDROXYQUINONE, V, 1011
2,5-Cyclohexadiene-1,4-dione, 2,3,5,6-tetrahydroxy- [319-89-1]
TETRAIODOPHTHALIC ANHYDRIDE, III, 796
1,3-Isobenzofurandione, 4,5,6,7-tetraiodo- [632-80-4]
TETRAKIS(TRIFLUOROMETHYL)-1,3-DITHIETANE, VII, 251
1,3-Dithietane, 2,2,4,4-tetrakis(trifluoromethyl)- [791-50-4]
TETRALIN HYDROPEROXIDE, IV, 895
Hydroperoxide, 1,2,3,4-tetrahydro-1-naphthalenyl- [771-29-9]
α-TETRALONE, II, 569; **III**, 798; **IV**, 898
1(2*H*)-Naphthalenone, 3,4-dihydro- [529-34-0]
β-TETRALONE, IV, 903
2(1*H*)-Naphthalenone, 3,4-dihydro- [530-93-8]
4,5,4',5'-TETRAMETHOXY-1,1'-BIPHENYL-2,2'-DICARBOXALDEHYDE, VIII, 586
[1,1'-Biphenyl]-2,2'-dicarboxaldehyde, 4,4',5,5'-tetramethoxy- [29237-14-7]
3,3,6,6-TETRAMETHOXY-1,4-CYCLOHEXADIENE, VI, 971
1,4-Cyclohexadiene, 3,3,6,6-tetramethoxy- [15791-03-4]
TETRAMETHYLAMMONIUM 1,1,2,3,3-PENTACYANOPROPENIDE, V, 1013
Methanaminium, *N,N,N*-trimethyl-, salt with 1-propene-1,1,2,3,3- pentacarbonitrile [53663-17-5]

TETRAMETHYLBIPHOSPHINE DISULFIDE, V, 1016; *Hazard* ; *Warning*, **VII**, 533
  Diphosphine, tetramethyl-, 1,2-disulfide [3676-97-9]
1,1,3,3-TETRAMETHYLBUTYL ISONITRILE, VI, 751
  Pentane, 2-isocyano-2,4,4-trimethyl- [14542-93-9]
N,N,N',N'-**TETRAMETHYLDIAMIDOPHOSPHOROCHLORIDATE, VII**, 66
  Phosphorodiamidic chloride, tetramethyl [1605-65-8]
**TETRAMETHYL 3,7-DIHYDROXYBICYCLO[3.3.0]OCTA-2,6-DIENE-2,4,6,8-
  TETRACARBOXYLATE, VII**, 50
  1,3,4,6-Pentalenetetracarboxylic acid, 1,3a,4,6a-tetrahydro-2,5-dihydroxy-,
    tetramethyl ester (1α,3aα,4α,6aα) [82416-04-4]
**TETRAMETHYL 3,7-DIHYDROXY-1,5-DIMETHYLBICYCLO[3.3.0]OCTA-
  2,6-DIENE-2,4,6,8-TETRACARBOXYLATE, VII**, 50
  1,3,4,6-Pentalenetetracarboxylic acid, 1,3a,4,6a-tetrahydro-2,5-dihydroxy-
    3a,6a-dimethyl-, tetramethyl ester [79150-94-0]
TETRAMETHYLENE CHLOROHYDRIN, II, 571
  1-Butanol, 4-chloro- [928-51-8]
TETRAMETHYL 1,1,2,2-ETHANETETRACARBOXYLATE, VII, 482
  1,1,2,2-Ethanetetracarboxylic acid, tetramethyl ester [5464-22-2]
2,3,4,6-TETRAMETHYL-D-GLUCOSE, III, 800
  D-Glucopyranose, 2,3,4,6-tetramethyl- [7506-68-5]
2,2,3,3-TETRAMETHYLIODOCYCLOPROPANE, VI, 974
  Cyclopropane, 3-iodo-1,1,2,2-tetramethyl- [39653-50-4]
2,2,6,6-TETRAMETHYLOLCYCLOHEXANOL, IV, 907
  1,1,3,3-Cyclohexanetetramethanol, 2-hydroxy- [5416-55-7]
**TETRAMETHYL [2.2]PARACYCLOPHANE-4,5,12,13-TETRACARBOXYLATE,
  VII**, 485
  Tricyclo[8.2.2.2^{4,7}]hexadeca-4,6,10,12,13,15-hexaene-5,6,11,12-
    tetracarboxylic acid, tetramethyl ester, stereoisomer [37437-90-4]
TETRAMETHYL-*p*-PHENYLENEDIAMINE, V, 1018
  1,4-Benzenediamine, *N,N,N',N'*-tetramethyl- [100-22-1]
2,3,4.5-TETRAMETHYLPYRROLE, V, 1022
  1*H*-Pyrrole, 2,3,4,5-tetramethyl- [1003-90-3]
(*R,R*)-(+)-*N,N,N',N'*-TETRAMETHYLTARTARIC ACID DIAMIDE, VII, 41
  Butanediamide, 2,3-dihydroxy-*N,N,N',N'*-tetramethyl-, [*R*-(*R*,R**)]- [26549-65-5]
2,2,5,5-TETRAMETHYLTETRAHYDRO-3-KETOFURAN, V, 1024
  3(2*H*)-Furanone, dihydro-2,2,5,5-tetramethyl- [5455-94-7]
α,α,α',α'-**TETRAMETHYLTETRAMETHYLENE GLYCOL, V**, 1026
  2,5-Hexanediol, 2,5-dimethyl- [110-03-2]
2,4,5,7-TETRANITROFLUORENONE, V, 1029
  9*H*-Fluoren-9-one, 2,4,5,7-tetranitro- [746-53-2]
(+)- and (−)-α-**(2,4,5,7-TETRANITRO-9-FLUORENYLIDENEAMINOOXY)PRO-
  PIONIC ACID, V**, 1031
  Propanoic acid, 2-[[(2,4,5,7-tetranitro-9*H*-fluoren-9-ylidene)amino]oxy]-, (*S*)-
    [50996-73-1]
  Propanoic acid, 2-[[(2,4,5,7-tetranitro-9*H*-fluoren-9-ylidene)amino]oxy]-, (*R*)-
    [50874-31-2]

TETRANITROMETHANE, III, 803
Methane, tetranitro- [509-14-8]
TETRAPHENYLARSONIUM CHLORIDE HYDROCHLORIDE, IV, 910
Arsonium, tetraphenyl-, (hydrogen dichloride) [21006-73-5]
TETRAPHENYLCYCLOPENTADIENONE, III, 806; **V,** 605
2,4-Cyclopentadien-1-one, 2,3,4,5-tetraphenyl- [479-33-4]
TETRAPHENYLETHYLENE, IV, 914
Benzene, 1,1',1'',1'''-(1,2-ethenediylidene)tetrakis- [632-51-9]
2,3,5,5-TETRAPHENYLISOXAZOLIDINE, V, 1126
Isoxazolidine, 2,3,5,5-tetraphenyl- [25116-92-1]
1,2,3,4-TETRAPHENYLNAPHTHALENE, V, 1037
Naphthalene, 1,2,3,4-tetraphenyl- [751-38-2]
TETRAPHENYLPHTHALIC ANHYDRIDE, III, 807
1,3-Isobenzofurandione, 4,5,6,7-tetraphenyl- [4741-53-1]
1,4,8,11-TETRATHIACYCLOTETRADECANE, VIII, 592
1,4,8,11-Tetrathiacyclotetradecane [24194-61-4]
TETROLIC ACID, V, 1043
2-Butynoic acid [590-93-2]
2-THENALDEHYDE, III, 811; **IV,** 915
2-Thiophenecarboxaldehyde [98-03-3]
3-THENALDEHYDE, IV, 918
3-Thiophenecarboxaldehyde [498-62-4]
3-THENOIC ACID, IV, 919
3-Thiophenecarboxylic acid [88-13-1]
3-THENYL BROMIDE, IV, 921
Thiophene, 3-(bromomethyl)- [34846-44-1]
cis-**8-THIABICYCLO[4.3.0]NONANE, VI,** 482
Benzo[c]thiophene, octahydro- [17739-77-4]
cis-**8-THIABICYCLO[4.3.0]NONANE 8,8-DIOXIDE, VI,** 482
Benzo[c]thiophene, 2,2-dioxide, octahydro-, *cis*- [57479-57-9]
4H-1,4-THIAZINE 1,1-DIOXIDE, VI, 976
4H-1,4-Thiazine, 1,1-dioxide [40263-61-4]
THIETANE 1,1-DIOXIDE, VII, 491
Thietane, 1,1-dioxide [5687-92-3]
THIETE 1,1-DIOXIDE, VII, 491
2H-Thiete, 1,1-dioxide [7285-32-7]
THIOBENZOIC ACID, IV, 924
Benzenecarbothioic acid [98-91-9]
THIOBENZOPHENONE, II, 573; **IV,** 927
Methanethione, diphenyl- [1450-31-3]
THIOBENZOYLTHIOGLYCOLIC ACID, V, 1046
Acetic acid, [(phenylthioxomethyl)thio]- [942-91-6]
m-**THIOCRESOL, III,** 809; *Warning,* **V,** 1050
Benzenethiol, 3-methyl- [108-40-7]
p-**THIOCYANODIMETHYLANILINE, II,** 574
Thiocyanic acid, 4-(dimethylamino)phenyl ester [7152-80-9]

β-**THIODIGLYCOL, II**, 576
  Ethanol, 2,2'-thiobis- [111-48-8]
2α-**THIOHOMOPHTHALIMIDE, V**, 1051
  3(2*H*)-Isoquinolinone, 1,4-dihydro-1-thioxo- [938-38-5]
THIOLACETIC ACID, IV, 928
  Ethanethioic acid [507-09-5]
7-**THIOMENTHOL, VIII**, 302
  Cyclohexanol, 2-(1-mercapto-1-methylethyl)-5-methyl-, [1*R*-(1α,2α,5α)]-
    [79563-68-1]; 1*R*-(1α,2β,5α)] [79563-59-0]; [1*S*-(1α,2α,5β)]-
    [79563-67-0]
2-**THIO-6-METHYLURACIL, IV**, 638
  4(1*H*)-Pyrimidinone, 2,3-dihydro-6-methyl-2-thioxo- [56-04-2]
THIOPHENE, II, 578
  Thiophene [110-02-1]
2-**THIOPHENEALDEHYDE, III**, 811, 812; **IV**, 915
  2-Thiophenecarboxaldehyde [98-03-3]
2-**THIOPHENETHIOL, VI**, 979
  2-Thiophenethiol [7774-74-5]
THIOPHENOL, I, 504; *Hazard*
  Benzenethiol [108-98-5]
THIOPHOSGENE, I, 506; *Hazard*
  Carbonothioic dichloride [463-71-8]
THIOSALICYLIC ACID, II, 580
  Benzoic acid, 2-mercapto- [147-93-3]
DL-**THREONINE, III**, 813
  DL-Threonine [80-68-2]
THYMOQUINONE, I, 511
  2,5-Cyclohexadiene-1,4-dione, 2-methyl-5-(1-methylethyl)- [490-91-5]
o-**TOLUALDEHYDE, III**, 818; **IV**, 932
  Benzaldehyde, 2-methyl- [529-20-4]
p-**TOLUALDEHYDE, II**, 583
  Benzaldehyde, 4-methyl- [104-87-0]
o-**TOLUAMIDE, II**, 586; *Warning*, **V**, 1054
  Benzamide, 2-methyl- [527-85-5]
p-**TOLUENESULFENYL CHLORIDE, IV**, 934
  Benzenesulfenyl chloride, 4-methyl- [933-00-6]
p-**TOLUENESULFINYL CHLORIDE, IV**, 937
  Benzenesulfinyl chloride, 4-methyl- [10439-23-3]
2-(*p*-**TOLUENESULFINYL)-2-CYCLOPENTENONE, (S)-(+)-, VII**, 495
  2-Cyclopenten-1-one, 2-[(4-methylphenyl)sulfinyl]-, (*S*)- [79681-26-8]
2-(*p*-**TOLUENESULFINYL)-2-CYCLOPENTENONE ETHYLENE KETAL**,
    (S)-(+)- **VII**, 495
  1,4-Dioxaspiro[4.4]non-6-ene, 6-[(4-methylphenyl)sulfinyl]-, (*S*)- [82136-15-0]
cis-2-(*p*-**TOLUENESULFONAMIDO)CYCLOHEXANOL, VII**, 375
  Benzenesulfonamide, *N*-(2-hydroxycyclohexyl)-4-methyl-, *cis*- [58107-40-7]
p-**TOLUENESULFONIC ANHYDRIDE, IV**, 940
  Benzenesulfonic acid, 4-methyl-, anhydride [4124-41-8]

p-TOLUENESULFONYLANTHRANILIC ACID, IV, 34
  Benzoic acid, 2-[(4-methylbenzenesulfonyl)amino]- [6311-23-5]
p-TOLUENESULFONYLHYDRAZIDE, V, 1055
  Benzenesulfonic acid, 4-methyl-, hydrazide [1576-35-8]
o-TOLUIC ACID, II, 588; III, 820
  Benzoic acid, 2-methyl- [118-90-1]
p-TOLUIC ACID, II, 589; III, 822
  Benzoic acid, 4-methyl- [99-94-5]
o-TOLUIDINESULFONIC ACID, III, 824
  Benzenesulfonic acid, 4-amino-3-methyl- [98-33-9]
o-TOLUNITRILE, I, 514
  Benzonitrile, 2-methyl- [529-19-1]
p-TOLUNITRILE, I, 514
  Benzonitrile, 4-methyl- [104-85-8]
p-TOLUYL-*o*-BENZOIC ACID, I, 517
  Benzoic acid, 2-(4-methylbenzoyl)- [85-55-2]
m-TOLYLBENZYLAMINE, III, 827
  Benzenemethanamine, *N*-(3-methylphenyl)- [5405-17-4]
p-TOLYL CARBINOL, II, 590
  Benzenemethanol, 4-methyl- [589-18-4]
1-*p*-TOLYLCYCLOPROPANOL, V, 1058
  Cyclopropanol, 1-(4-methylphenyl)- [40122-37-0]
o-TOLYL ISOCYANIDE, V, 1060
  Benzene, 1-isocyano-2-methyl- [10468-64-1]
p-TOLYLMERCURIC CHLORIDE, I, 519; *Hazard*
  Mercury, chloro(4-methylphenyl)- [539-43-5]
p-TOLYLSULFONYLDIAZOMETHANE, VI, 981
  Benzene, 1-[(diazomethyl)sulfonyl]-4-methyl- [1538-98-3]
2-(*p*-TOLYLSULFONYL)DIHYDROISOINDOLE, V, 1064
  1*H*-Isoindole, 2,3-dihydro-2-[(4-methylphenyl)sulfonyl]- [32372-83-1]
N-(*p*-TOLYLSULFONYLMETHYL)FORMAMIDE, VI, 987
  Formamide, *N*-[[(4-methylphenyl)sulfonyl]methyl]- [36635-56-0]
p-TOLYLSULFONYLMETHYL ISOCYANIDE, VI, 41, 987
  Benzene, 1-[(isocyanomethyl)sulfonyl]-4-methyl- [36635-61-7]
p-TOLYLSULFONYLMETHYLNITROSAMIDE, IV, 943
  Benzenesulfonamide, *N*,4-dimethyl-*N*-nitroso- [80-11-5]
p-TOLYLTHIOUREA, III, 77
  Thiourea, (4-methylphenyl)- [622-52-6]
p-TOLYL-2-(TRIMETHYLSILYL)ETHYNYL SULFONE, VIII, 281
  Silane, trimethyl[[(4-methylphenyl)sulfonyl]ethynyl]- [34452-56-7]
N-TOSYL-3-METHYL-2-AZABICYCLO[3.3.0]OCT-3-ENE, *cis*-, VII, 501
  Cyclopenta[*b*]pyrrole, 1,3a,4,5,6,6a-hexahydro-2-methyl-
    1-[(4-methylphenyl)sulfonyl]-, *cis*- [81097-07-6]
cis-*N*-TOSYL-2-(2-PROPENYL)CYCLOPENTYLAMINE, VII, 501
  Benzenesulfonamide, 4-methyl-*N*-[2-(2-propenyl)cyclopentyl], *cis*- [81097-06-5]
1,2,5-TRIACETOXYPENTANE, III, 833
  1,2,5-Pentanetriol, triacetate [5470-86-0]

1,3,5-TRIACETYLBENZENE, III, 829
  Ethanone, 1,1',1"-(1,3,5-benzenetriyl)tris- [779-90-8]
2,4,5-TRIAMINONITROBENZENE, V, 1067
  1,2,4-Benzenetriamine, 5-nitro- [6635-35-4]
1,2,4-TRIAZOLE, V, 1070
  1*H*-1,2,4-Triazole [288-88-0]
1,2,4-TRIAZOLE-3(5)-THIOL, V, 1070
  3*H*-1,2,4-Triazole-3-thione [36143-39-2]
TRIBIPHENYLCARBINOL, III, 831
  [1,1'-Biphenyl]-4-methanol, α,α-bis([1,1'-biphenyl]-4-yl)- [5341-14-0]
p,α,α-**TRIBROMOACETOPHENONE, IV**, 110
  Ethanone, 2,2-dibromo-1-(4-bromophenyl)- [13195-79-4]
sym-**TRIBROMOBENZENE, II**, 592
  Benzene, 1,3,5-tribromo- [626-39-1]
2,4,6-TRIBROMOBENZOIC ACID, IV, 947
  Benzoic acid, 2,4,6-tribromo- [633-12-5]
1,2,3-TRIBROMOPROPANE, I, 521
  Propane, 1,2,3-tribromo- [96-11-7]
TRI-*tert*-BUTYLCYCLOPROPENYL TETRAFLUOROBORATE, VI, 991
  Cyclopropenylium, tris(1,1-dimethylethyl)-, tetrafluoroborate, (1⁻) [60391-90-4]
TRIBUTYLETHYNYLSTANNANE, VIII, 268
  Stannane, tributylethynyl- [994-89-8]
**(Z)-TRIBUTYLSTANNYLMETHYLENE-4-ISOPROPYL-1,1-CYCLOPENTANE-
  DICARBOXYLIC ACID, DIMETHYL ESTER, VIII**, 381
  1,1-Cyclopentanedicarboxylic acid, 3-(1-methylethyl)-4-[(tributylstannyl)
  methylene]-, dimethyl ester [107473-15-4]
TRICARBALLYLIC ACID, I, 523
  1,2,3-Propanetricarboxylic acid [99-14-9]
TRICARBETHOXYMETHANE, II, 594
  Methanetricarboxylic acid, triethyl ester [6279-86-3]
2,3,6-TRICARBOETHOXYPYRIDINE, VIII, 597
  Pyridine-2,3,6-tricarboxylic acid, triethyl ester [122509-29-9]
TRICARBOMETHOXYMETHANE, II, 596
  Methanetricarboxylic acid, trimethyl ester [1186-73-8]
**TRICARBONYL[(2,3,4,5-η)-2,4-CYCLOHEXADIEN-1-ONE]IRON,
  VI**, 996
  Iron, tricarbonyl-[(2,3,4,5-η)-2,4-cyclohexadien-1-one]- [12306-92-2]
**TRICARBONYL[(1,2,3,4-η)-1- and 2-METHOXY-1,3-
  CYCLOHEXADIENE]IRON, VI**, 996
  Iron, tricarbonyl-[(1,2,3,4-η)-1-methoxycyclohexadiene]- [12318-18-2]
  Iron, tricarbonyl-[(1,2,3,4-η)-2-methoxycyclohexadiene]- [12318-19-3]
**TRICARBONYL[2-(2,3,4,5-η)-4-METHOXY-2,4-CYCLOHEXADIEN-1-YL]- 5,5-
  DIMETHYL-1,3-CYCLOHEXANEDIONE]IRON(1−), VI**, 1001
  Iron, tricarbonyl[2-[(2,3,4,5-η)-4-methoxy-2,4-cyclohexadien-1-yl]-5,5- dimethyl-
  1,3-cyclohexanedione]- [51539-52-7]

TRICARBONYL[(1,2,3,4,5-η)-2-METHOXY-2,4-CYCLOHEXADIEN-1-YL]IRON(1+) HEXAFLUOROPHOSPHATE(1−), VI, 996
Iron(1+), tricarbonyl[(1,2,3,4,5-η)-2-methoxy-2,4-cyclohexadien-1-yl]-hexafluorophosphate, (1−) [51508-59-9]

α,α,α-TRICHLOROACETANILIDE, V, 1074
Acetamide, 2,2,2-trichloro-N-phenyl- [2563-97-5]

1,1,1-TRICHLORO-4-ETHOXY-3-BUTEN-2-ONE, VIII, 254
3-Buten-2-one, 1,1,1-trichloro-4-ethoxy- [83124-74-7]

1,1,1-TRICHLORO-4-METHOXY-3-BUTEN-2-ONE, VIII, 238
3-Buten-2-one, 1,1,1-trichloro-4-methoxy- [138149-14-1]

2-(TRICHLOROMETHYL)BICYCLO[3.3.0]OCTANE, V, 93
Pentalene, octahydro-1-trichloromethyl- [18127-07-6]

TRICHLOROMETHYL CHLOROFORMATE, [503-38-8], **VI,** 715
Carbonochloridic acid, trichloromethyl ester [503-38-8]

TRICHLOROETHYL ALCOHOL, II, 598
Ethanol, 2,2,2-trichloro- [115-20-8]

TRICHLOROMETHYLPHOSPHONYL DICHLORIDE, IV, 950
Phosphonic dichloride, (trichloromethyl)- [21510-59-8]

1,1,3-TRICHLORONONANE, V, 1076
Nonane, 1,1,3-trichloro- [10575-86-7]

1,1,2-TRICHLORO-2,3,3-TRIFLUOROCYCLOBUTANE, V, 393
Cyclobutane, 1,1,2-trichloro-2,3,3-trifluoro- [697-17-6]

p-TRICYANOVINYL-N,N-DIMETHYLANILINE, IV, 953
Ethenetricarbonitrile, [4-(dimethylamino)phenyl]- [6673-15-0]

***endo*-TRICYCLO[4.4.0.0²,⁵]DECA-3,8-DIENE-7,10-DIONE, VI,** 1002
Tricyclo[4.4.0.0²,⁵]deca-3,8-diene-7,10-dione [34231-42-0]

TRICYCLO[4.3.1.0³,⁸]DECAN-4-ONE, VI, 958
2,5-Methano-1H-inden-7(4H)-one, hexahydro- [27567-85-7]

TRICYCLO[4.4.1.0¹,⁶]UNDECA-3,8-DIENE, VI, 731
Tricyclo[4.4.1.0¹,⁶]undeca-3,8-diene [27714-83-6]

TRIDEHYDRO[18]ANNULENE, VI, 68
1,3,7,9,13,15-Cyclooctadecahexaene-5,11,17-triyne [3891-75-6]

TRIETHYLCARBINOL, II, 602
3-Pentanol, 3-ethyl- [597-49-9]

TRIETHYL OXALYLSUCCINATE, III, 510; **V,** 687
1,2,3-Propanetricarboxylic acid, 1-oxo-, triethyl ester [42126-21-6]

TRIETHYLOXONIUM FLUOBORATE, V, 1080
Oxonium, triethyl-, tetrafluoroborate(1⁻) [368-39-8]

TRIETHYL PHOSPHITE, IV, 955
Phosphorous acid, triethyl ester [122-52-1]

TRIETHYL α-PHTHALIMIDOETHANE-α,α,β-TRICARBOXYLATE, IV, 55
1,1,2-Ethanetricarboxylic acid, 1-(1,3-dihydro-1,3-dioxo-2H-isoindol-2-yl)-, triethyl ester [76758-31-1]

TRIETHYL 1,2,4-TRIAZINE-3,5,6-TRICARBOXYLATE, VIII, 597
1,2,4-Triazine-3,5,6-tricarboxylic acid, triethyl ester [74476-38-3]

TRIETHYL *N*-TRICARBOXYLATE, III, 415
Nitridotricarbonic acid, triethyl ester [3206-31-3]
***N*-TRIFLUOROACETANILIDE, VI,** 1004
Acetamide, 2,2,2-trifluoro-*N*-phenyl- [404-24-0]
TRIFLUOROACETYL TRIFLATE, VII, 506
Acetic acid, trifluoro, anhydride with trifluoromethanesulfonic acid [68602-57-3]
***N*-TRIFLUOROACETYL-L-TYROSINE, VI,** 1004
L-Tyrosine, *N*-(trifluoroacetyl)- [350-10-7]
1,1,1-TRIFLUOROHEPTANE, V, 1082
Heptane, 1,1,1-trifluoro- [693-09-4]
***m*-TRIFLUOROMETHYLBENZENESULFONYL CHLORIDE, VII,** 508
Benzenesulfonyl chloride, 3-(trifluoromethyl)- [777-44-6]
***m*-TRIFLUOROMETHYL-*N,N*-DIMETHYLANILINE, V,** 1085
Benzenamine, *N,N*-dimethyl-3-(trifluoromethyl)- [329-00-0]
1,2,5-TRIHYDROXYPENTANE, III, 833
1,2,5-Pentanetriol [14697-46-2]
1,2,3-TRIIODO-5-NITROBENZENE, II, 604
Benzene, 1,2,3-triiodo-5-nitro- [53663-23-3]
3,4,5-TRIMETHOXYBENZALDEHYDE, VI, 1007
Benzaldehyde, 3,4,5-trimethoxy- [86-81-7]
TRIMETHYLACETIC ACID, I, 524
Propanoic acid, 2,2-dimethyl- [75-98-9]
TRIMETHYLAMINE, I, 528
Methanamine, *N,N*-dimethyl- [75-50-3]
TRIMETHYLAMINE HYDROCHLORIDE, I, 531
Methanamine, *N,N*-dimethyl-, hydrochloride [593-81-7]
4,6,8-TRIMETHYLAZULENE, V, 1088
Azulene, 4,6,8-trimethyl- [941-81-1]
TRIMETHYL-*p*-BENZOQUINONE, VI, 1010
2,5-Cyclohexadiene-1,4-dione, 2,3,5-trimethyl- [935-92-2]
2,5,5-TRIMETHYL-2-(2-BROMOETHYL)-1,3-DIOXANE, VII, 59
1,3-Dioxane, 2-(2-bromoethyl)-2,5,5-trimethyl- [87842-52-2]
**TRIMETHYL 2-CHLORO-2-CYCLOPROPYLIDENORTHOACETATE,
VIII,** 373
Cyclopropane, (1-chloro-2,2,2-trimethoxyethylidene)- [82979-34-8]
2,6,6-TRIMETHYL-2,4-CYCLOHEXADIENONE, V, 1092
2,4-Cyclohexadien-1-one, 2,6,6-trimethyl- [13487-30-4]
2,4,4-TRIMETHYLCYCLOPENTANONE, IV, 957
Cyclopentanone, 2,4,4-trimethyl- [4694-12-6]
2,2,4-TRIMETHYL-1,2-DIHYDROQUINOLINE (ACETONE ANIL), III, 329
Quinoline, 1,2-dihydro-2,2,4-trimethyl- [147-47-7]
TRIMETHYLENE BROMIDE, I, 30
Propane, 1,3-dibromo- [109-64-8]
TRIMETHYLENE CHLOROHYDRIN, I, 533
1-Propanol, 3-chloro- [627-30-5]

TRIMETHYLENE CYANIDE, I, 536
   Pentanedinitrile [544-13-8]
2,2-(TRIMETHYLENEDITHIO)CYCLOHEXANONE, VI, 1014
   1,5-Dithiaspiro[5.5]undecan-7-one [51310-03-3]
TRIMETHYLENE DITHIOTOSYLATE, VI, 1016
   Benzenesulfonothioic acid, 4-methyl-, *S,S'*-1,3-propanediyl ester [3866-79-3]
TRIMETHYLENE OXIDE, III, 835
   Oxetane [503-30-0]
TRIMETHYLETHYLENE DIBROMIDE, II, 409
   Butane, 2,3-dibromo-2-methyl- [594-51-4]
TRIMETHYLGALLIC ACID, I, 537
   Benzoic acid, 3,4,5-trimethoxy- [118-41-2]
3,4,5-TRIMETHYL-2,5-HEPTADIEN-4-OL, VIII, 505
   1,5-Heptadien-4-ol, 3,4,5-trimethyl- [64417-15-8]
TRIMETHYL (2-METHYLENE-4-PHENYL-3-BUTENYL)SILANE, VIII, 602
   Silane, trimethyl (2-methylene-4-phenyl-3-butenyl)- [80814-92-2]
TRIMETHYL O-4-BROMOBUTANOATE, VIII, 415
   Butane, 4-bromo-1,1,1-trimethoxy- [55444-67-2]
N-4,4-TRIMETHYL-2-OXAZOLINIUM IODIDE, VI, 65
   Oxazolium, 4,5-dihydro-3,4,4-trimethyl-, iodide [30093-97-1]
TRIMETHYLOXONIUM FLUOBORATE, V, 1096; **VI,** 1019
   Oxonium, trimethyl-, tetrafluoroborate (1⁻) [420-37-1]
TRIMETHYLOXONIUM 2,4,6-TRINITROBENZENESULFONATE, V, 1099
   Oxonium, trimethyl-, salt with 2,4,6-trinitrobenzenesulfonic acid (1:1)
      [13700-00-0]
3,5,5-TRIMETHYL-2-(2-OXOPROPYL)-2-CYCLOHEXEN-1-ONE, VI, 1024
   2-Cyclohexen-1-one, 3,5,5-trimethyl-2-(2-oxopropyl)- [61879-73-0]
2,2,4-TRIMETHYL-3-OXOVALERYL CHLORIDE, V, 1103
   Pentanoyl chloride, 2,2,4-trimethyl-3-oxo- [10472-34-1]
2,4,4-TRIMETHYLPENTANAL, VI, 1028
   Pentanal, 2,4,4-trimethyl- [17414-46-9]
2,4,6-TRIMETHYLPYRYLIUM PERCHLORATE, V, 1106
   Pyrylium, 2,4,6-trimethyl-, perchlorate [940-93-2]
2,4,6-TRIMETHYLPYRYLIUM TETRAFLUOROBORATE, V, 1112
   Pyrylium, 2,4,6-trimethyl-, tetrafluoroborate(1⁻) [773-01-3]
2,4,6-TRIMETHYLPYRYLIUM TRIFLUOROMETHANESULFONATE, V, 1114
   Pyrylium, 2,4,6-trimethyl-, salt with trifluoromethanesulfonic acid (1:1)
      [40927-60-4]
1-TRIMETHYLSILOXYCYCLOHEXENE, VI, 327; **VII,** 414; **VIII,** 461
   Silane, (1-cyclohexen-1-yloxy)trimethyl- [6651-36-1]
1-TRIMETHYLSILOXYCYCLOPENTENE, VII, 424
   Silane, (1-cyclopenten-1-yloxy)trimethyl- [19980-43-9]
2-[(TRIMETHYLSILOXY)METHYL]ALLYLTRIMETHYLSILANE, VII, 266
3-TRIMETHYLSILOXY-2-PENTENE, (Z)-, VII, 512
   Silane, [(1-ethyl-1-propenyl)oxy]trimethyl-, (Z)- [51425-54-8]

TRIMETHYLSILYLACETYLENE, VIII, 63, 606
Silane, ethynyltrimethyl- [1066-54-2]
TRIMETHYLSILYL AZIDE, VI, 1030
Silane, azidotrimethyl- [4648-54-8]
***O*-(TRIMETHYLSILYL)BENZOPHENONE CYANOHYDRIN, VII**, 20
Benzeneacetonitrile, α-phenyl-α-[(trimethylsilyl)oxy]- [40326-25-8]
3-TRIMETHYLSILYL-3-BUTEN-2-OL, VI, 1033
3-Buten-2-ol, 3-(trimethylsilyl)- [66374-47-8]
(Z)-4-(TRIMETHYLSILYL)-3-BUTEN-1-OL, VIII, 358, 609
3-Buten-1-ol, 4-(trimethylsilyl)-, (Z)- [87682-77-7]
3-TRIMETHYLSILYL-3-BUTEN-2-ONE, VI, 1033
3-Buten-2-one, 3-(trimethylsilyl)- [43209-86-5]
(Z)-4-(TRIMETHYLSILYL)-3-BUTENYL 4-METHYLBENZENESULFONATE,
VIII, 358
3-Buten-1-ol, 4-(trimethylsilyl)-, 4-methylbenzenesulfonate, (Z)- [87682-62-0]
4-(TRIMETHYLSILYL)-3-BUTYN-1-OL, VIII, 609
3-Butyn-1-ol, 4-(trimethylsilyl)- [2117-12-6]
TRIMETHYLSILYL CYANIDE, VII, 20, 294, 517
Silanecarbonitrile, trimethyl- [7677-24-9]
TRIMETHYLSILYLDIAZOMETHANE, VIII, 612
Silane, (diazomethyl)trimethyl- [18107-18-1]
***O*-TRIMETHYLSILYL-4-METHOXYMANDELONITRILE, VII**, 521
Benzeneacetonitrile, 4-methoxy-α-[(trimethylsilyl)oxy]- [66985-48-6]
1-TRIMETHYLSILYLOXYBICYCLO[4.1.0]HEPTANE, VI, 328
Silane, (bicyclo[4.1.0]hept-1-yloxy)trimethyl- [38858-74-1]
2-TRIMETHYLSILYLOXY-1,3-BUTADIENE, VI, 445
Silane, trimethyl[(1-methylene-2-propenyl)oxy]- [38053-91-7]
3-TRIMETHYLSILYL-2-PROPEN-1-OL, (*E*)-, VII, 524
2-Propen-1-ol, 3-(trimethylsilyl)-, (*E*)- [59376-64-6]
3-TRIMETHYLSILYL-2-PROPYN-1-OL, VII, 524
2-Propyn-1-ol, 3-(trimethylsilyl) [5272-36-3]
3-TRIMETHYLSILYL-2-PROPYN-1-YL METHANESULFONATE,
VIII, 471
2-Propyn-1-ol, 3-(trimethylsilyl)-, methanesulfonate [71231-17-0]
4,6,6-TRIMETHYL-2-TRIMETHYLSILOXYCYCLOHEXA-1,3-DIENE,
VII, 282
Silane, trimethyl[(3,3,5-trimethyl-1,5-cyclohexadien-1-yl)oxy]-
[54781-28-1]
TRIMETHYLVINYLTIN, VIII, 97
Stannane, ethenyltrimethyl- [754-06-3]
TRIMYRISTIN, I, 538
Tetradecanoic acid, 1,2,3-propanetriyl ester [555-45-3]
1,3,5-TRINITROBENZENE, I, 541
Benzene, 1,3,5-trinitro- [99-35-4]
2,4,6-TRINITROBENZOIC ACID, I, 543
Benzoic acid, 2,4,6-trinitro- [129-66-8]

2,4,7-TRINITROFLUORENONE, III, 837
9*H*-Fluoren-9-one, 2,4,7-trinitro- [129-79-3]
TRIPHENYLALUMINUM, V, 1116
Aluminum, triphenyl- [841-76-9]
TRIPHENYLAMINE, I, 544; *Hazard*
Benzenamine, *N,N*-diphenyl- [603-34-9]
2,4,6-TRIPHENYLANILINE, V, 1130
[1,1':3',1"-Terphenyl]-2'-amine, 5'-phenyl- [6864-20-6]
TRIPHENYLARSINE, IV, 910; *Hazard*
Arsine, triphenyl- [603-32-7]
TRIPHENYLARSINE OXIDE, IV, 911; *Hazard*
Arsine, oxide, triphenyl- [1153-05-5]
TRIPHENYLCARBINOL, III, 839
Benzenemethanol, α,α-diphenyl- [76-84-6]
TRIPHENYLCHLOROMETHANE, III, 841
Benzene, 1,1',1"-(chloromethylidyne)tris- [76-83-5]
TRIPHENYLENE, V, 1120
Triphenylene [217-59-4]
TRIPHENYLETHYLENE, II, 606
Benzene, 1,1',1"-(1-ethenyl-2-ylidene)tris- [58-72-0]
2,3,5-TRIPHENYLISOXAZOLIDINE, V, 1124
Isoxazolidine, 2,3,5-triphenyl- [13787-96-7]
TRIPHENYLMETHANE, I, 548
Benzene, 1,1',1"-methylidynetris- [519-73-3]
2,4,6-TRIPHENYLNITROBENZENE, V, 1128
1,1':3',1"-Terphenyl, 2'-nitro-5'-phenyl- [10368-47-5]
2,4,6-TRIPHENYLPHENOL, V, 1130
(*m*-Quaterphenyl)-2'-ol [23886-03-5]
2,4,6-TRIPHENYLPHENOXYL, V, 1130
2,5-Cyclohexadien-1-one, 2,4,6-triphenyl-4-(5'-phenyl[1,1':3,1"-terphenyl]-
2'-yloxy)- [10384-15-3]
α,β,β-TRIPHENYLPROPIONIC ACID, IV, 960
Benzenepropanoic acid, α,β-diphenyl- [53663-24-4]
α,α,β-TRIPHENYLPROPIONITRILE, IV, 962
Benzenepropanenitrile, α,α-diphenyl- [5350-82-3]
2,4,6-TRIPHENYLPYRYLIUM TETRAFLUOROBORATE, V, 1135
Pyrylium, 2,4,6-triphenyl-, tetrafluoroborate(1−) [448-61-3]
TRIPHENYLSELENONIUM CHLORIDE, II, 240; *Hazard*
Selenium, chlorotriphenyl-, (T-4)- [17166-13-1]
TRIPHENYLSTIBINE, I, 550; *Hazard*
Stibine, triphenyl- [603-36-1]
TRIPTYCENE, IV, 964
9,10[1',2']Benzenoanthracene, 9,10-dihydro- [477-75-8]
TRISAMMONIUM GERANYL PHOSPHATE, VIII, 616
Diphosphoric acid, mono(3,7-dimethyl-2,6-octadienyl) ester, triammonium salt, (*E*)-
[116057-55-7]

TRIS(DIETHYLAMINO)SULFONIUM DIFLUOROTRIMETHYLSILICATE, VIII, 329
Sulfiliminium, *S,S*-bis(diethylamino)-*N,N*-diethyl-, difluorotrimethylsilicate (1⁻) [59201-86-4]
TRIS(DIMETHYLAMINO)SULFONIUM DIFLUOROTRIMETHYLSILICATE, VII, 528
Sulfiliminium, *S,S*-bis(dimethylamino)-*N,N*-dimethyl-, difluorotrimethylsilicate (1–) [59218-87-0]
TRIS(TETRABUTYLAMMONIUM)HYDROGEN PYROPHOSPHATE TRIHYDRATE, VIII, 616
1-Butanaminium, *N,N,N*-tributyl-, (diphosphate) (3:1) [76947-02-9]
***N,N',N''*-TRIS-(*p*-TOLYLSULFONYL)DIETHYLENETRIAMINE, VI**, 652
Benzenesulfonamide, 4-methyl-*N,N*-bis[2-[[(4-methylphenyl]sulfonyl]amino]-ethyl]- [56187-04-3]
***N,N',N''*-TRIS(*p*-TOLYLSULFONYL)DIETHYLENETRIAMINE *N,N''*-DISODIUM SALT, VI**, 652
Benzenesulfonamide, 4-methyl-*N,N*-bis[2-[[(4-methylphenyl)sulfonyl]amino]ethyl]-, disodium salt [52601-80-6]
3,6,9-TRIS(*p*-TOLYLSULFONYL)-3,6,9-TRIAZAUNDECANE-1,11-DIOL, VI, 652
Benzenesulfonamide, 4-methyl-*N,N*-bis[2-[(2-hydroxyethyl)[(4-methylphenyl)sulfonyl]amino]ethyl]- [74461-29-3]
3,6,9-TRIS(*p*-TOLYLSULFONYL)-3,6,9-TRIAZAUNDECANE-1,11-DIMETHANESULFONATE, VI, 652
Benzenesulfonamide, 4-methyl-*N,N*-bis[2-[[(4-methylphenyl)sulfonyl][2-[(methylsulfonyl)oxy]ethyl]amino]ethyl- [74461-30-6]
***sym*-TRITHIANE, II**, 610
1,3,5-Trithiane [291-21-4]
TRITHIOCARBODIGLYCOLIC ACID, IV, 967
Acetic acid, 2,2'-[carbonothioylbis(thio)]bis- [6326-83-6]
TROPOLONE, VI, 1037
2,4,6-Cycloheptatrien-1-one, 2-hydroxy- [533-75-5]
TROPYLIUM FLUOBORATE, V, 1138
Cycloheptatrienylium, tetrafluoroborate(1–) [27081-10-3]
L-TRYPTOPHAN, II, 612
L-Tryptophan [73-22-3]
L-TYROSINE, II, 612
L-Tyrosine [60-18-4]

U

2,5-UNDECANEDIONE, VIII, 620
2,5-Undecanedione [7018-92-0]
UNDECYL ISOCYANATE, III, 846
Undecane, 1-isocyanato- [2411-58-7]
10-UNDECYNOIC ACID, IV, 969
10-Undecynoic acid [2777-65-3]
URAMIL, II, 617
2,4,6(1*H*,3*H*,5*H*)-Pyrimidinetrione, 5-amino- [118-78-5]

V

DL-VALINE, III, 848
DL-Valine [516-06-3]
L-VALINOL, VII, 530
1-Butanol, 2-amino-3-methyl-, (*S*)- [2026-48-4]
VANILLIC ACID, IV, 972
Benzoic acid, 4-hydroxy-3-methoxy- [121-34-6]
VERATRALDEHYDE, II, 619
Benzaldehyde, 3,4-dimethoxy- [120-14-9]
VERATRIC AMIDE, II, 44
Benzamide, 3,4-dimethoxy- [1521-41-1]
VERATRONITRILE, II, 622
Benzonitrile, 3,4-dimethoxy- [2024-83-1]
VINYLACETIC ACID, III, 851
3-Butenoic acid [625-38-7]
VINYL CHLOROACETATE, III, 853
Acetic acid, chloro, ethenyl ester [2549-51-1]
VINYL LAURATE and other VINYL ESTERS, IV, 977
Dodecanoic acid, ethenyl ester [2146-71-6]
2-VINYLTHIOPHENE, IV, 980
Thiophene, 2-ethenyl- [1918-82-7]
VINYLTRIMETHYLSILANE, VI, 1033
Silane, ethenyltrimethyl [754-05-2]
VINYLTRIPHENYLPHOSPHONIUM BROMIDE, V, 1145
Phosphonium, ethenyltriphenyl-, bromide [5044-52-0]

X

XANTHONE, I, 552
9*H*-Xanthen-9-one [90-47-1]
XANTHYDROL, I, 554
9*H*-Xanthen-9-ol [90-46-0]
***o*-XYLYLENE DIBROMIDE, IV**, 984
Benzene, 1,2-bis(bromomethyl)- [91-13-4]

CUMULATIVE CONTENTS INDEX

According to Chemical Abstracts Index Names

This index has an alphabetized arrangement of the CAS Index names first with their registry numbers followed by the more common names in capital letters corresponding to the names in Name of Compound Prepared index.

A

1,2-Acenaphthalenedione [82-86-0]
    ACENAPHTHENEQUINONE, **III**, 1
5-Acenaphthylenebutanoic acid, 1,2-dihydro-γ-oxo- [16294-60-3]
    β-(3-ACENAPHTHOYL)PROPIONIC ACID, **III**, 6
1-Acenaphthylenol, 1,2-dihydro- [6306-07-6]
    ACENAPHTHENOL-7, **III**, 3
1-Acenaphthylenol, 1,2-dihydro-, acetate [14966-36-0]
    ACENAPHTHENOL ACETATE, **III**, 3
Acetaldehyde, cyclohexylidene- [1713-63-9]
    CYCLOHEXYLIDENEACETALDEHYDE, **VI**, 358
Acetaldehyde, tribromo- [115-17-3]
    BROMAL, **II**, 87
Acetamide [60-35-5]
    ACETAMIDE, **I**, 3
Acetamide, *N*-[2-(2-acetyl-4,5-dimethoxyphenyl)ethyl]- [57621-03-1]
    2-(2-ACETAMIDOETHYL)-4,5-DIMETHOXYACETOPHENONE, **VI**, 1
Acetamide, *N*-[2,4-bis(1,3-diphenyl-2-imidazolidinyl)-5-hydroxyphenyl]- [67149-22-8]
    N-[2,4-BIS(1,3-DIPHENYLIMIDAZOLIDIN-2-YL)-5-
    HYDROXYPHENYL]ACETAMIDE, **VII**, 162
Acetamide, *N*-bromo- [79-15-2]
    N-BROMOACETAMIDE, **IV**, 104
Acetamide, *N*-(2-bromo-4-methylphenyl)- [614-83-5]
    3-BROMO-4-ACETAMINOTOLUENE, **I**, 111
Acetamide, 2-chloro- [79-07-2]
    CHLOROACETAMIDE, **I**, 153
Acetamide, *N*-[4-(chloroacetyl)phenyl]- [140-49-8]
    p-CHLOROACETYLACETANILIDE, **III**, 183

Acetamide, 2-cyano- [107-91-5]
   CYANOACETAMIDE, **I**, 179
Acetamide, 2,2-dichloro- [683-72-7]
   α,α-DICHLOROACETAMIDE, **III**, 260
Acetamide, *N*-(2,4-diformyl-5-hydroxyphenyl)- [67149-23-9]
   N-(2,4-DIFORMYL-5-HYDROXYPHENYL)ACETAMIDE, **VII**, 162
Acetamide, *N*-[2-(3,4-dimethoxyphenyl)ethyl]- [6275-29-2]
   N-[2-(3,4-DIMETHOXYPHENYL)ETHYL]ACETAMIDE, **VI**, 1
Acetamide, *N*-(hydroxymethyl)- [625-51-4]
   N-(HYDROXYMETHYL)ACETAMIDE, **VI**, 5
Acetamide, *N*-hydroxy-*N*-phenyl- [1795-83-1]
   N-ACETYL-*N*-PHENYLHYDROXYLAMINE, **VIII**, 16
Acetamide, *N*-(4-methoxy-2-nitrophenyl)- [119-81-3]
   2-NITRO-4-METHOXYACETANILIDE, **III**, 661
Acetamide, *N*-[(methylamino)carbonyl]- [623-59-6]
   ACETYL METHYLUREA, **II**, 462
Acetamide, *N*-(1-methyl-2-oxopropyl)- [6628-81-5]
   3-ACETAMIDO-2-BUTANONE, **IV**, 5
Acetamide, *N*-(1-nitro-2-naphthalenyl)- [5419-82-9]
   1-NITRO-2-ACETYLAMINONAPHTHALENE, **II**, 438
Acetamide, *N*-[4-[(4-nitrophenyl)sulfonyl]phenyl]- [1775-37-7]
   p-NITRO-*p*'-ACETYLAMINODIPHENYLSULFONE, **III**, 239
Acetamide, *N*-(2-oxopropyl)- [7737-16-8]
   ACETAMIDOACETONE, **V**, 27
Acetamide, *N*-[4-(4-oxo-2-thioxo-3-thiazolidinyl)phenyl]- [53663-36-8]
   N-(*p*-ACETYLAMINOPHENYL)RHODANINE, **IV**, 6
Acetamide, 2,2,2-trichloro-*N*-(1-ethenyl-1,5-dimethyl-4-hexenyl)- [51479-78-8]
   3,7-DIMETHYL-3-TRICHLOROACETAMIDO-1,6-OCTADIENE, **VI**, 507
Acetamide, 2,2,2-trichloro-*N*-phenyl- [2563-97-5]
   α,α,α-TRICHLOROACETANILIDE, **V**, 1074
Acetamide, 2,2,2-trifluoro-*N*-phenyl- [404-24-0]
   N-TRIFLUOROACETANILIDE, **VI**, 1004
Acetanilide, 4',4'''-azo-bis- [15446-39-6]
   4,4'-BIS(ACETAMIDO)AZOBENZENE, **V**, 341
Acetic acid, 1,1-dimethylethyl ester [540-88-5]
   tert-BUTYL ACETATE, **III**, 141; **IV**, 263
Acetic acid, (aminooxy)-, hydrochloride (2:1) [2921-14-4]
   CARBOXYMETHOXYLAMINE HEMIHYDROCHLORIDE, **III**, 172
Acetic acid, aminothioxo-, ethyl ester [16982-21-1]
   ETHYL THIOAMIDOOXALATE, **VIII**, 597
Acetic acid, anhydride with formic acid [2258-42-6]
   ACETIC FORMIC ANHYDRIDE, **VI**, 8
Acetic acid, arsono- [107-38-0]
   ARSONOACETIC ACID, **I**, 73
Acetic acid, (4-arsonophenoxy)- [53663-15-3]
   p-ARSONOPHENOXYACETIC ACID, **I**, 75

Acetic acid, azido-, 1,1-dimethylethyl ester [6367-36-8]
    tert-BUTYL AZIDOACETATE, **V**, 586
Acetic acid, bromo- [79-08-3]
    BROMOACETIC ACID, **III**, 381
Acetic acid, bromo, ethyl ester [105-36-2]
    ETHYL BROMOACETATE, **III**, 381
Acetic acid, 2-butyryl, ethyl ester [7152-15-0]
    ETHYL 2-BUTYRYLACETATE, **VII**, 213
Acetic acid, 2,2'-[carbonothioylbis(thio)]bis- [6326-83-6]
    TRITHIOCARBODIGLYCOLIC ACID, **IV**, 967
Acetic acid, chloro-, 1,1-dimethylethyl ester [107-59-5]
    tert-BUTYL CHLOROACETATE, **III**, 144
Acetic acid, chloro, ethenyl ester [2549-51-1]
    VINYL CHLOROACETATE, **III**, 853
Acetic acid, chloro-, ethyl ester [105-39-5]
    ETHYL CHLOROACETATE, **II**, 263
Acetic acid, chloro-, 5-methyl-2-(1-methyl-1-phenylethyl)cyclohexyl ester, [1*R*-
    (1α,2β,5α)]- [71804-27-8]
    (1*R*,2*S*,5*R*)-5-METHYL-2-[(1-METHYL-1-PHENYLETHYL)]CYCLOHEXYL
    CHLOROACETATE, **VIII**, 522
Acetic acid, chloro-, 2-phenylcyclohexyl ester, *trans*-(±)- [121906-81-8]
    trans-2-PHENYLCYCLOHEXYL CHLOROACETATE (±), (+)-, (–), **VIII**, 516
Acetic acid, chlorocyclopropylidene methyl ester [82979-45-1]
    METHYL 2-CHLORO-2-CYCLOPROPYLIDENACETATE, **VIII**, 373, 374
Acetic acid, chlorofluoro-, ethyl ester [401-56-9]
    ETHYL CHLOROFLUOROACETATE, **IV**, 423
Acetic acid, cyano, 1,1-dimethylethyl ester [1116-98-9]
    tert-BUTYL CYANOACETATE, **V**, 171
Acetic acid, cyano-, ethyl ester [105-56-6]
    ETHYL CYANOACETATE, **I**, 254
Acetic acid, cyanocyclohexylidene- [37107-50-9]
    CYCLOHEXYLIDENECYANOACETIC ACID, **IV**, 234
Acetic acid, cyclohexylidene, ethyl ester [1552-92-7]
    ETHYL CYCLOHEXYLIDENEACETATE, **V**, 547
Acetic acid, 2,2'-(1,2-diarsenediyl)bis- [544-27-4]
    ARSENOACETIC ACID, **I**, 73
Acetic acid, diazo-, 2-butenyl ester, (*E*)- [14746-03-3]
    CROTYL DIAZOACETATE, **V**, 258
Acetic acid, diazo-, 1,1-dimethylethyl ester [35059-50-8]
    tert-BUTYL DIAZOACETATE, **V**, 179
Acetic acid, diazo, ethyl ester [623-73-4]
    ETHYL DIAZOACETATE, **III**, 392; **IV**, 424
Acetic acid, dichloro- [79-43-6]
    DICHLOROACETIC ACID, **II**, 181
Acetic acid, diethoxy, ethyl ester [6065-82-3]
    ETHYL DIETHOXYACETATE, **IV**, 427

Acetic acid, ethoxy- [627-03-2]
ETHOXYACETIC ACID, **II**, 260
Acetic acid, ethoxy, ethyl ester [817-95-8]
ETHYL ETHOXYACETATE, **II**, 260
Acetic acid, 2-formylphenoxy- [6280-80-4]
o-FORMYLPHENOXYACETIC ACID, **V**, 251
Acetic acid, (hexahydro-2*H*-azepin-2-ylidene)-, ethyl ester, (Z)- [70912-51-5]
ETHYL α-(HEXAHYDROAZEPINYLIDENE-2)ACETATE, **VIII**, 263
Acetic acid, hydrazinoimino-, ethyl ester [53085-26-0]
ETHYL OXALAMIDRAZONATE, **VIII**, 597
Acetic acid, isocyano-, ethyl ester [2999-46-4]
ETHYL ISOCYANOACETATE, **VI**, 620
Acetic acid, [[5-methyl-2-(1-methylethyl)cyclohexyl]oxy]-, [1*R*-(1α,2β,5α)]-
[40248-63-3]
(−)-MENTHOXYACETIC ACID, **III**, 544
Acetic acid, [[(1-methylethylidene)amino]oxy]- [5382-89-8]
ACETONE CARBOXYMETHOXIME, **III**, 172
Acetic acid, nitro-, methyl ester [2483-57-0]
METHYL NITROACETATE, **VI**, 797
Acetic acid, oxo, butyl ester [6295-06-3]
BUTYL GLYOXYLATE, **IV**, 124
Acetic acid, [(phenylthioxomethyl)thio]- [942-91-6]
THIOBENZOYLTHIOGLYCOLIC ACID, **V**, 1046
Acetic acid, trifluoro, anhydride with trifluoromethanesulfonic acid [68602-57-3]
TRIFLUOROACETYL TRIFLATE, **VII**, 506
Acetic acid, trifluoro-, (Z)-4-chloro-4-hexenyl ester [28077-77-2]
(Z)-4-CHLORO-4-HEXENYL TRIFLUOROACETATE, **VI**, 273
Acetic acid, (trimethylsilyl)-, ethyl ester [4071-88-9]
ETHYL TRIMETHYLSILYLACETATE, **VII**, 512
Acetonitrile, amino-, sulfate [151-63-3]
AMINOACETONITRILE HYDROGEN SULFATE, **I**, 298
Acetonitrile, chloro- [107-14-2]
CHLOROACETONITRILE, **IV**, 144
Acetonitrile, cyclohexylidene- [4435-18-1]
CYCLOHEXYLIDENEACETONITRILE, **VII**, 108
Acetonitrile, dibromo- [3252-43-5]
DIBROMOACETONITRILE, **IV**, 254
Acetonitrile, (diethylamino)- [3010-02-4]
DIETHYLAMINOACETONITRILE, **III**, 275
Acetonitrile, hydroxy- [107-16-4]
GLYCOLONITRILE, **III**, 436
Acetonitrile, methoxy- [1738-36-9]
METHOXYACETONITRILE, **II**, 387
Acetonitrile, (methyleneamino)- [109-82-0]
METHYLENEAMINOACETONITRILE, **I**, 355

Acetyl chloride, [[5-methyl-2-(1-methylethyl)cyclohexyl]oxy]-, [1*R*-(1α,2β,5α)]- [15356-62-4]

    L-MENTHOXYACETYL CHLORIDE, **III**, 547

Acetyl isocyanate, chloro- [4461-30-7]

    α-CHLOROACETYL ISOCYANATE, **V**, 204

9-Acridinamine [90-45-9]

    9-AMINOACRIDINE, **III**, 53

Acridine, 9-butyl-1,2,3,4,5,6,7,8-octahydro- [99922-90-4]

    9-BUTYL-1,2,3,4,5,6,7,8-OCTAHYDROACRIDINE, **VIII**, 87

Acridine, 9-butyl-1,2,3,4,5,6,7,8-octahydro-, 10-oxide [136528-61-5]

    9-BUTYL-1,2,3,4,5,6,7,8-OCTAHYDROACRIDINE *N*-OXIDE, **VIII**, 87

Acridine, 9-chloro- [1207-69-8]

    9-CHLOROACRIDINE, **III**, 53

4-Acridinol, 9-butyl-1,2,3,4,5,6,7,8-octahydro- [99922-91-5]

    9-BUTYL-1,2,3,4,5,6,7,8-OCTAHYDROACRIDIN-4-OL, **VIII**, 87

9(10*H*)-Acridinone [578-95-0]

    ACRIDONE, **II**, 15

β-Alanine [107-95-9]

    β-ALANINE, **II**, 19; **III**, 34

DL-Alanine [302-72-7]

    DL-ALANINE, **I**, 21

β-Alanine, *N*-(3-ethoxy-3-oxopropyl)-*N*-methyl-, ethyl ester [6315-60-2]

    DI-β-CARBETHOXYETHYLMETHYLAMINE, **III**, 258

Alanine, 2-methyl- [62-57-7]

    α-AMINOISOBUTYRIC ACID, **II**, 29

L-Alanine, 3-sulfo-, hydrate [53643-49-5]

    CYSTEIC ACID MONOHYDRATE, **III**, 226

Aluminum, (cyano-C)diethyl- [5804-85-3]

    DIETHYLALUMINUM CYANIDE, **VI**, 436

Aluminum, 1-decenylbis(2-methylpropyl)-, (*E*)- [107441-86-1]

    (*E*)-1-DECENYLDIISOBUTYLALANE, **VIII**, 295

Aluminum, dimethyl-(2-methyl-1,3-butadienyl)- [96160-49-5]

    (*E*)-2-(METHYL-1,3-BUTADIENYL)DIMETHYLALANE, **VII**, 245

Aluminum, triphenyl- [841-76-9]

    TRIPHENYLALUMINUM, **V**, 1116

Androstan-17-ol, (5α,17β)- [1225-43-0]

    ANDROSTAN-17-OL, (5α,17β)-, **VI**, 62

Androstan-3-one, 17-(1,1-dimethylethoxy)-, (5α,17β)- [87004-41-9]

    17β-*tert*-BUTOXY-5α-ANDROSTAN-3-ONE, **VII**, 66

Androst-2-ene, 17-(1,1-dimethylethoxy)-, (5α,17β)- [87004-43-1]

    17β-*tert*-BUTOXY-5α-ANDROST-2-ENE, **VII** , 66

Androst-5-ene-17-carboxylic acid, 3-(acetyloxy)-, (3β,17β)- [7150-18-7]

    3β-ACETOXYETIENIC ACID, **V**, 8

Androst-5-ene-16,17-dione, 3-hydroxy-, 16-oxime, (3β)- [21242-37-5]

    16-OXIMINOANDROST-5-EN-3β-OL-17-ONE, **VI**, 840

Androst-5-en-17-one, 16-diazo-3-hydroxy-, (3β)- [26003-42-9]
   16-DIAZOANDROST-5-EN-3β-OL-17-ONE, **VI**, 840
Anthracene, 9-chloro- [716-53-0]
   9-CHLOROANTHRACENE, **V**, 206
Anthracene, 9,10-dibromo- [523-27-3]
   9,10-DIBROMOANTHRACENE, **I**, 207
Anthracene, 9,10-dihydro- [613-31-0]
   9,10-DIHYDROANTHRACENE, **V**, 398
Anthracene, 9-nitro- [602-60-8]
   9-NITROANTHRACENE, **IV**, 711
9-Anthracenecarboxaldehyde [642-31-9]
   9-ANTHRALDEHYDE, **III**, 98
9,10-Anthracenedione [84-65-1]
   ANTHRAQUINONE, **II**, 554
9,10-Anthracenedione, 1-bromo-4-(methylamino)- [128-93-8]
   1-METHYLAMINO-4-BROMOANTHRAQUINONE, **III**, 575
9,10-Anthracenedione, 1-chloro- [82-44-0]
   α-CHLOROANTHRAQUINONE, **II**, 128
9,10-Anthracenedione, 1,4-dihydroxy- [81-64-1]
   QUINIZARIN, **I**, 476
9,10-Anthracenedione, 2,3-dimethyl- [6531-35-7]
   2,3-DIMETHYLANTHRAQUINONE, **III**, 310
9,10-Anthracenedione, 2-methyl- [84-54-8]
   β-METHYLANTHRAQUINONE, **I**, 353
9,10-Anthracenedione, 1-(methylamino)- [82-38-2]
   1-METHYLAMINOANTHRAQUINONE, **III**, 573
9,10-Anthracenedione, 1,4,4a,9a-tetrahydro-2,3-dimethyl-, *cis*- [55538-11-9]
   2,3-DIMETHYL-1,4,5,6-TETRAHYDROANTHRAQUINONE, **III**, 310
1-Anthracenesulfonic acid, 9,10-dihydro-9,10-dioxo-, potassium salt [30845-78-4]
   POTASSIUM ANTHRAQUINONE-α-SULFONATE, **II**, 539
9(10*H*)-Anthracenone [90-44-8]
   ANTHRONE, **I**, 60
9(10*H*)-Anthracenone, 10-nitro- [6313-44-6]
   NITROANTHRONE, **I**, 390
D-Arabinose [10323-20-3]
   D-ARABINOSE, **III**, 101
L-Arabinose [5328-37-0]
   L-ARABINOSE, **I**, 67
L-Arginine, monohydrochloride [1119-34-2]
   L-ARGININE HYDROCHLORIDE, **II**, 49
Arsine, triphenyl- [603-32-7]
   TRIPHENYLARSINE, **IV**, 910
Arsine, oxide, triphenyl- [1153-05-5]
   TRIPHENYLARSINE OXIDE, **IV**, 911
Arsonic acid, [4-[(2-amino-2-oxoethyl)amino]phenyl]-, disodium salt [834-03-7]
   SODIUM *p*-ARSONO-*N*-PHENYLGLYCINAMIDE, **I**, 488

Arsonic acid, (4-aminophenyl)- [98-50-0]
 ARSANILIC ACID, **I**, 70
Arsonic acid, (4-hydroxyphenyl)-, sodium salt [53663-20-0]
 SODIUM *p*-HYDROXYPHENYLARSONATE, **I**, 490
Arsonic acid, (4-nitrophenyl)- [98-72-6]
 p-NITROPHENYLARSONIC ACID, **III**, 665
Arsonic acid, phenyl- [98-05-5]
 PHENYLARSONIC ACID, **II**, 494
Arsonium, tetraphenyl-, (hydrogen dichloride) [21006-73-5]
 TETRAPHENYLARSONIUM CHLORIDE HYDROCHLORIDE, **IV**, 910
DL-Aspartic acid [617-45-8]
 DL-ASPARTIC ACID, **IV**, 55
2-Azabicyclo[2.2.1]hept-5-ene, 2-(phenylmethyl)- [112375-05-0]
 N-BENZYL-2-AZANORBORNENE, **VIII**, 31
9-Azabicyclo[3.3.1]nonan-3-one, 9-methyl- [552-70-5]
 PSEUDOPELLETIERINE, **IV**, 816
1-Azabicyclo[2.2.2]octan-3-one, hydrochloride [1193-65-3]
 3-QUINUCLIDONE HYDROCHLORIDE, **V**, 989
2-Azabicyclo[2.2.2]octan-3-one [3306-69-2]
 3-ISOQUINUCLIDONE, **V**, 670
1*H*-Azepine, hexahydro-2-propyl-, (±)- [85028-29-1]
 2-PROPYL-1-AZACYCLOHEPTANE, **VIII**, 568
2*H*-Azepine, 3,4,5,6-tetrahydro-7-methoxy- [2525-16-8]
 O-METHYLCAPROLACTIM, **IV**, 588
2*H*-Azepin-2-one, 1,3-dihydro-3,5,7-trimethyl- [936-85-6]
 1,3-DIHYDRO-3,5,7-TRIMETHYL-2*H*-AZEPIN-2-ONE, **V**, 408
2*H*-Azepin-2-one, hexahydro- [105-60-2]
 2-KETOHEXAMETHYLENIMINE, **II**, 371
Azetidine [503-29-7]
 AZETIDINE, **VI**, 75
1-Azetidinepropanoic acid, ethyl ester [7730-42-9]
 ETHYL 3-(1-AZETIDINYL)PROPANOATE, **VI**, 75
Azetidinesulfonyl chloride, 2,2-dimethyl-4-oxo [17174-96-8]
 β-ISOVALEROLACTAM-*N*-SULFONYL CHLORIDE, **V**, 673
2-Azetidinone, 4-(acetyloxy)- [28562-53-0]
 4-ACETOXYAZETIDIN-2-ONE, **VIII**, 3
2-Azetidinone, 4,4-dimethyl- [4879-95-2]
 β-ISOVALEROLACTAM, **V**, 673
2-Azetidinone, 3-methoxy-1,3-dimethyl-4-phenyl- [82918-98-7]
 1,3-DIMETHYL-3-METHOXY-4-PHENYLAZETIDINONE, **VIII**, 216
1-Aziridinamine, 2,3-diphenyl-, *trans*- [28161-60-6]
 trans-1-AMINO-2,3-DIPHENYLAZIRIDINE, **VI**, 56, 679
1-Aziridinamine, 2-phenyl-, (±)- [19615-20-4]
 1-AMINO-2-PHENYLAZIRIDINE, **VI**, 56
1-Aziridinamine, 2-phenyl-, (±)-, monoacetate [37079-43-9]
 1-AMINO-2-PHENYLAZIRIDINIUM ACETATE, **VI**, 56

Aziridine [151-56-4]
 ETHYLENIMINE, **IV**, 433
Aziridine, 2,2-dimethyl- [2658-24-4]
 2,2-DIMETHYLETHYLENIMINE, **III**, 148
Aziridine, 1-ethyl-2-methylene- [872-39-9]
 N-ETHYLALLENIMINE, **V**, 541
Aziridine, 2-phenyl-3-(phenylmethyl)-, *cis*- [1605-08-9]
 cis-2-BENZYL-3-PHENYLAZIRIDINE, **V**, 83
2*H*-Azirine-2-carboxaldehyde, 3-phenyl- [42970-55-8]
 3-PHENYL-2*H*-AZIRINE-2-CARBOXALDEHYDE, **VI**, 893
2*H*-Azirine, 2-(dimethoxymethyl)-3-phenyl- [56900-68-6]
 2-(DIMETHOXYMETHYL)-3-PHENYL-2*H*-AZIRINE, **VI**, 893
2(1*H*)-Azocinone, hexahydro- [673-66-5]
 HEXAHYDRO-2-(1*H*)-AZOCINONE, **VII**, 254
Azulene [275-51-4]
 AZULENE, **VII**, 15
Azulene, 4,6,8-trimethyl- [941-81-1]
 4,6,8-TRIMETHYLAZULENE, **V**, 1088
1(2*H*)-Azulenone, 3,4-dihydro- [52487-41-9]
 3,4-DIHYDRO-1(2*H*)-AZULENONE, **VIII**, 196

B

Benzaldehyde, (phenylmethylene), hydrazone, (*E,E*)- [28867-76-7]
 BENZALAZINE, **II**, 395
Benzaldehyde, 2-amino [529-23-7]
 o-AMINOBENZALDEHYDE, **III**, 56
Benzaldehyde, 4-amino- [556-18-3]
 p-AMINOBENZALDEHYDE, **IV**, 31
Benzaldehyde, 3-bromo- [3132-99-8]
 m-BROMOBENZALDEHYDE, **II**,132
Benzaldehyde, 4-bromo- [1122-91-4]
 p-BROMOBENZALDEHYDE, **II**, 89, 442
Benzaldehyde, 2-bromo-4-methyl- [824-54-4]
 2-BROMO-4-METHYLBENZALDEHYDE, **V**, 139
Benzaldehyde, 3-chloro- [587-04-2]
 m-CHLOROBENZALDEHYDE, **II**, 130
Benzaldehyde, 4-chloro- [104-88-1]
 p-CHLOROBENZALDEHYDE, **II**, 133
Benzaldehyde, 3,4-dihydroxy- [139-85-5]
 PROTOCATECHUALDEHYDE, **II**, 549
Benzaldehyde, 3,4-dimethoxy- [120-14-9]
 VERATRALDEHYDE, **II**, 619
Benzaldehyde, 4,5-dimethoxy-2-nitro- [20357-25-9]
 6-NITROVERATRALDEHYDE, **IV**, 735
Benzaldehyde, 4-(dimethylamino)- [100-10-7]
 p-DIMETHYLAMINOBENZALDEHYDE, **I**, 214; **IV**, 331

Benzaldehyde, 2,4-dinitro- [528-75-6]
  2,4-DINITROBENZALDEHYDE, **II**, 223
Benzaldehyde, 3,5-dinitro- [14193-18-1]
  3,5-DINITROBENZALDEHYDE, **VI**, 529
Benzaldehyde, 4-ethoxy-3-hydroxy- [2539-53-9]
  4-ETHOXY-3-HYDROXYBENZALDEHYDE, **VI**, 567
Benzaldehyde, 3-hydroxy- [100-83-4]
  m-HYDROXYBENZALDEHYDE, **III**, 453
Benzaldehyde, 4-hydroxy-3,5-dimethoxy- [134-96-3]
  SYRINGIC ALDEHYDE, **IV**, 866
Benzaldehyde, 2-methoxy- [135-02-4]
  o-ANISALDEHYDE, **V**, 46; **VI**, 64
Benzaldehyde, 3-methoxy- [591-31-1]
  m-METHOXYBENZALDEHYDE, **III**, 564
Benzaldehyde, 2-methyl- [529-20-4]
  o-TOLUALDEHYDE, **III**, 818; **IV**, 932
Benzaldehyde, 4-methyl- [104-87-0]
  p-TOLUALDEHYDE, **II**, 583
Benzaldehyde, 2-nitro- [552-89-6]
  o-NITROBENZALDEHYDE, **III**, 641; **V**, 825
Benzaldehyde, 3-nitro- [99-61-6]
  m-NITROBENZALDEHYDE, **III**, 644; *Warning*, **III**, 645
Benzaldehyde, 4-nitro- [555-16-8]
  p-NITROBENZALDEHYDE, **II**, 441
Benzaldehyde, 3,4,5-trimethoxy- [86-81-7]
  3,4,5-TRIMETHOXYBENZALDEHYDE, **VI**, 1007
Benzaldehyde, 2,4,6-trimethyl- [487-68-3]
  MESITALDEHYDE, **III**, 549; **V**, 49
Benzamide, 3,4-dimethoxy- [1521-41-1]
  VERATRIC AMIDE, **II**, 44
Benzamide, *N*-hydroxy- [495-18-1]
  BENZOHYDROXAMIC ACID, **II**, 67
Benzamide, 2-hydroxy-*N*-(2-methylphenyl)- [7133-56-4]
  SALICYL-*o*-TOLUIDE, **III**, 765
Benzamide, 2-methyl- [527-85-5]
  o-TOLUAMIDE, **II**, 586; *Warning*, **V**, 1054
Benzamide, *N*-phenyl- [93-98-1]
  BENZANILIDE, **I**, 82
Benzamide, *N*-(2-phenylethyl)- [3278-14-6]
  N-(2-PHENYLETHYL)BENZAMIDE, **V**, 336
Benz[*a*]anthracene [56-55-3]
  BENZ[*a*]ANTHRACENE, **VII**, 18
7*H*-Benz[*de*]anthracen-7-one [82-05-31]
  BENZANTHRONE, **II**, 62
Benzenamine, 4,4'-azobis- [538-41-0]
  4,4'-DIAMINOAZOBENZENE, **V**, 341

Benzenamine, 4-bromo-*N,N*-dimethyl-3-(trifluoromethyl)- [51332-24-2]
 4-BROMO-*N,N*-DIMETHYL-3-(TRIFLUOROMETHYL)ANILINE, **VI**, 181
Benzenamine, 2-bromo-4-methyl- [583-68-6]
 3-BROMO-4-AMINOTOLUENE, **I**, 111
Benzenamine, 4-chloro-*N*-ethyl- [13519-75-0]
 N-ETHYL-*p*-CHLOROANILINE, **IV**, 420
Benzenamine, 2,6-dibromo- [608-30-0]
 2,6-DIBROMOANILINE, **III**, 262
Benzenamine, 2,6-dichloro- [608-31-1]
 2,6-DICHLOROANILINE, **III**, 262
Benzenamine, 4-(2,2-dichloroethenyl)-*N,N*-dimethyl- [6798-58-9]
 β,β-DICHLORO-*p*-DIMETHYLAMINOSTYRENE, **V**, 361
Benzenamine, *N,N*-diethyl-4-nitroso- [120-22-9]
Benzenamine, *N,N*-diethyl-4-nitroso-, hydrochloride [58066-98-1]
 p-NITROSODIETHYLANILINE and HYDROCHLORIDE, **II**, 224
Benzenamine, 2,6-diiodo-4-nitro- [5398-27-6]
 2,6-DIIODO-*p*-NITROANILINE, **II**, 196
Benzenamine, 3,4-dimethoxy- [6315-89-5]
 4-AMINOVERATROLE, **II**, 44
Benzenamine, 3-(dimethoxymethyl)- [53663-37-9]
 m-AMINOBENZALDEHYDE DIMETHYLACETAL, **III**, 59
Benzenamine, 3,4-dimethyl- [95-64-7]
 3,4-DIMETHYLANILINE, **III**, 307
Benzenamine, *N,N*-dimethyl-3-nitro- [619-31-8]
 m-NITRODIMETHYLANILINE, **III**, 658
Benzenamine, *N,N*-dimethyl-4-nitroso-, monohydrochloride [42344-05-8]
 p-NITROSODIMETHYLANILINE HYDROCHLORIDE, **I**, 410; **II**, 223
Benzenamine, *N,N*-dimethyl-2-(2-phenylethenyl)-, (*Z*)- [70197-43-2]
 β-[2-(*N,N*-DIMETHYLAMINO)PHENYL]STYRENE, (*Z*)-, **VII**, 172
Benzenamine, *N,N*-dimethyl-3-(trifluoromethyl)- [329-00-0]
 m-TRIFLUOROMETHYL-*N,N*-DIMETHYLANILINE, **V**, 1085
Benzenamine, 2,4-dinitro- [97-02-9]
 2,4-DINITROANILINE, **II**, 221
Benzenamine, 2,6-dinitro- [606-22-4]
 2,6-DINITROANILINE, **IV**, 364
Benzenamine, *N,N*-diphenyl- [603-34-9]
 TRIPHENYLAMINE, **I**, 544
Benzenamine, 4,4'-dithiobis- [722-27-0]
 p-AMINOPHENYL DISULFIDE, **III**, 86
Benzenamine, *N*-ethyl-3-methyl- [102-27-2]
 N-ETHYL-*m*-TOLUIDINE, **II**, 290
Benzenamine, *N*-hydroxy- [100-65-2]
 β-PHENYLHYDROXYLAMINE, **I**, 445; **VIII**, 16
Benzenamine, *N*-hydroxy-*N*-nitroso-, ammonium salt [135-20-6]
 CUPFERRON, **I**, 177
Benzenamine, 4-iodo- [540-37-4]
 p-IODOANILINE, **II**, 347

Benzenamine, *N,N'*-methanetetraylbis- [622-16-2]
 DIPHENYLCARBODIIMIDE (METHODS I and II), **V**, 501, 504
Benzenamine, 4-methoxy-2-nitro- [96-96-8]
 2-NITRO-4-METHOXYANILINE, **III**, 66
Benzenamine, 4-methoxy-*N*-[4-(trimethylsilyl)-3-butenyl]-, (Z)- [133157-30-9]
 N-(4-METHOXYPHENYL)-(Z)-4-(TRIMETHYLSILYL)-2-BUTENAMINE,
 VIII, 358
Benzenamine, 2-methyl-5-(1-methylethyl)- [2051-53-8]
 2-AMINO-*p*-CYMENE, **III**, 63
Benzenamine, 2-methyl-4-nitro- [99-52-5]
 2-AMINO-5-NITROTOLUENE, **IV**, 44
Benzenamine, 2-methyl-6-nitro- [570-24-1]
 2-AMINO-3-NITROTOLUENE, **IV**, 42
Benzenamine, *N*-methyl-*N*-nitroso- [614-00-6]
 N-NITROSOMETHYLANILINE, **II**, 460
Benzenamine, 2-nitro- [88-74-4]
 o-NITROANILINE, **I**, 388
Benzenamine, *N*-[(2-nitrophenyl)methylene]- [17064-77-6]
 o-NITROBENZALANILINE, **V**, 941
Benzenamine, *N*-(phenylmethylene)- [538-51-2]
 BENZALANILINE, **I**, 80
Benzenamine, *N*-(phenylmethylene)-*N*-oxide [1137-96-8]
 N,α-DIPHENYLNITRONE, **V**, 1124
Benzenamine, *N*-sulfinyl- [1122-83-4]
 N-SULFINYLANILINE, **V**, 504
Benzenamine, 4,4'-sulfonylbis- [80-08-0]
 4,4'-DIAMINODIPHENYLSULFONE, **III**, 239
Benzenamine, 4-(triphenylmethyl)- [22948-06-7]
 p-AMINOTETRAPHENYLMETHANE, **IV**, 47
Benzenaminium, *N,N,N*-trimethyl-, (tribromide) [4207-56-1]
 PHENYLTRIMETHYLAMMONIUM TRIBROMIDE, **VI**, 175
Benzene [71-43-2]
 REDUCTION OF ORGANIC HALIDES, CHLOROBENZENE TO BENZENE, **V**, 998
Benzene, azido- [622-37-7]
 PHENYL AZIDE, **III**, 710
Benzene, (1-azido-2-iodo-3,3-dimethoxypropyl)- [56900-66-4]
 (1-AZIDO-2-IODO-3,3-DIMETHOXYPROPYL)BENZENE, **VI**, 893
Benzene, 1-azido-2-nitro- [1516-58-1]
 o-NITROPHENYL AZIDE, **IV**, 75
Benzene, 1,2-bis(bromomethyl)- [91-13-4]
 o-XYLYLENE DIBROMIDE, **IV**, 984
Benzene, 1,2-bis(butylthio)- [53663-38-0]
 1,2-BIS(BUTYLTHIO)BENZENE, **V**, 107
Benzene, 1,2-bis(dibromomethyl)- [13209-15-9]
 α,α,α',α'-TETRABROMO-*o*-XYLENE, **IV**, 807
Benzene, 1,4-bis(dibromomethyl)- [1592-31-0]
 α,α,α',α'-TETRABROMO-*p*-XYLENE, **III**, 788

Benzene, 1,4-bis(4-phenyl-1,3-butadienyl)- [10162-88-6]
1,4-BIS(4-PHENYLBUTADIENYL)BENZENE, V, 985
Benzene, 1-bromo-2-chloro- [694-80-4]
o-CHLOROBROMOBENZENE, III, 185
Benzene, 4-bromo-1,2-dimethyl- [583-71-1]
4-BROMO-o-XYLENE, III, 138
Benzene, (2-bromoethoxy)- [589-10-6]
PHENOXYETHYL BROMIDE, I, 436
Benzene, [(2-bromoethyl)thio]- [4837-01-8]
1-PHENYLTHIO-2-BROMOETHANE (in solution), VII, 453
Benzene, (bromoethynyl)- [932-87-6]
PHENYLBROMOETHYNE, V, 921
Benzene, 1-bromo-2-fluoro- [1072-85-1]
1-BROMO-2-FLUOROBENZENE, V, 133
Benzene, 1-bromo-2-iodo- [583-55-1]
o-IODOBROMOBENZENE, V, 1120
Benzene, 1-(bromomethyl)-4-nitro- [100-11-8]
p-NITROBENZYL BROMIDE, II, 443
Benzene, 1-bromo-2-methyl- [95-46-5]
o-BROMOTOLUENE, I, 135
Benzene, 1-bromo-3-methyl- [591-17-3]
m-BROMOTOLUENE, I, 133
Benzene, 1-bromo-4-methyl- [106-38-7]
p-BROMOTOLUENE, I, 136
Benzene, 2-bromo-1-methyl-4-nitro- [7745-93-9]
2-BROMO-4-NITROTOLUENE, IV, 114
Benzene, 1-bromo-3-nitro- [585-79-5]
m-BROMONITROBENZENE, I, 123
Benzene, 1-bromo-4-(phenylmethyl)- [2116-36-1]
p-BROMODIPHENYLMETHANE, V, 130
Benzene, (3-bromo-1-propenyl)- [4392-24-9]
CINNAMYL BROMIDE, V, 249
Benzene, (3-bromopropoxy)- [588-63-6]
γ-PHENOXYPROPYL BROMIDE, I, 435
Benzene, 1-bromo-4-(2,2,2-trichloro-1-phenylethyl)- [39211-93-3]
1-p-BROMOPHENYL-1-PHENYL-2,2,2-TRICHLOROETHANE, V, 131
Benzene, 2-bromo-1,3,5-trimethyl- [576-83-0]
BROMOMESITYLENE, II, 95
Benzene, 1,1'-(1,3-butadiene-1,4-diyl)bis- [886-65-7]
1,4-DIPHENYL-1,3-BUTADIENE, II, 229; V, 499
Benzene, 1,3-butadienyl-, (E)- [16939-57-4]
trans-1-PHENYL-1,3-BUTADIENE, IV, 771
Benzene, 1-(1,3-butadienyl)-4-nitro- [20264-89-5]
1-(p-NITROPHENYL)-1,3-BUTADIENE, IV, 727
Benzene, 1,1'-(1,3-butadiyne-1,4-diyl)bis- [886-66-8]
DIPHENYLDIACETYLENE, V, 517

Benzene, 1-butoxy-2-nitro- [7252-51-9]
  o-BUTOXYNITROBENZENE, **III**, 140
Benzene, butyl- [104-51-8]
  BUTYLBENZENE, **III**, 157
Benzene, 1-chloro-4-(chloromethoxy)- [21151-56-4]
  p-CHLOROPHENOXYMETHYL CHLORIDE, **V**, 221
Benzene, (2-chloro-1,1-dimethylethyl)- [515-40-2]
  NEOPHYL CHLORIDE, **IV**, 702
Benzene, 2-chloro-1,3-dinitro- [606-21-3]
  1-CHLORO-2,6-DINITROBENZENE, **IV**, 160
Benzene, 1-chloro-3-ethenyl- [2039-85-2]
  m-CHLOROSTYRENE, **III**, 204
Benzene, 1,1'-[(2-chloroethoxy)methylene]bis- [32669-06-0]
  BENZHYDRYL β-CHLOROETHYL ETHER, **IV**, 72
Benzene, 1-chloro-4-isothiocyanato- [2131-55-7]
  p-CHLOROPHENYL ISOTHIOCYANATE, **I**, 165; **V**, 223
Benzene, (chloromethoxy)methyl- [3587-60-8]
  BENZYL CHLOROMETHYL ETHER, **VI**, 101
Benzene, 1,1',1"-(chloromethylidyne)tris- [76-83-5]
  TRIPHENYLCHLOROMETHANE, **III**, 841
Benzene, 2-(chloromethyl)-1,3,5-trimethyl- [1585-16-6]
  α²-CHLOROISODURENE, **III**, 557
Benzene, 1-chloro-2-methyl- [95-49-8]
  o-CHLOROTOLUENE, **I**, 170;
Benzene, 1-chloro-4-methyl- [106-43-4]
  p-CHLOROTOLUENE, **I**, 170;
Benzene, 1,1'-(1-chloro-3-methylcyclopropane-1,2-diyl)bis- (1α,2α,3β)- [80361-87-3];
    (1α,2β,3α)- [80408-28-2]
  1-CHLORO-1,2-DIPHENYL-3-METHYLCYCLOPROPANE, **VII**, 203
Benzene, 1-chloro-3-nitro- [121-73-3]
  m-CHLORONITROBENZENE, **I**, 162
Benzene, (1,5-cyclohexadien-1-ylsulfonyl)- [102860-22-0]
  2-(PHENYLSULFONYL)-1,3-CYCLOHEXADIENE, **VIII**, 540
Benzene, cyclohexyl- [827-52-1]
  CYCLOHEXYLBENZENE, **II** 151
Benzene, (1-cyclopenten-1-ylsulfonyl)- [64740-90-5]
  1-(PHENYLSULFONYL)CYCLOPENTENE, **VIII**, 38, 543
Benzene, cyclopropyl- [873-49-4]
  CYCLOPROPYLBENZENE, **V**, 328
  PHENYLCYCLOPROPANE, **V**, 929
Benzene, 1,1'-cyclopropylidenebis- [3282-18-6]
  1,1-DIPHENYLCYCLOPROPANE, **V**, 509
Benzene, diazomethyl [766-91-6]
  PHENYLDIAZOMETHANE, **VII**, 438
Benzene, 1,1'-(diazomethylene)bis- [883-40-9]
  DIPHENYLDIAZOMETHANE, **III**, 351

Benzene, 1-[(diazomethyl)sulfonyl]-4-methyl- [1538-98-3]
  p-TOLYLSULFONYLDIAZOMETHANE, **VI**, 981
Benzene, 1,1'-(2,2-dibromocyclopropylidene)bis [17343-74-7]
  1,1-DIBROMO-2,2-DIPHENYLCYCLOPROPANE, **VI**, 187
Benzene, 1,1'-(1,2-dibromoethylidene)bis- [40957-21-9]
  STILBENE DIBROMIDE, **III**, 350
Benzene, (1,3-dibromopropyl)- [17714-42-0]
  1,3-DIBROMO-1-PHENYLPROPANE, **V**, 328
Benzene, 1,2-dibutyl- [17171-73-2]
  1,2-DIBUTYLBENZENE, **VI**, 407
Benzene, (2,2-dichlorocyclopropyl)- [2415-80-7]
  1,1-DICHLORO-2-PHENYLCYCLOPROPANE, **VII**, 12
Benzene, 1,5-dichloro-2,4-dinitro- [3698-83-7]
  1,5-DICHLORO-2,4-DINITROBENZENE, **V**, 1067
Benzene, 1,1'-(dichloroethenylidene)bis[4-chloro- [72-55-9]
  1,1-DI-(p-CHLOROPHENYL)-2,2-DICHLOROETHYLENE, **III**, 270
Benzene, 2,4-dichloro-1-methoxy- [553-82-2]
  2,4-DICHLOROMETHOXYBENZENE, **VIII**, 167
Benzene, 1,3-dichloro-2-nitro- [601-88-7]
  2,6-DICHLORONITROBENZENE, **V**, 367
Benzene, [1-(diethoxymethyl)ethenyl]- [80234-04-4]
  ATROPALDEHYDE DIETHYL ACETAL, **VII**, 13
Benzene, (3,3-diethoxy-1-propynyl)- [6142-95-6]
  PHENYLPROPARGYLALDEHYDE DIETHYL ACETAL, **IV**, 801
Benzene, (2,2-difluoroethenyl)- [405-42-5]
  β,β-DIFLUOROSTYRENE, **V**, 390
Benzene, (difluoromethyl)- [455-31-2]
  α,α-DIFLUOROTOLUENE, **V**, 396
Benzene, [2,2-difluoro-1-(trifluoromethyl)ethenyl]- [1979-51-7]
  2-PHENYLPERFLUOROPROPENE, **V**, 949
Benzene, 1,3-dimethoxy-5-methyl- [4179-19-5]
  ORCINOL DIMETHYL ETHER, **VI**, 859
Benzene, 1-(dimethoxymethyl)-3-nitro- [3395-79-7]
  m-NITROBENZALDEHYDE DIMETHYLACETAL, **III**, 644
Benzene, (1,1-dimethylethoxy)- [6669-13-2]
  PHENYL tert-BUTYL ETHER (METHOD I), **V**, 924; (METHOD II), **V**, 926
Benzene, [(2,2-dimethylpropyl)thio]- [7210-80-2]
  NEOPENTYL PHENYL SULFIDE, **VI**, 833
Benzene, 1,2-dinitro- [528-29-0]
  o-DINITROBENZENE, **II**, 226
Benzene, 1,4-dinitro- [100-25-4]
  p-DINITROBENZENE, **II**, 225
Benzene, 1,1'-(1,1-ethanediyl)bis- [612-00-0]
  1,1-DIPHENYLETHANE, **VI**, 537
Benzene, 1,1'-(1,2-ethanediyl)bis[4-bromo- [19829-56-2]
  p,p'-DIBROMOBIBENZYL, **IV**, 257

Benzene, 1,1'-(1,2-ethanediyl)bis[4-nitro- [736-30-1]
  p,p'-DINITROBIBENZYL, **IV**, 367
Benzene, 1,1'-(1,2-ethenediyl)bis-, (*E*)- [103-30-0]
  trans-STILBENE, **III**, 786
Benzene, 1,1'-(1,2-ethenediyl)bis-, (*Z*)- [645-49-8]
  cis-STILBENE, **IV**, 857
Benzene, 1,1'-(1,2-ethenediyl)bis[4-methoxy-, (*E*)- [15638-14-9]
  trans-4,4'-DIMETHOXYSTILBENE, **V**, 428
Benzene, 1,1',1'',1'''-(1,2-ethenediylidene)tetrakis- [632-51-9]
  TETRAPHENYLETHYLENE, **IV**, 914
Benzene, ethenyl- [100-42-5]
  PHENYLETHYLENE (STYRENE), **I**, 440
Benzene, 1,1'-ethenylidenebis- [530-48-3]
  1,1-DIPHENYLETHYLENE, **I**, 226
Benzene, 1-ethenyl-2-nitro- [586-39-0]
  m-NITROSTYRENE, **IV**, 731
Benzene, 1,1'-[(ethenyloxy)methylene]bis- [23084-88-0]
  DIPHENYLMETHYL VINYL ETHER, **VI**, 552
Benzene, (ethenylsulfinyl)- [20451-53-0]
  PHENYL VINYL SULFOXIDE, **VII**, 453
Benzene, (ethenylsulfonyl)- [5535-48-8]
  PHENYL VINYL SULFONE **VII**, 453
Benzene, (ethenylthio)- [1822-73-7]
  PHENYL VINYL SULFIDE, **VII**, 453
Benzene, 1,1',1''-(1-ethenyl-2-ylidene)tris- [58-72-0]
  TRIPHENYLETHYLENE, **II**, 606
Benzene, 1-(2-ethoxycyclopropyl)-4-methyl-, *cis*- [40237-67-0]
  cis-1-ETHOXY-2-*p*-TOLYLCYCLOPROPANE, **VI**, 571
Benzene, 1-(2-ethoxycyclopropyl)-4-methyl-, *trans*- [40489-59-6]
  trans-1-ETHOXY-2-*p*-TOLYLCYCLOPROPANE, **VI**, 571
Benzene, 1,1'-ethylidenebis[4-methyl-] [530-45-0]
  1,1-DI-*p*-TOLYLETHANE, **I**, 229
Benzene, 1-ethyl-3-nitro- [7369-50-8]
  m-NITROETHYLBENZENE, **VII**, 393
Benzene, 1,1'-(1,2-ethynediyl)bis- [501-65-5]
  DIPHENYLACETYLENE, **III**, 350; **IV**, 377
Benzene, ethynyl- [536-74-3]
  PHENYLACETYLENE, **I**, 438; **IV**, 763
Benzene, 1-(ethynylsulfonyl)-4-methyl- [13894-21-8]
  ETHYNYL *p*-TOLYL SULFONE, **VIII**, 281
Benzene, fluoro- [462-06-6]
  FLUOROBENZENE, **II**, 295
Benzene, 1-(fluoromethyl)-4-nitro- [500-11-8]
  p-NITROBENZYL FLUORIDE, **VI**, 835
Benzene, hexamethyl- [87-85-4]
  HEXAMETHYLBENZENE, **II**, 248; **IV**, 520

Benzene, iodo- [591-50-4]
  IODOBENZENE, **I**, 323; **II**, 351
Benzene, (4-iodo-2-butenyl)-, (*E*)- [52534-83-5]
  (*E*)-1-IODO-4-PHENYL-2-BUTENE, **VI**, 704
Benzene, 4-iodo-1,2-dimethoxy- [5460-32-2]
  4-IODOVERATROLE, **IV**, 547
Benzene, 2-iodo-1,4-dimethyl- [1122-42-5]
  2-IODO-*p*-XYLENE, **VI**, 709
Benzene, 1-iodo-2,4-dinitro- [709-49-9]
  2,4-DINITROIODOBENZENE, **V**, 478
Benzene, 3-iodo-1,2,4,5-tetramethyl- [2100-25-6]
  IODODURENE, **VI**, 700
Benzene, iodosyl- [536-80-1]
  IODOSOBENZENE, **III**, 483; **V**, 658
Benzene, iodyl- [696-33-3]
  IODOXYBENZENE, **III**, 485; **V**, 665
Benzene, 1-isocyanato-4-nitro- [100-28-7]
  p-NITROPHENYL ISOCYANATE, **II**, 453
Benzene, (isocyanomethyl)- [10340-91-7]
  BENZYL ISOCYANIDE, **VII**, 27
Benzene, 1-isocyano-2-methyl- [10468-64-1]
  o-TOLYL ISOCYANIDE, **V**, 1060
Benzene, 1-[(isocyanomethyl)sulfonyl]-4-methyl- [36635-61-7]
  p-TOLYLSULFONYLMETHYL ISOCYANIDE, **VI**, 41, 987
Benzene, isothiocyanato- [103-72-0]
  PHENYL ISOTHIOCYANATE, **I**, 447
Benzene, methoxy- [100-66-3]
  ANISOLE, **I**, 58
Benzene, [(2-methoxy-1,3-butadienyl)thio]-, (*Z*)- [60466-66-2]
  (*Z*)-2-METHOXY-1-PHENYLTHIO-1,3-BUTADIENE, **VI**, 737
Benzene, 1,1'-[(2-methoxy-3-butenylidene)bis(thio)]bis- [60466-65-1]
  4,4-BIS(PHENYLTHIO)-3-METHOXY-1-BUTENE, **VI**, 737
Benzene, 1-methoxy-3,5-dinitro- [5327-44-6]
  3,5-DINITROANISOLE, **I**, 219
Benzene, 1-methoxy-2-(2-nitro-1-propenyl)- [6306-34-9]
  1-(*o*-METHOXYPHENYL)-2-NITRO-1-PROPENE, **IV**, 573
Benzene, 1-methoxy-2-phenoxy- [1695-04-1]
  2-METHOXYDIPHENYL ETHER, **III**, 566
Benzene, 1-methoxy-2-(2-propenyloxy)- [4125-43-3]
  GUAIACOL ALLYL ETHER, **III**, 418
Benzene, 1,1'-(3-methyl-1-cyclopropene-1,2-diyl)bis- [51425-87-7]
  1,2-DIPHENYL-3-METHYLCYCLOPROPENE, **VII**, 203
Benzene, 1,1'-methylenebis- [101-81-5]
  DIPHENYLMETHANE, **II**, 232
Benzene, 1,1'-[methylenebis(thio)]bis- [3561-67-9]
  BIS(PHENYLTHIO)METHANE, **VI**, 737

Benzene, 1,1'-methylenebis[2,4,6-trimethyl- [733-07-3]
    DIMESITYLMETHANE, **V**, 422
Benzene, 1,1',1"-methylidynetris- [519-73-3]
    TRIPHENYLMETHANE, **I**, 548
Benzene, 1-methyl-3-(1-methylethyl)- [535-77-3]
    m-CYMENE, **V**, 332
Benzene, 1-methyl-4-(methylethyl)-2-nitro- [943-15-7]
    2-NITRO-*p*-CYMENE, **III**, 653
Benzene, 1-methyl-4-(methylsulfinyl)- , (*S*)- [5056-07-5]
    (*S*)-(−)-METHYL *p*-TOLYL SULFOXIDE, **VIII**, 464
Benzene, 1-methyl-4-(methylsulfonyl)- [3185-99-7]
    METHYL-*p*-TOLYL SULFONE, **IV**, 674
Benzene, 1-methyl-3-nitro- [99-08-1]
    m-NITROTOLUENE, **I**, 415
Benzene, 2-methyl-4-nitro-1-nitroso- [57610-10-3]
    2-NITROSO-5-NITROTOLUENE, **III**, 334
Benzene, 2-methyl-2-propenyl- [3290-53-7]
    METHALLYLBENZENE, **VI**, 722
Benzene, (methylsulfinyl)- [1193-82-4]
    METHYL PHENYL SULFOXIDE, **V**, 791
Benzene, (2-nitroethenyl)- [102-96-5]
    β-NITROSTYRENE, **I**, 413
Benzene, (nitromethyl)- [622-42-4]
    PHENYLNITROMETHANE, **II**, 512
Benzene, [(nitromethyl)thio]- [60595-16-6]
    (PHENYLTHIO)NITROMETHANE, **VIII**, 550
Benzene, 1-nitro-2-phenoxy- [2216-12-8]
    o-NITRODIPHENYL ETHER, **II**, 446
Benzene, 1-nitro-4-phenoxy- [620-88-2]
    p-NITRODIPHENYL ETHER, **II**, 445
Benzene, nitroso- [586-96-9]
    NITROSOBENZENE, **III**, 668
Benzene, 1,3-octadienyl-, (*Z*,*E*)- [39491-66-2]
    (1*Z*,3*E*)-1-PHENYL-1,3-OCTADIENE, **VIII**, 532
Benzene, 1,3-pentadiynyl- [4009-22-7]
    1-PHENYL-1,3-PENTADIYNE, **VI**, 925
Benzene, 1,4-pentadiynyl- [6088-96-6]
    1-PHENYL-1,4-PENTADIYNE, **VI**, 925
Benzene, pentamethyl- [700-12-9]
    PENTAMETHYLBENZENE, **II**, 250
Benzene, pentyl [538-68-1]
    AMYLBENZENE, **II**, 47
Benzene, 1,1'-pentylidenebis- [1726-12-1]
    1,1-DIPHENYLPENTANE, **V**, 523
Benzene, (1-phenoxyethenyl)- [19928-57-5]
    1-PHENOXY-1-PHENYLETHENE, **VIII**, 512

Benzene, propyl- [103-65-1]
    PROPYLBENZENE, **I**, 471
Benzene, 1,1'-selenobis- [1132-39-4]
    DIPHENYL SELENIDE, **II**, 238
Benzene, 1,1'-[sulfonylbis(bromomethylene)]bis-, (*R*,R**)- (±)- [21966-50-7]
    α,α'-DIBROMODIBENZYL SULFONE, **VI**, 403
Benzene, 1,2,3,5-tetramethyl- [527-53-7]
    ISODURENE, **II**, 360
Benzene, 1,2,4,5-tetramethyl- [95-93-2]
    DURENE, **II**, 248; **VI**, 703
Benzene, 1,2,4,5-tetramethyl-3,6-dinitro- [5465-13-4]
    DINITRODURENE, **II**, 254
Benzene, 1,1'-thiobis- [139-66-2]
    DIPHENYL SULFIDE, **II**, 242
Benzene, 1,1'-[thiobis(methylene)]bis- [538-74-9]
    BENZYL SULFIDE , **VI**, 130
Benzene, 1,1'-thiobis[4-nitro- [1223-31-0]
    p-NITROPHENYL SULFIDE, **III**, 667
Benzene, 1,3,5-tribromo- [626-39-1]
    sym-TRIBROMOBENZENE, **II**, 592
Benzene, 1,2,3-triiodo-5-nitro- [53663-23-3]
    1,2,3-TRIIODO-5-NITROBENZENE, **II**, 604
Benzene, 1,2,3-trimethyl- [526-73-8]
    HEMIMELLITENE, **IV**, 508
Benzene, 1,3,5-trimethyl- [108-67-8]
    MESITYLENE, **I**, 341
Benzene, 1,3,5-trimethyl-2-nitro- [603-71-4]
    NITROMESITYLENE, **II**, 449
Benzene, 1,3,5-trinitro- [99-35-4]
    1,3,5-TRINITROBENZENE, **I**, 541
Benzeneacetaldehyde, α-methyl- [93-53-8]
    α-PHENYLPROPIONALDEHYDE, **III**, 733
Benzeneacetaldehyde, α-methylene- [4432-63-7]
    ATROPALDEHYDE, **VII**, 13
Benzeneacetaldehyde, α-methyl-α-2-propenyl- [24401-39-6]
    2-METHYL-2-PHENYL-4-PENTENAL, **VIII**, 451
Benzeneacetaldehyde, α-oxo- [1074-12-0]
    PHENYLGLYOXAL, **II**, 509; **V**, 937
Benzeneacetaldehyde, α-phenyl- [947-91-1]
    DIPHENYLACETALDEHYDE, **IV**, 375
Benzeneacetamide [103-81-1]
    PHENYLACETAMIDE, **IV**, 760
Benzeneacetamide, α-hydroxy- [4410-31-51]
    MANDELAMIDE, **III**, 536
Benzeneacetic acid [103-82-2]
    PHENYLACETIC ACID, **I**, 436; **IV**, 761

Benzeneacetic acid, ethyl ester [101-97-3]
   ETHYL PHENYLACETATE, **I**, 270
Benzeneacetic acid, α-acetyl-, ethyl ester [5413-05-8]
   ETHYL α-PHENYLACETOACETATE, **II**, 284
Benzeneacetic acid, α-(acetyloxy)- [5438-68-6]
   ACETYLMANDELIC ACID, **I**, 12
Benzeneacetic acid, α-amino-, (±)- [2835-06-5]
   DL-α-AMINOPHENYLACETIC ACID, **III**, 84
Benzeneacetic acid, 4-amino- [1197-55-3]
   p-AMINOPHENYLACETIC ACID, **I**, 52
Benzeneacetic acid, α-amino-α-methyl-, (±)- [6945-32-0]
   DL-α-AMINO-α-PHENYLPROPIONIC ACID, **III**, 88
Benzeneacetic acid, α-bromo- [4870-65-9]
   α-BROMOPHENYLACETIC ACID, **VI**, 403
Benzeneacetic acid, 4-bromo-α-hydroxy- [6940-50-7]
   p-BROMOMANDELIC ACID, **IV**, 110
Benzeneacetic acid, 2-carboxy- [89-51-0]
   HOMOPHTHALIC ACID, **III**, 449; **V**, 612
Benzeneacetic acid, α-chloro- [4755-72-0]
   α-CHLOROPHENYLACETIC ACID, **IV**, 169
Benzeneacetic acid, α-chloro-, ethyl ester [4773-33-5]
   ETHYL α-CHLOROPHENYLACETATE, **IV**, 169
Benzeneacetic acid, 4-chloro-α-(4-chlorophenyl)- [83-05-6]
   DI-(*p*-CHLOROPHENYL)ACETIC ACID, **III**, 270; *Correction*, **V**, 370
Benzeneacetic acid, α-cyano-, ethyl ester [4553-07-5]
   ETHYL PHENYLCYANOACETATE, **IV**, 461
Benzeneacetic acid, α-cyano-α-(2-cyanoethyl)-, ethyl ester [53555-70-7]
   α-PHENYL-α-CARBETHOXYGLUTARONITRILE, **IV**, 776
Benzeneacetic acid, α-cyano-2,3,4,5,6-pentafluoro-, ethyl ester [2340-87-6]
   ETHYL CYANO(PENTAFLUOROPHENYL)ACETATE, **VI**, 873
Benzeneacetic acid, 3,4-dimethoxy- [93-40-3]
   HOMOVERATRIC ACID, **II**, 333
Benzeneacetic acid, 3,4-dimethoxy-, methyl ester [15964-79-1]
   METHYL HOMOVERATRATE, **II**, 333
Benzeneacetic acid, 4-(dimethylamino)-, ethyl ester [17078-29-4]
   ETHYL *p*-DIMETHYLAMINOPHENYLACETATE, **V**, 552
Benzeneacetic acid, α-hydroxy- [90-64-2]
   MANDELIC ACID, **I**, 336; **III**, 538
Benzeneacetic acid, α-hydroxy-, ethyl ester [774-40-3]
   ETHYL MANDELATE, **IV**, 169
Benzeneacetic acid, α-hydroxy-2,5-dimethyl- [5766-40-5]
   2,5-DIMETHYLMANDELIC ACID, **III**, 326
Benzeneacetic acid, α-hydroxy-α-methyl- [515-30-0]
   ATROLACTIC ACID, **IV**, 58
Benzeneacetic acid, α-hydroxy-α-phenyl- [76-93-7]
   BENZILIC ACID, **I**, 89

Benzeneacetic acid, 4-nitro- [104-03-0]
  p-NITROPHENYLACETIC ACID, **I**, 406
Benzeneacetic acid, α-[(2-nitrophenylmethylene]-, (*E*)- [19319-35-8]
  trans-o-NITRO-α-PHENYLCINNAMIC ACID, **IV**, 730
Benzeneacetic acid, α-oxo- [611-73-4]
  BENZOYLFORMIC ACID, **III**, 114
Benzeneacetic acid, α-oxo-, ethyl ester [1603-79-8]
  ETHYL BENZOYLFORMATE, **I**, 241
Benzeneacetic acid, α-phenyl- [117-34-0]
  DIPHENYLACETIC ACID, **I**, 224
Benzeneacetic acid, α-(phenylmethylene)- [3368-16-9]
  α-PHENYLCINNAMIC ACID, **IV**, 777
Benzeneacetic acid, α,α'-thiobis-, (*R**,*R**)-(±)-[3442-23-7]; (*R**,*S**)- [2845-49-0]
  α,α'-DIPHENYLTHIODIGLYCOLIC ACID, **VI**, 403
Benzeneacetic acid, 2,4,6-trimethyl- [4408-60-0]
  MESITYLACETIC ACID, **III**, 557
Benzeneacetonitrile [140-29-4]
  BENZYL CYANIDE, **I**, 107
Benzeneacetonitrile, α-acetyl- [4468-48-8]
  α-PHENYLACETOACETONITRILE, **II**, 487
Benzeneacetonitrile, α-cyclohexyl- [3893-23-0]
  α-CYCLOHEXYLPHENYLACETONITRILE, **III**, 219
Benzeneacetonitrile, α-(dimethylamino)- [827-36-1]
  α-*N,N*-DIMETHYLAMINOPHENYLACETONITRILE, **V**, 437
Benzeneacetonitrile, α-[[[(1,1-dimethylethoxy)carbonyl]carbonyl]oxy]imino]-
  [58632-95-4]
  2-*tert*-BUTOXYCARBONYLOXYIMINO-2-PHENYLACETONITRILE, **VI**, 199,
  718
Benzeneacetonitrile, α-(diphenylmethylene)- [6304-33-2]
  α,β-DIPHENYLCINNAMONITRILE, **IV**, 387
Benzeneacetonitrile, α-ethenyl-α-ethyl- [13312-96-4]
  2-PHENYL-2-VINYLBUTYRONITRILE, **VI**, 940
Benzeneacetonitrile, α-ethyl- [769-68-6]
  2-PHENYLBUTYRONITRILE, **VI**, 897
Benzeneacetonitrile, α-(hydroxyimino)- [825-52-5]
  2-HYDROXYIMINO-2-PHENYLACETONITRILE, **VI**, 199
Benzeneacetonitrile, α-hydroxy-α-phenyl- [4746-48-9]
  BENZOPHENONE CYANOHYDRIN, **VII**, 20,
Benzeneacetonitrile, 4-methoxy- [104-47-2]
  p-METHOXYPHENYLACETONITRILE, **IV**, 576
Benzeneacetonitrile, 4-methoxy-α-[(trimethylsilyl)oxy]- [66985-48-6]
  O-TRIMETHYLSILYL-4-METHOXYMANDELONITRILE, **VII**, 521
Benzeneacetonitrile, 4-nitro- [555-21-5]
  p-NITROBENZYL CYANIDE, **I**, 396
Benzeneacetonitrile, α-oxo- [613-90-1]
  BENZOYL CYANIDE, **III**, 112

Benzeneacetonitrile, 2,3,4,5,6-pentafluoro- [653-30-5]
   (PENTAFLUOROPHENYL)ACETONITRILE, **VI**, 873
Benzeneacetonitrile, α-phenyl- [86-29-3]
   DIPHENYLACETONITRILE, **III**, 347
Benzeneacetonitrile, α-(phenylmethylene)- [2510-95-4]
   o-PHENYLCINNAMONITRILE, **III**, 715
Benzeneacetonitrile, α-phenyl-α-[(trimethylsilyl)oxy]- [40326-25-8]
   O-(TRIMETHYLSILYL)BENZOPHENONE CYANOHYDRIN, **VII**, 20
Benzeneacetonitrile, 2,4,6-trimethyl- [34688-71-6]
   MESITYLACETONITRILE, **III**, 557
Benzeneacetyl chloride, α-(acetyloxy)- [1638-63-7]
   ACETYLMANDELYL CHLORIDE, **I**, 12
Benzeneacetyl chloride, α-phenyl- [1871-76-7]
   DIPHENYLACETYL CHLORIDE, **VI**, 549
Benzeneacetyl isothiocyanate [29313-32-4]
   PHENYLACETYL ISOTHIOCYANATE, **V**, 1051
Benzenebutanenitrile, α-(4-chlorophenyl)-, β-oxo- [35741-47-0]
   α-(4-CHLOROPHENYL)-γ-PHENYLACETOACETONITRILE, **IV**, 174
Benzenebutanenitrile, γ-oxo-α-phenyl- [6268-00-4]
   α-PHENYL-β-BENZOYLPROPIONITRILE, **II**, 498
Benzenebutanoic acid [1821-12-1]
   γ-PHENYLBUTYRIC ACID, **II**, 499
Benzenebutanoic acid, ethyl ester [10031-93-3]
   ETHYL γ-PHENYLBUTYRATE, **II**, 196
Benzenebutanoic acid, α,α-dimethyl- [4374-44-1]
   2,2-DIMETHYL-4-PHENYLBUTYRIC ACID, **VI**, 517
Benzenebutanoic acid, β-ethyl-β-methyl- [53663-16-4]
   3-BENZYL-3-METHYLPENTANOIC ACID, **IV**, 93
Benzenebutanoic acid, β-oxo-, methyl ester [37779-49-0]
   METHYL PHENYLACETYLACETATE, **VII**, 359
Benzenebutanoic acid, γ-oxo- [2051-95-8]
   β-BENZOYLPROPIONIC ACID, **II**, 81
Benzenebutanoic acid, α-phenyl-, ethyl ester [53608-81-4]
   ETHYL 2,4-DIPHENYLBUTANOATE, **V**, 559
Benzenecarboperoxoic acid [93-59-4]
   PEROXYBENZOIC ACID, **I**, 431; **V**, 904; *Warning*, **V**, 900
Benezenecarboperoxoic acid, 2-carboxy- [2311-91-3]
   MONOPERPHTHALIC ACID, **III**, 619; **V**, 805
Benzenecarbothioic acid [98-91-9]
   THIOBENZOIC ACID, **IV**, 924
Benzenecarboximidamide, *N,N'*-diphenyl- [2556-46-9]
   N,N-DIPHENYLBENZAMIDINE, **IV**, 383
Benzenecarboximidamide, *N*-phenyl- [1527-91-9]
   N-PHENYLBENZAMIDINE, **IV**, 769
Benzenecarboximidoyl chloride, *N*-hydroxy- [698-16-8]
   BENZOHYDROXAMOYL CHLORIDE, **V**, 504

1,4-Benzenediacetic acid, 2,5-dihydroxy- [5488-16-4]
    2,5-DIHYDROXY-*p*-BENZENEDIACETIC ACID, **III**, 286
1,2-Benzenediamine [95-54-5]
    o-PHENYLENEDIAMINE, **II**, 501
1,3-Benzenediamine, 4,6-dinitro- [4987-96-6]
    1,5-DIAMINO-2,4-DINITROBENZENE, **V**, 1068
1,3-Benzenediamine, 4-methyl- [95-80-7]
    2,4-DIAMINOTOLUENE, **II**, 160
1,2-Benzenediamine, 4-nitro- [99-56-9]
    1,2-DIAMINO-4-NITROBENZENE, **III**, 242
1,4-Benzenediamine, *N,N,N',N'*-tetramethyl- [100-22-1]
    TETRAMETHYL-*p*-PHENYLENEDIAMINE, **V**, 1018
Benzenediazonium, 2-carboxy-, hydroxide, inner salt [1608-42-0]
    BENZENEDIAZONIUM-2-CARBOXYLATE, **V**, 54
1,2-Benzenedicarbonyl dichloride [88-95-9]
    sym-o-PHTHALYL CHLORIDE, **III**, 528
1,2-Benzenedicarboxaldehyde [643-79-8]
    o-PHTHALALDEHYDE, **IV**, 807
1,3-Benzenedicarboxaldehyde [626-19-7]
    ISOPHTHALALDEHYDE, **V**, 668
1,4-Benzenedicarboxaldehyde [623-27-8]
    TEREPHTHALALDEHYDE, **III**, 788
1,2-Benzenedicarboxylic acid, monooctyl ester [5393-19-1]
    sec-OCTYL HYDROGEN PHTHALATE, **I**, 418
1,2-Benzenedicarboxylic acid, 3-nitro- [603-11-2]
    3-NITROPHTHALIC ACID, **I**, 408
1,2-Benzenedicarboxylic acid, 4-nitro- [610-27-5]
    4-NITROPHTHALIC ACID, **II**, 457
1,3-Benzenedicarboxylic acid, 2-hydroxy- [606-19-9]
    2-HYDROXYISOPHTHALIC ACID, **V**, 617
1,4-Benzenedicarboxylic acid [100-21-0]
    TEREPHTHALIC ACID, **III**, 791
1,2-Benzenediol [120-80-9]
    CATECHOL, **I**, 149
1,3-Benzenediol, 4-bromo- [6626-15-9]
    4-BROMORESORCINOL, **II**, 100
1,4-Benzenediol, diacetate [1205-91-0]
    HYDROQUINONE DIACETATE, **III**, 452
1,2-Benzenediol, 3-methoxy- [934-00-9]
    PYROGALLOL 1-MONOMETHYL ETHER, **III**, 759
1,2-Benzenedithiol [17534-15-5]
    1,2-DIMERCAPTOBENZENE, **V**, 419
Benzeneethanamine [64-04-0]
    β-PHENYLETHYLAMINE, **III**, 720
Benzeneethanamine, α,α-dimethyl- [122-09-8]
    α,α-DIMETHYL-β-PHENETHYLAMINE, **V**, 471

Benzeneethanamine, *N,N*-dimethyl- [1126-71-2]
    β-PHENYLETHYLDIMETHYLAMINE, **III**, 723
Benzeneethanamine, *N*-methyl-α-phenyl- [53663-25-5]
    N-METHYL-1,2-DIPHENYLETHYLAMINE, **IV**, 605
Benzeneethanamine, *N*-methyl-α-phenyl-, hydrochloride [7400-77-3]
    N-METHYL-1,2-DIPHENYLETHYLAMINE HYDROCHLORIDE, **IV**, 605
Benzeneethanimidoyl chloride, *N*-hydroxy-α-oxo- [4937-87-5]
    ω-CHLOROISONITROSOACETOPHENONE, **III**, 191
Benzeneethanol, β-bromo-α-phenyl-, (*R**, *S**)- [10368-43-1]
    erythro-2-BROMO-1,2-DIPHENYLETHANOL, **VI**, 184
Benzenehexol [608-80-0]
    HEXAHYDROXYBENZENE, **V**, 595
Benzenemethanamine, 2,3-dimethoxy-*N*-methyl- [53663-28-8]
    N-METHYL-2,3-DIMETHOXYBENZYLAMINE, **IV**, 603
Benzenemethanamine, *N,N'*-[1,2-ethanediylbis(oxy-2,1-ethanediyl)]bis-
    [66582-26-1]
    1,10-DIBENZYL-4,7-DIOXA-1,10-DIAZADECANE, **VIII**, 152
Benzenemethanamine, α-methyl- [98-84-0]
    α-PHENYLETHYLAMINE, **II**, 503; **III**, 717
Benzenemethanamine, α-methyl-, (*R*)- [3886-69-9]
    α-PHENYLETHYLAMINE, *R*(+)-, **II**, 506
Benzenemethanamine, α-methyl-, (*S*)- [2627-86-3]
    α-PHENYLETHYLAMINE, *S*(−)-, **V**, 932
Benzenemethanamine, *N*-(3-methylphenyl)- [5405-17-4]
    m-TOLYLBENZYLAMINE, **III**, 827
Benzenemethanamine, *N*-phenyl- [103-32-2]
    BENZYLANILINE, **I**, 102; **IV**, 92
Benzenemethanamine, *N,N*,2-trimethyl- [4525-48-8]
    2-METHYLBENZYLDIMETHYLAMINE, **IV**, 585
Benzenemethanamine, *N*-[(trimethylsilyl)methyl]- [53215-95-5]
    N-BENZYL-*N*-[(TRIMETHYLSILYL)METHYL]AMINE, **VIII**, 231
Benzenemethanaminium, *N,N,N*-trimethyl-, ethoxide [27292-06-4]
    BENZYLTRIMETHYLAMMONIUM ETHOXIDE, **IV**, 98
Benzenemethanimine, α-phenyl- [1013-88-3]
    DIPHENYL KETIMINE, **V**, 520
Benzenemethanimine, α-phenyl-, hydrochloride [5319-67-5]
    DIPHENYLMETHANE IMINE HYDROCHLORIDE, **II**, 234
Benzenemethanol, 2-amino- [5344-90-1]
    o-AMINOBENZYL ALCOHOL, **III**, 60
Benzenemethanol, α-butyl-, (*R*)- [19641-53-3]
    (*R*)-1-PHENYL-1-PENTANOL, **VII**, 447
Benzenemethanol, 3-chloro-α-methyl- [6939-95-3]
    m-CHLOROPHENYLMETHYLCARBINOL, **III**, 200
Benzenemethanol, α-[2-cyclohexylimino)ethyl]-α-phenyl- [1235-46-7]
    N-(3-HYDROXY-3,3-DIPHENYLPROPYLIDENE)CYCLOHEXYLAMINE,
    VI, 901

Benzenemethanol, 2-(dimethylamino)-5-methyl-α,α-diphenyl- [23667-05-2]
    (2-DIMETHYLAMINO-5-METHYLPHENYL)DIPHENYLCARBINOL,
    VI, 478
Benzenemethanol, α,α-diphenyl- [76-84-6]
    TRIPHENYLCARBINOL, **III**, 839
Benzenemethanol, 4-iodo- [18282-51-4]
    p-IODOBENZYL ALCOHOL, **III**, 652
Benzenemethanol, 2-methyl- [89-95-2]
    o-METHYLBENZYL ALCOHOL, **IV**, 582
Benzenemethanol, 2-methyl-, acetate [17373-93-2]
    o-METHYLBENZYL ACETATE, **IV**, 582
Benzenemethanol, 4-methyl- [589-18-4]
    p-TOLYL CARBINOL, **II**, 590
Benzenemethanol, α-methyl-3-nitro- [5400-78-2[
    3'-NITRO-1-PHENYLETHANOL, **VIII**, 495
Benzenemethanol, 4-nitro- [619-73-8]
    p-NITROBENZYL ALCOHOL, **III**, 652
Benzenemethanol, 4-nitro-, acetate (ester) [619-90-9]
    p-NITROBENZYL ACETATE, **III**, 650
Benzenemethanol, α-phenyl- [91-01-0]
    BENZOHYDROL, **I**, 90
Benzenemethanol, α-trichloromethyl- [2000-43-3]
    1-PHENYL-2,2,2-TRICHLOROETHANOL, **V**, 130
Benzenepentanoic acid, δ-oxo- [1501-05-9]
    γ-BENZOYLBUTYRIC ACID, **II**, 82
Benzenepropanal [104-53-0]
    3-PHENYLPROPIONALDEHYDE, **VII**, 451
Benzenepropanal, α,α-dimethyl- [1009-62-7]
    2,2-DIMETHYL-3-PHENYLPROPIONALDEHYDE, **VI**, 526
Benzenepropanal, β-ethenyl- [939-21-9]
    3-PHENYL-4-PENTENAL, **VIII**, 536
Benzenepropanal, α-methyl- [5445-77-2]
    2-METHYL-3-PHENYLPROPANAL, **VII**, 361
    2-METHYL-3-PHENYLPROPIONALDEHYDE, **VI**, 815
Benzenepropanamide, β-oxo-*N*-phenyl- [959-66-0]
    BENZOYLACETANILIDE, **III**, 108; **IV**, 80
Benzenepropanenitrile, 2-chloro- [7315-17-5]
    o-CHLOROHYDROCINNAMONITRILE, **V**, 264
Benzenepropanenitrile, α,α-diphenyl- [5350-82-3]
    α,α,β-TRIPHENYLPROPIONITRILE, **IV**, 962
Benzenepropanoic acid [501-52-0]
    HYDROCINNAMIC ACID, **I**, 311
Benzenepropanoic acid, 4-acetyl-α-bromo- [18910-19-5]
    p-ACETYL-α-BROMOHYDROCINNAMIC ACID, **VI**, 21
Benzenepropanoic acid, α-acetyl-2,3-dimethoxy-, ethyl ester [53608-80-3]
    ETHYL α-ACETYL-β-(2,3-DIMETHOXYPHENYL)PROPIONATE, **IV**, 408

Benzenepropanoic acid, α-acetyl-β-oxo, ethyl ester [569-37-9]
ETHYL BENZOYLACETOACETATE, **II**, 266
Benzenepropanoic acid, β-amino-, (±)- [3646-50-2]
DL-β-AMINO-β-PHENYLPROPIONIC ACID, **III**, 91
Benzenepropanoic acid, α-bromo- [42990-49-8]
α-BROMO-β-PHENYLPROPIONIC ACID, **III**, 705
Benzenepropanoic acid, 2-carboxy- [776-79-4]
β-(*o*-CARBOXYPHENYL) PROPIONIC ACID, **IV**, 136
Benzenepropanoic acid, β-cyano-, ethyl ester [14025-83-3]
ETHYL β-PHENYL-β-CYANOPROPIONATE, **IV**, 804
Benzenepropanoic acid, β-cyano-α-oxo-, ethyl ester [6362-63-6]
ETHYL PHENYLCYANOPYRUVATE, **II**, 287
Benzenepropanoic acid, α,β-dibromo-, (*R*,S**)- [31357-31-0]
erythro-α,β-DIBROMO-β-PHENYLPROPIONIC ACID, **IV**, 961; **VII**, 173
Benzenepropanoic acid, α,β-dibromo-, ethyl ester [5464-70-0]
ETHYL α,β-DIBROMO-β-PHENYLPROPIONATE, **II**, 270
Benzenepropanoic acid, 3,4-dimethoxy-α-oxo- [2460-33-5]
3,4-DIMETHOXYPHENYLPYRUVIC ACID, **II**, 333, 335
Benzenepropanoic acid, α,α-dimethyl-β-oxo-, ethyl ester [25491-42-3]
ETHYL BENZOYLDIMETHYLACETATE, **II**, 268
Benzenepropanoic acid, α,β-diphenyl- [53663-24-4]
α,β,β-TRIPHENYLPROPIONIC ACID, **IV**, 960
Benzenepropanoic acid, β-ethenyl- [5703-57-1]
3-PHENYL-4-PENTENOIC ACID, **VII**, 164
Benzenepropanoic acid, β-hydroxy-, ethyl ester [5764-85-2]
ETHYL β-PHENYL-β-HYDROXYPROPIONATE, **III**, 408
Benzenepropanoic acid, β-hydroxy-α-methyl- (*2R*,3S**)- [14366-87-1]
(*2R*,3R**)-3-HYDROXY-3-PHENYL-2-METHYLPROPANOIC ACID, **VIII**, 339
Benzenepropanoic acid, 4-hydroxy-α-oxo- [156-39-8]
p-HYDROXYPHENYLPYRUVIC ACID, **V**, 627
Benzenepropanoic acid, β-hydroxy-β-phenyl-, ethyl ester [894-18-8]
ETHYL β-HYDROXY-β,β-DIPHENYLPROPIONATE, **V**, 564
Benzenepropanoic acid, α-oxo- [156-06-9]
PHENYLPYRUVIC ACID, **II**, 519
Benzenepropanoic acid, β-oxo, ethyl ester [94-02-0]
ETHYL BENZOYLACETATE, **II**, 266; **III**, 379; **IV**, 415
Benzenepropanoic acid, α-phenyl- [3333-15-1]
α,β-DIPHENYLPROPIONIC ACID, **V**, 526
Benzenepropanol [122-97-4]
3-PHENYL-1-PROPANOL, **IV**, 798
Benzenepropanol, β-amino-, (*S*)- [3182-95-4]
(*S*)-(+)-PHENYLALANOL, **VIII**, 528
Benzeneselenol [645-96-5]
SELENOPHENOL, **III**, 771
Benzeneselenenyl chloride [5707-04-0]
BENZENESELENENYL CHLORIDE, **VI**, 533

Benzenesulfenyl chloride [931-59-9]
  PHENYLSULFENYL CHLORIDE, **VIII**, 550
Benzenesulfenyl chloride, 2,4-dinitro- [528-76-7]
  2,4-DINITROBENZENESULFENYL CHLORIDE, **V**, 474
Benzenesulfenyl chloride, 4-methyl- [933-00-6]
  p-TOLUENESULFENYL CHLORIDE, **IV**, 934
Benzenesulfenyl chloride, 2-nitro- [7669-54-7]
  o-NITROPHENYLSULFUR CHLORIDE, **II**, 455
Benzenesulfinic acid, methyl ester [670-98-4]
  METHYL BENZENESULFINATE, **V**, 723
Benzenesulfinic acid, 4-(acetylamino)- [710-24-7]
  p-ACETAMINOBENZENESULFINIC ACID, **I**, 7
Benzenesulfinic acid, 2-amino-, monosodium salt [50827-53-7]
  SODIUM 2-AMINOBENZENESULFINATE, **V**, 61
Benzenesulfinic acid, 4-methyl-, sodium salt [824-79-3]
  SODIUM *p*-TOLUENESULFINATE, **I**, 492
Benzenesulfinic acid, 2-nitro- [13165-79-2]
  2-NITROBENZENESULFINIC ACID, **V**, 60
Benzenesulfinyl chloride, 4-methyl- [10439-23-3]
  p-TOLUENESULFINYL CHLORIDE, **IV**, 937
Benzenesulfonamide, 4-amino-3,5-dibromo- [39150-45-3]
  3,5-DIBROMOSULFANILAMIDE, **III**, 263
Benzenesulfonamide, 4-amino-3,5-dichloro- [22134-75-4]
  3,5-DICHLOROSULFANILAMIDE, **III**, 262
Benzenesulfonamide, *N*,4-dimethyl-*N*-nitroso- [80-11-5]
  p-TOLYLSULFONYLMETHYLNITROSAMIDE, **IV**, 943
Benzenesulfonamide, *N*-(1-ethyl-2-hydroxy-2-methylpropyl)-4-methyl- [87291-33-6]
  2-METHYL-3-(*p*-TOLUENESULFONAMIDO)-2-PENTANOL, **VII**, 375
Benzenesulfonamide, 4-formyl- [3240-35-5]
  4-FORMYLBENZENESULFONAMIDE, **VI**, 631
Benzenesulfonamide, *N*-(2-hydroxycyclohexyl)-4-methyl-, *cis*- [58107-40-7]
  cis-2-(*p*-TOLUENESULFONAMIDO)CYCLOHEXANOL, **VII**, 375
Benzenesulfonamide, 4-methyl-*N*,*N*-bis[2-[(2-hydroxyethyl)
    [(4-methylphenyl)sulfonyl]amino]ethyl]- [74461-29-3]
  3,6,9-TRIS(*p*-TOLYLSULFONYL)-3,6,9-TRIAZAUNDECANE-1,11-DIOL, **VI**, 652
Benzenesulfonamide, 4-methyl-*N*,*N*-bis[2-[[(4-methylphenyl]sulfonyl]amino]ethyl]-
    [56187-04-3]
  N,*N*′,*N*″-TRIS(*p*-TOLYLSULFONYL)DIETHYLENETRIAMINE, **VI**, 652
Benzenesulfonamide, 4-methyl-*N*,*N*-bis[2-[[(4-methylphenyl)sulfonyl]amino]ethyl]-,
    disodium salt [52601-80-6]
  N,*N*′,*N*″-TRIS(*p*-TOLYLSULFONYL)DIETHYLENETRIAMINE
    N,*N*″-DISODIUM SALT, **VI**, 652
Benzenesulfonamide, 4-methyl-*N*,*N*-bis[2-[[(4-methylphenyl)sulfonyl]
    [2-[(methylsulfonyl)oxy]ethyl]amino]ethyl]- [74461-30-6]
  3,6,9-TRIS(*p*-TOLYLSULFONYL)-3,6,9-TRIAZAUNDECANE-1,11-
    DIMETHANESULFONATE, **VI**, 652

Benzenesulfonamide, 4-methyl-*N*-[2-(2-propenyl)cyclopentyl]-, *cis*- [81097-06-5]
   cis-*N*-TOSYL-2-(2-PROPENYL)CYCLOPENTYLAMINE, **VII**, 501
Benzenesulfonamide, *N*-(phenylmethylene)- [13909-34-7]
   N-BENZYLIDENEBENZENESULFONAMIDE, **VIII**, 546
Benzenesulfonic acid, 2-amino- [88-21-1]
   ORTHANILIC ACID, **II**, 471
Benzenesulfonic acid, 4-amino-3-methyl- [99-33-9]
   o-TOLUIDINESULFONIC ACID, **III**, 824
Benzenesulfonic acid, 4-methyl-, anhydride [4124-41-8]
   p-TOLUENESULFONIC ANHYDRIDE, **IV**, 940
Benzenesulfonic acid, 4-methyl-, butyl ester [778-28-9]
   BUTYL *p*-TOLUENESULFONATE, **I**, 145
Benzenesulfonic acid, 4-methyl-, methyl ester [80-48-8]
   METHYL *p*-TOLUENESULFONATE, **I**, 146
Benzenesulfonic acid, 4-methyl-, hydrazide [1576-35-8]
   p-TOLUENESULFONYLHYDRAZIDE, **V**, 1055; **VI**, 63
Benzenesulfonic acid, 4-methyl-, hydrazide, phenylmethylene- [1666-17-7]
   BENZALDEHYDE TOSYLHYDRAZONE, **VII**, 438
Benzenesulfonic acid, 2,4,6-tris(1-methylethyl)-(1,7,7-trimethylbicyclo[2.2.1]hept-2-
   ylidene)hydrazide, 1*R*- [87068-34-6]
   D-CAMPHOR 2,4,6-TRIISOPROPYLBENZENESULFONYLHYDRAZONE, **VII**,
   77
Benzenesulfonoselenoic acid, Se-phenyl ester [60805-71-2]
   PHENYL BENZENESELENOSULFONATE, **VIII**, 543
Benzenesulfonothioic acid, 4-methyl-, *S,S'*-1,2-ethanediyl ester [2225-23-2]
   ETHYENE DITHIOTOSYLATE, **VI**, 1016
Benzenesulfonothioic acid, 4-methyl-, *S,S'*-1,3-propanediyl ester [3866-79-3]
   TRIMETHYLENE DITHIOTOSYLATE, **VI**, 1016
Benzenesulfonyl chloride [98-09-9]
   BENZENESULFONYL CHLORIDE, **I**, 84
Benzenesulfonyl chloride, 4-(acetylamino)- [121-60-8]
   p-ACETAMINOBENZENESULFONYL CHLORIDE, **I**, 8
Benzenesulfonyl chloride, 2-nitro- [1694-92-4]
   o-NITROBENZENESULFONYL CHLORIDE, **II**, 471
Benzenesulfonyl chloride, 3-(trifluoromethyl)- [777-44-6]
   m-TRIFLUOROMETHYLBENZENESULFONYL CHLORIDE, **VII**, 508
1,2,4,5-Benzenetetracarboxylic acid [89-05-4]
   PYROMELLITIC ACID, **II**, 551
Benzenethiol [108-98-5]
   THIOPHENOL, **I**, 504
Benzenethiol, 3-methyl- [108-40-7]
   m-THIOCRESOL, **III**, 809; *Warning*, **V**, 1050
1,2,4-Benzenetriamine, 5-nitro- [6635-35-4]
   2,4,5-TRIAMINONITROBENZENE, **V**, 1067
1,3,5-Benzenetriol [108-73-6]
   PHLOROGLUCINOL, **I**, 455

1,2,4-Benzenetriol, triacetate [613-03-6]
  HYDROXYHYDROQUINONE TRIACETATE, **I**, 317
9,10[1',2']Benzenoanthracene, 9,10-dihydro- [477-75-8]
  TRIPTYCENE, **IV**, 964
1*H*-Benzimidazole [51-17-2]
  BENZIMIDAZOLE, **II**, 65
2*H*-Benzimidazole-2-thione, 1,3-dihydro- [583-39-1]
  2-MERCAPTOBENZIMIDAZOLE, **IV**, 569
1*H*-Benz[*e*]indene-6-propanoic acid, dodecahydro-3-hydroxy-3a,6-dimethyl-7-oxo-,
  [3*S*-(3α,3aα,5aβ,6β,9aα,9bβ)]- [1759-35-9]
  17β-HYDROXY-5-OXO-3,5-*seco*-4-NORANDROSTANE-3-CARBOXYLIC
  ACID, **VI**, 690
2,1-Benzisoxazole, 1,3,3a,4,5,6,7,7a-octahydro-1,3,3,6-tetramethyl- [6603-39-0]
  HEXAHYDRO-1,3,3,6-TETRAMETHYL-2,1-BENZISOXAZOLINE, **VI**, 670
5*H*-Benzo[*a*]carbazole, 6,11-dihydro- [21064-49-3]
  1,2-BENZO-3,4-DIHYDROCARBAZOLE, **IV**, 885
Benzocarboperoxoic acid, 3-chloro- [937-14-4]
  m-CHLOROPERBENZOIC ACID, **VI**, 39, 276, 404,
1,3,2-Benzodioxaborole, 2-(1-hexenyl)-, (*E*)- [37490-22-5]
  (*E*)-1-HEXENYL-1,3,2-BENZODIOXABOROLE, **VIII**, 532,
1,3-Benzodioxole-5-carboxylic acid [94-53-1]
  PIPERONYLIC ACID, **II**, 538
1,3-Benzodioxol-2-one [2171-74-6]
  o-PHENYLENE CARBONATE, **IV**, 788
Benzofuran [271-89-6]
  COUMARONE, **V**, 251
Benzofuran, 3-methyl- [21535-97-7]
  3-METHYLCOUMARONE, **IV**, 590
2-Benzofurancarboxylic acid [496-41-3]
  COUMARILIC ACID, **III**, 209
Benzofurancarboxylic acid, 3-methyl- [24673-56-1]
  3-METHYLCOUMARILIC ACID, **IV**, 591
Benzofurancarboxylic acid, 3-methyl-, ethyl ester [22367-82-4]
  ETHYL 3-METHYLCOUMARILATE, **IV**, 590
4,7-Benzofurandione, 5,6-diethoxy- [138225-13-5]
  5,6-DIETHOXYBENZOFURAN-4,7-DIONE, **VIII**, 179
Benzofurazan, 1-oxide [480-96-6]
  BENZOFURAZAN OXIDE, **IV**, 74
Benzoic acid, anhydride [93-97-0]
  BENZOIC ANHYDRIDE, **I**, 91
Benzoic acid, 2-benzoylhydrazide [787-84-8]
  sym-DIBENZOYLHYDRAZINE, **II**, 208
Benzoic acid, hydrazide, 2-benzoyl-1,2-dimethyl- [1226-43-3]
  sym-DIBENZOYLDIMETHYLHYDRAZINE, **II**, 209
Benzoic acid, 2-phenylethyl ester [94-47-3]
  2-PHENYLETHYL BENZOATE, **V**, 336

Benzoic acid, phenylmethyl ester [120-51-4]
    BENZYL BENZOATE, **I**, 104
Benzoic acid, 4-acetyl-, methyl ester [3609-53-8]
    METHYL *p*-ACETYLBENZOATE, **IV**, 579
Benzoic acid, 2-(1-acetyl-2-oxopropyl)- [52962-26-2]
    2-(1-ACETYL-2-OXOPROPYL)BENZOIC ACID, **VI**, 36
Benzoic acid, 4-(acetyloxy)-, ethyl ester [13031-45-3]
    ETHYL 4-ACETOXYBENZOATE, **VI**, 576
Benzoic acid, 4-(acetyloxy)-, methyl ester [24262-66-6]
    METHYL 4-ACETOXYBENZOATE, **VI**, 576
Benzoic acid, 4-amino-, ethyl ester [94-09-7]
    ETHYL *p*-AMINOBENZOATE, **I**, 240
Benzoic acid, 2-amino-3,5-dichloro- [2789-92-6]
    3,5-DICHLORO-2-AMINOBENZOIC ACID, **IV**, 872
Benzoic acid, 2-amino-5-iodo- [5326-47-6]
    5-IODOANTHRANILIC ACID, **II**, 349
Benzoic acid, 4-amino-3-methyl-, ethyl ester [40800-65-5]
    ETHYL 4-AMINO-3-METHYLBENZOATE, **VI**, 560, 581
Benzoic acid, 4-amino-3-[(methylthio)methyl]- [50461-34-2]
    ETHYL 4-AMINO-3-(METHYLTHIOMETHYL)BENZOATE, **VI**, 581
Benzoic acid, 3-amino-2,4,6-tribromo- [6628-84-8]
    3-AMINO-2,4,6-TRIBROMOBENZOIC ACID, **IV**, 947
Benzoic acid, 5-bromo-2,4-dihydroxy- [7355-22-8]
    2,4-DIHYDROXY-5-BROMOBENZOIC ACID, **II**, 100
Benzoic acid, 2-bromo-3-methyl- [53663-39-1]
    2-BROMO-3-METHYLBENZOIC ACID, **IV**, 114
Benzoic acid, 2-bromo-3-nitro- [573-54-6]
    2-BROMO-3-NITROBENZOIC ACID, **I**, 125
Benzoic acid, 5-[(3-carboxy-4-hydroxyphenyl)(3-carboxy-4-oxo-2,5-cyclohexadien-
        1-ylidene)methyl]-2-hydroxy-, triammonium salt [569-58-4]
    AMMONIUM SALT OF AURIN TRICARBOXYLIC ACID, **I**, 54
Benzoic acid, 2-chloro- [118-91-2]
    o-CHLOROBENZOIC ACID, **II**, 135
Benzoic acid, 4-chloro-, anhydride [790-41-0]
    p-CHLOROBENZOIC ANHYDRIDE, **III**, 29
Benzoic acid, 2-(cyanomethyl)- [6627-91-4]
    o-CARBOXYPHENYLACETONITRILE, **III**, 174
Benzoic acid, 3,5-dichloro-4-hydroxy-, ethyl ester [17302-82-8]
    ETHYL 3,5-DICHLORO-4-HYDROXYBENZOATE, **III**, 267
Benzoic acid, 2-[(1,3-dihydro-1,3-dioxo-2*H*-isoindol-2-yl)methyl]-
        [53663-18-6]
    α-PHTHALIMIDO-*o*-TOLUIC ACID, **IV**, 810
Benzoic acid, 2,4-dihydroxy- [89-86-1]
    β-RESORCYLIC ACID, **II**, 557
Benzoic acid, 3,4-dihydroxy- [99-50-3]
    PROTOCATECHUIC ACID, **III**, 745

Benzoic acid, 3,5-dihydroxy- [99-10-5]
3,5-DIHYDROXYBENZOIC ACID, **III**, 288
Benzoic acid, 2-[[4-(dimethylamino)phenyl]azo]- [493-52-7]
METHYL RED, **I**, 374
Benzoic acid, 2,5-dinitro- [610-28-6]
2,5-DINITROBENZOIC ACID, **III**, 334
Benzoic acid, 3,5-dinitro- [99-34-3]
3,5-DINITROBENZOIC ACID, **III**, 337
Benzoic acid, 3,5-dinitro-, methyl ester [2702-58-1]
METHYL 3,5-DINITROBENZOATE, **V**, 798
Benzoic acid, 4-fluoro- [456-22-4]
p-FLUOROBENZOIC ACID, **II**, 299
Benzoic acid, 2-formyl- [119-67-5]
PHTHALALDEHYDIC ACID, **II**, 523; **III**, 737
Benzoic acid, 2-hydrazino-, hydrochloride [33906-30-8]
o-HYDRAZINOBENZOIC ACID HYDROCHLORIDE, **III**, 475
Benzoic acid, 4-hydroxy- [99-96-7]
p-HYDROXYBENZOIC ACID, **II**, 341
Benzoic acid, 2-hydroxy-, 4-chlorophenyl ester [2944-58-3]
p-CHLOROPHENYL SALICYLATE, **IV**, 178
Benzoic acid, 2-hydroxy-3,5-diiodo- [133-91-5]
2-HYDROXY-3,5-DIIODOBENZOIC ACID, **II**, 343
Benzoic acid, 4-hydroxy-3-methoxy- [121-34-6]
VANILLIC ACID, **IV**, 972
Benzoic acid, 3-iodo- [618-51-9]
m-IODOBENZOIC ACID, **II**, 353
Benzoic acid, 4-iodo- [619-58-9]
p-IODOBENZOIC ACID, **I**, 325
Benzoic acid, 2-mercapto- [147-93-3]
THIOSALICYLIC ACID, **II**, 580
Benzoic acid, 2-methyl- [118-90-1]
o-TOLUIC ACID, **II**, 588; **III**, 820
Benzoic acid, 4-methyl- [99-94-5]
p-TOLUIC ACID, **II**, 589; **III**, 822
Benzoic acid, 4-methyl-, 1,1-dimethylethyl ester [13756-42-8]
tert-BUTYL *p*-TOLUATE, **VI**, 259
Benzoic acid, 2-[(4-methylbenzenesulfonyl)amino]- [6311-23-5]
p-TOLUENESULFONYLANTHRANILIC ACID, **IV**, 34
Benzoic acid, 2-(4-methylbenzoyl)- [85-55-2]
p-TOLUYL-*o*-BENZOIC ACID, **I**, 517
Benzoic acid, 3-nitro- [121-92-6]
m-NITROBENZOIC ACID, **I**, 391
Benzoic acid, 3-nitro-, methyl ester [618-95-1]
METHYL *m*-NITROBENZOATE, **I**, 372
Benzoic acid, 4-nitro- [62-23-7]
p-NITROBENZOIC ACID, **I**, 392

Benzoic acid, pentachloro- [1012-84-6]
  PENTACHLORBENZOIC ACID, **V**, 890
Benzoic acid, 2-phenylamino- [91-40-7]
  N-PHENYLANTHRANILIC ACID, **II**, 15
Benzoic acid, 4-(phenylazo)- [1562-93-2]
  p-PHENYLAZOBENZOIC ACID, **III**, 711
Benzoic acid, 2-(1*H*- pyrrol-1-yl)-, methyl ester [10333-67-2]
  1-(2-METHOXYCARBONYLPHENYL)PYRROLE, **V**, 716
Benzoic acid, 2-sulfo-, monoammonium salt [6939-89-5]
  ACID AMMONIUM *o*-SULFOBENZOATE, **I**, 14
Benzoic acid, 2,4,6-tribromo- [633-12-5]
  2,4,6-TRIBROMOBENZOIC ACID, **IV**, 947
Benzoic acid, 3,4,5-trimethoxy- [118-41-2]
  TRIMETHYLGALLIC ACID, **I**, 537
Benzoic acid, 2,4,6-trimethyl- [480-63-7]
  MESITOIC ACID, **III**, 553; **V**, 706
Benzoic acid, 2,4,6-trinitro- [129-66-8]
  2,4,6-TRINITROBENZOIC ACID, **I**, 543
Benzonitrile, 2,4-dimethoxy- [4107-65-7]
  2,4-DIMETHOXYBENZONITRILE, **VI**, 465
Benzonitrile, 2,6-dimethoxy- [16932-49-3]
  2,6-DIMETHOXYBENZONITRILE, **III**, 293
Benzonitrile, 3,4-dimethoxy- [2024-83-1]
  VERATRONITRILE, **II**, 622
Benzonitrile, 2-methoxy-6-nitro- [38469-85-1]
  2-NITRO-6-METHOXYBENZONITRILE, **III**, 293
Benzonitrile, 2-methyl- [529-19-1]
  o-TOLUNITRILE, **I**, 514
Benzonitrile, 4-methyl- [104-85-8]
  p-TOLUNITRILE, **I**, 514
Benzonitrile, 2-methyl-3,5-dinitro- [948-31-2]
  3,5-DINITRO-*o*-TOLUNITRILE, **V**, 480
Benzonitrile, 4-nitro- [619-72-7]
  p-NITROBENZONITRILE, **III**, 646
2*H*-1-Benzopyran, 3,4-dihydro-2-methylene- [74104-13-5]
  3,4-DIHYDROMETHYLENE-2*H*-1-BENZOPYRAN, **VIII**, 512
2*H*-1-Benzopyran-3-carboxylic acid, 2-oxo-, ethyl ester [1846-76-0]
  3-CARBETHOXYCOUMARIN, **III**, 165
2*H*-1-Benzopyran-3-carboxylic acid, 5,6,7,8-tetrahydro-2-oxo-, methyl ester
  [85531-80-2]
  METHYL 2-OXO-5,6,7,8-TETRAHYDRO-2*H*-1-BENZOPYRAN-3-
    CARBOXYLATE, **VIII**, 444
1*H*-2-Benzopyran-4-carboxylic acid, octahydro-1,1,6-trimethyl-3-oxo-, methyl ester,
  (4α, 4aα, 6β, 8aβ)-, (±)- [138233-55-3]
  4-METHOXYCARBONYL-1,1,6-TRIMETHYL-1,4,4a,5,6,7,8,8a-OCTAHYDRO-
    2,3-BENZOPYRONE, (4*RS*,4a*RS*,6*RS*,8a*RS*), **VIII**, 353

1*H*-2-Benzopyran-4-carboxylic acid, octahydro-1,1,6-trimethyl-3-oxo-, methyl ester,
   [4*R*-(4α, 4aα, 6β, 8aβ)]- [121114-94-1]
      4-METHOXYCARBONYL-1,1,6-TRIMETHYL-1,4,4a,5,6,7,8,8a-OCTAHYDRO-
      2,3-BENZOPYRONE, (4*S*,4a*S*,6*S*,8a*S*), **VIII**, 353
1*H*-2-Benzopyran-4-carboxylic acid, octahydro-1,1,6-trimethyl-3-oxo-, methyl ester,
   [4*S*-(4α, 4aα, 6β, 8aβ)]- [138147-65-6]
      4-METHOXYCARBONYL-1,1,6-TRIMETHYL-1,4,4a,5,6,7,8,8a-OCTAHYDRO-
      2,3-BENZOPYRONE, (4*R*,4a*R*,6*R*,8a*R*), **VIII**, 353
1*H*-2-Benzopyran-1,3(4*H*)-dione [703-59-3]
      HOMOPHTHALIC ACID ANHYDRIDE, **III**, 449; **V**, 612
2*H*-1-Benzopyran-2-one, 8-acetyl-7-hydroxy-4-methyl- [2555-29-5]
      4-METHYL-7-HYDROXY-8-ACETYLCOUMARIN, **III**, 283
2*H*-1-Benzopyran-2-one, 7-acetyloxy-4-methyl- [2747-05-9]
      4-METHYL-7-ACETOXYCOUMARIN, **III**, 283
2*H*-1-Benzopyran-2-one, 3,4-dibromo-3,4-dihydro- [55077-11-7]
      COUMARIN DIBROMIDE, **III**, 209
2*H*-1-Benzopyran-2-one, 6,7-dihydroxy-4-methyl- [529-84-0]
      4-METHYLESCULETIN, **I**, 360
2*H*-1-Benzopyran-2-one, 7-hydroxy-4-methyl- [90-33-5]
      4-METHYL-7-HYDROXYCOUMARIN, **III**, 282
2*H*-1-Benzopyran-2-one, 4-methyl- [607-71-6]
      4-METHYLCOUMARIN, **III**, 581
3*H*-2-Benzopyran-3-one, 1,4-dihydro-6,7-dimethoxy- [16135-41-4]
      6,7-DIMETHOXY-3-ISOCHROMANONE, **VI**, 471
4*H*-1-Benzopyran-4-one, 2-ethyl- [14736-30-2]
      2-ETHYLCHROMONE, **III**, 387
4*H*-1-Benzopyran-4-one, 2-phenyl- [525-82-6]
      FLAVONE, **IV**, 478
1*H*,5*H*-Benzo[*ij*]quinolizine, 2,3,6,7-tetrahydro- [479-59-4]
      JULOLIDINE, **III**, 504
1,2,3-Benzothiadiazole, 1,1-dioxide [37150-27-9]
      1,2,3-BENZOTHIADIAZOLE 1,1-DIOXIDE, **V**, 60
2-Benzothiazolamine, 6-methyl- [2536-91-6]
      2-AMINO-6-METHYLBENZOTHIAZOLE, **III**, 76
Benzo[*c*]thiophene, octahydro-, *cis*- [17739-77-4]
      cis-8-THIABICYCLO[4.3.0]NONANE, **VI**, 482
Benzo[*c*]thiophene, 2,2-dioxide, octahydro- [57479-57-9]
      cis-8-THIABICYCLO[4.3.0]NONANE 8,8-DIOXIDE, **VI**, 482
Benzo[*c*]thiophene, 2,2-dioxide, octahydro-1,3-dimethyl- [60090-27-9]
      7,9-DIMETHYL-*cis*-8-THIABICYCLO[4.3.0]NONANE 8,8-DIOXIDE, **VI**, 482
1*H*-Benzotriazole [95-14-7]
      1,2,3-BENZOTRIAZOLE, **III**, 106
3*H*-2,1-Benzoxamercurol-3-one, 7-nitro- [53663-14-2]
      ANHYDRO-2-HYDROXYMERCURI-3-NITROBENZOIC ACID, **I**, 56
4*H*-1,3-Benzoxathiin, hexahydro-4,4,7-trimethyl- [59324-06-0]
      HEXAHYDRO-4,4,7-TRIMETHYL-4*H*-1,3-BENZOXATHIIN, **VIII**, 304

3*H*-2,1-Benzoxathiol-3-one, 1,1-dioxide [81-08-3]
   o-SULFOBENZOIC ANHYDRIDE, **I**, 495
3,1-Benzoxazepine [15123-59-8]
   3,1-BENZOXAZEPINE, **VII**, 23
2*H*-3,1-Benzoxazine-2,4(1*H*)-dione [118-48-9]
   ISATOIC ANHYDRIDE, **III**, 488
2(3*H*)-Benzoxazolone, 3-acetyl-3a,4,7,7a-tetrahydro-5,6-dimethyl- [65948-43-8]
   4-ACETYL-7,8-DIMETHYL-2-OXA-4-AZABICYCLO[4.3.0]NON-7-EN-3-ONE,
      VII, 5
Benzoyl azide, 3-nitro- [3532-31-8]
   m-NITROBENZAZIDE, **IV**, 715
Benzoyl chloride, 2-chloro- [609-65-4]
   o-CHLOROBENZOYL CHLORIDE, **I**, 155
Benzoyl chloride, 3-nitro- [121-90-4]
   m-NITROBENZOYL CHLORIDE, **IV**, 715
Benzoyl chloride, 4-nitro- [122-04-3]
   p-NITROBENZOYL CHLORIDE, **I**, 394
Benzoyl chloride, 4-pentyl [49763-65-7]
   4-PENTYLBENZOYL CHLORIDE, **VII**, 420
Benzoyl chloride, 4-(phenylazo)- [104-24-5]
   p-PHENYLAZOBENZOYL CHLORIDE, **III**, 712
Benzoyl chloride, 2,4,6-trimethyl- [938-18-1]
   MESITOYL CHLORIDE, **III**, 555
Benzoyl fluoride [455-32-3]
   BENZOYL FLUORIDE, **V**, 66
Bicyclo[1.1.0]butane [157-33-5]
   BICYCLO[1.1.0]BUTANE, **VI**, 133, 159
Bicyclo[2.2.1]hepta-2,5-diene, 7-(1,1-dimethylethoxy)- [877-06-5]
   7-*tert*-BUTOXYNORBORNADIENE, **V**, 151
Bicyclo[3.1.1]heptan-3-amine, 2,2,6-trimethyl- [17371-27-6]
   3-PINANAMINE, **VI**, 943
Bicyclo[4.1.0]heptane [286-08-8]
   NORCARANE, **V**, 855
Bicyclo[2.2.1]heptane, 1-chloro- [765-67-3]
   1-CHLORONORBORNANE, **VI** , 845
Bicyclo[2.2.1]heptane, 2,2-dichloro- [19916-65-5]
   2,2-DICHLORONORBORNANE, **VI**, 845
Bicyclo[2.2.1]heptane-1-carboxylic acid [18720-30-4]
   1-NORBORNANECARBOXYLIC ACID, **VI**, 845
Bicyclo[2.2.1]heptane-1-carboxylic acid, 7,7-dimethyl-2-oxo- [464-78-8]
   DL-KETOPINIC ACID, **V**, 689
Bicyclo[2.2.1]heptane-1-methanesulfonamide, 7,7-dimethyl-2-oxo- (1*S*)-
      [60933-63-3]
   (+)-(1*S*)-10-CAMPHORSULFONAMIDE, **VIII**, 104
Bicyclo[2.2.1]heptane-1-methanesulfonic acid, 7,7-dimethyl-2-oxo-, (±)- [5872-08-2]
   DL-10-CAMPHORSULFONIC ACID (REYCHLER'S ACID), **V**, 194

Bicyclo[2.2.1]heptane-1-methanesulfonyl chloride, 7,7-dimethyl-2-oxo-, (±)-
[6994-93-0]
DL-10-CAMPHORSULFONYL CHLORIDE, **V**, 196
Bicyclo[2.2.1]heptan-2-ol, formate, *exo-* [41498-71-9]
2-*exo*-NORBORNYL FORMATE, **V**, 852
Bicyclo[2.2.1]heptan-2-ol, 2-ethynyl-1,3,3-trimethyl-, (1*R-endo*)- [131062-94-7]
1-METHYL-2-ETHYNYL-*endo*-3,3-DIMETHYL-2-NORBORNANOL, **VIII**, 391
Bicyclo[3.1.1]heptan-3-ol, 6,6-dimethyl-2-methylene-, (1α,3β,5α)- [6712-79-4]
trans-PINOCARVEOL, **VI**, 946, 948
Bicyclo[3.1.1]heptan-3-ol, 2,6,6-trimethyl-, [1*R*-(1α,2β,3α,5α)]- [1196-00-5]
(−)-ISOPINOCAMPHEOL, **VI**, 719
Bicyclo[3.2.0]heptan-2-ol, 3,3-dimethyl- [71221-67-5]
3,3-DIMETHYL-*cis*-BICYCLO[3.2.0]HEPTAN-2-OL, **VII**, 177
Bicyclo[4.1.0]heptan-7-ol, (1α,6α,7α)- [13830-44-9]
exo-7-NORCARANOL, **V**, 859
Bicyclo [4.1.0]heptan-7-ol, (1α,6α,7β)- [931-31-7]
endo-7-NORCARANOL, **V**, 859
Bicyclo[2.2.1]heptan-2-one [497-38-1]
2-NORBORNANONE, **V**, 852; **VI**, 847
Bicyclo[2.2.1]heptan-2-one, 3-hydroxy-1,7,7-trimethyl-, *endo-* [21488-68-6]
3-HYDROXY-1,7,7-TRIMETHYLBICYCLO[2.2.1]HEPTAN-2-ONE, (endo
ISOMER), **VII**, 277
Bicyclo[3.2.0]heptan-6-one, 7,7-dichloro-1-methyl- [51284-43-6]
7,7-DICHLORO-1-METHYLBICYCLO[3.2.0]HEPTAN-6-ONE, **VIII**, 377
Bicyclo[3.2.0]heptan-2-one, 3,3-dimethyl- [71221-70-0]
3,3-DIMETHYL-*cis*-BICYCLO[3.2.0]HEPTAN-2-ONE, **VII**, 177
Bicyclo[2.2.1]heptan-2-one, 1-ethenyl-7,7-dimethyl-, (+)- [53585-70-9]
D-2-OXO-7,7-DIMETHYL-1-VINYLBICYCLO[2.2.1]HEPTANE, **V**, 877
Bicyclo[4.1.0]hepta-1,3,5-triene [4646-69-9]
BENZOCYCLOPROPENE, **VI**, 87
Bicyclo[2.2.1]hept-2-ene [498-66-8]
NORBORNYLENE, **IV**, 738
Bicyclo[2.2.1]hept-2-ene, 2-butyl-1,7,7-trimethyl- [87068-33-5]
2-BUTYLBORNENE, **VII**, 77,
Bicyclo[2.2.1]hept-2-ene, 7,7-dimethoxy- [875-04-7]
7,7-DIMETHOXYBICYCLO[2.2.1]HEPTENE, **V**, 424
Bicyclo[2.2.1]heptene, 1,2,3,4-tetrachloro-7,7-dimethoxy- [19448-78-3]
7,7-DIMETHOXY-1,2,3,4-TETRACHLOROBICYCLO[2.2.1]HEPTENE, **V**, 425
Bicyclo[2.2.1]hept-2-ene, 1,7,7-trimethyl- [464-17-5]
2-BORNENE, **VI**, 172
Bicyclo[3.1.1]hept-2-ene, 2-(2,2-dibromocyclopropyl)-6,6-dimethyl-, [1α, 2(*S**), 5α]-
[108404-80-4]; [1*R*-[1α, 2(*R**), 5α]]- [108450-07-3]
2-(2,2-DIBROMOCYCLOPROPYL)-6,6-DIMETHYLBICYCLO[2.2.1]HEPT-2-
ENE, **VIII**, 223
Bicyclo[3.1.1]hept-2-ene, 2-ethenyl-6,6-dimethyl- (1*R*)- [30293-06-2]
(1*R*)-NOPADIENE, **VIII** 223

Bicyclo[3.1.1]hept-2-ene, 2,6,6-trimethyl-, (1S)- [7785-26-4]
  PINENE, (–)-α-, **VIII**, 553
Bicyclo[4.1.0]hept-3-ene, 7,7-dibromo-1,6-dimethyl- [38749-43-8]
  7,7-DIBROMO-1,6-DIMETHYLBICYCLO[4.1.0]HEPT-3-ENE, **VII**, 200
Bicyclo[4.1.0]hept-3-ene, 7,7-dichloro- [16554-84-0]
  7,7-DICHLOROBICYCLO[4.1.0]HEPT-3-ENE, **VI**, 87
Bicyclo[3.1.1]hept-2-ene-2-ethanol, 6,6-dimethyl-, 4-methylbenzenesulfonate, (1R)-
  [81600-63-7]
  (1R)-NOPYL 4-METHYLBENZENESULFONATE, **VIII**, 223
Bicyclo[2.2.1]hept-2-en-7-one [694-71-3]
  BICYCLO[2.2.1]HEPTEN-7-ONE, **V**, 91
Bicyclo[3.2.0]hept-2-en-6-one, 7,7-dichloro- [5307-99-3]
  7,7-DICHLOROBICYCLO[3.2.0]HEPT-2-EN-6-ONE, **VI**, 1037
Bicyclo[2.2.0]hexa-2,5-diene, 1,2,3,4,5,6-hexamethyl- [7641-77-2]
  HEXAMETHYL DEWAR BENZENE, **VII**, 256
[1,1'-Bicyclohexyl]-1,1'-dicarbonitrile [18341-40-7]
  1,1'-DICYANO-1,1'-BICYCLOHEXYL, **IV**, 273
Bicyclo[6.1.0]nonane, 9,9-dibromo- [1196-95-8]
  9,9-DIBROMOBICYCLO[6.1.0]NONANE, **V**, 306
Bicyclo[3.3.1]nonan-9-one [17931-55-4]
  BICYCLO[3.3.1]NONAN-9-ONE, **VI**, 137
Bicyclo[3.3.1]nonan-9-one, 7-(iodomethyl)-, *endo*- [29817-49-0]
  endo-7-IODOMETHYLBICYCLO[3.3.1]NONAN-3-ONE, **VI**, 958
Bicyclo[4.2.0]octa-2,4-diene-7,8-diol, diacetate, (1α,6α,7α,8β)- [42301-50-8]
  trans-7,8-DIACETOXYBICYCLO[4.2.0]OCTA-2,4-DIENE, **VI**, 196
Bicyclo[4.2.0]octane-7-carbonitrile, 7-chloro-8-oxo-, (1α,6α,7β)- [89937-15-5]
  7-CHLORO-7-CYANOBICYCLO[4.2.0]OCTAN-8-ONE, **VIII**, 116
Bicyclo[2.2.2]octane-2-carboxylic acid, 1,3-dimethyl-5-oxo-, methyl ester,
  1α,2β,3α,4α- [121917-73-5]
  METHYL 1,3-DIMETHYL-5-OXOBICYCLO[2.2.2]OCTANE-2-CARBOXY-
  LATE, **VIII**, 219
Bicyclo[2.2.2]octane-2,6-dione, 4-methyl- [119986-98-0]
  4-METHYLBICYCLO[2.2.2]OCTANE-2,6-DIONE, **VIII**, 468
Bicyclo[3.2.1]octan-3-one [14252-05-2]
  BICYCLO[3.2.1]OCTAN-3-ONE, **VI**, 142
Bicyclo[4.2.0]octan-2-one, 7-(acetyloxy)-4,4,6,7-tetramethyl- [66016-89-5]
  7-ACETOXY-4,4,6,7-TETRAMETHYLBICYCLO[4.2.0]OCTAN-2-ONE, **VI**,
  1024
Bicyclo[4.2.0]octan-2-one, 7-hydroxy-4,4,6,7-tetramethyl- [61879-76-3]
  7-HYDROXY-4,4,6,7-TETRAMETHYLBICYCLO[4.2.0]OCTAN-2-ONE, **VI**, 1024
Bicyclo[4.2.0]octan-2-one, 6-methyl- [13404-66-5]
  6-METHYLBICYCLO[4.2.0]OCTAN-2-ONE, **VII**, 315
Bicyclo[4.2.0]octa-1,3,5-triene-7-carbonitrile [6809-91-2]
  1-CYANOBENZOCYCLOBUTENE, **V**, 263
Bicyclo[3.2.1]oct-2-ene, 3-chloro- [35242-17-2]
  3-CHLOROBICYCLO[3.2.1]OCT-2-ENE, **VI**, 142

Bicyclo[3.2.1]oct-2-ene, 3,4-dichloro-, *exo-* [2394-47-0]
  *exo-*3,4-DICHLOROBICYCLO[3.2.1]OCT-2-ENE, **VI**, 142
Bicyclo[4.2.0]oct-7-ene, 7,8-dimethyl-, *cis-* [53225-88-0]
  *cis-*7,8-DIMETHYLBICYCLO[4.2.0]OCT-7-ENE, **VI**, 482
Bi-2,4-cyclopentadien-1-yl [21423-86-9]
  9,10-DIHYDROFULVALENE, **VIII**, 298
Bicyclo[2.1.0]pentane [185-94-4]
  BICYCLO[2.1.0]PENTANE, **V**, 96
Bicyclo[2.1.0]pent-2-ene [5164-35-2]
  BICYCLO[2.1.0]PENT-2-ENE, **VI**, 145
Bicyclo[4.4.1]undeca-1,3,5,7,9-pentaene [2443-46-1]
  1,6-METHANO[10]ANNULENE, **VI**, 731
2,2'-Bi-1*H*-imidazole, 1,3-dibenzoyl-2,3-dihydro-, monohydrochloride [65276-00-8]
  2-(1,3-DIBENZOYLIMIDAZOLIN-2-YL)IMIDAZOLE HYDROCHLORIDE,
    VII, 287
2,2'-Bi-1*H*-imidazole, 1,1',3-tribenzoyl-2,3-dihydro- [62457-77-6]
  1-BENZOYL-2-(1,3-DIBENZOYL-4-IMIDAZOLIN-2-YL)IMIDAZOLE, **VII**, 287
1,1'-Binaphthalene, 2,2'-dibromo- [74866-28-7]
  2,2'-DIBROMO-1,1'-BINAPHTHYL, **VIII**, 57
[1,1'-Binaphthalene]-2,2'-diol, (±)- [602-09-5]; (*R*)- [18531-94-7]; (*S*)-(−)- [18531-99-2]
  1,1'-BINAPHTHALENE-2,2'-DIOL, **VIII**, 46, 50, 57
1,1'-Biphenyl [92-52-4]
  BIPHENYL, **VI**, 150, 490
1,1'-Biphenyl, 2-azido-4'-nitro- [14191-25-4]
  o-AZIDO-*p*'-NITROBIPHENYL, **V**, 830
1,1'-Biphenyl, 4-bromo- [92-66-0]
  p-BROMOBIPHENYL, **I**, 113
1,1'-Biphenyl, 4,4'-dibromo- [92-86-4]
  4,4'-DIBROMOBIPHENYL, **IV**, 256
1,1'-Biphenyl, 3,4-dichloro [2974-92-7]
  3,4-DICHLOROBIPHENYL, **V**, 51
1,1'-Biphenyl, 4,4'-difluoro- [398-23-2]
  4,4'-DIFLUOROBIPHENYL, **II**, 188
1,1'-Biphenyl, 3,3'-dimethoxy- [6161-50-8]
  3,3'-DIMETHOXYBIPHENYL, **III**, 295
1,1'-Biphenyl, 4,4'-dimethoxy- [2132-80-1]
  4,4'-DIMETHOXY-1,1'-BIPHENYL, **VI**, 468, 490,
1,1'-Biphenyl, 3,3'-dimethyl- [612-75-9]
  3,3'-DIMETHYLBIPHENYL, **III**, 295
1,1'-Biphenyl, 4,4'-dimethyl- [613-33-2]
  4,4'-DIMETHYLBIPHENYL, **VI**, 488
1,1'-Biphenyl, 2,2'-dinitro- [2436-96-6]
  2,2'-DINITROBIPHENYL, **III**, 339
1,1'-Biphenyl, 2-methyl- [643-58-3]
  2-METHYLBIPHENYL, **VI**, 747

1,1'-Biphenyl, 2-methyl-4'-nitro- [33350-73-1]
2-METHYL-4'-NITROBIPHENYL, **VIII**, 430

1,1'-Biphenyl, 3-nitro- [2113-58-8]
m-NITROBIPHENYL, **IV**, 718

[1,1'-Biphenyl]-2-amine [90-41-5]
o-AMINOBIPHENYL, **V**, 830

[1,1'-Biphenyl]-2-amine, 4'-nitro- [6272-52-2]
o-AMINO-*p*'-NITROBIPHENYL, **V**, 830

[1,1'-Biphenyl]-2-carboxylic acid, 2'-formyl- [6720-26-9]
DIPHENALDEHYDIC ACID, **V**, 493

[1,1'-Biphenyl]-4,4-diamine, *N*,*N*'-diethyl- [6290-86-4]
N,*N*'-DIETHYLBENZIDINE, **IV**, 283

[1,1'-Biphenyl]-2,2'-dicarboxaldehyde [1210-05-5]
DIPHENALDEHYDE, **V**, 489

[1,1'-Biphenyl]-2,2''dicarboxaldehyde, 4,4',5,5'-tetramethoxy- [29237-14-7]
4,5,4',5'-TETRAMETHOXY-1,1'-BIPHENYL-2,2'-DICARBOXALDEHYDE,
VIII, 586

[1,1'-Biphenyl]-2,2'-dicarboxylic acid [482-05-3]
DIPHENIC ACID, **I**, 222

[1,1'-Biphenyl]-2,2'-dicarboxylic acid, 4,4',6,6'-tetrachloro-, (±)- [53663-22-2]
DL-4,4',6,6'-TETRACHLORODIPHENIC ACID, **IV**, 872

[1,1'-Biphenyl]-3,3'-diol [612-76-0]
3,3'-DIHYDROXYBIPHENYL, **V**, 412

Biphenylene [259-79-0]
BIPHENYLENE, **V**, 54

[1,1'-Biphenyl]-4-methanol, α,α-bis([1,1'-biphenyl]-4-yl)- [5341-14-0]
TRIBIPHENYLCARBINOL, **III**, 831

2,2'-Bipyridine [366-18-7]
2,2'-BIPYRIDINE, **V**, 102

[5,5'-Bipyrimidine]-2,2',4,4',6,6'(1*H*,1'*H*,3*H*,3'*H*,5*H*,5'*H*)-hexone, 5,5'-dihydroxy-,
dihydrate [6011-27-4]
ALLOXANTIN DIHYDRATE, **III**, 42; **IV**, 25

9-Borabicyclo[3.3.1]nonane, 9-(2,6,6-trimethylbicyclo[3.1.1]hept-3-yl)- [73624-47-2]
B-3-PINANYL-9-BORABICYCLO[3.3.1]NONANE, **VII**, 402

9*b*-Boraphenalene, dodecahydro- [16664-33-8] *cis,cis*- (3aα,6aα,9aα)-
[1130-59-2]; *cis,trans*- (3aα,6aα,9aβ)- [2938-53-6]
PERHYDRO-9b-BORAPHENALENE, **VII**, 427

Boric acid (H$_3$BO$_3$), tributyl ester [688-74-4]
BUTYL BORATE, **II**, 106

Boric acid, tripentyl ester [621-78-3]
AMYL BORATE, **II**, 107

Boroxin, triphenyl- [3262-89-3]
BENZENEBORONIC ANHYDRIDE, **IV**, 68

1,3-Butadiene [106-99-0]
1,3-BUTADIENE, **II**, 102

1,3-Butadiene, 1-chloro-1,4,4-trifuoro- [764-14-7]
    1-CHLORO-1,4,4-TRIFLUOROBUTADIENE, **V**, 235
1,3-Butadiene, 2,3-dimethyl- [513-81-5]
    2,3-DIMETHYL-1,3-BUTADIENE, **III**, 312
1,3-Butadiene, 2,3-diphenyl- [2548-47-2]
    2,3-DIPHENYL-1,3-BUTADIENE, **VI**, 531
1,3-Butadiene-1,4-diol, diacetate, (*E*,*E*-) [15910-11-9]
    trans, *trans*-1,3-BUTADIENE-1,4-DIYL DIACETATE, **VI**, 196
2,3-Butadienoic acid, methyl ester [18913-35-4]
    METHYL BUTADIENOATE, **V**, 734
Butanal, 2-methyl-, (*S*)- [1730-97-8]
    2-METHYLBUTANAL, **VIII**, 367
Butanal-1-*d*, 2-methyl- [25132-57-4]
    2-METHYLBUTANAL, 1-*d*, **VI**, 751
Butanal, 2,2,3-trichloro- [76-36-8]
    BUTYRCHLORAL, **IV**, 130
Butanamide, 3-oxo-*N*-phenyl- [102-01-2]
    ACETOACETANILIDE, **III**, 10
1-Butanamine, *N*-methyl- [110-68-9]
    N-METHYLBUTYLAMINE, **V**, 736
1-Butanaminium, *N*,*N*,*N*-tributyl-, (diphosphate) (3:1) [76947-02-9]
    TRIS(TETRABUTYLAMMONIUM)HYDROGEN PYROPHOSPHATE
    TRIHYDRATE, **VIII**, 616
Butane, 1-bromo- [109-65-9]
    BUTYL BROMIDE, **I**, 28, 37
Butane, 2-bromo- [78-76-2]
    sec-BUTYL BROMIDE, **I**, 38
Butane, 1-bromo-3-methyl- [107-82-4]
    iso-AMYL BROMIDE, **I**, 27
Butane, 4-bromo-, 1,1,1-trimethoxy- [55444-67-2]
    TRIMETHYL ORTHO-4-BROMOBUTANOATE, **VIII**, 415
Butane, 1-chloro- [109-69-3]
    BUTYL CHLORIDE, **I**, 142
Butane, 2-chloro-2-methyl- [594-36-5]
    2-CHLORO-2-METHYLBUTANE, **I**, 144; **VII**, 425
Butane, 2,3-dibromo-2-methyl- [594-51-4]
    TRIMETHYLETHYLENE DIBROMIDE, **II**, 409
Butane, 1,4-diiodo- [628-21-7]
    1,4-DIIODOBUTANE, **IV**, 321
Butane, 1,4-dinitro- [4286-49-1]
    1,4-DINITROBUTANE, **IV**, 368
Butane, 1,1'-[(1-methylethylidene)bis(oxy)]bis- [141-72-0]
    ACETONE DIBUTYL ACETAL, **V**, 5
Butane, 1,1'-oxybis[4-chloro- [6334-96-9]
    4,4'-DICHLORODIBUTYL ETHER, **IV**, 266

Butanediamide, 2,3-dihydroxy-*N,N,N',N'*-tetramethyl-, [*R*-(*R**,*R**)]- [26549-65-5]
   (*R,R*)-(+)-*N,N,N',N'*-TETRAMETHYLTARTARIC ACID DIAMIDE,
      VII, 41
Butanediamide, 2,3-dimethoxy-*N,N,N',N'*-tetramethyl-, (*R,R*)- [26549-29-1]
   (*R,R*)-(+)-2,3-DIMETHOXY-*N,N,N',N'*-TETRAMETHYLSUCCINIC ACID
      DIAMIDE, **VII**, 41
1,4-Butanediamine, dihydrochloride [333-93-7]
   PUTRESCINE DIHYDROCHLORIDE, **IV**, 819
1,4-Butanediamine, 2,3-dimethoxy-*N,N,N',N'*-tetramethyl-, (*S,S*)-(+)- [26549-21-3]
   1,4-BIS(DIMETHYLAMINO)-2,3-DIMETHOXYBUTANE, (*S,S*)-(+)-, [DDB],
      VII, 41
Butanedinitrile, 2,3-diphenyl- [5424-86-2]
   2,3-DIPHENYLSUCCINONITRILE, **IV**, 392
Butanedioic acid, bis[5-methyl-2-(1-methylethyl)cyclohexyl] ester
      [1*R*-[1α(1*R**,2*S**,5*R**)], 2β,5α]]- [34212-59-4]
   (−)-DIMENTHYL SUCCINATE, **VIII**, 141
Butanedioic acid, diethyl ester [123-25-1]
   DIETHYL SUCCINATE, **V**, 993
Butanedioic acid, diphenyl ester [621-14-7]
   DIPHENYL SUCCINATE, **IV**, 390
Butanedioic acid, monomethyl ester [3878-55-5]
   METHYL HYDROGEN SUCCINATE, **III**, 169
Butanedioic acid, 2,3-dibromo- [526-78-3]
   α,β-DIBROMOSUCCINIC ACID, **II**, 177
Butanedioic acid, 2,2-difluoro- [665-31-6]
   2,2-DIFLUOROSUCCINIC ACID, **V**, 393
Butanedioic acid, 2,3-dihydroxy, (*R**,*R**)-(+)- [133-37-9]
   DL-TARTARIC ACID, **I**, 497
Butanedioic acid, dioxo-, diethyl ester [59743-08-7]
   DIETHYL DIOXOSUCCINATE, **VIII**, 597
Butanedioic acid, (diphenylmethylene)- ethyl ester [5438-22-2]
   β-CARBETHOXY-γ,γ-DIPHENYLVINYLACETIC ACID, **IV**, 132
Butanedioic acid, 2-ethyl-2-methyl- [631-31-2]
   α-ETHYL-α-METHYLSUCCINIC ACID, **V**, 572
Butanedioic acid, 2-hydroxy-3-(2-propenyl)-, (*S*-(*R**,*S**))-, diethyl ester
      [73837-97-5]
   DIETHYL (2*S*,3*R*)-(+)-3-ALLYL-2-HYDROXYSUCCINATE, **VII**, 153
Butanedioic acid, methyl- [498-21-5]
   METHYLSUCCINIC ACID, **III**, 615
Butanedioic acid, methylene- [97-65-4]
   ITACONIC ACID, **II**, 368
Butanedioic acid, 2-nitro-, dimethyl ester [28081-31-4]
   DIMETHYL NITROSUCCINATE, **VI**, 503
Butanedioic acid, (1-oxoheptyl)-, diethyl ester [41117-78-6]
   ETHYL ENANTHYLSUCCINATE, **IV**, 430

Butanedioic acid, 1-oxopropyl-, diethyl ester [4117-76-4]
  DIETHYL PROPIONYLSUCCINATE, **VI**, 615
Butanedioic acid, phenyl- [635-51-8]
  PHENYLSUCCINIC ACID, **I**, 451; **IV**, 804
Butanedioic acid, (3-phenyl-2-propenyl)- [5671-91-0]
  γ-PHENYLALLYLSUCCINIC ACID, **IV**, 766
2,3-Butanediol, monomethanesulfonate, (*R**,*S**)- [35405-98-2]
  erythro-2,3-BUTANEDIOL MONOMESYLATE, **VI**, 312
2,3-Butanediol, 1,4-bis(phenylmethoxy)-, [*S*-(*R**,*R**)]- [17401-06-8]
  1,4-DI-*O*-BENZYL-L-THREITOL, **VIII**, 155
2,3-Butanediol, 2,3-dimethyl-, hexahydrate [6091-58-3]
  PINACOL HYDRATE, **I**, 459
2,3-Butanedione, dioxime [95-45-4]
  DIMETHYLGLYOXIME, **II**, 204
2,3-Butanedione, monooxime [57-71-6]
  BIACETYL MONOXIME, **II**, 205
Butanedioyl dichloride, methylene- [1931-60-8]
  ITACONYL CHLORIDE, **IV**, 554
Butanenitrile, 2-carbonyl-3,3-dimethyl- [29342-22-1]
  tert-BUTYLCYANOKETENE, **VI**, 210
Butanenitrile, 4-chloro- [628-20-6]
  γ-CHLOROBUTYRONITRILE, **I**, 156
Butanenitrile, 2-(1-ethoxyethoxy)-, (*R**,*R**)- [72658-42-5] or (*R**,*S**)- [72658-43-6]
  2-[(1'-ETHOXY)-1-ETHOXY]BUTANENITRILE [72658-42-5], **VII**, 381
Butanenitrile, 2-hydroxy- [4476-02-2]
  2-HYDROXYBUTANENITRILE, **VII**, 381
1,2,3,4-Butanetetracarboxylic acid, tetraethyl ester [4373-15-3]
  TETRAETHYL 1,2,3,4-BUTANETETRACARBOXYLATE, **VII**, 479
Butanetetraylnaphthalene, 3,5,1,7-[1,2,3,4]-decahydro- [2292-79-7]
  DIAMANTANE, **VI**, 378
Butanethioic acid, *S*-(1-methyl-2-oxopropyl) ester [32425-98-2]
  S-(2-OXOBUT-3-YL) BUTANETHIOATE, **VI**, 776
Butanethioic acid, 3-oxo-, *S*-(1,1-dimethylethyl) ester [15925-47-0]
  S-*tert*-BUTYL ACETOTHIOACETATE, **VIII**, 71
Butanimidic acid, 4-bromo-, methyl ester, hydrochloride [21367-90-8]
  METHYL 4-BROMO-1-BUTANIMIDATE HYDROCHLORIDE, **VIII**, 415
Butanoic acid, butyl ester [109-21-7]
  BUTYL BUTYRATE, **I**, 138
Butanoic acid, 2-acetyl-3-methyl-, ethyl ester [1522-46-9]
  ETHYL α-ISOPROPYLACETOACETATE, **III**, 405
Butanoic acid, 2-acetyl-3-oxo-, ethyl ester [603-69-0]
  ETHYL DIACETYLACETATE, **III**, 390
Butanoic acid, 4-amino- [56-12-2]
  γ-AMINOBUTYRIC ACID, **II**, 25
Butanoic acid, 2-amino-2-ethyl- [2566-29-2]
  α-AMINODIETHYLACETIC ACID, **III**, 66

Butanoic acid, 4-bromo, ethyl ester [2969-81-5]
   ETHYL γ-BROMOBUTYRATE, **V**, 545
Butanoic acid, 2-bromo-3-methoxy-, *erythro*- [26839-93-0]; *threo*- [26839-91-8]
   α-BROMO-β–METHOXYBUTYRIC ACID, **III**, 813
Butanoic acid, 2-bromo-3-methyl- [565-74-2]
   α-BROMOISOVALERIC ACID, **II**, 93
Butanoic acid, 4-chloro-4-oxo, methyl ester [1490-25-1]
   β-CARBOMETHOXYPROPIONYL CHLORIDE, **III**, 169
Butanoic acid, 4-cyclohexyidene-4-(trimethylsilyl)oxy-, ethyl ester [65213-35-6]
   ETHYL 4-CYCLOHEXYLIDENE-4-(TRIMETHYLSILOXY)BUTANOATE, **VIII**, 578
Butanoic acid, 2-diazo-3-oxo-, 1,1-dimethylethyl ester [13298-76-5]
   tert-BUTYL α-DIAZOACETOACETATE, **V**, 180
Butanoic acid, 3-hydroxy-, (*S*)-, ethyl ester [56816-01-4]
   ETHYL (*S*)-(+)-3-HYDROXYBUTANOATE, **VII**, 215
Butanoic acid, 4-iodo-, ethyl ester [7425-53-8]
   ETHYL 4-IODOBUTYRATE, **VIII**, 274
Butanoic acid, 2-methyl-, (±)- [600-07-7]
   DL-METHYLETHYLACETIC ACID, **I**, 361
Butanoic acid, 3-methyl-, 1,1-dimethylethyl ester [16792-03-3]
   tert-BUTYL ISOVALERATE, **III**, 144
Butanoic acid, 2-nitro-, ethyl ester [2531-81-9]
   ETHYL α-NITROBUTYRATE, **IV**, 454
Butanoic acid, 3-oxo-, 1,1-dimethylethyl ester [1694-31-1]
   tert-BUTYL ACETOACETATE, **V**, 155
Butanoic acid, 3-oxo-, ethyl ester [141-97-9]
   ETHYL ACETOACETATE, **I**, 235
Butanoic acid, 3-oxo-, 1-ethenylheptyl ester [133538-60-0]
   (1-ETHENYL)HEPTANYL 3-KETOBUTANOATE, **VIII**, 235
Butanoic acid, 3-oxo-, 2-methoxy-1,1-dimethyl-2-oxoethyl ester [110451-07-5]
   1-CARBOMETHOXY-1-METHYLETHYL 3-OXOBUTANOATE, **VIII**, 71
1-Butanol, 2-amino-3-methyl-, (*S*)- [2026-48-4]
   VALINOL, L-, **VII**, 530,
1-Butanol, 4-chloro- [928-51-8]
   TETRAMETHYLENE CHLOROHYDRIN, **II**, 571
1-Butanol, 4-chloro-, benzoate [946-02-1]
   4-CHLOROBUTYL BENZOATE, **III**, 187
2-Butanol, 3-methyl- [598-75-4]
   METHYL ISOPROPYL CARBINOL, **II**, 406
1-Butanone, 3-hydroxy-3-methyl-1-phenyl- [43108-74-3]
   3-HYDROXY-3-METHYL-1-PHENYL-1-BUTANONE, **VIII**, 210, 323
1-Butanone, 1-(1-methyl-1*H*-pyrrol-3-yl)- [62128-46-5]
   3-BUTYROYL-1-METHYLPYRROLE, **VII**, 102
2-Butanone, 1-bromo-3-methyl [19967-55-6]
   1-BROMO-3-METHYL-2-BUTANONE, **VI**, 193
2-Butanone, 4-(4-chlorophenyl)- [3506-75-0]
   4-(4-CHLOROPHENYL)BUTAN-2-ONE, **VII**, 105

2-Butanone, 1-diazo-4-phenyl- [10290-42-3]
    1-DIAZO-4-PHENYL-2-BUTANONE, **VIII**, 196
2-Butanone, 1,3-dibromo- [815-51-0]
    1,3-DIBROMO-2-BUTANONE, **VI**, 711
2-Butanone, 3,4-dibromo-4-phenyl- [6310-44-7]
    BENZALACETONE DIBROMIDE, **III**, 105
2-Butanone, 4-(diethylamino)- [3299-38-5]
    1-DIETHYLAMINO-3-BUTANONE, **IV**, 281
2-Butanone, 3,3-dimethyl- [75-97-8]
    PINACOLONE, **I**, 462
2-Butanone, 3-methyl- [563-80-4]
    METHYL ISOPROPYL KETONE, **II**, 408
2-Butanone, 4-phenylmethoxy- [6278-91-7]
    4-BENZYLOXY-2-BUTANONE, **VII**, 386
2-Butenal, 2-chloro- [53175-28-3]
    α-CHLOROCROTONALDEHYDE, **IV**, 131
2-Butene, 2,3-dinitro-, (*E*)- [24335-43-1]; (*Z*)- [24335-44-2]
    2,3-DINITRO-2-BUTENE, **IV**, 374
2-Butenediamide, (*E*)- [627-64-5]
    FUMARAMIDE, **IV**, 486
2-Butenedinitrile, (*E*)- [764-42-1]
    FUMARONITRILE, **IV**, 486
2-Butenedinitrile, 2,3-diamino, (*Z*)- [1187-42-4]
    DIAMINOMALEONITRILE (HYDROGEN CYANIDE TETRAMER),
        V, 344
2-Butenedioic acid, (*E*)- [110-17-8]
    FUMARIC ACID, **II**, 302
2-Butenedioic acid, (*E*)-, 1,1-dimethylethyl ethyl ester [100922-16-5]
    tert-BUTYL ETHYL FUMARATE, **VII**, 93
2-Butenedioic acid, 2-methyl-, (*E*)- [498-24-8]
    MESACONIC ACID, **II**, 382
2-Butenedioic acid, 2-methyl-, (*Z*)- [498-23-7]
    CITRACONIC ACID, **II**, 140
2-Butene-1,4-dione, 1,4-diphenyl-, (*E*)- [959-28-4]
    trans-DIBENZOYLETHYLENE, **III**, 248
2-Butenedioyl dichloride, (*E*)- [627-63-4]
    FUMARYL CHLORIDE, **III**, 422
3-Butenenitrile [109-75-1]
    ALLYL CYANIDE, **I**, 46; **III**, 852
2-Butenoic acid, (*Z*)- [503-64-0]
    ISOCROTONIC ACID, **VI**, 711
2-Butenoic acid, (*Z*)-, ethyl ester [6776-19-8]
    ETHYL ISOCROTONATE, **VII**, 227
2-Butenoic acid, 1-methylpropyl ester, (*E*)- [10371-45-6]
    sec-BUTYL CROTONATE, **V**, 762
2-Butenoic acid, 2-cyano-3-phenyl-, ethyl ester [18300-89-5]
    ETHYL(1-PHENYLETHYLIDENE)CYANOACETATE, **IV**, 463

2-Butenoic acid, 2,3-dibromo-4-oxo-, (Z)- [488-11-9]
MUCOBROMIC ACID, **III**, 621; **IV**, 688
2-Butenoic acid, 4-hydroxy-, (E)-, ethyl ester [10080-68-9]
ETHYL 4-HYDROXYCROTONATE, **VII**, 221
2-Butenoic acid, 3-methyl- [541-47-9]
β,β-DIMETHYLACRYLIC ACID, **III**, 302
2-Butenoic acid, 4-(4-nitrophenyl)-4-oxo-, ethyl ester [131504-53-5]
ETHYL (E)-4-(4-NITROPHENYL)-4-OXO-2-BUTENOATE, **VIII**, 268
2-Butenoic acid, 4-oxo-4-phenyl- [583-06-2]
β-BENZOYLACRYLIC ACID, **III**, 109
2-Butenoic acid, 4-oxo-4-(phenylamino)- [37902-58-2]
MALEANILIC ACID, **V**, 944
2-Butenoic acid, 3-(phenylamino)-, ethyl ester [31407-07-5]
ETHYL β-ANILINOCROTONATE, **III**, 374
2-Butenoic acid, 3-(1-pyrrolidinyl)-, ethyl ester [54716-02-8]
ETHYL β-PYRROLIDINOCROTONATE, **VI**, 592
3-Butenoic acid [625-38-7]
VINYLACETIC ACID, **III**, 851
2-Buten-1-ol, 3-chloro- [40605-42-3]
3-CHLORO-2-BUTEN-1-OL, **IV**, 128
2-Buten-1-ol, 4-chloro-, acetate (E)- [34414-28-3]
1-ACETOXY-4-CHLORO-2-BUTENE, **VIII**, 9
2-Buten-1-ol, 4-(diethylamino)-, acetate (ester) [82736-47-8]
1-ACETOXY-4-DIETHYLAMINO-2-BUTENE, **VIII**, 9
2-Buten-1-ol, 4-[(phenylmethyl)amino]-, acetate (ester), (E)- [130892-14-7]
1-ACETOXY-4-BENZYLAMINO-2-BUTENE, **VIII**, 9,
3-Buten-1-ol, 2-chloro-, acetate [96039-67-7]
4-ACETOXY-3-CHLORO-1-BUTENE, **VIII**, 9,
3-Buten-1-ol, 4-(trimethylsilyl)-, (Z)- [87682-77-7]
(Z)-4-(TRIMETHYLSILYL)-3-BUTEN-1-OL, **VIII**, 358, 609
3-Buten-1-ol, 4-(trimethylsilyl)-, 4-methylbenzenesulfonate, (Z)- [87682-62-0]
(Z)-4-(TRIMETHYLSILYL)-3-BUTENYL 4-METHYLBENZENESULFONATE,
VIII, 358
3-Buten-2-ol, 4-phenyl- [17488-65-2]
trans-METHYLSTYRYLCARBINOL, **IV**, 773
3-Buten-2-ol, 3-(trimethylsilyl)- [66374-47-8]
3-TRIMETHYLSILYL-3-BUTEN-2-OL, **VI**, 1033
2-Buten-1-one, 1,3-diphenyl- [495-45-4]
DYPNONE, **III**, 367
2-Buten-1-one, 3-methyl-1-phenyl- [5650-07-7]
ISOPROPYLIDENACETOPHENONE, **VIII**, 210,
3-Buten-2-one, 3-bromo-4-phenyl- [31207-17-7]
α-BROMOBENZALACETONE, **III**, 125
3-Buten-2-one, 4-(2-furanyl)- [623-15-4]
2-FURFURALACETONE, **I**, 283
3-Buten-2-one, 4-phenyl- [122-57-6]
BENZALACETONE, **I**, 77

3-Buten-2-one, 1,1,1-trichloro-4-ethoxy- [83124-74-7]
    1,1,1-TRICHLORO-4-ETHOXY-3-BUTEN-2-ONE, **VIII**, 254
3-Buten-2-one, 1,1,1-trichloro-4-methoxy- [138149-14-1]
    1,1,1-TRICHLORO-4-METHOXY-3-BUTEN-2-ONE, **VIII**, 238
3-Buten-2-one, 3-(trimethylsilyl)- [43209-86-5]
    3-TRIMETHYLSILYL-3-BUTEN-2-ONE, **VI**, 1033
1-Buten-3-yne [689-97-4]
    MONOVINYLACETYLENE, **IV**, 683
1-Butyne, 1-ethoxy- [14272-91-4]
    1-ETHOXY-1-BUTYNE, **VI**, 564
2-Butynedioic acid [142-45-0]
    ACETYLENEDICARBOXYLIC ACID, **II**, 10
2-Butynedioic acid, diethyl ester [762-21-0]
    DIETHYL ACETYLENEDICARBOXYLATE, **IV**, 330
2-Butynedioic acid, dimethyl ester [762-42-5]
    DIMETHYL ACETYLENECARBOXYLATE, **IV**, 329
2-Butynoic acid [590-93-2]
    TETROLIC ACID, **V**, 1043
2-Butynoic acid, ethyl ester [4341-76-8]
    ETHYL TETROLATE, **VII**, 226
2-Butynoic acid, 4-chloro-, methyl ester [41658-12-2]
    METHYL 4-CHLORO-2-BUTYNOATE, **VIII**, 371
2-Butynoic acid, 4-hydroxy-, methyl ester [31555-05-2]
    METHYL 4-HYDROXY-2-BUTYNOATE, **VII**, 334
2-Butyn-1-ol [764-01-2]
    2-BUTYN-1-OL, **IV**, 128
3-Butyn-1-ol, 4-(trimethylsilyl)- [2117-12-6]
    4-(TRIMETHYLSILYL)-3-BUTYN-1-OL, **VIII**, 609
3-Butyn-2-ol, 2-methyl- [115-19-5]
    DIMETHYLETHYNYLCARBINOL, **III**, 320

C

Carbamic acid, butyl ester [592-35-8]
    BUTYL CARBAMATE, **I**, 140
Carbamic acid, 1,1-dimethylethyl ester [4248-19-5]
    tert-BUTYL CARBAMATE, **V**, 162
Carbamic acid, phenylmethyl ester [621-84-1]
    BENZYL CARBAMATE, **III**, 167
Carbamic acid, 1,3-butadienyl-, phenylmethyl ester, (*E*)- [71616-72-3]
    BENZYL *trans*-1,3-BUTADIENE-1-CARBAMATE, **VI**, 95
Carbamic acid, (chlorosulfonyl)-, methyl ester [36914-92-8]
    METHYL (CHLOROSULFONYL)CARBAMATE, **VI**, 788
Carbamic acid, dichloro-, methyl ester [16487-46-0]
    METHYL *N,N*-DICHLOROCARBAMATE, **VIII**, 427
Carbamic acid, (1-ethoxycyclopropyl)-, 1,1-dimethylethyl ester [28750-48-3]
    tert-BUTYL *N*-(1-ETHOXYCYCLOPROPYL)CARBAMATE, **VI**, 226

Carbamic acid, (1-formyl-3-methylbutyl)-, 1,1-dimethylethyl ester, (S)- [58521-45-2]
    N-*tert*-BUTOXYCARBONYL-L-LEUCINAL, **VIII**, 68
Carbamic acid, hexyl-, methyl ester [22139-32-8]
    METHYL N-HEXYLCARBAMATE, **VI**, 788
Carbamic acid, (2-hydroxy-1,2-diphenylethyl)-, (R*,R*)-, ethyl ester [73197-89-4]
    ETHYL threo-[1-(2-HYDROXY-1,2-DIPHENYL)ETHYL]CARBAMATE,
    VII, 223
Carbamic acid, [1-[(methoxymethylamino)carbonyl]-3-methylbutyl]-, 1,1-dimethylethyl
        ester, (S)- [87694-50-6]
    tert-BUTOXYCARBONYL-L-LEUCINE-N-METHYL-O-
    METHYLCARBOXAMIDE, **VIII**, 68
Carbamic acid, methyl-, ethyl ester [105-40-8]
    ETHYL N-METHYLCARBAMATE, **II**, 278
Carbamic acid, (2-methyl-2-butenyl)-, methyl ester [86766-65-6]
    METHYL N-(2-METHYL-2-BUTENYL)CARBAMATE, **VIII**, 427
Carbamic acid, methylnitroso-, ethyl ester [615-53-2]
    NITROSOMETHYLURETHANE, **II**, 464; *Warning*, **V**, 842
Carbamic acid, [[(4-methylphenyl)sulfonyl]methyl]-, ethyl ester [2850-26-2]
    ETHYL N-(p-TOLYLSULFONYLMETHYL)CARBAMATE, **VI**, 981
Carbamic acid, [[(4-methylphenyl)sulfonyl]methyl]nitroso-, ethyl ester [2951-53-3]
    ETHYL N-NITROSO-N-(p-TOLYLSULFONYLMETHYL)CARBAMATE, **VI**, 981
Carbamic acid, (phenylmethyl)-, ethyl ester [2621-78-5]
    ETHYL N-BENZYLCARBAMATE, **IV**, 780
Carbamic acid, (1,2,3,4-tetrahydro-2-iodo-1-naphthalenyl)-, methyl ester, *trans*-
        [1210-13-5]
    METHYL (*trans*-2-IODO-1-TETRALIN)CARBAMATE, **VI**, 795
Carbamimidic acid, methyl ester, monohydrochloride [5329-33-9]
    METHYLISOUREA HYDROCHLORIDE, **IV**, 645
Carbamimidothioic acid, methyl ester, sulfate (2:1) [867-44-7]
    S-METHYL ISOTHIOUREA SULFATE, **II**, 411
Carbamothioic acid, dimethyl-, O-2-naphthalenyl ester [2951-24-8]
    O-2-NAPHTHYL DIMETHYLTHIOCARBAMATE, **VI**, 824
Carbamothioic chloride, diethyl- [88-11-9]
    DIETHYLTHIOCARBAMYL CHLORIDE, **IV**, 307
1H-Carbazole, 2,3,4,9-tetrahydro- [942-01-8]
    1,2,3,4-TETRAHYDROCARBAZOLE, **IV**, 884
9H-Carbazole, 2-nitro- [14191-22-1]
    2-NITROCARBAZOLE, **V**, 829
[Carboethoxymethylene]triphenylphosphorane [1099-45-2]
    ETHYL (TRIPHENYLPHOSPHORANYLIDENE)ACETATE, **VII**, 232
Carbonazidic acid, 1,1-dimethylethyl ester [1070-19-5]
    tert-BUTYL AZIDOFORMATE, **V**, 157; **VI**, 207
Carbohydrazide, 1,5-diphenyl-3-thio- [622-03-7]
    DIPHENYLTHIOCARBAZIDE, **III**, 360
Carbonic acid, compound, with hydrazinecarboximidamide(1:1) [2582-30-1]
    AMINOGUANIDINE BICARBONATE, **III**, 73

Carbonic acid, 1,1-dimethylethyl phenyl ester [6627-89-0]
    tert-BUTYL PHENYL CARBONATE, **V**, 166
Carbonic acid, monoanhydride with diethyl phosphate, *tert*-butyl ester [14618-58-7]
    tert-BUTYLCARBONIC DIETHYLPHOSPHORIC ANHYDRIDE, **VI**, 207
Carbonic acid, thio-, *O-tert*-butyl *S*-methyl ester [29518-83-0]
    tert-BUTYL *S*-METHYLTHIOLCARBONATE, **V**, 168
Carbonochloridic acid, trichloromethyl ester [503-38-8]
    TRICHLOROMETHYL CHLOROFORMATE, **VI**, 715
Carbonothioic dichloride [463-71-8]
    THIOPHOSGENE, **I**, 506
Casein [9000-71-9]
    CASEIN, **II**, 120
Cholan-24-al [26606-02-0]
    CHOLAN-24-AL, **V**, 242
Cholane-24-thioic acid, 3,7,12-trihydroxy-, *S*-(1,1-dimethylethyl) ester, (3α,5β,7α,12α)-
    [58587-05-6]
    S-tert-BUTYL 3a,7a,12a-TRIHYDROXY-5β-CHOLANE-24-THIOATE,
    VII, 81
Cholan-24-oic acid, 3a,12a-dihydroxy-, methyl ester [20231-66-7]
    METHYL DESOXYCHOLATE, **III**, 237
Chol-23-ene-3,12-diol, 24,24-diphenyl-, diacetate, (3α,5β,12α)- [53608-88-1]
    3,12-DIACETOXY-bisnor-CHOLANYLDIPHENYLETHYLENE, **III**, 237
Cholesta-3,5-diene [747-90-0]
    CHOLESTA-3,5-DIENE, **VIII**, 126
Cholesta-3,5-dien-3-ol, trifluoromethanesulfonate [95667-40-6]
    CHOLESTA-3,5-DIEN-3-YL TRIFLUOROMETHANESULFONATE,
    VIII, 126
Cholestane, (5α)- [481-21-0]
    CHOLESTANE, **VI**, 289
Cholestane, 3-methoxy-, (3β,5α)- [1981-90-4]
    CHOLESTANYL METHYL ETHER, **V**, 245
Cholestane-5-carbonitrile, 3-(acetyloxy)-7-oxo-, (3β,5α)- [2827-02-3]
    3β-ACETOXY-5α-CYANOCHOLESTAN-7-ONE, **VI**, 14
Cholestan-3-ol, (3β,5α)- [80-97-7]
    DIHYDROCHOLESTEROL, **II**, 191
5α-Cholestan-3β-ol, 5,6β-dibromo- [1857-80-3]
    CHOLESTEROL DIBROMIDE, **IV**, 195
Cholestan-3-one, (5α)- [566-88-1]
    CHOLESTANONE, **II**, 139
Cholestan-3-one, 5,6-dibromo-, (5α,6β)- [2515-09-5]
    5α,6β-DIBROMOCHOLESTAN-3-ONE, **IV**, 197
Cholest-3-ene, (5β)- [13901-20-7]
    5β-CHOLEST-3-ENE, **VI**, 293
Cholest-3-ene, 5-methyl-, (5β)- [23931-38-6]
    5-METHYLCOPROST-3-ENE, **VI**, 762

Cholest-4-ene, 3-ethenyloxy-, (3β)- [56101-54-3]
   3β-CHOLEST-4-ENYL VINYL ETHER, **VI**, 298
Cholest-3-ene-5-acetaldehyde, (5β)- [56101-55-4]
   5β-CHOLEST-3-ENE-5-ACETALDEHYDE, **VI**, 298
Cholest-3-ene-5-acetamide, *N,N*-dimethyl-, (5β)- [56255-03-9]
   N,N-DIMETHYL-5β-CHOLEST-3-ENE-5-ACETAMIDE, **VI**, 491
Cholest-3-ene-5-acetic acid, ethyl ester, (5β)- [56101-56-5]
   ETHYL 5β-CHOLEST-3-ENE-5-ACETATE, **VI**, 584
Cholest-4-ene-3,6-dione [984-84-9]
   Δ⁴-CHOLESTEN-3,6-DIONE, **IV**, 189
Cholest-5-en-3-ol, (3β)- [57-88-5]
   CHOLESTEROL, **IV**, 195
Cholest-3-en-3-ol, 5-methyl-, diethyl hydrogen phosphate [23931-37-5]
   DIETHYL 5-METHYLCOPROST-3-EN-3-YL PHOSPHATE, **VI**, 762
Cholest-4-en-3-one [601-57-0]
   CHOLESTENONE, **III**, 207; Δ⁴-CHOLEST-3-ONE, **IV**, 192, 195
Cholest-5-en-3-one [601-54-7]
   Δ⁵-CHOLESTEN-3-ONE, **IV**, 195
Chromate(1–), diamminetetrakis(thiocyanato-*N*)-, ammonium, (OC-6-11)- [13573-16-5]
   REINECKE SALT, **II**, 555
Chromium, pentacarbonyl(1-methoxyethylidene)-, (OC-6-21)- [20540-69-6]
   [(METHYL)(METHOXY)CARBENE]PENTACARBONYL CHROMIUM(0),
     VIII, 216
Cobalt(1+), hexacarbonyl[*m*-[(2,3-η:2,3-η)-1-methyl-2-propynylium]]di-, (Co-Co), tetra-
   fluoroborate(1-) [62866-98-2]
   HEXACARBONYL (1-METHYL-2-PROPYNYLIUM)DICOBALT
     TETRAFLUOROBORATE, **VIII**, 460
Copper, tetrakis(pentafluorophenyl)tetra- [34077-61-7]
   PENTAFLUOROPHENYLCOPPER TETRAMER, **VI**, 875
Cyanamide, (2-chlorophenyl)- [45765-25-1]
   o-CHLOROPHENYLCYANAMIDE, **IV**, 172
Cyanamide, di-2-propenyl- [538-08-9]
   DIALLYLCYANAMIDE, **I**, 203
Cyanamide, methyl-1-naphthalenyl- [53663-33-5]
   N-METHYL-1-NAPHTHYLCYANAMIDE, **III**, 608
Cyanic acid, phenyl ester [1122-85-6]
   PHENYL CYANATE, **VII**, 435
Cyanogen bromide [506-68-3]
   CYANOGEN BROMIDE, **II**, 150
Cyclobuta-[*f*]-1,3-benzodioxole, 5,6-dihydro- [61099-23-8]
   4,5-METHYLENEDIOXYBENZOCYCLOBUTENE, **VII**, 326
Cyclobutanamine [2516-34-9]
   CYCLOBUTYLAMINE, **V**, 273
Cyclobutanamine hydrochloride [6291-01-6]
   CYCLOBUTYLAMINE HYDROCHLORIDE, **VIII**, 132

Cyclobutane, 1-bromo-3-chloro- [4935-03-9]
    1-BROMO-3-CHLOROCYCLOBUTANE, **VI**, 179, 272
Cyclobutane, 1,2-dibromo- [89033-70-5]
    1,2-DIBROMOCYCLOBUTANE, **VII**, 119
Cyclobutane, 1,2-diethenyl-, *cis*- [16177-46-1]
    cis-1,2-DIVINYLCYCLOBUTANE, **V**, 528
Cyclobutane, 1,2-diethenyl-, *trans*- [6553-48-6]
    trans-1,2-DIVINYLCYCLOBUTANE, **V**, 528
Cyclobutane, 1,1,2-trichloro-2,3,3-trifluoro- [697-17-6]
    1,1,2-TRICHLORO-2,3,3-TRIFLUOROCYCLOBUTANE, **V**, 393
Cyclobutanecarboxaldehyde [2987-17-9]
    CYCLOBUTANECARBOXALDEHYDE, **VI**, 312
Cyclobutanecarboxamide [1503-98-6]
    CYCLOBUTANECARBOXAMIDE, **VIII**, 132
Cyclobutanecarboxylic acid [3721-95-7]
    CYCLOBUTANECARBOXYLIC ACID, **III**, 213
Cyclobutanecarboxylic acid, 3-chloro- [35207-71-7]
    3-CHLOROCYCLOBUTANECARBOXYLIC ACID, **VI**, 271,
Cyclobutanecarboxylic acid, 1-methyl-2-[(methylsulfonyl)oxy]propyl ester, (*R**,*S**)-
    [35358-24-8]
    erythro-2-METHANESULFONYLOXY-2-BUTYL CYCLOBUTANE-
    CARBOXYLATE, **VI**, 312
Cyclobutanedicarbonitrile, 1-chloro- [3716-98-1]
    1-CHLORO-1,2-DICYANOCYCLOBUTANE, **VI**, 427
1,1-Cyclobutanedicarboxylic acid [5445-51-2]
    1,1-CYCLOBUTANEDICARBOXYLIC ACID, **III**, 213
1,1-Cyclobutanedicarboxylic acid, diethyl ester [3779-29-1]
    DIETHYL 1,1-CYCLOBUTANEDICARBOXYLATE, **IV**, 288
1,2-Cyclobutanedicarboxylic acid, 3-methylene-, dimethyl ester [53684-52-9]
    DIMETHYL 3-METHYLENECYCLOBUTANE-1,2-DICARBOXYLATE,
    V, 459
1,2-Cyclobutanediol, *cis*- [35358-33-9] and *trans*- [35358-34-0]
    1,2-CYCLOBUTANEDIOL, **VII**, 129
1,2-Cyclobutanedione [33689-28-0]
    1,2-CYCLOBUTANEDIONE, **VII**, 112
Cyclobutanol [2919-23-5]
    CYCLOBUTANOL, **VII**, 114, 117
Cyclobutanol, 4-methylbenzenesulfonate [10437-85-1]
    CYCLOBUTYL TOSYLATE, **VII**, 117
Cyclobutanone [1191-95-3]
    CYCLOBUTANONE, **VI**, 316, 320, 324; **VII**, 114
Cyclobutanone, 2-(1-ethoxycyclohexyl)-2-[(trimethylsilyl)oxy]- [69152-09-6]
    2-(1-ETHOXYCYCLOHEXYL)-2-(TRIMETHYLSILOXY)CYCLOBUTANONE,
    VIII, 579
Cyclobutanone, 3-hexyl- [138173-74-7]
    3-HEXYLCYCLOBUTANONE, **VIII**, 306

Cyclobutanone, 2-hydroxy- [17082-63-2]
    2-HYDROXYCYCLOBUTANONE, **VI**, 167
Cyclobutanone, 2-(1-propenyl)-, (*E*)- [63049-06-9]
    2-(1-PROPENYL)CYCLOBUTANONE, (*E*)-, **VIII**, 556
Cyclobutene [822-35-5]
    CYCLOBUTENE, **VII**, 117
Cyclobutene, 1-chloro-2,3,3-trifluoro- [694-62-2]
    1-CHLORO-2,3,3-TRIFLUOROCYCLOBUTENE, **V**, 394
Cyclobutene, 3,4-dichloro-, *cis*- [2957-95-1]
    cis-3,4-DICHLOROCYCLOBUTENE, **VI**, 422
Cyclobutene, 3,4-dichloro-1,2,3,4-tetramethyl- [1194-30-5]
    3,4-DICHLORO-1,2,3,4-TETRAMETHYLCYCLOBUTENE, **V**, 370
1-Cyclobutene-1,2-dicarbonitrile [3716-97-0]
    1,2-DICYANOCYCLOBUTENE, **VI**, 427
2-Cyclobuten-1-one, 3-butyl- [38425-48-8]
    3-BUTYLCYCLOBUTENONE, **VIII**, 82
2-Cyclobuten-1-one, 3-butyl-4,4-dichloro- [72284-70-9]
    3-BUTYL-4,4-DICHLOROCYCLOBUTENONE, **VIII**, 82
1,2-Cyclodecanediol [21014-77-7]
    1,2-CYCLODECANEDIOL, **IV**, 216
1,2-Cyclodecanedione [96-01-5]
    SEBACIL, **IV**, 838
Cyclododecanethiol [7447-11-2]
    CYCLODODECYL MERCAPTAN, **VII**, 124
Cyclodecanone [1502-06-3]
    CYCLODECANONE, **IV**, 218; **V**, 277
Cyclodecanone, 2-hydroxy- [96-00-4]
    SEBACOIN, **IV**, 840
Cyclododecanone, 2,12-dibromo- [24459-40-3]
    2,12-DIBROMOCYCLODODECANONE, **VI**, 368
Cyclododecene, (*E*)- [1486-75-5]
    CYCLODODECENE, (*E*)-, **VII**, 121
Cyclododecene, (*Z*)- [1129-89-1]
    cis-CYCLODODECENE, **V**, 281
Cycloheptanone [502-42-1]
    CYCLOHEPTANONE, **IV**, 221
Cycloheptanone, 2-phenyl- [14996-78-2]
    2-PHENYLCYCLOHEPTANONE, **IV**, 780
2,4,6-Cycloheptatrien-1-one, 2-hydroxy- [533-75-5]
    TROPOLONE, **VI**, 1037
Cycloheptatrienylium tetrafluoroborate(1−) [27081-10-3]
    TROPYLIUM FLUOBORATE, **V**, 1133
2-Cyclohepten-1-ol, acetate [826-13-1]
    2-CYCLOHEPTEN-1-YL ACETATE, **VIII**, 137
2-Cyclohepten-1-one [1121-66-0]
    2-CYCLOHEPTEN-1-ONE, **VI**, 327

1,3-Cyclohexadiene [592-57-4]
  1,3-CYCLOHEXADIENE, **V**, 285
1,4-Cyclohexadiene, 1,2-dimethyl- [17351-28-9]
  1,2-DIMETHYL-1,4-CYCLOHEXADIENE, **V**, 467
1,4-Cyclohexadiene, 1-methoxy- [2886-59-1]
  1-METHOXY-1,4-CYCLOHEXADIENE, **VI**, 996
1,4-Cyclohexadiene, 3,3,6,6-tetramethoxy- [15791-03-4]
  3,3,6,6-TETRAMETHOXY-1,4-CYCLOHEXADIENE, **VI**, 971
2,5-Cyclohexadiene-1-carbonitrile, 4-oxo-1-[(trimethylsilyl)oxy]- [40861-57-2]
  CYANOSILYLATION of BENZOQUINONE, **VII**, 517
2,5-Cyclohexadiene-1-carboxylic acid [4794-04-1]
  1,4-DIHYDROBENZOIC ACID, **V**, 400
3,5-Cyclohexadiene-1,2-dicarboxylic acid, *trans*- [5675-13-8]
  trans-1,2-DIHYDROPHTHALIC ACID, **VI**, 461
2,5-Cyclohexadiene-1,4-dione [106-51-4]
  QUINONE, **I**, 482; **II**, 553
2,5-Cyclohexadiene-1,4-dione, 2,6-bis(1,1-dimethylethyl)- [719-22-2]
  2,6-DI-*tert*-BUTYL-1,4-BENZOQUINONE, **VI**, 412
2,5-Cyclohexadiene-1,4-dione, 2-chloro- [695-99-8]
  CHLORO-*p*-BENZOQUINONE, **IV**, 148
2,5-Cyclohexadiene-1,4-dione, 3-chloro-2,5-bis(1,1-dimethylethyl)- [33611-70-0]
  3-CHLORO-2,5-DI-*tert*-BUTYL-1,4-BENZOQUINONE, **VI**, 210
2,5-Cyclohexadiene-1,4-dione, 2,5-diazido-3,6-bis(1,1-dimethylethyl)- [29342-21-0]
  3,6-DIAZIDO-2,5-DI-*tert*-BUTYL-1,4-BENZOQUINONE, **VI**, 211
2,5-Cyclohexadien-1-one, 2,6-dibromo-4-(chloroimino)- [537-45-1]
  2,6-DIBROMOQUINONE-4-CHLOROIMIDE, **II**, 175
2,5-Cyclohexadiene-1,4-dione, 2,5-dichloro-3,6-bis(1,1-dimethylethyl)- [33611-73-3]
  2,5-DI-*tert*-BUTYL-3,6-DICHLORO-1,4-BENZOQUINONE, **VI**, 210
2,5-Cyclohexadiene-1,4-dione, 2-methyl-5-(1-methylethyl)- [490-91-5]
  THYMOQUINONE, **I**, 511
2,5-Cyclohexadiene-1,4-dione, 2-(phenoxymethyl)- [7714-50-3]
  2-PHENOXYMETHYL-1,4-BENZOQUINONE, **VI**, 890
2,5-Cyclohexadien-1-one, 2,4,4,6-tetrabromo- [20244-61-5]
  2,4,4,6-TETRABROMO-2,5-CYCLOHEXADIEN-1-ONE, **VI**, 181
2,5-Cyclohexadiene-1,4-dione, 2,3,5,6-tetrahydroxy- [319-89-1]
  TETRAHYDROXYQUINONE, **V**, 1011
2,5-Cyclohexadiene-1,4-dione, 2,3,5,6-tetramethyl- [527-17-3]
  DUROQUINONE, **II**, 254
2,5-Cyclohexadiene-1,4-dione, 2,3,5-trimethyl- [935-92-2]
  TRIMETHYL-*p*-BENZOQUINONE, **VI**, 1010
3,5-Cyclohexadiene-1,2-dione, 4,5-dimethyl- [4370-50-7]
  4,5-DIMETHYL-*o*-BENZOQUINONE, **VI**, 480
2,4-Cyclohexadien-1-one, 2,3,4,5,6,6-hexamethyl- [3854-96-4]
  2,3,4,5,6,6-HEXAMETHYL-2,4-CYCLOHEXADIEN-1-ONE, **V**, 598
2,4-Cyclohexadien-1-one, 2,6,6-trimethyl- [13487-30-4]
  2,6,6-TRIMETHYL-2,4-CYCLOHEXADIENONE, **V**, 1092

2,5-Cyclohexadien-1-one, 2,4,6-triphenyl-4-(5'-phenyl[1,1':3,1"-terphenyl]-2'-yloxy)-
[10384-15-3]
2,4,6-TRIPHENYLPHENOXYL, **V**, 1130

Cyclohexanamine, *N*-[(2-bromo-4,5-dimethoxyphenyl)methylene]- [73252-55-8]
6-BROMO-3,4-DIMETHOXYBENZALDEHYDE CYCLOHEXYLIMINE, **VIII**, 586

Cyclohexanamine, *N,N*-dimethyl- [98-94-2]
N,N-DIMETHYLCYCLOHEXYLAMINE, **VI**, 499,

Cyclohexanamine, *N*-ethylidene- [1193-93-7]
N-ETHYLIDENECYCLOHEXYLAMINE, **VI**, 901

Cyclohexanamine, *N*-[(2-iodo-4,5-dimethoxyphenyl)methylene]- [61599-78-8]
6-IODO-3,4-DIMETHOXYBENZALDEHYDE CYCLOHEXYLIMINE, **VIII**, 586

Cyclohexanamine, 1-methyl- [6526-78-9]
1-AMINO-1-METHYLCYCLOHEXANE, **V**, 35

Cyclohexane, 1,2-bis(methylene)- [2819-48-9]
1,2-DIMETHYLENECYCLOHEXANE, **VIII**, 212

Cyclohexane, 1,1-bis(2-propenyloxy)- [53608-84-7]
CYCLOHEXANONE DIALLYL ACETAL, **V**, 292

Cyclohexane, 1-bromo-[(bromomethyl)sulfonyl]-1-methyl- [120696-44-8]
1-BROMO-1-METHYL-2-(BROMOMETHYLSULFONYL)CYCLOHEXANE,
VIII, 212

Cyclohexane, (2-bromo-2-propenyl)- [53608-85-8]
3-CYCLOHEXYL-2-BROMOPROPENE, **I**, 186

Cyclohexane, cyclohexylidene- [4233-18-5]
CYCLOHEXYLIDENECYCLOHEXANE, **V**, 297

Cyclohexane, 1,2-dibromo- [5401-62-7]
1,2-DIBROMOCYCLOHEXANE, **II**, 171

Cyclohexane, 1,2-dichloro-, *cis*- [10498-35-8]
cis-1,2-DICHLOROCYCLOHEXANE, **VI**, 424

Cyclohexane, 1,1-diethoxy- [1670-47-9]
CYCLOHEXANONE DIETHYL KETAL, **VIII**, 578, 579

Cyclohexane, iodo- [626-62-0]
IODOCYCLOHEXANE, **IV**, 324, 543; **VI**, 830

Cyclohexane, isocyano- [931-53-3]
CYCLOHEXYL ISOCYANIDE, **V**, 300

Cyclohexane, methoxy- [931-56-6]
CYCLOHEXYL METHYL ETHER, **VI**, 355

Cyclohexane, 1-methoxy-1-methyl-4-(1-methylethenyl)-, *cis*- [24655-72-9]
cis-p-MENTH-8-EN-1-YL METHYL ETHER, **VII**, 304

Cyclohexane, 1-methoxy-1-methyl-4-(1-methylethenyl)-, *trans*- [24655-71-8]
trans-p-MENTH-8-EN-1-YL METHYL ETHER, **VII**, 304

Cyclohexane, methylene- [1192-37-6]
METHYLENECYCLOHEXANE, **IV**, 612; **V**, 751

Cyclohexane, 4-methyl-2-methylene-1-(1-methylethyl)-, (1*R-cis*)- [122331-74-2]
3-METHYLENE-*cis-p*-MENTHANE, (+)-, **VIII**, 386

Cyclohexane, 2-propynyl- [17715-00-3]
3-CYCLOHEXYLPROPYNE, **I**, 191

Cyclohexaneacetic acid, 1-hydroxy-, ethyl ester [1127-01-1]
  ETHYL 1-HYDROXYCYCLOHEXYLACETATE, **VI**, 598
Cyclohexaneacetic acid, 1-methyl-3-oxo-, (±)- [119986-97-9]
  3-METHYLCYCLOHEXANONE-3-ACETIC ACID, **VIII**, 467
Cyclohexanebutanoic acid, γ-oxo-, ethyl ester [54966-52-8]
  ETHYL 4-CYCLOHEXYL-4-OXOBUTANOATE, **VIII**, 578
Cyclohexanecarbonitrile [766-05-2]
  CYCLOHEXANECARBONITRILE, **VI**, 305, 334
Cyclohexanecarbonitrile, 1,1'-azobis- [2094-98-6]
  1,1'-AZO-bis-1-CYCLOHEXANENITRILE, **IV**, 66
Cyclohexanecarbonitrile, 1,1'-hydrazobis- [17643-01-5]
  1,2-DI-1(1-CYANO)CYCLOHEXYLHYDRAZINE, **IV**, 274
Cyclohexanecarbonyl chloride [2719-27-9]
  CYCLOHEXANECARBOXYLIC ACID CHLORIDE, **VII**, 81
Cyclohexanecarbothioic acid, S-(1,1-dimethylethyl) ester [54829-37-7]
  S-*tert*-BUTYL CYCLOHEXYLMETHANETHIOATE, **VI**, 81, 87
Cyclohexanecarboxaldehyde [2043-61-0]
  CYCLOHEXANECARBOXALDEHYDE, **VI**, 272, 315, 338
Cyclohexanecarboxamide, N,N-dimethyl- [17566-51-7]
  N,N-DIMETHYLCYCLOHEXANECARBOXAMIDE, **VI**, 492
Cyclohexanecarboxylic acid, 1,1-dimethylethyl ester [16537-05-6]
  tert-BUTYL CYCLOHEXANECARBOXYLATE, **VII**, 87
Cyclohexanecarboxylic acid, 4-amino- [1776-53-0]
  4-AMINOCYCLOHEXANECARBOXYLIC ACID, cis–trans MIXTURE, **V**, 670
Cyclohexanecarboxylic acid, 1-cyano-2-methyl-, ethyl ester [5231-79-8]
  ETHYL 1-CYANO-2-METHYLCYCLOHEXANECARBOXYLATE, **VI**, 586
Cyclohexanecarboxylic acid, 1-(4-methoxyphenyl)-2-oxo-, ethyl ester [95793-86-5]
  ETHYL 1-(*p*-METHOXYPHENYL)-2-OXOCYCLOHEXANECARBOXYLATE,
    VII, 229
Cyclohexanecarboxylic acid, 1-methyl- [1123-25-7]
  1-METHYLCYCLOHEXANECARBOXYLIC ACID, **V**, 739
Cyclohexanecarboxylic acid, 2-oxo-, methyl ester [41302-34-5]
  METHYL 2-OXOCYCLOHEXANECARBOXYLATE, **VII**, 351
Cyclohexanecarboxylic acid, 3,4,5-trihydroxy-, (1α,3α,4α,5α)- [53796-39-7]
  HEXAHYDROGALLIC ACID, **V**, 591
Cyclohexanecarboxylic acid, 3,4,5-tris(acetyloxy)-, (1α,3α,4α,5α)- [53796-40-0]
  HEXAHYDROGALLIC ACID TRIACETATE, **V**, 591
1,2-Cyclohexanedicarboxylic acid, diethyl ester, *cis*- [17351-07-4]
  DIETHYL *cis*-HEXAHYDROPHTHALATE, **IV**, 304
1,3-Cyclohexanedicarboxylic acid, 2-oxo-, dimethyl ester [25928-05-6]
  DIMETHYL CYCLOHEXANONE-2,6-DICARBOXYLATE, **V**, 439
1,2-Cyclohexanedimethanol, dimethanesulfonate, *cis*- [66347-68-0]
  1,2-CYCLOHEXANEDIMETHANOL DIMETHANESULFONATE, *cis*-,
    VI, 482
cis-1,2-Cyclohexanediol [1792-81-0]
  cis-1,2-CYCLOHEXANEDIOL, **VI**, 342, 348

trans-1,2-Cyclohexanediol [1460-57-7]
  trans-1,2-CYCLOHEXANEDIOL, **III**, 217; **VI**, 348
1,3-Cyclohexanediol, 2-nitro-, (1α,2β,3α)- [38150-01-5]
  (1*R*,2*r*,3*S*)-2-NITROCYCLOHEXANE-1,3-DIOL, **VIII**, 332
1,3-Cyclohexanediol, 2-nitro-, 1-acetate, [1*S*-(1α,2β,3α)]- [108186-61-4]
  3-HYDROXY-2-NITROCYCLOHEXYL ACETATE, (1*S*,2*S*,3*R*), **VIII**, 332
(1*R*,2*r*,3*S*)-1,3-Cyclohexanediol, 2-nitro-, diacetate (ester), (1α,2β,3α)- [51269-14-8]
  (1*R*,2*r*,3*S*)-3-ACETOXY-2-NITRO-1-CYCLOHEXYL ACETATE, **VIII**, 332
1,2-Cyclohexanedione [765-87-7]
  1,2-CYCLOHEXANEDIONE, **IV**, 229
1,2-Cyclohexanedione, dioxime [492-99-9]
  1,2-CYCLOHEXANEDIONE DIOXIME, **IV**, 229
1,3-Cyclohexanedione [504-02-9]
  DIHYDRORESORCINOL, **III**, 278
1,3-Cyclohexanedione, 2,2-dimethyl- [562-13-0]
  2,2-DIMETHYL-1,3-CYCLOHEXANEDIONE, **VIII**, 312
1,3-Cyclohexanedione, 5,5-dimethyl- [126-81-8]
  5,5-DIMETHYL-1,3-CYCLOHEXANEDIONE, **II**, 200
1,3-Cyclohexanedione, 2-methyl- [1193-55-1]
  2-METHYL-1,3-CYCLOHEXANEDIONE, **V**, 743
1,3-Cyclohexanedione, 2-methyl-2-(3-oxobutyl)- [5073-65-4]
  2-METHYL-2-(3-OXOBUTYL)-1,3-CYCLOHEXANEDIONE, **VII**, 368
1,4-Cyclohexanedione [637-88-7]
  1,4-CYCLOHEXANEDIONE, **V**, 288
Cyclohexanemethanamine, *N,N*-dimethyl- [16607-80-0]
  N,N-DIMETHYLCYCLOHEXYLMETHYLAMINE, **IV**, 339; *Hazard*
Cyclohexanemethanol [100-49-2]
  CYCLOHEXYLCARBINOL, **I**, 188
Cyclohexanemethanol, 1-hydroxy- [15753-47-6]
  1-(HYDROXYMETHYL)CYCLOHEXANOL, **VIII**, 315
1,1,3,3-Cyclohexanetetramethanol, 2-hydroxy- [5416-55-7]
  2,2,6,6-TETRAMETHYLOLCYCLOHEXANOL, **IV**, 907
Cyclohexanimine, *N*-chloro- [6681-70-5]
  N-CHLOROCYCLOHEXYLIDENEIMINE, **V**, 208
Cyclohexanol [108-93-0]
  CYCLOHEXANOL, **VI**, 353
Cyclohexanol, 1-(aminomethyl)- [4000-72-0]
  1-(AMINOMETHYL)CYCLOHEXANOL, **IV**, 224
Cyclohexanol, 2-chloro- [1561-86-0]
  2-CHLOROCYCLOHEXANOL, **I**, 158
Cyclohexanol, 4-(1,1-dimethylethyl)-, *cis*- [937-05-3]
  cis-4-*tert*-BUTYLCYCLOHEXANOL, **VI**, 215
Cyclohexanol, 4-(1,1-dimethylethyl), *trans*- [21862-63-5]
  trans-4-*tert*-BUTYLCYCLOHEXANOL, **V**, 175
Cyclohexanol, 1-[[dimethyl(1-methoxyethoxy)silyl]methyl]- [138080-23-6]
  [(ISOPROPOXYDIMETHYLSILYL)METHYL]CYCLOHEXANOL, **VIII**, 315

Cyclohexanol, 1,1'-(1,2-ethynediyl)bis- [78-54-6]
    1,1'-ETHYNYLENE-BIS-CYCLOHEXANOL, **IV**, 471
Cyclohexanol, 1-ethynyl- [78-27-3]
    1-ETHYNYLCYCLOHEXANOL, **III**, 416
Cyclohexanol, 2-isocyano-, *trans*- [83152-97-0]
    2-ISOCYANOCYCLOHEXANOL, *trans*-, **VII**, 294
Cyclohexanol, 2-(1-mercapto-1-methylethyl)-5-methyl-, [1*R*-(1a,2a,5a)]- [79563-68-1];
    1*R*-(1a,2b,5a)] [79563-59-0]; [1*S*-(1a,2a,5b)]- [79563-67-0]
    2-(1-MERCAPTO-1-METHYLETHYL)-5-METHYLCYCLOHEXANOL
    (7-THIOMENTHOL), **VIII**, 304
Cyclohexanol, 1-methyl- [590-67-0]
    1-METHYLCYCLOHEXANOL, **VI**, 766
Cyclohexanol, 5-methyl-2-(1-methyl-1-phenylethyl)- [1*R*-(1α,2β,5α)]- [65253-04-5]
    8-PHENYLMENTHOL, (–)-, **VIII**, 522
Cyclohexanol, 1-nitromethyl- [3164-73-6]
    1-(NITROMETHYL)CYCLOHEXANOL, **IV**, 221, 224
Cyclohexanol, 2-phenyl-, (1*R-trans*)- [98919-68-7]
    PHENYLCYCLOHEXANOL, (–)-(1*R*,2*S*)-*trans*-2-, **VIII**, 516
Cyclohexanol, 2-phenyl-, (1*S-trans*)- [34281-92-0]
    PHENYLCYCLOHEXANOL, (+)-(1*S*,2*R*)-*trans*-2-, **VIII**, 516
Cyclohexanol, 2-(1-propenyl)-, [1α,2β(*E*)]- [76123-38-1]
    (*E*)-2-PROPENYLCYCLOHEXANOL, **VII**, 456
Cyclohexanone, 2-butyl-2-methyl- [1197-78-0]
    2-BUTYL-2-METHYLCYCLOHEXANONE, **V**, 187
Cyclohexanone, 2-chloro- [822-87-7]
    2-CHLOROCYCLOHEXANONE, **III**, 188
Cyclohexanone, 2-chloro-2-methyl- [10409-46-8]
    2-CHLORO-2-METHYLCYCLOHEXANONE, **IV**, 162
Cyclohexanone, 2-diazo- [3242-56-6]
    2-DIAZOCYCLOHEXANONE, **VI**, 389
Cyclohexanone, 4-(1,1-dimethylethyl)- [98-53-3]
    4-*tert*-BUTYLCYCLOHEXANONE, **VI**, 218, 220
Cyclohexanone, 2-ethenyl-3-methyl-, *trans*- [110222-94-1]
    trans-3-METHYL-2-VINYLCYCLOHEXANONE, **VIII**, 479
Cyclohexanone, 2-[(3-ethyl-5-methyl-4-isoxazolyl)methyl)]- [53984-03-5]
    2-(3-ETHYL-5-METHYL-4-ISOXAZOLYLMETHYL)CYCLOHEXANONE, **VI**,
    781
Cyclohexanone, 3-hydroxy-2,2-dimethyl-, (*S*)- [87655-21-8]
    (*S*)-(+)-3-HYDROXY-2,2-DIMETHYLCYCLOHEXANONE, **VIII**, 312
Cyclohexanone, 2-hydroxymethylene- [823-45-0]
    2-HYDROXYMETHYLENECYCLOHEXANONE, **IV**, 536, 537
Cyclohexanone, 5-methyl-2-(1-methylethyl)-, (2*R-cis*)- [1196-31-2]
    (+)-ISOMENTHONE, **VIII**, 386
Cyclohexanone, 5-methyl-2-(1-methylethyl)-, (2*S-trans*)- [14073-97-3]
    L-MENTHONE, **I**, 340

Cyclohexanone, 5-methyl-2-(1-methyl-1-phenylethyl)-, (2*S-cis*)- [65337-06-6]
  (2*R-trans*)- [57707-92-3]
  (2*RS*,5*R*)-5-METHYL-2-(1-METHYL-1-PHENYLETHYL)CYCLOHEXANONE,
  VIII, 523
Cyclohexanone, 5-methyl-2-[1-methyl-1-(phenylmethylthio)ethyl]-, (2*R-trans*)-
  [79563-58-9]; (2*S-cis*)- [79618-04-5]
  5-METHYL-2-[1-METHYL-1(PHENYLMETHYLTHIO)ETHYL]
  CYCLOHEXANONE, *cis*- and *trans*- (7-BENZYLTHIOMENTHONE), **VIII**, 304
Cyclohexanone, 2-methyl-2-(phenylmethyl)- [1206-21-9]
  2-BENZYL-2-METHYLCYCLOHEXANONE, **VI**, 121
Cyclohexanone, 2-methyl-6-(phenylmethyl)- [24785-76-0]
  2-BENZYL-6-METHYLCYCLOHEXANONE, **VI**, 121
Cyclohexanone, 3-methyl-2-(2-propenyl)- [56620-95-2]
  2-ALLYL-3-METHYLCYCLOHEXANONE, **VI**, 51
Cyclohexanone, 2-(1-methyl-2-propynyl)-, (*R**,*R**)-(±)- [130719-23-1]; (*R**,*S**)-(±)-
  [130719-24-3]
  2-(1-METHYL-2-PROPYNYL)CYCLOHEXANONE, **VIII**, 460
Cyclohexanone, 2-(2-oxopropyl)- [6126-53-0]
  2-(2-OXOPROPYL)CYCLOHEXANONE, **VII**, 414
Cyclohexanone, 2-(2-propenyl)- [94-66-6]
  2-ALLYLCYCLOHEXANONE, **III**, 44; **V**, 25
Cyclohexene [110-83-8]
  CYCLOHEXENE, **I**, 183; **II**, 152
Cyclohexene, 2-(1-buten-3-ynyl)-1,3,3-trimethyl-, (*E*)- [73395-75-2]
  BUTEN-3-YNYL-2,6,6-TRIMETHYL-1-CYCLOHEXENE, (*E*)-, **VII**, 63
Cyclohexene, 4-(1,1-dimethylethyl)-1-ethenyl- [33800-81-6]
  4-*tert*-BUTYL-1-VINYLCYCLOHEXENE, **VIII**, 97
Cyclohexene, 3-methyl- [591-48-0]
  3-METHYLCYCLOHEXENE, **VI**, 174, 769
1-Cyclohexene-1-acetonitrile [6975-71-9]
  1-CYCLOHEXENYLACETONITRILE, **IV**, 234
1-Cyclohexene-1-carboxaldehyde, 2-chloro- [1680-73-5]
  2-CHLORO-1-FORMYL-1-CYCLOHEXENE, **V**, 215
1-Cyclohexene-1-carboxaldehyde, 3-hydroxy- [67252-14-6]
  3-HYDROXY-1-CYCLOHEXENE-1-CARBOXALDEHYDE, **VIII**, 309
1-Cyclohexene-1-carboxylic acid, 2-[(diethoxyphosphinyl)oxy]-, methyl ester [71712-64-
  6]
  METHYL 2-(DIETHYLPHOSPHORYLOXY)-1-CYCLOHEXENE-
  1-CARBOXYLATE, **VII**, 351
1-Cyclohexene-1-carboxylic acid, 2-methyl-, methyl ester [25662-38-8]
  METHYL 2-METHYL-1-CYCLOHEXENE-1-CARBOXYLATE, **VII**, 351
2-Cyclohexene-1-carboxylic acid, 2,6-dimethyl-4-oxo-, ethyl ester [6102-15-4]
  3,5-DIMETHYL-4-CARBETHOXY-2-CYCLOHEXEN-1-ONE, **III**, 317
4-Cyclohexene-1,2-dicarboxylic acid, diethyl ester, *cis*- [4841-85-4]
  DIETHYL *cis*-Δ⁴-TETRAHYDROPHTHALATE, **IV**, 304

4-Cyclohexene-1,2-dicarboxylic acid, dimethyl ester, *cis-* [4841-84-3]
    DIMETHYL *cis*-Δ⁴-TETRAHYDROPHTHALATE, **IV**, 306

4-Cyclohexene-1,2-dicarboxylic acid, diethyl ester, *trans-* [5048-50-0]
    DIETHYL *trans*-Δ⁴-TETRAHYDROPHTHALATE, **VI**, 454

4-Cyclohexene-1,2-dicarboxylic acid, 4-[(trimethylsilyl)oxy]-, diethyl ester, *trans-*
    [61692-31-7]
    DIETHYL *trans*-4-TRIMETHYLSILYLOXY-4-CYCLOHEXENE-1,2-
    DICARBOXYLATE, **VI**, 445

2-Cyclohexene-1,4-dione, 5,6-dichloro-2,5-bis(1,1-dimethylethyl)- [33611-72-2]
    2,5-DI-*tert*-BUTYL-5,6-DICHLORO-2-CYCLOHEXENE-1,4-DIONE, **VI**, 210

2-Cyclohexene-1,4-dione, 3,5,6-trichloro-2,5-bis(1,1-dimethylethyl)-*cis-* [117257-59-7]
    trans- [117257-58-6]
    2,5-DI-*tert*-BUTYL-3,5,6-TRICHLORO-2-CYCLOHEXENE-1,4-DIONE,
    VI, 210

1-Cyclohexene-1-methanol, 4-(1-methylcyclopropyl)- [98678-72-9]
    1-HYDROXYMETHYL-4-(1-METHYLCYCLOPROPYL)-1-CYCLOHEXENE,
    VIII, 321

2-Cyclohexene-1-propanoic acid, 3-[(trimethylsilyl)oxy]-, ethyl ester [90147-64-1]
    ETHYL 3-[3-(TRIMETHYLSILOXY)CYCLOHEX-2-ENYL]PROPIONATE,
    VIII, 277

1-Cyclohexen-1-ol, 2-methyl-, acetate [75411-49-3]
    trans-2-METHYL-1-CYCLOHEXEN-1-YL ACETATE, **VI**, 121

2-Cyclohexen-1-ol, benzoate [3352-93-0]
    3-BENZOYLOXYCYCLOHEXENE, **V**, 70

2-Cyclohexen-1-ol, 4-chloro-, acetate, *cis-* [82736-39-8]
    cis-1-ACETOXY-4-CHLORO-2-CYCLOHEXENE, **VIII**, 6

2-Cyclohexen-1-ol, 1-(1,3-dithian-2-yl)- [53178-46-4]
    1-(1,3-DITHIAN-2-YL)-2-CYCLOHEXEN-1-OL, **VIII**, 309

2-Cyclohexen-1-ol, 3-methyl- [21378-21-2]
    3-METHYL-2-CYCLOHEXEN-1-OL, **VI**, 769

3-Cyclohexen-1-ol, 6-amino-3,4-dimethyl-, *cis-* [65948-45-0]
    6-AMINO-3,4-DIMETHYL-*cis*-3-CYCLOHEXEN-1-OL, **VII**, 5

2-Cyclohexen-1-one [930-68-7]
    2-CYCLOHEXENONE, **V**, 294

2-Cyclohexen-1-one, 2-acetyl- [52784-38-0]
    2-ACETYL-2-CYCLOHEXEN-1-ONE, **VI**, 23

2-Cyclohexen-1-one, 6-(1,2-dichloroethenyl)-3-ethoxy-6-methyl- [73843-25-1]
    6-[(*E*)-1,2-DICHLOROVINYL]-3-ETHOXY-6-METHYL-2-CYCLOHEXEN-1-
    ONE, **VII**, 241

2-Cyclohexen-1-one, 4-(1,2-dichloroethenyl)-4-methyl- [73843-27-3]
    4-[(*E*)-1,2-DICHLOROVINYL]-4-METHYL-2-CYCLOHEXEN-1-ONE, **VII**, 241

2-Cyclohexen-1-one, 3,5-dimethyl- [1123-09-7]
    3,5-DIMETHYL-2-CYCLOHEXEN-1-ONE, **III**, 317

2-Cyclohexen-1-one, 4,4-dimethyl- [1073-13-8]
    4,4-DIMETHYL-2-CYCLOHEXEN-1-ONE, **VI**, 496

2-Cyclohexen-1-one, 3-ethoxy- [5323-87-5]
   3-ETHOXY-2-CYCLOHEXENONE, **V**, 539
2-Cyclohexen-1-one, 3-ethoxy-6-methyl- [62952-33-4]
   3-ETHOXY-6-METHYL-2-CYCLOHEXEN-1-ONE, **VII**, 208, 241
2-Cyclohexen-1-one, 4-ethyl-4-(2-propenyl)-, (*R*)- [122444-62-6]
   4-ETHYL-4-ALLYL-2-CYCLOHEXEN-1-ONE, (*R*)-, **VIII**, 241
2-Cyclohexen-1-one, 2-heptyl- [87588-68-9]
   2-HEPTYL-2-CYCLOHEXENONE, **VII**, 249
2-Cyclohexen-1-one, 6-hydroxy-3,5,5-trimethyl- [61592-66-3]
   6-HYDROXY-3,5,5-TRIMETHYL-2-CYCLOHEXEN-1-ONE, **VII**, 282
2-Cyclohexen-1-one, 2-methyl- [1121-18-2]
   2-METHYL-2-CYCLOHEXENONE, **IV**, 162
2-Cyclohexen-1-one, 2-methyl-5-(1-methylethyl)- [43205-82-9]
   DIHYDROCARVONE, **VI**, 459
2-Cyclohexen-1-one, 3,5,5-trimethyl-2-(2-oxopropyl)- [61879-73-0]
   3,5,5-TRIMETHYL-2-(2-OXOPROPYL)-2-CYCLOHEXEN-1-ONE, **VI**, 1024
1,2-Cyclononadiene [1123-11-1]
   1,2-CYCLONONADIENE, **V**, 306
1,3,7,9,13,15-Cyclooctadecahexaene-5,11,17-triyne [3891-75-6]
   TRIDEHYDRO[18]ANNULENE, **VI**, 68
1,3,7,9,13,15-Cyclooctadecahexayne [7092-39-9]
   HEXADEHYDRO[18]ANNULENE , **VI**, 68
Cyclooctadecanonaene, 1,3,5,7,9,11,13,15,17- [2040-73-5]
   [18]ANNULENE, **VI**, 68
Cyclooctanecarboxylic acid, 2-oxo-, ethyl ester [4017-56-5]
   2-CARBETHOXYCYCLOOCTANONE, **V**, 198
Cyclooctanone [502-49-8]
   CYCLOOCTANONE, **V**, 310
Cyclooctanone, 2-chloro-, oxime [10499-33-9]
   2-CHLOROCYCLOOCTANONE OXIME, **V**, 266
Cyclooctanone, 2-methoxy-, oxime [10499-36-2]
   2-METHOXYCYCLOOCTANONE OXIME, **V**, 267
Cyclooctene, (*E*)- [931-89-5]
   trans-CYCLOOCTENE, **V**, 315
Cyclooctene, 1-nitro- [1782-03-2]
   1-NITROCYCLOOCTENE, **VI**, 837
1,3-Cyclopentadiene, 1,2,3,4,5-pentachloro-5-ethyl- [16177-48-3]
   1,2,3,4,5-PENTACHLORO-5-ETHYLCYCLOPENTADIENE, **V**, 893
1,3-Cyclopentadiene, 1,2,3,4,5-pentamethyl- [4045-44-7]
   1,2,3,4,5-PENTAMETHYLCYCLOPENTADIENE, **VII**, 505
1,3-Cyclopentadiene, 1,2,3,4-tetrachloro-5,5-dimethoxy- [2207-27-4]
   5,5-DIMETHOXY-1,2,3,4-TETRACHLOROCYCLOPENTADIENE, **V**, 424
1,3-Cyclopentadiene, 1,2,3,4-tetrachloro-5-(2,3,4,5-tetrachloro-2,4-cyclopentadien-
     1-ylidene)- [6298-65-3]
   PERCHLOROFULVALENE, **V**, 901

2,4-Cyclopentadiene-1-acetic acid, methyl ester [37455-98-4]
   METHYL 2,4-CYCLOPENTADIENE-1-ACETATE (in solution), **VII**, 339
2,4-Cyclopentadien-1-one, 2,3,4,5-tetraphenyl- [479-33-4]
   TETRAPHENYLCYCLOPENTADIENONE, **III**, 806
Cyclopentanamine, 1-phenyl- [17380-74-4]
   1-PHENYLCYCLOPENTYLAMINE, **VI**, 910
Cyclopentanamine, 2-(2-propenyl)-, *cis*- [81097-02-1]
   cis-2-(2-PROPENYL)CYCLOPENTYLAMINE, **VII**, 501
Cyclopentanecarboxaldehyde [872-53-7]
   CYCLOPENTANECARBOXALDEHYDE, **V**, 320
Cyclopentanecarboxaldehyde, 1-phenyl- [21573-69-3]
   1-PHENYLCYCLOPENTANECARBOXALDEHYDE, **VI**, 905
Cyclopentanecarboxylic acid, methyl ester [4630-80-2]
   METHYL CYCLOPENTANECARBOXYLATE, **IV**, 594
Cyclopentanecarboxylic acid, 2-ethenyl-5-oxo-, methyl ester [75351-19-8]
   2-CARBOMETHOXY-3-VINYLCYCLOPENTANONE, **VIII**, 112
Cyclopentanecarboxylic acid, 2-oxo-, ethyl ester [611-10-9]
   2-CARBETHOXYCYCLOPENTANONE, **II**, 116
Cyclopentanecarboxylic acid, 2-oxo-1-(phenylmethyl)-, methyl ester [10386-81-9]
   2-BENZYL-2-CARBOMETHOXYCYCLOPENTANONE, **V**, 75
1,1-Cyclopentanedicarboxylic acid, 3-methylene-4-(1-methylethyl)-, dimethyl ester
      [107473-16-5]
   3-METHYLENE-4-ISOPROPYL-1,1-CYCLOPENTANEDICARBOXYLIC ACID,
      DIMETHYL ESTER, **VIII**, 381
1,1-Cyclopentanedicarboxylic acid, 3-(1-methylethyl)-4-[(tributylstannyl)methylene]-,
      dimethyl ester [107473-15-4]
   (Z)-TRIBUTYLSTANNYLMETHYLENE-4-ISOPROPYL-1,1-CYCLOPENTANE
      DICARBOXYLIC ACID, DIMETHYL ESTER, **VIII**, 381
1,2-Cyclopentanedicarboxylic acid, 1-methyl-, *cis*-(±)- [70433-31-7]
   cis-1-METHYLCYCLOPENTANE-1,2-DICARBOXYLIC ACID, **VIII**, 377
1,3-Cyclopentanedione, 2-acetyl- [3859-39-0]
   2-ACETYL-1,3-CYCLOPENTANEDIONE, **VI**, 28
1,3-Cyclopentanedione, 2-methyl- [765-69-5]
   2-METHYLCYCLOPENTANE-1,3-DIONE, **V**, 747; **VI**, 30, 774
1,3-Cyclopentanedione, 2-methyl-2-(3-oxobutyl)- [25112-78-1]
   2-METHYL-2-(3-OXOBUTYL)-1,3-CYCLOPENTANEDIONE, **VII**, 363
Cyclopentaneglyoxylic acid, 3-methyl-2,4,5-trioxo-, ethyl ester [781-38-4]
   2-METHYL-4-ETHOXALYLCYCLOPENTANE-1,3,5-TRIONE, **V**, 747
Cyclopentane-1,2,4-trione, 3-methyl- [4505-54-8]
   2-METHYLCYCLOPENTANE-1,3,5-TRIONE HYDRATE, **V**, 748
Cyclopentanol, 2-chloro-, *trans*- [20377-80-4]
   trans-2-CHLOROCYCLOPENTANOL, **IV**, 157
Cyclopentanol, 2-(2-propenyl)-, *trans*- [74743-89-8]
   trans-2-(2-PROPENYL)CYCLOPENTANOL, **VII**, 501
Cyclopentanone [120-92-3]
   CYCLOPENTANONE, **I**, 192

Cyclopentanone, 2-(1,1-dimethylpropyl)- [25184-25-2]
2-*tert*-PENTYLCYCLOPENTANONE, **VII**, 424
Cyclopentanone, 2-ethenyl-2-methyl- [88729-76-4]
2-METHYL-2-VINYLCYCLOPENTANONE, **VII**, 486
Cyclopentanone, 2-(phenylmethyl)- [2867-63-2]
2-BENZYLCYCLOPENTANONE, **V**, 76
Cyclopentanone, 2,4,4-trimethyl- [4694-12-6]
2,4,4-TRIMETHYLCYCLOPENTANONE, **IV**, 957
Cyclopenta[*b*]pyrrole, 1,3a,4,5,6,6a-hexahydro-2-methyl-1-[(4-methylphenyl)sulfonyl]-,
    cis- [81097-07-6]
N-TOSYL-3-METHYL-2-AZABICYCLO[3.3.0]OCT-3-ENE, *cis*-, **VII**, 501
Cyclopentene, 3-chloro- [96-40-2]
3-CHLOROCYCLOPENTENE, **IV**, 238
2-Cyclopentene-1-acetic acid, 5-hydroxy-, methyl ester, (1*R-trans*)- [49825-99-2]
METHYL (1*R*,5*R*)-5-HYDROXY-2-CYCLOPENTENE-1-ACETATE, **VII**, 339
3-Cyclopentene-1,2-diol, *cis*- [694-29-1]
DIHYDROXYCYCLOPENTENE, **V**, 414
4-Cyclopentene-1,3-diol [4157-01-1]
3,5-DIHYDROXYCYCLOPENTENE, **V**, 414
4-Cyclopentene-1,3-diol, monoacetate, *cis*- [60410-18-6]
3-ACETOXY-5-HYDROXYCYCLOPENT-1-ENE, *cis*-, **VIII**, 134
4-Cyclopentene-1,3-dione [930-60-9]
2-CYCLOPENTENE-1,4-DIONE, **V**, 324
2-Cyclopenten-1-one [930-30-3]
CYCLOPENTENONE, **V**, 326
2-Cyclopenten-1-one, 4-(acetyloxy)- [768-48-9]
4-ACETOXY-2-CYCLOPENTEN-1-ONE, **VIII**, 15
2-Cyclopenten-1-one, 2-bromo- [10481-34-2]
2-BROMO-2-CYCLOPENTEN-1-ONE, **VII**, 271
2-Cyclopenten-1-one, 4,4-dimethyl- [22748-16-9]
4,4-DIMETHYL-2-CYCLOPENTEN-1-ONE, **VIII**, 208
2-Cyclopenten-1-one, 2,5-dimethyl-3-phenyl- [36461-43-5]
2,5-DIMETHYL-3-PHENYL-2-CYCLOPENTEN-1-ONE, **VI**, 520
2-Cyclopentenone, 2-hydroxymethyl- [68882-71-3]
2-HYDROXYMETHYL-2-CYCLOPENTENONE, **VII**, 271
2-Cyclopenten-1-one, 3-methyl-2-pentyl- [1128-08-1]
DIHYDROJASMONE; 3-METHYL-2-PENTYL-2-CYCLOPENTEN-1-ONE,
    VIII, 620
2-Cyclopenten-1-one, 2-[(4-methylphenyl)sulfinyl]-, (*S*)- [79681-26-8]
2-(*p*-TOLUENESULFINYL)-2-CYCLOPENTENONE, (*S*)-(+)-, **VII**, 495
Cyclopropane, 1,1-bis(bromomethyl)- [29086-41-7]
1,1-BIS(BROMOMETHYL)CYCLOPROPANE, **VI**, 153,
Cyclopropane, bromo- [4333-56-6]
BROMOCYCLOPROPANE, **V**, 126
Cyclopropane, 1-bromo-1-ethoxy- [95631-62-2]
1-BROMO-1-ETHOXYCYCLOPROPANE, **VIII**, 556

Cyclopropane, 1-chloro-1-(trichloroethenyl)- [82979-27-9]
    1-CHLORO-1-(TRICHLOROETHENYL)CYCLOPROPANE, **VIII**, 124, 373
Cyclopropane, (1-chloro-2,2,2-trimethoxyethylidene)- [82979-34-8]
    TRIMETHYL 2-CHLORO-2-CYCLOPROPYLIDENORTHOACETATE, **VIII**, 373
Cyclopropane, 3-iodo-1,1,2,2-tetramethyl- [39653-50-4]
    2,2,3,3-TETRAMETHYLIODOCYCLOPROPANE, **VI**, 974
Cyclopropanecarbonitrile [5500-21-0]
    CYCLOPROPYL CYANIDE: **III**, 223
Cyclopropanecarboxaldehyde [1489-69-6]
    CYCLOPROPANECARBOXALDEHYDE, **VII**, 129
Cyclopropanecarboxylic acid [1759-53-1]
    CYCLOPROPANECARBOXYLIC ACID, **III**, 221
Cyclopropanecarboxylic acid, 2-phenyl-, cis- [939-89-9]
    cis-2-PHENYLCYCLOPROPANECARBOXYLIC ACID, **VI**, 913
Cyclopropanecarboxylic acid, 2-phenyl-, trans- [939-90-2]
    trans-2-PHENYLCYCLOPROPANECARBOXYLIC ACID, **VI**, 915
1,1-Cyclopropanedicarboxylic acid [598-10-7]
    CYCLOPROPANE-1,1-DICARBOXYLIC ACID, **VII**, 411
1,2-Cyclopropanedicarboxylic acid, (1S-trans)- [14590-54-6]
    CYCLOPROPANE-1,2-DICARBOXYLIC ACID, (+)-(1S,2S)-, **VIII**, 141
1,2-Cyclopropanedicarboxylic acid, bis[5-methyl-2-(1-methylethyl)cyclohexyl] ester
    [1S-[1α(1S*,2S*,5R*)], 2β,5α]]- [96149-01-8]
    (−)-DIMENTHYL (1S,2S)-CYCLOPROPANE-1,2-DICARBOXYLATE, **VIII**, 142
Cyclopropanemethanol, 1-ethoxy-α-1-propenyl-, (E)- [130719-17-4]
    (E)-1-ETHOXY-1-(1-HYDROXY-2-BUTENYL)CYCLOPROPANE, **VIII**, 556
Cyclopropanol, 1-ethoxy- [13837-45-1]
    CYCLOPROPANONE ETHYL HEMIACETAL, **VII**, 131
Cyclopropanol, 1-(4-methylphenyl)- [40122-37-0]
    1-p-TOLYLCYCLOPROPANOL, **V**, 1058
Cyclopropene, 3,3-dimethoxy- [23529-83-1]
    3,3-DIMETHOXYCYCLOPROPENE, **VI**, 361
Cyclopropenone [2961-80-0]
    CYCLOPROPENONE, **VI**, 361
Cyclopropenone, 2,3-bis(1,1-dimethylethyl)- [19985-79-6]
    DI-tert-BUTYLCYCLOPROPENONE, **VI**, 991
2-Cyclopropen-1-one, 2,3-diphenyl- [886-38-4]
    DIPHENYLCYCLOPROPENONE, **V**, 514
Cyclopropenylium, 1,2,3-tris(1,1-dimethylethyl)-, tetrafluoroborate [60391-90-4]
    TRI-tert-BUTYLCYCLOPROPENYL TETRAFLUOROBORATE, **VI**, 991
Cycloundecanecarboxylic acid [4373-07-3]
    CYCLOUNDECANECARBOXYLIC ACID, **VII**, 135
Cycloundecanone [878-13-7]
    CYCLOUNDECANONE, **VI**, 368
1-Cycloundecene-1-carboxylic acid, methyl ester [3667-72-9]
    METHYL 1-CYCLOUNDECENECARBOXYLATE, **VI**, 368

L-Cysteine, *S*-[(acetylamino)methyl]-, monohydrochloride [28798-28-9]
　　S-ACETAMIDOMETHYL-L-CYSTEINE HYDROCHLORIDE, **VI**, 5
L-Cysteine, *S*-(1,1-dimethylethyl)-, 1,1-dimethylethyl ester [45157-84-4]
　　S-tert-BUTYL-L-CYSTEINE *tert*-BUTYL ESTER, **VI**, 255
L-Cysteine, *S*-(1,1-dimethylethyl)-*N*-[*O*-(1,1-dimethylethyl)-L-seryl]-, 1,1-dimethyl ester
　　[67942-95-4]
　　O-tert-BUTYL-L-SERYL-*S-tert*-BUTYL-L-CYSTEINE *tert*-BUTYL ESTER,
　　VI, 252
L-Cysteine, *N*-[*N*-(phenylmethoxy)carbonyl]-L-alanyl-, methyl ester [34804-78-3]
　　BENZYLOXYCARBONYL-L-ALANYL-L-CYSTEINE METHYL ESTER,
　　VII, 30
L-Cystine [56-89-3]
　　L-CYSTINE, **I**, 194

D

2,4-Decadienoic acid, ethyl ester, (*E*,*Z*)- [3025-30-7]
　　ETHYL (*E*,*Z*)-2,4-DECADIENOATE, **VIII**, 251
3,4-Decadienoic acid, ethyl ester [36186-28-4]
　　ETHYL 3,4-DECADIENOATE, **VIII**, 251
1-Decalol [529-32-8]
　　1-DECALOL, **VI**, 371
Decanal [112-31-2]
　　1-DECANAL, **VI**, 373
Decanal, 6-oxo- [63049-53-6]
　　6-OXODECANAL, **VIII**, 498
Decane [124-18-5]
　　DECANE, **VI**, 376
Decane, 1,10-dibromo- [4101-68-2]
　　DECAMETHYLENE BROMIDE, **III**, 227
1,10-Decanediamine [646-25-3]
　　DECAMETHYLENEDIAMINE, **III**, 229
Decanedinitrile [1871-96-1]
　　SEBACONITRILE, **III**, 768
Decanedioic acid, diethyl ester [110-40-7]
　　DIETHYL SEBACATE, **II**, 277
Decanedioic acid, dimethyl ester [106-79-6]
　　DIMETHYL DECANEDIOATE, **VII**, 177
Decanedioic acid, monoethyl ester [693-55-0]
　　ETHYL HYDROGEN SEBACATE, **II**, 276
1,10-Decanediol [112-47-0]
　　DECAMETHYLENE GLYCOL, **II**, 154
2,5-Decanediol, 2-methyl- [53731-34-3]
　　2-METHYL-2,5-DECANEDIOL, **IV**, 601
1,10-Decanedione, 1,10-bis(2-oxocyclohexyl)- [17343-93-0]
　　SEBACOYLDICYCLOHEXANONE, **V**, 533

Decanoic acid, 10-amino-10-oxo-, methyl ester [53663-35-7]
  METHYL SEBACAMATE, **III**, 613
Decanoic acid, 2-(methyldiphenylsilyl)-, ethyl ester [89638-16-4]
  ETHYL 2-(DIPHENYLSILYLMETHYL)DECANOATE, **VIII**, 474
Decanoic acid, 6-oxo- [4144-60-9]
  6-OXODECANOIC ACID, **VIII**, 499
Decanoic acid, 6-oxo-, methyl ester [61820-00-6]
  METHYL 6-OXODECANOATE, **VIII**, 441
2-Decanone [693-54-9]
  2-DECANONE, **VII**, 137
9-Decyn-1-ol [17643-36-6]
  9-DECYN-1-OL, **VIII**, 146
2,3-Diazabicyclo[2.2.1]heptane-2,3-dicarboxylic acid, diethyl ester [18860-71-4]
  DIETHYL 2,3-DIAZABICYCLO[2.2.1]HEPTANE-2,3-DICARBOXYLATE, **V**, 97
2,3-Diazabicyclo[2.2.1]hept-5-ene-2,3-dicarboxylic acid, diethyl ester [14011-60-0]
  DIETHYL 2,3-DIAZABICYCLO[2.2.1]HEPT-5-ENE-2,3-DICARBOXYLATE,
    V, 96
1,10-Diazacyclooctadecane [296-30-0]
  1,10-DIAZACYCLOOCTADECANE, **VI**, 382
1,2-Diazaspiro[2.5]oct-1-ene [930-82-5]
  3,3-PENTAMETHYLENEDIAZIRINE, **V**, 897
Diazene, diphenyl- [103-33-3]
  AZOBENZENE, **III**, 103
Diazene, diphenyl-, 1-oxide [495-48-7]
  AZOXYBENZENE, **II**, 57
Diazenecarbothioic acid, phenyl-, 2-phenylhydrazide [60-10-6]
  DITHIZONE, **III**, 360
Diazenecarboxylic acid, (1-cyanocyclohexyl)-, methyl ester [33670-04-1]
  2-(1-CYANOCYCLOHEXYL)DIAZENECARBOXYLIC ACID, METHYL
    ESTER, **VI**, 334
Diazenedicarboxylic acid, bis(1,1-dimethylethyl) ester [870-50-8]
  tert-BUTYL AZODIFORMATE, **V**, 160
Diazenedicarboxylic acid, bis(2,2,2-trichloroethyl) ester [38857-88-4]
  BIS(2,2,2-TRICHLOROETHYL) AZODICARBOXYLATE, **VII**, 56
Diazenedicarboxylic acid, diethyl ester [1972-28-7]
  ETHYL AZODICARBOXYLATE, **III**, 375; **IV**, 411; *Warning*, **V**, 544
Diazenedicarboxylic acid, dimethyl ester [2446-84-6]
  METHYL AZODICARBOXYLATE, **IV**, 411; *Warning*, **IV**, 412
3*H*-Diazirine, 3-chloro-3-phenyl- [4460-46-2]
  CHLOROPHENYLDIAZIRINE (in solution), **VI**, 276
Dibenz[*c,e*][1,2]azaborine, 5,6-dihydro-6-methyl- [15813-13-5]
  10-METHYL-10,9-BORAZAROPHENANTHRENE, **V**, 727
Dibenzo[*d,f*][1,2]-dioxocin, 5,8-dihydro-5,8-dimethoxy- [6623-54-7]
  3,8-DIMETHOXY-4,5,6,7-DIBENZO-1,2-DIOXACYCLOOCTANE, **V**, 493
Dibenzo[*b,k*][1,4,7,10,13,16]hexaoxacyclooctadecin, eicosahydro- [16069-36-6]
  DICYCLOHEXYL-18-CROWN-6 POLYETHER, **VI**, 395

Dibenzo[*b,k*][1,4,7,10,13,16]hexaoxacyclooctadecin, 6,7,9,10,17,18,20,21-octahydro-, [14187-32-7]
    DIBENZO-18-CROWN-6 POLYETHER, **VI**, 395
Dicarbonic acid, bis(1,1-dimethylethyl) ester [24424-99-5]
    DI-*tert*-BUTYL DICARBONATE, **VI**, 418
1,3-Dimethyl-5-oxobicyclo[2.2.2]octane-2-carboxylic acid, (1α,2β,3α,4α)- [121829-82-1]
    1,3-DIMETHYL-5-OXOBICYCLO[2.2.2]OCTANE-2-CARBOXYLIC ACID
    VIII, 219
1,3-Dimethyl-5-oxobicyclo[2.2.2]octane-2-carboxylic acid, methyl ester, (1α,2β,3α,4α)- [121917-73-5]
    1,3-DIMETHYL-5-OXOBICYCLO[2.2.2]OCTANE-2-CARBOXYLIC ACID, METHYL ESTER, **VIII**, 219
(*R*)-(−)- and (*S*)-(+)-Dinaphtho[2,1-*d*:1'2'-*f*][1,3,2]dioxaphosphepin, 4-hydroxy-4-oxide; (*R*)- [39648-67-4] and (*S*)- [35193-64-7]
    1,1'-BINAPHTHYL-2,2'-DIYL HYDROGEN PHOSPHATE, (*R*)-(−)- and (*S*)-(+)-, **VIII**, 46, 50
Dinaphtho[2,1-*d*:1'2'-*f*]dioxaphosphepin, 4-methoxy 4-oxide, (*R*)- [86334-02-3], 46
    (*R*)-(−)-METHYL 1,1'-BINAPHTHYL-2,2'-DIYL PHOSPHATE, **VIII**, 46
1,3-Dioxane, 2-(2-bromoethyl)- [33884-43-4]
    2-(2-BROMOETHYL)-1,3-DIOXANE, **VII**, 59
1,3-Dioxane, 2-(2-bromoethyl)-2,5,5-trimethyl- [87842-52-2]
    2,5,5-TRIMETHYL-2-(2-BROMOETHYL)-1,3-DIOXANE, **VII**, 59
1,3-Dioxane, 2-(bromomethyl)-2-(chloromethyl)- [60935-30-0]
    2-(BROMOMETHYL)-2-(CHLOROMETHYL)-1,3-DIOXANE, **VIII**, 173
1,3-Dioxane, 4-phenyl- [772-00-9]
    4-PHENYL-*m*-DIOXANE, **IV**, 786
1,4-Dioxane, 2,3-bis(1,1-dimethylethoxy)-, *cis*- [68470-78-0]
1,4-Dioxane, 2,3-bis(1,1-dimethylethoxy)-, *trans*- [68470-79-1]
    2,3-DI-*tert*-BUTOXY-1,4-DIOXANE, **VIII**, 161
1,4-Dioxane, 2,3-dichloro-, *trans*- [3883-43-0]
    trans-2,3-DICHLORO-1,4-DIOXANE, **VIII**, 161
1,3-Dioxane-5,5-dimethanol, 2-phenyl- [2425-41-4]
    MONOBENZALPENTAERYTHRITOL, **IV**, 679
1,3-Dioxane-4,6-dione, 5-(hexahydro-2*H*-azepin-2-ylidene)-2,2-dimethyl- [70912-54-8]
    ISOPROPYLIDENE α-(HEXAHYDROAZEPINYLIDENE-2)MALONATE, **VIII**, 263
1,4-Dioxaspiro[4.5]decane [177-10-6]
    1,4-DIOXASPIRO[4.5]DECANE, **V**, 303
1,4-Dioxaspiro[4.5]dec-6-ene, 8-ethynyl-8-methyl- [73843-26-2]
    8-ETHYNYL-8-METHYL-1,4-DIOXASPIRO[4.5]DEC-6-ENE, **VII**, 241
6,10-Dioxaspiro[4,5]dec-3-ene-1,1-dicarbonitrile, 2-phenyl- [88442-12-0]
    5,5-DICYANO-4-PHENYL-2-CYCLOPENTEN-1-ONE 1,3-PROPANEDIOL KETAL, **VIII**, 173
1,4-Dioxaspiro[4.4]non-6-ene, 6-bromo- [68241-78-1]
    2-BROMO-2-CYCLOPENTENONE ETHYLENE KETAL, **VII**, 271, 495

1,4-Dioxaspiro[4.4]non-6-ene, 6-[(4-methylphenyl)sulfinyl]-, (*S*)- [82136-15-0]
  2-(*p*-TOLUENESULFINYL)-2-CYCLOPENTENONE ETHYLENE KETAL,
  (*S*)-(+)-, **VII**, 495
1,4-Dioxaspiro[4.4]non-6-ene-6-methanol [80963-19-5]
  2-HYDROXYMETHYL-2-CYCLOPENTEN-1-ONE ETHYLENE KETAL,
  VII, 274
5,7-Dioxaspiro[2.5]octane-4,8-dione, 6,6-dimethyl- [5617-70-9]
  6,6-DIMETHYL-5,7-DIOXASPIRO[2.5]OCTANE-4,8-DIONE, **VII**, 411
4,8-Dioxaspiro[2.5]oct-1-ene [60935-21-9]
  CYCLOPROPENONE 1,3-PROPANEDIOL KETAL, **VIII**, 173
1*H*,6*H*-[1,3]-Dioxino[4,5-*c*][2]benzopyran-1-one, octahydro- 3,3,6,6,9-pentamethyl-,
  [6a*R*-(6aα,9a,10aβ)]- [78394-10-2]
  [6a*R*-(6aα,9a,10aβ)]-OCTAHYDRO-3,3,6,6,9-PENTAMETHYL-1*H*,6*H*-[1,3]-
  DIOXINO[4,5-*c*']BENZOPYRAN-1-ONE, **VIII**, 353
4*H*-1,3-Dioxin-4-one, 6-(chloromethyl)-2,2-dimethyl- [81956-31-2]
  6-CHLOROMETHYL-2,2-DIMETHYL-1,3-DIOXEN-4-ONE, **VIII**, 192
1,3-Dioxolane, 2-cyclobutyl-4,5-dimethyl-, (2α,4α,5β)- [36416-29-2]
  2-CYCLOBUTYL-*cis*-4-*trans*-5-DIMETHYL-1,3-DIOXOLANE, **VI**, 312
1,3-Dioxolane, 2,2-dimethyl-4,5-bis[(phenylmethoxy)methyl]-, (4*R*-*trans*)-
  [91604-40-9]; (4*S*-*trans*)- [68394-39-8]
  1,4-DI-*O*-BENZYL-2,3-DI-*O*-ISOPROPYLIDENE-L-THREITOL, **VIII**, 155
1,3-Dioxolane, 2-(4-ethoxy-3-methoxyphenyl)- [52987-93-6]
  4-ETHOXY-3-METHOXYBENZALDEHYDE ETHYLENE ACETAL, **VI**, 567
1,3-Dioxolane-4,5-dicarboxylic acid, 2,2-dimethyl-, bis(1-methylethyl) ester, (4*R*-*trans*)-
  [81327-47-1]
  DIISOPROPYL (2*S*,3*S*)-2,3-*O*-ISOPROPYLIDENETARTRATE, **VIII**, 201
1,3-Dioxolane-4,5-dicarboxylic acid, 2,2-dimethyl-, dimethyl ester, (4*R*-*trans*)-
  [37031-29-1] or (4*S*-*trans*)- [37031-30-4]
  DIMETHYL 2,3-*O*-ISOPROPYLIDENETARTRATE, **VIII**, 155
1,3-Dioxolane-4,5-dimethanol, 2,2-dimethyl-, (4*S*-*trans*)- [50622-09-8]
  2,3-DI-*O*-ISOPROPYLIDENE-L-THREITOL, **VIII**, 155
1,3-Dioxolane-4-methanol, 2,2-dimethyl-, (±)- [22323-83-7]
  DL-ISOPROPYLIDENEGLYCEROL, **III**, 502
1,3-Dioxolo[4,5-*g*]isoquinoline, 7,8-dihydro-5-phenyl- [55507-10-3]
  1-PHENYL-3,4-DIHYDRO-6,7-METHYLENEDIOXYISOQUINOLINE, **VII**, 326
Diphosphine, tetramethyl-, 1,2-disulfide [3676-97-9]
  TETRAMETHYLBIPHOSPHINE DISULFIDE, **V**, 1016; Warning, **VII**, 533
Diphosphoric acid, mono(3,7-dimethyl-2,6-octadienyl) ester, triammonium salt, (*E*)-
  [116057-55-7]
  TRISAMMONIUM GERANYL PHOSPHATE, **VIII**, 616
Diselenide, diphenyl- [1666-13-3]
  DIPHENYL DISELENIDE, **VI**, 533
Dispiro[5.1.5.1]tetradecane-7,14-dione [950-21-0]
  DISPIRO[5.1.5.1]TETRADECANE-7,14-DIONE, **V**, 297
Disulfide, bis(2-nitrophenyl) [1155-00-6]
  DI-*o*-NITROPHENYL DISULFIDE, **I**, 220

Disulfide, bis(3-nitrophenyl) [537-91-7]
    m-NITROPHENYL DISULFIDE, **V**, 843
Disulfide, dibenzoyl [644-32-6]
    BENZOYL DISULFIDE, **III**, 116
Disulfide, 1-methylethyl 1-methylpropyl [67421-86-7]
    sec-BUTYL ISOPROPYL DISULFIDE, **VI**, 235
1,3-Dithiane [505-23-7]
    1,3-DITHIANE, **VI**, 556
1,4-Dithiane [505-29-3]
    p-DITHIANE, **IV**, 396
1,3-Dithiane, 2-methylthio-2-phenyl- [34858-82-7]
    2-METHYLTHIO-2-PHENYL-1,3-DITHIANE, **VI**, 109
1,3-Dithiane, 2-phenyl- [5425-44-5]
    2-PHENYL-1,3-DITHIANE, **VI**, 109
1,4-Dithiaspiro[4.5]decan-6-one [27694-08-2]
    2,2-(ETHYLENEDITHIO)CYCLOHEXANONE, **VI**, 590
1,4-Dithiaspiro[4.11]hexadecane [16775-67-0]
    1,4-DITHIASPIRO[4.11]HEXADECANE, **VII**, 124
5,9-Dithiaspiro[3.5]nonane [15077-16-4]
    5,9-DITHIASPIRO[3.5]NONANE, **VI**, 316
1,5-Dithiaspiro[5.5]undecan-7-one [51310-03-3]
    2,2-(TRIMETHYLENEDITHIO)CYCLOHEXANONE, **VI**, 1014
1,3-Dithietane, 2,2,4,4-tetrakis(trifluoromethyl)- [791-50-4]
    TETRAKIS(TRIFLUOROMETHYL)-1,3-DITHIETANE, **VII**, 251
Dithioimidodicarbonic diamide [541-53-7]
    2,4-DITHIOBIURET, **IV**, 504
Docosanedioic acid [505-56-6]
    DOCOSANEDIOIC ACID, **V**, 533
Docosanedioic acid, 7,16-dioxo-, disodium salt [134507-60-1]
    7,16-DIKETODOCOSANEDIOIC ACID and DISODIUM SALT, **V**, 534, 536
13-Docosenoic acid, (Z)- [112-86-7]
    ERUCIC ACID, **II**, 258
2,6-Dodecadiene-1,11-diol, 10-bromo-3,7,11-trimethyl-, acetate, (*E,E*)- [54795-59-4]
    10-BROMO-11-HYDROXY-10,11-DIHYDROFARNESYL ACETATE, **VI**, 560
6,10-Dodecadien-2-yn-1-ol, 7,11-dimethyl-, (*E*)- [16933-56-5]
    (*E*)-7,11-DIMETHYL-6,10-DODECADIEN-2-YN-1-OL, **VIII**, 226
1-Dodecanamine, *N,N*-dimethyl-, *N*-oxide [1643-20-5]
    N,N-DIMETHYLDODECYLAMINE OXIDE, **VI**, 501
1-Dodecanamine, *N*-methyl- [7311-30-0]
    LAURYLMETHYLAMINE, **IV**, 564
Dodecane [112-40-3]
    DODECANE, **VI**, 376
Dodecane, 1-bromo- [143-15-7]
    DODECYL BROMIDE, **I**, 29; **II**, 246
1-Dodecanethiol [112-55-0]
    DODECYL MERCAPTAN, **III**, 363

Dodecanoic acid, ethenyl ester [2146-71-6]
  VINYL LAURATE and OTHER VINYL ESTERS, **IV**, 977
Dodecanoic acid, 2-methyl- [2874-74-0]
  2-METHYLDODECANOIC ACID, **IV**, 618
Dodecanoic acid, 2-methylene- [52756-21-5]
  2-METHYLENEDODECANOIC ACID, **IV**, 616
1-Dodecanol [112-53-8]
  LAURYL ALCOHOL, **II**, 372
1,3,6,10-Dodecatetraene, 3,7,11-trimethyl- [502-61-4]
  α-FARNESENE, **VII**, 245
2,6,10-Dodecatrien-1-ol, 3,7,11-trimethyl-, acetate, (*E,E*)- [4128-17-0]
  FARNESYL ACETATE, **VI**, 560
2-Dodecenoic acid, (*E*)- [32466-54-9]
  trans-2-DODECENOIC ACID, **IV**, 398
2-Dodecenoic acid, 2-methyl-, (*E*)- [53663-29-9]
  trans-2-METHYL-2-DODECENOIC ACID, **IV**, 608
5-Dodecen-2-one, (*E*)- [81953-05-1]
  5-DODECEN-2-ONE, (*E*)-, **VIII**, 235

E

4a,8a-Epoxynaphthalene, 2,3,6,7-tetrabromooctahydro- [16573-82-3]
  3,4,8,9-TETRABROMO-11-OXATRICYCLO[4.4.1.0^{1,6}]UNDECANE, **VI**, 862
4*a*,8*a*-Epoxynaphthalene, 1,4,5,8-tetrahydro- [16573-72-1]
  11-OXATRICYCLO[4.4.1.0^{1,6}]UNDECA-3,8-DIENE, **VI**, 862
1,3,5(10)-Estratrien-17-one, 16α-fluoro-3-methoxy- [2383-28-0]
  16α-FLUORO-3-METHOXY-1,3,5(10)-ESTRATRIEN-17-ONE, **VIII**, 288
Ethanamine, 2-bromo-, hydrobromide [2576-47-8]
  β-BROMOETHYLAMINE HYDROBROMIDE, **II**, 91
Ethanamine, 2-chloro-*N,N*-dimethyl-, hydrochloride [4584-46-7]
  β-DIMETHYLAMINOETHYL CHLORIDE HYDROCHLORIDE, **IV**, 333
Ethanamine, 2,2-diethoxy- [645-36-3]
  AMINOACETAL, **III**, 50
Ethanamine, *N*-methyl- [624-78-2]
  N-METHYLETHYLAMINE, **V**, 758
Ethanamine, *N*-(phenylmethylene)- [6852-54-6]
  N-BENZYLIDENEETHYLAMINE, **V**, 758
Ethanaminium, 2-(benzoyloxy)-*N,N,N*-trimethyl-, chloride [2964-09-2]
  BENZOYLCHOLINE CHLORIDE, **IV**, 84
Ethanaminium, 2-(benzoyloxy)-*N,N,N*-trimethyl-, iodide [17518-43-3]
  BENZOYLCHOLINE IODIDE, **IV**, 84
Ethanaminium, 2-chloro-*N,N,N*-trimethyl-2-oxo-, chloride [53684-57-4]
  N-CHLOROBETAINYL CHLORIDE, **IV**, 154
Ethanaminium, *N,N*-diethyl-*N*-[[methoxycarbonyl)amino]sulfonyl]-, inner salt
    [29684-56-8]
  METHYL(CARBOXYSULFAMOYL)TRIETHYLAMMONIUM HYDROXIDE,
    VI, 788

Ethanaminium, 2-hydrazino-*N,N,N*-trimethyl-2-oxo-, chloride [123-46-6]
    BETAINE HYDRAZIDE HYDROCHLORIDE, **II**, 85
Ethane, 1,2-bis(2-iodoethoxy)- [36839-55-1]
    1,2-BIS(2-IODOETHOXY)ETHANE, **VIII**, 153
Ethane, bromo- [74-96-4]
    ETHYL BROMIDE, **I**, 29, 36
Ethane, 2-bromo-1,1-diethoxy- [2032-35-1]
    BROMOACETAL, **III**, 123
Ethane, 1-bromo-2-ethoxy- [592-55-2]
    β-ETHOXYETHYL BROMIDE, **III**, 370
Ethane, 2-chloro-1-ethoxy-1,1,2-trifluoro- [310-71-4]
    2-CHLORO-1,1,2-TRIFLUOROETHYL ETHYL ETHER, **IV**, 184
Ethane, 1-chloro-2-(methythio)- [542-81-4]
    β-CHLOROETHYL METHYL SULFIDE, **II**, 136
Ethane, 1,2-dibromo-1-ethoxy- [2983-26-8]
    α,β-DIBROMOETHYL ETHYL ETHER, **IV**, 749
Ethane, 1,1-diethoxy [105-57-7]
    ACETAL, **I**, 1
Ethane, isocyano- [624-79-3]
    ETHYL ISOCYANIDE, **IV**, 438
Ethane,1,1',1",1'''-[methanetetrayltetrakis(oxy)]tetrakis- [78-09-1]
    ETHYL ORTHOCARBONATE, **IV**, 457
Ethane, 1,1',1"-[methylidynetris(oxy)]tris- [122-51-0]
    ETHYL ORTHOFORMATE, **I**, 258
1,2-Ethanediamine, *N,N*-dibutyl- [3529-09-7]
    β-DI-BUTYLAMINOETHYLAMINE, **III**, 254
Ethanedioic acid [144-62-7]
    OXALIC ACID (ANHYDROUS), **I**, 421
Ethanedioic acid, diethyl ester [95-92-1]
    ETHYL OXALATE, **I**, 261
Ethanedioic acid, dimethyl ester [553-90-2]
    METHYL OXALATE, **II**, 414
1,2-Ethanediol, 1-phenyl-, dimethanesulfonate [32837-95-9]
    STYRENE GLYCOL DIMESYLATE, **VI**, 56
1,2-Ethanediol, 1,1,2,2-tetraphenyl- [464-72-2]
    BENZOPINACOL, **II**, 71
Ethanedione, bis[4-(dimethylamino)phenyl]- [17078-27-2]
    4,4'-BIS(DIMETHYLAMINO)BENZIL, **V**, 111
Ethanedione, diphenyl- [134-81-6]
    BENZIL, **I**, 87
1,2-Ethanedisulfonic acid, 1,2-dihydroxy- [18381-20-9]
    GLYOXAL BISULFITE, **III**, 438
1,2-Ethanedithiol [540-63-6]
    ETHANEDITHIOL, **IV**, 401
Ethanesulfonic acid, 2-amino- [107-35-7]
    TAURINE, **II**, 563

Ethanesulfonic acid, 2-bromo-, sodium salt [4263-52-9]
   SODIUM 2-BROMOETHANESULFONATE, **II**, 558
1,1,2,2-Ethanetetracarboxylic acid, tetramethyl ester [5464-22-2]
   TETRAMETHYL 1,1,2,2-ETHANETETRACARBOXYLATE, **VII**, 482
Ethanethioic acid [507-09-5]
   THIOLACETIC ACID, **IV**, 928
Ethanethiol, 2,2-diethoxy- [53608-94-9]
   DIETHYL MERCAPTOACETAL, **IV**, 295
Ethanethiol, 2,2'-[1,3-propanediylbis(thio)]bis- [25676-62-4]
   3,7-DITHIANONANE-1,9-DITHIOL, **VIII**, 592
1,1,2-Ethanetricarboxylic acid, 1-(1,3-dihydro-1,3-dioxo-2*H*-isoindol-2-yl)-, triethyl
   ester [76758-31-1]
   TRIETHYL α-PHTHALIMIDOETHANE-α,α,β-TRICARBOXYLATE, **IV**, 55
Ethanimidamide, monohydrochloride [124-42-5]
   ACETAMIDINE HYDROCHLORIDE, **I**, 5
Ethanimidic acid, 2,2,2-trichloro-, 3,7-dimethyl-2,6-octadienyl ester, (*E*)-
   [51479-75-5]
   GERANIOL TRICHLOROACETIMIDATE, **VI**, 508
Ethanol, 2-bromo- [540-51-2]
   2-BROMOETHANOL, **I**, 117
Ethanol, 2-chloro-, benzoate [939-55-9]
   2-CHLOROETHYL BENZOATE, **IV**, 84
Ethanol, 2-(cyclohexyloxy)- [1817-88-5]
   2-CYCLOHEXYLOXYETHANOL, **V**, 303
Ethanol, 2,2-dichloro- [598-38-9]
   2,2-DICHLOROETHANOL, **V**, 271
Ethanol, 2-(diethylamino)- [100-37-8]
   β-DIETHYLAMINOETHYL ALCOHOL, **II**, 183
Ethanol, 2-iodo-, benzoate [39252-69-2]
   2-IODOETHYL BENZOATE, **IV**, 84
Ethanol, 2-[(1-methylethyl)amino]- [109-56-8]
   2-ISOPROPYLAMINOETHANOL, **III**, 501
Ethanol, 2-(methylthio)- [5271-38-5]
   β-HYDROXYETHYL METHYL SULFIDE, **II**, 345
Ethanol, 2-nitro- [625-48-9]
   2-NITROETHANOL, **V**, 833
Ethanol, 2,2'-[1,3-propanediylbis(thio)]bis- [16260-48-3]
   3,7-DITHIANONANE-1,9-DIOL, **VIII**, 592
Ethanol, 2,2'-thiobis- [111-48-8]
   β-THIODIGLYCOL, **II**, 576
Ethanol, 2,2,2-trichloro- [115-20-8]
   TRICHLOROETHYL ALCOHOL, **II**, 598
Ethanone, 2-(acetyloxy)-1,2-diphenyl- [574-06-1]
   BENZOIN ACETATE, **II**, 69
Ethanone, 2-amino-1-phenyl-, hydrochloride [5468-37-1]
   PHENACYLAMINE HYDROCHLORIDE, **V**, 909

Ethanone, 1-(9-anthracenyl)- [784-04-3]
    9-ACETYLANTHRACENE, **IV**, 8
Ethanone, 1,1',1''-(1,3,5-benzenetriyl)tris- [779-90-8]
    1,3,5-TRIACETYLBENZENE, **III**, 829
Ethanone, 1-[2-(benzoyloxy)phenyl]- [4010-33-7]
    o-BENZOYLOXYACETOPHENONE, **IV**, 478
Ethanone, 1,2-bis(4-methoxyphenyl)- [120-44-5]
    DEOXYANISOIN, **V**, 339
Ethanone, 2-bromo-1-(4-bromophenyl)- [99-73-0]
    p-BROMOPHENACYL BROMIDE, **I**, 127
Ethanone, 2-bromo-1-(6-methoxy-2-naphthalenyl)- [10262-65-4]
    2-BROMOACETYL-6-METHOXYNAPHTHALENE, **VI**, 175
Ethanone, 1-(3-bromophenyl)- [2142-63-4]
    3-BROMOACETOPHENONE, **V**, 117
Ethanone, 1-(4-bromophenyl)- [99-90-1]
    p-BROMOACETOPHENONE, **I**, 109
Ethanone, 2-bromo-1-phenyl- [70-11-1]
    PHENACYL BROMIDE, **II**, 480
Ethanone, 2-chloro-1,2-diphenyl- [447-31-4]
    DESYL CHLORIDE, **II**, 159
Ethanone, 1-(1-cyclohexen-1-yl)- [932-66-1]
    1-ACETYLCYCLOHEXENE, **III**, 22
Ethanone, 1-cyclohexyl- [823-76-7]
    CYCLOHEXYL METHYL KETONE, **V**, 775
Ethanone, 1-cyclopropyl- [765-43-5]
    METHYL CYCLOPROPYL KETONE, **IV**, 597
Ethanone, diazodiphenyl- [3469-17-8]
    PHENYLBENZOYLDIAZOMETHANE, **II**, 496
Ethanone, 2-diazo-1-phenyl- [3282-32-4]
    DIAZOACETOPHENONE, **VI**, 386
Ethanone, 2-diazo-1-naphthalenyl- [4372-76-3]
    1-(DIAZOACETYL)NAPHTHALENE, **VI**, 613,
Ethanone, 2,2-dibromo-1-(4-bromophenyl)- [13195-79-4]
    p,α,α-TRIBROMOACETOPHENONE, **IV**, 110
Ethanone, 2,2-dibromo-1-(6-methoxy-2-naphthalenyl)- [52997-56-5]
    2,2-DIBROMOACETYL-6-METHOXYNAPHTHALENE, **VI**, 175
Ethanone, 2,2-dichloro-1-phenyl- [2648-61-5]
    α,α-DICHLOROACETOPHENONE, **III**, 538
Ethanone, 1-(2',5'-dihydroxy[1,1'-biphenyl]-4-yl)- [3948-13-8]
    2-*p*-ACETYLPHENYLHYDROQUINONE, **IV**, 15
Ethanone, 1-(2,4-dihydroxyphenyl)- [89-84-9]
    RESACETOPHENONE, **III**, 761
Ethanone, 1-(2,5-dihydroxyphenyl)- [490-78-8]
    2,5-DIHYDROXYACETOPHENONE, **III**, 280
Ethanone, 1-(2,6-dihydroxyphenyl)- [699-83-2]
    2,6-DIHYDROXYACETOPHENONE, **III**, 281

Ethanone, 2,5-dimethoxyphenyl- [1201-38-3]
  QUINACETOPHENONE DIMETHYL ETHER, **IV**, 836
Ethanone, 1-(2,4-dimethyl-3-furanyl)- [32933-07-6]
  3-ACETYL-2,4-DIMETHYLFURAN, **VI**, 31
Ethanone, 1,2-diphenyl- [451-40-1]
  DESOXYBENZOIN, **II**, 156
Ethanone, 1-(9*H*-fluoren-2-yl)- [781-73-7]
  2-ACETYLFLUORENE, **III**, 23
Ethanone, 1-(1-hydroxycyclohexyl)- [1123-27-9]
  1-ACETYLCYCLOHEXANOL, **IV**, 13
Ethanone, 2-hydroxy-1,2-diphenyl- [119-53-9]
  BENZOIN, **I**, 94
Ethanone, 1-(2-hydroxy-5-methoxyphenyl)- [705-15-7]
  QUINACETOPHENONE MONOMETHYL ETHER, **IV**, 836
Ethanone, 2-hydroxy-2-(methylthio)-1-phenyl- [13603-49-1]
  PHENYLGLYOXAL HEMIMERCAPTAL, **V**, 937
Ethanone, 2-hydroxy-1-phenyl- [582-24-1]
  α-HYDROXYACETOPHENONE, **VII**, 263
Ethanone, 2-hydroxy-1-phenyl-, dimethylacetal [28203-05-6]
  α-HYDROXYACETOPHENONE DIMETHYL ACETAL, **VII**, 263
Ethanone, 1-(6-methoxy-2-naphthalenyl)- [3900-45-6]
  2-ACETYL-6-METHOXYNAPHTHALENE, **VI**, 34
Ethanone, 2-methoxy-1-phenyl- [4079-52-1]
  α-METHOXYACETOPHENONE, **III**, 562
Ethanone, 1-[2-methyl-5-(1-methylethyl)phenyl]- [1202-08-0]
  ACETO *p*-CYMENE, **III**, 3
Ethanone, 2'-methylsulfinyl-1-phenyl- [26524-93-6]
  2-(METHYLSULFINYL)ACETOPHENONE, **V**, 939
Ethanone, 1-(2-nitrophenyl)- [577-59-3]
  o-NITROACETOPHENONE, **IV**, 708
Ethanone, 1-(3-nitrophenyl)- [121-89-1]
  m-NITROACETOPHENONE, **II**, 434
Ethanone, 1-(1,2,3,4,5-pentamethyl-2,4-cyclopentadien-1-yl)- [15971-76-3]
  5-ACETYL-1,2,3,4,5-PENTAMETHYL-2,4-CYCLOPENTADIENE, **VI**, 39
Ethanone, 1-(9-phenanthrenyl)- [2039-77-2]
  9-ACETYLPHENANTHRENE, **III**, 26
Ethanone, 1-phenyl-, dimethylhydrazone [13466-32-5]
  ACETOPHENONE *N,N*-DIMETHYLHYDRAZONE, **VI**, 12
Ethanone, 1-phenyl-, hydrazone [13466-30-3]
  ACETOPHENONE HYDRAZONE, **VI**, 12
Ethanone, 1-(4-pyridinyl)-, oxime [1194-99-6]
  4-ACETYLPYRIDINE OXIME, **VII**, 149
Ethanone, 1-(4-pyridinyl)-, *O*-[(4-methylphenyl)sulfonyl]oxime [74209-52-2]
  4-ACETYLPYRIDINE OXIME TOSYLATE, **VII**, 149
Ethanone, tetraphenyl- [466-37-5]
  β-BENZOPINACOLONE, **II**, 73

Ethanone, 1-(2-thienyl)- [88-15-3]
 2-ACETOTHIENONE, **II**, 8; **III**, 14
Ethanone, 2,2,2-trichloro-1-(1*H*-pyrrol-2-yl)- [35302-72-8]
 PYRROL-2-YL TRICHLOROMETHYL KETONE, **VI**, 618
Ethanone, 1-(2,3,4-trihydroxyphenyl)- [528-21-2]
 GALLACETOPHENONE, **II**, 304
Ethanone, 1-(2,4,6-trihydroxyphenyl)- [480-66-0]
 PHLOROACETOPHENONE, **II**, 522
Ethenamine, 1,2,2-trichloro-*N,N*-diethyl- [686-10-2]
 N,N-DIETHYL-1,2,2-TRICHLOROVINYLAMINE, **V**, 387
Ethenaminium, 2-carboxy-*N,N,N*-trimethyl-, hydroxide, inner salt, (*E*)- [54299-83-1]
 (*E*)-(CARBOXYVINYL)TRIMETHYLAMMONIUM BETAINE, **VIII**, 536
Ethene, 1,1-dichloro-2,2-difluoro- [79-35-6]
 1,1-DICHLORO-2,2-DIFLUOROETHYLENE, **IV**, 268
Ethene, 1,1-diethoxy- [2678-54-8]
 KETENE DIETHYLACETAL, **III**, 506
Ethenesulfonic acid, 2-phenyl-, sodium salt [2039-44-3]
 SODIUM β-STYRENESULFONATE, **IV**, 846
Ethenesulfonyl chloride, 2-phenyl- [4091-26-3]
 β-STYRENESULFONYL CHLORIDE, **IV**, 846
Ethenetetracarbonitrile [670-54-2]
 TETRACYANOETHYLENE, **IV**, 877
Ethenetetracarboxylic acid, tetraethyl ester [6174-95-4]
 ETHYL ETHYLENETETRACARBOXYLATE, **II**, 273
Ethenetricarbonitrile, [4-(dimethylamino)phenyl]- [6673-15-0]
 p-TRICYANOVINYL-*N,N*-DIMETHYLANILINE, **IV**, 953
1,4-Ethenonaphthalene, 1,4-dihydro- [7322-47-6]
 BENZOBARRELENE, **VI**, 82
1,4-Ethenonaphthalene, 5,6,7,8-tetrachloro-1,4-dihydro- [13454-02-9]
 TETRACHLOROBENZOBARRELENE, **VI**, 82
Ethenone [463-51-4]
 KETENE, **I**, 330; **V**, 679
Ethenone, diphenyl- [525-06-4]
 DIPHENYLKETENE, **III**, 356; **VI**, 549
Ethyne, ethoxy- [927-80-0]
 ETHOXYACETYLENE, **IV**, 404
Ethyne, methoxy- [6443-91-0]
 METHOXYACETYLENE, **IV**, 406

F

Ferrate (2−), chloro[7,12-diethenyl-3,8,13,17-tetramethyl-21*H*,23*H*-porphine-
 2,18-dipropanoato(4−)-$N^{21},N^{22},N^{23},N^{24}$-, (SP-5-13)- [16009-13-5]
 HEMIN, **III**, 442
Ferrate(1−), dicarbonyl(η^5-2,4-cyclopentadien-1-yl)-, sodium [12152-20-4]
 SODIUM DICARBONYL(CYCLOPENTADIENYL)FERRATE, **VIII**, 479

Ferrocene [102-54-5]
FERROCENE, **IV**, 473
Ferrocene, carboxy- [1271-42-7]
FERROCENECARBOXLIC ACID, **VI**, 625
Ferrocene, (2-chlorobenzoyl)- [49547-67-3]
(2-CHLOROBENZOYL)FERROCENE, **VI**, 625
Ferrocene, (cyanomethyl)- [1316-91-2]
FERROCENYLACETONITRILE, **V**, 578
Ferrocene, (hydroxymethyl)- [1273-86-5]
HYDROXYMETHYLFERROCENE, **V**, 621
9*H*-Fluoren-2-amine [153-78-6]
2-AMINOFLUORENE, **II**, 447; **V**, 30
9*H*-Fluorene, 9-methyl- [2523-37-7]
9-METHYLFLUORENE, **IV**, 623
9*H*-Fluorene, 2-nitro- [607-57-8]
2-NITROFLUORENE, **II**, 447
9*H*-Fluorene-9-carboxylic acid [1989-33-9]
9-FLUORENECARBOXYLIC ACID, **IV**, 482
9*H*-Fluorene-2-carboxylic acid, 9-oxo- [784-50-9]
FLUORENONE-2-CARBOXYLIC ACID, **III**, 420
1*H*-Fluorene-2,3-dicarbonitrile, 4,4a,9,9a-tetrahydro- [52477-65-3]
2,3-DICYANO-1,4,4a,9a-TETRAHYDROFLUORENE, **VI**, 427
9*H*-Fluoren-9-one, 2,4,5,7-tetranitro- [746-53-2]
2,4,5,7-TETRANITROFLUORENONE, **V**, 1029
9*H*-Fluoren-9-one, 2,4,7-trinitro- [129-79-3]
2,4,7-TRINITROFLUORENONE, **III**, 837
Formamide, *N*-(4-chlorophenyl)-*N*-ethyl- [13519-67-0]
N-ETHYL-*p*-CHLOROFORMANILIDE, **IV**, 420
Formamide, *N*-[1-[(1,1-dimethylethoxy)-methyl]-2-methylpropyl]-, (*S*)- [90482-04-5]
(*S*)-*N*-FORMYL-*O*-*tert*-BUTYL-L-VALINOL, **VIII**, 204
Formamide, *N*-(2,2-dimethyl-1-phenylethyl)- [52117-13-2]
N-FORMYL-α,α-DIMETHYL-β-PHENETHYLAMINE, **V**, 471
Formamide, 1-hydroxymethyl-2-methylpropyl-, (*S*)- [89876-66-4]
(*S*)-*N*-FORMYL-L-VALINOL, **VIII**, 204
Formamide, *N*-methyl-*N*-phenyl- [93-61-8]
N-METHYLFORMANILIDE, **III**, 590
Formamide, *N*-[[(4-methylphenyl)sulfonyl]methyl]- [36635-56-0]
N-(*p*-TOLYLSULFONYLMETHYL)FORMAMIDE, **VI**, 987
Furan [110-00-9]
FURAN, **I**, 274
Furan, 2-(bromomethyl)tetrahydro- [1192-30-9]
TETRAHYDROFURFURYL BROMIDE, **III**, 793
Furan, 2-(chloromethyl)tetrahydro- [3003-84-7]
TETRAHYDROFURFURYL CHLORIDE, **III**, 698
Furan, 2,5-dihydro-2,5-dimethoxy- [332-77-4]
2,5-DIHYDRO-2,5-DIMETHOXYFURAN, **V**, 403

Furan, 3-methyl- [930-27-8]
  3-METHYLFURAN, **IV**, 628
Furan, tetrahydro- [109-99-9]
  TETRAHYDROFURAN, **II**, 566; *Warning* on PURIFICATION, **V**, 976
Furan, tetrahydro-2,2-dimethyl-5-pentyl- [53684-53-0]
  5,5-DIMETHYL-2-PENTYLTETRAHYDROFURAN, **IV**, 350
2-Furancarboxaldehyde [98-01-1]
  FURFURAL, **I**, 280
2-Furancarboxaldehyde, 5-methyl- [620-02-0]
  5-METHYLFURFURAL, **II**, 393
2-Furancarboxylic acid [88-14-2]
  2-FUROIC ACID, **IV**, 493
2-Furancarboxylic acid, 3-methyl- [4412-96-8]
  3-METHYL-2-FUROIC ACID, **IV**, 628
2-Furancarboxylic acid, 3-methyl-, methyl ester [6141-57-7]
  METHYL 3-METHYL-2-FUROATE, **IV**, 649
2-Furancarboxylic acid, tetrahydro-5-oxo-, (*S*)- [21461-84-7]
  γ-BUTYROLACTONE-γ-CARBOXYLIC ACID, (*S*)-(+)-, **VII**, 99
2,3-Furandione, dihydro-4,4-dimethyl- [13031-04-4]
  KETOPANTOYL LACTONE, **VII**, 417
2,5-Furandione, 3,4-bis(acetyloxy)dihydro-, (3*R-trans*)- [19523-83-2]
  DIACETYL D-TARTARIC ANHYDRIDE, **IV**, 242
2,5-Furandione, dihydro- [108-30-5]
  SUCCINIC ANHYDRIDE, **II**, 560
2,5-Furandione, dihydro-3-methylene- [2170-03-8]
  ITACONIC ANHYDRIDE, **II**, 368
2,5-Furandione, 3-methyl- [616-02-4]
  CITRACONIC ANHYDRIDE, **II**, 140
2-Furanmethanamine, *N,N*,5-trimethyl- [14496-35-6]
  5-METHYLFURFURYLDIMETHYLAMINE, **IV**, 626
2-Furanmethanethiol [98-02-2]
  2-FURFURYL MERCAPTAN, **IV**, 491
2-Furanmethanol [98-00-0]
  2-FURYLCARBINOL, **I**, 276
2-Furanmethanol, acetate [623-17-6]
  2-FURYLMETHYL ACETATE, **I**, 285
3-Furanol, tetrahydro- [453-20-3]
  3-HYDROXYTETRAHYDROFURAN, **IV**, 534
2(3*H*)-Furanone, 3-acetyl-5-(chloromethyl)dihydro- [3154-75-4]
  α-ACETYL-δ-CHLORO-γ-VALEROLACTONE, **IV**, 10
2(3*H*)-Furanone, 3-bromodihydro- [5061-21-2]
  α-BROMO-γ-BUTYROLACTONE, **V**, 255
2(3*H*)-Furanone, dihydro-3,4-dihydroxy-, [*R-(R*,S*)*]- [21730-93-8]
  D-ERYTHRONOLACTONE, **VII**, 297
2(3*H*)-Furanone, dihydro-3-hydroxy-4,4-dimethyl- [599-04-2]
  PANTOYL LACTONE, D-(−)-, **VII**, 417

2(3*H*)-Furanone, dihydro-5-(iodomethyl)-4-phenyl-, *cis*- [67279-70-3] and *trans*-
   [67279-69-0]
   4,5-DIHYDRO-5-IODOMETHYL-4-PHENYL-2(3*H*)-FURANONE, *cis*- and trans-
   VII, 164
2(3*H*)-Furanone, dihydro-5-octyl- [2305-05-7]
   γ-(OCTYL)-γ-BUTYROLACTONE, **VII**, 400
2(3*H*)-Furanone, dihydro-5-propyl- [105-21-5]
   γ-PROPYLBUTYROLACTONE, **III**, 742
2(3*H*)-Furanone, 5-phenyl-3-(phenylmethylene)- [4361-96-0]
   α-BENZYLIDENE-γ-PHENYL-Δ$^{\beta,\gamma}$-BUTENOLIDE, **V**, 80
2(5*H*)-Furanone [497-23-4]
   γ-CROTONOLACTONE, **V**, 255
2(5*H*)-Furanone, 3-acetyl-4-hydroxy-5,5-dimethyl- [22609-93-4]
   3-ACETYL-4-HYDROXY-5,5-DIMETHYLFURAN-2(5*H*)-ONE, **VIII**, 71
2(5*H*)-Furanone, 4-azido-3-chloro-5-(1-methylethoxy)- [126773-43-1]
   4-AZIDO-3-CHLORO-5-ISOPROPOXY-2(5*H*)-FURANONE, (in solution),
   VIII, 116
2(5*H*)-Furanone, 3,4-dichloro-5-(1-methylethoxy)- [29814-12-8]
   3,4-DICHLORO-5-ISOPROPOXY-2(5*H*)-FURANONE, **VIII**, 116
2(5*H*)-Furanone, 3-methyl- [22122-36-7]
   3-METHYL-2(5*H*)-FURANONE, **VIII**, 396
3(2*H*)-Furanone, dihydro-2,2,5,5-tetramethyl- [5455-94-7]
   2,2,5,5-TETRAMETHYLTETRAHYDRO-3-KETOFURAN, **V**, 1024
2-Furanpropanoic acid, tetrahydro- [935-12-6]
   β-(TETRAHYDROFURYL)PROPIONIC ACID, **III**, 742
Furazan, dimethyl- [4975-21-7]
   DIMETHYLFURAZAN, **IV**, 342
Furo[3,4-*d*]-1,3-dioxol-4(3a*H*)-one, dihydro-2,2-dimethyl-, (3a*R-cis*)- [25581-41-3]
   2,3-*O*-ISOPROPYLIDENE-D-ERYTHRONOLACTONE, **VII**, 297
2-Furoic acid [88-14-2]
   2-FURANCARBOXYLIC ACID, **I**, 276
Furo[3,2-*b*]pyridine, 2-phenyl- [18068-82-1]
   2-PHENYLFURO[3,2-*b*]PYRIDINE, **VI**, 916

G

Glucofuranose, 1,2:5,6-di-*O*-isopropylidene-, *S*-methyldithiocarbonate, α-D-
   [16667-96-2]
   1,2:5,6-DI-(*O*-ISOPROPYLIDENE-3-*O*-(*S*-METHYL DITHIOCARBONATE)-
   α-D-GLUCOFURANOSE, **VII**, 139
D-Gluconic acid, pentaacetate [17430-71-6]
   2,3,4,5,6-PENTA-*O*-ACETYL-D-GLUCONIC ACID, **V**, 887
D-Gluconic acid, 2,3,4,6-tetraacetate [61259-49-2]
   2,3,4,6-TETRA-*O*-ACETYL-D-GLUCONIC ACID MONOHYDRATE, **V**, 887
D-Gluconitrile, 2,3,4,5,6-pentaacetate [6272-51-1]
   PENTAACETYL D-GLUCONONITRILE, **III**, 690

D-Gluconyl chloride, pentaacetate [53555-69-4]
    2,3,4,5,6-PENTA-*O*-ACETYL-D-GLUCONYL CHLORIDE, **V**, 887
β-D-Glucopyranose, 1,2,3,4-tetraacetate [13100-46-4]
    β-D-GLUCOSE-1,2,3,4-TETRAACETATE, **III**, 432
β-D-Glucopyranose, 2,3,4,6-tetraacetate [3947-62-4]
    β-D-GLUCOSE-2,3,4,6-TETRAACETATE, **III**, 434
α-D-Glucopyranose, 4-*O*-(2,3,4,6-tetra-*O*-acetyl-β-D-glucopyranosyl)-, tetraacetate
    [5346-90-7]
    α-CELLOBIOSE OCTAACETATE, **II**, 124
β-D-Glucopyranose, 6-*O*-(2,3,4,6-tetra-*O*-acetyl-β-D-glucopyranosyl)-, tetraacetate
    [4613-78-9]
    β-GENTIOBIOSE OCTAACETATE, **III**, 428
D-Glucopyranose, 2,3,4,6-tetramethyl- [7506-68-5]
    2,3,4,6-TETRAMETHYL-D-GLUCOSE, **III**, 800
α-D-Glucopyranoside, methyl [97-30-3]
    α-METHYL D-GLUCOSIDE, **I**, 364
α-D-Glucopyranoside, methyl 6-bromo-6-deoxy-, 4-benzoate [10368-81-7]
    METHYL 4-*O*-BENZOYL-6-BROMO-6-DEOXY-α-D-GLUCOPYRANOSIDE,
    VIII, 363
α-D-Glucopyranoside, methyl 4,6-*O*-(phenylmethylene)- [3162-96-7]
    METHYL 4,6-*O*-BENZYLIDENE-α-D-GLUCOPYRANOSIDE, **VIII**, 363
α-D-Glucopyranosyl bromide, tetraacetate [572-09-8]
    ACETOBROMOGLUCOSE, **III**, 11
α-D-Glucopyranosyl chloride, 2-(acetylamino)-2-deoxy-, 3,4,6-triacetate [3068-34-6]
    2-ACETAMIDO-3,4,6-TRI-*O*-ACETYL-2-DEOXY-α-D-GLUCOPYRANOSYL
    CHLORIDE, **V**, **I**
D-Glucose, 2-amino-2-deoxy-, hydrochloride [66-84-2]
    D-GLUCOSAMINE HYDROCHLORIDE, **III**, 430
D-Glucose, 4-*O*-α-D-glucopyranosyl- [528-50-7]
    CELLOBIOSE, **II**, 122
D-Glutamic acid [6893-26-1]
    D-GLUTAMIC ACID, **I**, 286
Glutaramide, 2,4-dicyano-3-methyl- [5447-66-5]
    α,α'-DICYANO-β-METHYLGLUTARAMIDE, **III**, 591
D-Glycero-D-*ido*-nonononitrile, 4,8-anhydro-2,3-dideoxy-, 5,6,7,9-tetraacetate [86563-27-1]
    1-DEOXY-2,3,4,6-TETRA-*O*-ACETYL-1-(2-CYANOETHYL)-α-D-
    GLUCOPYRANOSE, **VIII**, 148
Glycine [56-40-6]
    GLYCINE, **I**, 298
Glycine, 1,1-dimethylethyl ester [6456-74-2]
    GLYCINE *tert*-BUTYL ESTER, **V**, 586
Glycine, ethyl ester, hydrochloride [623-33-6]
    GLYCINE ETHYL ESTER HYDROCHLORIDE, **II**, 310
Glycine, *N*-acetyl- [543-24-8]
    ACETYLGLYCINE, **II**, 11

Glycine, *N*-(aminoiminomethyl)- [352-97-6]
GUANIDOACETIC ACID, **III**, 440
Glycine, *N*-benzoyl- [495-69-2]
HIPPURIC ACID, **II**, 328
Glycine, *N*-(carboxymethyl)-*N*-methyl- [4408-64-4]
METHYLIMINODIACETIC ACID, **II**, 397
Glycine, *N*-formyl-, ethyl ester [3154-51-6]
N-FORMYLGLYCINE ETHYL ESTER, **VI**, 620
Glycine, *N*-[*N*-[3-hydroxy-1-[(phenylmethoxy)carbonyl]-L-prolyl]glycyl]-, ethyl ester
[57621-06-4]
N-CARBOBENZYLOXY-3-HYDROXY-L-PROLYLGLYCYLGLYCINE ETHYL
ESTER, **VI**, 263
Glycine, *N*-nitroso-*N*-phenyl- [6415-68-5]
N-NITROSO-*N*-PHENYLGLYCINE, **V**, 962
Glycine, *N*-(phenylmethoxy)carbonyl- [1138-80-3]
CARBOBENZOXYGLYCINE, **III**, 168
Glyoxylanilide, 2-oxime [1769-41-1]
ISONITROSOACETANILIDE, **I**, 327
Glyoxylic acid, *p*-toluenesulfonylhydrazone [14661-68-8]
Glyoxylic acid chloride, 2-(*p*-toluenesulfonylhydrazone) [14661-69-9]
GLYOXYLIC ACID (and ACID CHLORIDE)
p-TOLUENESULFONYLHYDRAZONE, **V**, 258
Guanidine, mononitrate [506-93-4]
GUANIDINE NITRATE, **I**, 302; *Warning*, **V**, 589
Guanidine, nitro- [556-88-7]
NITROGUANIDINE, **I**, 399
D-Gulonic acid, γ-lactone [6322-07-2]
D-GULONIC-γ-LACTONE, **IV**, 506

H

Heptacyclo[31.3.1.1^{3,7}.1^{9,13}.1^{15,19}.1^{21,25}.1^{27,31}]-
octatetraconta-1(37),3,5,7(42),9,11,13(41),15,17,19,(40),21,23,25(39),-
27,29,31(38),33,35-octadecaene-37,38,39,40,41,42-hexol,
5,11,17,23,29,35-hexakis (1,1-dimethylethyl)- [78092-53-2]
p-tert-BUTYLCALIX[6]ARENE, **VIII**, 77
1,6-Heptadien-3-ol, 4,4-dimethyl- [58144-16-4]
4,4-DIMETHYL-1,6-HEPTADIEN-3-OL, **VII**, 177
1,6-Heptadien-3-ol, 2-methyl- [53268-46-5]
2-METHYL-1,6-HEPTADIEN-3-OL, **VI**, 606
1,5-Heptadien-4-ol, 3,4,5-trimethyl- [64417-15-8]
3,4,5-TRIMETHYL-2,5-HEPTADIEN-4-OL, **VIII**, 505
Heptanal [111-71-7]
HEPTANAL, **VI**, 644, 650
Heptanal, oxime [629-31-2]
HEPTALDOXIME, **II**, 313

Heptanal, 2-bromo- [16486-84-3]
    α-BROMOHEPTALDEHYDE, **III**, 127
Heptanamide [628-62-6]
    HEPTAMIDE, **IV**, 513
1-Heptanamine [111-68-2]
    HEPTYLAMINE, **II**, 318
Heptane, 2-bromo-1,1-dimethoxy- [18207-18-6]
    α-BROMOHEPTALDEHYDE DIMETHYL ACETAL, **III**, 128
Heptane, 1-bromo-2-fluoro- [1786-32-9]
    1-BROMO-2-FLUOROHEPTANE, **V**, 136
Heptane, 1,1,1-trifluoro- [693-09-4]
    1,1,1-TRIFLUOROHEPTANE, **V**, 1082
Heptanedioic acid [111-16-0]
    PIMELIC ACID, **II**, 531
Heptanedioic acid, diethyl ester [2050-20-6]
    ETHYL PIMELATE, **II**, 536
Heptanedioic acid, 2-acetyl-3-oxo-, diethyl ester [61983-62-8]
    DIETHYL α-ACETYL-β-KETOPIMELATE, **V**, 384
Heptanedioic acid, 3-oxo-, diethyl ester [40420-22-2]
    DIETHYL β-KETOPIMELATE, **V**, 384
Heptanedioic acid, 4-oxo-, diethyl ester [6317-49-3]
    DIETHYL γ-OXOPIMELATE, **IV**, 302
2,5-Heptanedione [1703-51-1]
    2,5-HEPTANEDIONE, **VII**, 648
2,4-Heptanedione, 3-methyl- [13152-54-0]
    3-METHYL-2,4-HEPTANEDIONE, **VI**, 776
Heptanoic acid [111-14-8]
    HEPTANOIC ACID, **II**, 315
Heptanoic acid, anhydride [626-27-7]
    HEPTOIC ANHYDRIDE, **III**, 28
Heptanoic acid, 2-fluoro- [1578-58-1]
    2-FLUOROHEPTANOIC ACID, **V**, 580
Heptanoic acid, 3-methyl- [53663-30-2]
    3-METHYLHEPTANOIC ACID, **V**, 762
Heptanoic acid, 3-methyl-, 1-methylpropyl ester [16253-72-8]
    sec-BUTYL 3-METHYLHEPTANOATE, **V**, 763
Heptanoic acid, 6-oxo- [3128-07-2]
    δ-ACETYLVALERIC ACID, **IV**, 19
Heptanoic acid, 7-oxo-, methyl ester [35376-00-2]
    METHYL 7-OXOHEPTANOATE, **VI**, 807
1-Heptanol [111-70-6]
    HEPTYL ALCOHOL, **I**, 304
1-Heptanol, 2-fluoro-, acetate [1786-44-3]
    2-FLUOROHEPTYL ACETATE, **V**, 580
2-Heptanol [543-49-7]
    2-HEPTANOL, **II**, 317

2-Heptanone [110-43-0]
    METHYL AMYL KETONE, **I**, 351
2-Heptanone, 4-hydroxy-3-phenyl-, (*R*,R**)- [42052-62-0]
    threo-4-HYDROXY-3-PHENYL-2-HEPTANONE, **VI**, 692
2-Heptanone, 5-nitro- [42397-25-1]
    5-NITROHEPTAN-2-ONE, **VI**, 648
3-Heptanone, 5-hydroxy-2,4,6-trimethyl-2-[(trimethylsilyloxy)]-, (*R*,S**)-(±)-
        [64869-24-5]
    5-HYDROXY-2,4,6-TRIMETHYL-2-TRIMETHYLSILOXYHEPTAN-3-ONE,
        VII, 185
3-Heptanone, 4-methyl-, (*S*)- [51532-30-0]
    4-METHYL-3-HEPTANONE, (*S*)-(+)-, **VIII**, 403
4-Heptanone [123-19-3]
    4-HEPTANONE, **V**, 589
4-Heptanone, 3,5-dibromo-2,2,6,6-tetramethyl- [23438-05-3]
    α,α'-DIBROMODINEOPENTYL KETONE, **VI**, 991
4-Heptanone, 1,7-dichloro- [40624-07-5]
    1,7-DICHLORO-4-HEPTANONE, **IV**, 279
4-Heptanone, 2,2,6,6-tetramethyl- [4436-99-1]
    DINEOPENTYL KETONE, **VI**, 991
2-Heptene, 4-bromo- [22118-57-6]
    4-BROMO-2-HEPTENE, **IV**, 108
6-Heptenoic acid, 6-methyl-5-oxo-, ethyl ester [130892-17-0]
    ETHYL 5-OXO-6-METHYL-6-HEPTENOATE, **VIII**, 274
1-Hepten-1-ol, acetate (*E*)- [17574-85-5]; (*Z*)- [17574-86-6]
    HEPTALDEHYDE ENOL ACETATE, **III**, 127
1-Hepten-4-ol, 4,6-dimethyl- [32189-75-6]
    4,6-DIMETHYL-1-HEPTEN-4-OL, **V**, 452
1-Hepten-3-one, 1-chloro-6-methyl- [18378-90-0]
    β-CHLOROVINYL ISOAMYL KETONE, **IV**, 186
6-Hepten-2-one, 4-phenyl- [69492-29-1]
    4-PHENYL-6-HEPTEN-2-ONE, **VII**, 443
5-Heptynoic acid, 7-hydroxy-, methyl ester [50781-91-4]
    METHYL 7-HYDROXYHEPT-5-YNOATE, **VIII**, 415
1,4,7,10,13,16-Hexaazacyclooctadecane [296-35-5]
    1,4,7,10,13,16-HEXAAZACYCLOOCTADECANE, **VI**, 652
1,4,7,10,13,16-Hexaazacyclooctadecane, 1,4,7,10,13,16-
        hexakis[(methylphenyl)sulfonyl]- [52601-75-9]
    1,4,7,10,13,16-HEXAKIS(*p*-TOLYLSULFONYL)-1,4,7,10,13,16-
        HEXAAZACYCLOOCTADECANE, **VI**, 652
5,7-Hexadecadiene, (5Z,7E)- [86883-79-6]
    (5Z,7E)-5,7-HEXADECADIENE, **VIII**, 295
Hexadecane [544-76-3]
    HEXADECANE, **II**, 320
Hexadecane, 1-iodo- [544-77-4]
    HEXADECYL IODIDE, **II**, 322

Hexadecanoic acid [57-10-3]
    PALMITIC ACID, **III**, 605
Hexadecanoic acid, methyl ester [112-39-0]
    METHYL PALMITATE, **III**, 605
Hexadecanoic acid, 2-sulfo- [1782-10-1]
    α-SULFOPALMITIC ACID, **IV**, 862
1,5-Hexadiene [592-42-7]
    BIALLYL, **III**, 121
(Z,Z)-2,4-Hexadienedinitrile [1557-59-1]
    (Z,Z)-2,4-HEXADIENEDINITRILE, **VI**, 306, 662
2,4-Hexadienedioic acid, (Z,Z)- [505-70-4]
    MUCONIC ACID, **III**, 623
2,4-Hexadienoic acid, (E,E)- [110-44-1]
    SORBIC ACID, **III**, 783
2,4-Hexadienoic acid, monomethyl ester, (Z,Z)- [61186-96-7]
    cis,cis-MONOMETHYL MUCONATE, **VIII**, 490
1,5-Hexadien-3-ol [924-41-4]
    1,5-HEXADIEN-3-OL, **V**, 608
Hexanal [66-25-1]
    HEXALDEHYDE, **II**, 323
Hexanal, 6,6-dimethoxy- [55489-11-7]
    6,6-DIMETHOXYHEXANAL, **VII**, 168
Hexanal, 2,2-dimethyl-5-oxo- [13544-11-1]
    2,2-DIMETHYL-5-OXOHEXANAL, **VI**, 497
Hexane, 1,6-diiodo- [629-09-4]
    1,6-DIIODOHEXANE, **IV**, 323
Hexane, 1,6-diisocyanato- [822-06-0]
    HEXAMETHYLENE DIISOCYANATE, **IV**, 521
Hexane, 1-fluoro- [373-14-8]
    HEXYL FLUORIDE, **IV**, 525
Hexanedioic acid [124-04-9]
    ADIPIC ACID, **I**, 18
Hexanedioic acid, diethyl ester [141-28-6]
    DIETHYL ADIPATE, **II**, 264
Hexanedioic acid, 2,5-dibromo-, diethyl ester [869-10-3]
    DIETHYL α,δ-DIBROMOADIPATE, **III**, 623
1,6-Hexanediol [629-11-8]
    HEXAMETHYLENE GLYCOL, **II**, 325
2,5-Hexanediol, 2,5-dimethyl- [110-03-2]
    α,α,α',α'-TETRAMETHYLTETRAMETHYLENE GLYCOL, **V**, 1026
2,5-Hexanedione [110-13-4]
    ACETONYLACETONE, **II**, 219
2,5-Hexanedione, 3,4-diacetyl- [5027-32-7]
    TETRAACETYLETHANE, **IV**, 869
1,6-Hexanedione, 1,6-diphenyl- [3375-38-0]
    1,4-DIBENZOYLBUTANE, **II**, 169

Hexanenitrile, 2-ethyl- [4528-39-6]
  2-ETHYLHEXANONITRILE, **IV**, 436
Hexanoic acid, anhydride [2051-49-2]
  CAPROIC ANHYDRIDE, **III**, 164
Hexanoic acid, 2-acetyl-, ethyl ester [1540-29-0]
  ETHYL BUTYLACETOACETATE, **I**, 248
Hexanoic acid, 6-amino- [60-32-2]
  ε-AMINOCAPROIC ACID, **II**, 28; **IV**, 39
Hexanoic acid, 6-(benzoylamino)- [956-09-2]
  ε-BENZOYLAMINOCAPROIC ACID, **II**, 76
Hexanoic acid, 6-(benzoylamino)-2-bromo- [1700-05-6]
  ε-BENZOYLAMINO-α-BROMOCAPROIC ACID, **II**, 74
Hexanoic acid, 6-(benzoylamino)-2-chloro- [5107-15-3]
  ε-BENZOYLAMINO-α-CHLOROCAPROIC ACID, **VI**, 90
Hexanoic acid, 2-bromo- [616-05-7]
  α-BROMOCAPROIC ACID, **I**, 115; **II**, 95
Hexanoic acid, 2-cyano-, ethyl ester [7391-39-1]
  ETHYL BUTYLCYANOACETATE, **III**, 385
Hexanoic acid, 6,6-dimethoxy-, methyl ester [25176-55-0]
  METHYL 6,6-DIMETHOXYHEXANOATE, **VII**, 168
Hexanoic acid, 2-fluoro-, ethyl ester, (RS)-, (R)-, and (S)-; (RS)- [17841-31-5]; (R)-
    [124439-29-8]; (S)- [124439-31-2]
  ETHYL 2-FLUOROHEXANOATE, (RS)-, (R)-, and (S)-, **VIII**, 258
Hexanoic acid, 4-oxo-, ethyl ester [3249-33-0]
  ETHYL 4-OXOHEXANOATE, **VI**, 615
Hexanoic acid, 6-oxo-, methyl ester [6654-36-0]
  METHYL 6-OXOHEXANOATE, **VII**, 168
1-Hexanol [111-27-3]
  1-HEXANOL, **VI**, 919
  HEXYL ALCOHOL, **I**, 306
1-Hexanol, 6-chloro- [2009-83-8]
  HEXAMETHYLENE CHLOROHYDRIN, **III**, 446
1-Hexanol, 2-methyl- [624-22-6]
  2-METHYL-1-HEXANOL, **VI**, 786
2-Hexanone, 1,1-dimethoxy-5-methyl- [53684-58-51]
  β-KETOISOOCTALDEHYDE DIMETHYL ACETAL, **IV**, 558
Hexanoyl chloride, 2-bromo- [42768-46-7]
  2-BROMOHEXANOYL CHLORIDE, **VI**, 190
1,4,7,10,13,16-Hexaoxacyclooctadecane [17455-13-9]
  18-CROWN-6, **VI**, 301
1,2,4,5-Hexatetraene [29776-96-3]
  1,2,4,5-HEXATETRAENE (in solution), **VII**, 485
1,3,5-Hexatriene [2235-12-3]
  1,3,5-HEXATRIENE, **V**, 608
2-Hexenal, 4-hydroxy-, (E)- [17427-21-3]
  trans-4-HYDROXY-2-HEXENAL, **VI**, 683

3-Hexene, 3,4-dinitro- [53684-54-1]
    3,4-DINITRO-3-HEXENE, **IV**, 372
1-Hexene, 1-iodo-, (Z)- [16538-47-9]
    1-IODOHEXENE, (Z)-, **VII**, 290
2-Hexenedioic acid, dimethyl ester, (E)- [113327-79-0]
    DIMETHYL (E)-2-HEXENEDIOATE, **VIII**, 112
4-Hexen-1-ol, (E)- [6126-50-7]
    (E)-4-HEXEN-1-OL, **VI**, 675
5-Hexen-3-ol, 4,6-bis(methylthio)- [53107-07-6]
    3-BIS(METHYLTHIO)-1-HEXEN-4-OL, **VI**, 683
5-Hexen-2-one, 5-methyl- [3240-09-3]
    5-METHYL-5-HEXEN-2-ONE, **V**, 767
α-D-*ribo*-Hexofuranose, 3-deoxy-1,2:5,6-bis-*O*-(1-methylethylidene)- [4613-62-1]
    3-DEOXY-1,2:5,6-DI-*O*-ISOPROPYLIDENE-α-D-*ribo*-HEXOFURANOSE,
        VII, 139
α-D-*arabino*-Hexopyranose, 2-deoxy-, tetraacetate [16750-06-4]
    1,3,4,6-TETRA-*O*-ACETYL-2-DEOXY-α-D-GLUCOPYRANOSE, **VIII**, 583
5-Hexynal [29329-03-1]
    5-HEXYNAL, **VI**, 679
1-Hexyne [693-02-7]
    BUTYLACETYLENE, **IV**, 117
1-Hexyne, 3-ethyl- [76347-58-5]
    3-ETHYL-1-HEXYNE, **VI**, 595
2-Hexyne, 6-chloro- [28077-73-8]
    6-CHLORO-2-HEXYNE, **VI**, 273
2-Hexynoic acid, methyl ester [18937-79-6]
    METHYL 2-HEXYNOATE, **VI**, 791
L-Histidine, monohydrochloride [645-35-2]
    L-HISTIDINE MONOHYDROCHLORIDE, **II**, 330
Hydrazine, 1,1-dimethyl- [57-14-7]
    unsym-DIMETHYLHYDRAZINE , **II**, 213
Hydrazine, 1,1-dimethyl-, monohydrochloride [593-82-8]
    unsym-DIMETHYLHYDRAZINE HYDROCHLORIDE, **II**, 211
Hydrazine, 1,2-dimethyl-, dihydrochloride [306-37-6]
    sym-DIMETHYLHYDRAZINE DIHYDROCHLORIDE, **II**, 208
Hydrazine, (2,4-dinitrophenyl)- [119-26-6]
    2,4-DINITROPHENYLHYDRAZINE, **II**, 228
Hydrazine, methyl-, sulfate (1:1) [302-15-8]
    METHYLHYDRAZINE SULFATE, **II**, 395
Hydrazine, 1-methyl-1-phenyl- [618-40-6]
    α-METHYL-α-PHENYLHYDRAZINE, **II**, 418
Hydrazine, phenyl- [100-63-0]
    PHENYLHYDRAZINE, **I**, 442
Hydrazinecarboxamide, 2-(2-amino-1-methylethylidene)-, monohydrochloride
    [10469-70-2]
    AMINOACETONE SEMICARBAZONE HYDROCHLORIDE, **V**, 27

Hydrazinecarboxamide, *N*-phenyl- [537-47-3]
    4-PHENYLSEMICARBAZIDE, **I**, 450
Hydrazinecarboxylic acid, 1,1-dimethylethyl ester [870-46-2]
    tert-BUTYL CARBAZATE, **V**, 166
Hydrazinecarboxylic acid, ethyl ester [4114-31-2]
    ETHYL HYDRAZINECARBOXYLATE, **III**, 404
Hydrazinecarboxylic acid, 2-(1-cyanocyclohexyl)-, methyl ester [61827-29-0]
    METHYL 2-(1-CYANOCYCLOHEXYL)HYDRAZINECARBOXYLATE,
      VI, 334
Hydrazinecarboxylic acid, 2-[(phenylamino)carbonyl]-, ethyl ester [17696-94-5]
    4-PHENYL-1-CARBETHOXYSEMICARBAZIDE, **VI**, 936
1,2-Hydrazinedicarboxylic acid, diethyl ester [4114-28-7]
    DIETHYL HYDRAZODICARBOXYLATE, **III**, 375; **IV**, 411
1,2-Hydrazinedicarboxylic acid, bis(2,2,2-trichloroethyl) ester [38858-02-5]
    BIS(2,2,2-TRICHLOROETHYL) HYDRAZODICARBOXYLATE, **VII**, 56
Hydrazono, diethyl [38534-43-9]
    AZOETHANE, **VI**, 78
Hydroperoxide, 1,2,3,4-tetrahydro-1-naphthalenyl [771-29-9]
    TETRALIN HYDROPEROXIDE, **IV**, 895
Hydroperoxide, 1,1,3,3-tetramethylbutyl- [5809-08-5]
    tert-OCTYL HYDROPEROXIDE, **V**, 818
Hydroxylamine, *O*-(diphenylphosphinyl)- [72804-96-7]
    O-DIPHENYLPHOSPHINYLHYDROXYLAMINE, **VII**, 8
Hypochlorous acid, 1,1-dimethylethyl ester [507-40-4]
    tert-BUTYL HYPOCHLORITE, **IV**, 125; **V**, 184; *Warning*, **V**, 183

I

1*H*-Imidazole [288-32-4]
    IMIDAZOLE, **III**, 471
1*H*-Imidazole, 1,1'-carbonylbis- [530-62-1]
    1,1'-CARBONYLDIIMIDAZOLE, **V**, 201
1*H*-Imidazole-2-carboxaldehyde [10111-08-7]
    IMIDAZOLE-2-CARBOXALDEHYDE, **VII**, 287
1*H*-Imidazole-4,5-dicarboxylic acid [570-22-9]
    IMIDAZOLE-4,5-DICARBOXYLIC ACID, **III**, 471
1*H*-Imidazole-4-methanol, monohydrochloride [32673-41-9]
    4(5)-HYDROXYMETHYLIMIDAZOLE HYDROCHLORIDE, **III**, 460
2*H*-Imidazole-2-thione, 1,3-dihydro-1,3-dimethyl- [6596-81-2]
    1,3-DIMETHYLIMIDAZOLE-2-THIONE, **VII**, 195
Imidazolidine, 1,3-dibenzoyl-2(1*H*-imidazol-2-yl)-, monohydrochloride [65276-01-9]
    2-(1,3-DIBENZOYLIMIDAZOLIDIN-2-YL)IMIDAZOLE HYDROCHLORIDE,
      VII, 287
Imidazolidine, 2-(1,3-diphenyl-2-imidazolidinylidene)-1,3-diphenyl- [2179-89-7]
    BIS(1,3-DIPHENYLIMIDAZOLIDINYLIDENE-2), **V**, 115
2,4-Imidazolidinedione, 5,5-dimethyl- [77-71-4]
    5,5-DIMETHYLHYDANTOIN, **III**, 323

2,4-Imidazolidinedione, 5-[(4-hydroxyphenyl)methylene]- [80171-33-1]
5-(*p*-HYDROXYBENZAL)HYDANTOIN, **V**, 627
2-Imidazolidinethione [96-45-7]
ETHYLENE THIOUREA, **III**, 394
Imidazolidinetrione [120-89-8]
PARABANIC ACID, **IV**, 744
4-Imidazolidinone, 2-imino-1-methyl-5-vanillylidene- [29974-40-1]
5-(3-METHOXY-4-HYDROXYBENZAL)CREATININE, **III**, 587
1*H*-Imidazolium, 1,3-dimethyl-, iodide [4333-62-4]
1,3-DIMETHYLIMIDAZOLIUM IODIDE, **VII**, 195
4*H*-Imidazol-4-one, 2-amino-1,5-dihydro-1-methyl- [60-27-5]
CREATININE, **I**, 172
2*H*-Imidazol-2-one, 1,5-dihydro-4,5-diphenyl- [53684-56-3]
4,5-DIPHENYLGLYOXALONE, **II**, 231
Imidodicarbonic dihydrazide [4375-11-5]
DIAMINOBIURET, **III**, 404
1*H*-Indazole [271-44-3]
INDAZOLE, **III**, 475; **IV**, 536; **V**, 650
1*H*-Indazole, 3-chloro- [29110-74-5]
3-CHLOROINDAZOLE, **III**, 476
1*H*-Indazole, 5-nitro- [5401-94-5]
5-NITROINDAZOLE, **III**, 660
1*H*-Indazole, 4,5,6,7-tetrahydro- [2305-79-5]
4,5,6,7-TETRAHYDROINDAZOLE, **IV**, 536
2*H*-Indazole, 2-phenyl- [3682-71-1]
2-PHENYLINDAZOLE, **V**, 941
3*H*-3-Indazolone [5686-93-1]
INDAZOLONE, **III**, 476
1*H*-Indene, 1-chloro-2,3-dihydro- [35275-62-8]
α-CHLOROHYDRINDENE, **II**, 336
1*H*-Indene, 2,3-dihydro-1-methyl-3-phenyl- [6416-39-3]
1-METHYL-3-PHENYLINDANE, **IV**, 665
1*H*-Indene-2-carboxylic acid, 6,7-dimethoxy-3-methyl-, ethyl ester [5415-54-3]
ETHYL 6,7-DIMETHOXY-3-METHYLINDENE-2-CARBOXYLATE, **V**, 550
1*H*-Indene-1,5(4*H*)-dione, hexahydro-3a-hydroxy-7a-methyl-, (3a*S-cis*)- [33879-04-8]
(3a*S*,7a*S*)-(+)-2,3,3a,4,7,7a-HEXAHYDRO-3a-HYDROXY-7a-METHYL-1*H*-
INDENE-1,5(6*H*)-DIONE, **VII**, 363
1*H*-Indene-1,5(6*H*)-dione, 7a-methyl-2,3,7,7a-tetrahydro-, (*S*)- [17553-86-5]
(+)-(7a*S*)-7a-METHYL-2,3,7,7a-TETRAHYDRO-1*H*-INDENE-1,5(6*H*)-DIONE,
VII, 363
1*H*-Inden-1-one, 2,3-dihydro- [83-33-0]
α-HYDRINDONE, **II**, 336
1*H*-Inden-1-one, 2,3-dihydro-2-methyl- [17496-14-9]
2-METHYL-1-INDANONE, **V**, 624
1*H*-Inden-1-one, 2,3-diphenyl- [1801-42-9]
2,3-DIPHENYLINDONE (2,3-DIPHENYL-1-INDENONE), **III**, 353

2*H*-Inden-2-one, 1,3-dihydro- [615-13-4]
  2-INDANONE, **V**, 647
4*H*-Inden-4-one, 1,3a,5,6,7,7a-hexahydro-3,3a-dimethyl-6-(1-methylethenyl)-
    2-trimethylsilyl-, 3aα,6α,7aα)- [77494-23-6]
  cis-4-*exo*-ISOPROPENYL-1,9-DIMETHYL-8-(TRIMETHYLSILYL)
    BICYCLO[4.3.0]NON-8-EN-2-ONE, **VIII**, 347
5*H*-Inden-4-one, 1,2,3,3a,4,6-hexahydro- [131712-16-8]
  BICYCLO[4.3.0]NON-1-EN-4-ONE, **VIII**, 38
5*H*-Inden-5-one, 1,2,3,3*a*,4,7a-hexahydro-7a-(phenylsulfonyl)-, *cis*- [131712-15-7]
  4-OXO-1-(PHENYLSULFONYL)-*cis*-BICYCLO[4.3.0]NON-2-ENE, **VIII**, 39
1*H*-Indole [120-72-9]
  INDOLE, **III**, 479
1*H*-Indole, 1-methyl- [603-76-9]
  1-METHYLINDOLE, **V**, 769
1*H*-Indole, 2-methyl- [95-20-5]
  2-METHYLINDOLE, **III**, 597
1*H*-Indole, 4-nitro- [4769-97-5]
  4-NITROINDOLE, **VIII**, 493
1*H*-Indole, 2-phenyl- [948-65-2]
  2-PHENYLINDOLE, **III**, 725
1*H*-Indole, 3-(2-phenyl-1,3-dithian-2-yl)- [57621-00-8]
  3-(2-PHENYL-1,3-DITHIAN-2-YL)-1*H*-INDOLE, **VI**, 109
1*H*-Indole, 4-(phenylmethoxy)- [20289-26-3]
  4-BENZYLOXYINDOLE, **VII**, 34
1*H*-Indole, 1-(phenylmethyl)- [3377-71-7]
  1-BENZYLINDOLE, **VI**, 104, 106
1*H*-Indole, 3-(phenylmethyl)- [16886-10-5]
  3-BENZYLINDOLE, **VI**, 109
1*H*-Indole-3-acetic acid [87-51-4]
  INDOLE-3-ACETIC ACID, **V**, 654
1*H*-Indole-3-carbonitrile [5457-28-3]
  INDOLE-3-CARBONITRILE, **V**, 656
1*H*-Indole-3-carboxaldehyde [487-89-8]
  INDOLE-3-ALDEHYDE, **IV**, 539
1*H*-Indole-2-carboxylic acid, ethyl ester [3770-50-1]
  ETHYL INDOLE-2-CARBOXYLATE, **V**, 567
1*H*-Indole-5-carboxylic acid, 2-methyl-, ethyl ester [53600-12-7]
  ETHYL 2-METHYLINDOLE-5-CARBOXYLATE, **VI**, 601
1*H*-Indole-5-carboxylic acid, 2-methyl-3-(methylthio)-, ethyl ester
  ETHYL 2-METHYL-3-METHYLTHIOINDOLE-5-CARBOXYLATE, **VI**, 601
1*H*-Indole-2,3-dione [91-56-5]
  ISATIN, **I**, 327
1*H*-Indole-2,3-dione, 1-acetyl- [574-17-4]
  N-ACETYLISATIN, **III**, 456
2*H*-Indol-2-one, 3-acetyl-1,3-dihydro- [17266-70-5]
  3-ACETYLOXINDOLE, **V**, 12

2*H*-Indol-2-one, 1,3-dihydro-3-methyl- [1504-06-9]
   3-METHYLOXINDOLE, **IV**, 657
2*H*-Indol-2-one, 3-ethyl-1,3-dihydro-1-methyl- [2525-35-1]
   1-METHYL-3-ETHYLOXINDOLE, **IV**, 620
Iodine, bis(acetato-*O*)phenyl- [3240-34-4]
   IODOSOBENZENE DIACETATE, **V**, 660
Iodine, (dichlorophenyl)- [932-72-9]
   IODOBENZENE DICHLORIDE, **III**, 482
Iodine cyanide, I(CN) [506-78-5]
   CYANOGEN IODIDE, **IV**, 207
Iodonium, diphenyl, iodide [2217-79-0]
   DIPHENYLIODONIUM IODIDE, **III**, 355
Iron, tricarbonyl(η^4-1,3-cyclobutadiene)- [12078-17-0]
   CYCLOBUTADIENEIRON TRICARBONYL, **VI**, 310
Iron, tricarbonyl[(2,3,4,5-η)-2,4-cyclohexadien-1-one] [12306-92-2]
   TRICARBONYL[(2,3,4,5-η)-2,4-CYCLOHEXADIEN-1-ONE]IRON, **VI**, 996
Iron, tricarbonyl-[(1,2,3,4-η)-1-methoxycyclohexadiene] [12318-18-2]
Iron, tricarbonyl-[(1,2,3,4-η)-2-methoxycyclohexadiene] [12318-19-3]
   TRICARBONYLI(1,2,3,4-η)-1- and 2-METHOXY-1,3-
      CYCLOHEXADIENE]IRON, **VI**, 996
Iron, tricarbonyl[2-[(2,3,4,5-η)-4-methoxy-2,4-cyclohexadien-1-yl]-5,5- dimethyl-1,3-
   cyclohexanedione]- [51539-52-7]
   TRICARBONYL[2-(2,3,4,5-η)-4-METHOXY-2,4-CYCLOHEXADIEN-1-YL]-
      5,5-DIMETHYL-1,3-CYCLOHEXANEDIONE]IRON(1–), **VI**, 1001
Iron(1+), tricarbonyl[(1,2,3,4,5-η)-2-methoxy-2,4-cyclohexadien-1-yl]-
   hexafluorophosphate [51508-59-9]
   TRICARBONYL[(1,2,3,4,5-η)-2-METHOXY-2,4-CYCLOHEXADIEN-
      1-YL]IRON(1+) HEXAFLUOROPHOSPHATE(1–), **VI**, 996
1,3-Isobenzofurandione, 4-nitro- [641-70-3]
   3-NITROPHTHALIC ANHYDRIDE, **I**, 410
1,3-Isobenzofurandione, 3a,4,7,7a-tetrahydro-, *cis*- [935-79-5]
   cis-Δ^4-TETRAHYDROPHTHALIC ANHYDRIDE, **IV**, 890
1,3-Isobenzofurandione, 4,5,6,7-tetraiodo- [632-80-4]
   TETRAIODOPHTHALIC ANHYDRIDE, **III**, 796
1,3-Isobenzofurandione, 4,5,6,7-tetraphenyl- [4741-53-1]
   TETRAPHENYLPHTHALIC ANHYDRIDE, **III**, 807
1(3*H*)-Isobenzofuranone [87-41-2]
   PHTHALIDE, **II**, 526
1(3*H*)-Isobenzofuranone, 3-bromo- [6940-49-4]
   3-BROMOPHTHALIDE, **V**, 145
1(3*H*)-Isobenzofuranone, 3,3-dichloro- [601-70-7]
   unsym-o-PHTHALYL CHLORIDE, **II**, 528
1(3*H*)-Isobenzofuranone, hexahydro-, (3a*S-cis*)- [65376-02-5]
   8-OXABICYCLO[4.3.0]NONAN-7-ONE, (1*R*,6*S*)-(+)-, **VII**, 406
1(3*H*)-Isobenzofuranone, 3-(phenylmethylene)- [575-61-1]
   BENZALPHTHALIDE, **II**, 61

1,3,5(4*H*)-Isobenzofurantrione, tetrahydro-7-methoxy-, (3aα,7α,7aα)- [87334-37-0]
    5β-METHOXYCYCLOHEXAN-1-ONE-3β,4β-DICARBOXYLIC ACID
    ANHYDRIDE, **VII**, 312

1*H*-Isoindole, 2,3-dihydro- [496-12-8]
    1,3-DIHYDROISOINDOLE, **V**, 406

1*H*-Isoindole, 2,3-dihydro-2-[(4-methylphenyl)sulfonyl]- [32372-83-1]
    2-(TOLYLSULFONYL)DIHYDROISOINDOLE, **V**, 1064

2*H*-Isoindole-2-acetic acid, 1,3-dihydro-1,3-dioxo-α-(phenylmethyl)-, (*S*)- [5123-55-7]
    N-PHTHALYL-L-β-PHENYLALANINE, **V**, 973

1*H*-Isoindole-1,3(2*H*)-dione [85-41-6]
    PHTHALIMIDE, **I**, 457

1*H*-Isoindole-1,3(2*H*)-dione, 2-(2-bromoethyl)- [574-98-1]
    β-BROMOETHYLPHTHALIMIDE, **I**, 119; **IV**, 106

1*H*-Isoindole-1,3-(2*H*)-dione, 2-(bromomethyl)- [5332-26-3]
    N-BROMOMETHYLPHTHALIMIDE, **VIII**, 536

1*H*-Isoindole-1,3(2*H*)-dione, 2-(1,1-dimethylethyl)- [2141-99-3]
    tert-BUTYLPHTHALIMIDE, **III**, 152

1*H*-Isoindole-1,3(2*H*)-dione, *trans*-(±)-2-(2,3-diphenyl-1-aziridinyl- [37079-32-6]
    trans-2,3-DIPHENYL-1-PHTHALIMIDOAZIRIDINE, **VI**, 56

1*H*-Isoindole-1,3(2*H*)-dione, 5-nitro- [89-40-7]
    4-NITROPHTHALIMIDE, **II**, 459

1*H*-Isoindole-1,3(2*H*)-dione, 2-(phenylmethyl)- [2142-01-0]
    BENZYL PHTHALIMIDE, **II**, 83

DL-Isoleucine [443-79-8]
    DL-ISOLEUCINE, **III**, 495

Isoquinoline, 2-acetyl-1,2,3,4-tetrahydro-6,7-dimethoxy-1-methylene- [57621-04-2]
    2-ACETYL-6,7-DIMETHOXY-1-METHYLENE-1,2,3,4-
    TETRAHYDROISOQUINOLINE, **VI**, 1

Isoquinoline, 3,4-dihydro-6,7-dimethoxy-1-methyl- [4721-98-6]
    6,7-DIMETHOXY-1-METHYL-3,4-DIHYDROISOQUINOLINE, **VI**, 1

Isoquinoline 2-[[[1-[{1,1-dimethylethoxy)methyl]-2-methylpropyl]imino]methyl]-
    1,2,3,4-tetrahydro-6,7-dimethoxy-, (*S*)- [90482-03-4]
    6,7-DIMETHOXY-1,2,3,4-TETRAHYDRO-2-[(1-*tert*-BUTOXY-3-METHYL)-2-
    BUTYLIMINOMETHYL]ISOQUINOLINE, **VIII**, 573

Isoquinoline, 1-methyl- [1721-93-3]
    1-METHYLISOQUINOLINE, **IV**, 641

Isoquinoline, 1-(phenylmethyl)- [6907-59-1]
    1-BENZYLISOQUINOLINE, **VI**, 115

Isoquinoline, 1,2,3,4-tetrahydro-6,7-dimethoxy-1-methyl-, (S)- [493-48-1]
    SALSOLIDINE, (–)-, **VIII**, 573

1-Isoquinolinecarbonitrile, 2-benzoyl-1,2-dihydro- [844-25-7]
    1-CYANO-2-BENZOYL-1,2-DIHYDROISOQUINOLINE, **IV**, 641
    2-BENZOYL-1,2-DIHYDROISOQUINALDONITRILE, **VI**, 115

1-Isoquinolinecarbonitrile, 2-benzoyl-1,2-dihydro-1-methyl- [16576-32-2]
    2-BENZOYL-1-CYANO-1-METHYL-1,2-DIHYDROISOQUINOLINE, **IV**, 642

1-Isoquinolinecarbonitrile, 2-benzoyl-1,2-dihydro-1-(phenylmethyl)- [16576-37-2]
    1-BENZYL-2-BENZOYL-1,2-DIHYDROISOQUINALDONITRILE, **VI**, 118

1(2*H*)-Isoquinolinone, 2-hydroxy-3-methyl- [7114-79-6]
    2-HYDROXY-3-METHYLISOCARBOSTYRIL, **V**, 623
3(2*H*)Isoquinolinone, 1,4-dihydro-1-thioxo- [938-38-5]
    2α-THIOHOMOPHTHALIMIDE, **V**, 1051
Isoxazole, 4-chloromethyl-3-ethyl-5-methyl- [40500-39-8]
    4-CHLOROMETHYL-3-ETHYL-5-METHYLISOXAZOLE, **VI**, 781
Isoxazole, 3-(4-chlorophenyl)-5-(4-methoxyphenyl)- [24097-19-6]
    3-(4-CHLOROPHENYL)-5-(4-METHOXYPHENYL)ISOXAZOLE, **VI**, 278
4-Isoxazolecarboxyllc acid, 3-ethyl-5-methyl-, ethyl ester [53064-41-8]
    ETHYL 3-ETHYL-5-METHYL-4-ISOXAZOLECARBOXYLATE, **VI** , 592
4-Isoxazolemethanol, 3-ethyl-5-methyl- [53064-42-9]
    3-ETHYL-4-HYDROXYMETHYL-5-METHYLISOXAZOLE, **VI**, 781
Isoxazolidine, 2,3,5,5-tetraphenyl- [25116-92-1]
    2,3,5,5-TETRAPHENYLISOXAZOLIDINE, **V**, 1126
Isoxazolidine, 2,3,5-triphenyl- [13787-96-7]
    2,3,5-TRIPHENYLISOXAZOLIDINE, **V**, 1124
Isoxazolidine-4-carboxylic acid, 5-methyl-2,3-diphenyl-, ethyl ester [19744-10-6]
    ETHYL 2,3-DIPHENYL-5-METHYLISOXAZOLIDINE-4-CARBOXYLATE,
        V, 1126

L

DL-Leucine [328-39-2]
    DL-LEUCINE, **III**, 523
L-Leucine, *N*-[*N²*-[(phenylmethoxy)carbonyl]-L-asparaginyl]-, methyl ester [14317-83-0]
    N-CARBOBENZYLOXY-L-ASPARAGINYL-L-LEUCINE METHYL ESTER,
        VI, 263
Lithium, 1-methoxyethenyl- [42722-80-5]
    1-METHOXYVINYLLITHIUM, **VIII**, 19
Lithium, methyl [917-54-4]
    METHYLLITHIUM, LOW-HALIDE, **VII**, 346
DL-Lysine, dihydrochloride [617-68-5]
    DL-LYSINE DIHYDROCHLORIDE, **II**, 374
DL-Lysine, monohydrochloride [70-53-1]
    DL-LYSINE MONOHYDROCHLORIDE, **II**, 374
DL-Lysine, *N⁶*-benzoyl- [5107-18-6]
    DL-ε-BENZOYLLYSINE, **II**, 374

M

Magnesium, chloro[[dimethyl(1-methylethoxy)silyl]methyl]- [122588-50-5]
    (ISOPROPOXYDIMETHYLSILYL)METHYLMAGNESIUM CHLORIDE,
        VIII, 315
D-Mannose [3458-28-4]
    D-MANNOSE, **III**, 541
α-D-Mannoside, methyl [27939-30-6]
    α-METHYL MANNOSIDE, **I**, 371
Menthol, (−)-, (*S*)-*p*-toluenesulfinate, (−)- [1517-82-4]
    (*S*)-(−)-MENTHYL *p*-TOLUENESULFINATE, **VII**, 495

Mercurate (1–), (4-carboxylatophenyl)chloro-, hydrogen [90-03-9]
   p-CHLOROMERCURIBENZOIC ACID, **I**, 159
Mercury, bis(4-methylphenyl)- [537-64-4]
   DI-*p*-TOLYLMERCURY, **I**, 231
Mercury, chloro(2-hydroxyphenyl)- [90-03-9]
   o-CHLOROMERCURIPHENOL, **I**, 161
Mercury, chloro(4-methylphenyl)- [539-43-5]
   p-TOLYLMERCURIC CHLORIDE, **I**, 519
Mercury, chloro-2-naphthalenyl- [39966-41-1]
   β-NAPHTHYLMERCURIC CHLORIDE, **II**, 432
Mercury, chloro[2-(phenylsulfonyl)-3-cyclohexen-1-yl]-, *trans*- [102815-53-2]
   trans-2-(PHENYLSULFONYL)-4-(CHLOROMERCURI)CYCLOHEXENE, **VIII**, 540
Mercury, diphenyl- [587-85-9]
   DIPHENYLMERCURY, **I**, 228
Mercury, [2-hydroxybenzoato(2–), O^1, O^2]- [5970-32-1]
   ANHYDRO-2-HYDROXYMERCURIBENZOIC ACID, **I**, 57
Mercury, di-2-naphthalenyl- [19510-26-0]
   MERCURY DI-β-NAPHTHYL, **II**, 381
Mercury, [3-nitrobenzoato-)-C2, 01] [53663-14-2]
   ANHYDRO-2-HYDROXYMERCURI-3-NITROBENZOIC ACID, **I**, 56
Mercury, phenyl(trichloromethyl)- [3294-57-3]
   PHENYL(TRICHLOROMETHYL)MERCURY, **V**, 969
Methanamine, hydrochloride [593-51-1]
   METHYLAMINE HYDROCHLORIDE, **I**, 347
Methanamine, 1-(2,4-cyclopentadien-1-ylidene)-*N*,*N*-dimethyl- [696-68-4]
   6-(DIMETHYLAMINO)FULVENE, **V**, 431
Methanamine, *N*,*N*-dimethyl- [75-50-3]
   TRIMETHYLAMINE, **I**, 528
Methanamine, *N*,*N*-dimethyl-, hydrochloride [593-81-7]
   TRIMETHYLAMINE HYDROCHLORIDE, **I**, 531
Methanamine, *N*-hydroxy-*N*-methyl-, hydrochloride [16645-06-0]
   N,*N*-DIMETHYLHYDROXYLAMINE HYDROCHLORIDE, **IV**, 612
Methanamine, *N*-methyl-*N*-nitroso- [62-75-9]
   NITROSODIMETHYLAMINE, **II**, 211
Methanamine, *N*-(1-phenylethylidene)- [6907-71-7]
   N-(α-METHYLBENZYLIDENE)METHYLAMINE, **VI**, 818
Methanamine, *N*-(phenylmethylene)- [622-29-7]
   N-BENZYLIDENEMETHYLAMINE, **IV**, 605
Methanaminium, *N*-(chloromethylene)-*N*-methyl-, chloride [3724-43-4]
   N,*N*-DIMETHYL(CHLOROMETHYLENE)AMMONIUM CHLORIDE, **VIII**, 498
Methanaminium, *N*-(1-chloro-2-methylpropylidene)-*N*-methyl-, chloride [52851-35-1]
   1-CHLORO-*N*,*N*,2-TRIMETHYLPROPYLIDENIMINIUM CHLORIDE, **VI**, 282
Methanaminium, [[[(dimethylamino)methylene]amino]methylene]*N*-methyl-, chloride
   [20353-93-9]
   [3-(DIMETHYLAMINO)-2-AZAPROP-2-EN-1-YLIDENE]DIMETHYLAMMO-
   NIUM CHLORIDE, **VII**, 197

Methanaminium, *N*-5-[(dimethylamino)-2,4-pentadienylidene]-N-methyl-, chloride,
(*E,E*)- [70669-80-6]
[5-(DIMETHYLAMINO)-2-4-PENTADIENYLIDENE]DIMETHYLAMMONIUM
CHLORIDE (in solution), **VII**, 15

Methanaminium, 1-ferrocenyl-*N,N,N*-trimethyl-, iodide [12086-40-7]
N,N-DIMETHYLAMINOMETHYLFERROCENE METHIODIDE, **V**, 434

Methanaminium, *N,N,N*-trimethyl-, salt with 1-propene-1,1,2,3,3-pentacarbonitrile(1:1)
[53663-17-5]
TETRAMETHYLAMMONIUM 1,1,2,3,3-PENTACYANOPROPENIDE, **V**, 1013

Methane, chloromethoxy- [107-30-2]
MONOCHLOROMETHYL ETHER, **I**, 377

Methane-*d*$_2$, diazo- [14621-84-2]
DIDEUTERIODIAZOMETHANE, **VI**, 432

Methane, dibromo- [74-95-3]
METHYLENE BROMIDE, **I**, 357

Methane, dichloromethoxy- [4885-02-3]
DICHLOROMETHYL METHYL ETHER, **V**, 365

Methane, diiodo- [75-11-6]
METHYLENE IODIDE, **I**, 358

Methane, iodo- [75-11-6]
METHYL IODIDE, **II**, 399

Methane, isocyano- [593-75-9]
METHYL ISOCYANIDE, **V**, 772

Methane, isothiocyanato- [556-61-6]
METHYL ISOTHIOCYANATE, **III**, 599

Methane, nitro- [75-52-5]
NITROMETHANE, **I**, 401

Methane, oxybis[chloro- [542-88-1]
BISCHLOROMETHYL ETHER, **IV**, 101; (*Hazard Note*), **V**, 218

Methane, tetranitro- [509-14-8]
TETRANITROMETHANE, **III**, 803

Methanediamine, *N,N,N′,N′*-tetramethyl- [51-80-9]
BIS(DIMETHYLAMINO)METHANE, **VI**, 474

Methanediol, 2-furanyl, diacetate [613-75-2]
FURFURAL DIACETATE, **IV**, 489

Methanediol, (2-nitrophenyl)-, diacetate [6345-63-7]
o-NITROBENZALDIACETATE, **IV**, 713

Methanediol, (4-nitrophenyl)-, diacetate (ester) [2929-91-1]
p-NITROBENZALDIACETATE, **V**, 713

Methanesulfinyl chloride [676-85-7]
METHANESULFINYL CHLORIDE, **V**, 709

Methanesulfonic acid, trifluoro-, 3-butynyl ester [32264-79-2]
3-BUTYN-1-YL TRIFLUOROMETHANESULFONATE, **VI**, 324

Methanesulfonic acid, trifluoro, 4-(1,1-dimethylethyl)-1-cyclohexen-1-yl ester
[77412-96-5]
4-*tert*-BUTYLCYCLOHEXEN-1-YL TRIFLUOROMETHANESULFONATE, **VIII**, 97

Methanesulfonic acid, trifluoro-, 1,2-dimethyl-1-propenyl ester [28143-80-8]
    3-METHYL-2-BUTEN-2-YL TRIFLATE, **VI**, 757
Methanesulfonyl bromide, bromo- [54730-18-6]
    BROMOMETHANESULFONYL BROMIDE, **VIII**, 212
Methanesulfonyl chloride [124-63-0]
    METHANESULFONYL CHLORIDE, **IV**, 571
Methanesulfonyl cyanide [24225-08-9]
    METHANESULFONYL CYANIDE, **VI**, 727
Methanethioic acid, *O*-ethyl ester [29392-46-9]
    O-ETHYL THIOFORMATE, **VI**, 620
Methanethione, diphenyl- [1450-31-3]
    THIOBENZOPHENONE, **II**, 573; **IV**, 927
Methanetricarboxylic acid, triethyl ester [6279-86-3]
    TRICARBETHOXYMETHANE, **II**, 594
Methanetricarboxylic acid, trimethyl ester [1186-73-8]
    TRICARBOMETHOXYMETHANE, **II**, 596
Methanimidamide, monoacetate [3473-63-0]
    FORMAMIDINE ACETATE, **V**, 582
Methanimidamide, *N,N*-dimethyl-*N'*-(4-methylphenyl)- [7549-96-4]
    N,N-DIMETHYL-*N'*-*p*-TOLYLFORMAMIDINE, **VII**, 197
Methanimidamide, *N'*-[1-[(1,1-dimethylethoxy)methyl]-2-methylpropyl]-*N,N*-dimethyl-,
    (*S*)- [90482-06-7]
    N,N-DIMETHYL-*N'*-(1-*tert*-BUTOXY-3-METHYL-2-BUTYL)FORMAMIDINE,
    (*S*)- **VIII**, 204, 573
Methanimidic acid, *N*-(2-methyl-3-nitrophenyl)- ethyl ester [115118-93-9]
    ETHYL *N*-(2-METHYL-3-NITROPHENYL)FORMIMIDATE, **VIII**, 493
Methanimidic acid, *N*-phenyl-, ethyl ester [6780-49-0]
    ETHYL *N*-PHENYLFORMIMIDATE, **IV**, 464
3*H*-3a,6-Methano-2,1-benzisothiazole, hexahydro-8,8-dimethyl-, 2,2-dioxide,
    [3a*S*-(3aα,6α,7ab)]- [94594-90-8]
    (−)-D-2,10-CAMPHORSULTAM, **VIII**, 110
3*H*-3a,6-Methano-2,1-benzisothiazole, 4,5,6,7-tetrahydro-8,8-dimethyl-, 2,2-dioxide,
    (3a*S*)- [60886-80-8]
    (−)-(CAMPHORSULFONYL)IMINE, **VIII**, 104, 110
5,9-Methanobenzocycloocten-11-one, 10-butyldodecahydro-4a-hydroxy- [24133-22-0]
    8-BUTYL-2-HYDROXYTRICYCLO[7.3.1.0$^{2.7}$]TRIDECAN-13-ONE, **VIII**, 87
4,7-Methano-1*H*-indene, octahydro-, 3aα,4α,7α,7aα- [2825-83-4]
    endo-TETRAHYDRODICYCLOPENTADIENE, **V**, 16
4,6-Methano-2*H*-indene, 4,5,6,7-tetrahydro-5,5-dimethyl-, (4*R*)- [108404-79-1]
    (1*R*)-9,9-DIMETHYLTRICYCLO[6.1.1.0$^{2.6}$]DECA-2,5-DIENE, **VIII**, 223
2,5-Methano-1*H*-inden-7(4*H*)-one, hexahydro- [27567-85-7]
    4-PROTOADAMANTANONE; TRICYCLO[4.3.1.0$^{3.8}$]DECAN-4-ONE] **VI**, 958
Methanone, (2-aminophenyl)phenyl- [2835-77-0]
    2-AMINOBENZOPHENONE, **IV**, 34
Methanone, 1,2-cyclopropanediylbis[phenyl]-, *trans*- [38400-84-9]
    trans-1,2-DIBENZOYLCYCLOPROPANE, **VI**, 401

Methanone, dicyclopropyl- [1121-37-5]
DICYCLOPROPYL KETONE, **IV**, 278
Methanone, [4-(dimethylamino)phenyl]phenyl- [530-44-9]
p-DIMETHYLAMINOBENZOPHENONE, **I**, 217
Methanone, diphenyl- [119-61-9]
BENZOPHENONE, **I**, 95
Methanone, diphenyl-, oxime [574-66-3]
BENZOPHENONE OXIME, **II**, 70
Methanone, 1*H*-indol-3-ylphenyl- [15224-25-6]
3-BENZOYLINDOLE, **VI**, 109
Methanone, phenyl-3-pyridinyl- [5424-19-1]
3-BENZOYLPYRIDINE, **IV**, 88
Methanone, phenyl-4-pyridinyl- [14548-46-0]
4-BENZOYLPYRIDINE, **IV**, 89
Methanone, phenyl-2-thienyl- [135-00-2]
PHENYL THIENYL KETONE, **II**, 520
4*H*-4a,7-Methanooxazirino-[3,2-*i*][2,1]benzisothiazole, tetrahydro-9,9-dimethyl-,
3,3-dioxide, [4a*S*-(4aa,7a,8a*RR**)]- [104322-63-6]
(+)-(2*R*,8a*S*)-10-(CAMPHORSULFONYL)OXAZIRIDINE, **VIII**, 104
3,4,7-Metheno-7*H*-cyclopenta[*a*]pentalene-7,8-dicarboxylic acid, 3,3a,3b,4,6a,7a-
hexahydro- [61206-25-5]
3,3a,3b,4,6a,7a-HEXAHYDRO-3,4,7-METHENO-
7*H*-CYCLOPENTA[a]PENTALENE-7,8-DICARBOXYLIC ACID, **VIII**, 298
3,4,7-Metheno-7*H*-cyclopenta[*a*]pentalene-7,8-dicarboxylic acid,
3,3a,3b,4,6a,7a-hexahydro-, dimethyl ester [53282-97-6]
DIMETHYL 3,3a,3b,4,6a,7a-HEXAHYDRO-3,4,7-METHENO-7*H*-
CYCLOPENTA[*a*]PENTALENE-7,8-DICARBOXYLATE, **VIII**, 298
DL-Methionine [59-51-8]
DL-METHIONINE, **II**, 384
L-Methionine [63-68-3]
L-METHIONINE, **VI**, 253
Methylene, cyclopropyl [19527-12-9]
METHYLENECYCLOPROPANE, **VI**, 320
Molybdenum, (hexamethylphosphoric triamide-*O*)oxodiperoxy(pyridine)- [23319-63-3]
OXODIPEROXYMOLYBDENUM(PYRIDINE)(HEXAMETHYLPHOSPHORIC
TRIAMIDE), **VII**, 277
Morpholine, 4-chloro- [23328-69-0]
N-CHLOROMORPHOLINE, **VIII**, 167
Morpholine, 4-(1-cyclohexen-1-yl)- [670-80-4]
1-MORPHOLINO-1-CYCLOHEXENE, **V**, 808
Morpholine, 4-methyl-, 4-oxide- [7529-22-8]
N-METHYLMORPHOLINE *N*-OXIDE, **VI**, 342
Morpholine, 4-nitro- [4164-32-3]
N-NITROMORPHOLINE, **V**, 839
Morpholine, 4-(1-phenylethenyl)- [7196-01-2]
α-MORPHOLINOSTYRENE, **VI**, 520

N

1-Naphthalenamine, 4-nitro- [776-34-1]
    4-NITRO-1-NAPHTHYLAMINE, **III**, 664
2-Naphthalenamine, 1,2,3,4-tetrahydro- [2954-50-9]
    ac-TETRAHYDRO-β-NAPHTHYLAMINE, **I**, 499
Naphthalene [91-20-3]
    NAPHTHALENE, **VI**, 152, 821
Naphthalene, 1-bromo- [90-11-9]
    α-BROMONAPHTHALENE, **I**, 121
Naphthalene, 2-bromo- [580-13-2]
    2-BROMONAPHTHALENE, **V**, 142
Naphthalene, 1-(chloromethyl)- [86-52-2]
    1-CHLOROMETHYLNAPHTHALENE, **III**, 195
Naphthalene, 1,2-dihydro-4-phenyl- [7469-40-1]
    1-PHENYLDIALIN (1-PHENYL-3,4-DIHYDRONAPHTHALENE), **III**, 729
Naphthalene, 1,4-dinitro- [6921-26-2]
    1,4-DINITRONAPHTHALENE, **III**, 341
Naphthalene, 1-isothiocyanato- [551-06-4]
    α-NAPHTHYL ISOTHIOCYANATE, **IV**, 700
Naphthalene, 1,2,3,4,5,6,7,8-octahydro- [493-03-8]
    $\Delta^{9,10}$-OCTALIN, **VI**, 852
Naphthalene, 1-phenyl- [605-02-7]
    1-PHENYLNAPHTHALENE, **III**, 729
Naphthalene, 1,4,5,8-tetrahydro- [493-04-9]
    1,4,5,8-TETRAHYDRONAPHTHALENE, **VI**, 731
Naphthalene, 1,2,3,4-tetraphenyl- [751-38-2]
    1,2,3,4-TETRAPHENYLNAPHTHALENE, **V**, 1037
1-Naphthaleneacetic acid, ethyl ester [2122-70-5]
    ETHYL 1-NAPHTHYLACETATE, **VI**, 613
1-Naphthalenecarbonitrile [86-53-3]
    α-NAPHTHONITRILE, **III**, 631
Naphthalenecarbonitrile, 3,4-dihydro-6-methoxy- [6398-50-1]
    1-CYANO-6-METHOXY-3,4-DIHYDRONAPHTHALENE, **VI**, 307
1-Naphthalenecarboxaldehyde [66-77-3]
    1-NAPHTHALDEHYDE, **IV**, 690
1-Naphthalenecarboxaldehyde, 2-ethoxy- [19523-57-0]
    2-ETHOXY-1-NAPHTHALDEHYDE, **III**, 98
1-Naphthalenecarboxaldehyde, 2-hydroxy- [708-06-5]
    2-HYDROXY-1-NAPHTHALDEHYDE, **III**, 463
2-Naphthalenecarboxaldehyde [66-99-9]
    β-NAPHTHALDEHYDE, **III**, 626
1-Naphthalenecarboxylic acid [86-55-5]
    α-NAPHTHOIC ACID, **II**, 425
1-Naphthalenecarboxylic acid, ethyl ester [3007-97-4]
    ETHYL α-NAPHTHOATE, **II**, 282
2-Naphthalenecarboxylic acid [93-09-4]
    β-NAPHTHOIC ACID, **II**, 428

2-Naphthalenecarboxylic acid, 3-amino- [5959-52-4]
   3-AMINO-2-NAPHTHOIC ACID, III, 78
2-Naphthalenecarboxylic acid, 3,4-dihydro-6,7-dimethoxy- [53684-50-7]
   6,7-DIMETHOXY-3,4-DIHYDRO-2-NAPHTHOIC ACID, III, 300
2-Naphthalenecarboxylic acid, 1,3-dihydroxy- [3147-58-8]
   1,3-DIHYDROXY-2-NAPHTHOIC ACID, III, 638
2-Naphthalenecarboxylic acid, 1,3-dihydroxy-, ethyl ester [6843-89-6]
   ETHYL 1,3-DIHYDROXY-2-NAPHTHOATE, III, 638
2-Naphthalenecarboxylic acid, 5,6,7,8-tetrahydro-3-methoxy-, methyl ester [78112-34-2]
   6-METHOXY-7-METHOXYCARBONYL-1,2,3,4-
   TETRAHYDRONAPHTHALENE, VIII, 444
2,3-Naphthalenedicarboxylic acid [2169-87-1]
   2,3-NAPHTHALENEDICARBOXYLIC ACID, V, 810
2,6-Naphthalenedicarboxylic acid [1141-38-4]
   2,6-NAPHTHALENEDICARBOXYLIC ACID, V, 813
1,3-Naphthalenediol [132-86-5]
   NAPHTHORESORCINOL, III, 637
1,2-Naphthalenedione [524-42-5]
   1,2-NAPHTHOQUINONE, II, 430
1,4-Naphthalenedione [130-15-4]
   1,4-NAPHTHOQUINONE, I, 383; IV, 698
1,4-Naphthalenedione, 2-hydroxy- [83-72-7]
   2-HYDROXY-1,4-NAPHTHOQUINONE, III, 465
1,4-Naphthalenedione, 2-methoxy- [2348-82-5]
   2-METHOXY-1,4-NAPHTHOQUINONE, III, 465
1,6(2H,7H)-Naphthalenedione, 3,4,8,8a-tetrahydro-8a-methyl- [20007-72-1]
   8a-METHYL-3,4,8,8a-TETRAHYDRO-1,6(2H,7H)-NAPHTHALENEDIONE, (S)-
   [33878-99-8], VII, 368
1,6(2H,7H)-Naphthalenedione, 3,4,8,8a-tetrahydro-8a-methyl- [20007-72-1]
   1,6-DIOXO-8a-METHYL-1,2,3,4,6,7,8,8a-OCTAHYDRONAPHTHALENE, V, 486
1,3-Naphthalenedisulfonic acid, 6,6'-[(3,3'-dimethyl[1,1'-biphenyl]-4,4'-
   diyl)bis(azo)]bis[4-amino-5-hydroxy-, tetrasodium salt [314-13-6]
   COUPLING of o-TOLUIDINE and CHICAGO ACID, II, 145
1,5-Naphthalenedisulfonyl dichloride [1928-01-4]
   NAPHTHALENE-1,5-DISULFONYL CHLORIDE, IV, 693
1,5-Naphthalenedithiol [5325-88-2]
   1,5-NAPHTHALENEDITHIOL, IV, 695
1-Naphthalenesulfonic acid, 4-amino-3-hydroxy- [116-63-2]
   1-AMINO-2-NAPHTHOL-4-SULFONIC ACID, II, 42
1-Naphthalenesulfonic acid, 3,4-dihydro-3,4-dioxo-, ammonium salt [53684-60-9]
   1,2-NAPHTHOQUINONE-4-SULFONATE, AMMONIUM SALT, III, 633
1-Naphthalenesulfonic acid, 3,4-dihydro-3,4-dioxo-, potassium salt [5908-27-0]
   1,2-NAPHTHOQUINONE-4-SULFONATE POTASSIUM SALT, III, 633
(S)-1-Naphthalenethanamine, α-methyl- [10420-89-0]
   (S)-(−)-α-(1-NAPHTHYL)ETHYLAMINE, VI, 826, 828
2-Naphthalenethiol [91-60-1]
   2-NAPHTHALENETHIOL, VI, 824

1-Naphthalenol, 4-amino-, hydrochloride [5959-56-8]
    1,4-AMINONAPHTHOL HYDROCHLORIDE, **I**, 49; **II**, 39
1-Naphthalenol, decahydro- [529-32-8]
    1-DECALOL, **VI**, 371
1-Naphthalenol, 5,6,7,8-tetrahydro- [529-35-1]
    ar-TETRAHYDRO-α-NAPHTHOL, **IV**, 887
2-Naphthalenol, 1-amino-, hydrochloride [1198-27-2]
    1,2-AMINONAPHTHOL HYDROCHLORIDE, **II**, 33
2-Naphthalenol, 1-(aminophenylmethyl)- [481-82-3]
    β-NAPHTHOL PHENYLAMINOMETHANE, **I**, 381
2-Naphthalenol, 1-[α-(benzylideneamino)benzyl]-, (*E*)- [24609-80-1]
    1-[α-(BENZALAMINO)BENZYL]-2-NAPHTHOL, **I**, 381
2-Naphthalenol, 6-bromo- [15231-91-1]
    6-BROMO-2-NAPHTHOL, **III**, 132
2-Naphthalenol, 6-methoxy- [5111-66-0]
    PHENOLS: 6-METHOXY-2-NAPHTHOL, **V**, 918
2-Naphthalenol, 1-nitro- [550-60-7]
    1-NITRO-2-NAPHTHOL, **II**, 451
2-Naphthalenol, 1-nitroso- [131-91-9]
    NITROSO-β-NAPHTHOL, **I**, 411
1(2*H*)-Naphthalenone, 3,4-dihydro- [529-34-0]
    α-TETRALONE, **II**, 569; **III**, 798; **IV**, 898
1(2*H*)-Naphthalenone, 3,4-dihydro-2-methylene- [13203-73-1]
    2-METHYLENE-1-OXO-1,2,3,4-TETRAHYDRONAPHTHALENE, **VII**, 332
2(1*H*)-Naphthalenone, 3,4-dihydro- [530-93-8]
    β-TETRALONE, **IV**, 903
2(1*H*)-Naphthalenone, 3,4-dihydro-6-methoxy- [2472-22-2]
    6-METHOXY-β-TETRALONE, **VI**, 744
2(3*H*)-Naphthalenone, 1-butyl-4,4a,5,6,7,8-hexahydro-4a-methyl- [66252-93-5]
    1-BUTYL-10-METHYL-Δ$^{1(9)}$-2-OCTALONE, **VI**, 242
2(3*H*)-Naphthalenone, 4,4a,5,6,7,8-hexahydro-4a,5-dimethyl-, *cis*-, (±)- [20536-80-5]
    cis-4,4a,5,6,7,8-HEXAHYDRO-4a,5-DIMETHYL-2(3*H*)-
    NAPHTHALENONE, **VI**, 666
2(3*H*)-Naphthalenone, 4,4a,5,6,7,8-hexahydro- [1196-S5-0]
    Δ$^{1(9)}$-2-OCTALONE, **V**, 869
2(3*H*)-Naphthalenone, 4,4a,5,6,7,8-hexahydro-1-methyl- [5164-37-4]
    1-METHYL-4,4a,5,6,7,8-HEXAHYDRONAPHTHALEN-2(3*H*)-ONE, **VI**, 781
2(3*H*)-Naphthalenone, 4,4a,5,6,7,8-hexahydro-4a-methyl-, dimethylhydrazone, (*E*)-
    [66252-92-4]
    10-METHYL-Δ$^{1(9)}$-2-OCTALONE *N,N*-DIMETHYLHYDRAZONE, **VI**, 242
1*H*-Naphth[1,2-*b*]azirine, 1a,2,3,7b-tetrahydro- [1196-87-8]
    1,2,3,4-TETRAHYDRONAPHTHALENE(1,2)IMINE, **VI**, 967
Naphtho[1,2-*c*]furan-1,3-dione [5343-99-7]
    1,2-NAPHTHALIC ANHYDRIDE, **II**, 423
Naphtho[1,2-*c*]furan-1,3-dione, 4,5-dihydro- [37845-14-0]
    3,4-DIHYDRO-1,2-NAPHTHALIC ANHYDRIDE, **II**, 194

Naphtho[1,2-*d*]thiazol-2(1*H*)-imine, 1-methyl- [53663-31-3]
    1-METHYL-2-IMINO-β-NAPHTHOTHIAZOLINE, **III**, 595
Nitric acid, methyl ester [598-58-3]
    METHYL NITRATE, **II**, 412
Nitridotricarbonic acid, triethyl ester [3206-31-3]
    ETHYL *N*-TRICARBOXYLATE, **III**, 415
Nitrone, *N*-(*p*-dimethylaminophenyl)-α-(*o*-nitrophenyl)- [13664-79-4]
    N-(*p*-DIMETHYLAMINOPHENYL)-α-(*o*-NITROPHENYL)NITRONE, **V**, 826
Nitrous acid, butyl ester [544-16-1]
    BUTYL NITRITE, **II**, 108
Nitroxide, bis(1,1-dimethylethyl)- [2406-25-9]
    DI-*tert*-BUTYL NITROXIDE, **V**, 355
Nonacyclo[43.3.1.1^{3,7}.1^{9,13}.1^{15,19}.1^{21,25}.1^{27,31}.1^{33,37}.1^{39,43}]hexapentaconta-
    1(49),3,5,7(56),9,11,13(55),15,17,19(54),21,23,24(53),
    27,29,31(52),33,35,37(51),39,41,43(50),45,47-tetracosaene-
    49,50,51,52,53,54,55,56-octol, 5,11,17,23,29,35,41,47-
    octakis(1,1-dimethyethyl)- [68971-82-4]
    p-tert-BUTYLCALIX[8]ARENE, **VIII**, 80
Nonadecanoic acid [646-30-0]
    NONADECANOIC ACID, **VII**, 397
4,8-Nonadienoic acid, 4-methyl-, ethyl ester, (*E*)- [53359-96-9]
    ETHYL 4-METHYL-(*E*)-4,8-NONADIENOATE, **VI**, 606
1,2-Nonadien-4-ol [73229-28-4]
    4-HYDROXYNONA-1,2-DIENE, **VII**, 276
2,6-Nonadien-1-ol, 9-(3,3-dimethyloxiranyl)-3,7-dimethyl-, acetate, (*E,E*)- [50502-44-8]
    10,11-EPOXYFARNESYL ACETATE, **VI**, 560
3,7-Nonadien-1-ol, 4,8-dimethyl-, (*E*)- [459-88-1]
    HOMOGERANIOL, **VII**, 258
Nonane, 1,1,3-trichloro- [10575-86-7]
    1,1,3-TRICHLORONONANE, **V**, 1076
Nonanedinitrile [1675-69-0]
    AZELANITRILE, **IV**, 62
Nonanedioic acid [123-99-9]
    AZELAIC ACID, **II**, 53
2,4-Nonanedione [6175-23-1]
    2,4-NONANEDIONE, **V**, 848
4,6-Nonanedione, 2,8-dimethyl- [7307-08-6]
    DIISOVALERYLMETHANE, **III**, 291
Nonanoic acid [112-05-0]
    PELARGONIC ACID, **II**, 174
Nonanoic acid, 9-cyano- [5810-19-5]
    ω-CYANOPELARGONIC ACID, **III**, 768
Nonanoic acid, 9-cyano-, methyl ester [53663-26-6]
    METHYL ω-CYANOPELARGONATE, **III**, 584
5-Nonanol [623-93-8]
    DIBUTYLCARBINOL, **II**, 179

5-Nonanol, 5-(2-propenyl)- [76071-61-9]
  3-BUTYL-2-METHYL-1-HEPTEN-3-OL, **VI**, 240
Nona-1,3,6,8-tetraen-5-one, 1,9-diphenyl- [622-21-9]
  DICINNAMALACETONE, **VII**, 60,
1,3,7-Nonatriene, 4,8-dimethyl-, (*E*)- [19945-61-0]
  (*E*)-4,8-DIMETHYL-1,3,7-NONATRIENE, **VII**, 259
2-Nonene-1,4-diol, 3-pentyl- [138149-15-2]
  (*E*)-3-PENTYL-2-NONENE-1,4-DIOL, **VIII**, 507
1-Nonen-3-ol [21964-44-3]
  3-HYDROXY-1-NONENE, **VIII**, 235
D-Norandrost-5-ene-16-carboxylic acid, 3-hydroxy-, (3β,13α,16β)- [40013-51-2]
  D-NORANDROST-5-EN-3β-OL-16α-CARBOXYLIC ACID, **VI**, 840
D-Norandrost-5-ene-16-carboxylic acid, 3-hydroxy-, (3β,16β)- [50764-17-5]
  D-NORANDROST-5-EN-3β-OL-16β-CARBOXYLIC ACID, **VI**, 840
24-Norcholan-23-oic acid, 3,12-bis(acetyloxy)- (3α,5β,12α)- [63714-58-9]
  3,12-DIACETOXY-nor-CHOLANIC ACID, **III**, 234
24-Norcholan-23-oic acid, 3,12-dihydroxy-, (3α,5β,12α)- [53608-86-9]
  nor-DESOXYCHOLIC ACID, **III**, 234
L-Norleucine [5157-09-5]
  α-AMINOCAPROIC ACID, **I**, 48

O

9,12-Octadecadienoic acid, (*Z,Z*)- [60-33-3]
  LINOLEIC ACID, **III**, 526
9,12-Octadecadienoic acid, ethyl ester, (*Z,Z*)- [544-35-4]
  ETHYL LINOLEATE, **III**, 526
Octadecanedioic acid, diethyl ester [1472-90-8]
  ETHYL 1,16-HEXADECANEDICARBOXYLATE, **III**, 401
Octadecanedioic acid, dimethyl ester [1472-93-1]
  DIMETHYL OCTADECANEDIOATE, **V**, 463
Octadecanoic acid, 9,10-dihydroxy- [120-87-6]
  9,10-DIHYDROXYSTEARIC ACID, **IV**, 317
Octadecanoic acid, 9,10,12,13,15,16-hexabromo- [4167-08-2]
  9,10,12,13,15,16-HEXABROMOSTEARIC ACID, **III**, 531
Octadecanoic acid, 9,10,11,12-tetrabromo- [18464-04-5]
  TETRABROMOSTEARIC ACID, **III**, 526
9,12,15-Octadecatrienoic acid, (*Z,Z,Z*)- [463-40-1]
  LINOLENIC ACID, **III**, 531
9,12,15-Octadecatrienoic acid, ethyl ester, (*Z,Z,Z*)- [1191-41-9]
  ETHYL LINOLENATE, **III**, 532
9-Octadecenethioic acid, 12-hydroxy-, *S*-2-pyridyl ester, [*R*-(*E*)]- [100819-69-0]
  RICINELAIDIC ACID *S*-(2-PYRIDYL)CARBOTHIOATE, **VII**, 470
9-Octadecenoic acid, 12-hydroxy-, [*R*-(*E*)]- [540-12-5]
  RICINELAIDIC ACID, **VII**, 470

9-Octadecenoic acid, 12-hydroxy-, [(+)-(R)-*trans*]-, lactone [141-22-0]
Oxatridec-10-en-2-one, (E)- [79894-06-7]
    RICINELAIDIC ACID LACTONE, **VII**, 470
9-Octadecen-1-ol, (Z)- [143-28-2]
    OLEYL ALCOHOL, **II**, 468; **III**, 671
9-Octadecenoyl chloride, (Z)- [112-77-6]
    OLEOYL CHLORIDE, **IV**, 739
9-Octadecynoic acid [506-24-1]
    STEAROLIC ACID, **III**, 785; **IV**, 851
2,6-Octadienal, 3,7-dimethyl- [5392-40-5]
    GERANIAL, **VII**, 258
1,6-Octadien-1-amine, N,N-diethyl-3,7-dimethyl-, [R-(E)]- [67392-56-7]
    N,N-DIETHYL-(E)-CITRONELLALENAMINE, (R)-(−)-, **VIII**, 183
2,6-Octadien-1-amine, N,N-diethyl-3,7-dimethyl-, (E)- [40267-53-6]
    N,N-DIETHYLGERANYLAMINE, **VIII**, 183, 188
2,6-Octadien-1-amine, N,N-diethyl-3,7-dimethyl-, (Z)- [40137-00-6]
    N,N-DIETHYLNERYLAMINE, **VIII**, 183, 190
1,6-Octadien-3-amine, 3,7-dimethyl- [59875-02-4]
    3,7-DIMETHYL-1,6-OCTADIEN-3-AMINE, **VII**, 507
2,6-Octadiene, 1-chloro-3,7-dimethyl-, (E)- [5389-87-7]
    GERANYL CHLORIDE, **VI**, 634, 638
Octanal [124-13-0]
    OCTANAL, **V**, 872
Octane, 1-bromo- [111-83-1]
    OCTYL BROMIDE, **I**, 30
Octane, 2,7-dimethyl-2,7-dinitro- [53684-51-8]
    2,7-DIMETHYL-2,7-DINITROOCTANE, **V**, 445
Octane, 1-nitro- [629-37-8]
    1-NITROOCTANE, **IV**, 724
Octanenitrile, 8-oxo- [13050-09-4]
    7-CYANOHEPTANAL, **V**, 266
Octanoic acid, 7-methyl-4-oxo-, methyl ester [53663-32-4]
    METHYL 4-KETO-7-METHYLOCTANOATE, **III**, 601
Octanoic acid, 2-methyl-3-oxo-, 1-methylpropyl ester [53663-40-4]
    sec-BUTYL-α-CAPROYLPROPIONATE, **IV**, 120
Octanoic acid, 7-oxo-, methyl ester [16493-42-8]
    METHYL 7-OXOOCTANOATE, **VI**, 807,
1-Octanol [111-87-5]
    1-OCTANOL, **VI**, 919
2-Octanol [123-96-6]
    METHYL-*n*-HEXYLCARBINOL, **I**, 366
2-Octanol, (S)- [6169-06-8]
    L-OCTANOL-2, **I**, 418
2-Octanol, (R)- [5978-70-1]
    D-OCTANOL-2, **I**, 418

1-Octanone, 1-phenyl- [1674-37-9]
 PHENYL *n*-HEPTYL KETONE, **VI**, 919
4-Octanone, 5-hydroxy- [496-77-5]
 BUTYROIN, **II**, 114; **VII**, 95
4-Octanone, 3-hydroxy-2,2,3-trimethyl- [85083-71-2]
 3-HYDROXY-2,2,3-TRIMETHYLOCTAN-4-ONE, **VIII**, 343
6-Octenal, 3,7-dimethyl, (*R*)-(+)- [2385-77-5];.(*S*); (*RS*);
 CITRONELLAL, (*R*)-(+)-, **VIII**, 183
2-Octenoic acid, 4-ethyl-2-methyl- [6975-97-9]
 4-ETHYL-2-METHYL-2-OCTENOIC ACID, **IV**, 444
6-Octenoic acid, 2-cyano-, ethyl ester, (*E*)- [25143-86-6]
 ETHYL (*E*)-2-CYANO-6-OCTENOATE, **VI**, 586
(*Z*)-2-Octen-1-ol [26001-58-1]
 (*Z*)-2-OCTEN-1-OL, **VIII**, 508
(*Z*)-2-Octen-1-ol, 3-methyl- [30804-78-5]
 (*Z*)-3-METHYL-2-OCTEN-1-OL, **VIII**, 509
1-Octyn-3-ol, (*R*)- [32556-70-0]
 1-OCTYN-3-OL, (*R*)-(+)-, **VII**, 402
2-Oxabicyclo[4.1.0]heptane, 7,7-dichloro- [7556-13-0]
 2-OXA-7,7-DICHLORONORCARANE, **V**, 874
7-Oxabicyclo[4.1.0]heptane [286-20-4]
 CYCLOHEXENE OXIDE, **I**, 185
7-Oxabicyclo[4.1.0]heptane, 3,4-dibromo-1,6-dimethyl-, (1α,3α,4β,6α-), (±)-
  [137053-60-2]
 4,5-DIBROMO-1,2-DIMETHYL-1,2-EPOXYCYCLOHEXANE, **V**, 468
7-Oxabicyclo[4.1.0]heptan-2-one [6705-49-3]
 2,3-EPOXYCYCLOHEXANONE, **VI**, 679
7-Oxabicyclo[4.1.0]heptan-2-one, 4,4,6-trimethyl- [10276-21-8]
 ISOPHORONE OXIDE, **IV**, 552
7-Oxabicyclo[4.1.0]hept-2-ene, 1,6-dimethyl- [57338-10-0]
 1,2-DIMETHYL-1,2-EPOXYCYCLOHEX-4-ENE, **V**, 468
9-Oxabicyclo[6.1.0]nonane, *cis*- [4925-71-7]
 CYCLOOCTENE OXIDE, *cis*-, **VII**, 126
8-Oxabicyclo[3.2.1]oct-6-en-3-one, 2,4-dimethyl-, (*endo, endo*)- [37081-58-6]
 2α,4α-DIMETHYL-8-OXABICYCLO[3.2.1]OCT-6-EN-3-ONE, **VI**, 512
11-Oxabicyclo[4.4.1]undeca-1,3,5,7,9-pentaene [4759-11-9]
 1,6-OXIDO[10]ANNULENE, **VI**, 862
Oxacyclododecan-2-one [1725-03-7]
 11-HYDROXYUNDECANOIC LACTONE, **VI**, 698
1-Oxaspiro[2.5]octane [185-70-6]
 METHYLENECYCLOHEXANE OXIDE, **V**, 755
1-Oxaspiro[2.5]octane-2-carboxylic acid, ethyl ester [6975-17-3]
 ETHYL β,β-PENTAMETHYLENEGLYCIDATE, **IV**, 459
Oxaspiro[2.2]pentane [157-41-5]
 OXASPIROPENTANE, **VI**, 320

1,2-Oxathiane, 2,2-dioxide [1633-83-6]
  4-HYDROXY-1-BUTANESULFONIC ACID SULTONE, **IV**, 529
1,4-Oxathiane, 2,6-diethoxy-, 4,4-dioxide, *cis*- [40263-59-0]
  cis- and *trans*-2,6-DIETHOXY-1,4-OXATHIANE 4,4-DIOXIDE, **VI**, 976
Oxatridec-10-en-2-one, (*E*)- [79894-06-7]
  RICINELAIDIC ACID LACTONE, **VII**, 470
4*H*-1,3-Oxazine, 5,6-dihydro-4,4,6-trimethyl-2-(1-phenylcyclopentyl)-
  2-(1-PHENYLCYCLOPENTYL)-4,4,6-TRIMETHYL-
  5,6-DIHYDRO-1,3(4*H*)- OXAZINE, **VI**, 905
2*H*-1,3-Oxazine-, tetrahydro-4,4,6-trimethyl-2-(1-phenylcyclopentyl)-
  2-(1-PHENYLCYCLOPENTYL)-4,4,6-TRIMETHYLTETRAHYDRO-
  1,3-OXAZINE, **VI**, 905
Oxaziridine, 2-(1,1-dimethylethyl)-3-phenyl- [7731-34-2]
  2-*tert*-BUTYL-3-PHENYLOXAZIRANE, **V**, 191
Oxaziridine, 3-phenyl-2-(phenylsulfonyl)- [63160-13-4]
  (±)-*trans*-2-(PHENYLSULFONYL)-3-PHENYLOXAZIRIDINE, **VIII**, 546
2-Oxazolidinone [497-25-6]
  2-OXAZOLIDINONE, **VII**, 4
2-Oxazolidinone, 3-acetyl- [1432-43-5]
  3-ACETYL-2-OXAZOLIDINONE, **VII**, 5
2-Oxazolidinone, 3-acetyl-5-chloro- [60759-48-0]
  3-ACETYL-4- and 5-CHLORO-2-OXAZOLIDINONE, **VII**, 5
2-Oxazolidinone, 3-[3-hydroxy-2-methyl-1-oxo-3-phenylpropyl]-4-(phenylmethyl)-,
  [4*S*-[3(2*R**, 3*R**), 4*R**]]- [133467-37-5]
  (4*S*)-3-[(1-OXO-*syn*-2-METHYL-3-HYDROXY)-3-PHENYLPROPYL]-4-
  (PHENYLMETHYL)-2-OXAZOLIDINONE, **VIII**, 339
2-Oxazolidinone, 3-(1-oxopropyl)-4-(phenylmethyl)-, (*S*)- [101711-78-8]
  (*S*)-3-(1-OXOPROPYL)-4-(PHENYLMETHYL)-2-OXAZOLIDINONE,
  VIII, 339
2-Oxazolidinone, 4-(phenylmethyl)-, (*S*)- [90719-32-7]
  (*S*)-4-(PHENYLMETHYL)-2-OXAZOLIDINONE, **VIII**, 339, 528
Oxazolium, 4,5-dihydro-3,4,4-trimethyl-, iodide [30093-97-1]
  N,4,4-TRIMETHYL-2-OXAZOLINIUM IODIDE, **VI**, 65
2(3*H*)-Oxazolone, 3-acetyl- [60759-49-1]
  3-ACETYL-2(3*H*)-OXAZOLONE, **VII**, 4
5(4*H*)-Oxazolone, 4-[(3,4-dimethoxyphenyl)methylene]-2-phenyl-, (*E*)- [25349-38-6]
  (Z)- [25349-37-5]
  AZLACTONE of α-BENZOYLAMINO-β-(3,4-DIMETHOXYPHENYL)-
  ACRYLIC ACID, **II**, 55
5(4*H*)-Oxazolone, 2-phenyl- [1199-01-5]
  2-PHENYL-5-OXAZOLONE, **V**, 946
5(4*H*)-Oxazolone, 4-(phenylmethylene)-2-methyl- [881-90-3]
  AZLACTONE of α-ACETAMINOCINNAMIC ACID, **II**, 1
5(4*H*)-Oxazolone, 4-(phenylmethylene)-2-phenyl- [842-74-0]
  AZLACTONE of α-BENZOYLAMINOCINNAMIC ACID, **II**, 490

5*H*-Oxazolo[3,2-*a*]pyridin-5-one, 6-ethylhexahydro-3-(hydroxymethyl)-8a-methyl-2-
  phenyl-6-(2-propenyl)-, [2*S*-(2α,3β,6α, 8aβ)]- [122444-61-5]; [2*S*-(2α,3β,6β,
  8aβ)]- [122518-73-4]
  HEXAHYDRO-6-ETHYL-3-(HYDROXYMETHYL)-6-ALLYL-8a-METHYL-
  2-PHENYL[2*S*,3*S*,8a*R*]-5-OXO-5*H*-OXAZOLO[3,2-*a*]PYRIDINE,
  VIII, 241
5*H*-Oxazolo[3,2-*a*]pyridin-5-one, hexahydro-3-hydroxymethyl)-8a-methyl-2-phenyl-,
  [2*S*-(2α,3β,8aβ)]- [116950-01-7]
  HEXAHYDRO-3-(HYDROXYMETHYL)-8a-METHYL-2-PHENYL[2*S*,3*S*,8a*R*]-
  5-OXO-5*H*-OXAZOLO-[3,2-*a*]PYRIDINE, **VIII**, 241
2*H*-Oxecin-2-one, 3,4,5,8,9,10-hexahydro-9-(1-oxopropyl)-, (*E*)- [114633-68-0]
  8-PROPIONYL-(*E*)-5-NONENOLIDE, **VIII**, 562
Oxepin, 2,7-dimethyl- [1487-99-6]
  2,7-DIMETHYLOXEPIN, **V**, 467
Oxetane [503-30-0]
  TRIMETHYLENE OXIDE, **III**, 835
2-Oxetanone, 3,3-dimethyl-4-(1-methylethylidene)- [3173-79-3]
  DIMETHYLKETENE β-LACTONE DIMER, **V**, 456
2-Oxetanone, 4-methylene- [674-82-8]
  KETENE DIMER, **III**, 508
Oxirane, 2,3-bis(2-chlorophenyl)- [53608-92-7]
  2,2'-DICHLORO-α,α-EPOXYBIBENZYL, **V**, 358
Oxirane, (bromomethyl)- [3132-64-7]
  EPIBROMOHYDRIN, **II**, 256
Oxirane, (chloromethyl)- [106-89-8]
  EPICHLOROHYDRIN, **I**, 233; **II**, 256
Oxirane, 2,3-diphenyl-, *trans*- [1439-07-2]
  trans-STILBENE OXIDE, **IV**, 860
Oxirane, methyl- (*R*)- [15448-47-2]
  (*R*)-METHYLOXIRANE, **VIII**, 434
Oxirane, methyl-, (*S*)- [16088-62-3]
  METHYLOXIRANE, (*S*)-(–)-, **VII**, 356
Oxirane, phenyl- [96-09-3]
  STYRENE OXIDE, **I**, 494
Oxirane, [(phenylmethoxy)methyl]- [2930-05-4]
  BENZYL 2,3-EPOXYPROPYL ETHER, **VIII**, 33
Oxiranecarboxylic acid, 3-methyl-3-phenyl-, ethyl ester [77-83-8]
  PHENYLMETHYLGLYCIDIC ESTER, **III**, 727
Oxiranemethanol, 3-propyl-, 2*S*-*trans*- [89321-71-1]
  (2*S*,3*S*)-3-PROPYLOXIRANEMETHANOL, **VII**, 461
Oxiranetetracarbonitrile [3189-43-3]
  TETRACYANOETHYLENE OXIDE, **V**, 1007
Oxonium, triethyl-, tetrafluoroborate (1+) [368-39-8]
  TRIETHYLOXONIUM FLUOBORATE, **V**, 1080
Oxonium, trimethyl-, salt with 2,4,6-trinitrobenzenesulfonic acid (1:1) [13700-00-0]
  TRIMETHYLOXONIUM 2,4,6-TRINITROBENZENESULFONATE,
  V, 1099

Oxonium, trimethyl-, tetrafluoroborate(1-) [420-37-1]
   TRIMETHYLOXONIUM FLUOBORATE, **V**, 1096; **VI**, 1019
Ozone [10028-15-6]
   OZONE, **III**, 673

P

Pentacyclo[19.3.1.1^{3,7},1^{9,13},1^{15,19}]octacosa-1(25),3,5,7(28),9,11,13(27),15,17,19(26),21,23-dode-
   caene-25,26,27,28-tetraol, 5,11,17,23-tetrakis(1,1-dimethylethyl)- [60705-62-6]
   p-tert-BUTYLCALIX[4]ARENE, **VIII**, 75
3,5,1,7-[1,2,3,4]-Butanetetraylnaphthalene, decahydro- [2292-79-7]
   DIAMANTANE, **VI**, 378
Pentadecanal [2765-11-9]
   PENTADECANAL, **VI**, 869,
Pentadecane, 1,1-dimethoxy- [52517-73-4]
   PENTADECANAL DIMETHYL ACETAL, **VI**, 869
1,4-Pentadiene [591-93-5]
   1,4-PENTADIENE, **IV**, 746
2,3-Pentadienedioic acid, dimethyl ester [1712-36-3]
   DIMETHYL 2,3-PENTADIENEDIOATE, **VI**, 505
2,3-Pentadienoic acid, ethyl ester [131504-53-5]
   ETHYL 2,3-PENTADIENOATE, **VII**, 232
2,4-Pentadienoic acid, (*E*)- [21651-12-7]
   trans-2,4-PENTADIENOIC ACID, **VI**, 95
1,4-Pentadien-3-one, 1,5-diphenyl- [538-58-9]
   DIBENZALACETONE, **II**, 167
Pentalene, octahydro-1-trichloromethyl- [18127-07-6]
   2-(TRICHLOROMETHYL)BICYCLO[3.3.0]OCTANE, **V**, 93
1-Pentalenecarboxylic acid, octahydro-, (1α,3aα,6aα)- [18209-43-3]
   exo-cis-BICYCLO[3.3.0]OCTANE-2-CARBOXYLIC ACID, **V**, 93
1,4-Pentalenediol, octahydro- [17572-86-0]
   BICYCLO[3.3.0]OCTANE-2,6-DIOL, **VIII**, 43
1,4-Pentalenediol, octahydro-, diacetate [17572-85-9]
   2,6-DIACETOXYBICYCLO[3.3.0]OCTANE, **VIII**, 43
1,4-Pentalenedlone, hexahydro-, *cis*- [77483-80-8]
   BICYCLO[3.3.0]OCTANE-2,6-DIONE, *cis*-, **VIII**, 43
2,5-(1*H*,3*H*)-Pentalenedione, tetrahydro-, *cis*- [51716-63-3]
   BICYCLO[3.3.0]OCTANE-3,7-DIONE, *cis*-, **VII**, 50
2,5-(1*H*,3*H*)-Pentalenedione, tetrahydro-3a,6a-dimethyl-, *cis*- [21170-10-5]
   cis-1,5-DIMETHYLBICYCLO[3.3.0]OCTANE-3,7-DIONE, **VII**, 50
1,3,4,6-Pentalenetetracarboxylic acid, 1,3a,4,6a-tetrahydro-2,5-dihydroxy- tetramethyl
   ester, (1α,3aα,4α,6aα)- [82416-04-4]
   TETRAMETHYL 3,7-DIHYDROXYBICYCLO[3.3.0]OCTA-2,6-DIENE-2,4,6,8-
      TETRACARBOXYLATE, **VII**, 50
1,3,4,6-Pentalenetetracarboxylic acid, 1,3a,4,6a-tetrahydro-2,5-dihydroxy-
   3a,6a-dimethyl- tetramethyl ester [79150-94-0]
   TETRAMETHYL 3,7-DIHYDROXY-1,5-DIMETHYLBICYCLO[3.3.0]OCTA-2,6-
      DIENE- 2,4,6,8-TETRACARBOXYLATE, **VII**, 50

Pentanal, 2,2-dimethyl-4-oxo- [61031-76-3]
  2,2-DIMETHYL-4-OXOPENTANAL, **VIII**, 208
Pentanal, 5-hydroxy- [4221-03-8]
  5-HYDROXYPENTANAL, **III**, 470
Pentanal, 2,4,4-trimethyl- [17414-46-9]
  2,4,4-TRIMETHYLPENTANAL, **VI**, 1028
2-Pentanamine, *N*-(1,1-dimethylethoxy)-*N*-(1,1-dimethylethyl)-2,4,4-trimethyl- [90545-93-0]
  N-tert-BUTYL-*N-tert*-OCTYL-*O-tert*-BUTYLHYDROXYLAMINE, **VIII**, 93
2-Pentanamine, *N*-(1,1-dimethylethyl)-2,4,4-trimethyl- [90545-94-1]
  tert-BUTYL-*tert*-OCTYLAMINE, **VIII**, 93
2-Pentanamine, 4-methyl-4-(methylnitrosoamino)- [16339-21-2]
  N-NITROSO-β-METHYLAMINOISOBUTYL METHYL KETONE, **III**, 244
2-Pentanamine, 2,4,4-trimethyl-*N*-(2-methylethylidene-1-*d*)- [34668-70-7]
  N-(2-METHYLBUTYLIDENE-1-*d*)-1,1,3,3-TETRAMETHYLBUTYLAMINE, **VI**,
    751
Pentane [109-66-0]
  PENTANE, **II**, 478
Pentane, 1,5-dibromo- [111-24-0]
  PENTAMETHYLENE BROMIDE, **I**, 428; **III**, 692
Pentane, 1,5-dinitro- [6848-84-6]
  1,5-DINITROPENTANE, **IV**, 370
Pentane, 2-isocyano-2,4,4-trimethyl- [14542-93-9]
  1,1,3,3-TETRAMETHYLBUTYL ISONITRILE, **VI**, 751
Pentane, 2,2,4-trimethyl-4-nitro- [5342-78-9]
  4-NITRO-2,2,4-TRIMETHYLPENTANE, **V**, 845
Pentane, 2,2,4-trimethyl-4-nitroso- [31044-98-1]
  NITROSO-*tert*-OCTANE, **VIII**, 93
Pentanedial, 3-methyl- [6280-15-5]
  β-METHYLGLUTARALDEHYDE, **IV**, 661
Pentanedinitrile [544-13-8]
  TRIMETHYLENE CYANIDE, **I**, 536
Pentanedinitrile, 3-hydroxy- [13880-89-2]
  3-HYDROXYGLUTARONITRILE, **V**, 614
Pentanedioic acid [110-94-1]
  GLUTARIC ACID, **I**, 289; **IV**, 496
Pentanedioic acid, 2-acetyl-, ethyl ester [1501-06-0]
  ETHYL ACETOGLUTARATE, **II**, 263
Pentanedioic acid, 3,3-dimethyl- [4839-46-7]
  β,β-DIMETHYLGLUTARIC ACID, **IV**, 345
Pentanedioic acid, 3-ethyl-3-methyl- [5345-01-7]
  β-ETHYL-β-METHYLGLUTARIC ACID, **IV**, 441
Pentanedioic acid, 3-methyl- [626-51-7]
  β-METHYLGLUTARIC ACID, **III**, 591
Pentanedioic acid, 3-methyl-3-phenyl- [4160-92-3]
  β-METHYL-β-PHENYLGLUTARIC ACID, **IV**, 664
Pentanedioic acid, 2-oxo- [328-50-7]
  α-KETOGLUTARIC ACID, **III**, 510; **V**, 687

Pentanedioic acid, 2-oxo-, diethyl ester [5965-53-7]
ETHYL ETHOXALYLPROPIONATE, **II**, 272
Pentanedioic acid, 3-oxo- [542-05-2]
ACETONEDICARBOXYLIC ACID, **I**, 10
Pentanedioic acid, 3-oxo-, diethyl ester [105-50-0]
ETHYL ACETONEDICARBOXYLATE, **I**, 237
1,5-Pentanediol [111-29-5]
1,5-PENTANEDIOL, **III**, 693
1,5-Pentanediol, 3-methyl- [4457-71-0]
3-METHYL-1,5-PENTANEDIOL, **IV**, 660
1,5-Pentanedione, 3,3-dimethyl-1,5-diphenyl- [42052-44-8]
3,3-DIMETHYL-1,5-DIPHENYLPENTANE-1,5-DIONE, **VIII**, 210
2,4-Pentanedione [123-54-6]
ACETYLACETONE, **III**, 16
2,4-Pentanedione, 3-butyl- [1540-36-9]
3-BUTYL-2,4-PENTANEDIONE, **VI**, 245
2,4-Pentanedione, 3-methyl- [815-57-6]
3-METHYLPENTANE-2,4-DIONE, **V**, 785
2,4-Pentanedione, 1-phenyl- [3318-61-4]
1-PHENYL-2,4-PENTANEDIONE, **VI**, 928
1,2,5-Pentanetriol [14697-46-2]
1,2,5-TRIHYDROXYPENTANE, **III**, 833
1,2,5-Pentanetriol, triacetate [5470-86-0]
1,2,5-TRIACETOXYPENTANE, **III**, 833
1,3,5-Pentanetrione, 1-(4-methoxyphenyl)-5-phenyl- [1678-17-7]
1-(*p*-METHOXYPHENYL)-5-PHENYL-1,3,5-PENTANETRIONE, **V**, 718
Pentanoic acid, 2-bromo-3-methyl- [42880-22-8]
α-BROMO-β-METHYLVALERIC ACID, **II**, 95; **III**, 496
Pentanoic acid, 2-bromo-4-methyl- [49628-52-6]
α-BROMOISOCAPROIC ACID, **II**, 95; **III**, 523
Pentanoic acid, 5-bromo-, methyl ester [5454-83-1]
METHYL 5-BROMOVALERATE, **III**, 578
Pentanoic acid, 2,4-dioxo-, ethyl ester [615-79-2]
ETHYL ACETOPYRUVATE, **I**, 238
Pentanoic acid, 2-ethenyl-2-methyl-5-trimethylsilyl- [88729-72-0]
2-METHYL-5-TRIMETHYLSILYL-2-VINYLPENTANOIC ACID, **VIII**, 486
Pentanoic acid, 2-ethenyl-2-methyl-5-trimethylsilyl-, methyl ester [88729-80-0]
METHYL 2-METHYL-5-TRIMETHYLSILYL-2-VINYLPENTANOATE,
VIII, 486
Pentanoic acid, 3-hydroxy-2,4-dimethyl-, (*R**,*S**)-(±)- [64869-26-7]
2,4-DIMETHYL-3-HYDROXYPENTANOIC ACID, (2*SR*,3*RS*)-, **VII**, 185
Pentanoic acid, 3-hydroxy-2,4-dimethyl-, (*R**,*R**)-(±)- [73198-99-9]
2,4-DIMETHYL-3-HYDROXYPENTANOIC ACID, (2*SR*,3*SR*)-, **VII**, 190
Pentanoic acid, 3-hydroxy-2,4-dimethyl-, (*R**,*R**)-, 2,6-dimethylphenyl ester
[73198-92-2]
2',6'-DIMETHYLPHENYL (2*SR*,3*SR*)-2,4-DIMETHYL-3-HYDROXYPEN-
TANOATE, **VII**, 190

Pentanoic acid, 3-hydroxy-2-methyl-, methyl ester, (*R**,*R**)-(±)- [100992-75-4]
 METHYL *anti*-3-HYDROXY-2-METHYLPENTANOATE, **VIII**, 420
Pentanoic acid, 3-hydroxy-2-methylene-, methyl ester [18052-21-6]
 METHYL 3-HYDROXY-2-METHYLENEPENTANOATE, **VIII**, 420
Pentanoic acid, 3-methyl- [105-43-1]
 3-METHYLPENTANOIC ACID, **II**, 416
Pentanoic acid, 4-methyl-4-nitro-, methyl ester [16507-02-1]
 METHYL γ-METHYL-γ-NITROVALERATE, **IV**, 652
Pentanoic acid, 4-methyl-3-oxo-, ethyl ester [7152-15-0]
 ETHYL 2-BUTYRYLACETATE, **VII**, 213
Pentanoic acid, 4-oxo- [123-76-2]
 LEVULINIC ACID, **I**, 335
3-Pentanol, 3-ethyl- [597-49-9]
 TRIETHYLCARBINOL, **II**, 602
2-Pentanone, 4-amino-4-methyl-, ethanedioate (2:1) [53608-87-0]
 ACETONAMINE HYDROGEN OXALATE, **I**, 196
2-Pentanone, 5-chloro- [5891-21-4]
 5-CHLORO-2-PENTANONE, **IV**, 597
3-Pentanone, 2,4-dibromo- [815-60-1]
 2,4-DIBROMO-3-PENTANONE, **VI**, 512, 520
3-Pentanone, 1-(dimethylamino)-4-methyl- [5782-64-9]
 1-(DIMETHYLAMINO)-4-METHYL-3-PENTANONE, **VI**, 474
2-Pentanone, 4-hydroxy-4-methyl- [123-42-2]
 DIACETONE ALCOHOL, **I**, 199
3-Pentanone, 2-hydroxy-2-methyl- [2834-17-5]
 2-HYDROXY-2-METHYLPENTAN-3-ONE, **VII**, 381
3-Pentanone, 2-methyl-2-[(trimethylsilyl)oxy]- [72507-50-7]
 2-METHYL-2-TRIMETHYLSILOXYPENTAN-3-ONE, **VII**, 185, 381
Pentanoyl chloride, 2,2,4-trimethyl-3-oxo- [10472-34-1]
 2,2,4-TRIMETHYL-3-OXOVALERYL CHLORIDE, **V**, 1103
18-Pentatriacontanone [504-53-0]
 STEARONE, **V**, 854
4-Pentenal, 2,2-dimethyl- [5497-67-6]
 2,2-DIMETHYL-4-PENTENAL, **VII**, 177
2-Pentene [109-68-2]
 2-PENTENE, **I**, 430
1-Pentene, 5-bromo-4-ethoxy- [22089-55-0]
 α-ALLYL-β-BROMOETHYL ETHYL ETHER, **IV**, 750
2-Pentenedial, ion(1+), sodium [24290-36-6]
 GLUTACONALDEHYDE SODIUM SALT, **VI**, 640
2-Pentenedioic acid, 3-chloro-, dimethyl ester, (*E*)- [66016-87-3]; (*Z*)- [66016-88-4]
 DIMETHYL 3-CHLORO-2-PENTENEDIOATE, **VI**, 505
2-Pentenoic acid, 2-cyano-3-ethyl-, ethyl ester [868-04-2]
 ETHYL(1-ETHYLPROPYLIDENE)CYANOACETATE, **III**, 399
2-Pentenoic acid, 2-cyano-4-methyl-, ethyl ester [868-47-3]
 ETHYL *sec*-BUTYLIDENECYANOACETATE, **IV**, 94

3-Pentenoic acid, 2-cyano-3-ethyl-2-methyl-, ethyl ester [53608-83-6]
    ETHYL (1-ETHYLPROPENYL)METHYLCYANOACETATE, **III**, 397
3-Penten-2-ol [1569-50-2]
    3-PENTEN-2-OL, **III**, 696
4-Penten-1-ol [821-9-0]
    4-PENTEN-1-OL, **III**, 698
4-Penten-2-ol, 1-(phenylmethoxy)- [58931-16-11]
    1-BENZYLOXY-4-PENTEN-2-OL, **VIII**, 33
1-Penten-3-one, 4,4-dimethyl-1-phenyl- [538-44-3]
    BENZALPINACOLONE, **I**, 81
3-Penten-2-one, (*E*)- [3102-33-8]
    trans-3-PENTEN-2-ONE, **VI**, 883
3-Penten-2-one, 4-methoxy- [2845-83-2]
    4-METHOXY-3-PENTEN-2-ONE, **VIII**, 357
3-Penten-2-one, 4-methyl- [141-79-7]
    MESITYL OXIDE, **I**, 345
1-Penten-4-yn-3-ol, 1-phenyl- [14604-31-0]
    1-PHENYL-1-PENTEN-4-YN-3-OL, **IV**, 792
4-Pentyn-1-ol [5390-04-5]
    4-PENTYN-1-OL, **IV**, 755
Peroxide, bis(4-nitrobenzoyl)- [1712-84-1]
    p-NITROBENZOYL PEROXIDE, **III**, 649
9b*H*-Phenalen-9b-ol, dodecahydro-, (3aα,6aα,9aβ,9bα)- [16664-34-9]
    PERHYDRO-9*b*-PHENALENOL, **VII**, 427
Phenanthrene [85-01-8]
    PHENANTHRENE, **IV**, 313
Phenanthrene, 9-bromo- [573-17-1]
    9-BROMOPHENANTHRENE, **III**, 134
Phenanthrene, 9,10-dihydro- [776-35-2]
    9,10-DIHYDROPHENANTHRENE, **IV**, 313
Phenanthrene, 9-phenyl- [844-20-2]
    9-PHENYLPHENANTHRENE, **V**, 952
9-Phenanthrenecarbonitrile [2510-55-6]
    9-CYANOPHENANTHRENE, **III**, 212
9-Phenanthrenecarboxaldehyde [4707-71-5]
    PHENANTHRENE-9-ALDEHYDE, **III**, 701
1-Phenanthrenecarboxylic acid, 1,2,3,4,4a,4b,5,6,10,10a-decahydro-1,4-dimethyl-
    7-(1-methylethyl)-, [1*R*-(1α,4aβ,4bα,10aα)]- [514-10-3]
    ABIETIC ACID, **IV**, 1
1-Phenanthrenecarboxylic acid, 1,2,3,4,4a,4b,5,9,10,10a-decahydro-1,4-dimethyl-
    7-(1-methylethyl)-, [1*R*(1α,4aβ,4bα,10aα)]- [79-54-9]
    LEVOPIMARIC ACID, **V**, 699
4-Phenanthrenecarboxylic acid, 5-formyl- [5684-15-1]
    5-FORMYL-4-PHENANTHROIC ACID, **IV**, 484
9,10-Phenanthrenediol, 9,10-dihydro-, *trans*- [25061-61-4]
    trans-9,10-DIHYDRO-9,10-PHENANTHRENEDIOL, **VI**, 887

9,10-Phenanthrenedione [84-11-7]
    PHENANTHRENEQUINONE, **IV**, 757
2-Phenanthrenesulfonic acid [41105-40-2]
    2-PHENANTHRENESULFONIC ACID, **II**, 482
3-Phenanthrenesulfonic acid [2039-95-4]
    3-PHENANTHRENESULFONIC ACID, **II**, 482
Phenanthro[9,10-*b*]oxirene, 1*a*,9*b*-dihydro- [585-08-0]
    PHENANTHRENE-9,10-OXIDE, **VI**, 887
Phenazine, 1-methoxy- [2876-17-7]
    α-METHOXYPHENAZINE, **III**, 753
1-Phenazinol [528-71-2]
    α-HYDROXYPHENAZINE, **III**, 754
1(5*H*)-Phenazinone, 5-methyl- [85-66-5]
    PYOCYANINE, **III**, 753
Phenol, 2-amino-4-nitro- [99-57-0]
    2-AMINO-4-NITROPHENOL, **III**, 82
Phenol, 2-bromo- [95-56-7]
    o-BROMOPHENOL, **II**, 97
Phenol, 4-bromo- [106-41-2]
    p-BROMOPHENOL, **I**, 128
Phenol, 2-bromo-4-methyl- [6627-55-0]
    3-BROMO-4-HYDROXYTOLUENE, **III**, 130
Phenol, 2-(chloromethyl)-4-nitro- [2973-19-5]
    2-HYDROXY-5-NITROBENZYL CHLORIDE, **III**, 468
Phenol, 2,6-dibromo-4-nitro- [99-28-5]
    2,6-DIBROMO-4-NITROPHENOL, **II**, 173
Phenol, 2,6-dichloro- [87-65-0]
    2,6-DICHLOROPHENOL, **III**, 267
Phenol, 2-heptyl- [5284-22-0]
    o-HEPTYLPHENOL, **III**, 444
Phenol, 2-iodo- [533-58-4]
    o-IODOPHENOL, **I**, 326
Phenol, 4-iodo- [540-38-5]
    p-IODOPHENOL, **II**, 355
Phenol, 2-methoxy-4-methyl- [93-51-6]
    CREOSOL, **IV**, 203
Phenol, 3-methoxy-5-methyl- [3209-13-0]
    ORCINOL MONOMETHYL ETHER, **VI**, 859
Phenol, 2-methoxy-6-(2-propenyl)- [579-60-2]
    o-EUGENOL, **III**, 418
Phenol, 4-methyl- [106-44-5]
    p-CRESOL, **I**, 175
Phenol, 5-methyl-2-(1-methylethyl)-4-nitroso- [2364-54-7]
    NITROSOTHYMOL, **I**, 511
Phenol, 3-nitro- [554-84-7]
    m-NITROPHENOL, **I**, 404

Phenoxathiin [262-20-4]
   PHENOXTHIN, **II**, 485
DL-Phenylalanine [150-30-1]
   DL-PHENYLALANINE, **II**, 489; **III**, 705
Phenylalanine, *N*-acetyl- [2018-61-3]
   N-ACETYLPHENYLALANINE, **II**, 493
Phenylalanine, *N*-[(1,1-dimethylethoxy)carbonyl]-L- [13734-34-4]
   N-*tert*-BUTOXYCARBONYL-L-PHENYLALANINE, **VII**, 70, 75
Phosphine, [1,1'-binaphthalene]-2,2'-diylbis-[diphenyl-, (*R*)- [76189-55-4];
      (*S*)- [76189-56-5]
   2,2'-BIS(DIPHENYLPHOSPHINO)-1,1'-BINAPHTHYL (BINAP), (*R*)-(+)- and
      (*S*)-(–)-, **VIII**, 57, 183
Phosphine, (4-bromophenyl)diphenyl- [734-59-8]
   DIPHENYL-*p*-BROMOPHENYLPHOSPHINE, **V**, 496
Phosphine oxide, [1,1'-binaphthalene]-2,2'-diylbis-[diphenyl- [130164-89-5]
   (±)-2,2'-BIS(DIPHENYLPHOSPHINYL)-1,1'-BINAPHTHYL [(±)BINAPO],
      VIII, 58
Phosphinous chloride, bis(1-methylethyl)- [40244-90-4]
   CHLORODIISOPROPYLPHOSPHINE, **V**, 211
1*H*-Phosphole, 1,1-dichloro-1,1,2,3-tetrahydro-4-methyl-1-phenyl- [17154-12-0]
   3-METHYL-1-PHENYLPHOSPHOLENE 1,1-DICHLORIDE, **V**, 787
1*H*-Phosphole, 2,3-dihydro-4-methyl-1-phenyl-, 1-oxide [707-61-9]
   3-METHYL-1-PHENYLPHOSPHOLENE OXIDE, **V**, 787
Phosphonic acid, (aminomethyl)-, diethyl ester [50917-72-1]
   DIETHYL AMINOMETHYLPHOSPHONATE, **VIII**, 451
Phosphonic acid, [2-(cyclohexylamino)ethenyl]-, diethyl ester [20061-84-1]
   DIETHYL 2-(CYCLOHEXYLAMINO)VINYLPHOSPHONATE, **VI**, 448
Phosphonic acid, 2,2-diethoxyethyl-, diethyl ester [7598-61-0]
   DIETHYL 2,2-DIETHOXYETHYLPHOSPHONATE, **VI**, 448
Phosphonic acid, [(1,3-dihydro-1,3-dioxo-2*H*-isoindol-2-yl)methyl]-, diethyl ester
      [33512-26-4]
   DIETHYL PHTHALIMIDOMETHYLPHOSPHONATE, **VIII**, 451
Phosphonic acid, [(2,2-dimethyl-4-oxo-4*H*-1,3-dioxin-6-yl)methyl]-, diethyl ester
      [81956-28-7]
   6-DIETHYLPHOSPHONOMETHYL-2,2-DIMETHYL-1,3-DIOXEN-4-ONE,
      VIII, 192
Phosphonic acid, ethyl-, bis(1-methylethyl) ester [1067-69-2]
   DIISOPROPYL ETHYLPHOSPHONATE, **IV**, 326
Phosphonic acid, (hydroxymethyl)-, diethyl ester [3084-40-0]
   DIETHYL HYDROXYMETHYLPHOSPHONATE, **VII**, 160
Phosphonic acid, methyl, bis(1-methylethyl) ester [1445-75-6]
   DIISOPROPYL METHYLPHOSPHONATE, **IV**, 325
Phosphonic acid, (2-oxoethyl)-, diethyl ester [1606-75-3]
   DIETHYL FORMYLMETHYLPHOSPHONATE, **VI**, 448
Phosphonic acid, phenyl-, diethyl ester [1754-49-0]
   DIETHYL PHENYLPHOSPHONATE, **VI**, 451

Phosphonic acid, [[(phenylmethylene)amino]methyl]-, diethyl ester [50917-73-2]
  DIETHYL *N*-BENZYLIDENAMINOMETHYLPHOSPHONATE, **VIII**, 451
Phosphonic acid, [(tetrahydro-2*H*-pyran-2-yl)oxy]methyl]-, diethyl ester [71885-51-3]
  DIETHYL [(2-TETRAHYDROPYRANYLOXY)METHYL]PHOSPHONATE,
  VII, 160
Phosphonic dichloride, (2-phenylethenyl)- [4708-07-0]
  STYRYLPHOSPHONIC DICHLORIDE, **V**, 1005
Phosphonic dichloride, (trichloromethyl)- [21510-59-8]
  TRICHLOROMETHYLPHOSPHONYL DICHLORIDE, **IV**, 950
Phosphonium, ethenyltriphenyl-, bromide [5044-52-0]
  VINYL TRIPHENYLPHOSPHONIUM BROMIDE, **V**, 1145
Phosphonothioic dichloride, (chloromethyl)- [1983-27-3]
  CHLOROMETHYLPHOSPHONOTHIOIC DICHLORIDE, **V**,218
Phosphonous dichloride, phenyl- [644-97-3]
  PHENYLDICHLOROPHOSPHINE, **IV**, 784
Phosphoramidic acid, (cycloundecyl-1-pyrrolidinylmethylene)-, diphenyl ester
  [62914-02-7]
  DIPHENYL (CYCLOUNDECYL-1-PYRROLIDINYLMETHYLENE)-
  PHOSPHORAMIDATE, **VII**, 135
Phosphorane, (dichloromethylene)triphenyl- [6779-08-4]
  DICHLOROMETHYLENETRIPHENYLPHOSPHORANE, **V**, 361
Phosphorazidic acid, diphenyl ester [26386-88-9]
  DIPHENYL PHOSPHORAZIDATE, **VII**, 135, 206
Phosphoric acid, tributyl ester [126-73-8]
  BUTYL PHOSPHATE, **II**, 109
4-Phosphorinanone, 1-phenyl- [23855-87-0]
  1-PHENYL-4-PHOSPHORINANONE, **VI**, 932
3-Phosphorinocarbonitrile, 4-amino-1,2,5,6-tetrahydro-1-phenyl- [84819-76-1]
  4-AMINO-1,2,5,6-TETRAHYDRO-1-PHENYLPHOSPHORIN-
  3-CARBONITRILE, **VI**, 932
Phosphorodiamidic acid, tetramethyl-, 2-furanyl ester [105262-58-6]
  FURYL *N,N,N',N'*-TETRAMETHYLDIAMIDOPHOSPHATE, **VIII**, 396
Phosphorodiamidic chloride, tetramethyl [1605-65-8]
  N,N,N',N'-TETRAMETHYLDIAMIDOPHOSPHOROCHLORIDATE, **VII**, 66
Phosphorodichloridic acid, 2-furanyl ester [105262-70-2]
  FURYL PHOSPHORODICHLORIDATE, **VIII**, 396
Phosphorous acid, triethyl ester [122-52-1]
  TRIETHYL PHOSPHITE, **IV**, 955
Phosphorous triamide, hexamethyl [1608-26-0]
  HEXAMETHYLPHOSPHOROUS TRIAMIDE, **V**, 602
1,4-Phthalazinedione, 5-amino-2,3-dihydro- [521-31-3]
  5-AMINO-2,3-DIHYDRO-1,4-PHTHALAZINEDIONE, **III**, 69
1,4-Phthalazinedione, 2,3-dihydro-5-nitro- [3682-15-3]
  5-NITRO-2,3-DIHYDRO-1,4-PHTHALAZINEDIONE, **III**, 656
Piperazine, 1-(phenylmethyl)- [2759-28-6]
  1-BENZYLPIPERAZINE, **V**, 88

Piperidine, 1-benzoyl- [776-75-0]
BENZOYLPIPERIDINE, **I**, 99
Piperidine, 1-chloro- [2156-71-0]
N-CHLOROPIPERIDINE, **VI**, 968
Piperidine, 1-(cyclooctylidenemethyl)- [65226-88-2]
1-(CYCLOOCTYLIDENEMETHYL)PIPERIDINE, **VII**, 473
Piperidine, 1-ethyl- [766-09-6]
N-ETHYLPIPERIDINE, **V**, 575
Piperidine, 1-(3-hydroxy-2-methyl-1-oxo-3-phenylpropyl)-, (*R**,*R**)-(±)-
[99114-36-0]
1-(3-HYDROXY-2-METHYL-3-PHENYLPROPANOYL)PIPERIDINE, *erythro*,
VIII, 326
Piperidine, 1-(3-hydroxy-2-methyl-1-oxo-3-phenylpropyl)-, (*R**,*S**)-(±)- [116596-04-4]
1-(3-HYDROXY-2-METHYL-3-PHENYLPROPANOYL)PIPERIDINE, *threo*-,
VIII, 326
Piperidine, 1-(2-methyl-1,3-dioxo-3-phenylpropyl)-, (±)- [99114-34-8]
1-(2-BENZOYLPROPANOYL)PIPERIDINE, **VIII**, 326
Piperidine, 1-(2-naphthalenyl)- [5465-85-0]
N-β-NAPHTHYLPIPERIDINE, **V**, 816
Piperidine, 1-phenyl- [4096-20-2]
1-PHENYLPIPERIDINE, **IV**, 795
1-Piperidineacetic acid, 4-carboxy-, diethyl ester [1838-39-7]
1-CARBETHOXYMETHYL-4-CARBETHOXYPIPERIDINE, **V**, 989
3,5-Piperidinedicarbonitrile, 4-ethyl-4-methyl-2,6-dioxo- [1135-62-2]
α,α'-DICYANO-β-ETHYL-β-METHYLGLUTARIMIDE, **IV**, 441
3,5-Piperidinedicarbonitrile, 4-methyl-2,6-dioxo-4-phenyl- [6936-95-4]
β-METHYL-β-PHENYL-α,α'-DICYANOGLUTARIMIDE, **IV**, 662
2,6-Piperidinedione [1121-89-7]
GLUTARIMIDE, **IV**, 496
Plumbane, tris(acetyloxy)(4-methoxyphenyl)- [18649-43-9]
p-METHOXYPHENYLLEAD TRIACETATE, **VII**, 229
Pregn-5-ene-3,20-diol, 3-acetate, (3β,20*S*) [53603-96-6]
3β-ACETOXY-20β-HYDROXY-5-PREGNENE, **V**, 692
Pregn-5-en-18-oic acid, 3-(acetyloxy)-20-hydroxy-,γ-lactone, (3β, 20*R*)- [3020-10-8]
18,20-LACTONE OF 3β-ACETOXY-20β-HYDROXY-5-PREGNEN-18-OIC
ACID, **V**, 692
1,2-Propadiene [463-49-0]
ALLENE, **V**, 22
Propanal [123-38-6]
PROPIONALDEHYDE, **II**, 541
Propanal, 2,3-dihydroxy, (+)- [56-82-6]
DL-GLYCERALDEHYDE, **II**, 305
Propanal, 2-oxo-, 1-(phenylhydrazone) [5391-74-2]
METHYLGLYOXAL-ω-PHENYLHYDRAZONE, **IV**, 633
Propanamide, 2-hydroxy- [2043-43-8]
LACTAMIDE, **III**, 516

Propanamide, 2-methyl- [563-83-7]
   ISOBUTYRAMIDE, **III**, 490
2-Propanamine, *N*-hydroxy-2-methyl- [16649-50-6]
   N-tert-BUTYLHYDROXYLAMINE, **VI**, 804
2-Propanamine, 2-methyl- [75-64-9]
   tert-BUTYLAMINE, **III**, 148
2-Propanamine, 2-methyl, hydrochloride [10017-37-5]
   tert-BUTYLAMINE HYDROCHLORIDE, **III**, 153
2-Propanamine, 2-methyl-*N*-(2-methylpropylidene)- [6852-60-4]
   N-(2-METHYLPROPYLIDENE)-*tert*-BUTYLAMINE, **VI**, 526
2-Propanamine, 2-methyl-*N*-(phenylmethylene)- [6852-58-0]
   N-tert-BUTYLBENZALDIMINE, **V**, 191
1-Propanamine, 3,3'-(phenylphosphinidene)bis[*N,N*-dimethyl- [32357-32-7]
   BIS(3-DIMETHYLAMINOPROPYL)PHENYLPHOSPHINE, **VI**, 776
1-Propanaminium, 3-[(ethylcarbonimidoyl)amino]-*N,N*-diimethyl-, hydrochloride
   [22572-40-3]
   1-ETHYL-3-(3-DIMETHYLAMINO)PROPYLCARBODIIMIDE
   HYDROCHLORIDE, **V**, 555
Propane, 1-bromo- [106-94-5]
   PROPYL BROMIDE, **I**, 37; **II**, 359
Propane, 2-bromo- [75-26-3]
   ISOPROPYL BROMIDE, **I**, 37
Propane, 1-bromo-3-chloro-2,2-dimethoxy- [22089-54-9]
   1-BROMO-3-CHLORO-2,2-DIMETHOXYPROPANE, **VI**, 361
Propane, 1-bromo-2-methyl- [78-77-3]
   ISOBUTYL BROMIDE, **II**, 358
Propane, 2-bromo-2-methyl- [507-19-7]
   tert-BUTYL BROMIDE, **I** 38
Propane, 3-chloro-1,1-diethoxy- [35573-93-4]
   β-CHLOROPROPIONALDEHYDE ACETAL, **II**, 137
Propane, 2,2'-[(1-chloro-1,2-ethenediyl)bis(oxy)]bis[2-methyl-, (*E*)- [70525-93-8]
   (*E*)-1,2-DI-*tert*-BUTOXY-1-CHLOROETHENE, **VIII**, 161
Propane, 2-chloro-2-methyl- [507-20-0]
   tert-BUTYL CHLORIDE, **I**, 144
Propane, 2-diazo- [2684-60-8]
   2-DIAZOPROPANE, **VI**, 392
Propane, 2-diazo-1,1,1,3,3,3-hexafluoro- [684-23-1]
   BIS(TRIFLUOROMETHYL)DIAZOMETHANE, **VI**, 161
Propane, 1,3-dibromo- [109-64-8]
   TRIMETHYLENE BROMIDE, **I**, 30
Propane, 1,3-dibromo-2,2-bis(bromomethyl)- [3229-00-3]
   PENTAERYTHRITYL BROMIDE (or TETRABROMIDE), **II**, 476; **IV**, 753
Propane, 2,3-dibromo-1,1-diethoxy- [10160-86-8]
   2,3-DIBROMOPROPIONALDEHYDE DIETHYL ACETAL, **VI**, 954
Propane, 2,2'-[(1,2-dichloro-1,2-ethanediyl)]bis(oxy)bis[2-methyl-, (*R**,*R**)-(±)-
   [68470-80-4]; (*R**,*S**)-(±)- [68470-81-5]
   1,2-DI-*tert*-BUTOXY-1,2-DICHLOROETHANE, **VIII**, 161

Propane, 1,3-diiodo-2,2-bis(iodomethyl)- [1522-88-9]
   PENTAERYTHRITYL IODIDE, **II**, 476
Propane, 1,3-dinitro- [6125-21-9]
   1,3-DINITROPROPANE, **IV**, 370
Propane, 2,2'-[1,2-ethynediylbis(oxy)]bis[2-methyl- [66478-63-5]
   DI-*tert*-BUTOXYETHYNE, **VIII**, 161
Propane, 1,1,1,2,2,3,3-heptachloro- [594-89-8]
   unsym-HEPTACHLOROPROPANE, **II**, 312
Propane, 1-iodo-2,2-dimethyl- [15501-33-4]
   NEOPENTYL IODIDE, **VI**, 830
Propane, 2-isocyano-2-methyl- [7188-38-7]
   tert-BUTYL ISOCYANIDE, **VI**, 232
Propane, 2-methoxy-1,3-bis(methylthio)- [31805-84-2]
   1,3-BIS(METHYLTHIO)-2-METHOXYPROPANE, **VI**, 683
Propane, 2-methyl-2-nitro- [594-70-7]
   2-METHYL-2-NITROPROPANE, **VI**, 803
Propane, 2-methyl-2-nitroso- [917-95-3]
   2-METHYL-2-NITROSOPROPANE and DIMER, **VI**, 803
Propane, 1,1-thiobis- [111-47-7]
   PROPYL SULFIDE, **II**, 547
Propane, 1,2,3-tribromo- [96-11-7]
   1,2,3-TRIBROMOPROPANE, **I**, 521
Propane, 1,1,3-triethoxy- [7789-92-6]
   β-ETHOXYPROPIONALDEHYDE ACETAL, **III**, 371
Propanedial, nitro-, ion(1−), sodium, monohydrate [53821-72-0]
   SODIUM NITROMALONALDEHYDE MONOHYDRATE, **IV**, 844
      Warning, **V**, 1004
1,3-Propanediamine, *N,N*-dibutyl- [102-83-0]
   γ-DIBUTYLAMINOPROPYLAMINE, **III**, 256
1,3- Propanediamine, 2-(1,1-dimethylethyl)- [56041-82-8]
   2-*tert*-BUTYL-1,3-DIAMINOPROPANE, **VI**, 223
1,3-Propanediamine, *N'*-(ethylcarbonimidoyl)-*N,N*-dimethyl- [1892-57-5]
   1-ETHYL-3-(3-DIMETHYLAMINO)PROPYLCARBODIIMIDE, **V**, 555
1,3-Propanediamine, *N'*-(ethylcarbonimidoyl)-*N,N*-dimethyl-, monohydrochloride
      [25952-53-8]
   1-ETHYL-3-(3-DIMETHYLAMINO)PROPYLCARBODIIMIDE
      HYDROCHLORIDE, **V**, 555
Propanedinitrile [109-77-3]
   MALONONITRILE, **II**, 379; **III**, 535
Propanedinitrile, amino-, mono(4-methylbenzenesulfonate) [5098-14-6]
   AMINOMALONONITRILE *p*-TOLUENESULFONATE, **V**, 32
Propanedinitrile, 1,1-dimethylethyl- [4210-60-0]
   tert-BUTYLMALONONITRILE, **VI**, 223
Propanedinitrile, 1,3-dioxolan-2-ylidene- [5694-65-5]
   DICYANOKETENE ETHYLENE ACETAL, **IV**, 276
Propanedinitrile, (hydroxyimino)- [36568-05-5]
   OXIMINOMALONONITRILE, **V**, 32

Propanedinitrile, oxo- [1115-12-4]
  CARBONYL CYANIDE, **VI**, 268
Propanedioic acid [1141-82-2]
  MALONIC ACID, **II**, 376
Propanedioic acid, bis(1,1-dimethylethyl) ester [541-16-2]
  DI-*tert*-BUTYL MALONATE, **IV**, 261
Propanedioic acid, (1,1-dimethylethyl) ethyl ester [32864-38-3]
  ETHYL *tert*-BUTYL MALONATE, **IV**, 417
Propanedioic acid, acetyl-, diethyl ester [570-08-1]
  DIETHYL ACETOSUCCINATE, **II**, 262
Propanedioic acid, (acetylamino)-, diethyl ester [1068-90-2]
  DIETHYL ACETAMIDOMALONATE, **V**, 373
Propanedioic acid, [4-(acetyloxy)-2-cyclohexen-1-yl]-, dimethyl ester, *cis*- [82736-52-5]
  1-ACETOXY-4-(DICARBOMETHOXYMETHYL)-2-CYCLOHEXENE, *cis*-,
  VIII, 5
Propanedioic acid, [4-(acetyloxy)-2-cyclohexen-1-yl]-, dimethyl ester, *trans*-
  [82736-53-6]
  1-ACETOXY-4-(DICARBOMETHOXYMETHYL)-2-CYCLOHEXENE, *trans*-
  VIII, 5
Propanedioic acid, amino-, diethyl ester [6829-40-9]
  DIETHYL AMINOMALONATE, **V**, 376
Propanedioic acid, amino-, diethyl ester, hydrochloride [13433-00-6]
  DIETHYL AMINOMALONATE HYDROCHLORIDE, **V**, 376
Propanedioic acid, benzoyl-, diethyl ester [1087-97-4]
  DIETHYL BENZOYLMALONATE, **IV**, 285
Propanedioic acid, (benzoyloxy)ethyl-, diethyl ester [6259-78-5]
  DIETHYL [*O*-BENZOYL]ETHYLTARTRONATE, **V**, 379
Propanedioic acid, bis(hydroxymethyl)-, diethyl ester [20605-01-0]
  DIETHYL BIS(HYDROXYMETHYL)MALONATE, **V**, 381
Propanedioic acid, bromo-, diethyl ester [685-87-01]
  DIETHYL BROMOMALONATE, **I**, 245
Propanedioic acid, butyl, diethyl ester [133-08-4]
  DIETHYL BUTYLMALONATE, **I**, 250
Propanedioic acid, (chlorocyclopropylmethylene)-, diethyl ester [123844-18-8]
  DIETHYL 2-CHLORO-2-CYCLOPROPYLETHENE-1,1-DICARBOXYLATE,
  VIII, 247
Propanedioic acid, 2-cyclopenten-1-yl-, diethyl ester [53608-93-8]
  DIETHYL Δ²-CYCLOPENTENYLMALONATE, **IV**, 291
Propanedioic acid, (cyclopropylcarbonyl)-, diethyl ester [7394-16-3]
  DIETHYL CYCLOPROPYLCARBONYLMALONATE, **VIII**, 248
Propanedioic acid, diazo-, bis(1,1-dimethylethyl) ester [35207-75-1]
  DI-*tert*-BUTYL DIAZOMALONATE, **VI**, 414
Propanedioic acid, (1,3-dihydro-1,3-dioxo-2*H*-isoindol-2-yl)-, diethyl ester [5680-61-5]
  DIETHYL PHTHALIMIDOMALONATE, **I**, 271
Propanedioic acid, (1,1-dimethylethyl)-, diethyl ester [759-24-0]
  DIETHYL *tert*-BUTYLMALONATE, **VI**, 442

Propanedioic acid, (2,5-dimethylphenyl)hydroxy-, diethyl ester [83026-12-4]
DIETHYL 2,5-DIMETHYLPHENYLHYDROXYMALONATE, **III**, 326
Propanedioic acid, (ethoxymethylene)-, diethyl ester [87-13-8]
DIETHYL ETHOXYMETHYLENEMALONATE, **III**, 395
Propanedioic acid, ethylidene-, diethyl ester [1462-12-0]
DIETHYL ETHYLIDENEMALONATE, **IV**, 293
Propanedioic acid, hexadecyl-, diethyl ester [41433-81-2]
CETYLMALONIC ESTER, **IV**, 141
Propanedioic acid, (hydroxyimino)-, diethyl ester [6829-41-0]
DIETHYL OXIMINOMALONATE, **V**, 373
Propanedioic acid, methyl-, diethyl ester [609-08-5]
DIETHYL METHYLMALONATE, **II**, 279
Propanedioic acid, (3-methyl-2-butenyl)-, dimethyl ester [43219-18-7]
(3-METHYL-2-BUTENYL)PROPANEDIOIC ACID, DIMETHYL ESTER,
VIII, 381
Propanedioic acid, (3-methyl-2-butenyl)-2-propynyl-, dimethyl ester [107473-14-3]
(3-METHYL-2-BUTENYL)(2-PROPYNYL)PROPANEDIOIC ACID, DIMETHYL
ESTER, **VIII**, 381
Propanedioic acid, methylene-, bis(1,1-dimethylethyl) ester [86633-09-2]
DI-*tert*-BUTYL METHYLENEMALONATE, **VII**, 142
Propanedioic acid, methylene-, diethyl ester [3377-20-6]
DIETHYL METHYLENEMALONATE, **IV**, 298
Propanedioic acid, (1-methylethylidene)-, diethyl ester [6802-75-1]
DIETHYL ISOPROPYLIDENEMALONATE, **VI**, 442
Propanedioic acid, (1-methylpropyl)-, diethyl ester [83-27-2]
DIETHYL *sec*-BUTYLMALONATE, **III**, 495
Propanedioic acid, oxo-, diethyl ester [609-09-6]
ETHYL OXOMALONATE, **I**, 266
Propanedioic acid, phenyl-, diethyl ester [83-13-6]
DIETHYL PHENYLMALONATE, **II**, 288
Propanedioic acid, (phenylmethyl)-, diethyl ester [607-81-8]
DIETHYL BENZYLMALONATE, **III**, 705
Propanedioic acid, (phenylmethylene)-, diethyl ester [5292-53-5]
DIETHYL BENZALMALONATE, **III**, 377
1,2-Propanediol, (*R*)- [4254-14-2]
L-PROPYLENE GLYCOL, **II**, 545
1,2-Propanediol, (*S*)- [4254-15-3]
(*S*)-(+)-PROPANE-1,2-DIOL, **VII**, 356
1,3-Propanediol, 2,2-bis(hydroxymethyl)- [115-77-5]
PENTAERYTHRITOL, **I**, 425
1,3-Propanediol, 2-(bromomethyl)-2-(hydroxymethyl)- [19184-65-7]
MONOBROMOPENTAERYTHRITOL, **IV**, 681
1,2-Propanediol, 3-chloro- [96-24-2]
GLYCEROL α-MONOCHLOROHYDRIN, **I**, 294
1,2-Propanediol, 3,3-diethoxy-, (±)- [10487-05-5]
DL-GLYCERALDEHYDE DIETHYL ACETAL, **II**, 307

1,2-Propanediol, 3-phenoxy- [538-43-2]
α-GLYCERYL PHENYL ETHER, **I**, 296
1,2-Propanedione, 1-phenyl- [579-07-7]
ACETYLBENZOYL, **III**, 20
1,2-Propanedione, 1-phenyl-, 2-oxime [119-51-7]
ISONITROSOPROPIOPHENONE, **II**, 363
1,3-Propanedione, 2,2-dibromo-1,3-diphenyl- [16619-55-9]
DIBENZOYLDIBROMOMETHANE, **II**, 244
1,3-Propanedione, 1,3-diphenyl- [120-46-7]
DIBENZOYLMETHANE, **I**, 205; **III**, 251
1,3-Propanedione, 1-(2-hydroxyphenyl)-3-phenyl- [1469-94-9]
o-HYDROXYDIBENZOYLMETHANE, **IV**, 479
Propanedioyl dichloride [1663-67-8]
MALONYL DICHLORIDE, **IV**, 263
Propanenitrile, 3-amino- [151-18-8]
β-AMINOPROPIONITRILE, **III**, 93
Propanenitrile, 3-[(2-chlorophenyl)amino]- [94-89-3]
3-(o-CHLOROANILINO)PROPIONITRILE, **IV**, 146
Propanenitrile, 3-ethoxy- [2141-62-0]
β-ETHOXYPROPIONITRILE, **III**, 372
Propanenitrile, 3-hydroxy- [109-78-4]
ETHYLENE CYANOHYDRIN, **I**, 256
Propanenitrile, 3-[(2-hydroxyethyl)thio]- [15771-37-6]
(2-HYDROXYETHYLMERCAPTO)PROPIONITRILE, **III**, 458
Propanenitrile, 2-hydroxy-2-methyl- [75-86-5]
ACETONE CYANOHYDRIN, **II**, 7
Propanenitrile, 3,3'-iminobis- [111-94-4]
Bis-(β-CYANOETHYL)AMINE, **III**, 93
Propanenitrile, 2-methyl- [78-82-0]
ISOBUTYRONITRILE, **III**, 493
Propanenitrile, 2-methyl-2-nitrooxy- [40561-27-1]
ACETONE CYANOHYDRIN NITRATE, **V**, 839
Propanenitrile, 3-(phenylamino)- [1075-76-9]
N-2-CYANOETHYLANILINE, **IV**, 205
Propanenitrile, 3,3'-(phenylphosphinidene)bis- [15909-92-9]
BIS(2-CYANOETHYL)PHENYLPHOSPHINE, **VI**, 932
1,1,2,3-Propanetetracarboxylic acid, tetraethyl ester [635-03-0]
TETRAETHYL PROPANE-1,1,2,3-TETRACARBOXYLATE, **II**, 272
1,1,3,3-Propanetetracarboxylic acid, tetraethyl ester [2121-66-6]
TETRAETHYL PROPANE-1,1,3,3-TETRACARBOXYLATE, **I**, 290
1,2,3-Propanetricarboxylic acid [99-14-9]
TRICARBALLYLIC ACID, **I**, 523
1,2,3-Propanetricarboxylic acid, 1-oxo-, triethyl ester [42126-21-6]
TRIETHYL OXALYLSUCCINATE, **III**, 510; **V**, 687
Propanetrione, diphenyl- [643-75-4]
DIPHENYL TRIKETONE, **II**, 244

2-Propanimine, 1,1,1,3,3,3-hexafluoro- [1645-75-6]
 HEXAFLUOROACETONE IMINE, **VI**, 664
Propanoic acid, 1,1-dimethylethyl ester [51233-80-8]
 tert-BUTYL PROPIONATE, **III**, 143
Propanoic acid, 2,6-dimethylphenyl ester [51233-80-8]
 2',6'-DIMETHYLPHENYL PROPANOATE, **VII**, 190
Propanoic acid, 2-phenylhydrazide [20730-02-3]
 β-PROPIONYLPHENYLHYDRAZINE, **IV**, 657
Propanoic acid, 3-bromo- [590-92-1]
 β-BROMOPROPIONIC ACID, **I**, 131
Propanoic acid, 3-bromo-, ethyl ester [539-74-2]
 ETHYL β-BROMOPROPIONATE, **I**, 246
Propanoic acid, 3-bromo-, methyl ester [3395-91-3]
 METHYL β-BROMOPROPIONATE, **III**, 576
Propanoic acid, 3-bromo-2-(bromomethyl)- [41459-42-1]
 β,β'-DIBROMOISOBUTYRIC ACID, **VII**, 319
Propanoic acid, 3-bromo-2-(bromomethyl)-, methyl ester [22262-60-8]
 METHYL β,β-DIBROMOISOBUTYRATE, **VII**, 319
Propanoic acid, 2-bromo-3-methoxy- [65090-78-0]
 α-BROMO-β-METHOXYPROPIONIC ACID, **III**, 775
Propanoic acid, 2-bromo-3-methoxy-, methyl ester [27704-96-7]
 METHYL α-BROMO-β-METHOXYPROPIONATE, **III**, 774
Propanoic acid, 2-chloro-, (*S*)- [29617-66-1]
 (*S*)-2-CHLOROPROPANOIC ACID, **VIII**, 119, 434
Propanoic acid, 3-chloro- [107-94-8]
 β-CHLOROPROPIONIC ACID, **I**, 166
Propanoic acid, 3-chloro-2-oxo- [3681-17-2]
 CHLOROPYRUVIC ACID, **V**, 636
Propanoic acid, 3-chloro-2,2,3-trifluoro- [425-97-8]
 3-CHLORO-2,2,3-TRIFLUOROPROPIONIC ACID, **V**, 239
Propanoic acid, 3,3-diethoxy-, ethyl ester [10601-80-6]
 ETHYL 3,3-DIETHOXYPROPANOATE, **VIII**, 254
Propanoic acid, 3,3-diethoxy-, methyl ester [127581-78-6]
 METHYL 3,3-DIETHOXYPROPANOATE, **VIII**, 255
Propanoic acid, 2,2-dimethyl- [75-98-9]
 TRIMETHYLACETIC ACID, **I**, 524
Propanoic acid, 2-hydroxy-, 1-methylethyl ester [617-51-6]
 ISOPROPYL LACTATE, **II**, 365
Propanoic acid, 2-hydroxy-, 2-propenyl ester [5349-55-3]
 ALLYL LACTATE, **III**, 46
Propanoic acid, 3-hydroxy- [503-66-2]
 β-HYDROXYPROPIONIC ACID, **I**, 321
Propanoic acid, 2-iodo-3-nitro-, methyl ester
 METHYL 2-IODO-3-NITROPROPIONATE, **VI**, 799
Propanoic acid, 2-methyl-, 1,1-dimethylethyl ester [16889-72-8]
 tert-BUTYL ISOBUTYRATE, **III**, 143

Propanoic acid, 2-[[(1-methylethylidene)amino]oxy]-, (±)- [2009-90-7]; (S)- [2009-89-4]
  DL- and (+)- and (−)-α-(ISOPROPYLIDENEAMINOOXY)PROPIONIC ACID,
  V, 1032
Propanoic acid, 2-[[(1-methylethylidene)amino]oxy]- ethyl ester [54716-29-9]
  ETHYL α-(ISOPROPYLIDENEAMINOOXY)PROPIONATE, V, 1031
Propanoic acid, 2-oxo- [127-17-3]
  PYRUVIC ACID, I, 475
Propanoic acid, 2-oxo-, ethyl ester [617-35-6]
  ETHYL PYRUVATE, IV, 467
Propanoic acid, 2-oxo-, methyl ester [600-22-6]
  METHYL PYRUVATE, III, 610
Propanoic acid, 2-[[(2,4,5,7-tetranitro-9H-fluoren-9-ylidene)amino]oxy]-, (S)-
  [50996-73-1]
Propanoic acid, 2-[[(2,4,5,7-tetranitro-9H-fluoren-9-ylidene)amino]oxy]-, (R)-
  [50874-31-2]
  (+) and (−)-α-(2,4,5,7-TETRANITRO-9-FLUORENYLIDENEAMINOOXY)
  PROPIONIC ACID, V, 1031
Propanoic acid, 3,3'-thiobis-, dimethyl ester [4131-74-2]
  DIMETHYL β-THIODIPROPIONATE, IV, 669
2-Propanol, 1,3-bis(methylthio)- [31805-83-1]
  1,3-BIS(METHYLTHIO)-2-PROPANOL, VI, 683
2-Propanol, 1-bromo-, acetate, (S)- [39968-99-5]
  (S)-(−)-2-ACETOXY-1-BROMOPROPANE, VII, 356
1-Propanol, 2-chloro-, (S)-(+)- [19210-21-0]
  (S)-2-CHLOROPROPAN-1-OL, VIII, 434
1-Propanol, 3-chloro- [627-30-5]
  TRIMETHYLENE CHLOROHYDRIN, I, 533
1-Propanol, 3-chloro-, acetate [628-09-1]
  γ-CHLOROPROPYL ACETATE, III, 203
2-Propanol, 1,3-dibromo- [96-21-9]
  GLYCEROL α,γ-DIBROMOHYDRIN, II, 308
2-Propanol, 1,3-dichloro- [96-23-1]
  GLYCEROL α,γ-DICHLOROHYDRIN, I, 292
1-Propanol, 2,2-dimethyl- [75-84-3]
  NEOPENTYL ALCOHOL, V, 818
2-Propanol, 1,1,1,3,3,3-hexafluoro-2-phenyl-, potassium salt [37818-31-8]
  POTASSIUM 1,1,1,3,3,3-HEXAFLUORO-2-PHENYL-2-PROPANOLATE,
  VI, 163
1-Propanone, 2,3-dibromo-1,3-diphenyl- [611-91-6]
  BENZALACETOPHENONE DIBROMIDE, I, 205
1-Propanone, 3-(dimethylamino)-1-phenyl-, hydrochloride [879-72-1]
  β-DIMETHYLAMINOPROPIOPHENONE HYDROCHLORIDE, III, 305
1-Propanone, 2,2-dimethyl-1-phenyl- [938-16-9]
  tert-BUTYL PHENYL KETONE, VI, 248
1-Propanone, 1,3-diphenyl- [1083-30-3]
  BENZYLACETOPHENONE, I, 101

1-Propanone, 1-(3-hydroxy-2-methylphenyl)- [3338-15-6]
   3-HYDROXY-2-METHYLPROPIOPHENONE, **V**, 624
1-Propanone, 1-(2-hydroxyphenyl)- [610-99-1]
   o-PROPIOPHENOL, **II**, 543
1-Propanone, 1-(4-hydroxyphenyl)- [70-70-2]
   p-PROPIOPHENOL, **II**, 543
1-Propanone, 1,3,3-triphenyl- [606-86-0]
   β,β-DIPHENYLPROPIOPHENONE, **II**, 236
2-Propanone, hydrazone [5281-20-9]
   ACETONE HYDRAZONE, **VI**, 10
2-Propanone, oxime [127-06-0]
   ACETOXIME, **I**, 318
2-Propanone, 1-bromo- [598-31-2]
   BROMOACETONE, **II**, 88
2-Propanone, 1-chloro-3-phenyl- [937-38-2]
   BENZYL CHLOROMETHYL KETONE, **III**, 119
2-Propanone, 1-(4-chlorophenyl)-3-phenyl- [35730-03-1]
   1-(*p*-CHLOROPHENYL)-3-PHENYL-2-PROPANONE, **IV**, 176
2-Propanone, 1,3-dibromo-1,3-diphenyl- [958-79-2]
   α,α'-DIBROMODIBENZYL KETONE, **V**, 514
2-Propanone, 1,3-dichloro- [534-07-61
   α,γ-DICHLOROACETONE, **I**, 211
2-Propanone, 1,1-diphenyl- [781-35-1]
   α,α-DIPHENYLACETONE, **III**, 343
2-Propanone, 1,1,1,3,3,3-hexafluoro- [684-16-2]
   HEXAFLUOROACETONE, **VII**, 251
2-Propanone, 1,1,1,3,3,3-hexafluoro-, hydrazone [1588-41-6]
   HEXAFLUOROACETONE HYDRAZONE, **VI**, 161
2-Propanone, 1-hydroxy- [116-09-6]
   ACETOL, **II**, 5
2-Propanone, 1-(2-methoxyphenyl)- [5211-62-1]
   o-METHOXYPHENYLACETONE, **IV**, 573
2-Propanone, (1-methylethylidene)-, hydrazone [627-70-3]
   ACETONE AZINE, **VI**, 10
2-Propanone, 1-phenyl- [103-79-7]
   METHYL BENZYL KETONE, **II**, 389
Propanoyl bromide, 2-bromo-2-methyl- [20769-85-1]
   α-BROMOISOBUTYRYL BROMIDE, **IV**, 348
Propanoyl chloride, 3-isocyanato- [3729-19-9]
   3-ISOCYANATOPROPANOYL CHLORIDE, **VI**, 715
Propanoyl chloride, 2-oxo- [5704-66-5]
   PYRUVOYL CHLORIDE, **VII**, 467
2-Propenal [107-02-8]
   ACROLEIN, **I**, 15
2-Propenal, 2-bromo-3-phenyl- [5443-49-2]
   α-BROMOCINNAMIC ALDEHYDE, **III**, 731

2-Propenal, 3,3-diphenyl- [1210-39-5]
  β-PHENYLCINNAMALDEHYDE, **VI**, 901
2-Propenal, 3-(2-isothiocyanatophenyl)-, (*E*)- [19908-01-1]
  o-ISOTHIOCYANATO-(*E*)-CINNAMALDEHYDE, **VII**, 302
2-Propenal, 3-(2-nitrophenyl)- [1466-88-2]
  o-NITROCINNAMALDEHYDE, **IV**, 722
2-Propenamide, 2-methyl- [79-39-0]
  METHACRYLAMIDE, **III**, 560
2-Propenamide, *N*-(phenylmethyl)- [13304-62-6]
  N-BENZYLACRYLAMIDE, **V**, 73
1-Propen-1-amine, 1-chloro-*N*,*N*,2-trimethyl- [26189-59-3]
  1-CHLORO-*N*,*N*,2-TRIMETHYLPROPENYLAMINE, **VI**, 282; **VIII**, 441
1-Propen-1-amine, 2-phenyl-*N*-(phenylmethylene)- [64244-34-4]
  1,4-DIPHENYL-1-METHYL-3-AZA-1,4-BUTADIENE, **VIII**, 451, 455
2-Propen-1-amine [107-11-9]
  ALLYLAMINE, **II**, 24
2-Propen-1-amine, 2-bromo- [6943-51-7]
  2-BROMOALLYLAMINE, **V**, 121
2-Propen-1-amine, 2-bromo-*N*-ethyl- [871-23-8]
  N-(2-BROMOALLYL)ETHYLAMINE, **V**, 124
2-Propen-1-amine, *N*-2-propenyl- [124-02-7]
  DIALLYLAMINE, **I**, 201
1-Propene, 3-bromo- [106-95-6]
  ALLYL BROMIDE, **I**, 27
1-Propene, 2,3-dibromo- [513-31-5]
  2,3-DIBROMOPROPENE, **I**, 209
1-Propene, 3,3-diethoxy- [3054-95-3]
  ACROLEIN ACETAL, **II**, 17; **IV**, 21
1-Propene, 2-nitro- [4749-28-4]
  2-NITROPROPENE, **VII**, 396, 414
2-Propenenitrile, 3-(2-furanyl)- [7187-01-1]
  3-(2-FURYL)ACRYLONITRILE, **V**, 585
2-Propenenitrile, 3-phenyl- [4360-47-8]
  CINNAMONITRILE, **VI**, 304; **VII**, 108
1-Propene-1,2,3-tricarboxylic acid [499-12-7]
  ACONITIC ACID, **II**, 12
2-Propenoic acid [79-10-7]
  ACRYLIC ACID, **III**, 30
2-Propenoic acid, butyl ester [141-32-2]
  BUTYL ACRYLATE, **III**, 146
2-Propenoic acid, 2-(acetylamino)-3-phenyl- [5469-45-4]
  α-ACETAMINOCINNAMIC ACID, **II**, 1
2-Propenoic acid, 2-acetyl-3-(2,3-dimethoxyphenyl)-, ethyl ester [19411-81-5]
  ETHYL α-ACETYL-β-(2,3-DIMETHOXYPHENYL)ACRYLATE,
    IV, 408
2-Propenoic acid, 2-benzoylamino-3-phenyl- [1155-48-2]
  α-BENZOYLAMINOCINNAMIC ACID, **II**, 491

2-Propenoic acid, 2-(bromomethyl)- [72707-66-5]
α-(BROMOMETHYL)ACRYLIC ACID, **VII**, 210
2-Propenoic acid, 2-(bromomethyl)-, ethyl ester [17435-72-2]
ETHYL α-(BROMOMETHYL)ACRYLATE, **VII**, 210; **VIII**, 266
2-Propenoic acid, 2-(bromomethyl)-, methyl ester [4224-69-5]
METHYL α-(BROMOMETHYL)ACRYLATE, **VII**, 319
2-Propenoic acid, 3-(2-carboxyphenyl)- [612-40-8]
o-CARBOXYCINNAMIC ACID, **IV**, 136
2-Propenoic acid, 2-cyano-3-phenyl- [1011-92-3]
α-CYANO-β-PHENYLACRYLIC ACID, **I**, 181
2-Propenoic acid, 2-cyano-3-phenyl-, ethyl ester [2025-40-3]
ETHYL α-CYANO-β-PHENYLACRYLATE, **I**, 451
2-Propenoic acid, 3-(2,3-dimethoxyphenyl)- [7461-60-1]
2,3-DIMETHOXYCINNAMIC ACID, **IV**, 327
2-Propenoic acid, 3,3-diphenyl- [606-84-8]
β-PHENYLCINNAMIC ACID, **V**, 509
2-Propenoic acid, 2-formyl-3-hydroxy-, methyl ester [39947-70-1]
METHYL DIFORMYLACETATE, **VII**, 323
2-Propenoic acid, 3-(2-furanyl)- [539-47-9]
FURYLACRYLIC ACID, **III**, 425
2-Propenoic acid, 2-(hydroxymethyl)-, ethyl ester [10029-04-6]
ETHYL α-(HYDROXYMETHYL)ACRYLATE, **VIII**, 265
2-Propenoic acid, 3-nitro-, methyl ester, (*E*)- [52745-92-3]
METHYL (*E*)-3-NITROACRYLATE, **VI**, 799
2-Propenoic acid, 3-(3-nitrophenyl)- [555-68-0]
m-NITROCINNAMIC ACID, **I**, 398; **IV**, 731
2-Propenoic acid, 3-phenyl-, 1,1-dimethylethyl ester [14990-09-1]
tert-BUTYL CINNAMATE, **I**, 252
2-Propenoic acid, 3-phenyl-, ethyl ester [103-36-6]
ETHYL CINNAMATE, **I**, 252
2-Propenoic acid, 3-phenyl-, 5-methyl-2-(1-methylethyl)cyclohexyl ester,
[1*R*-(1α,2β,5α)]- [16205-99-5]
(−)-MENTHYL CINNAMATE, **VIII**, 350
2-Propenoic acid, 3-phenyl-, phenyl ester [2757-04-2]
PHENYL CINNAMATE, **III**, 714
2-Propenoic acid, 3-[(3-phenyl-2-propenyl)oxy]-, (*E,E*)- [88083-18-5]
(*E*)-3-[(*E*)-3-PHENYL-2-PROPENOXY]ACRYLIC ACID, **VIII**, 536
2-Propenoic acid, 3-(phenylsulfonyl)-, methyl ester, (*Z*)- [91077-67-7]
METHYL (*Z*)-3-(PHENYLSULFONYL)PROP-2-ENOATE, **VIII**, 458
2-Propenoic acid, 3-(tributylstannyl)-, ethyl ester, (*E*)- [106335-84-6]
ETHYL (*E*)-3-(TRIBUTYLSTANNYL)PROPENOATE, **VIII**, 268
1-Propen-2-ol, 1-phenyl, acetate, (*E*)- [19980-44-0]
2-ACETOXY-*trans*-1-PHENYLPROPENE, **VI**, 692
2-Propen-1-ol [107-18-6]
ALLYL ALCOHOL, **I**, 42
2-Propen-1-ol, 2-bromo-3,3-diphenyl-, acetate [14310-15-7]
2-BROMO-3,3-DIPHENYL-2-PROPEN-1-YL ACETATE, **VI**, 187

2-Propen-1-ol, 1-(trimethylsilyl)- [95061-68-0]
(1-HYDROXY-2-PROPENYL)TRIMETHYLSILANE, **VIII**, 501
2-Propen-1-ol, 3-(trimethylsilyl)-, (*E*)- [59376-64-6]
3-TRIMETHYLSILYL-2-PROPEN-1-OL, (*E*)-, **VII**, 524
2-Propen-1-ol, 2-[(trimethylsilyl)methyl]- [81302-80-9]
2-(HYDROXYMETHYL)ALLYLTRIMETHYLSILANE, **VII**, 266
1-Propen-1-one, 2-methyl- [598-26-5]
DIMETHYLKETENE, **IV**, 348
2-Propen-1-one, 3,3-bis(methylthio)-1-(2-pyridinyl)- [78570-34-0]
3,3-BIS(METHYLTHIO)-1-(2-PYRIDINYL)-2-PROPEN-1-ONE, **VII**, 476
2-Propen-1-one, 2-bromo-1,3-diphenyl- [6935-75-7]
α-BROMOBENZALACETOPHENONE, **III**, 126
2-Propen-1-one, 1-[4-(1,1-dimethylethyl)-1-cyclohexen-1-yl]- [92622-56-5]
1-(4-*tert*-BUTYLCYCLOHEXEN-1-YL)-2-PROPEN-1-ONE, **VIII**, 97
2-Propen-1-one, 1,3-diphenyl- [94-41-7]
BENZALACETOPHENONE, **I**, 78
2-Propynal [624-67-9]
PROPIOLALDEHYDE, **IV**, 813
2-Propynal, 3-phenyl- [2579-22-8]
PHENYLPROPARGYL ALDEHYDE, **III**, 731
1-Propyne, 3,3-diethoxy- [10160-87-9]
PROPIOLALDEHYDE DIETHYL ACETAL, **VI**, 954
2-Propynoic acid, 3-cyclopropyl-, ethyl ester [123844-20-2]
ETHYL CYCLOPROPYLPROPIOLATE, **VIII**, 247
2-Propynoic acid, 3-phenyl- [637-44-5]
PHENYLPROPIOLIC ACID, **II**, 515
2-Propyn-1-ol, 3-(trimethylsilyl)- [5272-36-3]
3-TRIMETHYLSILYL-2-PROPYN-1-OL, **VII**, 524
2-Propyn-1-ol, 3-(trimethylsilyl)-, methanesulfonate [71321-17-0]
3-TRIMETHYLSILYL-2-PROPYN-1-YL METHANESULFONATE, **VIII**, 471
1*H*-Purine-2,6-dione, 7-amino-3,7-dihydro-1,3-dimethyl- [81281-58-5]
7-AMINOTHEOPHYLLINE, **VII**, 8
1*H*-Purine-2,6-dione, 3,7-dihydro-1,3-dimethyl-7-[(phenylmethylene)amino]-
[81281-59-6]
7-BENZYLIDENEAMINOTHEOPHYLLINE, **VII**, 9
2*H*-Pyran, 3-chlorotetrahydro-2-methyl-, *cis*- [53107-04-3] and *trans*- [53107-05-4]
3-CHLORO-2-METHYLTETRAHYDROPYRAN, **VI**, 675
2*H*-Pyran, 2,3-dichlorotetrahydro- [5631-95-8]
2,3-DICHLOROTETRAHYDROPYRAN, **VI**, 675
2*H*-Pyran, 3,4-dihydro- [110-87-2]
2,3-DIHYDROPYRAN, **III**, 276
2*H*-Pyran, 3,4-dihydro-2-methoxy-4-methyl- [53608-95-0]
3,4-DIHYDRO-2-METHOXY-4-METHYL-2*H*-PYRAN, **IV**, 311
2*H*-Pyran, tetrahydro- [142-68-7]
TETRAHYDROPYRAN, **III**, 794

2*H*-Pyran, tetrahydro-2-(2-propynyloxy)- [6089-04-9]
    TETRAHYDRO-2-(2-PROPYNYLOXY)-2*H*-PYRAN, **VII**, 334
2*H*-Pyran-3-carboxylic acid, 4,6-dimethyl-2-oxo- [33953-26-3]
    ISODEHYDROACETIC ACID, **IV**, 549
2*H*-Pyran-5-carboxylic acid, 4,6-dimethyl-2-oxo-, ethyl ester [3385-34-0]
    ETHYL ISODEHYDROACETATE, **IV**, 549
2*H*-Pyran-5-carboxylic acid, 2-oxo- [500-05-0]
    COUMALIC ACID, **IV**, 201
2*H*-Pyran-5-carboxylic acid, 2-oxo-, methyl ester [6018-41-3]
    METHYL COUMALATE, **IV**, 532
4*H*-Pyran-2,6-dicarboxylic acid, 4-oxo- [99-32-1]
    CHELIDONIC ACID, **II**, 126
2*H*-Pyran-2,4(3*H*)-dione, 3-acetyl-6-methyl- [520-45-6]
    DEHYDROACETIC ACID, **III**, 231
2*H*-Pyran-2,6(3*H*)-dione, dihydro-4-methyl- [4166-53-4]
    β-METHYLGLUTARIC ANHYDRIDE, **IV**, 630
2*H*-Pyran-2,6(3*H*)-dione, dihydro-3-phenyl- [2959-96-8]
    α-PHENYLGLUTARIC ANHYDRIDE, **IV**, 790
2*H*-Pyran-2-one [504-31-4]
    α-PYRONE, **V**, 982
    2*H*-PYRAN-2-ONE, **VI**, 462
2*H*-Pyran-2-one, 5,6-dihydro- [3393-45-1]
    5,6-DIHYDRO-2*H*-PYRAN-2-ONE, **VI**, 462
2*H*-Pyran-2-one, 4,6-dimethyl- [675-09-2]
    4,6-DIMETHYLCOUMALIN, **IV**, 337
4*H*-Pyran-4-one, 2,6-dimethyl-3,5-diphenyl- [33731-54-3]
    2,6-DIMETHYL-3,5-DIPHENYL-4*H*-PYRAN-4-ONE, **V**, 450
4*H*-Pyran-4-one, 2-(4-methoxyphenyl)-6-phenyl- [14116-43-9]
    2-(*p*-METHOXYPHENYL)-6-PHENYL-4-PYRONE, **V**, 721
2*H*-Pyran-2-one-3-^{13}C, tetrahydro-4-hydroxy-4-methyl- [53771-22-5]
    MEVALONOLACTONE-2-^{13}C, (*R,S*)-, **VII**, 386
2*H*-Pyran-2-one, tetrahydro-4-methyl- [1121-84-2]
    β-METHYL-δ-VALEROLACTONE, **IV**, 677
2,3-Pyrazinedicarboxylic acid [89-01-0]
    2,3-PYRAZINEDICARBOXYLIC ACID, **IV**, 824
1*H*-Pyrazol-3-amine [1820-80-0]
    3(5)-AMINOPYRAZOLE, **V**, 39
3-Pyrazolamine, 4,5-dihydro-, sulfate [29574-26-3]
    3-AMINO-3-PYRAZOLINE SULFATE, **V**, 39
1*H*-Pyrazole, 3,5-dimethyl- [67-51-6]
    3,5-DIMETHYLPYRAZOLE, **IV**, 351
3-Pyrazoline, 3-amino-1-(*p*-tolylsulfonyl)- [1018-36-6]
    3-IMINO-1-(*p*-TOLYLSULFONYL)PYRAZOLIDINE, **V**, 40
2-Pyrazolin-5-one, 3-propyl- [29211-70-9]
    3-(1-PROPYL)-2-PYRAZOLIN-5-ONE, **VI**, 791

3*H*-Pyrazol-3-one, 5-amino-2,4-dihydro-2-phenyl- [4149-06-8]
    1-PHENYL-3-AMINO-5-PYRAZOLONE, **III**, 708
Pyrene, 1-bromo- [1714-29-0]
    3-BROMOPYRENE, **V**, 147
1-Pyrenol [5315-79-7]
    3-HYDROXYPYRENE, **V**, 632
3-Pyridinamine [462-08-8]
    3-AMINOPYRIDINE, **IV**, 45
2-Pyridinamine, 5-bromo- [1072-97-5]
    2-AMINO-5-BROMOPYRIDINE, **V**, 346
2-Pyridinamine, *N*-(phenylmethyl)- [6935-27-9]
    2-BENZYLAMINOPYRIDINE, **IV**, 91
Pyridine, 1-oxide [694-59-7]
    PYRIDINE- *N*-OXIDE, **IV**, 828
Pyridine, 2-amino-5-bromo-3-nitro- [6945-68-2]
    2-AMINO-5-BROMO-3-NITROPYRIDINE, **V**, 347
Pyridine, 2,6-bis(1,1-dimethylethyl)-4-methyl- [38222-83-2]
    2,6-DI-*tert*-BUTYL-4-METHYLPYRIDINE, **VII**, 144
Pyridine, 2-bromo- [109-04-6]
    2-BROMOPYRIDINE, **III**, 136
Pyridine, compound with hydrofluoric acid homopolymer [62778-11-4]
    PYRIDINIUM POLYHYDROGEN FLUORIDE, **VI**, 628
Pyridine, 2,6-dimethyl- [108-48-5]
    2,6-DIMETHYLPYRIDINE, **II**, 214
Pyridine, 4-ethyl- [536-75-4]
    4-ETHYLPYRIDINE, **III**, 410
Pyridine, 5-ethyl-2-methyl- [104-90-5]
    5-ETHYL-2-METHYLPYRIDINE, **IV**, 451
Pyridine, 3-methyl-4-nitro, 1-oxide [1074-98-2]
    3-METHYL-4-NITROPYRIDINE 1-OXIDE, **IV**, 654
Pyridine, 2-phenyl- [1008-89-5]
    2-PHENYLPYRIDINE, **II**, 517
Pyridine, 2,3,4,5-tetrahydro-, trimer [27879-53-4]
    2,3,4,5-TETRAHYDROPYRIDINE TRIMER, **VI**, 968
Pyridine, 1,2,3,6-tetrahydro-1-(4-methoxyphenyl)- [133157-31-0]
    1-(4-METHOXYPHENYL)-1,2,5,6-TETRAHYDROPYRIDINE, **VIII**, 358
Pyridine, 1,2,3,4-tetrahydro-1-methyl-6-phenyl- [30609-85-9]
    N-METHYL-2-PHENYL-Δ²-TETRAHYDROPYRIDINE, **VI**, 818
2-Pyridineacetic acid, ethyl ester [2739-98-2]
    ETHYL 2-PYRIDYLACETATE, **III**, 413
2-Pyridineacetic acid, 6-methyl-, ethyl ester [5552-83-0]
    ETHYL 6-METHYLPYRIDINE-2-ACETATE, **VI**, 611
3-Pyridinebutanenitrile, γ-oxo- [36740-10-0]
    4-OXO-4-(3-PYRIDYL)BUTYRONITRILE, **VI**, 866
3-Pyridinecarbonitrile [100-54-9]
    NICOTINONITRILE, **IV**, 706

3-Pyridinecarbonitrile, 2-chloro- [6602-54-6]
  2-CHLORONICOTINONITRILE, **V**, 166
3-Pyridinecarbonitrile, 1,2-dihydro-6-methyl-2-oxo- [4241-27-4]
  3-CYANO-6-METHYL-2(1)-PYRIDONE, **IV**, 210
2-Pyridinecarbonitrile, 6-methyl- [1620-75-3]
  2-CYANO-6-METHYLPYRIDINE, **IV**, 269
3-Pyridinecarboxamide, 1-oxide [1986-81-8]
  NICOTINAMIDE 1-OXIDE, **IV**, 704
2-Pyridinecarboxylic acid, hydrochloride [636-80-6]
  PICOLINIC ACID HYDROCHLORIDE, **III**, 740
3-Pyridinecarboxylic acid [59-67-6]
  NICOTINIC ACID, **I**, 385
3-Pyridinecarboxylic acid, anhydride [16837-38-0]
  NICOTINIC ANHYDRIDE, **V**, 822
3-Pyridinecarboxylic acid, 1,6-dihydro-6-oxo- [5006-66-6]
  6-HYDROXYNICOTINIC ACID, **IV**, 532
3-Pyridinecarboxylic acid, 5-methyl-2-(1-methylethyl)cyclohexyl ester, [1R-(1α,2β,5α)]-
  [133005-61-5]
  (–)-MENTHYL NICOTINATE, **VIII**, 350
2,3-Pyridinediamine [452-58-4]
  2,3-DIAMINOPYRIDINE, **V**, 346
2,3-Pyridinediamine, 5-bromo- [38875-53-5]
  2,3-DIAMINO-5-BROMOPYRIDINE, **V**, 346
3,5-Pyridinedicarboxylic acid, 1,4-dihydro-2,6-dimethyl-, diethyl ester [1149-23-1]
  1,4-DIHYDRO-3,5-DICARBETHOXY-2,6-DIMETHYLPYRIDINE,
    II, 214
3,5-Pyridinedicarboxylic acid, 2,6-dimethyl-, diethyl ester [1149-24-2]
  3,5-DICARBETHOXY-2,6-DIMETHYLPYRIDINE, **II**, 215
4-Pyridineethanamine, β,β-diethoxy- [74209-44-2]
  2,2-DIETHOXY-2-(4-PYRIDYL)ETHYLAMINE, **VII**, 149
2-Pyridineethanol, α-methyl- [5307-19-7]
  1-(α-PYRIDYL)-2-PROPANOL, **III**, 757
4-Pyridinesulfonic acid [5402-20-0]
  4-PYRIDINESULFONIC ACID, **V**, 977
Pyridine-2,3,6-tricarboxylic acid, triethyl ester [122509-29-9]
  2,3,6-TRICARBOETHOXYPYRIDINE, **VIII**, 597
Pyridinium, 1-amino-, iodide [6295-87-0]
  1-AMINOPYRIDINIUM IODIDE, **V**, 43
Pyridinium, 1-fluoro-, salt with trifluoromethanesulfonic acid (1:1) [107263-95-6]
  N-FLUOROPYRIDINIUM TRIFLATE, **VIII**, 287
Pyridinium, 1-methoxy-2-methyl-, methyl sulfate [55369-05-6]
  1-METHOXY-2-METHYLPYRIDINIUM METHYL SULFATE, **V**, 270
Pyridinium, 1-(2-nitrobenzyl)-, bromide [13664-80-7]
  o-NITROBENZYLPYRIDINIUM BROMIDE, **V**, 825
2(1H)-Pyridinone, 1-methyl- [694-85-9]
  1-METHYL-2-PYRIDONE, **II**, 419

1*H*-Pyrido[3,4-*b*]indole, 2,3,4,9-tetrahydro- [16502-01-5]
    1,2,3,4-TETRAHYDRO-β-CARBOLINE, **VI**, 965
2-Pyrimidinamine, *N,N*-dimethyl- [5621-02-3]
    2-(DIMETHYLAMINO)PYRIMIDINE, **IV**, 336
4-Pyrimidinamine, 2,6-dimethyl- [461-98-3]
    4-AMINO-2,6-DIMETHYLPYRIMIDINE, **III**, 71
Pyrimidine, 2-chloro- [1722-12-9]
    2-CHLOROPYRIMIDINE, **IV**, 182
Pyrimidine, 4-methyl- [3438-46-8]
    4-METHYLPYRIMIDINE, **V**, 794
5-Pyrimidinecarbonitrile, 1-heptyl-1,2,3,6-tetrahydro-6-imino-2-oxo- [53608-90-5]
    3-HEPTYL-5-CYANOCYTOSINE, **IV**, 515
5-Pyrimidinecarbonitrile, 4-hydroxy-2-mercapto- [23945-49-5]
    2-MERCAPTO-4-HYDROXY-5-CYANOPYRIMIDINE, **IV**, 566
5-Pyrimidinecarboxylic acid, 4-amino-1,2-dihydro-2-thioxo-, ethyl ester [774-07-2]
    2-MERCAPTO-4-AMINO-5-CARBETHOXYPYRIMIDINE, **IV**, 566
2,4(1*H*,3*H*)-Pyrimidinedione, 4,5-diamino-, monohydrochloride [53608-89-2]
    DIAMINOURACIL HYROCHLORIDE, **IV**, 247
2,4(1*H*,3*H*)-Pyrimidinedione, 6-methyl- [626-48-2]
    6-METHYLURACIL, **II**, 422
2(1*H*)-Pyrimidinethione [1450-85-7]
    2-MERCAPTOPYRIMIDINE, **V**, 703
2,4,6(1*H*,3*H*,5*H*)-Pyrimidinetrione [67-52-7]
    BARBITURIC ACID, **II**, 60
2,4,6(1*H*,3*H*,5*H*)-Pyrimidinetrione, 5-amino- [118-78-5]
    URAMIL, **II**, 617
2,4,6(1*H*,3*H*,5*H*)-Pyrimidinetrione, 5,5-dihydroxy- [3237-50-1]
    ALLOXAN MONOHYDRATE, **III**, 37; **IV**, 23
2,4,6(1*H*,3*H*,5*H*)-Pyrimidinetrione, 5-hydroxy- [444-15-5]
    DIALURIC ACID MONOHYDRATE, **IV**, 29
2,4,6(1*H*,3*H*,5*H*)-Pyrimidinetrione, 5-nitro- [480-68-2]
    NITROBARBITURIC ACID, **II**, 440
2,4,6(1*H*,3*H*,5*H*)-Pyrimidinetrione, 5-(phenylmethylene)- [27402-47-7]
    BENZALBARBITURIC ACID, **III**, 39
4(1*H*)-Pyrimidinone, 2,6-diamino- [56-06-4]
    2,4-DIAMINO-6-HYDROXYPYRIMIDINE, **IV**, 245
4(1*H*)-Pyrimidinone, 2,3-dihydro-6-methyl-2-thioxo- [56-04-2]
    2-THIO-6-METHYLURACIL, **IV**, 638
4(1*H*)-Pyrimidinone, 6-methyl- [3524-87-6]
    4-METHYL-6-HYDROXYPYRIMIDINE, **IV**, 638
1*H*-Pyrrole [109-97-7]
    PYRROLE, **I**, 473
1*H*-Pyrrole, 4,5-dihydro-1-methyl-2-(methylthio)- [25355-52-6]
    2-METHYLMERCAPTO-*N*-METHYL-Δ²-PYRROLINE, **V**, 780
1*H*-Pyrrole, 2,4-dimethyl- [625-82-1]
    2,4-DIMETHYLPYRROLE, **II**, 217

1*H*-Pyrrole, 2,5-dimethyl- [625-84-3]
    2,5-DIMETHYLPYRROLE, **II**, 219
1*H*-Pyrrole, 2,4-diphenyl- [3274-56-4]
    2,4-DIPHENYLPYRROLE, **III**, 358
1*H*-Pyrrole, 3-ethyl-2,4-dimethyl- [517-22-6]
    KRYPTOPYRROLE, **III**, 513
1*H*-Pyrrole, 2,3,4,5-tetramethyl- [1003-90-3]
    2,3,4.5-TETRAMETHYLPYRROLE, **V**, 1022
1*H*-Pyrrole-2-carboxaldehyde [1003-29-8]
    2-PYRROLEALDEHYDE, **IV**, 831
1*H*-Pyrrole-2-carboxylic acid, ethyl ester [2199-43-1]
    ETHYL PYRROLE-2-CARBOXYLATE, **VI**, 618
1*H*-Pyrrole-2-carboxylic acid, 4-acetyl-3,5-dimethyl-, ethyl ester [2386-26-7]
    2,4-DIMETHYL-3-ACETYL-5-CARBETHOXYPYRROLE, **III**, 513
1*H*-Pyrrole-2-carboxylic acid, 3,5-dimethyl-, ethyl ester [2199-44-2]
    2,4-DIMETHYL-5-CARBETHOXYPYRROLE, **II**, 198
1*H*-Pyrrole-2,4-dicarboxylic acid, 3,5-dimethyl-, diethyl ester [2436-79-5]
    2,4-DIMETHYL-3,5-DICARBETHOXYPYRROLE, **II**, 202
1*H*-Pyrrole-2,5-dione, 1-phenyl- [941-69-5]
    N-PHENYLMALEIMIDE, **V**, 944
1*H*-Pyrrole-3-methanol, 1-methyl-α-propyl- [70702-66-8]
    3-(1-HYDROXYBUTYL)-1-METHYLPYRROLE, **VII**, 102
1-Pyrrolidinamine, *N*-(1-ethyl-2-methylpentylidene)-2-methoxymethyl)-, [*S*-[*R**,*R**-(*Z*)]-
    [69943-24-4]
    4-METHYL-3-HEPTANONE SAMP-HYDRAZONE, (*S*)-(+)-, **VIII**, 403
1-Pyrrolidinamine, *N*-(1-ethylpropylidene)-2-(methoxymethyl)-, (*S*)- [59983-36-7]
    3-PENTANONE SAMP-HYDRAZONE, **VIII**, 403
1-Pyrrolidinamine, 2-(methoxymethyl)-, (*R*)-(+)- [72748-99-3]
    1-AMINO-2-METHOXYMETHYLPYRROLIDINE, (*R*)-(+)- (RAMP), **VIII**, 26
1-Pyrrolidinamine, 2-(methoxymethyl)-, (*S*)-(−)- [59983-30-0]
    1-AMINO-2-METHOXYMETHYLPYRROLIDINE, (*S*)-(−)- (SAMP) **VIII**, 26,
        403
Pyrrolidine, 1-butyl- [767-10-2]
    1-BUTYLPYRROLIDINE, **III**, 159
Pyrrolidine, 1-(1-cyclododecen-1-yl)- [25769-05-5]
    1-(*N*-PYRROLIDINO)-1-CYCLODODECENE, **VII**, 135
Pyrrolidine, 1-(1-cyclohexen-1-yl)- [1125-99-1]
    1-PYRROLIDINOCYCLOHEXENE, **VI**, 1014
Pyrrolidine, 2,2-dimethyl- [35018-15-6]
    2,2-DIMETHYLPYRROLIDINE, **IV**, 354
Pyrrolidine, 2-(methoxymethyl)-, (*S*)-(+)- [63126-47-6]
    (*S*)-(+)-2-METHOXYMETHYLPYRROLIDINE, **VIII**, 26
Pyrrolidine, 1-(2-methyl-1-propenyl)- [2403-57-8]
    1-(2-METHYLPROPENYL)PYRROLIDINE, **VI**, 496
Pyrrolidine, 1-[2-[2-nitro-6-(phenylmethoxy)phenyl]ethenyl]-, (*E*)- [99474-12-1]
    (*E*)-6-BENZYLOXY-2-NITRO-β-PYRROLIDINOSTYRENE, **VII**, 34

1-Pyrrolidineacetic acid, α-methyl-, ethyl ester [26846-86-6]
    ETHYL α-(1-PYRROLIDYL)PROPIONATE, **IV**, 466
1-Pyrrolidinecarboxaldehyde, 2-(hydroxymethyl)-, (S)-(−)- [55456-46-7]
    (S)-(−)-1-FORMYL-2-HYDROXYMETHYLPYRROLIDINE, **VIII**, 26
1-Pyrrolidinecarboxaldehyde, 2-(methoxymethyl)-, (S)-(−)- [63126-45-4]
    (S)-(−)-1-FORMYL-2-METHOXYMETHYLPYRROLIDINE, **VIII**, 26
1-Pyrrolidinecarboxamide, 2-(methoxymethyl)-, (S)- [95312-82-6]
    (S)-1-CARBAMOYL-2-METHOXYMETHYLPYRROLIDINE, **VIII**, 26
1-Pyrrolidinecarboxylic acid, methyl ester [56475-80-0]
    N-CARBOMETHOXYPYRROLIDINE, **VII**, 307
1-Pyrrolidinecarboxylic acid, 2-methoxy-, methyl ester [56475-88-8]
    2-METHOXY-N-CARBOMETHOXYPYRROLIDINE, **VII**, 307
3-Pyrrolidinecarboxylic acid, 2-oxo-1-phenyl- [56137-52-1]
    2-OXO-1-PHENYL-3-PYRROLIDINECARBOXYLIC ACID, **VII**, 411
1,2-Pyrrolidinedicarboxylic acid, 1-(1,1-dimethylethyl) ester, (S)- [15761-39-4]
    N-tert-BUTOXYCARBONYL-L-PROLINE, **VI**, 203
2,5-Pyrrolidinedione [123-56-8]
    SUCCINIMIDE, **II**, 562
1-Pyrrolidineethanol, β-methyl- [53663-19-7]
    2-(1-PYRROLIDYL)PROPANOL, **IV**, 834
2-Pyrrolidinemethanol, (S)-(+)- [23356-96-9]
    (S)-(+)-2-HYDROXYMETHYLPYRROLIDINE, **VIII**, 26
2-Pyrrolidinemethanol, α,α-diphenyl-, (±)- [22348-32-9]
    2-(DIPHENYLHYDROXYMETHYL) PYRROLIDINE **VI**, 542
2-Pyrrolidinethione, 1-methyl- [10441-57-3]
    N-METHYLTHIOPYRROLIDONE, **VII**, 372
2-Pyrrolidinone, 1,5-dimethyl- [5075-92-3]
    1,5-DIMETHYL-2-PYRROLIDONE, **III**, 328
2-Pyrrolidinone, 5,5-dimethyl- [5165-28-6]
    5,5-DIMETHYL-2-PYRROLIDONE, **IV**, 357
2H-Pyrrolo[3,4-d]isoxazole-4,6(3H,5H)-dione, dihydro-2,5-diphenyl-3-propyl-
    [53555-71-8]
    2-PHENYL-3-PROPYLISOXAZOLIDINE-4,5-cis-DICARBOXYLIC ACID
    N-PHENYLIMIDE, **V**, 957
2H-Pyrrol-2-one, 5-ethoxy-3,4-dihydro- [29473-56-1]
    2-ETHOXYPYRROLIN-5-ONE, **VI**, 226
Pyrrolo[3,4-c]pyrrole-1,3(2H,3aH)-dione, tetrahydro-2-phenyl-5-phenylmethyl)-, cis-
    [87813-00-1]
    2,6-DIOXO-1-PHENYL-4-BENZYL-1,4-DIAZABICYCLO[3.3.0]OCTANE,
    VIII, 231
Pyrylium, 2,6-bis(1,1-dimethylethyl)-4-methyl-, salt with trifluoromethanesulfonic acid
    [59643-43-5]
    2,6-DI-tert-BUTYL-4-METHYLPYRYLIUM TRIFLUOROMETHANESUL-
    FONATE, **VII**, 144
Pyrylium, 2,4,6-trimethyl-, perchlorate [940-93-2]
    2,4,6-TRIMETHYLPYRYLIUM PERCHLORATE, **V**, 1106

Pyrylium, 2,4,6-trimethyl-, salt with trifluoromethanesulfonic acid (1:1) [40927-60-4]
  2,4,6-TRIMETHYLPYRYLIUM TRIFLUOROMETHANESULFONATE, **V**, 1114
Pyrylium, 2,4,6-trimethyl-, tetrafluoroborate(1−) [773-01-3]
  2,4,6-TRIMETHYLPYRYLIUM TETRAFLUOROBORATE, **V**, 1112
Pyrylium, 2,4,6-triphenyl-, tetrafluoroborate(1−) [448-61-3]
  2,4,6-TRIPHENYLPYRYLIUM TETRAFLUOROBORATE, **V**, 1135

Q

(*m*-Quaterphenyl)-2'-ol [23886-03-5]
  2,4,6-TRIPHENYLPHENOL, **V**, 1130
2,4(1*H*,3*H*)-Quinazolinedione [86-96-4]
  BENZOYLENE UREA, **II**, 79
Quinoline [91-22-5]
  QUINOLINE, **I**, 478
Quinoline, 2-chloro-4-methyl- [634-47-9]
  2-CHLOROLEPIDINE, **III**, 194
Quinoline, 4,7-dichloro- [86-98-6]
  4,7-DICHLOROQUINOLINE, **III**, 272
Quinoline, 1,2-dihydro-2,2,4-trimethyl- [147-47-7]
  "ACETONE-ANIL"(2,2,4-TRIMETHYL-1,2-DIHYDROQUINOLINE), **III**, 329
Quinoline, 2,4-dimethyl- [1198-37-4]
  2,4-DIMETHYLQUINOLINE, **III**, 329
Quinoline, 6-methoxy-8-nitro- [85-81-4]
  6-METHOXY-8-NITROQUINOLINE, **III**, 568
Quinoline, 4-methyl- [491-35-0]
  LEPIDINE, **III**, 519
3-Quinolinecarboxylic acid, 7-chloro-4-hydroxy- [86-47-5]
  7-CHLORO-4-HYDROXY-3-QUINOLINECARBOXYLIC ACID, **III**, 273
4-Quinolinecarboxylic acid, 1,2-dihydro-2-oxo- [15733-89-8]
  2-HYDROXYCINCHONINIC ACID, **III**, 456
4-Quinolinecarboxylic acid, 3-hydroxy- [118-13-8]
  3-HYDROXYCINCHONINIC ACID, **V**, 636
3-Quinolinol [580-18-7]
  3-HYDROXYQUINOLINE, **V**, 635
4-Quinolinol, 2-methyl- [607-67-0]
  2-METHYL-4-HYDROXYQUINOLINE, **III**, 593
2(1*H*)-Quinolinone, 4-methyl- [607-66-9]
  4-METHYLCARBOSTYRIL, **III**, 580
Quinoxaline [91-19-0]
  QUINOXALINE, **IV**, 824
1,1':4',1":4",1"':4"',1""-Quinquephenyl [3073-05-0]
  p-QUINQUEPHENYL, **V**, 985

R

Ruthenocene [1287-13-4]
  RUTHENOCENE, **V**, 1001

S

Selenium, chlorotriphenyl-, (T-4)- [17166-13-1]
    TRIPHENYLSELONIUM CHLORIDE, **II**, 240
Selenium, dichlorodiphenyl- [2217-81-4]
    DIPHENYLSELENIUM DICHLORIDE, **II**, 240
Selenourea, *N,N*-diethyl- [15909-81-6]
    N,N-DIETHYLSELENOUREA, **IV**, 360
Selenourea, *N,N*-dimethyl- [5117-16-8]
    N,N-DIMETHYLSELENOUREA, **IV**, 359
Semicarbazide, 1-formyl-3-thio- [2302-84-3]
    1-FORMYL-3-THIOSEMICARBAZIDE, **V**, 1070
DL-Serine [302-84-1]
    DL-SERINE, **III**, 774
Silane, acetyltrimethyl- [13411-48-8]
    ACETYLTRIMETHYLSILANE, **VIII**, 19
Silane, azidotrimethyl- [4648-54-8]
    TRIMETHYLSILYL AZIDE, **VI**, 1030
Silane, (bicyclo[4.1.0]hept-1-yloxy)trimethyl- [38858-74-1]
    1-TRIMETHYLSILYLOXYBICYCLO[4.1.0]HEPTANE, **VI**, 328
Silane, (1-bromoethenyl)trimethyl- [13683-41-5]
    (1-BROMOETHENYL)TRIMETHYLSILANE, **VI**, 1034
Silane, 1,3-butadiyne-1,4-diylbis-[trimethyl- [4526-07-2]
    1,4-BIS(TRIMETHYLSILYL)BUTA-1,3-DIYNE, **VIII**, 63
Silane, (chloromethyl)isopropoxydimethyl- [18171-11-4]
    (ISOPROPOXYDIMETHYLSILYL)METHYL CHLORIDE, **VIII**, 317
Silane, (1-cyclobuten-1,2-ylenedioxy)bis[trimethyl- [17082-61-0]
    1,2-BIS(TRIMETHYLSILYLOXY)CYCLOBUTENE, **VI**, 167
Silane, (1-cyclohexen-1-yloxy)trimethyl- [6651-36-1]
    1-TRIMETHYLSILOXYCYCLOHEXENE, **VII**, 414; **VIII**, 460
Silane, (1-cyclopenten-1-yloxy)trimethyl- [19980-43-9]
    1-TRIMETHYLSILOXYCYCLOPENTENE, **VII**, 424
Silane, (diazomethyl)trimethyl- [18107-18-1]
    TRIMETHYLSILYLDIAZOMETHANE, **VIII**, 612
Silane, [(1-ethoxycyclopropyl)oxy]trimethyl- [27374-25-0]
    1-ETHOXY-1-TRIMETHYLSILOXYCYCLOPROPANE, **VII**, 131
Silane, [(1-ethyl-1-propenyl)oxy]trimethyl-, (Z)- [51425-54-8]
    3-TRIMETHYLSILOXY-2-PENTENE, (Z)-, **VII**, 512
Silane, ethynyltrimethyl- [1066-54-2]
    TRIMETHYLSILYLACETYLENE, **VIII**, 63, 606
Silane, (3-iodopropyl)trimethyl- [18135-48-3]
    1-IODO-3-TRIMETHYLSILYLPROPANE, **VIII**, 486
Silane, iodotrimethyl- [16029-98-4]
    IODOTRIMETHYLSILANE, **VI**, 353
Silane, [(2-isocyanocyclohexyl)oxy]trimethyl-, *trans*- [83152-87-8]
    [(*trans*-2-ISOCYANOCYCLOHEXYL)OXY]TRIMETHYLSILANE, **VII**, 294

Silane, [(3-methoxyestra-1,3,5(10),16-tetraen-17-yl)oxy]trimethyl- [115419-13-1]
  3-METHOXY-17-TRIMETHYLSILOXY-1,3,5(10)-ESTRATETRAENE, **VIII**, 286
Silane, (1-methoxyethenyl)trimethyl- [79678-01-6]
  1-(METHOXYVINYL)TRIMETHYLSILANE, **VIII**, 19
Silane, [(3-methoxy-1-methylene-2-propenyl)oxy]trimethyl- [59414-23-1]
  1-METHOXY-3-TRIMETHYLSILOXY-1,3-BUTADIENE, *trans-*, **VII**, 312
Silane, trimethyl(2-methylene-4-phenyl-3-butenyl)- [80814-92-2]
  TRIMETHYL (2-METHYLENE-4-PHENYL-3-BUTENYL)SILANE, **VIII**, 602
Silane, trimethyl[(1-methylene-2-propenyl)oxy]- [38053-91-7]
  2-TRIMETHYLSILYLOXY-1,3-BUTADIENE, **VI**, 445
Silane, trimethyl[(1-methylethenyl)oxy]- [1833-53-0]
  ACETONE TRIMETHYLSILYL ENOL ETHER, **VIII**, 1
Silane, trimethyl[[(4-methylphenyl)sulfonyl]ethynyl]- [34452-56-7]
  p-TOLYL-2-(TRIMETHYLSILYL)ETHYNYL SULFONE, **VIII**, 281
Silane, trimethyl(1-methyl-1,2-propadienyl)- [74542-82-8]
  1-METHYL-1-(TRIMETHYLSILYL)ALLENE, **VIII**, 471
Silane, trimethyl(1-oxo-2-propenyl)- [51023-60-0]
  (1-OXOPROPENYL)TRIMETHYLSILANE, **VIII**, 501
Silane, trimethyl[(1-phenylethenyl)oxy]- [13735-81-4]
  ACETOPHENONE TRIMETHYLSILYL ENOL ETHER, **VIII**, 324
Silane, trimethyl[(3,3,5-trimethyl-1,5-cyclohexadien-1-yl)oxy]- [54781-28-1]
  4,6,6-TRIMETHYL-2-TRIMETHYLSILOXYHEXA-1,3-DIENE, **VII**, 282
Silanecarbonitrile, trimethyl- [7677-24-9]
  TRIMETHYLSILYL CYANIDE, **VII**, 20, 294, 517
Spiro[4.5]decane-1,4-dione [39984-92-4]
  SPIRO[4.5]DECANE-1,4-DIONE, **VIII**, 578
Spiro[5,7]trideca-1,4-dien-3-one [41138-71-0]
  SPIRO[5,7]TRIDECA-1,4-DIEN-3-ONE, **VII**, 473
Spiro[5.7]tridec-1-en-3-one [60033-39-8]
  SPIRO[5.7]TRIDEC-1-EN-3-ONE, **VII**, 473
Stannane, dibutyldiethenyl- [7330-43-0]
  DIBUTYLDIVINYLTIN, **IV**, 258
Stannane, 1,2-ethenediylbis[dibutyl-, (*E*)- [14275-61-7]
  (*E*)-1,2-BIS(TRIBUTYLSTANNYL)ETHYLENE, **VIII**, 268
Stannane, ethenyltrimethyl- [754-06-3]
  TRIMETHYLVINYLTIN, **VIII**, 97
Stannane, tetraethyl- [597-64-8]
  TETRAETHYLTIN, **IV**, 881
Stannane, tributylethynyl- [994-89-8]
  TRIBUTYLETHYNYLSTANNANE, **VIII**, 268
Stannane, tributyl-2-propenyl- [24850-33-7]
  ALLYLTRIBUTYLTIN, **VIII**, 23
Stibine, triphenyl- [603-36-1]
  TRIPHENYLSTIBINE, **I**, 550
Sulfamide, *N,N'*-diethyl- [6104-21-8]
  N,N'-DIETHYLSULFAMIDE, **VI**, 78

Sulfide, benzyl 2,4-dinitrophenyl- [7343-61-5]
    2,4-DINITROPHENYL BENZYL SULFIDE, **V**, 474
Sulfiliminium, *S,S*-bis(diethylamino)-*N,N*-diethyl-, difluorotrimethylsilicate (1–)
    [59201-86-4]
    TRIS(DIETHYLAMINO)SULFONIUM DIFLUOROTRIMETHYLSILICATE,
    VIII, 329
Sulfiliminium, *S,S*-bis(dimethylamino)-*N,N* dimethyl-, difluorotrimethylsilicate (1–)
    [59218-87-0]
    TRIS(DIMETHYLAMINO)SULFONIUM DIFLUOROTRIMETHYLSILICATE,
    VII, 528
Sulfonium, cyclopropyldiphenyl-, tetrafluoroborate (1–) [33462-81-6]
    CYCLOPROPYLDIPHENYLSULFONIUM TETRAFLUOROBORATE, **VI**, 364
Sulfonium, dimethyl-2-propynyl-, bromide [23451-62-9]
    DIMETHYL-2-PROPYNYLSULFONIUM BROMIDE, **VI**, 31
Sulfur, bis[α,α-bis(trifluoromethyl)benzenemethanolato]diphenyl- [32133-82-7]
    BIS[2,2,2-TRIFLUORO-1-PHENYL-1-(TRIFLUOROMETHYL)ETHOXY]
    DIPHENYLSULFURANE, **VI**, 163
Sulfur, (diethylaminato)trifluoro- [38078-09-0]
    DIETHYLAMINOSULFUR TRIFLUORIDE, **VI**, 440
Sulfur, trifluorophenyl- [672-36-6]
    PHENYLSULFUR TRIFLUORIDE, **V**, 959
Sulfur (1+), tris(*N*-ethylethanaminato)-, difluorotrimethylsilicate (1–) [59201-86-4]
    TRIS(DIETHYLAMINO)SULFONIUM DIFLUOROTRIMETHYLSILICATE,
    VIII, 329
Sulfur(1+), tris(*N*-methylmethanaminato)-, difluorotrimethylsilicate(1–) [59218-87-0]
    TRIS(DIMETHYLAMINO)SULFONIUM DIFLUOROTRIMETHYLSILICATE
    VII, 528
Sulfur diimide, dicarboxy-, dimethyl ester [16762-82-6]
    N,N-BIS(METHOXYCARBONYL)SULFUR DIIMIDE, **VIII**, 427
Sulfuric acid, dibutyl ester [625-22-9]
    BUTYL SULFATE, **II**, 111
Sulfurous acid, dibutyl ester [626-85-7]
    BUTYL SULFITE, **II**, 112
Sulfuryl chloride isocyanate [1189-71-5]
    CHLOROSULFONYL ISOCYANATE, **V**, 226
Sydnone, 3-phenyl- [120-06-9]
    3-PHENYLSYDNONE, **V**, 962

T

Tartaric acid, (*R,R*)-(+)-, diethyl ester [608-84-4]
    DIETHYL (2*R*,3*R*)-(+)-TARTRATE, **VII**, 41, 461
Tartramide, *N,N,N',N'*-tetramethyl-, (+)- [26549-65-5]
    (*R,R*)-(+)-*N,N,N',N'*-TETRAMETHYLTARTARIC ACID DIAMIDE, **VII**, 41
1,1':3',1"-Terphenyl, 2'-nitro-5'-phenyl- [10368-47-5]
    2,4,6-TRIPHENYLNITROBENZENE, **V**, 1128
1,1':2',1"-Terphenyl, 3',4',5',6'-tetraphenyl- [992-04-1]
    HEXAPHENYLBENZENE, **V**, 604

[1,1':3',1"-Terphenyl]-2'-amine, 5'-phenyl- [6864-20-6]
  2,4,6-TRIPHENYLANILINE, **V**, 1130
2,2':6',2"-Terpyridine [1148-79-4]
  2,2':6',2"-TERPYRIDINE, **VII**, 476
2,2':6',2"-Terpyridine, 4'-(methylthio) [78570-35-1]
  4'-(METHYLTHIO)-2,2':6',2"-TERPYRIDINE, **VII**, 476
Tetracyclo[3.2.0.0^{2,7}.0^{4,6}]heptane [278-06-8]
  QUADRICYCLANE, **VI**, 962
Tetradecanoic acid [544-63-8]
  MYRISTIC ACID, **I**, 379; **III**, 605
Tetradecanoic acid, methyl ester [124-10-7]
  METHYL MYRISTATE, **III**, 605
Tetradecanoic acid, 1,2,3-propanetriyl ester [555-45-3]
  TRIMYRISTIN, **I**, 538
1,4,10,13-Tetraoxa-7,16-diazacyclooctadecane [23978-55-4]
  4,13-DIAZA-18-CROWN-6, **VIII**, 152
1,4,10,13-Tetraoxa-7,16-diazacyclooctadecane, 7,16-bis(phenylmethyl)- [69703-25-9]
  N,N'-DIBENZYL-4,13-DIAZA-18-CROWN-6, **VIII**, 152
21,22,23,24-Tetraoxapentacyclo[16.2.1.1^{3,6}.1^{8,11}.1^{3,16}]tetracosane-, 2,2,7,7,12,12,17,17-
    octamethyl- [22900-44-3]
  2,2,7,7,12,12,17,17-OCTAMETHYL-21,22,23,24- TETRAOXAPERHYDRO-
    QUATERENE, 856, **VI**
2,5,7,10-Tetraoxaundecane, 6-methylene- [5130-02-9]
  KETENE DI(2-METHOXYETHYL)ACETAL, **V**, 684
1,4,8,11-Tetrathiacyclotetradecane [24194-61-4]
  1,4,8,11-TETRATHIACYCLOTETRADECANE, **VIII**, 592
1H-Tetrazol-5-amine, N-(phenylmethyl)- [14832-58-7]
  5-BENZYLAMINOTETRAZOLE, **VII**, 27
1H-Tetrazole, 5-[(1,1'-biphenyl)-4-yloxy]-1-phenyl- [17743-27-0]
  4-(1-PHENYL-5-TETRAZOLYLOXY)BIPHENYL, **VI**, 150
7-Thiabicyclo[4.1.0]heptane [286-28-2]
  CYCLOHEXENE SULFIDE, **IV**, 232
4H-1,4-Thiazine, 1,1-dioxide [40263-61-4]
  4H-1,4-THIAZINE 1,1-DIOXIDE, **VI**, 976
2-Thiazolamine, 4-methyl- [1603-91-4]
  2-AMINO-4-METHYLTHIAZOLE, **II**, 31
Thiazole, 4,5-dihydro-2-[[1-(phenylmethyl)-2-propenyl]thio]- [52534-82-4]
  4,5-DIHYDRO-2-[(1-PHENYLMETHYL-2-PROPENYL)THIO]THIAZOLE,
    VI, 705
Thiazole, 4,5-dihydro-2-(2-propenylthio)- [3571-74-2]
  4,5-DIHYDRO-2-(2-PROPENYLTHIO)THIAZOLE, **VI**, 705
Thiazole, 2,4-dimethyl- [541-58-2]
  2,4-DIMETHYLTHIAZOLE, **III**, 332
4-Thiazolecarboxylic acid, ethyl ester [14527-43-6]
  ETHYL THIAZOLE-4-CARBOXYLATE, **VI**, 620
4-Thiazolidinone, 2-thioxo- [141-84-4]
  RHODANINE, **III**, 763

4(5*H*)-Thiazolone, 2-amino- [556-90-1]
  PSEUDOTHIOHYDANTOIN, **III**, 751
Thietane, 3-chloro-, 1,1-dioxide [15953-83-0]
  3-CHLOROTHIETANE-1,1-DIOXIDE, **VII**, 491
Thietane, 3,3-dichloro-, 1,1-dioxide [90344-85-7]
  3,3-DICHLOROTHIETANE-1,1-DIOXIDE, **VII**, 491
Thietane, 1,1-dioxide [5687-92-3]
  THIETANE 1,1-DIOXIDE, **VII**, 491
Thiete, 3-chloro-, 1,1-dioxide [90344-86-8]
  3-CHLOROTHIETE 1,1-DIOXIDE, **VII**, 491
Thiete, 1,1-dioxide [7285-32-7]
  THIETE 1,1-DIOXIDE, **VII**, 491
Thiirane [420-12-2]
  ETHYLENE SULFIDE, **V**, 562
Thiirane, chloro-, 1,1-dioxide [10038-13-8]
  2-CHLOROTHIIRANE 1,1-DIOXIDE, **V**, 231
Thiirene, diphenyl-, 1,1-dioxide [5162-99-2]
  2,3-DIPHENYLVINYLENE SULFONE, **VI**, 555
Thiocyanic acid, 4-(dimethylamino)phenyl ester [7152-80-9]
  p-THIOCYANODIMETHYLANILINE, **II**, 574
Thiocyanic acid, 1-methylethyl ester [625-59-2]
  ISOPROPYL THIOCYANATE, **II**, 366
Thioimidodicarbonic diamide[(H$_2$C(O)NHC(S)(NH$_2$)], *N'*-phenyl- [53555-72-9]
  1-PHENYL-2-THIOBIURET, **V**, 966
Thiophene [110-02-1]
  THIOPHENE, **II**, 578
Thiophene, 3-bromo- [872-31-1]
  3-BROMOTHIOPHENE, **V**, 149
Thiophene, 3-(bromomethyl)- [34846-44-1]
  3-THENYL BROMIDE, **IV**, 921
Thiophene, 2-(chloromethyl)- [765-50-4]
  2-CHLOROMETHYLTHIOPHENE, **III**, 197
Thiophene, 2-(1,1-dimethylethoxy)- [23290-55-3]
  2-*tert*-BUTOXYTHIOPHENE, **V**, 642
Thiophene, 2,5-dihydro-3-methyl-, 1,1-dioxide [1193-10-8]
  ISOPRENE CYCLIC SULFONE, **III**, 499
Thiophene, 2-ethenyl- [1918-82-7]
  2-VINYLTHIOPHENE, **IV**, 980
Thiophene, 2-iodo- [3437-95-4]
  2-IODOTHIOPHENE, **III**, 357; **IV**, 545
Thiophene, 3-methyl- [616-44-4]
  3-METHYLTHIOPHENE, **IV**, 671
Thiophene, 2-(methylthio)- [5780-36-9]
  METHYL 2-THIENYL SULFIDE, **IV**, 667
Thiophene, 2-nitro- [609-40-5]
  2-NITROTHIOPHENE, **II**, 466

Thiophene-2-ol [17236-58-7]
2-HYDROXYTHIOPHENE, **V**, 642
Thiophene, tetrahydro- [110-01-0]
TETRAHYDROTHIOPHENE, **IV**, 892
Thiophene, 2,2'-thiobis- [3988-99-6]
2,2'-DITHIENYL SULFIDE, **VI**, 558
2-Thiophenecarboxaldehyde [98-03-3]
2-THENALDEHYDE; 2-THIOPHENEALDEHYDE, **III**, 811; **IV**, 915
3-Thiophenecarboxaldehyde [498-62-4]
3-THENALDEHYDE, **IV**, 918
3-Thiophenecarboxylic acid [88-13-1]
3-THENOIC ACID, **IV**, 919
3,4-Thiophenedicarbonitrile, 2,5-diamino- [17989-89-8]
2,5-DIAMINO-3,4-DICYANOTHIOPHENE, **IV**, 243
2-Thiophenethiol [7774-74-5]
2-THIOPHENETHIOL, **VI**, 979
Thiourea, (aminoiminomethyl)- [2114-02-5]
GUANYLTHIOUREA, **IV**, 502
Thiourea, (2-chlorophenyl)- [5344-82-1]
o-CHLOROPHENYLTHIOUREA, **IV**, 180
Thiourea, (2,6-dimethylphenyl)- [6396-76-5]
2,6-DIMETHYLPHENYLTHIOUREA, **V**, 802
Thiourea, methyl- [598-52-7]
METHYLTHIOUREA, **III**, 617
Thiourea, *N*-methyl-*N*-1-naphthalenyl- [53663-34-6]
1-METHYL-1-(1-NAPHTHYL)-2-THIOUREA, **III**, 609
Thiourea, (4-methylphenyl)- [622-52-6]
p-TOLYLTHIOUREA, **III**, 77
Thiourea, phenyl- [103-85-5]
α-PHENYLTHIOUREA, **III**, 735
DL-Threonine [80-68-2]
DL-THREONINE, **III**, 813
Titanium, methyltris(2-propanolato)-, (T-4)- [18006-13-8]
METHYLTRIISOPROPOXYTITANIUM, **VIII**, 495
p-Toluenesulfonic acid, cholest-4-en-3-ylidenehydrazide [21301-41-7]
CHOLEST-4-EN-3-ONE *p*-TOLUENESULFONYLHYDRAZONE, **VI**, 293
1-Triazene, 3,3-dimethyl-1-(3-nitrophenyl)- [20241-06-9]
1-(*m*-NITROPHENYL)-3,3-DIMETHYLTRIAZENE, **IV**, 718
1-Triazene, 1,3-diphenyl [136-35-6]
DIAZOAMINOBENZENE, **II**, 163
1-Triazene, 1-methyl-3-(4-methylphenyl)- [21124-13-0]
1-METHYL-3-*p*-TOLYLTRIAZENE, **V**, 797
1,3,5-Triazine-2,4-diamine, 6-(chloromethyl-*N*-phenyl)- [30355-60-3]
2-AMINO-4-ANILINO-6-(CHLOROMETHYL)-*s*-TRIAZINE, **IV**, 29
1,3,5-Triazine-2,4-diamine, 6-phenyl- [91-76-9]
BENZOGUANAMINE, **IV**, 78

1,3,5-Triazine, hexahydro-1,3,5-tris(1-oxopropyl)- [30805-19-7]
    HEXAHYDRO-1,3,5-TRIPROPIONYL-*s*-TRIAZINE, **IV**, 518
1,2,4-Triazine-3,5,6-tricarboxylic acid, triethyl ester [74476-38-3]
    TRIETHYL 1,2,4-TRIAZINE-3,5,6-TRICARBOXYLATE, **VIII**, 597
1*H*-1,2,3-Triazol-5-amine, 1,4-diphenyl- [29704-63-0]
    1,4-DIPHENYL-5-AMINO-1,2,3-TRIAZOLE, **IV**, 380
1*H*-1,2,3-Triazol-4-amine, *N*,5-diphenyl- [53684-55-2]
    4-PHENYL-5-ANILINO-1,2,3-TRIAZOLE, **IV**, 380
1*H*-1,2,4-Triazol-3-amine [61-82-5]
    3-AMINO-1*H*-1,2,4-TRIAZOLE, **III**, 95
4*H*-1,2,4-Triazol-4-amine [584-13-4]
    4-AMINO-4*H*-1,2,4-TRIAZOLE, **III**, 96
1*H*-1,2,4-Triazole [288-88-0]
    1,2,4-TRIAZOLE, **V**, 1070
3*H*-1,2,4-Triazole-3,5(4*H*)-dione, 4-phenyl- [4233-33-4]
    4-PHENYL-1,2,4-TRIAZOLINE-3,5-DIONE, **VI**, 936
3*H*-1,2,4-Triazole-3-thione [36143-39-2]
    1,2,4-TRIAZOLE-3(5)-THIOL, **V**, 1070
Tricarbonic acid, bis(1,1-dimethylethyl)ester [24424-95-1]
    DI-*tert*-BUTYL TRICARBONATE, **VI**, 418
12-Tricosanone [540-09-0]
    LAURONE, **IV**, 560
Tricyclo[2.2.1.0^{2,6}]heptan-3-ol [695-04-5]
    NORTRICYCLANOL, **V**, 863
Tricyclo[2.2.1.0^{2,6}]heptan-3-ol, acetate [6555-48-2]
    NORTRICYCLYL ACETATE, **V**, 863
Tricyclo[2.2.1.0^{2,6}]heptanone [695-05-6]
    NORTRICYCLANONE, **V**, 866
Tricyclo[3.2.1.0^{2,7}]-octane-6,8-dione, 2-methyl- [119986-99-1]
    4-METHYLTRICYCLO[2.2.2.0^{3,5}]OCTANE-2,6-DIONE, **VIII**, 467
Tricyclo[3.3.1.1^{3,7}]decan-2-amine, 2-phenyl-, hydrochloride [73032-81-2]
    2-PHENYL-2-ADAMANTANAMINE HYDROCHLORIDE, **VII**, 433
Tricyclo[3.3.1.1^{3,7}]decane [281-23-2]
    ADAMANTANE, **V**, 16
Tricyclo[3.3.1.1^{3,7}]decane, 2-azido-2-phenyl- [65218-96-4]
    2-AZIDO-2-PHENYLADAMANTANE, **VII**, 433
Tricyclo[3.3.1.1^{3,7}]decane, 1-fluoro- [768-92-3]
    1-FLUOROADAMANTANE, **VI**, 628
Tricyclo[3.3.1.1^{3,7}]decane, tricyclo[3.3.1.1^{3,7}]decylidene- [30541-56-1]
    ADAMANTYLIDENEADAMANTANE, **VII**, 1
Tricyclo[3.3.1.1^{3,7}]decane-2-carbonitrile [35856-00-9]
    2-ADAMANTANECARBONITRILE, **VI**, 41
Tricyclo[3.3.1.1^{3,7}]decane-1-carboxylic acid [828-51-3]
    1-ADAMANTANECARBOXYLIC ACID, **V**, 20
Tricyclo[3.3.1.1^{3,7}]decan-1-ol [768-95-6]
    1-ADAMANTANOL, **VI**, 958

Tricyclo[3.3.1.1^{3,7}]decan-2-ol, 2-phenyl- [29480-18-0]
  2-PHENYL-2-ADAMANTANOL, **VII**, 433
Tricyclo[3.3.1.1^{3,7}]decanone [700-58-3]
  ADAMANTANONE, **VI**, 48
Tricyclo[4.1.0.0^{2,7}]hept-3-ene, 1,6-dimethyl- [61772-32-5]
  1,6-DIMETHYLTRICYCLO[4.1.0.0^{2,7}]HEPT-3-ENE, **VII**, 200
endo-Tricyclo[4.4.0.0^{2,5}]deca-3,8-diene-7,10-dione [34231-42-0]
  endo-TRICYCLO[4.4.0.0^{2,5}]DECA-3,8-DIENE-7,10-DIONE, **VI**, 1002
Tricyclo[4.4.1.0^{1,6}]undeca-3,8-diene [27714-83-6]
  TRICYCLO[4.4.1.0^{1,6}]UNDECA-3,8-DIENE, **VI**, 731
Tricyclo[4.4.1.0^{1,6}]undeca-3,8-diene, 11,11-dichloro- [39623-22-8]
  11,11-DICHLOROTRICYCLO[4.4.1.0^{1,6}]UNDECA-3,8-DIENE, **VI**, 731
Tricyclo[8.2.2.2^{4,7}]hexadeca-4,6,10,12,13,15-hexaene [1633-22-3]
  [2,2]PARACYCLOPHANE, **V**, 883
Tricyclo[8.2.2.2^{4,7}]hexadeca-4,6,10,12,13,15-hexaene-5,6,11,12-tetracarboxylic acid,
    tetramethyl ester [37437-90-4]
  TETRAMETHYL [2.2]PARACYCLOPHANE-4,5,12,13-TETRACARBOXY-
    LATE, **VII**, 485
Tridecanoic acid, ethyl ester [28267-29-0]
  ETHYL TRIDECYLATE, **II**, 292
Triphenylene [217-59-4]
  TRIPHENYLENE, **V**, 1120
1,3,5-Trithiane [291-21-4]
  sym-TRITHIANE, **II**, 610
1,3,5-Trithiane, 2-tetradecyl- [24644-07-3]
  2-TETRADECYL-*sym*-TRITHIANE, **VI**, 869
L-Tryptophan [73-22-3]
  L-TRYPTOPHAN, **II**, 612
L-Tyrosine [60-18-4]
  L-TYROSINE, **II**, 612
DL-Tyrosine, 3-hydroxy-*N*-methyl- [53663-27-7]
  N-METHYL-3,4-DIHYDROXYPHENYLALANINE, **III**, 586
L-Tyrosine, *N*-(trifluoroacetyl)- [350-10-7]
  N-TRIFLUOROACETYL-L-TYROSINE, **VI**, 1004

U

1,4-Undecadiene, 5-ethyl-, (*E*)- [107445-19-2]
  5-ETHYL-1,4-UNDECADIENE, (*E*)-, **VII**, 236
5,9-Undecadien-1-yne, 6,10-dimethyl-, (*E*)- [22850-55-1]
  (*E*)-6,10-DIMETHYL-5,9-UNDECADIEN-1-YNE, **VIII**, 226
Undecane, 1-isocyanato- [2411-58-7]
  UNDECYL ISOCYANATE, **III**, 846
Undecanedioic acid [1852-04-6]
  HENDECANEDIOIC ACID, **IV**, 510
Undecanedioic acid, monomethyl ester [3927-60-4]
  METHYL HYDROGEN HENDECANEDIOATE, **IV**, 635

Undecanedioic acid, 6-oxo- [3242-53-3]
    6-KETOHENDECANEDIOIC ACID, **IV**, 555
2,5-Undecanedione [7018-92-0]
    2,5-UNDECANEDIONE, **VIII**, 620
3,5,9-Undecatrien-2-one, 6,10-dimethyl- [141-10-6]
    PSEUDOIONONE, **III**, 747
2-Undecene, 2-methyl- [56888-88-1]
    2-METHYL-2-UNDECENE, **VIII**, 474
4-Undecenoic acid, 2-acetyl- [133538-61-1]
    3-CARBOXY-5-DODECEN-2-ONE, (*E*)-, **VIII**, 235
10-Undecynoic acid [2777-65-3]
    10-UNDECYNOIC ACID, **IV**, 969
Urea, (4-bromophenyl)- [1967-25-5]
    p-BROMOPHENYLUREA, **IV**, 49
Urea, *N*-cyano-*N'*-phenyl- [41834-91-7]
    1-CYANO-3-PHENYLUREA, **IV**, 213
Urea, cyclohexyl- [698-90-8]
    CYCLOHEXYLUREA, **V**, 801
Urea, *N,N*-dimethyl- [598-94-7]
    asym-DIMETHYLUREA, **IV**, 361
Urea, (1,1-dimethylethyl)- [1118-12-3]
    tert-BUTYLUREA, **III**, 256
Urea, (2,5-dioxo-4-imidazolidinyl)- [97-59-6]
    ALLANTOIN, **II**, 21
Urea, *N,N'*-diphenyl- [102-07-8]
    sym-DIPHENYLUREA, **I**, 453
Urea, (4-ethoxyphenyl)- [150-69-6]
    p-ETHOXYPHENYLUREA, **IV**, 52
Urea, heptyl- [42955-46-4]
    N-HEPTYLUREA, **IV**, 515
Urea, hydroxy- [127-07-1]
    HYDROXYUREA, **V**, 645
Urea, *N*-methyl-*N*-nitroso- [684-93-5]
    NITROSOMETHYLUREA, **II**, 461
Urea, nitro- [556-89-8]
    NITROUREA, **I**, 417
Urea, phenyl- [64-10-8]
    PHENYLUREA, **I**, 453

V

Valeric acid, 4-methyl-4-nitro- [32827-16-0]
    4-METHYL-4-NITROVALERIC ACID, **V**, 445
DL-Valine [516-06-3]
    DL-VALINE, **III**, 848
L-Valine, *N*-[*N*-[*N*-[(phenylmethoxy)carbonyl]-L-aspartyl]-L-phenylalanyl]-, 4-(1,1-
    dimethylethyl) 1-methyl ester [57850-41-6]

BENZYLOXYCARBONYL-L-ASPARTYL-(*tert*-BUTYL ESTER)-
L-PHENYLALANYL-L-VALINE METHYL ESTER, **VII**, 30

X

9*H*-Xanthen-9-ol [90-46-0]
 XANTHYDROL, **I**, 554
9*H*-Xanthen-9-one [90-47-1]
 XANTHONE, **I**, 552

Z

Zinc, diethyl- [557-20-0]
 DIETHYL ZINC, **II**, 184

SOLVENTS AND REAGENTS

Purification and/or Preparation

Organic Syntheses procedures frequently include notes describing the purification of solvents and reagents, and assay methods for both reagents and products. The preparation of starting materials is sometimes included in abbreviated form for simple transformations (e.g., esterification of an acid), as well as some techniques and qualitative tests. The information in these notes has been collected in this index. The CAS registry numbers have been included for compounds whose preparation has been described. In addition, many procedures, particularly in earlier volumes, described the preparation of materials that were employed as reagents rather than end products (e.g., diazomethane). Such preparations are found in this index rather than in the larger index, which is devoted to intermediates and final products. However, the distinction between these indices is an arbitrary choice of the editor, so both should be consulted in individual cases.

If any of the compounds in this index are listed in the *Toxic Substances List 1987*, they will have the word *Hazard* appended. Again, this listing is only to alert the user to potential problems in the handling and disposal of such compounds; details should be sought from appropriate OSHA bulletins.

At the present time many pure anhydrous solvents are available commercially. An increasing number of reagents are available as ACS Analytical Reagent Grades or as spectroscopically pure compounds. A great saving in time and effort can be made by use of these purified products, which give good results when applied according to *Organic Syntheses* procedures. However, care must be taken to use fresh, full containers of these compounds in those preparations specifying high purity.

A note of caution. Certain compounds, such as ethyl ether, dioxane, tetrahydrofuran, aldehydes, and unsaturated compounds, form explosive peroxides by reaction with oxygen. Partially filled containers of such compounds form peroxides when kept for a considerable amount of time. These peroxides should be destroyed by use of suitable reducing agents *before* using the purification procedures suggested.

Valuable criteria of purity, particularly of enantiomeric (ee) and diastereomeric (de) purity, are now available due to the development of chromatographic and spectroscopic (nuclear magnetic resonance, NMR) techniques. In particular, *Organic Syntheses* now regards such information as paramount, and while optical rotation data are still included, they are not regarded as sufficient for establishment of purity.

Acetic acid, drying, **VI**, 349
  purification of, **III**, 5; **IV**, 712
Acetic acid-2-^{13}C [1563-80-0], drying of,
  VII, 389
Acetic anhydride, *Hazard*, **VII**, 170
Acetone, *Hazard*, **I**, 239, 460; **III**, 17,
  208, 735, 830; **IV**, 211, 577,
  837; **VII**, 206, 381
Acetonitrile [75-05-8], *Hazard*, **I**, 6; **III**,
  72; **V**, 143, 250; **VI**, 514, 779,
  1015; **VII**, 87, 96, 108, 126,
  471, 522; **VIII**, 2, 7, 60, 232,
  482, 500, 618
Acetophenone, *Hazard*, **I**, 79
Acetophenone phenylhydrazone [583-11-
  9], preparation of, **III**, 726
4-Acetoxybenzoic acid [2345-34-8],
  preparation of, **VI**, 577
Acetylacetone, *Hazard*, **VI**, 37
Acetylene, *Hazard*, **IV**, 117, 187, 794;
  VI, 941; **VII**, 293; **VIII**, 607
N-Acetylhomoveratrylamine [6275-29-2],
  preparation of, **VI**, 2
2-Acetyl-2-phenylselenocyclohexanone
  [57205-12-6], *Hazard*,
  preparation of, **VI**, 23
Acid number, **IV**, 3
Active magnesium, **VI**, 846, 849
Adipyl chloride [111-50-2], *Hazard*,
  preparation of, **II**, 169; **III**,
  623
Adipyl hydrazide [1071-93-8],
  preparation of, **IV**, 819
β-Alanine, **IV**, 94
  hydrochloride, preparation of,
  VI, 717
Aldehydes, stabilization against
  autoxidation, **IV**, 447
Allenylmagnesium bromide, preparation
  of, **VII**, 485
Alizarin indicator, **IV**, 416
Allyl alcohol [107-18-6], *Hazard*, **VIII**, 502
  analysis of, **I**, 44
Allylamine, *Hazard*, nasal irritant, **II**, 25
Allyl bromide, *Hazard*, **VIII**, 244, 455
Allyllithium, preparation of, **V**, 452

Allylmagnesium bromide, preparation of,
  IV, 749; **VIII**, 23
Allylmagnesium chloride, preparation of,
  IV, 751
Allyltriphenyltin [76-63-1], *Hazard*,
  preparation of, **V**, 453
Alumina, activation of, **IV**, 520, 796
  chromatographic, activation of, **IV**,
  965
  deactivated, **VI**, 73
  preparation for chromatography, **IV**,
  818
Aluminum amalgam, **II**, 233; **III**, 49
Aluminum *tert*-butoxide, preparation of,
  III, 48
Aluminum ethoxide, preparation of, **II**,
  599
Aluminum isopropoxide, preparation of,
  IV, 193
Aluminum trichloride, *Hazard*, **VII**, 256
1-Amino-8-naphthol-2,4-disulfonic acid
  (Chicago acid), technical [82-
  47-3], **II**, 147
5-Aminotetrazole monohydrate [15454-
  54-3], **VII**, 27
Ammonia, handling and drying, **III**, 517;
  VI, 52, 998
Ammonium dithiocarbamate, preparation
  of, **III**, 763
Ammonium formate, **II**, 505
Aniline, *Hazard*, **V**, 769
Anisole, *Hazard*, **VI**, 998
Anthracene, *Hazard*, purification of, **IV**,
  9, 712
Antifoaming agent, **IV**, 138, 685, 856
Aqueous 50% sodium hydroxide, **VII**,
  412
Assay methods
  for "active bromine", **VI**, 176
  for arsonoacetic acid, **I**, 74
  for benzaldehyde, **VII**, 10
  for benzohydroxamic acid, **II**, 68
  for 1,2,3-benzothiadiazole 1,1-
  dioxide, **V**, 64
  for borane in tetrahydrofuran, **VI**,
  921

for *N*-bromoacetamide, **IV**, 105

for *tert*-butyl hydroperoxide [75-91-2], in toluene by NMR analysis, **VII**, 462

for *tert*-butyl hypochlorite, **V**, 185

for butyllithium, **VI**, 979; **VII**, 227, 284; **VIII**, 341
  with 2-butanol, menthol or 2-pentanol, 1,10-phenanthroline indicator, **VI**, 112, 138, 249, 479, 570, 687; **VII**, 389; **VIII**, 34, 143, 228, 379, 400, 431, 442, 455, 475, 502, or 2,2'-biquinoline indicator, **VII**, 174
  with 2,5-dimethoxybenzyl alcohol, **VIII**, 194, 392, 563
  with diphenylacetic acid, **VII**, 68, 155, 213, 389, 449, 497; **VIII**, 220, 238, 344, 405, 589
  with 1,3-diphenyl-2-propanone tosylhydrazone, **VIII**, 431, 502

for *sec*-butyllithium with diphenylacetic acid or by double titration with benzoic acid and allyl chloride, **VII**, 78, 268

for *tert*-butyllithium by double titration, **VI**, 250; **VIII**, 329, or tosylhydrazone method, **VIII**, 21
  with 2-butanol, 2,2'-bipydridyl indicator, **VI**, 250

for calcium cyanamide, **IV**, 646

for calcium hypochlorite, **VI**, 969

for chloride ion in methyllithium, **VII**, 349

for carbodiimide content of carbodiimide resin, **VI**, 953

for chlorine in chloromethylated polystyrene by potentiometric titration, **VI**, 953

for *m*-chloroperbenzoic acid [937-14-4], **VI**, 276, 277

for chromic acid, **IV**, 20

for chromium(II) perchlorate, **VI**, 822

for chromium(II) sulfate, **V**, 994

for 3-cyclopentenone in cyclopentadiene monoepoxide, **VIII**, 14

for cyclohexylmagnesium bromide, **I**, 187

for diazomethane, **II**, 166; **VI**, 387, 433, 614

for 2-diazopropane, **VI**, 393

for diborane solutions, **VII**, 261

for *N,N*-dimethyldodecylamine oxide, **VI**, 502

for dimethylketene, **IV**, 349

for 1,4-diphenyl-5-amino-1,2,3-triazole, **IV**, 381

for ethylmagnesium bromide with 2-butanol, 1,10-phenanthroline indicator, **VII**, 238; **VIII**, 611

for ethyl pyruvate, **IV**, 469

for fatty acid unsaturation, **III**, 528

for fluoroboric acid, **V**, 246

for formalin solutions, **I**, 378

for hydrogen peroxide, **IV**, 319; **VI**, 344

for hydroperoxide, **V**, 896

for hydroxylamine-*O*-sulfonic acid [2950-43-8], **V**, 898; **VI**, 945

for iodoso and iodoxy compounds, **III**, 484

for isopropylmagnesium chloride with 2-butanol, 1,10-phenanthroline indicator, **V**, 212

for ketene, **V**, 682

for lead tetraacetate, **V**, 724

for lithium aluminum hydride in ether, **VI**, 770

for lithium thiophenoxide, **VI**, 248

for methyllithium with 2-butanol, 1,10-phenanthroline or 2,2'-bipydridyl or 2,2'-biquinoline indicator, **V**, 777; **VI**, 125, 572, 667, 695, 764, 903; **VII**, 174, 201, 349, 354; **VIII**, 476
  by double titration, **VII**, 349; **VIII**, 482

for (4-methylphenyl)magnesium
bromide [4294-57-9], **VI**, 489
for oxaziridines, **V**, 192
for ozone concentration, **III**, 678;
VIII, 405
for peracetic acid, **V**, 661; **VI**,
863
for perbenzoic acid, **I**, 434; **V**,
906
for perphthalic acid, **V**, 806
for 4-phenyl-5-anilino-1,2,3-triazole,
IV, 382
for phenylmagnesium bromide,
VIII, 525
for phenyltrimethylammonium
tribromide, **VI**, 176
for *o*-phthalaldehyde, **IV**, 808
for "positive" chlorine, **V**, 37
for pyrrole in pyridine, **V**, 105
for reduction of Tl(III) to Tl(I),
VI, 793
for sodium β-styrenesulfonate,
IV, 849
for tetralin hydroperoxide, **IV**, 896
for tetralone, **IV**, 906
for thiophosgene [463-71-8], **VII**,
403
for trichloramine, **V**, 37
for trimethylsilyldiazomethane
[18107-18-1], titer of hexane
solution, **VIII**, 614
for 2,4,6-triphenylphenoxyl, **V**, 1133
for unsaturation, (bromate-bromide
titration), **IV**, 849
for vinyllithium with 2-butanol,
1,10-phenanthroline indicator,
VIII, 34
for water content, Karl Fischer
titration, **VI**, 344, 541
of deuterium content of
dideuteriodiazomethane
[14621-84-2], **VI**, 434
of diastereomeric purity by NMR
analysis, **VIII**, 112, 328, 334

of enantiomeric purity of alcohols by
NMR analysis of Mosher
esters, **VII**, 217; **VIII**, 314,
369, 519
of enantiomeric purity of amino
acids, **VIII**, 120
of enantiomeric purity by chiral
column HPLC, **VIII**, 47, 48,
61, 465
of enantiomeric purity of ethyl 2-
fluorohexanoate and 2-
fluorohexanoic acid, **VIII**, 261
of enantiomeric purity by GLC
analysis of Mosher ester, **VII**,
464; **VIII**, 530
of enantiomeric purity of (*S*)-(−)-2-
methyl-1-butanal, **VIII**, 368
of enantiomeric purity of (*S*)-(−)-2-
methyl-1-butanol [1565-80-6],
VIII, 367
of enantiomeric purity by NMR
analysis with shift reagent, **VI**,
721, 828; **VII**, 404, 463; **VIII**,
405, 423
of enantiomeric purity of 8-
oxabicyclo[4.3.0]nonan-7-one,
(1*R*,6*S*)-(+)-, **VII**, 408
of enantiomeric purity of L-(−)-
proline by GLC analysis, **VII**,
364
of enantiomeric purity of salsolidine
by chiral column HPLC of its
1-naphthamide, **VIII**, 575
of enantiomeric purity of Wieland–
Miescher ketone by NMR
analysis with shift reagent,
VII, 371
of enzymatic activity of horse liver
alcohol dehydrogenase, **VII**,
407
of proportions of structurally
isomeric benzylmethylcyclo-
hexanones by NMR analysis,
VI, 126

Azobisisobutyronitrile (AIBN) [78-67-1],
　　Hazard, **VIII**, 270, 383, 583

Baker's yeast, **VII**, 216; **VIII**, 313
Barium hydroxide in methanol, **IV**, 636
Beilstein test, use of, **IV**, 199, 725
Benzaldehyde, *Hazard*, **I**, 81, 95, 105,
　　253, 414; **II**, 29, 168; **III**, 208,
　　409, 645; **IV**, 778
Benzanilide [93-98-1], preparation of, **IV**,
　　384
Benzene, *Hazard*, **I**, 82; **III**, 208, 333;
　　IV, 556, 592, 702, 901; **V**, 199,
　　202, 737; **VI**, 219, 370, 384,
　　493, 522, 740, 938, 959; **VII**,
　　84, 174, 204, 373, 514; 275,
　　VIII, 564
Benzil monohydrazone [5344-88-7], **II**,
　　497; **III**, 357
　　dihydrazone [4702-78-7], **IV**, 377
Benzocaine, **VI**, 582
Benzoic carbonic anhydride, preparation
　　of, **IV**, 286
Benzophenone dichloride [2051-90-3],
　　preparation of, **II**, 573
Benzophenone hydrazone [5350-57-2],
　　preparation of, **III**, 352
p-Benzoquinone [106-51-4], **VI**, 892;
　　VII, 519
Benzoyl chloride, *Hazard*, **III**, 113; **VI**,
　　387
Benzyl alcohol, *Hazard*, **I**, 105, 106
Benzylamine, **VIII**, 11, 232
Benzylchlorobis(triphenylphosphine)
　　palladium(II) [22784-59-4],
　　VIII, 271
Benzyl chloroformate [501-53-1],
　　Hazard, preparation of, **III**, 167
Benzyl cyanide, *Hazard*, purification of,
　　I, 108; **IV**, 388
Benzyl disulfide, **VI**, 131
3-Benzyl-5-(2-hydroxyethyl)-4-methyl-
　　1,3-thiazolium chloride [4568-
　　71-2], **VII**, 95

Benzylidenearginine, in the isolation of
　　arginine, **II**, 49, 50
Benzyloxycarbonyl-L-aspartyl-(*tert*-butyl
　　ester)-L-phenylalanine [32771-
　　88-3], **VII**, 30
N$^\alpha$-Benzyloxycarbonyl-*O*-*tert*-butyl-L-
　　seryl-*S*-*tert*-butyl-L-cysteine
　　tert-butyl ester, preparation of,
　　VI, 256
Benzyltriethylammonium chloride [56-
　　37-1], **VI**, 233; **VII**, 411
Benzyltrimethylammonium iodide,
　　preparation of, **IV**, 585
2-Benzyl-4,4,6-trimethyl-5,6-dihydro-
　　1,3(4*H*)-oxazine [26939-22-0],
　　preparation of, **VI**, 907
(Bicyclo[2.2.1]heptadiene)(1,4-
　　bis(diphenylphosphino)butan-
　　erhodium
　　trifluoromethanesulfonate [122086-
　　63-9], preparation of, **VIII**, 420
Bicyclo[2.2.1]hepta-2,5-diene-2,4-
　　pentanedionatorhodium [32354-
　　50-0], preparation of, **VIII**, 420
1,1'-Binaphthol, **VIII**, 52
(*R*,*R*)-1,2-Bis(*o*-anisylphenylphosphino)
　　ethanerhodium trifluoromethane-
　　sulfonate, (DIPAMPRH$^+$),
　　preparation of, **VIII**, 423
Bis(2-carboethoxyethyl)zinc, preparation
　　of, **VIII**, 277
Bis[copper(I) trifluoromethanesulfonate]–
　　benzene complex [37234-97-2],
　　preparation of, **VI**, 739
[(+)-2,2'-Bis(diphenylphosphino)-1,1'-
　　binaphthyl]-(η^4-1,5-
　　cyclooctadiene)
　　rhodium(I) perchlorate [82822-45-
　　5], preparation of, **VIII**, 183
Bis[2-(2-methoxyethoxy)ethyl]ether (tetra-
　　glyme) [143-24-8], **VII**, 519
Bis(4-methoxyphenyl)tellurium
　　dichloride, preparation of
　　[4456-36-4], **VI**, 468

Bis(3-methyl-2-butyl)borane
(disiamylborane) [1069-54-1],
VI, 853, 919
1,3-Bis(methylthio)allyllithium,
preparation, **VI**, 683, 687
[Bis(trifluoroacetoxy)iodo]benzene [2712-
78-9], preparation of, **VIII**, 135
Boric anhydride-sulfuric acid scrubber,
for removing hydrogen
fluoride, **VII**, 394
Boron trichloride in xylene, *Hazard*,
V, 731
Boron trifluoride, *Hazard*, **V**, 334;
VII, 394
Boron trifluoride-acetic acid complex,
1:1, preparation of, **VI**, 246
Boron trifluoride etherate [109-63-7],
Hazard, **VI**, 112, 945; **VIII**,
311, 528, 578
Bromine, *Hazard*, **I**, 124, 246; **III**, 124,
524, 790; **IV**, 400, 619; **VI**,
404, 864
N-Bromoacetamide, determination of
purity, **IV**, 105
o-Bromobenzenediazonium
hexafluorophosphate,
preparation of, **V**, 133
2-Bromobenzoic acid, **VI**, 37
2-Bromo-2-butene (cis and trans
mixture), **VIII**, 505
α-Bromobutyryl chloride [22118-12-3],
preparation of, **IV**, 620
1-Bromo-3-chloropropane [109-70-6],
VI, 818
N-Bromoglutarimide [3699-18-1],
preparation of, **IV**, 498
6-Bromo-2-methoxynaphthalene [5114-
65-9], preparation of, **V**, 920
2-Bromo-4-methylbenzenediazonium
chloride, **V**, 140
4-Bromophenylmagnesium bromide,
preparation of, **V**, 496
N-Bromosuccinimide, **IV**, 922; **VI**, 464
Buffer solutions: ammonium bicarbonate
in aqueous isopropyl alcohol,
VIII, 618

ammonium bicarbonate in aqueous
isopropyl alcohol and
acetonitrile, **VIII**, 618
ammonium chloride-ammonium
hydroxide, **VI**, 667
disodium phosphate-monosodium
phosphate, **III**, 121
sodium acetate in aqueous acetic
acid, **III**, 121; **VI**, 244
Butadiene, *Hazard*, **II**, 105
trans-2-Butene oxide [21490-63-1],
preparation of, **VI**, 313
3-Buten-1-ol [627-27-0], preparation of,
VI, 608
Butyl alcohol, *Hazard*, **II**, 469, 475
tert-Butyl alcohol, *Hazard*, **III**, 49, 143,
145, 507; **IV**, 134, 460; **V**, 173,
938; **VI**, 72, 229, 259, 321, 764,
842; **VII**, 68, 145; **VIII**, 214
Butyl bromide, *Hazard*, **VII**, 78, 292
sec-Butyl α-bromopropionate [22710-17-
4], preparation of, **IV**, 122
4-*tert*-Butylcatechol [98-29-3], **VI**, 98
tert-Butylhydrazine, from the
hydrochloride, **VIII**, 95
tert-Butyl hydroperoxide [75-91-2], *Hazard*,
anhydrous in toluene, **VII**, 462
drying, **VI**, 502
Butyllithium, preparation, **VI**, 980;
VII, 292
Butylmagnesium bromide, preparation of,
VI, 407
sec-Butylmagnesium bromide,
preparation of, **VI**, 753
Butylmagnesium chloride, preparation of,
VIII, 606
preparation of unsolvated, **V**, 1141
Butyl oleate, **II**, 469
Butyryl chloride [141-75-3], *Hazard*,
preparation of, **I**, 147

Calcium hypochlorite [7778-54-3], **VI**, 969
Camphor tosylhydrazone [10224-32-5],
preparation of, **VI**, 173
Carbobenzyloxy-L-asparagine [2304-96-
3], **VI**, 265

N-Carbobenzyloxy-3-hydroxy-L-proline [57621-07-5], **VI**, 265
(3-Carboethoxypropyl)zinc iodide, **VIII**, 274
ω-Carbomethoxypelargonyl chloride [14065-32-8], preparation of, **III**, 613
Carbon dioxide, drying of, **VI**, 419
Carbon monoxide, *Hazard*, generation, **II**, 585
Carbon tetrachloride, *Hazard*, **IV**, 935; **VI**, 405, 634; **VIII**, 364
Caro's acid, preparation of, **III**, 334
δ-Carbomethoxyvaleryl chloride [35444-44-1], preparation of, **IV**, 556
Catalysts, enzyme
    Lipase (P-30), **VIII**, 258
    Lipase (*P. fluorescens*) [9001-62-1], **VIII**, 258
    Pig liver esterase, **VIII**, 258, 332
Catalysts, metal, *Hazard*
    chromium oxide–calcium carbonate, **IV**, 579
    copper chromite, **IV**, 216, 314, 324, 337, 678, 857
    Lindlar catalyst, **VII**, 226
        modified, **VIII**, 609
    Raney nickel W-2, **III**, 181; W-6, **III**, 176, 181; **IV**, 222, 229, 314, 357, 432, 603, 639, 660, 672; WJ-7, **V**, 102
    U.O.P. nickel, **III**, 278
    palladium, **III**, 685
        on carbon, **III**, 385, 520, 686, 687; **IV**, 306, 537, 886
        on barium carbonate, **III**, 688
        on barium sulfate, **III**, 685
        10% on calcium carbonate, **VI**, 73
        on calcium carbonate plus lead(Lindlar), **V**, 880
        recycling with copper(II) and oxygen, **VII**, 137
    palladium black, preparation of, **IV**, 612
    platinum black, **VI**, 254
    platinum oxide, **I**, 463; **IV**, 305, 306
    rhodium–palladium, preparation of, **V**, 671
Cation exchange resin, Dowex AG 50W-XB, ammonium form, **VIII**, 618
Celite (C-211), **VII**, 85
Cellulose chromatography, preparation of column, **VIII**, 619
Chlorine, anhydrous, *Hazard*, **III**, 483; **IV**, 935
    generation of, **II**, 46
4-Chloroacetophenone oxime [1956-39-4], preparation of, **VI**, 279
p-Chlorobenzenediazonium chloride, preparation of, **VII**, 105
o-Chlorobenzoic acid, technical, **II**, 16
m-Chlorobenzoyl chloride [618-46-2], preparation of, **VI**, 276
p-Chlorobenzoyl chloride [122-01-0], preparation of, **III**, 29
1-Chlorobutane [109-69-3], **V**, 1143; **VIII**, 607
4-Chloro-1-butene [563-52-0], preparation of, **VI**, 607
α-Chloroethyl ethyl ether [7081-78-9], preparation of, **IV**, 748
Chloroform, *Hazard*, **III**, 482; **IV**, 265; **V**, 536; **VI**, 228; **VII**, 230
1-Chloro-3-iodopropane [6940-76-7], preparation of, **VI**, 365
1-Chloromagnesioxy-3-octynylmagnesium chloride, **VIII**, 507
Chloromethanesulfonyl chloride [3518-65-8], preparation of, **V**, 232
Chloromethylated polystyrene, preparation of, **VI**, 952
N-Chlorosuccinimide [128-09-6], **VI**, 221; **VIII**, 617
Chlorosulfonyl isocyanate [1189-71-5], preparation of, **V**, 226; **VI**, 304, 465, 788
Chlorotrimethylsilane [75-77-4], **VII**, 132, 268, 514, 519, 521, 524; **VIII**, 21, 462, 502, 608, 610

debasifying glass apparatus, **VIII**, 406

drying of, **VI**, 169, 354, 446

Cholesterol, test for, **II**, 193

Chromic acid on silica gel, preparation of, **VIII**, 622

Chromic acid–sulfuric acid reagent, **V**, 310, 854, 866

Chromic acid solution, titration of, **IV**, 20

Chromium(II) perchlorate [13931-95-8], preparation of, **VI**, 822

Chromium (II) sulfate, preparation of, **V**, 993

Cinnamaldehyde dimethyl acetal [4364-06-1], preparation of, **VI**, 895

Citral [5392-40-5], **III**, 747, 749

Claisen's alkali (KOH in aq. methanol), **III**, 269, 663; **IV**, 191; **V**, 1095

Cobalt bromide-triphenylphosphine, preparation of, **VI**, 380

2,4,6-Collidine [108-75-8], **VIII**, 306

Copper bronze, **III**, 339

Copper powder, **II**, 446; **IV**, 337, 628, 732

Copper acetate, aq. soln, **III**, 391

Copper(I) bromide, **II**, 132; **III**, 186; **VI**, 22

dimethyl sulfide complex [54678-23-8], **VII**, 237, 290

Copper(I) chloride, *Hazard*, **I**, 162, 170; **II**, 585

tetramethylethylenediamine complex, **VIII**, 63

Copper(I) cyanide, *Hazard*, preparation of, **I**, 46

Copper(I) iodide [7681-65-4], **VI**, 250, 667; **VII**, 292, 353; **VIII**, 482

triethyl phosphite complex, preparation of, **VIII**, 587

Copper(I) phenylacetylide [13146-23-1], preparation of, **VI**, 916

Copper sulfite, preparation of, **III**, 341

18-Crown-6-potassium cyanide complex, **VII**, 519

Cuprous butylmercaptide, preparation of, **V**, 107

Cuprous oxide, preparation of, **V**, 108

Cuprous oxide–silver oxide catalyst, preparation of, **IV**, 494

Cuprous sulfate in aq. ammonia, preparation of, **IV**, 873

Cyanamide, preparation of, *Hazard*, **IV**, 645

4-Cyanobenzenesulfonamide [3119-02-6], preparation of, **VI**, 632

Cyanogen chloride [506-77-4], *Hazard*, preparation of, **VI**, 728

Cyclohexanone, anhydrous, *Hazard*, **IV**, 223

Cyclohexanone oxime methanesulfonate [80053-69-6], purification, **VIII**, 569

Cyclohexene, *Hazard*, **V**, 857; **VI**, 339

N-Cyclohexylformamide [766-93-8], preparation of, **V**, 301

Cyclohexyl methyl ether [931-56-6], preparation of, **VI**, 355

Cyclopentadiene [542-92-7], *Hazard*, preparation of, **IV**, 238; **V**, 415, 1002; **VI**, 147, 1038; **VII**, 339

drying and storage of, **VI**, 147

Cyclopentadiene monoepoxide, **VIII**, 14

Cyclopentadienylsodium, preparation of, **IV**, 474; **V**, 1088

1,3-Diaminopropane [109-76-2], *Hazard*, **VIII**, 146, 554

2,4-Di-*tert*-amylphenoxyacetyl chloride [88-34-6], preparation, **IV**, 742

Diazomethane, preparation of, *Hazard*, **II**, 165; **III**, 119, 244; **IV**, 225, 250; **V**, 351, 878; **VI**, 387, 432

solution in xylene, **V**, 1101

Dibenzyl ketoxime [1788-31-4], preparation of, **V**, 84

1,2-Dibromoethane [106-93-4], **VI**, 1040

Dibromomalononitrile–KBr complex, preparation of, **IV**, 877

Dibutyl ether, *Hazard*, **VI**, 998

Dibutylboron triflate [60669-69-4], **VIII**, 342

2,6-Di-*tert*-butyl-4-methylpyridine, recovery, **VIII**, 100

purification, **VIII**, 127
Dicarbonyl(cyclopentadienyl)(ethyl vinyl
    ether)iron tetrafluoroborate
    [75182-42-2], **VIII**, 479
Dicarbonyl(cyclopentadienyl)(*trans*-3-
    methyl-2-vinylcyclo-
    hexanone)iron tetrafluoroborate
    [121523-33-9], **VIII**, 479
Dichloroacetyl chloride [79-36-7],
    Hazard, preparation of, **VI**,
    1038
Dichloro[1,3-bis(diphenylphosphino)
    propane]nickel(II) [14647-23-
    5], preparation of, **VI**, 408
Dichlorobis(triphenylphosphine)palladium
    (II) [13965-03-2], preparation
    of, **VIII**, 533
1,2-Dichloroethane, *Hazard*, **VII**, 246;
    VIII, 307
Dichloroketene, from trichloroacetyl
    chloride, **VIII**, 84, 377
Dichloromethane (methylene chloride),
    Hazard, **VI**, 59, 284, 325, 374,
    646, 884; **VII**, 94, 169, 232,
    260, 285, 295, 421, 425, 444;
    VIII, 99, 127, 322, 324, 341,
    349, 388, 399, 405, 442, 462,
    465, 488, 499, 503, 579, 617
Dichloromethyl ethyl ether [13830-34-7],
    preparation of, **V**, 860
1,1-Di-(*p*-chlorophenyl)-2,2,2-
    trichloroethane (DDT) [50-29-
    3], *Hazard*, **III**, 271
Dicyclohexylcarbodiimide, **VII**, 94
Diethylamine, *Hazard*, **VI**, 949; **VIII**,
    189, 191
N,*N*-Diethylaminotrimethylsilane [996-50-
    9], preparation of, **VI**, 441
Diethyl ether, *Hazard*, **III**, 393, 555, 840;
    IV, 70, 564, 602, 822, 926; **VI**,
    8, 33, 52, 73, 164, 262, 299,
    329, 408, 607, 614, 749, 764,
    903, 949, 989; **VII**, 130, 132,
    237, 268, 328, 348, 353, 487,
    503, 514, 529; **VIII**, 83, 158,
    228, 278, 307, 372, 379, 399,

405, 418, 431, 482, 508, 525,
    558, 613
    removal of peroxides, **IV**
Diethyl malonate, **IV**, 287, 632
Diglyme [111-96-6], **V**, 391, 950; **VI**,
    530, 945; **VII**, 429
Dihydropyran, *Hazard*, **V**, 875
Diiron nonacarbonyl [15321-51-4],
    preparation of, **VI**, 311, 522
Diisopropylamine [108-18-9], *Hazard*,
    VI, 518, 544, 686, 819, 902;
    VII, 63, 66, 155, 187, 192, 208,
    243, 271, 276, 282, 351, 384,
    389; **VIII**, 193, 220, 244, 327,
    405, 445, 475, 563
N,*N*-Diisopropylethylamine [7087-68-5],
    VI, 97
(−)-2,3:4,6-Di-*O*-isopropylidene-2-keto-L-
    gulonic acid hydrate [(−)-
    DAG], recovery, **VI**, 827
Dilithioallene, preparation of, **VIII**, 226
2,2'-Dilithio-1,1'-binaphthyl, preparation
    of, **VIII**, 57
1,2-Dimethoxyethane [110-71-4],
    purification of, **IV**, 474; **V**, 355,
    719, 776, 970, 1002; **VI**, 122,
    626, 694; **VII**, 1, 284; **VIII**, 83
    a zinc chloride sequestering agent,
    VIII, 84
    potential hazard in purification, **V**,
    355
3,4-Dimethoxyphenylacetic acid [93-40-
    3], preparation of, **VI**, 471, 472
2-(3,4-Dimethoxyphenyl)-2-[4-(phenyl-
    methoxy)phenyl]ethanamine,
    resolution of, **VI**, 829
(*S*)-6,7-Dimethoxy-1,2,3,4-tetrahydro-
    isoquinoline [493-48-1], from
    hydrochloride, **VIII**, 574
6,7-Dimethoxy-1,2,3,4-tetrahydro-(3,4,5-
    trimethoxyphenyl) isoquinoline,
    resolution of, **VI**, 829
N,*N*-Dimethylacetamide [127-19-5],
    Hazard, **VI**, 504; **VIII**, 275, 306
Dimethylamine [124-40-3], *Hazard*, **IV**,
    336; **VI**, 854; **VII**, 63, 66

recovered from 25% aqueous
solution, **IV**, 336
recovered from 40% aqueous
solution, **VII**, 42
2-(*N,N*-Dimethylamino)phenyllithium
[22608-37-3], preparation of,
VII, 172
Dimethylaminosulfur trifluoride,
substitute for sulfur
tetrafluoride, **VII**, 529
N,N-Dimethylaniline [121-69-7], *Hazard*,
VII, 174
2,3-Dimethylbenzyltrimethylammonium
iodide, preparation of, **IV**, 331
Dimethyl ether [115-10-6], **VI**, 1019,
1020
N,N-Dimethylformamide, *Hazard*, **V**,
1065; **VI**, 329, 588, 724, 753,
784, 861, 867, 874; **VII**, 173,
198, 252, 323, 365; **VIII**, 137,
194, 462
N,N-Dimethylformamide–dimethyl
sulfate complex, **V**, 431
N,N-Dimethylformamide–phosphorus
oxychloride complex, **IV**, 331,
831; **V**, 215
N,N-Dimethylisobutyramide [21678-37-
5], preparation of, **VI**, 284
Dimethyl methoxymethylenemalonate
[22398-14-7], **VIII**, 445
4,4-Dimethyl-2-oxazoline, preparation of,
VI, 65
Dimethyloxosulfonium methylide,
preparation of, **V**, 755
Dimethyl sulfate, *Hazard*, handling of,
V, 271
Dimethyl sulfoxide, **V**, 243, 756, 938,
958; **VI**, 88, 209, 219, 532; **VII**,
119, 260, 370; **VIII**, 503
Dimethyl sulfoxide-methyl sulfide
reagent, **VI**, 221
N,N-Dimethylthiocarbamyl chloride
[16420-13-6], preparation of,
VI, 824
Dimsyl anion-dimethyl sulfoxide,
preparation of, **V**, 938; **VI**, 147

for removal of cyclopentadiene, **VI**,
147
3,5-Dinitrobenzoyl chloride [99-33-2],
VI, 530
Dinitrogen tetroxide [10544-72-6],
Hazard, **V**, 338; **VI**, 838
2,4-Dinitro-1-naphthol-7-sulfonic acid
(flavianic acid) [483-84-1],
preparation of, **II**, 51
Dioxane, *Hazard*, **IV**, 643, 848; **V**, 870;
VI, 716, 810; **VII**, 76, 198,
276; 475; **VIII**, 205
Diphenyldiselenide, *Hazard*, recovery
from benzeneseleninic acid,
VI, 25
Diphenyliodonium-2-carboxylate [1488-
42-2], preparation of, **V**, 1037
Diphenylmethylchlorosilane [144-79-6],
VIII, 474
Diphenylphosphine [829-85-6],
preparation of, **VI**, 569
(+)-Di-3-pinanylborane [21947-87-5],
VI, 944; **VII**, 339
Dipotassium salt of nitroacetic acid
[22742-08-1], **VI**, 797
Dipotassium nitrosodisulfonate [14293-
70-0], preparation of, **VI**, 481,
1012
Dipyridine chromium (VI) oxide [26412-
88-4], *Hazard*, **VI**, 644
2,2'-Dipyridyl disulfide [2127-03-9], **VII**,
471
Disodium hydroxylaminedisulfonate,
preparation of, **VI**, 1011
Disodium methylsuccinate, preparation
of, **IV**, 672
Disodium nitrosodisulfonate, **VI**, 1010
Disodium tetracarbonylferrate sesquidioxanate [57398-59-1],
VI, 807
1-Dodecyl tosylate [10157-76-3] ,
preparation of, **VI**, 377

Ethanethiol [75-08-1], *Hazard*, **VI**, 861
Ethanol, *Hazard*, **I**, 249, 251, 259; **II**,
155, 263, 286, 289, 319, 373,

488; **III**, 301, 398, 672, 830; **IV**, 11; **V**, 992; **VI**, 33; **VII**, 51, 212; **VIII**, 599

Ethanolamine, commercial, **II**, 92

(Ethoxycarbonylmethylene)triphenylphosphorane [1099-45-2], preparation of, **VII**, 234

Ethoxymagnesiummalonic ester, **II**, 594; **IV**, 285, 708

4-Ethoxy-3-methoxybenzaldehyde [120-25-2], **VI**, 569

Ethyl acetate, **I**, 227, 235, 253; **II**, 488; **III**, 19; **VI**, 938

Ethyl acetate-ethanol, preparation of, **I**, 235

Ethylamine, **VI**, 79, 854

Ethyl benzoate, purification of, **V**, 938

Ethyl bromide, **II**, 603; **VI**, 596

Ethyl bromoacetate, **III**, 409

Ethyl butyrate, **II**, 115

Ethyl carbonate, **II**, 283; **III**, 832

Ethyl α-chloroacetoacetate [609-15-4], preparation of, **IV**, 592

Ethyl copper complex, preparation of, **VII**, 236

Ethyl copper complex-1-octyne adduct, preparation of, **VII**, 236

Ethyl diazoacetate [623-73-4], preparation of, **VI**, 913

Ethyldiisopropylamine [7087-68-5], **VI**, 97, 740; **VII**, 321; **VIII**, 399

Ethylenediammonium diacetate (EDDA) [38734-69-9], **VIII**, 354

Ethylenediisothiouronium bromide [6943-65-3], preparation of, **IV**, 402

Ethylene glycol, anhydrous, *Hazard*, **IV**, 449

Ethylene glycol dimethyl ether, *Hazard*, purification of, **IV**, 475

Ethyl formate, *Hazard*, **II**, 6, 180; **III**, 830; **V**, 189; **VI**, 127

2-Ethylhexanamide [4164-92-5], preparation of, **IV**, 436

Ethyl iodide, **VIII**, 244

Ethyl isobutyrate, **II**, 269

Ethyl lithioacetate, **VI**, 598

Ethylmagnesium bromide [925-90-6], preparation of, **VI**, 526, 925

Ethyl nitrite, preparation of, **II**, 204; **VI**, 542

Ethyl oxalate, **I**, 239; **II**, 289

Ethyl phenylacetate, **II**, 289

Ethyl phenylacetylmalonate [20320-59-6], preparation of, **III**, 637

N-Ethyl-5-phenylisoxazolium-4'-sulfonate (Woodward's reagent K) [4156-16-5], **VI**, 265

Ethyl phthalimidomalonate sodium salt, preparation of, **II**, 384

Ethyl propionate, **III**, 389

Ethyl stearate [111-61-5], preparation of, **IV**, 142

Ethyl vinyl ether, **VI**, 299, 572

Ethynylmagnesium bromide, preparation of, **IV**, 792

Ethynylmagnesium chloride, preparation of, **VIII**, 606

Ferric chloride, ethanolic, **VIII**, 180 in N,N-dimethylformamide, preparation of, **VIII**, 470

Florisil, **VIII**, 194

Fluorine-containing organic compounds, potential poisons, *Hazard*, **V**, 393

10% Fluoroboric acid in acetic anhydride, **VI**, 994

Fluoroboric acid in ether, *Hazard*, **V**, 1136

Fluoroboric acid in ether-methylene chloride, *Hazard*, **V**, 246

Formaldoxime (10% aq. solution), **V**, 139

o-Formotoluide [94-69-9], preparation of, **III**, 480; **V**, 1062

Formyl fluoride, preparation with acetic formic anhydride, **VI**, 9

Fremy's salt, preparation of, **VI**, 1012

Furan, **V**, 404; **VI**, 514

Gallic acid, **V**, 593

Gilman's catalyst (Mg-Cu alloy + iodine), **II**, 361

Glycerol, **I**, 17; **III**, 503
Glycine, *N*-glycyl-, ethyl ester,
     monohydrochloride [623-33-6],
     VI, 265
Gold's reagent, **VII**, 197
Grignard reagents, test for presence of, **I**,
     189; **III**, 603

1,5-Hexadiyne [628-16-0], preparation of,
     VI, 72
Hexafluorophosphoric acid, handling of,
     Hazard, **V**, 135
1,1,1,3,3,3-Hexafluoro-2-phenyl-2-
     propanol [718-64-9],
     preparation of, **VI**, 163
Hexamethyldisiloxane [107-46-0],
     preparation of, **VI**, 354
Hexamethylenediammonium chloride,
     preparation of, **IV**, 522
Hexamethylphosphoric triamide (HMPA)
     [680-31-9], *Hazard*, **VI**, 66,
     377, 638; **VII**, 242; **VIII**, 277,
     278, 329
Hexane, *Hazard*, **V**, 610; **VI**, 116, 550,
     810; **VII**, 284; **VIII**, 158, 278,
     296, 418, 455, 569
1-Hexyne, **VIII**, 83
Hinsberg separation, procedure for, **III**,
     161
Hydrazine, anhydrous, *Hazard*, **VI**, 10, 11
     concentration of dilute solutions, **II**,
     86
Hydrazine hydrate, concentration of, **II**,
     86
     preparation of, **III**, 515
Hydrazine sulfate (1:1) [10034-93-2],
     Hazard, preparation of, **I**, 309
Hydrobromic acid, *Hazard*, preparation
     of, **I**, 26
     constant boiling, **III**, 524
8 *N* Hydrochloric acid, **VI**, 911
Hydrogen bromide, preparation of,
     Hazard, **II**, 246, 338; **III**, 13;
     V, 546
     removal of bromine, **III**, 577

Hydrogen chloride, anhydrous, *Hazard*,
     preparation and drying, **I**, 6,
     293, 534; **II**, 572; **IV**, 171, 607,
     751, 927; **V**, 378
     in ether, dry, **VI**, 771
     in absolute ethanol, **II**, 311; **IV**, 428
Hydrogen cyanide (anhydrous), *Hazard*,
     preparation of, **I**, 314
Hydrogen cyanide in tetrahydrofuran, **VI**,
     17
Hydrogen fluoride, *Hazard*, handling of,
     V, 66, 138, 482
Hydrogen peroxide, 90%, *Hazard*,
     handling of, **V**, 368, 599
Hydrogen selenide, *Hazard*, preparation
     of, **IV**, 360
o-Hydroxyacetophenone [118-93-4],
     preparation of, **III**, 389
Hydroxylamine, *Hazard*, **VII**, 10
Hydroxylamine hydrochloride [5470-11-
     1], preparation of, **I**, 318
Hydroxylamine-*O*-sulfonic acid, **V**, 44

Iodine, recovery from aqueous sodium
     iodide, **II**, 348
Iodine azide, *in situ* preparation of, **VI**, 893
Iodine isocyanate, preparation of, **VI**, 795
Iodine monochloride, *Hazard*, **II**, 197,
     344
Iodine number (Wijs titration), **IV**, 318,
     853, 978
Iodobenzene [591-50-4], **VI**, 452
N-Iodosuccinimide [516-12-1],
     preparation of, **V**, 664
Ion-exchange resin, treatment of, **IV**, 39,
     530
Isoamyl nitrite [110-46-3], preparation of,
     VI, 840
Isobutylmagnesium chloride, **VIII**, 508
Isobutyraldehyde [78-84-2], **VI**, 497
Isobutyryl chloride [79-30-1], preparation
     of, **III**, 490
Isopentane, *Hazard*, **VII**, 449
Isopropyl ether, *Hazard*, removal of
     peroxides, **III**, 13

Isopropyl nitrite, preparation of, **III**, 192

Jones' reagent, **VIII**, 45

Kerosene, purification of, *Hazard*, **III**, 72
β-Ketoesters, test for, **IV**, 122

Lauroyl chloride [112-16-3], preparation
    of, **IV**, 561
Laurylisothiourea hydrobromide,
    preparation of, **III**, 363
Lead tetraacetate, **VI**, 959; **VII**, 230
Lead thiocyanate, *Hazard*, preparation of,
    V, 1052
L-Leucine methyl ester hydrochloride
    [7517-19-3], **VI**, 265
Ligroin, **III**, 844
2-Lithiobornene, **VII**, 77
2-Lithio-2-butene, **VIII**, 505
2-Lithio-2-cyclopenten-1-one ethylene
    ketal, **VII**, 271, 495
6-Lithio-3,4-dimethoxybenzaldehyde
    cyclohexylimine, **VIII**, 586
1-Lithio-1-ethoxycyclopropane, **VIII**, 556
2-(2'-Lithio-4,5'-methylenedioxyphenyl)
    ethyl chloride, **VII**, 326
Lithium, handling of, **IV**, 888; **V**, 1094,
    1121
Lithium acetylide [1111-64-4], *Hazard*,
    VIII, 391
    compound with ethylenediamine
    [6867-30-7], **VIII**, 268
Lithium amide, *Hazard*, preparation of,
    V, 564
Lithium bis(trimethylsilyl)amide, **VI**,
    598
Lithium bromide, anhydrous, **VI**, 779
Lithium *tert*-butoxide, preparation of, **VI**,
    259
Lithium chloromethylacetylide, **VIII**, 371
Lithium dibutylcuprate (solution), **VII**, 290
Lithium diethylamide, *in situ* preparation
    of, **VI**, 948
Lithium di-[(Z)-hexenyl]cuprate
    (solution), **VII**, 290

Lithium diisopropylamide (LDA) [4111-
    54-0], preparation of, **VI**, 121,
    279, 542, 683, 818; **VII**, 63, 66,
    153, 208, 241, 277; **VIII**, 192,
    219, 235, 241, 326, 403, 444,
    467, 474, 486, 562
Lithium dimethylcuprate (solution)
    [15681-48-8], **VI**, 443; **VII**, 351
    preparation of, **VI**, 443, 762
Lithium 2,6-dimethylphenoxide,
    preparation of, **V**, 1092
Lithium divinylcyanocuprate, **VIII**, 112
Lithium iodide dihydrate, preparation of,
    V, 78
Lithium 1-lithio-3-propynoxide, **VIII**,
    415
Lithium perchlorate-1,2-dimethoxyethane
    complex, preparation of,
    VI, 857
Lithium phenylthio(*tert*-butyl)cuprate,
    VI, 248
Lithium 2,2,6,6-tetramethylpiperidide,
    VI, 571
Lithium thiophenoxide, preparation of,
    VI, 248
    titration of, **VI**, 248
Lithium triethylcarboxide, **VI**, 137
Lithium vinyl-2-thienylcyanocuprate,
    VIII, 33

Magnesium bromide [7786-30-3],
    anhydrous, preparation of,
    VI, 848
Magnesium ethoxide, anhydrous,
    preparation of, **VI**, 845
Magnesium methoxide, preparation of,
    V, 439
Magnesium methyl carbonate,
    preparation of, **V**, 440
L-Malic acid, recovery, **II**, 508
Malononitrile, *Hazard*, **V**, 34
Meldrum's acid, **VII**, 360
Mercuric sulfate, *Hazard*, regeneration
    from mercury and mercurous
    salts, **V**, 321

Mercury(II) acetate [1600-27-7], *Hazard*, **III**, 776; **VI**, 299

Mercury(II) trifluoroacetate [13257-51-7], *Hazard*, **VII**, 87

Methacryloyl chloride [920-46-7], *Hazard*, **VIII**, 275*d*

Methallyl bromide [1458-98-6], preparation of, **VI**, 723

π-Methallylnickel bromide, **VI**, 723

Methanesulfonyl chloride [124-63-0], **VII**, 472

Methanol, *Hazard*, **I**, 70, 220; **V**, 912; **VI**, 193; **VII**, 169, 264, 324; **VIII**, 383, 418

1-Methoxy-2-methylpyridinium methyl sulfate [55369-05-6], preparation of, **V**, 270

(4-Methoxyphenyl)acetyl chloride [4693-91-8], preparation of, **VI**, 745

2-Methoxyphenylmagnesium bromide, [16750-63-3], preparation of, **VI**, 64

Methylal [109-87-5], *Hazard*, preparation of, **III**, 469

Methylamine, *Hazard*, **IV**, 607

N-Methylaniline, recovery, *Hazard*, **IV**, 917

N-Methylanilinium trifluoroacetate [29885-95-8], **VII**, 332

(2-Methylbenzyl)ethyldimethylammonium bromide, preparation of, **IV**, 582

p-Methylbenzyltrimethylammonium bromide, preparation of, **V**, 884

3-Methyl-2-butanone [563-80-4], **VI**, 757

2-Methyl-2-butene, **VII**, 259

Methyl bromide generator, *Hazard*, **II**, 280

Methyl 6-bromohexanoate [14273-90-6], preparation of, **VI**, 811

3-Methyl-2-butanone [563-80-4], **VI**, 193

Methyl carbazate, preparation of, **VI**, 334

Methyl chloride generator, *Hazard*, **II**, 251

  transferring and metering, **VII**, 346

Methylcyclohexane, **V**, 1143

Methyl cyclohexanecarboxylate [4630-82-4], preparation of, **VI**, 493

2-Methylcyclohexanone [583-60-8], preparation of, **IV**, 164

Methyl *p*-ethylbenzoate [7364-20-7], preparation of, **IV**, 580

Methyl ethyl ketone, *Hazard*, **IV**, 86

2-Methylfuran, **IV**, 627

N-Methylhydroxylamine hydrochloride [4229-44-1], preparation of, **VI**, 670

Methyllithium [917-54-4], preparation of ether solution of, **V**, 860; **VI**, 903

Methyl mercaptan generator, *Hazard*, **II**, 345

Methyl methacrylate [80-62-6], **VI**, 241

N-Methylmorpholine, *Hazard*, **VIII**, 134

2-Methyl-2-propen-1-ol [513-42-8], **VII**, 268, 361

3-Methylpyridine-1-oxide [1003-73-2], preparation of, **IV**, 655

N-Methylpyrrolidinone [872-50-4], **VI**, 811

Methyltriphenylphosphonium iodide, **VII**, 260

Methyl vinyl ketone [78-94-4], **VI**, 446

Milas reagent, in the regeneration of osmium tetroxide, **VI**, 347

2-Naphthol [135-19-3], **VI**, 825

1-Naphthoyl chloride [879-18-5], preparation of, **VI**, 614

2-Naphthoyl chloride [2243-83-6], preparation of, **III**, 629

Nitrobenzene, *Hazard*, **VI**, 35

p-Nitrobenzenediazonium fluoborate, preparation of, **III**, 666

m-Nitrobenzoyl chloride [121-90-4], preparation of, **IV**, 715

Nitrogen, drying of, **IV**, 70, 134, 926; **VI**, 110, 420, 489, 1020

oxygen-free, preparation of, **IV**, 27;
  V, 81, 363, 1002, 1142;
  VI, 125, 810, 902, 1026
Nitromethane, *Hazard*, **IV**, 223; **V**, 1129
Nitrosyl chloride [2696-92-6], *Hazard*,
  preparation of, **VI**, 981
Nitronium tetrafluoroborate, preparation
  of, **V**, 480
Norbornene [498-66-8], preparation of,
  VI, 142

1-Octene [111-66-0], **VI**, 919
95% Orthophosphoric acid, *Hazard*,
  preparation of, **IV**, 543; **VI**, 219
1% Osmium tetroxide reagent, *Hazard*,
  VII, 223, 377
Ozone, *Hazard*, preparation and use of,
  III, 673

Palladium acetate [3375-31-3], **VII**, 362
Palladium(II) chloride-acetonitrile complex,
  preparation of, **VII**, 505
Palmitoyl chloride [112-67-4],
  preparation of, **IV**, 742
Paraformaldehyde, *Hazard*, **VIII**, 229
Pentaerythrityl tetrabromide [3229-00-3],
  VI, 158
Pentane, *Hazard*, **V**, 875; **VI**, 73, 164,
  450, 758, 787, 847; **VII**, 84,
  113, 234; **VIII**, 344, 515, 552
1,5-Pentanediol diacetate [6963-44-6],
  preparation of, **IV**, 748
2,4-Pentanedione [123-54-6], *Hazard*,
  VI, 929
Peroxides, test for, **IV**, 830; **V**, 644, 925
Peroxide test strips, **VI**, 344
Peroxyacetic acid, *Hazard*, handling of,
  V, 416
Peroxytrifluoroacetic acid [359-48-8],
  preparation of, **V**, 367, 598
Petroleum ether, *Hazard*, **IV**, 468; **VI**, 84;
  VII, 2, 201
Phenanthrene, *Hazard*, **III**, 134; **IV**, 313,
  758

9,10-Phenanthrenequinone [84-11-7],
  VI, 887
2-Phenyl-2-adamantanamine, **VII**, 434
L-Phenylalanine [63-91-2], **VIII**, 530
(4-Phenyl-2-butenyl)triphenylphosphonium
  iodide, preparation of, **VI**, 707
β-Phenyldithiocarbazic acid,
  phenylhydrazine salt, preparation
  of, **III**, 360
α-Phenylethylamine, resolution of,
  VI, 829
Phenylethynylmagnesium bromide,
  preparation of, **VI**, 925
Phenylnitroacetonitrile sodium salt,
  preparation of, **II**, 512
Phenyl propionate [637-27-4],
  preparation of, **II**, 544
1-Phenyl-5-tetrazolone, recovery,
  VI, 151
Phenyltrimethylammonium
  sulfomethylate [28001-58-3],
  VI, 175
Phosgene, purification, *Hazard*, **II**, 453;
  IV, 524
Phosphoric acid, anhydrous, *Hazard*, **VI**,
  219; **VII**, 68
Phosphorus oxychloride, *Hazard*,
  VIII, 379
Phosphorus pentachloride, *Hazard*, **I**,
  395; **II**, 529
Phosphorus trisulfide, *Hazard*,
  preparation, **II**, 579
Phthalide, **IV**, 811
Pinacol, **III**, 313
Pinacolone [75-97-8], **VIII**, 344
Pinene, (–)-α-, *Hazard*, purification by
  distillation from LiAlH$_4$, **VI**,
  721; **VII**, 402; **VIII**, 555
  from sodium, **VII**, 343
Pinene, (–)-β-, purification by distillation
  from LiAlH$_4$, **VIII**, 555
α-Pinene oxide [32162-27-9], preparation
  of, **VI**, 949
Piperidine, *Hazard*, **V**, 576

Polyphosphate ester as a drying agent, **VII**, 436

Polyphosphoric acid [8017-16-1], preparation of, **VIII**, 469

Potassium metal, *Hazard*, safe handling of, **IV**, 134

Potassium sand, *Hazard*, **II**, 195

Potassium amide [17242-52-3], preparation of, **IV**, 963; **V**, 12, 188; **VI**, 612; **VIII**, 173

Potassium 3-aminopropylamide, **VIII**, 553

Potassium azide, **VI**, 209

Potassium benzohydroxamate, preparation of, **II**, 67

Potassium *tert*-butoxide in *tert*-butyl alcohol, **IV**, 132, 399, 459; **V**, 306, 361, 1061; **VI**, 72

in heptane, **V**, 361

Potassium ethyl malonate, preparation of, **IV**, 417

Potassium fluoride, anhydrous, *Hazard*, **IV**, 526

Potassium hexacyanoferrate in 2 *N* sodium hydroxide, preparation of, **V**, 1133

Potassium 1,1,1,3,3,3-hexafluoro-2-phenyl-2-propanolate [37818-31-8], **VI**, 163

Potassium hydroxide, 10 *N*, *Hazard*, **VI**, 933

alcoholic, **I**, 121

30% methanolic, **IV**, 367

Potassium hypochlorite solution, preparation of, **IV**, 486; **VIII**, 30

Potassium methoxide, **III**, 71

Potassium pentoxide in 1-pentanol, **VII**, 356

Potassium permanganate [7722-64-7], preparation of, **VI**, 68, 731, 840

tert-butyl alcohol solution, preparation of, **VI**, 72

Potassium phthalimide [1074-82-4], preparation of, **I**, 119; **IV**, 812

purification of, **VI**, 953

Potassium thiotosylate [28519-50-8], **VI**, 1016

Potassium xanthate, preparation of, **III**, 668

2-Propanol, *Hazard*, **VIII**, 201, 526

Propargyl alcohol [107-19-7], *Hazard*, **VII**, 525; **VIII**, 418

Propargyl bromide, *Hazard*, **VII**, 487

Propionaldehyde [123-38-6], **VI**, 683

Propyl iodide, **VIII**, 405

Pyridine [110-86-1], *Hazard*, **II**, 518; **III**, 29, 411, 425, 632; **V**, 243; **VI**, 60, 72, 79, 374 562, 629, 663, 758; **VII**, 16, 119, 230, 280, 360, 504; **VIII**, 271, 360, 491, 500

removal of pyrrole from, **V**, 105

3-Pyridinecarboxaldehyde [500-22-1], purification of, **VI**, 867

Pyridinium poly(hydrogen fluoride) [32001-55-1], preparation of, **VI**, 628

solution in pyridine and tetrahydrofuran, **VIII**, 271

Pyridinium-1-sulfonate [42824-16-8], preparation of, **VI**, 641

Pyrrolidine [123-75-1], **VI**, 544

Quinoline, *Hazard*, **IV**, 733; **V**, 773

Quinoline-sulfur catalyst poison, **III**, 629

Raney nickel, *Hazard*, **VI**, 544

1 *M* Red-Al [22722-98-1] in toluene, **VIII**, 244

10% Rhodium-0.1% palladium catalyst, preparation of, **V**, 671

Ricinoleoyl chloride [7459-30-5], preparation of, **IV**, 742

Ruthenium trichloride, preparation of, **V**, 1002

Salicylaldehyde, **III**, 166

Sebacoyl chloride [111-19-3], preparation of, **V**, 536

Selenium dioxide, *Hazard*, preparation, **II**, 510

poison, **II**, 510
  recovery, **IV**, 231
Semicarbazide sulfate [17464-82-3],
    preparation of, **I**, 485
Shift reagents, chiral, **VII**, 217, 368,
    404
Silica gel, **VI**, 44, 131
Silver benzoate [532-31-0], preparation
    of, **VI**, 614
Silver carbonate, preparation of, **III**, 435
Silver chloride, preparation of, **IV**, 85
Silver cyanate [3315-16-0], preparation
    of, **VI**, 796
Silver nitrite, preparation of, **IV**, 725
Silver oxide, preparation of, **II**, 20; **IV**, 548,
    919, 972; **V**, 885; **VII**, 388
Silver perchlorate in toluene, **VII**, 470
Silver(I) trifluoroacetate [2966-50-9], **IV**,
    547; **VIII**, 73
Sodium sand, *Hazard*, **I**, 252, 297; **II**,
    195; **III**, 17, 387, 829; **V**, 386,
    1090; **VI**, 168, 677
3% Sodium amalgam, **I**, 554; **II**, 609; **III**,
    589; **VI**, 461; **VIII**, 479
  1%, **II**, 609
  5%, **IV**, 509
Sodium–potassium alloy, **VI**, 168, 170
Sodium acetylide, preparation of, **III**,
    416; **IV**, 117
Sodium acetylacetonate, preparation of,
    IV, 869
Sodium amide, *Hazard*, preparation of,
    III, 44, 219, 291; **IV**, 763, 970;
    VI, 273, 928; **VIII**, 165
Sodium benzenethiolate [930-69-8],
    preparation of, **VI**, 833
Sodium benzophenone ketyl, solution in
    diethyl ether, **VII**, 237
Sodium bisulfite in 40% aq. ethanol,
    III, 439
Sodium 2-butanethiolate [66782-99-1],
    VI, 235
Sodium chloride-sodium carbonate solu-
    tion, preparation of, **VIII**, 400
Sodium chlorodifluoroacetate,
    preparation of, **V**, 391, 950

Sodium cyanamide, aq. solution, **I**, 204
Sodium cyanoacetate, aq. solution, **I**, 181
Sodium cyanoborohydride, **VI**, 500
Sodium deuteroxide [14014-06-3],
    VI, 432
Sodium dimethyl malonate, **VIII**, 7
Sodium ethoxide solution, preparation of,
    VI, 228, 654; **VIII**, 533
Sodium ethylthiolate, **VI**, 859
Sodium hydride, *Hazard*, handling of,
    V, 718
  in oil, handling of, **V**, 198, 379, 753,
    755
Sodium hydrosulfide in alcohol, *Hazard*,
    II, 573
Sodium hydroxylamine monosulfonate,
    preparation of, **III**, 205
Sodium hypobromite, preparation of,
    IV, 45; **V**, 8; **VI**, 804; **VII**, 27
Sodium hypochlorite, *Hazard*,
    preparation of, **II**, 428; **IV**, 74,
    346; **VI**, 80
  reducing pH of, **VIII**, 367
Sodium iodide, **VIII**, 1, 152, 486
Sodium isopropyl thiosulfate [26726-19-
    2], preparation of, **VI**, 235
Sodium methanesulfinate, preparation of,
    VI, 727
Sodium methoxide in methanol, **IV**, 651;
    V, 912; **VI**, 775; **VII**, 439
Sodium methoxide, powdered, **IV**, 211;
    VI, 370
Sodium sulfide, *Hazard*, **VI**, 485
Sodium sulfite solution, saturated, **VI**,
    801
Sodium tetrafluoroborate [13755-29-8],
    preparation of, **VI**, 366
Sodium *p*-toluenesulfinate, preparation
    of, **IV**, 674; **VII**, 497
Sodium trichloroacetate, preparation and
    purification of, **V**, 970
Sodium trifluoroacetate [2923-18-4],
    preparation of, **VI**, 325
Sodium trifluoromethanesulfonate,
    preparation, **VIII**, 288
Stannous bromide, preparation of, **II**, 132

Stannous chloride, anhydrous, *Hazard*,
  III, 627
Styrene glycol [93-56-1], **VI**, 60
Succinimide silver salt [55047-82-0],
  V, 664; **VI**, 226
3-Sulfolene [77-79-2], **VI**, 455
Sulfur, *Hazard*, purification of, **IV**, 668
Sulfur (Lac), **VII**, 196
Sulfuric acid, 100%, preparation of,
  IV, 978; **VI**, 49
  fuming, preparation of, **III**, 289
  20%, preparation of, **VI**, 276
Sulfur trioxide, *Hazard*, distillation of,
  IV, 848; **V**, 228
  liquid, handling of, **IV**, 864

2,3,4,6-Tetra-*O*-acetyl-α-D-glucopyranosyl
  bromide [572-09-8],
  VIII, 583
Tetrabutylammonium bromide, **VI**, 158
Tetrabutylammonium fluoride,
  preparation of and drying,
  VII, 514
Tetrabutylammonium hydrogen sulfate,
  recovery, **VI**, 955
Tetrabutylammonium perchlorate [1923-
  70-2], **VII**, 480
Tetrachloroethylene, *Hazard*, **VIII**, 124
Tetrafluoroboric acid-acetic acid mixture,
  preparation of, **VIII**, 462
Tetrahydro-1,3-dimethyl-2(1*H*)-
  pyrimidinone (DMPU) [7226-
  23-5], **VII**, 66, 238
Tetrahydrofuran, *Hazard*, **IV**, 259, 356,
  474, 793; **V**, 84, 203, 511, 976,
  1090; **VI**, 15, 24, 66, 85, 139,
  208, 240, 249, 255, 279, 321,
  359, 366, 384, 420, 527, 544,
  599, 686, 706, 739, 764, 778,
  811, 819, 848, 870, 888, 907,
  921, 926, 953, 994, 1035; **VII**,
  43, 64, 85, 121, 130, 140, 155,
  187, 192, 205, 208, 213, 221,
  242, 246, 249, 260, 268, 280,
  333, 342, 352, 384, 389, 451,
  477, 497, 504, 514, 518, 525,

  532; **VIII**, 11, 14, 20, 29, 61,
  73, 100, 134, 143, 158, 180,
  192, 214, 220, 238, 244, 270,
  296, 300, 310, 317, 327, 341,
  344, 351, 379, 383, 387, 392,
  400, 431, 442, 445, 455, 472,
  475, 481, 488, 496, 500, 502,
  508, 514, 530, 537, 552, 563,
  575, 589, 603, 611
  removal of peroxides, **V**, 695,
  976
Tetralin, **IV**, 523, 896
β-Tetralone blue test, **IV**, 906
Tetrakis(triphenylphosphine)palladium
  [14221-01-3], **VII**, 246
N-(1,1,3,3-Tetramethylbutyl)formamide
  [10151-02-7], preparation of,
  VI, 753
N,N,N',N'-Tetramethyl-1,4-diamino-2,3-
  dimethoxybutane, (*S,S*)-(+)-
  [DDB], **VII**, 449
Tetramethylethylenediamine [110-18-9],
  VI, 479; **VII**, 66, 77, 268;
  VIII, 83
2,2,6,6-Tetramethylpiperidine [768-66-1],
  VI, 572
Thallium(III) nitrate [13746-98-0],
  Hazard, preparation of, **VI**, 792
Thallium(III) trifluoroacetate [23586-53-0],
  Hazard, preparation of, **VI**, 710
Theophylline, sodium salt, **VII**, 10
2-Thienylmethylhexamethyl-
  enetetrammonium chloride,
  preparation of, **III**, 811
Thiobutyric acid [3931-64-4], preparation
  of, **VI**, 779
Thionyl chloride, *Hazard*, **II**, 570
Thiophene [110-02-1], *Hazard*, **VIII**, 34
  recovery of, **II**, 9
Thiophosgene [463-71-8], analysis,
  VII, 303
  preparation of, **I**, 506
Thorium oxide, *Hazard*, **II**, 390
Titanium tetrachloride [7550-45-0],
  Hazard, **VII**, 424, 444;
  VIII, 324

solution in hexane, **VI**, 819

Titanium trichloride [7705-07-9], *Hazard*, **VII**, 105

Toluene, *Hazard*, **III**, 168, 220, 545; **IV**, 194; **VI**, 2, 168, 934; **VII**, 132, 140, 388, 493; **VIII**, 143, 446, 526

p-Toluenesulfonic acid monohydrate, dehydration, **VIII**, 360

p-Toluenesulfonyl azide, **V**, 179

p-Toluenesulfonyl chloride, **I**, 146; **V**, 181; **VI**, 1017; **VII**, 117, 151, 501, 504; **VIII**, 360

p-Toluoyl chloride [874-60-2], preparation of, **VI**, 262

N-o-Tolylformamide [94-69-9], preparation of, **III**, 480; **V**, 1062

1,2,3-Triazoles, 1,4-disubstituted-5-amino-, titration of, **IV**, 381

2,4,6-Tribromophenol, **VI**, 182

2,3,5-Tribromothiophene [3141-24-0], preparation of, **V**, 150

Tributylhexadecylphosphonium bromide [14937-45-2], preparation of, **VI**, 833

Trichloramine, preparation of, **V**, 35

Trichloroethylene, *Hazard*, **VII**, 243

1,1,1-Trichloro-3,3,3-trifluoroacetone [758-42-9], preparation of, **VI**, 1005

Triethylaluminum [97-93-8], procedures for handling, **VI**, 14, 436

in THF, **VI**, 17

Triethylaluminum-hydrogen cyanide, handling and storage of, **VI**, 14

preparation of, **VI**, 17

Triethylamine, **IV**, 557, 561; **V**, 515, 878, 1129; **VI**, 254, 265, 284, 329, 446, 464, 569, 739, 764, 790, 1015; **VII**, 170, 314, 321, 429; **VIII**, 1, 462, 503, 552, 621

Triethylbenzylammonium chloride [56-37-1], preparation of, **VII**, 411

Triethyl phosphite, *Hazard*, **V**, 894, 902

Triethylsilane [617-86-7], **VII**, 394

Trifluoroacetic acid, *Hazard*, **VI**, 274, 475

Trifluoromethanesulfonic anhydride [358-23-6], **VI**, 325; **VIII**, 127

Trihexylborane [1188-92-7], preparation of, **VI**, 919

Triisopropyl phosphite [116-17-6], preparation of, **IV**, 326, 955

3,4,5-Trimethoxybenzoyl chloride [4521-61-3], **VI**, 1008

Trimethylamine, *Hazard*, generation of, **I**, 528; **IV**, 86

Trimethylamine oxide [1184-78-7], preparation of, **V**, 872

Trimethyl borate, *Hazard*, **V**, 920

N,N,N-Trimethylcyclooctylammonium iodide and hydroxide, **V**, 315

Trimethyltin chloride [1066-45-1], *Hazard*, preparation of, **VIII**, 97

Triphenylcinnamylphosphonium chloride, **V**, 315

Triphenylmethylphosphonium bromide, **V**, 751

Triphenylmethyl tetrafluoroborate [341-02-6], preparation of, **VI**, 999

Triphenylphosphine–bromine complex, **V**, 142, 249

Tris(tetrabutylammonium) hydrogen pyrophosphate trihydrate [76947-02-9], preparation from disodium dihydrogen pyrophosphate, **VIII**, 618

Tris(triphenylphosphine)rhodium chloride [14694-95-2], preparation of, **VI**, 459

sym-Trithiane [291-24-4], **VI**, 869

6-Trityl-β-D-glucose-1,2,3,4-tetraacetate [37074-90-1], preparation of, **III**, 432

Vanadium pentoxide, *Hazard*, **II**, 302, 556

Veratraldehyde oxime, **II**, 622

Vinylmagnesium bromide, preparation of, **IV**, 258

Woodward's reagent K [4156-16-5],
    VI, 265

Xylene, *Hazard*, **III**, 19, 389; **IV**, 842;
    VI, 914; **VII**, 342; **VIII**, 39
Xylylene-bis(triphenylphosphonium
    chloride), preparation of,
    V, 985

Zinc, *Hazard*, **III**, 74, 409, 411; **VI**, 290;
    VIII, 387

Zinc amalgam, **II**, 499; **III**, 444, 786; **IV**,
    204, 696; **VI**, 769
Zinc chloride, *Hazard*, drying and
    powdering, **VII**, 313; **VIII**, 432
    ethereal solution, **VI**, 695
    recovery, **I**, 143
Zinc–copper couple [12019-27-1],
    preparation of, **II**, 185, 526;
    V, 330, 855, 857; **VIII**, 82,
    275, 378
Zinc dust, purification of, **III**, 74

TABLES INDEX

Scattered through the various procedures are tables that collect examples of other compounds that have been made by the procedures described. Editors of some Collective Volumes have included the entries in these tables in the indices; most have not. In this Cumulative Index an index to these tables, which have become more common in recent volumes, is provided, but individual entries have not been systematically included in the other indices.

Acetamidoethyl acetophenones, substituted, **VI**, 3

Arylureas, substituted, **IV**, 51

Acetylenic diethers, **VIII**, 166

α,β-Acetylenic esters from β-chloro-alkylidene malonates, **VIII**, 250

Acid chlorides with dilithio monoethyl malonate, RCOCl RCOCH$_2$CO$_2$C$_2$H$_5$, **VII**, 214

trans-1-*N*-Acylamino-1,3-butadienes, **VI**, 100

Acyl fluorides, **V**, 69

Acyloins by thiazolium ion-catalyzed condensation of aldehydes, **VII**, 97

Addition of monolithium acetylide to aldehydes and ketones, **VIII**, 395

tert-Alcohols from hydrocarbons with ozone on silica gel, **VI**, 46

Aldehydes, by oxidation of terminal olefins, **VI**, 1029

from aromatic nitriles, **VI**, 633

from 2-benzyl-4,4,6-trimethyl-5,6-dihydro-1,3(4*H*)oxazine, **VI**, 908

from ester-mesylates, **VI**, 315

from 2-lithio-1,3,5-trithiane and alkyl halides, **VI**, 871

from nitriles with Raney nickel in formic acid, **VI**, 633

from phenols and 1,1',3,3'-tetraphenyl-2,2'-biimidazo-lidinylidene, **VII**, 163

from 1,1,3,3-tetramethylbutyl isoni-trile and organometallic reagents, **VI**, 755

from *N*,4,4-trimethyl-2-oxazolinium iodides, **VI**, 66

Alkenyl carbinols, **VII**, 459

Alkenyl halide preparation by car-bocupration, **VII**, 293

Alkoxycarbonylating agents prepared from 2-hydroxyimino-2-phenyl-acetonitrile, **VI**, 202

α-Alkylated amines from oxime sul-fonates with trialkylaluminum-diisobutylaluminum hydride, **VIII**, 571

Alkylation, of aldimine and ketimine magnesium bromide salts, **VI**, 528

of indole sodium, **VI**, 107

of nitriles, **VI**, 899

Alkylation and cleavage of 2,4-pentane-dione, **V**, 769

Alkyl bromides from alcohols, **II**, 247

9-Alkylfluorenes from primary alcohols and fluorene, **IV**, 625

n-Alkyl fluorides from alkyl bromides, **IV**, 527

tert- and *sec*-Alkyl fluorides from alcohols with hydrogen fluoride-pyridine reagent, **VI**, 630

Alkyl mercaptans from alkyl halides, **III**, 365

(2*R*)-Alkyloxiranes, **VIII**, 439

5-Alkylpentachlorocyclopentadienes, **V**, 896

N-Alkylpyrrolidines from dialkylamines, **III**, 163

α-Allenic esters from the Wittig reaction, **VII**, 234

Allylic acetoxylation of olefins, **VIII**, 138–139

Amides from esters by aminolysis with dimethylaluminum amides, **VI**, 494

α-Amino acetals, **VII**, 152

4-Amino-2-alken-1-ol derivatives, **VIII**, 12

N-Amino compounds from imide sodium salts and DPH, **VII**, 11

α-Amino ketones, **V**, 915

Amines from *tert*-alcohols, **VII**, 434

Anodic syntheses of α-methoxycarbamates, **VII**, 311

Annulated hydroquinones and quinones from 4-aryl-4-hydroxy-2,3-dialkoxycyclobutenones, **VIII**, 182

[3 + 2]-Annulations employing allenylsilanes, **VIII**, 348

Aromatic hydrocarbons from aromatic ketones and aldehydes, **VI**, 539, 540

3-Arylcarbonyl compounds from allylic alcohols and "phenylpalladium acetate," **VI**, 816

2-Arylcycloheptanones from cyclohexanone, **IV**, 783

Asymmetric hydroboration of five-substituted cyclopentadienes, **VII**, 344

Asymmetric oxidation, of ketone-derived enolates using (+)-(2*R*, 8a*S*)-10-(camphorylsulfonyl)oxaziridine, **VIII**, 109

of lithium enolates of esters and amides using (+)-(2*R*, 8a*S*)-10-(camphorylsulfonyl)oxaziridine, **VIII**, 108

Asymmetric syntheses with (+)-DDB as a chiral auxiliary agent, **VII**, 44–48

Asymmetric synthesis, of cycloalkane dicarboxylates, **VIII**, 145

of optically active carbonyl compounds and amines using the SAMP/RAMP hydrazone method, **VIII**, 408–412

Aziridines, from reduction of oximes, **V**, 86

derivatives, **V**, 87

Benzaldehydes, substituted, from benzyl bromides, **IV**, 933

Benzobarrelenes, substituted, **VI**, 85

Benzanilides from phenyl benzoates and anilines, **III**, 766

O-Benzylidene acetals with *N*-bromosuccinimide, **VIII**, 365-366

Biaryls from the Ni- or Pd-catalyzed reaction of aryl metals with aryl halides, **VIII**, 433

Boc-protected amino acids, **VI**, 205; **VII**, 72–73

Bromohydrins from alkenes, **VI**, 186

Bromoketones and *cis*-α,β-unsaturated acids from methyl ketones, **VI**, 713

N-tert-Butoxycarbonyl-protected amino acids, **VI**, 202; **VII**, 72–73

Carbocyclic esters from aprotic double Michael addition, **VIII**, 221

3-Carbomethoxy-2-pyrones, **VIII**, 446–447

Carbonylation of alcohols to acids, **V**, 742

Chain elongation of alkenes, **VI**, 188

Chiral, 4,4-dialkylcyclohexenones, **VIII**, 246

tetrahydroisoquinolines, **VIII**, 577

nonracemic building blocks and natural products through alkylation of malic acid derivatives, **VII**, 158

Chlorination of aromatic compounds in trifluoroacetic acid, **VIII**, 170

(2S)-2-Chloroalkanoic acids, **VIII**, 122

(2S)-2-Chloroalkan-1-ols, **VIII**, 438

α-Chlorocarboxylic acids, **VI**, 93

α-Chloroenamines from amides, **VI**, 286, 287

Comparison of TMB with DDB used as cosolvents in the addition of butyllithium to various aldehydes, **VII**, 49

Condensation, of aryl esters with aldehydes, **VII**, 194

of dimethyl 1,3-acetonedicarboxylate with various 1,2-dicarbonyl compounds, **VII**, 54

of 2-methyl-2-(trimethylsiloxy)pentan-3-one with aldehydes, **VII**, 189

Conjugate additions of Zn(CH$_2$CH$_2$CO$_2$Et)$_2$, **VIII**, 280

Conjugate allylation of α,β-enones with allylsilanes, **VII**, 445

Conjugate reduction of α,β-unsaturated p-toluenesulfonylhydrazones to alkenes, **VI**, 295

Conversion, of allylic alcohols into rearranged trichloroacetamides, **VI**, 510

of epoxides into δ-acetoxy-trans-α,β-unsaturated aldehydes, **VI**, 689

of esters to allylsilanes, **VIII**, 605

of methyl ketones to terminal acetylenes via enol phosphates, **VII**, 65

of nitro groups to carbonyl groups with ozone, **VI**, 650

of RCH=C(OLi)R' into RCH(OH)COR', **VII**, 281

Cross-coupling of Grignard reagents with aryl and alkenyl halides, **VI**, 409, 410

Crossed aldols from enol silyl ethers, **VIII**, 325

Cyanates from hydroxy compounds, **VII**, 437

Cyanohydrins, **VII**, 22

Cyanosilylation of carbonyl compounds in situ, **VII**, 523

of ketones and aldehydes, **VII**, 520

Cycloaddition of N-benzyl-N-methoxymethyl-N-(trimethylsilyl)methylamine with electron-deficient dipolarophiles, **VIII**, 234

[4 + 2]-Cycloaddition reactions of 2,3-dicyanobutadiene, **VI**, 431

Cyclobutanones, from tert-amides and olefins, **VIII**, 308

from chlorocyanoketene, **VIII**, 118

via 1-bromo-1-ethoxycyclopropane, **VIII**, 560

Cyclobutenones from alkynes, **VIII**, 85

Cyclopentane-1,3-diones, **VIII**, 581

Cyclopentanone synthesis by intramolecular acylation of 5-trimethylsilylalkanoyl chlorides, **VIII**, 489

Cyclopropanes from alkyl halides, alkenes, and lithium 2,2,6,6-tetramethylpiperidide, **VI**, 573, 574

Cyclopropenone 1,3-propanediol ketal, with electron-deficient olefins, **VIII**, 177

2-Deoxy sugars via reductive rearrangement of glycosyl derivatives, **VIII**, 585

β-Dialkylaminoethyl bromide hydrobromides, **II**, 92

Diazo compounds, by pyrolysis of sodium salts of tosylhydrazones in ethylene glycol, **VII**, 442

by vacuum pyrolysis of sodium salts of tosylhydrazones, **VII**, 441

α-Diazo carbonyl compounds, 2-Diazocycloalkanones, **VI**, 391

α,ω-Dibromides from diols, **III**, 228

Diels–Alder reaction, of 3-carbomethoxy-2-pyrones, **VIII**, 449

  of 3-carbomethoxy-2-pyrones with 1,1-dimethoxyethylene, **VIII**, 448

  of triethyl 1,2,4-triazine-3,5,6-tricarboxylate, **VIII**, 601

Diene synthesis via the vinylogous Ramberg–Bäckland reaction, **VIII**, 215

1,4-Diketones from silyl enol ethers and nitroolefins, **VII**, 416

N,N-Dimethyl-N'-alkylformamidines, **VIII**, 207

4,4-Dimethyl-3-hydroxy-1-cyclopentene-1-carboxaldehyde, **VIII**, 311

α,ω-Diols by reduction of diesters, **II**, 155

cis-Diols by catalytic oxidation with osmium tetroxide, **VI**, 345, 346

m-Dioxanes from alkenes, **IV**, 787

Diphenylacetoacetonitriles, **IV**, 175

Directed aldol condensations with preformed lithium enolates, **VI**, 697

Directed homogeneous hydrogenation, **VIII**, 424

Direct nucleophilic acylation with the RLi/CO reagent, **VIII**, 345

Electrolytic reductive ring closure of dihalides, **VI**, 159

Electrophilic fluorinations with N-fluoropyridinium triflates, **VIII**, 292–293

Enantiomerically pure α-sulfinyl-α,β-enones from ethylene ketals of α-bromo-α,β-enones, **VII**, 499

Enantioselective yeast reduction of ketones to the corresponding alcohols, **VII**, 219

Endocyclic enamine synthesis, **VI**, 820

Esterification of carboxylic acids with trialkyloxonium fluoborates, **VI**, 579

O-Esters from thiol esters, **VII**, 89, 91

Esters prepared by alkoxide methods, **VI**, 260, 261

Fluoro-o-nitrobenzaldehydes, **V**, 828

Formamidines via N-formyl derivatives, **VIII**, 207

Furans via acetylenic sulfonium salts, **VI**, 32

Geminal acylation-alkylation of carbonyl compounds, **VIII**, 457

Glutaconaldehyde enol benzoates and 1-substituted 3-formyl-2(1H)-pyridinethiones, **VI**, 643

Hydrofluorination of alkenes with hydrogen fluoride-pyridine reagent, **VI**, 630

Hydroformylation of cyclohexene, **VI**, 340

Hydrogenation of aromatic nuclei, **VI**, 372

Hydrogenolysis of phenolic ethers, **VI**, 152

Hydromagnesiation reactions of acetylenes, **VIII**, 510

Indoles from 2-methylnitrobenzenes, **VII**, 37–39

Intramolecular 1,3-dipolar cycloadditions of nitrones to alkenes, **VI**, 671–673

Iodides from alcohols, **VI**, 832

Iodoarenes from the corresponding arenes, **VI**, 701, 702

Iodolactonization of olefinic acids, **VII**, 167

Iron carbonyl-promoted cyclopentenone synthesis, **VI**, 523, 524

Isocyanides, by the carbylamine reaction, **VI**, 234

  by oxidation of 5-aminotetrazoles, **VII**, 28

Isoxazoles derived from oximes, **VI**, 280

Ketene S,N-acetals, **V**, 784

β-Keto esters by acylation of cadmium reagents, **III**, 604

Ketone to butylalkene conversions, **VII**, 80

Ketone to nitrile transformation, **VI**, 336

Kinetic resolutions in acrylate hydrogenation, **VIII**, 425

γ-Lactones, from HLADH-catalyzed oxidations of meso-diols, **VII**, 409
from epoxides and ethyl acetoacetate, **IV**, 12

Macrocyclic diamines from terminal alkane ditosylates, **VI**, 655
yields of cyclization for various leaving groups, **VI**, 656
yields of macrocyclic amine-ethers, **VI**, 657-659
yields of macrocyclic polyamines, **VI**, 660-661

Mercaptans from ethylene thioketal cleavage/reduction, **VII**, 126

Methenylation of ketones with Zn/CH$_2$Br$_2$/TiCl$_4$, **VIII**, 389

Methyl 2-alkynoates from 2-pyrazolin-5-ones substituted at position 3, **VI**, 793

Mono- and dialkylation products with relative topicity *ul* of (*S*)-malic esters through alkoxide enolates, **VII**, 157

Mono-*N*-alkylation of primary aromatic amines, **IV**, 422

Monosubstituted dibenzyl ketones, **IV**, 177

Natural products from (*S*)- or (*R*)-ethyl 3-hydroxybutanoate, **VII**, 218

Nitriles prepared from chlorosulfonyl isocyanate, **VI**, 466, 467

Nitroalkanes from alkyl halides, **IV**, 726

Nitrodiol monoacetates prepared using PLE-powder instead of purified enzyme, **VIII**, 335-336

α-Nitro esters from α-bromo esters, **IV**, 456

Nucleophilic acylation and carboxylation of alkyl halides and tosylates with disodium tetracarbonylferrate, **VI**, 813

Nucleophilic hydroxymethylation of aldehydes, ketones, organic halides, alcohols, and epoxides, **VIII**, 319

Olefins from allylic alcohols, ethers, and acetates, **VI**, 772, 773

Organoselenium reagents, **VI**, 536

Oxidation, of primary alcohols to aldehydes, **VIII**, 370
of primary and secondary alcohols, **VI**, 222

Palladium-catalyzed, reaction of organolithium compounds and alkenyl halides, **VII**, 175
reduction of vinyl and aryl templates, **VIII**, 129, 130

[2.2]Paracyclophanes from Diels–Alder reaction of disubstituted acetylenes with 1,2,4,5-hexatriene, **VII**, 489

Pentafluorophenylcopper, complexes of, **VI**, 879
reactions of, **VI**, 880, 881

Peptide syntheses, **VI**, 267; **VII**, 32

Phase transfer-assisted permanganate oxidations of terminal alkenes, **VII**, 399

Phenylcyclopropanes from cinnamic esters, **V**, 513

Phenylmethylcarbinols, substituted, **III**, 201

Phenyl salicylates, substituted, **IV**, 179

2-(Phenylsulfonyl)1,3-dienes from 1,3-dienes, **VIII**, 542

Photoinduced selenosulfonation-elimination of olefins, **VIII**, 545

Preparation and irradiation of 2-ethoxypyrrolin-5-ones, **VI**, 230

Preparation and rearrangement of 3-allyloxyacrylic acids, **VIII**, 539

Preparation of 2-cycloalkenones and cycloalkane-1,3-diones, **VI**, 331, 332

Preparation of dialkyl and alkyl phenyl sulfides, **VI**, 834

Preparative-scale cycloadditions of enones to alkenes, **VII**, 318

Products of organotitanium triisopropoxides [RTi(O*i*Pr)$_3$] with functionalized carbonyl compounds, **VIII**, 497

Products of transesterification with titanate catalysis, **VIII**, 203

Quinolines from 2-chloroquinolines, **III**, 521

Quinones, substituted, derived by alkylation, **VI**, 891

Reaction of various halides with activated magnesium, **VI**, 851

Reaction of ketones and aldehydes with titanium trichloride-lithium, **VII**, 3

Reactions of various substrates with (vinyl)(2-thienyl)Cu(CN)Li$_2$, **VIII**, 37

Rearranged polycyclic ketones by fragmentation and recyclization of bridgehead alcohols, **VI**, 961

Rearrangement of β-allenic esters to (E,Z)-2,4-dienoic esters with alumina, **VIII**, 253

Reduction of aromatic acids to substituted benzenes, **VI**, 749

Reductions of alkynyl ketones with B-(3)-α-pinanyl-9-BBN, **VII**, 405

Reductive, aminations with sodium cyanoborohydride, **VI**, 500
    annulation of vinyl sulfones, **VIII**, 41
    arylation of electron-deficient olefins by arenediazonium salts, **VII**, 107
    ring contraction of α,α'-disubstituted sulfolane anions, **VI**, 486
    succinoylation, **VIII**, 582

Relative reactivity and selectivity of N-chlorinated amines in trifluoroacetic acid, **VIII**, 171

Removal of the Cbz-protecting group from sulfur-containing peptides, **VI**, 257

erythro-Selective reduction of α-substituted β-keto acid derivatives with PhMe$_2$SiH/H⁺, **VIII**, 330

threo-Selective reduction of β-keto amides with PhMe$_2$SiH/H⁺, **VIII**, 331

Small-ring β-enamino esters, **VIII**, 265

Styrenes, substituted, from phenylmethylcarbinols, **III**, 205, 206

α-Substituted sec-amines via electrophilic substitution, **VI**, 545–547

4-Substituted benzoyl chlorides from substituted benzenes, **VII**, 423

gem-Sulfonic acids–carboxylic acids, **IV**, 865

Sulfonyl cyanides from sulfonyl chlorides, **VI**, 728

2-Sulfonyloxaziridines, **VIII**, 549

Synthesis of polycyclic ketones, **VI**, 668

Synthetically useful L-threitol derivatives, **VIII**, 159

1,2,5,6-Tetrahydropyridines, substituted, **VIII**, 362

Thiazoles, from condensation of ethyl isocyanoacetate and α-isocyanoacetamide with thiono esters, **VI**, 623

Trialkylnitromethanes, R$_3$CNO$_2$, **V**, 847

Trichloroacetylamines, **V**, 1075

Trifluoroacetylamino compounds from 1,1,1-trifluoro-3,3,3-trichloroacetone, **VI**, 1006

Trifluoromethyl compounds from carboxylic acids, **V**, 1084

Trimethylsilyloxycyclohexenes, **VI**, 447

Trisubstituted alkenes, from addition of Grignard-derived alkylcopper complexes to acetylenes, followed by reaction with electrophilic reagents, **VII**, 239

Ullmann reaction, ambient temperature, **VIII**, 590-591

α,β-Unsaturated β-keto esters, β-diketones, and a β-keto sulfoxide, **VI**, 26

Unsaturated macrolides from cyclic enones, **VIII**, 566

γ,δ-Unsaturated methyl ketones, **VIII**, 239

Unsymmetrical dialkyl disulfides, **VI**, 238

Vinylation of enolates with vinyl cation equivalents, **VIII**, 484

Vinyl esters, **IV**, 979

TYPE OF REACTION INDEX

This index lists the preparations contained in the eight *Collective Volumes* in accordance with general types of reactions. Only those preparations that can be classified under the selected heading with some definiteness are included. The arrangement of types of preparations is alphabetical.

A valuable source of information concerning types of reactions described in *Organic Syntheses* is the *Organic Syntheses Reaction Guide*, edited by Dennis Liotta and Mark Volmer. The procedures are classified under 11 main headings with numerous subdivisions according to the types of functional groups undergoing change. Each preparation is summarized, in structural format, with solvents, catalysts, and experimental conditions over the arrow. Citations to *Collective Volumes I–VII* and *Annual Volumes 65–68* of *Organic Syntheses* are given.

ACETAL (and THIOACETAL) FORMATION

Acetone dibutyl acetal, **V**, 5

Acrolein acetal, **IV**, 21

Atropaldehyde diethyl acetal, **VII**, 13

Bromoacetal, **III**, 123

α-Bromocinnamic aldehyde acetal, **III**, 732

2-Bromo-2-cyclopenten-1-one ethylene ketal, **VII**, 271

α-Bromoheptaldehyde dimethyl acetal, **III**, 128

2-(Bromoethyl)-1,3-dioxane, **VII**, 59

2-(Bromomethyl)-2-(chloromethyl)-1,3-dioxane, **VIII**, 173

Cyclohexanone diallyl acetal, **V**, 292

2-Cyclohexyloxyethanol, **V**, 303

2,3-Di-*tert*-butoxy-1,4-dioxane, **VIII**, 161

8-[(*E*)-1,2-Dichlorovinyl]-8-methyl-1,4-dioxaspiro[4.5]dec-6-cene, **VII**, 241

Dicyanoketene ethylene acetal, **IV**, 276

cis- and *trans*-2,6-Diethoxy-1,4-oxathiane 4,4-dioxide, **VI**, 976

2,2-Diethoxy-2-(4-pyridyl)ethylamine, **VII**, 149

Diethyl [(2-tetrahydropyranyloxy)methyl]phosphonate, **VII**, 160

Diisopropylidene glycerol, **III**, 502

7,7-Dimethoxybicyclo[2.2.1]heptene, **V**, 424

6,6-Dimethoxyhexanal, **VII**, 168

5,5-Dimethoxy-1,2,3,4-tetrachloro-cyclopentadiene, **V**, 424

N,N-Dimethyl-5β-cholest-3-ene-5-acetamide, **VI**, 491

Dimethyl 2,3-*O*-isopropylidene-L-tartrate, **VIII**, 155

1,4-Dioxaspiro[4.5]decane, **V**, 303
1-(1,3-Dithian-2-yl)-2-cyclohexen-1-
ol, **VIII**, 309
1,4-Dithiaspiro[4.11]hexadecane,
VII, 124
2-[(1'-Ethoxy)-1-ethoxy]butyroni-
trile, **VII**, 381
4-Ethoxy-3-methoxybenzaldehyde
ethylene acetal, **VI**, 567
β-Ethoxypropionaldehyde acetal,
III, 371
Ethyl diethoxyacetate, **IV**, 427
8-Ethynyl-8-methyl-1,4-dioxa-
spiro[4.5]dec-6-ene, **VII**, 241
α-Hydroxyacetophenone dimethyl
acetal, **VII**, 263
2-Hydroxymethyl-2-cyclopentenone
ethylene ketal, **VII**, 271
2,3-*O*-Isopropylidene-D-erythrono-
lactone, **VII**, 297
Ketene diethylacetal, **III**, 506
β-Ketoisooctaldehyde dimethyl
acetal, **IV**, 558
Methyl 4,6-*O*-benzylidene-α-D-glu-
copyranoside, **VIII**, 36
Methyl 6,6-dimethoxyhexanoate,
VII, 168
Monobenzalpentaerythritol, **IV**, 681
m-Nitrobenzaldehyde dimethyl
acetal, **III**, 644
o-Nitrobenzaldiacetate, **III**, 641
p-Nitrobenzaldiacetate, **II**, 441
Phenylpropargylaldehyde, **III**, 731
Phenylpropargylaldehyde diethyl
acetal, **IV**, 801
2-Phenyl-1,3-dithiane, **VI**, 109
Tetrahydro-2-(2-propynyloxy)-2*H*-
pyran, **VII**, 334
3,3,6,6-Tetramethoxy-1,4-cyclohexa-
diene, **VI**, 971
(*S*)-(+)-2-(*p*-Toluenesulfinyl)-2-
cyclopentenone ethylene ketal,
VII, 495
2,5,5-Trimethyl-2-(2-bromoethyl)-
1,3-dioxane, **VII**, 60

**ACYLATION (including FORMYLA-
TION and FRIEDEL–
CRAFTS REACTIONS)
(see also Esterification)**
Acetamidoacetone, **V**, 27
3-Acetamido-2-butanone, **IV**, 5
Acetanilide, **I**, 332
β-(3-Acenaphthoyl)propionic acid,
III, 6
Acetic formic anhydride, **VI**, 8
Acetobromoglucose, **III**, 11
Aceto-*p*-cymene, **II**, 3
Acetophenone, **I**, 111
2-Acetothienone, **II**, 8; **III**, 14
p-Acetotoluidide, **I**, 111
3β-Acetoxyetienic acid, **V**, 8
(1*R*,2*r*,3*S*)-3-Acetoxy-2-nitrocyclo-
hexyl acetate, **VIII**, 332
2-Acetoxy-*trans*-1-phenylpropene,
VI, 692
9-Acetylanthracene, **IV**, 8
p-Acetyl-α-bromohydrocinnamic
acid, **VI**, 21
2-Acetyl-1,3-cyclopentanedione, **VI**,
28
2-Acetyl-6,7-dimethoxy-1-methyl-
ene-1,2,3,4-tetrahydroisoquino-
line, **VI**, 1
2-Acetylfluorene, **III** 23
Acetylglycine, **II**, 11
N-Acetylisatin, **III**, 456
Acetylmandelic acid, **I**, 12
2-Acetyl-6-methoxynaphthalene, **VI**,
34
3-Acetyl-2-oxazolidinone, **VII**, 5
N-Acetyl-*N*-phenylhydroxylamine,
VIII, 16
2-Aminobenzophenone, **IV**, 34
2-Amino-3-nitrotoluene, **IV**, 42
Benzanilide, **I**, 82
Benzohydroxamic acid, **II**, 67
Benzoin acetate, **II**, 69
Benzophenone, **I**, 95
Benzoylacetanilide, **IV**, 80
β-Benzoylacrylic acid, **III**, 109

ε-Benzoylaminocaproic acid, **II**, 76

γ-Benzoylbutyric acid, **II**, 82

Benzoylcholine chloride, **IV**, 84

1-Benzoyl-2-(1,3-dibenzoyl-4-imidazolin-2-yl)imidazole, **VII**, 287

o-Benzoyloxyacetophenone, **IV**, 478

Benzoylpiperidine, **I**, 99

1-(2-Benzoylpropanoyl)piperidine, **VIII**, 326

β-Benzoylpropionic acid, **II**, 81

3-Benzoylpyridine, **IV**, 88

Benzyl carbamate, **III**, 167

Benzyloxycarbonyl-L-alanyl-L-cysteine methyl ester, **VII**, 30

Benzyloxycarbonyl-L-aspartyl-(tert-butyl ester)-L-phenylalanyl-L-valine methyl ester, **VII**, 30

4,4'-Bis(dimethylamino)benzil, **V**, 111

Bis(2,2,2-trichloroethyl) hydrazodicarboxylate, **VII**, 56

p-Bromoacetophenone, **I**, 109

β-Bromoethylphthalimide, **IV**, 106

Boc-L-leucine-N-methyl-O-methylcarboxamide, **VIII**, 68

N-tert-Butoxycarbonyl-L-phenylalanine, **VII**, 70, 75

tert-Butyl acetate, **III**, 141

tert-Butyl azodiformate, **V**, 160

1-(4-tert-Butylcyclohexen-1-yl)-2-propen-1-one, **VIII**, 97

2-Butyl-2-methylcyclohexanone, **V**, 187

tert-Butyl S-methylthiolcarbonate, **V**, 166

S-tert-Butyl 3-oxobutanethioate, **VIII**, 71

3-Butyl-2,4-pentanedione, **VI**, 245

tert-Butyl phenylcarbonate, **V**, 166

tert-Butylphthalimide, **III**, 152

(S)-(−)-1-Carbamoyl-2-methoxymethylpyrrolidine, **VIII**, 26

2-Carbethoxycyclooctanone, **V**, 198

1-Carbomethoxy-1-methylethyl 3-oxobutanoate, **VIII**, 71

N-Carbomethoxypyrrolidine, **VII**, 307

1,1'-Carbonyldiimidazole, **V**, 201

α-Cellobiose octaacetate, **II**, 124

p-Chloroacetophenone, **I**, 111

p-Chloroacetylacetanilide, **III**, 183

α-Chloroacetyl isocyanate, **V**, 204

p-Chlorobenzoic anhydride, **III**, 29

(2-Chlorobenzoyl)ferrocene, **VI**, 625

2-Chloro-1-formyl-1-cyclohexene, **V**, 215

p-Chlorophenyl isothiocyanate, **V**, 223

p-Chlorophenyl salicylate, **IV**, 178

Cholesteryl acetate, **II**, 193

Cyclohexylurea, **V**, 801

Desoxybenzoin, **II**, 156

1-Diazo-4-phenyl-2-butanone, **VIII**, 196

1,4-Dibenzoylbutane, **II**, 169

trans-Dibenzoylethylene, **III**, 248

sym-Dibenzoylhydrazine, **II**, 208

Diethyl acetamidomalonate, **V**, 373

Diethyl α-acetyl-β-ketopimelate, **V**, 384

Diethyl benzoylmalonate, **IV**, 285

N-(2,4-Diformyl-5-hydroxyphenyl)-acetamide, **VII**, 162

1,2,5,6-Di-O-isopropylidene-3-O-(S-methyl dithiocarbonate)-α-D-glucofuranose, **VII**, 139

(−)-Dimenthyl succinate, **VIII**, 141

p-Dimethylaminobenzaldehyde, **IV**, 331

N,N-Dimethylcyclohexanecarboxamide, **IV**, 339

2,2-Dimethyl-5-phenylacetyl-1,3-dioxane-4,6-dione, **VII**, 359

2,6-Dimethylphenyl propanoate, **VII**, 190

2,6-Dimethylphenylthiourea, **V**, 802

α,α-Diphenylacetone, **III**, 343

Diphenylacetonitrile, **III**, 347

Diphenyl succinate, **IV**, 390

Diphenyl sulfide, **II**, 242

Durene, **II**, 248
10,11-Epoxyfarnesyl acetate, **VI**, 560
1-(Ethenyl)heptanyl 3-ketobutanoate, **VIII**, 235
Ethyl azodicarboxylate, **IV**, 411
Ethyl benzoylacetate, **II**, 266; **IV**, 415
Ethyl benzoylacetoacetate, **II**, 266
Ethyl 2-butyrylacetate, **VII**, 213
N-Ethyl-p-chloroformanilide, **IV**, 420
Ethyl diacetylacetate, **III**, 390
1-Ethyl-3-(3-dimethylamino)propylcarbodiimide hydrochloride and methiodide, **V**, 555
Ethyl 2,5-dimethylphenylhydroxymalonate, **III**, 326
Ethyl 2-fluorohexanoate, **VIII**, 258,
Ethyl 6-methylpyridine-2-acetate, **VI**, 611
Ethyl (E)-4-(4-nitrophenyl)-4-oxo-2-butenoate, **VIII**, 268
Ethyl N-tricarboxylate, **III**, 415
Ethyl 5-oxo-6-methylheptenoate, **VIII**, 274
Ethyl 2,3-pentadienoate, **VII**, 232
Ethyl tetrolate **VII**, 226
Farnesyl acetate, **VI**, 560
Flavone, **IV**, 478
(S)-N-Formyl-O-tert-butylvalinol, **VIII**, 204
(S)-(–)-1-Formyl-2-hydroxymethylpyrrolidine, **VIII**, 26
1-Formyl-3-thiosemicarbazide, **V**, 1070
Furfural diacetate, **IV**, 489
2-Furylmethyl acetate, **I**, 285
Fumaramide, **IV**, 486
Gallacetophenone, **II**, 304
β-Gentiobiose octaacetate, **III**, 428
β-D-Glucose-1,2,3,4-tetraacetate, **III**, 432
Heptaldehyde enol acetate, **III**, 127
Heptamide, **IV**, 513
Heptoic anhydride, **III**, 28

N-Heptylurea, **IV**, 515
Hexamethylene diisocyanate, **IV**, 521
Hexamethylphosphorous triamide, **V**, 602
Hippuric acid, **II**, 328
Hydroquinone diacetate, **III**, 452
Hydroxyhydroquinone triacetate, **I**, 317
2-Hydroxy-1-naphthaldehyde, **III**, 463
6-Hydroxynicotinic acid, **IV**, 532
Hydroxyurea, **V**, 645
Imidazole-2-carboxaldehyde, **VII**, 282
Indazole, **V**, 650
Indole-3-aldehyde, **IV**, 539
Isatoic anhydride, **III**, 488
α-Ketoglutaric acid, **V**, 687
Mesitoic acid, **V**, 706
p-Methoxyacetophenone, **I**, 111
1-(p-Methoxyphenyl)-5-phenyl-1,3,5-pentanetrione, **V**, 718
2-(p-Methoxyphenyl)-6-phenyl-4-pyrone, **V**, 721
p-Methylacetophenone, **I**, 111
4-Methyl-7-acetoxycoumarin, **III**, 283
4-Methylbicyclo[2.2.2]octane-2,6-dione, **VIII**, 467
3-Methyl-2-buten-2-yl triflate, **VI**, 757
Methyl 4-chloro-2-butynoate, **VIII**, 371
2-Methyl-1-cyclohexen-1-yl acetate, **VI**, 121
Methyl diformylacetate, **VII**, 323
Methyl hydrogen succinate, **III**, 169
Methyl 4-hydroxy-2-butynoate, **VII**, 334
trans-2-Methyl-2-dodecenoic acid, **IV**, 608
2-Methylenedodecanoic acid, **IV**, 616
N-Methyl-α-bromobutyranilide, **IV**, 620

(1*R*,2*S*,5*R*)-5-Methyl-2-(1-methyl-1-phenylethyl)cyclohexyl chloroacetate, **VIII**, 522
3-Methyloxindole, **IV**, 657
Methyl 2-oxocyclohexanecarboxylate, **VII**, 351
Methyl 6-oxodecanoate, **VIII**, 441
Methyl 7-oxoöctanoate, **VI**, 807
Methyl phenylacetylacetate, **VII**, 359
2-Methyl-2-vinylcyclopentanone, **VIII**, 486
o-Nitroacetophenone, **IV**, 708
o- and *p*-Nitrobenzaldiacetate, **IV**, 713
2-Nitro-4-methoxyaniline, **III**, 661
(*S*)-3-(1-Oxopropyl)-4-(phenylmethyl)-2-oxazolidinone, **VIII**, 339
Parabanic acid, **IV**, 744
Pentaerythrityl tetrabromide, **IV**, 753
trans-3-Penten-2-one, **VI**, 883
4-Pentylbenzoyl chloride, **VII**, 420
2-Phenylcycloheptanone, **IV**, 780
trans-2-Phenylcyclohexyl chloroacetate, **VIII**, 516
o-Phenylene carbonate, **IV**, 788
N-(2-Phenylethyl)benzamide, **V**, 336
Phenylglyoxal hemimercaptal, **V**, 937
N-Phenylmaleimide, **V**, 944
(*S*)-4-(Phenylmethyl)-2-oxazolidinone, **VIII**, 339
3-Phenylpropionaldehyde, **VII**, 451
Phenyl thienyl ketone, **II**, 520
1-Phenyl-2-thio-4-methylisobiuret, **V**, 966
1-Phenyl-2-thiobiuret, **V**, 966
Phloroacetophenone, **II**, 522
N-Phthalyl-L-β-phenylalanine, **V**, 973
β-Propionylphenylhydrazine, **IV**, 657
Putrescine dihydrochloride, **IV**, 819
2-Pyrrolealdehyde, **IV**, 831
Pyrrol-2-yl trichloromethyl ketone, **VI**, 618

Resacetophenone, **III**, 761
Sebacoyldicyclohexanone, **V**, 533
Syringic aldehyde, **IV**, 866
2,3,4,6-Tetra-*O*-acetyl-D-gluconic acid monohydrate, **V**, 887
α-Tetralone, **II**, 569; **IV**, 898
(*R*,*R*)-(+)-*N*,*N*,*N'*,*N'*-Tetramethyltartramide, **VII**, 41
2-Thenaldehyde, **IV**, 915
3-Thenaldehyde, **IV**, 918
2a-Thiohomophthalimide, **V**, 1051
DL-Threonine, **III**, 813
p-Toluenesulfonylhydrazide, **V**, 1055
p-Toluenesulfonylmethylnitrosamide, **IV**, 943
p-Toluyl-*o*-benzoic acid, **I**, 517
1,2,5-Triacetoxypentane, **III**, 833
1,2,4-Triazole, **V**, 1070
α,α,α-Trichloroacetanilide, **V**, 1074
1,1,1-Trichloro-4-ethoxy-3-buten-2-one, **VIII**, 254
Triethyl oxalylsuccinate, **V**, 687
Triethyl phosphite, **IV**, 955
N-Trifluoroacetanilide, **VI**, 1004
Trifluoroacetyl triflate, **VII**, 506
N-Trifluoroacetyl-L-tyrosine, **VI**, 1004
2,5-Undecanedione **VIII**, 620

ADDITION

  A. To C=C
1-Acetoxy-4-chloro-2-butene, **VIII**, 9
cis-1-Acetoxy-4-chloro-2-cyclohexene, **VIII**, 5
cis-3-Acetoxy-5-hydroxycyclopent-1-ene, **VIII**, 13
β-Aminopropionitrile, **III**, 93
[18]Annulene, **VI**, 68
(1-Azido-2-iodo-3,3-dimethoxypropyl)benzene, **VI**, 893
Benzalacetophenone dibromide, **I**, 205
exo-cis-Bicyclo[3.3.0]octane-2-carboxylic acid, **V**, 93
Bis(3-methyl-2-butyl)borane, **VI**, 919

Bromoacetal, **III**, 123

erythro-2-Bromo-1,2-diphenylethanol, **VI**, 184

1-Bromo-2-fluoroheptane, **V**, 136

α-Bromoheptaldehyde, **III**, 127

10-Bromo-11-hydroxy-10,11-dihydrofarnesyl acetate, **VI**, 560

1-Bromo-1-methyl-2-(bromomethylsulfonyl)cyclohexane, **VIII**, 212

17-β-*tert*-Butoxy-5α-androstan-3-one, **VII**, 66

Butyrchloral, **IV**, 130

2-Carbomethoxy-3-vinylcyclopentanone, **VIII**, 112

2-Chlorocyclooctanone oxime hydrochloride, **V**, 266

trans-2-Chlorocyclopentanol, **IV**, 157

1-Chloro-1,2-diphenyl-3-methylcyclopropane, **VII**, 203

2-Chlorocyclohexanol, **I**, 158

α-Chlorohydrindene, **II**, 336

4-(4-Chlorophenyl)butan-2-one, **VII**, 105

2-Chloro-1,1,2-trifluoroethyl ethyl ether, **IV**, 184

3-Chloro-2,2,3-trifluoropropionic acid, **V**, 239

Cholesterol, **IV**, 195

Cholesterol dibromide, **IV**, 195

Coumarin dibromide, **III**, 209

cis-1,2-Cyclohexanediol, **VI**, 343

Cyclohexylbenzene, **II**, 151

1,2-Cyclononadiene, **V**, 306

cis-Cyclooctene oxide, **VII**, 126

2-Decanone, **VII**, 137

1-Deoxy-2,3,4,6-*tetra*-*O*-acetyl-1-(2-cyanoethyl)-α-D-glucopyranose, **VIII**, 148

2,6-Diacetoxybicyclo[3.3.0]octane, **VIII**, 43

1,2-Dibromocyclohexane, **II**, 171

2-(2,2-Dibromocyclopropyl)-6,6-dimethylbicyclo[3.1.1]hept-2-ene, **VIII**, 223

7,7-Dibromo-1,6-dimethylbicyclo[4.1.0]hept-3-ene, **VII**, 200

4,5-Dibromo-1,2-dimethyl-1,2-epoxycyclohex-4-ene, **V**, 468

erythro-α,β-Dibromo-β-phenylpropionic acid, **IV**, 961; **VII**, 173

α,β-Dibromosuccinic acid, **II**, 177

2,5-Di-*tert*-butyl-5,6-dichloro-2-cyclohexene-1,4-dione, **VI**, 210

2,5-Di-*tert*-butyl-3,5,6-trichloro-2-cyclohexene-1,4-dione, **VI**, 211

Dicarbonyl(cyclopentadienyl)(*trans*-3-methyl-2-vinylcyclohexanone)iron tetrafluoroborate, **VIII**, 479

3,4-Dichloro-1,2,3,4-tetramethylcyclobutene, **V**, 370

2,3-Dichlorotetrahydropyran, **VI**, 675

Diethyl *tert*-butylmalonate, **VI**, 442

N,N-Diethylgeranylamine, **VIII**, 188

N,N-Diethylnerylamine, **VIII**, 190

Diethyl [(2-tetrahydropyranyloxy)methyl]phosphonate, **VII**, 160

Diethyl propionylsuccinate, **VI**, 615

Dihydrocarvone, **VI**, 459

4,5-Dihydro-5-iodomethyl-4-phenyl-2(3*H*)-furanone, *cis*- and *trans*-, **VII**, 164

9,10-Dihydroxystearic acid, **IV**, 317

3,3-Dimethyl-1,5-diphenylpentane-1,5-dione, **VIII**, 210

Dimethyl (*E*)-2-hexenedioate, **VIII**, 112

Diphenylacetylene, **III**, 350

trans-2,3-Diphenyl-1-phthalimidoaziridine, **VI**, 56

2-(1-Ethoxycyclohexyl)-2-(trimethylsiloxy)cyclobutanone, **VIII**, 579

2-[(1'-Ethoxy)-1-ethoxy]butyronitrile, **VII**, 381

Ethyl α,β-dibromo-β-phenylpropionate, **II**, 270

Ethyl enanthylsuccinate, **IV**, 430

Ethyl *threo*-[1-(2-hydroxy-1,2-diphenylethyl)]carbamate, **VII**, 223

Ethyl 3-[3-(trimethylsiloxy)cyclo-
hex-2-enyl]propionate, **VIII**,
277
unsym-Heptachloropropane, **II**, 312
Hexabromostearic acid, **III**, 531
Homogeraniol, **VII**, 258
β-(2-Hydroxyethylmercapto)propi-
onitrile, **III**, 458
1-Hydroxymethyl-4-(1-methylcyclo-
propyl)-1-cyclohexene, **VIII**,
321
2-Indanone, **V**, 647
Iodocyclohexane, **IV**, 543
(–)-Isopinocampheol, **VI**, 719
Linoleic acid, **III**, 526
Linolenic acid, **III**, 531
p-Menth-8-en-1-yl methyl ether,
VII, 304
6-Methylbicyclo[4.2.0]octan-2-one,
VII, 315
Methyl α-bromo-β-methoxypropionate, **III**, 774,
3-Methylcyclohexanone-3-acetic
acid, **VIII**, 467
Methyl 1,3-dimethyl-5-oxobicy-
clo[2.2.2]octane-2-carboxylate,
VIII, 219
2-Methyl-1-hexanol, **VI**, 786
Methyl (1*R*,5*R*)-5-hydroxy-2-
cyclopentene-1-acetate, **VII**,
339
Methyl 2-iodo-3-nitropropionate, **VI**,
799
Methyl (*trans*-2-iodo-1-tetralin)car-
bamate, **VI**, 795
(2*RS*,5*R*)-5-Methyl-2-(1-methyl-1-
phenylethyl)cyclohexanone,
VIII, 522
cis- and *trans*-5-Methyl-2-[1-methyl-
1-(phenylmethylthio)ethyl]-
cyclohexanone, **VIII**, 304
2-Methyl-3-(*p*-toluenesulfonamido)-
2-pentanol, **VII**, 375
2-*exo*-Norbornyl formate, **V**, 852
Nortricyclyl acetate, **V**, 863
γ-Octyl-γ-butyrolactone, **VII**, 400
tert-Octyl hydroperoxide, **V**, 818

4-Oxo-4-(3-pyridyl)butyronitrile, **VI**,
866
trans-3-Penten-2-one, **VI**, 883
Perhydro-9b-boraphenalene, **VII**,
427
4-Phenyl-*m*-dioxane, **IV**, 786
trans-3-(Phenylsulfonyl)-4-
(chloromercuri)cyclohexene,
VIII, 540
3-Pinanamine, **VI**, 943
8-Propionyl-(*E*)-5-nonenolide, **VIII**,
562
(2*S*,3*S*)-3-Propyloxiranemethanol,
VII, 461
Stearolic acid, **III**, 785; **IV**, 851
Stilbene dibromide, **III**, 350
Styrylphosphonic dichloride, **V**,
1005
Tetrabromostearic acid, **III**, 526
Tetrahydro-2-(2-propynyloxy)-2*H*-
pyran, **VII**, 334
2,2,5,5-Tetramethyltetrahydro-3-
ketofuran, **V**, 1024
cis-2-(*p*-Toluenesulfonamido)cyclo-
hexanol, **VII**, 375
1,2,3-Tribromopropane, **I**, 521
1,1,1-Trichloro-4-ethoxy-3-buten-2-
one, **VIII**, 254
1,1,3-Trichlorononane, **V**, 1076
Trihexylborane, **VI**, 919
2,5-Undecanedione, **VIII**, 620
10-Undecynoic acid, **IV**, 969
B. To C≡C
1-Acetylcyclohexanol, **IV**, 13
(*E*)-1,2-Bis(tributylstannyl)ethylene,
VIII, 268
(*E*)-(Carboxyvinyl)trimethylammo-
nium betaine, **VIII**, 536
β-Chlorovinyl isoamyl ketone, **IV**,
186
2,3-Diphenyl-1,3-butadiene, **VI**, 531
Ethyl 3,4-decadienoate, **VIII**, 251
Ethyl isocrotonate, **VII**, 226
(*E*)-5-Ethyl-1,4-undecadiene, **VII**,
236
Hexamethyl Dewar benzene, **VII**,
256

(*E*)-1-Hexenyl-1,3,2-benzodioxabo-
role, **VIII**, 532
Lithium di-(*Z*-hexenyl)cuprate (in
solution), **VII**, 290
(*E*)-(2-Methyl-1,3-butadienyl)-
dimethylalane, **VII**, 245
1-Methyl-1-(trimethylsilyl)allene,
VIII, 47
(*E*)-3-Pentyl-2-nonene-1,4-diol,
VIII, 507
2-Phenylfuro[3,2-*b*]pyridine, **VI**,
916
2-Phenyl-2-vinylbutyronitrile, **VI**,
940
Tetramethyl[2,2]paracyclophane-
4,5,12,13-tetracarboxylate, **VII**,
485
2,2,5,5-Tetramethyltetrahydro-3-
ketofuran, **V**, 1024
(*Z*)-4-(Trimethylsilyl)-3-buten-1-ol,
VIII, 471
(*E*)-Trimethylsilyl-2-propen-1-ol,
VII, 524
C. To C=N
1-Cyano-2-benzoyl-1,2-dihydroiso-
quinoline, **IV**, 641
1-Methyl-2-pyridone, **II**, 419
4-Nitroindole, **VIII**, 493
Trimethyl *ortho*-4-bromobutanoate,
VIII, 415
D. To C≡N
Acetamidine hydrochloride, **I**, 5
Benzamidine hydrochloride, **I**, 6
2-Benzoyl-1,2-dihydroisoquinal-
donitrile, **VI**, 115
sec-Butyl α-caproylpropionate, **IV**,
120
Diphenyl ketimine, **V**, 520
N,N-Dimethylselenourea, **IV**, 359
Ethyl α-phenylacetoacetate, **II**, 284
Ethyl thioamidooxalate, **VIII**, 597
N-Formyl-α,α-dimethyl-β-phenethyl-
amine, **V**, 471
Guanidine nitrate, **I**, 302
Guanylthiourea, **IV**, 502

Methyl 4-bromo-1-butanimidate
hydrochloride, **VIII**, 415
Methylisourea hydrochloride, **IV**,
645
N-Phenylbenzamidine, **IV**, 769
1-Phenyl-3,4-dihydro-6,7-methyl-
enedioxyisoquinoline, **VII**, 326
Phloroacetophenone, **II**, 522
E. To C=O
Acetone cyanohydrin, **II**, 7
DL-Alanine, **I**, 21
Atrolactic acid, **IV**, 58
Benzaldehyde tosylhydrazone, **VII**,
438
Benz[*a*]anthracene, **VII**, 18
Benzophenone cyanohydrin, **VII**, 20
p-Benzoquinone monotrimethylsilyl
cyanohydrin, **VII**, 517
N-Benzyl-2-azanorbornene, **VIII**, 31
Benzyl chloromethyl ether, **VI**, 101
N-Benzylidenebenzenesulfonamide,
VIII, 546
Benzyl isocyanide, **VII**, 27
5-Benzyloxy-3-hydroxy-3-
methylpentanoic-2-^{13}C acid,
VII, 386
1,3-Bis(methylthio)-1-hexen-4-ol,
VI, 684
2-Bromo-2-cyclopentenone ethylene
ketal, **VII**, 271
6-Bromo-3,4-dimethoxybenzalde-
hyde cyclohexylimine, **VIII**,
586
p-Bromophenylurea, **IV**, 49
S-tert-Butyl 3-oxobutanethioate,
VIII, 71
Butyroin, **VII**, 95
D-Camphor 2,4,6-triisopropylben-
zenesulfonylhydrazone, **VII**, 77
Cinnamonitrile, **VII**, 108
1-Cyano-3-phenylurea, **IV**, 213
1,2-Cyclobutanediol, **VII**, 129
Cyclodecanone, **V**, 277
1,2-Cyclohexanedione dioxime, **IV**,
229

Cyclohexylideneacetonitrile, **VII**, 108

Cyclohexyl methyl ketone, **V**, 775

Cyclohexylurea, **V**, 801

1-(Cyclooctylidenemethyl)piperi-
dine, **VII**, 473

Di-*tert*-butyl methylenemalonate,
VII, 142

1,2-Di-1-(1-cyano)cyclohexylhy-
drazine, **IV**, 274

Diethylaminoacetonitrile, **III**, 275

Diethyl *N*-benzylideneaminomethyl-
phosphonate, **VIII**, 451

Diethyl hydroxymethylphosphonate,
VII, 160

Dihydrojasmone, **VIII**, 620

3,4-Dihydro-2-methylene-2*H*-1-ben-
zopyran, **VIII**, 512

α-*N,N*-Dimethylaminophenylace-
tonitrile, **V**, 437

(*E*)-7,11-Dimethyl-6,10-dodecadien-
1-yne, **VIII**, 226

Dimethylethynylcarbinol, **III**, 320

4,4-Dimethyl-1,6-heptadien-3-ol,
VII, 177

(*E*)-4,8-Dimethyl-1,3,7-nonatriene,
VII, 259

2',6'-Dimethylphenyl (2*SR*,3*SR*)-2,4-
dimethyl-3-hydroxypentanoate,
VII, 190

Diphenylacetylene, **IV**, 377

2-(Diphenylhydroxymethyl)pyrroli-
dine, **VI**, 542

1,4-Diphenyl-1-methyl-3-aza-1,4-
butadiene, **VIII**, 451

N,α-Diphenylnitrone, **V**, 1124

1-(1,3-Dithian-2-yl)-2-cyclohexen-1-
ol, **VIII**, 309

1,4-Dithiaspiro[4.1.1]hexadecane,
VII, 124

(*E*)-1-Ethoxy-1-(1-hydroxy-2-
butenyl)cyclopropane, **VIII**,
556

p-Ethoxyphenylurea, **V**, 52

1-Ethyl-3-(3-dimethylamino)propyl-
carbodiimide, **V**, 555

(*S*)-(+)-Ethyl 3-hydroxybutanoate,
VII, 215

Ethyl 1-hydroxycyclohexylacetate,
VI, 598

Ethyl β-hydroxy-β,β-diphenylpropi-
onate, **V**, 564

Ethyl α-(hydroxymethyl)acrylate,
VIII, 265

Ethyl *N*-(2-methyl-3-
nitrophenyl)formimidate, **VIII**,
493

Ethyl 2,3-pentadienoate, **VII**, 232

1-Ethynylcyclohexanol, **III**, 416

1,1'-Ethynylene-bis-cyclohexanol,
IV, 471

Glycolonitrile, **III**, 436

Hexahydro-2-(1*H*)-azocinone, **VII**,
254

2-Hydroxybutanenitrile, **VII**, 381

3-(1-Hydroxybutyl)-1-methylpyr-
role, **VII**, 102

2-Hydroxymethyl-2-cyclopentenone,
VII, 271

2-Hydroxy-2-methylpentan-3-one,
VII, 381

3-Hydroxy-3-methyl-1-phenyl-1-
butanone, **VIII**, 323

4-Hydroxynona-1,2-diene, **VII**, 276

3-Hydroxy-1-nonene, **VIII**, 235

(2*S**,3*S**)-3-Hydroxy-3-phenyl-2-
methylpropanoic acid, **VIII**,
339

3-Hydroxy-2,2,3-trimethyloctan-4-
one, **VIII**, 343

5-Hydroxy-2,4,6-trimethyl-2-
trimethylsiloxyheptan-3-one,
VII, 185

1-[(Isopropoxydimethylsilyl)-
methyl]cyclohexanol, **VIII**, 315

Mandelic acid, **I**, 336

Methoxyacetonitrile, **II**, 387

N-Methyl-2,3-dimethoxybenzy-
lamine, **IV**, 603

Methyl 5,5-dimethoxy-3-methyl-2,3-
epoxypentanoate, **IV**, 649

N-Methyl-2-diphenylethylamine, **IV**, 605

(+)-Methylene-*cis*-*p*-menthane, **VIII**, 386

2-Methylene-1-oxo-1,2,3,4-tetrahydronaphthalene, **VII**, 332

1-Methyl-2-ethynyl-2-*endo*-3,3-dimethyl-2-norbornanol, **VIII**, 391

Methyl 3-hydroxy-2-methylenepentanoate, **VIII**, 420

Methyl (*trans*-2-iodo-1-tetralin)carbamate, **VI**, 795

[(Methyl)(methoxy)carbene]pentacarbonyl chromium(0), **VIII**, 216

3-Methyl-2-pentyl-2-cyclopenten-1-one, **VIII**, 620

2-Methyl-2-undecene, **VIII**, 474

1-Morpholino-1-cyclohexene, **V**, 808

m-Nitroethylbenzene, **VII**, 393

1-(Nitromethyl)cyclohexanol, **IV**, 221, 224

3'-Nitro-1-phenylethanol, **VIII**, 495

(*R*)-(+)-1-Octyn-3-ol, **VII**, 402

D-(−)-Pantoyl lactone, **VII**, 417

3-Pentanone SAMP-hydrazone, **VIII**, 403

(*E*)-3-Pentyl-2-nonene-1,4-diol, **VIII**, 507

1-Phenoxy-1-phenylethene, **VIII**, 512

(*R*)-(+)-Phenyl-1-pentanol, **VII**, 447

1-Phenyl-2,2,2-trichloroethanol, **V**, 130

1-(*N*-Pyrrolidino)-1-cyclododecene (in solution), **VII**, 135

1,2,3,4-Tetrahydrocarbazole, **IV**, 884

β-Tetralone, **IV**, 903

2,2,6,6-Tetramethylolcyclohexanol, **IV**, 907

Thiobenzophenone, **IV**, 927

Tri-*tert*-butylcyclopropenyl tetrafluoroborate, **VI**, 991

3,4,5-Trimethyl-2,5-heptadien-4-ol, **VIII**, 505

Trimethyl (2-methylene-4-phenyl-3-butenyl)silane, **VIII**, 602

O-(Trimethylsilyl)benzophenone cyanohydrin, **VII**, 20

O-Trimethylsilyl-4-methoxymandelonitrile, **VII**, 521

F. To C=S

Dithizone, **III**, 360

Ethylene thiourea, **III**, 394

Ethyl oxalamidrazonate, **VIII**, 597

Methyl isothiocyanate, **III**, 599

Rhodanine, **III**, 763

G. To C=C–C=C (see also Cyclization, B)

3-Chlorocyclopentene, **IV**, 238

2,5-Dihydrodimethoxy-2,5-furan, **V**, 403

2,3-Dimethylanthraquinone, **III**, 310

Isoprene cyclic sulfone, **III**, 499

1-(*p*-Nitrophenyl)-1,3-butadiene, **IV**, 727

Norbornylene, **IV**, 738

cis-Δ⁴-Tetrahydrophthalic anhydride, **IV**, 890

Tetraphenylphthalic anhydride, **III**, 807

Triptycene, **IV**, 964

H. To C=C–C=O and C=C–C≡N

2-*p*-Acetylphenylhydroquinone, **IV**, 15

DL-β-Amino-β-phenylpropionic acid, **III**, 91

Benzalacetone dibromide, **III**, 105

2-(2-Bromoethyl)-1,3-dioxane, **VII**, 59

3-(*o*-Chloroanilino)propionitrile, **IV**, 146

4-(4-Chlorophenyl)butan-2-one, **VII**, 105

β-Chloropropionic acid, **I**, 166

Coumarilic acid, **III**, 209

N-2-Cyanoethylaniline, **IV**, 205

β-Cyanoethylhydrazine, **V**, 39

Diacetonamine hydrogenoxalate, **I**, 196

DL-β-Carbethoxyethylmethylamine, **III**, 258

Diethyl 5-methylcoprost-3-en-3-yl phosphate, **VI**, 76

3,4-Dihydro-2-methoxy-4-methyl-2*H*-pyran, **IV**, 311

2,5-Dihydroxy-*p*-benzenediacetic acid, **III**, 286

5,5-Dimethyl-1,3-cyclohexanedione, **II**, 200

β,β-Diphenylpropiophenone, **II**, 236

β-Ethoxypropionaldehyde acetal, **III**, 371

Ethyl 3-benzyl-2-cyano-3-methylpentanoate, **IV**, 94

Ethyl α,β-dicyano-β-phenylpropionate, **I**, 452

Ethyl β-phenyl-β-cyanopropionate, **IV**, 804

Ethyl propane-1,1,2,3-tetracarboxylate, **I**, 272

cis-4,4a,5,6,7,8-Hexahydro-4a,5-dimethyl-2(3*H*)naphthalenone, **VI**, 666

Hydroxyhydroquinone triacetate, **I**, 317

6-Methylbicyclo[4.2.0]octan-2-one, **VII**, 118

Methyl β-bromopropionate, **III**, 576

β-Methylglutaric anhydride, **IV**, 630

Methyl γ-methyl-γ-nitrovalerate, **IV**, 652

2-Methyl-2-(3-oxobutyl)-1,3-cyclohexanedione, **VII**, 368

2-Methyl-2-(3-oxobutyl)-1,3-cyclopentanedione, **VII**, 363

β-Methyl-β-phenyl-α,α'-dicyanoglutarimide, **IV**, 662

Methylsuccinic acid, **III**, 615

Methyl β-thiodipropionate, **IV**, 669

γ-Phenylallylsuccinic acid, **IV**, 766

α-Phenyl-β-benzoylpropionitrile, **II**, 498

α-Phenyl-α-carbethoxyglutaronitrile, **IV**, 776

4-Phenyl-6-hepten-2-one, **VII**, 443

Phenylpropargylaldehyde, **III**, 731

Phenylsuccinic acid, **IV**, 804

DL-Serine, **III**, 774

Spiro[5.7]tridec-1-en-3-one, **VII**, 61, 129

Tetramethylammonium 1,1,2,3,3-pentacyanopropenide, **V**, 1013

DL-Threonine, **III**, 813

2,5,5-Trimethyl-2-(2-bromoethyl)-1,3-dioxane, **VII**, 59

I. To N=C=O

Benzoylene urea, **II**, 79

Methyl (chlorosulfonyl)carbamate, **VI**, 788

4-Phenyl-1-carbethoxysemicarbazide, **VI**, 936

Polymeric urea, **VI**, 951

J. To N=O

N-*tert*-Butyl-*N*-*tert*-octyl-*O*-*tert*-butylhydroxylamine, **VIII**, 93

N-(*p*-Dimethylaminophenyl)-α-(*o*-nitrophenyl)nitrone, **V**, 826

K. Miscellaneous

Acetoacetanilide, **III**, 10

2-Amino-6-methylbenzothiazole, **III**, 76

tert-Butyl carbamate, **V**, 162

o-Chlorophenylthiourea, **IV**, 180

2,5-Diamino-3,4-dicyanothiophene, **IV**, 243

2,3-Diphenylsuccinonitrile, **IV**, 392

[(*trans*-2-Isocyanocyclohexyl)oxy]-trimethylsilane, **VII**, 294

α-Phenylthiourea, **III**, 735

Vinyl chloroacetate, **III**, 853

ALKENYLATION

A. *C*-Alkenylation

(*E*)-6-Benzyloxy-2-nitro-β-pyrrolidinostyrene, **VII**, 34

2-Butylbornene, **VII**, 77

4-*tert*-Butyl-1-vinylcyclohexene, **VIII**, 97

4-[(*E*)-1,2-Dichlorovinyl]-4-methyl-2-cyclohexen-1-one, **VII**, 241

6,7-Dimethoxy-1,2,3,4-tetrahydro-2-[(1-*tert*-butoxy-3-methyl)-2-butyliminomethyl]isoquinoline, **VIII**, 573

(*Z*)-β-2-(*N*,*N*-Dimethylamino) phenyl] styrene, **VII**, 172

(*S*)-*N*,*N*-Dimethyl-*N'*-(1-*tert*-butoxy-3-methyl-2-butyl)formamidine, **VIII**, 204

Ethyl (*E*)-3-(tributylstannyl)propenoate, **VIII**, 268

(*E*)-5-Ethyl-1,4-undecadiene, **VII**, 236

α-Farnesene, **VII**, 245

(5*Z*,7*E*)-5,7-Hexadecadiene, **VIII**, 295

1,2,4,5-Hexatetraene, **VII**, 485

(1*Z*,3*E*)-1-Phenyl-1,3-octadiene, **VIII**, 532

(*E*)-2-Propenylcyclohexanol, **VII**, 456

B. Methylenation

3,4-Dihydro-2-methylene-2*H*-1-benzopyran, **VIII**, 512

(+)-3-Methylene-*cis*-*p*-menthane, **VIII**, 386

1-Phenoxy-1-phenylethene, **VIII**, 512

C. N-Alkenylation

cis-*N*-Tosyl-3-methyl-2-azabicyclo[3.3.0]oct-3-ene, **VII**, 501

ALKYLATION

A. *C*-Alkylation (including Friedel–Crafts Reactions)

cis- and *trans*-1-Acetoxy-4-(dicarbomethoxymethyl)-2-cyclohexene, **VIII**, 5

α-Acetyl-δ-chloro-γ-valerolactone, **IV**, 10

2-Allylcyclohexanone, **III**, 44

2-Allyl-3-methylcyclohexanone, **VI**, 51

p-Aminotetraphenylmethane, **IV**, 47

DL-Aspartic acid, **IV**, 55

1-[α-(Benzalamino)benzyl]-2-naphthol, **I**, 381

2-Benzoyl-1-cyano-1-methyl-1,2-dihydroisoquinoline, **IV**, 642

2-Benzylcyclopentanone, **V**, 76

2-Benzyl-2-methylcyclohexanone, **VI**, 121

2-Benzyl-6-methylcyclohexanone, **VI**, 121

1-Benzylisoquinoline, **VI**, 115

N-[2,4-Bis(1,3-diphenylimidazolidin-2-yl)-5-hydroxyphenyl]acetamide, **VII**, 162

p-Bromodiphenylmethane, **V**, 130

1-*p*-Bromophenyl-1-phenyl-2,2,2-trichloroethane, **V**, 131

Butylacetylene, **V**, 117

2-Butylbornene, **VII**, 77

2-Butyl-2-methylcyclohexanone, **V**, 187

tert-Butylmalononitrile, **VI**, 223

1-Butyl-10-methyl-$\Delta^{1(9)}$-2-octalone, **VI**, 242

α^{2}-Chloroisodurene, **III**, 557

1-Chloromethylnaphthalene, **III**, 195

2-Chloromethylthiophene, **III**, 197

1,1-Cyclobutanedicarboxylic acid, **III**, 213

α-Cyclohexylphenylacetonitrile, **III**, 219

Cyclopropanecarboxylic acid, **III**, 221

Cyclopropane-1,1-dicarboxylic acid, **VII**, 411

Cyclopropenone 1,3-propanediol ketal, **VIII**, 173

Cyclopropyl cyanide, **III**, 223

m-Cymene, **V**, 332

Dicarbonyl(cyclopentadienyl)(*trans*-3-methyl-2-vinylcyclohexanone)iron tetrafluoroborate, **VIII**, 479

α,α'-Dicyano-β-methylglutaramide, **III**, 591

Diethyl (2*S*,3*R*)-(+)-3-allyl-2-
hydroxysuccinate, **VII**, 153
Diethyl benzylmalonate, **III**, 705
Diethyl *sec*-butylmalonate, **III**, 495
Diethyl 1,1-cyclobutanedicarboxy-
late, **IV**, 288
Diethyl Δ²-cyclopentenylmalonate,
IV, 291
Diethyl β-ketopimelate, **V**, 384
9,10-Dihydrofulvalene **VIII**, 298
4,5-Dihydro-2-[(1-phenylmethyl-2-
propenyl)thio]thiazole, **VI**, 704
Dimesitylmethane, **V**, 422
2,2-Dimethylcyclohexane-1,3-dione,
VIII, 312
Dimethyl nitrosuccinate, **VI**, 503
2,2-Dimethyl-4-pentenal, **VII**, 177
2,2-Dimethyl-4-phenylbutyric acid,
VI, 517
2,2-Dimethyl-3-phenylpropionalde-
hyde, **VI**, 526
7,9-Dimethyl-*cis*-8-thiabicy-
clo[4.3.0]nonane 8,8-dioxide,
VI, 482
(*E*)-6,10-Dimethyl-5,9-undecadien-
1-yne, **VIII**, 226
3,4-Dinitro-3-hexene , **IV**, 372
1,6-Dioxo-8a-methyl-1,2,3,4,6,7,8a-
octahydronaphthalene, **V**, 486
1,1-Diphenylethane, **VI**, 537
Diphenylmethane, **II**, 232
1,1-Diphenylpentane, **V**, 523
Diphenylpropionic acid, **V**, 526
1-Ethoxy-1-butyne, **VI**, 564
2-(1-Ethoxycyclohexyl)-2-
(trimethylsiloxy)cyclobutanone,
VIII, 579
3-Ethoxy-6-methyl-2-cyclohexen-1-
one, **VII**, 208
Ethyl α-acetoglutarate, **II**, 263
Ethyl acetosuccinate, **II**, 262
Ethyl 4-amino-3-methylbenzoate,
VI, 581
Ethyl butylacetoacetate, **I**, 248
Ethyl butylmalonate, **I**, 250
Ethyl *sec*-butylmalonate, **II**, 417

Ethyl (*E*)-2-cyano-6-octenoate, **VI**,
586
Ethyl 4-cyclohexyl-4-oxobutanoate,
VIII, 578
Ethyl 2,4-diphenylbutanoate, **V**, 559
Ethyl (1-ethylpropenyl)methyl
cyanoacetate, **III**, 397
3-Ethyl-1-hexyne, **VI**, 595
Ethyl α-isopropylacetoacetate, **III**,
405
Ethyl isopropylmalonate, **II**, 94
2-(3-Ethyl-5-methyl-4-isoxazolyl-
methyl)cyclohexanone, **VI**, 781
Ethyl methylmalonate, **II**, 279
Ethyl 1-methylthiol-3-phthalimido-
propane-3,3-dicarboxylate, **II**,
384
(*E*)-5-Ethyl-1,4-undecadiene, **VII**, 236
α-Farnesene, **VII**, 245
2-Heptyl-2-cyclohexenone, **VII**, 249
Hexahydro-6-ethyl-3-(hydroxymethyl)-6-allyl-2-phenyl-
[2*S*,3*S*,6*S*,8a*R*]-5-oxo-5*H*-oxa-
zolo[3,2-*a*]pyridine, **VIII**, 241
Hexamethylbenzene, **II**, 248; **IV**,
520
2-Hydroxy-5-nitrobenzyl chloride,
III, 468
3-(1-Hydroxybutyl)-1-methylpyr-
role, **VII**, 102
Indole-3-acetic acid, **V**, 654
DL-Isoleucine, **III**, 495
Isopropylidene α-(hexahy-
droazepinylidene)malonate,
VIII, 263
(3-Methyl-2-butenyl)propanedioic
acid, dimethyl ester, **VIII**, 381
(3-Methyl-2-butenyl)(2-
propynyl)propanedioic acid,
dimethyl ester, **VIII**, 381
N-(2-Methylbutylidene-1-*d*)-1,1,3,3-
tetramethylbutylamine, **VI**, 751
3-Methylcyclohexanone-3-acetic
acid, **VIII**, 467
Methyl 2,4-cyclopentadiene-1-
acetate (in solution), **VII**, 339

4,5-Methylenedioxybenzo-
cyclobutene, **VII**, 326
1-Methyl-3-ethyloxindole, **IV**, 620
9-Methylfluorene, **IV**, 623
3-Methyl-2(5*H*)-furanone, **VIII**, 396
(*S*)-(+)-4-Methyl-3-heptanone, **VIII**,
403
5-Methyl-5-hexen-2-one, **V**, 767
Methyl 7-hydroxyhept-5-ynoate,
VIII, 415
1-Methylisoquinoline, **IV**, 641
Methyl *N*-(2-methyl-2-butenyl)car-
bamate, **VIII**, 68
Methyl 2-methyl-1-cyclohexene-1-
carboxylate, **VII**, 351
3-Methylpentane-2,4-dione, **V**, 785
2-Methyl-2-phenyl-4-pentenal, **VIII**,
451
N-Methyl-2-phenyl-Δ²-tetrahydropy-
ridine, **VI**, 818
2-(1-Methyl-2-propynyl)cyclohexa-
none, **VIII**, 460
2-Methyl-5-trimethylsilyl-2-
vinylpentanoic acid, **VIII**, 488
trans-3-Methyl-2-vinylcyclohexa-
none, **VIII**, 479
Neophyl chloride, **IV**, 702
2,4-Nonanedione, **V**, 848
Δ¹⁽⁹⁾-Octalone-2, **V**, 869
2-(2-Oxopropyl)cyclohexanone, **VII**,
414
Pelargonic acid, **II**, 474
1,2,3,4,5-Pentachloro-5-ethylcy-
clopentadiene, **V**, 893
Pentamethylbenzene, **II**, 248
1,2,3,4,5-Pentamethylcyclopenta-
diene, **VIII**, 505
2-*tert*-Pentylcyclopentanone, **VII**,
424
DL-Phenylalanine, **III**, 705
4-Phenyl-6-hepten-2-one, **VII**, 443
(Phenylthio)nitromethane, **VIII**, 550
2-Phenoxymethyl-1,4-benzoquinone,
VI, 890
2-Phenylbutyronitrile, **VI**, 897

2-(1-Phenylcyclopentyl)-4,4,6-
trimethyl-5,6-dihydro-
1,3(4*H*)oxazine, **VI**, 905
trans-2-(2-Propenyl)cyclopentanol,
VII, 501
2-Propyl-1-azacycloheptane, **VIII**,
568
Propylbenzene, **I**, 471
Salsolidine, **VIII**, 573
Tetraacetylethane, **IV**, 869
2-Tetradecyl-*sym*-trithiane, **VI**, 869
Tricarbonyl[2-[(2,3,4,5η)-4-
methoxy-2,4-cyclohexadien-1-
yl]-5,5-dimethyl-1,3-cyclohexa-
nedione]iron, **VI**, 1001
p-Tricyanovinyl-*N,N*-dimethylani-
line, **IV**, 953
Triethyl α-phthalimidoethane-α,α,β-
tricarboxylate, **IV**, 55
2,6,6-Trimethyl-2,4-cyclohexa-
dienone, **V**, 1092
Triphenylchloromethane, **III**, 842
Triphenylmethane, **I**, 548
α,β,β-Triphenylpropionic acid, **IV**,
960
Triphenylpropionitrile, **IV**, 962
2-Vinylthiophene, **IV**, 980
B. *N*-Alkylation
1-Acetoxy-4-diethylamino-2-butene,
VIII, 9
Benzoylcholine chloride, **IV**, 84
2-Benzylaminopyridine, **IV**, 91
Benzylaniline, **I**, 102
1-Benzylindole, **VI**, 104, 106
N-Benzyl-*N*-methoxymethyl-*N*-
[(trimethylsilyl)methyl]amine,
VIII, 231
Benzylphthalimide, **II**, 83
1-Benzylpiperazine, **V**, 88
Benzyltrimethylammonium ethoxide, **IV**, 98
N-Benzyl-*N*-[(trimethylsilyl)methyl]-
amine, **VIII**, 231
N-tert-Butyl-*N-tert*-octyl-*O-tert*-
butylhydroxylamine, **VIII**, 93

tert-Butylurea, **III**, 256
trans-Cycloocetene, **V**, 315
sym-Dibenzoyldimethylhydrazine,
    II, 209
γ-Dibutylaminopropylamine, **III**,
    256
N,N'-Diethylbenzidine, **IV**, 283
N,N-Dimethylaminomethylferrocene
    methiodide, **V**, 434
N,N'-Diphenylbenzamidine, **IV**, 383
Diphenylmethyl vinyl ether,
    VI, 552
Ethyl *N*-benzylcarbamate, **IV**, 780
N-Ethyl-*p*-chloroaniline, **IV**, 420
Ethyl *p*-dimethylaminophenyl-
    acetate, **V**, 552
1-Ethyl-3-(3-dimethylamino)propyl-
    carbodiimide methiodide,
    V, 555
Ethyl hydrazodicarboxylate, **IV**, 411
Ethyl *N*-(2-methyl-3-nitrophenyl)-
    formimidate, **VIII**, 493
Ethyl *N*-phenylformimidate, **IV**, 464
N-Ethylpiperidine, **V**, 575
Ethyl α-(1-pyrrolidyl)propionate,
    IV, 466
Ethyl-*m*-toluidine, **II**, 290
3-Heptylureidomethylenema-
    lononitrile, **IV**, 515
1,4,7,10,13,16-Hexakis(*p*-tolylsul-
    fonyl)-1,4,7,10,13,16-hexaaza-
    cyclooctadecane, **VI**, 652
1,3,5-Hexatriene, **V**, 608
Julolidine, **III**, 504
Methylamine hydrochloride, **I**, 347
o-Methylbenzyl alcohol, **IV**, 582
2-Methylbenzyldimethylamine, **IV**,
    585
N-Methylbutylamine, **V**, 736
N-Methylethylamine, **V**, 758
Methylhydrazine sulfate, **II**, 395
1-Methylindole, **V**, 769
o-Nitrobenzaldehyde, **V**, 825
o-Nitrobenzylpyridinium bromide,
    V, 825

2-Oxo-1-phenyl-3-pyrrolidinecar-
    boxylic acid, **VII**, 411
[2.2]Paracyclophane, **V**, 883
1-Phenyl-3,4-dihydro-6,7-methyl-
    enedioxyisoquinoline, **VII**, 326
1-Phenylpiperidine, **IV**, 795
Phenyltrimethylammonium sul-
    fomethylate, **VI**, 175
α-Phthalimido-*o*-toluic acid, **IV**, 810
Polymeric benzylamine, **VI**, 951
Pyocyanine, **III**, 753
3-Quinuclidone hydrochloride, **V**,
    989
Tetramethyl-*p*-phenylenediamine, **V**,
    1018
2-(*p*-Tolylsulfonyl)-dihydroisoin-
    dole, **V**, 1064
m-Trifluoromethyl-*N,N*-dimethylani-
    line, **V**, 1085
Trimethylamine, **I**, 528
Trimethylamine hydrochloride, **I**,
    531
3,6,9-Tris(*p*-tolylsulfonyl)-3,6,9-tri-
    azaundecane-1,11-diol, **VI**, 652
2-Vinylthiophene, **IV**, 980
C. *O*-Alkylation
Acetone carboxyoxime, **III**, 172
Alloxantin dihydrate, **IV**, 25
Anisole, **I**, 58
Benzyl chloromethyl ether, **VI**, 101
Benzyl 2,3-epoxypropyl ether, **VIII**,
    33
4-Benzyloxy-2-butanone, **VII**, 386
6-Benzyloxy-2-nitrotoluene, **VII**, 34
1,3-Bis(methylthio)-2-
    methoxypropane, **VI**, 683
4,4-Bis(phenylthio)-3-methoxy-1-
    butene, **VI**, 737
17-β-*tert*-Butoxy-5α-androstan-3-
    one, **VII**, 66
o-Butoxynitrobenzene, **III**, 140
N-tert-Butyl-*N-tert*-octyl-*O-tert*-
    butylhydroxylamine, **VIII**, 93
(S)-(+)-γ-Butyrolactone-γ-carboxylic
    acid, **VII**, 99

Carboxymethoxylamine hemihydrochloride, **III**, 172

Cholestanyl methyl ether, **V**, 245

Coumarone, **V**, 251

1,4-*O*-Dibenzyl-2,3-di-*O*-isopropyli-
dene-L-threitol, **VIII**, 155

3,4-Dichloro-5-isopropoxy-2(5*H*)-
furanone, **VIII**, 116

(*R,R*)-(+)-2,3-Dimethoxy-*N,N,N',N'*-
tetramethylsuccinic acid
diamide, **VII**, 41

2-Ethoxypyrrolin-5-one, **VI**, 226

Ethyl α-(isopropylideneaminooxy)
propionate, **V**, 1031

o-Eugenol, **III**, 418

(*S*)-*N*-Formyl-*O-tert*-butylvalinol,
VIII, 204

(*S*)-(–)-Formyl-2-methoxymethyl-
pyrrolidine, **VIII**, 26

D-Glucose-1,2,3,4-tetraacetate, **III**,
432

Guaiacol allyl ether, **III**, 418

Ketene di(2-methoxyethyl)acetal, **V**,
684

(–)-Menthoxyacetic acid, **III**, 544

Methoxyacetonitrile, **II**, 387

m-Methoxybenzaldehyde, **III**, 564

2-Methoxycyclooctanone oxime, **V**,
266

1-Methoxy-2-methylpyridinium
methyl sulfate, **V**, 270

4-Methoxy-3-penten-2-one, **VIII**,
357

Methyl *R*-(–)-1,1'-binaphthyl-2,2'-
diyl phosphate, **VIII**, 46

O-Methylcaprolactim, **IV**, 588

3-Methylcoumarone, **IV**, 590

α-Methyl-D-glucoside, **I**, 364

[(Methyl)(methoxy)carbene]pen-
tacarbonyl chromium(0), **VIII**,
216

Methyl β-naphthyl ether, **I**, 59

Orcinol dimethyl ether, **VI**, 859

Quinacetophenone monomethyl
ether, **V**, 836

2,3,4,6-Tetramethyl-D-glucose, **III**,
800

o-Tolualdehyde, **IV**, 932

Trimethylgallic acid, **I**, 537

Triethyloxonium fluoborate, **V**, 1080

Trimethyloxonium tetrafluoroborate,
V, 1096; **VI**, 1019

Trimethyloxonium 2,4,6-trinitroben-
zenesulfonate, **V**, 1099

Veratraldehyde, **II**, 619

D. *S*-Alkylation

S-Acetamidomethyl-L-cysteine
hydrochloride, **VI**, 5

1,2-Bis(butylthio)benzene, **V**, 107

4,5-Dihydro-2-(2-propenylthio)thia-
zole, **VI**, 704

p-Dithiane, **IV**, 396

3,7-Dithianonane-1,9-diol, **VIII**, 592

Dodecyl mercaptan, **III**, 363

Ethanedithiol, **IV**, 401

2-Furfurylmercaptan, **IV**, 491

Guanidoacetic acid, **III**, 440

S-Methyl isothiourea sulfate, **II**, 411

2-Methylmercapto-*N*-methyl-Δ²-
pyrroline, **V**, 780

Methyl *p*-tolyl sulfone, **IV**, 674

Neopentyl phenyl sulfide, **VI**, 833

S-(2-Oxobut-3-yl) butanethioate, **VI**,
776

1-Phenylthio-2-bromobutene (in
solution), **VII**, 453

Propyl sulfide, **II**, 547

Pseudothiohydantoin, **III**, 751

Rhodanine, **III**, 763

1,4,8,11-Tetrathiacyclotetradecane,
VIII, 592

E. Miscellaneous

Dicarbonyl(cyclopentadienyl)(ethyl
vinyl ether)iron tetrafluoroborate, **VIII**, 479

AMINATION

A. Direct

1-Amino-1-methylcyclohexane,
V, 35

1-Aminopyridinium iodide, **V**, 43

7-Aminotheophylline, **VII**, 10
7-Benzylideneaminotheophylline,
   VII, 9
B. Mannich Reaction
   1-Diethylamino-3-butanone, **IV**,
      281
   N,*N*-Dimethylaminomethylferrocene
      methiodide, **V**, 434
   β-Dimethylaminopropiophenone
      hydrochloride, **III**, 305
   5-Methylfurfuryldimethylamine,
      IV, 626
C. Reductive
   5-Benzylaminotetrazole, **VII**, 27
   α-*p*-Bromophenylethylamine,
      III, 505
   α-*p*-Chlorophenylethylamine,
      II, 505
   α-(β-Naphthyl)ethylamine, **II**, 505
   α-Phenylethylamine, **II**, 503
   α-*p*-Xenylethylamine, **II**, 505
   α-*p*-Tolylethylamine, **II**, 505
D. Miscellaneous
   Benzylacrylamide, **V**, 73
   Diethylaminosulfur trifluoride,
      VI, 440
   α,α-Dimethyl-β-phenethylamine,
      V, 471
   3-Pinanamine, **VI**, 943
AMINOMETHYLENATION
   (*E*)-6-Benzyloxy-2-nitro-β-pyrrolidi-
      nostyrene, **VII**, 34
   N,*N*-Dimethyl-*N'*-*p*-tolylformami-
      dine, **VII**, 197
ANNULATION
All ring-forming reactions organized
   as carbocyclic and heter-
   ocyclic by ring size. Bicyclic
   compounds are listed
   separately.
A. Carbocyclic-[3]
   1,1-Bis(bromomethyl)cyclopropane,
      VI, 153
   1-Chloro-1,2-diphenyl-3-methylcy-
      clopropane, **VII**, 203

exo/*endo*-7-(2-Chloroethoxy)bicy-
   clo[4.1.0]heptane, **V**, 859
Cyclopropanecarboxylic acid, **III**, 221
Cyclopropenone 1,3-propanediol
   ketal, **VIII**, 173
Cyclopropyl cyanide, **III**, 223
Cyclopropyldiphenylsulfonium tetrafluoroborate, **VI**, 364
trans-1,2-Dibenzoylcyclopropane,
   VI, 401
2-(2,2-Dibromocyclopropyl)-6,6-
   dimethylbicyclo[3.1.1]hept-2-
   ene, **VIII**, 223
7,7-Dibromo-1,6-dimethylbicy-
   clo[4.1.0]hept-3-ene, **VII**, 200
1,1-Dibromo-2,2-diphenylcyclo-
   propane, **VI**, 187
Di-*tert*-butylcyclopropenone, **VI**,
   991
1,1-Dichloro-2-phenylcyclopropane,
   VII, 12
11,11-Dichlorotricyclo[4.4.1.0^{1,6}]-
   undeca-3,8-diene, **VI**, 731
(–)-Dimenthyl (1*S*,2*S*)-cyclo-
   propane-1,2-dicarboxylate,
   VIII, 141
1,1-Diphenylcyclopropane, **V**, 509
Diphenylcyclopropenone, **V**, 514
1-Ethoxy-2-*p*-tolylcyclopropane, **VI**,
   571
Ethyl *cis*- and *trans*-2-phenylcyclo-
   propanecarboxylate, **VI**, 913
Methyl cyclopropyl ketone, **IV**, 597
Methylenecyclopropane, **VI**, 320
Norcarane, **V**, 855
2-Oxa-7,7-dichloronorcarane, **V**, 874
Phenylcyclopropane, **V**, 328, 929
2,2,3,3-Tetramethyliodocyclo-
   propane, **VI**, 974
1-*p*-Tolylcyclopropanol, **V**, 1058
1-Trimethylsilyloxybicyclo[4.1.0]-
   heptane, **VI**, 327
B. Carbocyclic-[4]
   7-Acetoxy-4,4,6,7-tetramethylbicy-
      clo[4.2.0]octan-2-one, **VI**, 1024

Biphenylene, **V**, 54
3-Butyl-4,4-dichlorocyclobutanone, **VIII**, 82
1-Chloro-1,4,4-trifluorocyclobutyl acetate, **V**, 235
1-Cyanobenzocyclobutene, **V**, 263
Cyclobutanone, **VI**, 320, 324
7,7-Dichlorobicyclo[3.2.0]hept-2-en-6-one, **VI**, 1037
7,7-Dichloro-1-methylbicyclo[3.2.0]heptan-6-one, **VIII**, 377
3,4-Dichloro-1,2,3,4-tetramethylcyclobutene, **V**, 370
Diethyl 1,1-cyclobutanedicarboxylate, **IV**, 213, 288
Dispiro[5.1.5.1]tetradecane-7,14-dione, **V**, 297
cis- and trans-1,2-Divinylcyclobutane, **V**, 528
6-Methylbicyclo[4.2.0]octan-2-one, **VII**, 315
3-Methylenecyclobutane-1,2-dicarboxylic anhydride, **V**, 459
1,1,2-Trichloro-2,3,3-trifluorocyclobutane, **V**, 393
C. Carbocyclic-[5]
2-Acetyl-1,3-cyclopentanedione, **VI**, 28
2-Carbethoxycyclopentanone, **II**, 116
2-Carbomethoxy-3-vinylcyclopentanone, **VIII**, 112
Cyclopentanone, **I**, 192
5,5-Dicyano-4-phenyl-2-cyclopenten-1-one 1,3-propanediol ketal, **VIII**, 173
Dihydrojasmone, **VIII**, 620
4,4-Dimethyl-2-cyclopenten-1-one, **VIII**, 208
(1R)-9,9-Dimethyltricyclo[6.1.1.0^{2,6}]-deca-2,5-diene, **VIII**, 223
2,3-Diphenylindone, **III**, 353
Ethyl 6,7-dimethoxy-3-methylindene-2-carboxylate, **V**, 550
Fluorenecarboxylic acid, **IV**, 482

cis-4-exo-Isopropenyl-1,9-dimethyl-8-(trimethylsilyl)bicyclo[4.3.0] non-8-en-2-one, **VIII**, 347
2-Methyl-1,3-cyclopentanedione, **V**, 747; **VI**, 774
2-Methyl-4-ethoxalylcyclopentane-1,3,5-trione, **V**, 747
1-Methyl-3-phenylindane, **IV**, 665
2-Methyl-2-vinylcyclopentanone, **VIII**, 486
1,2,3,4,5-Pentamethylcyclopentadiene, **VIII**, 505
2-(1-Phenylcyclopentyl)-4,4,6-trimethyl-5,6-dihydro-1,3(4H)-oxazine, **VI**, 905
4-Protoadamantanone, **VI**, 958
Tetraphenylcyclopentadienone, **III**, 806
D. Carbocyclic-[6]
4-Acetyl-7,8-dimethyl-2-oxa-4-azabicyclo[4.3.0]non-7-en-3-one, **VII**, 5
Benzanthrone, **II**, 62
2,3-Dicyano-1,4,4a,9a-tetrahydrofluorene, **VI**, 427
Diethyl trans-1,4-tetrahydrophthalate, **VI**, 454
Diethyl trans-4-trimethylsilyloxy-4-cyclohexene-1,2-dicarboxylate, **VI**, 445
3,4-Dihydro-1,2-naphthalic anhydride, **II**, 194
6,7-Dimethoxy-3,4-dihydro-2-naphthoic acid, **III**, 300
2,3-Dimethylanthraquinone, **III**, 310
3,5-Dimethyl-4-carbethoxy-2-cyclohexen-1-one, **III**, 318
5,5-Dimethyl-1,3-cyclohexanedione, **II**, 200
4,4-Dimethyl-2-cyclohexen-1-one, **VI**, 496
1,6-Dioxo-8a-methyl-1,2,3,4,6,7,8a-octahydronaphthalene, **V**, 486
Ethyl 1-cyano-2-methylcyclohexanecarboxylate, **VI**, 586

Ethyl 1,3-dihydroxy-2-naphthoate,
III, 638
cis-4,4a,5,6,7,8-Hexahydro-4a,5-
dimethyl-2(3H)-naphthalenone,
VI, 666
Hexahydro-1,3,3,6-tetramethyl-2,1-
benzisoxazoline, **VI**, 670
Hexaphenylbenzene, **V**, 604
Mesitylene, **I**, 341
5β-Methoxycyclohexan-1-one-
3β,4β-dicarboxylic acid anhydride, **VII**, 312
6-Methoxy-7-methoxycarbonyl-
1,2,3,4-tetrahydronaphthalene,
VIII, 444
6-Methoxy-β-tetralone, **VI**, 744
β-Methylanthraquinone, **I**, 353
1-Methyl-4,4a,5,6,7,8-hexahydro-
naphthalen-2(3H)-one, **VI**, 781
(8aS)-8a-Methyl-3,4,8,8a-tetrahydro-
1,6(2H,7H)-naphthalenedione,
VII, 368
$\Delta^{1(9)}$-Octalone-2, **V**, 869
Phenylphenanthrene, **V**, 952
Quinizarin, **I**, 476
Spiro[5.7]tridec-1-en-3-one, **VII**,
473
cis-Δ^4-Tetrahydrophthalic anhydride,
IV, 890
Tetrahydroxyquinone, **V**, 1011
α-Tetralone, **II**, 569; **III**, 798; **IV**,
898, 900
1,2,3,4-Tetraphenylnaphthalene, **V**,
1037
Tetraphenylphthalic anhydride, **III**,
807
1,3,5-Triacetylbenzene, **III**, 829
Triphenylene, **V**, 1120
2,4,6-Triphenylnitrobenzene, **V**,
1128
E. Carbocyclic-[>6]
Azulene, **VII**, 15
p-tert-Butylcalix[4]arene, **VIII**, 75
p-tert-Butylcalix[6]arene, **VIII**, 77
p-tert-Butylcalix[8]arene, **VIII**, 80
Cyclodecanone, **IV**, 218; **V**, 277

1,2-Cyclononadiene, **V**, 306
3,4-Dihydro-1(2H)-azulenone, **VIII**,
196
Hexadehydro[18]annulene, **VI**, 68
[2,2]Paracyclophane, **V**, 883
2-Phenylcycloheptanone, **IV**, 780
Sebacoin, **IV**, 840
Tetramethyl[2,2]paracyclophane-
4,5,12,13-tetracarboxylate, **VII**,
485
4,6,8-Trimethylazulene, **V**, 1088
F. Carbocyclic-[M,N]
Bicyclo[1.1.0]butane, **VI**, 133
Bicyclo[3.3.1]nonan-9-one, **VI**, 137
Bicyclo[3.3.0]octane-2,6-diol, **VIII**,
43
Bicyclo[3.3.0]octane-3,7-dione, cis-,
VII, 50
Bicyclo[2.1.0]pentane, **V**, 96
Bicyclo[2.1.0]pent-2-ene, **VI**, 145
trans-7,8-Diacetoxybicyclo[4.2.0]-
octa-2,4-diene, **VI**, 196
2,6-Diacetoxybicyclo[3.3.0]octane,
VIII, 43
Diamantane, **VI**, 378
Dicyclopropyl ketone, **IV**, 278
7,7-Dimethoxy-1,2,3,4-tetrachlorobi-
cyclo[2.2.1]heptene, **V**, 425
3,3-Dimethyl-cis-bicyclo[3.2.0]hep-
tan-2-ol, **VII**, 177
cis-1,5-Dimethylbicyclo[3.3.0]-
octane-3,7-dione, **VII**, 50
Dimethyl 3,3a,3b,4,6a,7a-hexahy-
dro-3,4,7-metheno-7H-cyclo-
penta[a]pentalene-7,8-dicar-
boxylate, **VIII**, 298
(1R)-9,9-Dimethyltricyclo[6.1.1.0^{2,6}]-
deca-2,5-diene, **VIII**, 223
1,6-Dimethyltricyclo[4.1.0.0^{2,7}]hept-
3-ene, **VII**, 200
Hexamethyl Dewar benzene, **VII**,
256
Methyl 1,3-dimethyl-5-oxobicy-
clo[2.2.2]octane-2-carboxylate,
VIII, 219
Norbornylene, **IV**, 738

Nortricyclyl acetate, **V**, 863
Perhydro-9b-phenalenol, **VII**, 427
Quadricyclane, **VI**, 962
p-Quinquephenyl, **V**, 985
Tetrachlorobenzobarrelene, **VI**, 82
2-(Trichloromethyl)-
 bicyclo[3.3.0]octane, **V**, 93
endo-Tricyclo[4.4.0.0^{2,5}]deca-3,8-
 diene-7,10-dione, **VI**, 1002
Triptycene, **IV**, 964
G. Heterocyclic-[3]
 1-Amino-2-phenylaziridine, **VI**, 56
 cis-2-Benzyl-3-phenylaziridine,
  V, 83
 2-*tert*-Butyl-3-phenyloxazirane, **V**,
  191
 (+)-(2*R*,8a*S*)-10-
  (Camphorylsulfonyl)-
  oxaziridine, **VIII**, 104
 Chlorophenyldiazirine, **VI**, 276
 2-Chlorothiirane 1,1-dioxide, **V**,
  231
 Cyclohexene oxide, **I**, 185
 Cyclohexene sulfide, **IV**, 232
 2,2'-Dichloro-α,α'-epoxybibenzyl,
  V, 358
 1,2-Dimethyl-1,2-epoxycyclohex-4-
  ene, **V**, 468
 2,2-Dimethylethylenimine, **III**, 148
 trans-2,3-Diphenyl-1-phthalimi-
  doaziridine, **VI**, 56
 2,3-Diphenylvinylene sulfone, **VI**,
  555
 Epibromohydrin, **II**, 256
 Epichlorohydrin, **I**, 233; **II**, 256
 2,3-Epoxycyclohexanone, **VI**, 679
 10,11-Epoxyfarnesyl acetate, **VI**,
  560
 N-Ethylallenimine, **V**, 541
 Ethylene sulfide, **V**, 562
 Ethylenimine, **IV**, 433
 Ethyl β,β-pentamethyleneglycidate,
  IV, 459
 Isophorone oxide, **IV**, 552
 Methyl 5,5-dimethoxy-3-methyl-2,3-
  epoxypentanoate, **IV**, 649

Methylenecyclohexane oxide, **V**,
 755
Methyl (3*RS*,4*RS*)-4,5-epoxy-3-
 phenylpentanoate, **VII**, 164
(*R*)-Methyloxirane, **VIII**, 434
Methyloxirane, (*S*)-(–)-, **VII**, 356
Oxaspiropentane, **VI**, 320
3,3-Pentamethylenediaziridine, **V**,
 897
3-Phenyl-2*H*-azirine-2-carboxalde-
 hyde, **VI**, 893
Phenylmethylglycidic ester, **III**, 727
(±)-*trans*-2-(Phenylsulfonyl)-3-
 phenyloxaziridine, **VIII**, 546
(2*S*,3*S*)-3-Propyloxiranemethanol,
 VII, 461
trans-Stilbene oxide, **IV**, 860
Styrene oxide, **I**, 494
Tetracyanoethylene oxide, **V**, 1007
1,2,3,4-Tetrahydronaphthalene-
 (1,2)imine, **VI**, 967
H. Heterocyclic-[4]
 4-Acetoxyazetidin-2-one, **VIII**, 3
 Dimethylketene β-lactone dimer, **V**,
  456
 1,3-Dimethyl-3-methoxy-4-phenyl-
  azetidinone, **VIII**, 216
 Ethyl 3-(1-azetidinyl)propanoate,
  VI, 75
 β-Isovalerolactam, **V**, 673
 β-Isovalerolactam-*N*-sulfonyl chloride, **V**, 673
 Ketene dimer, **III**, 508
 Trimethylene oxide, **III**, 835
I. Heterocyclic-[5]
 N-(*p*-Acetylamlnophenyl)rhodanlne,
  IV, 6
 α-Acetyl-δ-chloro-γ-valerolactone,
  IV, 10
 3-Acetyl-2,4-dimethylfuran, **VI**, 31
 3-Acetyl-4-hydroxy-5,5-dimethylfu-
  ran-2(5*H*)-one, **VIII**, 71
 3-Acetyloxindole, **V**, 12
 2-Amino-6-methylbenzothiazole,
  III, 76
 2-Amino-4-methylthiazole, **II**, 31

3-Amino-3-pyrazoline sulfate, **V**, 39
3-Amino-1*H*-1,2,4-triazole, **III**, 95
4-Amino-4*H*-1,2,4-triazole, **III**, 96
Anhydro-2-hydroxymercuri-3-
  nitrobenzoic acid, **I**, 56
Azlactone of α-Acetaminocinnamic
  acid, **II**, 1
Azlactone of α-Benzoylamino-β-
  (3,4-dimethoxyphenyl)acrylic
  acid, **II**, 55
Benzimidazole, **II**, 65
Benzofurazan oxide, **IV**, 74
1,2,3-Benzothiadiazole 1,1-dioxide,
  V, 60
1,2,3-Benzotriazole, **III**, 106
α-Benzylidene-γ-phenyl-Δ$^{\beta,\gamma}$-buteno-
  lide, **V**, 80
4-Benzyloxyindole, **VII**, 34
β-Bromoethylphthalimide, **I**, 119;
  IV, 106
tert-Butylphthalimide, **III**, 152
1-Butylpyrrollidine, **III**, 159
γ-Butyrolactone-γ-carboxylic acid,
  (*S*)-(+)-, **VII**, 99
(–)-(Camphorsulfonyl)imine, **VIII**,
  104
3-(4-Chlorophenyl)-5-(4-
  methoxyphenyl)isoxazole, **VI**,
  278
Coumarone, **V**, 251
Creatinine, **I**, 172
2-Cyclobutyl-*cis*-4-*trans*-5-dimethyl-
  1,3-dioxolane, **VI**, 312
Diacetyl D-tartaric anhydride, **IV**,
  242
2,5-Diamino-3,4-dicyanothiophene,
  IV, 243
4,5-Dihydro-5-iodomethyl-4-phenyl-
  2(3*H*)-furanone, *cis*- and *trans*-,
  VII, 164
3,4-Dihydro-1,2-naphthalic anhydride, **II**, 194
2,4-Dimethyl-3-acetyl-5-carb-
  ethoxypyrrole, **III**, 513
2,4-Dimethyl-3,5-dicarbethoxypyr-
  role, **II**, 202

Dimethylfurazan, **IV**, 342
5,5-Dimethylhydantoin, **III**, 323
5,5-Dlmethyl-2-pentyltetrahydrofu-
  ran, **IV**, 350
3,5-Dimethylpyrazole, **IV**, 351
2,5-Dimethylpyrrole, **II**, 219
1,5-Dimethyl-2-pyrrolidone, **III**, 328
5,5-Dimethyl-2-pyrrolidone, **IV**, 357
2,4-Dimethylthiazole, **III**, 332
2,6-Dioxo-1-phenyl-4-benzyl-1,4-
  diazabicyclo[3.3.0]octane, **VIII**,
  231
1,4-Diphenyl-5-amino-1,2,3-triazole,
  IV, 380
4,5-Diphenylglyoxalone, **II**, 231
2,4-Diphenyl-2-pyrroline, **III**, 358
2,2-(Ethylenedithio)cyclohexanone,
  VI, 590
Ethylene thiourea, **III**, 394
Ethyl 3-ethyl-5-methyl-4-isoxazole-
  carboxylate, **VI**, 592
Ethyl indole-2-carboxylate, **V**, 567
Ethyl 3-methylcoumarilate, **IV**, 590
Ethyl 2-methylindole-5-carboxylate,
  VI, 601
Ethyl thiazole-4-carboxylate, **VI**, 620
D-Gulonlc-γ-lactone, **IV**, 506
Hexahydro-1,3,3,6-tetramethyl-2,1-
  benzisoxazoline, **VI**, 670
4(5)-Hydroxymethylimidazole
  hydrochloride, **III**, 460
3-Hydroxytetrahydrofuran, **IV**, 534
Imidazole-4,5-dicarboxylic acid, **III**,
  471
Indazole, **III**, 475; **IV**, 536; **V**, 650
Indazolone, **III**, 476
Indole, **III**, 479
Isatin, **I**, 327
Isoprene cyclic sulfone, **III**, 499
Itaconic anhydride, **II** 368
18,20-Lactone of 3β-acetoxy-20β-
  hydroxy-5-pregnen-18-oic acid,
  V, 692
2-Mercaptobenzimidazole, **IV**, 569
1-(2-Methoxycarbonylphenyl)pyr-
  role, **V**, 716

1-Methyl-3-ethyloxindole, **IV**, 620
1-Methyl-2-imino-β-naphtho-
thiazoline, **III**, 595
2-Methylindole, **III**, 597
Methyl 3-methyl-2-furoate, **IV**, 649
3-Methyloxindole, **IV**, 657
3-Methyl-1-phenylphospholene
oxide, **V**, 787
3-Methylthiophene, **IV**, 671
5-Nitroindazole, **III**, 660
4-Nitroindole, **VIII**, 493
3-Nitrophthalic anhydride, **I**, 410
γ-Octyl-γ-butyrolactone, **VII**, 400
8-Oxabicyclo[4.3.0]nonan-7-one,
(1*R*,6*S*)-(+)-, **VII**, 406
2-Oxo-1-phenyl-3-pyrrolidinecar-
boxylic acid, **VII**, 411
Parabanic acid, **IV**, 744
1-Phenyl-3-amino-5-pyrazolone, **III**,
708
o-Phenylene carbonate, **IV**, 788
2-Phenylfuro[3,2-*b*]pyridine, **VI**,
916
2-Phenylindole, **III**, 725
4-Phenyl-1,2,4-triazoline-3,5-dione,
VI, 936
Phthalimide, **I**, 457
Phthalide, **II**, 526
3-(1-Propyl)-2-pyrazolin-5-one, **VI**,
791
Pseudothiohydantoin, **III**, 751
Pyrrole, **I**, 473
Rhodanine, **III**, 763
Succinic anhydride, **II**, 560
Succinimide, **II**, 562
o-Sulfobenzoic anhydride, **I**, 495
1,2,3,4-Tetrahydrocarbazole, **IV**, 884
4,5,6,7-Tetrahydroindazole, **IV**, 536
Tetrahydrothiophene, **IV**, 892
2,2,5,5-Tetramethyltetrahydro-3-
ketofuran, **V**, 1024
cis-8-Thiabicyclo[4.3.0]nonane, **VI**,
482
Thiophene, **II**, 578
2-(*p*-Tolylsulfonyl)dihydroisoindole,
V, 1064

N-Tosyl-3-methyl-2-azabicy-
clo[3.3.0]oct-3-ene, *cis*-, **VII**,
501
1,2,4-Triazole-3(5)-thiol, **V**, 1070
2,3,5-Triphenylisoxazolidine, **V**,
1124
J. Heterocyclic-[6]
"Acetone-anil"(2,2,4-trimethyl-1,2-
dihydroquinoline), **III**,
329
9-Aminoacridine, **III**, 53
2-Amino-4-anilino-6-(chloromethyl)-*s*-triazine, **IV**, 29
4-Amino-2,6-dimethylpyrimidine,
III, 71
4-Amino-1,2,5,6-tetrahydro-1-
phenylphosphorin-3-carboni-
trile, **VI**, 932
Barbituric acid, **II**, 60
Benzeneboronic anhydride, **IV**, 68
Benzoguanamine, **IV**, 78
Benzoylene urea, **II**, 79
9-Butyl-1,2,3,4,5,6,7,8-octahy-
droacridine, **VIII**, 87
3-Carboethoxycoumarin, **III**, 165
Chelidonic acid, **II**, 126
7-Chloro-4-hydroxy-3-quinolinecar-
boxylic acid, **III**, 273
Coumalic acid, **IV**, 201
3-Cyano-6-methyl-2(1*H*)-pyridone,
IV, 210
Dehydroacetic acid, **III**, 231
2,4-Diamino-6-hydroxypyrimidine,
IV, 245
Diaminouracil hydrochloride, **IV**,
247
2,6-Di-*tert*-butyl-4-methylpyridine,
VII, 144
3,4-Dihydro-2-methoxy-4-methyl-
2*H*-pyran, **IV**, 311
5,6-Dihydro-2*H*-pyran-2-one, **VI**,
462
6,7-Dimethoxy-3-isochromanone,
VI, 471
6,7-Dimethoxy-1-methyl-3,4-dihy-
droisoquinoline, **VI**, 1

2,6-Dimethyl-3,5-diphenyl-2*H*-
pyran-4-one, **V**, 450
2a,4a-Dimethyl-8-oxabicyclo[3.2.1]-
oct-6-en-3-one, **VI**, 512
2,6-Dimethylpyridine, **II**, 214
1,3-Dithiane, **VI**, 556
2-Ethylchromone, **III**, 387
Ethyl isodehydroacetate, **IV**, 549
Hexahydro-4,4,7-trimethyl-4*H*-1,3-
benzoxathiin, **VIII**, 304
Homophthalic acid anhydride, **III**,
449
4-Hydroxy-1-butanesulfonic acid
sultone, **IV**, 529
2-Hydroxycinchoninic acid, **III**, 456
3-Hydroxycinchoninic acid, **V**, 636
2-Hydroxy-3-methylisocarbostyril,
V, 623
6-Hydroxynicotinic acid, **IV**, 532
Isatoic anhydride, **III**, 488
Isodehydroacetic acid, **IV**, 549
Julolidine, **III**, 504
2-Mercapto-4-amino-5-car-
bethoxypyrimidine, **IV**, 566
2-Mercapto-4-hydroxy-5-cyanopy-
rimidine, **IV**, 566
2-Mercaptopyrimidine and
hydrochloride, **V**, 703
6-Methoxy-8-nitroquinoline, **III**, 568
α-Methoxyphenazine, **III**, 753
2-(*p*-Methoxyphenyl)-6-phenyl-4-
pyrone, **V**, 721
1-(4-Methoxyphenyl)-1,2,5,6-
tetrahydropyridine, **VIII**,
358
10-Methyl-10,9-borazarophenan-
threne, **V**, 727
4-Methylcarbostyril, **III**, 580
4-Methylcoumarin, **III**, 581
4-Methylesculetin, **I**, 360
β-Methylglutaric anhydride, **IV**,
630
4-Methyl-7-hydroxycoumarin, **III**,
282
2-Methyl-4-hydroxyquinoline, **III**,
593

Methyl 2-oxo-5,6,7,8-tetrahydro-2*H*-
1-benzopyran-3-carboxylate,
VIII, 444
β-Methyl-β-phenyl-α,α'-dicyanoglu-
tarimide, **IV**, 662
N-Methyl-2-phenyl-Δ²-tetrahydropy-
ridine, **VI**, 818
4-Methylpyrimidine, **V**, 794
6-Methyluracil, **II**, 422
β-Methyl-δ-valerolactone, **IV**, 677
Mevalonolactone-2-^{13}C, (*R,S*)-, **VII**,
386
5-Nitro-2,3-dihydro-1,4-phthalazine-
dione, **III**, 656
Phenoxthin, **II**, 485
N-Phenylanthranilic acid, **II**, 15
1-Phenyl-3,4-dihydro-6,7-methyl-
enedioxyisoquinoline, **VII**, 326
4-Phenyl-*m*-dioxane, **IV**, 786
α-Phenylglutaric anhydride, **IV**, 790
1-Phenylpiperidine, **IV**, 795
Quinoline, **I**, 478
Quinoxaline, **IV**, 824
1,2,3,4-Tetrahydro-β-carboline, **VI**,
965
4*H*-1,4-Thiazine 1,1-dioxide, **VI**,
976
2a-Thiohomophthalimide, **V**, 1051
2-Thio-6-methyluracil, **IV**, 638
Triethyl 1,2,4-triazine-3,5,6-tricar-
boxylate, **VIII**, 597
2,2-(Trimethylenedithio)cyclohexa-
none, **VI**, 1014
2,4,6-Trimethylpyrylium perchlorate,
V, 1106
2,4,6-Trimethylpyrylium tetrafluoroborate, **V**, 1112
2,4,6-Trimethylpyrylium trifluoromethanesulfonate, **V**, 1114
sym-Trithiane, **II**, 610
Xanthone, **I**, 552
K. Heterocyclic-[>6]
18-Crown-6, **VI**, 301
1,10-Diazacyclooctadecane, **VI**, 382
Dibenzo-18-crown-6 polyether,
VI, 395

1,4,7,10,13,16-Hexakis(p-tolylsulfonyl)-1,4,7,10,13,16-hexaaza-
cyclooctadecane, **VI**, 652
11-Hydroxyundecanoic lactone, **VI**,
698
2-Ketohexamethylenimine, **II**, 76,
371
2,2,7,7,12,12,17,17-Octamethyl-
21,22,23,24-tetraoxaperhydro-
quaterene, **VI**, 856
8-Propionyl-(E)-5-nonenolide, **VIII**,
562
Ricinelaidic acid lactone, **VII**, 470
1,4,8,11-Tetrathiacyclotetradecane,
VIII, 592
L. Heterocyclic-[M,N]
N-Benzyl-2-azanorbornene, **VIII**, 31
Bis(1,3-diphenylimidazolidinyli-
dene-2), **V**, 115
Diethyl 2,3-diazabicyclo[2.2.1]hept-
5-ene-2,3-dicarboxylate, **VI**, 96
Hexahydro-3-(hydroxymethyl)-8a-
methyl-2-phenyl[2S,3S,8aR]-5-
oxo-5H-oxazolo-[3,2-a]pyri-
dine, **VIII**, 241
3-Isoquinuclidone, **V**, 670
4'-(Methylthio)-2,2':6',2''-terpyridine,
VII, 476
Pseudopelletierine, **IV**, 816
3-Quinuclidone hydrochloride, **V**, 989
ARYLATION
p-Acetyl-α-bromohydrocinnamic
acid, **VI**, 21
2-(1-Acetyl-2-oxopropyl)benzoic
acid, **VI**, 36
2-p-Acetylphenylhydroquinone, **IV**,
15
4-(4-Chlorophenyl)butan-2-one, **VII**,
105
3,4-Dichlorobiphenyl, **V**, 51
(2-Dimethylamino-5-methylphenyl)-
diphenylcarbinol, **VI**, 478
(Z)-β-[2-(N,N-Dimethylamino)-
phenyl]styrene, **VII**, 172
2,4-Dinitrobenzenesulfenyl chloride,
V, 474

N-(2,4-Dinitrophenyl)pyridinium
chloride, **VII**, 15
Ethyl cyano(pentafluorophenyl)
acetate, **VI**, 873
Ethyl 2-(p-methoxyphenyl)-2-oxocy-
clohexanecarboxylate, **VII**, 229
2-Methoxydiphenyl ether, **III**, 566
2-Methyl-4'-nitrobiphenyl, **VIII**, 430
2-Methyl-3-phenylpropanal, **VII**,
361
2-Methyl-3-phenylpropionaldehyde,
VI, 815
m-Nitrobiphenyl, **IV**, 718
o-Nitrodiphenyl ether, **II**, 446
p-Nitrodiphenyl ether, **II**, 445
1-(p-Nitrophenyl)-1,3-butadiene, **IV**,
727
Phenyl tert-butyl ether (Method II),
V, 926
N-Phenylcarbazole, **I**, 547
1-Phenyl-2,4-pentanedione, **VI**, 928
2-Phenylpyridine, **II**, 517
4-Pyridinesulfonic acid, **V**, 977
Tetraphenylarsonium chloride
hydrochloride, **IV**, 910
4,5,4',5'-Tetramethoxy-1,1'-biphenyl-
2,2'-dicarboxaldehyde, **VIII**,
586
1,2,3,4-Tetraphenylnaphthalene, **V**,
1037
Triphenylamine, **I**, 544
Triphenylselenonium chloride, **II**,
241
CARBOMETALATION
Lithium di[(Z)-1-hexenyl)]cuprate
(in solution), **VII**, 290
(E)-(2-Methyl-1,3-butadienyl)-
dimethylalane, **VII**, 245
CARBONYLATION
(Carbonylamino)benzene, **VI**, 936
7-(Carbonylamino)-7-ethoxybicy-
clo[4.1.0]heptane, **VI**, 230
6-(Carbonylamino)-6-ethoxybicy-
clo[3.1.0]hexane, **VI**, 230
7-(Carbonylamino)-7-ethoxy-3-nor-
carene, **VI**, 230

6-(Carbonylamino)hexanoyl chloride, **VI**, 718
2-(Carbonylamino)propane, **VI**, 951
3-Hydroxy-2,2,3-trimethyloctan-4-
one, **VIII**, 343
Perhydro-9*b*-phenalenol, **VII**, 427
CARBOXYLATION
1-Adamantanecarboxylic acid, **V**,
20
tert-Butylcarbonic diethylphosphoric
anhydride, **VI**, 207
Di-*tert*-butyl tricarbonate, **VI**, 418
2-Carbethoxycyclooctanone, **V**,
198
Dimethyl cyclohexanone-2,6-dicar-
boxylate, **V**, 439
Ferrocenecarboxylic acid, **VI**, 625
Mesitoic acid, **V**, 706
1-Methylcyclohexanecarboxylic
acid, **V**, 739
1-Norbornanecarboxylic acid, **VI**,
845
β–Resorcylic acid, **II**, 557
Tetrolic acid, **V**, 1043
CLEAVAGE
A. Dealkylation
Catechol, **I**, 150
3,3'-Dihydroxybiphenyl, **V**, 412
4-Ethoxy-3-hydroxybenzaldehyde
ethylene acetal, **VI**, 567
α-Hydroxyphenazine, **III**, 754
(*R*,*S*)-Mevalonolactone-2-¹³C, **VII**,
386
Orcinol monomethyl ether, **VI**, 859
Phenyl α-methoxystyryl ketone, **I**,
205
Tetramethyl-*p*-phenylenediamine, **V**,
1018
B. Hydrolytic (see also Hydrolysis)
Dihydroxycyclopentene, **V**, 414
7,16-Diketodocosanedioic acid and
disodium salt, **V**, 534, 536
Ethyl γ-bromobutyrate, **V**, 545
3-Hydroxyglutaronitrile, **V**, 614
α-Ketoglutaric acid, **V**, 687
Neopentyl alcohol, **V**, 818

exo/endo-7-Norcaranol, **V**, 859
C. Ketonic
tert-Butyl diazoacetate, **V**, 179
2-Butyl-2-methylcyclohexanone, **V**,
187
Diethyl β-ketopimelate, **V**, 384
Docosanedioic acid, **V**, 533
Ferrocenecarboxylic acid, **VI**, 625
2-Methylcyclopentane-1,3,5-trione
hydrate, **V**, 748
5-Methyl-5-hexen-2-one, **V**, 767
D. Oxidative
o-Anisaldehyde, **V**, 46
Benzyl isocyanide , **VII**, 27
2-Cyclohepten-1-one, **VI**, 327
(2*SR*,3*RS*)-2,4-Dimethyl-3-hydrox-
ypentanoic acid, **VII**, 185
D-Erythronolactone, **VII**, 297
(*Z*,*Z*)-2,4-Hexadienedinitrile, **VI**,
662
1,2,4-Triazole, **V**, 1070
E. Reductive
O-tert-Butyl-L-seryl-*S-tert*-butyl-L-
cysteine *tert*-butyl ester, **VI**,
252
Cyclohexanol, **VI**, 353
2-Cyclohexyloxyethanol, **V**, 303
1,2-Dimercaptobenzene, **V**, 419
L-Methionine, **VI**, 252
5-Methylcoprost-3-ene, **VI**, 762
3-Methylcyclohexene, **VI**, 769
F. Miscellaneous
7-Cyanoheptanal, **V**, 266
2,4-Dinitrobenzenesulfenyl chloride,
V, 474
2,2,4-Trimethyl-3-oxovaleryl chloride, **V**, 1103
CONDENSATION
The term "condensation" is
used here in a restricted sense
and applies to those reactions
in which a carbon–carbon
bond is formed by the elimination of a simple molecule.
Cyclization and *Dehydration*
reactions are listed separately.

Many other reactions that produce a carbon–carbon bond are listed under other headings. (See *Addition; Alkylation, Diazotization; Friedel–Crafts Reaction; Grignard Reaction; Rearrangement; Reduction.*) The subheadings illustrate the types of reactions leading to the compounds listed.

A. Carbonyl–Active Methylene Condensations

2-Acetyl-1,3-cyclopentanedione, **VI**, 28

3-Acetyl-2,4-dimethylfuran, **VI**, 31

2-(1-Acetyl-2-oxopropyl)benzoic acid, **VI**, 36

Anisalacetone, **I**, 78

Azlactone of α-Acetaminocinnamic acid, **II**, 1

Azlactone of α-Benzoylamino-cinnamic acid, **II**, 490

Azlactone of α-Benzoylamino-β-(3,4-dimethoxyphenyl)acrylic acid, **II**, 55

Benzalacetone, **I**, 77

Benzalacetophenone, **I**, 78

Benzalbarbituric acid, **III**, 39

Benzalpinacolone, **I**, 81

α-Benzylidlne-γ-phenyl-Δ$^{\beta,\gamma}$-buteno-lide, **V**, 80

3-Benzyl-3-methylpentanoic acid, **IV**, 93

p-Bromodiphenylmethane, **V**, 130

3-Carbethoxycoumarin, **III**, 165

β-Carbethoxy-γ-diphenylvinylacetic acid, **IV**, 132

α-Cyano-β-phenylacrylic acid, **I**, 181

Cycloheptanone, **IV**, 221

Cyclohexylidenecyanoacetic acid, **IV**, 234

Diacetone alcohol, **I**, 199

Dibenzalacetone **II**, 167

α,α'-Dicyano-β-ethyl-β-methylglu-tarimide, **IV**, 441

Dicyclopropyl ketone, **IV**, 278

Diethyl bis(hydroxymethyl)mal-onate, **V**, 381

Diethyl ethylidenemalonate, **IV**, 293

1,4-Dihydro-3,5-dicarbethoxy-2,6-dimethylpyridine, **II**, 214

2,3-Dimethoxycinnamic acid, **IV**, 327

6-(Dimethylamino)fulvene **V**, 431

3,5-Dimethyl-2-cyclohexen-1-one, **III**, 317

1,4-Diphenylbutadiene, **II**, 229

Diphenylcinnamonitrile, **IV**, 387

2,3-Diphenylsuccinonitrile, **IV**, 392

Ethyl α-acetyl-β-(2,3-dimethoxyphenyl)acrylate, **IV**, 408

Ethyl benzalmalonate, **III**, 377

Ethyl butylcyanoacetate, **III**, 385

Ethyl *sec*-butylidenecyanoacetate, **IV**, 94

Ethyl cinnamate, **I**, 252

Ethyl (1-ethylpropylidene)cyanoace-tate, **III**, 399

α-Ethyl-α-methylsuccinic acid, **V**, 572

Ethyl *o*-nitrophenylpyruvate sodium salt, **V**, 567

Ethyl(1-phenylethylidene)cyanoace-tate, **IV**, 463

Ethyl propane-1,1,3,3-tetracarboxy-late, **I**, 290

2-Fururalacetone, **I**, 283

Furylacrylic acid, **III**, 425

3-(2-Furyl)acrylonitrile, **V**, 585

5-(*p*-Hydroxybenzal)hydantoin, **V**, 627

threo-4-Hydroxy-3-phenyl-2-hep-tanone, **VI**, 692

p-Hydroxyphenylpyruvic acid, **V**, 627

5-(3-Methoxy-4-hydroxybenzal)cre-atinine, **III**, 587

1-(*o*-Methoxyphenyl)-2-nitro-1-
   propene, **IV**, 573
N-Methyl-3,4-dihydroxyphenylala-
   nine, **III**, 586
β-Methylglutaric acid, **III**, 591
m-Nitrocinnamic acid, **I**, 398; **IV**,
   731
2-Nitroethanol, **V**, 833
trans-o-Nitro-α-phenylcinnamic
   acid, **IV**, 730
m-Nitrostyrene, **IV**, 731
β-Nitrostyrene, **I**, 413
trans-2,4-Pentadienoic acid, **VI**, 95
Pentaerythritol, **I**, 425
α-Phenylcinnamic acid, **IV**, 777
α-Phenylcinnamonitrile, **III**, 715
Sorbic acid, III, 783
B. Carbonyl–Ylide Condensation
   1,4-Bis(4-phenylbutadienyl)benzene,
      V, 985
   Cyclohexylideneacetaldehyde, **VI**,
      358
   Dichloromethylenetriphenylphosphor-
      ane, **V**, 361
   β,β-Difluorostyrene, **V**, 390
   (*E*)-4,8-Dimethyl-1,3,7-nonatriene,
      VII, 259
   1,4-Diphenyl-1,3-butadiene, **V**, 499
   1,1-Diphenylcyclopropane **V**, 509
   Ethyl cyclohexylideneacetate, **V**,
      547
   Ethyl 2,3-pentadienoate, **VII**, 232
   Ethyl (α-hydroxymethyl)acrylate,
      VIII, 265
   Methylenecyclohexane, **V**, 751
   β-Phenylcinnamic acid, **V**, 509
   2-Phenylperfluoropropene, **V**, 949
C. Ester–Active Methylene
      Condensation
   2-Carbethoxycyclopentanone, **II**, 116
   Diethyl isopropylidenemalonate, **VI**,
      442
   Ethyl ethoxyalylpropionate, **II**, 272
   Ethyl indole-2-carboxylate, **V**, 567
   Ethyl 2-ketocyclohexylglyoxalate,
      II, 531

Ethyl phenylcyanopyruvate, **II**,
   287
Ethyl phenylmalonate, **II**, 288
Ethyl thiazole-4-carboxylate, **VI**, 620
α-Phenylacetonitrile, **II**, 487
D. Ester-Ester Condensations
   Butyroin, **II**, 114
   Cetylmalonic ester, **IV**, 141
   3,4-Dihydro-1,2-naphthalic anhy-
      dride, **II**, 194
   6,7-Dimethoxy-3,4-dihydro-2-naph-
      thoic acid, **III**, 300
   Ethyl acetoacetate, **I**, 235
   Ethyl benzoylacetate, **III**, 379
   Ethyl oxalylsuccinate, **III**, 510
   Isobutyroin, **II**, 115
   α-Ketoglutaric acid, **III**, 510
   Pivaloin, **II**, 115
   Propionoin, **II**, 115
E. Ester–Ketone Condensations
   Acetylacetone, **III**, 17
   3-Cyano-6-methyl-2(1)-pyridone,
      IV, 210
   Dibenzoylmethane, **III**, 251
   Diisovalerylmethane, **III**, 291
   2-Ethylchromone, **III**, 387
   Ethyl β-hydroxy-β,β-diphenylpropi-
      onate, **V**, 564
   2-Hydroxymethylenecyclohexanone,
      IV, 536, 537
   Indazole, **IV**, 536
   1,3,5-Triacetylbenzene, **III**, 829
F. Miscellaneous
   Acetone azine, **VI**, 10
   Acetylacetone, **III**, 17
   Acridone, **II**, 15
   Aminoacetone semicarbazone
      hydrochloride, **V**, 27
   Benzalphthalide, **II**, 61
   Benzanthrone, **II**, 62
   Benzoin, **I**, 94
   2,2'-Bipyridine, **V**, 102
   Bis(1,3-diphenylimidazolidinyli-
      dene-2), **V**, 115
   3-(4-Chlorophenyl)-5-(4-meth-
      oxyphenyl)isoxazole, **VI**, 278

α-(4-Chlorophenyl)-γ-phenylace-
toacetonitrile, **IV**, 174
2-(1-Cyanocyclohexyl)hydrazinecar-
boxylic acid, methyl ester, **VI**,
334
Cyclopentanone, **I**, 192
Diaminomaleonitrile, **V**, 344
1,10-Diazacyclooctadecane-2,9-
dione, **VI**, 382
Diethyl 2-(cyclohexylamino)
vinylphosphonate, **VI**, 448
3,3-Dimethoxycyclopropene, **VI**,
361
p-Dimethylaminobenzophenone, **I**,
217
5,5-Dimethyl-1,3-cyclohexanedione,
II, 200
N,N-Dimethylcyclohexylamine, **VI**,
499
2,4-Dimethyl-3,5-dicarbethoxypyr-
role, **II**, 202
2,5-Dimethyl-3-phenyl-2-cyclopen-
ten-1-one, **VI**, 520
Dypnone, **III**, 367
Ethyl acetonedioxalate, **II**, 126
Ethyl acetopyruvate, **I**, 238
Ethyl ethylenetetracarboxylate, **II**,
273
Ethyl hydrazinecarboxylate, **VI**, 936
Ethyl phenylcyanoacetate, **IV**, 461
Ethyl propane-1,1,2,3-tetracarboxy-
late, **I**, 272
Ethyl β-pyrrolidinocrotonate, **VI**, 592
Furylacrylic acid, **III**, 426
4-Heptanone, **V**, 589
3-Heptylureidomethylenemalononi-
trile, **IV**, 515
5-Hexynal, **VI**, 679
Mesitylene, **I**, 341
p-Phenylazobenzoic acid, **III**, 711
Pseudoionone, **III**, 747
Tetracyanoethylene, **IV**, 877
Tetraphenylcyclopentadienone, **III**,
806
Tetraphenylethylene, **IV**, 914

p-Tolualdehyde, **II**, 583
m-Tolylbenzylamine, **III**, 827
COUPLING
α-Allyl-β-bromoethyl ethyl ether,
IV, 750
2,2'-Bipyridine, **V**, 102
4,4'-Diaminoazobenzene, **V**, 341
1,2-Dibutylbenzene, **VI**, 407
2,2'-Dichloro-α,α'-epoxybibenzyl,
V, 1026
4,4'-Dimethoxy-1,1'-biphenyl, **VI**,
468
4,4'-Dimethyl-1,1'-biphenyl, **VI**, 488
2,5-Dimethyl-3-phenyl-2-cyclopen-
ten-1-one, **VI**, 520
2,2'-Dinitrobiphenyl, **III**, 339
Diphenyldiacetylene, **V**, 517
Hexadehydro[18]annulene, **VI**, 68
2-Hydroxycyclobutanone, **VI**, 167
N-(3-Hydroxy-3,3-diphenylpropyli-
dene)cyclohexylamine, **VI**, 901
3-Isocyanatopropanoyl chloride, **VI**,
715
Methallylbenzene, **VI**, 722
3-Methyl-2,4-heptanedione, **VI**, 776
2-Methyl-4'-nitrobiphenyl, **VIII**, 430
1-(2-Methylpropenyl)pyrrolidine,
VI, 496
N-(2-Methylpropylidene)-tert-butyl-
amine, **VI**, 526
α-Morpholinostyrene, **VI**, 520
Nitroacetic acid, dipotassium salt,
VI, 797
1-Phenyl-1,4-pentadiyne, **VI**, 925
α,α,α',α'-Tetramethyltetramethylene
glycol, **V**, 1026
(E)-2-Undecene, **VI**, 1040
CYCLIZATION
A. Condensation
α-Acetyl-δ-chloro-γ-valerolactone,
IV, 10
3-Acetyl-2,4-dimethylfuran, **VI**, 31
3-Acetyl-4-hydroxy-5,5-dimethylfu-
ran-2(5H)-one, **VIII**, 71
Acridone, **II**, 15

3-Amino-3-pyrazoline sulfate, **V**, 39
4-Amino-1,2,5,6-tetrahydro-1-
phenylphosphorin-3-carboni-
trile, **VI**, 932
Azulene, **VII**, 15
Benzanthrone, **II**, 62
(*S*)-(+)- and (*R*)-(–)-1,1-Binaphthyl-
2,2'-diyl hydrogen phosphate,
VIII, 51
p-tert-Butylcalix[4]arene, **VIII**, 75
p-tert-Butylcalix[6]arene, **VIII**, 77
p-tert-Butylcalix[8]arene, **VIII**, 80
8-Butyl-2-hydroxytricyclo-
[7.3.1.0^{2,7}]tridecan-13-one,
VIII, 87
9-Butyl-1,2,3,4,5,6,7,8-octahy-
droacridine, **VIII**, 87
(–)-(Camphorsulfonyl)imine, **VIII**,
110
3-Carbethoxycoumarin, **III**, 165
2-Carbethoxycyclopentanone **II**, 116
2-Carbomethoxy-3-vinylcyclopen-
tanone, **VIII**, 112
Chelidonic acid, **II**, 126
7-Chloro-4-hydroxy-3-quinolinecar-
boxylic acid, **III**, 273
Coumalic acid, **IV**, 201
Coumarone, **V**, 251
3-Cyano-6-methyl-2(1*H*)-pyridone,
IV, 210
1,1-Cyclobutanedicarboxylic acid,
III, 213
Cyclobutanone, **VI**, 324
1,4-Cyclohexanedione, **V**, 288
Cyclopentanone, **I**, 192
Cyclopropanecarboxylic acid, **III**,
221
Cyclopropenone 1,3-propanediol
ketal, **VIII**, 173
Cyclopropyl cyanide, **III**, 223
Cyclopropyldiphenylsulfonium tetra-
fluoroborate, **VI**, 364
Dehydroacetic acid, **III**, 231
2,6-Diacetoxybicyclo[3.3.0]octane,
VIII, 43

1,10-Diazacyclooctadecane-2,9-
dione, **VI**, 382
trans-1,2-Dibenzoylcyclopropane,
VI, 401
N,N-Dibenzyl-4,13-diaza-18-crown-
6, **VIII**, 152
2,6-Di-*tert*-butyl-4-methylpyrylium
trifluoromethanesulfonate, **VII**,
144
2,2'-Dichloro-α,α'-epoxybibenzyl,
V, 358
α,α-Dicyano-β-ethyl-β-methylglu-
tarimide, **IV**, 441
Dicyclopropyl ketone, **IV**, 278
Diethyl 1,1-cyclobutanedicarboxy-
late, **IV**, 288
4,5-Dihydro-5-iodomethyl-4-
phenyl-2(3*H*)-furanone, *cis*-
and *trans*-, **VII**, 164
Dihydrojasmone, **VIII**, 620
3,4-Dihydro-1,2-naphthalic anhydride, **II**, 194
5,6-Dihydro-2*H*-pyran-2-one, **VIII**,
462
(–)-Dimenthyl (1*S*,2*S*)-cyclo-
propane-1,2-dicarboxylate,
VIII, 141
6,7-Dimethoxy-3,4-dihydro-2-naph-
thoic acid, **III**, 300
6,7-Dimethoxy-3-isochromanone,
VI, 471
6,7-Dimethoxy-1-methyl-3,4-dihy-
droisoquinoline, **VI**, 1
2,4-Dimethyl-3-acetyl-5-carb-
ethoxypyrrole, **III**, 513
3,5-Dimethyl-4-carbethoxy-2-cyclo-
hexen-1-one, **III**, 318
5,5-Dimethyl-1,3-cyclohexanedione,
II, 200
4,4-Dimethyl-2-cyclohexen-1-one,
VI, 496
4,4-Dimethyl-2-cyclopenten-1-one,
VIII, 208
2,4-Dimethyl-3,5-dicarbethoxypyr-
role, **II**, 202

2,2-Dimethylethylenimine, **III**, 148

2,6-Dimethyl-3,5-diphenyl-4H-pyran-4-one, **V**, 450

(1R)-9,9-Dimethyltricyclo[6.1.1.0^{2,6}]-deca-2,5-diene, **VIII**, 223

1,6-Dioxo-8a-methyl-1,2,3,4,6,7,8a-octahydronaphthalene, **V**, 486

Diphenylcyclopropenone, **V**, 514

2,3-Diphenylindone, **III**, 353

(R)-4-Ethyl-4-allyl-2-cyclohexen-1-one, **VIII**, 241

2-Ethylchromone, **III**, 387

Ethyl 1,3-dihydroxy-2-naphthoate, **III**, 638

Ethyl 6,7-dimethoxy-3-methylindene-2-carboxylate, **V**, 550

Ethyl isodehydroacetate, **IV**, 549

Ethyl 3-methylcoumarilate, **IV**, 590

Flavone, **IV**, 478; **IV**, 479

Fluorenecarboxylic acid, **IV**, 482

cis-4,4a,5,6,7,8-Hexahydro-4a,5-dimethyl-2(3H)-naphthalenone, **VI**, 666

Hexahydro-3-(hydroxymethyl)-8a-methyl-2-phenyl [2S,3S,8aR]-5-oxo-5H-oxazolo[3,2-a]pyridine, **VIII**, 241

Hexahydro-4,4,7-trimethyl-4H-1,3-benzoxathiin, **VIII**, 302

1,4,7,10,13,16-Hexakis(p-tolylsulfonyl)-1,4,7,10,13,16-hexaazacyclooctadecane, **VI**, 652

11-Hydroxyundecanoic lactone, **VI**, 698

Indazolone, **III**, 476

Isodehydroacetic acid, **IV**, 549

Julolidine, **III**, 504

Ketene dimer, **IV**, 508

Kryptopyrrole, **III**, 513

6-Methoxy-8-nitroquinoline, **III**, 568

α-Methoxyphenazine, **III**, 753

1-(4-Methoxyphenyl)-1,2,5,6-tetrahydropyridine, **VIII**, 358

β-Methylanthraquinone, **I**, 353

4-Methylcarbostyril, **III**, 580

4-Methylcoumarin, **III**, 581

2-Methyl-1,3-cyclopentanedione, **V**, 747; **VI**, 774

Methyl cyclopropyl ketone, **IV**, 597

Methyl 1,3-dimethyl-5-oxabicyclo[2.2.2]octane-2-carboxylate, **VIII**, 219

Methylenecyclopropane, **VI**, 320

Methylenecyclohexane oxide, **V**, 755

4-Methylesculetin, **I**, 360

2-Methyl-4-ethoxalylcyclopentane-1,3,5-trione, **V**, 747

1-Methyl-3-ethyloxindole, **IV**, 620

1-Methyl-4,4a,5,6,7,8-hexahydro-naphthalen-2(3H)-one, **VI**, 781

4-Methyl-7-hydroxycoumarin, **III**, 282

2-Methyl-4-hydroxyquinoline, **III**, 593

(R)-Methyloxirane, **VIII**, 434

Methyl 2-oxo-5,6,7,8-tetrahydro-2H-1-benzopyran-3-carboxylate, **VIII**, 444

1-Methyl-3-phenylindane, **IV**, 665

N-Methyl-2-phenyl-Δ²-tetrahydropyridine, **VI**, 818

(8aS)-8a-Methyl-3,4,8,8a-tetrahydro-1,6(2H,7H)-naphthalenedione, **VII**, 368

4'-(Methylthio)-2,2':6',2''-terpyridine, **VII**, 476

2-Methyl-2-vinylcyclopentanone, **VIII**, 486

Naphthoresorcinol, **III**, 637

(1R,2r,3S)-2-Nitrocyclohexane-1,3-diol, **VIII**, 332

4-Nitroindole, **VIII**, 493

Nortricyclanol, **V**, 863

Δ$^{1(9)}$-Octalone-2, **V**, 869

2,2,7,7,12,12,17,17-Octamethyl-21,22,23,24-tetraoxaquaterene, **VI**, 856

2-Oxazolidinone, **VII**, 4

2-Oxo-1-phenyl-3-pyrrolidinecarboxylic acid, **VII**, 411

1,2,3,4,5-Pentamethylcyclo-
pentadiene, **VIII**, 505
2-(1-Phenylcyclopentyl)-4,4,6-
trimethyl-5,6-dihydro-1,3(4*H*)-
oxazine, **VI**, 905
1-Phenyl-3,4-dihydro-6,7-methyl-
enedioxyisoquinoline, **VII**,
326
2-Phenylfuro[3,2-*b*]pyridine, **VI**,
916
1,2,3,4,5-Pentamethylcyclopenta-
diene, **VIII**, 505
(*S*)-4-(Phenylmethyl)-2-oxazolidi-
none, **VIII**, 528
Phenylphenanthrene, **V**, 952
4-Phenylurazole, **VI**, 936
8-Propionyl-(*E*)-5-nonenolide, **VIII**,
562
3-(1-Propyl)-2-pyrazolin-5-one, **VI**,
791
4-Protoadamantanone, **VI**, 958
Quinizarin, **I**, 476
Quinoline, **I**, 478
Quinoxaline, **IV**, 824
Ricinelaidic acid lactone, **VII**, 470
Sebacoin, **IV**, 840
Spiro[5.7]tridec-1-en-3-one, **VII**, 473
α-Tetralone, **II**, 569; **III**, 798; **IV**,
898; **IV**, 900
1,2,3,4-Tetrahydronaphthalen(1,2)-
imine, **VI**, 967
Tetramethyl 3,7-dihydroxybicy-
clo[3.3.0]octa-2,6-diene-2,4,
6,8-tetracarboxylate, **VII**, 50
Tetramethyl 3,7-dihydroxy-1,5-
dimethylbicyclo[3.3.0]octa-2,6-
diene-2,5,6,8-tetracarboxylate,
VII, 51
Tetraphenylcyclopentadienone, **III**,
806
1,4,8,11-Tetrathiacyclotetradecane,
VIII, 592
1-*p*-Tolylcyclopropanol, **V**, 1058
1,3,5-Triacetylbenzene, **III**, 829
1,2,4-Triazole-3(5)-thiol, **V**, 1070

(*Z*)-Tributylstannylmethylene-4-iso-
propyl-1,1-cyclopentanedicar-
boxylic acid, dimethyl ester,
VIII, 381
Triethyl 1,2,4-triazine-3,5,6-tricar-
boxylate, **VIII**, 597
Xanthone, **I**, 552
B. Cycloaddition
4-Acetoxyazetidin-2-one, **VIII**, 3
7-Acetoxy-4,4,6,7-tetramethyl-
bicyclo[4.2.0]octan-2-one,
VI, 1024
4-Acetyl-7,8-dimethyl-2-oxa-4-
azabicyclo[4.3.0]non-7-en-3-
one, **VII**, 5
Benzobarrelene, **VI**, 82
Binor-S, **VI**, 378,
Biphenylene, **V**, 54
3-Butyl-4,4-dichlorocyclobutanone,
VIII, 82
7-Chloro-7-cyanobicyclo[4.2.0]-
octan-8-one, **VIII**, 116
1-Chloro-1,4,4-trifluorobutadiene,
V, 235
trans, trans-1,4-Diacetoxy-1,3-buta-
diene, **VI**, 196
9,9-Dibromobicyclo[6.1.0]nonane,
V, 306
7,7-Dichlorobicyclo[3.2.0]hept-2-en-
6-one, **VI**, 1037
cis-3,4-Dichlorocyclobutene,
VI, 422
7,7-Dichloro-1-methylbicyclo-
[3.2.0]heptan-6-one, **VIII**, 377
3,4-Dichloro-1,2,3,4-tetramethylcy-
clobutene, **V**, 370
5,5-Dicyano-4-phenyl-2-cyclopen-
ten-1-one 1,3-propanediol ketal,
VIII, 173
2,3-Dicyano-1,4,4*a*,9*a*-tetrahydro-
fluorene, **VI**, 427
Diethyl 2,3-diazabicyclo[2.2.1]hept-
5-ene-2,3-dicarboxylate, **V**, 96
Diethyl *trans*-1,4-tetrahydrophtha-
late, **VI**, 454

Diethyl *trans*-4-trimethylsilyloxy-4-cyclohexene-1,2-dicarboxylate, **VI**, 445

3,4-Dihydro-2-methoxy-4-methyl-2*H*-pyran, **IV**, 311

7,7-Dimethoxy-1,2,3,4-tetrachlorobicyclo[2.2.1]heptene, **V**, 425

2,3-Dimethylanthraquinone, **III**, 310

3,3-Dimethyl-*cis*-bicyclo[3.2.0]heptan-2-ol, **VII**, 177

4,4-Dimethyl-2-cyclohexen-1-one, **VI**, 496

1,3-Dimethyl-3-methoxy-4-phenylazetidinone, **VIII**, 216

2α,4α-Dimethyl-8-oxabicyclo[3.2.1]oct-6-en-3-one, **VI**, 512

2,3-Dimethyl-1,4,5,6-tetrahydroanthraquinone, **III**, 310

2,6-Dioxo-1-phenyl-4-benzyl-1,4-diazabicyclo[3.3.0]octane, **VIII**, 231

Diphenylcarbodiimide (Method II), **V**, 504

1,1-Diphenylcyclopropane, **V**, 509

Dispiro[5.1.5.1]tetradecane-7,14-dione, **V**, 297

cis- and *trans*-1,2-Divinylcyclobutane, **V**, 528

Ethyl 3-ethyl-5-methyl-4-isoxazolecarboxylate, **VI**, 592

3,3a,3b,4,6a,7a-Hexahydro-3,4,7-metheno-7*H*-cyclopenta[*a*]pentalene-7,8-dicarboxylic acid, dimethyl ester, **VIII**, 298

Hexahydro-1,3,3,6-tetramethyl-2,1-benzisoxazoline, **VI**, 670

Hexamethyl Dewar benzene, **VII**, 256

Hexaphenylbenzene, **V**, 604

3-Hexylcyclobutanone, **VIII**, 306

Isoprene cyclic sulfone, **III**, 499

cis-4-*exo*-Isopropenyl-1,9-dimethyl-8-(trimethylsilyl)bicyclo[4.3.0]non-8-en-2-one, **VIII**, 347

β-Isovalerolactam-*N*-sulfonyl chloride and β-Isovalerolactam, **V**, 673

5β-Methoxycyclohexan-1-one-3β,4β-dicarboxylic acid anhydride, **VII**, 312

6-Methoxy-7-methoxycarbonyl-1,2,3,4-tetrahydronaphthalene, **VIII**, 444

6-Methylbicyclo[4.2.0]octan-2-one, **VII**, 315

3-Methylenecyclobutane-1,2-dicarboxylic anhydride, **V**, 459

3-Methyl-1-phenylphospholene 1,1-dichloride, **V**, 787

Norbornylene, **IV**, 738

[6a*R*-(6aα,9α,10aβ)]-Octahydro-3,3,6,6,9-pentamethyl-1*H*,6*H*-[1,3]dioxino-[4,5-*c'*][2]benzopyran-1-one, **VIII**, 353

4-Oxo-1-(phenylsulfonyl)-*cis*-bicyclo[4.3.0]non-2-ene, **VIII**, 39

[2,2]Paracyclophane, **V**, 883

1-Phenyl-3,4-dihydro-6,7-methylenedioxyisoquinoline, **VII**, 326

2-Phenyl-3-propylisoxazolidine-4,5-*cis*-dicarboxylic acid *N*-phenylimide, **V**, 957

p-Quinquephenyl, **V**, 985

Styrene oxide, **I**, 494

cis-Δ⁴-Tetrahydrophthalic anhydride, **IV**, 890

Tetramethyl [2,2]paracyclophane-4,5,12,13-tetracarboxylate, **VII**, 485

1,2,3,4-Tetraphenylnaphthalene, **V**, 1037

Tetraphenylphthalic anhydride, **III**, 807

2,3,6-Tricarboethoxypyridine, **VIII**, 597

1,1,2-Trichloro-2,3,3-trifluorocyclobutane, **V**, 393

Triphenylene, **V**, 1120

2,3,5-Triphenylisoxazolidine, **V**, 1124

C. Free Radical-Carbenoid/Nitrenoid

trans-1-Amino-2,3-diphenylaziridine, **VI**, 56

Benzocyclopropene, **VI**, 87

Bicyclo[1.1.0]butane, **VI**, 133

2-(Trichloromethyl)bicyclo[3.3.0]octane, **V**, 93

Bicyclo[2.1.0]pentane, **V**, 96

1-Chloro-1,2-diphenyl-3-methylcyclopropane, **VII**, 203

exo/*endo*-7-(2-Chloroethoxy)bicyclo[4.1.0]heptane, **V**, 859

Cyclopropylbenzene, **V**, 328

2-(2,2-Dibromocyclopropyl)-6,6-dimethylbicyclo[3.1.1]hept-2-ene, **VIII**, 223

7,7-Dibromo-1,6-dimethylbicyclo[4.1.0]hept-3-ene, **VII**, 200

1,1-Dibromo-2,2-diphenylcyclopropane, **VI**, 187

1,1-Dichloro-2-phenylcyclopropane, **VII**, 12

11,11-Dichlorotricyclo[4.4.1.0^{1,6}]-undeca-3,8-diene, **VI**, 731

3,4-Dihydro-1(2*H*)-azulenone, **VIII**, 196

1,6-Dimethyltricyclo[4.1 0.0^{2,7}]hept-3-ene, **VII**, 200

1-Ethoxy-2-*p*-tolylcyclopropane, **VI**, 571

Ethyl 1-cyano-2-methylcyclohexanecarboxylate, **VI**, 586

Ethyl *cis*- and *trans*-2-phenylcyclopropanecarboxylate, **VI**, 913

Methylenecyclohexane oxide, **V**, 755

2-Nitrocarbazole, **V**, 829

Norcarane, **V**, 855

exo/*endo*-7-Norcaranol, **V**, 859

γ-Octyl-γ-butyrolactone, **VII**, 99

2-Oxa-7,7-dichloronorcarane, **V**, 874

D-2-Oxo-7,7-dimethyl-1-vinylbicyclo[2.2.1]heptane, **V**, 877

2-Phenylcycloheptanone, **IV**, 780

Phenylindazole, **V**, 941

Pseudopelletierine, **IV**, 816

2,2,3,3-Tetramethyliodocyclopropane, **VI**, 974

endo-Tricyclo[4.4.0.0^{2,5}]deca-3,8-diene-7,10-dione, **VI**, 1002

1-Trimethylsilyloxybicyclo[4.1.0]heptane, **VI**, 327

Triptycene, **IV**, 964

D. Oxidative Cyclization

Benzofurazan oxide, **IV**, 74; **IV**, 75

(*S*)-(+)-γ-Butyrolactone-γ-carboxylic acid, **VII**, 99

trans-7,8-Diacetoxybicyclo[4.2.0]-octa-2,4-diene, **VI**, 196

4,5-Dihydro-5-iodomethyl-4-phenyl-2(3*H*)-furanone, **VII**, 164

18,20-Lactone of 3β-Acetoxy-20β-hydroxy-5-pregnen-18-oic acid, **V**, 692

(1*R*,6*S*)-(+)-8-Oxabicyclo[4.3.0]-nonan-7-one, **VII**, 406

Oxaspiropentane, **VI**, 320

Phenylchlorodiazirine (in solution), **VII**, 203

Pyromellitic acid, **II**, 551

trans-Stilbene oxide, **IV**, 860

Tetrahydroxyquinone, **V**, 1011

cis-*N*-Tosyl-3-methyl-2-azabicyclo[3.3.0]oct-3-ene, **VII**, 501

E. Reductive Cyclization

4-Benzyloxyindole, **VII**, 34

cis-2-Benzyl-3-phenylaziridine, **V**, 83

1,1-Bis(bromomethyl)cyclopropane, **VI**, 153

Cyclopropylbenzene (phenylcyclopropane), **V**, 328

Ethyl indole-2-carboxylate, **V**, 567

2-Hydroxycyclobutanone, **VI**, 167

(*R*,*S*)-Mevalonolactone-2-^{13}C, **VII**, 386

Phenacylamine hydrochloride, **V**, 909

2,3,4,5-Tetramethylpyrrole **V**, 1022

F. Miscellaneous

"Acetone-anil"(2,2,4-Trimethyl-1,2-dihydroquinoline), **III**, 329

3-Acetyloxindole, **V**, 12

3(5)-Aminopyrazole, **V**, 39

N-Acetylaminophenyl)rhodanine, **IV**, 6

2-Amino-4-anilino-6-(chloromethyl)-*s*-triazine, **IV**, 29

4-Amino-2,6-dimethylpyrimidine, **III**, 71

2-Amino-6-methylbenzothiazole, **III**, 76

2-Amino-4-methylthiazole, **II**, 31

3-Amino-1*H*-1,2,4-triazole, **III**, 95

4-Amino-4*H*-1,2,4-triazole, **III**, 96

Anhydro-2-hydroxymercuri-3-nitrobenzoic acid, **I**, 56

Azlactone of *o*-Acetaminocinnamic acid, **II**, 1

Azlactone of α-Benzoylamino-cinnamic acid, **II**, 490

Azlactone of α-Benzoylamino-β-(3,4-dimethoxyphenyl)acrylic acid, **II**, 55

Barbituric acid, **II**, 60

Benzeneboronic anhydride **IV**, 68

Benzimidazole, **II**, 65

Benzoguanamine, **IV**, 78

1,2,3-Benzothiadiazole 1,1-dioxide, **V**, 61

1,2,3-Benzotriazole, **III**, 106

Benzoylene urea, **II**, 79

α-Benzylidene-γ-phenyl-Δ^(β,γ)-buteno-lide, **V**, 80

Bis(1,3-diphenylimidazolidinyli-dene-2), **V**, 115

Di-*tert*-butylcyclopropenone, **VI**, 991

tert-Butylphthalimide, **III**, 152

1-Butylpyrrolidine, **III**, 159

ε-Caprolactam (2-ketohexamethy-lenimine), **II**, 77, 371

9-Chloroacridine, **III**, 53

7-Chloro-4-hydroxy-3-quinolinecar-boxylic acid, **III**, 273

Chlorothiirane 1,1-dioxide, **V**, 231

Coumarilic acid, **III**, 209

Creatinine, **I**, 172

1-Cyanobenzocyclobutene, **V**, 263

Cyclohexene oxide, **I**, 185

1,2-Cyclononadiene, **V**, 306

Diacetyl-D-tartaric anhydride, **IV**, 242

2,5-Diamino-3,4-dicyanothiophene, **IV**, 243

2,4-Diamino-6-hydroxypyrimidine, **IV**, 245

Diaminouracil hydrochloride, **IV**, 247

Dicyanoketene ethylene acetal, **IV**, 276

2,3-Dihydropyran, **III**, 276

2,4-Dimethyl-3-acetyl-5-carb-ethoxypyrrole, **III**, 513

2,4-Dimethyl-3,5-dicarbethoxypyr-role, **II**, 202

2,2-Dimethylenimine, **III**, 148

Dimethylfurazan, **IV**, 342

5,5-Dimethylhydantoin, **III**, 323

5,5-Dimethyl-2-pentyltetrahydrofu-ran, **IV**, 350

3,5-Dimethylpyrazole, **IV**, 351

2,6-Dimethylpyridine, **II**, 214

2,5-Dimethylpyrrole, **II**, 219

1,5-Dimethyl-2-pyrrolidone, **III**, 328

5,5-Dimethyl-2-pyrrolidone, **IV**, 357

2,4-Dimethylthiazole, **III**, 332

1,4-Diphenyl-5-amino-1,2,3-triazole, **IV**, 380

p-Dithiane, **IV**, 396

4,5-Diphenylglyoxalone, **II**, 231

2,4-Diphenylpyrrole, **III**, 358

Epibromohydrin, **II**, 257

Epichlorohydrin, **I**, 233; **II**, 256

N-Ethylallenimine, **V**, 541

Ethylene thiourea, **III**, 394

Ethylenimine, **IV**, 433

5-Ethyl-2-methylpyridine, **IV**, 451

Furfural, **I**, 280

Glutarimide, **IV**, 496

D-Gulonic-γ-lactone, **IV**, 506

3-Heptyl-5-cyanocytosine, **IV**, 515

Hexahydro-1,3,5-tripropionyl-*s*-tri-azine, **IV**, 518

Homophthalic anhydride, **III**, 449

4-Hydroxy-1-butanesulfonic acid sultone, **IV**, 529

2-Hydroxycinchoninic acid, **III**, 456

4(5)-Hydroxymethylimidazole
hydrochloride, **III**, 460
3-Hydroxytetrahydrofuran, **IV**, 534
Imidazole, **III**, 471
Indazole, **III**, 475; **IV**, 536; **V**, 650
Indole **III**, 479
Isatin, **I**, 327
Isatoic anhydride, **III**, 488
DL-Isopropylideneglycerol, **III**, 502
3-Isoquinuclidone, **V**, 670
Itaconic anhydride **II**, 368
2-Mercapto-4-amino-5-carb-
ethoxypyrimidine, **IV**, 566
Mercaptobenzimidazole, **IV**, 569
2-Mercapto-4-hydroxy-5-cyanopy-
rimidine, **IV**, 566
2-Mercaptopyrimidine, **V**, 703
Mesitylene, **I**, 341
1-(2-Methoxycarbonylphenyl)pyr-
role, **V**, 716
6-Methoxy-β-tetralone, **VI**, 744
10-Methyl-10,9-borazarophenan-
threne, **V**, 727
5-Methylfurfural, **II**, 393
β-Methylglutaric anhydride, **IV**, 630
4-Methyl-6-hydroxypyrimidine, **IV**,
638
1-Methyl-2-imino-β-naphthothiazo-
line, **III**, 595
2-Methylindole, **III**, 597
Methyl-3-methyl-2-furoate, **IV**, 649
3-Methyloxindole, **IV**, 657
β-Methyl-β-phenyl-α,α'-dicyanoglu-
tarimide, **IV**, 662
3-Methyl-1-phenylphospholene
1-oxide, **V**, 787
4-Methylpyrimidine, **V**, 794
3-Methylthiophene, **IV**, 671
6-Methyluracil, **II**, 422
β-Methyl-γ-valerolactone, **IV**, 677
Monobenzalpentaerythritol, **IV**, 679
5-Nitro-2,3-dihydro-1,4-phthalazine-
dione, **III**, 656
5-Nitroindazole, **III**, 660
3-Nitrophthalic anhydride, **I**, 410

Parabanic acid, **IV**, 744
3,3-Pentamethylenediaziridine, **V**,
897
3,3-Pentamethylenediazirine, **V**, 897
Phenoxthin, **II**, 485
1-Phenyl-3-amino-5-pyrazolone, **III**,
708
Phenylcyclopropane, **V**, 929
4-Phenyl-*m*-dioxane, **IV**, 786
o-Phenylene carbonate, **IV**, 788
α-Phenylglutaric anhydride, **IV**, 790
2-Phenylindole, **III**, 725
N-Phenylmaleimide, **V**, 944
Phenylmethylglycidic ester, **III**, 727
2-Phenyl-5-oxazolone, **V**, 946
3-Phenylsydnone, **V**, 962
Phthalhydrazide, **III**, 153
Phthalide, **II**, 526
Phthalimide, **I**, 457
unsym-Phthalyl chloride, **II**, 528
Pseudothiohydantoin, **III**, 751
Pyocyanine, **III**, 753
2,3-Pyrazinedicarboxylic acid, **IV**,
824
Pyrrole, **I**, 473
Quinoxaline, **IV**, 824
3-Quinuclidone hydrochloride, **V**,
989
Rhodanine, **III**, 763
Succinic anhydride, **II**, 560
Succinimide, **II**, 562
o-Sulfobenzoic anhydride, **I**, 495
1,2,3,4-Tetrahydrocarbazole, **IV**, 884
1,2,3,4-Tetrahydro-β-carboline, **VI**,
965
4,5,6,7-Tetrahydroindazole, **IV**, 536
Tetrahydrothiophene, **IV**, 892
2,2,5,5-Tetramethyltetrahydro-3-
ketofuran, **V**, 1024
2-Thiohomophthalimide, **V**, 1051
2-Thio-6-methyluracil, **IV**, 638
Thiophene, **II**, 578
2-(*p*-Tolylsulfonyl)dihydroisoindole,
V, 1064
1,2,4-Triazole, **V**, 1070

Trimethylene oxide, **III**, 835
2,4,6-Trimethylpyrylium perchlorate,
  V, 1106
2,4,6-Trimethylpyrylium tetrafluoroborate, **V**, 1112
2,4,6-Trimethylpyrylium trifluoromethanesulfonate, **V**, 1114
2,4,6-Triphenylnitrobenzene, **V**,
  1128
2,4,6-Triphenylpyrylium tetrafluoroborate, **V**, 1135
sym-Trithiane, **II**, 610

DECARBONYLATION

Cetylmalonic ester, **IV**, 141
Cyclohexylidenecyclohexane, **V**, 297
2,2-(Ethylenedithio)cyclohexanone,
  VI, 590
Methyl homoveratrate, **II**, 333
Tricarbonyl[(1,2,3,4-η)-1,3-cyclo-
  hexadien-1-one-] and [2-
  methoxy-1,3-cyclohexadien-1-
  yl]iron, **VI**, 996
2,4,4-Trimethylcyclopentanone, **IV**,
  957

DECARBOXYLATION

3-Acetamido-2-butanone, **IV**, 5
Acetonedicarboxylic acid, **I**, 10
Anhydro-*o*-hydroxymercuribenzoic
  acid, **I**, 57
Anhydro-2-hydroxymercuri-3-
  nitrobenzoic acid, **I**, 56
3-Benzyl-3-methylpentanenitrile, **IV**,
  95
cis-Bicyclo[3.3.0]octane-3,7-dione,
  VII, 50
α-Bromocaproic acid, **II**, 95
1-Bromo-3-chlorocyclobutane, **VI**,
  179
α-Bromoisocaproic acid, **II**, 95
α-Bromoisovaleric acid, **II**, 93
α-(Bromomethyl)acrylic acid, **VII**,
  210
α-Bromo-β-methylvaleric acid, **II**,
  95
4-Bromoresorcinol, **II**, 100
(Z)-β-Bromostyrene, **VII**, 172
Caproic acid, **II**, 417, 474

3-Chlorocyclobutanecarboxylic acid,
  VI, 271
5-Chloro-2-pentanone, **IV**, 597
Cinnamonitrile, **VI**, 304
Coumalic acid, **IV**, 201
Coumarone, **V**, 251
Cyclobutanecarboxylic acid, **III**, 213
Cyclodecanone, **V**, 277
Cyclohexanecarbonitrile, **VI**, 334
1,4-Cyclohexanedione, **V**, 288
1-Cyclohexenylacetonitrile, **IV**, 234
Cyclopentanone, **I**, 192
Diallylamine, **I**, 201
Dibromoacetonitrile, **IV**, 254
α,α'-Dibromodibenzyl sulfone, **VI**,
  403
Dibutylamine, **I**, 202
Di-*tert*-butyl dicarbonate, **VI**, 418
2,6-Dichlorophenol, **III**, 267
4,7-Dichloroquinoline, **III**, 272
Diethyl benzoylmalonate, **IV**, 285
cis-1,5-Dimethylbicyclo[3.3.0]-
  octane-3,7-dione, **VII**, 51
4,6-Dimethylcoumalin, **IV**, 337
3,5-Dimethyl-2-cyclohexen-1-one,
  III, 317
Dimethyl decanedioate, **VII**, 181
2,5-Dimethylmandelic acid, **III**, 326
2,6-Dimethylpyridine, **II**, 214
2,4-Dimethylpyrrole, **II**, 217
5-Dodecen-2-one, **VIII**, 235
Ethyl 2-butyrylacetate, **VII**, 213
Ethyl cyclopropylpropiolate, **VIII**,
  247
Ethyl α-(hexahydroazepinylidene)-
  acetate, **VIII**, 263
2-(3-Ethyl-5-methyl-4-isoxazolyl-
  methyl)cyclohexanone, **VI**, 781
α-Ethyl-α-methylsuccinic acid, **V**,
  572
Ethyl 4-oxohexanoate, **VI**, 615
Furan, **I**, 274
Heptamide, **IV**, 513
2-Heptyl-2-cyclohexenone, **VII**, 406
3-Hydroxyquinoline, **V**, 635
Imidazole, **III**, 471
DL-Isoleucine, **III**, 495

Itaconic anhydride, **II**, 368
α-Ketoglutaric acid, **III**, 510
Laurone, **IV**, 560
DL-Methionine, **II**, 384
Methyl amyl ketone, **I**, 351
Methyl benzyl ketone, **II**, 391
Methyl isobutyl ketone, **I**, 352
Methyl butyl ketone, **I**, 352
3-Methylcoumarone, **IV**, 590
3-Methylcyclohexanone-3-acetic
acid, **VIII**, 467
Methyl cyclopropyl ketone, **IV**, 597
Methyl diformylacetate, **VII**, 323
3-Methylfuran, **IV**, 628
4-Methylhexanone-2, **I**, 352
3-Methylpentanoic acid, **II**, 416
Methyl phenylacetylacetate, **VII**, 63,
198
β-Methyl-β-phenylglutaric acid, **IV**,
664
Naphthoresorcinol, **III**, 637
o-Nitroacetophenone, **IV**, 708
Nitromethane, **I**, 401
m-Nitrostyrene, **IV**, 731
Pelargonic acid, **II**, 474
trans-2,4-Pentadienoic acid, **VI**, 95
DL-Phenylalanine, **III**, 705
Phenylethylene, **I**, 440
Phenylnitromethane, **II**, 512
3-Phenyl-4-pentenal, **VIII**, 536
Phloroglucinol, **I**, 455
α-Pyrone, **V**, 982
Pyruvic acid, **I**, 475
cis-Stilbene, **IV**, 857
1,2,3,4-Tetrahydro-β-carboline, **VI**,
965
1,2,3,4-Tetraphenylnaphthalene, **V**,
1037
p-Tolylsulfonyldiazomethane, **VI**, 981
Tricarballylic acid, **I**, 523
1,3,5-Trinitrobenzene, **I**, 541
**DEOXYGENATION (see also
REDUCTION)**
Adamantylideneadamantane, **VII**, 1
Benz[*a*]anthracene, **VII**, 18
17β-*tert*-Butoxy-5α-androst-2-ene,
VII, 66

trans-Cyclododecene, **VII**, 121
3-Deoxy-1,2:5,6-di-*O*-isopropyli-
dene-α-D-ribo-hexofuranose,
VII, 139
m-Nitroethylbenzene, **VII**, 393
DESILYLATION
Benzophenone cyanohydrin,
VII, 20
1,2-Cyclobutanedione, **VII**, 112
Cyclopropanone ethyl hemiacetal,
VII, 131
2-(Hydroxymethyl)allyltrimethylsi-
lane, **VII**, 266
6-Hydroxy-3,5,5-trimethyl-2-cyclo-
hexen-1-one, **VII**, 282
trans-2-Isocyanocyclohexanol, **VII**,
294
3-Trimethylsilyl-2-propyn-1-ol, **VII**,
524
DESULFURIZATION
o-Chlorophenylcyanamide, **IV**, 172
Cyclododecyl mercaptan, **VII**, 124
Hexafluoroacetone, **VII**, 251
4-Methyl-6-hydroxypyrimidine, **IV**,
638
2,2′:6′,2″-Terpyridine, **VII**, 476
DIAZOTIZATION
A. Preparation of Diazonium Salts
(*Warning: Hazardous*. Observe
safety precautions)
p-Acetyl-α-bromohydrocinnamic
acid, **VI**, 21
Benzenediazonium-2-carboxylate, **V**,
54
4,4′-Biphenylene-bis-diazonium fluoborate, **II**, 188
o-Bromobenzenediazonium hexafluorophosphate, **V**, 133
2-Bromo-4-methylbenzenediazoni-
um chloride, **V**, 140
(S)-(+)-γ-Butyrolactone-γ-carboxylic
acid, **VII**, 99
p-Carbethoxybenzenediazonium fluoborate, **II**, 299
4-(4-Chlorophenyl)butan-2-one, **VII**,
105
1-(Diazoacetyl)naphthalene, **VI**, 613

16-Diazoandrost-5-en-3β-ol-17-one,
VI, 840

Naphthalenediazonium chloride-
mercuric chloride compound,
II, 432

o-Nitrobenzenediazonium fluoborate, II, 226

p-Nitrobenzenediazonium fluoborate, II, 225

m-Trifluoromethylbenzenesulfonyl
chloride, VII, 508

B. Coupling Reactions (Retention
of N₂)

Azo dye from o-toluidine and
Chicago acid, II, 145

4-Benzeneazo-1-naphthol, I, 49

1,2,3-Benzothiadiazole 1,1-dioxide,
V, 61

Diazoaminobenzene, II, 163

Methylglyoxal ω–phenylhydrazone,
IV, 633

Methyl Red, I, 374

1-Methyl-3-p-tolyltriazene, V, 797

1-(m-Nitrophenyl)-3,3-dimethyltri-
azene, IV, 718

Orange I, II, 39

Orange II, II, 36

C. Reduction of Azo and Diazonium
Groups

1,2-Aminonaphthol hydrochloride,
II, 35

1,4-Aminonaphthol hydrochloride, I,
49; II, 39

o-Hydrazinobenzoic acid hydrochloride, III, 475

Phenylhydrazine, I, 442

D. Replacement of Diazonium Group
(Loss of N₂)

1. By CHO
2-Bromo-4-methylbenzalde-
hyde, V, 139

2. By arsenic
p-Nitrophenylarsonic acid, III,
665

3. By aryl
p-Bromobiphenyl, I, 113

m-Nitrobiphenyl, IV, 718

4. By azide
o-Azido-p'-nitrobiphenyl, V, 830
2-Nitrocarbazole, V, 829
o-Nitrophenyl azide, IV, 75

5. By cyanide
o-Tolunitrile, I, 514
p-Tolunitrile, I, 514

6. By Halogen
(a) Br
m-Bromobenzaldehyde, II,
132
o-Bromotoluene, I, 135
p-Bromotoluene, I, 136
o-Chlorobromobenzene,
III, 185

(b) Cl
m-Chlorobenzaldehyde, II,
130
1-Chloro-2,6-dinitroben-
zene, IV, 160
m-Chloronitrobenzene, I,
162
2-Chloropyrimidine, IV,
182
o-Chlorotoluene, I, 170
p-Chlorotoluene, I, 170

(c) F
1-Bromo-2-fluorobenzene,
V, 133
4,4'-Difluorobiphenyl, II,
188
Fluorobenzene, II, 295
p-Fluorobenzoic acid, II,
299

(d) I
Iodobenzene, II, 351
o-Iodobromobenzene, V,
1120
p-Iodophenol, II, 355
1,2,3-Triodo-5-nitroben-
zene, II, 604

7. By hydrogen
m-Bromotoluene, I, 133
3,3'-Dimethoxybiphenyl, III,
295

3,3'-Dimethylbiphenyl, **III**, 295
m-Iodobenzoic acid, **II**, 353
sym-Tribromobenzene, **II**, 592
2,4,6-Tribromobenzoic acid,
  IV, 947
8. By hydroxyl
3-Bromo-4-hydroxytoluene, **III**,
  130
m-Hydroxybenzaldehyde, **III**,
  453, 564
3-Methoxy-5-nitrophenol, **I**,
  405
m-Nitrophenol, **I**, 404
3-Nitro-4,6-xylenol, **I**, 405
2,4,6-Triphenylphenol, **V**, 1130
9. By nitro
o-Dinitrobenzene, **II**, 226
p-Dinitrobenzene, **II**, 225
1,4-Dinitronaphthalene, **III**,
  341
m-Nitrotoluene, **I**, 415
10. By selenium
Diphenyl selenide, **II**, 238
Di-*m*-tolyl selenide, **II**, 239
Di-*o*-tolyl selenide, **II**, 239
Di-*p*-tolyl selenide, **II**, 239
11. By sulfhydryl
m-Thiocresol, **III**, 809
12. Miscellaneous
Biphenylene, **V**, 54
Diphenic acid, **I**, 222
Mercury di-β-naphthyl, **II**, 381
β-Naphthylmercuric chloride,
  II, 432
2-Nitrobenzenesulfinic acid, **V**,
  60
1-(*p*-Nitrophenyl)-4-chloro-2-
  butene, **IV**, 727
Phenylarsonic acid, **II**, 494
Thiosalicylic acid, **II**, 580
DL-4,4',6,6'-Tetrachlorodiphenic
  acid, **IV**, 872
DIAZO TRANSFER
tert-Butyl α-diazoacetoacetate, **V**,
  179
tert-Butyl diazoacetate, **V**, 179

2-Diazocyclohexanone, **VI**, 389
Di-*tert*-butyl diazomalonate, **VI**, 414
Trimethylsilyldiazomethane, **VIII**, 612
**DIELS–ALDER REACTIONS (includ-
  ing ENE REACTIONS)**
4-Acetyl-7,8-dimethyl-2-oxa-4-
  azabicyclo[4.3.0]non-7-en-3-
  one, **VII**, 4
N-Benzyl-2-azanorbornene, **VIII**, 31
7,7-Dimethoxy-1,2,3,4-tetrachlorobi-
  cyclo[2.2.1]heptene, **V**, 425
3,3a,3b,4,6a,7a-Hexahydro-3,4,7-
  metheno-7*H*-cyclopenta[a]pen-
  talene-7,8-dicarboxylic acid,
  VIII, 298
4-Methoxycarbonyl-1,1,6-trimethyl-
  1,4,4a,5,6,7,8,8*a*-octahydro-
  2,3-benzopyrone, **VIII**, 353
5β-Methoxycyclohexan-1-one-
  3β,4β-dicarboxylic acid anhydride, **VII**, 312
6-Methoxy-7-methoxycarbonyl-
  1,2,3,4-tetrahydronaphthalene,
  VIII, 444
Methyl *N*-(2-methyl-2-butenyl)car-
  bamate, **VIII**, 68
γ-Phenylallylsuccinic anhydride, **IV**,
  766
Tetramethyl [2.2]paracyclophane-
  4,5,12,13-tetracarboxylate, **VII**,
  485
2,3,6-Tricarboethoxypyridine, **VIII**,
  597
ELECTROLYTIC REACTIONS
Dimethyl decanedioate, **VII**, 181
2,7-Dimethyl-2,7-dinitrooctane, **V**, 445
Dimethyl octadecanedioate, **V**, 463
2-Methoxy-*N*-carbomethoxypyrroli-
  dine, **VII**, 307
Tetraethyl 1,2,3,4-butanetetracar-
  boxylate, **VII**, 479
Tetramethyl 1,1,2,2-ethanetetracar-
  boxylate, **VII**, 482
ELIMINATION
A. Dehalogenation
Allene, **V**, 22

Benzobarrelene, **VI**, 82
Bicyclo[1.1.0]butane, **VI**, 133
(Z)-β-Bromostyrene, **VII**, 172
3-Bromothiophene, **V**, 149
1-Chloro-2,3,3-trifluorocyclobutene, **V**, 394
Cholesterol, **IV**, 195
Cyclobutadieneiron tricarbonyl, **VI**, 310
1,1-Dichloro-2,2-difluoroethylene, **IV**, 268
7,7-Dimethoxybicyclo[2.2.1]heptene, **V**, 424
Dimethylketene, **IV**, 348
(1R)-9,9-Dimethyltricyclo[6.1.1.0^{2,6}]-deca-2,5-diene, **VIII**, 223
1,6-Dimethyltricyclo[4.1.0.0^{2,7}]hept-3-ene, **VII**, 200
Ethyl cyclopropylpropiolate, **VIII**, 247
Ethyl linoleate, **III**, 526
Ethyl linolenate, **III**, 532
8-Ethynyl-8-methyl-1,4-dioxaspiro[4.5]dec-6-ene, **VII**, 241
Perchlorofulvalene, **V**, 901
Triptycene, **IV**, 964
B. Dehydration
Acetamide, **I**, 3
Acetonedicarboxylic acid, **I**, 10
Aconitic acid, **II**, 12
Acrolein, **I**, 15
Allyl alcohol, **I**, 42
2-Amino-4-methylthiazole, **II**, 31
Azelanitrile, **IV**, 62
Benzalazine, **II**, 395
Benzalaniline, **I**, 80
Benzhydryl β-chloroethyl ether, **IV**, 72
Benzil hydrazone, **II**, 497
Benzimidazole, **II**, 65
Benzoic anhydride, **I**, 91
Benzophenone oxime, **II**, 70
Benzoylene urea, **II**, 79
Benzylidenearginine, **II**, 49
Bischloromethyl ether, **IV**, 101

α-(Bromomethyl)acrylic acid, **VII**, 210
Butyrchloral, **IV**, 130
Chelidonic acid, **II**, 126
Chloroacetonitrile, **IV**, 144
2-Chloronicotinonitrile, **IV**, 166
m-Chlorostyrene, **III**, 204
Coumalic acid, **IV**, 201
Creatinine, **I**, 172
1-Cyano-6-methoxy-3,4-dihydronaphthalene, **VI**, 307
Cyclohexanone oxime, **II**, 76, 314
Cyclohexene, **II**, 152; **I**, 183
2-Cyclohexenone, **V**, 294
Cyclohexyl isocyanide, **V**, 300
Cyclopentene, **II**, 153
2-Cyclopentenone, **V**, 326
Diacetyl-D-tartaric anhydride, **IV**, 242
4,4'-Dichlorodibutyl ether, **IV**, 266
Diethyl dioxosuccinate, **VIII**, 597
2,3-Dimethyl-1,3-butadiene, (Method I), **III**, 312; (Method II), **III**, 313
Dimethylfurazan, **IV**, 342
5,5-Dimethyl-2-pentyltetrahydrofuran, **IV**, 350
2,5-Dimethylpyrrole, **II**, 219
2,4-Dinitrobenzaldehyde, **II**, 223
1,1-Diphenylethylene, **I**, 226
4,5-Diphenylglyoxalone, **II**, 231
2,3-Diphenylindone, **III**, 353
1-Ethyl-3-(3-dimethylamino)propylcarbodiimide hydrochloride and methiodide, **V**, 555
2-Ethylhexanenitrile, **IV**, 436
Ethyl isocyanoacetate, **VI**, 620
Fumaronitrile, **IV**, 486
Furfural, **I**, 280
Glutarimide, **IV**, 496
Heptaldoxime, **II**, 313
Hexafluoroacetone hydrazone, **VI**, 161
Indole-3-carbonitrile, **V**, 656
Isobutyronitrile, **III**, 493
Isopropylideneacetophenone, **VIII**, 210

Itaconic anhydride, **II**, 368
Malononitrile, **II**, 379; **III**, 535
Mesityl oxide **I**, 345
Methacrylamide, **III**, 560
2-Methylbenzimidazole, **II**, 66
Methyl ω-cyanopelargonate, **III**, 584
β-Methylglutaric anhydride **IV**, 630
Methyl isocyanide, **V**, 772
(7a*S*)-(+)-7a-Methyl-2,3,7,7a-tetra-
    hydro-1*H*-indene-1,5(6*H*)-
    dione, **VII**, 363
6-Methyluracil, **II**, 422
Nicotinic anhydride, **V**, 822
Nicotinonitrile, **IV**, 706
3-Nitrophthalic anhydride, **I**, 410
2-Nitropropene, **VII**, 363
Oxalic acid, **I**, 421
Pentaacetyl-D-glucononitrile, **II**, 690
2-Pentene, **I**, 430
Phenanthrene-9,10-oxide, **VI**, 887
1-Phenylnaphthalene, **III**, 729
trans-1-Phenyl-1,3-butadiene, **IV**,
    771
α-Phenylglutaric anhydride, **IV**, 790
Pinacolone, **I**, 462
Polymeric carbodiimide, **VI**, 95
Pyrrole, **I**, 473
Pyruvic acid, **I**, 475
Sebaconitrile, **III**, 768
Succinic anhydride, **II**, 560
p-Toluenesulfonic anhydride, **IV**,
    940
o-Tolyl isocyanide, **V**, 1060
p-Tolylsulfonylmethyl isocyanide,
    VI, 987
Tri-*tert*-butylcyclopropenyl tetraflu-
    oroborate, **VI**, 99
Trifluoromethanesulfonic anhydride,
    VI, 757
Triphenylethylene, **II**, 606
sym-Trithiane, **II**, 610
Veratronitrile, **II**, 622
C. Dehydogenation
    2-Acetyl-2-cyclohexen-1-one, **VI**, 23
    2,2'-Bipyridine, **V**, 102

trans-4,4'-Dimethoxystilbene, **V**,
    428
2,4-Diphenylpyrrole, **III**, 358
Indazole, **IV**, 536
β-Methyl-δ-valerolactone, **IV**, 677
1,2-Naphthalic anhydride, **II**, 423
Phenoxthin, **II**, 485
Phenylbenzoyldiazomethane, **II**, 496
1-Phenylnaphthalene, **III**, 729
Spiro[5.7]trideca-1,4-dien-3-one,
    VII, 473
Tetraphenylphthalic anhydride, **III**,
    807
D. Dehydrohalogenation
    Acetylenedicarboxylic acid, **II**, 10
    3-Acetyl-2(3*H*)-oxazolone, **VII**, 5
    Acrolein acetal, **II**, 17
    Benzocyclopropene, **VI**, 87
    α-Bromobenzalacetone, **III**, 125
    tert-Butyl isocyanide, **VI**, 232
    2-Butyn-1-ol, **IV**, 128
    N-Chlorocyclohexylidenimine, **V**,
        208
    3-Chloro-2,5-di-*tert*-butyl-1,4-ben-
        zoquinone, **VI**, 210
    3-Chlorothiete 1,1-dioxide **VII**, 491
    1-Chloro-*N*,*N*,2-trimethylpropenyl-
        amine, **VI**, 282
    γ-Crotonolactone, **V**, 255
    2-Cyclohepten-1-one, **VI**, 327
    1,3-Cyclohexadiene, **V**, 285
    Cyclohexene oxide, **I**, 185
    3-Cyclohexylpropyne, **I**, 191
    1-Decyne, **I**, 192
    Dibenzoylmethane, **I**, 205
    2,3-Dibromopropene, **I**, 209
    (*E*)-1,2-Di-*tert*-butoxy-1-
        chloroethene, **VIII**, 161
    Di-*tert*-butoxyethyne, **VIII**, 161
    Di-*tert*-butylcyclopropenone, **VI**,
        991
    2,5-Di-*tert*-butyl-3,6-dichloro-1,4-
        benzoquinone, **VI**, 210
    Di-(*p*-chlorophenyl)acetic acid, **III**,
        270

1,1-Di-(*p*-chlorophenyl)-2,2-dichloroethylene, **III**, 270
1,2-Dicyanocyclobutene, **VI**, 427
3,3-Dimethoxycyclopropene, **VI**, 361
1,2-Dimethylenecyclohexane, **VIII**, 212
2,7-Dimethyloxepin, **V**, 467
Dimethyl 2,3-pentadienedioate, **VI**, 505
3,4-Dinitro-3-hexene, **IV**, 372
Diphenylacetylene **III**, 350
Diphenylcarbodiimide (Method II), **V**, 504
Diphenylcyclopropenone, **V**, 514
1,2-Diphenyl-3-methylcyclopropene, **VII**, 203
Diphenylketene, **VI**, 549
2,3-Diphenylvinylene sulfone, **VI**, 555
trans-2-Dodecenoic acid, **IV**, 398
Epibromohydrin, **II**, 257
Epichlorohydrin, **II**, 256; **I**, 233
Ethoxyacetylene, **IV**, 404
1-Ethoxy-1-butyne, **VI**, 564
Ketene diethylacetal, **III**, 506
6-Ketohendecanedioic acid, **IV**, 555
Laurone, **IV**, 560
Methyl α-(bromomethyl)acrylate, **VII**, 319
2-Methyl-2-cyclohexenone, **IV**, 162
trans-2-Methyl-2-dodecenoic acid, **IV**, 608
2-Methylenedodecanoic acid, **IV**, 616
Methyl *trans*-2-methyl-2-dodecenoic acid, **IV**, 608
Monovinylacetylene, **IV**, 683
Muconic acid, **III**, 623
9-Nitroanthracene, **IV**, 711
(1*R*)-Nopadiene, **VIII**, 223
4-Pentyn-1-ol, **IV**, 755
Phenylacetylene, **I**, 438; **IV**, 763
Phenacylamine hydrochloride, **V**, 909
4-Phenylbutyne-1, **I**, 192

Phenylpropargylaldehyde, **III**, 731; **III**, 731
Phenylpropiolic acid, **II**, 515
Phenyl vinyl sulfide, **VII**, 453
Propiolaldehyde diethyl acetal, **VI**, 954
2*H*-Pyran-2-one, **VI**, 462
Stearolic acid, **III**, 785; **IV**, 851
2,3,4,5-Tetrahydropyridine trimer, **VI**, 968
Thiete 1,1-dioxide, **VII**, 491
10-Undecynoic acid, **IV**, 969
E. Miscellaneous
2-Acetyl-2-cyclohexen-1-one, **VI**, 23
3(5)-Aminopyrazole, **V**, 39
o-Anisaldehyde, **V**, 46
Azetidine, **VI**, 75
(1-Azido-3, 3-dimethoxy-1-propenyl)benzene, **VI**, 893
Benzobarrelene, **VI**, 82
Bicyclo[2.1.0]pentane, **V**, 99
2-Bornene, **VI**, 172
2-Bromo-3,3-diphenyl-2-propen-1-ol acetate, **VI**, 187
1-(Bromovinyl)trimethylsilane, **VI**, 1033
o-Chlorophenylcyanamide, **IV**, 172
p-Chlorophenyl isothiocyanate, **V**, 223
Cinnamonitrile, **VI**, 304
Crotyl diazoacetate, **V**, 258
trans-Cyclooctene, **V**, 315
2-(Dimethoxymethyl)-3-phenyl-2*H*-azirine, **VI**, 893
Diphenylmethyl vinyl ether, **VI**, 552
1-Ethyl 3-(1-azetidinyl)propanoate, **VI**, 75
Ethyl 3-ethyl-5-methyl 4-isoxazole-carboxylate, **VI**, 592
1,3,5-Hexatriene, **V**, 608
(*E*)-4-Hexen-1-ol, **VI**, 675
(*Z*)-2-Methoxy-1-phenylthio-1,3-butadiene, **VI**, 737
Methyl 1-cycloundecenecarboxylate, **VI**, 368
1,6-Oxido[10]annulene, **VI**, 862

D-2-Oxo-7,7-dimethyl-1-vinylbicyclo[2.2.1]heptane, **V**, 877
1,4-Pentadiene, **IV**, 746; **IV**, 748
trans-3-Penten-2-one, **VI**, 883
β-Phenylcinnamaldehyde, **VI**, 901
Phenylcyclopropane, **V**, 929
Polymeric carbodiimide, **VI**, 951
Polymeric benzylamide, **VI**, 951
Propiolaldehyde diethyl acetal, **VI**, 954
2,3,4,5-Tetrahydropyridine trimer, **VI**, 968
1,1,3,3-Tetramethylbutylisonitrile, **VI**, 751
p-Tolylsulfonyldiazomethane, **VI**, 981
Vinyl triphenylphosphonium bromide, **V**, 1145

ENAMINE or IMINE FORMATION
Acetophenone *N,N*-dimethylhydrazone, **VI**, 12
Benzalazine, **II**, 395
Benzil dihydrazone, **IV**, 377
Benzylidenearginine, **II**, 49, 50
N-Benzylideneethylamine, **V**, 758
N-Benzylidenemethylamine, **IV**, 605
1,1-Bis(bromomethyl)cyclopropane, **VI**, 153
Bis(dimethylamino)methane, **V**, 435; **VI**, 474
N-*tert*-Butylbenzaldimine, **V**, 191
1-Chloro-*N,N*,2-trimethylpropenylamine, **VI**, 282
1-Chloro-*N,N*,2-trimethylpropylideniminium chloride, **VI**, 282
Cholest-4-en-3-one *p*-toluenesulfonylhydrazone, **VI**, 293
1-(Cyclooctylidenemethyl)piperidine, **VII**, 473
Glyoxylic acid *p*-toluenesulfonylhydrazone, **V**, 258
Hexafluoroacetone imine, **VI**, 664
N-(α-Methylbenzylidene)methylamine, **VI**, 818

10-Methyl-$\Delta^{1(9)}$-2-octalone *N,N*-dimethylhydrazone, **VI**, 242
1-Morpholino-1-cyclohexene, **V**, 808
o-Nitrobenzalaniline, **V**, 941
1-(*N*-Pyrrolidino)-1-cyclododecene, **VII**, 135
1-Pyrrolidinocyclohexene, **VI**, 1014
1,2,3,4-Tetrahydro-β-carboline, **VIII**, 965
3,3,6,6-Tetramethoxy-1,4-cyclohexadiene, **VI**, 971
4*H*-1,4-Thiazine 1,1-dioxide, **VI**, 976

ENZYMATIC REACTIONS
Ethyl (*R*)- and (*S*)-2-fluorohexanoate, **VIII**, 258
Ethyl (*S*)-(+)-3-hydroxybutanoate, **VII**, 215
(*S*)-(+)-3-Hydroxy-2,2-dimethylcyclohexanone, **VIII**, 312
(1*S*,2*S*,3*R*)-3-Hydroxy-2-nitrocyclohexyl acetate, **VIII**, 332
(1*R*,6*S*)-(+)-8-Oxabicyclo[4.3.0]-nonan-7-one, **VII**, 406
(−)-(1*R*,2*S*)- and (+)-(1*S*,2*R*)-*trans*-1-Phenylcyclohexanol, **VIII**, 516
L-Tryptophan, **II**, 612
L-Tyrosine, **II**, 612

EPOXIDE FORMATION
2-*tert*-Butyl-3-phenyloxaziridine, **V**, 191
cis-Cyclooctene oxide, **VII**, 126
Dihydroxycyclopentene, **V**, 414
1,2-Dimethyl-1,2-epoxycyclohex-4-ene, **V**, 468
2,3-Epoxycyclohexanone, **VI**, 679
10,11-Epoxyfarnesyl acetate, **VI**, 560
Ethyl β,β-pentamethyleneglycidate, **IV**, 459
Isophorone oxide, **IV**, 552
Methyl 5,5-dimethoxy-3-methyl-2,3-epoxypentanoate, **IV**, 649
Methyl (3*RS*,4*RS*)-4,5-epoxy-3-phenylpentanoate, **VII**, 166

(*S*)-(–)-Methyloxirane, **VII**, 356
(*R*)-Methyloxirane, **VIII**, 434
Oxaspiropentane, **VI**, 320
11-Oxatricyclo[4.4.1.0^{1,6}]undeca-3,8-
diene, **VI**, 862
Phenanthrene-9,10-oxide, **VI**, 887
Phenylmethylglycidic ester, **III**, 727
(2*S*,3*S*)-3-Propyloxiranemethanol,
VII, 461
trans-Stilbene oxide, **IV**, 860
Tetracyanoethylene oxide, **V**, 1007
ESTERIFICATION
A. Of Unsubstituted Monobasic Acids
(*S*)-(–)-2-Acetoxy-1-bromopropane,
VII, 356
(1*R*,2*r*,3*S*)-3-Acetoxy-2-nitrocyclo-
hexyl acetate, **VIII**, 332
7-Acetoxy-4,4,6,7-tetramethylbicy-
clo[4.2.0]octan-2-one, **VI**, 1024
Bicyclo[3.3.1]nonan-9-one, **VI**, 137
10-Bromo-11-hydroxy-10,11-dihy-
drofarnesyl acetate, **VI**, 560
Butyl butyrate, **I**, 138
sec-Butyl crotonate, **V**, 762
tert-Butyl cyclohexanecarboxylate,
VI, 87
tert-Butyl *p*-toluate, **VI**, 259
2-Chloroethyl benzoate, **IV**, 84
γ-Chloropropyl acetate, **III**, 203
trans-7,8-Diacetoxybicyclo[4.2.0]-
octa-2,4-diene, **VI**, 196
trans, *trans*-1,4-Diacetoxy-1,3-buta-
diene, **VI**, 196
2',6'-Dimethylphenyl propanoate,
VI, 190
10,11-Epoxyfarnesyl acetate, **VI**, 560
1-(Ethenyl)heptanyl 3-ketobu-
tanoate, **VIII**, 235
Ethyl 4-acetoxybenzoate, **VI**, 576
Ethyl 5β-cholest-3-ene-5-acetate,
VI, 584
Ethyl lithioacetate, **VI**, 598
Ethyl 4-methyl-(*E*)-4,8-nona-
dienoate, **VI**, 606
Ethyl *N*-tridecylate, **II**, 292
Ethyl phenylacetate, **I**, 270

Ethyl γ-phenylbutyrate, **II**, 96
Ethyl *cis*- and *trans*-2-phenylcyclo-
propanecarboxylate, **VI**, 913
Farnesyl acetate, **VI**, 560
2-Furylmethyl acetate, **I**, 285
Methyl 4-acetoxybenzoate, **VI**, 576
2-Methyl-1-cyclohexen-1-yl acetate,
VI, 121
Methyl 1-cycloundecenecarboxylate,
VI, 368
3-Methylheptanoic acid, **V**, 762
Methyl homoveratrate, **II**, 333
Methyl 2-hexynoate, **VI**, 791
2-*exo*-Norbornyl formate, **V**, 852
Nortricyclyl acetate, **V**, 863
2,3,4,5,6-Penta-*O*-acetyl-D-gluconic
acid and 2,3,4,5,6-penta-*O*-
acetyl-D-gluconyl chloride, **V**,
887
2,3,4,6-Tetra-*O*-acetyl-D-gluconic
acid monohydrate, **V**, 887
B. Of Substituted Monobasic Acids
2-Bromo-3,3-diphenyl-2-propen-1-ol
acetate, **VI**, 187
tert-Butyl acetoacetate, **V**, 155
tert-Butyl carbamate, **V**, 162
tert-Butyl carbazate, **V**, 166
tert-Butyl cyanoacetate, **V**, 171
tert-Butyl *S*-methylthiolcarbonate,
V, 166
tert-Butyl phenyl carbonate, **V**, 166
N-Carbobenzyloxy-L-asparaginyl-L-
leucine methyl ester, **VI**, 263
N-Carbobenzyloxy-3-hydroxy-L-pro-
lylglycylglycine ethyl ester, **VI**,
263
(*Z*)-4-Chloro-4-hexenyl trifluoroacetate, **VI**, 273
Crotyl diazoacetate, **V**, 258
2-(1-Cyanocyclohexyl)diazenecar-
boxylic acid methyl ester, **VI**,
334
2-(1-Cyanocyclohexyl)hydrazinecar-
boxylic acid methyl ester, **VI**,
334
Ethyl 4-acetoxybenzoate, **VI**, 586

Ethyl *p*-aminobenzoate, **II**, 300
Ethyl 4-amino-3-methylbenzoate, **VI**, 560, 581
Ethyl 3-(1-azetidinyl)propanoate, **VI**, 75
Ethyl benzoylformate, **I**, 241
Ethyl bromoacetate, **III**, 381
Ethyl α-(bromomethyl)acrylate, **VI**, 210
Ethyl β-bromopropionate, **I**, 246
Ethyl chloroacetate, **II**, 263
Ethyl α-chlorophenylacetate, **V**, 169
Ethyl cyanoacetate, **I**, 254
Ethyl 1-cyano-2-methylcyclohexanecarboxylate, **VI**, 586
Ethyl (*E*)-2-cyano-6-octenoate, **VI**, 586
Ethyl cyano(pentafluorophenyl)-acetate, **VI**, 873
Ethyl α-cyano-β-phenylacrylate, **I**, 451
Ethyl ethoxyacetate, **II**, 261
Ethyl 3-ethyl-5-methyl-4-isoxazolecarboxylate, **VI**, 592
Ethyl 1-hydroxycyclohexylacetate, **VI**, 598
Ethyl isocyanoacetate, **VI**, 620
Ethyl mandelate, **IV**, 169
Ethyl 2-methylindole-5-carboxylate, **VI**, 601
Ethyl 2-methyl-3-methylthioindole-5-carboxylate, **VI**, 601
Ethyl 6-methylpyridine-2-acetate, **VI**, 611
Ethyl 1-naphthylacetate, **VI**, 613
Ethyl 4-oxohexanoate, **VI**, 615
Ethyl 2-pyridylacetate, **III**, 413
Ethyl pyrrole-2-carboxylate, **VI**, 618
Ethyl β-pyrrolidinocrotonate, **VI**, 592
Ethyl thiazole-4-carboxylate, **VI**, 620
N-Formylglycine ethyl ester, **VI**, 620
Isopropyl lactate, **II**, 365
Methyl benzoylformate, **I**, 244

erythro-2-Methanesulfonyloxy-1-methyl-1-propyl cyclobutanecarboxylate, **VI**, 312
Methyl 4-acetoxybenzoate, **VI**, 576
Methyl coumalate, **IV**, 532
Methyl desoxycholate, **III**, 237
Methyl β,β-dibromoisobutyrate, **VII**, 319
Methyl 3,5-dinitrobenzoate, **V**, 797
3-Methyl-2-furoic acid, **IV**, 628
Methyl homoveratrate, **II**, 333
(1*R*,2*S*,5*R*)-5-Methyl 2-(1-methyl-1-phenylethyl)cyclohexyl chloroacetate, **VIII**, 522
Methyl nitroacetate, **VI**, 797
Methyl 7-oxoheptanoate, **VI**, 807
Methyl 7-oxooctanoate, **VI**, 807
Methyl pyruvate, **III**, 610
trans-2-Phenylcyclohexyl chloroacetate, **VIII**, 516
trans 1-Phenylpropen-2-yl acetate, **VI**, 692
Ricinelaidic acid lactone, **VII**, 470
C. Of Dicarboxylic Acids
tert-Butyl ethyl fumarate, **VII**, 93
Di-*tert*-butyl diazomalonate, **VI**, 414
Di-*tert*-butyl malonate, **IV**, 261
Diethyl *cis*-Δ⁴-tetrahydrophthalate, **IV**, 304
Diethyl *tert*-butylmalonate, **VI**, 442
Diethyl isopropylidenemalonate, **VI**, 442
Diethyl propionylsuccinate, **VI**, 615
Diethyl *trans*-Δ⁴-tetrahydrophthalate, **VI**, 454
Diethyl *trans*-4-trimethylsilyloxy-4-cyclohexene-1,2-dicarboxylate, **VI**, 445
(–)-Dimenthyl succinate, **VIII**, 141
Dimethyl acetylenedicarboxylate, **IV**, 329
Dimethyl 3-chloro-2-pentenedioate, **VI**, 505
6,6-Dimethyl-5,7-dioxaspiro[2,5]-octane-4,8-dione, **VII**, 411

Dimethyl 3-methylenecyclobutane-1,2-dicarboxylate, **V**, 459
Dimethyl nitrosuccinate, **VI**, 503
Dimethyl 2,3-pentadienedioate, **VI**, 505
Ethyl acetonedicarboxylate, **I**, 237
Ethyl adipate, **II**, 264
Ethyl *tert*-butyl malonate, **IV**, 417
Ethyl hydrogen adipate, **II**, 277
Ethyl hydrogen sebacate, **II**, 276
Ethyl oxalate, **I**, 261
Ethyl pimelate, **II**, 536
Ethyl sebacate, **II**, 277
Methyl hydrogen hendecanedioate, **IV**, 635
Methyl oxalate, **I**, 264; **II**, 414
sec-Octyl hydrogen phthalate, **I**, 418
D. Miscellaneous
Amyl borate, **II**, 107
Amyl phosphate, **II**, 110
Butyl borate, **II**, 106
S-tert-Butyl cholane-24-thioate, **VII**, 81
S-tert-Butyl cyclohexylmethanethioate, **VII**, 81
Butyl nitrite, **II**, 108
Butyl phosphate, **II**, 109
sec-Butyl phosphate, **II**, 110
Butyl sulfate, **II**, 111
Butyl sulfite, **II**, 112
Butyl *p*-toluenesulfonate, **I**, 145
1,2:5,6-Di-*O*-isopropylidene-3-*O*-(*S*-methyl dithiocarbonate)-α-D-glucofuranose, **VII**, 139
Dodecyl *p*-toluenesulfonate, **III**, 366
(*S*)-(–)-Menthyl *p*-toluenesulfinate, **VII**, 495
Methyl nitrate, **II**, 412
Methyl *p*-toluenesulfonate, **I**, 146
S-(2-Oxobut-3-yl) butanethioate, **VI**, 776
Propyl phosphate, **II**, 110
Propyl sulfate, **II**, 112
Ricinelaidic acid *S*-(2-pyridyl)carbothioate, **VII**, 470

ETHERIFICATION
Benzyl chloromethyl ether, **VI**, 101
1,3-Bis(methylthio)-2-methoxypropane, **VI**, 683
Cholestanyl methyl ether, **V**, 245
18-Crown-6, **VI**, 301
3-Ethoxy-2-cyclohexenone, **V**, 539
Ethyl α-(isopropylideneaminooxy)propionate, **V**, 1031
Ethyl 4-methyl-(*E*)-4,8-nonadienoate, **VI**, 606
o-Formylphenoxyacetic acid, **V**, 251
4-(1-Phenyl-5-tetrazolyloxy)biphenyl, **VI**, 150
(+)- and (–)-α-(2,4,5,7-Tetranitro-9-fluorenylideneaminooxy)propionic acid, **V**, 1031
2-Trimethylsilyloxy-1,3-butadiene, **VI**, 445
GRIGNARD and RELATED REACTIONS (Reformatsky, and so on)
9-Acetylphenanthrene, **III**, 26
Allylmagnesium chloride, **VIII**, 23
Allyltributyltin, **VIII**, 23
Amylbenzene, **II**, 47
o-Anisaldehyde, **VI**, 64
Benzeneboronic anhydride, **IV**, 68
Benzoic acid, **I**, 363
3-Benzyl-3-methylpentanoic acid, **IV**, 93
Biallyl, **III**, 121
Bis(3-dimethylaminopropyl)phenylphosphine, **VI**, 776
2-Bromo-1-decene, **I**, 187
2-*tert*-Butoxythiophene, **V**, 642
2-Butylbornene, **VII**, 77
sec-Butyl α-caproylpropionate, **IV**, 120
Butylmagnesium chloride, **VIII**, 606
3-Butyl-2-methyl-1-hepten-3-ol, **VI**, 240
tert-Butyl phenyl ketone, **VI**, 248
Chlorodiisopropylphosphine, **V**, 211

1-Chloromagnesioxy-3-octenylmag-
  nesium chloride, **VIII**, 507
3-Chloro-2-methyltetrahydropyran,
  VI , 675
m-Chlorophenylmethylcarbinol, **III**,
  200
Cyclohexanecarboxylic acid, **I**, 364
3-Cyclohexyl-2-bromopropene, **I**,
  186
Cyclohexylcarbinol, **I**, 188
Cyclohexyl methyl ketone, **V**, 775
3,12-Diacetoxy-bisnor-cholanyldi-
  phenylethylene, **III**, 237
1,2-Dibutylbenzene, **VI**, 407
Dibutyl carbinol, **II**, 179
Dibutyldivinyltin, **IV**, 258
Diethyl *tert*-butylmalonate, **VI**, 442
2-Dimethylamino-2'-methoxyben-
  zhydrol, **V**, 46
2,4-Dimethyl-5-carbethoxypyrrole,
  II, 198
(*E*)-7,11-Dimethyl-6,10-dodecadien-
  2-yn-1-ol, **VIII**, 226
4,4-Dimethyl-1,6-heptadien-3-ol,
  VII, 177
4,6-Dimethyl-1-hepten-4-ol, **V**, 452
2,2-Dimethyl-3-phenylpropionalde-
  hyde, **VI**, 526
Diphenyl-*p*-bromophenylphosphine,
  V, 496
1,1-Diphenylethylene, **I**, 226
2,3-Diphenylindone, **III**, 353
Diphenyl ketimine, **V**, 520
Dineopentyl ketone, **VI**, 991
Diphenyl diselenide, **VI**, 533
(*E*)-1-Ethoxy-1-(1-hydroxy-2-
  butenyl)cyclopropane, **VIII**,
  556
Ethyl 4-ethyl-2-methyl-3-hydroxyoc-
  tanoate, **IV**, 444
Ethyl phenylacetylmalonate, **III**, 637
Ethyl β-phenyl-β-hydroxypropi-
  onate, **III**, 408
Ethyl α-naphthoate, **II**, 282
Ethyl tetrolate, **VII**, 226

Ethynylmagnesium chloride, **VIII**,
  606
1,5-Hexadien-3-ol, **V**, 608
Hexaldehyde, **II**, 323
1,2,4,5-Hexatetraene (in solution),
  VII, 485
Hexyl alcohol, **I**, 306
2-Hydroxymethyl-2-cyclopentenone,
  VII, 271
1-Methylisoquinoline, **IV**, 641
2-Hydroxythiophene, **V**, 642
Isodurene, **II**, 360
1-[(Isopropoxydimethylsilyl)methyl]
  cyclohexanol, **VIII**, 315
2-Lithio-2-butene, **VIII**, 505
1-Lithio-3-chloropropyne, **VIII**, 371
1-Lithio-1-ethoxycyclopropane,
  VIII, 556
Lithium acetylide, **VIII**, 391
Lithium-1-lithiopropyn-3-oxide,
  VIII, 415
Mesitoic acid, **III**, 553
ω-Methoxyacetophenone, **III**, 562
6-Methoxy-2-naphthol, **V**, 918
10-Methyl-10,9-borazarophenan-
  threne, **V**, 727
Methyl 4-chloro-2-butynoate, **VIII**,
  371
2-Methyl-2,5-decanediol, **IV**, 601
N-Methyl-1,2-diphenylethylamine,
  IV, 605
4,5-Methylenedioxybenzocyclo-
  butene, **VII**, 326
DL-Methylethylacetic acid, **I**, 361
1-Methyl-2-ethynyl-*endo*-3,3-di-
  methyl-2-norbornanol, **VIII**, 391
2-Methyl-1,6-heptadien-3-ol, **VI**,
  606
3-Methylheptanoic acid, **V**, 762
Methyl 7-hydroxyhept-5-ynoate,
  VIII, 415
Methyl isopropyl carbinol, **II**, 406
Methyl 4-keto-7-methyloctanoate,
  III, 601
Methyllithium, low halide, **VII**, 346

Methyl 2-methyl-1-cyclohexene-1-carboxylate, **VII**, 351
(4-Methylphenyl)magnesium bromide, **VI**, 488
trans-Methylstyrylcarbinol, **IV**, 773
Methyl 4-(2-tetrahydropyranyloxy)-2-butynoate, **VII**, 334
Methyl 2-thienyl sulfide, **IV**, 667
α-Naphthoic acid, **II**, 425
1-Norbornanecarboxylic acid, **VI**, 845
Pentachlorobenzoic acid, **V**, 890
Pentafluorophenylcopper tetramer, **VI**, 875
Pentane, **II**, 478
3-Penten-2-ol, **III**, 696
(*E*)-3-Pentyl-2-nonene-1,4-diol, **VIII**, 507
Phenanthrene-9-aldehyde, **III**, 701
4-Phenyl-2-bromobutene-1, **I**, 187
trans-1-Phenyl-1,3-butadiene, **IV**, 771
Phenyl *tert*-butyl ether (Method I), **V**, 924
1-Phenyldialin (1-Hydroxy-1-phenyl-1,2,3,4-tetrahydronaphthalene), **III**, 729
1-Phenyl-3,4-dihydro-6,7-methylenedioxyisoquinoline, **VII**, 326
1-Phenylnaphthalene, **III**, 729
1-Phenyl-1,4-pentadiyne, **VI**, 925
(*R*)-(+)-1-Phenyl-1-pentanol, **VII**, 447
1-Phenyl-1-penten-4-yn-3-ol, **IV**, 792
3-Phenylpropionaldehyde, **VII**, 451
(*E*)-2-Propenylcyclohexanol, **VII**, 456
trans-2-(2-Propenyl)cyclopentanol, **VII**, 501
Propylbenzene, **I**, 471
Selenophenol, **III**, 771
Tetraethyltin, **IV**, 881
Tetramethylbiphosphine disulfide, **V**, 1016
Tetraphenylarsonium chloride hydrochloride, **IV**, 910
Tetrolic acid, **V**, 1043
p-Tolylcyclopropanol, **V**, 1058
Triamylcarbinol, **II**, 603
Tributylcarbinol, **II**, 603
Triethylcarbinol, **II**, 602
Triheptylcarbinol, **II**, 603
Trimethylacetic acid, **I**, 524
3-Trimethylsilyl-3-buten-2-ol, **VI**, 1033
3-Trimethylsilyl-2-propyn-1-ol, **VII**, 524
Triphenylcarbinol, **III**, 839
Triphenylethylene, **II**, 606
Triphenylstibine, **I**, 550
Tripropylcarbinol, **II** 603
Tri-*p*-tolylstibine, **I**, 551
Valeric acid, **I**, 363
Vinyltrimethylsilane, **VI**, 1033
HALOFORM REACTION
3β-Acetoxyetienic acid, **V**, 8
Ethyl 3,3-diethoxypropanoate, **VIII**, 254
Methyl 3,3-diethoxypropanoate, **VIII**, 255
β-Naphthoic acid, **II**, 428
HALOGENATION
A. Bromination
2-Amino-5-bromopyridine, **V**, 346
3-Amino-2,4,6-tribromobenzoic acid, **IV**, 947
ε-Benzoylamino-α-bromocaproic acid, **II**, 74
Bromal, **II**, 87
N-Bromoacetamide, **IV**, 104
3-Bromo-4-acetaminotoluene, **I**, 111
Bromoacetic acid, **III**, 381
Bromoacetone, **II**, 88
3-Bromoacetophenone, **V**, 117
2-Bromoacetyl-6-methoxynaphthalene, **VIII**, 175
p-Bromobenzaldehyde, **II**, 89
α-Bromo-γ-butyrolactone, **V**, 255
α-Bromocaproic acid, **I**, 115; **II**, 95

1-Bromo-3-chlorocyclobutane, **VI**, 179

1-Bromo-3-chloro-2,2-dimethoxy-propane, **VI**, 361

α-Bromocinnamic aldehyde, **III**, 731

2-Bromo-2-cyclopentenone, **VII**, 271

4-Bromo-*N*,*N*-dimethyl-3-(trifluoromethyl)aniline, **VI**, 181

erythro-2-Bromo-1,2-diphenylethanol, **VI**, 184

2-Bromohexanoyl chloride, **VI**, 190

4-Bromo-2-heptene, **IV**, 108

10-Bromo-11-hydroxy-10,11-dihydrofarnesyl acetate, **VI**, 561

α-Bromoisocaproic acid, **II**, 95; **III**, 523

α-Bromoisovaleric acid, **II**, 93; **III**, 848

p-Bromomandelic acid, **IV**, 110

Bromomesitylene, **II**, 95

Bromomethanesulfonyl bromide, **VIII**, 212

1-Bromo-3-methyl-2-butanone, **VIII**, 193

α-Bromo-β-methylvaleric acid, **II**, 95; **III**, 496

α-Bromonaphthalene, **I**, 121

6-Bromo-2-naphthol, **III**, 132

m-Bromonitrobenzene, **I**, 123

2-Bromo-4-nitrotoluene, **IV**, 114

p-Bromophenacyl bromide, **I**, 127

9-Bromophenanthrene, **III**, 134

o-Bromophenol, **II**, 97

p-Bromophenol, **I**, 128

α-Bromophenylacetic acid, **VI**, 403

α-Bromo-α-phenylacetone, **III**, 343

α-Bromo-α-phenylacetonitrile, **III**, 347

α-Bromo-β-phenylpropionic acid, **III**, 705

(2)3-Bromophthalide, **III**, 737; **V**, 145

3-Bromopyrene, **V**, 147

1-(Bromovinyl)trimethylsilane, **VI**, 1033

4-Bromo-*o*-xylene, **III**, 138

γ-Crotonolactone, **V**, 255

Cyanogen bromide, **II**, 150

Cyclopropylbenzene, **V**, 328

2,3-Diaminopyridine, **V**, 346

Dibenzoyldibromomethane, **II**, 244

Dibromoacetonitrile, **IV**, 254

2,2-Dibromoacetyl-6-methoxynaphthalene, **VIII**, 175

2,6-Dibromoaniline, **III**, 262

9,10-Dibromoanthracene, **I**, 207

4,4'-Dibromobiphenyl, **IV**, 256

1,3-Dibromo-2-butanone, **VI**, 711

2,12-Dibromocyclododecanone, **VI**, 368

α,α'-Dibromodibenzyl ketone, **V**, 514

α,α'-Dibromodibenzyl sulfone, **VI**, 403

α,α'-Dibromodineopentyl ketone, **VIII**, 991

1,1-Dibromo-2,2-diphenylcyclopropane, **VI**, 187

α,β-Dibromoethyl ethyl ether, **IV**, 749

2,4-Dibromo-3-pentanone, **VI**, 512, 520

1,3-Dibromo-1-phenylpropane, **V**, 328

2,3-Dibromopropionaldehyde diethyl acetal, **VI**, 954

2,6-Dibromo-4-nitrophenol, **II**, 173

3,5-Dibromosulfanilamide, **III**, 263

Diethyl α,δ-dibromoadipate, **III**, 623

2,4-Dihydroxy-5-bromobenzoic acid, **II**, 100

2,7-Dimethyloxepin, **V**, 467

Diphenylcyclopropenone, **V**, 514

trans-2-Dodecenoic acid, **IV**, 398

Ethyl bromoacetate, **III**, 381

Ethyl bromomalonate, **I**, 245

DL-Isoleucine, **III**, 495

DL-Leucine, **III**, 523

1-Methylamino-4-bromoanthraquinone, **III**, 575

Methyl 4-*O*-benzoyl-6-bromo-6-deoxy-α-D-glucopyranoside, **VIII**, 363
Methyl 5-bromovalerate, **III**, 578
trans-2-Methyl-2-dodecenoic acid, **IV**, 608
2-Methylenedodecanoic acid, **IV**, 616
Mucobromic acid, **III**, 621; **IV**, 688
Muconic acid, **III**, 623
o-Nitrobenzaldehyde, **V**, 825
p-Nitrobenzyl bromide, **II**, 443
Phenacyl bromide, **II**, 480
DL-Phenylalanine, **III**, 705
Phenylbromoethyne, **V**, 921
Phthalaldehydic acid, **III**, 737
2*H*-Pyran-2-one, **VI**, 462
DL-Serine, **III**, 774
Terephthalaldehyde, **III**, 788
2,4,4,6-Tetrabromo-2,5-cyclohexadien-1-one, **VI**, 181
3,4,8,9-Tetrabromo-11-oxatricyclo[4.4.1.0^{1,6}]undecane, **VI**, 862
α,α,α',α'-Tetrabromo-*o*-xylene, **IV**, 807
α,α,α',α'-Tetrabromo-*p*-xylene, **III**, 788
3-Thenyl bromide, **IV**, 921
DL-Threonine, **III**, 813
p,α,α-Tribromoacetophenone, **IV**, 110
sym-Tribromobenzene, **II**, 592
2,4,6-Tribromobenzoic acid, **IV**, 947
Trimethylethylene dibromide, **II**, 409
DL-Valine, **III**, 848
o-Xylylene dibromide, **IV**, 984
B. Chlorination
3-Acetyl-4- and 5-chloro-2-oxazolidinones, **VII**, 5
Benzohydroxamoyl chloride, **V**, 504
ε-Benzoylamino-α-chlorocaproic acid, **VI**, 90
Benzyl chloromethyl ether, **VI**, 101
Bicyclo[3.3.1]nonan-9-one, **VI**, 137
tert-Butyl hypochlorite, **IV**, 125; **V**, 184

9-Chloroanthracene, **V**, 206
α-Chloroanthraquinone, **II**, 128
p-Chlorobenzaldehyde, **II**, 133
o-Chlorobenzoyl chloride, **I**, 155
3-Chlorocyclobutanecarboxylic acid, **VI**, 271
2-Chlorocyclohexanone, **III**, 188
1-Chloro-1,2-dicyanocyclobutane, **VI**, 427
2-Chloro-1-formyl-1-cyclohexene, **V**, 215
(Z)-4-Chloro-4-hexenyl trifluoroacetate, **VI**, 273
2-Chloro-2-methylcyclohexanone, **IV**, 162
6-Chloromethyl-2,2-dimethyl-1,3-dioxen-4-one, **VIII**, 192
N-Chloromorpholine, **VIII**, 167
p-Chlorophenoxymethyl chloride, **V**, 221
N-Chloropiperidine, **VI**, 968
Chloropyruvic acid, **V**, 636
3-Chlorothietane 1,1-dioxide, **VII**, 491
2-Chlorothiirane 1,1-dioxide, **V**, 231
1,2-Di-*tert*-butoxy-1,2-dichloroethane, **VIII**, 161
α,α-Dichloroacetophenone, **III**, 538
3,5-Dichloro-2-aminobenzoic acid, **IV**, 872
2,6-Dichloroaniline, **III**, 262
cis-3,4-Dichlorocyclobutene, **VI**, 422
cis-1,2-Dichlorocyclohexane, **VI**, 424
trans-2,3-Dichloro-1,4-dioxane, **VIII**, 161
2,4-Dichloromethoxybenzene, **VIII**, 167
2,2-Dichloronorbornane, **VI**, 845
2,6-Dichlorophenol, **III**, 267
3,5-Dichlorosulfanilamide, **III**, 262
2,3-Dichlorotetrahydropyran, **VI**, 675
3,3-Dichlorothietane 1,1-dioxide, **VII**, 491

Diethyl 2-chloro-2-cyclopropyleth-
ene-1,1-dicarboxylate, **VIII**, 247
Diethylthiocarbamyl chloride, **IV**, 307
N,N-Dimethylchloromethylenam-
monium chloride, **VIII**, 498
Dimethyl 3-chloro-2-pentenedioate,
VI, 505
2,4-Dinitrophenylsulfur chloride, **II**,
456
Diphenylcarbodiimide (Method II),
V, 504
Ethyl 4-amino-3-(methylthiomethyl)benzoate, **VI**, 581
Ethyl *N*-chlorocarbamate, **VII**, 223
Ethyl 3,5-dichloro-4-hydroxyben-
zoate, **III**, 267
Ethyl 2-methyl-3-methylthioindole-
5-carboxylate, **VI**, 601
Geranyl chloride, **VI**, 634, 638
3-Hydroxyquinoline, **V**, 635
Mandelic acid, **III**, 538
3-Methylcoumarone, **IV**, 590
Methyl *N,N*-dichlorocarbamate,
VIII, 427
o-Nitrobenzenesulfonyl chloride, **II**,
471
o-Nitro-*p*-chlorophenylsulfur chloride, **II**, 456
o-Nitrophenylsulfur chloride, **II**, 455
Phenacylamine hydrochloride, **V**,
909
Phenylsulfenyl chloride, **VIII**, 550
DL-4,4',6,6'-Tetrachlorodiphenic
acid, **IV**, 872
p-Toluenesulfenyl chloride, **IV**, 934
Trichloromethyl chloroformate, **VI**,
715
C. Fluorination
1-Fluoroadamantane, **VI**, 628
16α-Fluoro-3-methoxy-1,3,5(10)-
estratrien-17-one, **VIII**, 286
N-Fluoropyridinium triflate, **VIII**,
286
p-Nitrobenzyl fluoride, **VI**, 348
Tris(dimethylamino)sulfonium difluorotrimethylsilicate, **VII**, 528

D. Iodination
Cyanogen iodide, **IV**, 207
cis-1,2-Cyclohexanediol, **VI**, 348
trans-1,2-Cyclohexanedlol, **VI**,
348
2,6-Diiodo-*p*-nitroaniline, **II**, 196
2-Hydroxy-3,5-diiodobenzoic acid,
II, 343
4-Hydroxy-3,5-diiodobenzoic acid,
II, 344
p-Iodoaniline, **II**, 347
5-Iodoanthranilic acid, **II**, 349
6-Iodo-3,4-dimethoxybenzaldehyde
cyclohexylimine, **VIII**, 586
Iododurene, **VI**, 700
(*Z*)-1-Iodohexene, **VII**, 290
endo-7-Iodomethylbicyclo[3.3.1]-
nonan-3-one, **VI**, 958
N-Iodosuccinimide, **V**, 663
2-Iodothiophene, **II**, 357; **IV**, 545
Iodotrimethylsilane, **VI**, 353
4-Iodoveratrole, **IV**, 547
2-Iodo-*p*-xylene, **VI**, 709
Tetraiodophthalic anhydride, **III**,
796
HYDROBORATION
(+)-Di-3-pinanylborane (in solution),
VII, 339
Homogeraniol, **VII**, 258
Methyl (1*R*,5*R*)-5-hydroxy-2-
cyclopentene-1-acetate, **VII**,
339
Perhydro-9*b*-boraphenalene, **VII**,
427
Perhydro-9*b*-phenalenol, **VII**, 427
β-3-Pinanyl-9-borabicyclo[3.3.1]-
nonane, **VII**, 402
**HYDROLYSIS (The subheadings indi-
cate the kinds of compounds
hydrolyzed)**
A. Acetals and Ketals
Atropaldehyde, **VII**, 13
Bicyclo[2.2.1]hepten-7-one, **V**, 91
1,4-Di-*O*-benzyl-L-threitol, **VIII**,
155
α-Hydroxyacetophenone, **VII**, 263

3-Hydroxy-1-cyclohexene-1-carbox-
aldehyde, **VIII**, 309
2-Hydroxymethyl-2-cyclopenten-1-
one, **VII**, 271
β-Methylglutaraldehyde, **IV**, 661
Methyl 4-hydroxy-2-butynoate,
VII, 334
(S)-(+)-2-(p-Toluenesulfinyl)-2-
cyclopenten-1-one, **VII**, 495
B. Acid Anhydrides
Citraconic acid, **II**, 141
Itaconic acid, **II**, 368
Mesaconic acid, **II**, 382
C. Acyl Peroxides
Perbenzoic acid, **I**, 431
D. Amides (including lactams)
ε-Aminocaproic acid, **II**, 28; **IV**, 39
L-Arginine hydrochloride, **II**, 49
Atrolactic acid, **IV**, 58
ε-Benzoylaminocaproic acid, **II**, 76
3-Bromo-4-aminotoluene, **I**, 111
tert-Butylamine, **III**, 154
O-tert-Butylvalinol, **VIII**, 204
L-Cystine, **I**, 194
Diazomethane, **V**, 351
1,3-Dihydroisoindole, **V**, 406
N-Ethyl-p-chloroaniline, **IV**, 420
D-Glutamic acid, **I**, 286
Glutaric acid, **IV**, 496
L-Histidine monohydrochloride, **II**,
330
Homophthalic acid, **V**, 612
(2S*,3S*)-3-Hydroxy-3-phenyl-2-
methylpropanoic acid, **VIII**,
339
DL-Lysine dihydrochloride, **II**, 375
DL-Methionine, **II**, 385
(S)-(+)-2-Methoxymethylpyrrolidine,
VIII, 26
1-Methylisoquinoline, **IV**, 641
N-Methyl-(3-methoxy-4-
hydroxy)alanine, **III**, 588
2-Nitro-4-methoxyaniline, **III**, 661
DL-β-Phenylalanine, **II**, 491
α-Phenylethylamine, **II**, 503
β-Valerolactam, **V**, 673

E. Amines
1-Nitro-2-naphthol, **II**, 451
Phenylpyruvic acid, **II**, 519
Phloroglucinol, **I**, 455
F. Azlactones
α-Acetaminocinnamic acid, **II**, 1
α-Benzoylaminocinnamic acid, **II**,
491
3,4-Dimethoxyphenylpyruvic acid,
II, 335
DL-β-Phenylalanine, **II**, 489
G. Enamines, Enol Ethers, and Imines
Acetyltrimethylsilane, **VIII**, 19
Aminoacetonitrile hydrochloride,
I, 298
D-Arginine hydrochloride, **II**, 50
(R)-(+)-Citronellal, **VIII**, 183
2,4-Dinitrobenzaldehyde, **II**, 223
Ethyl α-phenylacetoacetate, **II**, 284
Glycine, **I**, 298
Glycine ethyl ester hydrochloride,
II, 310
β-Naphthol phenylaminomethane,
I, 381
Phloroacetophenone, **II**, 522
o-Tolualdehyde, **III**, 818
H. Esters and Lactones
N-Benzyl-2-azanorbornene, **VIII**, 31
2-Benzylcyclopentanone, **V**, 76
3-Benzyl-3-methylpentanoic acid,
IV, 93
cis-Bicyclo[3.3.0]octane-3,7-dione,
VII, 51
Bicyclo[3.3.0]octene-2,6-diol,
VIII, 43
p-Bromobenzaldehyde, **II**, 442
α-Bromocaproic acid, **II**, 95
α-Bromoisocaproic acid, **II**, 95
α-Bromoisovaleric acid, **II**, 93
α-Bromo-β-methoxypropionic acid,
III, 775
α-Bromo-β-methylvaleric acid, **II**, 95
Caproic acid, **II**, 417, 474
(E)-(Carboxyvinyl)trimethylammon-
ium betaine, **VIII**, 536
Chelidonic acid, **II**, 126

2-Chloro-2-cyclopropylethene-1,1-
dicarboxylic acid, monoethyl
ester, **VIII**, 373
(+)-(1*S*,2*S*)-Cyclopropane-1,2-dicar-
boxylic acid, **VIII**, 141
cis-1,2-Cyclohexanediol, **VI**, 348
trans-Cyclohexanediol, **VI**, 348
2,6-Dichlorophenol, **III**, 267
1,3-Dihydroxy-2-naphthoic acid,
III, 638
3,4-Dimethoxyphenylpyruvic acid,
II, 333, 335
cis- 1,5-Dimethylbicyclo[3.3.0]-
octane-3,7-dione, **VII**, 51
(2*SR*,3*SR*)-2,4-Dimethyl-3-hydroxy-
pentanoic acid, **VII**, 190
2,5-Dimethylmandelic acid, **III**,
326
1,3-Dimethyl-5-oxobicyclo[2.2.2]
octane-2-carboxylic acid, **VIII**,
219
2,6-Dimethylpyridine, **II**, 214
2,4-Dimethylpyrrole, **II**, 217
10,11-Epoxyfarnesyl acetate, **VI**, 560
Erucic acid, **II**, 258
Ethyl benzoylacetate, **II**, 266
Ethyl *tert*-butyl malonate, **IV**, 417
Ethyl α-(hexahydroazepinylidene-
2)acetate, **VIII**, 263
2-(3-Ethyl-5-methyl-4-isoxazolyl-
methyl)cyclohexanone, **VI**, 781
Ethyl 4-oxohexanoate, **VI**, 615
2-Fluorohexanoic acid, **VIII**, 258
Glutaric acid, **I**, 291
3,3a,3b,4,6a,7a-Hexahydro-3,4,7-
metheno-7*H*-cyclopenta[*a*]-
pentalene-7,8-dicarboxylic acid,
VIII, 298
Homoveratric acid, **II**, 333
6-Hydroxynicotinic acid, **IV**, 532
(1*S*,2*S*,3*R*)-3-Hydroxy-2-nitrocyclo-
hexyl acetate, **VIII**, 332
7-Hydroxy-4,4,6,7-tetramethylbicy-
clo[4.2.0]octan-2-one, **VI**, 1024
Iodosobenzene, **V**, 658
DL-Isoleucine, **III**, 495

DL-α-(Isopropylideneaminooxy)pro-
pionic acid, **V**, 1032
α-Ketoglutaric acid, **III**, 510
Linoleic acid, **III**, 526
Linolenic acid, **III**, 531
Methyl amyl ketone, **I**, 351
o-Methylbenzyl alcohol, **IV**, 582
Methyl butyl ketone, **I**, 352
Methyl isobutyl ketone, **I**, 352
Methyl 2-chloro-2-cyclopropyliden-
acetate, **VIII**, 373
3-Methylcoumarilic acid, **IV**, 591
Methyl cyclopropyl ketone, **IV**, 597
Methyl diformylacetate, **VII**, 323
trans-2-Methyl-2-dodecenoic acid,
IV, 608
2-Methylenedodecanoic acid, **IV**, 616
Methyl (3*RS*,4*RS*)-4,5-epoxy-3-
phenylpentanoate, **VII**, 164
3-Methyl-2-furoic acid, **IV**, 628
β-Methylglutaric anhydride, **IV**, 630
Methylglyoxal ω-phenylhydrazone,
IV, 633
4-Methyl-2-hexanone, **I**, 352
Methyl hydrogen hendecanedioate,
IV, 635
Methyl 7-hydroxyhept-5-ynoate,
VIII, 415
4-Methyl-4-nitrovaleric acid, **V**, 445
3-Methylpentanoic acid, **II**, 416
1-Methylthio-3-phthalimido-
propane-3,3-dicarboxylic acid,
II, 385
2-Methyl-5-trimethylsilyl-2-vinyl-
pentanoic acid, **VIII**, 488
6-Methyluracil, **II**, 422
Myristic acid, **I**, 379
o-Nitroacetophenone, **IV**, 708
p-Nitrobenzaldehyde, **II**, 441
m-Nitrobenzoic acid, **I**, 391
p-Nitrobenzyl alcohol, **III**, 652
D- and L-Octanol-2, **I**, 418
6-Oxodecanoic acid, **VIII**, 498
Pelargonic acid, **II**, 474
(Pentafluorophenyl)acetonitrile,
VI, 873

DL-Phenylalanine, **III**, 705
(+)- and (–)-*trans*-2-Phenylcyclo-
  hexanol, **VIII**, 516
cis-2-Phenylcyclopropanecarboxylic
  acid, **VI**, 913
(–)-8-Phenylmenthol, **VIII**, 522
Phenylsuccinic acid, **I**, 451
1-Phenyl-2-thiobiuret, **V**, 966
Pimelic acid, **II**, 531
2,3,4,6-Tetra-*O*-acetyl-D-gluconic
  acid monohydrate, **V**, 887
(+)- and (–)-α-(2,4,5,7-Tetranitro-9-
  fluorenylideneaminooxy)propi-
  onic acid, **V**, 1031
Tricarballylic acid, **I**, 523
1,2,5-Trihydroxypentane, **III**, 833
I. Nitriles
β-Alanine, **III**, 34
DL-Alanine (HCI), **I**, 21
γ-Aminobutyric acid, **II**, 25
α-Aminodiethylacetic acid, **III**, 66
DL-α-Aminophenylacetic acid, **III**,
  84
DL-α-Amino-α-phenylpropionic
  acid, **III**, 88
Atrolactic acid, **IV**, 58
Benzoylformic acid, **III**, 114
3-Benzyl-3-methylpentanoic acid,
  IV, 93
2-Bromo-3-methylbenzoic acid, **IV**,
  414
β-Bromopropionic acid, **I**, 131
3-Chloro-2,2,3-trifluoropropionic
  acid, **V**, 239
Diallylamine, **I**, 201
Dibutylamine, **I**, 202
β-Ethyl-β-methylglutaric acid, **IV**,
  441
α-Ethyl-α-methylsuccinic acid, **V**,
  572
Ethyl *N*-tridecylate, **II**, 292
Glutaric acid, **I**, 289; **IV**, 496
Glycine, **I**, 298
Glycine ethyl ester hydrochloride,
  II, 310

D-Gulonic-γ-lactone, **IV**, 506
β-Hydroxypropionic acid, **I**, 321
Malonic acid **II**, 376
Mandelic acid, **I**, 336
Mesitylacetic acid, **III**, 557
Methyl benzyl ketone, **II**, 391
β-Methylglutaric acid, **III**, 591
β-Methyl-β-phenylglutaric acid, **IV**,
  664
p-Nitrophenylacetic acid, **I**, 406
Phenylacetamide, **IV**, 760
Phenylacetic acid, **I**, 436
α-Phenylglutaric anhydride, **IV**, 790
Phenylnitromethane, **II**, 512
Phenylsuccinic acid, **IV**, 804;
  I, 451
o-Toluamide, **II**, 586
o-Toluic acid, **II**, 588
p-Toluic acid, **II**, 589
Veratric amide, **II**, 44
Vinylacetic acid, **III**, 851
J. Miscellaneous
2-(2-Acetamidoethyl)-4,5-dimeth-
  oxyacetophenone, **VI**, 1
Acetonylacetone, **II**, 219
Acetylbenzoyl, **III**, 20
Allylamine, **II**, 24
2-Aminobenzophenone, **IV**, 34
6-Amino-3,4-dimethyl-*cis*-3-cyclo-
  hexen-1-ol, **VII**, 4
trans-1-Amino-2,3-diphenylaziri-
  dine, **VI**, 56
7-Aminotheophylline, **VII**, 10
o-Anisaldehyde, **VI**, 64
3-Benzoylindole, **VI**, 109
1-Benzylisoquinoline, **VI**, 115
exo-cis-Bicyclo[3.3.0]octane-2-car-
  boxylic acid, **V**, 93
Bicyclo[3.2.1]octan-3-one, **VI**, 142
Bicyclo[2.2.0]pentane, **V**, 96
o-Bromoheptaldehyde, **III**, 127
o-Bromophenol, **II**, 97
Cyclobutanecarboxaldehyde, **VI**, 312
Cyclobutanone, **VI**, 316
2-Cyclohexanone, **V**, 294

Cycloundecanecarboxylic acid, **VII**, 135

Diethyl formylmethylphosphonate, **VI**, 448

N-(2,4-Diformyl-5-hydroxyphenyl)-acetamide, **VII**, 162

4,4-Dimethyl-2-cyclohexen-1-one, **VI**, 496

sym-Dimethylhydrazine dihydrochloride, **II**, 208

2,6-Dinitroaniline, **IV**, 364

Docosanedioic acid, **V**, 533

4-Ethoxy-3-hydroxybenzaldehyde, **VI**, 567

Ethyl benzoylacetate, **IV**, 415

2-Furfuryl mercaptan, **IV**, 491

DL-Glyceraldehyde, **II**, 305

D-Glucosamine hydrochloride, **III**, 430

Glutaconaldehyde sodium salt, **VI**, 640

Glutaric acid, **IV**, 496

1,4,7,10,13,16-Hexaazacyclooctadecane, **VI**, 652

trans-4-Hydroxy-2-hexenal, **VI**, 683

Hydroxylamine hydrochloride, **I**, 318

5-Hydroxypentanal, **III**, 470

Imidazole-2-carboxaldehyde, **VII**, 287

6-Ketohendecanedioic acid, **IV**, 555

Laurone, **IV**, 560

D-Mannose, **III**, 541

2-Methoxy-1,4-naphthoquinone, **III**, 465

2-Methylbutyraldehyde-1-*d*, **VI**, 751

N-Methyl-3,4-dihydroxyphenylalanine, **III**, 586

3-Methylene-4-isopropyl-1,1-cyclopentanedicarboxylic acid, dimethyl ester, **VIII**, 381

1-Methyl-4,4a,5,6,7,8-hexahydronaphthalen-2(3*H*)-one, **VI**, 781

2-Methyl-1-hexanol, **VI**, 786

Methyl *N*-(2-methyl-2-butenyl)carbamate, **VIII**, 68

3-Methyl-1,5-pentanediol, **IV**, 660

2-Naphthalenethiol, **VI**, 824

o-Nitrobenzaldehyde, **V**, 825

Orthanilic acid, **II**, 471

Pentadecanal, **VI**, 869

γ-Phenylallylsuccinic acid, **IV**, 766

3-Phenyl-2*H*-azirine-2-carboxaldehyde, **VI**, 893

β-Phenylcinnamaldehyde, **VI**, 901

1-Phenylcyclopentanecarboxaldehyde, **VI**, 905

1-Phenylcyclopentylamine, **VI**, 910

Phenyl heptyl ketone, **VI**, 919

1-Phenyl-4-phosphorinanone, **VI**, 932

Phenylpropargylaldehyde, **III**, 731

Pseudopelletierine, **IV**, 816

Pyocyanine, **III**, 753

DL-Serine, **III**, 774

1,2,3,4-Tetrahydronaphthalen(1,2)-imine, **VI**, 967

β-Tetralone, **IV**, 903

2-Thiophenealdehyde, **III**, 811

Tricarbonyl[(2,3,4,5-η)-2,4-cyclohexadien-1-one]iron, **VI**, 996

2,2-(Trimethylenedithio)cyclohexanone, **VI**, 1014

HYDROXYLATION

α-Hydroxyacetophenone, **VII**, 263

3-Hydroxy-1,7,7-trimethylbicyclo[2.2.1]heptan-2-one, **VII**, 277

6-Hydroxy-3,5,5-trimethyl-2-cyclohexen-1-one, **VII**, 282

METALATION REACTIONS

A. Aluminum

(*E*)-1-Decenyldiisobutylalane, **VIII**, 295

Triphenylaluminum, **V**, 1116

B. Arsenic

Arsanilic acid, **I**, 70

p-Nitrophenylarsonic acid, **III**, 665

Sodium *p*-hydroxyphenylarsonate, **I**, 490

Triphenylarsine, **IV**, 910

C. Boron
  (*E*)-1-Hexenyl-1,3,2-benzodioxabo-
    role, **VIII**, 532
D. Chromium
  [(Methyl)(methoxy)carbene]pen-
    tacarbonyl chromium(0), **VIII**,
    216
E. Cobalt
  Hexacarbonyl(1-methyl-2-propynyli-
    um)dicobalt tetrafluoroborate,
    VIII, 460
F. Copper
  Lithium phenylthio(*tert*-butyl
    cuprate), **VI**, 248
  4,5,4',5'-Tetramethoxy-1,1'-biphenyl-
    2,2'-dicarboxaldehyde, **VIII**,
    586
G. Iron
  Dicarbonyl(cyclopentadienyl)(ethyl
    vinyl ether)iron tetrafluorobor-
    ate, **VIII**, 479
  Dicarbonyl(cyclopentadienyl)(*trans*-
    3-methyl-2-vinylcyclohexa-
    none)iron tetrafluoroborate,
    VIII, 479
  Ferrocene, **IV**, 473
H. Lead
  p-Methoxyphenyllead triacetate,
    VII, 229
I. Lithium
  Allyllithium, **V**, 452
  Benzobarrelene, **VI**, 82
  2-Benzyl-2-methylcyclohexanone,
    VI, 121
  2-Benzyl-6-methylcyclohexanone,
    VI, 121
  Dilithioallene, **VIII**, 226
  2,2'-Dilithio-1,1'-binaphthyl, **VIII**,
    57
  (2-Dimethylamino-5-methylphenyl)-
    diphenylcarbinol, **VI**, 478
  2-(*N*,*N*-Dimethylamino)phenyllithi-
    um, **VII**
  2,2-Dimethyl-4-phenylbutyric acid,
    VI, 517
  (1-Hydroxy-2-propenyl)trimethylsi-
    lane, **VIII**, 501

2-Lithiobornene, **VII**, 77
2-Lithio-2-butene, **VIII**, 505
1-Lithio-3-chloropropyne, **VIII**, 371
1-Lithio-1-ethoxycyclopropane,
    VIII, 556
2-(3-Lithio)furyl tetramethyldiami-
    dophosphate, **VIII**, 396
2-(2'-Lithio-4',5'-methylenedioxy-
    phenyl)ethyl chloride, **VII**, 326
Lithium acetylide, **VIII**, 391
Lithium 1-lithiopropyn-3-oxide,
    VIII, 415
Lithium phenylthio(*tert*-butyl
    cuprate), **VI**, 248
2-Methyl-1-hexanol, **VI**, 786
Methyllithium, **VII**, 346
Pentanoyllithium, **VIII**, 345
Phenyllithium, **III**, 413, 757
(*E*)-1-Propenyllithium, **VI**, 1040
2-Pyridylmethyllithium, **III**, 413, 757
o-Tolyllithium, **VIII**, 430
Tributyltinlithium, **VIII**, 562
J. Magnesium
  Allylmagnesium bromide, **VIII**, 23
  Butylmagnesium chloride, **VIII**, 606
  Butylmagnesium chloride, unsolvat-
    ed, **V**, 1141
  Ethynylmagnesium chloride, **VIII**,
    606,
  1-(Isopropoxydimethylsilyl)methyl-
    magnesium chloride, **VIII**, 315
  (*E*)-3-Pentyl-2-nonene-1,4-diol,
    VIII, 507
  Trimethylsilylmethylmagnesium
    chloride, **VIII**, 602, 612
K. Mercury
  o-Chloromercuriphenol, **I**, 161
  Cyclohexene-mercuric sulfate com-
    plex, **V**, 320
  Diphenylmercury, **I**, 228
  trans-3-(Phenylsulfonyl)-4-(chloro-
    mercuri)cyclohexene, **VIII**, 540
  Phenyl(trichloromethyl)mercury, **V**,
    969
L. Potassium
  Ethyl 6-methylpyridine-2-acetate,
    VI, 611

M. Ruthenium
  Ruthenocene, **V**, 1001
N. Silicon
  Ethyl trimethylsilylacetate, **VII**, 512
O. Silver
  Succinimide silver salt, **VI**, 226
P. Sodium
  Triphenylmethylsodium, **II**, 607
Q. Thallium
  2-Iodo-*p*-xylene, **VI**, 709
R. Tin
  Allyltributyltin, **VIII**, 23,
  Allyltriphenyltin, **V**, 453
  (*E*)-1,2-Bis(tributylstannyl)ethylene,
    VIII, 268
  Dibutyldivinyltin, **IV**, 258
  8-Propionyl-(*E*)-5-nonenolide, **VIII**,
    562
  Tetraethyltin, **IV**, 881
  Tributylethynylstannane, **VIII**, 268
  (*Z*)-3-Tributylstannylmethylene-4-
    isopropyl-1,1-dicarboxylic acid,
    dimethyl ester, **VIII**, 381
  Tributyltinlithium, **VIII**, 562
  Trimethylvinyltin, **VIII**, 97
S. Titanium
  Methyltriisopropoxytitanium, **VIII**,
    495
T. Zinc
  Bis(2-carboethoxyethyl)zinc, **VIII**,
    277
  3-Carboethoxyzinc iodide, **VIII**, 274
  Dibutyl zinc, **II**, 187
  Diethyl zinc, **II**, 184
  Diisoamyl zinc, **II**, 187
  Dipropyl zinc, **II**, 187
  o-Tolylzinc chloride, **VIII**, 430
NITRATION
  Acetone cyanohydrin nitrate, **V**, 839
  2-Amino-5-bromo-3-nitropyridine,
    V, 347
  o-Amino-*p*'-nitrobiphenyl, **V**, 830
  2-Amino-3-nitrotoluene, **IV**, 42
  1,5-Dichloro-2,4-dinitrobenzene, **V**,
    1067
  2,6-Dinitroaniline, **IV**, 364
  3,5-Dinitrobenzoic acid, **III**, 337

Dinitrodurene, **II**, 254
3,5-Dinitro-*o*-tolunitrile, **V**, 480
Ethyl *m*-nitrobenzoate, **I**, 373
Methyl *m*-nitrobenzoate, **I**, 372
3-Methyl-4-nitropyridine 1-oxide,
  IV, 654
m-Nitroacetophenone, **II**, 434
1-Nitro-2-acetylaminonaphthalene,
  II, 438
9-Nitroanthracene, **IV**, 711
Nitroanthrone, **I**, 390
Nitrobarbituric acid, **II**, 440
m-Nitrobenzaldehyde, **III**, 644
p-Nitrobenzyl cyanide, **I**, 396
o-Nitrocinnamaldehyde, **IV**, 722
2-Nitro-*p*-cymene, **III**, 653
m-Nitrodimethylaniline, **III**, 658
2-Nitrofluorene, **II**, 447
Nitroguanidine, **I**, 399
Nitromesitylene, **II**, 449
2-Nitro-4-methoxyacetanilide, **III**, 661
2-Nitro-4-methoxyaniline, **III**, 661
N-Nitromorpholine, **V**, 839
3-Nitrophthalic acid, **I**, 408
4-Nitrophthalimide, **II**, 459
2-Nitrothiophene, **II**, 466
Nitrourea, **I**, 417
6-Nitroveratraldehyde, **IV**, 735
2,4,5,7-Tetranitrofluorenone, **V**,
  1029
2,4,5-Triaminonitrobenzene, **V**, 1067
2,4,7-Trinitrofluorenone, **III**, 837
NITROSATION
  Aminomalononitrile *p*-toluenesul-
    fonate, **V**, 32
  Biacetyl monoxime, **II**, 205
  tert-Butyl azidoformate, **V**, 157
  ω-Chloroisonitrosoacetophenone,
    III, 191
  Cupferron, **I**, 177
  Diaminouracil hydrochloride, **IV**,
    247
  Diethyl acetamidomalonate, **V**, 373
  2,4-Dimethyl-3,5-dicarbethoxypyr-
    role, **II**, 202
  2-(Diphenylhydroxymethyl)pyrroli-
    dine, **VI**, 542

Ethyl *N*-nitroso-*N*-benzylcarbamate,
  IV, 780
Ethyl *N*-nitroso-*N*-(*p*-tolylsulfonyl-
  methyl)carbamate, **VI**, 981
N-Ethyl-*m*-toluidine, **II**, 290
2-Hydroxy-3-methylisocarbostyril,
  V, 623
Indazole, **V**, 650
2-Methyl-2-nitrosopropane, **VI**, 803
p-Nitrosodiethylaniline, **II**, 224
Nitrosodimethylamine, **II**, 211
p-Nitrosodimethylaniline hydro-
  chloride, **II**, 223
N-Nitroso-β-methylaminoisobutyl
  methyl ketone, **III**, 244
N-Nitrosomethylaniline, **II**, 460
Nitrosomethylurea, **II**, 461
Nitrosomethylurethane, **II**, 464
Nitroso-β-naphthol, **I**, 411
N-Nitroso-*N*-(2-phenylethyl)benza-
  mide, **V**, 336
N-Nitroso-*N*-phenylglycine, **V**, 962
Nitrosothymol, **I**, 511
Nortricyclanol, **V**, 863
3-Phenylsydnone, **V**, 962
p-Tolylsulfonylmethylnitrosamide,
  IV, 943
**OXIDATION (The subheadings indi-
  cate the types of oxidation)**
A. CH → C–C
  1,4-Bis(trimethysilyl)buta-1,3-diyne,
    VIII, 63
B. CH–CH → C=C
  1,6-Methano[10]annulene, **VI**, 731
C. CH₂ → CHOH and CHOR
  7-Acenaphthenol, **III**, 3
  Acenaphthenol acetate, **III**, 3
  1-Adamantanol, **VI**, 43
  3-Benzoyloxycyclohexene, **V**, 70
  7-*tert*-Butoxynorbornadiene, **V**, 151
  Diethyl (*O*-benzoyl)ethyltartronate,
    V, 379
  α-Hydroxyacetophenone, **VII**, 263
  3-Hydroxy-1,7,7-trimethylbicy-
    clo[2.2.1]heptan-2-one, **VII**,
    277

6-Hydroxy-3,5,5-trimethyl-2-cyclo-
  hexen-1-one, **VII**, 282
trans-Pinocarveol, **VI**, 946

$$\overset{O}{\overset{\|}{}}$$
D. CH=CH₂ → CCH₃
  2-Decanone, **VII**, 137
  2,2-Dimethyl-4-oxopentanal, **VIII**,
    208
E. C=C → 2CO₂H
  Azelaic acid, **II**, 53
  cis-1-Methylcyclopentane-1,2-dicar-
    boxylic acid, **VIII**, 377
  cis,cis-Monomethyl muconate, **VIII**,
    490
  Nonadecanoic acid, **VII**, 397
F. CH₂ → C=O
  Acenaphthenequinone, **III**, 1
  Adamantanone, **VI**, 48
  Alloxan monohydrate, **IV**, 23
  p-Bromobenzaldehyde, **II**, 442
  Δ⁴-Cholesten-3,6-dione, **IV**, 189
  1,2-Cyclohexanedione (SeO₂), **IV**,
    229
  3,12-Diacetoxy-nor-cholanic acid
    (CrO₃), **III**, 234
  Ethyl oxomalonate, **I**, 266
  Glyoxal bisulfite, **III**, 438
  Methyl *p*-acetylbenzoate, **IV**, 579
  o-Nitrobenzaldehyde, **V**, 825; **III**,
    641
  o- and *p*-Nitrobenzaldiacetate, **IV**,
    713
  p-Nitrobenzaldehyde, **II**, 441
  Phenylglyoxal, **II**, 509
  Propiolaldehyde, **IV**, 813
G. CH₃ → CO₂H
  o-Chlorobenzoic acid, **II**, 135
  p-Chloromercuribenzoic acid, **I**, 159
  2-Hydroxyisophthalic acid, **V**, 617
  2,3-Naphthalenedicarboxylic acid,
    V, 810
  p-Nitrobenzoic acid, **I**, 392
  Picolinic acid hydrochloride, **III**, 740
  o-Toluic acid, **III**, 820
  2,4,6-Trinitrobenzoic acid, **I**, 543

H. CH$_2$OH → CHO
1-Decanal, **VI**, 373
Heptanal, **VI**, 644
(S)-2-Methylbutanal, **VIII**, 367
Propionaldehyde, **II**, 541

I. CH$_2$OH → CO$_2$H
Butyl butyrate, **I**, 138
β-Chloropropionic acid, **I**, 168

J. CHO → CO$_2$H
β-Chloropropionic acid, **I**, 166
2-Furoic acid, **IV**, 493
Glutaric acid, **IV**, 496
Heptanoic acid, **II**, 315
Piperonylic acid, **II**, 538
3-Thenoic acid, **IV**, 919
Vanillic acid, **IV**, 972; **IV**, 974

K. CHOH → C=O
4-Acetoxy-2-cyclopenten-1-one, **VIII**, 15
Benzil, **I**, 87
Benzoylformic acid, **I**, 244
Bicyclo[3.3.0]octane-2,6-dione, **VIII**, 43
4-tert-Butylcyclohexanone, **VI**, 218, 220
3-Butyroyl-1-methylpyrrole, **VII**, 102
Cholan-24-al, **V**, 242
Cholestanone, **II**, 139
Δ^4-Cholesten-3,6-dione, **IV**, 189
Cholestenone, **III**, 207
Δ^4-Cholesten-3-one, **IV**, 192
Δ^5-Cholesten-3-one, **IV**, 195
Cyclobutanone, **VII**, 114
Cyclooctanone, **V**, 310
2-Cyclopentene-1,4-dione, **V**, 324
5α,6β-Dibromocholestanone, **IV**, 197
α,γ-Dichloroacetone, **I**, 211
3,3-Dimethyl-cis-bicyclo[3.2.0]heptan-2-one, **VII**, 177
Ethyl benzoylformate, **I**, 241
Ethyl pyruvate, **IV**, 467
Geranial, **VII**, 258
(+)-Isomenthone, **VIII**, 386
L-Menthone, **I**, 340

Methyl benzoylformate, **I**, 244
1-Methyl-2-pyridone, **II**, 419
2-Norbornanone, **V**, 852
Nortricyclanone, **V**, 866
(1-Oxo-2-propenyl)trimethylsilane, **VIII**, 501
Sebacil, **IV**, 838
3-Trimethylsilyl-3-buten-2-one, **VI**, 1033

L. COCH$_3$ → CO$_2$H
3β-Acetoxyetienic acid, **V**, 8
β,β-Dimethylacrylic acid, **III**, 302
Fluorenone-2-carboxylic acid, **III**, 420
β-Naphthoic acid, **II**, 428
Terephthalic acid, **III**, 791
Trimethylacetic acid, **I**, 524

M. CO$_2$H → CO$_3$H
m-Chloroperbenzoic acid, **VI**, 276
Monoperphthalic acid, **III**, 619; **V**, 805
Peroxybenzoic acid, **V**, 904

N. NH$_2$ → N=O
2-Nitroso-5-nitrotoluene, **III**, 334
Nitroso-tert-octane, **VIII**, 93

O. NH$_2$ → NO$_2$
2,6-Dichloronitrobenzene, **V**, 367
4-Nitro-2,2,4-trimethylpentane, **V**, 845
2-Methyl-2-nitropropane, **VI**, 803

P. R$_3$N → R$_3$NO
N,N-Dimethylcyclohexylmethylamine oxide, **IV**, 612
Nicotinamide 1-oxide, **IV**, 704
Pyridine N-oxide, **IV**, 828

Q. NH–NH → N=N
Azoethane, **VI**, 78
2,3-Diazabicyclo[2.2.1]hept-2-ene, **V**, 97
tert-Butyl azodiformate, **V**, 160
3,3-Pentamethylenediazirine, **V**, 897
Bis(2,2,2-trichloroethyl) azodicarboxylate, **VII**, 57
2-(1-Cyanocyclohexyl)diazenecarboxylic acid methyl ester, **VI**, 334
4-Phenyl-1,2,4-triazoline-3,5-dione, **VI**, 936

R. C=N → C–N
      \ /
       O

2-*tert*-Butyl-3-phenyloxazirane, **V**, 191
(+)-(2*R*,8a*S*)-10-(Camphorylsulfonyl)oxaziridine, **VIII**, 104
(±)-*trans*-2-(Phenylsulfonyl)-3-phenyloxaziridine, **VIII**, 546

S. S → SO
(S)-(–)-Methyl *p*-tolyl sulfoxide, **VIII**, 464
Phenyl vinyl sulfoxide, **VII**, 453

T. S → SO$_2$
Phenyl vinyl sulfone, **VII**, 453
Thietane 1,1-dioxide, **VII**, 491

U. Miscellaneous
Acenaphthenol acetate, **III**, 3
2-Acetyl-2-cyclohexen-1-one, **VI**, 23
δ-Acetylvaleric acid, **IV**, 19
Adipic acid, **I**, 18
Allantoin, **II**, 21
Alloxan monohydrate, **III**, 37; **III**, 39
Alloxantin dihydrate, **III**, 42
p-Aminobenzaldehyde, **IV**, 31
trans-1-Amino-2,3-diphenylaziridine, **VI**, 56
p-Aminophenyl disulfide, **III**, 86
Anthraquinone, **II**, 554
1,1'-Azo-bis-1-cyclohexanenitrile, **IV**, 66
Benzilic acid, **I**, 89
Benzofurazan oxide, **IV**, 74
Benzoyl disulfide, **III** 116
Benzyl isocyanide, **VII**, 27
Bicyclo[3.3.1]nonan-9-one, **VI**, 137
Bis(trifluoromethyl)diazomethane, **VI**, 161
Bis[2,2,2-trifluoro-1-phenyl-1-(trifluoromethyl)ethoxy]diphenylsulfurane, **VI**, 163
Bromomethanesulfonyl bromide, **VIII**, 212
Butyl glyoxylate, **IV**, 124
9-Butyl-1,2,3,4,5,6,7,8-octahydroacridine *N*-oxide, **VIII**, 87

o-Carboxycinnamic acid, **IV**, 136
Chloro-*p*-benzoquinone, **IV**, 148
N-Chloropiperidine, **VI**, 968
2-Chlorothiirane 1,1-dioxide, **V**, 231
1,2-Cyclobutanedione, **VII**, 112
cis-Cyclooctene oxide, **VII**, 124
cis-1,2-Cyclohexanediol, **VI**, 342, 348
trans-1,2-Cyclohexanediol, **III**, 217; **VI**, 348
Cysteic acid monohydrate, **III**, 226
nor-Desoxycholic acid, **III**, 234
4,4'-Diaminoazobenzene, **V**, 341
2-Diazopropane, **VI**, 392
α,α'-Dibromodibenzyl sulfone, **VI**, 403
2,6-Dibromoquinone-4-chloroimide, **II**, 175
2,6-Di-*tert*-butyl-*p*-benzoquinone, **VI**, 412
3,5-Dicarbethoxy-2,6-dimethylpyridine, **II**, 215
Diethyl oximinomalonate, **V**, 373
2,2-Difluorosuccinic acid, **V**, 393
6,6-Dimethoxyhexanal, **VII**, 168
4,5-Dimethyl-*o*-benzoquinone, **VI**, 480
Dimethyl decanedioate, **VII**, 181
N,*N*-Dimethyldodecylamine oxide, **VI**, 501
β,β-Dimethylglutaric acid, **IV**, 345
Diphenyl diselenide, **VI**, 533
(2*SR*,3*RS*)-2,4-Dimethyl-3-hydroxypentanoic acid, **VII**, 185
2,5-Dinitrobenzoic acid, **III**, 334
p,*p*'-Dinitrobibenzyl, **IV**, 367
Diphenylacetylene, **IV**, 377
Diphenyldiacetylene, **V**, 517
Diphenyldiazomethane, **III**, 351
Diphenylketene, **III**, 356
Disodium nitrosodisulfonate, **VI**, 1010
Duroquinone, **II**, 254
D-Erythronolactone, **VII**, 297
Ethyl azodicarboxylate, **IV**, 411
2-Fluoroheptanoic acid, **V**, 580
Fumaric acid, **II**, 302

2(5*H*)-Furanone, **VIII**, 396
Glutaric acid, **I**, 290
DL-Glyceraldehyde acetal, **II**, 307
Hexadehydro[18]annulene, **VI**, 68
(*Z,Z*)-2,4-Hexadienedinitrile, **VI**, 662
Hexafluoroacetone, **VII**, 251
2,3,4,5,6,6-Hexamethyl-2,4-cyclo-
    hexadien-1-one, **V**, 598
1-Hexanol, **VI**, 919
Homophthalic acid, **III**, 449
Homoveratric acid, **II**, 335
Hydrazine sulfate, **I**, 309
α-Hydrindone, **II**, 336
1-(Hydroxymethyl)cyclohexanol,
    VIII, 315
17β-Hydroxy-5-oxo-3,5-*seco*-4-
    norandrostane-3-carboxylic
    acid, **VI**, 690
2-Hydroxythiophene, **V**, 642
2-Indanone, **V**, 647
endo-7-Iodomethylbicyclo[3.3.1]-
    nonan-3-one, **VI**, 958
Iodosobenzene diacetate, **V**, 660
Iodoxybenzene, **V**, 665; **III**, 486
Isophthalaldehyde, **V**, 668
(–)-Isopinocampheol, **VI**, 719
DL-Ketopinic acid, **V**, 689
Methanesulfinyl chloride, **V**, 709
2-Methoxy-*N*-carbomethoxypyrroli-
    dine, **VII**, 307
6-Methoxy-2-naphthol, **V**, 918
Methyl benzenesulfinate, **V**, 723
Methyl 4-*O*-benzoyl-6-bromo-6-
    deoxy-α-D-glucopyranoside,
    VIII, 363
1-Methylcyclohexanol, **VI**, 766
Methyl 6,6-dimethoxyhexanoate,
    VII, 168
(*S*)-(+)-4-Methyl-3-heptanone, **VIII**,
    403
Methyl 2-hexynoate, **VI**, 791
2-Methyl-2-nitrosopropane, **VI**, 803
Methyl 6-oxohexanoate, **VII**, 168
Methyl phenyl sulfoxide, **V**, 791
4-Methyltricyclo[2.2.2.0³·⁵]octane-
    2,6-dione, **VIII**, 467

1-Naphthaldehyde, **IV**, 690
1,2-Naphthoquinone, **II**, 430
1,4-Naphthoquinone, **I**, 383; **IV**, 698
1,2-Naphthoquinone-4-sulfonic acid
    salts, **III**, 633
Neopentyl alcohol, **V**, 818
Nicotinic acid, **I**, 385
o-Nitrobenzenesulfonyl chloride,
    II, 471
p-Nitrobenzoyl peroxide, **III**, 649
Nitrosobenzene, **III**, 668
2-Nitroso-5-nitrotoluene, **III**, 334
1-Octanol, **VI**, 919
γ-Octyl-γ-butyrolactone, **VII**, 400
(1*R*,6*S*)-(+)-8-Oxabicyclo[4.3.0]-
    nonan-7-one, **VII**, 406
Oxaspiropentane, **VI**, 320
Oximinomalononitrile, **V**, 32
Pentadecanal dimethyl acetal, **VI**, 869,
Perhydro-9*b*-phenalenol, **VII**, 427
Phenyl benzeneselenosulfonate,
    VIII, 543
Phenyl *tert*-butyl ether, **V**, 924
Phenylchlorodiazirine, **VII**, 203
Phenylglyoxal, **V**, 937
1-(Phenylsulfonyl)cyclopentene,
    VIII, 543
Phenylsulfur trifluoride, **V**, 959
Phthalaldehydic acid, **II**, 523
8-Propionyl-(*E*)-5-nonenolide, **VIII**,
    562
2,3-Pyrazinedicarboxylic acid, **IV**, 824
Pyrogallol 1-monomethyl ether, **III**,
    759
Pyromellitic acid, **II**, 551
Quinone, **I**, 482; **II**, 553
Styrene oxide, **I**, 494
Tetralin hydroperoxide, **IV**, 895
3,3,6,6-Tetramethoxy-1,4-cyclohexa-
    diene, **VI**, 971
Tetramethyl 1,1,2,2-ethanetetracar-
    boxylate, **VII**, 482
Tetraphenylarsonium chloride
    hydrochloride, **IV**, 910
cis-8-Thiabicyclo[4.3.0]nonane 8,8-
    dioxide, **VI**, 482

Thymoquinone, **I**, 511
p-Toluic acid, **III**, 822
Tricarbonyl[(1,2,3,4,5-η)-2,4-cyclo-
  hexadien-1-one] and [2-meth-
  oxy-2,4-cyclohexadien-1-yl]-
  iron(1+) tetrafluoroborate (1–),
  VI, 996
3,5,5-Trimethyl-2-(2-oxopropyl)-2-
  cyclohexen-1-one, **VI**, 1024
Trimethyl-*p*-benzoquinone, **VI**,
  1010
2,4,4-Trimethylpentanal, **VI**,
  1028
Triphenylarsine oxide, **IV**, 911
2,4,6-Triphenylphenoxyl, **V**, 1130
Tropylium fluoborate, **V**, 1138
OZONOLYSIS
3,8-Dimethoxy-4,5,6,7-dibenzo-1,2-
  dioxacyclooctane, **V**, 493
6,6-Dimethoxyhexanal, **VII**, 168
Diphenaldehyde, **V**, 489
Diphenaldehydic acid, **V**, 493
5-Formyl-4-phenanthroic acid, **IV**,
  484
Methyl 6,6-dimethoxyhexanoate,
  VII, 168
Methyl 6-oxohexanoate, **VII**, 168
PALLADIUM-CATALYZED
  REACTIONS
1-Acetoxy-4-benzylamino-2-butene,
  VIII, 9
cis-1-Acetoxy-4-chloro-2-cyclohex-
  ene, **VIII**, 5
cis-1-Acetoxy-4-(dicarbomethoxy)-
  2-cyclohexene, **VIII**, 5
1-Acetoxy-4-diethylamino-2-butene,
  VIII, 9
cis-3-Acetoxy-5-hydroxycyclopent-
  1-ene, **VIII**, 13
1-(4-*tert*-Butylcyclohexen-1-yl)-2-
  propen-1-one, **VIII**, 97
4-*tert*-Butyl-1-vinylcyclohexene,
  VIII, 97
Cholesta-3,5-diene, **VIII**, 126
2-Cyclohepten-1-yl acetate, **VIII**, 137

4,13-Diaza-18-crown-6, **VIII**, 152
Dimethyl (*E*)-2-hexenedioate, **VIII**,
  112
2,2-Dimethyl-4-oxopentanal, **VIII**,
  208
Ethyl (*E*)-4-(4-nitrophenyl)-4-oxo-2-
  butenoate, **VIII**, 268
Ethyl 5-oxo-6-methyl-6-heptenoate,
  VIII, 274
(5*Z*,7*E*)-5,7-Hexadecadiene, **VIII**,
  295
2-Methyl-4'-nitrobiphenyl, **VIII**,
  430
(1*Z*,3*E*)-1-Phenyl-1,3-octadiene,
  VIII, 532
(*Z*)-4-(Trimethylsilyl)-3-buten-1-ol,
  VIII, 609
PHOSPHONATION (formation of
  C–P, N–P, and O–P bonds)
(*S*)-(+)- and (*R*)-(–)-1,1'-Binaphthyl-
  2,2'-diyl hydrogen phosphate,
  VIII, 51
(*E*)-Buten-3-ynyl-2,6,6-trimethyl-1-
  cyclohexene, **VII**, 63
17β-*tert*-Butoxy-5α-androst-2-ene,
  VII, 66
tert-Butyl thiocholate, **VII**, 81
6-Diethylphosphonomethyl-2,2-
  dimethyl-1,3-dioxen-4-one,
  VIII, 192
Diethyl phthalimidomethylphospho-
  nate, **VIII**, 451
Diisopropyl methylphosphonate, **IV**,
  325
O-Diphenylphosphinylhydroxylamine,
  VII, 9
Diphenyl phosphorazidate, **VII**, 206
Furyl phosphorodichloridate, **VIII**,
  396
Furyl *N,N,N',N'*-tetramethyldiami-
  dophosphate, **VIII**, 396
Methyl 2-(diethylphosphoryloxy)-1-
  cyclohexene-1-carboxylate,
  VII, 351
Phenyldichlorophosphine, **IV**, 784

Trisammonium geranyl diphosphate,
   VIII, 616
PHOTOCHEMICAL REACTIONS
7-Acetoxy-4,4,6,7-tetramethylbicy-
   clo[4.2.0]octan-2-one, **VI**, 1024
3-Acetyl-4- and 5-chloro-2-oxazo-
   lidinone, **VII**, 5
3,1-Benzoxazepine, **VII**, 23
Bicyclo[2.1.0]pent-2-ene, **VI**, 145
1-Bromo-1-methyl-2-(bromomethylsulfonyl)cyclohexane, **VIII**,
   212
tert-Butyl *N*-(1-ethoxycyclopropyl)-
   carbamate, **VI**, 226
3-Chlorothietane 1,1-dioxide, **VII**,
   491
Cyclohexylidenecyclohexane, **V**, 297
1-Deoxy-2,3,4,6-tetra-*O*-acetyl-1-(2-
   cyanoethyl)-α-D-glucopyranose,
   VIII, 148
3,3-Dichlorothietane 1,1-dioxide,
   VII, 491
Diethyl phenylphosphonate, **VI**, 451
Diethyl propionylsuccinate, **VI**, 615
3,3-Dimethyl-*cis*-bicyclo[3.2.0]hep-
   tan-2-ol, **VII**, 177
1,3-Dimethyl-3-methoxy-4-phenyl-
   azetidinone, **VIII**, 216
cis- and *trans*-1,2-Divinylcyclo-
   butane, **V**, 528
3-(1-Hydroxybutyl)-1-methylpyr-
   role, **VII**, 102
p-Menth-8-en-1-yl methyl ether,
   VII, 304
6-Methylbicyclo[4.2.0]octan-2-one,
   VII, 315
D-Norandrost-5-en-3β-ol-16-car-
   boxylic acids, **VI**, 840
9-Phenylphenanthrene, **V**, 952
1-(Phenylsulfonyl)cyclopentene,
   VIII, 543
Quadricyclane, **VI**, 962
Ricinelaidic acid, **VII**, 470
2,2,3,3-Tetramethyliodocyclopro-
   pane, **VI**, 974

Trichloromethyl chloroformate, **VI**,
   715
REARRANGEMENTS
1-Acetylcyclohexene, **III**, 22
Acetyl methylurea, **II**, 462
5-Acetyl-1,2,3,4,5-pentamethylcy-
   clopentadiene, **VI**, 39
Adamantane, **V**, 16
β-Alanine, **II**, 19
Allantoin, **II**, 21
2-Allylcyclohexanone, **V**, 25
(*S*)-(−)-1-Amino-2-methoxymethyl-
   pyrrolidine, **VIII**, 26
4-Aminoveratrole, **II**, 44
Azoethane, **VI**, 78
Benzilic acid, **I**, 89
β-Benzopinacolone, **II**, 73
3,1-Benzoxazepine, **VII**, 23
ε-Benzoylaminocaproic acid, **II**, 76
Benzyl *trans*-1,3-butadiene-1-carba-
   mate, **VI**, 95
Benzyl sulfide, **VI**, 130
Bicyclo[3.3.1]nonan-9-one, **VI**, 137
tert-Butyl *N*-(1-ethoxycyclopropyl)-
   carbamate, **VI**, 226
9-Butyl-1,2,3,4,5,6,7,8-octahy-
   droacridin-4-ol, **VIII**, 87
3-Carboxy-5-dodecen-2-one, **VIII**,
   235
Carroll rearrangement, **VIII**, 235
Chloro-*p*-benzoquinone, **IV**, 148
(*Z*)-4-Chloro-4-hexenyl trifluoroacetate, **VI**, 273
1-Chloronorbornane, **VI**, 846
5β-Cholest-3-ene, **VI**, 293
5β-Cholest-3-ene-5-acetaldehyde,
   VI, 298
Δ⁴-Cholesten-3-one, **IV**, 192
Citraconic anhydride, **II**, 140
7-Cyanoheptanal, **V**, 266
Cyclobutanol, **VII**, 114, 129
Cyclobutanone, **VI**, 320
Cyclobutylamine, **V**, 273
Cyclobutylamine hydrochloride,
   VIII, 132

1,2-Cyclononadiene, **V**, 306
Cyclopentanecarboxaldehyde, **V**, 320
2-Cyclopentenone, **V**, 326
Cyclopropanecarboxaldehyde, **VII**, 129
Cycloundecanecarboxylic acid, **VII**, 135
Cycloundecanone, **VI**, 368
m-Cymene, **V**, 332
9-Decyn-1-ol, **VII**, 146
trans-7,8-Diacetoxybicyclo-[4.2.0]octa-2,4-diene, **VI**, 196
trans, trans-1,4-Diacetoxy-1,3-butadiene, **VI**, 196
Diamantane, **VI**, 378
5,6-Diethoxybenzofuran-4,7-dione, **VIII**, 179
2,2-Diethoxy-2-(4-pyridyl)ethylamine, **VII**, 149
(*R*)-(–)-*N,N*-Diethyl-(*E*)-citronellalenamine, **VIII**, 183
N,N-Diethyl-1,2,2-trichlorovinylamine, **V**, 387
3,4-Dihydro-1(2*H*)-azulenone, **VIII**, 196
2,5-Dihydroxyacetophenone, **III**, 280
1-Dimethylamino-4-methyl-3-pentanone, **VI**, 474
N,N-Dimethyl-5β-cholest-3-ene-5-acetamide, **VI**, 491
1,2-Dimethylenecyclohexane, **VIII**, 212
Dimethylketene β-lactone dimer, **V**, 456
2,2-Dimethyl-4-pentenal, **VII**, 177
Diphenylacetaldehyde, **IV**, 375
Ethyl 4-amino-3-(methylthiomethyl)benzoate, **VI**, 581
Ethyl 5β-cholest-3-ene-5-acetate, **VI**, 584
Ethyl (*E,Z*)-2,4-decadienoate, **VIII**, 251
Ethyl 3,4-decadienoate, **VIII**, 251
Ethyl 2-methyl-3-methylthioindole-5-carboxylate, **VI**, 601

Ethyl 4-methyl-(*E*)-4,8-nonadienoate, **VI**, 606
Ethyl 1-naphthylacetate, **VI**, 613
o-Eugenol, **III**, 418
Flavone, **IV**, 478
Hexahydro-2-(1*H*)-azocinone, **VII**, 254
2,3,4,5,6,6-Hexamethyl-2,4-cyclohexadien-1-one, **V**, 598
p-Hydroxybenzoic acid, **II**, 341
3-Hydroxy-1-cyclohexene-1-carboxaldehyde, **VIII**, 309
o-Hydroxydibenzoylmethane, **IV**, 479
2-Hydroxy-3-methylisocarbostyril, **V**, 623
(1-Hydroxy-2-propenyl)trimethylsilane, **VIII**, 501
2-Indanone, **V**, 647
(*E*)-1-Iodo-4-phenyl-2-butene, **VI**, 704
Isocrotonic acid, **VI**, 711
2-Ketohexamethylenimine, **II**, 371
Mesaconic acid, **II**, 382
2-Methylbenzyldimethylamine, **IV**, 585
Methyl cyclopentanecarboxylate, **IV**, 594
Methylhexylcarbinol, **I**, 366
4-Methyl-7-hydroxy-8-acetylcoumarin, **III**, 283
Methyl isopropyl ketone, **II**, 408
Methyl *N*-(2-methyl-2-butenyl)carbamate, **VIII**, 427
Methyl 3-methyl-2-furoate, **IV**, 649
1-Methyl-2-pyridone, **II**, 419
2,6-Naphthalenedicarboxylic acid, **V**, 813
2-Naphthalenethiol, **VI**, 824
2-Oxo-1-phenyl-3-pyrrolidinecarboxylic acid, **VII**, 411
Phenacylamine hydrochloride, **V**, 909
4-Phenyl-5-anilino-1,2,3-triazole, **IV**, 380
1-Phenylcyclopentylamine, **VI**, 910

2-Phenylethyl benzoate, **V**, 336
Phenylglyoxal hemimercaptal, **V**, 937
1-Phenyl-1,3-pentadiyne, **VI**, 925
3-Phenyl-4-pentenal, **VIII**, 536
3-Phenyl-4-pentenoic acid, **VII**, 164
trans-3-(Phenylsulfonyl)-4-(chloromercuri)cyclohexene, **VIII**, 540
Phenylurea, **I**, 453
unsym-o-Phthalyl chloride, **II**, 528
Pinacolone, **I**, 462
(–)-β-Pinene, **VIII**, 553
trans-Pinocarveol, **VIII**, 948
(*E*)-2-(1-Propenyl)cyclobutanone, **VIII**, 556
(*E*)-2-Propenylcyclohexanol, **VII**, 456
o- and *p*-Propiophenol, **II**, 543
2-Propyl-1-azacycloheptane, **VIII**, 568
Ricinelaidic acid, **VII**, 470
Tetra-*O*-acetyl-2-deoxy-α-D-glucopyranose, **VIII**, 583
Trimethyl 2-chloro-2-cyclopropylidenorthoacetate, **VIII**, 373
2,4,4-Trimethylcyclopentanone, **IV**, 957
Xanthone, **I**, 552

REDUCTION

A. C≡C → C=C
Ethyl isocrotonate, **VII**, 226
(*E*)-3-Trimethylsilyl-2-propen-1-ol, **VII**, 524
(*Z*)-4-(Trimethylsilyl)-3-buten-1-ol, **VIII**, 609

B. C=C → HC–CH
N-Acetylhexahydrophenylalanine, **II**, 493
N-Acetylphenylalanine, **II**, 493
Adamantane, **V**, 16
4-Aminocyclohexanecarboxylic acid, *cis*–trans mixture, **V**, 670
Benzylacetophenone, **I**, 101
Bicyclo[2.1.0] pentane, **V**, 96
β-(*o*-Carboxyphenyl)propionic acid, **IV**, 136
Cyclodecanone, **V**, 277

cis-Cyclododecene, **V**, 281
Dicyclohexyl-18-crown-6-polyether, **VI**, 395
Diethyl 2,3-diazabicyclo[2.2.1]heptane-2,3-dicarboxylate, **V**, 97
Diethyl *cis*-hexahydrophthalate, **IV**, 304
Dihydrocarvone, **VI**, 459
trans-1,2-Dihydrophthalic acid, **VI**, 461
Diethyl methylenemalonate, **IV**, 298
Diethyl succinate, **V**, 993
9,10-Dihydroanthracene, **V**, 398
1,4-Dihydrobenzoic acid, **V**, 400
Dihydrocholesterol, **II**, 191
9,10-Dihydrophenanthrene, **IV**, 313
1,2-Dimethyl-1,4-cyclohexadiene, **V**, 467
Ethyl α-acetyl-β-(2,3-dimethoxyphenyl)propionate, **IV**, 408
Ethyl butylcyanoacetate, **III**, 385
β-Furylpropionic acid, **I**, 313
Hexahydrogallic acid and hexahydrogallic acid triacetate, **V**, 591
Hydrocinnamic acid, **I**, 311
Isopinocampheol, **VI**, 719
3-Isoquinuclidone, **V**, 670
5-(3-Methoxy-4-hydroxybenzyl)creatinine, **III**, 587
2-Methyl-1,3-cyclohexanedione, **V**, 743
Methyl *anti*-3-hydroxy-2-methylpentanoate, **VIII**, 420
$\Delta^{9,10}$-Octalin, **VI**, 852
2,2,7,7,12,12,17,17-Octamethyl-21,22,23,24-tetraoxaperhydroquaterene, **VI**, 856
DL-β-Phenylalanine, **II**, 489, 491
γ-Propylbutyrolactone, **III**, 742
3-Quinuclidone hydrochloride, **V**, 989
Succinic acid, **I**, 64
endo-Tetrahydrodicyclopentadiene, **V**, 16
Tetrahydrofuran, **II**, 566
β-(Tetrahydrofuryl)propionic acid, **III**, 742

1,4,5,8-Tetrahydronaphthalene, **VI**, 731
ar-Tetrahydro-α-naphthol, **IV**, 887
ac-Tetrahydro-β-naphthylamine, **I**, 499
Tetrahydropyran, **III**, 794
β-Tetralone, **IV**, 903

C. C=O → CH$_2$
Androstan-17β-ol, **VI**, 62
Anthrone, **I**, 60
Benz[*a*]anthracene, **VII**, 18
Cholestane, **VI**, 289
Creosol, **IV**, 203
1,10-Diazacyclooctadecane, **VI**, 382
Docosanedioic acid, **V**, 533
Hendecanedioic acid, **IV**, 510
o-Heptylphenol, **III**, 444
2-Methylcyclopentane-1,3-dione, **V**, 747
m-Nitroethylbenzene, **VII**, 393
γ-Phenylbutyric acid, **II**, 499
Phthalide, **II**, 526

D. C=O → CHOH
3β–Acetoxy-20β-hydroxy-5-pregnene, **V**, 692
Alloxantin dihydrate, **IV**, 25
Benzohydrol, **I**, 90
Benzyl alcohol, **II**, 591
cis-4-*tert*-Butylcyclohexanol, **VI**, 215
trans-4-*tert*-Butylcyclohexanol, **V**, 175
1,2-Cyclobutanediol, **VII**, 129
1,2-Cyclodecanediol, **IV**, 216
Cyclohexanone, **V**, 294
trans-9,10-Dihydro-9,10-phenanthrenediol, **VI**, 887
(*S*)-(+)-Ethyl 3-hydroxybutanoate, **VII**, 215
2-Heptanol, **II**, 317
Heptyl alcohol, **I**, 304
Hexahydroxybenzene, **V**, 595
(*S*)-(+)-3-Hydroxy-2,2-dimethylcyclohexanone, **VIII**, 312
erythro-1-(3-Hydroxy-2-methyl-3-phenylpropanoyl)piperidine, **VIII**, 326

threo-(3-Hydroxy-2-methyl-3-phenylpropanoyl)piperidine, **VIII**, 328
3-Methyl-2-cyclohexen-1-ol, **VI**, 769
(1*RS*,2*SR*,5*R*)-5-Methyl-2-(1-methyl-1-phenylethyl)cyclohexanol, **VIII**, 522
3-Methyl-1,5-pentanediol, **IV**, 660
(*R*)-(+)-1-Octyn-3-ol, **VII**, 402
D-(–)-Pantoyl lactone, **VII**, 417
Piperonyl alcohol, **II**, 591
L-Propylene glycol, **II**, 545
7-Thiomenthol, **VIII**, 302
p-Tolylcarbinol, **II**, 590
Trichloroethyl alcohol, **II**, 598
Xanthydrol, **I**, 554

E. CO$_2$R → CH$_2$OH
Cetyl alcohol, **II**, 374
Decamethylene glycol, **II**, 154
2,3-Di-*O*-isopropylidene-L-threitol, **VIII**, 155
Heptamethylene glycol, **II**, 155
Hexamethylene glycol, **II**, 325
Lauryl alcohol, **II**, 372
Myristyl alcohol, **II**, 374
Nonamethylene glycol, **II**, 155
Octadecamethylene glycol, **II**, 155
Oleyl alcohol, **II**, 468; **III**, 671
(*S*)-(+)-Propane-1,2-diol, **VII**, 356
2-(1-Pyrrolidyl)propanol, **IV**, 834
Tetradecamethylene glycol, **II**, 155
Tridecamethylene glycol, **II**, 155
Undecamethylene glycol, **II**, 155
Undecylenyl alcohol, **II**, 374

F. CO$_2$H → CH$_2$OH
(*S*)-2-Chloropropan-1-ol, **VIII**, 434
Ethyl 4-hydroxycrotonate, **VII**, 221
(*S*)-(+)-2-Hydroxymethylpyrrolidine, **VIII**, 26
(*S*)-Phenylalanol, **VIII**, 528
L-Valinol, **VII**, 530

G. COCl → CHO
6-Oxodecanal, **VIII**, 498

H. CONR$_2$ → CH$_2$NR$_2$

(S,S)-(+)-1,4-Bis(dimethylamino)-
  2,3-dimethoxybutane (DDB),
  VII, 41
N,N-Dimethylcyclohexylmethyl-
  amine, **IV**, 339
2,2-Dimethylpyrrolidine, **IV**, 354
Laurylmethylamine, **IV**, 564

I. CN → CH_2NH_2
Decamethylenediamine, **III**, 229
2,4-Diphenylpyrrole, **III**, 358
β-Phenylethylamine, **III**, 720

J. CX → CH
Benzene, **V**, 998
17β-*tert*-Butoxy-5α-androst-2-ene,
  VII, 66
3-Butylcyclobutenone, **VIII**, 82
3-Chlorobicyclo[3.2.1]oct-2-ene, **VI**,
  142
Cholesta-3,5-diene, **VIII**, 126
Decane, **VI**, 376
Deoxyanisoin, **V**, 339
3-Deoxy-1,2:5,6-di-O-isopropyli-
  dene-α-D-ribo-hexofuranose,
  VII, 139
2,3-Diaminopyridine, **V**, 346
7,7-Dimethoxybicyclo[2.2.1]hep-
  tene, **V**, 424
Diphenylacetic acid, **I**, 224
Dodecane, **VI**, 376
Hexadecane, **II**, 320
Indazole, **III**, 475
Lepidine, **III**, 519
Mesitaldehyde, **III**, 551
Methylene bromide, **I**, 357
Methylene iodide, **I**, 358
5-Methylfurfural, **II**, 393
β-Naphthaldehyde, **III**, 627
Naphthalene, **VI**, 821
1,3,4,6-Tetra-O-acetyl-2-deoxy-α-D-
  glucopyranose, **VIII**, 583
Thiophosgene, **I**, 509
o-Tolualdehyde, **III**, 818
Tricyclo[4.4.1.0^{1,6}]undeca-3,8-diene,
  VI, 731
3,4,5-Trimethoxybenzaldehyde, **VI**,
  1007

Triphenylmethane, **I**, 548

K. C=NOH → CH_2NH_2
Aminomalononitrile p-toluenesul-
  fonate, **V**, 32
Butylamine, **II**, 319
sec-Butylamine, **II**, 319
Cyclohexylamine, **II**, 319
Diethyl acetamidomalonate, **V**, 373
Diethyl aminomalonate hydrochlo-
  ride, **V**, 376
Heptylamlne, **II**, 318

L. NO → NH_2
1,2-Aminonaphthol hydrochloride,
  II, 33
2,4-Dimethyl-3,5-dicarbethoxypyr-
  role, **II**, 202
$unsym$-Dimethylhydrazine
  hydrochloride, **II**, 211
α-Methyl-α-phenylhydrazine, **II**, 418

M. NO_2 → NHOH
β-Phenylhydroxylamine, **I**, 445;
  VIII, 16

N. NO_2 → NH_2
o-Aminobenzaldehyde, **III**, 56
p-Aminobenzaldehyde, **IV**, 31
m-Aminobenzaldehyde dimethylace-
  tal, **III**, 59
o-Aminobiphenyl, **V**, 829
2-Amino-p-cymene, **III**, 63
5-Amino-2,3-dihydro-1,4-phthal-
  azinedione, **III**, 69
2-Aminofluorene, **V**, 30; **II**, 448
Aminoguanidine bicarbonate, **III**, 73
2-Amino-4-nitrophenol, **III**, 82
p-Aminophenylacetlc acid, **I**, 52
p-Aminophenyl disulfide, **III**, 86
m-Bromobenzaldehyde, **II**, 132
m-Chlorobenzaldehyde, **II**, 130
Cycloheptanone, **IV**, 221
4,4'-Diaminodiphenylsulfone, **III**, 239
Diaminodurene chlorostannate, **II**,
  255
1,2-Diamino-4-nitrobenzene, **III**,
  242
2,3-Diamino-5-bromopyridine, **V**,
  346

2,4-Diaminotoluene, **II**, 160
2,6-Dibromo-4-aminophenol
   chlorostannate, **II**, 175
5,5-Dimethyl-2-pyrrolidone, **IV**, 357
Ethyl *p*-aminobenzoate, **I**, 240
m-Hydroxybenzaldehyde, **III**, 453
Orthanilic acid, **II**, 471
o-Phenylenediamine, **II**, 501
Phloroglucinol, **I**, 455
Semicarbazide sulfate, **I**, 485
Sodium 2-aminobenzenesulfinate, **V**,
   61
2,4,5-Triaminonitrobenzene, **V**, 1067
2,4,6-Triphenylaniline, **V**, 1130
2,4,6-Triphenylphenoxyl, **V**, 1130
Uramil, **II**, 617
O. $NO_2 \rightarrow N=N$
Azobenzene, **III**, 103
P. $N=N \rightarrow NH_2$
1,2-Aminonaphthol hydrochloride,
   II, 35
1,4-Aminonaphthol hydrochloride,
   II, 39
Q. Miscellaneous
p-Acetaminobenzenesulfinic acid, **I**,
   7
Acetone hydrazone, **VI**, 10
Acetophenone hydrazone, **VI**, 12
Adamantylideneadamantane, **VII**, 1
Alloxantin dihydrate, **III**, 42
2-Allyl-3-methylcyclohexanone, **VI**,
   51
o-Aminobenzyl alcohol, **III**, 60
Aminomalononitrile *p*-toluenesul-
   fonate, **V**, 32
1,4-Aminonaphthol hydrochloride, **I**,
   49
DL-β-Amino-β-phenylpropionic acid,
   III, 91
Aminothymol, **I**, 512
Arsenoacetic acid, **I**, 74
Azoxybenzene, **II**, 57
Benzeneselenenyl chloride, **VI**, 533
Benzopinacol, **II**, 71
Benzoylpiperidine, **I**, 99

5-Benzylaminotetrazole, **VII**, 27
3-Benzylindole, **VI**, 109
4-Benzyloxyindole, **VII**, 34
Benzyl sulfide, **VI**, 130
Bicyclo[4.3.0]non-1-en-4-one, **VIII**,
   38
BINAP, **VIII**, 57
R-(+)-1,1'-Binaphthalene-2,2'-diol,
   VIII, 46
Biphenyl, **VI**, 150
Birch reduction of *o*-Anisic acid,
   VII, 249
1,1-Bis(bromomethyl)cyclopropane,
   VI, 153
17β-*tert*-Butoxy-5α-androst-2-ene,
   VII, 66
N-Boc-Leucinal, **VIII**, 68
tert-Butylamine, **III**, 148
2-*tert*-Butyl-1,3-diaminopropane,
   VI, 223
N-*tert*-Butylhydroxylamine, **VI**,
   803
tert-Butyl-*tert*-octylamine, **VIII**,
   93
(−)-(D)-2,10-Camphorsultam, **VIII**,
   110
Chloro-*p*-benzoquinone, **IV**, 148
Chloronicotinonitrile, **IV**, 166
4-(4-Chlorophenyl)butan-2-one, **VII**,
   105
5β-Cholest-3-ene, **VI**, 293
2-Cyclobutyl-*cis*-4-*trans*-5-dimethyl-
   1,3-dioxolane, **VI**, 312
Cyclodecanone, **IV**, 218
trans-Cyclododecene, **VII**, 121
Cyclododecylmercaptan, **VII**, 124
cis-1,2-Cyclohexanedimethanol, **VI**,
   482
2-Cyclohexyloxyethanol, **V**, 303
1-Decalol, **VI**, 371
2,5-Diamino-3,4-dicyanothiophene,
   IV, 243
Diaminouracil hydrochloride, **IV**,
   247
4,13-Diaza-18-crown-6, **VIII**, 152

2-(1,3-Dibenzoylimidazolidin-2-yl)-
  imidazole hydrochloride, **VII**,
  287
Di-*tert*-butyl nitroxide, **V**, 355
2,2-Dichloroethanol, **IV**, 271
4-[(*E*)-1,2-Dichlorovinyl]-4-methyl-
  2-cyclohexen-1-one, **VII**, 241
Diethyl mercaptoacetal, **IV**, 295
N,N-Diethyl-1,2,2-trichlorovinyl-
  amine, **V**, 387
Dihydroresorcinol, **III**, 278
N,N-Dimethylcyclohexylamine, **VI**,
  499
1,5-Dimethyl-2-pyrrolidone, **III**,
  328
3,5-Dinitrobenzaldehyde, **VI**, 529
1,1-Diphenylcyclopropane, **V**, 509
1,1-Diphenylethane, **VI**, 537
2-(Diphenylhydroxymethyl)pyrroli-
  dine, **VI**, 542
Disodium tetracarbonylferrate
  sesquidioxanate, **VI**, 807
1,3-Dithiane, **VI**, 556
Ethyl 4-amino-3-methylbenzoate,
  VI, 581
Ethyl *p*-dimethylaminophenylace-
  tate, **V**, 552
3-Ethyl-4-hydroxymethyl-5-
  methylisoxazole, **VI**, 781
Ethyl 2-methylindole-5-carboxylate,
  VI, 601
N-Ethyl-*m*-toluidine, **II**, 290
4-Formylbenzenesulfonamide, **VI**,
  631
Glycine *tert*-butyl ester, **V**, 586
Hemimellitene, **IV**, 508
2-Hydroxycyclobutanone, **VI**, 167
Methanesulfonyl cyanide, **VI**, 727
1-Methoxy-1,4-cyclohexadiene, **VI**,
  996
o-Methoxyphenylacetone, **IV**, 573
2-Methylbiphenyl, **VI**, 747
1-Methylcyclohexanol, **VI**, 766
N-Methyl-2,3-dimethoxybenzyl-
  amine, **IV**, 603

Methylene bromide, **I**, 357
Methylene iodide, **I**, 358
1-Methyl-4,4a,5,6,7,8-hexahydro-
  naphthalen-2(3*H*)-one, **VI**, 781
4-Methyl-6-hydroxypyrimidine, **IV**,
  638
(*R,S*)-Mevalonolactone-2-^{13}C, **VII**,
  386
β-Naphthaldehyde, **III**, 626
1,5-Naphthalenedithiol, **IV**, 695
m-Nitrophenyl disulfide, **V**, 843
Nitrobenzene, **III**, 668
Palladium catalyst for partial reduction of acetylenes, **V**, 887
1,5-Pentanediol, **III**, 693
2-Phenyl-2-adamantanamine
  hydrochloride, **VII**, 433
2-(1-Phenylcyclopentyl)-4,4,6-
  trimethyltetrahydro-1,3-
  oxazine, **VI**, 905
α-Phenylethylamine, **III**, 717
Phenylhydrazine, **I**, 442
3-Phenyl-1-propanol, **IV**, 798
Pimelic acid, **II**, 535
Pinacol hydrate, **I**, 459
(*E*)-2-Propenylcyclohexanol, **VII**, 456
γ-Propylbutyrolactone, **III**, 742
Sodium dicarbonyl(cyclopentadienyl)ferrate, **VIII**, 479
Sodium *p*-toluenesulfinate, **I**, 492
trans-Stilbene, **III**, 786
2,6':6',2''-Terpyridine, **VII**, 476
Tetrahydro-BINOR-S, **VI**, 378
ar-Tetrahydro-α-naphthol, **IV**, 887
2,2,6,6-Tetramethylolcyclohexanol,
  IV, 907
7-Thiomenthol, **VIII**, 302
Thiophenol, **I**, 504
Thiophosgene, **I**, 509
Thiosalicylic acid, **II**, 580
m-Tolylbenzylamlne, **III**, 827
7-Thiomenthol, **VIII**, 302
REPLACEMENT REACTIONS
  A. Alkoxy by Amino
    Benzoylacetanilide, **III**, 108

Lactamide, **III**, 516
Mandelamide, **III**, 536
Salicyl-*o*-toluide, **III**, 765
B. Alkoxy by Oxygen
Acrylic acid, **III**, 33
Allyl lactate, **III**, 46
Butyl acrylate, **III**, 146
Methyl myristate, **III**, 605
Methyl palmitate, **III**, 605
Nortricyclanol, **V**, 863
Vinyl laurate, **IV**, 977
C. Amino by Halogen
(*S*)-2-Chloropropanoic acid, **VIII**, 160
D. Amino by Oxygen
(S)-(+)-γ-Butyrolactone-γ-carboxylic
acid, **VII**, 99
2-Phenylethyl benzoate, **V**, 336
E. Ammonium by Oxygen
o-Methylbenzyl acetate, **IV**, 582
(*E*)-3-[(*E*)-3-Phenyl-2-propenoxy]-
acrylic acid, **VIII**, 536
F. Ether by Halogen
1-Bromo-1-ethoxycyclopropane,
VIII, 556
1,2-Di-*tert*-butoxy-1,2-dichloro-
ethane, **VIII**, 161
Methyl 4-*O*-benzoyl-6-bromo-6-
deoxy-α–D-glucopyranoside,
VIII, 363
G. Halogen by Amino
1-Acetoxy-4-benzylamino-2-butene,
VIII, 9
1-Acetoxy-4-diethylamino-2-butene,
VIII, 9
DL-Alanine, **I**, 23
Aminoacetal, **III**, 50
9-Aminoacridine, **III**, 53
2-Aminobenzophenone, **IV**, 34
α-Aminocaproic acid, **I**, 48
DL-ε-Benzoyllysine, **II**, 374
2-Bromoallylamine, **V**, 121
N-(2-Bromoallyl)ethylamine, **V**, 124
1,5-Diamino-2,4-dinitrobenzene, **V**,
1068
1,10-Diazacyclooctadecane-2,9-
dione, **VI**, 382

1,10-Dibenzyl-4,7-dioxa-1,10-
diazadecane, **VIII**, 152
β-Dibutylaminoethylamine, **III**, 254
β-Diethylaminoethyl alcohol, **II**, 183
N,*N*'-Diethylsulfamide, **VI**, 78
3,4-Dimethylaniline, **III**, 307
2,4-Dinitroaniline, **II**, 221
2,6-Dinitroaniline, **IV**, 364
2,4-Dinitrophenylhydrazine, **II**, 228
Ethyl *N*-methylcarbamate, **II**, 278
Glycine, **I**, 300
Hippuric acid, **II**, 328
Isobutyramide, **III**, 490
DL-Isoleucine, **III**, 495
DL-Leucine, **III**, 523
Methyliminodiacetic acid, **II**, 397
Methyl sebacamate, **III**, 613
DL-Phenylalanine, **III**, 705
N-Phenylanthranilic acid, **II**, 15
DL-Serine, **III**, 774
Taurine, **II**, 563
DL-Threonine, **III**, 813
2,4,5-Triaminonitrobenzene, **V**, 1067
Triphenylamine, **I**, 544
DL-Valine, **III**, 848
H. Halogen by Azide
4-Azido-3-chloro-5-isopropoxy-
2(5*H*)-furanone, **VIII**, 179
tert-Butyl azidoacetate, **V**, 586
Diphenyl phosphorazidate, **VII**, 206
I. Halogen by Carbon
2-(1-Acetyl-2-oxopropyl)benzoic
acid, **VI**, 36
J. Halogen by Cyanide
Allyl cyanide, **I**, 46
Benzoyl cyanide, **III**, 112
Benzyl cyanide, **I**, 107
γ-Chlorobutyronitrile, **I**, 156
9-Cyanophenanthrene, **III**, 212
β-Ethoxypropionitrile, **III**, 372
Ethyl cyanoacetate, **I**, 254
Ethylene cyanohydrin, **I**, 256
Ethyl tridecylate, **II**, 292
3-Hydroxyglutaronitrile, **V**, 614
Malonic acid, **II**, 376
Mesitylacetic acid, **III**, 557

p-Methoxyphenylacetonitrile, **IV**, 576
α-Naphthonitrile, **III**, 631
Sodium cyanoacetate, **I**, 181
Trimethylene cyanide, **I**, 536
Trimethylsilyl cyanide, **VII**, 517
K. Halogen by Diphosphate
Trisammonium geranyl diphosphate,
VIII, 616
L. Halogen by Halogen
Benzoyl fluoride, **V**, 66
1,2-Bis-(2-iodoethoxy)ethane, **VIII**,
153
2,4-Dinitroiodobenzene, **V**, 478
Ethyl 2-fluorohexanoate, **VIII**, 258
Ethyl 4-iodobutyrate, **VIII**, 274
2-Iodoethyl benzoate, **IV**, 84
1-Iodo-3-trimethylsilylpropane,
VIII, 486
M. Halogen by Nitro
Nitromethane, **I**, 401
N. Halogen by Oxygen
p-Arsonophenoxyacetlc acid, **I**, 75
2-Bromohexanoic acid, **VI**, 192
2-*tert*-Butoxycarbonyloxyimino-2-
phenylacetonitrile, **VI**, 206
3-Chloro-2-buten-1-ol, **IV**, 128
18-Crown-6, **VI**, 301
2-Cyclohepten-1-yl acetate, **VIII**,
137
Dibenzoylmethane, **I**, 205
Diphenylmethyl vinyl ether, **VI**, 552
Ethyl orthoformate, **I**, 258
2-Fluoroheptyl acetate, **V**, 580
β-D-Glucose-2,3,4,6-tetraacetate, **III**,
434
α-Glyceryl phenyl ether, **I**, 296
Hydroxylamine-*O*-sulfonic acid, **VI**,
943
3-Hydroxypyrene, **V**, 632
(Isopropoxydimethylsilyl)methyl
chloride, **VIII**, 315
erythro-3-Methanesulfonyloxy-2-
butyl cyclobutanecarboxylate,
VI, 312
2-Methoxycyclooctanone oxime, **V**,
267

β-Phenoxyethyl bromide, **I**, 436
γ-Phenoxypropyl bromide, **I**, 435
Phthalaldehydic acid, **III**, 737
Terephthalaldehyde, **III**, 788
Tropolone, **VI**, 1037
O. Halogen by Phosphine Oxide
(±)-2,2'-Bis(diphenylphosphinyl)-
1,1'-binaphthyl [(±)-BINAPO],
VIII, 57
P. Halogen by Phosphonate
6-Diethylphosphonomethyl-2,2-
dimethyl-1,3-dioxen-4-one,
VIII, 192
Q. Halogen by SO$_3$H
Sodium 2-bromoethanesulfonate, **II**,
558
Taurine, **II**, 564
R. Halogen by Various Groups
Acetol, **II**, 5
γ-Aminobutyric acid, **II**, 25
Arsonoacetic acid, **I**, 73
Benzophenone, **I**, 95
β-Bromoethylphthalimide, **I**, 119
Diallylcyanamide, **I**, 203
Dibutylcyanamide, **I**, 204
2,4-Dinitrophenyl benzyl sulfide, **V**,
474
Di-*o*-nitrophenyl disulfide, **I**, 220
Diphenylmethane, **II**, 232
Diphenyltriketone, **II**, 244
Ethoxyacetic acid, **II**, 260
Ethyl benzoylacetoacetate, **II**, 266
Ethyl benzoyldimethylacetate, **II**, 268
Ethyl butylacetoacetate, **I**, 248
Ethyl butylmalonate, **I**, 250
Ethyl isobutyrylisobutyrate, **II**, 270
Ethyl 1-methylthio-3-phthalimido-
propane-3,3-dicarboxylate, **II**,
384
Ethyl phthalimidomalonate, **I**, 271
α-Glyceryl phenyl ether, **I**, 296
β-Hydroxyethyl methyl sulfide, **II**,
345
Isopropyl thiocyanate, **II**, 366
p-Nitro-*p*'-acetylaminodiphenylsul-
fone, **III**, 239

Phenylacetyl isothiocyanate, **V**, 1051
2-Phenylpyridine, **II**, 517
Sodium *p*-arsono-*N*-phenylglycin-
    amide, **I**, 488
1,1,1',1'-Tetraethoxyethyl polysul-
    fide, **IV**, 295
Thiobenzophenone, **II**, 573
β-Thiodiglycol, **II**, 576
Tricarbethoxymethane, **II**, 594
Tricarbomethoxymethane, **II**, 596
S. Hydroxy by Amino
Cyclobutanecarboxamide, **VIII**, 132
cis-2-(2-Propenyl)cyclopentylamine,
    VII, 501
T. Hydroxy by Azide
2-Azido-2-phenyladamantane, **VII**,
    433
U. Hydroxy by Carbon
Ethyl 3,4-decadienoate, **VIII**, 251
V. Hydroxy by Halogen
2-Acetamido-3,4,6-tri-*O*-acetyl-2-
    deoxy-α-D-glucopyranosyl
    chloride, **V**, 1
(*S*)-(−)-2-Acetoxy-1-bromopropane,
    VII, 356
Acetylmandelyl chloride, **I**, 13
Adipic acid monomethyl ester
    monochloride, **VIII**, 441
Allyl bromide, **I**, 27
2-Aminobenzophenone, **IV**, 34
Amyl iodide, **II**, 402
Isoamyl bromide, **I**, 27
3-Benzoylpyridine, **IV**, 88
Benzyl chloromethyl ether, **VI**, 101
β-Bromoethylamine hydrobromide,
    II, 91
β-Bromoethylphthalimide, **IV**, 106
2-Bromohexanoyl chloride, **VI**, 190
α-(Bromomethyl)acrylic acid, **VII**,
    210
N-Bromomethylphthalimide, **VIII**,
    536
2-Bromonaphthalene, **V**, 142
β-Bromopropionic acid, **I**, 131
Butyl bromide, **I**, 28, 37
sec-Butyl bromide, **I**, 38; **II**, 359

tert-Butyl bromide, **I**, 38
Butyl chloride, **I**, 142
sec-Butyl chloride, **I**, 143
tert-Butyl chloride, **I**, 144
tert-Butyl cyanoacetate, **V**, 171
Butyl iodide, **II**, 402
Butyryl chloride, **I**, 147
DL-10-Camphorsulfonyl chloride,
    V, 196
β-Carbomethoxypropionyl chloride,
    III, 169
N-Chlorobetainyl chloride, **IV**, 154
β-Chloroethyl methyl sulfide, **II**,
    136
3-Chloroindazole, **III**, 476
2-Chlorolepidine, **III**, 194
4-Chloromethyl-3-ethyl-5-
    methylisoxazole, **VI**, 78
α-Chlorophenylacetic acid, **IV**, 169
Cinnamyl bromide, **V**, 249
Crotyl diazoacetate, **V**, 258
Cyclohexyl bromide, **II**, 247
Decamethylene bromide, **III**, 227
Desyl chloride, **II**, 159
2,2'-Dibromo-1,1'-binaphthyl, **VIII**,
    57
β,β-Dibromoisobutyric acid, **VII**,
    319
β-Dibutylaminoethyl bromide hydro-
    bromide, **II**, 92
1,7-Dichloro-4-heptanone, **IV**, 279
4,7-Dichloroquinoline, **III**, 272
β-Diethylaminoethyl bromide
    hydrochloride, **II**, 92
Diethyl 3-*N*-(3-chloropropyl)imin-
    odipropanoate, **VI**, 75
1,6-Diiodohexane, **IV**, 323
β-Dimethylaminoethyl bromide
    hydrobromide, **II**, 92
β-Dimethylaminoethyl chloride
    hydrochloride, **IV**, 333
N,*N*-Dimethylcyclohexylmethyl-
    amine, **IV**, 339
Diphenylacetyl chloride **VI**, 549
β-Dipropylaminoethyl bromide
    hydrobromide, **II**, 92

Dodecyl bromide, **I**, 29; **II**, 246
β-Ethoxyethyl bromide, **III**, 370
Ethyl α-chlorophenylacetate, **IV**, 169
Ethyl bromide, **I**, 29, 36
Ethyl iodide, **II**, 402
1-Fluoroadamantane, **VI**, 628
Geranyl chloride, **VI**, 634, 638;
  VIII, 616
Glycerol α,γ-dibromohydrin, **II**, 308
Glycerol α,γ-dichlorohydrin, **I**, 292
Glycerol α-monochlorohydrin, **I**,
  294
Heptyl bromide, **II**, 247
Hexadecyl iodide, **II**, 322
Hexamethylene chlorohydrin, **III**,
  446
1,3,5-Hexatriene, **V**, 608
Indazole, **III**, 475
Iodocyclohexane, **VI**, 830
Isoamyl iodide, **II**, 402
Isobutyl bromide, **II**, 358
Isobutyl iodide, **II**, 402
Isobutyramide, **III**, 490
Isopropyl bromide, **I**, 37; **II**, 359
Itaconyl chloride, **IV**, 554
Malonyl dichloride, **IV**, 263
L-Menthoxyacetyl chloride, **III**, 547
Mesitoyl chloride, **III**, 555
Methanesulfonyl chloride, **IV**, 571
p-Methoxyphenylacetonitrile, **IV**,
  576
1-Methyl-3-ethyloxindole, **IV**, 620
Methyl iodide, **II**, 399
Monobromopentaerythritol, **IV**, 681
Muconic acid, **III**, 623
Naphthalene-1,5-disulfonyl chloride,
  IV, 693
Neopentyl iodide, **VI**, 830
m-Nitrobenzazide, **IV**, 715
p-Nitrobenzoyl chloride, **I**, 394
p-Nitrobenzyl fluoride, **VI**, 835
Octadecyl bromide, **II**, 247
Octyl bromide, **I**, 30
Oleoyl chloride, **IV**, 739
2,3,4,5,6-Penta-O-acetyl-D-gluconic
  acid and 2,3,4,5,6-Penta-O-

acetyl-D-gluconyl chloride,
  V, 887
Pentaerythrityl bromide, **II**, 476
4-Penten-1-ol, **III**, 698
p-Phenylazobenzoyl chloride, **III**,
  712
Phenyl cinnamate, **III**, 714
Propyl bromide, **I**, 37; **II**, 359
Propyl chloride, **I**, 143
Propyl iodide, **II**, 402
Pyruvoyl chloride, **VII**, 467
β-Styrenesulfonyl chloride, **IV**, 846
Tetradecyl bromide, **II**, 247
Tetrahydrofurfuryl bromide, **III**, 793
Tetrahydrofurfuryl chloride, **III**, 698
p-Toluenesulfinyl chloride, **IV**, 937
Trimethylene bromide, **I**, 30
Trimethylene chlorobromide, **I**, 157
Trimethylene chlorohydrin, **I**, 533
Triphenylchloromethane, **III**, 841
W. Hydroxy by Thiol
  3,7-Dithianonane-1,9-dithiol, **VIII**,
    592
X. Sulfonate by Amino
  N-(4-Methoxyphenyl)-(Z)-4-
    (trimethylsilyl)-3-butenamine,
    VIII, 358
Y. Sulfonate by Carbon
  1-Methyl-1-(trimethylsilyl)allene,
    VIII, 471
Z. Sulfur by Hydrazine
  Ethyl oxalamidrazonate, **VIII**, 597
AA. Tin by Carbon
  Ethyl (E)-4-(4-nitrophenyl-4-oxo-2-
    butenoate, **VIII**, 268
  Ethyl (E)-3-(tributylstannyl)-
    propenoate, **VIII**, 268
BB. Miscellaneous
  1-Adamantanol, **VI**, 43
  3-Amino-2-naphthoic acid, **III**, 78
  1-Amino-2-phenylaziridine, **VI**, 56
  1-Amino-2-phenylaziridinium
    acetate, **VI**, 56
  3-Aminopyridine, **IV**, 45
  Benzoylcholine chloride, **IV**, 84
  Benzoyl disulfide, **III**, 116

Benzyl *trans*-1,3-butadiene-1-carba-
mate, **VI**, 95
Benzyltrimethylammonium ethoxide, **IV**, 98
Biphenyl, **VI**, 150
Bis(4-methoxyphenyl)tellurium
dichloride, **VIII**, 468
1,3-Bis(methylthio)-2-propanol, **VI**,
683
Bis(phenylthio)methane, **VI**, 737
Bromocyclopropane, **V**, 126
2-Bromo-3-methylbenzoic acid, **IV**,
114
2-Bromo-3-nitrobenzoic acid, **I**, 125
2-Bromopyridine, **III**, 136
tert-Butoxycarbonyl-L-proline, **VI**,
203
tert-Butyl azidoformate, **VI**, 207
tert-Butyl carbazate, **V**, 166
tert-Butylcarbonic diethylphosphoric
anhydride, **VI**, 207
sec-Butyl isopropyl disulfide, **VI**,
235
3-Butyn-1-yl trifluoromethanesulfonate, **VI**, 324
Carbobenzoxy chloride, **III**, 167
(Z)-4-Chloro-4-hexenyl trifluoroacetate, **VI**, 273
Chloromethylphosphonothioic
dichloride, **V**, 218
5-Chloro-2-pentanone, **IV**, 597
3-Chloropropyldiphenylsulfonium
tetrafluoroborate, **VI**, 364
2-(1-Cyanocyclohexyl)hydrazinecar-
boxylic acid methyl ester, **VI**,
334
2-Cyano-6-methylpyridine, **V**, 269
cis-1,2-Cyclohexanedimethanol
dimethanesulfonate, **VI**, 482
cis-1,2-Cyclohexanediol, **VI**, 348
trans-1,2-Cyclohexanediol, **VI**, 348
Cyclohexene sulfide, **IV**, 232
3,6-Diazido-5-di-*tert*-butyl-1,4-ben-
zoquinone, **VI**, 210
Diazoacetophenone, **VI**, 386

Dibenzo-18-crown-6 polyether, **VI**,
395
Dibenzyl sulfide, **VI**, 404
Di-*tert*-butyl diazomalonate, **VI**, 414
2,6-Di-*tert*-butyl-4-methylpyridine,
VII, 144
4,4'-Dichlorodibutyl ether, **IV**, 266
cis-1,2-Dichlorocyclohexane, **VI**,
424
Dichloromethyl methyl ether, **V**, 365
1,2-Di-1-(1-cyano)cyclohexylhy-
drazine, **IV**, 274
Dicyclopropyl ketone, **IV**, 278
Dideuteriodiazomethane, **VI**, 432
Diethylaluminum cyanide, **VI**, 436
Diethyl 2,2-diethoxyethylphospho-
nate, **VI**, 448
Diethyl mercaptoacetal, **IV**, 295
Diethyl phenylphosphonate, **VI**, 451
α,α-Difluorotoluene, **V**, 396
2,5-Dihydro-2,5-dimethoxyfuran, **V**,
403
3,5-Dihydroxybenzoic acid, **III**, 288
1,4-Diiodobutane, **IV**, 321
2,6-Dimethoxybenzonitrile, **III**, 293
α,*N*,*N*-Dimethylaminophenylaceto-
nitrile, **V**, 437
2-(Dimethylamino)pyrimidine, **IV**,
336
1,4-Dinitrobutane, **IV**, 368
N,*N*'-Diphenylbenzamidine, **IV**, 383
Diphenylcarbodiimide, **V**, 501
O-Diphenylphosphinylhydroxy-
lamine, **VII**, 8
Ethyl α-carbethoxy-β-(*m*-chloro-
anilino)acrylate, **III**, 272
Ethyl isocyanide, **IV**, 438
Ethyl α-nitrobutyrate, **IV**, 454
Ethyl orthocarbonate, **IV**, 457
Ferrocenylacetonitrile, **V**, 578
2-Fluoroheptanoic acid, **V**, 580
Formamidine acetate, **V**, 582
Fumaryl chloride, **III**, 422
Glycine *tert*-butyl ester, **V**, 586
Hexyl fluoride, **IV**, 525

4-Hydroxy-1-butanesulfonic acid sultone, **IV**, 529
Hydroxymethylferrocene, **V**, 621
2-Hydroxy-1,4-naphthoquinone, **III**, 465
Indole-3-carbonitrile, **V**, 656
p-Iodobenzoic acid, **I**, 325
2-Iodo-3-nitrobenzoic acid, **I**, 126
o-Iodophenol, **I**, 326
1-Methylaminonthraquinone, **III**, 573
o-Methylbenzyl alcohol, **IV**, 582
Methyl cyclopropyl ketone, **IV**, 597
Methyl iodide, **II**, 404
2-Methylmercapto-*N*-methyl-Δ^2-pyrroline, **V**, 780
N-Methyl-1-naphthylcyanamide, **III**, 608
Methyl *p*-tolyl sulfone, **IV**, 674
N-β-Naphthylpiperidine, **V**, 816
m-Nitrobenzazide, **IV**, 715
p-Nitrobenzyl acetate, **III**, 650
2-Nitro-6-methoxybenzonitrile, **III**, 293
1-Nitrooctane, **IV**, 724
p-Nitrophenyl sulfide, **III**, 667
Pentaerythrityl iodide, **II**, 477
Pentaerythrityl tetrabromide, **IV**, 753
sym-o-Phthalyl chloride, **II**, 528
Putrescine dihydrochloride, **IV**, 819
(+)- and (–)-α-(2,4,5,7-Tetranitro-9-fluorenylideneaminooxy)propionic acid, **V**, 1031
Thiobenzoic acid, **IV**, 924
Thiobenzoylthioglycolic acid, **V**, 1046
Thiolacetic acid, **IV**, 928
Trichloromethylphosphonyl dichloride, **IV**, 950
1,1,1-Trifluoroheptane, **V**, 1082
m-Trifluoromethylbenzenesulfonyl chloride, **VII**, 508

RESOLUTION
(*S*)-(+)- and (*R*)-(–)-1,1'-Binaphthyl-2,2'-diyl hydrogen phosphate, **VIII**, 51

Ethyl (*R*)- and (*S*)-2-fluorohexanoate, **VIII**, 258
(+)- and (–)-α-(Isopropylideneaminooxy)propionic acid, **V**, 1032
D- and L-Octanol-2, **I**, 418
D- and L-α-Phenylethylamine, **II**, 506
R-(+)- and *S*-(–)-α-Phenylethylamine, **V**, 932

RHODIUM-CATALYZED REACTIONS
(*R*)-(+)-Citronellal, **VIII**, 183
(*R*)-(–)-*N*,*N*-Diethyl-(*E*)-citronellalenamine, **VIII**, 183
3,4-Dihydro-1-(2*H*)-azulenone, **VIII**, 196
Methyl anti-3-hydroxy-2-methylpentanoate **VIII**, 420
Phenylhydroxylamine, **VIII**, 16

RING CONTRACTION REACTIONS
Bicyclo[2.1.0]pentane, **V**, 99
Cyclopropanecarboxaldehyde, **VII**, 129
Cycloundecanecarboxylic acid, **VII**, 135
Phenylcyclopropane, **V**, 929

RING EXPANSION REACTIONS
3,1-Benzoxazepine, **VII**, 23
Cyclobutanol, **VII**, 114, 117
Cyclodecanone, **V**, 277
Cycloheptanone, **IV**, 221
5,5-Dicyano-4-phenyl-2-cyclopenten-1-one 1,3-propanediol ketal, **VIII**, 173
5,6-Diethoxybenzofuran-4,7-dione, **VIII**, 179
1,3-Dihydro-2,3,7-trimethyl-2*H*-azepin-2-one, **V**, 408
2,7-Dimethyloxepin, **V**, 467
(1*R*)-9,9-Dimethyltricyclo[6.1.10^{2,6}]-deca-2,4-diene, **VIII**, 223
Hexahydro-2-(1*H*)-azocinone, **VII**, 254
3-Hydroxycinchoninic acid, **V**, 636
3-Hydroxyquinoline, **V**, 635

2-Oxo-1-phenyl-3-pyrrolidinecar-
boxylic acid, **VII**, 411
2-Phenylcycloheptanone, **IV**, 780
(E)-2-(1-Propenyl)cyclobutanone,
VIII, 556
2-Propyl-1-azacycloheptane, **VIII**,
568
Spiro[4.5]decane-1,4-dione, **VIII**, 578
4,6,8-Trimethylazulene, **V**, 1088
RING-OPENING REACTIONS
Atropaldehyde diethyl acetal, **VII**, 12
1-Benzyloxy-4-penten-2-ol, **VIII**, 33
(R)-(+)-1,1'-Binaphthalene-2,2'-diol,
VIII, 46
S-tert-Butyl 3-oxobutanoate, **VIII**,
71
Chlorocyanoketene, **VIII**, 118
1,4-Di-O-benzyl-L-threitol, **VIII**,
155
Bis(2-carboethoxyethyl)zinc, **VIII**,
277
6,6-Dimethoxyhexanal, **VII**, 168
3-(Dimethylamino)-2-azapropen-1-
ylidenedimethylammonium
chloride, **VII**, 197
5-Dimethylamino-2,4-pentadienyli-
denedimethyliminium salt (in
solution), **VII**, 15
(1-Ethenyl)heptanyl 3-ketobu-
tanoate, **VIII**, 235
(R)-4-Ethyl-4-allyl-2-cyclohexen-1-
one, **VIII**, 241
Ethyl 4-cyclohexyl-4-oxobutanoate,
VIII, 578
3-Hydroxy-1-cyclohexene-1-carbox-
aldehyde, **VIII**, 309
[(trans-2-Isocyanocyclohexyl)oxy]-
trimethylsilane, **VII**, 294
o-Isothiocyanato-(E)-cinnamalde-
hyde, **VII**, 302
cis-1-Methylcyclopentane-1,1-dicar-
boxylic acid, **VIII**, 377
Methyl 6,6-dimethoxyhexanoate,
VII, 168
Methyl 6-oxohexanoate, **VII**, 168

cis,cis-Monomethyl muconate, **VIII**,
490
trans-2-Phenylcyclohexanol, **VIII**,
516
trans-2-(2-Propenyl)cyclopentanol,
VII, 501
8-Propionyl-(E)-5-nonenolide, **VIII**,
562
SILYLATION
Acetone trimethylsilyl enol ether,
VIII, 1
Allyltrimethylsilyl ether, **VIII**, 501
Ethyl 2-(diphenylmethylsilyl)-
decanoate, **VIII**, 474,
1-Methoxy-3-trimethylsiloxy-1,3-
butadiene, **VII**, 312
Ethyl 3-[3-(trimethylsiloxy)cyclohex-
2-enyl]propionate, **VIII**, 277
(1-Hydroxy-2-propenyl)trimethylsi-
lane, **VIII**, 501
3-Methoxy-17-trimethylsiloxy-
1,3,5(10),16-estratetraene,
VIII, 286
2-Methyl-2-trimethylsiloxypentan-3-
one, **VII**, 381
1-Phenyl-1-trimethylsiloxyethylene,
VIII, 323
Trimethyl(2-methylene-4-phenyl-3-
butenyl)silane, **VIII**, 602
1-Trimethylsiloxycyclohexene, **VIII**,
460
1-Trimethylsiloxycyclopentene, **VII**,
424
2-(Trimethylsiloxymethyl)-
allyltrimethylsilane, **VII**, 266
3-Trimethylsiloxy-2-pentene, **VII**,
512
Trimethylsilylacetylene, **VIII**, 606
4-(Trimethylsilyl)-3-buten-1-ol,
VIII, 609
Trimethylsilyl cyanide, **VII**, 517
1-Trimethylsilyl-1-methoxyethene,
VIII, 19
3-Trimethylsilyl-2-propyn-1-ol, **VII**,
524

4,6,6-Trimethyl-2-trimethylsiloxycy-
     clohexa-1,3-diene, **VII**, 282
STRECKER SYNTHESIS
α-Aminodiethylacetic acid, **III**, 66
α-Aminoisobutyric acid, **II**, 29
DL-α-Aminophenylacetic acid, **III**, 84
DL-α-Amino-α-phenylpropionic
     acid, **III**, 88
SULFINYLATION
N-Sulfinylaniline, **V**, 504
(*S*)-(+)-2-(*p*-Toluenesulfinyl)-2-
     cyclopenten-1-one, **VII**, 495
SULFONATION
A. *C*-Sulfonation
p-Acetaminobenzenesulfonyl chlo-
     ride, **I**, 8
Barium 2-phenanthrenesulfonate, **II**,
     482
Benzenesulfonyl chloride, **I**, 85
o-Bromophenol, **II**, 97
4-*tert*-Butylcyclohexen-1-yl trifluo-
     romethanesulfonate, **VIII**, 97
DL-10-Camphorsulfonic acid, **V**, 194
Cholesta-3,5-dien-3-yl trifluo-
     romethanesulfonate, **VIII**, 126
Cyclohexanone oxime methanesul-
     fonate, **VIII**, 568
3,5-Dihydroxybenzoic acid, **III**, 288
2,6-Dinitroaniline, **IV**, 364
Methyl (Z)-3-(phenylsulfonyl)prop-
     2-enoate, **VIII**, 458
2-(Phenylsulfonyl)-1,3-cyclohexadi-
     ene, **VIII**, 540
1-(Phenylsulfonyl)cyclopentene,
     VIII, 543
Potassium anthraquinone-α-sul-
     fonate, **II**, 539
Potassium 3-phenanthrenesulfonate,
     II, 482
4-Pyridinesulfonic acid, **V**, 977
Sodium β-styrenesulfonate, **IV**, 846
α-Sulfopalmitic acid, **IV**, 862
o-Toluidinesulfonic acid, **III**, 824
p-Tolyl 2-(trimethylsilyl)ethynyl sul-
     fone, **VIII**, 281

(Z)-4-(Trimethylsilyl)-3-butenyl 4-
     methylbenzenesulfonate, **VIII**,
     358
3-Trimethylsilyl-2-propyn-1-yl
     methanesulfonate, **VIII**, 471
B. *N*-Sulfonation
(+)-(1*S*)-10-Camphorsulfonamide,
     VIII, 104
p-Toluenesulfonylanthranilic acid,
     IV, 34
cis-1-*N*-Tosyl-2-(2-propenyl)cyclo-
     pentylamine, **VII**, 501
C. *O*-Sulfonation
4-Acetylpyridine oxime tosylate,
     VII, 149
erythro-2,3-Butanediol monomesy-
     late, **VI**, 4
4-*tert*-Butylcyclohexen-1-yl trifluo-
     romethanesulfonate, **VIII**, 97
Butyl *p*-toluenesulfonate, **I**, 145
3-Butyn-1-yl trifluoromethanesul-
     fonate, **VI**, 324
Cholesta-3,5-dien-3-yl trifluo-
     romethanesulfonate, **VIII**,
     126
Cyclobutyl tosylate, **VII**, 117
Cyclohexanone oxime methanesul-
     fonate, **VIII**, 568
erythro-2-Methanesulfonyloxy-2-
     butyl cyclobutanecarboxylate,
     VI, 312
3-Methyl-2-butenyl triflate, **VI**, 757
Methyl *p*-toluenesulfonate, **I**, 146
(1*R*)-Nopyl 4-methylbenzenesul-
     fonate, **VIII**, 223
Styrene glycol dimesylate, **VI**, 56
(Z)-4-(Trimethylsilyl)-3-butenyl 4-
     methylbenzenesulfonate, **VIII**,
     358
3,6,9-Tris(*p*-tolylsulfonyl)-3,6,9-tri-
     azaundecane-1,11-dimethane-
     sulfonate, **VI**, 652
SULFURATION (THIATION)
*N*¹,*N*²-Bis(methoxycarbonyl)sulfur
     diimide, **VIII**, 427

1,3-Dimethylimidazole-2-thione,
VII, 195
N-Methylthiopyrrolidone, **V**, 235;
VII, 372
2,2,4,4-Tetrakis(trifluoromethyl)-1,3-
dithietane, **VII**, 251
**THERMAL DECOMPOSITION
REACTIONS (PYROLYTIC
REACTIONS)**
Acrylic acid, **III**, 30
1,3-Butadiene, **II**, 102
1-Chloro-1,4,4-trifluorobutadiene, **V**,
235
trans-Cyclooctene, **V**, 315
1,1'-Dicyano-1,1'-bicyclohexyl, **IV**,
273
Diethyl methylenemalonate, **IV**, 298
Dihydroxcyclopentene, **V**, 414
Diphenylcarbodiimide (Method II),
V, 504
Ethylene sulfide, **V**, 562
Ethyl 2-ketohexahydrobenzoate, **II**,
532
Ethyl methylmalonate, **II**, 279
Ethyl phenylmalonate, **II**, 288
Hexaphenylbenzene, **V**, 604

2-Hydroxythiophene, **V**, 642
Ketene, **I**, 330; **V**, 679
Methyl benzyl ketone, **II**, 389
Methyl butadienoate, **V**, 734
Methylenecyclohexane and *N*,*N*-
dimethylhydroxylamine
hydrochloride, **IV**, 612
α-Naphthyl isothiocyanate, **IV**, 700
1,4-Pentadiene, **IV**, 746
Putrescine dihydrochloride, **IV**, 819
Stearone, **IV**, 854
Succinimide, **II**, 562
2-Vinylthiophene, **IV**, 980
TRANSESTERIFICATION
1-Carbomethoxy-1-methylethyl 3-
oxobutanoate, **VIII**, 71
Diisopropyl (2*S*,3*S*)-2,3-*O*-isopropy-
lidenetartrate, **VIII**, 201
(–)-Menthyl cinnamate, **VIII**, 350
(–)-Menthyl nicotinate, **VIII**, 350
(4*RS*,4a*RS*,8a*RS*)-, (4*S*,4a*S*,6*S*,8a*S*)-,
and (4*R*,4a*R*,6*R*,8a*R*)-
4-Methoxycarbonyl-1,1,8-
trimethyl-1,4,4a,5,6,7,8,8a)
octahydro-2,3-benzopyrone,
VIII, 353

TYPE OF COMPOUND INDEX

Preparations are listed by functional groups or by ring systems. Many compounds are double listed, such as *m*-bromonitrobenzene, but some, such as substituted acyl halides, are not. This choice represents an arbitrary judgment by the editor as to the likely place a user would look for polyfunctional compounds. Salts are included with the corresponding acids and bases.

ACETALS (for THIOACETALS, see SULFUR COMPOUNDS)

Acetal, **I**, 1
Acetone dibutyl acetal, **V**, 5
Acrolein acetal, **II**, 17; **IV**, 21
Aminoacetal, **III**, 50
m-Aminobenzaldehyde dimethyl acetal, **III**, 59
Atropaldehyde diethyl acetal, VII, 13
Bromoacetal, **III**, 123
1-Bromo-3-chloro-2,2-dimethoxypropane, **VI**, 361
2-Bromo-2-cyclopenten-1-one ethylene ketal, **VII**, 271
2-(2-Bromoethyl)-1,3-dioxane, **VII**, 59
α-Bromoheptaldehyde dimethyl acetal, **III**, 128
2-(Bromomethyl)-2-(chloromethyl)-1,3-dioxane, **VIII**, 173
6-Chloromethyl-2,2-dimethyl-1,3-dioxen-4-one, **VIII**, 192
β-Chloropropionaldehyde acetal, **II**, 137
Cinnamaldehyde dimethyl acetal **VI**, 895
Crotonaldehyde diethyl acetal, **IV**, 22

2-Cyclobutyl-*cis*-4-*trans*-5-dimethyl-1,3-dioxolane, **VI**, 313
Cyclohexanone diallyl acetal, **V**, 292
Cyclohexanone diethyl ketal, **VIII**, 578, 579
Cyclopropanone ethyl hemiacetal, **VII**, 131
Cyclopropenone 1,3-propanediol ketal, **VIII**, 173
3-Deoxy-1,2:5,6-di-*O*-isopropylidene-α-D-ribo-hexofuranose, **VII**, 139
1,4-Di-*O*-benzyl-2,3-di-*O*-isopropylidene-L-threitol, **VIII**, 155
2,3-Dibromopropionaldehyde diethyl acetal, **VI**, 954
2,3-Di-*tert*-butoxy-1,4-dioxane, **VIII**, 161
Dicyanoketene ethylene acetal, **IV**, 276
5,5-Dicyano-4-phenyl-2-cyclopenten-1-one 1,3-propanediol ketal, **VIII**, 578, 579
cis and *trans*-2,6-Diethoxy-1,4-oxathiane 4,4-dioxide, **VI**, 976
2,2-Diethoxy-2-(4-pyridyl)ethylamine, **VII**, 149
Diethyl 2,2-diethoxyethylphosphonate, **VI**, 448

Diethyl mercaptoacetal, **IV**, 295
6-Diethylphosphonomethyl-2,2-
dimethyl-1,3-dioxen-4-one,
VIII, 192
Diethyl [(2-tetrahydropyranyloxy)
methylphosphonate, **VII**, 160
Diisopropyl (2*S*,3*S*)-2,3-*O*-isopropyl-
idenetartrate, **VIII**, 201
DL-lsopropylideneglycerol, **III**, 502
2,3-Di-*O*-isopropylidene-L-threitol,
VIII, 155
7,7-Dimethoxybicyclo[2.2.1]hep-
tene, **V**, 424
3,3-Dimethoxycyclopropene,
VI, 361
6,6-Dimethoxyhexanal, **VII**, 168
7,7-Dimethoxy-1,2,3,4-tetrachlorobi-
cyclo[2.2.1]heptene, **V**, 425
5,5-Dimethoxy-1,2,3,4-tetrachloro-
cyclopentadiene, **V**, 424
Dimethyl 2,3-*O*-isopropylidenetar-
trate, **VIII**, 155, 201
2-[(1'-Ethoxy)-1-ethoxy]butyroni-
trile, **VII**, 381
4-Ethoxy-3-methoxybenzaldehyde
ethylene acetal, **VI**, 567
β-Ethoxypropionaldehyde acetal,
III, 371
1-Ethoxy-1-trimethylsiloxycyclo-
propane, **VII**, 131
Ethyl diethoxyacetate, **IV**, 427
Ethyl 3,3-diethoxypropanoate,
VIII, 254
Ethyl orthocarbonate, **IV**, 457
Ethyl orthoformate, **I**, 258
8-Ethynyl-8-methyl-1,4-dioxa-
spiro[4.5]dec-6-ene, **VII**, 241
DL-Glyceraldehyde diethyl acetal,
II, 307
α-Hydroxyacetophenone dimethyl
acetal, **VII**, 263
2-Hydroxymethyl-2-cyclopenten-1-
one ethylene ketal, **VII**, 274
2,3-*O*-Isopropylidene-D-erythrono-
lactone, **VII**, 297
Ketene diethyl acetal, **III**, 506

Ketene di(2-methoxyethyl) acetal,
V, 684
β-Ketoisooctaldehyde dimethyl
acetal, **IV**, 558
Methylal, **III**, 469
Methyl 4,6-*O*-benzylidene-α-D-glu-
copyranoside, **VIII**, 363
Methyl 3,3-diethoxypropanoate,
VIII, 255
Methyl 6,6-dimethoxyhexanoate,
VII, 168
Monobenzalpentaerythritol, **IV**, 679
m-Nitrobenzaldehyde dimethyl
acetal, **III**, 644
Pentadecanal dimethyl acetal,
VI, 869
4-Phenyl-*m*-dioxane, **IV**, 786
Phenylpropargylaldehyde diethyl
acetal, **IV**, 801
Propiolaldehyde diethyl acetal, **VI**,
954
1,1,1',1'-Tetraethoxyethyl polysulfide, **IV**, 295
Tetrahydro-2-(2-propynyloxy)-2*H*-
pyran, **VII**, 334
Tiglylaldehyde diethyl acetal, **IV**, 22
(*S*)-(+)-2-(*p*-Toluenesulfinyl)-2-
cyclopenten-1-one ethylene
ketal, **VII**, 495
2,5,5-Trimethyl-2-(2-bromoethyl)-
1,3-dioxane, **VII**, 60

ACID ANHYDRIDES
Acetic formic anhydride, **VI**, 8
Benzeneboronic anhydride, **IV**, 68
Benzoic anhydride, **I**, 91
Benzoic carbonic anhydride, **IV**, 286
tert-Butylcarbonic diethylphosphoric
anhydride, **VI**, 207
Caproic anhydride, **III**, 164
p-Chlorobenzoic anhydride, **III**, 29
Citraconic anhydride, **II**, 140
Diacetyl-D-tartaric anhydride,
IV, 242
Di-*tert*-butyl dicarbonate, **VI**, 418
Di-*tert*-butyl tricarbonate, **VI**, 418

3,4-Dihydro-1,2-naphthalic anhydride, **II**, 194
Heptoic anhydride, **III**, 28
Homophthalic anhydride, **III**, 450
Isatoic anhydride, **III**, 488
Itaconic anhydride, **II**, 368
5β-Methoxycyclohexan-1-one-
   3β,4β-dicarboxylic acid
   anhydride, **VII**, 312
β-Methylglutaric anhydride, **IV**, 630
1,2-Naphthalic anhydride, **II**, 423
Nicotinic anhydride, **V**, 822
3-Nitrophthalic anhydride, **I**, 410
α-Phenylglutaric anhydride, **IV**, 790
Succinic anhydride, **II**, 560
o-Sulfobenzoic anhydride, **I**, 495
cis-Δ⁴-Tetrahydrophthalic anhydride,
   IV, 890
Tetraiodophthalic anhydride, **III**, 796
Tetraphenylphthalic anhydride, **III**,
   807
p-Toluenesulfonic anhydride, **IV**, 940
Trifluoroacetyl triflate, **VII**, 506
Trifluoromethanesulfonic anhydride,
   VI, 757

ACID HALIDES
Acetylmandelyl chloride, **I**, 13
p-Acetaminobenzenesulfonyl
   chloride, **I**, 8
Adipyl chloride, **II**, 169; **III**, 623
Benzenesulfinyl fluoride, **V**, 396
Benzenesulfonyl chloride, **I**, 84
Benzoyl fluoride, **V**, 66
Benzyl chloroformate, **III**, 167
2-Bromohexanoyl chloride, **VI**, 190
α-Bromobutyryl chloride, **IV**, 620
α-Bromoisobutyryl bromide, **IV**, 348
Butyryl chloride, **I**, 147
DL-10-Camphorsulfonyl chloride,
   V, 196
Carbobenzoxy chloride, **III**, 167
β-Carbomethoxypropionyl chloride,
   III, 169
δ-Carbomethoxyvaleryl chloride, **IV**,
   555

ω-Carbomethoxypelargonyl chloride, **III**, 613
o-Chlorobenzoyl chloride, **I**, 155
m-Chlorobenzoyl chloride, **VI**, 276,
Chlorobetainyl chloride, **IV**, 154
Chloromethanesulfonyl chloride, **V**,
   232
Chloromethylphosphonothioic
   dichloride, **V**, 218
Chlorosulfonyl isocyanate, **V**, 226
Cyclohexanecarboxylic acid chlor-
   ide, **VII**, 81
2,4-Di-tert-amylphenoxyacetyl
   chloride, **IV**, 742
Diethylthiocarbamyl chloride, **IV**, 307
N,N-Dimethylthiocarbamyl chloride,
   VI, 824
2,4-Dinitrobenzenesulfenyl chloride,
   V, 474
Diphenylacetyl chloride, **VI**, 549,
Fumaryl chloride, **III**, 422
Isobutyryl chloride, **III**, 490
3-Isocyanatopropanoyl chloride,
   VI, 715
β-Isovalerolactam-N-sulfonyl chlor-
   ide, **V**, 673
Itaconyl chloride, **IV**, 554
Lauroyl chloride, **IV**, 561
Malonyl dichloride, **IV**, 263
L-Menthoxyacetyl chloride, **III**, 547
Mesitoyl chloride, **III**, 555
Methanesulfinyl chloride, **V**, 709
Methanesulfonyl chloride, **IV**, 571
(4-Methoxyphenyl)acetyl chloride,
   VI, 745
Naphthalene-1,5-disulfonyl chloride,
   IV, 693
1-Naphthoyl chloride, **VI**, 614
2-Naphthoyl chloride, **III**, 629
o-Nitrobenzenesulfonyl chloride,
   II, 471
m-Nitrobenzoyl chloride, **IV**, 715
Oleoyl chloride, **IV**, 739
Palmitoyl chloride, **IV**, 742
2,3,4,5,6-Penta-O-acetyl-D-gluconyl
   chloride, **V**, 887

4-Pentylbenzoyl chloride, **VII**, 420
p-Phenylazobenzoyl chloride,
  III, 712
Phenyldichlorophosphine, **IV**, 784
sym- and *unsym-o-*Phthalyl chloride,
  II, 528
Pyruvoyl chloride, **VII**, 467
Ricinoleoyl chloride, **IV**, 742
Sebacoyl chloride, **V**, 536
β-Styrenesulfonyl chloride, **IV**, 846
Styrylphosphonic dichloride, **V**, 1005
Taconyl chloride, **IV**, 554
Thiophosgene, **I**, 506
p-Toluenesulfenyl chloride, **IV**, 934
p-Toluenesulfinyl chloride, **IV**, 937
p-Toluoyl chloride, **VI**, 262
Trichloromethyl chloroformate,
  VI, 715
Trichloromethylphosphonyl dichloride, **IV**, 950
m-Trifluoromethylbenzenesulfonyl
  chloride, **VII**, 508
2,2,4-Trimethyl-3-oxovaleryl chloride, **V**, 1103

ACIDS
  A. Unsubstituted
    1. Monobasic
      Abietic acid, **IV**, 1
      Acrylic acid, **III**, 30
      1-Adamantanecarboxylic acid, **V**, 20
      Benzoic acid, **I**, 363
      3-Benzyl-3-methylpentanoic acid, **IV**, 93
      exo-cis-Bicyclo[3.3.0]octane-2-carboxylic acid, **V**, 93
      Caproic acid, **II**, 417, 475
      Cyclobutanecarboxylic acid, **III**, 213
      Cyclohexanecarboxylic acid, **I**, 364
      Cyclopropanecarboxylic acid, **III**, 221

Cycloundecane carboxylic acid, **VII**, 135
1,4-Dihydrobenzoic acid, **V**, 400
Dimethylacrylic acid, **III**, 302
2,2-Dimethyl-4-phenylbutyric acid, **VI**, 517
Diphenylacetic acid, **I**, 224
α,β-Diphenylpropionic acid, **V**, 526
trans-2-Dodecenoic acid, **IV**, 398
Erucic acid, **II**, 258
3-Ethyl-3-methylhexanoic acid, **IV**, 97
4-Ethyl-2-methyl-2-octenoic acid, **IV**, 444
Ferrocenecarboxylic acid, **VI**, 625
9-Fluorenecarboxylic acid, **IV**, 482
Heptanoic acid, **II**, 315
Hydrocinnamic acid, **I**, 311
Isocrotonic acid, **VI**, 711
Levopimaric acid, **V**, 699
Linoleic acid, **III**, 526
Linolenic acid, **III**, 531
Mesitoic acid, **III**, 553; **V**, 706
Mesitylacetic acid, **III**, 557
1-Methylcyclohexanecarboxylic acid, **V**, 739
2-Methyldodecanoic acid, **IV**, 618
trans-2-Methyl-2-dodecenoic acid, **IV**, 608
2-Methylenedodecanoic acid, **IV**, 616
DL-Methylethylacetic acid, **I**, 361
3-Methylheptanoic acid, **V**, 762

3-Methylpentanoic acid,
II, 416
3-Methyl-3-phenylpen-
tanoic acid, IV, 97
Myristic acid, I, 379
α-Naphthoic acid, II, 425
β-Naphthoic acid, II, 428
Nonadecanoic acid,
VII, 397
1-Norbornanecarboxylic
acid, VI, 845
Palmitic acid, III, 605
Pelargonic acid, II, 474
trans-2,4-Pentadienoic
acid, VI, 95
Perbenzoic acid, I, 431;
V, 904
Phenylacetic acid, I, 436
γ-Phenylbutyric acid,
II, 499
α-Phenylcinnamic acid,
IV, 777
cis-2-Phenylcyclopro-
panecarboxylic acid,
VI, 913
3-Phenyl-4-pentenoic
acid, VII, 164
Phenylpropiolic acid,
II, 515
Sorbic acid, III, 783
Stearolic acid, III, 785;
IV, 851
Tetrolic acid, V, 1043
o-Toluic acid, II, 588;
III, 820
p-Toluic acid, II, 589;
III, 822
Trimethylacetic acid,
I, 524
α,β,β-Triphenylpropionic
acid, IV, 960
10-Undecynoic acid,
IV, 969
Valeric acid, I, 363
Vinylacetic acid, III,
851

2. Polybasic
Acetylenedicarboxylic
acid, II, 10
Aconitic acid, II, 12
Adipic acid, I, 18
Azelaic acid, II, 53
Calcium malonate, II, 376
o-Carboxycinnamic acid,
IV, 136
β-(o-Carboxyphenyl)pro-
pionic acid, IV, 136
3,4-seco-Δ⁵-Cholesten-
3,4-dioic acid, IV, 191
Citraconic acid, II, 140
1,1-Cyclobutanedicarbo-
xylic acid, III, 213;
VI, 272
Cyclopropane-1,1-dicar-
boxylic acid, VII, 411
(1S,2S)-(+)-Cyclopro-
pane-1,2-dicarboxylic
acid, VIII, 141
trans-1,2-Dihydro-
phthalic acid, VI, 461
β,β-Dimethylglutaric
acid, IV, 345
Diphenic acid, I, 222
Docosanedioic acid,
V, 533
β-Ethyl-β-methylglutaric
acid, IV, 441
α-Ethyl-α-methylsuc-
cinic acid, V, 572
Fumaric acid, II, 302
Glutaric acid, I, 289;
IV, 496
Hendecanedioic acid,
IV, 510
3,3a,3b,4,6a,7a-Hexa-
hydro-3,4,7-metheno-
7H-cyclopenta[a]pen-
talene-7,8-dicar-
boxylic acid,
VIII, 298
Homophthalic acid, III,
449; V, 612

Itaconic acid, **II**, 368
Malonic acid, **II**, 376
Mesaconic acid, **II**, 382
cis-1-Methylcyclo-
pentane-1,2-dicar-
boxylic acid,
VIII, 377
β-Methylglutaric acid,
III, 591
β-Methyl-β-phenylglu-
taric acid, **IV**, 664
Methylsuccinic acid,
III, 615
Monoperphthalic acid,
III, 619; **V**, 805
Muconic acid, **III**, 623
2,3-Naphthalenedicar-
boxylic acid, **V**, 810
2,6-Naphthalenedicar-
boxylic acid, **V**, 813
Oxalic acid (anhydrous),
I, 421
γ-Phenylallylsuccinic
acid, **IV**, 766
3-Phenyl-4-pentenoic
acid, **VII**, 164
Phenylsuccinic acid,
I, 451; **IV**, 804
Pimelic acid, **II**, 531
Pyromellitic acid, **II**, 551
Terephthalic acid,
III, 791
Tricarballylic acid, **I**, 523
Trithiocarbodiglycolic
acid, **IV**, 967
10-Undecynoic acid,
IV, 969
B.   Substituted
1.   Amino Acids
S-Acetamidomethyl-L-
cysteine hydrochlor-
ide, **VI**, 5
p-Acetaminobenzene-
sulfinic acid, **I**, 7
α-Acetaminocinnamic
acid, **II**, 1

Acetylglycine, **II**, 11
N-Acetylhexahydro-
phenylalanine, **II**, 493
N-Acetylphenylalanine,
II, 493
β-Alanine, **II**, 19; **III**,
34; **VI**, 717
DL-Alanine, **I**, 21
γ-Aminobutyric acid,
II, 25
α-Aminocaproic acid,
I, 48
ε-Aminocaproic acid,
II, 28; **IV**, 39
4-Aminocyclohexane-
carboxylic acid,
V, 670
α-Aminodiethylacetic
acid, **III**, 66
α-Aminoisobutyric acid,
II, 29
3-Amino-2-naphthoic
acid, **III**, 78
DL-α-Aminophenylacetic
acid, **III**, 84
p-Aminophenylacetic
acid, **I**, 52
DL-α-Amino-α-phenyl-
propionic acid, **III**, 88
DL-β-Amino-β-phenyl-
propionic acid, **III**, 91
3-Amino-2,4,6-tribromo-
benzoic acid, **IV**, 947
L-Arginine hydrochlor-
ide, **II**, 49
Arsanilic acid, **I**, 70
DL-Aspartic acid, **IV**, 55
ε-Benzoylamino-α-bro-
mocaproic acid, **II**, 74
ε-Benzoylamino-α-
chlorocaproic acid,
VI, 90
ε-Benzoylaminocaproic
acid, **II**, 76
α-Benzoylaminocin-
namic acid, **II**, 491

DL-ε-Benzoyllysine,
  II, 374
N-tert-Butoxycarbonyl-L-
  phenylalanine,
  VII, 70, 75
tert-Butoxycarbonyl-L-
  proline, **VI**, 203
O-tert-Butyl-L-seryl-*S-
  tert*-butyl-L-cysteine
  tert-butyl ester,
  VI, 252
Carbobenzyloxy-L-
  asparaginyl-L-leucine
  methyl ester, **VI**, 263
N-Carbobenzyloxy-3-hy-
  droxy-L-prolylglycine
  ethyl ester, **VI**, 263
Cysteic acid monohy-
  drate, **III**, 226
L-Cystine, **I**, 194
3,5-Dichloro-2-amino-
  benzoic acid, **IV**, 872
Guanidoacetic acid, **III**,
  440
D-Glutamic acid, **I**, 286
Glycine, **I**, 298
Hippuric acid, **II**, 328
L-Histidine monohy-
  drochloride, **II**, 330
6-Hydroxynicotinic acid,
  IV, 532
Indole-3-acetic acid,
  V, 654
5-Iodoanthranilic acid,
  II, 349
DL-Isoleucine, **III**, 495
DL-Leucine, **III**, 523
DL-Lysine dihydrochlo-
  ride, **II**, 374
DL-Lysine monohy-
  drochloride, **II**, 374
L-Methionine, **VI**, 253
DL-Methionine, **II**, 384
N-Methyl-3,4-dihydroxy-
  phenylalanine,
  III, 586

Methyliminodiacetic
  acid, **II**, 397
Nicotinic acid, **I**, 385
DL-Phenylalanine,
  III, 705
DL-β-Phenylalanine,
  II, 489
N-Phenylanthranilic acid,
  II, 15
N-Phthalyl-L-β-phenyl-
  alanine, **V**, 973
Picolinic acid hydrochlo-
  ride, **III**, 740
DL-Serine, **III**, 774
Taurine, **II**, 563
DL-Threonine, **III**, 813
N-Trifluoroacetyl-L-tyro-
  sine, **VI**, 1004
L-Tryptophan, **II**, 612
L-Tyrosine, **II**, 612
DL-Valine, **III**, 848

2. Haloacids
  p-Acetyl-α-bromohydro-
    cinnamic acid, **VI**, 21
  3-Amino-2,4,6-tribro-
    mobenzoic acid,
    IV, 947
  ε-Benzoylamino-α-bro-
    mocaproic acid, **II**, 74
  ε-Benzoylamino-α-
    chlorocaproic acid,
    VI, 90
  Bromoacetic acid,
    III, 381
  α-Bromocaproic acid, **I**,
    115; **II**, 95
  2-Bromohexanoic acid,
    VI, 192
  α-Bromoisocaproic acid,
    III, 523
  α-Bromoisovaleric acid,
    III, 848
  p-Bromomandelic acid,
    IV, 110
  α-Bromo-β-methoxypro-
    pionic acid, **III**, 775

α-(Bromomethyl)acrylic
acid, **VII**, 210

2-Bromo-3-methylben-
zoic acid, **IV**, 114

2-Bromo-3-nitrobenzoic
acid, **I**, 125

α-Bromophenylacetic
acid, **VI**, 403

α-Bromophenylpropionic
acid, **III**, 705

β-Bromopropionic acid,
I, 131

o-Chlorobenzoic acid,
II, 135

3-Chlorocyclobutane-
carboxylic acid,
VI, 271

7-Chloro-4-hydroxy-3-
quinolinecarboxylic
acid, **III**, 273

p-Chloromandelic acid,
IV, 112

α-Chlorophenylacetic
acid, **IV**, 169

(S)-2-Chloropropanoic
acid, **VIII**, 119

β-Chloropropionic acid,
I, 166

Chloropyruvic acid,
V, 636

3-Chloro-2,2,3-trifluoro-
propionic acid,
V, 239

Dibromohydrocinnamic
acid, **IV**, 961

β,β'-Dibromoisobutyric
acid, **VII**, 319

erythro-α,β-Dibromo-β-
phenylpropionic acid,
IV, 961; **VII**, 173

α,β-Dibromosuccinic
acid, **II**, 177

Dichloroacetic acid,
II, 181

3,5-Dichloro-2-amino-
benzoic acid, **IV**, 872

Di-(p-chlorophenyl)
acetic acid, **III**, 270

2,2-Difluorosuccinic
acid, **V**, 393

2,4-Dihydroxy-5-bromo-
benzoic acid, **II**, 100

p-Fluorobenzoic acid,
II, 299

2-Fluoroheptanoic acid,
V, 580

(S)-2-Fluorohexanoic
acid, **VIII**, 258

5-Iodoanthranilic acid,
II, 349

m-Iodobenzoic acid,
II, 353

p-Iodobenzoic acid,
I, 325

p-Iodomandelic acid,
IV, 112

2-Iodo-3-nitrobenzoic
acid, **I**, 126

Mucobromic acid, **III**,
621; **IV**, 688

Pentachlorobenzoic acid,
V, 890

DL-4,4',6,6'-Tetrachloro-
diphenic acid, **IV**, 872

2,4,6-Tribromobenzoic
acid, **IV**, 947

3. Hydroxy Acids
Atrolactic acid, **IV**, 58
Benzilic acid, **I**, 89
5-Benzyloxy-3-hydroxy-
3-methylpentanoic-2-
¹³C acid, **VII**, 386
nor-Desoxycholic acid,
III, 234
7-Chloro-4-hydroxy-3-
quinolinecarboxylic
acid, **III**, 273
5,5-Dihydroxybarbituric
acid, **IV**, 23
2,5-Dlhydroxy-p-ben-
zenediacetic acid,
III, 286

3,5-Dihydroxybenzoic
  acid, **III**, 288
2,4-Dihydroxy-5-bro-
  mobenzoic acid,
  II, 100
9,10-Dihydroxystearic
  acid, **IV**, 317
(2*SR*,3*RS*)-2,4-Dimethyl-
  3-hydroxypentanoic
  acid, **VII**, 185
(2*SR*,3*SR*)-2,4-Dimethyl-
  3-hydroxypentanoic
  acid, **VII**, 190
2,5-Dimethylmandelic
  acid, **III**, 326
Hexahydrogallic acid,
  V, 591
p-Hydroxybenzoic acid,
  II, 341
2-Hydroxycinchoninic
  acid, **III**, 456
2-Hydroxy-3,5-diiodo-
  benzoic acid, **II**, 343
2-Hydroxyisophthalic
  acid, **V**, 617
(2*S**,3*S**)-3-Hydroxy-2-
  methyl-3-phenyl-
  propanoic acid,
  VIII, 339
6-Hydroxynicotinic acid,
  IV, 532
17β-Hydroxy-5-oxo-3,5-
  seco-4-norandrostane-
  3-carboxylic acid,
  VI, 690
p-Hydroxyphenylpyruvic
  acid, **V**, 627
β-Hydroxypropionic
  acid, **I**, 321
Mandelic acid, **I**, 336;
  III, 538
Mucobromic acid, **III**,
  621; **IV**, 688
D-Norandrost-5-en-3β-ol-
  16-carboxylic acids,
  VI, 840

Protocatechuic acid,
  III, 745
β-Resorcylic acid, **II**, 557
Ricinelaidic acid,
  VII, 470
Sodium 2-phenyl-2-
  hydroxyethane-1-sul-
  fonate, **IV**, 850
DL-Tartaric acid, **I**, 497
Vanillic acid, **IV**, 972
4. Keto Acids
β-(1-Acenaphthoyl)pro-
  pionic acid, **III**, 9
β-(3-Acenaphthoyl)pro-
  pionic acid, **III**, 6
Acetonedicarboxylic
  acid, **I**, 10
2-(1-Acetyl-2-oxopropyl)
  benzoic acid, **VI**, 36
δ-Acetylvaleric acid,
  IV, 19
β-Benzoylacrylic acid,
  III, 109
γ-Benzoylbutyric acid,
  II, 82
Benzoylformic acid, **I**,
  244; **III**, 114
β-Benzoylpropionic acid,
  II, 81
3-Carboxy-5-dodecen-2-
  one, (*E*)-, **VIII**, 235
Chelidonic acid, **II**, 126
Chloropyruvic acid,
  V, 636
3,4-Dimethoxyphenyl-
  pyruvic acid, **IV**, 327
1,3-Dimethyl-5-oxobicy-
  clo[2.2.2]octane-2-
  carboxylic acid,
  VIII, 219
Fluorenone-2-carboxylic
  acid, **III**, 420
5-Formyl-4-phenanthroic
  acid, **IV**, 484
17β-Hydroxy-5-oxo-3,5-
  seco-4-norandrostane-

3-carboxylic acid,
VI, 690
p-Hydroxyphenylpyruvic
acid, **V**, 627
α-Ketoglutaric acid, **III**,
510; **V**, 687
6-Ketohendecanedioic
acid, **IV**, 555
DL-Ketopinic acid, **V**, 689
Levulinic acid, **I**, 335
3-Methylcyclohexanone-
3-acetic acid,
VIII, 467
γ-Oxocapric acid, **IV**, 432
6-Oxodecanoic acid,
VIII, 499
2-Oxo-1-phenyl-3-pyrro-
lidinecarboxylic acid,
VII, 411
Phenylpyruvic acid,
II, 519
Phthalaldehydic acid,
II, 523
Pyruvic acid, **I**, 475
p-Toluyl-o-benzoic acid,
I, 517

5. Nitro Acids
2-Bromo-3-nitrobenzoic
acid, **I**, 125
2,5-Dinitrobenzoic acid,
III, 334
3,5-Dinitrobenzoic acid,
III, 337
Nitroacetic acid, dipotassium salt, **VI**, 797
m-Nitrobenzoic acid,
I, 391
p-Nitrobenzoic acid,
I, 392
m-Nitrocinnamic acid,
I, 398; **IV**, 731
p-Nitrophenylacetic acid,
I, 406
trans-o-Nitro-α-phenyl-
cinnamic acid,
IV, 730

3-Nitrophthalic acid,
I, 408
4-Nitrophthalic acid,
II, 457
(+)- and (−)-α-(2,4,5,7-
Tetranitro-9-fluo-
renylideneaminooxy)
propionic acid,
V, 1031
2,4,6-Trinitrobenzoic
acid, **I**, 543
6. Miscellaneous
4-Acetoxybenzoic acid,
VI, 577
3β-Acetoxyetienic acid,
V, 8
Acetylmandelic acid,
I, 12
Acid ammonium *o*-sul-
fobenzoate, **I**, 14
Ammonium salt of aurin-
tricarboxylic acid,
I, 54
Arsenoacetic acid, **I**, 73
Arsonoacetic acid, **I**, 73
p-Arsonophenoxyacetic
acid, **I**, 75
Benzenediazonium-2-
carboxylate, **V**, 54
(*S*)-(+)-γ-Butyrolactone-
γ-carboxylic acid,
VII, 99
β-Carbethoxy-β,β-
diphenylvinylacetic
acid, **IV**, 132
Carbobenzoxyglycine,
III, 168
Carboxymethoxylamine
hemihydrochloride,
III, 172
o-Carboxyphenylaceto-
nitrile, **III**, 174
p-Chloromercuribenzoic
acid, **I**, 159
Coumalic acid, **IV**, 201
Coumarilic acid, **III**, 209

ω-Cyanopelargonic acid,
 III, 768
α-Cyano-β-phenylacrylic
 acid, **I**, 181
Cyclohexylidenecyano-
 acetic acid, **IV**, 234
1,3-Dihydroxy-2-naph-
 thoic acid, **III**, 638
7,16-Diketodocosanedi-
 oic acid and disodium
 salt, **V**, 534, 536
2,3-Dimethoxycinnamic
 acid, **IV**, 327
6,7-Dimethoxy-3,4-dihy-
 dro-2-naphthoic acid,
 III, 300
3,4-Dimethoxyphenyl-
 pyruvic acid, **II**, 335
Diphenaldehydic acid,
 V, 493
α,α'-Diphenylthiodigly-
 colic acid, **VI**, 403
Ethoxyacetic acid, **II**, 260
Ferrocenecarboxylic
 acid, **VI**, 625
2-Furancarboxylic acid,
 I, 276
2-Furoic acid, **IV**, 493
Furylacrylic acid, **III**, 425
Furylpropionic acid,
 I, 313
Hexahydrogalliċ acid tri-
 acetate, **V**, 591
Homoveratric acid,
 II, 333
o-Hydrazinobenzoic acid
 hydrochloride,
 III, 475
Indole-3-acetic acid,
 V, 654
Isodehydroacetic acid,
 IV, 549
L-Menthoxyacetic acid,
 III, 544
3-Methylcoumarilic acid,
 IV, 591

3-Methyl-2-furoic acid,
 IV, 628
Methyl hydrogen hende-
 canedioate, **IV**, 635
2-Methyl-5,6-pyrazinedi-
 carboxylic acid,
 IV, 827
Methyl Red, **I**, 374
2-Methyl-5-trimethylsi-
 lyl-2-vinylpentanoic
 acid, **VIII**, 488
cis,cis-Monomethyl
 muconate, **VIII**, 490
sec-Octyl hydrogen
 phthalate, **I**, 418
2-Oxo-1-phenyl-3-pyrro-
 lidinecarboxylic acid,
 VII, 411
Parabanic acid, **IV**, 744
2,3,4,5,6-Penta-O-acetyl-
 D-gluconic acid,
 V, 887
p-Phenylazobenzoic acid,
 III, 711
(E)-3-[(E)-3-Phenyl-2-
 propenoxy]acrylic
 acid, **VIII**, 536
α-Phthalimido-o-toluic
 acid, **IV**, 810
Piperonylic acid, **II**, 538
2,3-Pyrazinedicarboxylic
 acid, **IV**, 824
Sodium p-arsono-N-
 phenylglycinamide,
 I, 488
Sodium p-hydroxyphenyl-
 arsonate, **I**, 490
Sodium p-toluenesulfin-
 ate, **I**, 492
α-Sulfopalmitic acid,
 IV, 862
β-(Tetrahydrofuryl)propi-
 onic acid, **III**, 742
3-Thenoic acid, **IV**, 919
Thiobenzoic acid,
 IV, 924

Thiobenzoylthioglycolic
acid, **V**, 1046
Thiolacetic acid, **IV**, 928
Thiosalicylic acid, **II**, 580
p-Toluenesulfonylanthra-
nilic acid, **IV**, 34
Trimethylgallic acid,
I, 537
Trithiocarbodiglycolic
acid, **IV**, 967

ALCOHOLS
A.   Unsubstituted
Acenaphthenol-7, **III**, 3
1-Acetylcyclohexanol, **IV**, 13
1-Adamantanol, **VI**, 43
Allyl alcohol, **I**, 42
Androstan-17β-ol, **VI**, 62
Benzohydrol, **I**, 90
Benzopinacol, **II**, 71
Benzyl alcohol, **II**, 591
Bicyclo[3.3.0]octane-2,6-diol,
VIII, 43
3-Buten-1-ol, **VI**, 608
cis-4-*tert*-Butylcyclohexanol,
VI, 215
trans-4-*tert*-Butylcyclohexanol,
V, 175
3-Butyl-2-methylhept-1-en-3-
ol, **VI**, 240
2-Butyn-1-ol, **IV**, 128
Cetyl alcohol, **II**, 374
Cholesterol, **IV**, 195
1,2-Cyclobutanediol, **VII**, 129
Cyclobutanol, **VII**, 114, 117
1,2-Cyclodecanediol, **IV**, 216
cis-1,2-Cyclohexanedimethanol,
VI, 482,
cis-1,2-Cyclohexanediol, **VI**,
343, 348
trans-1,2-Cyclohexanediol, **III**,
217; **VI**, 348
Cyclohexanol, **VI**, 353
Cyclohexylcarbinol, **I**, 188
1-Decalol, **VI**, 371
Decamethylene glycol, **II**, 154

9-Decyn-1-ol, **VIII**, 146
Dibutyl carbinol, **II**, 179
Dihydrocholesterol, **II**, 191
trans-9,10-Dihydro-9,10-
phenanthrenediol, **VI**, 887
3,4-Dihydroxycyclopentene,
V, 414
3,5-Dihydroxycyclopentene,
V, 414
2,3-Dimethylbenzyl alcohol,
IV, 584
3,3-Dimethyl-*cis*-bicyclo[3.2.0]
heptan-2-ol, **VII**, 177
(*E*)-7,11-Dimethyl-6,10-
dodecadien-2-yn-1-ol,
VIII, 226
Dimethyl ethynyl carbinol,
III, 320
4,4-Dimethyl-1,6-heptadien-3-
ol, **VII**, 177
4,6-Dimethyl-1-hepten-4-ol,
V, 452
1-Ethynylcyclohexanol, **III**, 416
1,1'-Ethynylene-bis-cyclohexa-
nol, **IV**, 471
Heptamethylene glycol, **II**, 155
2-Heptanol, **II**, 317
Heptyl alcohol, **I**, 304
Hexamethylene glycol, **II**, 325
1-Hexanol, **I**, 306; **VI**, 919
(*E*)-4-Hexen-1-ol, **VI**, 675
Hexyl alcohol, **I**, 306; **VI**, 919
Homogeraniol, **VII**, 258
1-(Hydroxymethyl)cyclohexa-
nol, **VIII**, 315
1-Hydroxymethyl-4-(1-methyl-
cyclopropyl)-1-cyclohex-
ene, **VIII**, 321
4-Hydroxynona-1,2-diene,
VII, 276
3-Hydroxy-1-nonene, **VIII**, 235
(−)-Isopinocampheol, **VI**, 719,
Lauryl alcohol, **II**, 372
o-Methylbenzyl alcohol, **IV**,
582
1-Methylcyclohexanol, **VI**, 766

3-Methyl-2-cyclohexen-1-ol,
  VI, 769
2-Methyl-2,5-decanediol,
  IV, 601
1-Methyl-2-ethynyl-*endo*-2,3-
  dimethyl-2-norbornanol,
  VIII, 391
2-Methyl-1,6-heptadien-3-ol,
  VI, 606
2-Methyl-1-hexanol, **VI**, 786
Methyl hexyl carbinol, **I**, 366
Methyl isopropyl carbinol,
  II, 406
(1*RS*,2*SR*,5*R*)-5-Methyl-2-[1-
  methyl-1-(2-phenylethyl)]
  cyclohexanol, **VIII**, 522
(*Z*)-3-Methyl-2-octen-1-ol,
  VIII, 509
3-Methyl-1,5-pentanediol,
  IV, 660
trans-Methyl styryl carbinol,
  IV, 773
Myristyl alcohol, **II**, 374
Neopentyl alcohol, **V**, 818
Nonamethylene glycol, **II**, 155
exo/*endo*-7-Norcaranol, **V**, 859
Nortricyclanol, **V**, 863
Octadecamethylene glycol,
  II, 155
1-Octanol, **VI**, 919
D- and L-2-Octanol, **I**, 418
(*Z*)-2-Octen-1-ol, **VIII**, 508
(*R*)-(+)-1-Octyn-3-ol , **VII**, 402
Oleyl alcohol, **II**, 468; **III**, 671
Pentaerythritol, **I**, 425
1,5-Pentanediol, **III**, 693
1,2,5-Pentanetriol, **III**, 833
3-Penten-2-ol, **III**, 696
4-Penten-1-ol, **III**, 698
(*E*)-3-Pentyl-2-nonene-1,4-diol,
  VIII, 507
4-Pentyn-1-ol, **IV**, 755
Perhydro-9b-phenalenol,
  VII, 427
2-Phenyl-2-adamantanol,
  VII, 433

2-Phenylcyclohexanol, (±), (+)-
  and (–)-, **VIII**, 516
(–)-8-Phenylmenthol, **VIII**, 522
(*R*)-(+)-1-Phenyl-1-pentanol,
  VII, 447
1-Phenyl-1-penten-4-yn-3-ol,
  IV, 792
3-Phenyl-1-propanol, **IV**, 798
Pinacol hydrate, **I**, 459
trans-Pinocarveol, **VI**, 946, 948
(*S*)-(+)-Propane-1,2-diol,
  VII, 356
trans-2-(2-Propenyl)cyclopen-
  tanol, **VII**, 501
(*E*)-2-Propenylcyclohexanol,
  VII, 456
L-Propylene glycol, **II**, 545
Tetradecamethyleneglycol,
  II, 155
α,α,α′,α′-Tetramethyltetrameth-
  ylene glycol, **V**, 1026
p-Tolyl carbinol, **II**, 590
1-*p*-Tolylcyclopropanol,
  V, 1058
Triamyl carbinol, **II**, 603
Tribiphenyl carbinol, **III**, 831
Tributyl carbinol, **II**, 603
Tridecamethylene glycol, **II**, 155
Triethyl carbinol, **II**, 602
Triheptyl carbinol, **II**, 603
1,2,5-Trihydroxypentane,
  III, 833
3,4,5-Trimethyl-2,5-heptadien-
  4-ol, **VIII**, 505
Triphenyl carbinol, **III**, 839
Tripropyl carbinol, **II**, 603
Undecamethylene glycol,
  II, 155
Undecylenyl alcohol, **II**, 374
Xanthydrol, **I**, 554
B. Substituted (see also ACIDS B,
  ESTERS, KETONES, and
  LACTONES)
Acetol, **II**, 5
cis-3-Acetoxy-5-hydroxycy-
  clopent-1-ene, **VIII**, 13

1-Acetylcyclohexanol, **IV**, 13

3-Acetyl-4-hydroxy-5,5-dimethylfuran-2(5*H*)-one, **VIII**, 71

Aluminum *tert*-butoxide, **III**, 48

o-Aminobenzyl alcohol, **III**, 60

6-Amino-3,4-dimethyl-*cis*-3-cyclohexen-1-ol, **VII**, 4

1-(Aminomethyl)cyclohexanol, **IV**, 224

Atrolactic acid, **IV**, 58

Benzilic acid, **I**, 89

Benzoin, **I**, 94

Benzophenone cyanohydrin, **VII**, 20

5-Benzyloxy-3-hydroxy-3-methylpentanoic-2-^{13}C acid, **VII**, 386

1-Benzyloxy-4-penten-1-ol, **VIII**, 33

3-Bis(methylthio)-1-hexen-4-ol, **VI**, 683

1,3-Bis(methylthio)-2-propanol **VI**, 683,

erythro-2-Bromo-1,2-diphenylethanol, **VI**, 184

2-Bromoethanol, **I**, 117

10-Bromo-11-hydroxy-10,11-dihydrofarnesyl acetate, **VI**, 560

p-Bromomandelic acid, **IV**, 110

erythro-2,3-Butanediol monomesylate, **VI**, 4

8-Butyl-2-hydroxytricyclo[7.3.1.0^{2,7}]tridecan-13-one, **VIII**, 87

9-Butyl-1,2,3,4,5,6,7,8-octahydroacridin-4-ol, **VIII**, 87

Butyroin, **II**, 114; **VII**, 95

3-Chloro-2-buten-1-ol, **IV**, 128

2-Chlorocyclohexanol, **I**, 158

trans-2-Chlorocyclopentanol, **IV**, 157

m-Chlorophenyl methyl carbinol, **III**, 200

(*S*)-2-Chloropropan-1-ol, **VIII**, 434

Cholesterol dibromide, **IV**, 195

2-Cyclohexyloxyethanol, **V**, 303

Cyclopropanone ethyl hemiacetal, **VI**, 131

nor-Desoxycholic acid, **III**, 234

Diacetone alcohol, **I**, 199

16-Diazoandrost-5-en-3β-ol-17-one, **VI**, 840

1,4-Di-*O*-benzyl-L-threitol, **VIII**, 155

2,2-Dichloroethanol, **IV**, 271

Diethyl (2*S*,3*R*)-(+)-3-allyl-2-hydroxysuccinate, **VII**, 153

β-Diethylaminoethyl alcohol, **II**, 183

Diethyl bis(hydroxymethyl) malonate, **V**, 381

Diethyl 3-hydroxybutanoate, **VII**, 215

Diethyl hydroxymethylphosphonate, **VII**, 160

Diethyl (2*R*,3*R*)-(+)-tartrate, **VII**, 41, 461

9,10-Dihydroxystearic acid, **IV**, 317

2,3-Di-*O*-isopropylidene-L-threitol, **VIII**, 155

(2-Dimethylamino-5-methylphenyl)diphenylcarbinol, **VI**, 478

(2*SR*,3*RS*)-2,4-Dimethyl-3-hydroxypentanoic acid, **VII**, 185

(2*SR*,3*SR*)-2,4-Dimethyl-3-hydroxypentanoic acid, **VII**, 190

2,5-Dimethylmandelic acid, **III**, 326

2',6'-Dimethylphenyl (2*SR*,3*SR*)-2,4-dimethyl-3-hydroxypentanoate, **VII**, 190

2-(Diphenylhydroxymethyl)
    pyrrolidine, **VI**, 542
3,7-Dithianonane-1,9-diol,
    VIII, 592
1-(1,3-Dithian-2-yl)-2-cyclo-
    hexen-1-ol, **VIII**, 309
D-Erythronolactone, **VII**, 297
(*E*)-1-Ethoxy-1-(1-hydroxy-2-
    butenyl)cyclopropane,
    VIII, 556
Ethyl (*S*)-(+)-3-hydroxybu-
    tanoate, **VII**, 215
Ethyl 4-hydroxycrotonate, **VII**,
    221
Ethyl (1-hydroxycyclohexyl)
    acetate, **VI**, 598
Ethyl *threo*-[1-(2-hydroxy-1,2-
    diphenylethyl)]carbamate,
    VII, 223
Ethyl β-hydroxy-β,β-diphenyl-
    propionate, **V**, 564
Ethyl (α-hydroxymethyl)acry-
    late, **VIII**, 265
3-Ethyl-4-hydroxymethyl-5-
    methylisoxazole, **VI**, 781
Ethyl β-phenyl-β-hydroxypro-
    pionate, **III**, 408
(*S*)-(−)-1-Formyl-2-hydroxy-
    methylpyrrolidine,
    VIII, 26
(*S*)-*N*-Formyl-*O*-*tert*-butyl-L-
    valinol, **VIII**, 204
(*S*)-*N*-Formyl-L-valinol, **VIII**, 204
2-Furylcarbinol, **I**, 276
4-(2-Furyl)-4-hydroxy-2,3-
    diethoxycyclobutenone,
    VIII, 179
DL-Glyceraldehyde, **II**, 305
DL-Glyceraldehyde diethyl
    acetal, **II**, 307
Glycerol α,γ-dibromohydrin,
    II, 308
Glycerol α,γ-dichlorohydrin,
    I, 292
Glycerol α-monochlorohydrin,
    I, 294

α-Glyceryl phenyl ether, **I**, 296
Glycolonitrile, **III**, 436
Hexahydro-6-allyl-6-ethyl-3-
    (hydroxymethyl)-8a-
    methyl-2-phenyl
    [2*S*,3*S*,8a*R*]-5-oxo-5*H*-
    oxazolo[3,2-*a*]pyridine,
    VIII, 241
(3a*S*,7a*S*)-(+)-2,3,3a,4,7,7a-
    Hexahydro-3a-hydroxy-7a-
    methyl-1*H*-indene-
    1,5(6*H*)-dione, **VII**, 363
Hexahydro-3-(hydroxymethyl)-
    8a-methyl-2-phenyl
    [2*S*,3*S*,8a*R*]-5-oxo-5*H*-
    oxazolo[3,2-*a*]pyridine,
    VIII, 241
Hexamethylene chlorohydrin,
    III, 446
α-Hydroxyacetophenone,
    VII, 263
2-Hydroxybutanenitrile,
    VII, 381
3-(1-Hydroxybutyl)-1-
    methylpyrrole, **VII**, 102
2-Hydroxycyclobutanone, **VI**,
    167
3-Hydroxy-1-cyclohexene-1-
    carboxaldehyde, **VIII**, 309
(*S*)-(+)-3-Hydroxy-2,2-
    dimethylcyclohexanone,
    VIII, 312
N-(3-Hydroxy-3,3-diphenyl-
    propylidene)cyclohexy-
    lamine, **VI**, 901
Hydroxyethyl methyl sulfide,
    II, 345
3-Hydroxyglutaronitrile, **V**, 614
trans-4-Hydroxy-2-hexenal,
    VI, 683
N-(Hydroxymethyl)acetamide,
    VI, 5
2-(Hydroxymethyl)allyltrimeth-
    ylsilane, **VII**, 266
2-Hydroxymethyl-2-cyclopen-
    ten-1-one, **VII**, 271

Hydroxymethylferrocene, **V**, 621

4(5)-Hydroxymethylimidazole hydrochloride, **III**, 460

2-Hydroxy-2-methylpentan-3-one, **VII**, 381

3-Hydroxy-3-methyl-1-phenyl-1-butanone, **VIII**, 210, 323

(2*S**,3*S**)-3-Hydroxy-2-methyl-3-phenylpropanoic acid, **VIII**, 339

erythro-(3-Hydroxy-2-methyl-3-phenylpropanoyl)piperidine, **VIII**, 326

threo-(3-Hydroxy-2-methyl-3-phenylpropanoyl)piperidine, **VIII**, 326

(*S*)-(+)-2-Hydroxymethylpyrrolidine, **VIII**, 26

(1*S*,2*S*,3*R*)-3-Hydroxy-2-nitrocyclohexyl acetate, **VIII**, 332

17β-Hydroxy-5-oxo-3,5-*seco*-4-norandrostane-3-carboxylic acid, **VI**, 690

5-Hydroxypentanal, **III**, 470

threo-4-Hydroxy-3-phenyl-2-heptanone, **VI**, 692

(1-Hydroxy-2-propenyl)trimethylsilane, **VIII**, 501

β-Hydroxypropionic acid, **I**, 321

7-Hydroxy-4,4,6,7-tetramethylbicyclo[4.2.0]octan-2-one, **VI**, 1024

6-Hydroxy-3,5,5-trimethyl-2-cyclohexen-1-one, **VII**, 282

3-Hydroxy-2,2,3-trimethyloctan-4-one, **VIII**, 343

5-Hydroxy-2,4,6-trimethyl-2-trimethylsiloxyheptan-3-one, **VII**, 185

Isobutyroin, **II**, 115

trans-2-Isocyanocyclohexanol, **VII**, 294

1-[(Isopropoxydimethylsilyl)methyl]cyclohexanol, **VIII**, 315

2-Isopropylaminoethanol, **III**, 501

DL-Isopropylideneglycerol, **III**, 502

Mandelamide, **III**, 536

Mandelic acid, **I**, 336

2-(1-Mercapto-1-methylethyl)-5-methylcyclohexanol, **VIII**, 304

Methyl 4-*O*-benzoyl-6-bromo-6-deoxy-α-D-glucopyranoside, **VIII**, 363

Methyl 4,6-*O*-benzylidene-α-D-glucopyranoside, **VIII**, 363

2-Methyl-2,5-decanediol, **IV**, 601

Methyl 4-hydroxy-2-butynoate, **VII**, 334

Methyl (1*R*,5*R*)-5-hydroxy-2-cyclopentene-1-acetate, **VII**, 339

Methyl 7-hydroxyhept-5-ynoate, **VIII**, 415

Methyl 3-hydroxy-2-methylenepentanoate, **VIII**, 420

Methyl *anti*-3-hydroxy-2-methylpentanoate, **VIII**, 420

2-Methyl-3-(*p*-toluenesulfonamido)-2-pentanol, **VII**, 375

(*R*,*S*)-Mevalonolactone-2-^{13}C, **VII**, 386

Monobenzalpentaerythritol, **IV**, 679

Monobromopentaerythritol, **IV**, 681

p-Nitrobenzyl alcohol, **III**, 652

(1*R*,2*r*,3*S*)-2-Nitrocyclohexane-1,3-diol, **VIII**, 332

2-Nitroethanol, **V**, 833

3'-Nitro-1-phenylethanol, **VIII**, 495

D-Norandrost-5-en-3β-ol-16-
carboxylic acids, **VI**, 840
16-Oximinoandrost-5-en-3β-ol-
17-one, **VI**, 840
(4*S*)-3-[1-Oxo-(*syn*-2-methyl-3-
hydroxy)-3-phenylpropyl]-
4-benzyl-2-oxazolidinone
VIII, 339
D-(−)-Pantoyl lactone, **VII**, 417
(*S*)-(+)-Phenylalanol, **VIII**, 528
Piperonyl alcohol, **II**, 591
Pivaloin, **II**, 115
Potassium 1,1,1,3,3,3-hexafluo-
ro-2-phenyl-2-propanolate,
VI, 163
Propionoin, **II**, 115
(2*S*,3*S*)-3-Propyloxiranemeth-
anol, **VII**, 461
1(α-Pyridyl)-2-propanol,
III, 757
2-(1-Pyrrolidyl)propanol,
IV, 834
Ricinelaidic acid, **VII**, 470
Sebacoin, **IV**, 840
DL-Tartaric acid, **I**, 497
Tetramethyl 3,7-dihydroxybicy-
clo[3.3.0]octa-2,6-diene-
2,4,6,8-tetracarboxylate,
VII, 50
Tetramethyl 3,7-dihydroxy-1,5-
dimethylbicyclo[3.3.0]octa-
2,6-diene-2,4,6,8-tetracar-
boxylate, **VII**, 50
Tetramethylene chlorohydrin,
I, 571
2,2,6,6-Tetramethylolcyclo-
hexanol, **IV**, 907
β-Thiodiglycol, **II**, 576
7-Thiomenthol, **VIII**, 302,
cis-2-(*p*-Toluenesulfonamido)
cyclohexanol, **VII**, 375
Trichloroethyl alcohol, **II**, 598
Trimethylene chlorohydrin,
I, 533
3-Trimethylsilyl-3-buten-2-ol,
VI, 1033

(*Z*)-4-(Trimethylsilyl)-3-buten-
1-ol, **VIII**, 358, 609
4-(Trimethylsilyl)-3-butyn-1-ol,
VIII, 609
(*E*)-3-Trimethylsilyl-2-propen-
1-ol, **VII**, 524,
3-Trimethylsilyl-2-propyn-1-ol,
VII, 524
4,6,6-Trimethyl-2-trimethyl-
siloxyhexa-1,3-diene,
VII, 282
3,6,9-Tris(*p*-tolylsulfonyl)-
3,6,9-triazaundecane-1,11-
diol, **VI**, 652
Tropolone, **VI**, 1037
L-Valinol, **VII**, 530

ALDEHYDES

A. Aliphatic Aldehydes

Acrolein, **I**, 15
Atropaldehyde, **VII**, 13
Cholan-24-al, **V**, 242
5β-Cholest-3-ene-5-acetalde-
hyde, **VI**, 298
(*R*)-(+)-Citronellal, **VII**, 183
Cyclobutanecarboxaldehyde,
VI, 312
Cyclohexanecarboxaldehyde,
VI, 338
Cyclohexylideneacetaldehyde,
VI, 358
Cyclopentanecarboxaldehyde,
V, 320
Cyclopropanecarboxaldehyde,
VII, 129
1-Decanal, **VI**, 373
2,2-Dimethyl-4-pentenal,
VII, 177
2,2-Dimethyl-3-phenylpropi-
onaldehyde, **VI**, 526
Diphenylacetaldehyde, **IV**,
375
Geranial, **VII**, 258
Heptanal, **VI**, 644
Hexaldehyde, **II**, 323
5-Hexynal, **VI**, 679

(*S*)-(+)-2-Methylbutanal,
  VIII, 367
2-Methylbutyraldehyde-1-D,
  VI, 751
2-Methyl-4-phenyl-4-pentenal,
  VIII, 451
2-Methyl-3-phenylpropanal,
  VII, 361
2-Methyl-3-phenylpropionalde-
  hyde, **VI**, 815
Octanal, **V**, 872
Pentadecanal, **VI**, 869
β-Phenylcinnamaldehyde,
  VI, 901
1-Phenylcyclopentanecarbo-
  xaldehyde, **VI**, 905
3-Phenyl-4-pentenal, **VIII**,
  536
Phenylpropargyl aldehyde,
  III, 731
α-Phenylpropionaldehyde,
  III, 733
3-Phenylpropionaldehyde,
  VII, 451
Propiolaldehyde, **IV**, 813
Propionaldehyde, **II**, 541
2,4,4-Trimethylpentanal, **VI**,
  1028
B. Substituted Aliphatic Aldehydes
Bromal, **II**, 87
α-Bromocinnamic aldehyde,
  III, 731
α-Bromoheptaldehyde, **III**, 127
N-tert-Butoxycarbonyl-L-leuci-
  nal, **VIII**, 68
Butyl glyoxylate, **IV**, 124
Butyrchloral, **IV**, 130
α-Chlorocrotonaldehyde,
  IV, 131
2-Chloro-1-formyl-1-cyclohex-
  ene, **V**, 215
7-Cyanoheptanal, **V**, 266
Diethyl formylmethylphospho-
  nate, **VI**, 448
6,6-Dimethoxyhexanal,
  VII, 168

2,2-Dimethyl-4-oxopentanal,
  VIII, 208
DL-Glyceraldehyde, **II**, 305
3-Hydroxy-1-cyclohexene-1-
  carboxaldehyde, **VIII**, 309
trans-4-Hydroxy-2-hexenal,
  VI, 683
5-Hydroxypentanal, **III**, 470
o-Isothiocyanato-(E)-cin-
  namaldehyde, **VII**, 302
Methyl 7-oxoheptanoate,
  VI, 807
Methyl 6-oxohexanoate,
  VII, 168
o-Nitrocinnamaldehyde, **IV**,
  722
6-Oxodecanal, **VIII**, 498
3-Phenyl-2*H*-azirine-2-carbox-
  aldehyde, **VI**, 893
Phenylglyoxal, **II**, 509; **V**,937
Phenylpropargyl aldehyde,
  III, 731
C. Aromatic Aldehydes
p-Acetamidobenzaldehyde,
  IV, 32
o-Aminobenzaldehyde, **III**, 56
p-Aminobenzaldehyde, **IV**, 31
o-Anisaldehyde, **V**, 46; **VI**, 64
9-Anthraldehyde, **III**, 98
m-Bromobenzaldehyde, **II**, 132
p-Bromobenzaldehyde, **II**, 89,
  442
2-Bromo-4-methylbenzalde-
  hyde, **V**, 139
m-Chlorobenzaldehyde, **II**, 130
p-Chlorobenzaldehyde, **II**, 133
p-Dimethylaminobenzaldehyde,
  I, 214; **IV**, 331
2,4-Dinitrobenzaldehyde,
  II, 223
3,5-Dinitrobenzaldehyde,
  VI, 529
Diphenylacetaldehyde, **IV**, 375
Diphenaldehydic acid, **V**, 493
4-Ethoxy-3-hydroxybenzalde-
  hyde, **VI**, 567

2-Ethoxy-1-naphthaldehyde,
III, 98
4-Formylbenzenesulfonamide,
VI, 631
5-Formyl-4-phenanthroic acid,
IV, 484
m-Hydroxybenzaldehyde,
III, 453
2-Hydroxy-1-naphthaldehyde,
III, 463
Mesitaldehyde, **III**, 549; **V**,
49
m-Methoxybenzaldehyde,
III, 564
1-Naphthaldehyde, **IV**, 690
β-Naphthaldehyde, **III**, 626
o-Nitrobenzaldehyde, **III**, 641;
V, 825
m-Nitrobenzaldehyde, **III**, 644;
Warning, **III**, 645
p-Nitrobenzaldehyde, **II**, 441
o-Nitrocinnamaldehyde,
IV, 722
6-Nitroveratraldehyde, **IV**, 735
Phenanthrene-9-aldehyde,
III, 701
Phthalaldehydic acid, **III**, 737
Protocatechualdehyde, **II**, 549
Syringic aldehyde, **IV**, 866
o-Tolualdehyde, **III**, 818;
IV, 932
p-Tolualdehyde, **II**, 583
3,4,5-Trimethoxybenzaldehyde,
VI, 1007
Veratraldehyde, **II**, 619
D. Dialdehydes
N-(2,4-Diformyl-5-hydroxy-
phenyl)acetamide,
VII, 162
Diphenaldehyde, **V**, 489
Glutaconaldehyde sodium salt,
VI, 640
Glyoxal bisulfite, **III**, 438
Isophthalaldehyde, **V**, 668
Methyl diformylacetate,
VII, 323

β-Methylglutaraldehyde,
IV, 661
o-Phthalaldehyde, **IV**, 807
Sodium nitromalonaldehyde
monohydrate, **IV**, 844
Terephthalaldehyde, **III**, 788
4,5,4',5'-Tetramethoxy-1,1'-
biphenyl-2,2'-dicarbox-
aldehyde, **VIII**, 586
E. Heterocyclic Aldehydes
Furfural, **I**, 280
Imidazole-2-carboxaldehyde,
VII, 282
Indole-3-aldehyde, **IV**, 539
5-Methylfurfural, **II**, 393
2-Pyrrolealdehyde, **IV**, 831
2-Thenaldehyde, **IV**, 915 or
2-thiophenealdehyde,
III, 811
3-Thenaldehyde, **IV**, 918

**AMIDES, IMIDES, LACTAMS and
UREAS (see also HETE-
ROCYCLES for cyclic
amides)**
A. Unsubstituted
Acetamide, **I**, 3
p-Acetotoluide, **I**, 111
Acetyl methylurea, **II**, 462
Benzanilide, **I**, 82; **IV**, 384
Benzoylpiperidine, **I**, 99
N-Benzylacrylamide, **V**, 73
Benzylphthalimide, **II**, 83
tert-Butylphthalimide, **III**, 152
tert-Butylurea, **III**, 151
Cyclobutanecarboxamide,
VIII, 132
N-Cyclohexylformamide,
V, 301
Cyclohexylurea, **V**, 801
N,N-Dimethyl-5β-cholest-3-
ene-5-acetamide, **VI**, 491
N,N-Dimethylcyclohexanecar-
boxamide, **IV**, 339
N,N-Dimethylisobutyramide,
VI, 284

asym-Dimethylurea, **IV**, 361
sym-Diphenylurea, **I**, 453
2-Ethylhexanamide, **IV**, 436
o-Formotoluide, **III**, 480 or *N*-
 o-Tolylformamide,
  V, 1062
Fumaramide, **IV**, 486
Glutarimide, **IV**, 496
Heptamide, **IV**, 513
N-Heptylurea, **IV**, 515
Hexahydro-2-(1*H*)-azocinone,
 VII, 254
Isobutyramide, **III**, 490
Methacrylamide, **III**, 560
N-Methylformanilide, **III**, 590
Phenylacetamide, **IV**, 760
N-Phenylmaleimide, **V**, 944
N-Phenylimide, **V**, 957
Phenylurea, **I**, 453
Phthalimide, **I**, 457
Succinimide, **II**, 562
N-(1,1,3,3-Tetramethylbutyl)
 formamide, **VI**, 753
o-Toluamide, **II**, 586
B. Substituted
 3-Acetamido-2-butanone, **IV**, 5
 2-(2-Acetamidoethyl)-4,5-
  dimethoxyacetophenone,
   VI, 1
 2-Acetamido-3,4,6-tri-*O*-acetyl-
  2-deoxy-α-D-glucopyran-
   osyl chloride, **V**, 1
 S-Acetamidomethyl-L-cysteine
  hydrochloride, **VI**, 5
 p-Acetaminobenzenesulfinic
  acid, **I**, 7
 p-Acetaminobenzenesulfonyl
  chloride, **I**, 8
 α-Acetaminocinnamic acid,
  II, 1
 N-(*p*-Acetylaminophenyl)rhoda-
  nine, **IV**, 6
 Acetoacetanilide, **III**, 10
 Acetylglycine, **II**, 11
 Acetyl methylurea, **II**, 462
 N-Acetylphenylalanine, **II**, 493

N-Acetyl-*N*-phenylhydroxyl-
 amine, **VIII**, 16
N-Acetylhomoveratrylamine,
 VI, 2
Benzohydroxamic acid, **II**, 67
Benzoylacetanilide, **III**,108;
 IV, 80
ε-Benzoylamino-α-chloro-
 caproic acid, **VI**, 90
2-Benzoyl-1-cyano-1-methyl-
 1,2-dihydroisoquinoline,
  IV, 642
Benzoylene urea, **II**, 79
Benzoyl-2-methoxy-4-nitroacet-
 anilide, **IV**, 82
Benzoylpiperidine, **I**, 99
1-(2-Benzoylpropanoyl)piperi-
 dine, **VIII**, 326
N-Benzylacrylamide, **V**, 73
Benzyloxycarbonyl-L-alanyl-L-
 cysteine methyl ester,
  VII, 30
Benzyloxycarbonyl-L-aspartyl-
 (*tert*-butyl ester)-L-phenyl-
  alanyl-L-valine methyl
   ester, **VII**, 3
N-[2,4-Bis(1,3-diphenylimida-
 zolidin-2-yl)-5-hydroxy-
  phenyl]acetamide,
   VII, 162
N-Bromoacetamide, **IV**, 104
3-Bromo-4-acetaminotoluene,
 I, 111
β-Bromoethylphthalimide, **I**,
 119; **IV**, 106
N-Bromoglutarimide, **IV**, 498
N-Bromomethylphthalimide,
 VIII, 536
p-Bromophenylurea, **IV**, 49
tert-Butoxycarbonyl-L-leucinal,
 VIII, 68
tert-Butoxycarbonyl-L-leucine-
 N-methyl-*O*-methylcarbox-
  amide, **VIII**, 68
N-*tert*-Butylcarbonyl-L-phenyl-
 alanine, **VII**, 70, 75

(*S*)-1-Carbamoyl-2-methoxymethylpyrrolidine, **VIII**, 26

1,1'-Carbonyldiimidazole, **V**, 201

Chloroacetamide, **I**, 153

p-Chloroacetylacetanilide, **III**, 183

Cyanoacetamide, **I**, 179

1-Cyano-2-benzoyl-1,2-dihydroisoquinoline, **IV**, 641

1-Cyano-3-α-naphthylurea, **IV**, 215

1-Cyano-3-phenylurea, **IV**, 213

Diaminobiuret, **III**, 404

2,6-Dibromoquinone-4-chloroimide, **II**, 175

α,α-Dichloroacetamide, **III**, 260

α,α'-Dicyano-β-ethyl-β-methylglutarimide, **IV**, 441

α,α'-Dicyano-β-methylglutaramide, **III**, 591

Diethyl acetamidomalonate, **V**, 373

N-(2,4-Diformyl-5-hydroxyphenyl)acetamide, **VII**, 162

N-[2-(3,4-Dimethoxyphenyl)ethyl]acetamide, **VI**, 1

(*R,R*)-(+)-2,3-Dimethoxy-*N,N,N',N'*-tetramethylsuccinic acid diamide, **VII**, 41

N,N-Dimethylselenourea, **IV**, 359

2,6-Dioxo-1-phenyl-4-benzyl-1,4-diazabicyclo[3.3.0]octane, **VIII**, 231

2,4-Dithiobiuret, **IV**, 504

p-Ethoxyphenylurea, **IV**, 52

N-Ethyl-*p*-chloroformanilide, **IV**, 420

(*S*)-*N*-Formyl-*O*-*tert*-butyl-L-valinol, **VIII**, 204

N-Formylglycine ethyl ester, **VI**, 620

(*S*)-(−)-1-Formyl-2-hydroxymethylpyrrolidine, **VIII**, 26

(*S*)-(−)-1-Formyl-2-methoxymethylpyrrolidine, **VIII**, 26

N-Formyl-L-valinol, **VIII**, 204

Geraniol trichloroacetimidate, **VI**, 508

Guanylthiourea, **IV**, 502

3-Heptylureidomethylenemalononitrile, **IV**, 515

Hexahydro-6-allyl-6-ethyl-3-(hydroxymethyl)-8a-methyl-2-phenyl[2*S*,3*S*,6*S*,8a*R*]-5-oxo-5*H*-oxazolo[3,2-*a*]pyridine, **VIII**, 241

Hexahydro-3-(hydroxymethyl)-8a-methyl-2-phenyl-[2*S*,3*S*,8a*R*]-5-oxo-5*H*-oxazolo[3,2-*a*]pyridine, **VIII**, 241

Hexahydro-1,3,5-tripropionyl-*s*-triazine, **IV**, 518

Hexamethylphosphorus triamide, **V**, 602

N-(Hydroxymethyl)acetamide, **VI**, 5

erythro-(3-Hydroxy-2-methyl-3-phenylpropanoyl)piperidine, **VIII**, 326

threo-(3-Hydroxy-2-methyl-3-phenylpropanoyl)piperidine, **VIII**, 326

Hydroxyurea, **V**, 645

N-Iodosuccinimide, **V**, 663

Isatin, **I**, 327

Isonitrosoacetanilide, **I**, 327

Lactamide, **III**, 516

Mandelamide, **III**, 536

p-Methoxyphenylurea, **IV**, 53

Methylisourea hydrochloride, **IV**, 645

β-Methyl-β-phenyl-α,α'-dicyanoglutarimide, **IV**, 662

Methyl sebacamate, **III**, 613
Nicotinamide 1-oxide, **IV**, 704
1-Nitro-2-acetylaminonaphthalene, **II**, 438
2-Nitro-4-methoxyacetanilide, **III**, 661
4-Nitrophthalimide, **II**, 459
Nitrourea, **I**, 417
Nitrosomethylurea, **II**, 461
(4*S*)-3-[1-Oxo-(*syn*-methyl-3-hydroxy)-3-phenylpropyl]-4-benzyl-2-oxazolidinone, **VIII**, 339
2-Oxo-1-phenyl-3-pyrrolidinecarboxylic acid, **VII**, 411
(*S*)-3-(1-Oxopropyl)-4-(phenylmethyl)-2-oxazolidinone, **VIII**, 339
2-Phenyl-3-propylisoxazolidine-4,5-*cis*-dicarboxylic acid *N*-phenylimide, **V**, 957
4-Phenylsemicarbazide, **I**, 450
α-Phthalimido-*o*-toluic acid, **IV**, 810
N-Phthalyl-L-β-phenylalanine, **V**, 973
Salicyl-*o*-toluide, **III**, 765
Semicarbazide sulfate, **I**, 485
Sodium *p*-arsono-*N*-phenylglycinamide, **I**, 488
(*R*,*R*)-(+)-*N*,*N*,*N'*,*N'*-Tetramethyltartaric acid diamide, **VII**, 41
2α-Thiohomophthalimide, **V**, 1051
Triethyl α-phthalimidoethane-α,α,β-tricarboxylate, **IV**, 55
N-Trifluoroacetanilide, **VI**, 1004
N-Trifluoroacetyl-L-tyrosine, **VI**, 1004
Veratric amide, **II**, 44
C. Lactams
4-Acetoxyazetidin-2-one, **VIII**, 3
ε-Caprolactam (2-ketohexamethylenimine), **II**, 77, 371
3-Cyano-6-methyl-2(1*H*)-pyridone, **IV**, 210
1,3-Dihydro-3,5,7-trimethyl-2*H*-azepin-2-one, **V**, 408
1,3-Dimethyl-3-methoxy-4-phenylazetidinone, **VIII**, 216
1,5-Dimethyl-2-pyrrolidone, **III**, 328
5,5-Dimethyl-2-pyrrolidone, **IV**, 357
2-Ethoxypyrrolin-5-one, **VI**, 226
Hexahydro-6-allyl-6-ethyl-3-(hydroxymethyl)-8a-methyl-2-phenyl-[2*S*,3*S*,6*S*,8a*R*]-5-oxo-5*H*-oxazolo[3,2*a*]pyridine, **VIII**, 241
Hexahydro-2-(1*H*)-azocinone, **VII**, 254
Hexahydro-3-(hydroxymethyl)-8a-methyl-2-phenyl-[2*S*,3*S*,8a*R*]-5-oxo-5*H*-oxazolo[3,2-*a*]pyridine, **VIII**, 241
Isatin, **I**, 327
3-Isoquinuclidone, **V**, 670
β-Isovalerolactam, **V**, 673
β-Isovalerolactam-*N*-sulfonyl chloride, **V**, 673
2-Ketohexamethylenimine, **II**, 371
1-Methyl-3-ethyloxindole, **IV**, 620
3-Methyloxindole, **IV**, 657
1-Methyl-2-pyridone, **II**, 419

AMIDINES
Acetamidine hydrochloride, **I**, 5
Aminoguanidine bicarbonate, **III**, 73
Benzamidine hydrochloride, **I**, 6
Creatinine, **I**, 172

6,7-Dimethoxy-1,2,3,4-tetrahydro-2-
[(1-*tert*-butoxy-3-methyl)-2-
butyliminomethyl]isoquinoline,
VIII, 573
[3-(Dimethylamino)-2-azaprop-2-en-
1-ylidene]dimethylammonium
chloride, **VII**, 197
(*S*)-*N*,*N*-Dimethyl-*N*'-(1-*tert*-butoxy-
3-methyl-2-butyl)formamidine,
VIII, 204, 573
N,*N*-Dimethyl-*N*'-*p*-tolylformami-
dine, **VII**, 197
N,*N*'-Diphenylbenzamidine, **IV**,
383
Formamidine acetate, **V**, 582
Guanidine nitrate, **I**, 302
Guanidoacetic acid, **III**, 440
Nitroguanidine, **I**, 399
N-Phenylbenzamidine, **IV**, 769

**AMINES (and SALTS including QUA-
TERNARY AMMONIUM
COMPOUNDS)**
A. Aliphatic
Acetonamine hydrogen oxalate,
I, 196
1-Acetoxy-4-benzylamino-2-
butene, **VIII**, 9
1-Acetoxy-4-diethylamino-2-
butene, **VIII**, 9
Allylamine, **II**, 24
Aminoacetal, **III**, 50
Aminoacetone semicarbazone
hydrochloride, **V**, 27
6-Amino-3,4-dimethyl-*cis*-3-
cyclohexen-1-ol, **VII**, 4
trans-1-Amino-2,3-diphenyl-
aziridine, **VI**, 56
Aminomalononitrile *p*-toluene-
sulfonate, **V**, 32
1-Amino-1-methylcyclohexane,
V, 35
1-(Aminomethyl)cyclohexanol,
IV, 224
1-Amino-2-phenylaziridine,
VI, 56

1-Amino-2-phenylaziridinium
acetate, **VI**, 56
β-Aminopropionitrile, **III**, 93
N-Benzyl-*N*-(methoxymethyl)-
N-[(trimethylsilyl)methyl]
amine, **VIII**, 231
N-Benzyl-*N*-[(trimethylsilyl)
methyl]amine, **VIII**, 231
Bis(β-cyanoethyl)amine, **III**, 93
(*S*,*S*)-(+)-1,4-Bis(dimethylamino)-2,3-dimethoxybu-
tane (DDB), **VII**, 41
Bis(dimethylamino)methane, **V**,
435; **VI**, 474
Bis(3-dimethylaminopropyl)
phenylphosphine, **VI**, 776
2-Bromoallylamine, **V**, 121
N-(2-Bromoallyl)ethylamine,
V, 124
Bromoethylamine hydrobro-
mide, **II**, 91
α-(*p*-Bromophenyl)ethylamine,
II, 505
Butylamine, **II**, 319
sec-Butylamine, **II**, 319
tert-Butylamine, **III**, 148
tert-Butylamine hydrochloride,
III, 153
2-*tert*-Butyl-1,3-diamino-
propane, **VI**, 223
tert-Butyl-*tert*-octylamine,
VIII, 93
Carboxymethoxylamine hemi-
hydrochloride, **III**, 172
α-(*p*-Chlorophenyl)ethylamine,
II 505
Cyclobutylamine, **V**, 273
Cyclobutylamine hydrochlo-
ride, **VIII**, 132
Cyclohexylamine, **II**, 319
Decamethylenediamine,
III, 229
Diacetonamine hydrogen
oxalate, **I**, 196
Diallylamine, **I**, 201
Diaminomaleonitrile, **V**, 344

4,13-Diaza-18-crown-6, **VIII**, 152

1,10-Diazacyclooctadecane, **VI**, 382

N,*N*'-Dibenzyl-4,13-diaza-18-crown-6, **VIII**, 152

1,10-Dibenzyl-4,7-dioxa-1,10-diazadecane, **VIII**, 152

Dibutylamine, **I**, 202

β-Dibutylaminoethylamine, **III**, 254

γ-Dibutylaminopropylamine, **III**, 256

Di-β-carbethoxyethylmethyl-amine, **III**, 258

2,2-Diethoxy-2-(4-pyridyl)ethylamine, **VII**, 149

Diethylaminoacetonitrile, **III**, 275

1-Diethylamino-3-butanone, **IV**, 281

β-Diethylaminoethyl alcohol, **II**, 183

Diethyl aminomalonate hydrochloride, **V**, 376

Diethyl aminomethylphosphonate, **VIII**, 451

Diethylaminosulfur trifluoride, **VI**, 440

Diethyl 2-(cyclohexylamino)vinylphosphonate, **VI**, 448

N,*N*-Diethylgeranylamine, **VIII**, 183, 188

N,*N*-Diethylnerylamine, **VIII**, 183, 190

N,*N*-Dimethylcyclohexylamine, **VI**, 499

N,*N*-Diethyl-1,2,2-trichlorovinylamine, **V**, 387

β-Dimethylaminoethyl chloride hydrochloride, **IV**, 333

6-(Dimethylamino)fulvene, **V**, 431

1-(Dimethylamino)-4-methyl-3-pentanone, **VI**, 474

α-*N*,*N*-Dimethylaminophenylacetonitrile, **V**, 437

β-Dimethylaminopropio-phenone hydrochloride, **III**, 305

2,3-Dimethylbenzyldimethyl-amine, **IV**, 587

N,*N*-Dimethylcyclohexyl-methylamine, **IV**, 339

3,7-Dimethyl-1,6-octadien-3-amine, **VII**, 507

α,α-Dimethyl-β-phenethyl-amine, **V**, 471

1-Ethyl-3-(3-dimethylamino)propylcarbodiimide hydrochloride, **V**, 555

Glycine *tert*-butyl ester, **V**, 586

Heptylamine, **II**, 318

2-Isopropylaminoethanol, **III**, 501

Laurylmethylamine, **IV**, 564

N-(4-Methoxyphenyl)-(*Z*)-4-(trimethylsilyl)-2-butenamine, **VIII**, 358

Methylamine hydrochloride, **I**, 347

2-Methylbenzyldimethylamine, **IV**, 585

N-Methylbutylamine, **V**, 736

N-Methyl-2,3-dimethoxybenzyl-amine, **IV**, 603

N-Methyl-1,2-diphenylethyl-amine, **IV**, 605

Methylenaminoacetonitrile, **I**, 355

N-Methylethylamine, **V**, 758

5-Methylfurfuryldimethyl-amine, **IV**, 626

Methyliminodiacetic acid, **II**, 397

N-Methyl-2-phenyl-Δ^2-tetrahydropyridine, **VI**, 818

β-Naphthol phenylaminomethane, **I**, 381

(*S*)-(−)-α-(1-Naphthyl)ethyl-amine, **VI**, 826, 828

α-(β-Naphthyl)ethylamine, **II**, 505

Phenacylamine hydrochloride,
V, 909
2-Phenyl-2-adamantanamine
hydrochloride, VII, 433
(S)-Phenylalanol, VIII, 528
1-Phenylcyclopentylamine,
VI, 910
β-Phenylethylamine, III, 720
DL-α-Phenylethylamine, II,
503; III, 717
R(+)- and S(−)-α-Phenylethyl-
amine, II, 506; V, 932
β-Phenylethyldimethylamine,
III, 723
3-Pinanamine, VI, 943
Polymeric benzylamine, VI, 951
cis-2-(2-Propenyl)cyclopentyl-
amine, VII, 501
Putrescine dihydrochloride,
IV, 819
ac-Tetrahydro-β-naphthyl-
amine, I, 499
α-p-Tolylethylamine, II, 505
Trimethylamine, I, 528
Trimethylamine hydrochloride,
I, 531
α-p-Xenylethylamine, II, 505
L-Valinol, VII, 530
B. Aromatic Amines
o-Aminobenzaldehyde, III, 56
p-Aminobenzaldehyde, IV, 31
m-Aminobenzaldehyde
dimethyl acetal, III, 59
2-Aminobenzophenone, IV, 34
o-Aminobenzyl alcohol, III, 60
o-Aminobiphenyl, V, 830
2-Amino-p-cymene, III, 63
2-Aminofluorene, II, 448; V, 30
3-Amino-2-naphthoic acid,
III, 78
1,2-Aminonaphthol hydrochloride, II, 33
1,4-Aminonaphthol hydrochloride, I, 49; II, 39
1-Amino-2-naphthol-4-sulfonic
acid, II, 42

o-Amino-p'-nitrobiphenyl,
V, 830
2-Amino-4-nitrophenol, III, 82
2-Amino-3-nitrotoluene, IV, 42
2-Amino-5-nitrotoluene, IV, 44
p-Aminophenyl disulfide,
III, 86
p-Aminotetraphenylmethane,
IV, 47
Aminothymol, I, 512
4-Aminoveratrole, II, 44
Arsanilic acid, I, 70
N-Benzylaniline, I, 102; IV, 92
4,4'-Bis(dimethylamino)benzil,
V, 111
3-Bromo-4-aminotoluene, I, 111
4-Bromo-N,N-dimethyl-3-(trifluoromethyl)aniline, VI,
181
3-(o-Chloroanilino)propioni-
trile, IV, 146
N-2-Cyanoethylaniline, IV, 205
4,4'-Diaminoazobenzene,
V, 341
1,5-Diamino-2,4-dinitroben-
zene, V, 1068
4,4'-Diaminodiphenyl sulfone,
III, 239
Diaminodurene chlorostannate,
II, 255
1,2-Diamino-4-nitrobenzene,
III, 242
2,4-Diaminotoluene, II, 160
N,N'-Dibenzyl-p-phenylenedi-
amine, IV, 92
2,6-Dibromo-4-aminophenol
chlorostannate, II, 175
2,6-Dibromoaniline, III, 262
3,5-Dibromosulfanilamide,
III, 263
2,6-Dichloroaniline, III, 262
Dichloro-p-dimethylamino-
styrene, V, 361
N,N'-Diethylbenzidine, IV, 283
2,6-Diiodo-p-nitroaniline,
II, 196

p-Dimethylaminobenzaldehyde,
IV, 331
(2-Dimethylamino-5-methyl-
phenyl)diphenylcarbinol,
VI, 478
(Z)-β-[2-(N.N-Dimethylamino)
phenyl]styrene, **VII**, 172
3,4-Dimethylaniline, **III**, 307
2,4-Dinitroaniline, **II**, 221
2,6-Dinitroaniline, **IV**, 364
Ethyl p-aminobenzoate, **I**, 240
Ethyl 4-amino-3-methyl-
benzoate, **VI**, 560, 581
Ethyl 4-amino-3-(methyl-
thiomethyl)benzoate,
VI, 581
Ethyl β-anilinocrotonate,
III, 374
N-Ethyl-p-chloroaniline,
IV, 420
Ethyl p-dimethylaminophenyl-
acetate, **V**, 552
N-Ethyl-m-toluidine, **II**, 290
p-Iodoaniline, **II**, 347
1-(4-Methoxyphenyl)-1,2,5,6-
tetrahydropyridine,
VIII, 358
N-(4-Methoxyphenyl)-(Z)-
(trimethylsilyl)-3-buten-
amine, **VIII**, 358
1-Methylaminoanthraquinone,
III, 573
4'-Methyl-2-aminobenzophe-
none, **IV**, 38
1-Methylamino-4-bromoan-
thraquinone, **III**, 575
β-Naphthol phenylamino-
methane, **I**, 381
N-β-Naphthylpiperidine, **V**, 816
o-Nitroaniline, **I**, 388
m-Nitrodimethylaniline,
III, 658
2-Nitro-4-methoxyaniline,
III, 661
4-Nitro-1-naphthylamine,
III, 664

Orthanilic acid, **II**, 471
o-Phenylenediamine, **II**, 501
1-Phenylpiperidine, **IV**, 795
Sodium p-arsono-N-phenyl-
glycinamide, **I**, 488
ac-Tetrahydro-β-naphthyl-
amine, **I**, 499
Tetramethyl-p-phenylenediamine, **V**, 1018
p-Thiocyanodimethylaniline,
II, 574
o-Toluidinesulfonic acid, **III**, 824
m-Tolylbenzylamine, **III**, 827
2,4,5-Triaminonitrobenzene,
V, 1067
p-Tricyanovinyl-N,N-dimethyl-
aniline, **IV**, 953
m-Trifluoromethyl-N,N-dimethyl-
aniline, **V**, 1085
Triphenylamine, **I**, 544
C.   Heterocyclic
9-Aminoacridine, **III**, 53
2-Amino-4-anilino-6-
(chloromethyl)-s-triazine,
IV, 29
2-Amino-5-bromo-3-nitropyri-
dine, **V**, 347
2-Amino-5-bromopyridine, **V**,
346
5-Amino-2,3-dihydro-1,4-
phthalazinedione, **III**, 69
4-Amino-2,6-dimethylpyrimi-
dine, **III**, 71
trans-1-Amino-2,3-diphenyl-
aziridine, **VI**, 56
(S)-(−)- and (R)-(+)-1-Amino-2-
methoxymethylpyrrolidine,
VIII, 26
2-Amino-6-methylbenzothia-
zole, **III**, 76
2-Amino-4-methylthiazole,
II, 31
1-Amino-2-phenylaziridine,
VI, 56
1-Amino-2-phenylaziridinium
acetate, **VI**, 56

3(5)-Aminopyrazole, **V**, 39
4-Amino-1,2,5,6-tetrahydro-1-
    phenylphosphorin-3-car-
    bonitrile, **VI**, 932
7-Aminotheophylline, **VII**, 10
2-Amino-6-methylbenzothia-
    zole, **III**, 76
3(5)-Aminopyrazole, **V**, 39
3-Aminopyridine, **IV**, 45
1-Aminopyridinium iodide,
    V, 43
7-Aminotheophylline, **VII**, 10
3-Amino-1H-1,2,4-triazole,
    III, 95
4-Amino-4H-1,2,4-triazole,
    III, 96
Azetidine, **VI**, 75
Benzoguanamine, **IV**, 78
3,1-Benzoxazepine, **VII**, 23
2-Benzylaminopyridine, **IV**,
    91
5-Benzylaminotetrazole,
    VII, 27
N-Benzyl-2-azanorbornene,
    VIII, 31
4-Benzyloxyindole, **VII**, 34
cis-2-Benzyl-3-phenylaziridine,
    V, 83
$tert$-Butoxycarbonyl-L-proline,
    VI, 203
2,6-Di-$tert$-butyl-4-methylpyri-
    dine, **VII**, 144
9-Butyl-1,2,3,4,5,6,7,8-octahy-
    droacridine, **VIII**, 87
9-Butyl-1,2,3,4,5,6,7,8-octahy-
    droacridine N-oxide,
    VIII, 87
9-Butyl-1,2,3,4,5,6,7,8-octahy-
    droacridin-4-ol, **VIII**, 87
1-Butylpyrrolidine, **III**, 159
3-Butyroyl-1-methylpyrrole,
    VII, 102
2-Chloromethyl-4,6-diamino-s-
    triazine, **IV**, 30
N-Chloropiperidine, **VI**, 968
Creatinine, **I**, 172

2,3-Diamino-5-bromopyridine,
    V, 346
2,5-Diamino-3,4-dicyanothio-
    phene, **IV**, 243
2,4-Diamino-6-hydroxypyrimi-
    dine, **IV**, 245
2,4-Diamino-6-phenyl-s-tri-
    azine, **IV**, 78
2,3-Diaminopyridine, **V**, 346
Diaminouracil hydrochloride,
    IV, 247
2,2-Diethoxy-2-(4-pyridyl)ethyl-
    amine, **VII**, 149
2-(Dimethylamino)pyrimidine,
    IV, 336
2,2-Dimethylpyrrolidine,
    IV, 354
2,6-Dioxo-1-phenyl-4-benzyl-
    1,4-diazabicyclo[3.3.0]
    octane, **VIII**, 231
1,4-Diphenyl-5-amino-1,2,3-tri-
    azole, **IV**, 380
N-Ethylallenimine, **V**, 541
Ethyl 3-(1-azetidinyl)
    propanoate, **VI**, 75
Ethyl p-dimethylaminophenyl-
    acetate, **V**, 552
Ethylenimine, **IV**, 433
N-Ethylpiperidine, **V**, 575
1,4,7,10,13,16-Hexaazacyclo-
    octadecane, **VI**, 652
3-(1-Hydroxybutyl)-1-
    methylpyrrole **VII**, 102,
(S)-(+)-2-Hydroxymethylpyrro-
    lidine, **VIII**, 26
Imidazole-1-carboxaldehyde,
    VII, 282
2-Mercapto-4-amino-5-carb-
    ethoxypyrimidine, **IV**, 566
(S)-(+)-2-Methoxymethylpyrro-
    lidine, **VIII**, 26
1-(4-Methoxyphenyl)-1,2,5,6-
    tetrahydropyridine,
    VIII, 358
N-Methylaminopyrimidine,
    IV, 336

1-Phenyl-3-amino-5-pyra-
zolone, **III**, 708
4-Phenyl-5-anilino-1,2,3-tri-
azole, **IV**, 380
3-Phenyl-2*H*-azirine-2-carbox-
aldehyde, **VI**, 893
N-Phenylcarbazole, **I**, 547
1-Phenyl-3,4-dihydro-6,7-meth-
ylenedioxyisoquinoline,
VII, 326
2-Propyl-1-azacycloheptane,
VIII, 568
Pseudopelletierine, **IV**, 816
Quinoline, **I**, 478
3-Quinuclidone hydrochloride,
V, 989
(–)-Salsolidine, **VIII**, 573
1,2,3,4-Tetrahydro-β-carboline,
VI, 965
1,2,3,4-Tetrahydronaphthalene
(1,2)imine, **VI**, 967
2,3,4,5-Tetrahydropyridine
trimer, **VI**, 968
Uramil, **II**, 617
D. Miscellaneous
6-Amino-3,4-dimethyl-*cis*-3-
cyclohexen-1-ol, **VII**, 2
6-Bromo-3,4-dimethoxyben-
zaldehyde cyclohexylim-
ine, **VIII**, 586
N-tert-Butyl-*N-tert*-octyl-*O-*
tert-butylhydroxylamine,
VIII, 93
(*E*)-(Carboxyvinyl)trimethylam-
monium betaine, **VIII**, 536
Diethyl *N*-benzylidenamino-
methylphosphonate,
VIII, 451
[3-(Dimethylamino)-2-azaprop-
2-en-1-ylidene]dimethyl-
ammonium chloride,
VII, 197
[5-(Dimethylamino)-2,4-penta-
dienylidene]dimethylam-
monium chloride (in solu-
tion), **VII**, 15

N,N-Dimethyl(chloromethyl-
ene)ammonium chloride,
VIII, 498
N,N-Dimethyldodecylamine
oxide, **VI**, 501
1,4-Diphenyl-1-methyl-3-aza-
1,4-butadiene, **VIII**, 451,
455
O-Diphenylphosphinylhy-
droxylamine, **VII**, 9
6-Iodo-3,4-dimethoxybenzalde-
hyde cyclohexylimine,
VIII, 586
(*S*)-(+)-4-Methyl-3-heptanone
SAMP hydrazone,
VIII, 403
3-Pentanone SAMP-hydrazone,
VIII, 403
Phenylhydroxylamine, **VIII**,
16
E. Quaternary Ammonium
Compounds
1-Aminopyridinium iodide,
V, 43
Benzoylcholine chloride, **IV**, 84
Benzoylcholine iodide, **IV**, 84
Benzyltrimethylammonum
ethoxide, **IV**, 98
Benzyltrimethylammonium
iodide, **IV**, 585
Betaine hydrazide hydrochlo-
ride, **II**, 85
(*E*)-(Carboxyvinyl)trimethylam-
monium betaine, **VIII**, 536
N-Chlorobetainyl chloride,
IV, 154
N,N-Dimethylaminomethylfer-
rocene methiodide, **V**, 434
2,3-Dimethylbenzyltrimethyl-
ammonium iodide, **IV**, 331
N,N-Dimethyl(chloromethylene)ammonium chloride,
VIII, 498
1-Ethyl-3-(3-dimethylamino)
propylcarbodiimide
methiodide, **V**, 555

(2-Methylbenzyl)ethyldimethyl-
ammonium bromide,
IV, 582
p-Methylbenzyltrimethyl-
ammonium bromide,
V, 884
Methyl(carboxysulfamoyl)tri-
ethylammonium hydrox-
ide, **VI**, 788
Phenyltrimethylammonium sulfomethylate, **VI**, 175
Phenyltrimethylammonium tribromide, **VI**, 175
Tetrabutylammonium fluoride,
VII, 514
Tetramethylammonium
1,1,2,3,3-pentacyanopro-
penide, **V**, 1013
2-Thienylmethylhexamethyl-
enetetrammonium chlo-
ride, **III**, 811
Tris(tetrabutylammonium)
hydrogen pyrophosphate
trihydrate, **VIII**, 616

AZIDES
4-Azido-3-chloro-5-isopropoxy-
2(5*H*)-furanone (in solution),
VIII, 116
(1-Azido-2-iodo-3,3-dimethoxy-
propyl)benzene, **VI**, 893
2-Azido-2-phenyladamantane,
VII, 433
tert-Butyl azidoformate, **V**, 157
Diphenyl phosphorazidate, **VII**,
135
m-Nitrobenzazide, **IV**, 715
o-Nitrophenyl azide, **IV**, 75
Phenyl azide, **III**, 710
Trimethylsilyl azide, **VI**, 1030

AZO and AZOXY COMPOUNDS
Azobenzene, **III**, 103
1,1'-Azo-bis-1-cyclohexanenitrile,
IV, 66
2,2'-Azo-bis-isobutyronitrile, **IV**, 67

Azo dye from *o*-toluidine and
Chicago acid, **II**, 145
Azoethane, **VI**, 78
Azoxybenzene, **II**, 57
Bis(2,2,2-trichloroethyl) azodicar-
boxylate, **VII**, 56
tert-Butyl azodiformate, **V**, 160
2-(1-Cyanocyclohexyl)diazenecar-
boxylic acid methyl ester,
VI, 334
4,4'-Diaminoazobenzene, **V**, 341
Diazoaminobenzene, **II**, 163
Diethyl azodicarboxylate, **III**, 375;
IV, 411
Dimethyl azodicarboxylate, **IV**, 414
Methyl Red, **I**, 374
1-Methyl-3-*p*-tolyltriazene, **V**, 797
1-(*m*-Nitrophenyl)-3,3-dimethyltri-
azene, **IV**, 718
p-Phenylazobenzoic acid, **III**, 711
p-Phenylazobenzoyl chloride,
III, 712

CARBAMATES
Benzyl *trans*-1,3-butadiene-1-carba-
mate, **VI**, 95
Benzyl carbamate, **III**, 168
Benzyloxycarbonyl-L-alanine-L-cys-
teine methyl ester, **VII**, 30
Benzyloxycarbonyl-L-aspartyl-(*tert*-
butyl ester)-L-phenylalanyl-L-
valine methyl ester, **VII**, 30
N-*tert*-Butoxycarbonyl-L-leucinal,
VIII, 68
N-*tert*-Butoxycarbonyl-L-phenylala-
nine, **VII**, 70, 75
tert-Butoxycarbonyl-L-proline,
VI, 203
Butyl carbamate, **I**, 140
tert-Butyl carbamate, **V**, 162
tert-Butyl *N*-(1-ethoxycyclopropyl)
carbamate, **VI**, 226
Carbobenzoxyglycine, **III**, 168
Carbobenzyloxy-L-asparaginyl-L-
leucine methyl ester,
VI, 263

N-Carbomethoxypyrrolidine,
VII, 307
N,N'-Dicarbethoxyputrescine,
IV, 822
Ethyl *N*-benzylcarbamate, **IV**, 780
Ethyl *threo*-[1-(2-hydroxy-1,2-
diphenylethyl)]carbamate,
VII, 223
Ethyl *N*-methylcarbamate, **II**, 278
Ethyl *N*-nitroso-*N*-(*p*-tolylsulfonyl-
methyl)carbamate, **VI**, 981
Ethyl *N*-(*p*-tolylsulfonylmethyl)car-
bamate, **VI**, 981
2-Methoxy-*N*-carbomethoxypyrroli-
dine, **VII**, 307
Methyl(carboxysulfamoyl)triethyl-
ammonium hydroxide, **VI**, 788
Methyl (chlorosulfonyl)carbamate,
VI, 788
Methyl *N,N*-dichlorocarbamate,
VIII, 427
Methyl *N*-hexylcarbamate, **VI**, 788
Methyl (*trans*-2-iodo-1-tetralin)car-
bamate, **VI**, 795
Methyl *N*-(2-methyl-2-butenyl)car-
bamate, **VIII**, 427
O-2-Naphthyl dimethylthiocarba-
mate, **VI**, 824
Nitrosomethylurethane, **II**, 464
(*S*)-3-(1-Oxopropyl)-4-(phenyl-
methyl)-2-oxazolidinone,
VIII, 339
(*S*)-4-(Phenylmethyl)-2-oxazolidi-
none, **VIII**, 339, 528

CARBODIIMIDES
Diphenylcarbodiimide (Method I),
V, 501
Diphenylcarbodiimide (Method II),
V, 504
1-Ethyl-3-(3-dimethylamino)propyl-
carbodiimide hydrochloride,
V, 555
1-Ethyl-3-(3-dimethylamino)propyl-
carbodiimide methiodide,
V, 555

CARBOHYDRATE DERIVATIVES
2-Acetamido-3,4,6-tri-*O*-acetyl-2-
deoxy-α-D-glucopyranosyl
chloride, **V**, 1
Acetobromoglucose, **III**, 11
L-Arabinose, **I**, 67
D-Arabinose, **III**, 101
Cellobiose, **II**, 122
α-Cellobiose octaacetate, **II**, 124
3-Deoxy-1,2:5,6-di-*O*-isopropyli-
dene-α-D-ribo-hexofuranose,
VII, 139
1-Deoxy-2,3,4,6-tetra-*O*-acetyl-1-(2-
cyanoethyl)-α-D-glucopyranose
VIII, 148
2,3-Di-*O*-benzyl-2,3-di-*O*-isopropyl-
idene-L-threitol, **VIII**, 155
1,2:5,6-Di-(*O*-isopropylidene-3-*O*-
(*S*-methyl dithiocarbonate)-α-D-
glucofuranose, **VII**, 139
Diisopropyl (2*S*,3*S*)-2,3-*O*-isopro-
pylidenetartrate, **VIII**, 201
Dimethyl 2,3-*O*-isopropylidenetar-
trate, **VIII**, 155, 201
β-Gentiobiose octaacetate, **III**, 428
D-Glucosamine hydrochloride,
III, 430
β-D-Glucose-1,2,3,4-tetraacetate,
III, 432
β-D-Glucose-2,3,4,6-tetraacetate,
III, 434
D-Gulonic-γ-lactone, **IV**, 506
D-Mannose, **III**, 541
Methyl 4-*O*-benzoyl-6-bromo-6-
deoxy-α-D-glucopyranoside,
VIII, 363,
Methyl 4,6-*O*-benzylidene-α-D-glu-
copyranoside, **VIII**, 363
α-Methyl D-glucoside, **I**, 364
α-Methyl mannoside, **I**, 371
Pentaacetyl D-glucononitrile, **III**,
690
2,3,4,5,6-Penta-*O*-acetyl-D-gluconic
acid, **V**, 887
2,3,4,5,6-Penta-*O*-acetyl-D-gluconyl
chloride, **V**, 887

1,3,4,6-Tetra-*O*-acetyl-2-deoxy-α-D-
glucopyranose, **VIII**, 583
2,3,4,6-Tetramethyl-D-glucose,
III, 800
6-Trityl-β-D-glucose-1,2,3,4-tetraac-
etate, **III**, 432

CHIRAL AUXILIARIES
(*S*)-(–)- and (*R*)-(+)-1-Amino-2-
methoxymethylpyrrolidine
(SAMP/RAMP), **VIII**, 26, 403
1,1'-Binaphthalene-2,2'-diol,
VIII, 46
(*R,R*)-1,2-Bis(*o*-anisylphenylphos-
phino)ethanerhodium trifluo-
romethanesulfonate,
(DIPAMPRh⁺), **VIII**, 423
(*R*)-(+)- and (*S*)-(–)-2,2'-Bis
(diphenylphosphino)-1,1'-
binaphthyl (BINAP), **VIII**,
57, 183
D-(–)-2,10-Camphorsultam **VIII**, 110
1,4-Di-*O*-benzyl-L-threitol, **VIII**, 155
Diethyl (2*S*,3*S*)-tartrate, **VIII**, 464
(*S*)-*N,N*'-Dimethyl-*N*'-(1-*tert*-butoxy-
3-methyl-2-butyl)formamidine,
VIII, 204, 573
Hexahydro-3-(hydroxymethyl)-8a-
methyl-2-phenyl[2*S*,3*S*,8a*R*]-5-
oxo-5*H*-oxazolo[3,2-*a*]pyridine,
VIII, 241
Hexahydro-4,4,7-trimethyl-4*H*-1,3-
benzoxathiin, **VIII**, 302
(–)-Menthol, **VIII**, 141, 350
trans-2-Phenylcyclohexanol,
(1*R*,2*S*)-(–)- and (1*S*,2*R*)-(+)-,
VIII, 516
(–)-8-Phenylmenthol, **VIII**, 522
(*S*)-4-(Phenylmethyl)-2-oxazolidi-
none, **VIII**, 339, 528
(–)-α-Pinene, **VIII**, 553

CYANAMIDES
o-Chlorophenylcyanamide, **IV**, 172
Diallylcyanamide, **I**, 203
Dibutylcyanamide, **I**, 204

N-Methyl-1-naphthylcyanamide, **III**,
608
α-Naphthycyanamide, **IV**, 174

DIAZO and DIAZONIUM COMPOUNDS
Benzenediazonium-2-carboxylate,
V, 54
4,4'-Biphenylene-bis-diazonium fluoborate, **II**, 188
Bis(trifluoromethyl)diazomethane,
VI, 161
o-Bromobenzenediazonium hexafluorophosphate, **V**, 133
2-Bromo-4-methylbenzenediazoni-
um chloride, **V**, 140
tert-Butyl diazoacetate, **V**, 179
tert-Butyl diazoacetoacetate, **V**, 180
p-Carbethoxybenzenediazonium
fluoborate, **II**, 299
p-Chlorobenzenediazonium chloride,
VII, 105
Crotyl diazoacetate, **V**, 258
Diazoacetophenone, **VI**, 386
1-(Diazoacetyl)naphthalene, **VI**, 613
16-Diazoandrost-5-en-3β-ol-17-one,
VI, 840
2-Diazocyclohexanone, **VI**, 389
Diazomethane, **II**, 165; **III**, 244; **IV**,
225, 250; **V**, 351, 878
1-Diazo-4-phenyl-2-butanone, **VIII**,
196
2-Diazopropane, **VI**, 392
Di-*tert*-butyl diazomalonate, **VI**, 414
Dideutereodiazomethane, **VI**, 432
Diphenyldiazomethane, **III**, 351
Ethyl diazoacetate, **III**, 392; **IV**, 424
β-Naphthalenediazonium chloride–
mercuric chloride compound,
II, 432
o-Nitrobenzenediazonium fluoborate, **II**, 226
p-Nitrobenzenediazonium fluoborate, **II**, 225; **III**, 666
Phenylbenzoyldiazomethane, **II**, 496
Phenyldiazomethane, **VII**, 438

p-Tolylsulfonyldiazomethane, **VIII**, 981

m-Trifluoromethylbenzenediazonium chloride, **VII**, 508

Trimethylsilyldiazomethane, **VIII**, 612

ENAMINES AND IMINES

1-[α-(Benzalamino)benzyl]-2-naphthol, **I**, 381

Benzalaniline, **I**, 80

7-Benzylidenaminotheophylline, **VII**, 9

Benzylidenearginine, **II**, 49

N-Benzylidenebenzenesulfonamide, **VIII**, 546

N-Benzylideneethylamine, **V**, 758

N-Benzylidenemethylamine, **IV**, 605

1-Benzylindole, **VI**, 104

3-Benzylindole, **VI**, 109

(*E*)-6-Benzyloxy-2-nitro-β-pyrrolidinostyrene, **VII**, 34

6-Bromo-3,4-dimethoxybenzaldehyde cyclohexylimine, **VIII**, 586

2-*tert*-Butoxycarbonyloxyimino-2-phenylacetonitrile, **VI**, 199

N-Chlorocyclohexylidenimine, **V**, 208

1-Chloro-*N,N*,2-trimethylpropenylamine, **VI**, 282; **VIII**, 441

1-Chloro-*N,N*,2-trimethylpropylideniminium chloride, **VI** , 282

1-(Cyclooctylidenemethyl)piperidine, **VII**, 473

2,6-Dibromoquinone-4-chloroimide, **II**, 175

Diethyl *N*-benzylidenaminomethylphosphonate, **VIII**, 451

(*R*)-(−)-*N,N*-Diethyl-(*E*)-citronellalenamine, **VIII**, 183

Diethyl 2-(cyclohexylamino)vinylphosphonate, **VI**, 448

N,N-Diethyl-1,2,2-trichlorovinylamine, **V**, 387

Diphenyl ketimine, **V**, 520

Diphenylmethanimine hydrochloride, **II**, 234

Ethyl α-(hexahydroazepinylidene-2)acetate, **VIII**, 263

N-Ethylidenecyclohexylamine, **VI**, 901

Hexafluoroacetone imine, **VI**, 664

N-(3-Hydroxy-3,3-diphenylpropylidene)cyclohexylamine, **VI**, 901

6-Iodo-3,4-dimethoxybenzaldehyde cyclohexylimine, **VIII**, 586

Isopropylidene α-(hexahydroazepinylidene-2)malonate, **VIII**, 263

N-(α-Methylbenzylidene)methylamine, **VI**, 818

N-(2-Methylbutylidene-1-*d*)-1,1,3,3-tetramethylbutylamine, **VI**, 751

O-Methylcaprolactim, **IV**, 588

Methylenaminoacetonitrile, **I**, 355

N-Methyl-2-phenyl-Δ²-tetrahydropyridine, **VI**, 818

1-(2-Methylpropenyl)pyrrolidine, **VI**, 496

N-(2-Methylpropylidene)-*tert*-butylamine, **VI**, 526

1-Morpholino-1-cyclohexene, **V**, 808

α-Morpholinostyrene, **VI**, 520

1-(*N*-Pyrrolidino)-1-cyclododecene, **VII**, 135

1-Pyrrolidinocyclohexene, **VI**, 1014

cis-*N*-Tosyl-3-methyl-2-azabicyclo[3.3.0]oct-3-ene, **VII**, 501

ESTERS (including ortho esters)

A. Of Unsubstituted Monobasic Acids

Acenaphthenol acetate, **III**, 3

2-Acetamido-3,4,6-tri-*O*-acetyl-2-deoxy-α-D-glucopyranosyl chloride, **V**, 13

4-Acetoxy-2-azetidin-2-one, **VIII**, 3

4-Acetoxybenzoic acid, **VI**, 577

1-Acetoxy-4-benzylamino-2-butene, **VIII**, 9

(*S*)-(−)-2-Acetoxy-1-bromo-
    propane, **VII**, 356
1-Acetoxy-4-chloro-2-butene,
    VIII, 9
4-Acetoxy-3-chloro-1-butene,
    VIII, 9
3β-Acetoxy-5α-cyano-
    cholestan-7-one, **VI**, 14
4-Acetoxy-2-cyclopenten-1-
    one, **VIII**, 15
1-Acetoxy-4-diethylamino-2-
    butene, **VIII**, 9
3β-Acetoxyetienic acid, **V**, 8
cis-3-Acetoxy-5-hydroxycy-
    clopent-1-ene, **VIII**, 13
(1*R*,2*r*,3*S*)-3-Acetoxy-2-nitro-1-
    cyclohexyl acetate,
    VIII, 332
2-Acetoxy-*trans*-1-phenyl-
    propene, **VI**, 692
7-Acetoxy-4,4,6,7-tetramethyl-
    bicyclo[4.2.0]octan-2-one,
    VI, 1024
Acetylmandelic acid, **I**, 12
Acetylmandelyl chloride, **I**, 13
2-*p*-Acetylphenylhydroquinone
    diacetate, **IV**, 16
Benzoin acetate, **II**, 69
Benzoylcholine chloride, **IV**, 84
Benzoylcholine iodide, **IV**, 84
o-Benzoyloxyacetophenone,
    IV, 478
3-Benzoyloxycyclohexene,
    V, 70
Benzyl benzoate, **I**, 104
p-Bromobenzaldiacetate, **II**, 442
2-Bromo-3,3-diphenyl-2-
    propen-1-ol acetate,
    VI, 187
10-Bromo-11-hydroxy-10,11-
    dihydrofarnesyl acetate,
    VI, 560
trans, *trans*-1,3-Butadiene-1,4-
    diyl diacetate, **VI**, 196
tert-Butyl acetate, **III**, 141;
    IV, 263

Butyl acrylate, **III**, 146
Butyl butyrate, **I**, 138
sec-Butyl crotonate, **V**, 762
tert-Butyl cyclohexanecarboxy-
    late, **VII**, 87
Butyl oleate, **II**, 469
tert-Butyl *p*-toluate, **VI**, 259
4-Chlorobutyl benzoate,
    III, 187
2-Chloroethyl benzoate, **IV**, 84
γ-Chloropropyl acetate, **III**, 203
Cholesteryl acetate, **II**, 193
2-Cyclohepten-1-yl acetate,
    VIII, 137
1-Deoxy-2,3,4,6-tetra-*O*-acetyl-
    1-(2-cyanoethyl)-α-D-glu-
    copyranose, **VIII**, 148
trans-7,8-Diacetoxybicyclo
    [4.2.0]octa-2,4-diene,
    VI, 196
2,6-Diacetoxybicyclo[3.3.0]
    octane, **VIII**, 43
trans, *trans*-1,4-Diacetoxy-1,3-
    butadiene, **VI**, 196
3,12-Diacetoxy-bisnor-
    cholanyldiphenylethylene,
    III, 237
Diacetyl-D-tartaric anhydride,
    IV, 242
2,3-Dimethylbenzyl acetate,
    IV, 584
2',6'-Dimethylphenyl propa-
    noate, **VII**, 190
10,11-Epoxyfarnesyl acetate,
    VI, 560
Ethyl 5β-cholest-3-ene-5-
    acetate, **VI**, 584
Ethyl cinnamate, **I**, 252
Ethyl cyclohexylideneacetate,
    V, 547
Ethyl cyclopropylpropiolate,
    VIII, 247
Ethyl (2*E*,4*Z*)-2,4-decadienoate,
    VIII, 251
Ethyl 3,4-decadienoate,
    VIII, 251

Ethyl 2,4-diphenylbutanoate,
  V, 559
Ethyl isocrotonate, **VII**, 227
Ethyl 4-methyl-(*E*)-4,8-nona-
  dienoate, **VI**, 606
Ethyl α-naphthoate, **II**, 282
Ethyl 1-naphthylacetate,
  VI, 613
Ethyl 2,3-pentadienoate,
  VII, 232
Ethyl phenylacetate, **I**, 270
Ethyl γ-phenylbutyrate, **II**, 196
Ethyl *cis* and *trans*-2-phenylcy-
  clopropanecarboxylate,
  VI, 913
Ethyl stearate, **IV**, 142
Ethyl tetrolate, **VII**, 226
Ethyl tridecylate, **II**, 292
Farnesyl acetate, **VI**, 560
Furfural diacetate, **IV**, 489
2-Furylmethyl acetate, **I**, 285
Hexahydrogallic acid triacetate,
  V, 591
Hydroquinone diacetate,
  III, 452
Hydroxyhydroquinone triac-
  etate, **I**, 317
(1*S*,2*S*,3*R*)-3-Hydroxy-2-nitro-
  1-cyclohexyl acetate,
  VIII, 332
2-Iodoethyl benzoate, **IV**, 84
18,20-Lactone of 3β-acetoxy-
  20β-hydroxy-5-pregnene-
  18-oic acid, **V**, 692
(−)-Menthyl cinnamate,
  VIII, 350
erythro-2-Methanesulfonyloxy-
  2-butyl cyclobutanecar-
  boxylate, **VI**, 312
Methyl 4-*O*-benzoyl-6-bromo-
  6-deoxy-α-D-glucopyra-
  noside, **VIII**, 363
o-Methylbenzyl acetate, **IV**,
  582
Methyl butadienoate, **V**, 734
Methyl cyclohexanecarboxylate, **VI**, 493

trans-2-Methyl-1-cyclohexen-1-
  yl acetate, **VI**, 121
Methyl 2,4-cyclopentadiene-1-
  acetate (in solution),
  VII, 339
Methyl cyclopentanecarboxy-
  late, **IV**, 594
Methyl 1-cycloundecenecar-
  boxylate, **VI**, 368
Methyl *p*-ethylbenzoate,
  IV, 580
Methyl 2-hexynoate, **VI**, 791
Methyl 2-methyl-1-cyclohex-
  ene-1-carboxylate,
  VII, 351
Methyl myristate, **III**, 605
Methyl palmitate, **III**, 605
o- and *p*-Nitrobenzaldiacetate,
  IV, 713
p-Nitrobenzyl acetate, **III**, 650
1,5-Pentanediol diacetate,
  IV, 748
2,3,4,5,6-Penta-*O*-acetyl-D-glu-
  conic acld, **V**, 887
2,3,4,5,6-Penta-*O*-acetyl-D-glu-
  conyl chlorlde, **V**, 887
Phenyl cinnamate, **III**, 714
2-Phenylethyl benzoate, **V**, 336
1-Phenylpropen-2-yl acetate
  VI, 692,
Phenyl propionate, **II**, 544
1,3,4,6-Tetra-*O*-acetyl-2-deoxy-
  α-D-glucopyranose,
  VIII, 583
Trimyristin, **I**, 538
Vinyl laurate, **IV**, 977
B. Of Substituted Monobasic
    Acids
1. Cyano Esters
    tert-Butyl cyanoacetate,
      V, 171
    2-(1-Cyanocyclohexyl)
      diazenecarboxylic
      acid, methyl ester,
      VI, 334
    Ethyl butylcyanoacetate,
      III, 385

Ethyl *sec*-butylidene-
cyanoacetate, **IV**, 94
Ethyl cyanoacetate, **I**, 254
Ethyl 1-cyano-2-methyl-
cyclohexanecarboxy-
late, **VI**, 586
Ethyl (*E*)-2-cyano-6-
octenoate, **VI**, 586
Ethyl cyano(pentafluoro-
phenyl)acetate, **VI**, 873
Ethyl α-cyano-β-phenyl-
acrylate, **I**, 451
Ethyl (1-ethylpropenyl)
methylcyanoacetate,
III, 397
Ethyl (1-ethylpropyli-
dene)cyanoacetate,
III, 399
Ethyl phenylcyanoace-
tate, **IV**, 461
Ethyl β-phenyl-β-
cyanopropionate,
IV, 804
Ethyl phenylcyanopyru-
vate, **II**, 287
Ethyl (1-phenylethyli-
dene)cyanoacetate,
IV, 463
Methyl 2-(1-cyanocyclo-
hexyl)hydrazinecar-
boxylate, **VI**, 334
Methyl ω-cyanopelar-
gonate, **III**, 584
α-Phenyl-α-carbethoxy-
glutaronitrile, **IV**, 776
2. Halogen Esters
tert-Butyl bromoacetate,
IV, 263
sec-Butyl α-bromopropi-
onate, **IV**, 122
tert-Butyl chloroacetate,
IV, 263
tert-Butyl α-chloropropi-
onate, **IV**, 263
(*Z*)-4-Chloro-4-hexenyl
trifluoroacetate,
VI, 273

6-Chloromethyl-2,2-
dimethyl-1,3-dioxen-
4-one, **VIII**, 192
Ethyl bromoacetate,
III, 381
Ethyl γ-bromobutyrate,
V, 545
Ethyl α-(bromomethyl)
acrylate, **VII**, 210;
VIII, 266
Ethyl β-bromopropi-
onate, **I**, 246
Ethyl chloroacetate,
II 263
Ethyl α-chloroacetoac-
etate, **IV**, 592
Ethyl chlorofluoroace-
tate, **IV**, 423
Ethyl α-chlorophenylace-
tate, **IV**, 169
Ethyl α,β-dlbromo-β-
phenylpropionate,
II, 270
Ethyl 2-fluorohexanoate,
(±)-, (+)-, and (−)-,
VIII, 258
Ethyl 4-iodobutyrate,
VIII, 274
Methyl 4-bromo-1-butan-
imidate hydrochlo-
ride, **VIII**, 415
Methyl α-(bromomethyl)
acrylate **VII**, 319
Methyl β-bromopropi-
onate, **III**, 576
Methyl 5-bromovalerate,
III, 578
Methyl 4-chloro-2-
butynoate, **VIII**, 371
Methyl 2-chloro-2-cyclo-
propylidenacetate,
VIII, 373, 374
Methyl β,β-dibro-
moisobutyrate,
VII, 319
Methyl *N*,*N*-dichlorocar-
bamate, **VIII**, 427

Methyl 2-iodo-3-nitro-
propionate, **VI**, 799

(1*R*,2*S*,5*R*)-5-Methyl-2-
[(1-methyl-1-phenyl-
ethyl)]cyclohexyl
chloroacetate,
VIII, 522

trans-2-Phenylcyclo-
hexyl chloroacetate
(±)-, (+)-, (−)-,
VIII, 516

Trimethyl 2-chloro-2-
cyclopropylidenortho-
acetate, **VIII**, 373

Trimethyl *ortho*-4-bromo-
butanoate, **VIII**, 415

Vinyl chloroacetate,
III, 853

3. Hydroxy Esters
Allyl lactate, **III**, 46

p-Chlorophenyl salicy-
late, **IV**, 178

2',6'-Dimethylphenyl
(2*SR*,3*SR*)-2,4-
dimethyl-3-hydroxy-
pentanoate, **VII**, 190

Ethyl 4-ethyl-2-methyl-
3-hydroxyoctanoate,
IV, 447

Ethyl (*S*)-(+)-3-hydroxy-
butanoate, **VII**, 215

Ethyl 4-hydroxycroto-
nate, **VII**, 221

Ethyl 1-hydroxycyclo-
hexylacetate **VI**, 598

Ethyl β-hydroxy-β,β-
diphenylpropionate,
V, 564

Ethyl α-(hydroxymethyl)
acrylate, **VIII**, 265

Ethyl mandelate, **IV**, 169

Ethyl β-phenyl-β-
hydroxypropionate,
III, 408

Isopropyl lactate, **II**, 365

Methyl 4-hydroxy-2-
butynoate, **VII**, 334

Methyl (1*R*,5*R*)-5-
hydroxy-2-cyclopen-
tene-1-acetate, **VII**,
339

Methyl 7-hydroxyhept-5-
ynoate, **VIII**, 415

Methyl 3-hydroxy-2-
methylenepentanoate,
VIII, 420

Methyl *anti*-3-hydroxy-
2-methylpropanoate,
VIII, 420

4. Keto Esters
tert-Butyl acetoacetate,
V, 155

tert-Butyl *o*-benzoylben-
zoate, **IV**, 263

sec-Butyl α-caproylpro-
pionate, **IV**, 120

Butyl glyoxylate, **IV**, 124

2-Carbethoxycycloocta-
none, **V**, 198

2-Carbethoxycyclo-
pentanone, **II**, 116

1-Carbomethoxy-1-
methylethyl 3-oxobu-
tanoate, **VIII**, 71

2-Carbomethoxy-3-vinyl-
cyclopentanone,
VIII, 112

Diethyl dioxosuccinate,
VIII, 597

3,5-Dimethyl-4-car-
bethoxy-2-cyclohex-
en-1-one, **III**, 317

1,3-Dimethyl-5-oxobicy-
clo[2.2.0]octane-2-
carboxylic acid,
methyl ester,
VIII, 219

(1-Ethenyl)heptanyl 3-
ketobutanoate,
VIII, 235

Ethyl acetoacetate, **I**, 235
Ethyl α-acetoglutarate,
 II, 263
Ethyl acetonedioxalate,
 II, 126
Ethyl acetopyruvate,
 I, 238
Ethyl α-acetyl-β-(2,3-
 dimethoxyphenyl)
 acrylate, **IV**, 408
Ethyl α-acetyl-β-(3,4-
 dimethoxyphenyl)
 acrylate, **IV**, 410
Ethyl α-acetyl-β-(2,3-
 dimethoxyphenyl)
 propionate, **IV**, 408
Ethyl α-acetyl-β-(3,4-
 dimethoxyphenyl)
 propionate, **IV**, 410
Ethyl benzoylacetate,
 II, 266; **III**, 379;
 IV, 415
Ethyl benzoylacetoac-
 ctate, **II**, 266
Ethyl benzoyldimethyl-
 acetate, **II**, 268
Ethyl benzoylformate,
 I, 241
Ethyl butylacetoacetate,
 I, 248
Ethyl 2-butyrylacetate,
 VII, 213
Ethyl α-chloroacetoace-
 tate, **IV**, 592
Ethyl 4-cyclohexyl-4-
 oxobutanoate,
 VIII, 578
Ethyl diacetylacetate,
 III, 390
Ethyl isobutyrylisobu-
 tyrate, **II**, 270
Ethyl α-isopropylace-
 toacetate, **III**, 405
Ethyl 2-ketocyclohexyl-
 glyoxalate, **II**, 531

Ethyl 2-ketohexahy-
 drobenzoate, **II**, 532
Ethyl 1-(p-methoxy-
 phenyl)-2-oxocyclo-
 hexanecarboxylate,
 VII, 229
Ethyl (E)-(4-nitro-
 phenyl)-4-oxo-2-
 butanoate, **VIII**, 268
Ethyl 4-oxohexanoate,
 VI, 615
Ethyl 5-oxo-6-methyl-6-
 heptenoate, **VIII**, 274
Ethyl α-phenylacetoace-
 tate, **II**, 284
Ethyl pyruvate, **IV**, 467
Methyl p-acetylbenzoate,
 IV, 579
Methyl benzoylformate,
 I, 244
Methyl diformylacetate,
 VII, 323
Methyl 1,3-dimethyl-5-
 oxobicyclo[2.2.2]oct-
 ane-2-carboxylate,
 VIII, 219
Methyl-4-keto-7-methyl-
 octanoate, **III**, 601
Methyl 2-oxocyclohex-
 anecarboxylate,
 VII, 351
Methyl 6-oxodecanoate,
 VIII, 441
Methyl 7-oxoheptanoate,
 VI, 807
Methyl 6-oxohexanoate,
 VII, 168
Methyl 7-oxooctanoate,
 VI, 807
Methyl 2-oxo-5,6,7,8-
 tetrahydro-2H-1-ben-
 zopyran-3-carboxy-
 late, **VIII**, 444
Methyl phenylacetylace-
 tate, **VII**, 359

Methyl pyruvate, **III**, 610

5. Lactones

α-Acetyl-δ-chloro-γ-valerolactone, **IV**, 10

3-Acetyl-4-hydroxy-5,5-dimethylfuran-2(5*H*)-one, **VIII**, 71

4-Azido-3-chloro-5-isopropoxy-2(5*H*)-furanone, **VIII**, 116

Benzalphthalide, **II**, 61

α-Benzylidene-γ-phenyl-Δ$^{\beta,\gamma}$-butenolide, **V**, 80

3-Bromophthalide, **V**, 145

(*S*)-(+)-γ-Butyrolactone-γ-carboxylic acid, **VII**, 99

γ-Caprilactone, **IV**, 432

3-Carbethoxycoumarin, **III**, 165

Coumalic acid, **IV**, 201

γ-Crotonolactone, **V**, 255

Dehydroacetic acid, **III**, 231

3,4-Dichloro-5-isopropoxy-2(5*H*)-furanone, **VIII**, 116

cis-4,5-Dihydro-5-iodomethyl-4-phenyl-2(3*H*)-furanone, **VII**, 164

trans-4,5-Dihydro-5-iodomethyl-4-phenyl-2(3*H*)-furanone, **VII**, 164

5,6-Dihydro-2*H*-pyran-2-one, **VI**, 562

6,7-Dimethoxy-3-isochromanone, **VI**, 471

4,6-Dimethylcoumalin, **IV**, 337

Dimethylketene β-lactone dimer, **V**, 456

D-Erythronolactone, **VII**, 297

4-Ethyl-4-hydroxy-2-methyloctanoic acid γ-lactone, **IV**, 447

Ethyl isodehydroacetate, **IV**, 549

2(5*H*)-Furanone, **VIII**, 396

D-Gulonic-γ-lactone, **IV**, 506

11-Hydroxyundecanoic lactone, **VI**, 69

Isodehydroacetic acid, **IV**, 549

2,3-*O*-Isopropylidene-D-erythronolactone, **VII**, 297

Ketene dimer, **III**, 508

Ketopantoyl lactone, **VII**, 417

18,20-Lactone of 3β-Acetoxy-20β-hydroxy-5-pregnene-18-oic acid, **V**, 692

4-Methoxycarbonyl-1,1,6-trimethyl-1,4,4a,5,6,7,8,8a-octahydro-2,3-benzopyrone (±)-, (+)-, (−)-, **VIII**, 353

Methyl coumalate, **IV**, 532

3-Methyl-2(5*H*)-furanone, **VIII**, 396

Methyl 2-oxo-5,6,7,8-tetrahydro-2*H*-1-benzopyran-3-carboxylate, **VIII**, 444

β-Methyl-δ-valerolactone, **IV**, 677

(*R,S*)-Mevalonolactone-2-^{13}C, **VII**, 386

γ-Octyl-γ-butyrolactone, **VII**, 400

(1*R*,6*S*)-(+)-8-Oxabicyclo[4.3.0] nonan-7-one, **VII**, 406

D-(−)-Pantoyl lactone,
  VII, 417
Phthalide, **II**, 526
unsym-o-Phthalyl chloride, **II**, 528
8-Propionyl-(*E*)-5-nonenolide, **VIII**, 562
γ-Propylbutyrolactone,
  III, 742
2*H*-Pyran-2-one, **VI**, 462
α-Pyrone, **V**, 982
Ricinelaidic acid lactone,
  VII, 470
6. Miscellaneous
Benzyloxycarbonyl-L-alanyl-L-cysteine methyl ester, **VII**, 30
Benzyloxycarbonyl-L-aspartyl-(*tert*-butyl ester)-L-phenylalanine-L-valine methyl ester, **VII**, 30
tert-Butyl azidoformate,
  V, 157
tert-Butyl azodiformate,
  V, 157
tert-Butyl carbazate,
  V, 166
S-tert-Butyl-L-cysteine *tert*-butyl ester,
  VI, 255
tert-Butyl diazoacetate,
  V, 179
tert-Butyl diazoacetoacetate, **V**, 180
Butyl glyoxylate, **IV**, 124
3-Carbethoxycoumarin,
  III, 165
Carbobenzyloxy-L-asparaginyl-L-leucine methyl ester, **VI**, 263
N-Carbobenzyloxy-3-hydroxy-L-prolylglycylglycine ethyl ester,
  VI, 263

3-Carbomethoxy-3-methyl-6-trimethylsilyl-1-hexene,
  VIII, 487
N-Carbomethoxypyrrolidine, **VII**, 307
Crotyl diazoacetate, **V**, 258
Di-β-carbethoxyethylmethylamine, **III**, 258
6-Diethylphosphonomethyl-2,2-dimethyl-1,3-dioxen-4-one,
  VIII, 192
2,4-Dimethyl-3-acetyl-5-carbethoxypyrrole,
  II, 198
2,4-Dimethyl-5-carbethoxypyrrole, **II**, 198
Ethyl 4-acetoxybenzoate,
  VI, 586
Ethyl *p*-aminobenzoate,
  II, 300
Ethyl 4-amino-3-methylbenzoate, **VI**,
  560, 581
Ethyl 4-amino-3-(methylthiomethyl) benzoate, **VI**, 581
Ethyl β-anilinocrotonate,
  III, 374
Ethyl 3-(1-azetidinyl) propanoate, **VI**, 75
Ethyl chlorofluoroacetate, **IV**, 423
Ethyl 4-cyclohexylidene-4-(trimethylsiloxy) butanoate, **VIII**, 578
Ethyl diazoacetate, **III**,
  392; **IV**, 424
Ethyl diethoxyacetate,
  IV, 427
Ethyl 3,3-diethoxypropanoate, **VIII**, 254
Ethyl 6,7-dimethoxy-3-methylindene-2-carboxylate, **V**, 550

Ethyl *p*-dimethyl-
aminophenylacetate,
V, 552
Ethyl 2-(diphenylsilyl-
methyl) decanoate,
VIII, 474
Ethyl ethoxyacetate,
II, 261
Ethyl 3-ethyl-5-methyl-
4-isoxazolecarboxy-
late, **VI**, 592
Ethyl α-(hexahydroaze-
pinylidene-2)acetate,
VIII, 263
Ethyl hydrazinecarboxy-
late, **III**, 404
Ethyl indole-2-carboxy-
late, **V**, 567
Ethyl isocyanoacetate,
VI, 620
Ethyl isodehydroacetate,
IV, 549
Ethyl lithioacetate,
VI, 598
Ethyl 3-methylcoumari-
late, **IV**, 590
Ethyl 2-methylindole-5-
carboxylate, **VI**, 601
Ethyl 2-methyl-3-
methylthioindole-5-
carboxylate, **VI**, 601
Ethyl *N*-(2-methyl-3-
nitrophenyl)formimi-
date, **VIII**, 493
Ethyl 6-methylpyridine-
2-acetate, **VI**, 611
Ethyl α-nitrobutyrate,
IV, 454
Ethyl (*E*)-4-(4-nitro-
phenyl)-4-oxo-2-
butenoate, **VIII**, 268
Ethyl *N*-nitroso-*N*-ben-
zylcarbamate, **IV**, 780
Ethyl β,β-pentamethyl-
eneglycidate, **IV**, 459

Ethyl *N*-phenylformimi-
date, **IV**, 464
Ethyl 2-pyridylacetate,
III, 413
Ethyl pyrrole-2-carboxy-
late, **VI**, 618,
Ethyl β-pyrrolidinocroto-
nate, **VI**, 592
Ethyl α-(1-pyrrolidyl)
propionate, **IV**, 466
Ethyl thiazole-4-carboxy-
late, **VI**, 620
Ethyl (*E*)-3-(tributylstan-
nyl) propenoate,
VIII, 268
Ethyl 3-[3-(trimethyl-
siloxy) cyclohex-2-
enyl]propionate,
VIII, 277
Ethyl trimethylsilylace-
tate, **VII**, 512
Ethyl (triphenylphospho-
ranylidene)acetate,
VII, 232
N-Formylglycine ethyl
ester, **VI**, 620
Glycine *tert*-butyl ester,
V, 586
Glycine ethyl ester
hydrochloride, **II**, 310
(–)-Menthyl nicotinate,
VIII, 350
2-Mercapto-4-amino-5-
carbethoxypyrimi-
dine, **IV**, 566
2-Methoxy-*N*-carbometh-
oxypyrrolidine,
VII, 307
1-(2-Methoxycarbonyl-
phenyl)pyrrole,
V, 716
6-Methoxy-7-methoxy-
carbonyl-1,2,3,4-
tetrahydronaphtha-
lene, **VIII**, 444

Methyl 4-acetoxyben-
zoate, **VI**, 576
Methyl coumalate,
IV, 532
Methyl 3,3-diethoxypro-
panoate, **VIII**, 255
Methyl 2-(diethylphos-
phoryloxy)-1-cyclo-
hexene-1-carboxylate,
VII, 351
Methyl 6,6-dimethoxy-
hexanoate, **VII**, 168
Methyl 5,5-dimethoxy-3-
methyl-2,3-epoxypen-
tanoate, **IV**, 649
Methyl (3*RS*,4*RS*)-4,5-
epoxy-3-phenylpen-
tanoate, **VII**, 164
Methyl homoveratrate,
II, 333
Methyl 3-methyl-2-
furoate, **IV**, 649
Methyl γ-methyl-γ-
nitrovalerate, **IV**, 652
Methyl nitroacetate,
VI, 797
Methyl (*E*)-3-nitroacry-
late, **VI**, 799
Methyl *m*-nitrobenzoate,
I, 372
Methyl (*Z*)-3-(phenylsul-
fonyl)prop-2-enoate,
VIII, 458
Phenylmethylglycidic
ester, **III**, 727
C.   Of Dibasic Acids
cis- and *trans*-1-Acetoxy-4-
(dicarbomethoxymethyl)-
2-cyclohexene, **VIII**, 5
Bis-(2,2,2-trichloroethyl)azodi-
carboxylate, **VII**, 56
Bis-(2,2,2-trichloroethyl) hydra-
zodicarboxylate, **VII**, 56
tert-Butyl azodiformate,
V, 160

tert-Butyl ethyl fumarate,
VII, 93
tert-Butyl phenyl carbonate,
V, 166
β-Carbethoxy-γ,γ-dlphenylvinyl-
acetic acid, **IV**, 132
1-Carbethoxymethyl-4-car-
bethoxypiperidine, **V**, 989
Carbomethoxypropionyl chlor-
ide, **III**, 169
Cetylmalonic ester, **IV**, 141
2-Chloro-2-cyclopropylethene-
1,1-dicarboxylic acid,
monoethyl ester, **VIII**, 247
Di-*tert*-butyl dicarbonate,
VI, 418
Di-*tert*-butyl diazomalonate,
VI, 414
Di-*tert*-butyl β,β-dimethylglu-
tarate, **IV**, 263
Di-*tert*-butyl glutarate, **IV**, 263
Di-*tert*-butyl malonate, **IV**, 261
Di-*tert*-butyl methylene-
malonate, **VII**, 152
Di-*tert*-butyl succinate, **IV**, 263
3,5-Dicarbethoxy-2,6-
dimethylpyridine, **II**, 216
Diethyl acetamidomalonate,
V, 373
Diethyl acetonedicarboxylate,
I, 237
Diethyl acetosuccinate, **II**, 262
Diethyl acetylenedicarboxylate,
IV, 330
Diethyl α-acetyl-β-ketopime-
late, **V**, 384
Diethyl adipate, **II**, 264
Diethyl (2*S*,3*R*)-(+)-3-allyl-2-
hydroxysuccinate, **VII**, 153
Diethyl aminomalonate
hydrochloride, **V**, 376
Diethyl azodicarboxylate, **III**,
375; **IV**, 411
Diethyl benzalmalonate,
III, 377

Diethyl (*O*-benzoyl)ethyl-
tartronate, **V**, 379
Diethyl benzoylmalonate,
IV, 285
Diethyl benzylmalonate,
III, 705
Diethyl bis(hydroxymethyl)
malonate, **V**, 381
Diethyl bromomalonate, **I**, 245
Diethyl butylmalonate, **I**, 250
Diethyl *sec*-butylmalonate, **II**,
417; **III**, 495
Diethyl *tert*-butylmalonate, **IV**,
417; **VI**, 442
Diethyl 2-chloro-2-cyclopropy-
lethene-1,1-dicarboxylate,
VIII, 247
Diethyl 1,1-cyclobutanedicar-
boxylate, **IV**, 288
Diethyl cyclohexanone-2,6-
dicarboxylate, **V**, 439
Diethyl Δ²-cyclopentenyl-
malonate, **IV**, 291
Diethyl cyclopropylcarbonyl-
malonate, **VIII**, 248
Diethyl α,δ-dibromoadipate,
III, 623
Diethyl dioxosuccinate,
VIII, 597
Diethyl enanthylsuccinate,
IV, 430
Diethyl ethoxalylpropionate,
II, 272
Diethyl ethoxymethylene-
malonate, **III**, 395
Diethyl ethylidenemalonate,
IV, 293
Diethyl ethylphosphonate,
IV, 326
Diethyl 1,16-hexadecanedicar-
boxylate, **III**, 401
Diethyl *cis*-hexahydrophthalate,
IV, 304
Diethyl hydrazodicarboxylate,
III, 375; **IV**, 411

Diethyl isopropylidene-
malonate, **VI**, 442
Diethyl isopropylmalonate,
II, 94
Diethyl β-ketopimelate, **V**, 384
Diethyl methylenemalonate,
IV, 298
Diethyl methylmalonate, **II**,
279
Diethyl methylphosphonate,
IV, 326
Diethyl oxalate, **I**, 261
Diethyl oximinomalonate,
V, 573
Diethyl γ-oxopimelate, **IV**,302
Diethyl phenylacetylmalonate,
III, 637
Diethyl phenylmalonate,
II, 288
Diethyl phthalimidomalonate,
I, 271
Diethyl pimelate, **II**, 536
Diethyl propionylsuccinate,
VI, 615
Diethyl sebacate, **II**, 277
Diethyl succinate, **V**, 993
Diethyl (2*R*,3*R*)-(+)-tartrate,
VII, 41, 461
Diethyl *cis*-Δ⁴-4-tetrahydroph-
thalate, **IV**, 304
Diethyl *trans*-Δ⁴-tetrahydroph-
thalate, **VI**, 454
Diethyl *trans*-4-trimethylsilyl-
oxy-4-cyclohexene-1,2-
dicarboxylate, **VI**, 445
1,4-Dihydro-3,5-dicarbethoxy-
2,6-dimethylpyridine,
II, 216
Diisopropyl ethylphosphonate,
IV, 326
Diisopropyl (2*S*,3*S*)-2,3-*O*-iso-
propylidenetartrate,
VIII, 201
Diisopropyl methylphospho-
nate, **IV**, 325

(–)-Dimenthyl (1*S*,2*S*)-cyclo-
propane-1,2-dicarboxylate,
VIII, 141
(–)-Dimenthyl succinate,
VIII, 141
Dimethyl acetylenedicarboxy-
late, **IV**, 329
Dimethyl azodicarboxylate,
IV, 414
Dimethyl 3-chloro-2-pentene-
dioate, **VI**, 505
Dimethyl decanedioate,
VII, 181
2,4-Dimethyl-3,5-dicarb-
ethoxypyrrole, **II**, 202
6,6-Dimethyl-5,7-dioxa-
spiro[2.5]octane-4,8-dione,
VII, 411
Dimethyl 3,3a,3b,4,6a,7a-hexa-
hydro-3,4,7-metheno-7*H*-
cyclopenta[*a*]pentalene-
7,8-dicarboxylate,
VIII, 298
Dimethyl *cis*-hexahydrophtha-
late, **IV**, 306
Dimethyl (*E*)-2-hexenedioate,
VIII, 112
Dimethyl hydrazodicarboxylate,
IV, 413
Dimethyl 2,3-*O*-isopropylidene-
tartrate, **VIII**, 155, 201
Dimethyl 3-methylenecyclobu-
tane-1,2-dicarboxylate,
V, 459
Dimethyl nitrosuccinate,
VI, 503
Dimethyl octadecanedioate,
V, 463
Dimethyl oxalate, **I**, 264; **II**, 414
Dimethyl 2,3-pentadienedioate,
VI, 505
Dimethyl *cis*-Δ⁴-tetrahydroph-
thalate, **IV**, 306
Dimethyl β-thiodipropionate,
IV, 669

Diphenyl succinate, **IV**, 390
Ethoxymagnesium malonic
ester, **IV**, 285
Ethyl *tert*-butyl malonate,
IV, 417
Ethyl hydrogen adipate, **II**, 277
Ethyl hydrogen sebacate, **II**,
276
Isopropylidene α-(hexahydro-
azepinylidene-2)malonate,
VIII, 263
(3-Methyl-2-butenyl)propane-
dioic acid, dimethyl ester,
VIII, 381
(3-Methyl-2-butenyl)(2-propy-
nyl)propanedioic acid,
dimethyl ester, **VIII**, 381
3-Methylene-4-isopropyl-1,1-
cyclopentanedicarboxylic
acid, dimethyl ester,
VIII, 381
Methyl hydrogen hendecanedioate, **IV**, 635
Methyl sebacamate, **III**, 613
cis,*cis*-Monomethyl muconate,
VIII, 490
sec-Octyl hydrogen phthalate,
I, 418
o-Phenylene carbonate, **IV**, 788
(*Z*)-Tributylstannylmethylene-
4-isopropyl-1,1-cyclopen-
tanedicarboxylic acid,
dimethyl ester, **VIII**, 381
Triethyl orthoformate, **I**, 258
D. Miscellaneous
Ethyl oxalamidrazonate, **VIII**,
597
Ethyl thioamidooxalate,
VIII, 597
S-(2-Oxo-3-butyl)
butanethioate, **VI**, 776
Tetraethyl 1,2,3,4-butanete-
tracarboxylate, **VII**, 479
Tetraethyl ethylenetetracar-
boxylate, **II**, 273

Tetraethyl orthocarbonate,
IV, 457
Tetraethyl propane-1,1,2,3-
tetracarboxylate, **I**, 272
Tetraethyl propane-1,1,3,3-
tetracarboxylate, **I**, 290
Tetramethyl 3,7-dihydroxybicy-
clo[3.3.0]octa-2,6-diene-
2,4,6,8-tetracarboxylate,
VII, 50
Tetramethyl 3,7-dihydroxy-1,5-
dimethylbicyclol3.3.0]octa-
2,6-diene-2,4,6,8-tetracar-
boxylate, **VII**, 50
Tetramethyl ethane-1,1,2,2-
tetracarboxylate, **VII**, 482
Tetramethyl [2,3]paracyclo-
phane-4,5,12,13-tetracar-
boxylate, **VII**, 485
Tricarbethoxymethane, **II**, 594
2,3,6-Tricarboethoxypyridine,
VIII, 597
Tricarbomethoxymethane,
II, 596
Triethyl α-phthalimidoethane-
α,α,β-tricarboxylate,
IV, 55
Triethyl 1,2,4-triazine-3,5,6-tri-
carboxylate, **VIII**, 597
Triethyl *N*-tricarboxylate,
III, 415

ETHERS
A. Epoxides
Benzyl 2,3-epoxypropyl ether,
VIII, 33
trans-2-Butene oxide, **VI**, 313
Cyclohexene oxide, **I**, 185;
cis-Cyclooctene oxide, **VII**,
126
2,2'-Dichloro-α,α'-epoxybiben-
zyl, **V**, 358
Epibromohydrin, **II**, 257
Epichlorohydrin, **I**, 233;
II, 256

2,3-Epoxycyclohexanone,
VI, 679
10,11-Epoxyfarnesyl acetate,
VI, 560
Ethyl β,β-pentamethyleneglyci-
date, **IV**, 459
Isophorone oxide, **IV**, 552
Methyl 5,5-dimethoxy-3-
methyl-2,3-epoxypen-
tanoate, **IV**, 649
Methylenecyclohexane oxide,
V, 755
Methyl (3*RS*,4*RS*)-4,5-epoxy-3-
phenylpentanoate, **VII**, 166
(*R*)-Methyloxirane, **VIII**, 434
(*S*)-(−)-Methyloxirane, **VII**, 356
Oxaspiropentane, **VI**, 320
11-Oxatricyclo[4.4.1.0^{1,6}]unde-
ca-3,8-diene, **VI**, 862
Phenanthrene-9,10-oxide,
VI, 887
Phenylmethylglycidic ester,
III, 727
α-Pinene oxide, **VI**, 949
(2*S*,3*S*)-3-Propyloxiranemeth-
anol, **VII**, 461
trans-Stilbene oxide, **IV**, 860
Styrene oxide, **I**, 494
3,4,8,9-Tetrabromo-11-oxatricy-
clo[4.4.1.0^{1,6}]undecane,
VI, 862
Tetracyanoethylene oxide,
V, 1007
B. Silyl
Acetone trimethylsilyl enol
ether, **VIII**, 1
Acetophenone trimethylsilyl
enol ether, **VIII**, 210, 323
Benzoquinone *O*-trimethylsilyl
cyanohydrin, **VII**, 517
1,2-Bis(trimethylsilyloxy)
cyclobutene, **VI**, 167
Diethyl *trans*-trimethylsilyloxy-
4-cyclohexene-1,2-dicar-
boxylate, **VI**, 445

2-(1-Ethoxycyclohexyl)-2-
   (trimethylsiloxy)cyclobu-
   tanone, **VIII**, 579
1-Ethoxy-1-trimethylsiloxycy-
   clopropane, **VII**, 131
Ethyl 4-cyclohexylidene-4-
   (trimethylsiloxy)butanoate,
   VIII, 578
Ethyl 3-[3-(trimethylsiloxy)
   cyclohex-2-enyl]propi-
   onate, **VIII**, 277
5-Hydroxy-2,4,6-trimethyl-2-
   trimethylsiloxyheptan-3-
   one, **VII**, 185
[(*trans*-2-Isocyanocyclohexyl)
   oxytrimethylsilane,
   VII, 294
trans-1-Methoxy-3-trimethyl-
   siloxy-1,3-butadiene,
   VII, 312
3-Methoxy-17-trimethylsiloxy-
   1,3,5(10)-estratetraene,
   VIII, 286
2-Methyl-2-(trimethylsiloxy)
   pentan-3-one, **VII**,
   185, 381
4-Oxo-1-[(trimethylsilyl)oxy]-
   2,5-cyclohexadiene-1-car-
   bonitrile, **VII**, 517
1-Trimethylsiloxycyclohexene,
   VIII, 460,
1-Trimethylsiloxycyclopentene,
   VII, 424
2-[(Trimethylsiloxy)methyl]
   allyltrimethylsilane,
   VII, 266
(*Z*)-3-Trimethylsiloxy-2-pen-
   tene, **VII**, 512,
O-(Trimethylsilyl)benzophe-
   none cyanohydrin,
   VII, 20
O-Trimethylsilyl-4-methoxy-
   mandelonitrile, **VII**, 521
2-Trimethylsilyloxy-1,3-butadi-
   ene, **VI**, 445

4,6,6-Trimcthyl-2-trimethyl-
   siloxycyclohexa-1,3-diene,
   VII, 282
C. Miscellaneous
2-(2-Acetamidoethyl)-4,5-
   dimethoxyacetophenone,
   VI, 1
2-Acetyl-6,7-dimethoxy-1-
   methylene-1,2,3,4-tetrahy-
   droisoquinoline, **VI**, 1
2-Acetyl-6-methoxynaphtha-
   lene, **VI**, 34
α-Allyl-β-bromoethyl ethyl
   ether, **IV**, 750
(*S*)-(−)-1-Amino-2-methoxy-
   methylpyrrolidine, **VIII**,
   26, 403
4-Aminoveratrole, **II**, 44
o-Anisaldehyde, **V**, 46; **VI**, 64
Anisole, **I**, 58
p-Arsonophenoxyacetic acid,
   I, 75
(1-Azido-2-iodo-3,3-dime-
   thoxypropyl)benzene,
   VI, 893
Benzhydryl β-chloroethyl ether,
   IV, 72
Benzyl chloromethyl ether,
   VI, 101
Benzyl 2,3-epoxypropyl ether,
   VIII, 33
4-Benzyloxy-2-butanone, **VII**,
   386
5-Benzyloxy-3-hydroxy-3-
   methylpentanoic-2-[13]C
   acid, **VII**, 386
4-Benzyloxyindole, **VII**, 34
6-Benzyloxy-2-nitrotoluene,
   VII, 34
1-Benzyloxy-4-penten-2-ol,
   VIII, 33
Bis(chloromethyl) ether, **IV**, 101
(*S,S*)-(+)-1,4-Bis(dimethylamino)-2,3-dimethoxybu-
   tane (DDB), **VII**, 41

1,2-Bis(2-iodoethoxy)ethane,
VIII, 152, 153

1,3-Bis(methylthio)-2-methoxy-
propane, **VI**, 683

4,4-Bis(phenylthio)-3-methoxy-
1-butene, **VI**, 737

Bis[2,2,2-trifluoro-1-phenyl-1-
(trifluoromethyl)ethoxy]
diphenylsulfurane,
VI, 163

2-Bromoacetyl-6-methoxynaph-
thalene, **VI**, 175

6-Bromo-3,4-dimethoxybenz-
aldehyde cyclohexylimine,
VIII, 586

1-Bromo-1-ethoxycyclo-
propane, **VIII**, 556

α-Bromo-β-methoxypropionic
acid, **III**, 775

17β-*tert*-Butoxy-5α-androstan-
3-one, **VII**, 66

17β-*tert*-Butoxy-5α-androst-2-
ene, **VII**, 66

o-Butoxynitrobenzene, **III**, 140

7-*tert*-Butoxynorbornadiene,
V, 151

2-*tert*-Butoxythiophene, **V**, 642

tert-Butyl *N*-(1-ethoxycyclo-
propyl)carbamate, **VI**, 226

α-Chloroethyl ethyl ether,
IV, 748

3-Chloro-2-methyltetrahydropy-
ran, **VI**, 675

p-Chlorophenoxymethyl chloride, **V**, 221

3-(4-Chlorophenyl)-5-(4-meth-
oxyphenyl)isoxazole,
VI, 278

2-Chloro-1,1,2-trifluoroethyl
ethyl ether, **IV**, 1 84

Cholestanyl methyl ether, **V**, 245

Creosol, **IV**, 203

18-Crown-6, **VI**, 301

1-Cyano-6-methoxy-3,4-dihy-
dronaphthalene, **VI**, 307

Cyclohexyl methyl ether,
VI, 355

2-Cyclohexyloxyethanol,
V, 303

Cyclopropanone ethyl hemiacetal, **VI**, 131

Deoxyanisoin, **V**, 339

4,13-Diaza-18-crown-6,
VIII, 152

Dibenzo-18-crown-6 polyether,
VI, 395

N,*N*-Dibenzyl-4,13-diaza-18-
crown-6, **VIII**, 152

1,4-Di-*O*-benzyl-2,3-di-*O*-iso-
propylidene-L-threitol,
VIII, 155

1,10-Dibenzyl-4,7-dioxa-1,10-
diazadecene, **VIII**, 152

1,4-Di-*O*-benzyl-L-threitol,
VIII, 155

2,2-Dibromoacetyl-6-methoxy-
naphthalene, **VI**, 175

Dibromoethyl ethyl ether,
IV. 749

(*E*)-1,2-Di-*tert*-butoxy-1-
chloroethene, **VIII**, 161

1,2-Di-*tert*-butoxy-1,2-
dichloroethane, **VIII**, 161

2,3-Di-*tert*-butoxy-1,4-dioxane,
VIII, 161

Di-*tert*-butoxyethyne, **VIII**, 161

4,4'-Dichlorodibutyl ether,
IV, 266

trans-2,3-Dichloro-1,4-dioxane,
VIII, 161

3,4-Dichloro-5-isopropoxy-
2(5*H*)-furanone, **VIII**, 116

2,4-Dichloromethoxybenzene,
VIII, 167

Dichloromethyl methyl ether,
V, 365

2,3-Dichlorotetrahydropyran,
VI, 675

Dicyclohexyl-18-crown-6 polyether, **VI**, 395

5,6-Diethoxybenzofuran-4,7-
dione, **VIII**, 179
2,5-Dihydro-2,5-dimethoxyfu-
ran, **V**, 403
3,4-Dihydro-2-methoxy-4-
methyl-2*H*-pyran, **IV**, 311
3,4-Dihydro-2-methylene-2*H*-1-
benzopyran, **VIII**, 512
2,3-Dihydropyran, **III**, 276
2,4-Dimethoxybenzonitrile,
VI, 465
2,6-Dimethoxybenzonitrile,
III, 293
3,3'-Dimethoxybiphenyl, **III**, 295
4,4'-Dimethoxy-1,1'-biphenyl
VI, 468,
2,3-Dimethoxycinnamic acid,
IV, 327
6,7-Dimethoxy-3,4-dihydro-2-
naphthoic acid, **III**, 300
6,7-Dimethoxy-3-isochro-
manone, **VI**, 471
6,7-Dimethoxy-1-methyl-3,4-
dihydroisoquinoline, **VI**, 1
3,4-Dimethoxyphenylpyruvic
acid, **IV**, 327
trans-4,4'-Dimethoxystilbene,
V, 428
(*R,R*)-(+)-2,3-Dimethoxy-
N,N,N',N'-tetramethylsuc-
cinic acid diamide, **VII**,
41
(*S*)-*N,N*-Dimethyl-*N'*-(1-*tert*-
butoxy-3-methyl-2-
butyl)formamidine, **VIII**,
204, 573
1,3-Dimethyl-3-methoxy-4-
phenylazetidinone,
VIII, 216
2α,4α-Dimethyl-8-oxabicy-
clo[3.2.1]oct-3-en-3-one,
VI, 512
3,5-Dinitroanisole, **I**, 219
Diphenylmethyl vinyl ether,
VI, 552

Ethoxyacetic acid, **II**, 260
Ethoxyacetylene, **IV**, 404
1-Ethoxy-1-butyne, **VI**, 564
3-Ethoxy-2-cyclohexenone,
V, 539
β-Ethoxyethyl bromide,
III, 370
4-Ethoxy-3-hydroxybenzalde-
hyde, **VI**, 567
(*E*)-1-Ethoxy-1-(1-hydroxy-2-
butenyl)cyclopropane,
VIII, 556
4-Ethoxy-3-methoxybenzalde-
hyde ethylene acetal,
VI, 567
3-Ethoxy-6-methyl-2-cyclohex-
en-1-one, **VII**, 208
2-Ethoxy-1-naphthaldehyde,
III, 98
p-Ethoxyphenylurea, **IV**, 52
β-Ethoxypropionitrile, **III**, 372
1-Ethoxy-2-*p*-tolylcyclo-
propane, *cis*- and *trans*-,
VI, 571
Ethyl α-acetyl-β-(2,3-dimeth-
oxyphenyl)acrylate,
IV, 408
Ethyl α-acetyl-β-(3,4-dimeth-
oxyphenyl)acrylate,
IV, 410
Ethyl α-acetyl-β-(2,3-dimeth-
oxyphenyl)propionate, **IV**,
408
Ethyl α-acetyl-β-(3,4-dimeth-
oxyphenyl)propionate, **IV**,
410
Ethyl diethoxyacetate, **IV**, 427
Ethyl 6,7-dimethoxy-3-methyl-
indene-2-carboxylate,
V, 550
Ethyl ethoxyacetate, **II**, 261
Ethyl 1-(*p*-methoxyphenyl)-2-
oxocyclohexanecarboxy-
late, **VII**, 229
o-Eugenol, **III**, 418

16α-Fluoro-3-methoxy-
1,3,5(10)-estratrien-17-
one, **VIII**, 287
(*S*)-*N*-Formyl-*O*-*tert*-butyl-L-
valinol, **VIII**, 204
(*S*)-(−)-1-Formyl-2-methoxy-
methylpyrrolidine,
VIII, 26
α-Glyceryl phenyl ether, **I**, 296
Homoveratric acid, **II**, 333
4-Iodoveratrole, **IV**, 547
(Isopropoxydimethylsilyl)
methyl chloride, **VIII**, 315
p-Menth-8-en-1-yl methyl
ether, **VII**, 304
L-Menthoxyacetic acid, **III**, 544
L-Menthoxyacetyl chloride,
III, 547
Methoxyacetonitrile, **II**, 387
ω-Methoxyacetophenone,
III, 562
Methoxyacetylene, **IV**, 406
m-Methoxybenzaldehyde, **III**,
564
2-Methoxy-*N*-carbomethoxy-
pyrrolidine, **VII**, 307
1-Methoxy-1,4-cyclohexadiene,
VI, 996
5β-Methoxycyclohexan-1-one-
3β,4β-dicarboxylic acid
anhydride, **VII**, 312
2-Methoxydiphenyl ether, **III**,
566
6-Methoxy-7-methoxycarbonyl-
1,2,3,4-tetrahydronaphtha-
lene, **VIII**, 444
(*S*)-(+)-2-Methoxymethylpyr-
rolidine, **VIII**, 26
6-Methoxy-2-naphthol, **V**, 918
6-Methoxy-8-nitroquinoline,
III, 568
4-Methoxy-3-penten-2-one,
VIII, 357
o-Methoxyphenylacetone,
IV, 573

p-Methoxyphenylacetonitrile,
IV, 576
p-Methoxyphenyllead triacetate, **VII**, 229
1-(*o*-Methoxyphenyl)-2-nitro-1-
propene, **IV**, 573
1-(*p*-Methoxyphenyl)-5-phenyl-
1,3,5-pentanetrione, **V**, 718
2-(*p*-Methoxyphenyl)-6-phenyl-
4-pyrone, **V**, 721
1-(4-Methoxyphenyl)-1,2,5,6-
tetrahydropyridine,
VIII, 358
p-Methoxyphenylurea, **IV**, 53
(*Z*)-2-Methoxy-1-phenylthio-
1,3-butadiene, **VI**, 737
N-(4-Methoxyphenyl)-(*Z*)-
(trimethylsilyl)-3-buten-
amine, **VIII**, 358
6-Methoxy-β-tetralone, **VI**, 744
O-Methylcaprolactim, **IV**, 588
N-Methyl-2,3-dimethoxybenzyl-
amine, **IV**, 603
Methyl 5,5-dimethoxy-3-
methyl-2,3-epoxypen-
tanoate, **IV**, 649
Methylisourea hydrochloride,
IV, 645
Methyl β-naphthyl ether, **I**, 59
Monochloromethyl ether, **I**, 377
o-Nitrodiphenyl ether, **II**, 446
p-Nitrodiphenyl ether, **II**, 445
2-Nitro-4-methoxyacetanilide,
III, 661
2-Nitro-4-methoxyaniline,
III, 661
6-Nitroveratraldehyde, **IV**, 735
2,2,7,7,12,12,17,17-Octa-
methyl-21,22,23,24-tetra-
oxaperhydroquaterene,
VI, 856
Orcinol dimethyl ether,
VI, 859
Orcinol monomethyl ether,
VI, 859

2-Oxa-7,7-dichloronorcarane,
  V, 874
1,6-Oxido[10]annulene, **VI**, 862
β-Phenoxyethyl bromide, **I**, 436
2-Phenoxymethyl-1,4-benzo-
  quinone, **VI**, 890
1-Phenoxy-1-phenylethene,
  VIII, 512
γ-Phenoxypropyl bromide,
  I, 435
Phenyl *tert*-butyl ether (Method
  I), **V**, 924
Phenyl *tert*-butyl ether (Method
  II), **V**, 926
(*E*)-3-[(*E*)-3-Phenyl-2-
  propenoxy]acrylic acid,
  VIII, 536
4-(1-Phenyl-5-tetrazolyloxy)
  biphenyl, **VI**, 150
Pyrogallol 1-monomethyl ether,
  III, 759
Quinacetophenone dimethyl
  ether, **IV**, 837
Quinacetophenone monomethyl
  ether, **IV**, 836
Syringic aldehyde, **IV**, 866
Tetrahydro-2-(2-propynyloxy)-
  2*H*-pyran, **VII**, 334
Tetrahydropyran, **III**, 794
4,5,4',5'-Tetramethoxy-1,1'-
  biphenyl-2,2'-dicarbox-
  aldehyde, **VIII**, 586
3,3,6,6-Tetramethoxy-1,4-
  cyclohexadiene, **VI**, 971
2,3,4,6-Tetramethyl-D-glucose,
  III, 800
1,1,1-Trichloro-4-ethoxy-3-
  buten-2-one, **VIII**, 254
1,1,1-Trichloro-4-methoxy-3-
  buten-2-one, **VIII**, 238
3,4,5-Trimethoxybenzaldehyde,
  VI, 1007
Trimethylgallic acid, **I**, 537
O-Trimethylsilyl-4-methoxy-
  mandelonitrile, **VII**, 521

Vanillic acid, **IV**, 972
Veratraldehyde, **II**, 619
Veratronitrile, **II**, 622

HALOGENATED COMPOUNDS

A. Bromo Compounds

Acetobromoglucose, **III**, 11
(*S*)-(−)-2-Acetoxy-1-bromo-
  propane, **VII**, 356
p-Acetyl-α-bromohydrocin-
  namic acid, **VI**, 21
Allyl bromide, **I**, 27
α-Allyl-β-bromoethyl ethyl
  ether, **IV**, 750
2-Amino-5-bromo-3-nitropyri-
  dine, **V**, 347
2-Amino-5-bromopyridine,
  V, 346
3-Amino-2,4,6-tribromobenzoic
  acid, **IV**, 947
Benzalacetone dibromide,
  III, 105
Benzalacetophenone dibromide,
  I, 205
ε-Benzoylamino-α-bromo-
  caproic acid, **II**, 74
1,1-Bis(bromomethyl)cyclo-
  propane, **VI**, 153
Bromal, **II**, 87
Bromoacetal, **III**, 123, 506
Bromoacetamide, **IV**, 104
3-Bromo-4-acetaminotoluene,
  I, 111
Bromoacetic acid, **III**, 381
Bromoacetone, **II**, 88
3-Bromoacetophenone, **V**, 117
p-Bromoacetophenone, **I**, 109
2-Bromoacetyl-6-methoxynaph-
  thalene, **VI**, 175
2-Bromoallylamine, **V**, 121
(2-Bromoallyl)ethylamine,
  V, 124
3-Bromo-4-aminotoluene, **I**, 111
α-Bromobenzalacetone, **III**, 125
m-Bromobenzaldehyde, **II**, 132

p-Bromobenzaldehyde, **II**, 89, 442

p-Bromobiphenyl, **I**, 113

α-Bromo-γ-butyrolactone, **V**, 255

α-Bromobutyryl chloride, **IV**, 620

α-Bromocaproic acid, **I**, 115; **II**, 95

α-Bromocinnamic aldehyde, **III**, 731

1-Bromo-3-chlorocyclobutane, **VI**, 179

1-Bromo-3-chloro-2,2-dimethoxypropane, **VI**, 361

2-Bromo-2-cyclopenten-1-one, **VII**, 271

2-Bromo-2-cyclopenten-1-one ethylene ketal, **VII**, 271; **VII**, 498

Bromocyclopropane, **V**, 126

2-Bromodecene-1, **I**, 187

6-Bromo-3,4-dimethoxybenzaldehyde cyclohexylimine, **VIII**, 586

4-Bromo-*N*,*N*-dimethyl-3-(trifluoromethyl)aniline, **VI**, 181

erythro-2-Bromo-1,2-diphenylethanol, **VI**, 184

p-Bromodiphenylmethane, **V**, 130

2-Bromo-3,3-diphenyl-2-propen-1-ol acetate, **VI**, 187

2-Bromoethanol, **I**, 117

1-Bromo-1-ethoxycyclopropane, **VIII**, 556,

Bromoethylamine hydrobromide, **II**, 91

2-(2-Bromoethyl)-1,3-dioxane, **VII**, 59

β-Bromoethylphthalimide, **I**, 119; **IV**, 106

(1-Bromoethynyl)trimethylsilane, **VI**, 1034

1-Bromo-2-fluorobenzene, **V**, 133

1-Bromo-2-fluoroheptane, **V**, 136

Bromoglutarimide, **IV**, 498

α-Bromoheptaldehyde, **III**, 127

α-Bromoheptaldehyde dimethyl acetal, **III**, 128

4-Bromo-2-heptene, **IV**, 108

2-Bromohexanoic acid, **VI**, 192

2-Bromohexanoyl chloride, **VI**, 190

10-Bromo-11-hydroxy-10,11-dihydrofarnesyl acetate, **VI**, 560

3-Bromo-4-hydroxytoluene, **III**, 130

α-Bromoisobutyryl bromide, **IV**, 348

α-Bromoisocaproic acid, **II**, 95

α-Bromoisovaleric acid, **II**, 93

p-Bromomandelic acid, **IV**, 110

Bromomesitylene, **II**, 95

Bromomethanesulfonyl bromide, **VIII**, 212

6-Bromo-2-methoxynaphthalene, **V**, 920

α-Bromo-β-methoxypropionic acid, **III**, 775

α-(Bromomethyl)acrylic acid, **VII**, 210

2-Bromo-4-methylbenzaldehyde, **V**, 139

2-Bromo-3-methylbenzoic acid, **IV**, 114

1-Bromo-1-methyl-2-(bromomethylsulfonyl)cyclohexane, **VIII**, 21

1-Bromo-3-methyl-2-butanone, **VI**, 193

2-(Bromomethyl)-2-(chloromethyl)-1,3-dioxane, **VIII**, 173

N-Bromomethylphthalimide, **VIII**, 536

α-Bromo-β-methylvaleric acid, **II**, 95

α-Bromonaphthalene, **I**, 121

2-Bromonaphthalene, **V**, 142
6-Bromo-2-naphthol, **III**, 132
m-Bromonitrobenzene, **I**, 123
2-Bromo-3-nitrobenzoic acid,
    I, 125
2-Bromo-4-nitrotoluene,
    IV, 114
p-Bromophenacyl bromide,
    I, 127
9-Bromophenanthrene, **III**, 134
o-Bromophenol, **II**, 97
p-Bromophenol, **I**, 128
α-Bromophenylacetic acid,
    VI, 403
α-Bromo-α-phenylacetone,
    III, 343
α-Bromo-α-phenylacetonitrile,
    III, 347
1-*p*-Bromophenyl-1-phenyl-
    2,2,2-trichloroethane,
    V, 131
α-Bromophenylpropionic acid,
    III, 705
p-Bromophenylurea, **IV**, 49
3-Bromophthalide, **V**, 145
β-Bromopropionic acid, **I**, 131
3-Bromopyrene, **V**, 147
2-Bromopyridine, **III**, 136
4-Bromoresorcinol, **II**, 100
(*Z*)-β-Bromostyrene, **VII**, 172
3-Bromothiophene, **V**, 149
o-Bromotoluene, **I**, 135
m-Bromotoluene, **I**, 133
p-Bromotoluene, **I**, 136
1-(Bromovinyl)trimethylsilane,
    VI, 1033
4-Bromo-*o*-xylene, **III**, 138
Butyl bromide, **I**, 28, 37
sec-Butyl bromide, **I**, 38; **II**,
    359
tert-Butyl bromide, **I**, 38
o-Chlorobromobenzene,
    III, 185
Cholesterol dibromide, **IV**, 195
Cinnamyl bromide, **V**, 249
Coumarin dibromide, **III**, 209

Cyanogen bromide, **II**, 150
Cyclohexyl bromide, **II**, 247
3-Cyclohexyl-2-bromopropene,
    I, 186
Decamethylene bromide,
    III, 227
2,3-Diamino-5-bromopyridine,
    V, 346
Dibenzoyldibromomethane,
    II, 244
Dibromoacetonitrile, **IV**, 254
2,2-Dibromoacetyl-6-methoxy-
    naphthalene, **VI**, 175
2,6-Dibromoaniline, **III**, 262
9,10-Dibromoanthracene, **I**, 207
4,4'-Dibromobibenzyl, **IV**, 257
9,9-Dibromobicyclo[6.1.0]
    nonane, **V**, 306
2,2'-Dibromo-1,1'-binaphthyl,
    VIII, 57
4,4'-Dibromobiphenyl, **IV**, 256
1,3-Dibromo-2-butanone,
    VI, 711
5α,6β-Dibromocholestan-3-
    one, **IV**, 197
1,2-Dibromocyclohexane, **II**, 171
2,12-Dibromocyclododecanone,
    VI, 368
1,2-Dibromocyclobutane,
    VII, 119
2-(2,2-Dibromocyclopropyl)-
    6,6-dimethylbicyclo[2.2.1]
    hept-2-ene, **VIII**, 223
α,α'-Dibromodibenzyl sulfone,
    VI, 403
7,7-Dibromo-1,6-dimethylbicy-
    clo[4.1.0]hept-3-ene,
    VII, 200
α,α'-Dibromodineopentyl
    ketone, **VI**, 991
1,1-Dibromo-2,2-diphenylcy-
    clopropane, **VI**, 187
Dibromoethyl ethyl ether,
    IV, 749
β,β-Dibromoisobutyric acid,
    VII, 319

2,6-Dibromo-4-nitrophenol,
II, 173
2,4-Dibromo-3-pentanone, **VI**,
512, 520
1,3-Dibromo-1-phenylpropane,
V, 328
2,3-Dibromopropene, **I**, 209
2,3-Dibromopropionaldehyde
diethyl acetal, **VI**, 954
erythro-α,β-Dibromo-β-phenyl-
propionic acid, **IV**, 961;
VII, 173
2,6-Dibromoquinone-4-
chloroimide, **II**, 175
Dibromosuccinic acid, **II**, 177
3,5-Dibromosulfanilamide,
III, 263
β-Dibutylaminoethyl bromide
hydrobromide, **II**, 92
β-Diethylaminoethyl bromide
hydrobromide, **II**, 92
Diethyl α,δ-dibromoadipate,
III, 623
2,4-Dihydroxy-5-bromobenzoic
acid, **II**, 100
β-Dimethylaminoethyl bromide
hydrobromide, **II**, 92
Diphenyl-*p*-bromophenylphos-
phine, **V**, 496
β-Dipropylaminoethyl bromide
hydrobromide, **II**, 92
Dodecyl bromide, **I**, 29; **II**,
246
Epibromohydrin, **II**, 257
β-Ethoxyethyl bromide, **III**, 370
Ethyl bromide, **I**, 29, 36
Ethyl bromoacetate, **III**, 381
Ethyl γ-bromobutyrate, **V**, 545
Ethyl bromomalonate, **I**, 245
Ethyl α-(bromomethyl)acrylate,
VII, 210; **VIII**, 266
Ethyl α,β-dibromo-β-phenyl-
propionate, **II**, 270
Glycerol α,γ-dibromohydrin,
II, 308
Heptyl bromide, **II**, 247

Isoamyl bromide, **I**, 27
Isobutyl bromide, **II**, 358
Isopropyl bromide, **I**, 37; **II**, 359
Methallyl bromide [1458-98-6],
VI, 723
1-Methylamino-4-bromoan-
thraquinone, **III**, 575
Methyl 4-*O*-benzoyl-6-bromo-
6-deoxy-α-D-glucopyra-
noside, **VIII**, 363
Methyl 4-bromo-1-butanimidate
hydrochloride, **VIII**, 415
Methyl α-(bromomethyl)acry-
late, **VII**, 319
Methyl β-bromopropionate,
III, 576
Methyl β,β-dibromoisobutyrate,
VII, 319
Methylene bromide, **I**, 357
Monobromopentaerythritol,
IV, 681
p-Nitrobenzyl bromide, **II**, 443
Octadecyl bromide, **II**, 247
Octyl bromide, **I**, 30
Pentaerythrityl bromide. **II**, 476
Pentaerythrityl tetrabromide,
IV, 753
Pentamethylene bromide, **I**,
428; **III**, 692
Phenacyl bromide, **II**, 480
β-Phenoxyethyl bromide,
I, 436
γ-Phenoxypropyl bromide,
I, 435
4-Phenyl-2-bromobutene-1
I, 187
Phenylbromoethyne, **V**, 921
1-Phenylthio-2-bromoethane (in
solution), **VII**, 453
Phenyltrimethylammonium tri-
bromide, **VI**, 175
Propyl bromide, **I**, 37; **II**, 359
Sodium 2-bromoethanesul-
fonate, **II**, 558
2,4,4,6-Tetrabromo-2,5-cyclo-
hexadien-1-one, **VI**, 181

3,4,8,9-Tetrabromo-11-oxatricy-
clo[4.4.1.0^{1,6}]undecane,
VI, 862
α,α,α',α'-Tetrabromo-*o*-xylene,
IV, 807
Tetradecyl bromide, **II**, 247
Tetrahydrofurfuryl bromide,
III, 793
3-Thenyl bromide, **IV**, 921
sym-Tribromobenzene, **II**, 592
2,4,6-Tribromobenzoic acid,
IV, 947
2,3,5-Tribromothiophene,
V, 150
1,2,3-Tribromopropane, **I**, 521
Trimethyl ortho-4-bromobu-
tanoate, **VIII**, 415
2,5,5-Trimethyl-2-(2-bro-
moethyl)-1,3-dioxane,
VII, 60
Trimethylene bromide, **I**, 30
Trimethylethylene dibromide,
II, 419
Vinyltriphenylphosphonium
bromide, **V**, 1145
o-Xylylene dibromide, **IV**, 984
B. Chloro Compounds
2-Acetamido-3,4,6-tri-*O*-acetyl-
2-deoxy-α-D-glucopyran-
osyl chloride, **V**, 1
p-Acetaminobenzenesulfonyl
chloride, **I**, 8
1-Acetoxy-4-chloro-2-butene,
VIII, 9
4-Acetoxy-3-chloro-1-butene,
VIII, 9
cis-1-Acetoxy-4-chloro-2-
cyclohexene, **VIII**, 5
α-Acetyl-δ-chloro-γ-valerolactone, **IV**, 10
2-Amino-4-anilino-6-
(chloromethyl)-*s*-triazine,
IV, 29
4-Azido-3-chloro-5-isopro-
poxy-2(5*H*)-furanone,
VIII, 116

Benzeneselenenyl chloride,
VI, 533
Benzenesulfonyl chloride,
I, 84
Benzhydryl β-chloroethyl ether,
IV, 72
Benzohydroxamoyl chloride,
V, 504
Benzophenone dichloride, **II**, 573
ε-Benzoylamino-α-chloro-
caproic acid, **VI**, 90
Benzoylcholine chloride, **IV**, 84
Benzyl chloromethyl ether,
VI, 101
Benzyl chloromethyl ketone,
III, 119
Bis(chloromethyl) ether,
IV, 101
Bis(2,2,2-trichloroethyl) azodi-
carboxylate, **VII**, 56
α-Bromobutyryl chloride,
IV, 620
1-Bromo-3-chlorocyclobutane,
VI, 179
1-Bromo-3-chloro-2,2-dimeth-
oxypropane, **VI**, 361
2-Bromohexanoyl chloride,
VI, 190
2-(Bromomethyl)-2-(chloro-
methyl)-1,3-dioxane,
VIII, 173
1-*p*-Bromophenyl-1-phenyl-
2,2,2-trichloroethane,
V, 131
Butyl chloride, **I**, 142
sec-Butyl chloride, **I**, 143
tert-Butyl chloride, **I**, 144
3-Butyl-4,4-dichlorocy-
clobutenone, **VIII**, 82
tert-Butyl hypochlorite, **IV**, 125;
V, 184; *Warning*, **V**, 183
Butyrchloral, **IV**, 130
Chloroacetamide, **I**, 153
Chloroacetonitrile, **IV**, 144
4-Chloroacetophenone oxime
VI, 279

p-Chloroacetylacetanilide,
     III, 260
α-Chloroacetyl isocyanate,
     V, 204
9-Chloroacridine, **III**, 54
3-(o-Chloroanilino)propioni-
     trile, **IV**, 146
9-Chloroanthracene, **V**, 206
α-Chloroanthraquinone, **II**, 128
m-Chlorobenzaldehyde. **II**, 130
p-Chlorobenzaldehyde, **II**, 133
o-Chlorobenzoic acid, **II**, 135
p-Chlorobenzoic anhydride,
     III, 29
Chloro-p-benzoquinone,
     IV, 148
o-Chlorobenzoyl chloride,
     I, 155
(2-Chlorobenzoyl)ferrocene,
     VI, 625
N-Chlorobetainyl chloride,
     IV, 154
3-Chlorobicyclo[3.2.1]oct-2-
     ene, **VI**, 142
o-Chlorobromobenzene,
     III, 185
4-Chloro-1-butene, **VI**, 607
3-Chloro-2-buten-1-ol, **IV**, 128
4-Chlorobutyl benzoate,
     III, 187
γ-Chlorobutyronitrile, **I**, 156
α-Chlorocrotonaldehyde,
     IV, 130
7-Chloro-7-cyanobicyclo[4.2.0]
     octan-8-one, **VIII**, 116
3-Chlorocyclobutanecarboxylic
     acid, **VI**, 271
2-Chlorocyclohexanol, **I**, 158
2-Chlorocyclohexanone,
     III, 188
N-Chlorocyclohexylidenimine,
     V, 208
trans-2-Chlorocyclopentanol,
     IV, 157
3-Chlorocyclopentene, **IV**, 239

3-Chloro-2,5-di-tert-butyl-p-
     benzoquinone, **VI**, 210
1-Chloro-1,2-dicyanocyclobu-
     tane, **VI**, 427
Chlorodiisopropylphosphine,
     V, 211
1-Chloro-2,6-dinitrobenzene,
     IV, 160
1-Chloro-1,2-diphenyl-3-
     methylcyclopropane,
     VII, 203
α-Chloroethyl ethyl ether,
     IV, 748
β-Chloroethyl methyl sulfide,
     II, 136
2-Chloro-1-formyl-1-cyclohex-
     ene, **V**, 215
(Z)-1-Chloro-1-hexenyl trifluo-
     roacetate, **VI**, 273
6-Chloro-2-hexyne, **VI**, 273
α-Chlorohydrindene, **II,**, 336
3-Chloroindazole, **III**, 476
1-Chloro-3-iodopropane,
     VI, 365
ω-Chloroisonitrosoaceto-
     phenone, **III**, 191
2-Chlorolepidine, **III**, 194
2-Chloro-2-methylbutane, **I**,
     144; **VII**, 425
2-Chloro-2-methylcyclohexa-
     none, **IV**, 162
2-Chloromethyl-4,6-diamino-s-
     triazine, **IV**, 30
6-Chloromethyl-2,2-dimethyl-
     1,3-dioxen-4-one,
     VIII, 192
4-Chloromethyl-3-ethyl-5-
     methylisoxazole,
     VI, 781
1-Chloromethylnaphthalene,
     III, 195
Chloromethylphosphonothioic
     dichloride, **V**, 218
3-Chloro-2-methyltetrahydropy-
     ran, **VI**, 675

2-Chloromethylthiophene,
III, 197
N-Chloromorpholine, **VIII**, 167
2-Chloronicotinonitrile, **IV**, 166
m-Chloronitrobenzene, **I**, 162
1-Chloronorbornane, **VI**, 845
5-Chloro-2-pentanone, **IV**, 597
m-Chloroperbenzoic acid,
VI, 276
p-Chlorophenoxymethyl chloride, **V**, 221
α-Chlorophenylacetic acid,
IV, 169
4-(4-Chlorophenyl)butan-2-one,
VII, 105
o-Chlorophenylcyanamide,
IV, 172
3-(4-Chlorophenyl)-5-(4-meth-
oxyphenyl)isoxazole,
VI, 278
p-Chlorophenyl isothiocyanate,
V, 223
m-Chlorophenyl methyl
carbinol, **III**, 200
α-(4-Chlorophenyl)-γ-phenyl-
acetoacetonitrile, **IV**, 174
1-(*p*-Chlorophenyl)-3-phenyl-2-
propanone, **IV**, 176
p-Chlorophenyl salicylate,
IV, 178
o-Chlorophenylthiourea,
IV, 180
N-Chloropiperidine, **VI**, 968
(*S*)-2-Chloropropanoic acid,
VIII, 119, 434
(*S*)-2-Chloropropan-1-ol,
VII, 434
β-Chloropropionaldehyde
acetal, **II**, 137
β-Chloropropionic acid, **I**, 166
γ-Chloropropyl acetate,
III, 203
3-Chloropropyldiphenyl-
sulfonium tetrafluorobo-
rate, **VI**, 364

2-Chloropyrimidine, **IV**, 182
Chloropyruvic acid, **V**, 636
m-Chlorostyrene, **III**, 204
Chlorosulfonyl isocyanate,
V, 226
2-Chlorothiirane 1,1-dioxide,
V, 231
o-Chlorotoluene, **I**, 170
p-Chlorotoluene, **I**, 170
3-Chlorotoluquinone, **IV**, 152
1-Chloro-1-(trichloroethenyl)
cyclopropane, **VIII**,
124, 373
1-Chloro-1,4,4-trifluorobutadi-
ene, **V**, 235
1-Chloro-2,3,3-trifluorocy-
clobutene, **V**, 394
2-Chloro-1,1,2-trifluoroethyl
ethyl ether, **IV**, 184
3-Chloro-2,2,3-trifluoropropi-
onic ac!d, **V**, 239
1-Chloro-*N,N*,2-trimethyl-
propenylamine, **VI**, 282
1-Chloro-*N,N*,2-trimethylpropyl-
ideniminium chloride,
VI, 282
β-Chlorovinyl isoamyl ketone,
IV, 186
Cholesterol dibromide, **IV**, 195
Desyl chloride, **II**, 159
2,6-Dibromoquinone-4-
chloroimide, **II**, 175
(*E*)-1,2-Di-*tert*-butoxy-1-
chloroethene, **VIII**, 161
1,2-Di-*tert*-butoxy-1,2-
dichloroethane, **VIII**, 161
2,5-Di-*tert*-butyl-5,6-dichloro-
2-cyclohexene-1,4-dione,
VI, 210
2,5-Di-*tert*-butyl-3,5,6-
trichloro-2-cyclohexene-
1,4-dione, **VI**, 210
α,α-Dichloroacetamide,
III, 260
Dichloroacetic acid, **II**, 181

α,γ-Dichloroacetone, **I**, 211
Dichloroacetonitrile, **IV**, 255
Dichloroacetophenone, **III**, 538
2,6-Dichloroaniline, **III**, 262
7,7-Dichlorobicyclo[4.1.0]hept-3-ene, **VI**, 87
7,7-Dichlorobicyclo[3.2.0]hept-2-en-6-one, **VI**, 1037
exo-3,4-Dichlorobicyclo[3.2.1]oct-2-ene, **VI**, 142
3,4-Dichlorobiphenyl, **V**, 51
cis-3,4-Dichlorocyclobutene, **VI**, 422
cis-1,2-Dichlorocyclohexane, **VI**, 424
4,4'-Dichlorodibutyl ether, **IV**, 266
1,1-Dichloro-2,2-difluoroethylene, **IV**, 268
β,β-Dichloro-p-dimethylaminostyrene, **V**, 361
1,5-Dichloro-2,4-dinitrobenzene, **V**, 1067
trans-2,3-Dichloro-1,4-dioxane, **VIII**, 161
2,2'-Dichloro-α,α'-epoxybibenzyl, **V**, 358
2,2-Dichloroethanol, **IV**, 271
3,4-Dichloro-5-isopropoxy-2(5H)-furanone, **VIII**, 116
2,4-Dichloromethoxybenzene, **VIII**, 167
7,7-Dichloro-1-methylbicyclo[3.2.0]heptan-6-one, **VIII**, 377
Dichloromethylenetriphenylphosphorane, **V**, 361
Dichloromethyl methyl ether, **V**, 365
2,6-Dichloronitrobenzene, **V**, 367
2,2-Dichloronorbornane, **VI**, 845

2,6-Dichlorophenol, **III**, 267
Di-(p-chlorophenyl)acetic acid, **III**, 270
1,1-Dichloro-2-phenylcyclopropane, **VII**, 12
1,1-Di-(p-chlorophenyl)-2,2-dichloroethylene, **III**, 270
4,7-Dichloroquinoline, **III**, 272
2,5-Dichloroquinone, **IV**, 152
3,5-Dichlorosulfanilamide, **III**, 262
2,3-Dichlorotetrahydropyran, **VI**, 675
3,4-Dichloro-1,2,3,4-tetramethylcyclobutene, **V** 370
1,11-Dichlorotricyclo[4.4.1.0^{1,6}]undeca-3,8-diene, **VI**, 731
6-[(E)-1,2-Dichlorovinyl]-3-ethoxy-6-methyl-2-cyclohexen-1-one, **VII**, 241
4-[(E)-1,2-Dichlorovinyl]-4-methyl-2-cyclohexen-1-one, **VII**, 241
Diethyl 2-chloro-2-cyclopropylethene-1,1-dicarboxylate, **VIII**, 247
N,N-Diethyl-1,2,2-trichlorovinylamine, **V**, 387
7,7-Dimethoxy-1,2,3,4-tetrachlorobicyclo[2.2.1]heptene, **V**, 425
5,5-Dimethoxy-1,2,3,4-tetrachlorocyclopentadiene, **V**, 424
β-Dimethylaminoethyl chloride hydrochloride, **IV**, 333
Dimethyl 3-chloro-2-pentenedioate, **VI**, 505
2,4-Dinitrobenzenesulfenyl chloride, **V**, 474
Epichlorohydrin, **I**, 233; **II**, 256
Ethyl α-chloroacetoacetate, **IV**, 592
N-Ethyl-p-chloroaniline, **IV**, 420

Ethyl chlorofluoroacetate, **IV**, 423
N-Ethyl-*p*-chloroformanilide,
      IV, 420
Geraniol trichloroacetimidate,
      VI, 508
Geranyl chloride, **VI**, 634, 638;
      VIII, 226, 616
Glycerol α,γ-dichlorohydrin,
      I, 292
Glycerol α-monochlorohydrin,
      I, 294
unsym-Heptachloropropane,
      II, 312
Hexamethylene chlorohydrin,
      III, 446
2-Hydroxy-5-nitrobenzyl chlor-
      ide, **III**, 468
Iodobenzene dichloride, **III**, 482
(Isopropoxydimethyl-
      silyl)methyl chloride,
      VIII, 315
Mesitoyl chloride, **III**, 555
Methanesulfinyl chloride,
      V, 709
Methyl 4-chloro-2-butynoate,
      VIII, 371
Methyl 2-chloro-2-cyclopropyl-
      idenacetate, **VIII**, 373
Methyl (chlorosulfonyl)carba-
      mate, **VI**, 788
Methyl *N,N*-dichlorocarbamate,
      VIII, 427
(1*R*,2*S*,5*R*)-5-Methyl-2-[(1-
      methyl-1-phenylethyl)]
      cyclohexyl chloroacetate,
      VIII, 522
Monochloromethyl ether, **I**, 377
Neophyl chloride, **IV**, 702
m-Nitrobenzoyl chloride,
      IV, 715
p-Nitrobenzoyl chloride, **I**, 394
1-(1-Nitrophenyl)-4-chloro-2-
      butene, **IV**, 727
o-Nitrophenylsulfur chloride,
      II, 455

2-Oxa-7,7-dichloronorcarane,
      V, 874
Pentachlorobenzoic acid, **V**, 890
1,2,3,4,5-Pentachloro-5-ethyl-
      cyclopentadiene, **V**, 893
4-Pentylbenzoyl chloride,
      VII, 420
Perchlorofulvalene, **V**, 901
Phenylchlorodiazirine, **VII**, 203
trans-2-Phenylcyclohexyl
      chloroacetate (±)-, (+)-,
      (−)-, **VIII**, 516
Phenyldichlorophosphine,
      IV, 784
Phenylsulfenyl chloride,
      VIII, 550
Phenyl(trichloromethyl)mer-
      cury, **V**, 969
sym- and *unsym*-*o*-Phthalyl
      chloride, **II**, 528
Propyl chloride, **I**, 143
Pyrrol-2-yl trichloromethyl
      ketone, **VI**, 618
Pyruvoyl chloride, **VII**, 467
β-Styrenesulfonyl chloride,
      IV, 846
Styrylphosphonic dichloride,
      V, 1005
Tetrachlorobenzobarrelene, **VI**, 82
DL-4,4',6,6'-Tetrachlorodiphenic
      acid, **IV**, 872
Tetramethylene chlorohydrin,
      I, 571
Tetraphenylarsonium chloride
      hydrochloride, **IV**, 910
p-Toluenesulfenyl chloride,
      IV, 934
p-Toluenesulfinyl chloride,
      IV, 937
α,α,α-Trichloroacetanilide,
      V, 1074
1,1,1-Trichloro-4-ethoxy-3-
      buten-2-one, **VIII**, 254
1,1,1-Trichloro-4-methoxy-3-
      buten-2-one, **VIII**, 255

Trichloroethyl alcohol, **II**, 598
Trichloromethyl chloroformate,
 VI, 715
Trichloromethylphosphonyl
 dichloride, **IV**, 950
1,1,3-Trichlorononane, **V**, 1076
m-Trifluoromethylbenzene-
 sulfonyl chloride, **VII**, 508
Trimethyl 2-chloro-2-cyclo-
 propylidenorthoacetate,
 VIII, 373
Trimethylene chlorobromide,
 I, 157
Trimethylene chlorohydrin,
 I, 533
2,4,6-Trimethylpyrylium per-
 chlorate. **V**, 1106
Triphenylchloromethane, **III**, 841
Vinyl chloroacetate, **III**, 853
C. Fluorine Compounds
Benzoyl fluoride, **V**, 66
Bis(trifluoromethyl)dia-
 zomethane, **VI**, 161
Bis[2,2,2-trifluoro-1-phenyl-1-
 (trifluoromethyl)ethoxy]di-
 phenylsulfurane, **VI**, 163
4-Bromo-*N*,*N*-dimethyl-3-(trifluoromethyl)aniline,
 VI, 181
1-Bromo-2-fluorobenzene,
 V, 133
1-Bromo-2-fluoroheptane,
 V, 136
(Z)-1-Chloro-1-hexenyl trifluoroacetate, **VI**, 273
1-Chloro-1,4,4-trifluorobutadi-
 ene, **V**, 235
1-Chloro-2,3,3-trifluorocy-
 clobutene, **V**, 394
2-Chloro-1,1,2-trifluoroethyl
 ethyl ether, **IV**, 184
3-Chloro-2,2,3-trifluoropropi-
 onic ac!d, **V**, 239
Cholesta-3,5-dien-3-yl trifluoromethanesulfonate,
 VIII, 126

1,1-Dichloro-2,2-difluoroethyl-
 ene, **IV**, 268
Diethylaminosulfur trifluoride,
 VI, 440
4,4'-Difluorobiphenyl, **II**, 188
β,β-Difluorostyrene, **V**, 390
2,2-Difluorosuccinic acid,
 V, 393
α,α-Difluorotoluene, **V**, 396
Ethyl chlorofluoroacetate,
 IV, 423
Ethyl cyano(pentafluorophenyl)
 acetate, **VI**, 873
Ethyl 2-fluorohexanoate, (±)-,
 (+)-, and (−)-, **VIII**, 258
1-Fluoroadamantane, **VI**, 628
Fluorobenzene, **II**, 295
p-Fluorobenzoic acid, **II**, 299
2-Fluoroheptanoic acid, **V**,
 580
2-Fluorohexanoic acid,
 VIII, 259
16α-Fluoro-3-methoxy-
 1,3,5(10)-estratrien-17-
 one, **VIII**, 286
N-Fluoropyridinium triflate,
 VIII, 286
Hexafluoroacetone, **VII**, 251
Hexyl fluoride, **IV**, 525
p-Nitrobenzyl fluoride, **VI**, 835
(Pentafluorophenyl)acetonitrile,
 VI, 873
Pentafluorophenylcopper
 tetramer, **VI**, 875
2-Phenylperfluoropropene, **V**, 949
Phenylsulfur trifluoride, **V**, 959
Potassium 1,1,1,3,3,3-hexafluo-
 ro-2-phenyl-2-propanolate,
 VI, 163
Pyridinium polyhydrogen fluor-
 ide, **VI**, 628
Tetrakis(trifluoromethyl)-1,3-
 dithietane, **VII**, 251
Triethyloxonium fluoborate,
 V, 1080
N-Trifluoroacetanilide, **VI**, 1004

Trifluoroacetyl triflate, **VII**, 506

m-Trifluoromethylbenzene-
sulfonyl chloride, **VII**, 508

m-Trifluoromethyl-*N,N*-
dimethylaniline, **V**, 1085

1,1,1-Trifluoroheptane, **V**, 1082

m-Trifluoromethylbenzene-
sulfonyl chloride, **VII**, 508

Trimethyloxonium fluoborate,
V, 1096

2,4,6-Trimethylpyrylium tetrafluoroborate, **V**, 1112

2,4,6-Trimethylpyrylium trifluoromethanesulfonate, **V**, 1114

2,4,6-Triphenylpyrylium tetrafluoroborate, **V**, 1135

Tris(dimethylamino)sulfonium
difluorotrimethylsilicate,
VII, 528

D. Iodine Compounds

Amyl iodide, **II**, 402

(1-Azido-2-iodo-3,3-
dimethoxypropyl)benzene,
VI, 893

1,2-Bis(2-iodoethoxy)ethane,
VIII, 152, 153

Benzoylcholine iodide, **IV**, 84

Benzyltrimethylammonium
iodide, **IV**, 585

[Bis(trifluoroacetoxy)iodo]ben-
zene, **VIII**, 135

Butyl iodide, **II**, 402; **IV**, 322

tert-Butyl iodide, **IV**, 324

1-Chloro-3-iodopropane,
VI, 365

Cyanogen iodide, **IV**, 207

cis-4,5-Dihydro-5-iodomethyl-
4-phenyl-2(3*H*)-furanone,
VII, 164

trans-4,5-Dihydro-5-
iodomethyl-4-phenyl-
2(3*H*)-furanone, **VII**, 164

1,4-Diiodobutane, **IV**, 321

1,6-Diiodohexane, **IV**, 323

2,6-Diiodo-*p*-nitroaniline,
II, 196

N,N-Dimethylaminomethylfer-
rocene methiodide, **V**, 434

2,4-Dinitroiodobenzene, **V**, 478

Diphenyliodonium-2-carboxy-
late, **V**, 1037

Diphenyliodonium iodide,
III, 355

1-Ethyl-3-(3-dimethylamino)
propylcarbodimide methio-
dide, **V**, 555

Ethyl iodide, **II**, 402

Ethyl 4-iodobutyrate, **VIII**, 274

Hexadecyl iodide, **II**, 322

2-Hydroxy-3,5-diiodobenzoic
acid, **II**, 343

4-Hydroxy-3,5-diiodobenzoic
acid, **II**, 344

p-Iodoaniline, **II**, 347

5-Iodoanthranilic acid, **II**, 349

Iodobenzene, **I**, 323; **II**, 351

Iodobenzene dichloride,
III, 482

m-Iodobenzoic acid, **II**, 353

p-Iodobenzoic acid, **I**, 325

Iodocyclohexane, **IV**, 324, 543;
VI, 830

6-Iodo-3,4-dimethoxybenzalde-
hyde cyclohexylimine,
VIII, 586

Iododurene, **VI**, 700

(*Z*)-1-Iodohexene, **VII**, 290

endo-7-Iodomethylbicyclo
[3.3.1]nonan-3-one,
VI, 958

2-Iodo-3-nitrobenzoic acid,
I, 126

o-Iodophenol, **I**, 326

p-Iodophenol, **II**, 355

(*E*)-1-Iodo-4-phenyl-2-butene,
VI, 704

Iodosobenzene, **III**, 483; **V**, 658

Iodosobenzene diacetate, **V**, 660

N-Iodosuccinimide, **V**, 663

2-Iodothiophene, **II**, 357;
IV 545

Iodotrimethylsilane, **VI**, 353

1-Iodo-3-trimethylsilylpropane,
VIII, 486
4-Iodoveratrole, **IV**, 547
Iodoxybenzene, **III**, 485; **V**, 665
2-Iodo-*p*-xylene, **VI**, 709
Isoamyl iodide, **II**, 402
Isobutyl iodide, **II**, 402; **IV**, 324
Isopropyl iodide, **IV**, 322
Methylene iodide, **I**, 358
Methyl iodide, **II**, 399
Methyl 2-iodo-3-nitropropi-
onate, **VI**, 799
Methyl (*trans*-2-iodo-1-
tetralin)carbamate, **VI**, 795
Neopentyl iodide, **VI**, 830
Pentaerythrityl iodide, **II**, 477
Propyl iodide, **II**, 402; **IV**, 324
Tetraiodophthalic anhydride,
III, 796
2,2,3,3-Tetramethyliodocyclo-
propane, **VI**, 974
1,2,3-Triiodo-5-nitrobenzene,
II, 604

HETEROCYCLES
A.   Three-Membered, One
Nitrogen
trans-1-Amino-2,3-diphenyl-
aziridine, **VI**, 56
1-Amino-2-phenylaziridine,
VI, 56
1-Amino-2-phenylaziridinium
acetate, **VI**, 56
cis-2-Benzyl-3-phenylaziridine,
V, 83
2-(Dimethoxymethyl)-3-phenyl-
2H-azirine, **VI**, 893
trans-2,3-Diphenyl-1-phthalim-
idoaziridine, **VI**, 56
N-Ethylallenimine, **V**, 541
Ethylenimine, **IV**, 433
3-Phenyl-2*H*-azirine-2-carbox-
aldehyde, **VI**, 893
1,2,3,4-Tetrahydronaphthalen
(1,2)imine, **VI**, 967

B.   Three-Membered, Two
Nitrogen Atoms
3,3-Pentamethylene (−) diaziri-
dine, **V**, 897
3,3-Pentamethylenediazirine,
V, 897
Phenylchlorodiazirine,
VII, 203
C.   Three-Membered, One
Nitrogen, One Oxygen
2-*tert*-Butyl-3-phenyloxazirane,
V, 191
(+)-(2*R*,8a*S*)-10-(Camphor-
sulfonyl)oxaziridine,
VIII, 104
(±)-*trans*-2-(Phenylsulfonyl)-3-
phenyloxaziridine,
VIII, 546
D.   Three-Membered, One Oxygen
Benzyl 2,3-epoxypropyl ether,
VIII, 33
Cyclohexene oxide, **I**, 185
cis-Cyclooctene oxide, **VII**, 126
2,2'-Dichloro-α,α'-epoxybiben-
zyl, **V**, 358
Epibromohydrin, **II**, 257
Epichlorohydrin, **I**, 233; **II**, 256
2,3-Epoxycyclohexanone,
VI, 679
10,11-Epoxyfarnesyl acetate,
VI, 560
Ethyl β,β-pentamethyleneglyci-
date, **IV**, 459
Isophorone oxide, **IV**, 552
Methyl 5,5-dimethoxy-3-
methyl-2,3-epoxypen-
tanoate, **IV**, 649
Methylenecyclohexane oxide,
V, 755
Methyl (3*RS*,4*RS*)-4,5-epoxy-3-
phenylpentanoate,
VII, 164
(*R*)-Methyloxirane, **VIII**, 434
(*S*)-(−)-Methyloxirane, **VII**, 356
Oxaspiropentane, **VI**, 320

11-Oxatricyclo[4.4.1.0^{1,6}]unde-
ca-3,8-diene, **VI**, 862
Phenanthrene-9,10-oxide, **VI**, 887
Phenylmethylglycidic ester,
  III, 727
(2*S*,3*S*)-3-Propyloxiranemeth-
anol, **VII**, 461
trans-Stilbene oxide, **IV**, 860
Styrene oxide, **I**, 494
3,4,8,9-Tetrabromo-11-oxatricy-
clo[4.4.1.0^{1,6}]undecane,
  VI, 862
Tetracyanoethylene oxide, **V**,
  1007

E. Three-Membered, One Sulfur
2-Chlorothiirane 1,1-dioxide,
  V, 231
Cyclohexene sulfide, **IV**, 232
Ethylene sulfide, **V**, 562
2,3-Diphenylene sulfone, **VI**, 555

F. Four-Membered, One Nitrogen
4-Acetoxyazetidin-2-one,
  VIII, 3
Azetidine, **VI**, 75
1,3-Dimethyl-3-methoxy-4-
phenylazetidinone,
  VIII, 216
Ethyl 3-(1-azetidinyl)pro-
panoate, **VI**, 75
β-Isovalerolactam, **V**, 673
β-Isovalerolactam-*N*-sulfonyl
chloride, **V**, 673

G. Four-Membered, One Oxygen
Dimethylketene β-lactone
dimer, **V**, 456
Ketene dimer, **III**, 508
Trimethylene oxide, **III**, 835

H. Four-Membered, Two
Phosphorus, Two Sulfur
2,4-Bis-(4-methoxyphenyl)-
1,3,2,4-dithiaphosphetane-
2,4-disulfide, **VII**, 372

I. Four-Membered, Sulfur
3-Chlorothietane 1,1-dioxide,
  VII, 491

3-Chlorothiete 1,1-dioxide,
  VII, 491
3,3-Dichlorothietane 1,1-diox-
ide, **VII**, 491
Thietane 1,1-dioxide, **VII**, 491
Thiete 1,1-dioxide, **VII**, 491

J. Four-Membered, Two Sulfur
2,2,4,4-Tetrakis(trifluo-
romethyl)-1,3-dithietane,
  VII, 251

K. Five-Membered, One Nitrogen
N-Acetylisatin, **III**, 456
3-Acetyloxindole, **V**, 12
(*S*)-(−)-1-Amino-2-methoxy-
methylpyrrolidine, **VIII**,
  26, 403
1,2-Benzo-3,4-dihydrocar-
bazole, **IV**, 885
3-Benzoylindole, **VI**, 109
1-Benzylindole, **VI**, 104, 106
3-Benzylindole, **VI**, 109
4-Benzyloxindole, **VII**, 34
β-Bromoethylphthalimide,
  I, 119
tert-Butoxycarbonyl-L-proline,
  VI, 203
1-Butylpyrrolidine, **III**, 159
3-Butyroyl-1-methylpyrrole,
  VII, 102
(*S*)-(−)-1-Carbamoyl-2-meth-
oxymethylpyrrolidine,
  VIII, 26
N-Carbomethoxypyrrolidine,
  VII, 307
1,3-Dihydroisoindole, **V**, 406
2,4-Dimethyl-3-acetyl-5-carb-
ethoxypyrrole, **II**, 198
2,4-Dimethyl-5-carbethoxypyr-
role, **II**, 198
2,4-Dimethyl-3,5-dicarb-
ethoxypyrrole, **II**, 202
2,4-Dimethylpyrrole, **II**, 217
2,5-Dimethylpyrrole, **II**, 219
2,2-Dimethylpyrrolidine,
  IV, 354

1,5-Dimethyl-2-pyrrolidone,
III, 328
5,5-Dimethyl-2-pyrrolidone,
IV, 357
2-(Diphenylhydroxymethyl)
pyrrolidine, **VI**, 542
2,4-Diphenylpyrrole, **III**, 358
2-Ethoxypyrrolin-5-one,
VI, 226
Ethyl indole-2-carboxylate,
V, 567
Ethyl 2-methylindole-5-carb-
oxylate, **VI**, 601
Ethyl 2-methyl-3-methylthioin-
dole-5-carboxylate,
VI, 601
Ethyl pyrrole-2-carboxylate
VI, 618
Ethyl β-pyrrolidinocrotonate,
VI, 592
Ethyl α-(1-pyrrolidyl)propio-
nate, **IV**, 466
(S)-(−)-1-Formyl-2-hydroxy-
methylpyrrolidine,
VIII, 26
(S)-(−)-1-Formyl-2-methoxy-
methylpyrrolidine,
VIII, 26
3-(1-Hydroxybutyl)-1-methyl-
pyrrole, **VII**, 102
(S)-(+)-2-Hydroxymethylpyr-
rolidine, **VIII**, 26
Indole, **III**, 479
Indole-3-acetic acid, **V**, 654
Indole-3-aldehyde, **IV**, 539
Indole-3-carbonitrile, **V**, 656
Isatin, **I**, 327
Kryptopyrrole, **III**, 513
2-Methoxy-N-carbomethoxy-
pyrrolidine, **VII**, 307
1-(2-Methoxycarbonylphenyl)
pyrrole, **V**, 716
(S)-(+)-2-Methoxymethylpyr-
rolidine, **VIII**, 26
1-Methyl-3-ethyloxindole,
IV, 620

(S)-(+)-4-Methyl-3-heptanone
SAMP hydrazone, **VIII**, 403
1-Methylindole, **V**, 769
2-Methylindole, **III**, 597
2-Methylmercapto-N-methyl-
Δ^2-pyrroline, **V**, 780
3-Methyloxindole, **IV**, 657
1-(2-Methylpropenyl)pyrroli-
dine, **VI**, 496
N-Methylthiopyrrolidone,
VII, 372
2-Nitrocarbazole, **V**, 829
4-Nitroindole, **VIII**, 493
2-Oxo-1-phenyl-3-pyrroli-
dinecarboxylic acid,
VII, 411
3-Pentanone SAMP hydrazone
VIII, 403
3-(2-Phenyl-1,3-dithian-2-yl)-
1H-indole, **VI**, 109
2-Phenylindole, **III**, 725
N-Phenylmaleimide, **V**, 944
2-Phenyl-3-N-propylisoxazoli-
dine-4,5-cis-dicarboxylic
acid N-phenylimide,
V, 957
Phthalimide, **I**, 457
Pyrrole, **I**, 473
2-Pyrrolealdehyde, **IV**, 831
1-(N-Pyrrolidino)-1-cyclodo-
decene, **VII**, 135
1-Pyrrolidinocyclohexene,
VI, 1014
2-(1-Pyrrolidyl)propanol,
IV, 834
Pyrrol-2-yl trichloromethyl
ketone, **VI**, 618
1,2,3,4-Tetrahydrocarbazole,
IV, 884
2,3,4,5-Tetramethylpyrrole,
V, 1022
2-(p-Tolylsulfonyl)dihy-
droisoindole, **V**, 1064
cis-N-Tosyl-3-methyl-2-azabi-
cyclo[3.3.0]oct-3-ene,
VII, 501

L. Five-Membered, Two Nitrogen
   Atoms
   Allantoin, **II**, 21
   3(5)-Aminopyrazole, **V**, 39
   Benzimidazole, **II**, 65
   1-Benzoyl-2-(1,3-dibenzoyl-4-
      imidazolin-2-yl)imidazole,
      VII, 287
   N-[2,4-Bis(1,3-diphenylimida-
      zolidin-2-yl)-5-hydroxy-
      phenylacetamide, **VII**, 162
   Bis(1,3-diphenylimidazo-
      lidinylidene-2), **V**, 115
   1,1'-Carbonyldiimidazole, **V**, 201
   3-Chloroindazole, **III**, 476
   Creatinine, **I**, 172
   2-(1,3-Dibenzoylimidazolidin-
      2-yl)imidazole hydrochlo-
      ride, **VII**, 287
   2-(1,3-Dibenzoylimidazolin-2-
      yl)imidazole hydrochlo-
      ride, **VII**, 287
   5,5-Dimethylhydantoin,
      III, 323
   1,3-Dimethylimidazole-2-
      thione, **VII**, 195
   3,5-Dimethylpyrazole, **IV**, 351
   4,5-Diphenylglyoxalone, **II**, 231
   Ethylene thiourea, **III**, 394
   4(5)-Hydroxymethylimidazole
      hydrochloride, **III**, 460
   Imidazole, **III**, 471
   Imidazole-2-carboxaldehyde,
      VII, 287
   Indazole, **III**, 475; **IV**, 536;
      V, 650
   Indazolone, **III**, 476
   2-Mercaptobenzimidazole,
      IV, 569
   2-Methylbenzimidazole, **II**,
      66
   5-Nitroindazole, **III**, 660
   Parabanic acid, **IV**, 744
   1-Phenyl-3-amino-5-pyra-
      zolone, **III**, 708
   2-Phenylindazole, **V**, 941

3-(1-Propyl)-2-pyrazolin-5-one,
   VI, 791
4,5,6,7-Tetrahydroindazole,
   IV, 537

M. Five-Membered Ring, Three
   Nitrogen Atoms
   3-Amino-1H-1,2,4-triazole,
      III, 95
   4-Amino-4H-1,2,4-triazole,
      III, 96
   1,2,3-Benzotriazole, **III**, 106
   1,4-Diphenyl-5-amino-1,2,3-tri-
      azole, **IV**, 380
   4-Phenyl-5-anilino-1,2,3-tri-
      azole, **IV**, 380
   4-Phenyl-1,2,4-triazoline-3,5-
      dione, **VI**, 936
   4-Phenylurazole, **VI**, 936
   1,2,4-Triazole, **V**, 1070

N. Five-Membered, Four Nitrogen
   Atoms
   5-Benzylaminotetrazole,
      VII, 27
   4-(1-Phenyl-5-tetrazolyloxy)
      biphenyl, **VI**, 150

O. Five-Membered, One Nitrogen,
   One Oxygen
   3-Acetyl-4- and 5-chloro-2-oxa-
      zolidinone, **VII**, 5,
   4-Acetyl-7,8-dimethyl-2-
      oxa-4-azabicyclo
      [4.3.0]non-7-en-
      3-one, **VII**, 5
   3-Acetyl-2-oxazolidinone,
      VII, 5
   3-Acetyl-2(3H)-oxazolone,
      VII, 5
   Azlactone of α-acetaminocin-
      namic acid, **II**, 3
   Azlactone of α-benzoylamino-
      cinnamic acid, **II**, 490
   Azlactone of α-benzoylamino-
      β-(3,4-dimethoxyphenyl)
      acrylic acid, **II**, 55
   4-Chloromethyl-3-ethyl-5-
      methylisoxazole, **VI**, 781

3-(4-Chlorophenyl)-5-(4-meth-oxyphenyl)isoxazole, **VI**, 278
4,4-Dimethyl-2-oxazoline, **VI**, 65
Ethyl 3-ethyl-5-methyl-4-isoxa-zolecarboxylate, **VI**, 592
3-Ethyl-4-hydroxymethyl-5-methylisoxazole, **VI**, 781
2-(3-Ethyl-5-methyl-4-isoxa-zolylmethyl)cyclohexa-none, **VI**, 781
Hexahydro-1,3,3,6-tetramethyl-2,1-benzisoxazoline, **VI**, 670
2-Mercaptobenzoxazole, **IV**, 570
2-Oxazolidinone, 4, **VII**
(4*S*)-3-[(1-Oxo-*syn*-2-methyl-3-hydroxy)-3-phenylpropyl]-4-(phenylmethyl)-2-oxazo-lidinone, **VIII**, 339
(*S*)-3-(1-Oxopropyl)-4-(phenyl-methyl)-2-oxazolidinone, **VIII**, 339
(*S*)-4-(Phenylmethyl)-2-oxazo-lidinone, **VIII**, 339, 528
2-Phenyl-5-oxazolone, **V**, 946
2-Phenyl-3-propylisoxazoli-dine-4,5-*cis*-dicarboxylic acid *N*-phenylimide, **V**, 957
N,4,4-Trimethyloxazolinium iodide, **VI**, 65
2,3,5-Triphenylisoxazolidine, **V**, 1124

P. Five-Membered, One Nitrogen, One Sulfur
N-(*p*-Acetylaminophenyl)rhoda-nine, **IV**, 6
2-Amino-6-methylbenzothia-zole, **III**, 76
2-Amino-4-methylthiazole, **II**, 31
(−)-(Camphorsulfonyl)imine, **VIII**, 104, 110

(−)-2,10-Camphorsultam, **VIII**, 110
4,5-Dihydro-2-[(1-phenyl-methyl-2-propenyl) thio]thiazole, **VI**, 705
4,5-Dihydro-2-(2-propenylthio) thiazole, **VI**, 705
2,4-Dimethylthiazole, **III**, 332
Ethyl thiazole-4-carboxylate, **VI**, 620
1-Methyl-2-imino-β-naphtho-thiazoline, **III**, 595
Pseudothiohydantoin, **III**, 751
Rhodanine, **III**, 763

Q. Five-Membered, Two Nitrogen, One Oxygen
Benzofurazan oxide, **IV**, 74
Dimethylfurazan, **IV**, 342
3-Phenylsydnone, **V**, 962

R. Five-Membered, Two Nitrogen, One Sulfur
1,2,3-Benzothiadiazole 1,1-dioxide, **V**, 61

S. Five-Membered, One Oxygen
3-Acetyl-2,4-dimethylfuran, **VI**, 31
3-Acetyl-4-hydroxy-5,5-dimethylfuran-2(5*H*)-one, **VIII**, 71
4-Azido-3-chloro-5-isopropoxy-2(5*H*)-furanone, **VIII**, 116
α-Acetyl-δ-chloro-γ-valerolac-tone, **IV**, 10
Benzalphthalide, **II**, 61
α-Benzylidene-γ-phenyl-Δ^{β,γ}-butenolide, **V**, 80
α-Bromo-γ-butyrolactone, **V**, 255
3-Bromophthalide, **V**, 145
(*S*)-(+)-γ-Butyrolactone-γ-car-boxylic acid, **VII**, 99
Coumarilic acid, **III**, 209
Coumarone, **V**, 251
γ-Crotonolactone, **V**, 255
3-Deoxy-1,2:5,6-di-*O*-isopropyl-idene-D-ribo-hexofuranose, **VII**, 139

3,4-Dichloro-5-isopropoxy-
2(5*H*)-furanone, **VIII**, 116
unsym-o-Phthalyl chloride,
II, 528
5,6-Diethoxybenzofuran-4,7-
dione, **VIII**, 179
cis-4,5-Dihydro-5-iodomethyl-
4-phenyl-2(3*H*)-furanone,
VII, 164
trans-4,5-Dihydro-5-
iodomethyl-4-phenyl-
2(3*H*)-furanone, **VII**, 164
2,5-Dihydro-2,5-dimethoxyfu-
ran, **V**, 403
5,5-Dimethyl-2-pentyltetrahy-
drofuran, **IV**, 350
D-Erythronolactone, **VII**, 297
Ethyl 3-methylcoumarilate,
IV, 590
Furan, **I**, 274
2-Furancarboxylic acid, **I**, 276
2(5*H*)-Furanone, **VIII**, 396
2-Furoic acid, **IV**, 493
Furfural, **I**, 280
2-Furfuralacetone, **I**, 283
Furfural diacetate, **IV**, 489
2-Furfuryl mercaptan, **IV**, 491
Furylacrylic acid, **III**, 425
3-(2-Furyl)acrylonitrile,**V**, 585
2-Furylcarbinol, **I**, 276
4-(2-Furyl)-4-hydroxy-2,3-
diethoxycyclobutenone,
VIII, 179
2-Furylmethyl acetate, **I**, 285
Furyl phosphorodichloridate,
VIII, 396
β-Furylpropionic acid, **I**, 313
Furyl *N,N,N',N'*-tetramethyl-
diamidophosphate, **VIII**,
396
3-Hydroxytetrahydrofuran, **IV**,
534
2,3-*O*-Isopropylidene-D-ery-
thronolactone, **VII**, 297
3-Methylcoumarilic acid,
IV, 591

3-Methylcoumarone, **IV**,590
3-Methylfuran, **IV**, 628
3-Methyl-2(5*H*)-furanone,
VIII, 396
5-Methylfurfural, **II**, 393
5-Methylfurfuryldimethyl-
amine, **IV**, 626
3-Methyl-2-furoic acid, **IV**, 628
Methyl 3-methyl-2-furoate,
IV, 649
2,2,7,7,12,12,17,17-Octa-
methyl-21,22,23,24-tetra-
oxaperhydroquaterene,
VI, 856
2,2,7,7,12,12,17,17-Octa-
methyl-21,22,23,24-tetra-
oxaquaterene, **VI**, 856
γ-Octyl-γ-butyrolactone,
VII, 400
(1*R*,6*S*)-(+)-8-Oxabicyclo
[4.3.0]nonan-7-one,
VII, 406
D-(−)-Pantoyl lactone, **VII**, 417
Phthalide, **II** , 526
γ-Propylbutyrolactone, **III**, 742
Tetrahydrofuran, **II**, 566
Tetrahydrofurfuryl bromide,
III, 793
β-(Tetrahydrofuryl)propionic
acid, **III**, 742
2,2,5,5-Tetramethyltetrahydro-
3-ketofuran, **V**, 1024
T. Five-Membered, Two Oxygen
2-Bromo-2-cyclopenten-1-one
ethylene ketal, **VII**,
271, 495
2-Cyclobutyl-*cis*-4-*trans*-5-
dimethyl-1,3-dioxolane,
VI, 313
1,4-Di-*O*-benzyl-2,3-di-*O*-iso-
propylidene-L-threitol,
VIII, 155
DL-Isopropylideneglycerol,
III, 502
Dimethyl 2,3-*O*-isopropylidene-
tartrate, **VIII**, 155, 201

4-Ethoxy-3-methoxybenzalde-
hyde ethylene acetal,
VI, 567
8-Ethynyl-8-methyl-1,4-dioxa-
spiro[4.5]dec-3-ene,
VII, 241
Diisopropyl (2*S*,3*S*)-2,3-*O*-iso-
propylidenetartrate,
VIII, 201
2,3-*O*-Isopropylidene-L-threitol,
VIII, 155
Dimethyl 2,3-*O*-isopropylidene-
tartrate, **VIII**, 155, 201
2-Hydroxymethyl-2-cyclopen-
ten-1-one ethylene ketal,
VII, 271
4,5-Methylenedioxybenzo-
cyclobutene, **VII**, 326
o-Phenylene carbonate, **IV**, 788
Piperonylic acid, **II**, 538
(*S*)-(+)-2-*p*-Toluenesulfinyl)-2-
cyclopenten-1-one ethyl-
ene ketal, **VII**, 495
U. Five-Membered, One Boron,
Two Oxygen
(*E*)-1-Hexenyl-1,3,2-benzodiox-
aborole, **VIII**, 532
V. Five-Membered, One Sulfur,
VI
2-Acetothienone, **III**, 14
3-Bromothiophene, **V**, 149
2-*tert*-Butoxythiophene, **V**, 642
2-Chloromethylthiophene, **III**,
197
2,5-Diamino-3,4-dicyanothio-
phene, **IV**, 243
7,9-Dimethyl-*cis*-8-thiabicy-
clo[4.3.0]nonane 8,8-diox-
ide, **VI**, 482
2,2'-Dithienyl sulfide, **VI**, 558
2-Hydroxythiophene, **V**, 642
2-Iodothiophene, **IV**, 545
Isoprene cyclic sulfone, **III**, 499
Methyl 2-thienyl sulfide, **IV**, 667
3-Methylthiophene, **IV**, 671
2-Nitrothiophene, **II**, 466

Phenyl thienyl ketone, **II**, 520
Tetrahydrothiophene, **IV**, 892
2-Thenaldehyde, **IV**, 915 or
2-Thiophenealdehyde,
III, 811
3-Thenaldehyde, **IV**, 918
3-Thenoic acid, **IV**, 919
3-Thenyl bromide, **IV**, 921
cis-8-Thiabicyclo[4.3.0]nonane,
VI, 482
cis-8-Thiabicyclo[4.3.0]nonane
8,8-dioxide, **VI**, 482
Thiophene, **II**, 578
2-Thiophenethiol, **VI**, 979
2,3,5-Tribromothiophene, **V**, 150
2-Vinylthiophene, **IV**, 980
W. Five-Membered, Two Sulfur
Atoms
1,4-Dithiaspiro[4.11]hexade-
cane, **VII**, 124
5,9-Dithiaspiro[3.5]nonane,
VI, 316
2,2-(Ethylenedithio)cyclohexa-
none, **VI**, 590
X. Six-Membered, One Boron
Perhydro-9*b*-boraphenalene,
VII, 427
Y. Six-Membered, One Boron,
One Nitrogen
10-Methyl-10,9-borazarophen-
anthrene, **V**, 727
Z. Six-Mmbered, One Nitrogen
Acetone anil, **III**, 329
2-Acetyl-6,7-dimethoxy-1-
methylene-1,2,3,4-tetrahy-
droisoquinoline, **VI**, 1
4-Acetylpyridine oxime, **VII**, 149
4-Acetylpyridine oxime tosylate, **VII**, 149
Acridone, **II**, 15
9-Aminoacridine, **III**, 53
3-Aminopyridine, **IV**, 45
1-Aminopyridinium iodide, **V**, 43
2-Benzoyl-1-cyano-1-methyl-
1,2-dihydroisoquinoline,
IV, 642

2-Benzoyl-1,2-dihydroiso-
quinaldonitrile, **IV**, 641;
VI, 115
Benzoylpiperidine, **I**, 99
3-Benzoylpyridine, **IV**, 88
4-Benzoylpyridine, **IV**, 89
2-Benzylaminopyridine, **IV**, 91
1-Benzylisoquinoline, **VI**, 115
3,3-Bis(methylthio)-1-(2-pyri-
dinyl)-2-propen-1-one,
VII, 476
2,2'-Bipyridine, **V**, 102
2-Bromopyridine, **III**, 136
9-Butyl-1,2,3,4,5,6,7,8-octahy-
droacridine, **VIII**, 87
9-Butyl-1,2,3,4,5,6,7,8-octahydro-
acridine *N*-oxide, **VIII**, 87
9-Butyl-1,2,3,4,5,6,7,8-octahy-
droacridin-4-ol, **VIII**, 87
9-Chloroacridine, **III**, 54
7-Chloro-4-hydroxy-3-quino-
linecarboxylic acid,
III, 273
2-Chlorolepidine, **III**, 194
2-Chloronicotinonitrile,
IV, 166
N-Chloropiperidine, **VI**, 968
1-Cyano-2-benzoyl-1,2-dihy-
droisoquinoline, **IV**, 641;
VI, 115
2-Cyano-6-methylpyridine,
V, 269
3-Cyano-6-methyl-2(1H)-pyri-
done, **IV**, 210
1-(Cyclooctylidenemethyl)
piperidine, **VII**, 473
2,3-Diamino-5-bromopyridine,
V, 346
2,3-Diaminopyridine, **V**, 346
2,6-Di-*tert*-butyl-4-methylpyri-
dine, **VII**, 144
3,5-Dicarbethoxy-2,6-dimethyl-
pyridine, **II**, 216
4,7-Dichloroquinoline, **III**, 272
2,2-Diethoxy-2-(4-pyridyl)
ethylamine, **VII**, 149

1,4-Dihydro-3,5-dicarbethoxy-
2,6-dimethylpyridine,
II, 216
6,7-Dimethoxy-1-methyl-3,4-
dihydroisoquinoline,
VI, 1
(*S*)-6,7-Dimethoxy-1-methyl-
1,2,3,4-tetrahydroisoquino-
line, **VIII**, 57
6,7-Dimethoxy-1,2,3,4-tetrahy-
dro-2-[(1-*tert*-butoxy-3-
methyl)-2-butylimino-
methyl]isoquinoline,
VIII, 573
2,6-Dimethylpyridine, **II**, 214
2,4-Dimethylquinoline, **III**, 329
Dipyridine chromium (**VI**)
oxide, **VI**, 644
5-Ethyl-2-methylpyridine, **IV**,
451
Ethyl 6-methylpyridine-2-
acetate, **VI**, 611
N-Ethylpiperidine, **V**, 575
4-Ethylpyridine, **III**, 410
Ethyl 2-pyridylacetate, **III**, 413
N-Fluoropyridinium triflate,
VIII, 286
2-Hydroxycinchoninic acid,
III, 456
2-Hydroxy-3-methylisocar-
bostyril, **V**, 623
erythro-(3-Hydroxy-2-methyl-
3-phenylpropanoyl)piperi-
dine, **VIII**, 326
threo-(3-Hydroxy-2-methyl-3-
phenylpropanoyl)piperi-
dine, **VIII**, 326
6-Hydroxynicotinic acid,
IV, 532
3-Hydroxyquinoline, **V**, 635
3-Isoquinuclidone, **V**, 670
Julolidine, **III**, 504
Lepidine, **III**, 519
(−)-Menthyl nicotinate, **VIII**, 350
1-Methoxy-2-methylpyridinium
methylsulfate, **V**, 270

6-Methoxy-8-nitroquinoline, **III**, 568
1-(4-Methoxyphenyl)-1,2,5,6-tetrahydropyridine, **VIII**, 358
4-Methylcarbostyril, **III**, 580
2-Methyl-4-hydroxyquinoline, **III**, 593
1-Methylisoquinoline, **IV**, 641
3-Methyl-4-nitropyridine 1-oxide, **IV**, 654
N-Methyl-2-phenyl-Δ²-tetrahydropyridine, **VI**, 818
3-Methylpyridine 1-oxide, **IV**, 655
1-Methyl-2-pyridone, **II**, 419
4'-(Methylthio)-2,2':6',2"-terpyridine, **VII**, 476
N-β-Naphthylpiperidine, **V**, 816
Nicotinamide 1-oxide, **IV**, 704
Nicotinic acid, **I**, 385
Nicotinic acid anhydride, **V**, 822
Nicotinonitrile, **IV**, 706
4-Oxo-4-(3-pyridyl)butyronitrile, **VI**, 866
1-Phenyl-3,4-dihydro-6,7-methylenedioxyisoquinoline, **VII**, 326
1-Phenylpiperidine, **IV**, 795
2-Phenylpyridine, **II**, 517
Picolinic acid hydrochloride, **III**, 740
Pyridine 1-oxide, **IV**, 828
4-Pyridinesulfonic acid, **V**, 977
Pyridinium polyhydrogen fluoride, **VI**, 628
1-(α-Pyridyl)-2-propanol, **III**, 757
Quinoline, **I**, 478
3-Quinuclidone hydrochloride, **V**, 989
(−)-Salsolidine, **VIII**, 57
2,2':6',2"-Terpyridine, **VII**, 476
1,2,3,4-Tetrahydro-β-carboline, **VI**, 965

2,3,4,5-Tetrahydropyridine trimer, **VI**, 968
2α-Thiohomophthalimlde, **V**, 1051
2,3,6-Tricarboethoxypyridine, **VIII**, 597
AA. Six-Membered, Two Nitrogen
Alloxan monohydrate, **III**, 37; **IV**, 23
Alloxantin dihydrate, **III**, 42; **IV**, 25
5-Amino-2,3-dihydro-1,4-phthalazinedione, **III**, 69
4-Amino-2,6-dimethylpyrimidine, **III**, 71
Barbituric acid, **II**, 60
Benzalbarbituric acid, **III**, 39
Benzoylene urea, **II**, 79
1-Benzylpiperazine, **V**, 88
2-Chloropyrimidine, **IV**, 182
2,4-Diamino-6-hydroxypyrimidine, **IV**, 245
Diaminouracil hydrochloride, **IV**, 247
2-(Dimethylamino)pyrimidine, **IV**, 336
3-Heptyl-5-cyanocytosine, **IV**, 515
2-Mercapto-4-amino-5-carbethoxypyrimidine, **IV**, 566
2-Mercapto-4-hydroxy-5-cyanopyrimidine, **IV**, 566
2-Mercaptopyrimidine, **V**, 703
N-Methylaminopyrimidine, **IV**, 336
4-Methyl-6-hydroxypyrimidine, **IV**, 638
2-Methyl-5,6-pyrazinedicarboxylic acid, **IV**, 827
4-Methylpyrimidine, **V**, 794
6-Methyluracil, **II**, 422
Nitrobarbituric acid, **II**, 440
5-Nitro-2,3-dihydro-1,4-phthalazinedione, **III**, 656
Pyocyanine, **III**, 753

2,3-Pyrazinedicarboxylic acid,
  IV, 824
Quinoxaline, **IV**, 824
2-Thio-6-methyluracil, **IV**, 638
Uramil, **II**, 617
BB. Six-Membered, Three Nitrogen
  2-Amino-4-anilino-6-(chloro-
    methyl)-s-triazine, **IV**, 29
  Benzoguanamine, **IV**, 78
  2-Chloromethyl-4,6-diamino-s-
    triazine, **IV**, 30
  2,4-Diamino-6-phenyl-s-tri-
    azine, **IV**, 78
  Hexahydro-1,3,5-tripropionyl-*s*-
    triazine, **IV**, 518
  Triethyl 1,2,4-triazine-3,5,6-tri-
    carboxylate, **VIII**, 597
CC. Six-Membered Ring, One
    Nitrogen, One Oxygen
  2-Benzyl-4,4,6-trimethyl-5,6-
    dihydro-1,3(4*H*)-oxazine,
    VI, 907
  N-Chloromorpholine, **VIII**,
    167
  Isatoic anhydride, **III**, 488
  N-Methylmorpholine *N*-oxide,
    VI, 342
  1-Morpholino-1-cyclohexene,
    V, 808
  α-Morpholinostyrene, **VI**, 520
  N-Nitromorpholine, **V**, 839
  2-(1-Phenylcyclopentyl)-4,4,6-
    trimethyl-5,6-dihydro-
    1,3(4*H*)-oxazine, **VI**, 905
  2-(1-Phenylcyclopentyl)-4,4,6-
    trimethyltetrahydro-1,3-
    oxazine, **VI**, 905
DD. Six-Membered Ring, One
    Nitrogen, One Sulfur
  4*H*-1,4-Thiazine 1,1-dioxide,
    VI, 976
EE. Six-Membered, One Oxygen
  3-Carbethoxycoumarin, **III**,165
  Chelidonic acid, **II**, 126
  Coumalic acid, **IV**, 201
  Coumarin dibromide, **III**,209

Dehydroacetic acid, **III**, 231
1-Deoxy-2,3,4,6-tetra-*O*-acetyl-
  1-(2-cyanoethyl)-α-D-glu-
  copyranose, **VII**, 148
2,6-Di-*tert*-butyl-4-methylpyry-
  lium trifluoromethanesulfonate, **VII**, 144
Diethyl [(2-tetrahydropyranyl-
  oxy)methyl]phosphonate,
  VII, 160
3,4-Dihydro-2-methoxy-4-
  methyl-2*H*-pyran, **IV**, 311
3,4-Dihydro-2-methylene-2*H*-
  benzopyran, **VIII**, 512
2,3-Dihydropyran, **III**, 276
5,6-Dihydro-2*H*-pyran-2-one,
  VI, 462
4,6-Dimethylcoumalin,
  IV,337
2,6-Dimethyl-3,5-diphenyl-4*H*-
  pyran-4-one, **V**, 450
2α,4α-Dimethyl-8-oxabicy-
  clo[3.2.1]oct-6-en-3-one,
  VI, 512
2-Ethylchromone, **III**, 387
Ethyl isodehydroacetate,
  IV, 549
Flavone, **IV**, 478
Isodehydroacetic acid, **IV**, 549
2-(*p*-Methoxyphenyl)-6-phenyl-
  4-pyrone, **V**, 721
Methyl 4-*O*-benzoyl-6-bromo-
  6-deoxy-α-D-glucopyra-
  noside, **VIII**, 363
Methyl coumalate, **IV**, 532
4-Methylcoumarin, **III**, 581
4-Methylesculetin, **I**, 360
Methyl 2-oxo-5,6,7,8-tetrahy-
  dro-2*H*-1-benzopyran-3-
  carboxylate, **VIII**, 444
(*R*,*S*)-Mevalonolactone-2-¹³C,
  VII, 386
2-Oxa-7,7-dichloronorcarane,
  V, 874
2*H*-Pyran-2-one, **VI**, 462
α-Pyrone, **V**, 982

Tetrahydro-2-(2-propynyloxy)-2H-pyran, **VII**, 334
Tetrahydropyran, **III**, 794
2,4,6-Trimethylpyrylium perchlorate, **V**, 1106
2,4,6-Trimethylpyrylium tetrafluoroborate, **V**,1112
2,4,6-Trimethylpyrylium trifluoromethanesulfonate, **V**, 1114
2,4,6-Triphenylpyrylium tetrafluoroborate, **V**, 1135
Xanthone, **I**, 552
Xanthydrol, **I**, 554

FF. Six-Membered, Two Oxygen
2-Bromo-2-cyclopenten-1-one ethylene ketal, **VII**, 271
2-(2-Bromoethyl)-1,3-dioxane, **VII**, 59,
2-(Bromomethyl)-2-(chloromethyl)-1,3-dioxane, **VIII**, 173
6-Chloromethyl-2,2-dimethyl-1,3-dioxen-4-one, **VIII**, 192
Cyclopropenone 1,3-propanediol ketal, **VIII**, 173
2,3-Di-*tert*-butoxy-1,4-dioxane, **VIII**, 161
trans-2,3-Dichloro-1,4-dioxane, **VIII**, 161
5,5-Dicyano-4-phenyl-2-cyclopenten-1-one 1,3-propanediol ketal, **VIII**, 173
6-Diethylphosphonomethyl-2,2-dimethyl-1,3-dioxen-4-one, **VIII**, 192
8-Ethynyl-8-methyl-1,4-dioxaspiro[4.5]dec-6-ene, **VII**, 241
4-Methoxycarbonyl-1,1,6-trimethyl-1,4,4a,5,6,7,8,8a-octahydro-2,3-benzopyrone, (rac), (+), (−), **VIII**, 353
Monobenzalpentaerythritol, **IV**, 679

4-Phenyl-*m*-dioxane, **IV**, 786
(*S*)-(+)-(*p*-Toluenesulfinyl)-2-cyclopenten-1-one ethylene ketal, **VII**, 495
2,5,5-Trimethyl-2-(2-bromoethyl)-1,3-dioxane, **VII**, 59

GG. Six-Membered Ring, One Oxygen, One Sulfur
cis- and *trans*-2,6-Diethoxy-1,4-oxathiane 4,4-dioxide, **VI**, 976
Hexahydro-4,4,7-trimethyl-4H-1,3-benzoxathiin, **VIII**, 302
Phenoxthin, **II**, 485

HH. Six-Membered Ring, One Phosphorus
4-Amino-1,2,5,6-tetrahydro-1-phenylphosphorin-3-carbonitrile, **VI**, 932
1-Phenyl-4-phosphorinanone, **VI**, 932

II. Six-Membered, Two Sulfur
1,3-Dithiane, **VI**, 556
1,4-Dithiane, **IV**, 396
1-(1,3-Dithian-2-yl)-2-cyclohexen-1-ol, **VIII**, 309
2-Methylthio-2-phenyl-1,3-dithiane, **VI**, 109
2-Phenyl-1,3-dithiane, **VI**, 109
2,2-(Trimethylenedithio)cyclohexanone, **VI**, 1014

JJ. Six-Membered, Three Sulfur Atoms
sym-Trithiane, **II**, 610
2-Tetradecyl-*sym*-trithiane, **VI**, 869

KK. Seven-Membered, One Nitrogen
1,3-Dihydro-3,5,7-trimethyl-2H-azepin-2-one, **V**, 408
Ethyl α-(hexahydroazepinylidene-2)acetate, **VIII**, 263
Isopropylidene α-(hexahydroazepinylidene-2)malonate, **VIII**, 263

2-Ketohexamethylenimine,
   II, 371
O-Methylcaprolactim, **IV**, 588
2-Propyl-1-azacycloheptane
   VIII, 568
LL.   Seven-Membered, One
        Nitrogen, One Oxygen
        3,1-Benzoxazepine, **VII**, 23
MM. Seven-Membered, One
        Oxygen
        2,7-Dimethyloxepin, **V**, 467
NN. Eight-Membered, One Nitrogen
        Hexahydro-2-(1*H*)azocinone,
           VII, 254
OO. Thirteen-Membered, One
        Oxygen
        Ricinelaidic acid lactone,
           VII, 470
PP.   Fourteen-Membered, Four
        Sulfur Atoms
        1,4,8,11-Tetrathiacyclotetra-
           decane, **VIII**, 592
QQ. Sixteen-Membered, Four
        Oxygen Atoms
        2,2,7,7,12,12,17,17-Octa-
           methyl-21,22,23,24-tetra-
           oxaperhydroquaterene,
           VI, 856
RR. Eighteen-Membered Ring, Two
        Nitrogen Atoms
        1,10-Diazacyclooctadecane,
           VI, 382
SS.   Eighteen-Membered Ring, Six
        Nitrogen Atoms
        1,4,7,10,13,16-Hexaazacyclo-
           octadecane, **VI**, 652
        1,4,7,10,13,16-Hexakis(*p*-tolyl-
           sulfonyl)-1,4,7,10,13,16-
           hexaazacyclooctadecane,
           VI, 652
TT.   Eighteen-Membered Ring, Two
        Nitrogen Atoms, Four
        Oxygen Atoms
        4,13-Diaza-18-crown-6,
           VIII, 152
        N,N-Dibenzyl-4,13-diaza-18-
           crown-6, **VIII**, 152

UU. Eighteen-Membered Ring, Six
        Oxygen Atoms
        18-Crown-6, **VI**, 301
        Dibenzo-18-crown-6 polyether,
           VI, 395
        Dicyclohexyl-18-crown-6 poly-
           ether, **VI**, 395
VV. Miscellaneous
        7-Aminotheophylline,
           VII, 10
        Anhydro-*o*-hydroxymercuriben-
           zoic acid, **I**, 57
        Anhydro-2-hydroxymercuri-3-
           nitrobenzoic acid, **I**, 56
        N-Benzyl-2-azanorbornene,
           VIII, 31
        7-Benzylidenaminotheo-
           phylline, **VII**, 9
        6-Dioxo-1-phenyl-4-benzyl-1,4-
           diazabicyclo[3.3.0]octane,
           VIII, 231
        Hemin, **III**, 443
        Hexahydro-6-allyl-6-ethyl-3-
           (hydroxymethyl)-8a-methyl-
           2-phenyl-[2*S*,3*S*,6*S*,8a*R*]-5-
           oxo-5*H*-oxazolo[3,2a]pyri-
           dine, **VIII**, 241
        Hexahydro-3-(hydroxymethyl)-
           8a-methyl-2-phenyl
           [2*S*,3*S*,8a*R*]-5-oxo-5*H*-
           oxazolo[3,2-a]pyridine,
           VIII, 241
        2,3-*O*-Isopropylidene-D-ery-
           thronolactone, **VII**, 297
        3-Isoquinuclidone, **V**, 670
        [6a*R*-(6aα,9a,10aβ)]-Octa-
           hydro-3,3,6,6,9-penta-
           methyl-1*H*,6*H*-[1,3]-diox-
           ino-[4,5-*c*]benzopyran-1-
           one, **VIII**, 353
        2α,4α-Dimethyl-8-oxabicy-
           clo[3.2.1]oct-6-en-3-one,
           VI, 512
        2-Phenyfuro[3,2-*b*]pyridine,
           VI, 916
        B-3-Pinanyl-9-borabicy-
           clo[3.3.1]nonane, **VII**, 402

Pseudopelletierine, **IV**, 816
3-Quinuclidone hydrochloride,
  V, 989
o-Sulfobenzoic anhydride, **I**, 495

HYDRAZINES and DERIVATIVES
Acetone azine, **VI**, 10
Acetone hydrazone, **VI**, 10
Acetophenone *N*,*N*-dimethylhydra-
  zone, **VI**, 12
Acetophenone hydrazone, **VI**, 12
Acetophenone phenylhydrazone,
  III, 726
Adipyl hydrazide, **IV**, 819
Aminoacetone semicarbazone
  hydrochloride, **V**, 27
Aminoguanidine bicarbonate, **III**, 73
trans-1-Amino-2,3-diphenylaziri-
  dine, **VI**, 56
1-Amino-2-phenylaziridine, **VI**, 56
1-Amino-2-phenylaziridinium
  acetate, **VI**, 56
Benzalazine, **II**, 395
Benzil dihydrazone, **IV**, 377
Benzil hydrazone, **II**, 497; **III**, 357
Benzophenone hydrazone, **III**, 352
Betaine hydrazide hydrochloride,
  II, 85
tert-Butyl carbazate, **V**, 166
Camphor tosylhydrazone, **VI**, 173
Cholest-4-en-3-one *p*-toluenesul-
  fonylhydrazone, **VI**, 293
Diaminobiuret, **III**, 404
sym-Dibenzoyldimethylhydrazine,
  II, 209
1,2-Di-1-(1-cyano)cyclohexylhy-
  drazine, **IV**, 274
1,2-Di-2-(2-cyano)propylhydrazine,
  IV, 275
Diethyl hydrazodicarboxylate,
  IV, 411
unsym-Dimethylhydrazine, **II**, 213
sym-Dimethylhydrazine dihydrochloride, **II**, 208
unsym-Dimethylhydrazine hydrochloride, **II**, 211

2,4-Dinitrophenylhydrazine, **II**, 228
trans-2,3-Diphenyl-1-phthalimi-
  doaziridine, **VI**, 56
Ethyl hydrazinecarboxylate, **VI**, 936
Hexafluoroacetone hydrazone,
  VI, 161
Hydrazine sulfate, **I**, 309
Methyl 2-(1-cyanocyclohexyl)hydra-
  zinecarboxylate, **VI**, 334
Methylglyoxal ω-phenylhydrazone,
  IV, 633
4-Methyl-3-heptanone SAMP-hydra-
  zone, (*S*)-(+)-, **VIII**, 403
Methylhydrazine sulfate, **II**, 395
Methyl hydrazodicarboxylate,
  IV, 413
10-Methyl-$\Delta^{1(9)}$-2-octalone *N*,*N*-
  dimethylhydrazone,
  VI, 242
α-Methyl-α-phenylhydrazine,
  II, 418
3-Pentanone SAMP-hydrazone,
  VIII, 403
4-Phenyl-1-carbethoxysemicar-
  bazide, **VI**, 936
Phenylhydrazine, **I**, 442
4-Phenylsemicarbazide, **I**, 450
β-Propionylphenylhydrazine,
  IV, 657
Semicarbazide sulfate, **I**, 485
p-Toluenesulfonylhydrazide, **V**, 1055

HYDROCARBONS
Adamantane, **V**, 16
Adamantylideneadamantane, **VII**, 1
Allene, **V**, 22
Amylacetylene, **IV**, 119
Amylbenzene, **II**, 47
[18]Annulene, **VI**, 68
Azulene, **VII**, 15
Benz[*a*]anthracene, **VII**, 18
Benzene, **V**, 998
Benzobarrelene, **VI**, 82,
Benzocyclopropene, **VI**, 87
Biallyl, **III**, 121
Bicyclo[1.1.0]butane, **VI**, 133, 159

Bicyclo[2.1.0]pentane, **V**, 96
Bicyclo[2.1.0]pent-2-ene, **VI**, 145
Binor-*S*, **VI**, 378
Biphenyl, **VI**, 150
Biphenylene, **V**, 54
1,4-Bis-(4-phenylbutadienyl)ben-
    zene, **V**, 985
2-Bornene, **VI**, 172,
1,3-Butadiene, **II**, 102
(*E*)-Buten-3-ynyl-2,6,6-trimethyl-1-
    cyclohexene, **VII**, 63
Butylacetylene, **IV**, 117
Butylbenzene, **III**, 157
2-Butylbornene, **VII**, 77
4-*tert*-Butyl-1-vinylcyclohexene,
    VIII, 97
Cholesta-3,5-diene, **VIII**, 126
Cholestane, **VI**, 289
5β-Cholest-3-ene, **VI**, 293
Cyclobutene, **VII**, 117
cis-Cyclododecene, **V**, 281
trans-Cyclododecene, **VII**, 121
1,3-Cyclohexadiene, **V**, 285
Cyclohexene, **I**, 183; **II**, 152
Cyclohexylbenzene, **II**, 151
Cyclohexylidenecyclohexane, **V**, 297
3-Cyclohexylpropyne, **I**, 191
1,2-Cyclononadiene, **V**, 306
trans-Cyclooctene, **V**, 315
Cyclopentadiene, **IV**, 238
Cyclopentene, **II**, 153
Cyclopropylbenzene, **V**, 328, 929
m-Cymene, **V**, 332
Decane, **VI**, 376
1-Decyne, **I**, 192
Diamantane, **VI**, 378
1,2-Dibutylbenzene, **VI**, 407
9,10-Dihydroanthracene, **V**, 398
9,10-Dihydrofulvalene, **VIII**, 298
9,10-Dihydrophenanthrene, **IV**, 313
Dimesitylmethane, **V**, 422
cis-7,8-Dimethylbicyclo[4.2.0]oct-7-
    ene, **VI**, 482
3,3'-Dimethylbiphenyl, **III**, 295
4,4'-Dimethyl-1,1'-biphenyl, **VI**, 488
2,3-Dimethyl-1,3-butadiene, **III**, 312

1,2-Dimethyl-1,4-cyclohexadiene,
    V, 467
1,2-Dimethylenecyclohexane,
    VIII, 212
(*E*)-4,8-Dimethyl-1,3,7-nonatriene,
    VII, 258
1,6-Dimethyltricyclo[4.1.0.0^{2,7}]hept-
    3-ene, **VII**, 200
(1*R*)-9,9-Dimethyltricyclo[6.1.1.0^{2,6}]
    deca-2,5-diene, **VIII**, 223
(*E*)-6,10-Dimethyl-5,9-undecadien-
    1-yne, **VIII**, 226
Diphenylacetylene, **III**, 350; **IV**, 377
1,4-Diphenyl-1,3-butadiene, **II**, 229;
    V, 499
2,3-Diphenyl-1,3-butadiene, **VI**, 531
1,1-Diphenylcyclopropane, **V**, 509
Diphenyldiacetylene, **V**, 517
1,1-Diphenylethane, **VI**, 537
1,1-Diphenylethylene, **I**, 226
Diphenylmethane, **II**, 232
1,2-Diphenyl-3-methylcyclopropene,
    VII, 203
1,1-Diphenylpentane, **V**, 523
1,1-Di-*p*-tolylethane, **I**, 229
cis- and *trans*-1,2-Divinylcyclo-
    butane, **V**, 528
Dodecane, **VI**, 376
Durene, **II**, 248
3-Ethyl-1-hexyne, **VI**, 595
(*E*)-5-Ethyl-1,4-undecadiene,
    VII, 236
α-Farnesene, **VII**, 245
Hemimellitene, **IV**, 508
(5*Z*,7*E*)-5,7-Hexadecadiene,
    VIII, 295
Hexadecane, **II**, 320
Hexadehydro[18]annulene, **VI**, 68
Hexamethylbenzene, **II**, 248;
    IV, 520
Hexamethyl Dewar benzene,
    VII, 256
Hexaphenylbenzene, **V**, 604
1,2,4,5-Hexatetraene (in solution),
    VII, 485
1,3,5-Hexatriene, **V**, 608

Hexylacetylene, **IV**, 119
Isoamylacetylene, **IV**, 119
Isodurene, **II**, 360
Mesitylene, **I**, 341
Methallylbenzene, **VI**, 722
1,6-Methano[10]annulene, **VI**, 731
2-Methylbiphenyl, **VI**, 747
5-Methylcoprost-3-ene, **VI**, 762
3-Methylcyclohexene **VI**, 76,
Methylenecyclohexane, **IV**, 612;
    V, 751
Methylenecyclopropane, **VI**, 320
(+)-3-Methylene-*cis-p*-menthane,
    VIII, 386
9-Methylfluorene, **IV**, 623
1-Methyl-3-phenylindane, **IV**, 665
2-Methyl-2-undecene, **VIII**, 474
Monovinylacetylene, **IV**, 683
Naphthalene, **VI**, 821
(1*R*)-Nopadiene, **VIII**, 223
Norbornylene, **IV**, 738
Norcarane, **V**, 855
$\Delta^{9,10}$-Octalin, **VI**, 852
[2.2]Paracyclophane, **V**, 883
1,4-Pentadiene, **IV**, 746
1,2,3,4,5-Pentamethylcyclopenta-
    diene, **VIII**, 505
Pentamethylbenzene, **II**, 248
Pentane, **II**, 478
2-Pentene, **I**, 430
Phenanthrene, **IV**, 313
Phenylacetylene, **I**, 438; **IV**, 763
trans-1-Phenyl-1,3-butadiene,
    IV, 771
4-Phenylbutyne-1, **I**, 192
Phenylcyclopropane, **V**, 328, 929
Phenylethylene, **I**, 440
1-Phenylnaphthalene, **III**, 729
(1*Z*,3*E*)-1-Phenyl-1,3-octadiene,
    VIII, 532
1-Phenyl-1,3-pentadiyne, **VI**, 925
1-Phenyl-1,4-pentadiyne, **VI**, 925
9-Phenylphenanthrene, **V**, 952
α-Pinene, **VIII**, 553
Propylacetylene, **IV**, 119
Propylbenzene, **I**, 471

Quadricyclane, **VI**, 962
p-Quinquephenyl, **V**, 985
cis-Stilbene, **IV**, 857
trans-Stilbene, **III**, 786
Tetrahydro-Binor-S, **VI**, 378
1,4,5,8-Tetrahydronaphthalene,
    VI, 731
Tetraphenylethylene, **IV**, 914
1,2,3,4-Tetraphenylnaphthalene,
    V, 1037
Tricyclo[4.4.1.0^{1,6}]undeca-3,8-diene,
    VI, 731
Tridehydro[18]annulene, **VI**, 68
4,6,8-Trimethylazulene, **V**, 1088
1,1,3-Trimethyl-3-phenylindane,
    IV, 666
Triphenylene, **V**, 1120
Triphenylethylene, **II**, 606
Triphenylmethane, **I**, 548
Triptycene, **IV**, 964
(*E*)-2-Undecene, **VI**, 1040

HYDROXYLAMINES, OXIMES, and DERIVATIVES

Acetoxime, **I**, 318
N-Acetyl-*N*-phenylhydroxylamine,
    VIII, 16
4-Acetylpyridine oxime, **VII**, 149
4-Acetylpyridine oxime tosylate,
    VII, 149
Benzohydroxamic acid, **II**, 67
Benzohydroxamoyl chloride, **V**, 504
Benzophenone oxime, **II**, 70
Biacetyl monoxime, **II**, 205
2-*tert*-Butoxycarbonyloxyimino-2-
    phenylacetonitrile, **VI**, 199
N-tert-Butylhydroxylamine, **VI**, 804
N-tert-Butyl-*N-tert*-octyl-*O-tert*-butyl-
    hydroxylamine, **VIII**, 93
Carboxymethoxylamine hemihy-
    drochloride, **III**, 172
4-Chloroacetophenone oxime,
    VI, 279
ω-Chloroisonitrosoacetophenone,
    III, 191
Cupferron, **I**, 177

1,2-Cyclohexanedione dioxime,
    IV, 229
Cyclohexanone oxime, **II**, 76, 314
Cyclohexanone oxime methanesul-
    fonate, **VIII**, 568
Dibenzyl ketoxime, **V**, 84
Di-*tert*-butyl nitroxide, **V**, 355
Diethyl oximinomalonate, **V**, 573
Dimethylglyoxime, **II**, 204
N,N-Dimethylhydroxylamine
    hydrochloride, **IV**, 612
O-Diphenylphosphinylhydroxyl-
    amine, **VII**, 8
Heptaldoxime, **II**, 313
Hydroxylamine hydrochloride, **I**, 318
2-Hydroxyimino-2-phenylacetoni-
    trile, **VI**, 199
2-Hydroxy-3-methylisocarbostyril,
    V, 623
Hydroxyurea, **V**, 645
Isonitrosoacetanilide, **I**, 327
Isonitrosopropiophenone, **II**, 363
16-Oximinoandrost-5-en-3β-ol-17-
    one, **VI**, 840
β-Phenylhydroxylamine, **I**, 445;
    VIII, 16
Potassium benzohydroxamate, **II**, 67
(+)- and (−)-α-(2,4,5,7-Tetranitro-9-
    fluorenylideneaminooxy)
    propionic acid, **V**, 1031
Veratraldoxime, **II**, 622

ISOCYANATES
α-Chloroacetyl isocyanate, **V**, 204
Chlorosulfonyl isocyanate, **V**, 226
Hexamethylene diisocyanate,
    IV, 521
3-Isocyanatopropanoyl chloride,
    VI, 715
p-Nitrophenyl isocyanate, **II**, 453
Undecyl isocyanate, **III**, 846

ISONITRILES
Benzyl isocyanide, **VII**, 27
tert-Butyl isocyanide, **VI**, 232
Cyclohexyl isocyanide, **V**, 300

Ethyl isocyanide, **IV**, 438
trans-2-Isocyanocyclohexanol,
    VII, 294
[(*trans*-2-Isocyanocyclohexyl)oxy-
    trimethylsilane, **VII**, 294
Methyl isocyanide, **V**, 772
1,1,3,3-Tetramethylbutyl isonitrile,
    VI, 751
o-Tolyl isocyanide, **V**, 1060
p-Tolylsulfonylmethyl isocyanide,
    VI, 41, 987

KETONES
A. Monoketones
    1. Unsubstituted
        Aceto-*p*-cymene, **II**, 3
        Acetophenone, **I**, 111
        9-Acetylanthracene,
            IV, 8
        1-Acetylcyclohexene,
            III, 22
        2-Acetylfluorene, **III**,
            23
        5-Acetyl-1,2,3,4,5-penta-
            methylcyclopentadi-
            ene, **VI**, 39
        9-Acetylphenanthrene,
            III, 26
        Adamantanone, **VI**, 48
        2-Allylcyclohexanone,
            III, 44; **V**, 25
        2-Allyl-3-methylcyclo-
            hexanone, **VI**, 51
        Anthrone, **I**, 60
        Benzalacetone, **I**, 77
        Benzalacetophenone,
            I, 78
        Benzalpinacolone, **I**, 81
        Benzanthrone, **II**, 62
        Benzophenone, **I**, 95
        β-Benzopinacolone,
            II, 73
        Benzylacetophenone,
            I, 101
        2-Benzylcyclopentanone,
            V, 76

2-Benzyl-2-methylcyclo-
hexanone, **VI**, 121
2-Benzyl-6-methylcyclo-
hexanone, **VI**, 121
Bicyclo[2.2.1]hepten-7-
one, **V**, 91
Bicyclo[3.3.1]nonan-9-
one, **VI**, 137
Bicyclo[4.3.0]non-1-en-
4-one, **VIII**, 38
Bicyclo[3.2.1]octan-3-
one, **VI** , 142
3-Butylcyclobutenone,
VIII, 82
4-*tert*-Butylcyclohexa-
none **VI**, 218, 220
1-(4-*tert*-Butylcyclo-
hexen-1-yl)-2-propen-
1-one, **VIII**, 97
2-Butyl-2-methylcyclo-
hexanone, **V**, 187
1-Butyl-10-methyl-$\Delta^{1(9)}$-
2-octalone, **VI**, 242
tert-Butyl phenyl ketone,
VI, 248
Cholestanone, **II**, 139
Cholestenone, **III**, 207
Δ^4-Cholesten-3-one,
IV, 192
Δ^5-Cholesten-3-one,
IV, 195
Cyclobutanone, **VI**, 316,
320, 324; **VII**, 114
Cyclodecanone, **IV**, 218;
V, 277
Cycloheptanone, **IV**, 221
2-Cyclohepten-1-one,
VI, 327
2-Cyclohexenone, **V**, 294
Cyclohexyl methyl
ketone, **V**, 775
Cyclooctanone, **V**, 310
Cyclopentanone, **I**, 192
2-Cyclopentenone,
V, 326

Cyclopropenone, **VI**, 361
Cycloundecanone, **VI**, 368
2-Decanone, **VII**, 137
Desoxybenzoin, **II**, 156
Dibenzalacetone, **II**, 167
Di-*tert*-butylcycloprope-
none, **VI** , 991
Dicinnamalacetone,
VII, 60
Dicyclopropyl ketone,
IV, 278
3,4-Dihydro-1(2*H*)-azu-
lenone, **VIII**, 196
Dihydrocarvone, **VI**, 459
Dihydrojasmone,
VIII, 620
3,3-Dimethyl-*cis*-bicy-
clo[3.2.0]heptan-2-
one, **VII**, 177
3,5-Dimethyl-2-cyclo-
hexen-1-one, **II**, 317
4,4-Dimethyl-2-cyclo-
hexen-1-one, **VI**, 496
4,4-Dimethyl-2-cyclo-
penten-1-one, **VIII**,
208
Di-(2-methylcyclopropyl)
ketone, **IV**, 280
2,5-Dimethyl-3-phenyl-
2-cyclopenten-1-one,
VI, 520
Dineopentyl ketone,
VI, 991
Diphenylacetone, **III**, 343
Diphenylcyclopro-
penone, **V**, 514
2,3-Diphenylindone,
II, 353
1,9-Diphenylnona-
1,3,6,8-tetraen-5-one,
VII, 60
Diphenylpropiophenone,
II, 236
5-Dodecen-2-one,
VIII, 235

Dypnone, **III**, 367
(*R*)-4-Ethyl-4-allyl-2-
    cyclohexen-1-one,
    VIII, 241
4-Heptanone, **V**, 589
2-Heptyl-2-cyclohexen-
    1-one, **VII**, 249
cis-4,4a,5,6,7,8-
    Hexahydro-4a,5-
    dimethyl-2(3*H*)naph-
    thalenone, **VI**, 666
2,3,4,5,6,6-Hexamethyl-
    2,4-cyclohexadien-1-
    one, **V**, 598
3-Hexylcyclobutanone,
    VIII, 306
α-Hydrindone, **II**, 336
2-Indanone, **V**, 647
(+)-Isomenthone,
    VIII, 386
Isopropylidenaceto-
    phenone, **VIII**, 210
Laurone, **IV**, 560
L-Menthone, **I**, 340
Mesityl oxide, **I**, 345
p-Methylacetophenone,
    I, 111
Methyl amyl ketone,
    I, 351
Methyl benzyl ketone,
    II, 389
6-Methylbicyclo[4.2.0]
    octan-2-one,
    VII, 315
Methyl butyl ketone,
    I, 352
2-Methylcyclohexanone,
    IV, 164
2-Methyl-2-cyclohexen-
    one, **IV**, 162
Methyl cyclopropyl
    ketone, **IV**, 597
2-Methylene-1-oxo-
    1,2,3,4-tetrahydro-
    naphthalene, **VII**, 332

(*S*)-(+)-4-Methyl-3-hep-
    tanone, **VIII**, 403
1-Methyl-4,4a,5,6,7,8-
    hexahydronaphthalen-
    2(3*H*)-one, **VI**, 781
4-Methylhexanone-2,
    I, 352
5-Methyl-5-hexen-2-one,
    V, 767
Methyl isobutyl ketone,
    I, 352
Methyl isopropyl ketone,
    II, 408
(2*RS*,5*R*)-5-Methyl-2-[1-
    methyl-1-(2-phenyl-
    ethyl)]cyclohexanone,
    VIII, 522
3-Methyl-2-pentyl-2-
    cyclopenten-1-one,
    VIII, 620
2-(1-Methyl-2-propynyl)
    cyclohexanone **VIII**,
    460
trans-3-Methyl-2-vinyl-
    cyclohexanone,
    VIII, 479
2-Methyl-2-vinylcyclo-
    pentanone, **VIII**, 486
2-Norbornanone,
    V, 852
Nortricyclanone, **V**, 866
$\Delta^{1(9)}$-Octalone-2, **V**, 869
(+)-2-Oxo-7,7-dimethyl-
    1-vinylbicyclo[2.2.1]
    heptane, **V**, 877
trans-3-Penten-2-one,
    VI, 883
2-*tert*-Pentylcyclopenta-
    none, **VII**, 424
2-Phenylcycloheptanone,
    IV, 780
4-Phenyl-6-hepten-2-one,
    VII, 443
Phenyl heptyl ketone,
    VI, 919

Pinacolone, **I**, 462
(*E*)-2-(1-Propenyl)cyclo-
butanone, **VIII**, 556
4-Protoadamantanone,
VI, 958
Pseudoionone, **III**, 747
Spiro[5.7]trideca-1,4-
dien-3-one, **VII**, 473
Spiro[5.7]tridec-1-en-3-
one, **VII**, 473
Stearone, **IV**, 854
α-Tetralone, **II**, 569; **III**,
798; **IV**, 898
β-Tetralone, **IV**, 903
Tetraphenylcyclopen-
tadienone, **III**, 806
2,6,6-Trimethyl-2,4-
cyclohexadienone,
V, 1092
2,4,4-Trimethylcyclopen-
tanone, **IV**, 957
2. Amino and amido ketones
3-Acetamido-2-butanone,
IV, 5
2-(2-Acetamidoethyl)-
4,5-dimethoxyace-
tophenone, **VI**, 1
Acetoacetanilide, **III**, 10
Acetonamine hydrogen
oxalate, **I**, 196
2-Acetyl-6,7-dimethoxy-
1-methylene-1,2,3,4-
tetrahydroisoquino-
line, **VI**, 1
Acridone, **II**, 15
2-Aminobenzophenone,
IV,34
Benzoylacetanilide, **III**,
108; **IV**, 80
3-Benzoylindole, **VI**, 109
Benzoyl-2-methoxy-4-
nitroacetanilide,
IV, 82
1-(2-Benzoylpropanoyl)
piperidine, **VIII**, 326

3-Benzoylpyridine, **IV**, 88
4-Benzoylpyridine, **IV**, 89
3,3-Bis(methylthio)-1-(2-
pyridinyl)-2-propen-
1-one, **VII**, 476
3-Butyroyl-1-methylpyr-
role, **VII**, 102
p-Chloroacetylacetani-
lide, **III**, 183
1-Diethylamino-3-
butanone, **IV**, 281
p-Dimethylaminoben-
zophenone, **I**, 217
1-(Dimethylamino)-4-
methyl-3-pentanone,
VI, 474
3-Dimethylaminopro-
piophenone hydro-
chloride, **III**, 305
4'-Methyl-2-aminoben-
zophenone, **IV**, 38
4-Oxo-4-(3-pyridyl)buty-
ronitrile, **VI**, 866
Phenacylamine hydro-
chloride, **V**, 909
Phenyl 4-pyridyl ketone,
IV, 89
Pseudopelletierine,
IV, 816
Pyrrol-2-yl trichloro-
methyl ketone,
VIII, 618
3-Quinuclidone hydro-
chloride, **V**, 989
3. Halogen substituted
p-Acetyl-α-bromohy-
drocinnamic acid,
VI, 21
α-Acetyl-δ-chloro-γ-
valerolactone, **IV**, 10
3-Acetyl-4- and 5-
chloro-2-oxazolidi-
none **VII**, 5
Benzalacetone dibro-
mide, **III**, 105

Benzyl chloromethyl
  ketone, **III**, 119
Bromoacetone, **II**, 88
3-Bromoacetophenone,
  V, 117
p-Bromoacetophenone,
  I, 109
2-Bromoacetyl-6-meth-
  oxynaphthalene,
  VI, 175
α-Bromobenzalacetone,
  III, 125
2-Bromo-2-cyclopenten-
  1-one, **VII**, 271
1-Bromo-3-methyl-2-
  butanone, **VI**, 193
p-Bromophenacyl bro-
  mide, **I**, 127
α-Bromo-α-phenylace-
  tone, **III**, 343
3-Butyl-4,4-dichlorocy-
  clobutenone,
  VIII, 82
p-Chloroacetophenone,
  I, 111
p-Chloroacetylacetani-
  lide, **III**, 183
(2-Chlorobenzoyl)fer-
  rocene, **VI**, 625
7-Chloro-7-cyanobicy-
  clo[4.2.0]octan-8-one,
  VIII, 116
2-Chlorocyclohexanone,
  III, 188
ω-Chloroisonitrosoace-
  tophenone, **III**, 191
2-Chloro-2-methylcyclo-
  hexanone, **IV**, 162
5-Chloro-2-pentanone,
  IV, 597
4-(4-Chlorophenyl)
  butan-2-one, **VII**, 105
α-(4-Chlorophenyl)-γ-
  phenylacetoacetoni-
  trile, **IV**, 174

1-(p-Chlorophenyl)-3-
  phenyl-2-propanone,
  IV, 176
Chloropyruvic acid, **V**, 636
β-Chlorovinyl isoamyl
  ketone, **IV**, 186
Desyl chloride, **II**, 159
2,2-Dibromoacetyl-6-
  methoxynaphthalene,
  VI, 175
1,3-Dibromo-2-butanone,
  VI, 711
5α,6β-Dibromocholes-
  tan-3-one, **IV**, 197
2,12-Dibromocyclodo-
  decanone, **VI**, 368
α,α'-Dibromodibenzyl
  ketone, **V**, 514
α,α'-Dibromodineopen-
  tyl ketone, **VI**, 991
2,4-Dibromo-3-pen-
  tanone, **VI**, 512, 520
α,γ-Dichloroacetone,
  I, 211
Dichloroacetophenone,
  III, 538
7,7-Dichlorobicyclo
  [3.2.0]hept-2-en-6-
  one, **VI**, 1037
1,7-Dichloro-4-hep-
  tanone, **IV**, 279
7,7-Dichloro-1-methylbi-
  cyclo[3.2.0]heptan-6-
  one, **VIII**, 377
6-[(E)-1,2-Dichloro-
  vinyl]-3-ethoxy-6-
  methyl-2-cyclohexen-
  1-one, **VII**, 241
4-[(E)-1,2-Dichloro-
  vinyl]-4-methyl-2-
  cyclohexen-1-one,
  VII, 241
16α-Fluoro-3-methoxy-
  1,3,5(10)-estratrien-
  17-one, **VIII**, 287

Hexafluoroacetone,
VII, 251
endo-7-Iodomethyl-
bicyclo[3.3.1]nonan-
3-one, **VI**, 958
Phenacyl bromide, **II**, 480
Pyrrol-2-yl trichloro-
methyl ketone,
VIII, 618
2,4,4,6-Tetrabromo-2,5-
cyclohexadien-1-one,
VI, 181
p,α,α-Tribromoaceto-
phenone, **IV**, 110
1,1,1-Trichloro-4-ethoxy-
3-buten-2-one,
VIII, 254
1,1,1-Trichloro-4-
methoxy-3-buten-2-
one, **VIII**, 238

4. Hydroxy ketones
Acetol, **II**, 5
1-Acetylcyclohexanol,
IV, 13
3-Acetyl-4-hydroxy-5,5-
dimethylfuran-2(5H)-
one, **VIII**, 71
2-p-Acetylphenylhydro-
quinone, **IV**, 15
Benzoin, **I**, 94
8-Butyl-2-hydroxytricy-
clo[7.3.1.0^{2,7}]tridecan-
13-one, **VIII**, 87
Butyroin, **II**, 114; **VII**, 95
Diacetone alcohol, **I**, 199
16-Diazoandrost-5-en-
3β-ol-17-one, **VI**, 840
2,5-Dihydroxyaceto-
phenone, **III**, 280
2,6-Dihydroxyaceto-
phenone, **III**, 281
4-(2-Furyl)-4-hydroxy-
2,3-diethoxycy-
clobutenone,
VIII, 179
Gallacetophenone, **II**, 304

α-Hydroxyacetophenone,
VII, 263
o-Hydroxyacetophenone,
III, 389
2-Hydroxycyclobuta-
none, **VI**, 167
(S)-(+)-3-Hydroxy-2,2-
dimethylcyclohexa-
none, **VIII**, 312
2-Hydroxymethyl-2-
cyclopenten-1-one,
VII, 271
2-Hydroxymethylene-
cyclohexanone, **IV**,
536; **VI**, 390, 590
2-Hydroxy-2-methyl-
pentan-3-one,
VII, 381
3-Hydroxy-3-methyl-1-
phenyl-1-butanone,
VIII, 210, 323
17β-Hydroxy-5-oxo-3,5-
$seco$-4-norandrostane-
3-carboxylic acid,
VI, 690
$threo$-4-Hydroxy-3-
phenyl-2-heptanone,
VI, 692
p-Hydroxyphenylpyruvic
acid, **V**, 627
7-Hydroxy-4,4,6,7-tetra-
methylbicyclo[4.2.0]
octan-2-one, **VI**, 1024
3-Hydroxy-1,7,7-trimeth-
ylbicyclo[2.2.1] hep-
tan-2-one, **VII**, 277
6-Hydroxy-3,5,5-tri-
methyl-2-cyclohexen-
1-one, **VII**, 282
3-Hydroxy-2,2,3-tri-
methyloctan-4-one,
VIII, 343
5-Hydroxy-2,4,6-tri-
methyl-2-trimethyl-
siloxyheptan-3-one,
VII, 185

(4*S*)-3-[(1-Oxo-*syn*-2-
  methyl-3-hydroxy)-3-
  phenylpropyl]-4-
  (phenylmethyl)-2-
  oxazolidinone,
  VIII, 339
Phloroacetophenone,
  II, 522
o- and *p*-Propiophenol,
  II, 543
Quinacetophenone
  monomethyl ether,
  IV, 836
Sebacoin, **IV**, 840
Tropolone, **VI**, 1037

5. Keto acids
p-Acetyl-α-bromohydro-
  cinnamic acid, **VI**, 21
2-(1-Acetyl-2-oxopropyl)
  benzoic acid, **VI**, 36
δ-Acetylvalerlc acid,
  IV, 19
β-Benzoylacrylic acid,
  III, 109
γ-Benzoylbutyrlc acid,
  II, 82
Benzoylformic acid, **I**,
  244; **III**, 114
β-Benzoylpropionic acid,
  II, 81
3-Carboxy-5-dodecen-2-
  one, **VIII**, 235
Chloropyruvic acid, **V**, 636
3,4-Dimethoxyphenyl-
  pyruvic acid, **IV**, 327
1,3-Dimethyl-5-oxobicy-
  clo[2.2.2]octane-2-
  carboxylic acid,
  VIII, 219
Fluorenone-2-carboxylic
  acid, **III**, 420
α-Ketoglutaric acid, **III**,
  510; **V**, 687
6-Ketohendecanedioic
  acid, **IV**, 555
DL-Ketopinic acid, **V**, 689

Levulinic acid, **I**, 335
3-Methylcyclohexanone-
  3-acetic acid,
  VIII, 467
6-Oxodecanoic acid,
  VIII, 499
Phenylpyruvic acid,
  II, 519
p-Toluyl-*o*-benzoic acid,
  I, 517

6. Keto esters
3β-Acetoxy-5α-cyanoc-
  holestan-7-one, **VI**, 14
4-Acetoxy-2-cyclopen-
  ten-1-one, **VIII**, 15
7-Acetoxy-4,4,6,7-tetra-
  methylbicyclo[4.2.0]
  octan-2-one, **VI**, 1024
Benzoin acetate, **II**, 69
o-Benzoyloxyaceto-
  phenone, **IV**, 478
tert-Butyl acetoacetate,
  V, 155
sec-Butyl α-caproylpro-
  pionate, **IV**, 120
S-*tert*-Butyl 3-oxobu-
  tanethioate, **VIII**, 71
2-Carbethoxycycloocta-
  none, **V**, 198
2-Carbethoxycyclopen-
  tanone, **V**, 198
1-Carbomethoxy-1-
  methylethyl 3-oxobu-
  tanoate, **VIII**, 71
2-Carbomethoxy-3-vinyl-
  cyclopentanone,
  VIII, 112
Diethyl acetonedicar-
  boxylate, **I**, 237
Diethyl acetosuccinate,
  II, 262
Diethyl benzoylmalo-
  nate, **IV**, 285
Diethyl cyclohexanone-
  2,6-dicarboxylate,
  V, 439

Diethyl cyclopropylcar-
bonylmalonate,
VIII, 248
Diethyl enanthylsucci-
nate, **IV**, 430
Diethyl β-ketopimelate,
V, 384
Diethyl oxomalonate,
I, 266
Diethyl γ-oxopimelate,
IV, 302
Diethyl phenylacetyl-
malonate, **III**, 637
Diethyl propionylsucci-
nate, **VI**, 615
3,5-Dimethyl-4-car-
bethoxy-2-cyclohex-
en-1-one, **III**, 317
Dimethyl cyclohexa-
none-2,6-dicarboxy-
late, **V**, 439
1,3-Dimethyl-5-oxobicy-
clo[2.2.0]octane-2-car-
boxylic acid, methyl
ester, **VIII**, 219
(1-Ethenyl)heptanyl 3-
ketobutanoate,
VIII, 235
Ethyl acetoacetate, **I**, 235
Ethyl acetopyruvate, **I**, 238
Ethyl α-acetyl-β-(2,3-
dimethoxyphenyl)pro-
pionate, **IV**, 408
Ethyl benzoylacetate, **II**,
266; **III**, 379; **IV**, 415
Ethyl benzoyldimethyl-
acetate, **II**, 268
Ethyl benzoylformate,
I, 241
Ethyl butylacetoacetate,
I, 248
Ethyl 2-butyrylacetate,
VII, 213
Ethyl 4-cyclohexyl-4-
oxobutanoate,
VIII, 578

Ethyl ethoxalylpropi-
onate, **II**, 272
Ethyl α-isopropylace-
toacetate, **III**, 405
Ethyl 1-(p-methoxy-
phenyl)-2-oxocyclo-
hexanecarboxylate,
VII, 229
Ethyl 4-oxohexanoate,
VI, 615
Ethyl (E)-4-(4-nitro-
phenyl)-4-oxo-2-
butenoate, **VIII**, 268
Ethyl 5-oxo-6-methyl-6-
heptenoate, **VIII**,
274
Ethyl α-phenylacetoace-
tate, **II**, 284
Ethyl pyruvate, **IV**, 467
Methyl p-acetylbenzoate,
IV, 579
Methyl benzoylformate,
I, 244
Methyl-4-keto-7-methyl-
octanoate, **III**, 601
Methyl 2-oxocyclohex-
anecarboxylate,
VII, 351
Methyl 6-oxodecanoate,
VIII, 441
Methyl 6-oxohexanoate,
VII, 168
Methyl 7-oxooctanoate,
VI, 807
Methyl phenylacetylace-
tate, **VII**, 359
Methyl pyruvate, **III**, 610
S-(2-Oxobut-3-yl)
butanethioate **VI**, 776
7. Miscellaneous
2-Acetothienone, **II**, 8;
III, 14
3-Acetyl-2,4-dimethylfu-
ran, **VI**, 31
4-Acetyl-7,8-dimethyl-2-
oxa-4-azabicyclo

[4.3.0]non-7-en-3-
one, **VII**, 5
2-Acetyl-6-methoxy-
naphthalene, **VI**, 34
3-Acetyl-2-oxazolidi-
none, **VII**, 5
3-Acetyl-2(3*H*)-oxa-
zolone, **VII**, 5
3-Acetyloxindole, **V**, 12
2-*p*-Acetylphenylhydro-
quinone diacetate,
IV, 17
2-*p*-Acetylphenyl-
quinone, **IV**, 16
Acetyltrimethylsilane,
VIII, 19
Anisalacetone, **I**, 78
4-Benzyloxy-2-butanone,
386, **VII**
Biacetyl monoxime,
II, 205
17β–Butoxy-5α-andro-
stan-3-one, **VII**, 66
(+)-(1S)-10-Camphorsul-
fonamide, **VIII**, 104
DL-10-Camphorsulfonic
acid, **V**, 194
DL-10-Camphorsulfonyl
chloride, **V**, 196
(−)-(Camphorsulfonyl)
imine, **VIII**, 104, 110
(+)-(2*R*,8a*S*)-10-(Cam-
phorsulfonyl)oxaziri-
dine, **VIII**, 104
(−)-D-2,10-Camphor-
sultam, **VIII**, 110
Carbonyl cyanide, **VI**, 268
Deoxyanisoin, **V**, 339
Diazoacetophenone,
VI, 386
1-(Diazoacetyl)naphtha-
lene, **VI**, 613
2-Diazocyclohexanone,
VI, 389
1-Diazo-4-phenyl-2-
butanone, **VIII**, 196

2α,4α-Dimethyl-8-
oxabicyclo[3.2.1]oct-
3-en-3-one, **VI**, 512
2,2-Dimethyl-4-oxopen-
tanal, **VIII**, 208
2,3-Epoxycyclohexa-
none, **VI**, 679
3-Ethoxy-2-cyclo-
hexenone, **V**, 539
2-(1-Ethoxycyclohexyl)-
2-(trimethylsiloxy)
cyclobutanone,
VIII, 579
3-Ethoxy-6-methyl-2-
cyclohexen-1-one,
VII, 208
2-Ethylchromone,
III, 387
2,2-(Ethylenedithio)
cyclohexanone,
VI, 590
2-(3-Ethyl-5-methyl-4-
isoxazolylmethyl)-
cyclohexanone,
VI, 781
Ethyl phenylcyanopyru-
vate, **II**, 287
Flavone, **IV**, 478
2-Furfuralacetone,
I, 283
Isonitrosopropiophenone,
II, 363
Isophorone oxide,
IV, 552
cis-4-*exo*-Isopropenyl-
1,9-dimethyl-8-
(trimethylsilyl)bicy-
clo[4.3.0]non-8-en-2-
one, **VIII**, 347
β-Ketoisooctaldehyde
dimethyl acetal,
IV, 558
Ketopantoyl lactone,
VII, 417
ω-Methoxyaceto-
phenone, **III**, 562

p-Methoxyacetophenone,
 I, 111
5β-Methoxycyclohexan-
 1-one-3β,4β-dicar-
 boxylic acid anhy-
 dride **VII**, 312
4-Methoxy-3-penten-2-
 one, **VIII**, 357
o-Methoxyphenyl-
 acetone, **IV**, 573
6-Methoxy-β-tetralone,
 VI, 744
Methyl 2-(diethylphos-
 phoryloxy)-1-cyclo-
 hexene-1-carboxylate,
 VII, 351
Methylthio-2-propanone,
 VI, 601
2-Methyl-2-trimethyl-
 siloxypentan-3-one,
 VII, 185, 381
o-Nitroacetophenone,
 IV, 708
m-Nitroacetophenone,
 II, 434
Nitroanthrone, **I**, 390
5-Nitroheptan-2-one,
 VI, 648
16-Oximinoandrost-5-en-
 3β-ol-17-one, **VI**, 840
6-Oxodecanal, **VIII**, 498
4-Oxo-1-(phenylsul-
 fonyl)-*cis*-bicy-
 clo[4.3.0]non-2-ene,
 VIII, 39
(1-Oxo-2-propenyl)tri-
 methylsilane,
 VIII, 501
(*S*)-3-(1-Oxopropyl)-4-
 (phenylmethyl)-2-
 oxazolidinone,
 VIII, 339
4-Oxo-1-[(trimethylsi-
 lyl)oxy]-2,5-cyclo-
 hexadiene-1-carboni-
 trile, **VII**, 517

α-Phenylacetoaceto-
 nitrile, **II**, 487
Phenylbenzoyldi-
 azomethane, **II**, 496
α-Phenyl-β-benzoylpropionitrile, **II**, 498
Phenylglyoxal, **V**, 937
1-Phenyl-4-phosphori-
 nanone, **VI**, 932
Phenyl thienyl ketone,
 II, 520
8-Propionyl-(*E*)-5-nonen-
 olide, **VIII**, 562
o- and *p*-Propiophenol,
 II, 543
3-(1-Propyl)-2-pyrazolin-
 5-one, **VI**, 791
Pyruvic acid, **I**, 475
Quinacetophenone
 dimethyl ether,
 IV, 837
Resacetophenone,
 III, 761
2,2,5,5-Tetramethyltetra-
 hydro-3-ketofuran,
 V, 1024
2,4,5,7-Tetranitrofluoren-
 one, **V**, 1029
(*S*)-(+)-2-(*p*-Toluene-
 sulfinyl)-2-cyclopen-
 tenone, **VII**, 495
2,2-(Trimethylenedithio)
 cyclohexanone,
 VI, 1014
2,2,4-Trimethyl-3-oxo-
 valeryl chloride,
 V, 1103
2,4,7-Trinitrofluorenone,
 III, 837
Xanthone, **I**, 552
B.   Diketones
 Acetonylacetone, **II**, 219
 Acetylacetone, **III**, 16
 Acetylbenzoyl, **III**, 20
 2-Acetyl-2-cyclohexen-
  1-one, **VI**, 23

2-(1-Acetyl-2-oxopropyl)
benzoic acid **VI**, 36,
2-Acetyl-2-phenylseleno-
cyclohexanone, **VI**, 23
Benzil, **I**, 87
Bicyclo[3.3.0]octane-2,6-
dione, **VIII**, 43
cis-Bicyclo[3.3.0]octane-
3,7-dione, **VII**, 51
4,4'-Bis(dimethylamino)
benzil, **V**, 111
3-Butyl-2,4-pentane-
dione, **VI**, 245
3-Chloro-2,5-di-*tert*-
butyl-1,4-benzo-
quinone, **VI**, 210
Δ^4-Cholesten-3,6-dione,
IV, 189
1,2-Cyclobutanedione,
VII, 112
1,2-Cyclohexanedione,
IV, 229
1,4-Cyclohexanedione,
V, 288
2-Cyclopentene-1,4-
dione, **V**, 324
Dehydroacetic acid,
III, 231
1,4-Dibenzoylbutane,
II, 169
trans-1,2-Dibenzoylcy-
clopropane, **VI**, 401
Dibenzoyldibromo-
methane, **II**, 244
trans-Dibenzoylethylene,
II, 248
Dibenzoylmethane, **I**,
205; **III**, 251
2,5-Di-*tert*-butyl-5,6-
dichloro-2-cyclohex-
ene-1,4-dione,
VI, 210
2,5-Di-*tert*-butyl-3,5,6-
trichloro-2-cyclohex-
ene-1,4-dione,
VI, 210

5,6-Diethoxybenzo-
furan-4,7-dione,
VIII, 179
Diethyl α-acetyl-β-keto-
pimelate, **V**, 384
Diethyl dioxosuccinate,
VIII, 597
Dihydroresorcinol,
III, 278
Diisovalerylmethane,
III, 291
7,16-Diketodocosane-
dioic acid and disodi-
um salt, **V**, 534, 536
cis-1,6-Dimethylbicyclo
[3.3.0]octane-3,7-
dione, **VII**, 51
2,2-Dimethyl-1,3-cyclo-
hexanedione,
VIII, 312
5,5-Dimethyl-1,3-cyclo-
hexanedione, **II**, 200
3,3-Dimethyl-1,5-di-
phenylpentane-1,5-
dione, **VIII**, 210
1,6-Dioxo-8a-methyl-
1,2,3,4,6,7,8a-octahy-
dronaphthalene,
V, 486
Ethyl diacetylacetate,
III, 390
2,5-Heptanedione,
VI, 648
(3a*S*,7a*S*)-(+)-
2,3,3a,4,7,7a-
Hexahydro-3a-
hydroxy-7a-methyl-
1*H*-indene-1,5(6*H*)-
dione, **VII**, 363
o-Hydroxydibenzoyl-
methane, **IV**, 479
2-(Hydroxymethylene)
cyclohexanone, **IV**,
536; **VI**, 390, 590
α-Ketoglutaric acid,
V, 687

4-Methylbicyclo[2.2.2]
octane-2,6-dione,
VIII, 467
2-Methyl-1,3-cyclohex-
anedione, **V**, 743
2-Methylcyclopentane-
1,3-dione, **V**, 747;
VI, 774
3-Methyl-2,4-heptane-
dione, **VI**, 776
3-Methylpentane-2,4-
dione, **V**, 785
(7a*S*)-(+)-7a-Methyl-
2,3,7,7a-tetrahydro-
1*H*-indene-1,5(6*H*)-
dione **VII**, 363
(8a*S*)-8a-Methyl-
3,4,8,8a-tetrahydro-
1,6(2*H*,7*H*)-naph-
thalenedione,
VII, 368
4-Methyltricyclo
[2.2.2.0^{3,5}]octane-3,5-
dione, **VIII**, 467
2,4-Nonanedione,
V, 848
2-(2-Oxopropyl)cyclo-
hexanone, **VII**, 414
1-Phenyl-2,4-pentane-
dione, **VI**, 928
4-Phenyl-1,2,4-triazo-
line-3,5-dione,
VI, 936
4-Phenylurazole **VI**, 936,
Sebacil, **IV**, 838
Spiro[4.5]decane-1,4-
dione, **VIII**, 578
Tricarbonyl[2-[(2,3,4,5-
η)-4-methoxy-2,4-
cyclohexadien-1-yl]-
5,5-dimethyl-1,3-
cyclohexanedione]
iron, **VI**, 1001
endo-Tricyclo[4.4.0.0.^{2,5}]
deca-3,8-diene-7,10-
dione, **VI**, 1002

3,5,5-Trimethyl-2-(2-
oxopropyl)-2-cyclo-
hexen-1-one **VI**, 1024
3,5-Undecanedione,
VIII, 620
C. Polyketones
2-Acetyl-1,3-cyclopen-
tanedione, **VI**, 28
Diphenyl triketone,
II, 244
Diphenyl triketone
hydrate, **II**, 244
1-(*p*-Methoxyphenyl)-5-
phenyl-1,3,5-pen-
tanetrione, **V**, 718
2-Methyl-2-(3-oxobutyl)-
1,3-cyclohexane-
dione, **VII**, 368
2-Methyl-2-(3-oxobutyl)-
1,3-cyclopentane-
dione, **VII**, 363
1,3,5-Triacetylbenzene,
III, 829
Tetraacetylethane,
IV, 869

NITRILES
Acetone cyanohydrin, **II**, 7, 29;
III, 324
Acetone cyanohydrin nitrate,
V, 839
3β-Acetoxy-5α-cyanocholestan-7-
one, **VI**, 14
2-Adamantanecarbonitrile, **VI**, 41
Allyl cyanide, **I**, 46; **III**, 852
Aminomalononitrile *p*-toluenesul-
fonate, **V**, 32
β-Aminopropionitrile, **III**, 93
4-Amino-1,2,5,6-tetrahydro-1-
phenylphosphorin-3-car-
bonitrile, **VI**, 932
Azelanitrile, **IV**, 62
1,1'-Azo-bis-1-cyclohexanenitrile,
IV, 66
2,2'-Azo-bis-isobutyronitrile, **IV**, 67
Benzophenone cyanohydrin, **VII**, 20

Benzoquinone (*O*-trimethylsilyl) cyanohydrin, **VII**, 517
Benzoyl cyanide, **III**, 112
2-Benzoyl-1-cyano-1-methyl-1,2-dihydroisoquinoline, **IV**, 642
2-Benzoyl-1,2-dihydroisoquinaldonitrile, **IV**, 641; **VI**, 115
Benzyl cyanide, **I**, 107
3-Benzyl-3-methylpentanenitrile, **IV**, 95
Bis(β-cyanoethyl)amine, **III**, 93
Bis(2-cyanoethyl)phenylphosphine, **VI**, 932
α-Bromo-α-phenylacetonitrile, **III**, 347
2-*tert*-Butoxycarbonyloxyimino-2-phenylacetonitrile, **VI**, 199
tert-Butyl cyanoacetate, **V**, 171
tert-Butyl cyanoketene, **VI**, 210
tert-Butylmalononitrile, **VI**, 223
Carbonyl cyanide, **VI**, 268
o-Carboxyphenylacetonitrile, **III**, 174
Chloroacetonitrile, **IV**, 144
3-(*o*-Chloroanilino)propionitrile, **IV**, 146
γ-Chlorobutyronitrile, **I**, 156
7-Chloro-7-cyanobicyclo[4.2.0]octan-8-one, **VIII**, 116
1-Chloro-1,2-dicyanocyclobutane, **VI**, 427
2-Chloronicotinonitrile, **IV**, 166
α-(4-Chlorophenyl)-γ-phenylacetoacetonitrile, **IV**, 174
Cinnamonitrile, **VI**, 304; **VII**, 108
Cyanamide, **IV**, 645
Cyanoacetamide, **I**, 179
4-Cyanobenzenesulfonamide, **VI**, 632
1-Cyanobenzocyclobutene, **V**, 263
1-Cyano-2-benzoyl-1,2-dihydroisoquinoline, **IV**, 641; **VI**, 115
2-(1-Cyanocyclohexyl)diazenecarboxylic acid methyl ester, **VI**, 334

2-(1-Cyanocyclohexyl)hydrazinecarboxylic acid methyl ester, **VI**, 334
N-2-Cyanoethylaniline, **IV**, 205
Cyanogen bromide, **II**, 150
Cyanogen iodide, **IV**, 207
7-Cyanoheptanal, **V**, 266
1-Cyano-6-methoxy-3,4-dihydronaphthalene, **VI**, 307
2-Cyano-6-methylpyridine, **V**, 269
3-Cyano-6-methyl-2(1*H*)-pyridone, **IV**, 210
1-Cyano-3-α-naphthylurea, **IV**, 215
ω-Cyanopelargonic acid, **II**, 768
9-Cyanophenanthrene, **III**, 212
α-Cyano-β-phenylacrylic   acid, **I**, 181
1-Cyano-3-phenylurea, **IV**, 213
Cyclohexanecarbonitrile, **VI**, 334
1-Cyclohexenylacetonitrile, **IV**, 234
Cyclohexylideneacetonitrile, **VII**, 108
Cyclohexylidenecyanoacetic acid, **IV**, 234
α-Cyclohexylphenylacetonitrile, **III**, 219
Cyclopropyl cyanide, **III**, 223
1-Deoxy-2,3,4,6-tetra-*O*-acetyl-1-(2-cyanoethyl)-α-D-glucopyranose, **VIII**, 148
2,5-Diamino-3,4-dicyanothiophene, **IV**, 243
Diaminomaleonitrile, **V**, 344
Dibromoacetonitrile, **IV**, 254
Dichloroacetonitrile, **IV**, 255
1,1-Dicyano-1,1'-bicyclohexyl, **IV**, 273
1,2-Dicyanocyclobutene, **VI**, 427
1,2-Di-1-(1-cyanocyclohexyl)hydrazine, **IV**, 274
α,α'-Dicyano-β-ethyl-β-methylglutarimide, **IV**, 441
α,α'-Dicyano-β-methylglutaramide, **III**, 591
Dicyanoketene ethylene acetal, **IV**, 276

5,5-Dicyano-4-phenyl-2-cyclopen-
ten-1-one 1,3-propanediol
ketal, **VIII**, 173
1,2-Di-2-(2-cyanopropyl)hydrazine,
IV, 275
2,3-Dicyano-4,4a,9a-tetrahydrofluor-
ene, **VI**, 427
Diethylaluminum cyanide, **VI**, 436
Diethylaminoacetonitrile, **III**, 275
2,4-Dimethoxybenzonitrile, **VI**, 465
2,6-Dimethoxybenzonitrile, **III**, 293
α-*N*,*N*-Dimethylaminophenyl-
acetonitrile, **V**, 437
3,5-Dinitro-*o*-tolunitrile, **V**, 480
Diphenylacetonitrile, **III**, 347
α,β-Diphenylcinnamonitrile, **IV**, 387
2,3-Diphenylsuccinonitrile, **IV**, 392
2-[1'-Ethoxy)-1-ethoxy]butyronitrile,
VII, 381
β-Ethoxypropionitrile, **III**, 372
Ethyl butylcyanoacetate, **III**, 385
Ethyl cyanoacetate, **I**, 254
Ethyl 1-cyano-2-methylcyclohex-
anecarboxylate, **VI**, 586
Ethyl (*E*)-2-cyano-6-octenoate,
VI, 586
Ethyl cyano(pentafluorophenyl)
acetate, **VI**, 873
Ethyl α,β-dicyano-β-phenylpropi-
onate, **I**, 452
Ethylene cyanohydrin, **I**, 256
Ethyl (1-ethylpropenyl)methylcyano-
acetate, **III**, 397
Ethyl (1-ethylpropylidene)cyanoace-
tate, **III**, 399
2-Ethylhexanonitrile, **IV**, 436
3-Ethyl-3-methylhexanenitrile,
IV, 97
Ethyl phenylcyanoacetate, **IV**, 461
Ethyl phenylcyanopyruvate, **II**, 287
Ethyl (1-phenylethylidene)cyanoace-
tate, **IV**, 463
Ferrocenylacetonitrile, **V**, 578
Fumaronitrile, **IV**, 486
3-(2-Furyl)acrylonitrile, **V**, 585
Glycolonitrile, **III**, 436

3-Heptyl-5-cyanocytosine, **IV**, 515
3-Heptylureidomethylenemalono-
nitrile, **IV**, 515
(*Z*,*Z*)-2,4-Hexadienedinitrile,
VI, 662
2-Hydroxybutyronitrile, **VII**, 381
β-(2-Hydroxyethylmercapto)propio-
nitrile, **III**, 458
3-Hydroxyglutaronitrile, **V**, 614
2-Hydroxyimino-2-phenylacetoni-
trile, **VI**, 199
Indole-3-carbonitrile, **V**, 656
Isobutyronitrile, **III**, 493
Malononitrile, **II**, 379; **III**, 535
2-Mercapto-4-hydroxy-5-cyanopy-
rimidine, **IV**, 566
Methanesulfonyl cyanide, **VI**, 727
Methoxyacetonitrile, **II**, 387
p-Methoxyphenylacetonitrile,
IV, 576
Methyl 2-(1-cyanocyclohexyl)hydra-
zinecarboxylate, **VI**, 334
Methyl ω-cyanopelargonate, **III**, 584
Methyleneaminoacetonitrile, **I**, 355
β-Methyl-β-phenyl-α,α'-dicyanoglu-
tarimide, **IV**, 662
3-Methyl-3-phenylpentanenitrile,
IV, 97
α-Naphthonitrile, **III**, 631
Nicotinonitrile, **IV**, 706
p-Nitrobenzonitrile, **III**, 646
p-Nitrobenzyl cyanide, **I**, 396
m-Nitrophenol, **I**, 404
4-Oxo-4-(3-pyridyl)butyronitrile,
VI, 866
4-Oxo-1-[(trimethylsilyl)oxy]-2,5-
cyclohexadiene-1-carboni-
trile, **VII**, 517
Palmitonitrile, **IV**, 437
Pentaacetyl-D-gluconitrile,
III, 690
(Pentafluorophenyl)acetonitrile,
VI, 873
α-Phenylacetoacetonitrile, **II**, 487
α-Phenyl-β-benzoylpropionitrile,
II, 498

2-Phenylbutyronitrile, **VI**, 897

α-Phenyl-α-carbethoxyglutaronitrile,
IV, 776

α-Phenylcinnamonitrile, **III**, 715

2-Phenyl-2-vinylbutyronitrile,
VI, 940

Sebaconitrile, **III**, 768

Tetracyanoethylene, **IV**, 877

Tetracyanoethylene oxide, **V**, 1007

Tetramethylammonium 1,1,2,3,3-
pentacyanopropenide,
V, 1013

Tetramethylsuccinonitrile, **IV**, 273

o-Tolunitrile, **I**, 514

p-Tolunitrile, **I**, 514

p-Tricyanovinyl-N,N-dimethylani-
line, **IV**, 953

Trimethylene cyanide, **I**, 536

O-(Trimethylsilyl)benzophenone
cyanohydrin, **VII**, 20

O-(Trimethylsilyl)-4-methoxyman-
delonitrile, **VII**, 521

Trimethylsilyl cyanide, **VII**, 20,
294, 517

α,α,β-Triphenylpropionitrile,
IV, 962

Veratronitrile, **II**, 622

NITRO COMPOUNDS

(1R,2r,3S)-3-Acetoxy-2-nitrocyclo-
hexyl acetate, **VIII**, 332

2-Amino-5-bromo-3-nitropyridine,
V, 347

o-Amino-p'-nitrobiphenyl, **V**, 830

2-Amino-4-nitrophenol, **III**, 82

2-Amino-3-nitrotoluene, **IV**, 42

2-Amino-5-nitrotoluene, **IV**, 44

Anhydro-2-mercuri-3-nitrobenzoic
acid, **I**, 56

Benzoyl-2-methoxy-4-nitroacet-
anilide, **IV**, 82

(E)-6-Benzyloxy-2-nitro-3-pyrrolidi-
nostyrene, **VII**, 34

m-Bromonitrobenzene, **I**, 123

2-Bromo-3-nitrobenzoic acid, **I**, 125

2-Bromo-4-nitrotoluene, **IV**, 114

o-(Butoxy)nitrobenzene, **III**, 140

1-Chloro-2,6-dinitrobenzene,
IV, 160

m-Chloronitrobenzene, **I**, 162

1,5-Diamino-2,4-dinitrobenzene,
V, 1068

1,2-Diamino-4-nitrobenzene,
III, 242

2,6-Dibromo-4-nitrophenol, **II**, 173

1,5-Dichloro-2,4-dinitrobenzene,
V, 1067

2,6-Dichloronitrobenzene, **V**, 367

2,6-Diiodo-p-nitroaniline, **II**, 196

2,7-Dimethyl-2,7-dinitrooctane, **V**, 445

Dimethyl nitrosuccinate, **VI**, 503

2,4-Dinitroaniline, **II**, 221

2,6-Dinitroaniline, **IV**, 364

3,5-Dinitroanisole, **I**, 219

2,4-Dinitrobenzaldehyde, **II**, 223

3,5-Dinitrobenzaldehyde, **VI**, 529

o-Dinitrobenzene, **II**, 226

p-Dinitrobenzene, **II**, 225

2,4-Dinitrobenzenesulfenyl chloride,
V, 474

2,5-Dinitrobenzoic acid, **III**, 334

3,5-Dinitrobenzoic acid, **III**, 337

p,p'-Dinitrobibenzyl, **IV**, 367

2,2'-Dinitrobiphenyl, **III**, 339

1,4-Dinitrobutane, **IV**, 368

2,3-Dinitro-2-butene, **IV**, 374

Dinitrodurene, **II**, 254

1,6-Dinitrohexane, **IV**, 370

3,4-Dinitro-3-hexene, **IV**, 372

2,4-Dinitroiodobenzene, **V**, 478

1,4-Dinitronaphthalene, **III**, 341

1,5-Dinitropentane, **IV**, 370

2,4-Dinitrophenylhydrazine, **II**, 228

Di-o-nitrophenyl disulfide, **I**, 220

1,3-Dinitropropane, **IV**, 370

3,5-Dinitro-o-tolunitrile, **V**, 480

Ethyl N-(2-methyl-3-nitrophenyl)
formimidate, **VIII**, 493

Ethyl m-nitrobenzoate, **I**, 373

Ethyl α-nitrobutyrate, **IV**, 454

Ethyl (E)-4-(4-nitrophenyl)-4-oxo-2-
butenoate, **VIII**, 268

2-Hydroxy-5-nitrobenzyl chloride,
  III, 468
(1*S*,2*S*,3*R*)-3-Hydroxy-2-nitrocyclo-
  hexyl acetate, **VIII**, 332
3-Methoxy-5-nitrophenol, **I**, 405
6-Methoxy-8-nitroquinoline, **III**, 568
1-(*o*-Methoxyphenyl)-2-nitro-1-
  propene, **IV**, 573
Methyl 2-iodo-3-nitropropionate,
  VI, 799
Methyl γ-methyl-γ-nitrovalerate,
  IV, 652
Methyl nitroacetate, **VI**, 797
Methyl (*E*)-3-nitroacrylate, **VI**, 799
Methyl *m*-nitrobenzoate, **I**, 372
2-Methyl-4'-nitrobiphenyl, **VIII**,
  430
2-Methyl-2-nitropropane, **VI**, 803
3-Methyl-4-nitropyridine 1-oxide,
  IV, 654
Nitroacetic acid, dipotassium salt,
  VI, 797
o-Nitroacetophenone, **IV**,708
m-Nitroacetophenone, **II**, 434
1-Nitro-2-acetylaminonaphthalene,
  II, 438
o-Nitroaniline, **I**, 388
9-Nitroanthracene, **IV**, 711
Nitroanthrone, **I**, 390
Nitrobarbituric acid, **II**, 440
o-Nitrobenzaldehyde, **III**, 641,
  V, 825
m-Nitrobenzaldehyde, **III**, 644;
  Warning, **III**, 645
p-Nitrobenzaldehyde, **II**, 441
m-Nitrobenzaldehyde dimethyl
  acetal, **III**, 644
o- and *p*-Nitrobenzaldiacetate,
  IV, 713
m-Nitrobenzazide, **IV**, 715
m-Nitrobenzoic acid, **I**, 391
p-Nitrobenzoic acid, **I**, 392
p-Nitrobenzonitrile, **III**, 646
m-Nitrobenzoyl chloride, **IV**, 715
p-Nitrobenzoyl chloride, **I**, 394
p-Nitrobenzoyl peroxide, **III**, 649

p-Nitrobenzyl acetate, **III**, 650
p-Nitrobenzyl alcohol, **III**, 652
p-Nitrobenzyl bromide, **II**, 443
p-Nitrobenzyl cyanide, **I**, 396
p-Nitrobenzyl fluoride, **VI**, 835
m-Nitrobiphenyl, **IV**, 718
2-Nitrocarbazole, **V**, 829
o-Nitrocinnamaldehyde, **IV**, 722
m-Nitrocinnamic acid, **I**, 398;
  IV, 731
(1*R*,2*r*,3*S*)-2-Nitrocyclohexane-1,3-
  diol, **VIII**, 332
1-Nitrocyclooctene, **VI**, 837
2-Nitro-*p*-cymene, **III**, 653
5-Nitro-2,3-dihydro-1,4-phthalazine-
  dione, **III**, 656
m-Nitrodimethylaniline, **III**, 658
o-Nitrodiphenyl ether, **II**, 446
p-Nitrodiphenyl ether, **II**, 445
2-Nitroethanol, **V**, 833
m-Nitroethylbenzene, **VII**, 393
2-Nitrofluorene, **II**, 447
Nitroguanidine, **I**, 399
5-Nitroheptan-2-one, **VI**, 648
5-Nitroindazole, **III**, 660
4-Nitroindole, **VIII**, 493
Nitromesitylene, **II**, 449
Nitromethane, **I**, 401
2-Nitro-4-methoxyacetanilide,
  III, 661
2-Nitro-4-methoxyaniline, **III**, 661
N-Nitromorpholine, **V**, 839
1-Nitro-2-naphthol, **II**, 451
4-Nitro-1-naphthylamine, **III**, 664
1-Nitrooctane, **IV**, 724
p-Nitrophenylacetic acid, **I**, 406
p-Nitrophenylarsonic acid, **III**, 665
o-Nitrophenyl azide, **IV**, 75
1-(*p*-Nitrophenyl)-1,3-butadiene,
  IV, 727
1-(*p*-Nitrophenyl)-4-chloro-2-
  butene, **IV**, 727
trans-o-Nitro-α-phenylcinnamic
  acid, **IV**, 730
1-(*m*-Nitrophenyl)-3,3-dimethyltri-
  azene, **IV**, 718

m-Nitrophenyl disulfide, **V**, 843
3'-Nitro-1-phenylethanol, **VIII**, 495
p-Nitrophenyl isocyanate, **II**, 453
p-Nitrophenyl sulfide, **III**, 667
o-Nitrophenylsulfur chloride, **II**, 455
3-Nitrophthalic acid, **I**, 408
4-Nitrophthalic acid, **II**, 457
3-Nitrophthalic anhydride, **I**, 410
4-Nitrophthalimide, **II**, 459
2-Nitropropene, **VII**, 396
β-Nitrostyrene, **I**, 413
m-Nitrostyrene, **IV**, 731
2-Nitrothiophene, **II**, 466
m-Nitrotoluene, **I**, 415
4-Nitro-2,2,4-trimethylpentane,
    V, 845
Nitrourea, **I**, 417
6-Nitroveratraldehyde, **IV**, 735
Phenylnitromethane, **II**, 512
(Phenylthio)nitromethane, **VIII**, 550
Sodium nitromalonaldehyde mono-
    hydrate, **IV**, 844
2,4,5,7-Tetranitrofluorenone, **V**, 1029
(+)- and (−)-α-(2,4,5,7-Tetranitro-9-
    fluorenylideneaminooxy)
    propionic acid, **V**, 1031
Tetranitromethane, **III**, 803
2,4,5-Triaminonitrobenzene, **V**, 1067
1,2,3-Triiodo-5-nitrobenzene, **II**, 604
Trimethyloxonium 2,4,6-trinitroben-
    zenesulfonate, **V**, 1099
1,3,5-Trinitrobenzene, **I**, 541
2,4,6-Trinitrobenzoic acid, **I**, 543
2,4,7-Trinitrofluorenone, **III**, 837
2,4,6-Triphenylnitrobenzene,
    V, 1128

NITROSO COMPOUNDS

Cupferron, **I**, 177
Disodium nitrosodisulfonate,
    VI, 1010
Ethyl *N*-nitroso-*N*-benzylcarbamate,
    IV, 780
Ethyl *N*-nitroso-*N*-(*p*-tolylsulfonyl-
    methyl)carbamate, **VI**, 981
2-Methyl-2-nitrosopropane, **VI**, 803

Nitrosobenzene, **III**, 668
p-Nitrosodiethylaniline, **II**, 224
Nitrosodimethylamine, **III**, 211
p-Nitrosodimethylaniline hydrochlo-
    ride, **III**, 223
N-Nitrosomethylaniline, **II**, 460
Nitrosomethylurea, **II**, 461
Nitrosomethylurethane, **II**, 464
Nitroso-β-naphthol, **I**, 411
Nitroso-*tert*-octane, **VIII**, 93
Nitrosothymol, **I**, 511
p-Tolylsulfonylmethylnitrosamide,
    IV, 943

ORGANOMETALLIC COMPOUNDS

A. Aluminum
    (*E*)-1-Decenyldiisobutylalane,
        VIII, 295
    Diethylaluminum cyanide,
        VI, 436
    (*E*)-2-(Methyl-1,3-butadienyl)
        dimethylalane, **VII**, 245
    Triphenylaluminum, **V**, 1116
B. Antimony
    Triphenylstibine, **I**, 550
    Tri-*p*-tolylstibine, **I**, 551
C. Arsenic
    Arsanilic acid, **I**, 70
    Arsenoacetic acid, **I**, 74
    Arsonoacetic acid, **I**, 73
    p-Arsonophenoxyacetic acid,
        I, 75
    p-Nitrophenylarsonic acid, **III**,
        665
    Phenylarsonic acid, **II**, 494
    Sodium *p*-arsono-*N*-phenyl-
        glycinamide, **I**, 488
    Sodium *p*-hydroxyphenylarson-
        ate, **I**, 490
    Tetraphenylarsonium chloride
        hydrochloride, **IV**, 910
    Triphenylarsine, **IV**, 910
    Triphenylarsine oxide, **IV**, 911
D. Boron
    Benzeneboronic anhydride,
        IV, 68

Bis(3-methyl-2-butyl)borane
(Disiamylborane), **VI**,
853, 919

Butyl borate, **II**, 106

(+)-Di-3-pinanylborane, **VII**, 339

10-Methyl-10,9-borazarophen-
anthrene, **V**, 727

(*E*)-1-Hexenyl-1,3,2-benzodioxa-
borole, **VIII**, 532

Perhydro-9b-boraphenalene,
VII, 427

B-3-Pinanyl-9-borabicyclo
[3.3.1]nonane, **VII**, 402

E. Cerium
Trimethylsilylmethyl-
magnesium chloride-
cerium chloride, **VIII**, 602

F. Chromium
[(Methyl)(methoxy)
carbene]pentacarbonyl
chromium(0), **VIII**, 216

Reinecke salt, **II**, 555

G. Cobalt
Hexacarbonyl (1-methyl-2-
propynylium)dicobalt tetrafluoroborate, **VIII**, 460

H. Copper
Copper(I) phenylacetylide,
VI, 916

Ethyl copper complex, **VII**, 236

Ethyl copper complex-1-octyne
adduct, **VII**, 236

Lithium dibutylcuprate (solution), **VII**, 290

Lithium di-[(*Z*)-hexenyl]cuprate
(solution) **VII**, 290

Lithium dimethylcuprate (solution), **VII**, 351

Lithium divinylcyanocuprate,
VIII, 112

Lithium phenylthio(*tert*-butyl)
cuprate, **VI**, 248

Lithium vinyl-2-thienyl-
cyanocuprate, **VIII**, 33

Pentafluorophenylcopper
tetramer, **VI**, 875

I. Iron
2-Chlorobenzoyl)ferrocene,
VI, 625

Cyclobutadieneiron tricarbonyl,
VI, 310

Dicarbonyl(cyclopenta-
dienyl)(ethyl vinyl
ether)iron tetrafluorobo-
rate, **VIII**, 479

Dicarbonyl(cyclopentadienyl)
(*trans*-3-methyl-2-vinylcy-
clohexanone) iron tetrafluoroborate, **VIII**, 479

N,N-Dimethylaminomethyl-
ferrocene methiodide,
V, 434

Disodium tetracarbonylferrate
sesquidioxanate, **VI**, 807

Ferrocene, **IV**, 473

Ferrocenecarboxylic acid,
VI, 625

Ferrocenylacetonitrile, **V**, 578

Hemin, **III**, 442

Hydroxymethylferrocene, **V**, 621

Sodium dicarbonyl(cyclopentadienyl)ferrate, **VIII**, 479

Tricarbonyl[(2,3,4,5-η)-2,4-
cyclohexadien-1-one]iron,
VI, 996

Tricarbonyl[(1,2,3,4-η)-1- and
2-methoxy-1,3-cyclohexa-
diene]iron, **VI**, 996

Tricarbonyl[2-[(2,3,4,5-η)-4-
methoxy-2,4-cyclohexadi-
en-1-yl]-5,5-dimethyl-1,3-
cyclohexanedione]iron,
VI, 1001

Tricarbonyl[(1,2,3,4,5-η)-2-
methoxy-2,4-cyclohexadi-
en-1-yl]iron(+1) hexafluo-
rophosphate(1-), **VI**, 996

Tricarbonyl[(1,2,3,4,5-η)-1-
and 2-methoxy-2,4-cyclo-
hexadien-1-yl]iron(+1)
tetrafluoroborate(1−),
VI, 996

J. Lead
    p-Methoxyphenyllead triace-
      tate, **VII**, 229
K. Lithium
    Allyllithium, **V**, 452
    1,3-Bis(methylthio)allyllithium,
      VI, 683, 687
    Dilithioallene **VIII**, 226,
    2,2'-Dilithio-1,1'-binaphthyl,
      VIII, 57
    2-(*N,N*-Dimethylamino)phenyl-
      lithium, **VII**, 172
    2-Lithiobornene, **VII**, 77
    2-Lithio-2-butene, **VIII**, 505
    2-Lithio-2-cyclopenten-1-one
      ethylene ketal, **VII**,
      271, 495
    6-Lithio-3,4-dimethoxybenz-
      aldehyde cyclohexylimine,
      VIII, 586
    1-Lithio-1-ethoxycyclopropane,
      VIII, 556
    2-(2'-Lithio-4,5'-methylene-
      dioxyphenyl)ethyl chlo-
      ride, **VII**, 326
    Lithium acetylide, **VIII**, 391
    Lithium chloromethylacetylide,
      VIII, 371
    Lithium 1-lithio-3-propynoxide,
      VIII, 415
    1-Methoxyvinyllithium, **VIII**, 19
    Methyllithium, low halide, **VII**,
      346
    1,3-Bis(methylthio)allyllithium,
      VI, 683, 687
    (*E*)-1-Propenyllithium, **VI**, 1040
    Propynyllithium, **VII**, 226
    o-Tolyllithium, **VIII**, 430
    Tributyltinlithium, **VIII**, 562
L. Magnesium
    Allenylmagnesium bromide,
      VII, 485
    Allylmagnesium bromide, **IV**,
      749; **VIII**, 23
    Allylmagnesium chloride,
      IV, 751
    4-Bromophenylmagnesium bro-
      mide, **V**, 496
    Butylmagnesium bromide,
      VI, 407
    sec-Butylmagnesium bromide,
      VI, 753
    Butylmagnesium chloride,
      VIII, 606
      unsolvated, **V**, 1141
    1-Chloromagnesioxy-3-octynyl-
      magnesium chloride,
      VIII, 507
    Ethynylmagnesium bromide,
      IV, 792
    Ethynylmagnesium chloride,
      VIII, 606
    Isobutylmagnesium chloride,
      VIII, 508
    (Isopropoxydimethylsilyl)
      methylmagnesium chlo-
      ride, **VIII**, 315
    2-Methoxyphenylmagnesium
      bromide, **VI**, 64
    (4-Methylphenyl)magnesium
      bromide, **VI**, 488
    Phenylethylmagnesium chlo-
      ride, **VII**, 451
    Phenylethynylmagnesium bro-
      mide **VI**, 925,
    Phenylmagnesium bromide,
      VIII, 516, 522
    Propynylmagnesium bromide,
      VII, 456
    3-(2-Tetrahydropyranyloxy)
      propynylmagnesium bro-
      mide, **VII**, 334
    Trimethylsilylmethylmag-
      nesium chloride, **VIII**,
      602, 612
    Vinylmagnesium bromide,
      VII, 177
M. Mercury
    Anhydro-2-mercuri-3-nitroben-
      zoic acid, **I**, 56
    p-Chloromercuribenzoic acid,
      I, 159

o-Chloromercuriphenol, **I**, 161

Diphenylmercury, **I**, 228

Mercury di-β-naphthyl, **II**, 381

β-Naphthylmercuric chloride,
  II, 432

trans-2-(Phenylsulfonyl)-4-
  chloromercuri)cyclohex-
  ene, **VIII**, 540

Phenyl(trichloromethyl)mer-
  cury, **V**, 969

Di-*p*-tolylmercury, **I**, 231

p-Tolylmercuric chloride, **I**, 519

N. Nickel

π-Methallylnickel bromide,
  VI, 723

O. Rhodium

(Bicyclo[2.2.1]heptadiene)(1,4-
  bis(diphenylphosphino)bu-
  tanerhodium trifluoro-
  methanesulfonate, **VIII**, 420

(*R*,*R*)-1,2-Bis(*o*-ani-
  sylphenylphosphino)ethane
  rhodium trifluoromethanesulfonate, (DIPAMPRH⁺),
  VIII, 423

[(+)-2,2'-Bis(diphenylphosphino)-1,1'-binaphthyl]-(η⁴-
  1,5-cyclooctadiene)rhodi-
  um(I) perchlorate,
  VIII, 183

P. Ruthenium

Ruthenocene, **V**, 1001

Q. Selenium

2-Acetyl-2-phenylselenocyclo-
  hexanone, **VI**, 23

Benzeneselenenyl chloride,
  VI, 533

N,N-Dimethylselenourea,
  IV, 359

Diphenyl diselenide, **VI**, 533

Diphenyl selenide, **II**, 238

Diphenylselenium dichloride,
  II. 240

Di-*m*-tolyl selenide, **II**, 239

Di-*o*-tolyl selenide, **II**, 239

Di-*p*-tolyl selenide, **II**, 239

Phenyl benzeneselenosulfonate,
  VIII, 543

Selenophenol, **III**, 771

Triphenylselenonium chloride,
  II, 241

R. Silicon

Acetyltrimethylsilane, **VIII**, 19

N-Benzyl-*N*-(methoxymethyl)-
  N-[(trimethylsilyl)methyl]
  amine, **VIII**, 231

N-Benzyl-*N*-[(trimethylsilyl)
  methyl]amine, **VIII**, 231

1,4-Bis(trimethylsilyl)buta-1,3-
  diyne, **VIII**, 63

(1-Bromoethynyl)trimethylsi-
  lane, **VI**, 1034

Ethyl 2-(diphenylmethylsilyl)
  decanoate, **VIII**, 474

Ethyl trimethylsilylacetate,
  VII, 512

2-(Hydroxymethyl)allyltri-
  methylsilane, **VII**, 266

(1-Hydroxy-2-propenyl)tri-
  methylsilane, **VIII**, 501

Iodotrimethylsilane **VI**, 353,

1-Iodo-3-trimethylsilylpropane,
  VIII, 486

cis-4-*exo*-Isopropenyl-1,9-
  dimethyl-8-(trimethylsilyl)bicyclo[4.3.0]non-8-en-
  2-one, **VIII**, 347

(Isopropoxydimethyl-
  silyl)methyl chloride,
  VIII, 315

1-[(Isopropoxydimethyl-
  silyl)methyl]cyclohexanol,
  VIII, 315

N-(4-Methoxyphenyl)-(*Z*)-
  (trimethylsilyl)-3-buten-
  amine, **VIII**, 358

1-(Methoxyvinyl)trimethylsi-
  lane, **VIII**, 19

1-Methyl-1-(trimethylsilyl)
  allene, **VIII**, 471

(1-Oxo-2-propenyl)trimethylsi-
  lane, **VIII**, 501

p-Tolyl-2-(trimethylsilyl)ethyn-
   yl sulfone, **VIII**, 281
Trimethyl-(2-methylene-4-
   phenyl-3-butenyl)silane,
   VIII, 602
2-(Trimethylsiloxymethyl)
   allyltrimethylsilane,
   VII, 266
Trimethylsilylacetylene, **VIII**,
   63, 606
(Z)-4-(Trimethylsilyl)-3-buten-
   1-ol, **VIII**, 358, 609
3-Trimethylsilyl-3-buten-2-one,
   VI, 1033
(Z)-4-(Trimethylsilyl)-3-butenyl
   4-methylbenzenesulfonate,
   VIII, 358
4-(Trimethylsilyl)-3-butyn-1-ol,
   VIII, 609
6-(Trimethylsilyl)-3-car-
   bomethoxy-3-methyl-1-
   hexene, **VIII**, 486
6-(Trimethylsilyl)-3-carboxy-3-
   methyl-1-hexene,
   VIII, 486
Trimethylsilyldiazomethane,
   VIII, 612
1-Trimethylsilyloxybicyclo
   [4.1.0]heptane, **VI**, 328
(E)-3-Trimethylsilyl-2-propen-
   1-ol, **VII**, 524
3-Trimethylsilyl-2-propyn-1-ol,
   VII, 524,
3-(Trimethylsilyl)-2-propyn-1-
   yl methanesulfonate,
   VIII, 471
S. Sodium
Ethyl sodium phthalimidomalonate, **II**, 384
Sodium cyclopentadienide, **IV**,
   474; **V**, 1088; **VII**, 15,
   339
Sodium phenyl-aci-nitro-
   methane, **II**, 512
Triphenylmethylsodium,
   II, 607

T. Tellurium
Bis(4-methoxyphenyl)tellurium
   dichloride, **VI**, 468
U. Tin
Allyltributyltin, **VIII**, 23
Allyltriphenyltin, **V**, 453
(E)-1,2-Bis(tributylstannyl)eth-
   ylene, **VIII**, 268
Dibutyldivinyltin, **IV**, 258
Ethyl (E)-3-(tributylstannyl)
   propenoate, **VIII**, 268
Tetraethyltin, **IV**, 881
Tributylethynylstannane
   VIII, 268
(Z)-3-(Tributylstannyl)methyl-
   ene-4-isopropyl-1,1-cyclo-
   pentanedicarboxylic acid,
   dimethyl ester, **VIII**, 381
Tributyltinlithium, **VIII**, 562
Trimethyltin chloride, **VIII**, 97
Trimethylvinyltin, **VIII**, 97
V. Titanium
Methyltriisopropoxytitanium,
   VIII, 495
W. Zinc
Bis(2-carboethoxyethyl)zinc,
   VIII, 277
(3-Carboethoxypropyl)zinc
   iodide, **VIII**, 274,
Dibutyl zinc, **II**, 187
Diethyl zinc, **II**, 184
Diisoamyl zinc, **II**, 187
Dipropyl zinc, **II**, 187
o-Tolylzinc chloride, **VIII**, 430

OXONIUM COMPOUNDS
Triethyloxonium fluoborate, **V**, 1080
Trimethyloxonium fluoborate,
   V, 1096
Trimethyloxonium 2,4,6-trinitroben-
   zenesulfonate, **V**, 1099
2,4,6-Trimethylpyrylium tetrafluoroborate, **V**, 1112
2,4,6-Trimethylpyrylium trifluoromethanesulfonate,
   V, 1114

2,4,6-Trimethylpyrylium perchlorate,
  V, 1106
2,4,6-Triphenylpyrylium tetrafluoroborate, **V**, 1135

PEROXY COMPOUNDS
Bis-3,4-dichlorobenzoyl peroxide,
  V, 52
m-Chloroperbenzoic acid, **VI**, 39,
  276, 404
Monoperphthalic acid, **V**, 805
p-Nitrobenzoyl peroxide, **III**, 649
Peroxybenzoic acid, **I**, 431; **V**, 900,
  904
Tetralin hydroperoxide, **V**, 895

PHENOLS
2-*p*-Acetylphenylhydroquinone,
  IV, 15
1,2-Aminonaphthol hydrochloride,
  II, 33
1,4-Aminonaphthol hydrochloride, **I**,
  49; **II**, 39
1-Amino-2-naphthol-4-sulfonic acid,
  II, 42
2-Amino-4-nitrophenol, **III**, 82
Ammonium salt of aurin tricarboxylic acid, **I**, 54
1-[α-(Benzalamino)benzyl]-2-naph-
  thol, **I**, 381
(*R*)-(+)-1,1'-Binaphthalene-2,2'-diol,
  VIII, 46
N-[2,4-Bis(1,3-diphenylimidazo-
  lidin-2-yl)-5-hydroxy-
  phenyl]acetamide,
  VII, 162
3-Bromo-4-hydroxytoluene, **III**, 130
6-Bromo-2-naphthol, **III**, 132
o-Bromophenol, **II**, 97
p-Bromophenol, **I**, 128
4-Bromoresorcinol, **II**, 100
p-tert-Butylcalix[4]arene, **VIII**, 75
p-tert-Butylcalix[6]arene, **VIII**, 77
p-tert-Butylcalix[8]arene, **VIII**, 80
Catechol, **I**, 149
o-Chloromercuriphenol, **I**, 161

p-Chlorophenyl salicylate, **IV**, 178
Creosol, **IV**, 203
p-Cresol, **I**, 175
2,6-Dibromo-4-nitrophenol, **II**, 173
2,6-Dichlorophenol, **III**, 267
N-(2,4-Diformyl-5-hydroxyphenyl)
  acetamide, **VII**, 162
2,5-Dihydroxyacetophenone,
  III, 280
2,6-Dihydroxyacetophenone,
  III, 281
2,5-Dihydroxy-*p*-benzenediacetic
  acid, **III**, 286
3,5-Dihydroxybenzoic acid, **III**, 288
3,3'-Dihydroxybiphenyl, **V**, 412
2,4-Dihydroxy-5-bromobenzoic acid,
  II, 100
1,3-Dihydroxy-2-naphthoic acid,
  III, 638
4-Ethoxy-3-hydroxybenzaldehyde,
  VI, 567
o-Eugenol, **III**, 418
Gallacetophenone, **II**, 304
o-Heptylphenol, **III**, 444
Hexahydrogallic acid, **V**, 591
Hexahydroxybenzene, **V**, 595
o-Hydroxyacetophenone, **III**, 389
m-Hydroxybenzaldehyde, **III**, 453
p-Hydroxybenzoic acid, **II**, 341
2-Hydroxy-3,5-diiodobenzoic acid,
  II, 343
2-Hydroxyisophthalic acid, **V**, 617
2-Hydroxy-1-naphthaldehyde,
  III, 463
2-Hydroxy-1,4-naphthoquinone,
  III, 465
2-Hydroxy-5-nitrobenzyl chloride,
  III, 468
p-Hydroxyphenylpyruvic acid,
  V, 627
3-Hydroxypyrene, **V**, 632
3-Hydroxyquinoline, **V**, 635
2-Hydroxythiophene, **V**, 642
o-Iodophenol, **I**, 326
p-Iodophenol, **II**, 355
6-Methoxy-2-naphthol, **V**, 918

3-Methoxy-5-nitrophenol, **I**, 405
N-Methyl-3,4-dihydroxyphenylala-
    nine, **III**, 586
4-Methylesculetin, **I**, 360
2-Methyl-4-hydroxyquinoline,
    III, 593
β-Naphthol phenylaminomethane,
    I, 381
Naphthoresorcinol, **III**, 637
1-Nitro-2-naphthol, **II**, 451
m-Nitrophenol, **I**, 404
3-Nitro-4,6-xylenol, **I**, 405
Nitroso-β-naphthol, **I**, 411
Orcinol monomethyl ether, **VI**, 859
Phloroglucinol, **I**, 455
o- and *p*-Propiophenol, **II**, 543
Protocatechualdehyde, **II**, 549
Protocatechuic acid, **III**, 745
Pyrogallol 1-monomethyl ether,
    III, 759
Quinacetophenone monomethyl
    ether, **IV**, 836
Quinizarin, **I**, 476
Resacctophenone, **III**, 761
β-Resorcylic acid, **II**, 557
Salicyl-*o*-toluide, **III**, 765
Sodium *p*-hydroxyphenylarsonate,
    I, 490
Syringic aldehyde, **IV**, 866
ar-Tetrahydro-α-naphthol, **IV**, 887
Tetrahydroxyquinone, **V**, 1011
Vanillic acid, **IV**, 972

PHOSPHORUS COMPOUNDS

4-Amino-1,2,5,6-tetrahydro-1-
    phenylphosphorin-3-car-
    bonitrile, **VI**, 932
1,1'-Binaphthyl-2,2'-diyl hydrogen
    phosphate, (*S*)-(+)- and
    (*R*)-(−)-, **VIII**, 46, 50
Bis(2-cyanoethyl)phenylphosphine,
    VI, 932
Bis(3-dimethylaminopropyl) phenyl-
    phosphine, **VI**, 776
2,2'-Bis(diphenylphosphino)-1,1'-
    binaphthyl (BINAP), (*R*)-

(+)- and (*S*)-(−)-, **VIII**,
    57, 183
(±)-2,2'-Bis(diphenylphosphinyl)-
    1,1'-binaphthyl [(±)-
    BINAPO], **VIII**, 57
2,4-Bis(4-methoxyphenyl)-1,3,2,4-
    dithiaphosphetane 2,4-
    disulfide, **VII**, 372
17β-*tert*-Butoxy-5α-androstan-3-
    enyl-*N,N,N',N'*-tetra-
    methylphosphoroamidate,
    VII, 66
tert-Butylcarbonic diethylphosphoric
    anhydride, **VI**, 207
Butyl phosphate, **II**, 109
Chlorodiisopropylphosphine, **V**,
    211
Chloromethylphosphonothioic di-
    chloride, **V**, 218
Dichloromethylenetriphenylphos-
    phorane, **V**, 361
Diethyl aminomethylphosphonate,
    VIII, 451
Diethyl *N*-benzylidenaminomethyl-
    phosphonate, **VIII**, 451
Diethyl 2-(cyclohexylamino)vinyl-
    phosphonate, **VI**, 448
Diethyl 2,2-diethoxyethylphospho-
    nate, **VI**, 448
Diethyl ethylphosphonate, **IV**, 326
Diethyl formylmethylphosphonate,
    VI, 448
Diethyl hydroxymethylphosphonate,
    VII, 160
Diethyl 5-methylcoprost-3-en-3-yl
    phosphate, **VI**, 762
Diethyl methylphosphonate, **IV**, 326
Diethyl phenylphosphonate, **VI**, 451
Diethylphosphonomethyl-2,2-
    dimethyl-1,3-dioxen-4-
    one, **VIII**, 192
(*E*)-[3-(Diethylphosphoryloxy)buta-
    dienyl]-2,6,6-trimethyl-1-
    cyclohexene, **VII**, 63
Diethyl phthalimidomethylphospho-
    nate, **VIII**, 451

Diethyl [(2-tetrahydropyranyloxy)
methyl]phosphonate,
VII, 160
Diisopropyl ethylphosphonate,
IV, 326
Diisopropyl methylphosphonate,
IV, 325
Diphenyl-*p*-bromophenylphosphine,
V, 496
Diphenyl (cycloundecyl-1-pyrro-
lidinylmethylene)phospho-
ramidate, **VII**, 135
Diphenylphosphine, **VI**, 569
O-Diphenylphosphinylhydroxyl-
amine, **VII**, 8
Diphenyl phosphorazidate, **VII**,
135, 206
Ethyl (triphenylphosphoranylidene)
acetate, **VII**, 232
Furyl phosphorodichloridate,
VIII, 396
Furyl *N,N,N',N'*-tetramethyldiami-
dophosphate, **VIII**, 396
Hexamethylphosphorus triamide,
V, 602
(*R*)-(−)-Methyl 1,1'-binaphthyl-2,2'-
diyl phosphate, **VIII**, 46
Methyl 2-(diethylphosphoryloxy)-1-
cyclohexene-1-carboxy-
late, **VII**, 232
3-Methyl-1-phenylphospholene
oxide, **V**, 787
(4-Phenyl-2-butenyl)triphenylphos-
phonium iodide, **VI**, 707
Phenyldichlorophosphine, **IV**, 784
1-Phenyl-4-phosphorinanone,
VI, 932
Styrylphosphonic dichloride, **V**, 1005
Tetramethylbiphosphine disulfide,
V, 1016
N,N,N',N'-Tetramethyldiamidophos-
phorochloridate, **VII**, 66
Tributylhexadecylphosphonium bromide, **VI**, 833
Trichloromethylphosphonyl dichloride, **IV**, 950

Triethyl phosphite, **IV**, 955
Triphenylcinnamylphosphonium
chloride, **V**, 315
Triphenylmethylphosphonium bro-
mide, **V**, 751
Triphenylphosphine dibromide,
VIII, 57
Trisammonium geranyl phosphate,
VIII, 616
Tris(tetrabutylammonium) hydrogen
pyrophosphate trihydrate,
VIII, 616
Vinyltriphenylphosphonium bro-
mide, **V**, 1145
Xylylene-bis(triphenylphosphonium
chloride), **V**, 985

QUINONES
Acenaphthenequinone, **III**, 1
p-Acetylphenylquinone, **IV**, 16
Anthraquinone, **II**, 554
p-Benzoquinone, **IV**, 152
α-Chloroanthraquinone, **II**, 128
Chloro-*p*-benzoquinone, **IV**, 148
3-Chloro-2,5-di-*tert*-butyl-*p*-benzo-
quinone, **VI**, 210
3-Chlorotoluquinone, **IV**, 152
2,6-Dibromoquinone-4-chloroimide,
II, 175
2,6-Di-*tert*-butyl-*p*-benzoquinone,
VI, 412
2,5-Dichloroquinone, **IV**, 152
5,6-Diethoxybenzofuran-4,7-dione,
VIII, 179
2,3-Dimethylanthraquinone, **III**,
310
4,5-Dimethyl-*o*-benzoquinone,
VI, 480
Duroquinone, **II**, 254
2-Hydroxy-1,4-naphthoquinone,
III, 465
Methoxyquinone, **IV**, 153
1-Methylaminoanthraquinone,
III, 573
1-Methylamino-4-bromoanthra-
quinone, **III**, 575

Methylanthraquinone, **I**, 353
1,2-Naphthoquinone, **II**, 430
1,4-Naphthoquinone, **I**, 383; **IV**, 698
1,2-Naphthoquinone-4-sulfonic acid,
　salts of, **III**, 633
Phenanthrenequinone, **IV**, 757
2-Phenoxymethyl-1,4-benzoquinone,
　VI, 890
Phloroacetophenone, **II**, 522
Potassium anthraquinone-α-sul-
　fonate, **II**, 539
Quinizarin, **I**, 476
Quinone, **I**, 482; **II**, 553
Tetrahydroxyquinone, **V**, 1011
Thymoquinone, **I**, 511
Trimethyl-*p*-benzoquinone, **VI**, 1010

RESOLVING AGENTS
　(*S*)-(+)- and (*R*)-(–)-1,1'-Binaphthyl-
　2,2'-diyl hydrogen phos-
　phate (BNP acid), **VIII**, 50
　(+)-Cinchonine, **VIII**, 50
　2,3-Di-*O*-benzoyltartaric acids, (–)-
　　L- or (+)-D-, **VIII**, 57

SULFUR COMPOUNDS
　A.　Disulfides
　　p-Aminophenyl disulfide, **III**, 86
　　Benzoyl disulfide, **III**, 116
　　2,4-Bis(4-methoxyphenyl)-
　　　1,3,2,4-dithiadiphosphe-
　　　tane 2,4-disulfide, **VII**, 372
　　sec-Butyl isopropyl disulfide,
　　　VI, 235
　　L-Cystine, **I**, 194
　　Dibenzhydryl disulfide, **II**, 574
　　Di-*o*-nitrophenyl disulfide, **I**, 220
　　m-Nitrophenyl disulfide, **V**, 843
　　Tetramethylbiphosphine disul-
　　　fide, **V**, t016
　B.　Sulfones
　　1-Bromo-1-methyl-2-(bro-
　　　momethylsulfonyl)cyclo-
　　　hexane, **VIII**, 212
　　3-Chlorothietane 1,1-dioxide,
　　　VII, 491

3-Chlorothiete 1,1-dioxide,
　VII, 491
4,4'-Diaminodiphenylsulfone,
　III, 239
α,α'-Dibromodibenzyl sulfone,
　VI, 403
3,3-Dichlorothietane 1,1-diox-
　ide, **VII**, 491
7,9-Dimethyl-*cis*-8-thiabicy-
　clo[4.3.0]nonane 8,8-diox-
　ide, **VI**, 482
Ethyl *N*-nitroso-*N*-(*p*-tolylsul-
　fonylmethyl)carbamate,
　VI, 981
Ethyl *N*-(*p*-tolylsulfonylmethyl)
　carbamate, **VI**, 981
Ethynyl *p*-tolyl sulfone, **VIII**, 281
Isoprene cyclic sulfone,
　III, 499
Methyl (*Z*)-3-(phenylsulfonyl)
　prop-2-enoate, **VIII**, 458
Methyl *p*-tolyl sulfone, **IV**, 674
4-Oxo-1-(phenylsulfonyl)-*cis*-
　bicyclo[4.3.0]non-2-ene,
　VIII, 39
trans-3-(Phenylsulfonyl)-4-
　(chloromercuri)cyclohex-
　ene, **VIII**, 540
3-(Phenylsulfonyl)-1,3-cyclo-
　hexadiene, **VIII**, 540
1-(Phenylsulfonyl)cyclopen-
　tene, **VIII**, 38, 543
Phenyl vinyl sulfone, **VII**, 453
cis-8-Thiabicyclo[4.3.0]nonane
　8,8-dioxide, **VI**, 482
Thietane 1,1-dioxide, **VII**, 491
Thiete 1,1-dioxide, **VII**, 491
p-Tolylsulfonyldiazomethane,
　VIII, 981
p-Tolylsulfonylmethyl iso-
　cyanide, **VI**, 41, 987
p-Tolyl 2-(trimethylsilyl)
　ethynyl sulfone, **VIII**, 281
　C.　Sulfonic Acids and Derivatives
　　p-Acetaminobenzenesulfonyl
　　　chloride, **I**, 8

4-Acetylpyridine oxime tosylate, **VII**, 149

Acid ammonium *o*-sulfobenzoate, **I**, 14

1-Amino-2-naphthol-4-sulfonic acid, **II**, 42

Barium 2-phenanthrenesulfonate, **II**, 482

Benzenesulfonyl chloride, **I**, 84

N-Benzylidenebenzenesulfonamide, **VIII**, 546

Bromomethanesulfonyl bromide, **VIII**, 212

erythro-2,3-Butanediol monomesylate, **VI**, 4

4-*tert*-Butylcyclohexen-1-yl trifluoromethanesulfonate, **VIII**, 97

Butyl *p*-toluenesulfonate, **I**, 145

3-Butyn-1-yl trifluoromethanesulfonate, **VI**, 324

(+)-(1*S*)-10-Camphorsulfonamide, **VIII**, 104

DL-10-Camphorsulfonic acid. **V**, 194

DL-10-Camphorsulfonyl chloride, **V**, 196

(−)-(Camphorsulfonyl)imine, **VIII**, 104, 110

(+)-(2*R*,8a*S*)-10-(Camphorsulfonyl)oxaziridine, **VIII**, 104

(−)-D-2,10-Camphorsultam, **VIII**, 110

Chloromethanesulfonyl chloride, **V**, 232

Chlorosulfonyl isocyanate, **V**, 226

Cholesta-3,5-dien-3-yl trifluoromethanesulfonate, **VIII**, 126

Coupling of *o*-toluidine and Chicago acid, **II**, 145

4-Cyanobenzenesulfonamide, **VI**, 632

Cyclobutyl tosylate, **VII**, 117

Cyclohexanone oxime methanesulfonate, **VIII**, 568

Cysteic acid monohydrate, **III**, 226

3,5-Dibromosulfanilamide, **III**, 263

3,5-Dichlorosulfanilamide, **III**, 262

N,*N*'-Diethylsulfamide, **VI**, 78

Disodium nitrosodisulfonate, **VI**, 1010

Dodecyl *p*-toluenesulfonate, **III**, 366; **VI**, 377

4-Formylbenzenesulfonamide, **VI**, 631

Glyoxal bisulfite, **III**, 438

4-Hydroxy-1-butanesulfonic acid sultone, **IV**, 529

3-Isovalerolactam-*N*-sulfonyl chloride, **V**, 673

Methanesulfonyl chloride, **IV**, 571

erythro-2-Methanesulfonyloxy-2-butyl cyclobutanecarboxylate, **VI**, 312

3-Methyl-2-butenyl triflate, **VI**, 757

Methyl(carboxysulfamoyl)triethylammonium hydroxide, **VI**, 788

2-Methyl-3-(*p*-toluenesulfonamido)-2-pentanol, **VII**, 375

Methyl *p*-toluenesulfonate, **I**, 146

Naphthalene-1,5-disulfonyl chloride, **IV**, 693

1,2-Naphthoquinone-4-sulfonic acid, salts, **III**, 633

(1*R*)-Nopyl 4-methylbenzenesulfonate, **VIII**, 223

Orthanilic acid, **II**, 471

2- and 3-Phenanthrenesulfonic acids, **II**, 482

trans-2-(Phenylsulfonyl)-3-phenyloxaziridine, **VIII**, 546

Potassium anthraquinone-α-sulfonate, **II**, 539

Potassium 3-phenanthrenesulfonate, **II**, 482

4-Pyridinesulfonic acid, **V**, 977

Pyridinium-1-sulfonate, **VI**, 641

Sodium 2-bromoethanesulfonate, **II**, 558

Sodium 2-phenyl-2-hydroxyethane-1-sulfonate, **IV**, 850

Sodium β-styrenesulfonate, **IV**, 846

Styrene glycol dimesylate, **VI**, 56

β-Styrenesulfonyl chloride, **IV**, 846

o-Sulfobenzoic anhydride, **I**, 495

α-Sulfopalmitic acid, **IV**, 862

Taurine, **II**, 563

cis-2-(*p*-Toluenesulfonamido) cyclohexanol, **VII**, 375

p-Toluenesulfonic anhydride, **IV**, 940

p-Toluenesulfonylanthranilic acid, **IV**, 34

p-Toluenesulfonylhydrazide, **V**, 1055

o-Toluidinesulfonic acid, **III**, 824

2-(*p*-Tolylsulfonyl)dihydroisoindole, **V**, 1064

p-Tolylsulfonylmethylnitrosamide, **IV**, 943

cis-*N*-Tosyl-3-methyl-2-azabicyclo[3.3.0]oct-3-ene, **VII**, 501

cis-*N*-Tosyl-2-(2-propenyl) cyclopentylamine, **VII**, 501

m-Trifluoromethylbenzenesulfonyl chloride, **VII**, 508

Trimethyloxonium 2,4,6-trinitrobenzenesulfonate, **V**, 1099

2,4,6-Trimethylpyrylium trifluoromethanesulfonate, **V**, 1114

(Z)-4-(Trimethylsilyl)-3-butenyl 4-methylbenzenesulfonate, **VIII**, 358

3-Trimethylsilyl-2-propyn-1-yl methanesulfonate, **VIII**, 471

N,N'N''-Tris(*p*-tolylsulfonyl) diethylenetriamine *N,N''*-disodium salt, **VI**, 652

3,6,9-Tris(*p*-tolylsulfonyl)-3,6,9-triazaundecane-1,11-dimethanesulfonate, **VI**, 652

3,6,9-Tris(*p*-tolylsulfonyl)-3,6,9-triazaundecane-1,11-diol, **VI**, 652

cis-2-(*p*-Toluenesulfonamido) cyclohexanol, **VII**, 375

D. Sulfoxides

Methyl phenyl sulfoxide, **V**, 791

(S)-(−)-Methyl *p*-tolyl sulfoxide, **VIII**, 464

Phenyl vinyl sulfoxide, **VII**, 453

(S)-(+)-2-(*p*-Toluenesulfinyl)-2-cyclopenten-1-one, **VII**, 495

(S)-(+)-(*p*-Toluenesulfinyl)-2-cyclopenten-1-one ethylene ketal, **VII**, 495

E. Thioacetals

3,3-Bis(methylthio)-1-(2-pyridinyl)-2-propen-1-one, **VII**, 476

Bis(phenylthio)methane, **VI**, 737

4,4-Bis(phenylthio)-3-methoxy-1-butene, **VI**, 737

1,3-Dithiane, **VI**, 556

1-(1,3-Dithian-2-yl)-2-cyclohexen-1-ol, **VIII**, 309

1,4-Dithiaspiro[4.11]hexade-
cane, **VII**, 124
5,9-Dithiaspiro[3.5]nonane,
VI, 316
2,2-(Ethylenedithio)cyclohexa-
none, **VI**, 590
2-Methylthio-2-phenyl-1,3-
dithiane, **VI**, 109
2-Phenyl-1,3-dithiane, **VI**, 109
3-(2-Phenyl-1,3-dithian-2-yl)-
1*H*-indole, **VI**, 109
2,2-(Trimethylenedithio)cyclo-
hexanone, **VI**, 1014
sym-Trithiane, **II**, 610
2-Tetradecyl-*sym*-trithiane,
VI, 869
F. Thioesters
S-tert-Butyl acetothioacetate,
VIII, 71
S-tert-Butyl 3α,7α,12α-trihy-
droxy-5β-cholane-24-
thioate, **VII**, 81
S-tert-Butyl cyclohexanecarbothioate, **VII**, 81, 87
1,2:5,6-Di-(*O*-isopropylidene-3-
O-(*S*-methyl dithiocarbonate)-α-D-glucofuranose,
VII, 139
Ricinelaidic acid *S*-(2-pyridyl)
carbothioate, **VII**, 470
G. Thioethers (Sulfides)
S-Acetamidomethyl-L-cysteine
hydrochloride, **VI**, 5
Benzyl sulfide, **VI**, 130
7-Benzylthiomenthone,
VIII, 302
1,2-Bis(butylthio)benzene,
V, 107
1,3-Bis(methylthio)-1-hexen-4-
ol, **VI**, 683
1,3-Bis(methylthio)-2-methoxy-
propane, **VI**, 683
1,3-Bis(methylthio)-2-propanol,
VI, 683
S-tert-Butyl-L-cysteine, *tert*-
butyl ester, **VI**, 255

β-Chloroethyl methyl sulfide,
II, 136
Dimethyl β-thiodipropionate,
IV, 669
Diphenyl sulfide, **II**, 242
3,7-Dithianonane-1,9-diol,
VIII, 592
3,7-Dithianonane-1,9-dithiol,
VIII, 592
2,2'-Dithienyl sulfide, **VI**, 558
Ethyl 4-amino-3-(methylthio-
methyl)benzoate,
VI, 581
Ethyl 2-methyl-3-methylthioin-
dole-5-carboxylate,
VI, 601
Ethyl 1-methylthio-3-phthalimi-
dopropane-3,3-dicarboxy-
late, **II**, 384
β-(2-Hydroxyethylmercapto)
propionitrile, **III**, 458
Hydroxyethyl methyl sulfide,
II, 345
DL-Methionine, **II**, 384
(*Z*)-2-Methoxy-1-phenylthio-
1,3-butadiene, **VI**, 737
2-Methylmercapto-*N*-methyl-
Δ²-pyrroline, **V**, 780
Methyl 2-thienyl sulfide,
IV, 667
4'-(Methylthio)-2,2':6',2"-ter-
pyridine, **VII**, 476
Neopentyl phenyl sulfide,
VI, 833
p-Nitrophenyl sulfide,
III, 667
1-Phenylthio-2-bromoethane (in
solution) **VII**, 453,
(Phenylthio)nitromethane,
VIII, 550
Phenyl vinyl sulfide, **VII**, 453
Propyl sulfide, **II**, 547
1,1,1',1'-Tetraethoxyethyl poly-
sulfide, **IV**, 295
1,4,8,11-Tetrathiacyclotetra-
decane, **VIII**, 592

cis-8-Thiabicyclo[4.3.0]nonane,
  VI, 482
β-Thiodiglycol, **II**, 576
H. Thiols
  Cyclododecyl mercaptan
    VII, 124
  Diethyl mercaptoacetal, **IV**, 295
  1,2-Dimercaptobenzene,
    V, 419
  3,7-Dithianonane-1,9-dithiol,
    VIII, 592
  Dodecyl mercaptan, **III**, 363
  Ethanedithiol, **IV**, 401
  2-Furfuryl mercaptan, **IV**, 491
  2-Mercapto-4-amino-5-carb-
    ethoxypyrimidine, **IV**, 566
  2-Mercaptobenzimidazole,
    IV, 569
  2-Mercaptobenzoxazole, **IV**, 570
  2-Mercapto-4-hydroxy-5-cyano-
    pyrimidine, **IV**, 566
  2-(1-Mercapto-1-methylethyl)-
    5-methylcyclohexanol,
    VIII, 304
  2-Mercaptopyrimidine, **V**, 703
  1,5-Naphthalenedithiol, **IV**, 695
  2-Naphthalenethiol, **VI**, 824
  Thallium(I) 2-methylpropane-2-
    thiolate, **VII**, 81
  m-Thiocresol, **III**, 809
  7-Thiomenthol, **VIII**, 302
  2-Thio-6-methyluracil, **IV**, 638
  2-Thiophenethiol, **VI**, 979
  Thiophenol, **I**, 504
  Thiosalicylic acid, **II**, 580
I. Miscellaneous
  p-Acetaminobenzenesulfinic
    acid, **I**, 7
  N-(*p*-Acetylaminophenyl)rhoda-
    nine, **IV**, 6
  Benzenesulfinyl fluoride,
    V, 396
  7-Benzylthiomenthone,
    VIII, 304
  N,N'-Bis(methoxycarbonyl)sul-
    fur diimide, **VIII**, 427

Bis[2,2,2-trifluoro-1-phenyl-1-
    (trifluoromethyl)ethoxy]
    diphenylsulfurane,
    VI, 163
m-Bromophenyl isothiocyanate,
    I, 448
p-Bromophenyl isothiocyanate,
    I, 448
tert-Butyl *S*-methylthiolcarbon-
    ate, **V**, 166
Butyl sulfate, **II**, 111
Butyl sulfite, **II**, 112
Chloromethylphosphonothioic
    dichloride, **V**, 218
p-Chlorophenyl isothiocyanate,
    I, 165; **V**, 223
o-Chlorophenylthiourea, **IV**, 180
Cyclopropyldiphenylsulfonium
    tetrafluoroborate, **VI**, 364
Diethylaminosulfur trifluoride,
    VI, 440
Diethylthiocarbamyl chloride,
    IV, 307
1,2:5,6-Di-(*O*-isopropylidene-3-
    O-(*S*-methyl dithiocarbon-
    ate)-α-D-glucofuranose,
    VII, 139
p-Dimethylaminophenyliso-
    thiocyanate, **I**, 448
1,3-Dimethylimidazole-2-
    thione, **VII**, 195
2,6-Dimethylphenylthiourea,
    V, 802
Dimethyl-2-propynylsulfonium
    bromide, **VI**, 31
Dimethyloxosulfonium
    methylide, **V**, 755
2,4-Dinitrobenzenesulfenyl
    chloride, **V**, 474
2,4-Dinitrophenylsulfur chlor-
    ide, **II**, 456
Diphenylthiocarbazide, **III**, 361
2,4-Dithiobiuret, **IV**, 504
Dithizone, **III**, 360
Ethylenediisothiuronium bro-
    mide, **IV**, 402

Ethylene dithiotosylate,
VI, 1016
Ethylene thiourea, III, 394
Ethyl thioamidooxalate,
VIII, 597
O-Ethyl thioformate, VI, 620
Guanylthiourea, IV, 502
Hexahydro-4,4,7-trimethyl-4H-
1,3-benzoxathiin,
VIII, 302
p-Iodophenyl isothiocyanate,
II, 366
p-Isopropylphenyl isothio-
cyanate, I, 448
o-Isothiocyano-(E)-cinnamalde-
hyde, VII, 302
Laurylisothiourea hydrobro-
mide, III, 363
Lithium phenylthio(tert-butyl)
cuprate, VI, 248
(S)-(−)-Menthyl p-toluenesulfi-
nate, VII, 495
Methanesulfinyl chloride,
V, 709
Methanesulfonyl cyanide,
VI, 727
Methyl benzenesulfinate,
V, 723
2-Methyl-4-isopropylphenyl
isothiocyanate, I, 448
Methyl isothiocyanate,
III, 599
S-Methyl isothiourea sulfate,
II, 411
1-Methyl-1-(1-naphthyl)-2-
thiourea, III, 609
N-Methylthiopyrrolidone,
VII, 372
Methylthiourea, III, 617
Morland salt, II, 555
O-2-Naphthyl dimethylthiocar-
bamate, VI, 824
α-Naphthyl isothiocyanate,
IV, 700
β-Naphthyl isothiocyanate,
I, 448

o-Nitro-p-chlorophenylsulfur
chloride, II, 456
o-Nitrophenylsulfur chloride,
II, 455
S-(2-Oxobut-3-yl) butanethioate
VI, 776,
Phenyl benzeneselenosulfonate,
VIII, 543
β-Phenyldithiocarbazic acid,
phenylhydrazine salt,
III, 360
Phenyl isothiocyanate, I, 447
Phenylsulfenyl chloride,
VIII, 550
Phenylsulfur trifluoride,
V, 959
1-Phenyl-2-thiobiuret, V,966
α-Phenylthiourea, III, 735
Reinecke salt, II, 555
Sodium p-toluenesulfinate,
I, 492
2,2,4,4-Tetrakis(trifluoro-
methyl)-1,3-dithietane,
VII, 251
Tetramethylbiphosphine disul-
fide, V, 1016
Thiobenzoic acid, IV, 924
Thiobenzophenone, II, 573;
IV, 927
Thiobenzoylthioglycolic acid,
V, 1046
Thiobutyric acid, VI, 779
Thiocarbonyl perchloride,
I, 506
p-Thiocyanodimethylaniline,
II, 574
2a-Thiohomophthalimide,
V, 1051
Thiolacetic acid, IV, 928
Thiophosgene, I, 509
p-Toluenesulfenyl chloride,
IV, 934
p-Toluenesulfinyl chloride,
IV, 937
Trifluoroacetyl triflate,
VII, 506

FORMULA INDEX

All preparations listed in the Cumulative Contents and Solvents and Regents Indices are recorded in this index. The system of indexing is that used by *Chemical Abstracts*. The essential principles involved are as follows. (1) The arrangement of symbols in formulas is alphabetical except that in carbon compounds C always comes first, followed immediately by H if hydrogen is also present. (2) The arrangement of formulas is also alphabetical except that the number of atoms of any specific kind influences the order of compounds, for example, all formulas with one carbon atom precede those with two carbon atoms; thus, CH_2I_2, CH_3NO_2, CH_5N, $C_2H_2O_3$. (3) The arrangement of entries under any heading is strictly alphabetical according to the names of the isomers. (4) Inorganic salts of organic acids and inorganic addition compounds of organic compounds are listed under the formula of the compounds from which they are derived.

The names in this index are the common names of the compounds prepared and the reagents employed. To obtain the *Chemical Abstracts Index* name, the Cumulative Contents Index should be consulted.

CBrN Cyanogen bromide, **II**, 150
CClNO$_3$S Chlorosulfonyl isocyanate, **V**, 226
CCl$_2$S Thiophosgene, **I**, 509
CCl$_4$S Thiocarbonyl perchloride, **I**, 506
CCl$_5$OP Trichloromethylphosphonyl dichloride, **IV**, 950
CD$_2$N$_2$ Dideuteriodiazomethane, **VI**, 432
CIN Cyanogen iodide, **IV**, 207
CHN Hydrogen cyanide (anhydrous), **I**, 314
CH$_2$Br$_2$ Methylene bromide, **I**, 357
CH$_2$Br$_2$O$_2$S Bromomethanesulfonyl bromide, **VIII**, 212
CH$_2$Cl$_2$O$_2$S Chloromethanesulfonyl chloride, **V, 232**
CH$_2$Cl$_3$PS Chloromethylphosphonothioic dichloride, **V**, 218

CH$_2$I$_2$ Methylene iodide, **I**, 358
CH$_2$N$_2$ Diazomethane, **II**, 165; **III**, 244; **IV**, 250; **V**, 351
CH$_3$ClOS Methanesulfinyl chloride, **V**, 709
CH$_3$ClO$_2$S Methanesulfonyl chloride, **IV**, 571
CH$_3$I Methyl iodide, **II**, 399
CH$_3$Li Methyllithium, **VII**, 346
CH$_3$NO$_2$ Methyl nitrite, **II**, 363
Nitromethane, **I**, 401
CH$_3$NO$_3$ Methyl nitrate, **II**, 412
CH$_3$N$_3$O$_3$ Nitrourea, **I**, 417
CH$_4$N$_2$O$_2$ Hydroxyurea, **V**, 645
CH$_4$N$_4$O$_2$ Nitroguanidine, **I**, 399
CH$_4$S Methyl mercaptan, **II**, 345
CH$_5$N·HCl Methylamine hydrochloride, **I**, 347

$CH_5NO \cdot HCl$ *N*-Methylhydroxylamine hydrochloride, **VI**, 670

$CH_5N_3O \cdot H_2O_4S$ Semicarbazide sulfate, **I**, 485

$CH_6N_2 \cdot H_2O_4S$ Methylhydrazine sulfate, **II**, 395

$CH_6N_4 \cdot CH_2O_3$ Aminoguanidine bicarbonate, **III**, 73

CN_4O_8 Tetranitromethane, **III**, 803

$C_2Cl_2F_2$ 1,1-Dichloro-2,2-difluoroethylene, **IV**, 268

$C_2Cl_4O_2$ Trichloromethyl chloroformate, **VI**, 715

$C_2F_6O_5S_2$ Trifluoromethanesulfonic anhydride, **VI**, 324, 757

C_2HBr_2N Dibromoacetonitrile, **IV**, 254

C_2HBr_3O Bromal, **II**, 87

C_2HBrMg Ethynylmagnesium bromide, **IV**, 792

C_2HClMg Ethynylmagnesium chloride, **VIII**, 606

C_2HCl_3O Dichloroacetyl chloride, **VI**, 1038

$C_2HF_3O_3$ Peroxytrifluoroacetic acid, **V**, 367, 598

C_2HLi Lithium acetylide, **VIII**, 391

$C_2HNO_4K_2$ Dipotassium salt of nitroacetic acid, **VI**, 797

C_2HN_3S 1,2,4-Triazole-3(5)-thiol, **V**, 1070

C_2H_2ClN Chloroacetonitrile, **IV**, 144

$C_2H_2Cl_2O_2$ Dichloroacetic acid, **II**, 181

C_2H_2O Ketene, **I**, 330; **V**, 679

$C_2H_2O_2$ Glyoxal, **III**, 438

$C_2H_2O_4$ Oxalic acid (anhydrous), **I**, 421

C_2H_3BrMg Vinylmagnesium bromide, **IV**, 258

$C_2H_3BrO_2$ Bromoacetic acid, **III**, 381

$C_2H_3ClO_2S$ 2-Chlorothiirane 1,1-dioxide, **V**, 231

$C_2H_3Cl_2NO$ α,α-Dichloroacetamide, **III**, 260

$C_2H_3Cl_2NO_2$ Methyl *N,N*-dichlorocarbamate, **VIII**, 427

$C_2H_3Cl_3O$ Trichloroethyl alcohol, **II**, 598

C_2H_3N Methyl isocyanide, **V**, 772

C_2H_3NO Glycolonitrile, **III**, 436

$C_2H_3NO_2S$ Methanesulfonyl cyanide, **VI**, 727

C_2H_3NS Methyl isothiocyanate, **III**, 599

$C_2H_3N_3$ 1,2,4-Triazole, **V**, 1070

C_2H_4BrNO *N*-Bromoacetamide, **IV**, 104

C_2H_4ClNO Chloroacetamide, **I**, 153

$C_2H_4ClNO_4S$ Methyl (chlorosulfonyl)carbamate, **VI**, 788

$C_2H_4Cl_2O$ Bis(chloromethyl) ether, **IV**, 101

2,2-Dichloroethanol, **IV**, 271

Dichloromethyl methyl ether, **V**, 365

$C_2H_4N_4$ 3-Amino-1*H*-1,2,4-triazole, **III**, 95

4-Amino-4*H*-1,2,4-triazole, **III**, 96

C_2H_4OS Thiolacetic acid, **IV**, 928

C_2H_4S Ethylene sulfide, **V**, 562

$C_2H_5AsO_5$ Arsonoacetic acid, **I**, 73

C_2H_5Br Ethyl bromide, **I**, 29, 36

C_2H_5BrMg Ethylmagnesium bromide, **VI**, 526, 925

C_2H_5BrO 2-Bromoethanol, **I**, 117

$C_2H_5BrO_3S \cdot Na$ Sodium 2-bromoethanesulfonate, **II**, 558

C_2H_5ClO Monochloromethyl ether, **I**, 377

$C_2H_5ClO_2S$ 2-Chlorothiirane 1,1-dioxide, **V**, 231

C_2H_5I Ethyl iodide, **II**, 402

C_2H_5N Ethylenimine, **IV**, 433

C_2H_5NO Acetamide, **I**, 3

$C_2H_5NO_2$ Glycine, **I**, 298

Ethyl nitrite, **II**, 204; **VI**, 542

$C_2H_5NO_3$ Carboxymethoxylamine hemihydrochloride, **III**, 172

2-Nitroethanol, **V**, 833

$C_2H_5N_3$ 3(5)-Aminopyrazole, **V**, 39

$C_2H_5N_3O_2$ Nitrosomethylurea, **II**, 461

$C_2H_5N_3S$ 1-Formyl-3-thiosemicarbazide, **V**, 1070

$C_2H_5N_3S_2$ Dithioimidodicarbonic diamide, **IV**, 504

$C_2H_6BrN \cdot HBr$ β-Bromoethylamine hydrobromide, **II**, 91

$C_2H_6N_2$ Acetamidine hydrochloride, **I**, 5

Azoethane, **VI**, 78

$C_2H_6N_2O \cdot HCl$ Methylisourea hydrochloride, **IV**, 645

Nitrosodimethylamine, **II**, 211

$C_2H_6N_2O_2$ Methyl carbazate, **VI**, 334

$C_2H_6N_2O_2S$ *N,N'*-Dimethylsulfamide, **VI**, 78

$C_2H_6N_2S$ *S*-Methyl isothiourea sulfate, **IV**, 645

Methylthiourea, **III**, 617

$C_2H_6N_4S$ Guanylthiourea, **IV**, 502

$C_2H_6O_8S_2$ Glyoxal bisulfite, **III**, 438;

$C_2H_6S_2$ Ethanedithiol, **IV**, 401

$C_2H_7NO \cdot HCl$ *N,N*-Dimethylhydroxylamine hydrochloride, **IV**, 612

$C_2H_7NO_3S$ Taurine, **II**, 563

$C_2H_7N_3O_2$ Diaminobiuret, **III**, 404

$C_2H_8N_2$ *sym*-Dimethylhydrazine dihydrochloride, **II**, 208

unsym-Dimethylhydrazine, **II**, 213

unsym-Dimethylhydrazine hydrochloride, **II**, 212

$C_3Cl_3F_3O$ 1,1,1-Trichloro-3,3,3-trifluoroacetone, **VI**, 1005

$C_3F_6N_2$ Bis(trifluoromethyl)diazomethane, **VI**, 161

C_3F_6O Hexafluoroacetone, **VII**, 251

$C_3F_6O_4S$ Trifluoroacetyl triflate, **VII**, 506

C_3HCl_7 *unsym*-Heptachloropropane, **II**, 312

C_3HF_6N Hexafluoroacetone imine, **VI**, 161, 664

C_3HN_3O Oximinomalononitrile, **V**, 32

$C_3H_2ClF_3O_2$ 3-Chloro-2,2,3-trifluoropropionic acid, **V**, 239

$C_3H_2ClNO_2$ α-Chloroacetyl isocyanate, **V**, 204

$C_3H_2Cl_2O_2$ Malonyl dichloride, **IV**, 263

$C_3H_2F_6N_2$ Hexafluoroacetone hydrazone, **VI**, 161

$C_3H_2Li_2$ Dilithioallene, **VIII**, 226,

$C_3H_3NO_4H_2O \cdot Na$ Sodium nitromalonaldehyde monohydrate, **IV**, 844

$C_3H_2N_2$ Malononitrile, **II**, 379; **III**, 535

$C_3H_2N_2O_3$ Parabanic acid, **IV**, 744

C_3H_2O Cyclopropenone, **VI**, 361

Propiolaldehyde, **IV**, 813

$C_3H_3ClO_2$ Pyruvoyl chloride, **VII**, 467*

$C_3H_3ClO_2S$ 3-Chlorothiete 1,1-dioxide, **VII**, 491

$C_3H_3ClO_3$ Chloropyruvic acid, **V**, 636

$C_3H_3NOS_2$ Rhodanine, **III**, 763

$C_3H_3NO_2$ Sodium cyanoacetate, **I**, 181

C_3H_4 Allene, **V**, 22

$C_3H_4Br_2$ 2,3-Dibromopropene, **I**, 209

$C_3H_4Cl_2O$ α,γ-Dichloroacetone, **I**, 211

$C_3H_4Cl_2O_2S$ 3,3-Dichlorothietane-1,1-dioxide, **VII**, 491

$C_3H_4N_2$ Imidazole, **III**, 471

Methyleneaminoacetonitrile, **I**, 355

$C_3H_4N_2O$ Cyanoacetamide, **I**, 179

$C_3H_4N_2OS$ Pseudothiohydantoin, **III**, 751

C_3H_4O Acrolein, **I**, 15

$C_3H_4O_2$ Acrylic acid, **III**, 30

$C_3H_4O_2S$ Thiete 1,1-dioxide, **VII**, 491

$C_3H_4O_3$ Acetic formic anhydride, **VI**, 8

Pyruvic acid, **I**, 475

$C_3H_4O_4$ Calcium malonate, **II**, 376

Malonic acid, **II**, 376

C_3H_5Br Allyl bromide, **I**, 27

Bromocyclopropane, **V**, 126

C_3H_5BrMg Allylmagnesium bromide, **VIII**, 23

C_3H_5BrO Bromoacetone, **II**, 88

Epibromohydrin, **II**, 257

$C_3H_5BrO_2$ β-Bromopropionic acid, **I**, 131

$C_3H_5Br_3$ 1,2,3-Tribromopropane, **I**, 521

C_3H_5ClMg Allylmagnesium chloride, **IV**, 751

C_3H_5ClO Epichlorohydrin, **I**, 233; **II**, 256

$C_3H_5ClO_2$ (*S*)-2-Chloropropanoic acid, **VIII**, 119, 434

β-Chloropropionic acid, **I**, 166

$C_3H_5ClO_2S$ 3-Chlorothietane-1,1-dioxide, **VII**, 491

C_3H_5Li Allyllithium, **V**, 452

C_3H_5LiO 1-Methoxyvinyllithium, **VIII**, 19

C_3H_5N Ethyl isocyanide, **IV**, 438

C_3H_5NO Ethylene cyanohydrin, **I**, 256

Methoxyacetonitrile, **II**, 387

$C_3H_5NO_2$ 2-Nitropropene, **VII**, 396

2-Oxazolidinone, **VII**, 4

$C_3H_5NO_4$ Methyl nitroacetate, **VI**, 797

$C_3H_5N_3$ 3(5)-Aminopyrazole, **V**, 39

C_3H_6BrCl Trimethylene chlorobromide, **I**, 156

C_3H_6BrN 2-Bromoallylamine, **V**, 121
$C_3H_6Br_2$ Trimethylene bromide, **I**, 30
$C_3H_6Br_2O$ Glycerol α,γ-dibromohydrin, **II**, 308
C_3H_6ClI 1-Chloro-3-iodopropane, **VI**, 365
C_3H_6ClNS N,N-Dimethylthiocarbamyl chloride, **VI**, 825
$C_3H_6Cl_2O$ Dichloromethyl ethyl ether, **V**, 860
  Glycerol α,γ-dichlorohydrin, **I**, 292
$C_3H_6N_2$ β-Aminopropionitrile, **III**, 93
  2-Diazopropane, **VI**, 392
$C_3H_6N_2S$ Ethylene thiourea, **III**, 394
C_3H_6O Allyl alcohol, **I**, 42
  (R)-Methyloxirane, **VIII**, 434
  (S)-(−)-Methyloxirane, **VII**, 356
  Propionaldehyde, **II**, 541
  Trimethylene oxide, **III**, 835
C_3H_6OS O-Ethyl thioformate, **VI**, 620
$C_3H_6O_2$ Acetol, **II**, 5
$C_3H_6O_2S$ Thietane 1,1-dioxide, **VII**, 491
$C_3H_6O_3$ DL-Glyceraldehyde, **II**, 305
  β-Hydroxypropionic acid, **I**, 321
$C_3H_6S_3$ sym-Trithiane, **II**, 610
C_3H_7Br Isopropyl bromide, **I**, 37; **II**, 359
  Propyl bromide, **I**, 37; **II**, 359
C_3H_7Cl Propyl chloride, **I**, 143
$C_3H_7ClN·Cl$ N,N-Dimethylchloromethylen-ammonium chloride, **VIII**, 498
C_3H_7ClO (S)-2-Chloropropan-1-ol, **VIII**, 434
  Trimethylene chlorohydrin, **I**, 533
$C_3H_7ClO_2$ Glycerol α-monochlorohydrin, **I**, 294
C_3H_7ClS β-Chloroethyl methyl sulfide, **II**, 136
C_3H_7I Propyl iodide, **II**, 402
C_3H_7N Allylamine, **II**, 24
  Azetidine, **VI**, 75
C_3H_7NO Acetoxime, **I**, 318
$C_3H_7NO_2$ β-Alanine, **II**, 19; **III**, 34
  DL-Alanine, **I**, 21
  N-(Hydroxymethyl)acetamide, **VI**, 5
  Isopropyl nitrite, **III**, 192
  Lactamide, **III**, 516
$C_3H_7NO_3$ DL-Serine, **III**, 774

$C_3H_7NO_5S$ Cysteic acid monohydrate, **III**, 226
$C_3H_7N_3$ 3-Amino-3-pyrazoline sulfate, **V**, 39
$C_3H_7N_3O_2$ Guanidoacetic acid, **III**, 440
$C_3H_7O_3S_2Na$ Sodium isopropyl thiosulfate or Sodium 1-methylethyl thio-sulfate, **VI**, 235
$C_3H_8N_2$ Acetone hydrazone, **VI**, 10
$C_3H_8N_2O$ asym-Dimethylurea, **IV**, 361
$C_3H_8N_2O_2$ Formamidine acetate, **V**, 582
  Ethyl hydrazinecarboxylate, **III**, 404; **VI**, 936
$C_3H_8N_2Se$ N,N-Dimethylselenourea, **IV**, 359
C_3H_8OS Dimethyloxosulfonium methylide, **V**, 755
  β-Hydroxyethyl methyl sulfide, **II**, 345
$C_3H_8O_2$ Methylal, **III**, 469
  L-Propylene glycol, **II**, 545
C_3H_9ISi Iodotrimethylsilane, **VI**, 353
C_3H_9N N-Methylethylamine, **V**, 758
  Trimethylamine, **I**, 528
  Trimethylamine hydrochloride, **I**, 531
$C_3H_9N_3Si$ Trimethylsilyl azide, **VI**, 1030
$C_3H_9OBF_4$ Trimethyloxonium tetrafluo-roborate, **V**, 1096; **VI**, 1019
C_3N_2O Carbonyl cyanide, **VI**, 268
$C_4FeNa_2O_4·dioxane$ Disodium tetracar-bonylferrate sesquidioxanate, **VI**, 807
C_4HBr_3S 2,3,5-Tribromothiophene, **V**, 150
$C_4H_2Br_2O_3$ Mucobromic acid, **III**, 621; **IV**, 688
$C_4H_2ClF_3$ 1-Chloro-1,4,4-trifluorobutadi-ene, **V**, 235
  1-Chloro-2,3,3-trifluorocyclobutene, **V**, 394
$C_4H_2Cl_3F_3$ 1,1,2-Trichloro-2,3,3-trifluoro-cyclobutane, **V**, 393
$C_4H_2F_2O_4$ 2,2-Difluorosuccinic acid, **V**, 393
$C_4H_2N_2$ Fumaronitrile, **IV**, 486
$C_4H_2N_2O_4$ Alloxan monohydrate, **III**, 37; **IV**, 23

$C_4H_2N_2S$ 4-Cyanothiazole, **VI**, 623
$C_4H_2O_4$ Acetylenedicarboxylic acid, **II**, 10
C_4H_3BrS 3-Bromothiophene, **V**, 149
$C_4H_3ClN_2$ 2-Chloropyrimidine, **IV**, 182
$C_4H_3Cl_2O_3P$ Furyl phosphorodichloridate, **VIII**, 396
C_4H_3IS 2-Iodothiophene, **II**, 357; **IV**, 545
$C_4H_3NO_2S$ 2-Nitrothiophene, **II**, 466
$C_4H_3N_3O_5$ Nitrobarbituric acid, **II**, 440
C_4H_4 Monovinylacetylene, **IV**, 683
$C_4H_4Br_2O_4$ α,β-Dibromosuccinic acid, **II**, 177
$C_4H_4ClNO_2$ 3-Isocyanatopropanoyl chloride, **VI**, 715
$C_4H_4Cl_2$ *cis*-3,4-Dichlorocyclobutene, **VI**, 422
$C_4H_4INO_2$ *N*-Iodosuccinimide, **V**, 663
$C_4H_4NO_2Ag$ Succinimide silver salt, **VI**, 226
$C_4H_4N_2O$ Imidazole-2-carboxaldehyde, **VII**, 287
$C_4H_4N_2O_3$ Barbituric acid, **II**, 60
$C_4H_4N_2O_4$ Dialuric acid monohydrate, **IV**, 29
$C_4H_4N_2O_5$ Alloxan monohydrate, **III**, 37; **IV**, 23
$C_4H_4N_2S$ 2-Mercaptopyrimidine, **V**, 703
$C_4H_4N_4$ Diaminomaleonitrile, **V**, 344
C_4H_4O Furan, **I**, 274
C_4H_4OS 2-Hydroxythiophene, **V**, 642
$C_4H_4O_2$ γ-Crotonolactone, **V**, 255
1,2-Cyclobutanedione, **VII**, 112
2(5*H*)-Furanone, **VIII**, 396
Ketene dimer, **III**, 508
Tetrolic acid, **V**, 1043
$C_4H_4O_3$ Succinic anhydride, **II**, 560
$C_4H_4O_4$ Fumaric acid, **II**, 302
C_4H_4S Thiophene, **II**, 578
$C_4H_4S_2$ 2-Thiophenethiol, **VI**, 979
$C_4H_5BrO_2$ α-Bromo-γ-butyrolactone, **V**, 255
C_4H_5ClO α-Chlorocrotonaldehyde, **IV**, 131
$C_4H_5ClO_2$ Vinyl chloroacetate, **III**, 853
$C_4H_5Cl_3O$ Butyrchloral, **IV**, 130

C_4H_5N Allyl cyanide, **I**, 46; **III**, 852
Cyclopropyl cyanide, **III**, 223
Pyrrole, **I**, 473
$C_4H_5NO_2$ Succinimide, **II**, 562
$C_4H_5NO_2S$ 4*H*-1,4-Thiazine 1,1-dioxide, **VI**, 976
$C_4H_5NO_4$ Methyl (*E*)-3-nitroacrylate, **VI**, 799
$C_4H_5N_3O_3$ Uramil, **II**, 617
C_4H_6 Bicyclo[1.1.0]butane, **VI**, 133, 59
1,3-Butadiene, **II**, 102
Cyclobutene, **VII**, 117
Methylenecyclopropane, **VI**, 320
$C_4H_6As_2O_4$ Arsenoacetic acid, **I**, 74
C_4H_6BrCl 1-Bromo-3-chlorocyclobutane, **VI**, 179
C_4H_6BrClO α-Bromobutyryl chloride, **IV**, 620
$C_4H_6Br_2$ 1,2-Dibromocyclobutane, **VII**, 119
$C_4H_6Br_2O$ α-Bromoisobutyryl bromide, **IV**, 348
1,3-Dibromo-2-butanone, **VI**, 711
$C_4H_6Br_2O_2$ β,β'-Dibromoisobutyric acid, **VII**, 319
$C_4H_6ClFO_2$ Ethyl chlorofluoroacetate, **IV**, 423
$C_4H_6ClF_3O$ 2-Chloro-1,1,2-trifluoroethyl ethyl ether, **IV**, 184
C_4H_6ClN γ-Chlorobutyronitrile, **I**, 156
$C_4H_6Cl_2O_2$ *trans*-2,3-Dichloro-1,4-dioxane, **VIII**, 161
$C_4H_6CrN_6S_4 \cdot H_4N$ Reinecke salt, **II**, 555;
$C_4H_6INO_4$ Methyl 2-iodo-3-nitropropionate, **VI**, 799
$C_4H_6N_2$ 4(5)-Hydroxymethylimidazole hydrochloride, **III**, 460
$C_4H_6N_2O$ Dimethylfurazan, **IV**, 342
4(5)-Hydroxymethylimidazole hydrochloride, **III**, 460
$C_4H_6N_2O_2$ Ethyl diazoacetate, **III**, 392; **IV**, 424; **VI**, 913
$C_4H_6N_2O_3$ Acetone cyanohydrin nitrate, **V**, 839
$C_4H_6N_2O_4$ Dimethyl azodicarboxylate, **IV**, 411; *Warning*, **IV**, 412

$C_4H_6N_2O_4S$ N^1,N^2-Bis(methoxycarbonyl)-sulfur diimide, **VIII**, 427

$C_4H_6N_2S$ 2-Amino-4-methylthiazole, **II**, 31

$C_4H_6N_4O$ 2,4-Diamino-6-hydroxypyrimidine, **IV**, 245

$C_4H_6N_4O_2$·HCl Diaminouracil hydrochloride, **IV**, 247

$C_4H_6N_4O_3$ Allantoin, **II**, 21

C_4H_6O trans-2-Butene oxide, **VI**, 312
2-Butyn-1-ol, **IV**, 128
Cyclobutanone, **VI**, 316, 320, 324; **VII**, 114
Cyclopropanecarboxaldehyde, **VI**, 315
Dimethylketene, **IV**, 348
Ethoxyacetylene, **IV**, 404
Oxaspiropentane, **VI**, 320

$C_4H_6O_2$ Cyclopropanecarboxylic acid, **III**, 221
2-Hydroxycyclobutanone, **VI**, 167
Isocrotonic acid, **VI**, 711
Vinylacetic acid, **III**, 851

$C_4H_6O_3$ Methyl pyruvate, **III**, 610

$C_4H_6O_4$ D-Erythronolactone, **VII**, 297
Methyl oxalate, **I**, 264; **II**, 414

$C_4H_6O_6$ DL-Tartaric acid, **I**, 497

C_4H_7Br Methallyl bromide, **VI**, 723

$C_4H_7BrO_2$ Ethyl bromoacetate, **III**, 381
Methyl β-bromopropionate, **III**, 576

$C_4H_7BrO_3$ α-Bromo-β-methoxypropionic acid, **III**, 775

C_4H_7Cl 4-Chloro-1-butene, **VI**, 607

C_4H_7ClO Butyryl chloride, **I**, 147
3-Chloro-2-buten-1-ol, **IV**, 128
Isobutyryl chloride, **III**, 490

$C_4H_7ClO_2$ Ethyl chloroacetate **II**, 263

C_4H_7N Isobutyronitrile, **III**, 493

C_4H_7NO Acetone cyanohydrin, **II**, 7
Methacrylamide, **III**, 560

$C_4H_7NO_2$ Biacetyl monoxime, **II**, 205

$C_4H_7NO_2S$ Ethyl thioamidooxalate, **VIII**, 597

$C_4H_7NO_3$ Acetylglycine, **II**, 11

$C_4H_7NO_4$ DL-Aspartlc acid, **IV**, 55

C_4H_7NS Isopropyl thiocyanate, **II**, 366

$C_4H_7N_3O$ Creatinine, **I**, 172

$C_4H_7O_2Li$ Ethyl lithioacetate, **VI**, 598

$C_4H_8Br_2O$ α,β-Dibromoethyl ethyl ether, **IV**, 749

C_4H_8ClNO N-Chloromorpholine, **VIII**, 167

$C_4H_8I_2$ 1,4-Diiodobutane, **IV**, 321

$C_4H_8N_2O_2$ Acetyl methylurea, **II**, 462
Dimethylglyoxime, **II**, 204
Sodium dimethylglyoximate, **II**, 206

$C_4H_8N_2O_3$ Nitrosomethylurethane, **II**, 464
N-Nitromorpholine, **V**, 839

$C_4H_8N_2O_4$ 1,4-Dinitrobutane, **IV**, 368

C_4H_8O trans-2-Butene oxide, **VI**, 313
3-Buten-1-ol, **VI**, 608
Cyclobutanol, **VII**, 114, 117
Tetrahydrofuran, **II**, 566

C_4H_8OS Methylthio-2-propanone, **VI**, 601
Thiobutyric acid, **VI**, 779

$C_4H_8O_2$ 3-Hydroxytetrahydrofuran, **IV**, 534
1,2-Cyclobutanediol, **VII**, 129

$C_4H_8O_3$ Ethoxyacetic acid, **II**, 260

$C_4H_8O_3S$ 4-Hydroxy-1-butanesulfonic acid sultone, **IV**, 529

C_4H_8S Tetrahydrothiophene, **IV**, 892

$C_4H_8S_2$ 1,3-Dithiane **VI**, 556,
p-Dithiane, **IV**, 396

C_4H_9Br Butyl bromide, **I**, 28, 37
sec-Butyl bromide, **I**, 38; **II**, 359
tert-Butyl bromide, **I**, 38
Isobutyl bromide, **II**, 358

C_4H_9BrMg Butylmagnesium bromide, **VI**, 407
sec-Butylmagnesium bromide, **VI**, 753

C_4H_9BrO β-Ethoxyethyl bromide, **III**, 370

C_4H_9Cl Butyl chloride, **I**, 142
sec-Butyl chloride, **I**, 143
tert-Butyl chloride, **I**, 144

C_4H_9ClMg Butylmagnesium chloride, **V**, 1141; **VIII**, 606

C_4H_9ClO t-Butyl hypochlorite, **IV**, 125; **V**, 184
α-Chloroethyl ethyl ether, **IV**, 748
Tetramehylene chlorohydrin, **II**, 571

C_4H_9I Butyl iodide, **II**, 402
  Isobutyl iodide, **II**, 402
C_4H_9Li Butyllithium, **VI**, 980; **VII**, 292
C_4H_9N Cyclobutylamine, **V**, 273
  2,2-Dimethylethylenimine, **III**, 148
C_4H_9NHCl Cyclobutylamine hydrochloride, **VIII**, 132
C_4H_9NO Isobutyramide, **III**, 490
  2-Methyl-2-nitrosopropane, **VI**, 803
$C_4H_9NO_2$ γ-Aminobutyric acid, **II**, 25
  α-Aminoisobutyric acid, **II**, 29
  Butyl nitrite, **II**, 108
  Ethyl N-methylcarbamate, **II**, 278
  Glycine ethyl ester hydrochloride, **II**, 310
  2-Methyl-2-nitropropane, **VI**, 803,
$C_4H_9NO_3$ DL-Threonine, **III**, 813
C_4H_9NSi Cyanotrimethylsilane or Trimethylsilylcarbonitrile or Trimethylsilyl cyanide, **VII**, 517
$C_4H_9N_3O_2$ Ethyl oxalamidrazonate, **VIII**, 597
C_4H_9SNa Sodium 2-butanethiolate, **VI**, 236
$C_4H_{10}BrN·HBr$ β-Dimethylaminoethyl bromide hydrobromide, **II**, 92
$C_4H_{10}ClN·HCl$ β-Dimethylaminoethyl chloride hydrochloride, **IV**, 333
$C_4H_{10}F_3NS$ Diethylaminosulfur trifluoride, **VI**, 440, 835
$C_4H_{10}N_2$ Azoethane, **VI**, 78
$C_4H_{10}N_2Si$ Trimethylsilyldiazomethane, **VIII**, 612
$C_4H_{10}N_4O·HCl$ Aminoacetone semicarbazone hydrochloride, **V**, 27
$C_4H_{10}N_4S_2·2HBr$ Ethylenediisothiuronium bromide, **IV**, 402
$C_4H_{10}O·1/3 Al$ Aluminum tert-butoxide, **III**, 48
$C_4H_{10}O_2S$ β-Thiodiglycol, **II**, 576
$C_4H_{10}Zn$ Diethylzinc, **II**, 184
$C_4H_{11}ClSiMg$ (Trimethylsilyl)methylmagnesium chloride, **VIII**, 602, 612
$C_4H_{11}N$ Butylamine, **II**, 319
  sec-Butylamine, **II**, 319

tert-Butylamine, **II**, 148
tert-Butylamine hydrochloride, **III**, 153
$C_4H_{11}NO$ N-tert-Butylhydroxylamine, **VI**, 804
$C_4H_{12}ClN_2OP$ N,N,N',N'-Tetramethyldiamidophosphorochloridate, **VII**, 66,
$C_4H_{12}N_2$ Putrescine dihydrochloride, **IV**, 819
$C_4H_{12}N_2O_2S$ N,N'-Diethylsulfamide, **VI**, 78
$C_4H_{12}P_2S_2$ Tetramethylbiphosphine disulfide, **V**, 1016
$CN_2·C_2H_4O_2$ Formamidine acetate, **V**, 582
$C_5H_2N_2Se$ N,N-Diethylselenourea, **IV**, 360
$C_5H_3N_3OS$ 2-Mercapto-4-hydroxy-5-cyanopyrimidine, **IV**, 566
C_5H_4BrN 2-Bromopyridine, **III**, 136
$C_5H_4N_2O_4$ Imidazole-4,5-dicarboxylic acid, **III**, 471;
$C_5H_4BrN_3O_2$ 2-Amino-5-bromo-3-nitropyridine, **V**, 347
$C_5H_4Cl_4$ 1-Chloro-1-(trichloroethenyl)-cyclopropane, **VIII**, 124, 373
$C_5H_4N_2O_4$ Imidazole-4,5-dicarboxylic acid, **III**, 471
C_5H_4OS 2-Thenaldehyde, **IV**, 915
  3-Thenaldehyde, **IV**, 918
  2-Thiophenealdehyde, **III**, 811
$C_5H_4O_2$ 2-Cyclopentene-1,4-dione, **V**, 324
  Furfural, **I**, 280
  2H-Pyran-2-one, **VI**, 462 or α-Pyrone, **V**, 982
$C_5H_4O_2S$ 3-Thenoic acid, **IV**, 919
$C_5H_4O_3$ Citraconic anhydride, **II**, 140
  2-Furancarboxylic acid, **I**, 276
  2-Furoic acid, **IV**, 493
  Itaconic anhydride, **II**, 368
$C_5H_4O_6$ α-Ketoglutaric acid, **V**, 687
$C_5H_5BrN_2$ 2-Amino-5-bromopyridine, **V**, 346
C_5H_5BrO 2-Bromo-2-cyclopenten-1-one, **VII**, 271
$C_5H_5BrO_2$ 5-Bromo-5,6-dihydro-2H-pyran-2-one, **VI**, 462

C_5H_5BrS 3-Thenyl bromide, **IV**, 921

$C_5H_5ClO_2$ Methyl 4-chloro-2-butynoate, **VIII**, 371

C_5H_5ClS 2-Chloromethylthiophene, **III**, 197

$C_5H_5F_3O_3S$ 3-Butyn-1-yl trifluoromethane sulfonate or 3-Butyn-1-yl triflate, **VI**, 324

C_5H_5NO Pyridine *N*-oxide, **IV**, 828
2-Pyrrolealdehyde, **IV**, 831

$C_5H_5NO_3$ 3-Acetyl-2(3*H*)-oxazolone, **VII**, 5

$C_5H_5NO_3S$ 4-Pyridinesulfonic acid, **V**, 977
Pyridinium-1-sulfonate, **VI**, 641

$C_5H_5N_3O$ *trans*-2,4-Pentadienoic acid azide, **VI**, 95

C_5H_5Na Cyclopentadienylsodium, **IV**, 474; **V**, 1088

$C_5H_5O_2\cdot Na$ Glutaconaldehyde sodium salt, **VI**, 641

C_5H_6 Bicyclo[2.1.0]pent-2-ene, **VI**, 145
Cyclopentadiene, **IV**, 238; **V**, 415, 1002; **VI**, 147, 1038; **VII**, 339

$C_5H_6BrNO_2$ *N*-Bromoglutarimide, **IV**, 498

$C_5H_6BrN_3$ 2,3-Diamino-5-bromopyridine, **V**, 346;

$C_5H_6ClNO_3$ 3-Acetyl-4- and 5-chloro-2-oxazolidinone, **VII**, 5

$C_5H_6N_2$ 3-Aminopyridine, **IV**, 45
3-Hydroxyglutaronitrile, **V**, 614
4-Methylpyrimidine, **V**, 794
Trimethylene cyanide, **I**, 536

$C_5H_6N_2O$ 3-Hydroxyglutaronitrile, **V**, 614
4-Methyl-6-hydroxypyrimidine, **IV**, 638

$C_5H_6N_2O_2$ 6-Methyluracil, **II**, 422

$C_5H_6N_2OS$ 2-Thio-6-methyluracil, **IV**, 638

C_5H_6O 2-Cyclopentenone, **V**, 326
3-Methylfuran, **IV**, 628

C_5H_6OS 2-Furfuryl mercaptan, **IV**, 491

$C_5H_6O_2$ 5,6-Dihydro-2*H*-pyran-2-one, **VI**, 462

Cyclopentadiene, **IV**, 238

2-Furyl carbinol, **I**, 276

3-Methyl-2(5*H*)-furanone, **VIII**, 396

trans-2,4-Pentadienoic acid, **VI**, 95

$C_5H_6O_3$ Methyl 4-hydroxy-2-butynoate, **VII**, 334

$C_5H_6O_4$ (*S*)-(+)-γ-Butyrolactone-γ-carboxylic acid, **VII**, 99
Citraconic acid, **II**, 141
Cyclopropane-1,1-dicarboxylic acid, **VII**, 411
(+)-(1*S*,2*S*)-Cyclopropane-1,2-dicarboxylic acid, **VIII**, 141
Itaconic acid, **II**, 368
Mesaconic acid, **II**, 382
Methyl diformylacetate, **VII**, 323

$C_5H_6O_4S_3$ Trithiocarbodiglycolic acid, **IV**, 967

$C_5H_6O_5$ Acetonedicarboxylic acid, **I**, 10
Ethyl oxomalonate, **I**, 266
α-Ketoglutaric acid, **III**, 510

C_5H_6S 3-Methylthiophene, **IV**, 671

$C_5H_6S_2$ Methyl 2-thienyl sulfide, **IV**, 667

$C_5H_7BrO_2$ Methyl α-(bromomethyl)acrylate, **VII**, 319

C_5H_7Cl 3-Chlorocyclopentene, **IV**, 238

$C_5H_7ClO_2$ 3-Chlorocyclobutanecarboxylic acid, **VI**, 179, 271

$C_5H_7ClO_3$ β-Carbomethoxypropionyl chloride, **III**, 169

$C_5H_7N_2\cdot I$ 1-Aminopyridinium iodide, **V**, 43

$C_5H_7NO_2$ Ethyl cyanoacetate, **I**, 254
Ethyl isocyanoacetate or Ethyl (carbylamino)acetate, **VI**, 620
Glutarimide, **IV**, 496

$C_5H_7NO_3$ 4-Acetoxyazetidin-2-one, **VIII**, 3
3-Acetyl-2-oxazolidinone, **VII**, 5

C_5H_7NS 2,4-Dimethylthiazole, **III**, 332

$C_5H_7N_3$ 2,3-Diaminopyridine, **V**, 346

C_5H_8 Bicyclo[2.1.0]pentane, **V**, 96
1,4-Pentadiene, **IV**, 746

$C_5H_8Br_2$ 1,1-Bis(bromomethyl)cyclopropane, **VI**, 153

C₅H₈Br₂O 2,4-Dibromo-3-pentanone, **VI**, 512, 520

C₅H₈Br₂O₂ Methyl β,β-dibromoisobutyrate, **VII**, 319

C₅H₈Br₄ Pentaerythrityl bromide, **II**, 476; **IV**, 753

C₅H₈ClNO₃S β-Isovalerolactam-*N*-sulfonyl chloride, **V**, 673

C₅H₈Cl₂O 2,3-Dichlorotetrahydropyran, **VI**, 675

C₅H₈F₃NO₂ Dimethyl(methylene)ammonium trifluoroacetate **VI**, 474

C₅H₈I₄ Pentaerythrityl iodide, **II**, 477

C₅H₈N 2,3,4,5-Tetrahydropyridine, **VI**, 970

C₅H₈N₂ 3,5-Dimethylpyrazole, **IV**, 351

C₅H₈N₂O₂ 5,5-Dimethylhydantoin, **III**, 323

C₅H₈N₂S 1,3-Dimethylimidazole-2-thione, **VII**, 195

C₅H₈O Cyclobutanecarboxaldehyde, **VI**, 312

Cyclopentanone, **I**, 192

2,3-Dihydropyran, **III**, 276

Dimethylethynylcarbinol, **III**, 320

Methyl cyclopropyl ketone, **IV**, 597

trans-3-Penten-2-one, **VI**, 883

4-Pentyn-1-ol, **IV**, 755

C₅H₈O₂ Acetylacetone, **III**, 16

Cyclobutanecarboxylic acid, **III**, 213

Dihydroxycyclopentene, **V**, 414

3,5-Dihydroxycyclopentene, **V**, 414;

3,3-Dimethoxycyclopropene, **VI**, 361

β,β-Dimethylacrylic acid, **III**, 302

C₅H₈O₂S Isoprene cyclic sulfone, **III**, 499

C₅H₈O₃ Ethyl pyruvate, **IV**, 467

Levulinic acid, **I**, 335

C₅H₈O₄ Glutaric acid, **I**, 289; **IV**, 496

Methylsuccinic acid, **III**, 615

C₅H₉BrO 1-Bromo-1-ethoxycyclopropane, **VIII**, 556

1-Bromo-2-methyl-2-butanone, **VI**, 193

Tetrahydrofurfuryl bromide, **III**, 793

C₅H₉BrO₂ (*S*)-(−)-2-Acetoxy-1-bromopropane, **VII**, 356

α-Bromoisovaleric acid, **II**, 93; **III**, 848

Ethyl β-bromopropionate, **I**, 246

C₅H₉BrO₃ *erythro*-α-Bromo-β-methoxybutyric acid, **III**, 813

Methyl α-bromo-β-methoxypropionate, **III**, 774

C₅H₉S·Br Dimethyl-2-propynylsulfonium bromide, **VI**, 31

C₅H₉ClO *trans*-2-Chlorocyclopentanol, **IV**, 157

5-Chloro-2-pentanone, **IV**, 597

C₅H₉ClO₂ γ-Chloropropyl acetate, **III**, 203

C₅H₉DO 2-Methylbutanal-1-*d*, **VI**, 751

C₅H₉I 1-Iodo-*cis*, *trans*-2,3-dimethylcyclopropane, **VI**, 975

C₅H₉N *tert*-Butyl isocyanide or 2-(carbylamino)-2,2-dimethylpropane, **VI**, 234

N-Ethylallenimine, **V**, 541

(C₅H₉N)₃ 2,3,4,5-Tetrahydropyridine trimer, **VI**, 968

C₅H₉NO 5-Aminopentanal, **VI**, 970

Cyclobutanecarboxamide, **VIII**, 132

4,4-Dimethyl-2-oxazoline, **VI**, 65

β-Ethoxypropionitrile, **III**, 372

β-Isovalerolactam, **V**, 673

C₅H₉NOS β-(2-Hydroxyethylmercapto)-propionitrile, **III**, 458

C₅H₉NO₂ Acetamidoacetone, **V**, 27

Isoamyl nitrite, **VI**, 840

C₅H₉NO₃ Acetone carboxymethoxime, **III**, 172

N-Formylglycine ethyl ester, **VI**, 620

C₅H₉NO₄ D-Glutamic acid, **I**, 286

Methyliminodiacetic acid, **II**, 397

C₅H₉NS *N*-Methylthiopyrrolidone, **V**, 780; **VII**, 372

C₅H₉N₃ 4-Amino-2,6-dimethylpyrimidine, **III**, 71

C₅H₉N₃O₂ *tert*-Butyl azidoformate, **V**, 157; **VI**, 207

$C_5H_9LiS_2$ 1,3-Bis(methylthio)allyllithium, **VI**, 683, 687

C_5H_{10} 2-Pentene, **I**, 430

$C_5H_{10}AlN$ Diethylaluminum cyanide, **VI**, 436

$C_5H_{10}BrClO_2$ 1-Bromo-3-chloro-2,2-dimethoxypropane, **VI**, 361

$C_5H_{10}BrN$ N-(2-Bromoallyl)ethylamine, **V**, 124

$C_5H_{10}Br_2$ Pentamethylene bromide, **I**, 428; **III**, 692

Trimethylethylene dibromide, **II**, 409

$C_5H_{10}ClN$ N-Chloropiperidine, **VI**, 968

$C_5H_{10}ClNS$ Diethylthiocarbamyl chloride, **IV**, 307

$C_5H_{10}I_2$ 1,3-Diiodo-2,2-dimethylpropane, **VI**, 832

$C_5H_{10}N_2O_4$ 1,5-Dinitropentane, **IV**, 370

$C_5H_{10}O$ (S)-(+)-2-Methylbutanal, **VIII**, 367

Methyl isopropyl ketone, **II**, 408

3-Penten-2-ol, **III**, 696

4-Penten-1-ol, **III**, 698

Tetrahydropyran, **III**, 794

$C_5H_{10}O_2$ Cyclopropanone ethyl hemiacetal, **VII**, 131

5-Hydroxypentanal, **III**, 470

DL-Methylethylacetic acid, **I**, 361

Trimethylacetic acid, **I**, 524

Valeric acid, **I**, 363

$C_5H_{10}O_5$ D-Arabinose, **III**, 101

L-Arabinose, **I**, 67

$C_5H_{10}Si$ Trimethylsilylacetylene, **VIII**, 63, 606

$C_5H_{11}Br$ Isoamyl bromide, **I**, 27

$C_5H_{11}BrClNO$ Methyl 4-bromo-1-butanimidate hydrochloride, **VIII**, 415

$C_5H_{11}BrO_3$ Monobromopentaerythritol, **IV**, 681

$C_5H_{11}BrSi$ (1-Bromoethenyl)trimethylsilane, **VI**, 1033

$C_5H_{11}ClNO·Cl$ N-Chlorobetainyl chloride, **IV**, 154

$C_5H_{11}I$ Amyl iodide, **II**, 402

Isoamyl iodide, **II**, 402

Neopentyl iodide or 1-Iodo-2,2-dimethylpropane, **VI**, 830

$C_5H_{11}NO$ (S)-(+)-2-Hydroxymethylpyrrolidine, **VIII**, 26

$C_5H_{11}NO_2$ Butyl carbamate, **I**, 140

tert-Butyl carbamate, **V**, 162

N-Methylmorpholine N-oxide, **VI**, 342

DL-Valine, **III**, 848

$C_5H_{11}NO_2S$ DL-Methionine, **II**, 384

C_5H_{12} Pentane, **II**, 478

$C_5H_{12}CrN_9S_4$ Morland salt, **II**, 555

$C_5H_{12}N_2O$ tert-Butylurea, **III**, 151

$C_5H_{12}N_2O_2$ tert-Butyl carbazate, **V**, 166

$C_5H_{12}O$ Methyl isopropyl carbinol, **II**, 406

Neopentyl alcohol, **V**, 818

$C_5H_{12}OS_2$ 1,3-Bis(methylthio)-2-propanol, **VI**, 683

$C_5H_{12}OSi$ Acetyltrimethylsilane, **VIII**, 19

$C_5H_{12}O_2$ 1,5-Pentanediol, **III**, 693

$C_5H_{12}O_3$ 1,2,5-Trihydroxypentane, **III**, 833

$C_5H_{12}O_4$ Pentaerythritol, **I**, 425

$C_5H_{12}O_4S$ erythro-2,3-Butanediol monomesylate, **VI**, 312

$C_5H_{12}Si$ Vinyltrimethylsilane, **VI**, 1033

$C_5H_{12}Sn$ Trimethylvinyltin, **VIII**, 97

$C_5H_{13}Cl_2N$ 3-Chloro-N,N-dimethyl-1-propylamine, **VI**, 776

$C_5H_{13}N$ N-Methylbutylamine, **V**, 736

$C_5H_{13}NO$ 2-Isopropylaminoethanol, **III**, 501

L-Valinol, **VII**, 530

$C_5H_{13}O_4P$ Diethyl hydroxymethylphosphonate, **VII**, 160

$C_5H_{14}NO_3P$ Diethyl aminomethylphosphonate, **VIII**, 451

$C_5H_{14}N_2$ Bis(dimethylamino)methane, **VI**, 474

$C_5H_{14}N_3O·Cl$ Betaine hydrazide hydrochloride, **II**, 85

C_6BrF_5 Bromopentafluorobenzene, **VI**, 875

C₆Cl₆ Hexachlorobenzene, **VI**, 82

(C₆CuF₅)₄ Pentafluorophenylcopper tetramer, **VI**, 875

C₆F₆ Hexafluorobenzene, **VI**, 873

C₆F₉O₆Tl Thallium(III) trifluoroacetate, **VI**, 709

C₆F₁₂S₂ Tetrakis(trifluoromethyl)-1,3-dithietane, **VII**, 251

C₆H₂Br₂ClNO 2,6-Dibromoquinone-4-chloroimide, **II**, 175

C₆H₂Br₄O 2,4,4,6-Tetrabromo-2,5-cyclohexadien-1-one, **VI**, 181

C₆H₂I₃NO₂ 1,2,3-Triiodo-5-nitrobenzene, **II**, 604

C₆H₂N₃O₉S·C₃H₉O Trimethyloxonium 2,4,6-trinitrobenzenesulfonate, **V**, 1099; **VI**, 1021

C₆H₃Br₂NO₃ 2,6-Dibromo-4-nitrophenol, **II**, 173

C₆H₃Br₃ sym-Tribromobenzene, **II**, 592

C₆H₃Br₃O 2,4,6-Tribromophenol, **VI**, 181

C₆H₃ClNO₂S o-Nitro-p-chlorophenylsulfur chloride, **II**, 456

C₆H₃ClN₂ 2-Chloronicotinonitrile, **IV**, 166

C₆H₃ClN₂O₄ 1-Chloro-2,6-dinitrobenzene, **IV**, 160

C₆H₃ClN₂O₄S 2.4-Dinitrobenzenesulfenyl chloride, **V**, 474
  2,4-Dinitrophenylsulfur chloride, **II**, 456

C₆H₃ClO₂ Chloro-p-benzoquinone, **IV**, 148

C₆H₃Cl₂NO₂ 2,6-Dichloronitrobenzene, **V**, 367

C₆H₃IN₂O₄ 2,4-Dinitroiodobenzene, **V**, 478

C₆H₄N₃O₂·BF₄ o-Nitrobenzenediazonium fluoborate, **II**, 226

C₆H₄N₃O₆ 1,3,5-Trinitrobenzene, **I**, 541
  p-Nitrobenzenediazonium fluoborate, **II**, 225

C₆H₄BrCl o-Chlorobromobenzene, **III**, 185

C₆H₄BrClMg 4-Chlorophenylmagnesium bromide, **VI**, 851

C₆H₄BrF 1-Bromo-2-fluorobenzene, **V**, 133

C₆H₄BrI o-Iodobromobenzene, **V**, 1120

C₆H₄BrMg·Br 4-Bromophenylmagnesium bromide, **V**, 496

C₆H₄BrN 2-Bromopyridine, **III**, 136

C₆H₄BrNO₂ m-Bromonitrobenzene, **I**, 123

C₆H₄BrN₂·PF₆ o-Bromobenzenediazonium hexafluorophosphate, **V**, 133

C₆H₄Br₂Mg₂ 1,4-Phenylenedimagnesium dibromide or phenylene-1,4-dimagnesium dibromide, **VI**, 851

C₆H₄ClNO₂ m-Chloronitrobenzene, **I**, 162

C₆H₄ClNO₂S o-Nitrophenylsulfur chloride, **II**, 455

C₆H₄ClNO₄S o-Nitrobenzenesulfonyl chloride, **II**, 471

C₆H₄ClN₂· Cl p-Chlorobenzenediazonium chloride, **VII**, 105

C₆H₄Cl₂ 1,2-Dichlorobenzene, **VI**, 407

C₆H₄Cl₂O p-Chlorobenzoyl chloride, **III**, 29,
  2,6-Dichlorophenol, **III**, 267

C₆H₄Cl₃NO 2-Pyrrolyl trichloromethyl ketone or 2-(Trichloroacetyl)pyrrole, **VI**, 618

C₆H₄Cl₆N₂O₄ Bis(2,2,2-trichloroethyl)azodicarboxylate, **VII**, 56

C₆H₄I₂N₂O₂ 2,6-Diiodo-p-nitroaniline, **II**, 196

C₆H₄N₂ 1,2-Dicyanocyclobutene or 1-Cyclobutene-1,2-dicarbonitrile, **VI**, 427
  (Z,Z)-2,4-Hexadienedinitrile, **VI**, 662
  Nicotinonitrile, **IV**, 706

C₆H₄N₂O 2-(Cyanoformyl)pyrrole or 2-Pyrrolecarbonyl cyanide, **VI**, 270

C₆H₄N₂O₂ Benzofurazan oxide, **IV**, 74
  Dicyanoketene ethylene acetal, **IV**, 276

C₆H₄N₂O₂S 1,2,3-Benzothiadiazole 1,1-dioxide, **V**, 61

$C_6H_4N_2O_4$ o-Dinitrobenzene, **II**, 226
  p-Dinitrobenzene, **II**, 225
  2,3-Pyrazinedicarboxylic acid, **IV**,
    824
$C_6H_4N_3O_2 \cdot BF_4$ p-Nitrobenzenediazonium
  fluoborate, **III**, 666
$C_6H_4N_4O_2$ o-Nitrophenyl azide, **IV**, 75
$C_6H_4N_4S$ 2,5-Diamino-3,4-dicyanothio-
  phene, **IV**, 243
$C_6H_4O_2$ 1,4-Benzoquinone **VI**, 890 or
  p-Benzoquinone, **I**, 482; **II**, 553; **VI**,
    100
$C_6H_4O_4$ Coumalic acid, **IV**, 201
$C_6H_4O_6$ Tetrahydroxyquinone, **V**, 1011
$C_6H_5BO_2$ Catecholborane, **VI**, 293
C_6H_5Br Bromobenzene, **VI**, 452, 722
C_6H_5BrClN 4-Bromo-2-chloroaniline, **VI**,
    183
  4-Bromo-3-chloroaniline, **VI**, 183
$C_6H_5BrN_2O_2$ 4-Bromo-2-nitroaniline, **VI**,
    183
  4-Bromo-3-nitroaniline, **VI**, 183
C_6H_5BrO o-Bromophenol, **II**, 97
  p-Bromophenol, **I**, 128
$C_6H_5BrO_2$ 4-Bromoresorcinol, **II**, 100
C_6H_5BrSe Benzeneselenenyl bromide, **VI**,
    24, 535
$C_6H_5Br_2N$ 2,4-Dibromoaniline, **VI**, 183
  2,6-Dibromoaniline, **III**, 262
  3,4-Dibromoaniline, **VI**, 183
$C_6H_5Br_2NO$ 2,6-Dibromo-4-aminophenol
  chlorostannate, **II**, 175
C_6H_5ClHgO o-Chloromercuriphenol, **I**,
    161
$C_6H_5ClHgO_2$ p-Chloromercuribenzoic
  acid, **I**, 159
$C_6H_5ClN_2$ 1-Chloro-1,2-dicyanocyclobu-
  tane or 1-Chlorocyclobutane-
  1,2-dicarbonitrile, **VI**, 427
C_6H_5ClO Benzoyl chloride, **VI**, 116
C_6H_5ClOSe Benzeneseleninyl chloride,
    VI, 536
$C_6H_5ClO_2S$ Benzenesulfonyl chloride, **I**,
    84
C_6H_5ClS Phenylsulfenyl chloride, **VIII**,
    550

C_6H_5ClSe Benzeneselenenyl chloride, **VI**,
    23
$C_6H_5Cl_2I$ Iodobenzene dichloride, **III**, 482
$C_6H_5Cl_2N$ 2,6-Dichloroaniline, **III**, 262
$C_6H_5Cl_2P$ Dichlorophenylphosphine, **VI**,
    776 or Phenyldichlorophos-
    phine, **IV**, 784
C_6H_5F Fluorobenzene, **II**, 295
$C_6H_5FN \cdot CF_3O_3S$ N-Fluoropyridinium tri-
  flate, **VIII**, 286
C_6H_5FOS Benzylsulfinyl fluoride, **V**,
    396
$C_6H_5F_3S$ Phenylsulfur trifluoride, **V**, 959
C_6H_5I Iodobenzene, **I**, 323; **II**, 351; **VI**,
    451
C_6H_5IO o-Iodophenol, **I**, 326
  p-Iodophenol, **II**, 355
  Iodosobenzene, **III**, 483; **V**, 658
$C_6H_5IO_2$ Iodoxybenzene, **III**, 485; **V**,
    665
C_6H_5NO Nitrosobenzene, **III**, 668
  3-Pyridinecarboxaldehyde, **VI**, 866
C_6H_5NOS N-Sulfinylaniline, **V**, 504
$C_6H_5NO_2$ Nicotinic acid, **I**, 385
  Picolinic acid hydrochloride, **III**,
    740
$C_6H_5NO_3$ 6-Hydroxynicotinic acid, **IV**,
    532
  m-Nitrophenol, **I**, 404
$C_6H_5NO_3S$ 4-Pyridinesulfonic acid, **V**,
    977
$C_6H_5NO_4S$ 2-Nitrobenzenesulfinic acid,
    V, 60
$C_6H_5N_3$ 1,2,3-Benzotriazole, **III**, 106
  Phenyl azide, **III**, 710
$C_6H_5N_3O_4$ 2,4-Dinitroaniline, **II**, 221
  2,6-Dinitroaniline, **IV**, 364
C_6H_5SLi Lithium thiophenoxide, **VI**,
    248
C_6H_5SNa Sodium benzenethiolate, **VI**,
    833
C_6H_5SeNa Sodium selenophenoxide or
  Sodium benzeneselenolate, **VI**,
    535
C_6H_6 Benzene, **V**, 998; **VI**, 150
  Bicyclo[2.2.0]hexa-2,5-diene, **VI**, 462

1,5-Hexadiyne, **VI**, 68
1,2,4,5-Hexatetraene, **VII**, 485
$C_6H_6AsNO_3$ p-Nitrophenylarsonic acid,
 III, 665
C_6H_6BrN 4-Bromoaniline, **VI**, 183
$C_6H_6Br_2N_2O_2S$ 3,5-Dibromosulfanilamide,
 III, 263
$C_6H_6Cl_2N_2O_2S$ 3,5-Dichlorosulfanilamide,
 III, 262
$C_6H_6Cl_6N_2O_4$ Bis(2,2,2-trichloroethyl)
 hydrazodicarboxylate, **VII**, 56
C_6H_6IN p-Iodoaniline, **II**, 347
$C_6H_6N_2$ 1,2-Dicyanocyclobutane or
 1,2-Cyclobutanedicarbonitrile, **VI**,
 427
$C_6H_6N_2O_2$ Nicotinamide 1-oxide, **IV**,
 704
 o-Nitroaniline, **I**, 388
 N-Nitroso-N-phenylhydroxylamine,
 I, 177
$C_6H_6N_2O_3$ 2-Amino-4-nitrophenol, **III**, 82
 3-Methyl-4-nitropyridine 1-oxide,
 IV, 654
$C_6H_6N_4O_4$ 2,4-Dinitrophenylhydrazine, **II**,
 228
C_6H_6OS 2-Acetothienone, **III**, 14; **II**, 8
$C_6H_6O_2$ Catechol, **I**, 149; **VI**, 395,
 Hydroquinone, **VI**, 427, 454, 620,
 947
 5-Methylfurfural, **II**, 393
$C_6H_6O_3$ 2-Methylcyclopentane-1,3,5-tri-
 one hydrate, **V**, 748
 3-Methyl-2-furoic acid, **IV**, 628
 Phloroglucinol, **I**, 455
$C_6H_6O_4$ Dimethyl acetylenedicarboxylate,
 IV, 329; **VI**, 196, 422
 Muconic acid, **III**, 623
$C_6H_6O_6$ Aconitic acid, **II**, 12
 Hexahydroxybenzene, **V**, 595
C_6H_6S Thiophenol or benzenethiol, **I**,
 504; **VI**, 248, 737, 833, 980
 2-Vinylthiophene, **IV**, 980
$C_6H_6S_2$ 1,2-Dimercaptobenzene, **V**, 419
C_6H_6Se Selenophenol, **III**, 771 or
 Benzeneselenol, **VI**, 535
$C_6H_7AsO_3$ Phenylarsonic acid, **II**, 494

$C_6H_7AsO_4 \cdot Na$ Sodium p-hydroxyphenyl-
 arsonate, **I**, 490
$C_6H_7BrN_2$ 3-Amino-4-bromoaniline, **VI**,
 183
 2-Amino-4-bromo-6-methylpyridine
 VI, 183,
$C_6H_7ClO_2$ Methyl 2-chloro-2-cyclopropyl-
 idenacetate, **VIII**, 373, 374
$C_6H_7Cl_3O_2$ 1,1,1-Trichloro-4-ethoxy-3-
 buten-1-one, **VIII**, 254
C_6H_7N Aniline, **VI**, 152, 1004
 2,4-Hexadienenitrile, **VI**, 306
C_6H_7NO Isopropylcyanoketene, **VI**, 215
 3-Methylpyridine 1-oxide, **IV**, 655
 1-Methyl-2-pyridone, **II**, 419
 Phenylhydroxylamine, **I**, 445; **VIII**,
 16
$C_6H_7NO_2S$ Ethyl thiazole-4-carboxylate,
 VI, 620
$C_6H_7NO_2SNa$ Sodium 2-aminobenzene-
 sulfinate, **V**, 61
$C_6H_7NO_3S$ Orthanilic acid, **II**, 471
$C_6H_7N_3O_2$ 1,2-Diamino-4-nitrobenzene,
 III, 242
C_6H_7P Phenylphosphine, **VI**, 932
C_6H_8 Bicyclo[2.2.0]hex-2-ene, **VI**, 462
 1,3-Cyclohexadiene, **V**, 285
 1,4-Cyclohexadiene, **VI**, 87
 1,3,5-Hexatriene, **V**, 608
$C_6H_8AsNO_3$ Arsanilic acid, **I**, 70
$C_6H_8Br_2$ 1,2-Dibromo-2-cyclohexene, **VI**,
 188
$C_6H_8Cl_2O$ 2-Oxa-7,7-dichloronorcarane,
 V, 874
$C_6H_8Cl_2O_2$ Adipyl chloride, **II**, 169; **III**,
 623
$C_6H_8F_3NO_3$ N-Trifluoroacetyl-L-proline,
 VI, 1006
$C_6H_8N_2$ 1,2-Benzenediamine, **VI**, 662
 Phenylhydrazine, **I**, 442
 o-Phenylenediamine, **II**, 501
 o-Phenylenediamine dihydrochloride, **II**, 502
$C_6H_8N_2O$ 2-Diazocyclohexanone, **VI**,
 389
$C_6H_8N_2O_2$ Crotyl diazoacetate, **V**. 258

$C_6H_8N_2O_3$ Ethyl 2-diazoacetoacetate, **VI**, 416

$C_6H_8N_4O_2$ 2,4,5-Triaminonitrobenzene, **V**, 1067

C_6H_8O 2-Cyclohexen-1-one, **V**, 294; **VI**, 331, 679

    5-Hexynal, **VI**, 679

    2-Vinylcyclopropanecarboxalde-hyde, **VI**, 647

$C_6H_8O_2$ 1,2-Cyclohexanedione, **IV**, 229

    1,4-Cyclohexanedione, **V**, 288

    Cyclopropenone 1,3-propanediol ketal, **VIII**, 173

    Dihydroresorcinol, **III**, 278

    2,3-Epoxycyclohexanone, **VI**, 679

    Ethyl tetrolate, **VII**, 226

    2-Hydroxymethyl-2-cyclopentenone, **VII**, 271

    2-Methyl-1,3-cyclopentanedione, **V**, 747; **VI**, 30

    Methyl 2-pentynoate, **VI**, 793,

    Sorbic acid, **III**, 783

$C_6H_8O_3$ Ketopantoyl lactone, **VII**, 417

    β-Methylglutaric anhydride, **IV**, 630

$C_6H_8O_4$ 1,1-Cyclobutanedicarboxylic acid, **III**, 213; **VI**, 271

    Dimethyl fumarate, **VI**, 579

    Dimethyl maleate, **VI**, 579

$C_6H_8O_6$ Tricarballylic acid, **I**, 523

$C_6H_8O_7$ Citric acid, **VI**, 419

C_6H_9Br 3-Bromo-2-methyl-1,3-pentadi-ene, **VI**, 88

C_6H_9BrO 2-Bromo-2-cyclohexen-1-ol, **VI**, 188

$C_6H_9BrO_2$ Ethyl α-(bromomethyl)acry-late, **VII**, 210

C_6H_9Cl 6-Chloro-2-hexyne, **VI**, 273

C_6H_9ClO 2-Chlorocyclohexanone, **III**, 188

$C_6H_9ClO_2$ 1-Acetoxy-3-chloro-1-butene, **VIII**, 9

    1-Acetoxy-4-chloro-2-butene, **VIII**, 9

$C_6H_9ClO_3$ Ethyl α-chloroacetoacetate, **IV**, 592

C_6H_9DO Cyclopentanecarboxaldehyde-1-*d*, **VI**, 755

$C_6H_9F_3O_3S$ 3-Methyl-1-buten-2-yl triflate, **VI**, 757

    3-Methyl-2-buten-2-yl triflate, **VI**, 757

C_6H_9N 2,4-Dimethylpyrrole, **II**, 217

    2,5-Dimethylpyrrole, **II**, 219

$C_6H_9NO_2$ 2-Ethoxypyrrolin-5-one, **VI**, 226

    Ethyl 3-cyanopropionate, **VI**, 305

$C_6H_9NO_6$ Dimethyl nitrosuccinate, **VI**, 503

$C_6H_9NS_2$ 4,5-Dihydro-2-(2-propenylthio)-thiazole, **VI**, 704

$C_6H_9N_3$ 4-Amino-2,6-dimethylpyrimidine, **III**, 71

    bis(β-Cyanoethyl)amine, **III**, 93

    2-(Dimethylamino)pyrimidine, **IV**, 336

$C_6H_9N_3O_2$ 4-*tert*-Butyl-1,2,4-triazoline-3,5-dione, **VI**, 939

    L-Histidine monohydrochloride, **II**, 330

C_6H_{10} Butylacetylene, **IV**, 117

    Biallyl, **III**, 121

    Cyclohexene, **I**, 183; **II**, 152; **VI**, 338

    1,3-Dimethyllbicyclo[1.1.0]butane, **VI**, 159

    2,3-Dimethyl-1,3-butadiene, **III**, 312

    1-Hexyne, **VI**, 595

$C_6H_{10}BrClO$ 2-Bromohexanoyl chloride, **VI**, 190

$C_6H_{10}BrClO_2$ 2-(Bromomethyl)-2-(chloro-methyl)-1,3-dioxane, **VIII**, 173

$C_6H_{10}Br_2$ 1,2-Dibromocyclohexane, **II**, 171

$C_6H_{10}Br_2O$ 1,3-Dibromo-2-hexanone, **VI**, 713

    1,3-Dibromo-4-methyl-2-pentanone, **VI**, 713

$C_6H_{10}ClN$ N-Chlorocyclohexylidenimine, **V**, 208

$C_6H_{10}Cl_2$ *cis*-1,2-Dichlorocyclohexane, **VI**, 424

$C_6H_{10}Cl_3N$ N,N-Diethyl-1,2,2-trichloro-
vinylamine, **V**, 387

$C_6H_{10}F_3NO_3$ N-Trifluoroacetyl-L-valine,
VI, 1006

$C_6H_{10}N_2$ 3,3-Pentamethylenediazirine, **V**,
897

$C_6H_{10}N_2O$ 3-(1-Propyl)-2-pyrazolin-5-one,
VI, 791

$C_6H_{10}N_2O_2$ *tert*-Butyl diazoacetate, **V**, 179;
VI, 416

  1,2-Cyclohexanedione dioxime, **IV**,
229

$C_6H_{10}N_2O_4$ 3,4-Dinitro-3-hexene, **IV**, 372
  Ethyl azodicarboxylate, **III**, 375; **IV**,
411

$C_6H_{10}O$ *exo*-Bicyclo[2.2.0]hexan-2-ol, **VI**,
462

  Cyclohexene oxide, **I**, 185
  Cyclopentanecarboxaldehyde, **V**,
320

  Cyclohexanone, **VI**, 327, 334, 358,
499, 598, 1014

  1,2-Epoxycyclohexane, **VI**, 424
  2-Ethoxy-1,3-butadiene, **VI**, 447
  1-Ethoxy-1-butyne **VI**, 564,
  1,5-Hexadien-3-ol, **V**, 608
  Mesityl oxide, **I**, 345

$C_6H_{10}O_2$ Acetonylacetone, **II**, 219
  Ethyl isocrotonate, **VII**, 226
  cis-2-Hexenoic acid, **VI**, 713,
  trans-4-Hydroxy-2-hexenal, **VI**,
683

  trans-4-Hydroxy-4-methyl-2-penten-
al, **VI**, 688

  4-Methoxy-3-penten-2-one, **VIII**,
357

  β-Methylglutaraldehyde, **IV**, 661
  3-Methyl-2,4-pentanedione, **V**, 785;
VI, 27, 247

  cis-4-Methyl-2-pentenoic acid, **VI**,
713

  β-Methyl-δ-valerolactone, **IV**, 677

$C_6H_{10}O_3$ Allyl lactate, **III**, 46
  Butyl glyoxylate, **IV**, 124
  2,5-Dihydro-2,5-dimethoxyfuran, **V**,
403

  Ethyl acetoacetate, **I**, 235; **VI**, 592
  Ethyl 4-hydroxycrotonate, **VII**,
221

  Ethyl (α-hydroxymethyl)acrylate,
VIII, 265

  Mevalonolactone-2-^{13}C, **VII**, 386
  D-(−)-Pantoyl lactone, **VII**, 417

$C_6H_{10}O_4$ Adipic acid, **I**, 18
  Conduritol-D, **VI**, 198
  Ethyl oxalate, **I**, 261
  β-Methylglutaric acid, **III**, 591

$C_6H_{10}O_6$ D-Gulonic-γ-lactone, **IV**, 506

$C_6H_{10}S$ Cyclohexene sulfide, **IV**, 232

$C_6H_{11}Br$ Cyclohexyl bromide, **II**, 247

$C_6H_{11}BrO_2$ α-Bromocaproic acid, **I**, 115;
II, 95

  2-(2-Bromoethyl)-1,3-dioxane, **VII**,
59

  α-Bromoisocaproic acid, **II**, 95;
III,523

  α-Bromo-β-methylvaleric acid, **II**,
95

  Ethyl γ-bromobutyrate, **V**, 545
  Methyl 5-bromovalerate, **III**,
578

$C_6H_{11}ClO$ *tert*-Butylacetyl chloride, or
3,3-Dimethylbutanoyl chloride,
VI, 991

  2-Chlorocyclohexanol, **I**, 158
  3-Chloro-2-methyltetrahydropyran,
VI, 675

$C_6H_{11}ClO_2$ *tert*-Butyl chloroacetate, **III**,
144

  2-Chloro-2-methylpentanoic acid,
VI, 93,

  2-Chloro-3-methylpentanoic acid,
VI, 93

  2-Chloro-4-methylpentanoic acid,
VI, 93

$C_6H_{11}F$ Cyclohexyl fluoride or fluorocy-
clohexane, **VI**, 630

$C_6H_{11}FO_2$ 2-Fluorohexanoic acid, **VIII**,
258

$C_6H_{11}I$ Cyclohexyl iodide or Iodocyclohexane, **IV**, 543; **VI**, 830

  (Z)-1-Iodo-1-hexene, **VII**, 290

$C_6H_{11}IO_2Zn$ (3-Carboethoxypropyl)zinc iodide, **VIII**, 274

$C_6H_{11}N$ Diallylamine, **I**, 201

$C_6H_{11}NO$ Cyclohexanone oxime, **II**, 76,314

1,5-Dimethyl-2-pyrrolidone, **III**,328

5,5-Dimethyl-2-pyrrolidone, **IV**,357

2-Ketohexamethylenimine, **II**, 371

$C_6H_{11}NO_2$ 3-Acetamido-2-butanone, **IV**,5

N-Carbomethoxypyrrolidine, **VII**, 307

(E)-(Carboxyvinyl)trimethylammonium betaine **VIII**, 536,

(S)-(−)-1-Formyl-2-hydroxymethyl-pyrrolidine, **VIII**, 26

$C_6H_{11}NO_3$ DL-, (+)- and (−)-α-(Isopropylideneaminooxy)propionic acid, **V**, 1032

$C_6H_{11}NO_4$ Ethyl α-nitrobutyrate, **IV**,454

4-Methyl-4-nitrovaleric acid, **V**, 445

(1R,2r,3S)-2-Nitrocyclohexane-1,3-diol, **VIII**, 332

$C_6H_{11}NS$ 2-Methylmercapto-N-methyl-Δ²-pyrroline, **V**, 780

$C_6H_{11}N_3O_2$ tert-Butyl azidoacetate, **V**, 586

4-tert-Butylurazole, **VI**, 938

$C_6H_{11}O$ 2-Ethylcyclobutanone, **VI**, 326

$C_6H_{11}OP$ 1-Methyl-4-phosphorinanone, **VI**, 934

C_6H_{12} 2,3-Dimethyl-2-butene, **VI**, 974

1-Hexene, **VI**, 919

4-Methyl-2-pentene, **VI**, 295

$C_6H_{12}BrN$ 1-Bromo-N,N,2-trimethyl-propenylamine, **VI**, 287

$C_6H_{12}ClN$ 1-Chloro-N,N,2-trimethyl-propenylamine, **VI**, 282

$C_6H_{12}Cl_2O_2$ 1,2-Bis(2-chloroethoxy)-ethane, **VI**, 301

$C_6H_{12}F_2$ 2,2-Difluorohexane, **VI**, 630

3,3-Difluorohexane, **VI**, 630

$C_6H_{12}NO·I$ N,4,4-Trimethyl-2-oxazolinium iodide, **VI**, 64

$C_6H_{12}I_2$ 1,6-Diiodohexane, **IV**, 323

$C_6H_{12}NO_3P$ Diethyl cyanomethylphosphonate, **VI**, 450

$C_6H_{12}N_2$ Acetone azine, **VI**, 10

1,4-Diazabicyclo[2.2.2]octane, **VI**, 419

Diethylaminoacetonitrile, **III**, 275

$C_6H_{12}N_2O_3·HCl$ S-Acetamidomethyl-L-cysteine hydrochloride, **VI**, 5

$C_6H_{12}N_2O_4$ Diethyl hydrazodicarboxylate, **III**, 375

$C_6H_{12}N_2O_4S_2$ L-Cystine, **I**, 194

$C_6H_{12}O$ Cyclohexanol, **VI**, 353, 830

Hexaldehyde, **II**, 323 or Hexanal, **VI**, 315, 871

(E)-4-Hexen-1-ol, **VI**, 586, 675

Methyl butyl ketone, **I**, 352

Methyl isobutyl ketone, **I**, 352

(S)-(−)-3-Methylpentanal, **VI**, 871

Pinacolone, **I**, 462

$C_6H_{12}OSi$ (1-Oxo-2-propenyl)trimethylsilane, **VIII**, 501

3-Trimethylsilyl-2-propyn-1-ol, **VII**, 524

$C_6H_{12}O_2$ tert-Butyl acetate, **III**, 141; **IV**, 263

Caproic acid, **II**, 417, 475

cis-1,2-Cyclohexanediol, **VI**, 342, 348

trans-1,2-Cyclohexanediol, **III**, 217; **VI**, 348

1,3-Cyclohexanediol, **VI**, 372

1,4-Cyclohexanediol, **VI**, 372

Diacetone alcohol, **I**, 199

Hexanoic acid, **VI**, 190

2-Hydroxy-2-methylpentan-3-one, **VII**, 381

Ketene diethylacetal, **III**, 506

3-Methylpentanoic acid, **II**, 416

(2S,3S)-3-Propyloxirane methanol, **VII**, 461

$C_6H_{12}O_2S$ tert-Butyl S-methylthiolcarbonate, **V**, 168

$C_6H_{12}O_3$ Ethyl ethoxyacetate, **II**, 261

Ethyl (S)-3-hydroxybutanoate, **VII**, 215

DL-Isopropylideneglycerol, **III**, 502

Isopropyl lactate, **II**, 365

$C_6H_{12}O_6$ D-Mannose, **III**, 541

$C_6H_{12}S_2$ Allyl isopropyl disulfide, **VI**, 238
1,3-Bis(methylthio)-1-butene, **VI**, 687

$C_6H_{13}BrO$ 2-Bromo-3,3-dimethyl-1-butanol, **VI**, 186

$C_6H_{13}BrO_2$ Bromoacetal, **III**, 123 or Bromoacetaldehyde diethyl acetal, **VI**, 448

$C_6H_{13}ClO$ Hexamethylene chlorohydrin, **III**, 446

$C_6H_{13}ClO_2$ Chloroacetaldehyde diethyl acetal, **VI**, 564

$C_6H_{13}ClN\cdot Cl$ 1-Chloro-N,N,2-trimethylpropylideniminium chloride, **VI**, 282

$C_6H_{13}F$ Hexyl fluoride, **IV**, 525

$C_6H_{13}I$ Hexyl iodide or 1-Iodohexane, **VI**, 832

$C_6H_{13}N$ 1,2-Benzo-3,4-dihydrocarbazole, **IV**, 885
Cyclohexylamine, **II**, 319; **VI**, 448, 500, 901
2,2-Dimethylpyrrolidine, **IV**, 354
N-Methylpiperidine, **VI**, 500

$C_6H_{13}NO$ Acetonamine hydrogen oxalate, **I**, 196
N,N-Dimethylisobutyramide, **VI**, 282
(S)-(+)-2-Methoxymethylpyrrolidine, **VIII**, 26

$C_6H_{13}NO_2$ α-Aminocaproic acid, **I**, 48
ε-Aminocaproic acid, **II**, 28; **IV**, 39
α-Aminodiethylacetic acid, **III**, 66
N-tert-Butylaminoacetic acid, **VI**, 545
(S)-N-Formyl-L-valinol, **VIII**, 204
Glycine tert-butyl ester, **V**, 586
DL-Isoleucine, **III**, 495
DL-Leuclne, **III**, 523

$C_6H_{13}NO_4$ Nitroacetaldehyde diethyl acetal, **VI**, 978

$C_6H_{13}NO_5\cdot HCl$ D-Glucosamine hydrochloride, **III**, 430

$C_6H_{13}N_2O_3\cdot Cl$ Glycylglycine ethyl ester hydrochloride, **VI**, 263

$C_6H_{13}N_2O_3S\cdot Cl$ S-Acetamidomethyl-L-cysteine hydrochloride, **VI**, 5

$C_6H_{13}O_4P$ Diethyl formylmethylphosphonate, **VI**, 448

$C_6H_{14}BrN$ β-Diethylaminoethyl bromide hydrobromide, **II**, 92

$C_6H_{14}ClP$ Chlorodiisopropylphosphine, **V**, 211

$C_6H_{14}N_2$ Azopropane or Dipropyldiazene, **VI**, 80

$C_6H_{14}N_2O$ (S)-(−)-1-Amino-2-methoxymethylpyrrolidine (SAMP), **VIII**, 26, 403

$C_6H_{14}N_2O_2$ DL-Lysine dihydrochloride, **II**, 375; **VI**, 94
DL-Lysine monohydrochloride, **II**, 375

$C_6H_{14}N_3Cl$ [3-(Dimethylamino)-2-azaprop-2-en-1-ylidene]dimethylammonium chloride, **VII**, 197

$C_6H_{14}N_4O_2$ Adipyl hydrazide, **IV**, 819
D-Arginine hydrochloride, **II**, 49

$C_6H_{14}O$ Hexyl alcohol, **I**, 306 or 1-Hexanol, **VI**, 788, 919

$C_6H_{14}OS_2$ 1,3-Bis(methylthio)-2-methoxypropane, **VI**, 683

$C_6H_{14}OSi$ Acetone trimethylsilyl enol ether, **VIII**, 1
(1-Hydroxy-2-propenyl)trimethylsilane, **VIII**, 501
1-(Methoxyvinyl)trimethylsilane, **VIII**, 19
(E)-3-Trimethylsilyl-2-propen-1-ol, **VII**, 524

$C_6H_{14}O_2$ Acetal, **I**, 1
Hexamethylene glycol, **II**, 325
3-Methyl-1,5-pentanediol, **IV**, 660
Pinacol hydrate, **I**, 459
Triethylene glycol, **VI**, 301

$C_6H_{14}O_2S$ Diethyl mercaptoacetal, **IV**, 295

$C_6H_{14}O_4S$ Propyl sulfate, **II**, 112

$C_6H_{14}S$ Propyl sulfide, **II**, 547

$C_6H_{14}S_2$ Ethyl sec-butyl disulfide, **VI**, 238
Ethyl tert-butyl disulfide, **VI**, 238
Ethyl isobutyl disulfide **VI**, 238

$C_6H_{14}Zn$ Dipropylzinc, **II**, 187

$C_6H_{15}Al$ Triethylaluminum, **VI**, 436

$C_6H_{15}ClOSi$ (Isopropoxydimethylsilyl) methyl chloride, **VIII**, 315

$C_6H_{15}O\cdot BF_4$ Triethyloxonium tetrafluoroborate **V**, 1080; **VI**, 576, 1019

$C_6H_{15}ISi$ 1-Iodo-3-trimethylsilylpropane, **VIII**, 486

$C_6H_{15}N$ Diisopropylamine, **VI**, 121, 517, 542, 648, 684, 818, 901

$C_6H_{15}NO$ β-Diethylaminoethyl alcohol, **II**, 183

$C_6H_{15}NO_2$ Aminoacetal, **III**, 50
N,N-Dimethylacetamide dimethyl acetal, **VI**, 491

$C_6H_{15}O_3P$ Triethyl phosphite, **IV**, 955; **VI**, 448

$C_6H_{16}N_2$ N,N,N'N'-Tetramethyl-1,2-ethanediamine, **VI**, 786 or N,N,N'N'-Tetramethylethylenediamine, **VI**, 478

$C_6H_{18}N_3OP$ Hexamethylphosphoric triamide, **VI**, 64, 106, 242, 376, 638, 704

$C_6H_{18}N_3P$ Hexamethylphosphorous triamide, **V**, 602; **VI**, 130

$C_6H_{18}N_3\cdot C_3H_9F_2Si$ Tris(dimethylamino) sulfonium difluorotrimethylsilicate, **VII**, 528

$C_6H_{18}OSi_2$ Hexamethyldisiloxane, **VI**, 353

$C_6H_{19}NSiLi$ Lithium bis(trimethylsilyl) amide, **VI**, 598

$C_6H_{19}NSi_2$ Hexamethyldisilazane, **VI**, 598

C_6N_4 Tetracyanoethylene, **IV**, 877

C_6N_4O Tetracyanoethylene oxide, **V**, 1007; **VI**, 268

$C_7HCl_5O_2$ Pentachlorobenzoic acid, **V**, 890

$C_7H_3Br_3O_2$ 2,4,6-Tribromobenzoic acid, **IV**, 947

$C_7H_3ClN_2O_5$ 3,5-Dinitrobenzoyl chloride, **VI**, 529

$C_7H_3F_5$ 2,3,4,5,6-Pentafluorotoluene, **VI**, 880

$C_7H_3HgNO_4$ Anhydro-2-hydroxymercuri-3-nitrobenzoic acid, **I**, 56

$C_7H_3N_3O_8$ 2,4,6-Trinitrobenzoic acid, **I**, 543

$C_7H_4BrNO_4$ 2-Bromo-3-nitrobenzoic acid, **I**, 125

$C_7H_4Br_3NO_2$ 3-Amino-2,4,6-tribromobenzoic acid, **IV**, 947

C_7H_4BrNS m-Bromophenyl isothiocyanate, **I**, 448
p-Bromophenyl isothiocyanate, **I**, 448

C_7H_4ClNO 2-Chlorobenzoxazole, **VI**, 151

$C_7H_4ClNO_2S$ 4-Chlorobenzenesulfonyl cyanide, **VI**, 729

$C_7H_4ClNO_3$ 3-Nitrobenzoyl chloride, **IV**, 715
p-Nitrobenzoyl chloride, **I**, 394

C_7H_4ClNS p-Chlorophenyl isothiocyanate, **I**, 165; **V**, 223

$C_7H_4Cl_2O$ o-Chlorobenzoyl chloride, **I**, 155 or 2-Chlorobenzoyl chloride, **VI**, 625
m-Chlorobenzoyl chloride, **VI**, 276

$C_7H_4F_3ClO_2S$ m-Trifluoromethylbenzenesulfonyl chloride, **VII**, 508

$C_7H_4FeO_3$ Cyclobutadieneiron tricarbonyl, **VI**, 310, 1002

$C_7H_4HgO_2$ Anhydro-o-hydroxymercuribenzoic acid, **I**, 57

$C_7H_4INO_2$ 2-Iodo-3-nitrobenzoic acid, **I**, 126

C_7H_4INS p-Iodophenyl isothiocyanate, **I**, 448

$C_7H_4I_2O_3$ 2-Hydroxy-3,5-diiodobenzoic acid, **II**, 343
4-Hydroxy-3,5-diiodobenzoic acid, **II**, 344

$C_7H_4N_2O_2$ Benzenediazonium-2-carboxylate, **V**, 54
p-Nitrobenzonitrile **III**, 646

$C_7H_4N_2O_3$ p-Nitrophenyl isocyanate, **II**, 453

$C_7H_4N_2O_5$ 2,4-Dinitrobenzaldehyde, **II**, 223
3,5-Dinitrobenzaldehyde, **VI**, 529

$C_7H_4N_2O_4S$ 4-Nitrobenzenesulfonyl cyanide, **VI**, 728

$C_7H_4N_2O_6$ 2,5-Dinitrobenzoic acid, **III**, 334

3,5-Dinitrobenzoic acid, **III**, 337

$C_7H_4N_4O_3$ m-Nitrobenzazide, **IV**, 715

$C_7H_4O_3$ o-Phenylene carbonate, **IV**, 788

$C_7H_4O_4S$ o-Sulfobenzoic anhydride, **I**, 495

$C_7H_4O_6$ Chelidonic acid, **II**, 126

$C_7H_5BrF_3N$ 4-Bromo-2-(trifluoromethyl) aniline, **VI**, 183

C_7H_5BrO m-Bromobenzaldehyde, **II**, 132

p-Bromobenzaldehyde, **II**, 89 442

$C_7H_5BrO_2$ 2-Bromobenzoic acid, **VI**, 36

$C_7H_5BrO_4$ 2,4-Dihydroxy-5-bromobenzoic acid, **II**, 100

$C_7H_5ClHgO_2$ p-Chloromercuribenzoic acid, **I**, 159

$C_7H_5ClN_2$ o-Chlorophenylcyanamide, **IV**, 172

3-Chloro-3-phenyldiazirine, **VII**, 203

3-Chloroindazole, **III**, 476

$C_7H_5ClN_4$ 1-Phenyl-5-chlorotetrazole, **VI**, 150

C_7H_5ClO Benzoyl chloride, **VI**, 248, 386

2-Chlorobenzaldehyde, **VI**, 633

m-Chlorobenzaldehyde, **II**, 130

p-Chlorobenzaldehyde, **II**, 133 or 4-Chlorobenzaldehyde, **VI**, 633

$C_7H_5ClO_2$ o-Chlorobenzoic acid, **II**, 135

4-Chlorobenzoic acid, **VI**, 851

$C_7H_5ClO_3$ m-Chloroperbenzoic acid or 3-Chloroperbenzoic acid, **VI**, 39, 276, 404

$C_7H_5Cl_3Hg$ Phenyl(trichloromethyl)mercury, **V**, 969

$C_7H_5Cl_5$ 1,2,3,4,5-Pentachloro-5-ethylcyclopentadiene, **V**, 893

C_7H_5DO Benzaldehyde-formyl-d, **VI**, 755

$C_7H_5DO_2$ Benzoic acid-O-d, **VI**, 434

C_7H_5FO Benzoyl fluoride, **V**, 66

$C_7H_5FO_2$ p-Fluorobenzoic acid, **II**, 299

$C_7H_5FeO_2 \cdot Na$ Sodium dicarbonyl(cyclopentadienyl)ferrate, **VIII**, 479

$C_7H_5IO_2$ p-Iodobenzoic acid, **I**, 325

m-Iodobenzoic acid, **II**, 353

C_7H_5N Phenyl isocyanide or (Carbylamino)benzene, **VI**, 936

C_7H_5NO 3-(2-Furyl)acrylonitrile, **V**, 585

Phenyl cyanate, **VII**, 435

Phenyl isocyanate or (Carbonylamino)benzene **VI**, 936

$C_7H_5NO_3$ o-Nitrobenzaldehyde, **III**, 641; **V**, 825

p-Nitrobenzaldehyde, **II**, 441 or 4-Nitrobenzaldehyde, **VI**, 646

$C_7H_5NO_4$ m-Nitrobenzoic acid, **I**, 391

p-Nitrobenzoic acid, **I**, 392

$C_7H_5NO_5$ 4-Nitroperbenzoic acid, **VI**, 320

C_7H_5NS Phenyl isothiocyanate, **I**, 447

$C_7H_5N_3O$ Indazolone, **III**, 476

$C_7H_5N_3O_2$ 5-Nitroindazole, **III**, 660

$C_7H_5O_2Ag$ Silver benzoate, **VI**, 613

C_7H_6 Benzocyclopropene, **VI**, 87, 736

$C_7H_6BrNO_2$ 2-Amino-5-bromobenzoic acid, **VI**, 183

p-Nitrobenzyl bromide, **II**, 443

$C_7H_6BrN_2 \cdot Cl$ 2-Bromo-4-methylbenzenediazonium chloride, **V**, 140

$C_7H_6ClNO_3$ 2-Hydroxy-5-nitrobenzyl chloride, **III**, 468

$C_7H_6Cl_2O$ p-Chlorophenoxymethyl chloride, **V**, 221

7,7-Dichlorobicyclo[3.2.0]hept-2-en-6-one, **VI**, 1037

2,4-Dichloromethoxybenzene or 2,4-Dichloroanisole, **VIII**, 167

$C_7H_6Cl_4O$ 5,5-Dimethoxy-1,2,3,4-tetrachlorocyclopentadiene, **V**, 424

$C_7H_6FNO_2$ p-Nitrobenzyl fluoride, **VI**, 835

$C_7H_6F_2$ α,α-Difluorotoluene, **V**, 396

$C_7H_6INO_2$ 5-Iodoanthranilic acid, **II**, 349

$C_7H_6N_2$ Benzimidazole, **II**, 65

2-Cyano-6-methylpyridine, **V**, 269

Indazole, **III**, 475; **IV**, 536; **V**, 650

Phenyldiazomethane, **VII**, 438

$C_7H_6NO \cdot K$ Potassium benzohydroxamate, **II**, 67

$C_7H_6N_2O$ 3-Cyano-6-methyl-2(1H)-pyridone, **IV**, 210

2-Hydroxyimino-2-phenylacetoni-
trile, **VI**, 199

Indazolone, **III**, 476

$C_7H_6N_2O_2S$ 4-Cyanobenzenesulfonamide,
VI, 631

$C_7H_6N_2O_3$ 2-Nitroso-5-nitrotoluene, **III**,
334

$C_7H_6N_2O_5$ 3,5-Dinitroanisole, **I**, 219

$C_7H_6N_2S$ 2-Mercaptobenzimidazole, **IV**,
569

$C_7H_6N_4O$ 1,1-Carbonyldiimidazole, **V**,
201

C_7H_6O Benzaldehyde, **VI**, 66, 222, 315,
633, 650, 897

C_7H_6OS Thiobenzoic acid, **IV**, 924

$C_7H_6O_2$ Benzoic acid, **I**, 363; **VI**, 432,
851

m-Hydroxybenzaldehyde, **III**, 453
564

Tropolone, **VI**, 1037

$C_7H_6O_2S$ Thiosalicylic acid, **II**, 580

$C_7H_6O_3$ Furylacrylic acid, **III**, 425

p-Hydroxybenzoic acid, **II**, 341

Perbenzoic acid, **I**, 431 or
Peroxybenzoic acid, **V**, 904

Protocatechualdehyde, **II**, 549

$C_7H_6O_4$ 3,5-Dihydroxybenzoic acid, **III**,
288

Methyl coumalate, **IV**, 532

Protocatechuic acid, **III**, 745

β-Resorcylic acid, **II**, 557

$C_7H_6O_5S·H_3N$ Acid ammonium o-sul-
fobenzoate, **I**, 14

$C_7H_7·BF_4$ Tropylium fluoborate, **V**,
1138

C_7H_7Br Benzyl bromide, **VI**, 104, 121,
637, 704

m-Bromotoluene, **I**, 133

o-Bromotoluene, **I**, 135

p-Bromotoluene, **I**, 136; **VI**, 749 or
4-Bromotoluene, **VI**, 488

C_7H_7BrMg (4-Methylphenyl)magnesium
bromide, **VI**, 488

C_7H_7BrMgO (2-Methoxyphenyl)magne-
sium bromide, **VI**, 64

$C_7H_7BrN_2O$ p-Bromophenylurea, **IV**, 49

C_7H_7BrO 3-Bromo-4-hydroxytoluene, **III**,
130

C_7H_7Cl Benzyl chloride, **VI**, 106, 115,
404, 526

o-Chlorotoluene, **I**, 170

p-Chlorotoluene, **I**, 170; **VI**, 749

C_7H_7ClHg p-Tolymercuric chloride, **I**,
519

$C_7H_7ClN_2S$ o-Chlorophenylthiourea, **IV**,
180

C_7H_7ClOS p-Toluenesulfinyl chloride, **IV**,
937

$C_7H_7ClO_2S$ p-Toluenesulfonyl chloride,
VI, 586, 638, 652, 951, 1016

C_7H_7ClS p-Toluenesulfenyl chloride, **IV**,
934

C_7H_7ClZn o-Tolylzinc chloride, **VIII**,
430

C_7H_7IO p-Iodobenzyl alcohol, **III**, 652

C_7H_7NO o-Aminobenzaldehyde, **III**, 56

p-Aminobenzaldehyde, **IV**, 31

C_7H_7NOS N-Methyl-3-formyl-2(1H)-
pyridinethione **VI**, 643

$C_7H_7NO_2$ Benzohydroxamic acid, **II**, 67

m-Nitrotoluene, **I**, 415

Phenylnitromethane, **II**, 512

Potassium benzohydroxamate, **II**, 67

$C_7H_7NO_2S$ (Phenylthio)nitromethane,
VIII, 550

$C_7H_7NO_3$ p-Nitrobenzyl alcohol, **III**, 652;
VI, 835

$C_7H_7NO_3S$ 4-Formylbenzenesulfonamide,
VI, 631

$C_7H_7NO_4$ 3-Methoxy-5-nitrophenol, **I**, 405

$C_7H_7N_3O_2S$ p-Toluenesulfonyl azide, **VI**,
389, 414

$C_7H_7O_2S_2K$ Potassium thiotosylate **VI**,
1016

C_7H_8 Norbornadiene or 2,5-Norborna-
diene, **VI**, 378, 962

Quadricyclane, **VI**, 962

Toluene, **VI**, 749

C_7H_8BrN 3-Bromo-4-aminotoluene, **I**, 111

4-Bromo-N-methylaniline, **VI**, 183

4-Bromo-2-methylaniline, **VI**, 183

4-Bromo-3-methylaniline, **VI**, 183

C₇H₈BrNO 4-Bromo-2-methoxyaniline, **VI**, 183

4-Bromo-3-methoxyaniline, **VI**, 183

6-Bromo-3-methoxyaniline, **VI**, 183

C₇H₈Br₂ *exo*-5-*anti*-7-Dibromonorbornene, **VI**, 964

2,6-Dibromonortricyclene, **VI**, 964

C₇H₈ClN₃O₃ 4-Azido-3-chloro-5-isopropoxy-2(5*H*)-furanone, **VIII**, 16

C₇H₈Cl₂ 7,7-Dichlorobicyclo[4.1.0]hept-3-ene, **VI**, 87

C₇H₈Cl₂O₃ 3,4-Dichloro-5-isopropoxy-2(5*H*)-furanone, **VIII**, 116

C₇H₈N₂ Benzamidine hydrochloride, **I**, 6

C₇H₈N₂O 4-Acetylpyridine oxime, **VII**, 149

N-Nitrosomethylaniline, **II**, 460

3-(2-Oxopyrrolidinyl)propenenitrile, **VI**, 466

Phenylurea, **I**, 453

C₇H₈N₂O₂ 2-Amino-3-nitrotoluene, **IV**, 42

2-Amino-5-nitrotoluene, **IV**, 44

o-Hydrazinobenzoic acid hydrochloride, **III**, 475

C₇H₈N₂O₃ 2-Nitro-4-methoxyaniline, **III**, 661

C₇H₈N₂S α-Phenylthiourea, **III**, 735

C₇H₈N₂S₂·C₆H₈N₂ β-Phenyldithiocarbazic acid, phenylhydrazine salt, **III**, 360

C₇H₈O Anisole, **I**, 58; **VI**, 152, 468, 996

Benzyl alcohol, **II**, 591; **VI**, 95, 101

Bicyclo[2.2.1]hepten-7-one, **V**, 91

p-Cresol, **I**, 175

Nortricyclanone, **V**, 866

C₇H₈O₂ Bicyclo[2.2.1]heptane-2,5-dione, **VI**, 962

1,4-Dihydrobenzoic acid, **V**, 400

4,6-Dimethylcoumalin, **IV**, 337

Guaiacol, **VI**, 861

4-Methoxyphenol, **VI**, 861

Orcinol, **VI**, 861

Tropolone, **VI**, 1037

C₇H₈O₃ 4-Acetoxy-2-cyclopenten-1-one, **VIII**, 15

2-Acetyl-1,3-cyclopentanedione, **VI**, 28

2-Furylmethyl acetate, **I**, 285

α-Furylpropionic acid, **I**, 313

Methyl 3-methyl-2-furoate, **IV**, 649

Methyl 5-oxo-1-cyclopentene-1-carboxylate, **VI**, 26

Pyrogallol 1-monomethyl ether, **III**, 759; **VI**, 861

C₇H₈OS Methyl phenyl sulfoxide, **V**, 791

C₇H₈O₂S Methyl benzenesulfinate, **V**, 723

Sodium *p*-toluenesulfinate, **I**, 492

C₇H₈O₃S *p*-Toluenesulfonic acid, **VI**, 245, 308

C₁₀H₈O₃S·C₃H₃N₃ Aminomalononitrile *p*-toluenesulfonate, **V**, 32

C₇H₈O₄ Dimethyl 2,3-pentadienedioate, **VI**, 505

cis,cis-Monomethyl muconate, **VIII**, 490

C₇H₈S *m*-Thiocresol, **III**, 809

C₇H₉BrN₂ 4-Bromo-2-(dimethylamino)pyridine, **VI**, 183

4-Bromo-3-(dimethylamino)pyridine, **VI**, 183

C₇H₉BrO₂ 2-Bromo-2-cyclopentenone ethylene ketal, **VII**, 271, 495

C₇H₉ClO 2-Chloro-1-formyl-1-cyclohexene, **V**, 215

C₇H₉ClO₂ γ-Chloropropyl acetate, **III**, 203

C₇H₉ClO₃ α-Acetyl-δ-chloro-γ-valerolactone, **IV**, 10

6-Chloromethyl-2,2-dimethyl-1,3-dioxen-4-one, **VIII**, 192

C₇H₉ClO₄ Dimethyl 3-chloro-2-pentenedioate, **VI**, 505

C₇H₉N 2,6-Dimethylpyridine, **II**, 214 or 2,6-Lutidine, **VI**, 611

4-Ethylpyridine, **III**, 410

C₇H₉NO *o*-Aminobenzyl alcohol, **III**, 60

tert-Butylcyanoketene, **VI**, 210

C₇H₉NO₂ Ethyl pyrrole-2-carboxylate, **VI**, 618

C₇H₉NO₂S Ethyl 5-methylthiazole-4-carboxylate, **VI**, 623

$C_7H_9NO_3S$ *o*-Toluidinesulfonic acid, **III**, 824

$C_7H_9N_3O$ 4-Phenylsemicarbazide, **I**, 450

$C_7H_9N_3O_2S$ 2-Mercapto-4-amino-5-carbethoxypyrimidine, **IV**, 566

$C_7H_9N_5O_2$ 7-Aminotheophylline, **VII**, 8

C_7H_{10} Norbornylene, **IV**, 738 or Norbornene, **VI**, 142

$C_7H_{10}ClNO$ 4-Chloromethyl-3-ethyl-5-methylisoxazole, **VI**, 781

$C_7H_{10}ClNO_2$ 6-Isocyanatohexanoyl chloride or 6-(Carbonylamino)hexanoyl chloride **VI**, 718,

$C_7H_{10}Cl_2$ 2,2-Dichloronorbornane, **VI**, 845

$C_7H_{10}NO\cdot CH_3O_4S$ 1-Methoxy-2-methylpyridinium methyl sulfate, **V**, 270

$C_7H_{10}N_2$ Diallylcyanamide, **I**, 203
2,4-Diaminotoluene, **II**, 160
2,4-Diaminotoluene sulfate, **II**, 160
tert-Butylmalononitrile, **VI**, 223
α-Methyl-α-phenylhydrazine, **II**, 418

$C_7H_{10}N_2O$ 2-Diazocycloheptanone, **VI**, 391

$C_7H_{10}N_2O_2S$ *p*-Toluenesulfonylhydrazide, **V**, 1055; **VI**, 293 or Tosylhydrazide, **VI**, 62

$C_7H_{10}O$ 2-Cyclohepten-1-one, **VI**, 327
3-Cyclohexenecarboxaldehyde, **VI**, 315
Dicyclopropyl ketone, **IV**, 278
4,4-Dimethyl-2-cyclopenten-1-one, **VIII**, 208
1-Methoxy-1,4-cyclohexadiene, **VI**, 996
2-Methyl-2-cyclohexen-1-one, **IV**, 162; **VI**, 666
2-Norbornanone, **V**, 852 or Norcamphor, **VI**, 845
Nortricyclanol, **V**, 863
(*E*)-2-(1-Propenyl)cyclobutanone, **VIII**, 556

$C_7H_{10}O_2$ 2-Acetylcyclopentanone, **VI**, 247
1,3-Cycloheptenedione, **VI**, 332

2-Ethyl-1,3-cyclopentanedione, **VI**, 775

Ethyl 2,3-pentadienoate, **VII**, 232

3-Hydroxy-1-cyclohexene-1-carboxaldehyde, **VIII**, 309

2-(Hydroxymethylene)cyclohexanone, **VI**, 389

2-Methyl-1,3-cyclohexanedione, **V**, 743

Methyl 2-hexynoate, **VI**, 791

endo-5-Norbornene-2,3-diol, **VI**, 346

exo-5-Norbornene-2,3-diol, **VI**, 346

$C_7H_{10}O_3$ *cis*-3-Acetoxy-5-hydroxy-1-cyclopentene **VIII**, 13

$C_7H_{10}O_4$ Ethyl acetopyruvate, **I**, 238
2,3-*O*-Isopropylidene-D-erythronolactone, **VII**, 297

$C_7H_{10}O_5$ Ethyl oxomalonate, **I**, 266
Shikimic acid, **VI**, 198,

$C_7H_{10}O_6$ Tricarbomethoxymethane, **II**, 596

$C_7H_{10}S_2$ 5,9-Dithiaspiro[3.5]nonane, **VI**, 316

$C_7H_{11}Br$ 3-Bromo-2,4-dimethyl-1,3-pentadiene, **VI**, 188

$C_7H_{11}BrO_2$ 2-Bromo-3-methyl-2-buten-1-yl acetate, **VI**, 188
3-Bromo-2-methyl-3-buten-2-yl acetate **VI**, 188
3-Bromo-3-penten-2-yl acetate, **VI**, 188

$C_7H_{11}BrO_4$ Ethyl bromomalonate, **I**, 245

$C_7H_{11}Cl$ 1-Chloronorbornane, **VI**, 845

$C_7H_{11}ClO$ 2-(2-Chloroethoxy)-1-ethenylcyclopropane, **VI**, 574
2-Chloro-2-methylcyclohexanone, **IV**, 162
Cyclohexanecarboxylic acid chloride, **VII**, 81

$C_7H_{11}ClO_2$ 1-Chlorocyclohexanecarboxylic acid, **VI**, 93
3-Chloropropiolaldehyde diethyl acetal, **VI**, 956

$C_7H_{11}ClO_3$ δ-Carbomethoxyvaleryl chloride, **IV**, 556

$C_7H_{11}Cl_2N$ *N*-(1,2-Dichloro-1-propen-1-yl)
  pyrrolidine, **VI**, 286
$C_7H_{11}F$ 2-Fluoronorbornane, **VI**, 630
$C_7H_{11}F_3O_3S$ 2,2-Dimethyl-1-methylene-
  propyl triflate, **VI**, 759
  1-Methylenepentyl triflate, **VI**, 759
  1-Ethyl-1-butenyl triflate, **VI**, 759
$C_7H_{11}N$ Cyclohexyl isocyanide, **V**, 300;
  VI, 234
  Cyclohexanecarbonitrile, **VI**, 305,
  334
$C_7H_{11}NO$ (*E*)-4,6-Heptadienamide, **VI**,
  494
  trans-2-Isocyanocyclohexanol, **VII**,
  294
  3-Isoquinuclidone, **V**, 670
  Hydrochloride, **V**, 989
$C_7H_{11}NO_2$ *tert*-Butyl cyanoacetate, **V**, 171
  6-Ethoxy-4,5-dihydro-2(3*H*)-pyri-
  done, **VI**, 231
  3-Ethyl-4-hydroxymethyl-5-methyl-
  isoxazole or 3-Ethyl-5-methyl-
  4-isoxazolylmethanol, **VI**, 781
$C_7H_{11}NO_2S$ Cyclohexanesulfonyl cyanide,
  VI, 728
$C_7H_{11}NO_5$ Diethyl oximinomalonate, **V**,
  373
$C_7H_{11}O_3S$ *p*-Toluenesulfonic acid, **VI**,
  242
C_7H_{12} 1-Methylcyclohexene, **VI**, 409, 766
  3-Methylcyclohexene, **VI**, 121
  Methylenecyclohexane, **IV**, 612; **V**,
  751
  Norcarane, **V**, 855
$C_7H_{12}Br_2O$ 1,3-Dibromo-4,4-dimethyl-2-
  pentanone, **VI**, 713
  1,3-Dibromo-2-heptanone, **VI**, 713
$C_7H_{12}Cl_2O$ 1,7-Dichloro-4-heptanone, **IV**,
  279
$C_7H_{12}F_3NO_3$ *N*-Trifluoroacetyl-L-leucine,
  VI, 1006
$C_7H_{12}NO_3$ *trans*-2,2,5-Trimethyl-1,3-diox-
  olane-4-carboxamide, **VI**, 494
$C_7H_{12}O$ Cycloheptanone, **IV**, 221
  Cyclohexanecarboxaldehyde, **VI**,
  222, 315, 338

2-Isopropylcyclobutanone or 2-(1-
  Methylethyl)cyclobutanone, **VI**,
  326
1-Methoxy-2-phenylcyclopropane or
  (2-Methoxy-1-cyclopropyl)ben-
  zene, **VI**, 573
2-Methylcyclohexanone, **VI**, 121
3-Methyl-2-cyclohexen-1-ol, **VI**,
  769
Methylenecyclohexane oxide, **V**,
  755
5-Methyl-4-hexenal, **VI**, 813
5-Methyl-5-hexen-2-one, **V**, 767
exo/endo-7-Norcaranol, **V**, 859
$C_7H_{12}O_2$ Butyl acrylate, **III**, 146
  Cyclohexanecarboxylic acid, **I**, 364
  3,4-Dihydro-2-methoxy-4-methyl-
  2*H*-pyran, **IV**, 311
  2,2-Dimethyl-4-oxopentanal, **VIII**,
  208
  3,3-Dimethyl-2,4-pentanedione, **VI**,
  247
  cis-4,4-Dimethyl-2-pentenoic acid,
  VI, 713
  3-Ethyl-2,4-pentanedione, **VI**, 247
  2,5-Heptanedione, **VI**, 648
  cis-2-Heptenoic acid, **VI**, 713
  Methyl cyclopentanecarboxylate, **IV**,
  594
  3-Methyl-2,4-hexanedione, **VI**, 247
  Propiolaldehyde diethyl acetal, **VI**,
  954
  γ-Propylbutyrolactone, **III**, 742
$C_7H_{12}O_3$ δ-Acetylvaleric acid, **IV**, 19
  3-Hydroxycyclohexanecarboxylic
  acid, **VI**, 372
  Methyl 3-hydroxy-2-methylenepen-
  tanoate, **VIII**, 420
  Methyl 6-oxohexanoate **VII**, 168
  β-(Tetrahydrofuryl)propionic acid,
  III, 742
$C_7H_{12}O_4$ Diethyl malonate, **VI**, 442
  β,β-Dimethylglutaric acid, **IV**, 345
  α-Ethyl-α-methylsuccinic acid, **V**,
  572
  Pimelic acid, **II**, 531

$C_7H_{12}O_5$ Hexahydrogallic acid, **V**, 591

$C_7H_{12}S_2$ 5,9-Dithiaspiro[3.5]nonane, **VI**, 316

$C_7H_{13}Br$ 4-Bromo-2-heptene, **IV**, 108

$C_7H_{13}Al$ (*E*)-2-(Methyl-1,3-butadienyl) dimethylalane, **VII**, 245

$C_7H_{13}BrO$ α-Allyl-β-bromoethyl ethyl ether, **IV**, 750

α-Bromoheptaldehyde, **III**, 127

$C_7H_{13}BrO_2$ *sec*-Butyl α-bromopropionate, **IV**, 122

Methyl 6-bromohexanoate, **VI**, 807

$C_7H_{13}ClO_2$ 7-Chloroheptanoic acid, **VI**, 813

$C_7H_{13}DO$ Heptanal-1-*d*, **VI**, 755

$C_7H_{13}F$ Cycloheptyl fluoride or Fluorocycloheptane, **VI**, 630

$C_7H_{13}FO_2$ 2-Fluoroheptanoic acid, **V**, 580

$C_7H_{13}F_3$ 1,1,1-Trifluoroheptane, **V**, 1082

$C_7H_{13}I$ 2,2,3,3-Tetramethyliodocyclopropane or 3-Iodo-1,1,2,2-tetramethylcyclopropane, **VI**, 974

$C_7H_{13}NO$ *N*-Cyclohexylformamide, **V**, 301

Heptamide, **IV**, 513

Hexahydro-2(1*H*)-azocinone, **VII**, 254

O-Methylcaprolactim, **IV**, 588

cis-*N*-Methyl-3-oxa-2-azabicyclo[3.3.0]octane or *cis*-*N*-Methylhexahydro-1H-cyclopent[*c*]isoxazole, **VI**, 671

$C_7H_{13}NO_2$ 4-Aminocyclohexanecarboxylic acid, *cis–trans* mixture, **V**, 670

(*S*)-(−)-1-Formyl-2-methoxymethylpyrrolidine, **VIII**, 26

Methyl *N*-(2-methyl-2-butenyl)carbamate, **VIII**, 68

$C_7H_{13}NO_3$ 2-Methoxy-*N*-carbomethoxypyrrolidine, **VII**, 307

5-Nitroheptan-2-one, **VI**, 648

1-(Nitromethyl)cyclohexanol, **IV**, 221, 224

$C_7H_{13}NO_4$ Diethyl aminomalonate and hydrochloride, **V**, 376

$C_7H_{13}NO_3S$ Cyclohexanone oxime methanesulfonate, **VIII**, 568

$C_7H_{13}NO_4$ Methyl γ-methyl-γ-nitrovalerate, **IV**, 652

$C_7H_{13}OP$ 1-Ethyl-4-phosphorinanone, **VI**, 934

$C_7H_{13}O_2$ 4-Methoxycyclohexanol, **VI**, 372

$C_7H_{14}BrF$ 1-Bromo-2-fluoroheptane, **V**, 136

$C_7H_{14}Br_2O_2$ 2,3-Dibromopropionaldehyde diethyl acetal, **VI**, 954

$C_7H_{14}ClNO_4$ Diethyl aminomalonate hydrochloride, **V**, 376

$C_7H_{14}N_2O$ Cyclohexylurea, **V**, 801

$C_7H_{14}N_2O_2$ (*S*)-(−)-1-Carbamoyl-2-methoxymethylpyrrolidine, **VIII**, 26

N-Nitroso-β-methylaminoisobutyl methyl ketone, **III**, 244

$C_7H_{14}O$ Cyclohexylcarbinol, **I**, 188

Cyclohexyl methyl ether, **VI**, 353

1,1-Dimethyl-2-ethoxycyclopropane, **VI**, 573

Heptanal, **VI**, 644, 650

4-Heptanone, **V**, 589

Methyl amyl ketone, **I**, 351

1-Methylcyclohexanol, **VI**, 46, 766

2-Methylhexanal, **VI**, 528

4-Methylhexanone-2, **I**, 352

$C_7H_{14}OSi$ 3-Trimethylsilyl-3-buten-2-one, **VI**, 666, 1033

2-Trimethylsilyloxy-1,3-butadiene, **VI**, 445

4-(Trimethylsilyl)-3-butyn-1-ol, **VIII**, 609

$C_7H_{14}O_2$ Acrolein acetal, **II**, 17; **IV**, 21

tert-Butyl propionate, **III**, 143

Heptanoic acid, **II**, 315; **VI**, 813

2-Heptanone, **VI**, 245

1-(Hydroxymethyl)cyclohexanol, **VIII**, 315

$C_7H_{14}O_3$ 2,4-Dimethyl-3-hydroxypentanoic acid, (2*SR*,3*RS*), **VII**, 185, and (2*SR*,3*SR*), **VII**, 190

Methyl *anti*-3-hydroxy-2-methylpentanoate, **VIII**, 420

C$_7$H$_{14}$O$_3$SSi 3-Trimethylsilyl-2-propynyl
  methanesulfonate, **VIII**, 471
C$_7$H$_{14}$O$_4$ 2,3-Di-*O*-isopropylidene-L-thre-
  itol, **VIII**, 155
C$_7$H$_{14}$O$_6$ α-Methyl D-glucoside, **I**, 364
  α-Methyl mannoside, **I**, 371
C$_7$H$_{14}$S$_2$ Allyl *sec*-butyl disulfide, **VI**,
  238
  Allyl *tert*-butyl disulfide, **VI**, 238
  Allyl isobutyl disulfide, **VI**, 238
C$_7$H$_{14}$Si 1-Methyl-1-(trimethylsilyl)allene,
  VIII, 471
C$_7$H$_{15}$Br Heptyl bromide, **II**, 247
C$_7$H$_{15}$BrO *erythro*-3-Bromo-4,4-dimethyl-
  2-pentanol, **VI**, 186
  threo-3-Bromo-4,4-dimethyl-2-pen-
  tanol, **VI**, 186
  1-Bromo-2,3,3-trimethyl-2-butanol,
  VI, 186,
C$_7$H$_{15}$BrO$_3$ Trimethyl *ortho*-4-bromo-
  butanoate, **VIII**, 415
C$_7$H$_{15}$ClO$_2$ β-Chloropropionaldehyde
  acetal, **II**, 137
C$_7$H$_{15}$F 3-Ethyl-3-fluoropentane, **VI**, 630
C$_7$H$_{15}$N 1-Amino-1-methylcyclohexane,
  V, 35
  N-Ethylpiperidine, **V**, 575
  N-Methylcyclohexylamine, **VI**, 500
C$_7$H$_{15}$NO 1-(Aminomethyl)cyclohexanol,
  IV, 224
  Heptaldoxime, **II**, 313
  Heptamide, **IV**, 513
  2-(1-Pyrrolidyl)propanol, **IV**, 834
C$_7$H$_{15}$OLi Lithium triethylcarboxide or
  Lithium 3-ethyl-3-pentoxide,
  VI, 137
C$_7$H$_{16}$INO$_2$ L-Leucine methyl ester hydro-
  chloride, **VI**, 263
C$_7$H$_{16}$O 3-Ethyl-3-pentanol, **VI**, 137
  Heptyl alcohol, **I**, 304 or 1-
  Heptanol, **VI**, 644
  2-Heptanol, **II**, 317
  2-Methyl-1-hexanol, **VI**, 786
  Triethyl carbinol, **II**, 602
C$_7$H$_{16}$OSi 2-(Hydroxymethyl)allyltri-
  methylsilane, **VII**, 266

3-Trimethylsilyl-3-buten-2-ol, **VI**,
  1033
(*Z*)-4-(Trimethylsilyl)-3-buten-1-ol,
  VIII, 358, 609
C$_7$H$_{16}$O$_2$ Heptamethylene glycol, **II**, 155
C$_7$H$_{16}$O$_2$S$_2$ 3,7-Dithianonane-1,9-diol,
  VIII, 592
C$_7$H$_{16}$O$_2$Si Ethyl trimethylsilylacetate,
  VII, 512
C$_7$H$_{16}$O$_3$ Ethyl orthoformate, **I**, 258 or
  Triethyl orthoformate, **VI**, 620,
  954
C$_7$H$_{16}$O$_4$ DL-Glyceraldehyde diethyl
  acetal, **II**, 307
C$_7$H$_{16}$S$_2$ *sec*-Butyl isopropyl disulfide, **VI**,
  235
  tert-Butyl isopropyl disulfide, **VI**,
  238
  Isobutyl isopropyl disulfide, **VI**, 238
C$_7$H$_{16}$S$_4$ 3,7-Dithianonane-1,9-dithiol,
  VIII, 592
C$_7$H$_{17}$N Heptylamine, **II**, 318
C$_7$H$_{17}$O$_3$P Diisopropyl methylphospho-
  nate, **IV**, 325
C$_7$H$_{18}$N$_2$ 2-*tert*-Butyl-1,3-diaminopropane,
  VI, 223,
C$_7$H$_{19}$NSi *N*,*N*-Diethylaminotrimethyl-
  silane, **VI**, 440,
C$_8$Cl$_2$N$_2$O$_2$ 2,3-Dichloro-5,6-dicyano-1,4-
  benzoquinone or
  4,5-Dichloro-3,6-dioxocyclohexane-
  1,2-dicarbonitrile, **VI**, 731
C$_8$O$_8$Co$_2$ Dicobalt octacarbonyl, **VI**, 339
C$_8$F$_8$ Perfluorostyrene, **VI**, 881
C$_8$H$_2$F$_5$N (Pentafluorophenyl)acetonitrile,
  VI, 873
C$_8$H$_3$F$_4$N 4,5,6,7-Tetrafluoroindole, **VI**,
  874
C$_8$H$_3$F$_5$OP 2,3,4,5,6-Pentafluoroacetophe-
  none, **VI**, 881
C$_8$H$_4$Cl$_2$O$_2$ *sym*- and *unsym*-*o*-Phthalyl
  chloride, **II**, 528
C$_8$H$_4$F$_3$NS *m*-(Trifluoromethyl)phenyl
  isothiocyanate, **I**, 448
C$_8$H$_4$NO$_2$K Potassium phthalimide, **VI**,
  951

$C_8H_3NO_5$ 3-Nitrophthalic anhydride, **I**, 410

$C_8H_4N_2O_2S$ 4-Cyanobenzenesulfonyl cyanide or 4-Cyanophenylsulfonyl cyanide, **VI**, 728

$C_8H_4N_2O_4$ 4-Nitrophthalimide, **II**, 459

$C_8H_4N_4O_4$ 4-(4-Nitrophenyl)-1,2,4-triazoline-3,5-dione, **VI**, 938

C_8H_5Br Phenylbromoethyne, **V**, 921

$C_8H_5BrO_2$ 3-Bromophthalide, **V**, 145

C_8H_5BrMg Phenylethynylmagnesium bromide, **VI**, 925

$C_8H_5Br_3O$ p,α,α-Tribromoacetophenone, **IV**, 110

C_8H_5Cu Copper(I) phenylacetylide, **VI**, 916

$C_8H_5F_3O_2Se$ Benzeneselenenyl trifluoroacetate, **VI**, 535

C_8H_5NO Benzoyl cyanide, **III**, 112

$C_8H_5NO_2$ Isatin, **I**, 327
Phthalimide, **I**, 457; **VI**, 59

$C_8H_5NO_3$ Isatoic anhydride, **III**, 488

$C_8H_5NO_6$ 3-Nitrophthalic acid, **I**, 408
4-Nitrophthalic acid, **II**, 457

$C_8H_5N_2O_2Na$ Phenylnitroacetonitrile sodium salt, **II**, 512

$C_8H_5N_3O_2$ 4-Phenyl-1,2,4-triazoline-3,5-dione, **VI**, 936,

$C_8H_5N_3O_4$ 3,5-Dinitro-o-tolunitrile, **V**, 480
5-Nitro-2,3-dihydro-1,4-phthalazinedione, **III**, 656

C_8H_6 Phenylacetylene, **I**, 438; **IV**, 763; **VI**, 916, 926

C_8H_6BrN α-Bromo-α-phenylacetonitrile (in solution), **III**, 347

$C_8H_6Br_2O$ p-Bromophenacyl bromide, **I**, 127

$C_8H_6Br_4$ $\alpha,\alpha,\alpha',\alpha'$-Tetrabromo-$o$-xylene, **IV**, 807
$\alpha,\alpha,\alpha',\alpha'$-Tetrabromo-$p$-xylene, **III**, 788

$C_8H_6ClNO_2$ ω-Chloroisonitrosoacetophenone, **III**, 191

$C_8H_6Cl_2N_2$ 1,2-Dicyano-4,5-$trans$-dichlorocyclohexene or 4,5-$trans$-Dichloro-1,2-cyclohexenedicarbonitrile, **VI**, 431

$C_8H_6Cl_2O$ Dichloroacetophenone, **III**, 538

$C_8H_6Cl_3NO$ α,α,α-Trichloroacetanilide, **V**, 1074

$C_8H_6CrO_6$ [(Methyl)(methoxy)carbene] pentacarbonyl chromium(0), **VIII**, 216

$C_8H_6F_2$ β,β-Difluorostyrene, **V**, 390

$C_8H_6F_3NO$ N-Trifluoroacetanilide, **VI**, 1004

$C_8H_6F_6O_6S_2Cu_2$ Bis[copper(I) trifluoromethanesulfonate]–benzene complex, **VI**, 737

$C_8H_6N_2$ Quinoxaline, **IV**, 824

$C_8H_6N_2O$ Diazoacetophenone, **VI**, 386, 919

$C_8H_6N_2O_2$ N-Aminophthalimide **VI**, 56
Benzoylene urea, **II**, 79
p-Nitrobenzyl cyanide, **I**, 396
4-Nitroindole, **VIII**, 493
3-Phenylsydnone, **V**, 962

$C_8H_6N_2O_3$ 2-Nitro-6-methoxybenzonitrile, **III**, 293

$C_8H_6N_2O_6$ Methyl 3,5-dinitrobenzoate, **V**, 798

$C_8H_6N_4O_4$ 4-(4-Nitrophenyl)urazole, **VI**, 938

$C_8H_6N_4O_8$ Alloxantin dihydrate, **III**, 42; **IV**, 25

C_8H_6O Coumarone, **V**, 251

$C_8H_6O_2$ Isophthalaldehyde, **V**, 668
Phenylglyoxal, **II**, 509; **V**, 937
o-Phthalaldehyde, **IV**, 807
Phthalide, **II**, 526
Terephthalaldehyde, **III**, 788

$C_8H_6O_3$ Benzoylformic acid, **I**, 244; **III**, 114
Bicyclo[2.2.0]hex-5-ene-2,3-dicarboxylic anhydride, **VI**, 462
cis-1,2-Dihydrophthalic anhydride, **VI**, 462
Phthalaldehydic acid, **II**, 513; **III**, 737

$C_8H_6O_4$ Phthalic acid, **VI**, 461
Piperonylic acid, **II**, 538
Terephthalic acid, **III**, 791

$C_8H_6O_5$ 2-Hydroxyisophthalic acid, **V**, 617

Monoperphthalic acid, **III**, 619; **V**, 805; **VI**, 482

$C_8H_6S_3$ 3,3'-Dithienyl sulfide, **VI**, 559

2,2'-Dithienyl sulfide, **VI**, 558

2,3'-Dithienyl sulfide, **VI**, 559

C_8H_7BrO 3-Bromoacetophenone, **V**, 117

p-Bromoacetophenone, **I**, 109

2-Bromo-4-methylbenzaldehyde, **V**, 139

Phenacyl bromide, **II**, 480

$C_8H_7BrO_2$ 2-Bromo-3-methylbenzoic acid, **IV**, 114

α-Bromophenylacetic acid, **VI**, 403

$C_8H_7BrO_3$ p-Bromomandelic acid, **IV**, 110

C_8H_7Cl m-Chlorostyrene, **III**, 204

4-Chlorostyrene, **VI**, 410

$C_8H_7ClF_5N$ 2-(Pentafluorophenyl)ethyl-amine hydrochloride, **VI**, 874

C_8H_7ClO p-Chloroacetophenone, **I**, 111

p-Toluoyl chloride, **VI**, 259

$C_8H_7ClO_2$ Carbobenzoxy chloride, **III**, 167

2-(2-Chloroethyl)-1,4-benzoquinone, **VI**, 891

α-Chlorophenylacetic acid, **IV**, 169

$C_8H_7ClO_2S$ β-Styrenesulfonyl chloride, **IV**, 846

$C_8H_7Cl_2OP$ Styrylphosphonic dichloride, **V**, 1005

$C_8H_7Cl_3O$ 1-Phenyl-2,2,2-trichloroethanol, **V**, 130

$C_8H_7DO_2$ 2-Methoxybenzaldehyde-*formyl-d*, **VI**, 66

C_8H_7N Benzyl cyanide, **I**, 107; **VI**, 199 or Phenylacetonitrile, **VI**, 305, 897

Benzyl isocyanide or α-(Carbyla-mino)toluene, **VI**, 234; **VII**, 27

Indole, **III**. 479; **VI**, 104, 106, 109

o-Tolunitrile, **I**, 514

p-Tolunitrile, **I**, 514

o-Tolyl isocyanide, **V**, 1060

$C_8H_7NO_2$ β-Nitrostyrene, **I**, 413

m-Nitrostyrene, **IV**, 731

$C_8H_7NO_2S$ Benzylsulfonyl cyanide, **VI**, 728

4-Methylphenylsulfonyl cyanide, **VI**, 728

$C_8H_7NO_3$ 3-Methoxy-4-nitrobenzalde-hyde, **VI**, 1009

o-Nitroacetophenone, **IV**, 708

m-Nitroacetophenone, **II**, 434

$C_8H_7NO_3S$ 4-Methoxybenzenesulfonyl cyanide or 4-Methoxyphenyl-sulfonyl cyanide, **VI**, 728

$C_8H_7NO_4$ Methyl m-nitrobenzoate, **I**, 372

p-Nitrophenylacetic acid, **I**, 406

$C_8H_7N_3O$ 1-Cyano-3-phenylurea, **IV**, 213

$C_8H_7N_3O_2$ 5-Amino-2,3-dihydro-1,4-phthalazinedione, **III**, 69

4-Phenylurazole, **VI**, 936

C_8H_8 Cubane, **VI**, 1003

Cyclooctatetraene, **VI**, 196, 422

Phenylethylene, **I**, 440 or Styrene, **VI**, 913

Tetracyclo[4.2.0.0^{2,4}.0^{3,5}]oct-7-ene, **VI**, 574

$C_8H_8Br_2$ o-Xylylene dibromide, **IV**, 984

$C_8H_8ClNO_3S$ p-Acetaminobenzenesul-fonyl chloride, **I**, 8

N-Hydroxy-1-(p-tolylsulfonyl) methanimidoyl chloride, **VI**, 985

C_8H_8ClNO 4-Chloroacetophenone oxime, **VI**, 278

$C_8H_8I_2$ 1,2-Diiodo-4,5-dimethylbenzene, **VI**, 701

$C_8H_8N_2$ 1,2-Dicyanocyclohexene or Cyclohexene-1,2-dicarbonitrile, **VI**, 431

2-Methylbenzimidazole, **II**, 66

$C_8H_8N_2O_2$ Isonitrosoacetanilide, **I**, 327

$C_8H_8N_2O_2S$ p-Tolylsulfonyldiazomethane, **VI**, 981

$C_8H_8N_2O_3$ N-Nitroso-N-phenylglycine, **V**, 962

$C_8H_8N_2S$ 2-Amino-6-methylbenzothia-zole, **III**, 76

C_8H_8O Acetophenone, **I**, 111; **VI**, 12, 520, 818, 962

2-Methylbenzaldehyde, **VI**, 633 or o-Tolualdehyde, **III**, 818; **IV**, 932

Phenylacetaldehyde, **VI**, 66, 871

Styrene oxide, **I**, 494

p-Tolualdehyde, **II**, 583

C_8H_8OS Phenyl vinyl sulfoxide **VII**, 453,

$C_8H_8O_2$ *o*-Anisaldehyde, **V**, 46; **VI**, 64 or
2-Methoxybenzaldehyde, **VI**,
633, 646

   syn-Bishomoquinone, **VI**, 973

   4,5-Dimethyl-*o*-benzoquinone, **VI**,
   480

   2-Furfuralacetone, **I**, 283

   α-Hydroxyacetophenone, **VII**, 263

   o-Hydroxyacetophenone, **III**, 389

   m-Methoxybenzaldehyde, **III**, 564

   4-Methoxybenzaldehyde, **VI**, 633

   Methyl benzoate, **VI**, 434, 579,

   Phenylacetic acid, **I**, 436; **VI**, 403

   o-Toluic acid, **II**, 588; **III**, 820

   p-Toluic acid, **II**, 589; **III**, 822

$C_8H_8O_2S$ Phenyl vinyl sulfone, **VII**, 453

$C_8H_8O_2Se$ Benzeneselenenyl acetate, **VI**,
535

$C_8H_8O_3$ 2,5-Dihydroxyacetophenone, **III**,
280

   2,6-Dihydroxyacetophenone, **III**, 281

   Mandelic acid, **I**, 336; **III**, 538

   Phenoxyacetic acid, **VI**, 890

   Piperonyl alcohols, **II**, 591

   Resacetophenone, **III**, 761

   cis-Δ⁴-Tetrahydrophthalic anhydride,
   IV, 890

$C_8H_8O_3S \cdot Na$ Sodium β-styrenesulfonate,
IV, 846

$C_8H_8O_4$ Dehydroacetic acid, **III**, 231

   trans-1,2-Dihydrophthalic acid, **VI**,
   461

   2,6-Dimethoxy-*p*-benzoquinone, **VI**,
   413

   Gallacetophenone, **II**, 304

   Isodehydroacetic acid, **IV**, 549

   Phloroacetophenone, **II**, 522

   Vanillic acid, **IV**, 972

$C_8H_8O_7$ Diacetyl-D-tartaric anhydride, **IV**,
242

$C_8H_9AsO_6$ *p*-Arsonophenoxyacetic acid, **I**,
75

C_8H_9Br (2-Bromoethyl)benzene, **VI**, 517

   4-Bromo-*o*-xylene, **III**, 138

C_8H_9BrO 2-Bromo-1-phenylethanol, **VI**,
186

   β-Phenoxyethyl bromide, **I**, 436

C_8H_9BrS 1-Phenylthio-2-bromoethane (in
solution), **VII**, 453

C_8H_9Cl α-Chloro-*p*-xylene, **VI**, 571

C_8H_9ClO Benzyl chloromethyl ether, **VI**,
101

   m-Chlorophenylmethylcarbinol, **III**,
   200

$C_8H_9ClO_2$ 4-Iodoveratrole, **IV**, 547

$C_8H_9ClO_6S$ *p*-Tolylsulfonylmethyl perchlorate, **VI**, 985

C_8H_9F 1-Fluoro-1-phenylethane, **VI**, 630

   1-Fluoro-2-phenylethane, **VI**, 836

C_8H_9I 2-Iodo-1-phenylethane, **VI**, 832

   2-Iodo-*p*-xylene, **VI**, 709

C_8H_9N N-Benzylidenemethylamine, **IV**,
605

   1,3-Dihydroisoindole, **V**, 406

C_8H_9NO Acetanilide, **VI**, 494

   4-Aminoacetophenone, **VI**, 21

   o-Formotoluide, **III**, 480 or *N-o*-
   Tolylformamide, **V**, 1062

   N-Methylformanilide, **III**, 590

   Phenacylamine hydrochloride, **V**,
   909

   Phenylacetamide, **IV**, 760

   o-Toluamide, **II**, 586

C_8H_9NOS N-Ethyl-3-formyl-2(1*H*)-
pyridinethione, **VI**, 643

$C_8H_9NO_2$ N-Acetyl-N-phenylhydroxyl-
amine, **VIII**, 16

   DL-α-Aminophenylacetic acid, **III**,
   84

   p-Aminophenylacetic acid, **I**, 52

   Benzyl carbamate, **III**, 168

   Mandelamide, **III**, 536

   m-Nitroethylbenzene, **VII**, 393

$C_8H_9NO_3$ 3'-Nitro-1-phenylethanol, **VIII**,
495

   3-Nitro-4,6-xylenol, **I**, 405

$C_8H_9NO_3S$ *p*-Acetaminobenzenesulfinic
acid, **I**, 7

$C_8H_9N_3OS$ 1-Phenyl-2-thiobiuret, **V**, 966
$C_8H_9N_5$ 5-Benzylaminotetrazole, **VII**, 27
C_8H_{10} *m*-Xylene, **VI**, 749
    o-Xylene, **VI**, 749
    p-Xylene, **VI**, 709, 749
$C_8H_{10}BrN$ 4-Bromo-*N*,*N*-dimethylaniline,
    VI, 183
    4-Bromo-*N*,2-dimethylaniline, **VI**,
    183
    4-Bromo-2,3-dimethylaniline, **VI**,
    183
    4-Bromo-2,5-dimethylaniline **VI**,
    183
    4-Bromo-3,5-dimethylaniline **VI**,
    183
    α-(*p*-Bromophenyl)ethylamine, **II**,
    505
$C_8H_{10}ClF_3O_2$ 4-Chloro-4-hexenyl trifluo-
    roacetate, **VI**, 273
$C_8H_{10}ClN$ α-(*p*-Chlorophenyl)ethylamine,
    II, 505
    N-Ethyl-*p*-chloroaniline, **IV**, 420
$C_8H_{10}Cl_2$ *exo*-3,4-Dichlorobicyclo[3.2.1]
    oct-2-ene, **VI**, 142
$C_8H_{10}Cl_2O$ 3-Butyl-4,4-dichlorocy-
    clobutenone, **VIII**, 82
    7,7-Dichloro-1-mcthylbicyclo[3.2.0]
    heptan-6-one, **VIII**, 377
$C_8H_{10}Cl_3NO$ 3-Trichloroacetamido-1-
    cyclohexene or *N*-(2-Cyclo-
    hexen-1-yl)trichloroacetamide,
    VI, 510
$C_8H_{10}LiN$ 2-(*N*,*N*-Dimethylamino)phenyl-
    lithium, **VII**, 172
$C_8H_{10}N_2$ Acetophenone hydrazone, **VI**, 12
    1-Amino-2-phenylaziridine, **VI**, 56
    monoacetate, **VI**, 56
$C_8H_{10}N_2O$ *p*-Nitrosodimethylaniline
    hydrochloride, **II**, 223
$C_8H_{10}N_2O_2$ *m*-Nitrodimethylaniline, **III**,
    658
$C_8H_{10}N_2O_3S$ *p*-Tolylsulfonylmethylnitrosa-
    mide, **IV**, 943
$C_8H_{10}N_2S$ *p*-Tolylthiourea, **III**, 77
$C_8H_{10}N_4O_2$ α,α'-Dicyano-β-methylglu-
    taramide, **III**, 591

1-(*m*-Nitrophenyl)-3,3-dimethyltri-
    azene, **IV**, 718
$C_8H_{10}O$ 2,7-Dimethyloxepin, **V**, 467
    2,6-Dimethylphenol, **VI**, 137
    3,4-Dimethylphenol, **VI**, 480
    o-Methylbenzyl alcohol, **IV**, 582
    p-Tolyl carbinol, **II**, 590
$C_8H_{10}OS$ (*S*)-(−)-Methyl *p*-tolyl sulfoxide,
    VIII, 464
$C_8H_{10}O_2$ 2-Acetyl-2-cyclohexen-1-one,
    VI, 23
    3-Acetyl-2,4-dimethylfuran, **VI**,
    31
    Bicyclo[3.3.0]octane-2,6-dione, *cis*-,
    VIII, 43
    cis-Bicyclo[3.3.0]octane-3,7-dione,
    VII, 50
    Creosol, **IV**, 203
    1,4-Dimethoxybenzene, **VI**, 971
    Ethyl cyclopropylpropiolate, **VIII**,
    247
    Methyl 2,4-cyclopentadiene-1-
    acetate (in solution), **VII**, 339
    Orcinol monomethyl ether, **VI**, 859
    Resorcinol dimethyl ether, **VI**, 465
    Styrene glycol, **VI**, 56
$C_8H_{10}O_2S$ Methyl *p*-tolyl sulfone, **IV**,
    674
$C_8H_{10}O_3$ 2-Acetyl-1,3-cyclohexanedione,
    VI, 30
    2-Acetyl-4-methyl-1,3-cyclopentane-
    dione, **VI**, 30
    cis-1,2-Cyclohexanedicarboxylic
    anhydride, **VI**, 482
    4,4-Dimethoxy-2,5-cyclohexa-
    dienone, **VI**, 973
    Ethyl 5-oxo-1-cyclopentene-1-car-
    boxylate, **VI**, 26
    Methyl *p*-toluenesulfonate, **I**, 146
    Phloroglucinol dimethyl ether, **VI**,
    861
$C_8H_{10}O_3S$ Methyl *p*-toluenesulfonate, **I**,
    146
$C_8H_{10}O_4$ 3-Acetyl-4-hydroxy-5,5-
    dimethyl-2(5*H*)-furanone, **VIII**,
    71

trans, trans-1,3-Butadiene-1,4-diyl diacetate, **VI**, 196

6,6-Dimethyl-5,7-dioxaspiro[2.5] octane-4,8-dione, **VII**, 411

$C_8H_{10}O_6$ Diethyl dioxosuccinate, **VIII**, 597

$C_8H_{11}AsN_2O_4 \cdot 2Na$ Sodium *p*-arsono-*N*-phenylglycinamide, **I**, 488

$C_8H_{11}Cl$ 3-Chlorobicyclo[3.2.1]oct-2-ene, **VI**, 142

$C_8H_{11}ClO_2$ *cis*-1-Acetoxy-4-chloro-2-cyclohexene, **VIII**, 5

$C_8H_{11}N$ 1-Cyclohexenylacetonitrile, **IV**, 234

Cyclohexylideneacetonitrile, **VII**, 108

N,N-Dimethylaniline, **VI**, 175, 199

6-(Dimethylamino)fulvene, **V**, 431

3,4-Dimethylaniline, **III**, 307

5-Ethyl-2-methylpyridine, **IV**, 451

α-Phenylethylamine, **II**, 506; **III**, 717; **VI**, 500

DL-α-Phenylethylamine, **II**, 503

D-α-Phenylethylamine, **II**, 506

R(+) and *S*(−)-α-Phenylethylamine, **V**, 932

β-Phenylethylamine, **III**, 720

$C_8H_{11}N \cdot C_6H_6O_5$ D-α-Phenylethylamine acid L-malate, **II**, 506

$C_8H_{11}N \cdot C_6H_6O_6$ L-α-Phenylethylamine acid D-tartrate, **II**, 507

$C_8H_{11}NO$ *tert*-Pentylcyanoketene, **VI**, 215

1-(α-Pyridyl)-2-propanol, **III**, 757

$C_8H_{11}NO_2$ 4-Aminoveratrole, **II**, 44

$C_8H_{11}NSe$ *N,N*-Dimethylbenzeneselenenamide, **VI**, 536

$C_8H_{11}N_3$ 1-Methyl-3-*p*-tolyltriazene, **V**, 797

$C_8H_{11}O \cdot BF_4$ 2,4,6-Trimethylpyrylium tetrafluoroborate, **V**, 1112

$C_8H_{11}O \cdot ClO_4$ 2,4,6-Trimethylpyrylium perchlorate, **V**, 1106

$C_8H_{11}O \cdot CF_3O_3S$ 2,4,6-Trimethylpyrylium trifluoromethanesulfonate, **V**, 1114

C_8H_{12} 1,2-Dimethyl-1,4-cyclohexadiene, **V**, 467

1,2-Dimethylenecyclohexane, **VIII**, 212

cis-1,2-Divinylcyclobutane, **V**, 528; *trans*-1,2-Divinylcyclobutane, **V**, 528

$C_8H_{12}Br_2O$ 4,5-Dibromo-1,2-dimethyl-1,2-epoxycyclohexane, **V**, 468

$C_8H_{12}Cl_2$ 3,4-Dichloro-1,2,3,4-tetramethylcyclobutene, **V**, 370

$C_8H_{12}Cl_2O_2$ Suberyl dichloride, **VI**, 382

$C_8H_{12}Cl_3NO$ 3-Trichloroacetamido-1-hexene or *N*-(1-Hexen-1-yl) trichloroacetamide, **VI**, 510

$C_8H_{12}N_2O$ 2-Diazocyclooctanone, **VI**, 391

Diazomethyl cyclohexyl ketone, **VI**, 388

$C_8H_{12}N_2O_2$ Hexamethylene diisocyanate, **IV**, 521

$C_8H_{12}N_2O_3$ *tert*-Butyl diazoacetoacetate, **VI**, 416

$C_8H_{12}O$ 1-Acetylcyclohexene, **III**, 22

Bicyclo[3.2.1]octan-3-one, **VI**, 142

3-Butylcyclobutenone, **VIII**, 82

Cyclohexylideneacetaldehyde, **VI**, 358

cis-Cyclooctene oxide, **VII**, 126

2-Cycloocten-1-one, **VI**, 331

3,5-Dimethyl-2-cyclohexen-1-one, **III**, 317

4,4-Dimethyl-2-cyclohexen-1-one, **VI**, 496

1,2-Dimethyl-1,2-epoxycyclohex-4-ene, **V**, 468

1-Ethynylcyclohexanol, **III**, 416

7-Methylbicyclo[3.2.0]heptan-6-one, **VI**, 326

3-Methyl-2-cyclohepten-1-one, **VI**, 331

2-Methyl-3-vinylcyclopentanone, **VIII**, 486

$C_8H_{12}OS$ 2-tert-Butoxythiophene, **V**, 642

$C_8H_{12}OS_2$ 2,2-(Ethylenedithio)cyclohexanone, **VI**, 590

$C_8H_{12}O_2$ 2-Acetylcyclohexanone, **VI**, 23, 247

Bicyclo[3.3.0]octane-2,6-dione, **VIII**, 43

2,2-Dimethylcyclohexane-1,3-dione, **II**, 200; **VIII**, 312

5,5-Dimethyl-1,3-cyclohexanedione, **VI**, 1001

Dimethylketene β-lactone dimer, **V**, 456

3-Ethoxy-2-cyclohexenone, **V**, 539

2-Methylcycloheptane-1,3-dione, **VI**, 332

Methyl 5-methyl-2-hexynoate, **VI**, 793

1-Norbornanecarboxylic acid, **VI**, 845

2-*exo*-Norbornyl formate, **V**, 852

(1*R*,6*S*)-(+)-8-Oxabicyclo[4.3.0] nonen-7-one, **VII**, 406

Tetrahydro-2-(2-propynyloxy)-2*H*-pyran, **VII**, 334

$C_8H_{12}O_3$ 2-Carbethoxycyclopentanone, **II**, 116

trans-1-Ethyl-4-oxo-2-butenyl acetate, **VI**, 689

2-Hydroxymethyl-2-cyclopenten-1-one ethylene ketal, **VII**, 274

Methyl (1*R*,5*R*)-5-hydroxy-2-cyclopentene-1-acetate, **VII**, 339

Methyl 7-hydroxyhept-5-ynoate, **VIII**, 415

$C_8H_{12}O_4$ Diethyl fumarate, **VI**, 445, 454, 579

Diethyl maleate, **VI**, 579, 615

Diethyl methylenemalonate, **IV**, 298

Dimethyl (*E*)-2-hexenedioate, **VIII**, 112

Ethyl diacetylacetate, **III**, 390

cis-1-Methylcyclopentane-1,2-dicarboxylic acid, **VIII**, 377

$C_8H_{12}O_8Pb$ Lead tetraacetate, **VI**, 56

$C_8H_{13}BrO_2$ 2-Bromo-1,3-dimethyl-2-butenyl acetate, **VI**, 188

$C_8H_{13}ClO$ 6-(2-Chloroethoxy)bicyclo[3.1.0]hexane, **VI**, 573

β-Chlorovinyl isoamyl ketone, **IV**, 186

$C_8H_{13}ClO_2$ 2,2,4-Trimethyl-3-oxovaleryl chloride, **V**, 1103

$C_8H_{13}ClO_3$ Trimethyl 2-chloro-2-cyclopropylidenorthoacetate, **VIII**, 373

$C_8H_{13}N$ Kryptopyrrole, **III**, 513

2-Methylcyclohexanecarbonitrile, **VI**, 336

2,3,4,5-Tetramethylpyrrole, **V**, 1022

$C_8H_{13}NO$ Bicyclo[3.2.1]octan-3-one oxime, **VI**, 144

7-Cyanoheptanal, **V**, 266

5-Methylfurfuryldimethylamine, **IV**, 626

$C_8H_{13}NO_2$ 2-Ethenylcyclopropyl *N,N*-dimethylcarbamate, **VI**, 573

Ethyl (*trans, trans*)-1,3-pentadiene-1-carbamate, **VI**, 100

1-Nitrocyclooctene, **VI**, 837

6-Propoxy-4,5-dihydro-2(3*H*)-pyridone, **VI**, 231

$C_8H_{13}NO_5$ (1*S*,2*S*,3*R*)-3-Hydroxy-2-nitrocyclohexyl acetate, **VIII**, 332

3-Nitrophthalic anhydride, **I**, 410

C_8H_{14} Cyclooctene, **VI**, 837

trans-Cyclooctene, **V**, 315

3-Ethyl-1-hexyne **VI**, 595

Ethylidenecyclohexane, **VI**, 295

2,3-Octadiene, **VI**, 296

Spiro[2.5]octane, **VI**, 136

$C_8H_{14}Br_2O$ 1,3-Dibromo-6-methyl-2-heptanone, **VI**, 713

1,3-Dibromo-2-octanone, **VI**, 713

$C_8H_{14}Br_2O_2S$ 1-Bromo-1-methyl-2-(bromomethylsulfonyl)cyclohexane, **VIII**, 212

$C_8H_{14}ClN$ 2-Chloro-1,3-dimethyl-4,5,6,7-tetrahydro-1*H*-azepine, **VI**, 286

$C_8H_{14}ClNO$ 2-Chlorocyclooctanone oxime, **V**, 266

$C_8H_{14}Cl_2OMg_2$ 1-Chloromagnesioxy-3-octynylmagnesium chloride, **VIII**, 507

$C_8H_{14}O$ Cyclooctanone, **V**, 310

Cyclohexyl methyl ketone, **V**, 775

2,3-Dimethylcyclohexanone, **VI**, 53

2-Methyl-1,6-heptadien-3-ol, **VI**, 606

trans-2-Octenal, **VI**, 688

(*R*)-(+)-1-Octyn-3-ol, **VII**, 402

2,4,4-Trimethylcyclopentanone, **IV**, 957

C₈H₁₄O₂ 1-Acetylcyclohexanol, **IV**, 13
Bicyclo[3.3.0]octane-2,6-diol, **VIII**, 43

1,4-Dioxaspiro[4.5]decane, **V**, 303

2-Hydroxycyclooctanone, **VI**, 171

(*S*)-(+)-3-Hydroxy-2,2-dimethylcyclohexanone, **VIII**, 312

3-Isopropyl-2,4-pentanedione, **VI**, 27

Methyl cyclohexanecarboxylate, **VI**, 492

1-Methylcyclohexanecarboxylic acid, **V**, 739

2-Methyl-1,6-heptadien-3-ol, **VI**, 606

3-Methyl-2,4-heptanedione, **VI**, 776

cis-6-Methyl-2-heptenoic acid , **VI**, 713

cis-2-Octenoic acid, **VI**, 713

2,2,5,5-Tetramethyltetrahydro-3-ketofuran, **V**, 1024

C₈H₁₄O₂S S-*tert*-Butyl 3-oxobutanethioate, **VIII**, 71

S-(2-Oxobut-3-yl) butanethioate or S-(1-Methyl-2- oxopropyl)butanethioate, **VI**, 776

cis-8-Thiabicyclo[4.3.0]nonane 8,8-dioxide, **VI**, 482

C₈H₁₄O₃ *tert*-Butyl acetoacetate, **V**, 155
Ethyl 2-butyrylacetate, **VII**, 213
Ethyl 3-oxohexanoate, **VI**, 791
Ethyl 4-oxohexanoate, **VI**, 615, 774
Methyl 7-oxoheptanoate, **VI**, 807
7-Oxooctanoic acid, **VI**, 814

C₈H₁₄O₄ Diethyl methylmalonate, **II**, 2 79
Diethyl succinate, **V**, 993; **VI**, 167
Ethyl hydrogen adipate, **II**, 277
β-Ethyl-β-methylglutaric acid, **IV**, 441

C₈H₁₄O₄S Methyl β-thiodipropionate, **IV**, 669

C₈H₁₄O₆ Diethyl (2*R*,3*R*)-(+)-tartrate, **VII**, 41, 461

C₈H₁₄S *cis*-8-Thiabicyclo[4.3.0]nonane, **VI**, 482

C₈H₁₅B 9-Borabicyclo[3.3.1]nonane, **VI**, 137

C₈H₁₅F Fluorocyclooctane or Cyclooctyl fluoride, **VI**, 836

C₈H₁₅FO₂ Ethyl 2-fluorohexanoate, **VIII**, 258

C₈H₁₅N 2-Ethylhexanonitrile, **IV**, 436
N-Ethylidenecyclohexylamine, **VI**, 901

1-(2-Methylpropenyl)pyrrolidine, **VI**, 496

cis-2-(2-Propenyl)cyclopentylamine, **VII**, 501

C₈H₁₅NO 6-Amino-3,4-dimethyl-*cis*-3-cyclohexen-1-ol, **VII**, 5

1,6a-Dimethylhexahydro-1*H*-cyclopent[*c*]isoxazole, **VI**, 672

C₈H₁₅NO₂ Ethyl 3-(1-azetidinyl)propionate, **VI**, 75

C₈H₁₅NO₃ Ethyl α-(isopropylideneaminooxy)propionate, **V**, 1031

C₈H₁₅N₂O₃P Furyl *N*,*N*,*N'*,*N'*-tetramethyldiamidophosphate, **VIII**, 396

C₈H₁₆ 1-Octene, **VI**, 919
2,4,4-Trimethyl-1-pentene, **VI**, 1028

C₈H₁₆ClN 1-Chloro-*N*,*N*,3,3-tetramethyl-1-buten-1-amine, **VI**, 286

C₈H₁₆ClNO *N*-*tert*-Butyl-4-chlorobutanamide, **VI**, 494
4-Chloro-*N*-(1-methylpropyl)butanamide, **VI**, 494

C₈H₁₆Cl₂O 4,4'-Dichlorodibutyl ether, **IV**, 266

C₈H₁₆N₂O₄ (*R*,*R*)-(+)-*N*,*N*,*N'*,*N'*-Tetramethyltartaric acid diamide, **VII**, 41

C₈H₁₆O *cis*-and *trans*-1,4-Dimethylcyclohexanol, **VI**, 46
2,2-Dimethylhexanal, **VI**, 528
(*S*)-(+)-4-Methyl-3-heptanone, **VIII**, 403
Octanal, **V**, 872; **VI**, 222, 871

(Z)-2-Octen-1-ol, **VIII**, 508

2,4,4-Trimethylpentanal, **VI**, 1028

C$_8$H$_{16}$OS$_2$ 1,3-Bis(methylthio)-1-hexen-4-ol, **VI**, 683

C$_8$H$_{16}$OSi 1-Trimethylsiloxycyclopentene, **VII**, 424

C$_8$H$_{16}$O$_2$ Butyl butyrate, **I**, 138

tert-Butyl isobutyrate, **III**, 143

Butyroin, **II**, 114; **VII**, 95

cis-1,2-Cyclohexanedimethanol, **VI**, 482

cis-1,2-Cyclooctanediol, **VI**, 345

2-Cyclohexyloxyethanol, **V**, 303

Isobutyroin, **II**, 115

3-Methylheptanoic acid, **V**, 762

C$_8$H$_{16}$O$_2$Si *trans*-1-Methoxy-3-trimethylsilyloxy-1,3-butadiene, **VII**, 312

C$_8$H$_{16}$O$_3$ 6,6-Dimethoxyhexanal, **VII**, 168

C$_8$H$_{16}$O$_3$S *cis*-and *trans*-2,6-Diethoxy-1,4-oxathiane 4,4-dioxide, **VI**, 976

C$_8$H$_{16}$O$_4$ Ethyl diethoxyacetate, **IV**, 427

Ketene di(2-methoxyethyl) acetal, **V**, 684

Methyl 3,3-diethoxypropanoate, **VIII**, 255

C$_8$H$_{17}$Br Octyl bromide, **I**, 30

C$_8$H$_{17}$F 3-Fluoro-3-methylheptane, **VI**, 630

4-Fluoro-3-methylheptane, **VI**, 630

1-Fluorooctane or *n*-Octyl fluoride, **VI**, 836

C$_8$H$_{17}$N 1-Butylpyrrolidine, **III**, 159

N,N-Dimethylcyclohexylamine, **VI**, 499

N-(2-Methylpropylidene)-*tert*-butylamine, **VI**, 526

C$_8$H$_{17}$NO 1-Diethylamino-3-butanone, **IV**, 281

1-(Dimethylamino)-4-methyl-3-pentanone, **VI**, 474

2-Ethylhexanamide, **IV**, 436

Nitroso-*tert*-octane, **VIII**, 93

cis-6-Propyl-3-piperidinol, **VI**, 547

C$_8$H$_{17}$NO$_2$ Methyl *N*-hexylcarbamate, **VI**, 788

1-Nitrooctane, **IV**, 724

4-Nitro-2,4,4-trimethylpentane, **V**, 845

C$_8$H$_{17}$N$_3$ 1-Ethyl-3-(3-dimethylamino)propylcarbodiimide and hydrochloride, **V**, 555

C$_8$H$_{17}$N$_3$O$_3$ 4-*tert*-Butyl-1-carbethoxysemicarbazide or Ethyl 2-[*N*-1,1-dimethylethylcarbamoyl]hydrazine-1-carboxylate, **VI**, 938

C$_8$H$_{18}$ 3-Ethylhexane, **VI**, 596

C$_8$H$_{18}$BrN β-Dipropylaminoethyl bromide hydrobromide, **II**, 92

C$_8$H$_{18}$ClN$_3$ 1-Ethyl-3-(3-dimethylamino)propylcarbodiimide hydrochloride, **V**, 555

C$_8$H$_{18}$NO Di-*tert*-butyl nitroxide, **V**, 355

C$_8$H$_{18}$N$_2$ Azobutane or dibutyldiazene, **VI**, 80

Azo-*tert*-butane or bis(1,1-dimethylethyl)diazene, **VI**, 80

Azoisobutane or bis(2-methylpropyl)diazene, **VI**, 80

C$_8$H$_{18}$N$_2$O *N*-Heptylurea, **IV**, 515

C$_8$H$_{18}$O 3-Methyl-4-heptanol, **VI**, 241

Methyl hexyl carbinol, **I**, 366

1-Octanol, **VI**, 919

D- and L-2-Octanol, **I**, 418

C$_8$H$_{18}$OSi (Z)-3-Trimethylsiloxy-2-pentene, **VII**, 512

C$_8$H$_{18}$O$_2$ *tert*-Octyl hydroperoxide, **V**, 818

α,α,α',α'-Tetramethyltetramethylene glycol, **V**, 1026

C$_8$H$_{18}$O$_2$Si 1-Ethoxy-1-trimethylsiloxycyclopropane, **VII**, 131

C$_8$H$_{18}$O$_3$ Ethyl orthoacetate, **VI**, 610, or Triethyl orthoacetate, **VI**, 584

C$_8$H$_{18}$O$_3$S Butyl sulfite, **II**, 112

C$_8$H$_{18}$O$_4$S Butyl sulfate, **II**, 111

C$_8$H$_{18}$S Butyl sulfide or Dibutyl sulfide, **VI**, 268

C$_8$H$_{18}$Zn Dibutylzinc, **II**, 187

C$_8$H$_{19}$N Dibutylamine, **I**, 202

N,N-Diisopropylethylamine, **VI**, 95, 576, 737

C$_8$H$_{19}$N$_2$O$_4$S Methyl (carboxysulfamoyl)triethylammonium hydroxide, inner salt, **VI**, 788

$C_8H_{20}N \cdot BF_4$ Tetraethylammonium tetrafluoroborate, **VI**, 153

$C_8H_{20}N_2$ 1,8-Diaminooctane, **VI**, 382

$C_8H_{20}Sn$ Tetraethyltin, **IV**, 881

$C_8I_4O_3$ Tetraiodophthalic anhydride, **III**, 796

$C_8N_5 \cdot C_4H_{12}N$ Tetramethylammonium 1,1,2,3,3-pentacyanopropenide, **V**, 1013

$C_9Fe_2O_9$ Diiron nonacarbonyl, **VI**, 310, 515, 520

$C_9H_5Cl_2N$ 4,7-Dichloroquinoline, **III**, 272

$C_9H_5F_5$ 2-Phenylperfluoropropene, **V**, 949

$C_9H_5MnO_5$ Carboxycyclopentadienyltricarbonylmanganese, **VI**, 627

$C_9H_6BrNO_2$ N-Bromomethylphthalimide, **VIII**, 536

$C_9H_6Br_2O_2$ Coumarin dibromide, **III**, 209

$C_9H_6ClF_3O_3S$ α-(3-Chlorostyryl) triflate, **VI**, 760

  α-(4-Chlorostyryl) triflate, **VI**, 760

$C_9H_6F_3NO_5S$ α-(4-Nitrostyryl) triflate, **VI**, 760

$C_9H_6F_6O$ 1,1,1,3,3,3-Hexafluoro-2-phenyl-2-propanol and potassium salt, **VI**, 163

$C_9H_6FeO_4$ Tricarbonyl[(2,3,4,5-η)-2,4-cyclohexadien-1-one]iron, **VI**, 996

$C_9H_6N_2$ Indole-3-carbonitrile, **V**, 656

$C_9H_6N_4O_2$ 4-Benzalamino-1,2,4-triazoline-3,5-dione, **VI**, 939

C_9H_6O Phenylpropargyl aldehyde, **III**, 731 or Phenylpropynal, **VI**, 66

$C_9H_6O_2$ Phenylpropiolic acid, **II**, 515

$C_9H_6O_3$ Coumarilic acid, **III**, 209

  Homophthalic anhydride, **III**, 450

C_9H_7BrO α-Bromocinnamic aldehyde, **III**, 731

$C_9H_7F_3O_3S$ α-Styryl triflate, **VI**, 758

C_9H_7N Cinnamonitrile, **VI**, 304; **VII**, 108

  1-Cyanobenzocyclobutene, **V**, 263

  Isoquinoline, **VI**, 115

  Quinoline, **I**, 478; **VI**, 883,

C_9H_7NO 3,1-Benzoxazepine, **VII**, 23

  3-Hydroxyquinoline, **V**, 635

Indole-3-aldehyde, **IV**, 539

3-Phenyl-2H-azirine-2-carboxaldehyde, **VI**, 893

C_9H_7NOS 2α-Thiohomophthalimide, **V**, 1051

Phenylacetyl isothiocyanate, **V**, 1051

$C_9H_7NO_2$ o-Carboxyphenylacetonitrile, **III**, 174

2-Phenyl-5-oxazolone, **V**, 946

$C_9H_7NO_3$ o-Nitrocinnamaldehyde, **IV**, 722

$C_9H_7NO_4$ m-Nitrocinnamic acid, **1**, 398

C_9H_7NS 2-(2-Thienyl)pyridine, **VI**, 410

C_9H_8 Indene, **VI**, 427

C_9H_8BrClO 2-Bromo-3-phenylpropanoyl chloride, **VI**, 192

C_9H_8ClN o-Chlorohydrocinnamonitrile, **V**, 264

$C_9H_8Cl_2O_3$ Ethyl 3,5-dichloro-4-hydroxybenzoate, **III**, 267

$C_9H_8N_2O$ 1-Diazo-1-phenylacetone, **VI**, 416

  4-(3-Pyridyl)-4-oxobutyronitrile, **VI**, 866

$C_9H_8N_4O_2$ 4-Benzalaminourazole, **VI**, 938

C_9H_8O Atropaldehyde, **VII**, 12,

  Cinnamaldehyde, **VI**, 66, 315, 633

  2-Indanone, **V**, 647

  α-Hydrindone, **II**, 336

  3-Methylcoumarone, **IV**, 590

$C_9H_8O_2$ Acetylbenzoyl, **III**, 20

  Cinnamic acid, **VI**, 304

  4,5-Methylenedioxybenzocyclobutene, **VII**, 326

$C_9H_8O_2S$ Ethynyl p-tolyl sulfone, **VIII**, 281

$C_9H_8O_2S_2$ Thiobenzoylthioglycolic acid, **V**, 1046

$C_9H_8O_3$ Methyl benzoylformate, **I**, 244

  Phenylpyruvic acid, **II**, 519

$C_9H_8O_4$ 2-Acetoxybenzoic acid, **VI**, 577

  4-Acetoxybenzoic acid, **VI**, 576

  o-Formylphenoxyacetic acid, **V**, 251

  Homophthalic acid, **III**, 449; **V**, 612

  p-Hydroxyphenylpyruvic acid, **V**, 627

C_9H_9Br Cinnamyl bromide, **V**, 249

$C_9H_9BrF_3N$ 4-Bromo-*N,N*-dimethyl-3-(trifluoromethyl)aniline, **VI**, 181

C_9H_9BrO α-Bromo-α-phenylacetone (in
   solution), **III**, 343

C_9H_9Cl α-Chlorohydrindene, **II**, 336

$C_9H_9ClN_2$ 3-(*o*-Chloroanilino)propioni-
   trile, **IV**, 146

C_9H_9ClO Benzyl chloromethyl ketone,
   III, 119

$C_9H_9ClO_2$ 2-Chloroethyl benzoate, **IV**, 84
   (4-Methoxyphenyl)acetyl chloride,
   VI, 744

C_9H_9DO 3-Phenylpropionaldehyde-1-*d*,
   VI, 755

$C_9H_9IO_2$ 2-Iodoethyl benzoate, **IV**, 84

C_9H_9N 1-Methylindole, **V**, 769; **VI**, 107
   2-Methylindole, **III**, 597

C_9H_9NO Cinnamamide, **VI**, 494
   p-Methoxyphenylacetonitrile, **IV**,
   576
   3-Methyloxindole, **IV**, 657

$C_9H_9NO_2$ 2,4-Dimethoxybenzonitrile, **VI**,
   465
   2,6-Dimethoxybenzonitrile, **III**, 293
   Isonitrosopropiophenone, **II**, 363
   Veratronitrile, **II**, 622

$C_9H_9NO_2 \cdot BF_4$ *p*-Carbethoxybenzenediazo-
   nium fluoborate, **II**, 299

$C_9H_9NO_2S$ *p*-Tolylsulfonylmethyl isocyanide or 4-Methyl-1-[((isocyanomethyl)sulfonyl)]benzene
   or 1-Methyl-4-[(carbylamino)
   methylsulfonyl]benzene, **VI**,
   41, 987

$C_9H_9NO_3$ Hippuric acid, **II**, 328

$C_9H_9NO_4$ Ethyl *m*-nitrobenzoate, **I**, 373
   p-Nitrobenzyl acetate, **III**, 650

$C_9H_9NO_5$ 6-Nitroveratraldehyde, **IV**, 735

$C_9H_9N_3O$ 1-Phenyl-3-amino-5-pyrazolone,
   III, 708

$C_9H_9N_5$ Benzoguanamine, **IV**, 78

C_9H_{10} Allylbenzene, **VI**, 174, 295
   Cyclopropylbenzene, **V**, 328 or
   Phenylcyclopropane, **V**, 929; **VI**,
   159

3-Phenyl-1-propene, **VI**, 772

$C_9H_{10}BrNO$ 3-Bromo-4-acetaminotoluene,
   I, 111

$C_9H_{10}ClN$ *N*-(1-Chloroethenyl)-*N*-methyl-
   aniline, **VI**, 287

$C_9H_{10}ClNO$ 7-Chloro-7-cyanobicy-
   clo[4.2.0]octan-8-one, **VIII**,
   116
   N-Ethyl-*p*-chloroformanilide, **IV**,
   420

$C_9H_{10}Cl_2O$ 4-[(*E*)-1,2-Dichlorovinyl]-4-
   methyl-2-cyclohexen-1-one,
   VII, 241

$C_9H_{10}F_3N$ *N,N*-Dimethyl-3-(trifluoromethyl)aniline, **VI**, 181 or
   m-Trifluoromethyl-*N,N*-
   dimethylaniline, **V**, 1085

$C_9H_{10}IO$ 2-Allyl-6-iodophenol, **VI**, 351

$C_9H_{10}N_2$ Myosmine, **VI**, 868
   N-2-Cyanoethylaniline, **IV**, 205

$C_9H_{10}N_2O$ Methylglyoxal ω-phenylhydrazone, **IV**, 633

$C_9H_{10}N_2O_4$ 2-Nitro-4-methoxyacetanilide,
   III, 661

$C_9H_{10}N_2S$ *p*-Dimethylaminophenyl isothiocyanate, **I**, 448
   p-Thiocyanodimethylaniline, **II**, 574

$C_9H_{10}O$ 3,4-Dimethylbenzaldehyde, **VI**,
   1009
   p-Methylacetophenone, **I**, 111
   Methyl benzyl ketone, **II**, 389
   Phenylacetone, **VI**, 692
   2-Phenylpropanal, **VI**, 1029 or 2-
   Phenylpropionaldehyde, **VI**, 908
   or α-Phenylpropionaldehyde,
   III, 733
   3-Phenylpropanal, **VI**, 813; **VII**, 451
   or 3-Phenylpropionaldehyde,
   VI, 816

$C_9H_{10}O_2$ 3,4-Dimethylbenzoic acid, **VI**, 627
   Ethyl benzoate, **VI**, 152, 579
   Hydrocinnamic acid, **I**, 311
   p-Methoxyacetophenone, **I**, 111
   α-Methoxyacetophenone, **III**, 562
   4-Methyltricyclo[2.2.2.0^{3,5}]octane-
   2,6-dione, **VIII**, 467

Nortricyclyl acetate, **VI**, 964

Phenyl propionate, **II**, 544

o- and *p*-Propiophenol, **II**, 543

Trimethyl-*p*-benzoquinone, **VI**, 1010

$C_9H_{10}O_2S$ 2-(Methylsulfinyl)acetophenone, **V**, 939

Phenylglyoxal hemimercaptal, **V**, 937

$C_9H_{10}O_3$ Atrolactic acid, **IV**, 58

2,6-Dimethoxybenzaldehyde, **VI**, 633

4-Ethoxy-3-hydroxybenzaldehyde, **VI**, 567

Methyl anisate or Methyl 4-methoxybenzoate, **VI**, 278

Quinacetophenone monomethyl ether, **IV**, 836

Veratraldehyde, **II**, 619

$C_9H_{10}O_4$ 2,4-Dimethoxybenzoic acid, **VI**, 627

3,4-Dimethoxybenzoic acid, **VI**, 627

Syringic aldehyde, **IV**, 866

$C_9H_{10}O_5$ Furfural diacetate, **IV**, 489

5β-Methoxycyclohexan-1-one-3β,4β-dicarboxylic acid anhydride, **VII**, 312

$C_9H_{10}S$ 1-Phenylthio-1-propene, **VI**, 742

$C_9H_{11}Br$ Bromomesitylene, **II**, 95

$C_9H_{11}BrO$ 1-Bromo-2-phenyl-2-propanol, **VI**, 186

erythro-2-Bromo-1-phenyl-1-propanol, **VI**, 186

threo-2-Bromo-1-phenyl-1-propanol, **VI**, 186

3-Bromo-1-phenyl-2-propanol, **VI**, 186

γ-Phenoxypropyl bromide, **I**, 435

$C_9H_{11}ClO_4$ 2-Chloro-2-cyclopropylethene-1,1-dicarboxylic acid, monomethyl ester, **VIII**, 247

$C_9H_{11}N$ *N*-Benzylideneethylamine, **V**, 758

N-(α-Methylbenzylidene)methyl-amine, **VI**, 818

$C_9H_{11}NO$ *p*-Acetotoluidide, **I**, 111

p-Dimethylaminobenzaldehyde, **I**, 214; **IV**, 331

Ethyl *N*-phenylformimidate, **IV**, 464

$C_9H_{11}NOS$ *N*-Isopropyl-3-formyl-2(1*H*)-pyridinethione, **VI**, 643

Formyl-*N*-propyl-2(1*H*)-pyridine-thione, **VI**, 643

$C_9H_{11}NO_2$ DL-α-Amino-α-phenylpropionic acid, **III**, 88

DL-β-Amino-β-phenylpropionic acid, **III**, 91

Cyclohexylidenecyanoacetic acid, **IV**, 234

Ethyl *p*-aminobenzoate, **I**, 240; **II**, 300 or Ethyl 4-aminobenzoate, **VI**, 581, 601

Ethyl 2-pyridylacetate, **III**, 413

Nitromesitylene, **II**, 449

DL-β-Phenylalanine, **II**, 489 or DL-Phenylalanine, **III**, 705

$C_9H_{11}NO_3$ L-(−)-Tyrosine, **II**, 613; **VI**, 1004

Veratric amide, **II**, 44

$C_9H_{11}NO_3S$ *N*-(*p*-Tolylsulfonylmethyl)for-mamide, **VI**, 987

$C_9H_{11}NO_4$ Methyl 6-nitro-2-norbornene-5-carboxylate, **VI**, 802

m-Nitrobenzaldehyde dimethyl acetal, **III**, 644

$C_9H_{11}N_3O_2$ 7-Aminotheophylline, **VII**, 8

C_9H_{12} 1,6-Dimethyltricyclo[4.1.0 $0^{2,7}$]hept-3-ene, **VII**, 200

Hemimellitene, **IV**, 508

Mesitylene, **I**, 341; **VI**, 317, 749

Propylbenzene, **I**, 471 or 1-Phenylpropane, **VI**, 409,

2-Phenylpropane, **VI**, 409

$C_9H_{12}BrN$ 4-Bromo-*N*,*N*,2-trimethylani-line, **VI**, 183

4-Bromo-*N*,*N*,3-trimethylaniline, **VI**, 183

$C_9H_{12}Br_2$ 7,7-Dibromo-1,6-dimethylbicy-clo[4.1.0]hept-3-ene, **VII**, 200

$C_9H_{12}N_2$ *N*,*N*-Dimethyl-*N'*-*p*-tolylfor-mamidine, **VII**, 197

$C_9H_{12}N_2O$ β-Propionylphenylhydrazine, **IV**, 657

$C_9H_{12}N_2O_2$ *p*-Ethoxyphenylurea, **IV**, 52

$C_9H_{12}N_2O_4$ [R-(R^*,R^*)]-2-Amino-1-(4-nitrophenyl)-1,3-propanediol, VI, 829

$C_9H_{12}N_2S$ 2,6-Dimethylphenylthiourea, V, 802

$C_9H_{12}O$ Bicyclo[4.3.0]non-1-en-4-one, VIII, 38,
   3-Phenyl-1-propanol, IV, 798
   Tricyclo[3.3.1.0$^{2.7}$]nonan-3-one, VI, 961
   2,6,6-Trimethyl-2,4-cyclohexadienone, V, 1092
   2,3,6-Trimethylphenol, VI, 1010

$C_9H_{12}O_2$ 2-Acetyl-2-cyclohepten-1-one, VI, 27
   2α,4α-Dimethyl-8-oxabicyclo[3.2.1]oct-6-en-3-one, VI, 512
   4-Methylbicyclo[2.2.2]octane-2,6-dione, VIII, 467
   Methyl p-ethylbenzoate, IV, 580
   Methyl 2-octynoate, VI, 793
   Nortricyclyl acetate, V, 863
   Orcinol dimethyl ether, VI, 860

$C_9H_{12}O_3$ 2-Acetyl-1,3-cycloheptanedione, VI, 30
   2-Acetyl-4,4-dimethyl-1,3-cyclopentanedione, VI, 30
   2-Carbomethoxy-3-vinylcyclopentanone, VIII, 112
   Ethyl 2,4-dimethylfuran-3-carboxylate, VI, 33
   Ethyl 6-oxo-1-cyclohexene-1-carboxylate, VI, 26
   α-Glyceryl phenyl ether, I, 296

$C_9H_{12}O_4$ Dimethyl 3-methylenecyclobutane-1,2-dicarboxylate, V, 459
   Methyl butadienoate, V, 734

$C_9H_{13}BrO_2$ 2-Bromo-2-cyclohepten-1-ol acetate VI, 188

$C_9H_{13}Cl_3$ 2-(Trichloromethyl)bicyclo[3.3.0]octane, V, 93

$C_9H_{13}F_3O_3S$ 2,6-Dimethyl-1-cyclohexen-1-yl triflate, VI, 760

$C_9H_{13}N$ 2-Butylpyridine, VI, 409
   N,N-Dimethyl-p-toluidine, VI, 478

Ethyl-m-toluidine, II, 290
   N-Methylphenethylamine, VI, 500
   α-p-Tolylethylamine, II, 505

$C_9H_{13}NO$ 3-Butyroyl-1-methylpyrrole, VII, 102
   1,3-Dihydro-3,5,7-trimethyl-2H-azepin-2-one, V, 408
   2-(Methylamino)-1-phenylethanol, VI, 545
   (S)-Phenylalanol, VIII, 528

$C_9H_{13}NO_2$ m-Aminobenzaldehyde dimethylacetal, III, 59
   6-(Carbonylamino)-6-ethoxybicyclo[3.1.0]hexane, VI, 230
   2,4-Dimethyl-5-carbethoxypyrrole, II, 198
   Ethyl sec-butylidenecyanoacetate, IV, 94

$C_9H_{13}NO_2S$ Ethyl 5-propylthiazole-4-carboxylate, VI, 623

$C_9H_{13}NO_3$ Ethyl 3-ethyl-5-methyl-4-isoxazolecarboxylate, VI, 592, 782,

$C_9H_{13}N_3O_2$ Methyl 2-(1-cyanocyclohexyl)diazenecarboxylate, VI, 334

C_9H_{14} 3-Cyclohexylpropyne, I, 191
   1,2-Cyclononadiene, V, 306

$C_9H_{14}Br_2$ 9,9-Dibromobicyclo[6.1.0]nonane, V, 306

$C_9H_{14}Cl_3NO$ (E)-1-Trichloroacetamido-2-heptene or (E)-N-(2-Hepten-1-yl)trichloroacetamide, VI, 510

$C_9H_{14}N·Br_3$ Phenyltrimethylammonium tribromide, VI, 175

$C_9H_{14}N_2$ Azelanitrile, IV, 62

$C_9H_{14}N_2O$ $trans$-N-(1,3-Butadienyl)-1-pyrrolidine-1-carboxamide, VI, 100
   2-Diazocyclononanone, VI, 391

$C_9H_{14}O$ 2-Allylcyclohexanone, III, 44; V, 25
   Bicyclo[3.2.2]nonan-3-one, VI, 144
   Bicyclo[3.3.1]nonan-9-one, VI, 137
   3,3-Dimethylbicyclo[3.2.0]heptan-2-one, VII, 177
   6-Ethoxy-6-methyl-2-cyclohexen-1-one, VII, 208

Isophorone or 3,5,5-Trimethyl-2-
cyclohexen-1-one, **VI**, 1024
1-Methylbicyclo[3.2.1]octan-3-one,
VI, 144
6-Methylbicyclo[4.2.0]octan-2-one,
VII, 315
trans-3-Methyl-2-vinylcyclohexa-
none, **VIII**, 479
2,4,4,5-Tetramethyl-2-cyclopenten-
1-one, **VI**, 523
$C_9H_{14}OS_2$ 2,2-(Trimethylenedithio)cyclo-
hexanone, **VI**, 1014
$C_9H_{14}O_2$ *exo-cis*-Bicyclo[3.3.0]octane-2-
carboxylic acid, **V**, 93
2-Cyclohepten-1-yl acetate, **VIII**,
137
1,3-Cyclononanedione, **VI**, 332
7,7-Dimethoxybicyclo[2.2.1]hep-
tene, **V**, 424
3-Ethoxy-6-methyl-2-cyclohexen-1-
one, **VII**, 208, 241
6-Hydroxy-3,5,5-trimethyl-2-cyclo-
hexen-1-one, **VII**, 282
Isophorone oxide, **IV**, 552
2-Methyl-1-cyclohexen-1-yl acetate,
VI, 121
Methyl 2-methyl-1-cyclohexene-1-
carboxylate, **VII**, 351
2-Oxa-1-adamantanol, **VI**, 961
2-(2-Oxopropyl)cyclohexanone, **VII**,
414
$C_9H_{14}O_3$ Ethyl 2-cyclohexanonecarboxy-
late, **VI**, 782
Ethyl 2-ketohexahydrobenzoate, **II**,
532
3-Methylcyclohexanone-3-acetic
acid, **VIII**, 467
$C_9H_{14}O_4$ Diethyl ethylidenemalonate, **IV**,
293
$C_9H_{14}O_5$ 1-Carbomethoxy-1-methylethyl 3-
oxobutanoate, **VIII**, 71
Dimethyl 4-oxoheptanedioate, **VI**,
650
Ethyl acetonedicarboxylate, **I**, 237
Ethyl acetosuccinate, **II**, 262
Ethyl ethoxalylpropionate, **II**, 272

$C_9H_{14}O_6$ Dimethyl 2,3-*O*-isopropylidene-
L-tartrate, **VIII**, 155, 201
$C_9H_{14}SCuLi$ Lithium phenylthio(*tert*-
butyl)cuprate, **VI**, 248
$C_9H_{15}Br$ 3-Cyclohexyl-2-bromopropene,
I, 186
$C_9H_{15}ClOSiMg$ (Isopropoxydimethylsilyl)
methylmagnesium chloride,
VIII, 315
$C_9H_{15}N$ 2,3,6,7,8,9-Hexahydro-4*H*-quino-
lizine, **VI**, 820
$C_9H_{15}NO$ 3-(1-Hydroxybutyl)-1-
methylpyrrole **VII**, 102,
N-Methyl-2-oxa-3-azabicyclo
[8.1^{4,7}.0^{1,5}]decane, **VI**, 673
Pseudopelletierine, **IV**, 816
$C_9H_{15}NO_2$ Ethyl butylcyanoacetate, **III**,
385
tert-Butyl *trans*-1,3-butadiene-1-car-
bamate, **VI**, 100
$C_9H_{15}NO_5$ Diethyl acetamidomalonate, **V**,
373
$C_9H_{15}N_3O$ Bicyclo[3.2.1]octan-3-one
semicarbazone, **VI**, 143
$C_9H_{15}N_3O_2$ Methyl 2-(1-cyanocyclo-
hexyl)hydrazinecarboxylate,
VI, 334
$C_9H_{15}NO_5$ Diethyl acetamidomalonate, **V**,
373
$C_9H_{15}NO_6$ Dimethyl 2-nitropentanedioate,
VI, 504
Ethyl *N*-tricarboxylate, **III**, 415
C_9H_{16} 3,5,5-Trimethylcyclohexene, **VI**,
295
$C_9H_{16}Br_2O$ 1,3-Dibromo-2-nonanone, **VI**,
713
$C_9H_{16}ClN$ *N*-(1-Chloro-2-methyl-1-
propen-1-yl)piperidine, **VI**,
286
$C_9H_{16}O$ 2-Butylcyclopentanone, **VI**, 528
3,3-Dimethyl-*cis*-bicyclo[3.2.0]hep-
tan-2-ol, **VII**, 177
4,4-Dimethyl-1,6-heptadien-3-ol,
VII, 177
4-Hydroxynona-1,2-diene, **VII**, 276
2-Isopropylcyclohexanone, **VI**, 528

cis-2-Methylhexahydrobenzofuran, **VI**, 372

cis-2-Nonenoic acid, **VI**, 713

(*E*)-2-Propenylcyclohexanol, **VII**, 456

C₉H₁₆O₂ 3-Butyl-2,4-pentanedione, **VI**, 245

2-Cyclobutyl-*cis*-4-*trans*-5-dimethyl-1,3 dioxolane, **VI**, 312

(*E*)-1-Ethoxy-1-(1-hydroxy-2-butenyl)cyclopropane, **VIII**, 556

Heptaldehyde enol acetate, **III**, 127

2-Hydroxycyclononanone, **VI**, 171

2,4-Nonanedione, **V**, 848

C₉H₁₆O₃ Ethyl 1-hydroxycyclohexylacetate, **VI**, 598

Ethyl α-isopropylacetoacetate, **III**, 405

Methyl 7-oxooctanoate, **VI**, 807

C₉H₁₆O₄ Azelaic acid, **II**, 53

Ethyl *tert*-butyl malonate, **IV**, 417

1,5-Pentanediol diacetate, **IV**, 748

C₉H₁₆O₆ Diethyl bis(hydroxymethyl)malonate, **V**, 446

C₉H₁₇BrO₂ 2,5,5-Trimethyl-2-(bromoethyl)-1,3-dioxane, **VII**, 59

C₉H₁₇ClO₂ (*E*)-6-Chloro-4-methyl-4-hexenal dimethyl acetal, **VI**, 639

(*Z*)-6-Chloro-4-methyl-4-hexenal dimethyl acetal, **VI**, 639

C₉H₁₇Cl₃ 1,1,3-Trichlorononane, **V**, 1076

C₉H₁₇FO₂ 2-Fluoroheptyl acetate, **V**, 580

C₉H₁₇N 1,1,3,3-Tetramethylbutyl isonitrile or 2-(carbylamino)-2,4,4-trimethylpentane, **VI**, 751

C₉H₁₇NO *N,N*-Dimethylcyclohexanecarboxamide, **VI**, 492

cis-1,N-Dimethyl-3-oxa-2-azabicyclo[3.3.0]octane or *cis*-1,7a-Dimethylhexahydro-1H-cyclopent[c]isoxazole, **VI**, 672

cis,cis-N-Ethyl-4-methyl-3-oxa-2-azabicyclo[3.3.0]octane or *cis,cis*-N-Ethyl-3-methylhexahydro-1H- cyclopent[c]isoxazole, **VI**, 671

C₉H₁₇NO₂ ω-Cyanopelargonic acid, **III**, 768

Ethyl α-(1-pyrrolidyl)propionate, **IV**, 466

2-Methoxycyclooctanone oxime, **V**, 267

C₉H₁₇NO₃ Methyl *N*-(*cis*-2,3-dimethyl-1-ethoxycyclopropyl)carbamate, **VI**, 230

Methyl *N*-(*trans*-2,3-dimethyl-1-ethoxycyclopropyl)carbamate, **VI**, 230

C₉H₁₇N₂Cl [5-(Dimethylamino)-2-4-pentadienylidene]dimethylammonium chloride, **VII**, 15

C₉H₁₈NLi Lithium 2,2,6,6-tetramethylpiperidide, **VI**, 571

C₉H₁₈N₂ Dibutylcyanamide, **I**, 204

C₉H₁₈O 4,6-Dimethyl-1-hepten-4-ol, **V**, 452

3-Hydroxy-1-nonene, **VIII**, 235

2-Methyloctanal, **VI**, 813

(*Z*)-3-Methyl-2-octen-1-ol, **VIII**, 509

2-Nonanone, **VI**, 222

4-Nonen-1-ol, **VI**, 678

cis-2,4,4-Trimethylcyclohexanol, **VI**, 217

cis-3,5,5-Trimethylcyclohexanol, **VI**, 217

C₉H₁₈OSi 1-Trimethylsilyloxycyclohexene, **VI**, 327; **VIII**, 460

C₉H₁₈O₂ *tert*-Butyl trimethylacetate, **VI**, 261

Pelargonic acid, **II**, 474

C₉H₁₈O₃ β-Ketoisooctaldehyde dimethyl acetal, **IV**, 558

C₉H₁₈O₄ Ethyl 3,3-diethoxypropanoate, **VIII**, 254

Malonaldehydic acid diethyl acetal, ethyl ester, **VI**, 978,

Methyl 6,6-dimethoxyhexanoate, **VII**, 168

C₉H₁₉BrO₂ α-Bromoheptaldehyde dimethyl acetal, **III**, 128

C₉H₁₉N *N,N*-Dimethylcyclohexylmethylamine, **IV**, 339

2-Propyl-1-azacycloheptane, **VIII**, 568,

C$_9$H$_{19}$NO *N*-(1,1,3,3-Tetramethylbutyl)formamide, **VI**, 751

C$_9$H$_{19}$O$_6$P *tert*-Butylcarbonic diethylphosphoric anhydride, **VI**, 207

C$_9$H$_{20}$N$_3$·I 1-Ethyl-3-(3-dimethylamino)propylcarbodiimide methiodide, **V**, 555

C$_9$H$_{20}$O Dibutyl carbinol, **II**, 179 or 5-Nonanol, **VI**, 240

C$_9$H$_{20}$O$_2$ Nonamethylene glycol, **II**, 155

C$_9$H$_{20}$O$_2$Si 2-Methyl-2-(trimethylsiloxy)pentan-3-one, **VII**, 381

C$_9$H$_{20}$O$_3$ β-Ethoxypropionaldehyde acetal, **III**, 371

C$_9$H$_{20}$O$_4$ Ethyl orthocarbonate, **IV**, 457

C$_9$H$_{21}$N Tripropylamine, **VI**, 747

C$_9$H$_{21}$O$_3$P Triisopropyl phosphite, **IV**, 326, 955

C$_9$H$_{21}$O$_4$P Propyl phosphate, **II**, 110

C$_{10}$Cl$_8$ Perchlorfulvalene, **V**, 901

C$_{10}$H$_5$Co$_2$O$_6$·BF$_4$ Hexacarbonyl (1-methyl-2-propynylium)dicobalt tetrafluoroborate, **VIII**, 460

C$_{10}$H$_5$F$_6$IO$_4$ [Bis(trifluoroacetoxy)iodo]benzene, **VIII**, 135

C$_{10}$H$_6$ClNO$_3$ 7-Chloro-4-hydroxy-3-quinolinecarboxylic acid, **III**, 273

C$_{10}$H$_6$Cl$_2$O$_4$S$_2$ Naphthalene-1,5-disulfonyl chloride, **IV**, 693

C$_{10}$H$_6$N$_2$O$_4$ 1,4-Dinitronaphthalene, **III**, 341

C$_{10}$H$_6$O$_2$ 1,2-Naphthoquinone, **II**, 430
1,4-Naphthquinone, **I**, 383; **IV**, 698

C$_{10}$H$_6$O$_3$ 2-Hydroxy-1,4-naphthoquinone, **III**, 465

C$_{10}$H$_6$O$_5$S 1,2-Naphthoquinone-4-sulfonic acid salts, **III**, 633

C$_{10}$H$_6$O$_8$ Pyromellitic acid, **II**, 551

C$_{10}$H$_7$Br α-Bromonaphthalene, **I**, 121 or 1-Bromonaphthalene, **VI**, 821
2-Bromonaphthalene, **V**, 142

C$_{10}$H$_7$BrHg β-Naphthylmercuric bromide, **II**, 434

C$_{10}$H$_7$BrO 6-Bromo-2-naphthol, **III**, 132

C$_{10}$H$_7$ClHg β-Naphthylmercuric chloride, **II**, 432

C$_{10}$H$_7$ClO$_2$ Methyl 3-(4-chlorophenyl)-2-propynoate **VI**, 793

C$_{10}$H$_7$F$_5$O$_2$ Ethyl (pentafluorophenyl)acetate, **VI**, 880

C$_{10}$H$_7$F$_6$O$_3$S α-(4-Trifluoromethyl)styryl triflate, **VI**, 760

C$_{10}$H$_7$FeO· PF$_6$ Tricarbonyl[1,2,3,4,5-η)-2-methoxy-2,4-cyclohexadien-1-yl] iron(1+) hexafluorophosphate, **VI**, 1001

C$_{10}$H$_7$NOS *o*-Isothiocyanatocinnamaldehyde, (*E*)- and (*Z*)-, **VII**, 302

C$_{10}$H$_7$NO$_2$ α-Cyano-β-phenylacrylic acid, **I**, 181
Nitroso-β-naphthol, **I**, 411
N-Phenylmaleimide, **V**, 944

C$_{10}$H$_7$NO$_3$ *N*-Acetylisatin, **III**, 456
2-Hydroxycinchoninic acid, **III**, 456
3-Hydroxycinchoninic acid, **V**, 636
1-Nitro-2-naphthol, **II**, 451

C$_{10}$H$_7$N$_2$· Cl-HgCl$_2$ β-Naphthalenediazonium chloride-mercuric chloride compound, **II**, 432

C$_{10}$H$_7$N$_3$S 2-(4-Thiazolyl)benzimidazole or thiabendazole, **VI**, 624

C$_{10}$H$_7$O$_3$ β-Benzoylacrylic acid, **III**, 109

C$_{10}$H$_8$ Azulene, **VII**, 15
Naphthalene, **VI**, 152, 731, 821, 852

C$_{10}$H$_8$BrN 1-Amino-4-bromonaphthalene or 4-Bromo-α-naphthylamine, **VI**, 183

C$_{10}$H$_8$BrNO$_2$ β-Bromoethylphthalimide, **I**, 120; **IV** 106

C$_{10}$H$_8$ClN 2-Chlorolepidine, **III**, 194

C$_{10}$H$_8$ClNO 4-Oxo-4-(4-chlorophenyl)butyronitrile, **VI**, 867

C$_{10}$H$_8$N$_2$ 2,2'-Bipyridine, **V**, 102 or 2,2'-Bipyridyl, **VI**, 122, 693, 764

C$_{10}$H$_8$N$_2$O$_2$ 4-Nitro-1-naphthylamine, **III**, 664

C$_{10}$H$_8$N$_2$O$_3$ 6-Methoxy-8-nitroquinoline, **III**, 568

C$_{10}$H$_8$O 1-Naphthol, **VI**, 371

2-Naphthol, **VI**, 824

1,6-Oxido[10]annulene, **VI**, 736, 862

$C_{10}H_8O_2$ 2,5-Dihydroxyacetophenone, **III**, 280

2,6-Dihydroxyacetophenone, **III**, 281

4-Methylcoumarin, **III**, 581

Methyl 3-phenyl-2-propynoate, **VI**, 793

Naphthoresorcinol, **III**, 637

endo-Tricyclo[4.4.0.0^{2,5}]deca-3,8-diene-7,10-dione, **VI**, 1002

cis, syn, cis-Tricyclo[5.3.0.0^{2,6}]deca-4,8-diene-3,10-dione, **VI**, 1003

$C_{10}H_8O_3$ 3-Methylcoumarilic acid, **IV**, 591

4-Methyl-7-hydroxycoumarin, **III**, 282

$C_{10}H_8O_4$ *o*-Carboxycinnamic acid, **IV**, 136

4-Methylesculetin, **I**, 360

$C_{10}H_8S$ 2-Naphthalenethiol, **VI**, 824

$C_{10}H_8S_2$ 1,5-Naphthalenedithiol, **IV**, 695

$C_{10}H_9FeO_3 \cdot PF_6$ Tricarbonyl[(1,2,3,4,5-η)-1-methoxy-2,4- cyclohexadien-1-yl]iron(1+) hexafluorophosphate, **VI**, 996

Tricarbonyl[(1,2,3,4,5-η)-2-methoxy-2,4-cyclohexadien-1-yl]iron(1+) hexafluorophosphate, **VI**, 996

$C_{10}H_9BrO$ α-Bromobenzalacetone, **III**, 125

$C_{10}H_9ClO_3$ Acetylmandelyl chloride, **I**, 13

$C_{10}H_9N$ 1,6-Imino[10]annulene, **VI**, 865

Lepidine, **III**, 519

1-Methylisoquinoline, **IV**, 641

$C_{10}H_9NO$ 1,2-Aminonaphthol hydrochloride, **II**, 33

1,4-Aminonaphthol hydrochloride, **I**, 49; **II**, 39

4-Methylcarbostyril, **III**, 580

2-Methyl-4-hydroxyquinoline, **III**, 593

4-Oxo-4-phenylbutyronitrile, **VI**, 867

α-Phenylacetoacetonitrile, **II**, 487

$C_{10}H_9NO_2$ 1-Acetoxy-4-diethylamino-2-butene, **VIII**, 9

3-Acetyloxindole, **V**, 12

2-Hydroxy-3-methylisocarbostyril, **V**, 623

Indole-3-acetic acid, **V**, 654

1-(p-Nltrophenyl)-1,3-butadiene, **IV**, 727 or 1-(4-Nitrophenyl)-1,3-butadiene, **VI**, 22

$C_{10}H_9NO_3$ Maleanilic acid, **V**, 944

$C_{10}H_9NO_4$ (2,4-Diformyl-5-hydroxyphenyl)acetamide, **VII**, 162

$C_{10}H_9NO_4S$ 1-Amino-2-naphthol-4-sulfonic acid, **II**, 42

$C_{10}H_{10}$ 9,10-Dihydrofulvalene, **VIII**, 298

1,2-Dihydronaphthalene, **VI**, 795

1-Phenyl-1,2-butadiene, **VI**, 296

trans-1-Phenyl-1,3-butadiene, **IV**, 771

4-Phenylbutyne-1, **I**, 192

Tetracyclo[5.3.0.0^{2,5}.0^{3,10}]deca-4,8-diene, **VI**, 1003

p-Tolylacetylene, **VI**, 956

Tricyclo[4.4.0.0^{2,5}]deca-3,7,9-triene, **VI**, 1003

$C_{10}H_{10}Br_2O$ Benzalacetone dibromide, **III**, 105

$C_{10}H_{10}ClNO_2$ p-Chloroacetylacetanilide, **III**, 183

$C_{10}H_{10}ClN_5$ 2-Amino-4-anilino-6-(chloromethyl)-s-triazine, **IV**, 29

$C_{10}H_{10}F_3NO_3$ N-(Trifluoroacetyl)-L-phenylalanine **VI**, 1006

N-(Trifluoroacetyl)-DL-phenylalanine, **VI**, 1006

$C_{10}H_{10}Fe$ Ferrocene, **IV**, 473; **VI**, 625

$C_{10}H_{10}N_2O$ 1-Diazo-4-phenyl-2-butanone, **VI**, 388; **VIII**, 196

$C_{10}H_{10}N_2O_3Cr$ Dipyridine chromium(VI) oxide, **VI**, 644

$C_{10}H_{10}N_4O_4$ Bis-(N-methyl-N-nitroso)terephthalamide, **VI**, 432

$C_{10}H_{10}O$ Benzalacetone, **I**, 77

3,4-Dihydro-1(2H)-azulenone, **VIII**, 196

3,4-Dihydro-2-methylene-2*H*-1-ben-
zopyran, **VIII**, 512
2-Methyl-1-indanone, **V**, 624
1-Phenylcyclopropanecarboxalde-
hyde, **VI**, 908
α-Tetralone, **II**, 569; **III**, 798; **IV**,
898
β-Tetralone, **IV**, 903
$C_{10}H_{10}O_2$ Benzoylacetone, **VI**, 247
cis-2-Phenylcyclopropanecarboxylic
acid, **VI**, 913
trans-2-Phenylcyclopropane-
carboxylic acid, **VI**, 915
$C_{10}H_{10}O_3$ α-Benzoylpropionic acid, **II**, 81
Ethyl benzoylformate, **I**, 241
Methyl *p*-acetylbenzoate, **IV**, 579
$C_{10}H_{10}O_4$ Acetylmandelic acid, **I**, 12
Benzoic carbonic anhydride, **IV**, 286
β-(*o*-Carboxyphenyl)propionic acid,
IV, 136
Hydroquinone diacetate, **III**, 452
Methyl 2-acetoxybenzoate, **VI**, 576
Methyl 4-acetoxybenzoate, **VI**, 576
Phenylsuccinic acid, **I**, 451; **IV**, 804
$C_{10}H_{10}O_4$Fe Tricarbonyl[(1,2,3,4-η)-1-
methoxy-1,3-cyclohexa-
diene]iron, **VI**, 996
Tricarbonyl[(1,2,3,4-η)-2-methoxy-
1,3-cyclohexadiene]iron, **VI**,
996
$C_{10}H_{10}O_4$S Methyl (*Z*)-3-(phenylsulfonyl)
prop-2-enoate, **VIII**, 458
$C_{10}H_{10}O_6$ 2,5-Dihydroxy-*p*-benzenedi-
acetic acid, **III**, 286
2-Methyl-4-ethoxalylcyclopentane-
1,3,5-trione, **V**, 747
$C_{10}H_{10}Ru$ Ruthenocene, **V**, 1001
$C_{10}H_{10}S$ (*Z*)-1-Phenylthio-1,3-butadiene,
VI, 742
(*Z*)-2-Phenylthio-1,3-butadiene, **VI**,
742
$C_{10}H_{11}Br$ 4-Phenyl-2-bromobutene-1, **I**,
187
$C_{10}H_{11}ClO$ 4-(4-Chlorophenyl)-2-
butanone, **VII**, 105
Mesitoyl chloride, **III**, 555

$C_{10}H_{11}ClO_2$ Ethyl α-chlorophenylacetate,
IV, 169
$C_{10}H_{11}ClO_4$ 3,4,5-Trimethoxybenzoyl
chloride, **VI**, 1007
$C_{10}H_{11}Cl_2N$ β,β-Dichloro-*p*-dimethyl-
aminostyrene, **V**, 361
$C_{10}H_{11}I$ (*E*)-1-Iodo-4-phenyl-2-butene, **VI**,
704
$C_{10}H_{11}IO_2$ 3-Iodo-2,4,6-trimethylbenzoic
acid, **VI**, 702
$C_{10}H_{11}IO_4$ Iodobenzene diacetate, **V**, 660
$C_{10}H_{11}N$ 1-Ethylindole, **VI**, 107
2-Phenylbutyronitrile **VI**, 897, 940
1,2,3,4-Tetrahydronaphthalen(1,2)
imine, **VI**, 967
2,4,6-Trimethylbenzonitrile, **VI**, 466
$C_{10}H_{11}NO$ *N*-Benzylacrylamide, **V**, 73
$C_{10}H_{11}NOS_2$ 3,3-Bis(methylthio)-1-(2-
pyridinyl)-2-propen-1-one, **VII**,
476
$C_{10}H_{11}NO_2$ Acetoacetanilide, **III**, 10
(*S*)-4-(Phenylmethyl)-2-oxazolidi-
none, **VIII**, 339, 528
$C_{10}H_{11}NO_3$ 1-(*o*-Methoxyphenyl)-2-nitro-
1-propene, **IV**, 573
$C_{10}H_{11}NO_4$ Carbobenzoxyglycine, **III**,
168
$C_{10}H_{11}NS$ *p*-Isopropylphenyl isothiocyanate, **I**, 448
$C_{10}H_{11}N_3O_2$ α,α'-Dicyano-β-ethyl-β-
methylglutarimide, **IV**, 441
$C_{10}H_{12}$ Dicyclopentadiene **VI**, 147,
Methallylbenzene, **VI**, 722
1-Methylindane, **VI**, 539
$\Delta^{9,10}$-Octalin, **VI**, 852
(*S*)-(+)-3-Phenyl-1-butene, **VI**, 409
1-Phenyl-2-butene, **VI**, 295
1,4,5,8-Tetrahydronaphthalene, **VI**,
731, 862
$C_{10}H_{12}BrN$ 4-Bromo-1-(dimethylamino)
naphthalene or 4-Bromo-*N*,*N*-
dimethyl-1-naphthylamine, **VI**,
183
$C_{10}H_{12}Br_4O$ 3,4,8,9-Tetrabromo-11-oxatri-
cyclo[4.4.1.0^{1,6}]undecane, **VI**,
862

$C_{10}H_{12}ClN$ N-(1-Chloro-1-propen-1-yl)-N-methylaniline, **VI**, 286

$C_{10}H_{12}I_2$ 1,2-Diiodo-3,4,5,6-tetramethylbenzene, **VI**, 702

$C_{10}H_{12}N_2$ α-N,N-Dimethylaminophenylacetonitrile, **V**, 437

$C_{10}H_{12}N_2O_3$ Ethyl N-(2-methyl-3-nitrophenyl)formimidate, **VIII**, 493

$C_{10}H_{12}N_2O_4$ Dinitrodurene, **II**, 254

$C_{10}H_{12}N_4O_5$ 4-(4-Nitrophenyl)-1-carbethoxysemicarbazide or Ethyl 2-[N-(4-nitrophenyl)carbamoyl]hydrazine-1-carboxylate, **VI**, 938

$C_{10}H_{12}O$ Mesitaldehyde, **III**, 549; **V**, 49
2-Methyl-2-phenylpropionaldehyde, **VI**, 908
2-Methyl-3-phenylpropionaldehyde, **VI**, 815 or 2-Methyl-3-phenylpropanal, **VII**, 361
trans-Methylstyrylcarbinol, **IV**, 773
11-Oxatricyclo[4.4.1.0^{1,6}]undeca-3,8-diene, **VI**, 862
3-Phenylbutyraldehyde, **VI**, 815
ar-Tetrahydro-α-naphthol, **IV**, 887
1-p-Tolylcyclopropanol, **V**, 1058

$C_{10}H_{12}O_2$ Benzyl 2,3-epoxypropyl ether, **VIII**, 33
2-tert-Butyl-1,4-benzoquinone, **VI**, 891
Duroquinone, **II**, 254
Ethyl phenylacetate, **I**, 270
o-Eugenol, **III**, 418
Guaiacol allyl ether, **III**, 418
3-Hydroxy-2-methylpropiophenone, **V**, 624
Mesitoic acid, **III**, 553; **V**, 706
o-Methoxyphenylacetone, **IV**, 573
(7aS)-(+)-7a-Methyl-2,3,7,7a-tetrahydro-1H-indene-1,5(6H)-dione, **VII**, 363
γ-Phenylbutyric acid, **II**, 499
4-Phenyl-m-dioxane, **IV**, 786
Tetralin hydroperoxide, **IV**, 895
Thymoquinone, **I**, 512

2,4,6-Trimethylbenzoic acid, **VI**, 627

$C_{10}H_{12}O_3$ 2,5-Dimethylmandelic acid, **III**, 326
Ethyl mandelate, **IV**, 169
4-Ethoxy-3-methoxybenzaldehyde, **VI**, 567
(2S*,3S*)-3-Hydroxy-3-phenyl-2-methylpropanoic acid, **VIII**, 339
Quinacetophenone dimethyl ether, **IV**, 836

$C_{10}H_{12}O_4$ 3,4-Dimethoxyphenylacetic acid, **VI**, 471
Ethyl isodehydroacetate, **IV**, 549
Homoveratric acid, **II**, 333
1,4,9,12-Tetraoxadispiro[4.2.4.2]tetradeca-6,13-diene, **VI**, 972
3,4,5-Trimethoxybenzaldehyde, **VI**, 1007

$C_{10}H_{12}O_5$ Trimethylgallic acid, **I**, 537

$C_{10}H_{12}S$ 2-Methyl-1-phenylthio-1-propene, **VI**, 742

$C_{10}H_{12}S_2$ 2-Phenyl-1,3-dithiane, **VI**, 109

$C_{10}H_{13}Cl$ α^2-Chloroisodurene, **III**, 557
Neophyl chloride, **IV**, 702

$C_{10}H_{13}ClN_2$ Tryptamine hydrochloride, **VI**, 965

$C_{10}H_{13}ClO_2$ 4-Chlorobutyl benzoate, **III**, 187

$C_{10}H_{13}ClO_3S$ 1-(1-Chloro-1-ethoxymethylsulfonyl)-4-methylbenzene or 4-(1-Chloro-1-ethoxymethylsulfonyl)-1-methylbenzene, **VI**, 985

$C_{10}H_{13}I$ Iododurene, **VI**, 700

$C_{10}H_{13}N$ ac-Tetrahydro-β-naphthylamine, **I**, 499

$C_{10}H_{13}NO_2$ 7-(Carbonylamino)-7-ethoxy-3-norcarene, **VI**, 230
Ethyl 4-amino-3-methylbenzoate, **VI**, 581
Ethyl N-benzylcarbamate, **IV**, 780
Ethyl 6-methylpyridine-2-acetate, **VI**, 611
2-Nitro-p-cymene, **III**, 653
Nitrosothymol, **I**, 511

$C_{10}H_{13}NO_2Si$ p-Benzoquinone monotrimethylsilyl cyanohydrin, **VII**, 517

$C_{10}H_{13}NO_3$ N-Methyl-3,4-dihydroxyphenylalanine, **III**, 586

o-Butoxynitrobenzene, **III**, 140

$C_{10}H_{13}NO_4$ N-Methyl-3,4-dihydroxyphenylalanine, **III**, 586

2-Nitrophenylacetaldehyde dimethyl acetal, **VI**, 978

$C_{10}H_{13}N_3O_2S$ 3-Imino-1-(p-tolylsulfonyl)pyrazolidine, **V**, 40

$C_{10}H_{13}N_3O_3$ 4-Phenyl-1-carbethoxysemicarbazide or Ethyl 2-(N-phenylcarbamoyl)hydrazine-1-carboxylate, **VI**, 936

$C_{10}H_{14}$ Butylbenzene, **III**, 157; **VI**, 409

m-Cymene, **V**, 332

p-Cymene or 4-Methyl-(1-methylethyl)benzene, **VI**, 152

Durene, **II**, 248; **VI**, 700

Isodurene, **II**, 360

$C_{10}H_{14}BrN$ 4-Bromo-N,N-diethylaniline, **VI**, 183

$C_{10}H_{14}CuLiS$ Lithium phenylthio(tert-butyl)cuprate, **VI**, 248

$C_{10}H_{14}NO$ tert-Butylphenylnitrone **VI**, 806

$C_{10}H_{14}N_2$ Acetophenone N,N-dimethylhydrazone, **VI**, 12

N,N-Dimethyl-N'-p-tolylformamidine, **VII**, 197

$C_{10}H_{14}N_2O$ p-Nitrosodiethylaniline, **II**, 224

$C_{10}H_{14}N_2O_2$ 1-Amino-2-phenylaziridinium acetate, **VI**, 56

$C_{10}H_{14}O$ Adamantanone, **VI**, 41

Carvone, **VI**, 459

1,3-Dimethyl-4,5,6,6a-tetrahydro-2(1H)-pentalenone, **VI**, 523

2,4-Dimethylbicyclo[3.2.1]oct-6-en-3-one, **VI**, 515

2,4-Dimethylbicyclo[3.3.0]oct-1-en-3-one, **VI**, 523

2-(1-Methyl-2-propynyl)cyclohexanone, **VIII**, 460

$\Delta^{1(9)}$-2-Octalone, **V**, 869

Phenyl tert-butyl ether(Method I), **V**, 924; (Method II), **V**, 926

4-Protoadamantanone, **VI**, 958

Tricyclo[4.3.1.0^{3,8}]decan-4-one, **VI**, 958

Tricyclo[5.2.1.0^{4,8}]decan-3-one, **VI**, 961

Tricyclo[5.2.1.0^{3,8}]decan-4-one, **VI**, 961

$C_{10}H_{14}O_2$ 4-tert-Butylcatechol, **VI**, 96, 455

cis-1,5-Dimethylbicyclo[3.3.0]octane-3,7-dione, **VII**, 50

Spiro[4.5]decane-1,4-dione, **VIII**, 578

$C_{10}H_{14}O_3$ trans-3-(trans-2-Acetoxy-1-cyclopentyl)-2-propenal or trans-2-(trans-3-Oxo-1-propen-1-yl)-1-cyclopentyl acetate, **VI**, 689

2-Acetyl-5,5-dimethyl-1,3-cyclohexanedione, **VI**, 30

Ethyl 4-ethyl-2-methylfuran-3-carboxylate, **VI**, 32

Ethyl-7-oxo-1-cycloheptene-1-carboxylate, **VI**, 26

(3aS,7aS)-(+)-2,3,3a,4,7,7a-Hexahydro-3a-hydroxy-7a-methyl-1H-indene-1,5(6H)-dione, **VII**, 363

α-Hydroxyacetophenone dimethylacetal, **VII**, 263

DL-Ketopinic acid, **V**, 689

2-Methyl-2-(3-oxobutyl)-1,3-cyclopentanedione, **VII**, 363

$C_{10}H_{14}O_4$ Dimethyl trans-Δ^4-tetrahydrophthalate, **VI**, 457

Ethyl 2-ketocyclohexylglyoxalate, **II**, 531

Tetraacetylethane, **IV**, 869

$C_{10}H_{14}O_5$ Diethyl acetone-1,3-dicarboxylate or Diethyl 3-oxopentanedioate, **VI**, 505

Dimethyl cyclohexanone-2,6-dicarboxylate, **V**, 439

$C_{10}H_{14}O_6S_2$ Styrene glycol dimesylate, **VI**, 56

$C_{10}H_{14}O_8$ Tetramethyl 1,1,2,2-ethanete-
tracarboxylate, **VII**, 482

$C_{10}H_{15}ClO_3S$ DL-10-Camphorsulfonyl
chloride, **V**, 196

$C_{10}H_{15}F$ 1-Fluoroadamantane, **VI**, 630

$C_{10}H_{15}IO$ endo-7-Iodomethylbicyclo
[3.3.1]nonan-3-one, **VI**, 958

$C_{10}H_{15}N$ 2-Amino-p-cymene, **III**, 63

α,α-Dimethyl-β-phenylethylamine,
V, 471

N-Isobutylaniline, **VI**, 500

2-Methylbenzyldimethylamine, **IV**,
585

β-Phenylethyldimethylamine, **III**, 723

$C_{10}H_{15}NO$ Aminothymol, **I**, 512

$C_{10}H_{15}NO_2$ 7-Ethoxybicyclo[4.1.0]hept-7-
yl isocyanate or 7-(Carbonyla-
mino)-7-ethoxybicyclo[4.1.0]
heptane **VI**, 230

Ethyl (1-ethylpropylidene)
cyanoacetate, **III**, 399

N-Methyl-2,3-dimethoxybenzyl-
amine, **IV**, 603

$C_{10}H_{15}NO_2S$ (−)-(Camphorsulfonyl)imine,
VIII, 104, 110

$C_{10}H_{15}NO_3S$ (+)-(2R,8aS)-10-(Camphorsulfonyl)oxaziridine, **VIII**, 104

$C_{10}H_{15}NO_6$ (1R,2r,3S)-3-Acetoxy-2-nitro-
cyclohexyl acetate, **VIII**, 332

$C_{10}H_{15}O_3P$ Diethyl phenylphosphonate,
VI, 451

$C_{10}H_{16}$ Adamantane, **V**, 16; **VI**, 43, 48

2-Bornene, **VI**, 172

cis-7,8-Dimethylbicyclo[4.2.0]oct-7-
ene, **VI**, 482

1-Methyl-4-(1-methylethenyl)-1-
cyclohexene, **VI**, 772

$Δ^{1,9}$-Octalin, **VI**, 852

$Δ^{9,10}$-Octalin, **VI**, 852

1,2,3,4,5-Pentamethylcyclopenta-
diene, **VIII**, 505

(±)-α-Pinene **VI**, 944,

(+)-α-Pinene, **VI**, 720; **VIII**, 553

β-Pinene, **VI**, 947

endo-Tetrahydrodicyclopentadiene,
V, 16

$C_{10}H_{16}Br_2O_4$ Diethyl α,δ-dibromoadipate,
III, 623

$C_{10}H_{16}ClN$ Benzyltrimethylammonium
chloride, **VI**, 119

$C_{10}H_{16}Cl_2O_2$ Sebacoyl chloride, **V**, 536

$C_{10}H_{16}CoO_2$ Cobalt(II) acetylacetonate,
VI, 339

$C_{10}H_{16}F$ 1-Fluoroadamantane **VI**, 628,

$C_{10}H_{16}N_2$ Diaminodurene chlorostannate,
II, 255

Sebaconitrile, **III**, 768 or Sebacic
acid dinitrile or Decanedinitrile,
VI, 305

Tetramethyl-p-phenylenediamine, **V**,
1018

$C_{10}H_{16}N_2O$ trans,trans-N-(1,3-Penta-
dienyl)pyrrolidine-1-carboxam-
ide, **VI**, 100

2-Diazocyclodecanone, **VI**, 391

$C_{10}H_{16}O$ 1-Adamantanol, **VI**, 43, 628, 958

2-Allyl-3-methylcyclohexanone, **VI**,
51

Dihydrocarvone, **VI**, 459

Geranial, **VII**, 258

9-Methylbicyclo[5.2.0]nonan-8-one,
VI, 326

Perhydro-4,7-methanoinden-3a-ol,
VI, 46

α-Pinene oxide, **VI**, 949

trans-Pinocarveol, **VI**, 946, 948

$C_{10}H_{16}OS_2$ 1-(1,3-Dithianyl)-2-cyclohex-
en-1-ol, **VIII**, 309

$C_{10}H_{16}O_2$ Adamantane-1,3-diol, **VI**, 47

Ethyl cyclohexylideneacetate, **V**,
547

3-Hydroxy-1,7,7-trimethylbicy-
clo[2.2.1]heptan-2-one, endo
isomer, **VII**, 277

Methyl 2-nonynoate **VI**, 793,

Sebacil, **IV**, 838

$C_{10}H_{16}O_3$ Ethyl 5-oxo-6-methyl-6-hep-
tenoate, **VIII**, 274

Ethyl β,β-pentamethyleneglycidate,
IV, 459

$C_{10}H_{16}O_4$ Diethyl 1,1-cyclobutanedicar-
boxylate, **IV**, 288; **VI**, 272

Diethyl isopropylidenemalonate, **VI**, 442

Ethyl *tert*-butyl fumarate, **VII**, 93

(3-Methyl-2-butenyl)propanedioic acid, dimethyl ester, **VIII**, 381

3,3,6,6-Tetramethoxy-1,4-cyclohexadiene, **VI**, 971

$C_{10}H_{16}O_4S$ DL-10-Camphorsulfonic acid, **V**, 194

$C_{10}H_{16}O_5$ Diethyl acetosuccinate, **II**, 262

Diethyl ethoxymethylenemalonate, **III**, 395

$C_{10}H_{16}O_6$ Tricarbethoxymethane, **II**, 594

$C_{10}H_{17}Cl$ Geranyl chloride, **VI**, 634, 638; **VIII**, 226

Linalyl chloride, **VI**, 635

Neryl chloride, **VI**, 639

$C_{10}H_{17}N$ 1-Pyrrolidinocyclohexene or 1-(1-Cyclohexenyl)pyrrolidine, **VI**, 1014

$C_{10}H_{17}NO$ 1-Morpholino-1-cyclohexene, **V**, 808

$C_{10}H_{17}NO_2$ ω-Cyanopelargonic acid, **III**, 768

Ethyl α-(hexahydroazepinylidene-2) acetate, **VIII**, 263

Ethyl β-pyrrolidinocrotonate, **VI**, 592

$C_{10}H_{17}NO_2S$ (−)-D-2,10-Camphorsultam, **VIII**, 110

$C_{10}H_{17}NO_3S$ (+)-(1S)-10-Camphorsulfonamide, **VIII**, 104

$C_{10}H_{17}NO_4$ *tert*-Butoxycarbonyl-L-proline, **VI**, 203

$C_{10}H_{17}NO_4S$ Phenyltrimethylammonium sulfomethylate, **VI**, 175

$C_{10}H_{18}$ 1-*tert*-Butylcyclohexene, **VI**, 855

Decyne-1, **I**, 192

Hexamethylbicyclo[1.1.0]butane, **VI**, 136

1-Isopropyl-4-methylcyclohexene, **VI**, 296

$C_{10}H_{18}N_2O$ 1-Diazo-2-decanone, **VI**, 388

$C_{10}H_{18}N_2O_2$ N-(5-Ethoxybicyclo[2.1.0] pent-5-yl)-N',N'-dimethylurea, **VI**, 230

$C_{10}H_{18}N_2O_4$ *tert*-Butyl azodiformate, **V**, 160

$C_{10}H_{18}O$ 2-Butylcyclohexanone, **VI**, 528

4-*tert*-Butylcyclohexanone, **VI**, 215, 218, 220

(R)-(+)-Citronellal, **VIII**, 183

Cyclodecanone, **IV**, 218; **V**, 277; **VI**, 370

1-Decalol, **VI**, 371

cis-1-Decalol, **VI**, 47

trans-9-Decalol, **VI**, 46,

2-Decalol, **VI**, 372,

9-Decyn-1-ol, **VIII**, 146

cis-2,2-Dimethylhexahydrobenzofuran, **VI**, 372

3,7-Dimethyl-6-octenal, **VI**, 670

Geraniol, **VI**, 507, 634, 638

3-Hexylcyclobutanone, **VIII**, 306

Isomenthone, **VIII**, 386

(−)-Isopinocampheol, **VI**, 719,

L-Menthone, **I**, 340

2-*tert*-Pentylcyclopentanone, **VII**, 424

3,4,5-Trimethyl-2,5-heptadien-4-ol, **VIII**, 505

$C_{10}H_{18}O_2$ Di-*tert*-butoxyethyne, **VIII**, 161

2-Hydroxycyclodecanone, **VI**, 171

6-Oxodecanal, **VIII**, 498,

Sebacoin, **IV**, 840

$C_{10}H_{18}O_2S$ 7,9-Dimethyl-*cis*-8-thiabicyclo[4.3.0]nonane 8,8-dioxide, **VI**, 482,

$C_{10}H_{18}O_3$ Ethyl butylacetoacetate, **I**, 248

Ethyl 5,5-dimethyl-4-oxohexanoate, **VI**, 251

Ethyl isobutyrylisobutyrate, **II**, 270

Methyl 4-keto-7-methyloctanoate, **III**, 601

6-Oxodecanoic acid, **VIII**, 499

$C_{10}H_{18}O_4$ Ethyl adipate, **II**, 264

Ethyl isopropylmalonate, **II**, 94

(±)-Nonactic acid, **VI**, 515

$C_{10}H_{18}O_4Zn$ Bis(2-carboethoxyethyl)zinc, **VIII**, 277

$C_{10}H_{18}O_5$ Di-*tert*-butyl dicarbonate, **VI**, 418

$C_{10}H_{18}O_5S$ *erythro*-3-Methanesulfonyloxy-1-methyl-1-propyl cyclobutanecarboxylate, **VI**, 312

$C_{10}H_{18}O_8$ Hexahydrogallic acid triacetate, **V**, 591

$C_{10}H_{18}Si_2$ 1,4-Bis(trimethylsilyl)buta-1,3-diyne, **VIII**, 63

$C_{10}H_{19}Br$ 2-Bromodecene-1, **I**, 187

$C_{10}H_{19}ClO_2$ 1,2-Di-*tert*-butoxy-1-chloroethene, **VIII**, 161

$C_{10}H_{19}N$ 3-Pinanamine, **VI**, 943
3,7-Dimethyl-1,6-octadien-3-amine, **VI**, 507

$C_{10}H_{19}NOSi$ [(*trans*-2-Isocyanocyclohexyl)oxy]trimethylsilane, **VII**, 294

$C_{10}H_{19}NO_2$ 1-Acetoxy-4-diethylamino-2-butene, **VIII**, 9

$C_{10}H_{19}NO_3$ *tert*-Butyl N-(1-ethoxycyclopropyl)carbamate **VI**, 226

$C_{10}H_{20}Br_2$ Decamethylene bromide, **III**, 227

$C_{10}H_{20}Cl_2O_2$ 1,2-Di-*tert*-butoxy-1,2-dichloroethane, **VIII**, 161

$C_{10}H_{20}I$ Iododecane or Decyl iodide, **VI**, 376

$C_{10}H_{20}N_2O_4$ 2,7-Dimethyl-2,7-dinitrooctane, **V**, 445

$C_{10}H_{20}O$ *cis*-3-*tert*-Butylcyclohexanol, **VI**, 215
cis-4-*tert*-Butylcyclohexanol, **VI**, 215
trans-4-*tert*-Butylcyclohexanol, **V**, 175; **VI**, 218
1-Decanal, **VI**, 373
2-Decanone, **VII**, 137
3-Methyl-2-oxononane **VI**, 813,

$C_{10}H_{20}OS$ 7-Thiomenthol, **VIII**, 302

$C_{10}H_{20}OSi$ 1-Trimethylsilyloxybicyclo[4.1.0]heptane, **VI**, 327

$C_{10}H_{20}O_2$ 1,2-Cyclodecanediol, **IV**, 216
Cyclohexanone diethyl ketal, **VIII**, 579
Pivaloin, **II**, 115

$C_{10}H_{20}O_3$ β-Ketoisooctaldehyde dimethyl acetal, **IV**, 558

Methyl 6-oxodecanoate, **VIII**, 441

$C_{10}H_{20}O_5$ 2,2,6,6-Tetramethylolcyclohexanol, **IV**, 907

$C_{10}H_{20}O_6$ 2,3,4,6-Tetramethyl-D-glucose, **III**, 800

$C_{10}H_{20}O_6S_2$ *cis*-1,2-Cyclohexanedimethanyl dimethanesulfonate, **VI**, 482

$C_{10}H_{20}O_7P\cdot3 H_3N$ Trisammonium geranyl diphosphate, **VIII**, 616

$C_{10}H_{20}S_4$ 1,4,8,11-Tetrathiacyclotetradecane, **VIII**, 592

$C_{10}H_{21}NO_2$ (*S*)-N-Formyl-*O*-*tert*-butylvalinol, **VIII**, 204

$C_{10}H_{21}O_5P$ Diethyl [(2-tetrahydropyranyloxy)methyl]phosphonate, **VII**, 160

$C_{10}H_{22}$ Decane, **VI**, 376

$C_{10}H_{22}BrN$ β-Dibutylaminoethyl bromide hydrobromide, **II**, 92

$C_{10}H_{22}O$ 1-Decanol, **VI**, 373
Tripropyl carbinol, **II**, 603

$C_{10}H_{22}O_2$ Decamethylene glycol, **II**, 154

$C_{10}H_{22}O_2Si_2$ 1,2-Bis(trimethylsilyloxy)cyclobutene, **VI**, 167

$C_{10}H_{22}S$ Dineopentyl sulfide, **VI**, 834
Ethyl octyl sulfide or Ethyl 1-octyl sulfide, **VI**, 834
Ethyl 2-octyl sulfide or Ethyl 1-methyl-1-heptyl sulfide, **VI**, 834

$C_{10}H_{22}Zn$ Diisoamylzinc, **II**, 187

$C_{10}H_{23}B$ Bis(3-methyl-2-butyl)borane or Bis(1,2-dimethylpropyl)borane **VI**, 852, 919,

$C_{10}H_{23}O_5P$ Diethyl 2,2-diethoxyethylphosphonate, **VI**, 448

$C_{10}H_{24}N_2$ Decamethylenediamine, **III**, 229
β-Dibutylaminoethylamine, **III**, 254

$C_{10}H_{24}N_2O_2$ N,N,N,N-Tetramethyl-1,4-diamino-2,3-dimethoxybutane (DDB), **VII**, 41

$C_{10}H_{24}O$ 4-*tert*-Butylcyclohexanone, **VI**, 220

$C_{10}H_{24}O_3Ti$ Methyltriisopropoxytitanium, **VIII**, 495

$C_{11}H_6F_5NO_2$ Ethyl cyano(pentafluorophenyl)acetate, **VI**, 873,

$C_{11}H_7ClO$ 1-Naphthoyl chloride, **VI**, 613, 2-Naphthoyl chloride, **III**, 629

$C_{11}H_7Cl_3N_2O_3$ 2-(2,2,2-Trichloroethoxy) carbonyloxyimino-2-phenylacetonitrile, **VI**, 202

$C_{11}H_7N$ α-Naphthonitrile, **III**, 631 or 1-Naphthalenecarbonitrile, **VI**, 305

$C_{11}H_7NS$ α-Naphthyl isothiocyanate, **IV**, 700
β-Naphthyl isothiocyanate, **I**, 448

$C_{11}H_8$ 1-Phenyl-1,3-pentadiyne, **VI**, 925
1-Phenyl-1,4-pentadiyne **VI**, 925

$C_{11}H_8N_2O_3$ Benzalbarbituric acid, **III**, 39

$C_{11}H_8O$ 1-Naphthaldehyde, **IV**, 690
β-Naphthaldehyde, **III**, 626 or 2-Formylnaphthalene, **VI**, 633

$C_{11}H_8OS$ Phenyl thienyl ketone, **II**, 520

$C_{11}H_8O_2$ 2-Hydroxy-1-naphthaldehyde, **III**, 463
α-Naphthoic acid, **II**, 425 or 1-Naphthoic acid, **VI**, 614
β-Naphthoic acid, **II**, 428

$C_{11}H_8O_3$ 2-Acetyl-1,3-indanedione, **VI**, 30
2-Methoxy-1,4-naphthoquinone, **III**, 465

$C_{11}H_8O_4$ 1,3-Dihydroxy-2-naphthoic acid, **III**, 638

$C_{11}H_9BrO$ 6-Bromo-2-methoxynaphthalene, **V**, 920

$C_{11}H_9Cl$ 1-Chloromethylnaphthalene, **III**, 195

$C_{11}H_9N$ tert-Butyl isocyanide or 2-(Carbylamino)-2-methylpropane, **VI**, 232
2-Phenylpyridine, **II**, 517

$C_{11}H_9NO_2$ 3-Amino-2-naphthoic acid, **III**, 78
Azlactone of α-acetaminocinnamic acid, **II**, 1

$C_{11}H_9NO_4S$ 4-(4-Carboxyphenyl)-4H-1,4-thiazine 1,1-dioxide, **VI**, 978

$C_{11}H_{10}$ 1,6-Methano[10]annulene, **VI**, 731, 865

$C_{11}H_{10}Cl_3NO$ 3-Phenyl-3-trichloroacetamido-1-propene or N-(1-Phenyl-2-propen-1-yl)trichloroacetamide, **VI**, 510

$C_{11}H_{10}F_3NO_4$ N-Trifluoroacetyl-L-tyrosine, **VI**, 1004

$C_{11}H_{10}FeO_2$ Ferrocenecarboxylic acid, **VI**, 625

$C_{11}H_{10}N_2O_2S_2$ N-(p-Acetylphenyl)rhodanine, **IV**, 6

$C_{11}H_{10}O$ 5,6-Dihydrobenzocyclohepten-7-one, **VI**, 331
2-Methoxynaphthalene, **VI**, 34
2-Methylene-1-oxo-1,2,3,4-tetrahydronaphthalene, **VII**, 332
Methyl β-naphthyl ether, **I**, 59
1-Phenyl-1-penten-4-yn-3-ol, **IV**, 792

$C_{11}H_{10}O_2$ 2-Ethylchromone, **III**, 387
6-Methoxy-2-naphthol, **V**, 918
2-Methylene-1-phenyl-1,3-butanedione, **VI**, 63

$C_{11}H_{10}O_3$ α-Phenylglutaric anhydride, **IV**, 790

$C_{11}H_{11}BrO_2$ 2-Bromo-3-phenyl-2-propen-1-yl acetate, **VI**, 188

$C_{11}H_{11}BrO_3$ p-Acetyl-α-bromohydrocinnamic acid, **VI**, 21

$C_{11}H_{11}BrO_4$ p-Bromobenzaldiacetate, **II**, 442

$C_{11}H_{11}ClO_2$ 1-(4-Chlorophenyl)-2,4-pentanedione, **VI**, 930

$C_{11}H_{11}DO$ 2-Phenyl-4-pentenal-1-d, **VI**, 908

$C_{11}H_{11}F_3O_3S$ α,β-Dimethylstyryl triflate, **VI**, 760

$C_{11}H_{11}I$ (E)-1-Iodo-4-phenyl-2-butene, **VI**, 704

$C_{11}H_{11}IO_2$ 4,5-Dihydro-5-iodomethyl-4-phenyl-2(3H)-furanone, cis and trans, **VII**, 164

$C_{11}H_{11}N$ 1-Allylindole, **VI**, 107
2,4-Dimethylquinoline, **III**, 329

$C_{11}H_{11}NO$ 2-Methyl-4-oxo-4-phenylbutyronitrile, **VI**, 867

3-Methyl-4-oxo-4-phenylbutyroni-
trile, **VI**, 867

$C_{11}H_{11}NOS$ *S*-Phenyl *trans*-1,3-butadiene-
1-thiocarbamate, **VI**, 100

$C_{11}H_{11}NO_2$ Ethyl indole-2-carboxylate, **V**,
567

Ethyl phenylcyanoacetate, **IV**, 461

Phenyl *trans*-1,3-butadiene-1-carba-
mate, **VI**, 100

$C_{11}H_{11}NO_3$ α-Acetaminocinnamic acid, **II**,
1

2-Oxo-1-phenyl-3-pyrrolidinecar-
boxylic acid, **VII**, 411

$C_{11}H_{11}NO_4S$ *N*-Ethyl-5-phenylisoxazoli-
um-3'-sulfonate, **VI**, 263

N-Ethyl-5-phenylisoxazolium-4'-sul-
fonate, **VI**, 265

$C_{11}H_{11}NO_6$ *o*-Nitrobenzaldiacetate, **IV**,
713

p-Nitrobenzaldiacetate, **II**, 441; **IV**,
713

$C_{11}H_{12}Br_2O_2$ Ethyl α,β-dibromophenylpro-
pionate, **II**, 270

$C_{11}H_{12}Cl_2$ 11,11-Dichlorotricyclo[4.4.1.0$^{1.6}$]
undeca-3,8-diene, **VI**, 732

$C_{11}H_{12}FeO$ Hydroxymethylferrocene, **V**,
621

$C_{11}H_{12}N_2$ 3,4-Dicyanobicyclo[4.3.0]non-3-
ene or

Bicyclo[4.3.0]non-3-ene-3,4-dicar-
bonitrile, **VI**, 431

1,2,3,4-Tetrahydro-α-carboline, **VI**,
965

$C_{11}H_{12}N_2O_2$ L-Tryptophan, **II**, 612

$C_{11}H_{12}N_4O_4$ Cyclobutanecarboxaldehyde
2,4-dinitrophenylhydrazone,
VI, 314

$C_{11}H_{12}O$ Isopropylidenacetophenone,
VIII, 210

4-Oxahexacyclo[5.4.0.0^{2,6}.0^{3,10}.0^{5,9}.0^{8,11}]
undecane, **VI**, 1003

1-Phenylcyclobutanecarboxalde-
hyde, **VI**, 908

(α-Phenylcyclopropyl)methyl
ketone, **VI**, 899

3-Phenyl-4-pentenal, **VIII**, 536

$C_{11}H_{12}OS$ 1-Methoxy-4-phenylthio-1,3-
butadiene, **VI**, 743

2-Methoxy-3-phenylthio-1,3-butadi-
ene, **VI**, 743

(Z)-2-Methoxy-1-phenylthio-1,3-
butadiene, **VI**, 737

$C_{11}H_{12}O_2$ 2-Acetoxy-*trans*-1-phenyl-
propene or 1-Methyl-2-phenyl-
1-ethenyl acetate, **VI**, 692

Anisalacetone, **I**, 78

Cyclopropylmethyl benzoate, **VI**,
261

Ethyl cinnamate, **I**, 252

6-Methoxy-1-tetralone or 6-
Methoxy-α-tetralone, **VI**, 307

6-Methoxy-β-tetralone or 6-
Methoxy-2-tetralone, **VI**, 744

1-Phenyl-2,4-pentanedione, **VI**, 928

3-Phenyl-2,4-pentanedione, **VI**, 247

$C_{11}H_{12}O_2S$ 2-(Methylsulfinyl)-1-phenyl-1-
buten-1-one, **VI**, 26,

1-(Phenylsulfonyl)cyclopentene,
VIII, 38, 543

$C_{11}H_{12}O_3$ γ-Benzoylbutyric acid, **II**, 82

Ethyl benzoylacetate, **II**, 266; **III**,
379; **IV**, 415

Methyl phenylacetylacetate, **VII**,
359

$C_{11}H_{12}O_4$ 2,3-Dimethoxycinnamic acid,
IV, 327

6,7-Dimethoxy-3-isochromanone,
VI, 471

Ethyl 2-acetoxybenzoate **VI**, 579

Ethyl 4-acetoxybenzoate, **VI**, 576

Methyl 2-oxo-5,6,7,8-tetrahydro-2*H*-
1-benzopyran-3-carboxylate,
VIII, 444

$C_{11}H_{12}O_5$ 3,4-Dimethoxyphenylpyruvic
acid, **II**, 335

$C_{11}H_{13}BF_4O_3Fe$ Dicarbonyl(cyclopentadienyl)(ethyl vinyl ether) iron
tetrafluoroborate, **VIII**, 479

$C_{11}H_{13}N$ Mesitylacetonitrile, **III**, 557

$C_{11}H_{13}NO$ 2-(Methylenemethoxy)-2-
phenylpropionitrile, **VI**, 899

1-Methyl-3-ethyloxindole, **IV**, 620

$C_{11}H_{13}NO_2$ 2-(Dimethoxymethyl)-3-
phenyl-2H-azirine, **VI**, 893
Ethyl 6-methylpyridine-2-acetate,
VI, 611
cis-N-Methyl-1,3a,4,9b-tetrahydro-
3H-[1]benzopyrano[3,4-c] isoxazole, **VI**, 670
$C_{11}H_{13}NO_3$ N-Acetylphenylalanine, **II**, 493
$C_{11}H_{13}NS$ 2-Methyl-4-isopropylphenyl
isothiocyanate, **I**, 448
$C_{11}H_{13}N_3O_2$ (1-Azido-3,3-dimethoxy-1-
propenyl)benzene, **VI**, 893
$C_{11}H_{13}OP$ 3-Methyl-1-phenylphospholene
1-oxide, **V**, 787
1-Phenyl-4-phosphorinanone, **VI**,
932
$C_{11}H_{14}$ 3,4-Dimethyltricyclo[4.2.1.0$^{2.5}$]
nona-3,7-diene, **VI**, 486
1-Methyl-1,2,3,4-tetrahydronaphtha-
lene, **VI**, 539
Tricyclo[4.4.1.0$^{1.6}$]undeca-3,8-diene,
VI, 731
$C_{11}H_{14}ClN$ 1-Chloro-N,2-dimethyl-N-
phenyl-1-propen-1-amine, **VI**,
286
1-Chloro-N,N-dimethyl-2-phenyl-1-
propen-1-amine, **VI**, 286
$C_{11}H_{14}Cl_2O_2$ 6-[(E)-1,2-Dichlorovinyl]-3-
ethoxy-6-methyl-2-cyclohexen-
1-one, **VII**, 241
$C_{11}H_{14}IN_3O_2$ (1-Azido-2-iodo-3,3-
dimethoxypropyl)benzene, **VI**,
893
$C_{11}H_{14}N_2$ 4-(Diethylamino)benzonitrile
VI, 466,
$C_{11}H_{14}N_2O_5S$ Ethyl N-nitroso-N-(p-tolyl-
sulfonylmethyl)carbamate, **VI**,
981
$C_{11}H_{14}O$ sec-Butyl phenyl ketone, **VI**, 251
tert-Butyl phenyl ketone, **VI**, 248
2,2-Dimethyl-3-phenylpropionalde-
hyde **VI**, 526,
(R)-4-Ethyl-4-allyl-2-cyclohexen-1-
one, **VIII**, 241
2-Phenylpentanal, **VI**, 908
4-Phenyl-2-pentanone, **VI**, 816

$C_{11}H_{14}O_2$ tert-Butyl benzoate **VI**, 26,
Cinnamaldehyde dimethyl acetal,
VI, 893
2',6'-Dimethylphenyl propanoate,
VII, 190
1,6-Dioxo-8α-methyl-1,2,3,4,6,7,8a-
octahydronaphthalene, **V**, 486
8-Ethynyl-8-methyl-1,4-dioxas-
piro[4.5]dec-3-ene, **VII**, 241
3-Hydroxy-3-methyl-1-phenyl-1-
butanone, **VIII**, 210, 323,
Mesitylacetic acid, **III**, 557
(8aS)-Methyl-3,4,8,8a-tetrahydro-
1,6(2H,7H)-naphthalene-
dione, **VII**, 368; and 1,6-Dioxo-
8a-methyl-1,2,3,4,6,7,8,8a-
octahydronaphthalene, **V**, 486
Methyl 2,4,6-trimethylbenzoate, **VI**,
579
$C_{11}H_{14}O_3$ tert-Butyl phenyl carbonate, **VI**,
203
Ethyl β-phenyl-β-hydroxypropionate, **III**, 408
$C_{11}H_{14}O_3S$ Cyclobutyl tosylate, **VII**, 117
$C_{11}H_{14}O_4$ Methyl homoveratrate, **II**, 333
$C_{11}H_{14}O_7$ Ethyl acetonedicarboxylate, **II**,
126
$C_{11}H_{14}S$ 2-Methyl-3-phenylthio-2-butene,
VI, 742
$C_{11}H_{14}S_3$ 2-Methylthio-2-phenyl-1,3-dithi-
ane **VI**, 109,
$C_{11}H_{15}ClO_4$ Diethyl 2-chloro-2-cyclo-
propylethene-1,1-dicarboxylate,
VIII, 247
$C_{11}H_{15}F_3N_2O_4$ N-Trifluoroacetyl-L-prolyl-
glycine ethyl ester, **VI**. 1006
$C_{11}H_{15}N$ 2-Adamantanecarbonitrile, **VI**, 41
N-tert-Butylbenzaldimine, **V**, 191
1-Phenylcyclopentylamine, **VI**, 910
1-Phenylpiperidine, **IV**, 795
$C_{11}H_{15}NO$ 2-tert-Butyl-3-phenyloxazi-
rane, **V**, 191
β-Dimethylaminopropiophenone
hydrochloride, **III**, 305
N-Formyl-α,α-dimethyl-β-phenethyl-
amine, **V**, 471

$C_{11}H_{15}NO_2S$ Ethyl 4-amino-3-(methylthiomethyl)benzoate, **VI**, 581

$C_{11}H_{15}NO_3$ 4-Acetyl-7,8-dimethyl-2-oxa-4-
azabicyclo[4.3.0]non-7-en-3-
one, **VII**, 5

2,4-Dimethyl-3-acetyl-5-car-
bethoxypyrrole, **III**, 513

$C_{11}H_{15}NO_4S$ Ethyl N-(p-tolylsulfonyl-
methyl)carbamate, **VI**, 981

$C_{11}H_{16}$ Amylbenzene, **II**, 47

(1R)-Nopadiene, **VIII**, 223

Pentamethylbenzene, **II**, 248

$C_{11}H_{16}Cl_3NO$ 3,5,5-Trimethyl-3-
trichloroacetamido-1-cyclohex-
ene or N-(1,5,5-Trimethyl-2-
cyclohexen-1-yl)trichloroac-
etamide, **VI**, 510

$C_{11}H_{16}N_2$ 1-Benzylpiperazine, **V**, 88

$C_{11}H_{16}N_2O_4$ Diethyl 2,3-diazabicyclo
[2.2.1]hept-5-ene-2,3-dicar-
boxylate, **V**, 96

$C_{11}H_{16}O$ 1-Adamantanol, **VI**, 958

7-$tert$-Butoxynorbornadiene, **V**, 151

1,3-Dimethyl-1,4,5,6,7,7a-hexahy-
dro-2-indenone, **VI**, 523

cis-1,7a-Dimethyl-1,2,3,6,7,7a-hexa-
hydro-5-indenone, **VI**, 668

1-Methyl-4,4a,5,6,7,8-hexahydro-
naphthalen-2(3H)-one, **VI**, 781

10-Methyl-$\Delta^{1(9)}$-2-octalone, **VI**, 242

D-2-Oxo-7,7-dimethyl-1-vinylbicy-
clo[2.2.1]heptane, **V**, 877

(R)-(+)-1-Phenylpentanol, **VII**, 447

Tricyclo[4.3.1.1^{3,8}]undecan-4-one,
VI, 961

Tricyclo[4.3.1.1^{5,8}]undecan-4-one,
VI, 961

Tricyclo[4.3.2.0^{3,8}]undecan-4-one,
VI, 961

$C_{11}H_{16}OSi$ Acetophenone trimethylsilyl
enol ether, **VIII**, 210, 323

$C_{11}H_{16}O_2$ 1-Adamantanecarboxylic acid,
V, 20

$C_{11}H_{16}O_3$ 3,5-Dimethyl-4-carbethoxy-2-
cyclohexen-1-one, **III**, 317

1,3-Dimethyl-5-oxobicyclo[2.2.2]
octane-2-carboxylic acid, **VIII**,
219

Ethyl 8-oxo-1-cyclooctene-1-car-
boxylate, **VI**, 26

2-Methyl-2-(3-oxobutyl)-1,3-cyclo-
hexanedione, **VII**, 368

$trans$-2-(3-Oxo-1-propen-1-yl)-1-
cyclohexyl acetate, **VI**, 688

$C_{11}H_{16}O_3S$ Butyl p-toluenesulfonate, **I**,
145

$C_{11}H_{16}O_5$ Diethyl cyclopropylcarbonylmalonate, **VIII**, 248

$C_{11}H_{16}S$ Neopentyl phenyl sulfide, **VI**,
833

$C_{11}H_{17}F_3O_3S$ 4-$tert$-Butylcyclohexen-1-yl
trifluoromethanesulfonate,
VIII, 97

$C_{11}H_{17}N$ N-Isopropylphenethylamine, **VI**,
545

$C_{11}H_{17}NO$ 2-(Ethylamino)-1-phenyl-1-
propanol, **VI**, 546

N-(4-Methyl-2,4-hexadienoyl)pyrro-
lidine, **VI**, 494

$C_{11}H_{17}NO_2$ Ethyl 1-cyano-2-methylcyclo-
hexanecarboxylate, **VI**, 586

Ethyl (1-ethylpropenyl)methylcyano-
acetate, **III**, 397

Ethyl (E)-2-cyano-6-octenoate, **VI**,
586

Methyl 3-(2-cyano-1-cyclohexyl)
propionate, **VI**, 336

$C_{11}H_{18}$ (E)-4,8-Dimethyl-1,3,7-nonatriene,
VII, 259

$C_{11}H_{18}N_2O$ 2-Diazocycloundecanone, **VI**,
391

$C_{11}H_{18}N_2O_2$ 2,2-Diethoxy-2-(4-pyridyl)eth-
ylamine, **VII**, 149

$C_{11}H_{18}N_2O_3$ Diethyl 2,3-diazabicyclo
[2.2.1]heptane-2,3-dicarboxy-
late, **V**, 97

$C_{11}H_{18}N_2O_4$ Di-$tert$-butyl diazomalonate,
VI, 414

$C_{11}H_{18}O$ 3-(3-Butenyl)-2-methylcyclo-
hexanone, **VI**, 54

Di-$tert$-butylcyclopropenone, **VI**,
991

Dihydrojasmone, **VIII**, 620

4-Ethyl-4-allyl-2-cyclohexen-1-one, (*R*)-, **VIII**, 241

1-Hydroxymethyl-4-(1-methylcyclopropyl)-1-cyclohexene, **VIII**, 321

$C_{11}H_{18}O_2$ 10-Undecynoic acid, **IV**, 969

$C_{11}H_{18}O_3$ 2-Carbethoxycyclooctanone, **V**, 198

2-Oxabicyclo[4.1.0]hept-7-yl trimethylacetate, **VI**, 574

$C_{11}H_{18}O_5$ Diethyl (2*S*,3*R*)-(+)-3-allyl-2-hydroxysuccinate, **VII**, 153

Diethyl β-ketopimelate, **V**, 384

Diethyl γ-oxopimelate, **IV**, 302

Diethyl propionylsuccinate, **VI**, 615

Ethyl α-acetoglutarate, **II**, 263

6-Ketohendecanedioic acid, **IV**, 555

$C_{11}H_{18}O_5S_2$ Di-*tert*-butyl dithiotricarbonate, **VI**, 421

$C_{11}H_{18}O_6$ 1,2,5-Triacetoxypentane, **III**, 833

$C_{11}H_{18}O_7$ Di-*tert*-butyl tricarbonate, **VI**, 418

$C_{11}H_{19}ClO_3$ N-Acetylhexahydrophenylalanine, **II**, 493

ω-Carbomethoxypelargonyl chloride, **III**, 613

$C_{11}H_{19}N$ 4-*tert*-Butylcyclohexanecarbonitrile, **VI**, 336

2-Phenylpyridine, **II**, 517

$C_{11}H_{19}NO_2$ 8-(*N*,*N*-Dimethylcarbamoyl)bicyclo[5.1.0]octane, **VI**, 574

Methyl ω-cyanopelargonate, **III**, 584

$C_{11}H_{19}NO_3$ N-Acetylhexahydrophenylalanine, **II**, 493

$C_{11}H_{19}NSi$ N-Benzyl-N-[(trimethylsilyl)methyl]amine, **VIII**, 231

$C_{11}H_{19}O_6P$ 6-Diethylphosphonomethyl-2,2-dimethyl-1,3-dioxen-4-one, **VIII**, 192

$C_{11}H_{20}$ 3-Butyl-1-heptyne, **VI**, 596

2-Butyl-2-methylcyclohexanone, **V**, 187

(+)-Methylene-*cis-p*-menthane, **VIII**, 386

$C_{11}H_{20}Br_2O$ α,α'-Dibromodineopentyl ketone, **VI**, 991

$C_{11}H_{20}ClN$ (α-Chlorocyclohexylidenemethyl)diethylamine, **VI**, 286

$C_{11}H_{20}O$ 2-Butylcycloheptanone, **VI**, 528

2-Butyl-2-methylcyclohexanone, **V**, 187;

Homogeraniol, **VII**, 288

p-Menth-8-en-1-yl methyl ether, **VII**, 304

$C_{11}H_{20}OS$ S-*tert*-Butyl cyclohexylmethanethioate, **VII**, 81, 87

Hexahydro-4,4,7-trimethyl-4*H*-1,3-benzoxathiin, **VIII**, 302

$C_{11}H_{20}O_2$ *tert*-Butyl cyclohexanecarboxylate, **VII**, 87

Diisovalerylmethane, **III**, 291

4-Ethyl-2-methyl-2-octenoic acid, **IV**, 444

2-Hydroxycycloundecanone, **VI**, 171

11-Hydroxyundecanoic lactone, **VI**, 698

2,3,3-Trimethylcyclopropyl trimethylacetate, **VI**, 574

2,5-Undecanedione, **VIII**, 620

$C_{11}H_{20}O_2Si$ 4-Trimethylsilyloxy-4-cyclohexenylethanone, **VI**, 447

$C_{11}H_{20}O_3$ Methyl 6-oxodecanoate, **VIII**, 441

$C_{11}H_{20}O_4$ Di-*tert*-butyl malonate, **IV**, 261; **VI**, 414

Diethyl butylmalonate, **I**, 250

Diethyl *sec*-butylmalonate, **II**, 417

Diethyl *tert*-butylmalonate, **VI**, 442

Diethyl pimelate, **II**, 536

Hendecanedioic acid, **IV**, 510

$C_{11}H_{21}BrO_2$ 11-Bromoundecanoic acid, **VI**, 698

$C_{11}H_{21}NO$ Hexahydro-1,3,3,6-tetramethyl-2,1-benzisoxazoline, **VI**, 670

1-(2-Piperidyl)cyclohexanol, **VI**, 546

$C_{11}H_{21}NO_3$ N-*tert*-Butoxycarbonyl-L-leucinal, **VIII**, 68

Methyl sebacamate, **III**, 613

$C_{11}H_{21}NO_4$ Di-β-carbethoxyethylmethylamine, **III**, 258

C₁₁H₂₂ (E)-2-Undecene, **VI**, 1040

$C_{11}H_{22}MoN_4O_6P$ MoOPH, **VII**, 277

$C_{11}H_{22}N_2O$ 3-Pentanone SAMP- hydrazone, **VIII**, 403

$C_{11}H_{22}O$ 2-Butylheptanal, **VI**, 528

   3-Butyl-2-heptanone, **VI**, 596

   5,5-Dimethyl-2-pentyltetrahydrofuran, **IV**, 350

   Dineopentyl ketone, **VI**, 991

   3-Oxoundecane, **VI**, 813

   Undecanal, **VI**, 871

   Undecylenyl alcohol, **II**, 374

$C_{11}H_{22}O_2$ Acetone dibutyl acetal, **V**, 5

   Ethyl nonanoate, **VI**, 813

   5-Hydroxy-1,1,6,6-tetramethyl-3-heptanone, **VI**, 697

   3-Hydroxy-2,2,3-trimethyloctan-4-one, **VIII**, 343

   2-Methyl-2,5-decanediol, **IV**, 601

   Undecamethylene glycol, **II**, 155

$C_{11}H_{22}O_2Si$ 2-(1-Methylethoxy)-1-(trimethylsilyloxy)-1-ethenylcyclopropane, **VI**, 574

   2-Methyl-5-trimethylsilyl-2-vinylpentanoic acid, **VIII**, 486, 468

$C_{11}H_{23}MoN_4O_6P$ Oxodiperoxymolybdenum(pyridine)(hexamethylphosphoric triamide), **VII**, 277

$C_{11}H_{23}NO_2S$ S-tert-Butyl-L-cysteine **VI**, 252

$C_{11}H_{24}O_2$ 2-Methyl-2,5-decanediol, **IV**, 601

$C_{11}H_{24}O_4$ Malonaldehyde bis(diethyl acetal), **VI**, 978

$C_{11}H_{26}N_2$ γ-Dibutylaminopropylamine, **III**, 256

$C_{12}F_{10}$ 2,2',3,3',4,4',5,5',6,6'-Decafluorobiphenyl **VI**, 880,

$C_{12}H_4F_5NO_2$ 2-(Pentafluorophenyl)nitrobenzene, **VI**, 880

   3-(Pentafluorophenyl)nitrobenzene, **VI**, 880

   4-(Pentafluorophenyl)nitrobenzene, **VI**, 880

$C_{12}H_4F_6$ 3-(Pentafluorophenyl)fluorobenzene, **VI**, 880

$C_{12}H_5F_5$ (Pentafluorophenyl)benzene, **VI**, 879

$C_{12}H_5N_3O_7$ 2,4,7-Trinitrofluorenone, **III**, 837

$C_{12}H_6Cl_4$ Tetrachlorobenzobarrelene, **VI**, 82

$C_{12}H_6O_2$ Acenaphthenequinone, **III**, 1

$C_{12}H_6O_3$ 1,2-Naphthalic anhydride, **II**, 423

$C_{12}H_8$ Biphenylene, **V**, 54

$C_{12}H_8B_2F_3N_4$ 4,4'-Biphenylene-bis-diazonium fluoborate, **II**, 188

$C_{12}H_8Br_2$ 4,4'-Dibromobiphenyl, **IV**, 256

$C_{12}H_8Br_2O_3$ 2,4-Dibromoglutaconaldehyde enol benzoate, **VI**, 643

$C_{12}H_8Cl_2$ 3,4-Dichlorobiphenyl, **V**, 51

   4,4'-Dichloro-1,1-biphenyl, **VI**, 490

$C_{12}H_8Cl_2O_3$ 2,4-Dichloroglutaconaldehyde enol benzoate, **VI**, 643

$C_{12}H_8FNOS$ N-(3-Fluorophenyl)-2(1H)-pyridinethione, **VI**, 643

   N-(4-Fluorophenyl)-2(1H)-pyridinethione, **VI**, 643

$C_{12}H_8F_2$ 4,4'-Difluoro-1,1'-biphenyl, **II**, 188; **VI**, 490

$C_{12}H_8IN$ 3-Iodocarbazole, **VI**, 702

$C_{12}H_8INO_2$ Methyl (trans-2-iodo-1-tetralin)carbamate, **VI**, 795

$C_{12}H_8I_2$ 4,4'-Diiodobiphenyl, **VI**, 701

$C_{12}H_8NO_3$ Nicotinic anhydride, **V**, 822

$C_{12}H_8N_2O$ 1-(Diazoacetyl)naphthalene, **VI**, 613

   α-Hydroxyphenazine, **III**, 754

$C_{12}H_8N_2O_2$ 2-Nitrocarbazole, **V**, 829

$C_{12}H_8N_2O_3$ Nicotinic anhydride, **V**, 822

$C_{12}H_8N_2O_4$ 2,2'-Dinitrobiphenyl, **III**, 339

$C_{12}H_8N_2O_4S$ p-Nitrophenyl sulfide, **III**, 667

$C_{12}H_8N_2O_4S_2$ Di-o-nitrophenyl disulfide, **I**, 220

   m-Nitrophenyl disulfide, **V**, 843

$C_{12}H_8N_4O_2$ o-Azido-p'-nitrobiphenyl, **V**, 830

$C_{12}H_8N_4O_4$ Azobis(4-nitrobenzene), **VI**, 80

$C_{12}H_8OS$ Phenoxthin, **II**, 485

$C_{12}H_8O_2$ 2-Benzyl-1,4-benzoquinone, **VI**, 891

$C_{12}H_8O_3$ 3,4-Dihydro-1,2-naphthalic anhydride, **II**, 194

$C_{12}H_8O_4$ 2,3-Naphthalenedicarboxylic acid, **V**, 810
2,6-Naphthalenedicarboxylic acid, **V**, 813

$C_{12}H_9Br$ p-Bromobiphenyl, **I**, 113

$C_{12}H_9BrO_3$ 2-Bromoglutaconaldehyde enol benzoate, **VI**, 643

$C_{12}H_9ClO_3$ 2-Chloroglutaconaldehyde enol benzoate, **VI**, 643

$C_{12}H_9I$ 5-Iodoacenaphthene, **VI**, 701

$C_{12}H_9IO_3$ 2-Iodoglutaconaldehyde enol benzoate, **VI**, 643

$C_{12}H_9NO$ 3-Benzoylpyridine, **IV**, 88
4-Benzoylpyridine, **IV**, 89
2-Methoxynaphthalene-1-carbonitrile, **VI**, 467
4-Methoxynaphthalene-1-carbonitrile, **VI**, 467

$C_{12}H_9NOS$ N-Phenyl-3-formyl-2(1H)-pyridinethione, **VI**, 643

$C_{12}H_9NO_2$ m-Nitrobiphenyl, **IV**, 718

$C_{12}H_9NO_3$ o-Nitrodiphenyl ether, **II**, 446
p-Nitrodiphenyl ether, **II**, 445

$C_{12}H_9O_4$ 3-Carbethoxycoumarin, **III**, 165

$C_{12}H_{10}$ Benzobarrelene, **VI**, 82
Biphenyl, **VI**, 150, 490

$C_{12}H_{10}BrN$ N-(4-Bromophenyl)aniline, **VI**, 183

$C_{12}H_{10}ClI$ Diphenyliodonium chloride, **VI**, 928

$C_{12}H_{10}Cl_2$ Benzophenone dichloride, **II**, 573

$C_{12}H_{10}Cl_2Se$ Diphenylselenium dichloride, **II**, 240

$C_{12}H_{10}Cl_2Te$ Diphenyltellurium dichloride, **VI**, 470

$C_{12}H_{10}Hg$ Diphenylmercury, **I**, 228

$C_{12}H_{10}I\cdot I$ Diphenyliodonium iodide, **III**, 355

$C_{12}H_{10}NS$ 1-Methyl-2-imino-β-naphthothiazoline, **III**, 595

$C_{12}H_{10}N_2$ Azobenzene, **III**, 103
N-Methyl-1-naphthylcyanamide, **III**, 608

$C_{12}H_{10}N_2O$ Azoxybenzene, **II**, 57

$C_{12}H_{10}N_2O_3$ 1-Nitro-2-acetylaminonaphthalene, **II**, 438

$C_{12}H_{10}N_2S$ 1-Methyl-2-amino-β-naphthothiazoline, **III**, 595

$C_{12}H_{10}N_3O_2P$ Diphenyl phosphorazidate, **VII**, 135, 206

$C_{12}H_{10}O$ Acenaphthenol-7, **III**, 3
4-Phenylphenol, **VI**, 150

$C_{12}H_{10}O_2$ 3,3'-Dihydroxybiphenyl, **V**, 412

$C_{12}H_{10}O_2SSe$ Phenyl benzeneselenosulfonate, **VIII**, 543

$C_{12}H_{10}O_3$ Glutaconaldehyde enol benzoate, **VI**, 643
2-(Methoxymethyl)-1,4-naphthoquinone, **VI**, 891

$C_{12}H_{10}O_4$ 4-Methyl-7-acetoxycoumarin, **III**, 283
4-Methyl-7-hydroxy-8-acetylcoumarin, **III**, 283

$C_{12}H_{10}O_3Se_2$ Benzeneseleninic anhydride, **VI**, 536

$C_{12}H_{10}O_4$ 3-Carbethoxycoumarin, **III**, 165

$C_{12}H_{10}S$ Diphenyl sulfide, **II**, 242; **VI**, 163, 364

$C_{12}H_{10}Se$ Diphenyl selenide, **II**, 238

$C_{12}H_{10}Se_2$ Diphenyl diselenide, **VI**, 533

$C_{12}H_{11}$ Bicyclo[5.4.1]dodecapentaenylium ion, **VI**, 736

$C_{12}H_{11}FeN$ Ferrocenylacetonitrile, **V**, 578

$C_{12}H_{11}I$ 2,3-Dimethyl-1-iodonaphthalene, **VI**, 701

$C_{12}H_{11}N$ Diphenylamine, **VI**, 1000

$C_{12}H_{11}NO$ 1-Cyano-6-methoxy-3,4-dihydronaphthalene or 3,4-Dihydro-6-methoxynaphthalene-1-carbonitrile, **VI**, 307

$C_{12}H_{11}NO_2$ Ethyl α-cyano-β-phenylacrylate, **I**, 451
1-(2-Methoxycarbonylphenyl)-pyrrole, **V**, 716

$C_{12}H_{11}NO_2S$ Ethyl 5-phenylthiazole-4-carboxylate, **VI**, 623

$C_{12}H_{11}NO_3$ Ethyl phenylcyanopyruvate, **II**, 287

$C_{12}H_{11}NO_5$ Ethyl (E)-4-(4-nitrophenyl)-4-oxo-2-butenoate, **VIII**, 268

$C_{12}H_{11}N_2O_2$ Tetrahydro-β-carboline-1-carboxylic acid, **VI**, 965

$C_{12}H_{11}N_2O_2$·Br o-Nitrobenzylpyridinium bromide, **V**, 825

$C_{12}H_{11}N_3$ Diazoaminobenzene, **II**, 163

$C_{12}H_{11}P$ Diphenylphosphine, **VI**, 568

$C_{12}H_{12}$ trans-Stilbene, **III**, 786

$C_{12}H_{12}NO_2P$ O-Diphenylphosphinylhydroxylamine, **VII**, 9

$C_{12}H_{12}N_2$ Benzophenone hydrazone, **III**, 352

2-Benzylaminopyridine, **IV**, 91

$C_{12}H_{12}N_2O_2S$ 4,4'-Diaminodiphenylsulfone, **III**, 239

$C_{12}H_{12}N_2S$ 1-Methyl-1-(1-naphthyl)-2-thiourea, **III**, 609

$C_{12}H_{12}N_2S_2$ p-Aminophenyl disulfide, **III**, 86

$C_{12}H_{12}N_4$ 4,4'-Diaminoazobenzene, **V**, 341

$C_{12}H_{12}O_2S$ 2-(Phenylsulfonyl)-1,3-cyclohexadiene, **VIII**, 540

2-(p-Toluenesulfinyl)-2-cyclopenten-1-one, **VII**, 495

$C_{12}H_{12}O_3$ Ethyl 3-methylcoumarilate, **IV**, 590

(E)-3-[(E)-3-Phenyl-3-propenoxy]acrylic acid, **VIII**, 536

1,3,5-Triacetylbenzene, **III**, 829

$C_{12}H_{12}O_4$ 2-(1-Acetyl-2-oxopropyl)benzoic acid, **VI**, 36

$C_{12}H_{12}O_5$ 5,6-Diethoxybenzofuran-4,7-dione, **VIII**, 179

$C_{12}H_{12}O_6$ Hydroxyhydroquinone triacetate, **I**, 317

$C_{12}H_{13}ClHgO_2S$ trans-3-(Phenylsulfonyl)-4-(chloromercuri)cyclohexene, **VIII**, 540

$C_{12}H_{13}N$ α-(β-Naphthyl)ethylamine, **II**, 505

(S)-(−)-α-(1-Naphthyl)ethylamine, **VI**, 826

1-Phenylcyclopentanecarbonitrile, **VI**, 899

2-Phenyl-2-vinylbutyronitrile, **VI**, 940

1,2,3,4-Tetrahydrocarbazole, **IV**, 884

$C_{12}H_{13}NOS$ S-Phenyl trans, trans-1,3-pentadiene-1-thiocarbamate, **VI**, 100

$C_{12}H_{13}NO_2$ Benzyl trans-1,3-butadiene-1-carbamate, **VI**, 95

tert-Butylphthalimlde, **III**, 152

Ethyl 2-methylindole-5-carboxylate, **VI**, 601

Ethyl β-phenyl-β-cyanopropionate, **IV**, 804

Phenyl trans, trans-1,3-pentadiene-1-carbamate, **VI**, 100

$C_{12}H_{13}N_2P$ 4-Amino-1,2,5,6-tetrahydro-1-phenylphosphorin-3-carbonitrile, **VI**, 932

Bis(2-cyanoethyl)phenylphosphine, **VI**, 932

$C_{12}H_{13}N_3O_3$ 5-(3-Methoxy-4-hydroxybenzal)creatinine, **III**, 587

$C_{12}H_{14}$ 1-Phenylcyclohexene, **VI**, 771

$C_{12}H_{14}INO_2$ Methyl (trans-2-iodo-1-tetralin)carbamate, **VI**, 967

$C_{12}H_{14}NO_2$ Ethyl β-anilinocrotonate, **III**, 374

$C_{12}H_{14}N_2$ 3-Methyl-1,2,3,4-tetrahydrocarboline, **VI**, 966

9-Methyl-1,2,3,4-tetrahydro-β-carboline, **VI**, 966

$C_{12}H_{14}N_2O$ 6-Methoxy-1,2,3,4-tetrahydro-β-carboline, **VI**, 966

$C_{12}H_{14}N_2O_5$ Carbobenzyloxy-L-asparagine, **VI**, 263

$C_{12}H_{14}N_4O_5$ trans-4-Hydroxy-2-hexenal 2,4-dinitrophenylhydrazone, **VI**, 685

$C_{12}H_{14}O$ 2-Benzylcyclopentanone, **V**, 76

2-Methyl-2-phenyl-4-pentenal, **VIII**, 451

1-Phenylcyclopentanecarboxaldehyde, **VI**, 905

$C_{12}H_{14}OS$ (Z)-2-Methoxy-3-methyl-1-phenylthio-1,3-butadiene, **VI**, 737

$C_{12}H_{14}O_2$ Ethyl cis-2-phenylcyclopropanecarboxylate, **VI**, 913

Ethyl *trans*-2-phenylcyclo-
propanecarboxylate, **VI**, 913
1-Phenylcyclopentanecarboxylic
acid, **VI**, 48
$C_{12}H_{14}O_3$ Ethyl α-phenylacetoacetate, **II**,
284
Phenylmethylglycidic ester, **III**, 727
$C_{12}H_{14}O_4$ (1α,6α,7α,8β)- or *trans*-
Bicyclo[4.2.0]octa-2,4-dien-7,8-
diyl diacetate, or (1α,6α,7α,8β)-
or *trans*-7,8-Diacetoxybicyclo-
[4.2.0]octa- 2,4-diene, **VI**, 196
β-Methyl-β-phenylglutaric acid, **IV**,
664
$C_{12}H_{14}S$ 1-Phenylthiocyclohexene, **VI**, 742
$C_{12}H_{15}BO_2$ (*E*)-1-Hexenyl-1,3,2-benzodi-
oxaborole, **VIII**, 532
$C_{12}H_{15}ClO$ 4-Pentylbenzoyl chloride, **VII**,
420
$C_{12}H_{15}N$ "Acetone-anil"(2,2,4-trimethyl-
1,2-dihydroquinoline), **III**, 329
Julolidine, **III**, 504
N-Methyl-2-phenyl-Δ²-tetrahydropy-
ridine, **VI**, 818
$C_{12}H_{15}NO$ Benzoylpiperidine, **I**, 99
1-(4-Methoxyphenyl)-1,2,5,6-
tetrahydropyridine, **VIII**, 358
α-Morpholinostyrene, **VI**, 520
N-(1-Phenylethenyl)morpholine, **VI**,
523
$C_{12}H_{15}NO_2$ 6,7-Dimethoxy-1-methyl-3,4-
dihydroisoquinoline, **VI**, 1
1,3-Dimethyl-3-methoxy-4-phenyl-
azetidinone, **VIII**, 216
Ethyl β-anilinocrotonate, **III**, 374
$C_{12}H_{15}O_2Rh$ Bicyclo[2.2.1]hepta-2,5-
diene-2,4-pentanedionatorhodi-
um, **VIII**, 532
$C_{12}H_{15}N_3O_6$ Triethyl 1,2,4-triazine-3,5,6-
tricarboxylate, **VIII**, 597
$C_{12}H_{16}$ Cyclohexylbenzene, **II**, 151
(1*R*)-9,9-Dimethyltricyclo[6.1.1.0²,⁶]
deca-2,5-diene, **VIII**, 223
$C_{12}H_{16}Br_2$ 2-(2,2-Dibromocyclopropyl)-
6,6-dimethylbicyclo[2.2.1]hept-
2-ene, **VIII**, 223

$C_{12}H_{16}ClNO_2$ 6,7-Dimethoxy-1-methyl-
3,4-dihydroisoquinoline
hydrochloride, **VI**, 2
$C_{12}H_{16}Cl_2NO_4P$ 6,7-Dimethoxy-1-methyl-
3,4-dihydroisoquinoline
dichlorophosphate, **VI**, 1
$C_{12}H_{16}N_2O$ 1,2-Dicyano-4-butoxycyclo-
hexene or 4-Butoxy-1-cyclohex-
ene-1,2-dicarbonitrile, **VI**, 431
$C_{12}H_{16}O$ Aceto-*p*-cymene, **II**, 3
1-Ethoxy-2-*p*-tolylcyclopropane, **VI**,
571
4-Methyl-4-phenyl-2-pentanone, **VI**,
816
cis-1-Ethenyl-1,2,3,6,7,7a-hexahy-
dro-7a-methyl-5-indenone, **VI**,
668
trans-2-Phenylcyclohexanol,
(1R,2S)- and (1S, 2R)-, **VIII**,
516
$C_{12}H_{16}O_2$ 1-Benzyloxy-4-penten-2-ol,
VIII, 33
tert-Butyl phenylacetate, **VI**, 262
tert-Butyl-p-toluate, **VI**, 259
2,2-Dimethyl-4-phenylbutyric acid,
VI, 517
2,2-Dimethylpropyl benzoate or
Neopentyl benzoate, **VI**, 260
Ethyl γ-phenylbutyrate, **II**, 196
Ethyl 2,4,6-trimethylbenzoate, **VI**,
579
$C_{12}H_{16}O_2SSi$ *p*-Tolyl-2-(trimethylsilyl)-
ethynyl sulfone, **VIII**, 281
$C_{12}H_{16}O_3S$ DL-1-(4-Methoxyphenyl-
sulfonyl)-*trans*-2,3-dimethylcy-
clopropane, **VI**, 985
$C_{12}H_{16}O_4$ 4-Ethoxy-3-methoxybenzalde-
hyde ethylene acetal, **VI**, 567
Monobenzalpentaerythritol, **IV**, 679
$C_{12}H_{17}ClO_3S$ 1-(1-*tert*-Butoxy-1-chloro-
methylsulfonyl)-4-methyl-
benzene or 4-(1-*tert*-Butoxy-1-
chloromethylsulfonyl)-1-methyl-
benzene, **VI**, 985
$C_{12}H_{17}I$ 5-*tert*-Butyl-2,3-dimethyliodoben-
zene, **VI**, 701

$C_{12}H_{17}NO_2$ Ethyl p-dimethylaminophenyl-
acetate, **V**, 552
(–)-Salsolidine **VIII**, 573,
$C_{12}H_{17}NO_2Si$ O-(Trimethylsilyl)-4-
methoxymandelonitrile, **VII**,
521
$C_{12}H_{17}NO_3$ N-Acetylhomoveratrylamine or
N-[2-(3,4-Dimethoxyphenyl)
ethyl]acetamide, **VI**, 1
$C_{12}H_{17}NO_4$ 2,4-Dimethyl-3,5-dicarb-
ethoxypyrrole, **II**, 202
Isopropylidene α-(hexahy-
droazepinylidene-2)malonate,
VIII, 263
$C_{12}H_{18}$ 1-$tert$-Butyl-4-ethylbenzene, **VI**,
539
Hexamethylbenzene, **II**, 248; **IV**,
520
Hexamethyl Dewar benzene, **VI**, 39;
VII, 256
cis-1,6,7,8-Tetramethylbicyclo[4.2.0]
octa-3,7-diene, **VI**, 486
$C_{12}H_{18}Cl_3NO$ 3,7-Dimethyl-3-trichloroac-
etamido-1,6-octadiene or N-
(1,5-Dimethyl-1-ethenyl-4-hex-
enyl)trichloroacetamide, **VI**,
507
Geraniol trichloroacetimidate, **VI**,
507
$C_{12}H_{18}NO_2 \cdot Cl$ Benzoylcholine chloride,
IV, 84
$C_{12}H_{18}NO_2 \cdot I$ Benzoylcholine iodide, **IV**,
84
$C_{12}H_{18}NO_3P$ Diethyl benzylidenamino-
methylphosphonate, **VIII**, 451
$C_{12}H_{18}N_4O$ 3-Heptyl-5-cyanocytosine, **IV**,
515
$C_{12}H_{18}O$ 5-Acetyl-1,2,3,4,5-pentamethyl-
cyclopentadiene, **VI**, 39
8,10-Dimethylbicyclo[5.3.0]dec-7-
en-9-one, **VI**, 524
cis-4,4a,5,6,7,8-Hexahydro-4a,5-
dimethyl-2(3H)-naphthalenone,
VI, 666
2,3,4,5,6,6-Hexamethyl-2,4-cyclo-
hexadien-1-one, **V**, 598

1-Methyl-2-ethynyl-$endo$-3,3-
dimethyl-2-norbornanol, **VIII**,
391
$C_{12}H_{18}O_2$ 3,5,5-Trimethyl-2-(2-oxo-
propyl)-2-cyclohexen-1-one,
VI, 1024
$C_{12}H_{18}O_3$ Ethyl 1-hydroxycyclohexylace-
tate, **VI**, 599
Methyl 1,3-dimethyl-5-oxobicy-
clo[2.2.2]octane-2- carboxylate,
VIII, 219
8-Propionyl-(E)-5-nonenolide, **VIII**,
562
$C_{12}H_{18}O_4$ 2,6-Diacetoxybicyclo[3.3.0]
octane, **VIII**, 43
Diethyl Δ²-cyclopentenylmalonate,
IV, 291
Diethyl cis-Δ⁴-tetrahydrophthalate,
IV, 304; **VI**, 456
Diethyl $trans$-4-tetrahydrophthalate,
VI, 454
$C_{12}H_{18}O_7$ Triethyl oxalylsuccinate, **III**,
510; **V**, 687
$C_{12}H_{19}N$ 1,2,6,7,7a,8,9,10-Octahydro-
3H,5H-benzo[i,j]quinolizine,
VI, 820
$C_{12}H_{20}$ 4-$tert$-Butyl-1-vinylcyclohexene,
VIII, 97
Cyclohexylidenecyclohexane, **V**,
297
$C_{12}H_{20}Br_2O$ 2,12-Dibromocyclododec-
anone, **VI**, 368
$C_{12}H_{20}N_2O$ 2-Diazocyclodecanone, **VI**,
391
$C_{12}H_{20}N_2O_4$ 4,13-Diaza-18-crown-6, **VIII**,
152
$C_{12}H_{20}O_2$ Cyclohexanone diallyl acetal, **V**,
292
Ethyl (E,Z)-2,4-decadienoate, **VIII**,
251
Ethyl 3,4-decadienoate, **VIII**, 251
Ethyl 4-methyl-(E)-4,8-nona-
dienoate, **VI**, 606
7-Hydroxy-4,4,6,7-tetramethylbicy-
clo[4.2.0]octan-2-one, **VI**,
1024

$C_{12}H_{20}O_3$ Ethyl 4-cyclohexyl-4-oxobu-
tanoate, **VIII**, 578

$C_{12}H_{20}O_4$ Di-*tert*-butyl methylenemalonate, **VII**, 142

Diethyl *cis*-hexahydrophthalate, **IV**,
304

$C_{12}H_{20}O_5$ 3-Deoxy-1,2:5,6-di-*O*-isopropyl-
idene-α-D-ribo-hexofuranose,
VII, 139

$C_{12}H_{20}O_8$ (–)-2,3:4,6-Di-*O*-isopropylidene-
2-keto-L-gulonic acid hydrate,
VI, 826

$C_{12}H_{21}B$ Perhydro-9b-boraphenalene, **VII**,
427

$C_{12}H_{21}ClO_2$ L-Menthoxyacetyl chloride,
III, 547

$C_{12}H_{21}NO$ Benzyltrimethylammonium
ethoxide, **IV**, 98

$C_{12}H_{21}NO_4$ 1-Carbethoxymethyl-4-car-
bethoxypiperidine, **V**, 989

$C_{12}H_{21}N_3O_3$ Hexahydro-1,3,5-tripropionyl-
s-triazine, **IV**, 518

$C_{12}H_{21}O_6P$ Methyl 2-(diethylphosphoryloxy)-1-cyclohexene-1-carboxy-
late, **VII**, 351

$C_{12}H_{22}$ *cis*-Cyclododecene, **V**, 281
trans-Cyclododecene, **VII**, 121

$C_{12}H_{22}Br_3N_3O_3$ Pyrrolidine hydrotribro-
mide or Pyrrolidinium tribro-
mide, **VI**, 178

$C_{12}H_{22}N_2$ Azocyclohexane, **VI**, 80

$C_{12}H_{22}O$ Cyclododecanone, **VI**, 368
5-Dodecen-2-one, **VIII**, 235

$C_{12}H_{22}OSi$ 4,6,6-Trimethyl-2-trimethyl-
siloxyhexa-1,3-diene, **VII**, 282

$C_{12}H_{22}O_2$ Cycloundecanecarboxylic acid,
VII, 135
trans-2-Dodecenoic acid, **IV**, 398
2-Hydroxycyclododecanone, **VI**, 171
γ-Octyl-γ-butyrolactone, **VII**, 400

$C_{12}H_{22}O_3$ Caproic anhydride, **III**, 164
L-Menthoxyacetic acid, **III**, 54

$C_{12}H_{22}O_4$ Dimethyl decanedioate, **VII**, 181
Ethyl hydrogen sebacate, **II**, 276
Methyl hydrogen hendecanedioate,
IV, 635

$C_{12}H_{22}O_{11}$ Cellobiose, **II**, 122

$C_{12}H_{23}ClO$ Lauroyl chloride, **IV**, 561

$C_{12}H_{23}NO$ Undecyl isocyanate, **III**, 846

$C_{12}H_{24}$ 2-Methyl-2-undecene, **VIII**, 474

$C_{12}H_{24}N_2O_2S$ *N,N'*-Dicyclohexylsulfamide,
VI, 79

$C_{12}H_{24}NO_3P$ Diethyl 2-(cyclohexylamino)vinylphosphonate or
Diethyl 2-(cyclohexylamino)
ethenylphosphonate, **VI**, 358,
448

$C_{12}H_{24}O$ 3-Butyl-2-methyl-1-hepten-3-ol,
VI, 240
4,4-Dimethyl-2-neopentylpentanal or
4,4-Dimethyl-2-(2,2-dimethyl-
1-propyl)pentanal, **VI**, 1029

$C_{12}H_{24}O_2$ 1,1-Diethylpropyl trimethylacetate, **VI**, 261

$C_{12}H_{24}O_2Si$ 2-Methyl-5-trimethylsilyl-2-
vinylpentanoic acid, methyl
ester, **VIII**, 486

$C_{12}H_{24}O_4$ 2,3-Di-*tert*-butoxy-1,4-dioxane,
VIII, 161

$C_{12}H_{24}S$ Cyclododecyl mercaptan, **VII**,
124

$C_{12}H_{24}O_6$ 18-Crown-6 polyether, **VI**, 301

$C_{12}H_{24}Sn$ Dibutyldivinyltin, **IV**, 258

$C_{12}H_{25}Br$ Dodecyl bromide, **I**, 29; **II**,
246

$C_{12}H_{25}N_3O$ 3-Butyl-2-heptanone semicarbazone, **VI**, 596

$C_{12}H_{26}$ Dodecane, **VI**, 376

$C_{12}H_{26}N_2O$ (*S*)-*N,N*-Dimethyl-*N'*-(1-*tert*-
butoxy-3-methyl-2-butyl)for-
mamidine, **VIII**, 204, 573

$C_{12}H_{26}N_2O_4$ 4,13-Diaza-18-crown-6, **VIII**,
152

$C_{12}H_{26}O$ Lauryl alcohol, **II**, 372

$C_{12}H_{26}O_2Si$ [(Isopropoxydimethylsilyl)
methyl]cyclohexanol, **VIII**, 315

$C_{12}H_{26}S$ Dodecyl mercaptan, **III**, 363

$C_{12}H_{27}N$ *tert*-Butyl-*tert*-octylamine, **VIII**,
93

$C_{12}H_{27}O_4P$ Butyl phosphate, **II**, 109

$C_{12}H_{30}N_3P$ Hexaethylphosphorous triamide, **VI**, 131

$C_{12}H_{30}N_3S \cdot C_3H_9F_2Si$ Tris(dimethylamino) sulfonium difluorotrimethylsilicate, **VII**, 528;

$C_{12}H_{30}N_6$ 1,4,7,10,13,16-Hexaazacyclooctadecane, **VI**, 652

$C_{12}H_{27}AlO_3$ Aluminum tert-butoxide, **III**, 48

$C_{12}H_{27}BO_3$ Butyl borate, **II**, 106

$C_{12}H_{27}O_4P$ Butyl phosphate, **II**, 109
    sec-Butyl phosphate, **II**, 110

$C_{13}F_{10}O$ Perfluorobenzophenone, **VI**, 881

$C_{13}H_4N_4O_9$ 2,4,5,7-Tetranitrofluorenone, **V**, 1029

$C_{13}H_5F_5O$ 2,3,4,5,6-Pentafluorobenzophenone, **VI**, 881

$C_{13}H_5N_3O_7$ 2,4,7-Trinitrofluorenone, **III**, 837;

$C_{13}H_7F_5$ Benzylpentafluorobenzene, **VI**, 880

$C_{13}H_8ClN$ 9-Chloroacridine, **III**, 54

$C_{13}H_8O_2$ Xanthone, **I**, 552

$C_{13}H_9ClN_2O$ p-Phenylazobenzoyl chloride, **III**, 712

$C_{13}H_9ClO_3$ p-Chlorophenyl salicylate, **IV**, 178

$C_{13}H_9I$ 2-Iodofluorene, **VI**, 701

$C_{13}H_9NO$ Acridone, **II**, 15
    2-Phenylfuro[3,2-b]pyridine, **VI**, 916

$C_{13}H_9NO_2$ 2-Nitrofluorene, **II**, 447

$C_{13}H_{10}Br_2O_2$ 2,2-Dibromoacetyl-6-methoxynaphthalene, **VI**, 175

$C_{13}H_{10}Cl_2$ Benzophenone dichloride, **II**, 573

$C_{13}H_{10}IO_2$ Diphenyliodonium-2-carboxylate, **V**, 1037

$C_{13}H_{10}N_2$ 9-Aminoacridine, **III**, 53
    Diphenylcarbodiimide(Method I), **V**, 501
    Diphenylcarbodiimide(Method II), **V**, 504
    Diphenyldiazomethane, **III**, 351; **VI**, 393
    2-Phenylindazole, **V**, 941

$C_{13}H_{10}N_2O$ α-Methoxyphenazine, **III**, 753
    Pyocyanine, **III**, 753

$C_{13}H_{10}N_2O_2$ p-Phenylazobenzoic acid, **III**, 711

$C_{13}H_{10}N_2O_4S$ 2,4-Dinitrophenyl benzyl sulfide, **V**, 474

$C_{13}H_{10}N_4$ p-Tricyanovinyl-N,N-dimethylaniline, **IV**, 953

$C_{13}H_{10}O$ Benzophenone, **I**, 95; **VI**, 478, 537, 542, 807, 901

$C_{13}H_{10}O_2$ Biphenyl-4-carboxylic acid or 4-Biphenylcarboxylic acid, **VI**, 627
    2-Biphenylcarboxylic acid, **VI**, 747
    Xanthydrol, **I**, 554

$C_{13}H_{10}O_3$ 2-Phenoxymethyl-1,4-benzoquinone, **VI**, 890

$C_{13}H_{10}S$ Thiobenzophenone, **II**, 573; **IV**, 927

$C_{13}H_{11}Br$ p-Bromodiphenylmethane, **V**, 130

$C_{13}H_{11}BrO_2$ 2-Bromoacetyl-6-methoxynaphthalene, **VI**, 175

$C_{13}H_{11}N$ 2-Aminofluorene, **II**, 448; **V**, 30
    Benzalaniline, **I**, 80
    Diphenyl ketimine, **V**, 520 and Diphenylmethanimine hydrochloride, **II**, 234

$C_{13}H_{11}NO$ 2-Aminobenzophenone, **IV**, 34
    Benzanilide, **I**, 82; **IV**, 384
    Benzophenone oxime, **II**, 70
    N,α-Diphenylnitrone, **V**, 1124

$C_{13}H_{11}NO_2$ N-Phenylanthranilic acid, **II**, 15

$C_{13}H_{11}NO_2$ 2-Methyl-4'-nitrobiphenyl, **VIII**, 430
    N-Phenylanthranilic acid, **II**, 15

$C_{13}H_{11}NO_2S$ N-Benzylidenebenzenesulfonamide, **VIII**, 546

$C_{13}H_{11}NO_3S$ trans-2-(Phenylsulfonyl)-3-phenyloxaziridine, **VIII**, 546

$C_{13}H_{11}N_3$ 1,2-Dicyano-4-(2-pyridyl)cyclohexene or 4-(2-Pyridyl)cyclohexene-1,2-dicarbonitrile, **VI**, 43

$C_{13}H_{12}$ Diphenylmethane, **II**, 232
    2-Methylbiphenyl, **VI**, 747
    (Methylethenyl)naphthalene, **VI**, 410

$C_{13}H_{12}BN$ 10-Methyl-10,9-borazarophenanthrene, **V**, 727

$C_{13}H_{12}N_2$ 4,5-Dicyanotricyclo [6.2.1.0$^{2.7}$] undeca-4,9-diene or Tricyclo [6.2.1.0$^{2.7}$]undeca-4,9-diene-4,5-dicarbonitrile, **VI**, 431

N-Phenylbenzamidine, **IV**, 769

$C_{13}H_{12}N_2O$ Benzil hydrazone, **II**, 497

sym-Diphenylurea, **I**, 453

$C_{13}H_{12}N_2O_2$ Ethyl α,β-dicyano-β-phenyl-propionate, **I**, 452

$C_{13}H_{12}N_4S$ Dithizone, **III**, 360

$C_{13}H_{12}O$ Benzhydrol, **I**, 90; **VI**, 241

1-Methoxybenzobarrelene, **VI**, 85

$C_{13}H_{12}O_2$ 2-Acetyl-6-methoxynaphthalene, **VI**, 34, 176

2-Ethoxy-1-naphthaldehyde, **III**, 98

Ethyl α-naphthoate, **II**, 282

2-Methoxydiphenyl ether, **III**, 566

$C_{13}H_{12}O_3$ 4-Methylglutaconaldehyde enol benzoate, **VI**, 642

$C_{13}H_{12}O_4$ 4-Methoxyglutaconaldehyde enol benzoate, **VI**, 642

$C_{13}H_{12}S_2$ Bis(phenylthio)methane, **VI**, 737

$C_{13}H_{13}N$ Benzylaniline, **I**, 102

$C_{13}H_{13}NOS$ *O*-2-Naphthyldimethylthio-carbamate, **VI**, 824

$C_{13}H_{13}NO_2$ Ethyl (1-phenylethylidene) cyanoacetate, **IV**, 463

$C_{13}H_{13}NO_2S$ Ethyl 5-benzylthiazole-4-car-boxylate, **VI**, 623

$C_{13}H_{14}$ Spiro[cyclopentane-1,1'-indene], **VI**, 899

4,6,8-Trimethylazulene, **V**, 1088

$C_{13}H_{14}NO_5$ *N*-Carbobenzyloxy-3-hydroxy-L-proline, **VI**, 264

$C_{13}H_{14}N_2O_3$ 2-*tert*-Butoxycarbonyloxyi-mino-2-phenylacetonitrile, **VI**, 199, 718

$C_{13}H_{14}N_4O_4$ Cyclohexanecarboxaldehyde 2,4-dinitrophenylhydrazone, **VI**, 339

$C_{13}H_{14}O$ 2,5-Dimethyl-3-phenyl-2-cyclo-penten-1-one, **VI**, 520

α-Phenylcyclopentylideneacetalde-hyde, **VI**, 908

$C_{13}H_{14}O_2$ 3-Benzoyloxycyclohexane, **V**, 70

$C_{13}H_{14}O_3S$ 2,4-Dimethyl-3-(*p*-toluenesul-fonyl)furan, **VI**, 32

$C_{13}H_{14}O_4$ 6,7-Dimethoxy-3,4-dihydro-2-naphthoic acid, **III**, 300

Ethyl benzoylacetoacetate, **II**, 266

γ-Phenylallylsuccinic acid, **IV**, 766

Sclerin, 472, **VI**

$C_{13}H_{15}N$ N-Benzyl-2-azanorbornene, **VIII**, 31

3-Methyl-2,3,4,5-tetrahydro-1*H*-benz[*e*]indole, **VI**, 820

$C_{13}H_{15}NO_2S$ Ethyl 2-methyl-3-methylthio-indole-5-carboxylate, **VI**, 601

$C_{13}H_{15}NO_3$ (*S*)-3-(1-Oxopropyl)-4-(phenylmethyl)-2-oxazolidi-none, **VIII**, 339

$C_{13}H_{15}NO_4$ 2-(2-Acetamidoethyl)-4,5-methylenedioxyacetophenone or *N*-[2-(4,5-Methylenedioxy-2-oxoethylphenyl)ethyl] acetamide, **VI**, 3

1,3a,4,9b-Tetrahydro-3-methoxycar-bonyl-*cis,cis*-N-methyl-3*H*-ben-zopyrano[4,3-*c*]isoxazole **VI**, 673

$C_{13}H_{15}NS_2$ 4,5-Dihydro-2-[(1-phenyl-methyl-2-propenyl)thio]thia-zole, **VI**, 704

$C_{13}H_{16}$ Cyclohexylidenephenylmethane **VI**, 772,

$C_{13}H_{16}BrNO_3$ ε-Benzoylamino-α-bromo-caproic acid, **VI**, 94

$C_{13}H_{16}ClNO_3$ ε-Benzoylamino-α-chloro-caproic acid, **VI**, 90

$C_{13}H_{16}NO_5P$ Diethyl phthalim-idomethylphosphonate, **VIII**, 451

$C_{13}H_{16}N_4S$ Diphenylthiocarbazide, **III**, 361

$C_{13}H_{16}O$ Benzalpinacolone, **I**, 81

2-Benzylcyclohexanone, **VI**, 526

2-Phenylcycloheptanone, **IV**, 780

4-Phenyl-6-hepten-2-one, **VII**, 443

$C_{13}H_{16}O_2$ *tert*-Butyl cinnamate, **VI**, 261

threo-2-(1-Hydroxy-1-phenylmethyl) cyclohexanone, **VI**, 697

1-Phenyl-2,4-heptanedione, **VI** , 930

2-Phenyl-3,5-heptanedione, **VI**, 930

Phenylpropargylaldehyde diethyl acetal, **IV**, 801

$C_{13}H_{16}O_3$ Ethyl benzoyldimethylacetate, **II**, 268

6-Methoxy-7-methoxycarbonyl-1,2,3,4-tetrahydronaphthalene, **VIII**, 444

$C_{13}H_{16}O_4$ Diethyl phenylmalonate, **II**, 288

$C_{13}H_{16}O_7Pb$ p-Methoxyphenyllead triacetate, **VII**, 229

$C_{13}H_{16}S$ Phenylthiomethylidenecyclohexane, **VI**, 742

$C_{13}H_{17}NO$ N-Methyl-5-phenyl-6-oxa-7-azabicyclo[3.2.1]octane, **VI**, 671

cis-N-Methyl-5-phenyl-3-oxa-2-azabicyclo[3.3.0]octane or cis-N-Methyl-3a-phenylhexahydro-1H-cyclopent[c]isoxazole, **VI**, 671

$C_{13}H_{17}NO_2$ 1-Acetoxy-4-benzylamino-2-butene, **VIII**, 9

$C_{13}H_{17}NO_3$ 2-(2-Acetamidoethyl)-4-methoxyacetophenone or N-[2-(4-Methoxy-2-oxoethylphenyl)ethyl]acetamide, **VI**, 3

ε-Benzoylaminocaproic acid, **II**, 76; **VI**, 90

Methyl 2-(N,N-diethylcarbamoyl)benzoate, **VI**, 579

$C_{13}H_{17}NO_4$ 3,5-Dicarbethoxy-2,6-dimethylpyridine, **II**, 215

$C_{13}H_{17}NO_4S$ N-Benzyloxycarbonyl-L-methionine, **VI**, 253

$C_{13}H_{18}$ (E)-Buten-2-ynyl-2,6,6-trimethyl-1-cyclohexene, **VII**, 63

1-Butylindane or 1-butyl-2,3-dihydro-1H-indene, **VI**, 539

$C_{13}H_{18}N_2O_3$ DL-ε-Benzoyllysine, **II**, 374

$C_{13}H_{18}N_4O_2$ Benzylidenearginine, **II**, 49

$C_{13}H_{18}O$ Spiro[5.7]trideca-1,4-dien-3-one, **VII**, 473

$C_{13}H_{18}O_2$ 3-Benzyl-3-methylpentanoic acid, **IV**, 93

1-Ethyl-1-methylpropyl benzoate, **VI**, 260

threo-4-Hydroxy-3-phenyl-2-heptanone, **VI**, 692

tert-Butyl β-phenylpropionate, **VI**, 261

1,2,2-Trimethylpropyl benzoate, **VI**, 260

$C_{13}H_{18}O_3S$ (E)-4-Hexen-1-yl p-toluenesulfonate, **VI**, 586

$C_{13}H_{18}O_4$ (3-Methyl-2-butenyl)(2-propynyl)propanedioic acid, dimethyl ester, **VIII**, 381

$C_{13}H_{18}O_6$ cis- and trans-1-Acetoxy-4-(dicarbomethoxymethyl)-2-cyclohexene, **VIII**, 5

$C_{13}H_{18}O_8$ 5,6-Dihydroxy-7-(methoxycarbonylmethyl)-2-methylnorbornane-2,3-dicarboxylic acid, **VI**, 346

Hexahydrogallic acid triacetate, **V**, 591

$C_{13}H_{19}I$ 6-tert-Butyl-2,3,4-trimethyliodobenzene, **VI**, 701

$C_{13}H_{19}NO_2$ 2-(3-Ethyl-5-methyl-4-isoxazolylmethyl)cyclohexanone, **VI**, 781

$C_{13}H_{19}NO_3$ 2-(tert-Butylamino-1-(3,4-methylenedioxyphenyl)ethanol, **VI**, 545

$C_{13}H_{19}NO_3S$ cis-2-(p-Toluenesulfonamido)cyclohexanol, **VII**, 375

$C_{13}H_{19}NO_4$ 1,4-Dihydro-3,5-dicarbethoxy-2,6-dimethylpyridine, **II**, 214

$C_{13}H_{20}$ (E)-6,10-Dimethyl-5,9-undecadien-1-yne, **VIII**, 226

$C_{13}H_{20}O$ 1-(4-tert-Butylcyclohexen-1-yl)-2-propen-1-one, **VIII**, 97

o-Heptylphenol, **III**, 444

Pseudoionone, **III**, 747

Spiro[5.7]tridec-1-en-3-one, **VII**, 473

$C_{13}H_{20}O_4$ 3-Methylene-4-isopropyl-1,1-cyclopentanedicarboxylic acid, dimethyl ester, **VIII**, 381

$C_{13}H_{20}O_6$ Diethyl α-acetyl-β-ketopimelate, **V**, 384

$C_{13}H_{20}S$ Thiobenzophenone, **II**, 573; **IV**, 927

$C_{13}H_{21}NO_3S$ 2-Methyl-3-(p-toluenesulfon-amido)-2-pentanol, **VII**, 375

$C_{13}H_{22}ClN$ Benzyltriethylammonium chloride, **VI**, 232, 897 940

$C_{13}H_{22}N_2$ 10-Methyl-$\Delta^{1(9)}$-2-octalone N,N-dimethylhydrazone, **VI**, 242

$C_{13}H_{22}O$ trans-2-Cyclotridecen-1-one, **VI**, 331

2-Heptyl-2-cyclohexenone, **VII**, 249

Perhydro-9b-phenalenol, **VII**, 427

$C_{13}H_{22}O_2$ Methyl 1-cycloundecene-1-carboxylate, **VI**, 368

$C_{13}H_{22}O_3$ 3-Carboxy-5-dodecen-2-one, **VIII**, 234

(1-Ethenyl)heptanyl 3-ketobutanoate, **VIII**, 235

$C_{13}H_{22}O_5Si$ Dimethyl trans-4-trimethylsi-lyloxy-4-cyclohexene-1,2-dicarboxylate, **VI**, 447

$C_{13}H_{22}O_6$ Diisopropyl (2S,3S)-2,3-O-iso-propylidenetartrate, **VIII**, 201

$C_{13}H_{23}NO$ N-(1-Methylethyl)-1,8,8-trimethyl-2-oxa-3-azabicy-clo[7.1^{4,7}.0^{1,5}]nonane, **VI**, 672

N-(1-Methylethyl)-5,6,6-trimethyl-2-oxa-3-azabicyclo[7.1^{4,7}.0^{1,5}] nonane, **VI**, 672

$C_{13}H_{23}NOSi$ N-Benzyl-N-methoxymethyl-N-[(trimethylsilyl)methyl] amine, **VIII**, 231

$C_{13}H_{24}$ (E)-5-Ethyl-1,4-undecadiene, **VII**, 236

$C_{13}H_{24}O_2$ 2-Hydroxycyclotridecanone, **VI**, 171

trans-2-Methyl-2-dodecenoic acid, **IV**, 608

2-Methylenedodecanoic acid, **IV**, 616

$C_{13}H_{24}O_3$ sec-Butyl α-caproylpropionate, **IV**, 120

$C_{13}H_{25}N$ Dodecyl isocyanide or 1-(Carbyl-amino)dodecane, **VI**, 234

$C_{13}H_{25}NO_5$ Diethyl 3-N-(3-hydroxypropyl) iminodipropionate, **VI**, 75

$C_{13}H_{26}DN$ N-(2-Methylbutylidene-1-d)-1,1,3,3-tetramethylbutylamine, **VI**, 752

$C_{13}H_{26}N_2O_4$ Boc-L-leucine N-methyl-O-methylcarboxamide, **VIII**, 68

$C_{13}H_{26}O_2$ 1,1-Diethylpropyl 3,3-dimethyl-butyrate, **VI**, 261

2-Methyldodecanoic acid, **IV**, 618

$C_{13}H_{28}N_2S\cdot HBr$ Laurylisothiourea hydro-bromide, **III**, 363

$C_{13}H_{28}O$ Tributylcarbinol or 5-Butyl-5-nonanol, **II**, 603; **VI**, 241

$C_{13}H_{28}O_2$ Tridecamethylene glycol, **II**, 155

$C_{13}H_{29}N$ Laurylmethylamine, **IV**, 564

$C_{14}F_{10}$ Di-(pentafluorophenyl)acetylene, **VI**, 881

$C_{14}H_6Cl_4O_4$ DL-4,4',6,6'-Tetrachloro-diphenic acid, **IV**, 872

$C_{14}H_7ClO_2$ α-Chloroanthraquinone, **II**, 128

$C_{14}H_8Br_2$ 9,10-Dibromoanthracene, **I**, 207

$C_{14}H_8Cl_2O_3$ p-Chlorobenzoic anhydride, **III**, 29

$C_{14}H_8N_2O_8$ p-Nitrobenzoyl peroxide, **III**, 649

$C_{14}H_8O_2$ Anthraquinone, **II**, 554

9,10-Phenanthrenequinone, **IV**, 757; **VI**, 887

$C_{14}H_8O_3$ Fluorenone-2-carboxylic acid, **III**, 420

$C_{14}H_8O_4$ Quinizarin, **I**, 476

$C_{14}H_8O_5SK$ Potassium anthraquinone-α-sulfonate, **II**, 539

$C_{14}H_9Br$ 9-Bromophenanthrene, **III**, 134

$C_{14}H_9Cl$ 9-Chloroanthracene, **V**, 206

$C_{14}H_9F_5O_2$ Ethyl 4-(pentafluorophenyl) benzoate, **VI**, 880

$C_{14}H_9NO_2$ 9-Nitroanthracene, **IV**, 711

$C_{14}H_9NO_3$ Nitroanthrone, **I**, 390

$C_{14}H_{10}$ Diphenylacetylene, **III**, 350; **IV**, 377; **VI**, 531

Phenanthrene, **IV**, 313

$C_{14}H_{10}Cl_2O$ 2,2'-Dichloro-α,α'-epoxy-bibenzyl, **V**, 358

$C_{14}H_{10}Cl_2O_2$ Di-(p-chlorophenyl)acetic acid, **III**, 270

$C_{14}H_{10}F_5N$ 4-(Pentafluorophenyl)-*N,N*-dimethylaniline, **VI**, 880

$C_{14}H_{10}N_2O$ Phenylbenzoyldiazomethane, **II**, 496

$C_{14}H_{10}O$ Anthrone, **I**, 60
   Diphenylketene, **III**, 356; **VI**, 549
   Phenanthrene-9,10-oxide, **VI**, 887

$C_{14}H_{10}O_2$ Benzil, **I**, 87
   Diphenaldehyde, **V**, 489
   9-Fluorenecarboxylic acid, **IV**, 482

$C_{14}H_{10}O_2S$ 2,3-Diphenylvinylene sulfone
   or Diphenylthiirene 1,1-dioxide,
   VI, 404, 555

$C_{14}H_{10}O_2S_2$ Benzoyl disulfide, **III**, 116

$C_{14}H_{10}O_3$ Benzoic anhydride, **I**, 91
   Diphenaldehydic acid, **V**, 493

$C_{14}H_{10}O_3S$ 2-Phenanthrenesulfonic acid,
   II, 482
   3-Phenanthrenesulfonic acid, **II**, 482

$C_{14}H_{10}O_4$ Diphenic acid, **I**, 222

$C_{14}H_{11}ClO$ Desyl chloride, **II**, 159
   Diphenylacetyl chloride, **VI**, 549

$C_{14}H_{11}N$ Diphenylacetonitrile, **III**, 347
   2-Phenylindole, **III**, 725

$C_{14}H_{11}NO$ Benzophenone cyanohydrin,
   VII, 20

$C_{14}H_{11}NO_2$ 3-Cyano-4-methyl-6-(4-methylphenyl)-2*H*-pyran-2-one,
   VI, 466

$C_{14}H_{11}N_3O_2$ β-Methyl-β-phenyl-α,α'-dicyanoglutarimide, **IV**, 662

$C_{14}H_{12}$ 9,10-Dihydroanthracene, **V**, 398
   9,10-Dihydrophenanthrene, **IV**, 313;
   V, 398
   1,1-Diphenylethylene, **I**, 226; **VI**, 187
   9-Methylfluorene, **IV**, 623
   (*E*)-Stilbene, **VI**, 56, 184
   (*Z*)-Stilbene, **IV**, 857; **VI**, 410

$C_{14}H_{12}Br_2$ Stilbene dibromide, **III**, 350

$C_{14}H_{12}Br_2O_2S$ α,α'-Dibromodibenzyl sulfone, **VI**, 403, 555

$C_{14}H_{12}N_2$ Benzalazine, **II**, 395

$C_{14}H_{12}N_2O_2$ *sym*-Dibenzoylhydrazine, **II**, 208

$C_{14}H_{12}N_2O_4$ *p,p'*-Dinitrobibenzyl, **IV**, 367

$C_{14}H_{12}N_4$ 1,4-Diphenyl-5-amino-1,2,3-triazole, **IV**, 380
   4-Phenyl-5-anilino-1,2,3-triazole, **IV**, 380

$C_{14}H_{12}O$ Desoxybenzoin, **II**, 156
   Diphenylacetaldehyde, **IV**, 375; **VI**, 1029
   1-Phenoxy-1-phenylethene, **VIII**, 512
   trans-Stilbene oxide, **IV**, 860

$C_{14}H_{12}O_2$ Acenaphthenol acetate, **III**, 3
   Benzoin, **I**, 94
   Benzyl benzoate, **I**, 104
   trans-9,10-Dihydro-9,10-phenanthrenediol, **VI**, 887
   Diphenylacetic acid, **I**, 224; **VI**, 549
   p-Cresyl benzoate, **VI**, 261
   2-Cyclopropyl-3-methyl-1,4-benzoquinone, **VI**, 891

$C_{14}H_{12}O_3$ Benzilic acid, **I**, 89
   2-*p*-Acetylphenylhydroquinone, **IV**, 15

$C_{14}H_{12}O_4$ 3,3a,3b,4,6a,7a-Hexahydro-3,4,7-metheno-7*H*-cyclopenta [*a*]pentalene-7,8-dicarboxylic acid, **VIII**, 298

$C_{14}H_{12}S$ α-Phenylthiostyrene, **VI**, 742

$C_{14}H_{13}BrO$ *erythro*-2-Bromo-1,2-diphenylethanol, **VI**, 184
   threo-2-Bromo-1,2-diphenylethanol, **VI**, 185

$C_{14}H_{13}FO_2$ Ethyl α-fluoronaphthaleneacetate, **VI**, 836

$C_{14}H_{13}F_5$ 1-Pentafluorophenylbicyclo [2.2.2]octane, **VI**, 880

$C_{14}H_{13}N$ 10,11-Dihydro-5H-dibenzo[*b,f*]azepine, **VI**, 105

$C_{14}H_{13}NO$ *N*-Benzylbenzamide, **VI**, 494

$C_{14}H_{13}NO_2$ Salicyl-*o*-toluide, **III**, 765

$C_{14}H_{13}N_5O_2$ 7-Benzylideneaminotheophylline, **VII**, 9

$C_{14}H_{14}$ [14]Annulene, **VI**, 74
   3,3'-Dimethyl-1,1'-biphenyl, **III**, 295; **VI**, 490
   4,4'-Dimethyl-1,1'-biphenyl, **VI**, 488
   VI, 490
   1,1-Diphenylethane, **VI**, 537

$C_{14}H_{14}Cl_2O_2Te$ Bis(4-methoxyphenyl)tellurium dichloride, **VI**, 468

$C_{14}H_{14}Hg$ Di-*p*-tolylmercury, **I**, 231

$C_{14}H_{14}N_2$ Acetophenone phenylhydrazone, **III**, 726

trans-1-Amino-2,3-diphenylaziridine, **VI**, 56, 679

$C_{14}H_{14}N_2O_2$ α-Phenyl-α-carbethoxyglutaronitrile, **IV**, 776

$C_{14}H_{14}N_2O_2S$ Benzaldehyde tosylhydrazone, **VII**, 438

$C_{14}H_{14}N_2O_3S$ 4-Acetylpyridine oxime tosylate, **VII**, 149

$C_{14}H_{14}N_4$ Benzil dihydrazone, **IV**, 377

$C_{14}H_{14}O$ 1-Methoxy-2-methylbenzobarrelene, **VI**, 85, 679

$C_{14}H_{14}O_2$ 3,3'-Dimethoxybiphenyl, **III**, 295

4,4'-Dimethoxy-1,1'-biphenyl, **VI**, 468, 488

Ethyl 1-naphthylacetate, **VI**, 613

$C_{14}H_{14}O_2P_2S_4$ Lawesson's reagent [2,4-bis(4-methoxyphenyl]-1,3,2,4-dithiaphosphetane]2,4-disulfide, **VII**, 372

$C_{14}H_{14}O_5S_2$ *p*-Toluenesulfonic anhydride, **IV**, 940

$C_{14}H_{14}S$ Benzyl sulfide, **VI**, 130,

$C_{14}H_{14}S_2$ Benzyl disulfide, **VI**, 130,

$C_{14}H_{14}Se$ Di-*m*-tolyl selenide, **II**, 239

Di-*o*-tolyl selenide, **II**, 239

Di-*p*-tolyl selenide, **II**, 239

$C_{14}H_{15}ClN_2O$ 1-(4-Chlorophenyl)-3-(4'-methyl-1-piperazinyl)-2-propyn-1-one, **VII**, 30

$C_{14}H_{15}F_5Si$ [(Pentafluorophenyl)ethynyl]triethylsilane, **VI**, 881

$C_{14}H_{15}N$ *m*-Tolylbenzylamine, **III**, 827

α-*p*-Xenylethylamine, **II**, 505

$C_{14}H_{15}NO_5S$ 1-Methylthiol-3-phthalamidopropane-3,3-dicarboxylic acid, **II**, 385

$C_{14}H_{16}$ Binor-S, **VI**, 378

cis-2a,3,4,8b-Tetrahydro-1,2-dimethylcyclobuta[*a*]naphthalene, **VI**, 486

$C_{14}H_{16}O_2$ 2-(Butylidene)-1-phenyl-1,3-butanedione, **VI**, 26

Propyl 1-naphthylacetate, **VI**, 614

$C_{14}H_{16}O_2Se$ 2-Acetyl-2-phenylselenocyclohexanone, **VI**, 23

$C_{14}H_{16}O_3S$ 2-(*p*-Toluenesulfinyl)-2-cyclopentenone ethylene ketal, (*S*)-(+)-, **VII**, 495

$C_{14}H_{16}O_4$ Diethyl benzalmalonate, **III**, 377

$C_{14}H_{16}O_5$ Diethyl benzoylmalonate, **IV**, 285

$C_{14}H_{17}BrO_6$ Methyl 4-*O*-benzoyl-6-bromo-6-deoxy-α-D-glucopyranoside, **VIII**, 363

$C_{14}H_{17}ClO_2$ *trans*-2-Phenylcyclohexyl chloroacetate, **VIII**, 516

$C_{14}H_{17}N$ α-Cyclohexylphenylacetonitrile, **III**, 219

1,2,3,4,5,6-Hexahydro-4-methylbenzo[*f*]quinoline, **VI**, 820

$C_{14}H_{17}NO$ *N*-Phenyl-2-oxa-3-aza-bicyclo[9.0^{1.5}.0^{4.5}]decane, **VI**, 672

$C_{14}H_{17}NO_3$ 2-Acetyl-6,7-dimethoxy-1-methylene-1,2,3,4-tetrahydroisoquinoline, **VI**, 1

$C_{14}H_{17}NO_6$ 2,3,6-Tricarboethoxypyridine, **VIII**, 597

$C_{14}H_{17}O_3$ 2-Acetyl-6,7-dimethoxy-1-methylene-1,2,3,4-tetrahydroisoquinoline, **VI**, 1

$C_{14}H_{18}$ 11,12-Dimethyltricyclo[4.4.2.0^{1.6}]dodeca-2,4,11-triene, **VI**, 486

11,12-Dimethyltricyclo[4.4.2.0^{1.6}]dodeca-3,7,11-triene, **VI**, 486

(1*Z*,3*E*)-1-Phenyl-1,3-octadiene, **VIII**, 532

$C_{14}H_{18}Cl_2O_2$ 2,5-Di-*tert*-butyl-3,6-dichloro-1,4-benzoquinone, **VI**, 210

$C_{14}H_{18}N_6O_2$ 3,6-Diazido-2,5-di-*tert*-butyl-1,4-benzoquinone, **VI**, 210

$C_{14}H_{18}O$ 2-Benzyl-2-methylcyclohexanone, **VI**, 121

2-Benzyl-6-methylcyclopropane, **VI**, 121

$C_{14}H_{18}O_4$ Diethyl benzylmalonate, **III**, 705

$C_{14}H_{18}O_6$ Methyl 4,6-O-benzylidene-α-D-glucopyranoside, **VIII**, 363

$C_{14}H_{19}BrO_9$ Acetobromoglucose, **III**, 11

$C_{14}H_{19}ClO_2$ 3-Chloro-2,5-di-*tert*-butyl-1,4-benzoquinone, **VI**, 210

$C_{14}H_{19}Cl_3O_2$ 2,5-Di-*tert*-butyl-3,5,6-trichloro-2-cyclohexene-1,4-dione, **VI**, 210

$C_{14}H_{19}NO$ 2-Benzyl-4,4,6-trimethyl-5,6-dihydro-1,3(4*H*)-oxazine, **VI**, 905

N-Methyl-6-phenyl-7-oxa-8-azabicyclo[4.2.1]nonane, **VI**, 671

$C_{14}H_{19}NO_3$ Ethyl 2-(*N,N*-diethylcarbamoyl)benzoate, **VI**, 579

$C_{14}H_{19}NO_4$ 2-(2-Acetamidoethyl)-4,5-dimethoxyacetophenone or *N*-[2-(4,5-Dimethoxy-2-oxoethylphenyl)ethyl] acetamide, **VI**, 1

N-(*tert*-Butoxycarbonyl)-L-phenylalanine, **VII**, 70, 75

$C_{14}H_{20}$ 1-Butyl-1,2,3,4-tetrahydronaphthalene, **VI**, 539

Diamantane, **VI**, 378

11,12-Dimethyltricyclo[4.4.2.0^1,6] dodeca-3,11-diene, **VI**, 486

Tetrahydro-Binor-S, **VI**, 378

$C_{14}H_{20}ClNO_8$ 2-Acetamido-3,4,6-tri-*O*-acetyl-2-deoxy-α-D-glycopyranosyl chloride, **V**, 1

$C_{14}H_{20}Cl_2O_2$ 3,6-Di-*tert*-butyl-5,6-dichloro-2-cyclohexene-1,4-dione, **VI**, 210

$C_{14}H_{20}Cl_3NO_3$ 3-(4,4-Ethylenedioxybutyl)-3-trichloroacetamido-1-cyclohexene or *N*-[1-(4,4-Ethylenedioxy-1-butyl)-2-cyclohexen-1-yl]trichloroacetamide, **VI**, 510

$C_{14}H_{20}FeN·I$ *N,N*-Dimethylaminomethylferrocene methiodide, **V**, 434

$C_{14}H_{20}N_2$ 1,1'-Dicyano-1,1'-bicyclohexyl, **IV**, 273

$C_{14}H_{20}N_4$ 1,1'-Azo-bis-1-cyclohexanenitrile, **IV**, 66

$C_{14}H_{20}O$ *cis*-4,4a,5,6,7,8-Hexahydro-4a-methyl-5-(1-methylethenyl)-

2(3*H*)-naphthalenone, **VI**, 668

Phenyl heptyl ketone, **VI**, 919

$C_{14}H_{20}O_2$ 2,5-Di-*tert*-butyl-1,4-benzoquinone, **VI**, 210

2,6-Di-*tert*-butyl-1,4-benzoquinone, **VI**, 412

1,1-Diethylpropyl benzoate, **VI**, 260

Dispiro[5.1.5.1]tetradecane-7,14-dione, **V**, 297

3-(Hydroxyphenylmethyl)-2-heptanone, **VI**, 697

$C_{14}H_{20}O_3S$ 1-(4-Methoxyphenylsulfonyl)-2,2,3,3-tetramethylcyclopropane, **VI**, 985

$C_{14}H_{20}O_8$ Ethyl ethylenetetracarboxylate, **II**, 273

$C_{14}H_{20}O_9$ 1,3,4,6-Tetra-*O*-acetyl-2-deoxy-α-D-glucopyranose, **VIII**, 583

$C_{14}H_{20}O_{10}$ β-D-Glucose-1,2,3,4-tetraacetate, **III**, 432

β-D-Glucose-2,3,4,6-tetraacetate, **III**, 434

$C_{14}H_{20}O_{11}$ 2,3,4,6-Tetra-*O*-acetyl-D-gluconic acid monohydrate, **V**, 887

$C_{14}H_{20}Si$ Trimethyl(2-methylene-4-phenyl-3-butenyl)silane **VIII**, 602,

$C_{14}H_{22}$ 1,2-Dibutylbenzene, **VI**, 407

4-(1-Methylethyl)-1-pentylbenzene, **VI**, 539

$C_{14}H_{22}N_4$ 1,2-Di-(1-cyano)cyclohexylhydrazine, **IV**, 274

$C_{14}H_{22}O$ 2,6-Di-*tert*-butylphenol, **VI**, 412

(*E*)-7,11-Dimethyl-6,10-dodecadien-2-yn-1-ol, **VIII**, 226

trans-4,4a,5,6,7,8-Hexahydro-5-methyl-8-(1-methylethyl)-2(3*H*)-naphthalenone, **VI**, 668

$C_{14}H_{22}O_2$ 1,1'-Ethynylene-bis-cyclohexanol, **IV**, 471

$C_{14}H_{22}O_3SSi$ (*Z*)-4-(Trimethylsilyl)-3-butenyl tosylate, **VIII**, 358

$C_{14}H_{22}O_3$ 7-Acetoxy-4,4,6,7-tetramethylbicyclo[4.2.0]octan-2-one, **VI**, 1024

$C_{14}H_{22}O_4$ (4R,4aR,6R,8aR)-4-Methoxy-carbonyl-1,1,6-trimethyl-1,4,4a,5,6,7,8,8a-octahydro-2,3-benzopyrone, **VIII**, 353

$C_{14}H_{22}O_6S_2$ 1,2:5,6-Di-(O-isopropylidene-3-O-(S-methyl dithiocarbonate) α-D-glucofuranose, **VII**, 139

$C_{14}H_{22}S$ Phenyl octyl sulfide, **VI**, 834
Phenyl 2-octyl sulfide, **VI**, 834

$C_{14}H_{22}S_2$ 1,2-Bis(butylthio)benzene, **V**, 107

$C_{14}H_{23}N$ 2,6-Di-tert-butyl-4-methylpyridine, **VII**, 144

$C_{14}H_{23}NOSi$ N-(4-Methoxyphenyl)-(Z)-4-(trimethylsilyl)-2-butenamine, **VIII**, 358

$C_{14}H_{23}O·CF_3O_3S$ 2,6-Di-tert-butyl-4-methylpyrylium trifluoromethanesulfonate, **VII**, 144

$C_{14}H_{24}$ 2-Butylbornene, **VII**, 77

$C_{14}H_{25}N$ 1-(Cyclooctylidenemethyl)piperidine, **VII**, 473

$C_{14}H_{26}O_2$ 2-Hydroxycyclotetradecanone, **VI**, 171
Vinyl laurate, **IV**, 977

$C_{14}H_{26}O_3$ Heptoic anhydride, **III**, 28

$C_{14}H_{26}O_3Si$ Ethyl 3-[3-(trimethylsiloxy) cyclohex-2-enyl]propionate, **VIII**, 277

$C_{14}H_{26}O_4$ Diethyl sebacate, **II**, 277

$C_{14}H_{26}S_2$ 1,4-Dithiaspiro[4.11]hexadecane, **VII**, 124

$C_{14}H_{26}Si_2$ 1,3-Bis(Trimethylsilylmethyl) benzene, **VI**, 409

$C_{14}H_{27}N$ (R)-(−)-N,N-Diethyl-(E)-citronellalenamine, **VIII**, 183
N,N-Diethylgeranylamine, **VIII**, 188
N,N-Diethylnerylamine, **VIII**, 190

$C_{14}H_{28}N_2O$ (S)-(+)-4-Methyl-3-heptanone SAMP-hydrazone, **VIII**, 403

$C_{14}H_{28}O_2$ Myristic acid, **I**, 379
(E)-3-Pentyl-2-nonene-1,4-diol, **VIII**, 507

$C_{14}H_{28}Sn$ Tributylethynyltin, **VIII**, 268

$C_{14}H_{29}Br$ 1-Bromotetradecane or Tetradecyl bromide, **II**, 247; **VI**, 869

$C_{14}H_{30}O$ Myristyl alcohol, **II**, 374

$C_{14}H_{30}O_2$ Tetradecamethylene glycol, **II**, 155

$C_{14}H_{31}N$ N,N-Dimethyldodecylamine, **VI**, 501

$C_{14}H_{31}NO$ N,N-Dimethyldodecylamine oxide, **VI**, 501

$C_{15}H_9N$ 9-Anthracenecarbonitrile, **VI**, 467
9-Cyanophenanthrene, **III**, 212

$C_{15}H_{10}BrNO_2$ 1-Methylamino-4-bromoanthraquinone, **III**, 575

$C_{15}H_{10}Br_2O_2$ Dibenzoyldibromomethane, **II**, 244

$C_{15}H_{10}ClNO$ 3-(4-Chlorophenyl)-5-phenylisoxazole, **VI**, 280
5-(4-Chlorophenyl)-3-phenylisoxazole, **VI**, 280

$C_{15}H_{10}O$ 9-Anthraldehyde, **III**, 98
Diphenylcyclopropenone, **V**, 514
Phenanthrene-9-aldehyde, **III**, 701

$C_{15}H_{10}O_2$ Benzalphthalide, **II**, 61
Flavone, **IV**, 478
β-Methylanthraquinone, **I**, 353

$C_{15}H_{10}O_3$ Diphenyl triketone and hydrate, **II**, 244

$C_{15}H_{11}BrO$ α-Bromobenzalacetophenone, **III**, 126

$C_{15}H_{11}N$ α-Phenylcinnamonitrile, **III**, 715

$C_{15}H_{11}NO$ 3-Benzoylindole, **VI**, 109
3,5-Diphenylisoxazole, **VI**, 280

$C_{15}H_{11}NO_2$ Benzylphthalimide, **II**, 83
1-Methylaminoanthraquinone, **III**, 573

$C_{15}H_{11}NO_4$ trans-o-Nitro-α-phenylcinnamic acid, **IV**, 730

$C_{15}H_{11}N_3$ Terpyridine, **VII**, 476

$C_{15}H_{12}$ Diphenylacetylene, **VI**, 956

$C_{15}H_{12}Br_2$ 1,1-Dibromo-2,2-diphenylcyclopropane, **VI**, 187

$C_{15}H_{12}Br_2O$ Benzalacetophenone dibromide, **I**, 205
α,α'-Dibromodibenzyl ketone, **V**, 514

$C_{15}H_{12}ClNO_2$ 3-(4-Chlorophenyl)-5-(4-methoxyphenyl)isoxazole, **VI**, 278

$C_{15}H_{12}N_2$ 2,3-Dicyano-1,4,4a,9a-tetrahy-
drofluorene or 1,4,4a,9a-
Tetrahydro-2,3-fluorenedicar-
bonitrile, **VI**, 427

$C_{15}H_{12}N_2O$ 4,5-Diphenylglyoxalone, **II**,
231

$C_{15}H_{12}N_2O_2$ 2-(4-Nitrophenyl)-2-phenyl-
propionitrile, **VI**, 899

$C_{15}H_{12}O$ 2-Acetylfluorene, **III**, 23
Benzalacetophenone, **I**, 78
α-Phenylcinnamaldehdye, **VI**, 908
β-Phenylcinnamaldehyde, **VI**, 901

$C_{15}H_{12}O_2$ Dibenzoylmethane, **I**, 205; **III**,
251
α-Phenylcinnamlc acid, **IV**, 777
β-Phenylcinnamic acid, **V**, 509
Phenyl cinnamate, **III**, 714

$C_{15}H_{12}O_3$ o-Benzoyloxyacetophenone, **IV**,
478
o-Hydroxydibenzoylmethane, **IV**,
479
Methyl 2-benzoylbenzoate, **VI**, 576
p-Toluyl-o-benzoic acid, **I**, 517

$C_{15}H_{13}ClO$ 1-(p-Chlorophenyl)-3-phenyl-
2-propanone, **IV**, 176

$C_{15}H_{13}N$ 1-Benzylindole, **VI**, 104, 106
3-Benzylindole, **VI**, 109

$C_{15}H_{13}NO$ 4-Benzyloxyindole, **VII**, 34

$C_{15}H_{13}NO_2$ Benzoylacetanilide, **III**, 108;
IV, 80

$C_{15}H_{13}OP$ 2,3-Dihydro-1-phenyl-4(1H)-
phosphinolinone, **VI**, 935

$C_{15}H_{14}$ 1,1-Diphenylcyclopropane, **V**, 509
1-Phenylindane or 2,3-Dihydro-1-
phenyl-1H-indene, **VI**, 540

$C_{15}H_{14}ClN$ N-(1-Chloro-2-phenylethenyl)-
N-methylaniline, **VI**, 287

$C_{15}H_{14}NO_6Na$ Ethyl sodium phthalimidomalonate, **II**, 384

$C_{15}H_{14}O$ Benzylacetophenone, **I**, 101
α,α-Diphenylacetone, **III**, 343
Diphenylmethyl vinyl ether, **VI**, 552
α-Methylbenzoin, **VI**, 222

$C_{15}H_{14}O_2$ α,β-Diphenylpropionic acid, **V**,
526
2-Phenylethyl benzoate, **V**, 336

$C_{15}H_{14}O_4$ 2-(ω-Carboxybutyl)-1,4-naph-
thoquinone or 4-(1,4-Dioxo-2-
naphthyl)pentanoic acid, **VI**,
891

$C_{15}H_{15}ClO$ Benzhydryl β-chloroethyl
ether, **IV**, 72

$C_{15}H_{15}ClO_2S$ Chloro-p-tolylsulfonyl-p-
tolylthiomethane, **VI**, 985

$C_{15}H_{15}N$ cis-2-Benzyl-3-phenylazlridine,
V, 83

$C_{15}H_{15}NO$ Dibenzyl ketoxime, **V**, 84
p-Dimethylaminobenzophenone, **I**,
217
N-(2-Phenylethyl)benzamide, **V**,
336

$C_{15}H_{15}NO_2S$ 2-(p-Tolylsulfonyl)dihy-
droisoindole, **V**, 1064

$C_{15}H_{15}NO_6$ Ethyl phthalimidomalonate, **I**,
271

$C_{15}H_{15}N_3O_2$ Methyl Red, **I**, 374

$C_{15}H_{15}N_3O_3$ N-(p-Dimethylaminophenyl)-
α-(o-nitrophenyl)nitrone, **V**,
826

$C_{15}H_{15}S·BF_4$ Cyclopropyldiphenylsulfonium tetrafluoroborate, **VI**, 364

$C_{15}H_{16}ClS·BF_4$ 3-Chloropropyldiphenyl-
sulfonium tetrafluoroborate, **VI**,
364

$C_{15}H_{16}O$ 2,9-Dimethyl-1-methoxybenzo-
barrelene, **VI**, 85
2,10-Dimethyl-1-methoxybenzobar-
relene, **VI**, 85

$C_{15}H_{16}O_2$ 1-Ethynyl-1-cyclohexyl benzoate, **VI**, 260

$C_{15}H_{16}O_3$ Ethyl 4-benzyl-2-methylfuran-3-
carboxylate, **VI**, 32

$C_{15}H_{16}O_3S$ 4-Oxo-1-(phenylsulfonyl)-cis-
bicyclo[4.3.0]non-2-ene, **VIII**,
39

$C_{15}H_{16}S_2$ 1,1-Bis(phenylthio)propane, **VI**,
742

$C_{15}H_{17}N$ N-Methyl-1,2-diphenylethyl-
amine, **IV**, 605
N-Methyl-1,2-diphenylethylamine
hydrochloride, **IV**, 605
N-β-Naphthylpiperidine, **V**, 816

$C_{15}H_{18}N_4O_4$ Bicyclo[3.3.1]nonan-9-one 2,4-dinitrophenylhydrazone, **VI**, 140

$C_{15}H_{18}O_4$ Ethyl 6,7-dimethoxy-3-methylindene-2-carboxylate, **V**, 550

$C_{15}H_{18}O_5$ Ethyl α-acetyl-β-(2,3-dimethoxyphenyl)acrylate, **IV**, 408
Ethyl phenylacetylmalonate, **III**, 637

$C_{15}H_{19}N$ 2-Ethenyl-2-phenylheptanenitrile, **VI**, 942

$C_{15}H_{19}NO_2$ 1-(2-Benzoylpropanoyl)piperidine, **VIII**, 326

$C_{15}H_{19}NO_2S$ cis-N-Tosyl-3-methyl-2-azabicyclo[3.3.0]oct-3-ene, **VII**, 501

$C_{15}H_{19}NO_3$ Hexahydro-3-(hydroxymethyl)-8a-methyl-2-phenyl[2S,3S,8aR]-5-oxo-5H-oxazolo[3,2-a]pyridine, **VIII**, 241

$C_{15}H_{20}BrNO_2$ 6-Bromo-3,4-dimethoxybenzaldehyde cyclohexylimine, **VIII**, 586

$C_{15}H_{20}INO_2$ 6-Iodo-3,4-dimethoxybenzaldehyde cyclohexylimine, **VIII**, 586

$C_{15}H_{20}N_2O_2$ N-[2-Phenyl-1-(1-pyrrolidinecarbonyl)ethyl]acetamide, **VI**, 494

$C_{15}H_{20}N_2O_5S$ Benzyloxycarbonyl-L-alanyl-L-cysteine methyl ester **VII**, 30,

$C_{15}H_{20}O_2$ 1-Phenyl-2,4-nonanedione, **VI**, 930

$C_{15}H_{20}O_5$ Ethyl α-acetyl-β-(2,3-dimethoxyphenyl)propionate, **IV**, 408

$C_{15}H_{21}IO$ 2-(2,2-Dimethyl-1-oxopropyl)-3,4,5,6-tetramethyliodobenzene or 2,2-Dimethyl-1-(2,3,4,5-tetramethyl-6-iodophenyl)-1-propanone, **VI**, 702

$C_{15}H_{21}NO_2$ erythro-1-(3-Hydroxy-2-methyl-3-phenylpropanoyl) piperidine, **VIII**, 326
threo-1-(3-Hydroxy-2-methyl-3-phenylpropanoyl)piperidine, **VIII**, 326

$C_{15}H_{21}NO_5$ 2-(2-Acetamidoethyl)-4,5,6-trimethoxyacetophenone or N-[2-(1-Oxoethyl-3,4,5-trimethoxyphenyl)ethyl]acetamide, **VI**, 3

$C_{15}H_{22}$ 4-Methyldiamantane, **VI**, 381

$C_{15}H_{22}O$ Aristolone, **VI**, 669

$C_{15}H_{22}O_2$ 1,1-Diethylpropyl p-toluate, **VI**, 260

$C_{15}H_{22}O_3$ 2',6'-Dimethylphenyl (2SR,3SR)-2,4-dimethyl-3-hydroxypentanoate, **VII**, 190

$C_{15}H_{23}N$ 16-Azabicyclo[10.3.1]hexadeca-12,14,16(1)-triene, **VI**, 410

$C_{15}H_{24}$ α-Farnesene, **VII**, 245

$C_{15}H_{24}O$ 1-Butyl-10-methyl-$\Delta^{1(9)}$-2-octalone, **VI**, 242
3,3a,4,5,6,7-Hexahydro-2,4-bis(1-methylethyl)-2-indenone, **VI**, 524
Fukinone, **VI**, 669

$C_{15}H_{24}O_4$ 3a,5a-cis-Hexahydro-4-(1,2-dihydroxyheptyl)cyclopropa[3,4]-cyclopenta[1,2-b]furan-2(3H)-one, **VI**, 345

$C_{15}H_{24}O_8$ Ethyl propane-1,1,2,3-tetracarboxylate, **I**, 272
Ethyl propane-1,1,3,3-tetracarboxylate, **I**, 290

$C_{15}H_{25}Br$ Farnesyl bromide, **VI**, 637

$C_{15}H_{26}O$ Farnesol, **VI**, 560

$C_{15}H_{26}O_5$ Ethyl enanthylsuccinate, **IV**, 430

$C_{15}H_{26}O_5Si$ Diethyl cis-4-trimethylsilyloxy-4-cyclohexene-1,2-dicarboxylate, **VI**, 447
Diethyl trans-4-trimethysilyloxy-4-cyclohexene-1,2-dicarboxylate, **VI**, 445

$C_{15}H_{27}\cdot BF_4$ Tri-tert-butylcyclopropenyl tetrafluoroborate, **VI**, 991

$C_{15}H_{28}O_2$ 15-Hydroxypentadecanoic acid lactone, **VI**, 699

$C_{15}H_{28}O_3Si$ 2-(1-Ethoxycyclohexyl)-2-(trimethylsiloxy)cyclobutanone, **VIII**, 579
Ethyl 4-cyclohexylidene-4-trimethylsiloxybutanoate, **VIII**, 578

$C_{15}H_{30}O$ Pentadecanal, **VI**, 869

$C_{15}H_{30}O_2$ 1,1-Diethylpropyl diisopropylacetate, **VI**, 261

Ethyl tridecylate, **II**, 292

Methyl myristate, **III**, 605

$C_{15}H_{32}Sn$ Allyltributyltin, **VIII**, 23

$C_{15}H_{33}BO_3$ Amyl borate, **II**, 107

$C_{15}H_{33}O_4P$ Amyl phosphate, **II**, 110

$C_{16}H_4F_{10}O_2$ 1,4-Di(perfluorophenyl)-1,4-butanedione, **VI**, 881

$C_{16}H_9Br$ 3-Bromopyrene, **V**, 147

$C_{16}H_9N_5O_{11}$ (+)- and (−)-α-(2,4,5,7-Tetranitro-9-fluorenylideneaminooxy)propionic acid, **V**, 1031

$C_{16}H_{10}$ Diphenyldiacetylene, **V**, 517

3-Hydroxypyrene, **V**, 632

$C_{16}H_{10}O_3$ 5-Formyl-4-phenanthroic acid, **IV**, 484

$C_{16}H_{11}NO_2$ Azlactone of α-benzoylaminocinnamic acid, **II**, 490

$C_{16}H_{11}NO_4$ α-Phthalimido-*o*-toluic acid, **IV**, 810

$C_{16}H_{12}$ 1-Phenylnaphthalene, **III**, 729

$C_{16}H_{12}ClNO$ α-(4-Chlorophenyl)-γ-phenylacetoacetonitrile, **IV**, 174

$C_{16}H_{12}ClNO_2$ 5-(4-Chlorophenyl)-3-(4-methoxyphenyl)isoxazole, **VI**, 280

$C_{16}H_{12}NO_2S$ 3,5-Diphenyl-4*H*-1,4-thiazine 1,1-dioxide, **VI**, 978

$C_{16}H_{12}N_2$ 2,3-Diphenylsuccinonitrile, **IV**, 392

$C_{16}H_{12}N_2O_3$ 2-Benzyloxycarbonyloxyimino-2-phenylacetonitrile, **VI**, 202

$C_{16}H_{12}O$ 9-Acetylanthracene, **IV**, 8

9-Acetylphenanthrene, **III**, 26

$C_{16}H_{12}O_2$ *trans*-Dibenzoylethylene, **III**, 248

2,3-Dimethylanthraquinone, **III**, 310

$C_{16}H_{13}N$ 1-Benzylisoquinoline, **VI**, 115

2,4-Diphenylpyrrole, **III**, 358

$C_{16}H_{13}NO$ α-Phenyl-β-benzoylpropionitrile, **II**, 498

3,4-Diphenyl-4-oxobutyronitrile, **VI**, 867

$C_{16}H_{13}NO_2$ 5-(4-Methoxyphenyl)-3-phenylisoxazole, **VI**, 280

1-Phenyl-3,4-dihydro-6,7-methylenedioxyisoquinoline, **VII**, 326

$C_{16}H_{13}NO_3$ α-Benzoylaminocinnamic acid, **II**, 490

$C_{16}H_{13}N_3S$ 4'-(Methylthio)-2,2':6',2''-terpyridine, **VII**, 476

$C_{16}H_{14}$ 1,4-Diphenylbutadiene, **II**, 229; **V**, 499

2,3-Diphenyl-1,3-butadiene, **VI**, 531

1,2-Diphenyl-3-methylcyclopropene, **VII**, 203

1-Phenyldialin (1-phenyl-3,4-dihydronaphthalene), **III**, 729

$C_{16}H_{14}BrN$ 4-Bromo-2,2-diphenylbutyronitrile, **VI**, 899

$C_{16}H_{14}N_2O_2$ 5,5-Dicyano-4-phenyl-2-cyclopenten-1-one 1,3-propanediol ketal, **VIII**, 173

$C_{16}H_{14}O$ Dypnone, **III**, 367

$C_{16}H_{14}O_2$ 1,4-Diphenyl-1,3-butanedione, **VI**, 930

$C_{16}H_{14}O_3$ β-(1-Acenaphthoyl)propionic acid, **III**, 9

β-(3-Acenaphthoyl)propionic acid, **III**, 6

Benzoin acetate, **II**, 69

Ethyl 2-benzoylbenzoate, **VI**, 579

$C_{16}H_{14}O_4$ Diphenyl succinate, **IV**, 390

$C_{16}H_{15}F_5$ 1-(Pentafluorophenyl)adamantane, **VI**, 882

$C_{16}H_{15}N$ 1,4-Diphenyl-1-methyl-3-aza-1,4-butadiene, **VIII**, 451, 455

$C_{16}H_{16}$ [16]Annulene, **VI**, 74

1-Methyl-3-phenylindane, **IV**, 665

[2.2]Paracyclophane, **V**, 883

$C_{16}H_{16}N_2O_2$ Dibenzoyldimethylhydrazine, **II**, 209

$C_{16}H_{16}N_4O_2$ 4,4'-Bis(acetamido)azobenzene, **V**, 341

$C_{16}H_{16}O_2$ *trans*-4,4'-Dimethoxystilbene, **V**, 428

2,3-Dimethyl-1,4,5,6-tetrahydroanthraquinone, **III**, 310

$C_{16}H_{16}O_3$ Deoxyanisoin, **V**, 339

$C_{16}H_{16}O_4$ 3-Benzyloxy-4,5-dimethoxybenz-aldehyde, **VI**, 1009

Dibenzo-12-crown-4 polyether, **VI**, 399

3,8-Dimethoxy-4,5,6,7-dibenzo-1,2-dioxacyclooctane, **V**, 493

Dimethyl 3,3a,3b,4,6a,7a-hexahy-dro-3,4,7-metheno-7H-cyclo-penta[a]pentalene-1,2- and 7,8-dicarboxylate, **VIII**, 298

$C_{16}H_{17}N$ (Z)-β-[2-(N,N-Dimethylamino)phenyl]styrene, **VII**, 172

$C_{16}H_{17}NO$ α,α-Diphenyl-2-azetidine-methanol, **VI**, 546

$C_{16}H_{18}$ 1,1-Di-p-tolylethane, **I**, 229

4-(1-Methylethyl)-1-(phenylmethyl)benzene, **VI**, 540

3,3',4,4'-Tetramethyl-1,1'-biphenyl, **VI**, 490

$C_{16}H_{18}O$ 2-Methoxy-2',4',6'-trimethyl-biphenyl, **VI**, 410

$C_{16}H_{18}O_4S_4$ Ethylene dithiotosylate, **VI**, 590, 1016

$C_{16}H_{18}O_{10}$ Tetramethyl 3,7-dihydroxybicy-clo[3.3.0]octa-2,6-diene-2,4,6,8-tetracarboxylate, **VII**, 50

$C_{16}H_{18}S_2$ 1,1-Bis(phenylthio)-3-methyl-propane, **VI**, 742,

$C_{16}H_{19}N_3$ 2-Azido-2-phenyl-2-adamantane, **VII**, 433

$C_{16}H_{19}FeO_3 \cdot BF_4$ Dicarbonyl(cyclopentadi-enyl)(trans-3-methyl-2-vinylcy-clohexanone)iron tetrafluorobo-rate, **VIII**, 479

$C_{16}H_{20}N_2$ N,N'-Diethylbenzidine, **IV**, 283

$C_{16}H_{20}N_6O_{10}S$ Arginine dinitronaphtholsul-fonate (flavianate), **II**, 50

$C_{16}H_{20}O$ 2-Phenyl-2-adamantanol, **VII**, 433

$C_{16}H_{20}O_4$ Ethyl 1-(p-methoxyphenyl)-2-oxocyclohexanecarboxylate, **VII**, 229

$C_{16}H_{20}O_6$ Diethyl [O-benzoyl]ethyl-tartronate, **V**, 379

$C_{16}H_{21}ClO_{11}$ 2,4,4,5,6-Penta-O-acetyl-D-gluconyl chloride, **V**, 887

$C_{16}H_{21}N$ 2-Phenyl-2-adamantamine (and HCl salt), **VII**, 433

$C_{16}H_{21}NO_{10}$ Pentaacetyl D-glucononitrile, **III**, 690

$C_{16}H_{22}O$ (2RS,5R)-5-Methyl-2-(1-methyl-1-phenylethyl)cyclohexanone, **VIII**, 304

$C_{16}H_{22}O_2$ 1-(4-Methylphenyl)-2,4-nonane-dione, **VI**, 930

$C_{16}H_{22}O_4$ sec-Octyl hydrogen phthalate, **I**, 418

$C_{16}H_{22}O_{12}$ 2,3,4,5,6-Penta-O-acetyl-D-glu-conic acid, **V**, 887

$C_{16}H_{23}NO_2$ (−)-Menthyl nicotinate, **VIII**, 350

$C_{16}H_{23}NO_9$ 1-Deoxy-2,3,4,6-tetra-O-acetyl-1-(2-cyanoethyl)-α-D-gluco-pyranose, **VIII**, 148

$C_{16}H_{24}O$ 8-Phenylmenthol, (−)-, **VIII**, 522

$C_{16}H_{24}O_4$ [6aR-(6aα,9a,10aβ)]-Octahydro-3,3,6,6,9-pentamethyl-1H,6H[1,3]dioxino[4,5-c'][2]benzopy-ran-1-one, **VIII**, 353

$C_{16}H_{26}O_8$ Tetraethyl 1,2,3,4-butanetet-racarboxylate, **VII**, 479

$C_{16}H_{28}$ cis-1-Cyclohexyl-3,7-dimethyl-2,6-octadiene, **VI**, 725

trans-1-Cyclohexyl-3,7-dimethyl-2,6-octadiene, **VI**, 725

$C_{16}H_{29}N$ 1-(N-Pyrrolidino)-1-cyclodo-decene, **VII**, 135

$C_{16}H_{29}N_2P$ Bis(3-dimethylaminopropyl)phenylphosphine, **VI**, 776

$C_{16}H_{30}$ (5Z,7E)-5,7-Hexadecadiene, **VIII**, 295

$C_{16}H_{30}N_2O_2$ 1,10-Diazacyclooctadecane-2,9-dione, **VI**, 382

$C_{16}H_{31}ClO$ Palmitoyl chloride, **IV**, 742

$C_{16}H_{32}O_2$ Palmitic acid, **III**, 605;

$C_{16}H_{32}O_5S$ α-Sulfopalmitic acid, **IV**, 862

$C_{16}H_{33}I$ Hexadecyl iodide, **II**, 322

$C_{16}H_{34}$ Hexadecane, **II**, 320

$C_{16}H_{34}N_2$ 1,10-Diazacyclooctadecane, **VI**, 382

$C_{16}H_{34}O$ Cetyl alcohol, **II**, 374

Triamyl carbinol, **II**, 603

$C_{16}H_{34}S$ Dioctyl sulfide, **VI**, 834
Di-2-octyl sulfide, **VI**, 833

$C_{16}H_{35}NO$ N-tert-Butyl-N-tert-octyl-O-tert-butylhydroxylamine, **VIII**, 93

$C_{16}H_{36}NBr$ Tetra-n-butylammonium bromide, **VI**, 153

$C_{16}H_{36}NHO_4S$ Tetrabutylammonium hydrogen sulfate, **VI**, 955

$C_{17}H_{10}O$ Benzanthrone, **II**, 62

$C_{17}H_{12}N_2O$ 2-Benzoyl-1,2-dihydroisoquinaldonitrile, **VI**, 115

$C_{17}H_{12}O_2$ α-Benzylidene-γ-phenyl-$\Delta^{\beta,\gamma}$-butenolide, **V**, 80

$C_{17}H_{13}ClFeO$ (2-Chlorobenzoyl)ferrocene, **VI**, 625

$C_{17}H_{13}NO_4$ N-Phthalyl-L-β-phenylalanine, **V**, 973

$C_{17}H_{14}N_2O_4$ 2-(4-Methoxybenzyloxy)carbonyloxyimino-2-phenylacetonitrile, **VI**, 202

$C_{17}H_{14}O$ Dibenzalacetone, **II**, 167

$C_{17}H_{14}O_2$ trans-1,2-Dibenzoylcyclopropane, **VI**, 401

$C_{17}H_{15}BrO_2$ 2-Bromo-3,3-diphenyl-2-propen-1-yl acetate, **VI**, 187

$C_{17}H_{15}NO$ β-Naphthol phenylaminomethane, **I**, 381

$C_{17}H_{16}N_2$ 9-Phenyl-1,2,3,4-tetrahydro-β-carboline, **VI**, 966

$C_{17}H_{16}O_2$ 1,3-Dibenzoylpropane or 1,5-Diphenyl-1,5-pentanedione, **VI**, 401
1,3-Diphenyl-2,4-pentanedione, **VI**, 247
4-(4-Methylphenyl)-1-phenyl-1,3-butanedione, **VI**, 930

$C_{17}H_{16}O_4$ 2-Acetoxy-3-(γ,γ-dimethylallyl)-1,4-naphthoquinone or 3-(3-Methyl-3-butenyl)-1,4-dioxo-2-naphthyl acetate, **VI**, 891

$C_{17}H_{17}N$ 2-Benzyl-2-phenylbutyronitrile, **VI**, 899

$C_{17}H_{18}N_4O_4$ 2,2-Dimethyl-3-phenylpropionaldehyde 2,4-dinitrophenylhydrazone, **VI**, 527

$C_{17}H_{18}OS_2$ 4,4-Bis(phenylthio)-3-methoxy-1-butene, **VI**, 737

$C_{17}H_{18}O_3$ Ethyl β-hydroxy-β,β-diphenylpropionate, **V**, 564

$C_{17}H_{19}NO$ 2-(Diphenylhydroxymethyl)pyrrolidine or Diphenyl-2-pyrrolidinemethanol, **VI**, 542

$C_{17}H_{19}NO_3$ Ethyl threo-[1-(2-hydroxy-1,2-diphenylethyl)] carbamate, **VII**, 223

$C_{17}H_{20}$ 4-(1,1-Dimethylethyl)-1-(phenylmethyl)benzene, **VI**, 540
1,1-Diphenylpentane, **V**, 523
(1-Phenylpentyl)benzene, **VI**, 539

$C_{17}H_{20}O_4S_4$ Trimethylene dithiotosylate, **VI**, 1014, 1016

$C_{17}H_{20}S_2$ 3,3-Bis(phenylthio)-2-methylbutane, **VI**, 742

$C_{17}H_{21}NO$ 2-(Diphenylmethoxy)-N,N-dimethylethylamine or Dimethyl(diphenylmethoxyethyl)amine **VI**, 552
α-Hydroxy-α-phenyl-β-(2-propylamino)ethylbenzene or 2-(1-Methylethylamino)-1,1-diphenylethanol, **VI**, 545

$C_{17}H_{21}NO_2$ tert-Butyl 3-cyano-3-phenylpentanoate, **VI**, 899

$C_{17}H_{22}O$ 2,5-Diisopropyl-3-phenyl-2-cyclopenten-1-one, **VI**, 523

$C_{17}H_{22}O_2$ 1,7,7-Trimethyl-2-norbornyl benzoate, **VI**, 260

$C_{17}H_{23}NO_9$ 1-Deoxy-2,3,4,6-tetra-O-acetyl-1-(2-cyanoethyl)-α-D-glucopyranose, **VIII**, 148;

$C_{17}H_{24}N_2O_2S$ Camphor tosylhydrazone, **VI**, 172

$C_{17}H_{24}OS$ 7-Benzylthiomenthone, **VIII**, 302

$C_{17}H_{25}N$ 9-Butyl-1,2,3,4,5,6,7,8-octahydroacridine, **VIII**, 87

$C_{17}H_{25}NO$ 9-Butyl-1,2,3,4,5,6,7,8-octahydroacridine-N-oxide, **VIII**, 87
9-Butyl-1,2,3,4,5,6,7,8-octahydroacridin-4-ol, **VIII**, 87

$C_{17}H_{26}O_2S$ (S)-(-)-Menthyl p-toluenesulfinate, **VII**, 495

$C_{17}H_{28}O$ 13,15-Dimethylbicyclo[10.3.0]pentadec-12-en-14-one, **VI**, 524

$C_{17}H_{28}OSi$ cis-4-exo-Isopropenyl-1,9-dimethyl-8-(trimethylsilyl)bicyclo[4.3.0]non-8-en-2-one, **VIII**, 347

$C_{17}H_{28}O_2$ 8-Butyl-2-hydroxytricyclo[7.3.1.0^{2,7}]tridecan-13-one, **VIII**, 87

Farnesyl acetate, **VI**, 560

$C_{17}H_{28}O_3$ 10,11-Epoxyfarnesyl acetate, **VI**, 560

cis-4-(4,4-Dimethoxybutyl)-1,7a-dimethyl-1,2,3,6,7,7a-hexahydro-5-indenone, **VI**, 668

$C_{17}H_{29}BrO_3$ 10-Bromo-11-hydroxy-10,11-dihydrofarnesyl acetate, **VI**, 560

$C_{17}H_{32}O_2$ 17-Hydroxyheptadecanoic lactone, **VI**, 699

$C_{17}H_{34}O$ Heptadecanal, **VI**, 871

$C_{17}H_{34}O_2$ 1,1-Diethylpropyl di-tert-butyl-acetate, **VI**, 261

Methyl palmitate, **III**, 605

$C_{17}H_{34}O_2Sn$ Ethyl (E)-3-(tributylstannyl)propenoate, **VIII**, 268

$C_{17}H_{34}S_3$ 2-Tetradecyl-sym-trithiane, **VI**, 869

$C_{17}H_{36}O_2$ Pentadecanal dimethyl acetal, **VI**, 869

$C_{18}H_{12}$ Benz[a]anthracene, **VII**, 18

Hexadehydro[18]annulene, **VI**, 68

Tridehydro[18]annulene, **VI**, 68

Triphenylene, **V**, 1120

$C_{18}H_{12}N_2$ 6b,7,10,10a-Tetrahydro-8,9-fluoranthenedicarbonitrile, **VI**, 431

$C_{18}H_{12}O_2$ 2,6-Diphenyl-p-benzoquinone, **VI**, 413

$C_{18}H_{13}N$ N-Phenylcarbazole, **I**, 547

$C_{18}H_{14}BrP$ Diphenyl-p-bromophenylphosphine, **V**, 496

$C_{18}H_{14}N_2O$ 2-Benzoyl-1-cyano-1-methyl-1,2-dihydroisoquinoline, **IV**, 642

$C_{18}H_{14}N_4O$ 4-(1-Phenyl-5-tetrazolyloxy)biphenyl, **VI**, 150

$C_{18}H_{14}O_2$ 3-Benzoyl-4-methyl-2-phenylfuran, **VI**, 32

$C_{18}H_{14}O_3$ 2-(p-Methoxyphenyl)-6-phenyl-4-pyrone, **V**, 721

$C_{18}H_{15}Al$ Triphenylaluminum, **V**, 1116

$C_{18}H_{15}As$ Triphenylarsine, **IV**, 910

$C_{18}H_{15}AsO$ Triphenylarsine oxide, **IV**, 911

$C_{18}H_{15}B_3O_3$ Benzeneboronic anhydride, **IV**, 68

$C_{18}H_{15}ClSe$ Triphenylselenonium chloride, **II**, 241

$C_{18}H_{15}N$ Triphenylamine, **I**, 544

$C_{18}H_{15}NO_4$ Azlactone of α-benzoylamino-β-(3,4-dimethoxyphenyl)-acrylic acid, **II**, 55

$C_{18}H_{15}P$ Triphenylphosphine, **VI**, 424, 634

$C_{18}H_{15}O_3P$ Triphenyl phosphite, **VI**, 830

$C_{18}H_{15}Sb$ Triphenylstibine, **I**, 550

$C_{18}H_{16}O_4$ 1-(p-Methoxyphenyl)-5-phenyl-1,3,5-pentanetrione, **V**, 718

$C_{18}H_{17}NO$ 1,1-Diphenyl-2-ethoxy-3-butenenitrile, **VI**, 942

$C_{18}H_{17}NS_2$ 3-(2-Phenyl-1,3-dithian-2-yl)indole, **VI**, 109

$C_{18}H_{18}$ [18]Annulene, **VI**, 68

$C_{18}H_{18}O_2$ 1,4-Dibenzoylbutane or 1,6-Diphenyl-1,6-hexanedione, **II**, 169; **VI**, 401

$C_{18}H_{18}O_6$ 4,5,4',5'-Tetramethoxy-1,1'-biphenyl-2,2'-dicarboxaldehyde, **VIII**, 586

$C_{18}H_{19}FeO_6$ Tricarbonyl[2-[(2,3,4,5-η)-4-methoxy-2,4-cyclohexadien-1-yl]-dimethyl-1,3-cyclohexanedione]iron, **VI**, 1001

$C_{18}H_{19}NO_4$ 1,2,3,4-Tetrahydro-6,7-dimethoxy-1-[3,4-(methylenedioxy)phenyl]isoquinoline, **VI**, 829

$C_{18}H_{20}N_2O_2$ 4,4'-Bis(dimethylamino)benzil, **V**, 111

$C_{18}H_{20}O_2$ Ethyl 2,4-diphenylbutanoate, **V**, 559

$C_{18}H_{20}S_2$ 1,1-Bis(phenylthio)cyclohexene, **VI**, 732

$C_{18}H_{21}NO_6S$ Diethyl 1-methylthiol-3-phthalimidopropane-3,3-dicarboxylate, **II**, 384

$C_{18}H_{22}O_4$ 1,4-Di-O-benzyl-L-threitol, **VIII**, 155

3a,5a-cis-Hexahydro-4-(1,2-dihydroxy-4-phenylbutyl)cyclopropa[3,4]cyclopenta[1,2-b]furan-2(3H)-one, **VI**, 345

$C_{18}H_{22}O_{10}$ Tetramethyl 3,7-dihydroxy-1,5-dimethylbicyclo[3.3.0]octa-2,6-diene-2,4,6,8-tetracarboxylate, **VII**, 50

$C_{18}H_{24}INO$ 2-(Diphenylmethoxy)-N,N-dimethylethylamine methiodide or Dimethyl(diphenylmethoxyethyl)amine methiodide, **VI**, 552

$C_{18}H_{25}ClO_2$ (1R,2S,5R)-5-Methyl-2-(1-methyl-1-phenylmethyl)cyclohexyl chloroacetate, **VIII**, 522

$C_{18}H_{25}NO$ 2-(1-Phenylcyclopentyl)-4,4,6-trimethyl-5,6-dihydro-1,3(4H)-oxazine, **VI**, 905

$C_{18}H_{26}N_3O_6$ Carbobenzoxy-L-asparaginyl-L-leucine methyl ester, **VI**, 263

$C_{18}H_{26}O_2$ $threo$-5-$tert$-Butyl-2-(hydroxyphenylmethyl)-2-methylcyclohexanone, **VI**, 697

$C_{18}H_{27}ClO_2$ 2,4-Di-$tert$-amylphenoxyacetyl chloride, **IV**, 742

$C_{18}H_{27}NO$ 2-(1-Phenylcyclopentyl)-4,4,6-trimethyltetrahydro-1,3-oxazine, **VI**, 905

$C_{18}H_{28}O$ cis-1,2,3,6,7,7a-Hexahydro-7a-methyl-1-(1-methyl-4-hexen-1-yl)-5-indenone, **VI**, 668

$C_{18}H_{28}O_4$ 17-Hydroxy-5-oxo-3,5-$seco$-4-norandrostane-3-carboxylic acid, **VI**, 690

$C_{18}H_{29}I$ 3-Iodo-1,2,4,5-tetrakis(1-methylethyl)benzene, **VI**, 701

$C_{18}H_{30}O_2$ Linolenic acid, **III**, 531

$C_{18}H_{32}Br_4O_2$ Tetrabromostearic acid, **III**, 526

$C_{18}H_{32}ClO$ Oleoyl chloride, **IV**, 739

$C_{18}H_{32}O_2$ Linoleic acid, **III**, 526
Recinelaidic acid lactone, **VII**, 470
Stearolic acid, **III**, 785; **IV**, 851

$C_{18}H_{33}ClO$ Oleoyl chloride, **IV**, 739

$C_{18}H_{33}ClO_2$ Ricinoleoyl chloride, **IV**, 742

$C_{18}H_{33}NO_2$ 1-Nitro-1-octadecene, **VI**, 839

$C_{18}H_{34}O_3$ Ricinelaidic acid, **VII**, 470

$C_{18}H_{36}Cl_2$ $threo$-7,8-Dichlorooctadecane, **VI**, 425

$C_{18}H_{36}O$ Oleyl alcohol, **II**, 468; **III**, 671

$C_{18}H_{36}O_4$ 9,10-Dihydroxystearic acid, **IV**, 317

$C_{18}H_{37}Al$ (E)-1-Decenyldiisobutylalane, **VIII**, 295

$C_{18}H_{37}Br$ Octadecyl bromide, **II**, 247

$C_{18}H_{38}O_2$ Octadecamethylene glycol, **II**, 155

$C_{18}H_{39}B$ Trihexylborane, **VI**, 919

$C_{19}H_{14}N_4O$ 4-(1-Phenyl-5-tetrazolyloxy)biphenyl, **VI**, 150

$C_{19}H_{15}\cdot BF_4$ Triphenylmethyl tetrafluoroborate, **VI**, 999

$C_{19}H_{15}Cl$ Triphenylchloromethane, **III**, 841

$C_{19}H_{15}Cl_2P$ Dichloromethylenetriphenylphosphorane, **V**, 361

$C_{19}H_{15}Na$ Triphenylmethylsodium, **II**, 607

$C_{19}H_{16}$ Triphenylmethane, **I**, 548; **VI**, 540, 638

$C_{19}H_{16}N_2$ N,N'-Diphenylbenzamidine, **IV**, 383

$C_{19}H_{16}O$ Triphenylcarbinol, **III**, 839

$C_{19}H_{16}O_2$ 2,6-Dimethyl-3,5-diphenyl-4H-pyran-4-one, **V**, 450

$C_{19}H_{18}N_2O_2$ 2,6-Dioxo-1-phenyl-4-benzyl-1,4-diazabicyclo[3.3.0]octane, **VIII**, 231

$C_{19}H_{18}O_4$ $trans$-1,2-Bis(4-methoxybenzoyl)cyclopropane, **VI**, 402
β-Carbethoxy-γ,γ-diphenylvinylacetic acid, **IV**, 132

$C_{19}H_{19}N$ 2,4-Diphenyl-2-(1-methylethyl)-3-butenenitrile, **VI**, 942

$C_{19}H_{20}N_2O_3$ (E)-6-Benzyloxy-2-nitro-β-pyrrolidinostyrene, **VII**, 34

$C_{19}H_{20}N_4$ Bis[4-(dimethylamino)phenyl] malononitrile, **VI**, 270

$C_{19}H_{20}O_2$ 3,3-Dimethyl-1,5-diphenylpen-
tane-1,5-dione, **VIII**, 210

$C_{19}H_{21}NO_3$ Ethyl 2,3-diphenyl-5-
methylisoxazolidine-4-carboxy-
late, **V**, 1126

$C_{19}H_{22}S_2$ [Bis(phenylthio)methyl]cyclo-
hexane, **VI**, 742

$C_{19}H_{23}FO_2$ 16α-Fluoroestrone 3-methyl
ether, **VIII**, 286,

$C_{19}H_{23}NO$ α,α-Diphenylperhydro-2-
azepinemethanol, **VI**, 547

$C_{19}H_{24}$ Dimesitylmethane, **V**, 422

$C_{19}H_{25}N_3O_7$ N-Carbobenzyloxy-3-
hydroxy-L-prolylglycylglycine
ethyl ester,**VI**, 263

$C_{19}H_{26}N_2O_2$ 16-Diazoandrost-5-en-3β-ol-
17-one, **VI**, 841

$C_{19}H_{26}O_2$ (−)-Menthyl cinnamate, **VIII**,
350

$C_{19}H_{27}NO_3$ 16-Oximinoandrost-5-en-3β-
ol-17-one, **VI**, 840

$C_{19}H_{27}N_3O_6$ N-Carbobenzyloxy-L-
asparaginyl-L-leucine methyl
ester, **VI**, 263

$C_{19}H_{28}O_2$ Dehydroisoandrosterone, **VI**,
840

1-Phenyl-2,4-tridecanedione, **VI**,
930

$C_{19}H_{28}O_3$ D-Norandrost-5-en-3β-ol-16α-
and 16β-carboxylic acids, **VI**,
840

$C_{19}H_{30}$ Androst-16-ene, **VI**, 174

$C_{19}H_{30}O_2$ 5α-Androstan-17β-ol-3-one, **VI**,
62

$C_{19}H_{32}O$ Androstan-17-ol or (5α,17β)-
Androst-2-an-17-ol, **VI**, 62

$C_{19}H_{32}O_3S$ 1-Dodecyl tosylate or Dodecyl
p-toluenesulfonate, **III**, 366;
VI, 376

$C_{19}H_{33}NO$ 2-(Hexylamino)-1-phenyl-1-
heptanol, **VI**, 546

$C_{19}H_{38}O_2$ Nonadecanoic acid, **VII**, 397

$C_{20}H_{12}Br_2$ (R)-(+)-2,2'-Dibromo-1,1'-
binaphthyl, **VIII**, 57

$C_{20}H_{12}Li_2$ 2,2'-Dilithio-1,1'-binaphthyl,
VIII, 57

$C_{20}H_{13}O_4P$ 1,1'-Binaphthyl-2,2'-diyl
hydrogen phosphate, **VIII**,
46, 50

$C_{20}H_{14}$ 9-Phenylphenanthrene, **V**, 952
Triptycene, **IV**, 964

$C_{20}H_{14}Hg$ Mercury di-β-naphthyl, **II**,
381

$C_{20}H_{14}O_2$ (R)-(+)-1,1'-Binaphthalene-2,2'-
diol, **VIII**, 46

$C_{20}H_{16}$ Triphenylethylene, **II**, 606

$C_{20}H_{16}N_2$ 4,5-Diphenylcyclohexene-1,2-
dicarbonitrile, **VI**, 431

$C_{20}H_{16}N_2O_3$ DL-Desoxycamptothecin, **VI**,
473

$C_{20}H_{17}OP$ Formylmethylenetriphenylphos-
phorane, **VI**, 450

$C_{20}H_{18}P·Br$ Vinyltriphenylphosphonium
bromide, **V**, 1145

$C_{20}H_{18}ClNO_4$ Pseudoberberine, **VI**, 473

$C_{20}H_{18}N_4O_2$ 2-(1,3-Dibenzoylimidazolidin-
2-yl)imidazole hydrochloride,
VII, 287

$C_{20}H_{18}S_2$ 1,1-Bis(phenythio)-1-phenyl-
ethane, **VI**, 742

$C_{20}H_{20}$ [20]Annulene, **VI**, 74

$C_{20}H_{20}N_2$ 1-Methyl-15,16,17,18,19,20-
hexahydroyohimban, **VI**, 473

$C_{20}H_{21}NO_4$ (4S)-3-[1-Oxo-(syn-2-methyl-
3-hydroxy)-3-phenylpropyl]-2-
oxazolidinone, **VIII**, 339

$C_{20}H_{20}N_2O_3$ 2-Phenyl-3-propylisoxazoli-
dine-4,5-cis-dicarboxylic acid
N-phenylimide, **V**, 957

$C_{20}H_{24}O_6$ Dibenzo-18-crown-6-polyether,
VI, 395

$C_{20}H_{25}NO_5$ 1,2,3,4-Tetrahydro-6,7-
dimethoxy-1-(3,4,5-trimethoxy-
phenyl)isoquinoline, **VI**, 829

$C_{20}H_{26}N_2O_2$ 1,10-Dibenzyl-4,7-dioxa-1,10-
diazadecane, **VIII**, 152

$C_{20}H_{26}N_2O_5S_2$ 4,7-Bis(p-tolylsulfonyl)-1-
oxa-4,7-diazacyclononane, **VI**,
657

$C_{20}H_{26}N_3O_8$ N-Carbobenzoxy-3-hydroxy-
L-prolylglycylglycine ethyl
ester, **VI**, 263

C$_{20}$H$_{27}$NO$_3$ Hexahydro-6-ethyl-3-(hydroxymethyl)-6-allyl-2-phenyl-
[2S,3S,6S,8aR]-5-oxo-5H-oxa-
zolo[3,2-a]pyridine, **VIII**, 241

C$_{20}$H$_{28}$ Adamantylideneadamantane, **VII**,
1

C$_{20}$H$_{30}$O$_2$ Abietic acid, **IV**, 1
Levopimaric acid, **V**, 699

C$_{20}$H$_{30}$O$_3$ Methyl D-norandrost-5-en-3-ol-
16-carboxylate, **VI**, 842

C$_{20}$H$_{34}$O$_2$ Ethyl linolenate, **III**, 532

C$_{20}$H$_{35}$B (−)-Diisopinocampheylborane,
VI, 719
Di-3-pinanylborane, **VI**, 944

C$_{20}$H$_{36}$O$_2$ Ethyl linoleate, **III**, 526

C$_{20}$H$_{36}$O$_6$ Dicyclohexyl-18-crown-6 polyether, **VI**, 395

C$_{20}$H$_{38}$O$_4$ Dimethyl octadecanedioate, **V**,
463

C$_{20}$H$_{40}$O$_2$ Ethyl stearate, **IV**, 142

C$_{21}$H$_{14}$O 2,3-Diphenylindone, **III**, 353

C$_{21}$H$_{15}$·ClO$_4$ Triphenylcyclopropenyl perchlorate, **VI**, 994

C$_{21}$H$_{15}$N α,β-Diphenylcinnamonitrile, **IV**,
387

C$_{21}$H$_{15}$O$_4$P (R)-(−)-Methyl 1,1'-binaphthyl-
2,2'-diyl phosphate, **VIII**, 46

C$_{21}$H$_{17}$N 2,3,3-Triphenylpropionitrile, **IV**,
962; **VI**, 899

C$_{21}$H$_{18}$O Dicinnamalacetone, **VII**, 60
β,β-Diphenylpropiophenone, **II**, 236

C$_{21}$H$_{18}$O$_2$ 2,2-Diphenylethyl benzoate, **VI**,
260
α,β,β-Triphenylpropionic acid, **IV**,
960

C$_{21}$H$_{19}$NO 2,3,5-Triphenylisoxazolidine,
V, 1124

C$_{21}$H$_{19}$NO$_3$S p-Toluidine 2-phenanthrene-
sulfonate, **II**, 483
p-Toluidine 3-phenanthrenesul-
fonate, **II**, 483

C$_{21}$H$_{20}$Br$_3$O$_2$P 2-Carboxyethyltriphenyl-
phosphonium perbromide, **VI**,
178

C$_{21}$H$_{20}$O$_3$S 2,2-Diphenylethyl p-toluene-
sulfonate, **VI**, 261

C$_{21}$H$_{20}$Sn Allyltriphenyltin, **V**, 453

C$_{21}$H$_{21}$Sb Tri-p-tolylstibine, **I**, 551

C$_{21}$H$_{22}$O$_4$ $trans$-1,2-Bis(2-methoxy-5-
methylbenzoyl)cyclopropane,
VI, 402

C$_{21}$H$_{24}$N$_2$O$_3$ Tetrahydroalstonine, **VI**, 105

C$_{21}$H$_{25}$NO N-(3-Hydroxy-3,3-diphenyl-
propylidene)cyclohexylamine
or 3-(Cyclohexylimino)-1,1-
diphenylethanol, **VI**, 901

C$_{21}$H$_{26}$O$_4$ 1,4-Di-O-benzyl-2,3-di-O-iso-
propylidene-L-threitol, **VIII**, 155

C$_{21}$H$_{28}$N$_2$O$_5$S$_2$ 4,8-Bis(p-tolylsulfonyl)-1-
oxa-4,8-diazacyclodecane, **VI**,
657

C$_{21}$H$_{30}$F$_3$NO N-(Trifluoroacetyl)dehydro-
abietylamine, **VI**, 1006

C$_{21}$H$_{30}$O$_3$ Testosterone acetate, **VI**, 690

C$_{21}$H$_{34}$N$_2$O$_3$ 6,7-Dimethoxy-1,2,3,4-
tetrahydro-2-[(1-$tert$-butoxy-
3-methyl)-2-butylimino-
methyl]isoquinoline, **VIII**, 573

C$_{21}$H$_{41}$NO$_2$ Glutaconaldehyde tetrabutylammonium salt, **VI**, 642

C$_{22}$H$_{14}$O$_9$·3NH$_3$ Ammonium salt of aurin
tricarboxylic acid, **I**, 54

C$_{22}$H$_{16}$N$_2$O$_2$ $trans$-2,3-Diphenyl-1-phthal-
imidoaziridine, **VI**, 56

C$_{22}$H$_{21}$O$_2$P Ethyl (triphenylphosphoranylidene)acetate, **VII**, 232,

C$_{22}$H$_{22}$ [22]Annulene, **VI**, 74

C$_{22}$H$_{22}$S$_3$ 1,1,3-Tris(phenylthio)butane, **VI**,
742
1,3,3-Tris(phenylthio)butane, **VI**,
742

C$_{22}$H$_{23}$NO (2-Dimethylamino-5-methyl-
phenyl)diphenylcarbinol, **VI**, 478

C$_{22}$H$_{30}$N$_2$O$_5$S$_2$ 5,8-Bis(p-tolylsulfonyl)-1-
oxa-5,8-diazacycloundecane,
VI, 657

C$_{22}$H$_{30}$N$_2$O$_6$S$_2$ 4,10-Bis(p-tolylsulfonyl)-
1,7-dioxa-4,10-diazacyclodode-
cane, **VI**, 657
7,10-Bis(p-tolylsulfonyl)-1,4-dioxa-
7,10-diazacyclododecane, **VI**,
657

$C_{22}H_{30}O_2S$ 3-*tert*-Butyl-4-hydroxy-5-methylphenyl sulfide, **VI**, 738

$C_{22}H_{32}O_2Si$ 3-Methoxy-17-trimethylsiloxy-1,3,5(10)-estratetraene, **VIII**, 286

$C_{22}H_{32}O_4$ 3β-Acetoxyetienic acid, **V**, 8

$C_{22}H_{34}O_4$ Sebacoyldicyclohexanone, **V**, 533

$C_{22}H_{38}O_6$·2 Na 7,16-Diketodocosanedioic acid and disodium salt, **V**, 534, 536

$C_{22}H_{42}O$ Triheptyl carbinol, **II**, 603

$C_{22}H_{42}O_2$ Butyl oleate, **II**, 469
Erucic acid, **II**, 258

$C_{22}H_{42}O_4$ Dimethyl octadecanedioate, **V**, 463
Docosanedioic acid, **V**, 533
Ethyl 1,16-hexadecanedicarboxylate, **III**, 401

$C_{23}H_{17}OBF_4$ 2,4,6-Triphenylpyrylium tetrafluoroborate, **V**, 1135

$C_{23}H_{25}NO_3$ 2-[4-(Phenylmethoxy)phenyl]2-(3,4-dimethoxyphenyl)ethylamine, **VI**, 829

$C_{23}H_{28}O$ 17β-*tert*-Butoxy-5α-androst-2-ene, **VII**, 66

$C_{23}H_{32}N_2O_5S_2$ 5,9-Bis(*p*-tolylsulfonyl)-1-oxa-5,9-diazacyclododecane, **VI**, 657

$C_{23}H_{32}N_2O_6S_2$ 7,11-Bis(*p*-tolylsulfonyl)-1,4-dioxa-7,11-diazacyclotridecane, **VI**, 657

$C_{23}H_{32}O_2Si$ 3-Methoxy-17-trimethylsiloxy-1,3,5(10)-estratetraene, **VIII**, 286

$C_{23}H_{32}O_4$ 18,20-Lactone of 3β-acetoxy-20β-hydroxy-5-pregnene-18-oic acid, **V**, 692

$C_{23}H_{34}O_6$ (11β,20*S*)-21-Acetoxy-11,17,20-trihydroxypregn-4-en-3-one, **VI**, 346

$C_{23}H_{36}O_3$ 3β-Acetoxy-20β-hydroxy-5-pregnene, **V**, 692

$C_{23}H_{37}NO_2S$ Ricinelaidic acid *S*-(2-pyridyl)carbothioate, **VII**, 470

$C_{23}H_{38}O$ 17β-*tert*-Butoxy-5α-androst-2-ene, **VII**, 66;

$C_{23}H_{38}O_2$ 17β-*tert*-Butoxy-5α-androstan-3-one, **VII**, 66

$C_{23}H_{38}O_4$ nor-Desoxycholic acid, **III**, 234

$C_{23}H_{44}O_4$ Cetylmalonic ester, **IV**, 141

$C_{23}H_{46}O$ Laurone, **IV**, 560

$C_{24}H_{17}NO_2$ 2,4,6-Triphenylnitrobenzene, **V**, 1128

$C_{24}H_{17}O$ 2,4,6-Triphenylphenoxyl, **V**, 1130

$C_{24}H_{18}N_2O$ 1-Benzyl-2-benzoyl-1,2-dihydroisoquinaldonitrile, **VI**, 118

$C_{24}H_{18}O$ 2,4,6-Triphenylphenol, **V**, 1130

$C_{24}H_{19}N$ 2,4,6-Triphenylaniline, **V**, 1130

$C_{24}H_{19}NO$ 1-[α-(Benzalamino)benzyl]-2-naphthol, **I**, 381

$C_{24}H_{20}AsCl_2$ Tetraphenylarsonium chloride hydrochloride, **IV**, 910

$C_{24}H_{24}$ [24]Annulene, **VI**, 74

$C_{24}H_{24}O_8$ Tetramethyl [2.2]paracyclophane-4,5,12,13-tetracarboxylate, **VII**, 485

$C_{24}H_{32}O_8$ Dibenzo-24-crown-8 polyether, **VI**, 399

$C_{24}H_{34}N_2O_7S_2$ 7,13-Bis(*p*-tolylsulfonyl)-1,4,10-trioxa-7,13-diazacyclopentadecane, **VI**, 658
10,13-Bis(*p*-tolylsulfonyl)-1,4,7-trioxa-10,13-diazacyclopentadecane, **VI**, 658

$C_{24}H_{40}O$ Cholan-24-al, **V**, 242

$C_{24}H_{42}O_4$ (−)-Dimenthyl succinate, **VIII**, 141
Methyl desoxycholate, **III**, 237

$C_{25}H_{21}N$ *p*-Aminotetraphenylmethane, **IV**, 47

$C_{25}H_{27}NO_4$ 1,2,3,4-Tetrahydro-6,7-dimethoxy-1-[3-methoxy-4-(phenylmethoxy)phenyl]isoquinoline, **VI**, 829

$C_{25}H_{31}N_3O_6S_3$ *N,N′,N″*-Tris(*p*-tolylsulfonyl)diethylenetriamine and *N,N″*-disodium salt, **VI**, 652

$C_{25}H_{36}N_2O_7S_2$ 10,14-Bis(*p*-tolylsulfonyl)-1,4,7-trioxa-10,14-diazacyclohexadecane, **VI**, 658

$C_{25}H_{36}O_2Si$ Ethyl 2-(diphenylmethylsilyl)decanoate, **VIII**, 474

$C_{25}H_{40}N_2O_2S$ D-Camphor 2,4,6-triiso-
propylbenzenesulfonylhydra-
zone, **VII**, 77

$C_{25}H_{42}O_4$ (–)-Dimenthyl (1S,2S)-cyclo-
propane-1,2-dicarboxylate,
VIII, 141

$C_{25}H_{46}O_4Sn$ (Z)-Tributylstannylmethylene-
4-isopropyl-1,1-cyclopentane-
dicarboxylic acid, dimethyl
ester, **VIII**, 381

$C_{25}H_{54}ClN$ Methyltrioctylammonium chloride, **VI**, 414

$C_{26}H_{20}$ Tetraphenylethylene, **IV**, 914

$C_{26}H_{20}O$ β-Benzopinacolone, **II**, 73

$C_{26}H_{22}O_2$ Benzopinacol, **II**, 71

$C_{26}H_{22}S_2$ Dibenzohydryl disulfide, **II**, 574

$C_{26}H_{30}IP$ 2-Octenyltriphenylphosphonium
iodide, **VI**, 707

$C_{26}H_{38}N_2O_4$ N,N'-Dibenzyl-4,13-diaza-18-
crown-6, **VIII**, 152

$C_{26}H_{38}N_2O_8S_2$ 7,16-Bis(p-tolylsulfonyl)-
1,4,10,13-tetraoxa-7,16- diaza-
cyclooctadecane, **VI**, 659

10,16-Bis(p-tolylsulfonyl)-1,4,7,13-
tetraoxa-10,16-diazacyclooc-
tadecane, **VI**, 695

$C_{26}H_{42}N_2O_6S$ N-Benzyloxycarbonyl-O-
tert-butyl-L-seryl-S-tert-butyl-L-
cysteine tert-butyl ester, **VI**,
252

$C_{26}H_{56}Sn_2$ (E)-1,2-Bis(tributylstannyl)eth-
ylene, **VIII**, 268

$C_{27}H_{23}NO$ 2,3,5,5-Tetraphenylisoxazoli-
dine, **V**, 1126

$C_{27}H_{26}Cl_2NiP_2$ Dichloro[1,3-bis(diphenylphosphino)propane]nickel(II),
VI, 407

$C_{27}H_{27}NS$ 4-(Butylthio)-3-(diphenylmethyl)-2-phenyl-3-buteneni-
trile, **VI**, 942

$C_{27}H_{33}N_3O_6S_3$ 1,4,7-Tris(p-tolylsulfonyl)-
1,4,7-triazacyclononane, **VI**,
655

$C_{27}H_{42}O_2$ Δ⁴-Cholesten-3,6-dione, **IV**, 189

$C_{27}H_{42}O_6$ 3,12-Diacetoxy-nor-cholanic
acid, **III**, 234

$C_{27}H_{44}$ Cholesta-3,5-diene, **VIII**, 126

$C_{27}H_{44}Br_2O$ 5α,6β-Dibromocholestan-3-
one, **IV**, 197

$C_{27}H_{44}O$ Cholestan-3α-ol, **VI**, 217
Cholestenone, **III**, 207
Cholest-4-en-3-one, **VI**, 293, 762
Δ⁵-Cholesten-3-one, **IV**, 195

$C_{27}H_{46}$ 5β-Cholest-3-ene, **VI**, 293
Cholest-6-ene, **VI**, 174

$C_{27}H_{46}Br_2$ 3α,4β-Dibromo-5β-cholestane,
VI, 294

$C_{27}H_{46}Br_2O$ Cholesterol dibromide, **IV**,
195

$C_{27}H_{46}O$ Cholestan-3-one, **II**, 139; **VI**,
289
Cholest-4-en-3β-ol, **VI**, 298, 491,
584
Cholesterol, **IV**, 195

$C_{27}H_{48}$ Cholestane, **VI**, 289,

$C_{27}H_{48}O$ Cholestan-3α-ol, **VI**, 217
Dihydrocholesterol, **II**, 191

$C_{28}H_{29}P·I$ (E)-4-Phenyl-2-butenyltri-
phenylphosphonium iodide,
VI, 707

$C_{28}H_{32}O_4$ 2,2,7,7,12,12,17,17-Octamethyl-
21,22,23,24-tetraoxaquaterene,
VI, 856

$C_{28}H_{35}N_3O_6S_3$ 1,4,7-Tris(p-tolylsulfonyl)-
1,4,7-triazacyclodecane, **VI**,
655

$C_{28}H_{38}O_{19}$ α-Cellobiose octaacetate, **II**,
124
β-Gentiobiose octaacetate, **III**, 428

$C_{28}H_{40}O_{10}$ Dibenzo-30-crown-10 polyether, **VI**, 399

$C_{28}H_{43}F_3O_3S$ Cholesta-3,5-dien-3-yl trifluoromethanesulfonate, **VIII**, 126

$C_{28}H_{45}NO_2$ 3β-Hydroxy-5α-cyano-
cholestan-7-one, **VI**, 20

$C_{28}H_{48}$ 5-Methylcoprost-3-ene, **VI**, 762

$C_{28}H_{48}O_4$ 2,2,7,7,12,12,17,17-Octamethyl-
21,22,23,24-tetraoxaperhydro-
quaterene, **VI**, 856

$C_{28}H_{48}O_4S$ S-tert-Butyl 3α,7α,12α-trihy-
droxy-5β-cholane-24-thioate,
VII, 81

$C_{28}H_{50}O$ Cholestanyl methyl ether, **V**, 245

$C_{28}H_{60}P \cdot Br$ Tributylhexadecylphosphoniu m bromide, **VI**, 883

$C_{29}H_{20}O$ Tetraphenylcyclopentadienone, **III**, 806

$C_{29}H_{28}IOP$ (4-Methoxyphenyl-2-butenyl) triphenylphosphonium iodide, **VI**, 707

$C_{29}H_{37}N_3O_6S_3$ 1,4,7-Tris(*p*-tolylsulfonyl)-1,4,7-triazacycloundecane, **VI**, 655

1,4,8-Tris(*p*-tolylsulfonyl)-1,4,8-triazacycloundecane, **VI**, 660

$C_{29}H_{37}N_3O_7S_3$ 4,7,10-Tris(*p*-tolylsulfonyl)-1-oxa-4,7,10-triazacyclododecane, **VI**, 657

$C_{29}H_{39}N_3O_8$ Benzyloxycarbonyl L-aspartyl-(*tert*-butyl ester)-L-phenylalanyl-L-valine methyl ester, **VII**, 30

$C_{29}H_{39}N_3O_{8S3}$ 3,6,9-Tris(p-tolylsulfonyl)-3,6,9-triazaundecane-1,11-diol, **VI**, 652

$C_{29}H_{46}O_3$ 3β-Acetoxycholest-5-en-7-one, **VI**, 14

$C_{29}H_{48}O$ 5β-Cholest-3-ene-5-acetaldehyde, **VI**, 298

$C_{29}H_{48}O_2$ Cholesteryl acetate, **II**, 193

$C_{30}H_{20}F_{12}O_2S$ Bis[2,2,2-trifluoro-1-phenyl-1-(trifluoromethyl)ethoxy]-diphenylsulfurane, **VI**, 163

$C_{30}H_{22}$ *p*-Quinquephenyl, **V**, 985

$C_{30}H_{28}N_4$ Bis(1,3-diphenylimidazolidinylidene-2), **V**, 115

$C_{30}H_{39}N_3O_6S_3$ 1,4,7-Tris(*p*-tolylsulfonyl)-1,4,7-triazacyclododecane, **VI**, 655

1,5,9-Tris(*p*-tolylsulfonyl)-1,5,9-triazacyclododecane, **VI**, 660

$C_{30}H_{39}N_3O_7S_3$ 4,7,10-Tris(*p*-tolylsulfonyl)-1-oxa-4,7,10-triazacyclotridecane, **VI**, 658

$C_{30}H_{47}NO_3$ β-Acetoxy-5α-cyanocholestan-7-one, **VI**, 14

$C_{30}H_{48}O$ Squalene-2,3-oxide, **VI**, 563

$C_{30}H_{48}OS_2$ 2,2-(Trimethylenedithio)

cholest-4-en-3-one, **VI**, 1015

$C_{31}H_{33}N_3O_6S_3$ 1,4,7-Tris(*p*-tolylsulfonyl) benzo[*h*]1,4,7-triazacyclononane, **VI**, 660

$C_{31}H_{41}N_3O_6S_3$ 1,4,7-Tris(*p*-tolylsulfonyl)-1,4,7-triazacyclotridecane, **VI**, 655

$C_{31}H_{41}N_3O_7S_3$ 4,8,12-Tris(*p*-tolylsulfonyl)-1-oxa-4,8,12-triazacyclotetradecane, **VI**, 657

$C_{31}H_{41}N_3O_8$ Benzyloxycarbonyl-L-aspartyl-(*tert*-butyl ester)-L-phenylalanyl-L-valine methyl ester, **VII**, 30

$C_{31}H_{41}N_3O_8S_3$ 7,10,13-Tris(*p*-tolylsulfonyl)-1,4-dioxa-7,10,13-triazacyclopentadecane, **VI**, 658

$C_{31}H_{43}N_3O_{12}S_3$ 3,6,9-Tris(*p*-tolylsulfonyl)-3,6,9-triazaundecane-1,11-dimethanesulfonate, **VI**, 652

$C_{31}H_{52}O_2$ Ethyl 5β-cholest-3-ene-5-acetate, **VI**, 584

$C_{31}H_{53}NO$ *N,N*-Dimethyl-5β-cholest-3-ene-5-acetamide, **VI**, 491

$C_{32}H_{20}O_3$ Tetraphenylphthalic anhydride, **III**, 807

$C_{32}H_{30}N_4O_5Co_2$ Salcomine, **VI**, 412

$C_{32}H_{57}O_4P$ Diethyl 5-methylcoprost-3-en-3-yl phosphate, **VI**, 762

$C_{33}H_{34}O_{10}$ 6-Trityl-β-D-glucose-1,2,3,4-tetraacetate, **III**, 432

$C_{33}H_{45}N_3O_7S_3$ 5,9,13-Tris(*p*-tolylsulfonyl)-1-oxa-5,9,13-triazacyclohexadecane, **VI**, 658

$C_{33}H_{45}N_3O_8S_3$ 7,11,15-Tris(*p*-tolylsulfonyl)-1,4-dioxa-7,11,15-triazacycloheptadecane, **VI**, 658

$C_{33}H_{45}N_3O_9S_3$ 10,13,16-Tris(*p*-tolylsulfonyl)-1,4,7-trioxa-10,13,16-triazacyclooctadecane, **VI**, 659

$C_{34}H_{24}$ 1,2,3,4-Tetraphenylnaphthalene, **V**, 1037

$C_{34}H_{28}N_6O_{14}S_4$ Coupling product from *o*-toluidine and Chicago acid, **II**, 145

$C_{34}H_{34}FeN_4O_4$ Hemin, **III**, 443

$C_{34}H_{52}N_2O_2S$ Cholest-4-en-3-one p-toluenesulfonylhydrazone, **VI**, 293

$C_{35}H_{36}P_2Rh \cdot CF_3O_3S$ (Bicyclo[2.2.1]heptadiene)(1,4-bis(diphenylphosphino)butanerhodium trifluoromethanesulfonate, **VIII**, 420

$C_{35}H_{49}N_3O_9S_3$ 10,14,18-Tris(p-tolylsulfonyl)-1,4,7-trioxa-10,14,18-triazacycloeicosane, **VI**, 659

$C_{35}H_{70}O$ Stearone, **IV**, 854

$C_{36}H_{36}F_3O_3P_2SRh$ Bicyclo[2.2.1]hepta-2,5-diene 1,4-bis(diphenylphosphino)butanerhodium trifluoromethanesulfonate, **VIII**, 420

$C_{36}H_{36}N_2O_2S$ N,N'-Dispiro[cyclohexane-1,9'-fluoren]-4-ylsulfamide, **VI**, 80

$C_{36}H_{50}IP$ 2-Octadecenyltriphenylphosphonium iodide, **VI**, 707

$C_{37}H_{28}O$ Tribiphenylcarbinol, **III**, 831

$C_{37}H_{46}N_4O_8S_4$ 1,4,7,10-Tetrakis(p-tolylsulfonyl)-1,4,7,10-tetraazacyclotridecane, **VI**, 660

$C_{38}H_{37}N_5O_2$ N-[2,4-Bis(1,3-diphenylimidazolidin-2-yl)-5-hydroxyphenyl]acetamide, **VII**, 162

$C_{38}H_{48}N_4O_8S_4$ 1,4,7,11-Tetrakis(p-tolylsulfonyl)-1,4,7,11-tetraazacyclotetradecane, **VI**, 661
1,4,8,11-Tetrakis(p-tolylsulfonyl)-1,4,8,11-tetraazacyclotetradecane, **VI**, 66

$C_{38}H_{48}N_4O_9S_4$ 4,7,10,13-Tetrakis(p-tolylsulfonyl)-1-oxa-4,7,10,13-tetraazacyclopentadecane, **VI**, 658

$C_{40}H_{52}N_4O_8S_4$ 1,5,9,13-Tetrakis(p-tolylsulfonyl)-1,5,9,13-tetraazacyclohexadecane, **VI**, 661

$C_{40}H_{52}N_4O_9S_4$ 5,8,11,14-Tetrakis(p-tolylsulfonyl)-1-oxa-5,8,11,14-tetraazacycloheptadecane, **VI**, 658

$C_{40}H_{52}N_4O_{10}S_4$ 7,10,13,16-Tetrakis(p-tolylsulfonyl)-1,4-dioxa-7,10,13,16-tetraazacyclooctadecane, **VI**, 652

$C_{40}H_{52}O_4$ 3,12-Diacetoxy-bisnor-cholanyldiphenylethylene, **III**, 237

$C_{40}H_{64}O_{12}$ Nonactin, **VI**, 515,

$C_{42}H_{30}$ Hexaphenylbenzene, **V**, 604

$C_{42}H_{56}N_4O_{11}S_4$ 10,13,16,19-Tetrakis(p-tolylsulfonyl)-1,4,7-trioxa-10,13,16,19-tetraazacycloheneicosane, **VI**, 659

$C_{44}H_{32}O_2P_2$ 2,2'-Bis(diphenylphosphinyl)-1,1'-binaphthyl, **VIII**, 57

$C_{44}H_{32}P_2$ $S(-)$-2,2'-Bis(diphenylphosphino)-1,1'-binaphthyl, **VIII**, 57

$C_{44}H_{38}P_2 \cdot Cl_2$ Xylylene-bis-(triphenylphosphonium chloride), **V**, 985

$C_{44}H_{56}O_4$ p-tert-Butylcalix[4]arene, **VIII**, 75

$C_{45}H_{55}N_5O_{10}S_5$ 1,4,7,10,13-Pentakis(p-tolylsulfonyl)-1,4,7,10,13-pentaazacyclopentadecane, **VI**, 661

$C_{45}H_{86}O_6$ Trimyristin, **I**, 538

$C_{46}H_{57}N_5O_{10}S_5$ 1,4,7,10,13-Pentakis(p-tolylsulfonyl)-1,4,7,10,13-pentaazacyclohexadecane, **VI**, 661

$C_{52}H_{44}Rh \cdot ClO_4$ [(+)-2,2'-Bis(diphenylphosphino)-1,1'-binaphthyl]-(η^4-1,5-cyclooctadiene)rhodium(I) perchlorate, **VIII**, 183

$C_{54}H_{66}N_6O_{12}S_6$ 1,4,7,10,13,16-Hexakis (p-tolylsulfonyl)-1,4,7,10,13,16-hexaazacyclooctadecane, **VI**, 652

$C_{63}H_{77}N_7O_{14}S_7$ 1,4,7,10,13,16,19-Heptakis (p-tolylsulfonyl)-1,4,7,10,13,16,19-heptaazacycloheneicosane, **V**, 661

$C_{66}H_{84}O_6$ p-tert-Butylcalix[6]arene, **VIII**, 77

$C_{72}H_{88}N_8O_{16}S_8$ 1,4,7,10,13,16,19,22-Octakis(p-tolylsulfonyl)-1,4,7,10,13,16,19,22-octaazacyclotetracosane, **VI**, 661

$C_{88}H_{112}O_8$ p-tert-Butylcalix[8]arene, **VIII**, 80

$(FH)_x \cdot x\ C_5H_5N$ Pyridinium polyhydrogen fluoride, **VI**, 628

APPARATUS INDEX

A number of the procedures in *Organic Syntheses* describe the use of special or less common pieces of apparatus and equipment. References to many of them are recorded in this index. Illustrations are indicated by page numbers in boldface type.

Addition funnels, **VIII, 593**
  Büchner, cooling jacket for, **IV,**
      373, 928
    slit-sieve, **VI,** 240
  calibrated for benzene solution of
      phosgene, **V,** 201
  electrically heated, **IV,** 811
  for solids, **VI,** 354
  Hershberg, **II, 129; IV,** 70, **747,** 844,
      964; **V,** 891; **VI,** 167
  plastic, **V,** 55
  pressure, **VI,** 875
  pressure-equalizing, **VI,** 23, 39,
      51, 602
  steam-jacketed, **IV,** 561, 856
  with needle valve adaptors, **VI, 383**
  with temperature-control jacket, **V,**
      675; **VI,** 604; **VII,** 106, 327
Apparatus, for acyloin cyclization,
      IV, 840
  for addition of butadiene to maleic
      anhydride, **IV, 891**
  for addition of potassium perman-
      ganate, **VI,** 993
  for addition of reagents, **VIII, 491**
    under inert atmosphere, **VI, 317**
  for assay of butylmagnesium chlor-
      ide, **V,** 1143
  for bromination of *m*-aminobenzoic
      acid, **IV, 947**
  for bromination or chlorination using
      aluminum chloride cata-
      lyst, **V,** 117, 120

for carbonation of magnesium
    methoxide, **V,** 439
for catalytic hydrogenation, **I,** 61;
    VI, 543; VII, 390
  in liquid ammonia, **VI, 253**
for caustic fusion of vanillin,
    IV, 975
for chlorination, **VI, 91,** 212
  with phosgene, **VI, 283**
for collecting volatile reaction pro-
    ducts, **VIII,** 436
for condensing liquid ammonia,
    VIII, 304
for continuous chromatography,
    V, 730
for distillation of diethyl methylene-
    malonate, **IV, 300**
for evacuating and introducing nitro-
    gen into reaction vessels,
    IV, 133, 459
for extraction of liquid ammonia
    from commercial cylinder,
    eductor tube, **VI,** 52
for filtration in an inert atmosphere,
    VI, 876
for handling boron trifluoride,
    V, 600
for handling fluorine gas, **VIII, 287**
for handling hydrogen fluoride, **V,**
    66, 136, 332
for handling hydrogen fluoride and
    boron trifluoride, **V,** 480
for high dilution, **VIII, 198**

for introduction of chlorine gas,
V, 371
for maintaining an inert atmosphere,
VI, **125**, 695; VII, **154**;
VIII, 351
for measuring volume of acetylene,
VII, 293
for measuring volume of liquid
allene, VIII, 228
for measuring volume of liquid
methanethiol, VI, 686
for measuring volume of liquid methyl
chloride, VII, **346**, **347**
for measuring volume of and the
addition of sulfur tetrafluoride,VII, **528**
for monitoring carbon dioxide evolution, V, 127
for oxidation with molecular oxygen, VIII, **491**
for observing gas evolution, VI,
565, 616
for preparation of acrolein, I, **15**
of alkyl bromides, I, **39**
of alloxantin dihydrate, IV, **26**
of allyl vinyl ethers and subsequent Claisen rearrange-
ment, VI, **585**, **608**
of bromoacetal, III, **123**
of 2-bromophthalide, III, **738**
of 1-butylpyrrolidine, III, **160**
of β-chloropropionaldehyde
acetal, II, **138**
of chromous sulfate, V, 994
of cyclopentadiene, VII, **340**
of cyclopropyl cyanide, III, **223**
of diazomethane, V, 351
of hexamethylene chlorohydrin,
III, **446**
of isonitrosopropiophenone,
II, **363**
of methyl iodide, II, **401**
of p-nitrophenyl isocyanate,
II, **453**
of potassium salt of dimethyl
sulfoxide, V, 938

of Raney nickel catalyst, III,
177; V, 103
of sodium amide, III, 779; IV,
117
of thiophosgene, I, **508**
of p-tolualdehyde, II, **584**
of tribromoaniline, II, **592**
of trimethyloxonium 2,4,6-trini-
trobenzenesulfonate using
diazomethane, V, **1100**
of zinc–copper couple, V, 857
for purification of boron trifluoride
gas, VII, 394
for reaction of ethanol with chloro-
trifluoroethylene, IV, **184**
for reactions in liquid ammonia,
V, 12
for reactions of cyanogen chloride
and sulfur trioxide, V, 227
for removal of supernatant solvent
under inert atmosphere,
VI, **808**
for solution transfer in inert atmosphere, VI, 139; VII, 290,
347, 487; VIII, **327**, **523**
for transfer of Grignard reagents,
V, 1058
for transferring triethylaluminum
from a lecture bottle, VI,
16, **18**, **19**, **437**
stopcock used in this apparatus,
VI, **17**
for transferring reaction mixture
under inert atmosphere to
acid quench, VI, **308**
for trapping volatile product, VI,
318; VIII, **436**
resealable glass tube for pressure
reactions, VIII, 446
to prepare solution of acetylene in
THF, VIII, **393**
Autoclaves, Pressure Reactors; (see also
Hydrogenation Equipment)
autoclave, VI, 1008; VIII, 124
for Diels–Alder reaction, VI,
455; VII, 7

glass, **VIII**, 190
Hastelloy-C vessels for fluorine
   compounds, V, 1083
   for alkaline rearrangements,
      V, 814
high-pressure reactors, **IV**, 311,
   530, 930
pressure vessel, for carbonylation,
   VII, 430
steel pressure vessel, **III**, 517
stirred or rocking autoclaves for
   cycloaddition reactions, **V**,
   237, 393, 459
use of hydrogenation bombs as reactors, **IV**, 294, 452, 623,
   738, 930; **V**, 811

Bath, cooling,
   cooling, large scale, **V**, 312
   Dry Ice and acetone, **IV**, 69, 75,
      268, 747, 751
   Dry Ice–ethanol, **IV**, 20, 385
   Dry Ice and ethylene glycol
      monomethyl ether, **IV**, 240
   Dry Ice and methylene chloride,
      IV, 369
   Dry Ice and trichloroethylene, **IV**,
      60, 387, 404
   refrigerated, **VI**, 241, 530; **VII**, 238,
      447, **448**; **VIII**, 372; 418
Bath, heating
   fused salt (potassium nitrate + sodi-
      um nitrite), **IV**, 498
   hot water circulator, **VII**, 342
   wax, **IV**, 488, 683; **V**, 619
   Wood's metal, **IV**, 488, 592,
      768, 854
Beaker, stainless-steel, for caustic fusion,
   IV, 974; **V**, 617
Blender, Waring-type, **IV**, 111
Bottle, pressure, **IV**, 261, 418
Bubbler, Nujol, **VI**, 82
   mercury, **VI**, 309, 318, 526, 598
Buret, plastic, for dispensing methylene
   chloride solution of dia-
   zomethane, **V**, 247

Capillary, sealed, **IV**, 209, 211, 941
Capillary tube, evacuated, **IV**, 199
Chromatography column, water-jacketed,
   VII, **147**
Circulating bath, refrigerated, **VII**, 178,
   365, 468
   pump, **VI**, 97; **VIII**, 214
   warm water, **VI**, 69
Condenser, air-cooled, **VI**, 40
   bulb, **VI**, 404
   Dewar, **VI**, 51, 161, 1019
   Dimroth, for refluxing high-boiling
      liquids, **VI**, 92
   Dry Ice–acetone, **IV**, 118, 119, 258,
      269, 296, 755, 851, 970;
      VI, 97, 148, 361
      for low-boiling liquids, **I**, **117**, **529**;
      VI, 526
   Friedrichs, **VI**, 273, 517
   Fries condenser, **VI**, 1026
   Liebig condenser, **VI**, 774, 1034
   low temperature, **V**, 947; **VI**, 274
   partial, Hahn type, **III**, **231**
Continuous reactors, **IV**, **81**, **740**

Dean–Stark water separator, **VI**, 243,
   497, 1014; **VIII**, 548
Distillation
   apparatus, short path, **VI**, **88**, **438**
      for dehydration, **I**, **422**
      for azeotropic esterification,
         I, **262**
      for evaporative distillation, **VI**,
         764
      for laurone, **IV**, **562**
      for low-boiling liquids, **V**, 23, **682**
      for solids, **III**, **133**; **IV**, 830
      for sulfur trioxide, **IV**, 848
      for steam distillation, **I**, **282**, **479**,
         507; **III**, **7**, **65**; **V**, 1029
      of solids, **IV**, **397**
      rapid, **V**, 1020
   Kugelrohr distillation oven **VI**,
      24, 447, 106, 538, 577,
      649, 667, 739
      of volatile aldehydes, **VI**, 754

columns
  Crismer column, **VI**, 677
  Dufton column, **VI**, 512, 770
  low temperature, **IV**, 686
  for methyl esters, Clarke–
    Rahrs, **III**, **611**
  Podbielniak still, **VI**, 664
  spinning-band, **VI**, 40, 161
  vacuum-jacketed, **VI**, 317
  vacuum-jacketed, silvered,
    VI, 456
  Vigreux, vacuum-jacketed, **VI**,
    374, 376
  Widmer, **III**, **423**; **VI**, 273,
    712
  flasks, special types, **I**, **130**, **501**;
    II, **18**, 227; **III**, 65,
    520
  steam, **I**, **282**, **479**, **507**; **III**, 7, 65;
    VI, 49
Dry-box, in the preparation of 2-hydroxy-
    cyclobutanone, **VI**, 167

Electrolysis apparatus,
  cell, **VII**, **309**
    electrode clamp, **VI**, 972
    electrodes, **VI**, 972; **VII**, 308
    power supply for, **VII**, 309,
      480, 483
    for electrohydrodimerization,
      VII, 479
    electrode assembly for,
      VII, **183**
    for Kolbe's synthesis, **V**, 363, 446,
      464, **466**
    for reduction, **I**, **486**; **III**, **61**; **IV**,
      149; **VI**, 154, **155**
    Teflon anode support, **VI**, **156**
    reference electrode and salt
      bridges, **VI**, **157**
Evaporator, reduced pressure, **I**, **427**
  for solvent stripping, **V**, 974
  rotary, **IV**, 614
  still, evaporative, **IV**, 794
    special for thiophosgene, **I**,
      507, **508**

Extractors, continuous,
  of solutions, **I**, **277**; **II**, **378**, **615**;
    III, **539**; **IV**, 20, 46, 580;
    V, 630, 615,835; **VI**, 78,
    384, 1039
  of solids, **I**, **375**, **539**; **IV**, 167, 548,
    564, 645

Filter aids, **IV**, 100, 115, 217, 283, 358,
    365, 581, 660, 838, 909
Filtration, inverted, **II**, **610**
Flasks
  for addition of solids, **III**, **550**;
    IV, 649
  Claisen, modified, **IV**, 109
  copper, **IV**, 95, 446
  creased, **IV**, 523, 529, 840; **VI**, 459
    Morton, **IV**, 68
  Dewar, **IV**, 724
  equipped with stopcock at the bot-
    tom, **VI**, **69**
  jacketed, **IV**, 431
  sausage, **IV**, 766
  stainless-steel, **IV**, 95, 446, 449
Flow meter, **IV**, 64, 126
Flow-rate-indicating fluid (arochlors), **IV**,
    126

Gas bag for adding gaseous reactant,
    VIII, 101, 271
Gas dispersion tube, **IV**, 184, 523, 669, 895
Gas generation apparatus
  for bromine vapors, **II**, **340**
  for ethyl nitrite, **II**, **20**
  for hydrogen bromide, **II**, **339**;
    V, 546
  for hydrogen cyanide, **I**, **314**
  for hydrogen sulfide, **IV**, 928
  for methyl nitrite, **II**, **363**
  for nitrogen oxides, **I**, **266**; **V**, 651
Gas inlet, polyolefin, **VI**, 628
Gasometer, acetylene, **I**, **230**
Gas storage tube, **VI**, **135**
Gas traps
  absorption, **I**, **97**; **II**, **4**; **IV**, 44, 63,
    157, 162, 178, 558

bubbler, mercury, **IV**, 966
carbon dioxide collection, **V**, 949
flow regulator, **IV**, 580
Glass wool filter, **VII**, 348,
Glove bag, **VI**, 165
Gooch tube, **VI**, 383

Hopper for additions, **I**, **233**
Hydrogenation apparatus, **I**, **63**, **66**; **IV**,
217, 222, 299, 305, 357,
409, 603, 660, 888, 930;
V, 591, 743, 881, 989,
1132; **VI**, 459
Parr, **VI**, 151, 371, 783
modified for temperature obser-
vation, **V**, 240
bombs, catalytic, **VI**, 455

Infrared heat lamp for controlled heating,
V, 327
Ion exchange chromatography, **VIII**, 616
Ion-exchange column, **IV**, **40**

Labjack, pneumatically operated, **VI**, 241
Lubricant for stirrer shaft, **VIII**, 300
Lyophilization, **VIII**, 616

Oscillating motor, for rotating flask under
vacuum, **VIII**, 454
Oven, Kugelrohr, **VI**, 300
Ozonizer and accessories, **III**, **674**, **675**,
676, **677**, **682**; **IV**, 485; **V**,
489; **VI**, 45, 977
vessel, **VI**, 43

pH controller, **VIII**, 519
Photochemical apparatus, **VI**, **146**, 229,
452, 716, 963, 974, **1025**;
VII, 23, 24, 103, 178,
305, 316, **317**, 471; **VIII**,
148, 214
immersion well, quartz, **VI**, 842, 974
quartz, waterjacketed, **VI**,
145, 227
Lamps, photochemical, **IV**, 807,
984; **VII**, 5, 24, 470

Mercury-vapor, **VI**, 715
Vitalites, **VIII**, 218
Photolysis, apparatus for, **V**, 298,
529, 952
Polarograph, **VI**, 158
Pressure reactions, use of resealable glass
tubes, **VIII**, 175, 446
Pressure vessel, **VI**, 338, 454; **VII**, 6, 7
Pyrolysis apparatus
for condensation of tetrahydropyran
with aniline, **IV**, **795**
for decarboxylation of coumalic
acid, **V**, 982
for decomposition of diazonium
hexafluorophosphates,
V, 134
for elimination of acetate group, **IV**,
746; **V**, 236
for Hofmann decomposition, **V**, 316
for preparation of,
acrylic acid, **III**, **30**
azelanitrile, **IV**, **63**
benzyl methyl ketone, **II**, **389**
bicyclo[2.1.0]pentane, **V**, 99
butadiene, **II**, **103**
ketene from acetone, **I**, **331**
from diketene, **V**, 679
methyl butadienoate, **V**, 734
2-octanol, **I**, **368**
for thermal cycloaddition, **V**, 425

Rubber dam, **IV**, 155, 332, 662, 864,
954

Separator, water,
automatic, for liquids, **I**, **151**, **262**,
277, **282**, **368**; **III**,
382, **503**
Barrett-type, **IV**, 94
Dean and Stark, **IV**, 235, 573,
580, 606
Separatory funnel, polyolefin, **VI**, 629
Shaking machine, **I**, **63**
Parr Instrument Co., **I**, **66**
Sodium, device for holding, **IV**, 117, 763
Sonicator, **VIII**, 233

Stirrers
  air-driven, **VII**, 457
  centrifugal, **IV**, 649
  Duplex-dispersator, **IV**, 70
  gas dispersion, **IV**, 158
  gas-tight bushing, **III**, **178**
  glass propeller, **I**, **34**
  glycerin-sealed, **IV**, 291, **546**
  Hershberg, **II**, **117**; **VI**, 11, 76, 751
  high-speed, **IV**, 70, 840; **V**, 40, 1027
  magnetic, **IV**, 252, 308, 339, 340,
      532, 662, 794, 870
  mercury-sealed, **I**, **33**; **VI**, 320, 488
  metal, for caustic fusion, **IV**, **975**
  sealed, **IV**, 89, 368, 650, 862
  Stir-O-Vac, **IV**, 70
  sweep-type, **IV**, 573, 598
  Teflon bearing, **VIII**, 106
  Trubore, **IV**, 89, 683, 724, 807, 814,
      829, 947
  Vibromixer, **VI**, 168, 677; **VIII**, 64, 288
    for making sodium sand, **V**,
      76, 1090
  wire, **II**, **534**; **V**, 819
Stirring bar, magnetic, glass-covered,
      VI, 122
Stopcock to allow simultaneous addition
      of liquid reagents and inert
      gases, **VII**, **237**

Sublimation apparatus, **IV**, 153, 880;
      VII, 329
Syringe filter, **VI**, 255
Syringe pump, **VIII**, 344
Syringe, gas-tight, **VI**, 329
Syringe needle, double-ended, for han-
      dling and transferring air-
      sensitive solutions,
      VI, 138

Tower, soda-lime, **IV**, 99, 387
Tangential continuous reactors, **IV**,
      81, **740**
Tape, heating, **IV**, 465, 741
Thermometer, Pt, **VII**, 154
Trap, for hydrogen bromide, **VI**, 520
Tube, sealed (Carius), **V**, 1076
  Pyrex bomb, **V**, 1093

Ultrasound catalysis, **VIII**, 231

Vacuum "dumb-bell" desiccator (for dry-
      ing large amounts of mate-
      rials), **III**, **845**
Vacuum manifold system, **VI**, **134**, **135**
Vibromixer, **VI**, 168, 677; **VIII**, 64, 288
  for making sodium sand, **V**, 76,
      1090
Voltmeter, **VI**, 158

HAZARD AND WASTE DISPOSAL INDEX

This index includes specific warnings concerning the hazards involved in preparing and handling the compounds listed in Indices 1 and 6. These warnings and cautions are specific to the procedures in *Organic Syntheses* and are usually in addition to those found, for example, in the *Toxic Substance List for 1987*. As far as possible, the specific hazard involved is listed but the user should be aware there may be additional hazards. In all cases the appropriate OSHA bulletins should be consulted.

Acetic acid, **VII**, 142

Acetone cyanohydrin nitrate, potential explosive, **V**, 839, 841

Acetyl chloride, irritant, **V**, 712; **VIII**, 18

Acetylene, purifying and handling, **VIII**, 391, 393, 606
   explosion hazard, **IV**, 187

Acetylenic compounds, potential explosives, **VII**, 335; **VIII**, 65

N-Acetyl-*N*-phenylhydroxylamine, potential carcinogen, **VIII**, 16

Acrolein toxicity, **I**, 17, 167; **II**, 138; **V**, 608

Acrylonitrile toxicity, **III**, 93; **IV**, 777; **V**, 39, 73; **VIII**, 148

Active magnesium, **VI**, 845

Adipyl azide, explosive, **IV**, 822

Air sensitive solutions, handling **VIII**, 20, 271, 461, 553, 557, 558,

Alkane thiols, **VI**, 236; **VII**, 124

Alkyl halides, toxicity, **VIII**, 29, 405

Aluminum alkyls, **VIII**, 295, 321, 512, 568

Aluminum amalgam, **VI**, 804

Amine vapors, **VI**, 852

2-Aminobiphenyl, carcinogen, **V**, 831
   also may be contaminated with carcinogenic 4-isomer, **V**, 731

Ammonia, **IV**, 62, 117, 128, 297, 387, 404, 585, 755, 867, 962;
969; **V**, 12, 187, 400, 420, 467, 523, 526, 541, 559, 564, 629, 769, 848; **VI**, 51, 252, 320, 361, 611, 664, 928; **VII**, 249

Arsenic compounds, poison, **IV**, 910

Azides, **VI**, 893
   acyl, **VI**, 96

Azoalkanes, **VI**, 78

Benzalacetone, skin irritant, **I**, 77

Benz[*a*]anthracene, **VII**, 18

Benzene, **VI**, 57, 251, 274, 305, 562; **VII**, 18, 23, 82, 203, 256, 273, 312, 372, 400, 485, 495, 512; **VIII**, 14, 48, 68, 82, 130, 150, 196, 199, 210, 218, 220, 242, 271, 307, 310, 338, 358, 368, 371, 382, 395 403, 418, 424, 427, 436, 442, 451, 459, 468, 488, 492, 503, 520, 529, 542, 549, 555, 558, 567, 571, 590, 592, 662, 704, 722, 727, 737, 776, 781, 788, 797, 824, 826, 837, 840, 859, 875, 887, 897, 913, 936, 940, 946, 951, 958, 976, 987, 1004, 1014, 1024, 1037

Benzeneboronic acid and anhydride, irritant, toxic, **IV**, 68
Benzenediazonium-2-carboxylate, shock-sensitive, **V**, 54, 57
Benzocyclopropene, **VI**, 87
1,2,3-Benzothiadiazole 1,1-dioxide, shock-sensitive, **V**, 62
3,1-Benzoxazepine, **VII**, 23
β-Benzoylacrylic acid, nose irritant, **III**, 111
Benzoyl fluoride, irritant, toxic, **V**, 68
Benzoyl peroxide, shock-sensitive, **I**, 432; **V**, 379
Benzyl bromide, **VI**, 122; **VII**, 386
Benzyl chloromethyl ether, **VI**, 101
Benzyl isocyanide, **VII**, 27
Bis(chloromethyl)ether, carcinogenic, irritant, **IV**, 101; **V**, 218
Bis(dimethylamino)methane, lachrymator, **V**, 434
Bis(trifluoromethyl)diazomethane, **VI**, 161
9-Borabicyclo[3.3.1]nonane, **VI**, 139
Borane–THF, **VII**, 428
Boron tribromide, **V**, 413
Boron trichloride, **V**, 727
Boron trifluoride. **III**, 406; **VII**, 393
    disposal, **V**, 1136
Bromine, **VI**, 153, 181, 190; **VII**, 172, 435
Bromoacetal, lachrymator, **III**, 124
Bromoacetone, lachrymator, **VI**, 21
2-Bromoallylamine, irritant, **V**, 121
N-(2-Bromoallyl)ethylamine, irritant, **V**, 124, 541
α-Bromobenzalacetone, skin irritant, **III**, 125
p-Bromobenzal bromide, irritant, **II**, 90
α-Bromo-γ-butyrolactone, irritant, **V**, 255
α-Bromoheptaldehyde, lachrymator, **III**, 127
Bromoketones, **II**, 88; **VI**, 21, 175, 193, 992
1-Bromo-3-methyl-2-butanone, lachrymator, skin irritant, **VI**, 193

α-Bromo-α-phenylacetonitrile, lachrymator, **III**, 348
3-Bromophthalide, skin irritant, **V**, 146
3-Bromopyrene, potential carcinogen, **V**, 147, 632
N-Bromosuccinimide, skin irritant, **V**, 328
α-Bromo-*p*-xylene, lachrymator, irritant, **V**, 883
Butane, **VI**, 598
tert-Butyl azidoacetate, potential explosive, **V**, 587
tert-Butyl azidoformate, potential explosive, headaches, **V**, 158; **VI**, 207
    substitutes, **VI**, 207
sec-Butyl α-bromopropionate, lachrymator, **IV**, 122
tert-Butyl hypochlorite, sensitive to light, **IV**, 127
    need for temperature control during preparation, **V**, 183
Butyllithium (concd), **VII**, 121, 125, 187, 384
tert-Butyllithium, **VII**, 329; **VIII**, 430, 502, 556, 573
Butyl nitrite, heart stimulant, headaches, **II**, 108

Carbon disulfide, **I**, 110; **III**, 23, 183; **IV**, 243; **VII**, 139
Carbon monoxide, **I**, 11; **III**, 808; **IV**, 143; **V**, 20, 706, 739; **VI**, 232, 807, 997, 1002; **VII**, 144, 420
Carbonyl cyanide, **VI**, 268
Chloramine reaction, **V**, 408, 409
Chlorine, **V**, 712; **VI**, 90, 675; **VIII**, 161, 427
Chlorine-containing organic compounds, toxic, **V**, 424
Chloroacetaldehyde diethyl acetal, irritant, mutagen, **VIII**, 479
Chloroacetamide, **I**, 154
Chloroacetone, lachrymator, **VI**, 601
Chloroacetyl chloride, lachrymator, **III**, 183; **VIII**, 516

9-Chloroacridine, irritant, **III**, 54
α-Chlorocrotonaldehyde, lachrymator,
     IV, 131
α-Chloroethyl ethyl ether, lachrymator,
     IV, 751
β-Chloroethyl methyl sulfide, vesicant,
     II, 136
p-Chloromercuribenzoic acid, irritant,
     I, 325
Chloromethanesulfonyl chloride, lachry-
     mator, **V**, 231
Chloromethylmethylether, **VI**,.952
1-Chloromethylnaphthalene, lachrymator,
     skin irritant, **III**, 195;
     IV, 690
2-Chloromethylthiophene, lachrymator,
     skin irritant, **III**, 197
     storage hazard, **III**, 198
N-Chloromorpholine, decomposition,
     VIII, 168
p-Chlorophenyl isothiocyanate, skin irri-
     tant, **V**, 223
Chloropicrin, lachrymator, skin irritant,
     IV, 458
N-Chlorosuccinimide, carcinogen,
     VIII, 167
Chlorosulfonic acid, **VI**, 90
Chlorosulfonyl isocyanate, **V**, 226, 673;
     VI, 304, 465, 788
     violent reaction with water, **VI**, 228
Chlorotrimethylsilane, in acyloin reac-
     tions, **VI**, 169
β-Chlorovinyl isoamyl ketone, irritant,
     IV, 187
Cholestererol dibromide, skin irritant, **IV**,
     199
Chromic acid–methanol, inflammable,
     III, 235
Chromium trioxide, mixing with acetic
     anhydride, **IV**, 714
     mixing with pyridine, **VI**, 644
Copper cyanide, poison, **VIII**, 33, 112
     disposal, **III**, 113; **VIII**, 114
Crotyl diazoacetate, **V**, 261
Crown ethers, **VI**, 301
Cyanide salts, **VI**, 115

disposal, **VI**, 117
Cyanogen, **I**, 47
Cyanogen bromide, poison, potential
     explosive, **II**, 136; **III**,
     608; **VII**, 435
Cyanogen chloride, **V**, 226; **VI**, 727
Cyanogen iodide, poison, **IV**, 207
Cyanuric chloride, **VII**, 197
Cyclohexyl isocyanide, stench, toxic, **V**, 300
     decontamination, **V**, 300

1,3-Diaminopropane, irritant, **VIII**,
     146, 553
Diazirines, potential explosives, **V**, 899
Diazoacetic esters, explosive, toxic, **V**,
     179, 261; **VI**, 914
Diazoacetophenone, **VI**, 386
1-(Diazoacetyl)naphthalene, skin irritant,
     VI, 614
Diazo compounds, **VII**, 438; **VIII**, 162
2-Diazocyclohexanone, **VI**, 389
Diazomalonic esters, **VI**, 414
Diazomethane, explosive, toxic, **II**, 165;
     III, 119, 245; **IV**, 225,
     250; **V**, 231, 247, 351,
     352, 877, 1099; **VI**, 386,
     432, 613; **VIII**, 196
Diazonium salt solutions and metallic
     polysulfides, potentially
     explosive, **III**, 810
Diazonium xanthates, explosive, **III**, 809;
     V, 1050
2-Diazopropane, **VI**, 392
2,2-Dibromoacetyl-6-methoxynaphtha-
     lene, skin irritant, **VI**, 177
1,3-Dibromo-2-butanone, lachrymator,
     skin irritant, **VI**, 711
α,α'-Dibromodibenzyl ketone, skin irri-
     tant, **V**, 515
Dibromodineopentyl ketone, **VI**, 992
α,β-Dibromoethyl ethyl ether, lachryma-
     tor, **IV**, 751
Dibromomalononitrile–KBr complex,
     irritant, **IV**, 877
2,4-Dibromo-3-pentanone, lachrymator,
     VI, 512, 520

2,3-Dibromopropene, lachrymator, **I**, 209;
V, 121, 124

α,γ-Dichloroacetone, irritant, **I**, 212

Dichloroacetophenone, irritant, **III**, 538

1,5-Dichloro-2,4-dinitrobenzene, skin
irritant, **V**, 1069

α,α-Dichloromethyl methyl ether,
VII, 468

Dichlorophenylphosphine, **VI**, 776

Dicyclohexylcarbodiimide, **VII**, 93

Dicyclohexyl-18-crown-6, irritant, toxic,
VI, 396

Diethylaminosulfur trifluoride, **VI**,
440, 835

Diethyl squarate, skin irritant, **VIII**,
179, 180

Diethylzinc, pyrophoric, **VI**, 327
disposal, **VI**, 329

1,4-Dihydrobenzoic acid, stench, **V**, 401

Diisopropyl ether, explosive peroxides,
VI, 344

Diketene, irritant, poison, **V**, 155

Dimethyl acetylenedicarboxylate, lachrymator, vesicant, **IV**, 329

Dimethylaniline, poison, **VIII**, 516

Dimethylketene β-lactone dimer, lachrymator, **V**, 456, 1103

2,6-Dimethylphenol, **VII**, 192

Dimethyl sulfate, poison, **I**, 58, 538; **II**,
210, 362, 387, 396, 404,
411, 620; **IV**, 675; **V**, 269,
737, 1018; **VI**, 175; **VII**,
41; **VIII**, 46
ammonia antidote, **I**, 537; **II**, 210,
362, 387, 396, 404, 411,
620; **IV**, 675

Dimethyl sulfide, **VI**, 31; **VII**, 236, 531

Dimsyl anion-dimethyl sulfoxide, **VI**, 145

2,4-Dinitrobenzenesulfenyl chloride,
potential explosive, **V**, 475

2,3-Dinitro-2-butene, explosive, **VI**, 374

2,4-Dinitrochlorobenzene, skin irritant, **V**,
474, 478

Dinitrogen tetroxide, toxicity, **V**, 336,
338; **VI**, 837

3,4-Dinitro-3-hexene, explosive, **VI**, 374

1,1-Dinitropropane, potassium salt,
explosive, **VI**, 373

Diphenylcyclopropenone, skin irritant,
V, 515

Diphenylmercury, poison, **I**, 228

Diphenylphosphine, pyrophoric, **VI**, 569

Diphenyl phosphorazidate, **VII**, 136

Distillation residues, exposure to air, **I**,
359, 405; **III**, 370; **V**, 389,
536, 603, 835, 865, 922,
930; **VII**, 519, 526; **VIII**,
399, 558, 608

Ethoxyacetylene, sodium salt, pyrophoric,
IV, 405

N-Ethylallenimine, skin irritant, **V**, 541
sensitivity to acids, **V**, 543

Ethyl azodicarboxylate, explosive, **IV**,
412; **V**, 544

Ethyl (α-(bromomethyl)acrylate,
VII, 210

Ethyl bromoacetate, lachrymator, **III**,
383, 409; **VII**, 155

Ethyl α-bromopropionate, lachrymator,
IV, 447

Ethyl chloroacetate, **I**, 512

Ethyl α-chloroacetoacetate, lachrymator,
IV, 592

Ethyl diazoacetate, toxic, potential explosive, **III**, 393; **IV**, 412

Ethylene oxide, toxicity, **I**, 118

Ethylenimine, toxic, skin irritant, **IV**, 435

3-Ethyl-1-hexyne, irritant, **VI**, 596

Ethyl isocyanide, stench, potential explosive, **IV**, 438

Ethyl nitrite, **II**, 296

Ethyl N-nitroso-N-benzylcarbamate, skin
irritant, potential explo-
sive, **IV**, 782

Fluorine, handling, **VIII**, 286

1-Fluoroalkanes, toxic, **IV**, 525

Fluoroboric acid, **VI**, 991

Formaldehyde, **VII**, 142

Formic acid, skin irritant, **V**, 854

2-Furfuryl mercaptan, stench, **IV**, 492

Fumaronitrile, lachrymator, skin irritant, **IV**, 486, 488

Hexafluoroacetone, **VI**, 664; **VII**, 251
Hexafluoroacetone imine, **VI**, 161
65% Hexafluorophosphoric acid, corrosive, **V**, 135
    treatment of burns, **V**, 135
Hexafluoropropene, **VII**, 251
Hexamethyl Dewar benzene, **VII**, 257
Hexamethylenediamine, toxic, **IV**, 521
Hexamethylene diisocyanate, toxic, **IV**, 521
Hexamethylphosphoric triamide (HMPA), toxicity, **VI**, 64, 106, 242, 376, 638, 704; **VII**, 66, 236, 241, 278; **VIII**, 277, 326, 486
Hexamethylphosphorous triamide, irritant, **VI**, 131
Hexyl fluoride, toxic, **IV**, 526
HMPA, see Hexamethylphosphoric triamide
Hydrazine, anhydrous, **VII**, 504
Hydrazine toxicity, **IV**, 274; **V**, 30, 39, 166, 168, 281, 929, 1055; **VI**, 10, 56, 161; **VIII**, 16, 573, 597
Hydrazines, **VI**, 12
Hydrazoic acid, **IV**, 106; **V**, 273; **VI**, 96, 368, 1030
Hydrofluoric acid, 48%, **II**, 226, 297, 301; **V**, 1136; **VII**, 514
    60%, **II**, 189
Hydrogenation catalysts, **VII**, 289, 388, 424, 434
Hydrogen azide, **V**, 832; **VII**, 433
Hydrogen chloride, **VI**, 1, 28, 190, 505, 747, 883, 976; **VII**, 1, 28, 420
    water trap, **VIII**, 415
Hydrogen cyanide, **I**, 22, 257; **III**, 275, 436, 550, 615; **IV**, 59, 115, 274, 276, 496, 641, 804, 877, 953; **V**, 240, 264, 269, 344, 438, 471, 841, 1013; **VI**, 14, 334

antidotes, **VI**, 335
possible formation from chloral-hydroxylamine mixtures, for example, **I**, 327 (Chemistry in Britain, April, 1988, letters)
gas absorption trap, **I**, 97
Hydrogen fluoride, **IV**, 359; **V**, 66, 136, 138, 332, 480, 482; **VI**, 628
    treatment of burns, **V**, 66, 67, 138, 333, 482; **VI**, 629
Hydrogen gas, **V**, 14, 40, 439, 775, 1064; **VI**, 675, 692, 783, 804, 860, 945; **VII**, 530
Hydrogen peroxide, **V**, 368, 599, 904; **VI**, 137; **VII**, 403
90% H_2O_2 in methylene chloride, shock sensitive, **V**, 369
Hydrogen selenide, poison, **III**, 771; **IV**, 359
Hydrogen sulfide, toxicity, **IV**, 243, 669, 671, 928; **V**, 780; **VI**, 621, 1016; **VII**, 372; **VIII**, 597
    trapping, **III**, 117, 610; **V**, 782
2-Hydroxybutanenitrile, **VII**, 383
2-Hydroxy-3-methoxybenzaldehyde, nose irritant, **III**, 760
3-Hydroxypyrene, potential carcinogen, **V**, 632

Iodosobenzene, heat-sensitive, **V**, 658
Iodoxybenzene, heat-sensitive, **V**, 665
Iron carbonyl compounds, **VI**, 520
Iron pentacarbonyl, **VI**, 807, 997
Isonitriles (Isocyanides), **V**, 300, 772, 1060; **VI**, 232, 753
    decontamination, **V**, 300, 1062

Ketene, lung irritant, toxic, **I**, 333; **V**, 679
    potentially explosive under pressure, **V**, 679

Lead hydroxide, **III**, 68
Lithium, dispersion, **VII**, 346

Lithium aluminum hydride, **IV**, 356, 835;
 V, 305; **VII**, 328
 disposal, **V**, 178; **VIII**, 155
 distillation of solutions containing
 LiAlH₄ **IV**, 356

Macrocyclic polyethers, **VI**, 395
Magnesium, highly reactive, **VI**, 845
Mercaptans, stench, **VIII**, 71, 302, 350, 597
 trapping with permanganate trap,
 III, 365
Metal carbonyls, toxicity, **VIII**, 217, 461
Methanesulfinyl chloride, irritant, **V**, 712
 not to be stored under pressure,
 V, 715
Methanethiol, stench, **VI**, 601
Methanol, **VII**, 278
1-Methoxy-2-methylpyridinium methyl
 sulfate, potential explo-
 sive, **V**, 271
1-(o-Methoxyphenyl)-2-nitro-1-propene,
 lachrymator, skin irritant,
 IV, 573
Methyl benzenesulfinate, skin irritant,
 V, 724
Methyl bromide, **VI**, 675
Methyl bromoacetate, **VII**, 339
Methyl α-(bromomethyl)acrylate,
 VII, 319
Methyl chloride, **VII**, 346
Methyl 4-chloro-2-butynoate, irritant,
 VIII, 371
Methyl chloroformate, irritant, **VIII**, 371
Methyl chlorothiolformate, stench,
 V, 166
Methyl N,N-dichlorocarbamate, skin irri-
 tant, **VIII**, 427
Methyl 4-hydroxy-2-butynoate, **VII**, 335
Methyl iodide, **V**, 1093; **VI**, 64, 442, 483,
 501, 552, 683, 704, 807,
 830; **VII**, 139
Methyl isocyanide, poison, stench,
 V, 772
Methyllithium, ethereal solution of, **VI**,
 122 **VIII**, 112, 192, 216,
 223, 268, 371, 479, 486

Methylmercuric iodide, poison, **II**, 403
Methyl nitrate, potential explosive,
 II, 413
Methyl (E)-3-nitroacrylate, **VI**, 801
4-Methyl-4-nitrovaleric acid, decomposi-
 tion during distillation,
 V, 448
Methyloxirane, carcinogen, **VIII**, 434
Methyl propiolate, lachrymator, **V**, 278;
 VIII, 458
1-Methyl-3-p-tolyltriazene, potential
 explosive, **V**, 799
Methyl vinyl ketone, **VII**, 59, 105, 363,
 368, 473
MoOPH, **VII**, 279

β-Naphthalenediazonium chloride–mercuric chloride double salt,
 potential explosive, **II**, 433
Nickel carbonyl, poison, **VI**, 722
 antidote, **VI**, 723
Nitric acid-acetic acid-acetic anhydride
 mixtures, **II**, 450; **V**, 839
m-Nitrobenzaldehyde, **III**, 645
Nitrobenzene, poison, **III**, 669; **VIII**,
 16, 17
o-Nitrobenzyl bromide, lachrymator, skin
 irritant, **V**, 827
p-Nitrobenzyl bromide, skin irritant,
 II, 444
Nitro compounds, sodium salts: potential
 explosives when dry,
 IV, 223
Nitrogen dioxide, **VI**, 799
Nitrogen oxides, **V**, 1071; **VII**, 57, 99
2-Nitropropene, **VII**, 396, 414
Nitrosamines, poison, **VIII**, 405
 disposal, **VIII**, 405
Nitrosobenzene, poison, **III**, 669
N-Nitroso-β-methylaminoisobutyl methyl
 ketone, *Warning*, **III**, 245
Nitrosomethylurea, *Warning*, **II**, 165, 461
Nitrosomethylurethane, *Warning*, **II**, 461;
 III, 119; **V**, 842
 skin irritant, **II**, 465
N-Nitrosopyrrolidine, **VIII**, 542

β-Nitrostyrene, irritant, **I**, 414
2-Nitrothiophene, poison, vesicant,
    II, 467
Nitrosyl chloride, **V**, 864; **VIII**, 981

Osmium tetroxide, **VI**, 342; **VII**, 375
Oxalyl chloride, toxicity, **V**, 111, 205;
    VIII, 499
Oxaziridines, potential explosives, **V**,
    191, 193
Ozone, **VI**, 43, 44; **VIII**, 43
Ozonides, explosive, **III**, 680; **VIII**, 405
Ozonolysis reactions, **VII**, 168

Palladium black, **VI**, 252
Palladium catalysts, pyrophoric, **III**, 689;
    V, 30; **VI**, 1008
3,3-Pentamethylenediazirine, **V**, 899;
Peracetic acid, handling, use, and dispos-
    al, **V**, 416, 660, 661, 665
Peracids, **V**, 367, 416, 805; **VI**, 39
Perbenzoic acid, *Warning*, **I**, 431; **V**,
    900, 906
Perchloric acid, **VI**, 122
Perhydro-9b-boraphenalene, **VII**, 428, 429
Peroxides, **V**, 70, 379, 818; **VII**, 126, 168
    detection, **IV**, 830; **V**, 644, 925;
        VIII, 316, 318, 399
    disposal, **VIII**, 318,
    removal, **III**, 499, 500; **V**, 644
Peroxytrifluoroacetic acid, **V**, 367, 598
Phenacyl bromide, lachrymator, skin irri-
    tant, **II**, 481
Phenacyl chloride, lachrymator, skin irri-
    tant, **III**, 192
Phenol, corrosive, **I**, 162; **V**, 1145
    treatment of burns, **I**, 162
Phenyl azide, potential explosive, **III**, 711
Phenylchlorodiazirine, **VII**, 203
Phenylhydrazine, poison, **I**, 445
Phenylhydroxylamine, carcinogen, **III**,
        669; **V**, 957, 1124; **VIII**,16
2-Phenylperfluoropropene, potential poi-
    son, **V**, 949;
2-(Phenylsulfonyl)-3-phenyloxaziridine,
    decomposition, **VIII**, 546

Phenylsulfur trifluoride, toxic, skin irri-
    tant, **V**, 396, 959
1-Phenylthio-2-bromoethane, **VII**, 453
(Phenylthio)nitromethane, disposal,
    VIII, 550
Phosgene, **II**, 454; **III**, 168, 488; **IV**, 521,
        788; **V**, 201, 822; **VI**, 199,
        282, 418; **VII**, 420
    trapping with 20% NaOH, **II**, 454 or
        toluene, **III**, 168
Phosphines, aromatic, odor, **V**, 789
    disposal, **V**, 789
Phosphorus (yellow), **I**, 39, 40; **II**, 403
Phosphorus oxychloride, irritant, **IV**, 167;
    VIII, 50
Phosphorus pentafluoride, poison, **V**, 134
Phosphorus tribromide and trichloride,
    possible formation of
    explosive phosphine
    byproducts, **III**, 370
Piperidine, **VII**, 473
Potassium, **III**, 71; **IV**, 134, 399, 616,
        962; **V**, 14, 188, 567,
        1061; **VI**, 14, 612, 845;
        VII, 151
    disposal, **III**, 511; **IV**, 134
Potassium amide, inflammable with
    water, **V**, 14
Potassium cyanide, **VII**, 521
Potassium hydride, fire hazard, **VII**, 351;
    VIII, 553
Potassium selenide, odor, **II**, 239
Pressure reactions, explosive, **V**, 528;
    VIII, 446
Propargyl bromide, **VI**, 31; **VII**, 485
Propargyl chloride, irritant, **VIII**, 371
Propene, **VI**, 883
Propiolaldehyde, lachrymator, **IV**, 813
    sensitivity to base, **IV**, 814
Pyrene, potential carcinogen, **V**, 147
Pyridine, poison, **VI**, 644; **VIII**, 50
Pyridinium polyhydrogen fluoride, **VI**, 628

Raney nickel catalyst, **III**, 329, 721, 743;
        IV, 284, 640; **V**, 102; **VI**,
        544, 581, 601

disposal, **III**, 743; **IV**, 138
Rhodium-on-alumina, pyrophoric,
  VI, 372

Selenium compounds, toxicity, **III**, 771;
  VI, 23, 533, 946
disposal, **III**, 773
Silver difluoride, corrosive, **V**, 959
Sodamide, potential explosive, **III**, 782;
  V, 263
hazards of old samples, **I**, 191
Sodium, **IV**, 297, 315; **VII**, 133
disposal, **IV**, 315
Sodium amalgam, fire hazard, **VIII**, 479
Sodium cyanide, disposal, **VIII**, 588
poison, **VI**, 866; **VII**, 383, 435;
  VIII, 587
Sodium ethoxide, **VIII**, 783
Sodium hydride, disposal, **VIII**, 161
handling, **V**, 718; **VIII**, 27, 300
Sodium methanenitronate, explosive,
  VIII, 55
Sodium nitromalonaldehyde monohy-
  drate, shock sensitive,
  explosive, **IV**, 844;
  V, 1004
Sodium–potassium alloy, **VI**, 170
Stannic chloride, **VII**, 414
Styrene dibromide, skin irritant, **IV**, 763
Sulfinyl chlorides, aromatic, do not distil,
  V, 715
Sulfur, odor, **VII**, 196
Sulfur compounds, **VI**, 109
Sulfur dichloride, poison, **VIII**, 427
Sulfur dioxide, **IV**, 334, 937; **V**, 60, 259;
  VI, 48; **VII**, 495
gas absorption trap, **I**, 97
Sulfur dioxide–bromine, **I**, 34
Sulfur tetrafluoride, poison, **V**, 1082; **VI**,
  440; **VII**, 528
disposal, **V**, 1083
Sulfur trioxide, **IV**, 864; **V**, 226
reaction with chlorinated solvents,
  IV, 864
Sulfuryl chloride, poison, **VIII**, 550

Tellurium compounds, **VI**, 468
Tetrachloroethane, toxic, **III**, 550; **V**, 127
Tetracyanoethylene, skin irritant, **IV**, 243,
  877, 953; **V**, 1007, 1014
source of hydrogen cyanide, **IV**,
  877; **V**, 1007
Tetracyanoethylene oxide, source of
  hydrogen cyanide, **V**, 1007
Tetrahydrofuran, purification, **V**, 34, *976*;
  VII, 85, 130, 155, 389,
  451, 477
Tetramethylbiphosphine disulfide, **VII**,
  533
Tetramethyl-*p*-phenylenediamine, skin
  irritant, **V**, 1018
Tetranitromethane, explosive, toxic,
  III, 804
Thallium, **VI**, 791; **VII**, 82
3-Thenyl bromide, lachrymator, skin irri-
  tant, **IV**, 923
m-Thiocresol, stench, skin irritant,
  III, 809
Thiolacetic acid, toxic, stench, **IV**, 928
Thiophosgene, **I**, 165; **VII**, 302
Tin alkyls, disposal, **VIII**, 114
poison, **VIII**, 97, 114, 268
Titanium tetrachloride, disposal,
  VIII, 347
o-Toluamide, potential explosion hazard,
  II, 586
p-Tolylsulfonylmethylnitrosamide, poten-
  tial explosion hazard,
  IV, 945;
p,α,α-Trlbromoacetophenone, lachryma-
  tor, **IV**, 112
Tributylphosphine, reaction with air,
  V, 389
Tributyltin hydride, **VII**, 139
Trichloroacetonitrile, **VI**, 507
Trichloroethylene, **V**, 215; **VII**, 241
Trichloromethyl chloroformate, **VI**, 715
Trichlorosilane, **VI**, 747
Triethylaluminum, pyrophoric, **VI**, 14,
  15, 436
handling procedures, **VI**, 15

Trifluoroacetic acid, irritant, **V**, 162
Trifluoroacetyl triflate, **VII**, 506
Trifluoromethylbenzenesulfonyl chloride,
    VII, 508
Trimethylaluminum, **VII**, 245
Trimethylamine, **V**, 578, 621, 883
Trimethylpyrylium perchlorate, explo-
        sive, shock sensitive, **V**,
        1089, 1106, 1107, 1108
Trimethysilyl cyanide, **VII**, 294, 517
Triphenylaluminum, reactive with air and
        water, **V**, 1089

Vinyl chloroacetate, lachrymator, **III**, 853
Vinyltriphenylphosphonium bromide,
        nose irritant, **V**, 1145

o-Xylylene dibromide, lachrymator, skin
        irritant, **IV**, 112, 984;
        V, 1064
    decontamination, **IV**, 984

Zinc dust, inflammability, **I**, 91; **IV**, 21

CUMULATIVE AUTHOR INDEX

Each preparation was carried out and written directions submitted by the chemists whose names are shown just beneath the equations for each synthesis. Their names were also listed in the front of Annual Volumes 1-69 as "Contributors".

Abbott, T. W., **I**, 440; **II**, 10, 270, 515
Abell, R. D., **I**, 205
Abramovitch, B., **III**, 142
Abrams, S. R., **VIII**, 146
Ackerman, J. H., **V**, 668; **VI**, 529
Adams, A. C., **IV**, 415
Adams, E. W., **I**, 459
Adams J. T., **III**, 291, 405
Adams, Roger, **I**, 10, 61, 84, 94, 101,
    107, 109, 128, 214, 237, 240, 250,
    270, 280, 309, 341, 355, 358, 366,
    394, 436, 459, 463, 504, 528, 531;
    II, 287; **V**, 107
Adger, B. M., **VIII**, 16
Adkins, Homer, **I**, 1, 15; **II**, 262, 206;
    III, 17, 176, 217, 367, 671, 785
Advani, B. G., **V**, 504
Aebi. J., **VII**, 153
Aguilar, V. N., **V**, 572
Ainsworth, C., **IV**, 536; **V**, 1070
Åkermark, B., **VIII**, 137
Akutagawa, S., **VIII**, 57, 183
Albert, Adrien, **III**, 53
Alberti, C. G., **III**, 371
Albertson, Noel F., **VI**, 618
Albini, A., **VII**, 23
Albright, Charles F., **IV**, 68
Alderdice, M., **VII**, 351
Alexakis, A., **VII**, 290
Alexander, Elliot, **III**, 385
Alexander, Kliem, **IV**, 266
Allen, C. F. H., **I**, 156, 205, 226; **II**, 3,

15, 62, 81, 128, 156, 219, 221, 228,
    336, 387, 498, 517, 539, 580; **III**, 1,
    28, 63, 76, 78, 96, 136, 140, 203,
    275, 310, 312, 353, 358, 377, 394,
    404, 415, 418, 573, 578, 597, 621,
    710, 727, 731, 733, 765, 783, 796,
    824, 827, 846; **IV**, 45, 80, 433, 739,
    804, 866
Allen, C. Freeman, **IV**, 398, 608, 616
Allen, George A., **III**, 59
Allen, Lloyd, **VI**, 968
Allen P., Jr., **I**, 77
Allen, R. E., **III**, 200
Alles, Gorden A., **III**, 644, 723, 763
Allinger, Norman L., **IV**, 840
Almond, M. R., **VIII**, 132
Alonso, M. E., **VI**, 235
Alt, G. H., **V**, 208
Althausen, Darrell, **II**, 270
Alvardo, A. M., **I**, 3
Amend, W. J., **II**, 325
Amin, G. C., **III**, 280
Amin, J. H., **V**, 36, 580
Amstutz, E. D., **II**, 462
Amundsen, Lawrence H., **III**, 254, 256
Andell, O. S., **VIII**, 540
Anderson, Albert G., **VII**, 144
Anderson, Arthur G., Jr., **IV**, 221
Anderson, B. C., **V**, 235
Anderson, Ernest, **I**, 67
Anderson, F., **VII**, 208
Andreades, S., **V**, 679

Andrews, S. A., **VII**, 197
Andrews, S. D., **VI**, 392
Andrist, A. Harry, **VI**, 145
Andrus, D. W., **III**, 276, 692, 794
Anelli, P. L., **VIII**, 367
Anello, L. B., **VII**, 251
Angyal, S. J., **IV**, 690
Anker, H. S., **III**, 172
Ansbacher, S., **III**, 138,307
Anslow, W. K., **I**, 298
Anspon, Harry D., **III**, 711, 712
Aoyama, T., **VIII**, 612
Aragona, H., **VI**, 235
Archer, W. L., **IV**, 331
Arduengo, A. J., **VII**, 111, 195
Arens, J. F., **V**, 211
Arhart, R. J., **VI**, 163
Armendt, B. F., **I**, 417
Arndt, F., **II**, 165, 461; **III**, 231
Arnold, H. R., **II**, 142
Arnold, Richard T., **II**, 10
Arora, S. K., **VII**, 263
Arsenijevic, L., **VI**, 34
Arsenijevic, V., **VI**, 34
Arters, A. A., **III**, 833
Arvanaghi, M., **VII**, 451
Ashcroft, M. R., **VI**, 512
Ashley, W. C., **III**, 725
Ashworth, P. J., **IV**, 128
Asirvatham, E., **VIII**, 562
Aslam, M., **VIII**, 212
Asscher, M., **V**, 1076
Aston, J. G., **III**, 538
Atkins, T. J., **VI**, 652
Atkinson, Edward R., **IV**, 872
Atkinson, E. R., **I**, 222
Audrieth, L. F., **III**, 516, 536
Auerbach, Robert A., **VI**, 692
Augelli, C. E., **VIII**, 235
Augustine, Robert L., **V**, 869
Ault, Addison, **V**, 932
Aumiller, James C., **VI**, 133, 179, 271
Axelrod, L. R., **VI**, 690
Ayers, Gilbert B., **I**, 23

Baba, S., **VIII**, 295
Babasinian, V. S., **II**, 466

Babcock, G. S, **I**, 85
Babcock, S. H., **II**, 328
Babers, Frank H., **II**, 412, 512
Bach, Robert D., **V**, 315, 1145; **VII**, 126
Bachman, G. Bryant, **II**, 323
Bachmann, W. E., **I**, 113; **II**, 71, 73; **III**, 839, 841
Backer, H. J., **IV**, 225, 250, 943
Bäckvall, J. E., **VIII**, 5, 9, 540
Baclawski, Leona M., **VI**, 611
Bader, Alfred R., **IV**, 898
Badhwar, I. C., **II**, 304
Bailey, C. F., **II**, 563
Bailey, Denis M. **VI**, 618,
Bailey, Philip S., **V**, 489, 493
Baizer, Manuel, **III**, 323
Bak, B., **IV**, 207
Baker, Robert H., **III**, 141
Baker, Wilson, **I**, 181, 451
Bal, B., **VII**, 185
Balaban, A. T., **V**, 1106, 1112, 1114
Balasubramanian, M., **V**, 439
Balasubrahmanyam, S. N., **V**, 439
Balcom, Don M., **IV**, 603
Balderman, D., **VII**, 433
Baldwin, John E., **VI**, 145
Bales, Stephen E., **VI**, 845
Ballard, Donald A., **I**, 89
Ballard, Murel M., **II**, 97
Ballesteros, P., **VII**, 142
Baltzly, R., **III**, 239
Ban, Y., **VI**, 462
Banks, Charles V., **IV**, 229
Bankston, D., **VIII**, 490
Bannard, R. A. B., **IV**, 393
Banner, B. L., **VII**, 297
Bantjes, A., **IV**, 534
Bao, L. Q., **VI**, 190
Bare, Thomas M., **V**, 775
Barger, G., **II**, 384
Barker, W. E., **II**, 156
Barnes, R. P., **III**, 551,555
Barnier, J. P., **VII**, 129
Barrett, A. G. M., **VIII**, 550
Barrett, J. H., **V**, 467
Barthel, John W., **IV**, 218
Bartlett, P. A., **VII**, 164

Bartlett, Paul D., **V**, 194, 196, 689
Basha, Anwer, **VI**, 492
Bass, K. C., **V**, 398
Bast, Klaus, **V**, 650
Bastiaansen, L. A. M., **VII**, 287
Batcho, A. D., **VII**, 34
Baude, Frederic J., **V**, 567
Bauer, M., **VII**, 435
Baum, A. A., **V**, 797
Baumgarten, Henry E., **V**, 909
Bavin, P. M. G., **V**, 30
Baxter, J. F., **III**, 154
Beaber, N. J., **I**, 145
Beadle, J. R., **VII**, 108
Beal, G. D., **I**, 379, 538
Beam, Charles F., Jr., **VI**, 278
Bean, H. J., **II**, 29
Beaujon, Jan H. R., **V**, 376
Beavers, Ellington M., **V**, 269
Becher, Jan, **VI**, 640
Beck, A. K., **VI**, 316, 869
Becker, Ernest I., **IV**, 174, 176, 623, 657,
   771; **V**, 111, 552
Bedell, S. F., **IV**, 810
Bedoukian, Paul, Z., **III**, 127
Behr, Letha Davies, **II**, 19, 562
Behr, Lyell C., **IV**, 342
Bell, Alan, **III**, 78, 96, 310, 312, 404,
   415, 846
Bell, E. V., **II**, 223
Bell, E. W., **IV**, 125
Bell, T. W., **VIII**, 87
Bellis, M. P., **IV**, 722,903
Bellus, D., **VI**, 427
Benac, B. L., 195, **VII**
Benaim, J., 242, **VI**
Bender, J. A., **I**, 410
Bender, Myron L., **IV**, 932
Benkeser, Robert A., **VI**, 747, 852
Bennett, G. M., **II**, 223
Bennett, Frank, **IV**, 359
Benson, Richard, **III**, 105
Benson, Richard E., **IV**, 588, 746
Benson, W. L., **II**, 273
Benson, W. R., **V**, 663
Bercaw, J. E., **VIII**, 505
Berchet, G. J., **II**, 397

Bere, C. M., **I**, 7
Berenbaum, M. B., **IV**, 66, 273, 274
Berger, Alfred, **III**, 786
Bergman, J., **VI**, 468; **VIII**, 493
Bergman, R. G., **VIII**, 505
Bergstrom, F. W., **III**, 778
Berlin, K. D., **VI**, 932; **VII**, 210
Berlinguet, L., **IV**, 496
Berndt, A., **V**, 1128, 1130
Berney, D., **VIII**, 241
Berthold, Robert V., **V**, 703
Bertz, R. T., **IV**, 489
Bertz, S. H., **VII**, 50
Betti, M., **I**, 381
Bettinetti, G. F., **VII**, 23
Beverung, W. N., **VI**, 371
Bey, P., **VI**, 459
Beyer, E., **V**, 365
Bigelow, H. E., **II**, 57; **III**, 103
Bigelow, Lucius A., **I**, 80, 133, 135, 136,
   476; **II**, 244
Biggs, B. S., **III**, 229, 768
Bill, J. C., **IV**, 807
Billek, Gerhard, **V**, 627
Billica, Harry R., **III**, 176
Billig, Franklin A., **IV**, 68
Billman, John H., **III**, 360
Billups, W. E., **VI**, 87
Bingham, E. M., **VI**, 440, 835
Birch, A. J., **VI**, 996, 1001
Bircher, L. J., **I**, 485
Birkofer, L., **VI**, 1030
Bisgrove, D. E., **IV**, 372
Bishop, W. S., **III**, 229, 584, 613, 768
Bissing, D. E., **V**, 361
Bistline, R. G., Jr., **IV**, 862
Bithos, Joe J. **V**, 591
Black, Alvin P., **II**, 412, 512
Blacklock, Thomas, **VI**, 893; **VII**, 203
Blackwood, Robert K., **IV**, 454
Blake, J., **IV**, 327
Blakeney, A. J., **VI**, 87
Blanchard, K. C., **I**, 453
Blankley, C. John, **V**, 258
Blardinelli, Albert J., **IV**, 311
Blaszczak, L. C., **VIII**, 23
Blatter, H. M., **V**, 656

Blegen, James, R., **III**, 116
Bloch, Henery S., **III**, 637
Block, E., **VIII**, 212
Block, Paul, Jr., **V**, 381, 1031
Blomquist, A. T., **IV**, 216, 838
Bloomfield, Jordan J., **VI**, 167
Blues, E. T., **V**, 1141
Blum, David M., **VI**, 666, 1033
Boatman, S., **V**, 187, 767
Boden, H., **V**, 1029
Bodesheim, Ferdinand, **V**, 300
Boeckman, Robert K., Jr., **VI**, 101, 666, 1033; **VIII**, 192
Boehme, Werner R., **IV**, 590
Boekelheide, V., **IV**, 298, 641
Boes, M., **VIII**, 204, 573
Bogdanowicz, Mitchell J., **VI**, 364
Boger, D. L., **VIII**, 173, 444, 597
Bogert, Marston T., **I**, 220
Bohm, I., **VII**, 485
Böhme, Horst, **III**, 619
Boisselle, A. P., **V**, 406
Bollinger, K. M., **I**, 473
Bond, F. T., **VII**, 77
Borch, Richard F., **VI**, 499
Bordwell, Frederick G., **III**, 141; **IV**, 846
Bornstein J., **IV**, 329, 810; **V**, 406, 1064
Borum, O.H., IV,5
Bose, Ajay K., **V**, 973
Bost, R. W., **II**, 547, 610; **III**, 98
Botteghi, C., **VI**, 338
Bottini, Albert., **V**, 121, 124, 541
Bottorff, E. M., **V**, 687
Bou, A., **VIII**, 161
Boulton, A. J., **V**, 1112, 1114
Bourgel, M., **I**, 186, 191, 209
Bourgeois, R. C., **IV**, 918
Bourns, A. N., **IV**, 795
Bousquet, E. W., **I**, 526; **II**, 313
Boutwell, P. E., **III**, 605
Bowden, Everett, **I**, 424; **II**, 414
Bowen, Douglas M., **III**, 553
Boyd, Thomas, **I**, 36
Boyer, J. H., **IV**, 75, 532; **V**, 1067
Bramann, G. M., **I**, 81
Branca, S. J., **VII**, 271

Brand, E., **II**, 49; **III**, 440
Brand, F. C., **III**, 440
Brasen, W. R., **IV**, 508, 582, 585
Braude, E. A., **IV**, 698, 927
Braun, Charles E., **III**, 430; **IV**, 8, 711; **V**, 887
Braun, Géza, **I**, 431; **II**, 122, 124, 256, 308; **III**, 101
Braxmeier, H., **VIII**, 427
Bredereck, H., **V**, 794
Bremer, Keith, **IV**, 777
Brener, L., **VI**, 1002
Brenninger, W., **V**, 533, 808, 1018
Brent, John T., **IV**, 342
Breslow, David, **III**, 408
Breslow, R., **V**, 514; **VI**, 361
Brethen, M. R., **I**, 121, 162; **II**, 278
Breuer, A., **VI**, 731, 862
Brewster, R. Q., **I**, 323, 447; **II**, 347, 445, 574; **III**, 337
Bridson, John N., **VI**, 386
Brinkmeyer, R. S., **VI**, 64
Broach, Vicky, **VI**, 293
Brod, Albert O., **VI**, 576
Brooks, L. A., **II**, 171, 531; **III**, 471, 698
Brossi, A., **VI**, 1, 471
Brotherton, C. E., **VIII**, 173
Brotherton, T. K., **V**, 816
Brown, B. K., **I**, 309, 528
Brown, C. A., **VIII**, 553
Brown, E. A., **IV**, 960
Brown, George Bosworth, **III**, 615
Brown, Herbert C., **V**, 977; **VI**, 137; **VII**, 427
Brown, J. B., **IV**, 851
Brown, J. F., Jr., **IV**, 372
Brown, J. M., **VIII**, 420
Brown, M. **V**, 1043
Brown, M. J., **VIII**, 609
Brubaker, M. M., **I**, 485
Bruce, William F., **II**, 12, 139, 191; **IV**, 788
Bruggink, Alle, **VI**, 36
Brun, P., **VI**, 142
Brüning, Ingrid, **V**, 957, 1124
Bryce-Smith, D., **V**, 993, 1141

Bryson, T. A., **VI**, 505
Bublitz, D. E., **V**, 1001
Buc, Saul R., **III**, 93; **IV**, 101
Buchdahl, M. R., **IV**, 106
Büchi, G. H., **VIII**, 536
Buchner, B., **IV**, 784
Buchschacher, P., **VII**, 368
Buck, Allen, **III**, 195, 270, 449
Buck, J. S., **II**, 44, 55, 130, 333, 549, 619, 622, 290; **III**, 239
Buckles, Robert E., **IV**, 256, 722, 777, 857, 914
Budde, W. M., **IV**, 31
Budewitz, E. P., **IV**, 748
Buehler, C. A., **II**, 341; **III**, 468
Bui Khac, Trung, **VI**, 474
Bulford, R. H., **II**, 494
Bunnelle, W. H., **VIII**, 602
Bunnett, Joseph F., **IV**, 114; **V**, 12, 478, 816; **VI**, 451
Buntin, S. A., **VII**, 361
Burckhalter, J. H., **IV**, 333
Bürger, W., **V**, 221
Burgess, Edward M., **VI**, 788; **VII**, 195
Burgstahler, Albert W., **V**, 251, 591
Buriks, R. S., **V**, 1067
Burks, R. E., Jr., **III**, 785
Burness, D. M., **IV**, 628, 649; **V**, 403
Burnett, Robert E., **II**, 246, 338
Burns. Timothy P., **VI**, 845
Burpitt, R. D., **V**, 277
Burt, Pauline, **I**, 494
Burton, Donald J., **V**, 949
Buse, C. T., **VII**, 185, 381
Buswell, R. J., **II**, 276
Buter, J., **VIII**, 592
Butler, John Mann, **IV**, 486
Byers, J. R., **II**, 31, 322, 451
Byers, J. R., Jr., **IV**, 739

Cabezas, J., **VIII**, 226
Cacchi, S., **VIII**, 126
Cadogan, J. I. G., **V**, 941
Cadogan, M., **VI**, 744
Caesar, P. D., **IV**, 693, 695
Caglioti, L., **VI**, 62

Cahiez, G., **VII**, 290
Caine, Drury, **VI**, 51
Cairncross, Allan, **VI**, 875
Cairns, Theodore L., **II**, 604; **IV**, 588
Calder, A., **VI**, 803
Caldwell, W. E., **VII**, 315
Calkins, A. E., **III**, 544, 547
Callen, Joseph E., **III**, 26, 134, 212
Calvery, H. O., **I**, 228, 533; **II**, 287
Calzada, Jose G., **VI**, 634; **VIII**, 226
Cambie, R. C., **VI**, 348
Campaigne, E., **IV**, 31, 331, 918, 919, 921
Campbell, Barbara, K., **III**, 148; **IV**, 117, 763
Campbell, I. D., **V**, 517
Campbell, Kenneth N., **III**, 148, 446; **IV**, 117, 763
Campbell, Tod W., **V**, 499, 501, 985
Cannon, George W., **IV**, 597
Caputo, Joseph A., **V**, 869
Carabateas, C. D., **VI**, 529
Carboni, R. A., **IV**, 877
Cargill, R. L., **VII**, 315
Carhart, Homer W., **IV**, 436
Carlsmith, Allan, **IV**, 828
Carlson, Bruce A., **VI**, 137
Carlson, H. D., **V**, 679; **VI**, 664
Carlson, Robert M., **VI**, 196
Carlsson, R., **VI**, 468
Carmack, Marvin, **IV**, 669
Carozza, L., **VII**, 297
Carpenter, Wayne R., **V**, 288
Carpino, Barbara A., **V**, 157, 166
Carpino, Louis A., **V**, 157, 166; **VI**, 403, 555
Carr, R. V. C., **VII**, 453
Carter, H. E., **III**, 167, 774, 813; **V**, 946
Carter, K. N., **III**, 154
Carter, P. H., **V**, 339
Casanova, Joseph, Jr., **V**, 273, 772
Cason, James, **III**, 3, 169, 601, 627; **IV**, 510, 555, 630, 635
Cassady, J. M., **VII**, 319
Castle, J. E., **V**, 393
Castro, C. E., **V**, 993; **VI**, 821, 916

Casy, G., **VIII**, 415
Cate, W. A., **II**, 341
Catlin, W. E., **I**, 186, 471
Caudle, E. C., **I**, 224
Cava, M. P., **V**, 944
Cejka, L., **I**, 283
Celerier, J. P., **VIII**, 263
Cha, D. D., **VIII**, 82
Cha, D. Y., **VI**, 342
Chabala, John C., **VI**, 106
Chadha, N. K., **VII**, 339
Chadhary, Sohan S., **V**, 35
Chadwell, H. M., **I**, 78
Chalmers, William, **II**, 598
Chamberlain, K. B., **VI**, 996, 1001
Chamberlin, A. R., **VII**, 77
Champion, J., **VI**, 320; **VII**, 112, 129
Chan, D. M. T., **VII**, 266
Chan, W. K., **VII**, 81, 87
Chan, Yihlin, **VI**, 496
Chanan, Henry, **III**, 121
Chang, T. C. T., **VIII**, 479
Chang, V. S., **VII**, 397
Chao, Sam T., **VI**, 51
Chao, Tai Siang, **IV**, 380
Chard, S. J., **IV**, 520
Charles, Robert G., **IV**, 869
Charleson, D. A., **VII**, 282
Chastrette, Francine, **VI**, 856
Chastrette, Maurice, **VI**, 856
Chatziiosifidis, I., **VII**, 424
Chaux, R., **I**, 166
Chaykovsky, Michael, **V**, 171,755
Cheetham, H. C., **I**, 70
Chiles, H. M., **I**, 10, 237
Chinn, Leland J., **IV**, 459
Chiu, I-C., **VII**, 249
Cho, Y.-M., **VIII**, 87
Chow, W. Y., **VI**, 87
Christiansen, W. G., **I**, 490
Chu, Vera, **VI**, 615, 774
Chudd, C. C., **IV**, 748
Ciabattoni, J., **VI**, 991
Ciganek, Engelbert, **IV**, 339, 612
Citterio, A., **VII**, 105
Clapp, L. B., **IV**, 372

Clark, John R., **II**, 181; **III**, 260
Clark, Peter S., **VI**, 533
Clark, R. Donald, **IV**, 674; **V**, 456
Clarke, H. T., **I**, 14, 85, 87, 91, 115, 121,
    150, 153, 233, 261, 272, 304, 374,
    415, 421, 455, 478, 495, 514, 523,
    541, 543; **II**, 19, 29, 135, 562, 588;
    III, 172, 226, 690
Claus, R. E., **VII**, 168
Clausen, K., **VII**, 372
Claxton, George P., **VI**, 968
Cleland, Elizabeth S., **III**, 360
Cleland, George H., **VI**, 21
Clemens, David H., **IV**, 463, 662, 664
Cleveland, E. A., **III**, 60, 227, 339, 465
Clibbens, D. A., **II**, 557
Closson, R. D., **V**, 575
Clizbe, Lane A., **VI**, 507
Coan, Stephen B., **IV**, 174, 176
Coburn, E. R., **III**, 313, 696
Cockerille, F. O., **II**, 468
Cocolas, George, **V**, 376
Coffman, Donald D., **III**, 320
Coghlan, M. J., **VIII**, 543
Cohen, F. L. **I**, 240
Cohen, Martin L., **VI**, 533
Cohen, N., **VII**, 297
Cohen, Theodore, **VI**, 737
Cohen, Zvi, **VI**, 43
Cohn, E. J., **II**, 120
Coleman, George H., **I**, 3, 158, 183, 442;
    II, 89, 179, 443, 583, 592; **III**, 60,
    134, 159, 212, 221, 434, 668, 701,
    712
Coleman, Gerald H., **I**, 214
Collington, E. W., **VI**, 64
Collins, David, **V**, 166
Collins, J. C., **VI**, 644
Collins, Peter M., **V**, 598
Colon, Ismael, **VI**, 401
Colonge, J., **IV**, 350, 601
Conant, J. B., **I**, 49, 199, 211, 292, 294,
    345; **II**, 33
Conard, Charles R., **II**, 167
Conez, Cecilia, **VI**, 887
Conia, J. M., **VI**, 320; **VII**, 112, 129

Conn, M. W., **II**, 547
Conner, R. M., **V**, 478
Connor, D. S., **VI**, 101
Connor, Ralph, **II**, 181, 441; **III**, 260, 363
Conrow, Kenneth, **V**, 1138
Constable, E. W., **II**, 610
Converse, S., **I**, 226
Cook, Clinton D., **IV**, 711; **V**, 887
Cook, Fred L., **VI**, 301
Cook, J. M., **VII**, 50
Cook, William B., **III**, 71
Cooke, Giles B., **II**, 406
Cookson, R. C., **VI**, 936
Cooper, F. C., **IV**, 769
Cope, Arthur C., **II**, 181; **III**, 219, 385,
    397, 399, 501; **IV**, 62, 218, 234,
    304, 339, 377, 612, 816, 890; **V**, 315
Copenhaver, J. E., **I**, 142
Corbin, Thomas F., **I**, 328
Corey, E. J., **V**, 755; **VI**, 220, 556;
    VIII, 93
Cornell, John H., Jr., **V**, 422
Cornforth, J. W., **IV**, 467
Corson, B. B., **I**, 49, 179, 241, 336; **II**,
    33, 69, 151, 229, 231, 273, 378, 434,
    596; **IV**, 884
Cortese, Frank, **II**, 9, 564
Cotter, Robert J., **IV**, 62, 377
Cowan, Dorotha, **V**, 890
Cowper, R. M., **II**, 480
Cox, Gerald J., **II**, 612
Cox, Lucile, **III**, 286
Cox, R. F. B., **II**, 7, 272, 279, 453
Cox, W. M., **II**, 468
Coxon, J. M., **VI**, 946
Coyle, Christopher, **VI**, 190
Coyner, Eugene C., **IV**, 727
Crabbe, P., **VII**, 276
Cragoe, Edward J., Jr., **V**, 635
Craig, David, **II**, 179, 583
Craig, W. E., **IV**, 130
Cram, Donald J., **III**, 125; **V**, 926;
    VI, 301
Crandall, J. K, **V**, 863, 866; **VI**, 786, 948
Crass, G., **VII**, 41
Crawley, L. C., **VI**, 948

Creary, X., **VII**, 438
Creegan, Francis J., **V**, 623
Creger, P. L., **VI**, 517
Cressman, Homer, W. J., **III**, 595, 608,
    609, 796
Criegee, R., **V**, 370
Cripps, H. N., **V**, 22, 459
Crisp, G. T. **VIII**, 97,
Crocker, Richard E., **IV**, 278
Crockett, Geoffrey C., **VI**, 226
Cromwell, Norman H., **III**, 105, 125
Crosby, Donald G., **V**, 654, 703
Crossland, I., **VII**, 12
Crossley, Frank, **II**, 363; **III**, 599, 617
Crovetti, Aldo J., **IV**, 166, 654, 704
Crow, R., **VIII**, 33
Crowley, Paul., **V**, 157, 160
Crowther, G. P., **VI**, 259
Croxall, W. J., **IV**, 98
Cruickshank, Philip A., **V**, 555
Crumrine, David S., **VI**, 692
Culhane, Paul J., **I**, 56, 125, 408
Cullinane, N. M., **IV**, 520
Cupas, C., **V**, 16
Curphey, T. J., **VI**, 1019
Curtin, David Y., **V**, 1092
Curtis, Omer E., Jr., **IV**, 278
Cymerman-Craig, J., **IV**, 205, 667, 700;
    V, 88,339
Czyzewski, J., **VI**, 940

Dabovich, Thomas C., **III**, 39
D'Addieco, Alfred A., **IV**, 234
Daeniker, H. U., **V**, 989
Daignault, Ronald A., **V**, 303
Dains, F. B., **I**, 323, 447; **II**, 355
Dakin, H. D., **I**, 149; **II**, 555
Dalton, D. R., **VI**, 184
Dalton, J. R., **VII**, 315
Damerell, V. R., **II**, 204
Damschroder, R. E., **III**, 107
Danheiser, R. L., **VIII**, 82, 347, 471, 501
Daniels, J. J., **VI**, 719
Danishefsky, S., **VII**, 312, 411
Dansted, E., **VI**, 946
Darby, William J., **III**, 460

Darling, S. D., **V**, 918
Daub, Guido H., **IV**, 390
Dauben, Hyp J., Jr., **IV**, 221
Davidsen, S. K., **VIII**, 451
Davidson, David, **II**, 590
Davidson, L. H., **II**, 480
Davis, Anne W., **I**, 261, 421, 478
Davis, F. A., **VIII**, 104, 110, 546
Davis, R. B., **IV**, 392
Davis, Robert, **V**, 589
Davis, Tenny L., **I**, 140, 302, 399, 453
Dawkins, C. W. C., **IV**, 520
Dawson, D. J., **VI**, 298, 491, 584
Day, A. C., **VI**, 10, 392
Dayan, J. E., **IV**, 499
Deacon, B. D., **IV**, 569
Dean, F. H., **V**, 136,580
Deana, A. A., **V**, 944
Deardorff, D. R., **VIII**, 13
Deatherage, F. E., **IV**, 851
DeBoer, Charles D., **V**, 528
de Boer, Th. J., **IV**, 225, 250, 943
Deebel, George F., **III**, 468; **IV**, 579
Degering, E. F., **I**, 36
Deghenghi, R., **V**, 645
Dehesch, T., **VI**, 324
Dehn, William M., **I**, 89
Dekoker, A., **VI**, 282
Deloisy-Marchalant, E., **VIII**, 263
Delonge, C. R. H. I., **VI**, 412
DeMaster, Robert D., **VI**, 976
de Meijere, A., **VIII**, 124, 373
Denekas, M. O.,**III**, 317
Denis, J. M., **VII**, 112
Denmark, S. E., **VII**, 524
Dent, W., **VIII**, 231
Deprés, J.-P., **VIII**, 377
DePuy, C. H., **V**, 324, 326, 1058
Dessau, R. M.,,.**VII**, 400
Dessy, R. E., **IV**, 484
de Stevens, G., **V**, 656
DeTar, D. F., **IV**, 34, 730
Dev, Vasu, **V**, 121
de Vries, G., **V**, 223
Dewar, M. J. S., **V**, 727
Dewar, R. B. K., **V**, 727

DeWitt, C. C., **II**, 25
Dhanak, D., **VIII**, 550
Dhawan, B., **VIII**, 77
Diamanti, Joseph, **V**, 198
DiBiase, S. A., **VII**, 108
Dickel, Geraldine B., **IV**, 317
Dickey, J. B., **II**, 21, 31, 60, 173, 175,
    242, 322, 451, 494; **III**, 573
Dickinson, C. L., **IV**, 276
Dickman, D. A., **VII**, 530; **VIII**, 204, 573
Diebold, James L., **IV**, 254
Diehl, Harvey, **III**, 370, 372; **IV**, 229
Dimming, D. A., **III**, 165
Dimroth, K., **V**, 1128, 1130, 1135
Dinsmore, R., **II**, 358
Dittmer, D. C., **VII**, 491
Dobson, A. G., **IV**, 854
Dodge, Ruth A., **I**, 241, 336
Doering, W. E., **III**, 50
Dolak, T. M., **VI**, 505
Dolan, L. A., **VI**, 1
Dolliver, Morris A., **II**, 167
Domeier, L. A., **VI**, 818
Donahoe, Hugh B., **IV**, 157
Donaldson, M. M., **V**, 16
Donleavy, John J., **II**, 422
Dorn, H.,**V**,39
Dornfeld, Clinton A., **III**, 26, 134,
    212, 701
Dorsey, James E., **VI**, 276
Dorsky, Julian, **III**, 538
Dougherty, Charles M., **VI**, 571
Douglas, W. M., **VI**, 39
Douglass, Irwin B., **V**, 709, 710
Doumani, Thomas, F., **III**, 653
Dowbenko, R., **V**, 93
Dowd, S. R., **VI**, 526
Doyle, T. W., **VI**, 215
Dox, A. W., **I**, 5, 266
Drake, N. L., **I**, 77; **II**, 406
Drake, W. V., **II**, 196
Dreger, E. E., **I**, 14, 87, 238, 258, 304,
    306, 357, 495; **II**, 150
Drummond, P. E., **IV**, 810
Dryden, Hugh L., Jr., **IV**, 816
Dubois, Richard H., **VI**, 1028

Dubuis, R., **V**, 880
Dueber, Thomas E., **VI**, 757
Duncan, J. H., **VI**, 172
Duncan, M. P., **VIII**, 43
Duncan, Warren G., **V**, 390; **VI**, 210
Dunham, M., 273, **VI**
Dunn, M. S., **IV**, 55
Dunnavant, W. R., **IV**, 962; **V**, 526, 564
Dupuis, J., **VIII**, 148
Durham, Lois J., **IV**, 510, 555, 635
Dutton, G. R., **II**, 109
du Vigneaud, V., **I**, 48; **II**, 93
Dyer, Morgan C. D., **VI**, 278
Dyson, G. Malcolm, **I**, 165, 506

Earl, R. A., **VII**, 334
Easley, William K., **IV**, 892
Eastham, Jerome F., **IV**, 192
Eatough, Harry, **I**, 80
Eberle, M., **VIII**, 332
Eberly, Floyd, **II**, 355
Eck, J. C., **II**, 28, 74, 76, 371, 374
Edens, C. O., **III**, 394, 731
Edgar, Graham, **I**, 172
Efraty, A., **VI**, 39
Egan, Richard, **III**, 270, 449
Eglinton, G., **IV**, 404, 755; **V**, 517
Ehler, David F., **VI**, 747
Ehrenfeld, Louis, **I**, 388
Ehrhart, Wendell A., **V**, 582
Ehrlich, P., **II**, 457
Eilers, K. L., **V**, 326
Eisenbraun, E. J., **V**, 8, 310
Eissenstat, M. A., **VI**, 334
Eitel, D. E., **V**, 797
Elam, Edward U., **V**, 1103
Eliel, E. L., **IV**, 58, 169, 626; **V**, 175, 303; **VI**, 215, 442, **VIII**, 302
Ellingboe, E. K., **IV**, 928
Ellis, B. A., **I**, 18
Ellis, M. K., **VII**, 356
Ellis, Ray C., **IV**, 597
Ellison, David L., **VI**, 692
Elphimoff-Felkin, I., **VI**, 769
Elser, W., **V**, 780

Elsinger, Fritz, **V**, 76
Elsom, L. F., **VI**, 488
Embleton, H. W., **IV**, 795
Emerson, Oliver, H., **III**, 151
Emerson, William S., **III**, 831; **IV**, 302, 311, 579, 660, 677, 980
Emmick, Thomas L., **V**, 450
Emmons, W. D., **V**, 191, 367, 547
Enders, D., **VI**, 542; **VIII**, 26, 403
Endler, Abraham S., **IV**, 657
Endres, Leland, **V**, 285
England, D. C., **V**, 239
Engler, A., **VIII**, 219
English, J., Jr., **IV**, 499
Englund, Bruce, **IV**, 184, 423
Epstein, William W., **VI**, 496
Erickson, Bruce W., **VI**, 683
Erickson, Floyd B., **IV**, 430
Erickson, Ronald E., **V**, 489, 493
Eschenmoser, A., **VI**, 56, 679
Escudero, J., **VIII**, 306
Estes, Leland L., **IV**, 62
Evanega, George, **V**, 269
Evans, D. A., **VI**, 818; **VIII**, 339, 528
Evans, P. L., 420, **VII**
Evans, J. C. W., **II**, 517
Evans, Russell F., **V**, 977
Evans, Wm. Lloyd, **II**, 17, 137, 305, 307; **III**, 432
Evers, W. L., **II**, 408
Exon, C. M., **VII**, 461

Faber, E. M., **II**, 576
Fadel, A., **VII**, 117
Fahlbusch, K.-G., **VIII**, 353
Fahrenholtz, Susan R., **V**, 151
Fahrni, H. P., **VII**, 30
Falck, J. R., **VI**, 737
Falkoff, M. M., **III**, 350
Falmagne, J. B., **VIII**, 306
Fanta, Paul E., **III**, 267, 661; **IV**, 844
Farah, Basil Said, **V**, 709
Farlow, Mark W., **II**, 312; **IV**, 521
Farmer, H. H., **IV**, 441
Farrand, R., **VII**, 302
Fatiadi, A. J., **V**, 595, 1011

Fawcett, F. S., **V**, 883
Fawcett, J. S., **IV**, 698
Feely, Wayne, **IV**, 298; **V**, 269
Fegley, Marion F., **III**, 488; **IV**, 98
Fehnel, Edward A., **IV**, 669
Feiring, A. E., **VI**, 991
Feldkamp, R. F., **IV**, 671
Feldman, A. M., **V**, 355
Felix, Arthur M., **VI**, 252
Felix, Dorothee, **VI**, 56, 679
Ferretti, Aldo, **V**, 107, 419
Ferris, J. P., **V**, 32, 344
Ferry Clayton W., **II**, 290; **III**, 239
Fetscher, Charles A., **IV**, 735
Feuer, Henry, **IV**, 368, 554
Feutrill, G. I., **VI**, 859
Fey, P., **VIII**, 26, 403
Field, Lamar, **IV**, 674; **V**, 723
Field, R. F., **III**, 317
Fieser, Louis F., **I**, 353, 383, 517; **II**, 35,
    39, 42, 145, 194, 423, 430, 482, 560,
    569; **III**, 6, 98, 465, 590, 633; **IV**,
    189, 195; **V**, 604, 1037
Filler, Robert, **V**, 80; **VI**, 873
Findlay, J. W. A., **V**, 428
Finelli, A. F., **IV**, 461, 776, 790
Fink, D. M., **VIII**, 347, 471, 501
Finke, Richard G., **VI**, 807
Finkel, J.M., **V**, 332
Finkelstein, J., **VI**, 471
Firestone, A., **VIII**, 87
Fischer, Hans, **II**, 198, 202, 217; **III**,
    442, 513
Fischer, Nikolaus, **V**, 877
Fish, M. S., **IV**, 327, 408
Fishbein, P. L., **VIII**, 116
Fishel, D. L., **VI**, 12; **VII**, 420
Fisk, Milton T., **IV**, 169, 626
Fitch, Howard, M., **III**, 658
Flann, C. J., **VIII**, 358
Fleming, Ian, **VI**, 799
Fleming, M. P., **VII**, 1
Flood, D. T., **II**, 295
Flosdorf, E. W., **I**, 462
Floyd, Don E., **IV**, 141
Fluckiger, E., 776, **VI**

Fludzinski, P., **VII**, 208, 221,241
Fogel, Elaine R., **VI**, 737
Fones, William S., **IV**, 293
Fonken, Gunther S., **IV**, 261
Forbus, T. R., Jr., **VII**, 506
Ford, Jared H., **III**, 34
Ford-Moore, A. H., **IV**, 84, 325, 955
Ford, S. G., **II**, 140, 368, 372, 382
Formo, M. W., **III**, 483, 486
Forrester, A. R., **VI**, 803
Forth, M. A., **VIII**, 16
Fortunak, J., **VI**, 799
Foster, Duncan G., **III**, 771
Foster, G. L., **II**, 330
Foster, H. M., **IV**, 638
Foster, L. S., **III**, 157
Fouquey, C., **VIII**, 50
Fowler, K. W., **VIII**, 586
Fox, B. A., **V**, 346
Fox, C. M. J., **VIII**, 71
Fox, G. J., **VI**, 181
Frank, Robert, L., **III**, 116, 167, 328, 410,
    499, 735, 829; **IV**, 451
Frankenfield, E., **V**, 1018
Franz, J. A., **VI**, 163
Franzen, Volker, **V**, 872
Fraser, W. A., **I**, 475
Freeborn, Emeline, **II**, 231
Freeman, Fillmore, **VI**, 1028
Freeman, Jeremiah P., **IV**, 58; **V**,
    839
Freeman, R. C., **V**, 387
Freudenberger, H., **II**, 573
Freudenberg, W., **III**, 541
Friedman, Lester, **III**, 510; **V**, 54,
    810, 1055
Frisell, C., **V**, 642, 924
Fry, J. L. **VII**, 393,
Frye, John R., **III**, 282
Frye, L. L., **VII**, 495
Frye, S. V., **VIII**, 302
Fujii, Shotaro, **VI**, 327
Fujisawa, T., **VIII**, 441, 498
Fujita, M., **VIII**, 326
Fujita, T., **VIII**, 188
Fujiwara, Kunio, **IV**, 72

Fukushima D. K., **III**, 809
Fung, A. P., **VII**, 254
Fuqua, Samuel A., **V**, 390
Furrow, C. L., Jr., **IV**, 242
Furst, A., **VII**, 368
Furuta, K., **VIII**, 141
Fuson, Reynold C., **II**, 169, 260; **III**, 209, 339, 549, 557

Gadwood, R. C., **VIII**, 556,
Gage, J. R. **VIII**, 339, 528,
Gale, D. M., **VI**, 161
Gaibel, Z. L. F., **V**, 727
Gall, Martin, **VI**, 121
Galli, C., **VI**, 698
Galun, A. B., **V**, 130
Gander, Robert, **III**, 200
Ganem, Bruce, **VI**, 666, 1033
Gannon, Walter F., **V**, 294, 539
Garbisch, E. W., Jr., **VI**, 368
Gardner, J. H., **II**, 523, 526
Gariano, Anne L., **VI**, 576
Gariano, Patrick, Jr., **VI**, 576
Garrison, W. E., Jr., **V**, 463
Gassman, P. G., **V**, 91, 96, 424; **VI**, 432, 581, 601; **VII**, 20, 294
Gates, John W., Jr., **III**, 140, 353, 363, 418; **VI**, 150
Gatto, V. J., **VIII**, 152
Gaudry, Michel, **VI**, 193, 474
Gaudry, Roger, **III**, 436; **IV**, 496
Gauerke, C. G., **I**, 123
Gault, R., **VI**, 371
Gawish, A., **VII**, 50
Gawley, R. E., **VII**, 530
Gaylord, Norman G., **IV**, 178
Gelblum, E., **V**, 355
Gellerson, Hilda E., **IV**, 903
Geluk, H. W., **VI**, 48
Georg, G. I., **VIII**, 173
Georgiadis, G. M., **VII**, 124
Gerhart, H. L., **II**, 111, 112
Gerlach, H., **VII**, 470
Gerold, Corinne, **IV**, 104
Gershbein, Leon, **III**, 458
Ghosez, L., **VI**, 282; **VII**, 306

Ghosh, S., **VII**, 177
Giese, B., **VIII**, 148, 583
Giesecke, H., **VII**, 162
Gilbert, E. C., **VI**, 215
Gilbert, Leonard S., **V**, 904
Gillespie, H. B., **II**, 489
Gillespie, R. H., **III**, 671
Gillis, Richard G., **IV**, 396
Gilman, Henry, **I**, 145, 188, 361, 471; **II**, 47, 425
Gilow, H. M., 102, **VII**
Ginsberg, Helen, **III**, 742
Girard, André, **II**, 85
Girard, F. H., **V**, 572
Giza, Chester A., **V**, 157, 166
Glass, D. B., **III**, 504
Gleason, John G., **VI**, 190
Glickman, Samuel A., **IV**, 234
Godefroi, E. F., **VII**, 287
Godt, H. C., Jr., **V**, 893
Goel, O. P., **VIII**, 68
Gofton, B. F., **IV**, 927
Goheen, D. W., **IV**, 594
Gokel, George W., **VI**, 232, 301; **VII**, 108; **VIII**, 152
Golding, B. T., **VII**, 356
Goldstein, Albert, **IV**, 216, 838
Gollis, Morton H., **V**, 422
Gollman, H., **VII**, 323
Gomberg, M., **I**, 113
Gompper, R., **V**, 780
Gonzales, F. B., 164, **VII**
Goodman, Marjorie M., **V**, 162
Goodwin, G. B. T., **VIII**, 1
Gorgues, A., 954, **VI**
Gortner, R. A., **I**, 194
Goshorn, R. H., **I**, 36; **IV**, 307
Gösl, R., **V**, 43
Gott, P. Glenn, **V**, 1103
Gottfried, Sidney, **III**, 381
Göwecke, Siegfried, **V**, 166
Graboski, G. G., **VIII**, 550
Gradsten, M. A., **IV**, 518
Graf, Roderich, **V**, 226, 673
Graham, R. S., **VII**, 402
Granito, Charles, **V**, 589

Gras, J. L., **VII**, 332
Grashey, Rudolf, **V**, 957, 1124
Gray, A. R., **II**, 60, 509
Greene, A. E., **VIII**, 377
Greenlee, W. J. **VI**, 432,
Greenwood, F. L., **III**, 673; **IV**, 108
Gregson, Michael, **VI**, 638
Greull, Gerhard, **V**, 320
Grieco, Paul A., **VI**, 638; **VIII**, 31
Griffin, Gary W., **VI**, 401
Griffin, K. P., **III**, 242
Gringas, Leopold, **III**, 95
Gringore, O. H., **VII**, 99
Grisar, J. Martin, **VI**, 968
Grob, C. A., **V**, 989
Groening, Theodore, **II**, 445
Gröninger, K. S., **VIII**, 583
Gronowitz, S., **V**, 149
Gronwall, Susanne, **V**, 155
Gross, A. W., **VIII**, 93
Gross, H., **V**, 49, 221, 365
Gruber, J. M., **VII**, 282
Gruetzmacher, G., **VI**, 581
Grummitt, Oliver, **III**, 195, 270, 449,
    806, 807, 833; **IV**, 748, 771; **V**, 320
Grunewald, Gary L., **V**, 982
Guaciaro, M. A., **VII**, 271
Guggenheim, T. L., **VII**, 294
Guha, Maya, **V**, 384
Guha, P. C., **III**, 623
Guida, Wayne C., **VI**, 576
Gulati, K. C., **II**, 522
Gulen, R., **IV**, 679
Gumprecht, W. H., **V**, 147, 632
Gund, Tamara M., **VI**, 378
Gunn, B. P., **VII**, 249
Gunsher, J., **VI**, 142
Gunstone, F. D., **IV**, 160
Gupte, S. S., **VI**, 936
Gupton, J. T., **VII**, 197
Gurien, H., **VI**, 1007
Gutenstein, Morris, **V**, 822
Guthrie, J. L., **IV**, 513
Gutsche, C. D., **IV**, 780, 887; **VIII**, 75,
    77, 80
Gutzwiller, J., **VII**, 368

Haaf, W., **V**, 20, 739
Hach, Clifford C., **IV**, 229
Haddadin, Makhluf J., **V**, 1037
Haese, W., **VIII**, 620
Haeseler, P. R., **I**, 196
Hafner, K., **V**, 431, 1088, 1106; **VII**, 15
Hageman, H. J., **VI**, 412
Hager, F. D., **I**, 58, 224, 248, 351, 544
Hagiwara, Daijiro, **VI**, 199
Hahn, Roger C., **V**, 328
Hain, B. E., **III**, 605
Hajos, Z. G., **VII**, 363
Hales, Neil J., **VI**, 82
Hall, Luther A. R., **IV**, 333
Hall, Stan S., **VI**, 537
Hallas, G., **VI**, 181
Halligan, N. G., **VIII**, 23
Halvey, Neil, **VI**, 1033
Hamada, Y., **VII**, 135
Hamersak, Zdenko, **VI**, 958
Hamilton, Cliff S., **IV**, 950
Hamilton, Frances H., **I**, 231, 492, 519
Hamilton, Robert W., **IV**, 576
Hammond, George S., **V**, 528
Hampton, K. Geraid, **V**, 838; **VI**, 928
Hamrick, Phillip J., **V**, 523
Hanack, M., **VI**, 324
Hancock, Evelyn M., **III**, 219, 397,
    399, 501
Handrick, R. G., **III**, 471
Hanessian, S., **VIII**, 363
Hankin, Hilda, **III**, 742
Hansen, H.-J., **VII**, 232
Hanslick, Roy S., **II**, 244; **IV**, 788
Hanson, E. R., **I**, 161, 326
Hansson, S., **VIII**, 137
Hansuld, Mary K., **IV**, 795
Hanzlik, R. P., **VI**, 560
Hargrove, W. W., **V**, 117
Harman, R. E., **IV**, 148
Harpp, David N., **VI**, 130, 190
Harris, Arlo D., **V**, 736
Harris, Charles H., **III**, 125
Harris, G. C., **IV**, 1
Harris, Henry P., **VI**, 301
Harris, S. A., **I**, 241, 336

Harris, Thomas M., **V**, 187, 718, 848; **VI**, 478, 928
Harrison, George C., **III**, 370, 372
Harrison, J. G., **V**, 946
Harrisson, R. J., **IV**, 493
Hart, Harold, **IV**, 278; **V**, 598
Hartfiel, U., **VIII**, 254
Harting, Walter H., **III**, 191
Hartley, W. M., **VI**, 342
Hartman, G. D., **VI**, 620
Hartman, R. J., **IV**, 93
Hartmann, W., **VII**, 4
Hartman, W. W., **I**, 162, 175, 233, 357, 455, 541, 543; **II**, 21, 150, 163, 173, 175, 183, 232, 242, 278, 322, 404, 418, 438, 440, 451, 460, 464, 617; **III**, 20, 37, 82, 650, 652
Hartough, Howard, D., **III**, 14
Hartshorn, M. P., **VI**, 946
Hartung, Walter H., **I**, 15; **II**, 363, 543; **V**, 376
Hartwell, J. L., **II**, 145; **III**, 98, 185
Harvey, Ronald G., **VI**, 887; **VII**, 18
Hasek, Robert H., **V**, 456, 1103
Hasek, W. R., **V**, 1082
Hashimoto, K., **VII**, 512
Hass, Henry, **II**, 17, 137, 305, 307; **IV**, 932
Hassner, A., **VI**, 795, 967
Hatch, Lewis F., **VI**, 454
Hatt, H. H., **II**, 208, 211, 395; **IV**, 854
Hauck, Hans, **V**, 957, 1124
Hauser, Charles R., **II**, 67, 268, 607; **III**, 142, 291, 374, 405, 408, 593, 842; **IV**, 508, 582, 585, 708, 962; **V**, 187, 434, 437, 523, 564, 578, 621, 718, 721, 767, 848; **VI**, 245, 259, 278, 928
Haveaux, R., **VI**, 282,
Hawks, George H., **VI**, 549
Hay, J. **V**., **VI**, 478
Hayakawa, Y., **VI**, 520
Hayase, Yoshio, **VI**, 358, 448
Hayashi, J., **VII**, 443
Hays, High R., **V**, 562
Hazen, R. K., **I**, 241; **II**, 434

Healy, K., **VIII**, 87
Heaney, Harry, **V**, 1120; **VI**, 82, 104
Hearst, Peter J., **IV**, 571
Heathcock, C. H., **VI**, 795, 967; **VII**, 1
Heator, J. S., **I**, 207
Heck, R. F., **VI**, 815; **VII**, 361
Hedrick, Glen W., **V**, 699
Hegedus, L. S., **VII**, 501; **VIII**, 216
Heiba, E. I., **VII**, 400
Heidelberger, M., **I**, 153, 488
Heidtke, W. J., **IV**, 960
Heilbron, I. M., **I**, 207
Heilman, S. M., **VII**, 521
Heimbach, H., **VII**, 424
Heine, H.-G., **VII**, 4
Heininger, S. A., **IV**, 146
Heinzelman, R. V., **IV**, 573
Heisig, G. B., **I**, 54
Helferich, B. **I**, 147, 364; **III**, 428
Helmkamp, G. K., **V**, 1099
Helquist, Paul M., **VI**, 722; **VII**, 236, 326
Hemperly, S. B., **VIII**, 155
Henderson, A., **V**, 355
Hendry, J. L., **II**, 120
Henery, J., **VI**, 310, 422
Hengartner, U., **VI**, 774
Hennion, G. F., **IV**, 683
Hepburn, S. P., 803, **VI**
Hepworth, John D., **V**, 27; **VI**, 181
Herbst, R. M., **II**, 1, 11, 389, 491, 519
Hering, H., **IV**, 203
Herkes, Frank E., **V**, 949
Herranz, E., **VII**, 223, 375
Herrick, Elbert C., **IV**, 304,
Hershberg, E. B., **I**, 168; **II**, 111, 627
Herzog, Hershel L., **IV**, 753
Hess, W. W., **VI**, 644
Hesse, A., **VI**, 901
Hessler, John C., **I**, 438
Hetzel, Frederick W., **VI**, 824
Hetzner, H. P., **III**, 839
Heumann, A., **VIII**, 137
Heusler, K., **V**, 692
Hexner, Peter E., **IV**, 351
Hey, D. H., **V**, 51
Heyne, Hans-Ulrich, **VI**, 78

Hibbert, Harold, **I**, 494
Hidber, A., **VII**, 447
Hiers, G. S., **I**, 58, 117, 327
Higgins, J. G., **V**, 142, 249
Hill, D. T., **VI**, 142
Hill,G.A., **I**, 81,462
Hill, J. G., **VII**, 461
Hill, Julian W., **II**, 53, 325
Hill, Richard K., **VI**, 196
Hill, Robert M., **I**, 300
Hillebert A., **IV**, 207
Hinegardner, W. S., **I**, 172
Hinga, F. M., **V**, 215
Hino, T., **VI**, 462
Hirai, Koichi, **VI**, 704
Hirata, Yoshimasa, **VI**, 289
Hirschmann, Ralph, **VI**, 5
Hirst, G. C., **VIII**, 458
Hixon, R. M., **I**, 511; **II**, 566, 571
Hiyama, T., **VIII**, 326
Ho, Beng T., **VI**, 965
Hobbs, F. W., Jr., **VIII**, 112
Hocker, J., **V**, 179; **VII**, 162,
Hodgson, H. H., **III**, 341
Hoehn, H. H., **III**, 356
Hoenben, G., **VI**, 412
Hoffman, Joseph, **V**, 818
Hoffman, R. V., **VII**, 508
Hoffman, W. F., **I**, 194
Hoffmann, A. K., **V**, 355
Hoffmann, H. M. R., **VI**, 512
Hoffmann, R. W., **V**, 61
Hofle, G., **VII**, 27
Höft, E., **V**, 49, 365
Holden, M. S., **VII**, 501
Holden, Raymond F., **III**, 800
Holleman, A. F., **I**, 497, 552, 554
Hollinshead, John H., **VI**, 82
Holly, F. W., **VII**, 75
Holmes, A. B., **VIII**, 63, 606
Holmes, H. L., **II**, 428; **III**, 300
Holmgren, A. V., **IV**, 23
Holt, D. A., **VII**, 456
Honeywell, G. E., **II**, 89, 443
Hontz, Arthur C., **IV**, 383
Hoogenboom, B. E., **VI**, 987

Hooz, John, **VI**, 386, 634, 919; **VIII**, 226
Hopf, H., **VII**, 485
Horan, J. E., **V**, 647
Horeau, A., **VI**, 34
Hormi, O., **VIII**, 247
Horne, W. H., **II**, 453
Horning, E. C., **III**, 165, 317, 549; **IV**, 144, 408, 461, 620, 776, 790
Horning, M. G., **III**, 165
Horovitch, Sharon, **VI**, 190
Horton, Derek, **V**, 1
Hosomi, A., **VII**, 443
Hou, K.-C., **VII**, 263
House, Herbert O., **IV**, 367, 375, 552, 860, 957; **V**, 258, 294, 324, 539, 775; **VI**, 121, 692; **VII**, 346
Howard, John C., **IV**, 42
Howard, J. W., **I**, 475; **II**, 87
Howard, W. L., **V**, 5, 25, 292
Howe. Eugene E., **V**, 373
Howell, Charles F., **IV**, 816
Howes, P. D., **VI**, 31
Howie, G. A., **VII**, 319
Howk, B. W., **IV**, 801
Hrutfiord, B. F., **V**, 12
Hsiao, C.-N., **VIII**, 3
Hsueh, C. M., **II**, 416
Huang, E. P. Y., **IV**, 93
Huang, Pao-Tung, **IV**, 66, 274
Huang, Z., **VIII**, 616
Hubacher, Max H., **II**, 453
Hubner, F., **VII**, 424
Hudack, N. J., **IV**, 738
Hudlicky, T., **VIII**, 208, 309
Hudnall, Phillip M., **VIII**, 845
Hudrlik, O., **III**, 673
Hudson, B. E., **III**, 142
Hudson, Boyd E., Jr., **III**, 842
Hudson, C. S., **I**, 371
Hufferd, R. W., **I**, 128, 341
Hui, R., **VIII**, 343
Huisgen, Rolf, **V**, 650, 957, 1124
Hulce, M., **VII**, 495
Hull, R., **VII**, 302
Hummel, K., **VI**, 324
Humphlett, W. J., **IV**, 80, 739

Hunig, S., **V**, 533, 808, 1018
Hunt, Richard H., **IV**, 459
Huntress, E. H., **I**, 168; **II**, 457, 459; **IV**, 329
Hurd, Charles D., **I**, 217, 330; **II**, 541; **III**, 458
Huston, Ralph C., **II**, 97
Hutchins, Robert O., **VI**, 215, 223, 293, 376, 442
Hutchinson, C. R., **VII**, 323
Hwa, Jesse C. H., **V**, 608
Hwang Dorothy,, **VI**, 670
Hymans, W. E., **V**, 863, 866
Hyre, John E., **V**, 684

Iacono, S., **VII**, 139
Icke, Roland N., **III**, 59, 644, 723, 763
Ide, Junya, **VI**, 531
Ide, W. S., **II**, 44, 55, 130, 333, 622
Imwinkelried, R., **VIII**, 201, 495
Inaishi, Morio, **VI**, 90
Ingersoll, A. W., **I**, 311, 417, 485; **II**, 328, 503, 506
Inglis, J. K. H., **I**, 235, 254
Insalaco, Michael A., **VI**, 207
Ipatieff, V. N., **II**, 151
Iqbal, M., **VIII**, 15
Ireland, Robert E., **V**, 171; **VI**, 298, 459, 491, 567, 584, 606; **VII**, 66
Irvine, James, **III**, 326
Isbell, Neville, **II**, 262
Ishida, N., **VIII**, 315
Issidorides, C. H., **IV**, 679, 681
Ito, Yoshihiko, **VI**, 327; **VIII**, 315
Itoh, Masumi, **VI**, 199
Iwai, Issei, **VI**, 531
Iwakura, Yoshio, **VI**, 715
Iwanaga, K., **VIII**, 14
Iyer, R. S., **VII**, 236

Jackson, H. L., **IV**, 438
Jacobs, Thomas L., **III**, 456
Jacobs, W. A., **I**, 153, 488
Jacobsen, Niels, **VI**, 890
Jakobsen, Preben, **V**, 70
Jacques, J., **VI**, 34, 175; **VIII**, 50

Jadhav, P. K., **VIII**, 553
Jaekel, F., **VIII**, 124
Jakiela, D. J., **VII**, 326
Jakovac, I. J., **VII**, 406
James, A. P., **VIII**, 420
James, Philip N., **IV**, 539
Janssen, Donald E., **IV**, 547
Jasor, Yves, **VI**, 474
Jaul, E., **IV**, 307
Jawdosiuk, M., **VI**, 940
Jefferson, George D., **II**, 284, 487
Jefford, C. W., **VI**, 142
Jenkins, D. M., **III**, 538
Jenkins, R. L., **I**, 347, 394
Jenner, E. J., **V**, 1026
Jerkunica, J. M., **VI**, 766
Jessup, Peter J., **VI**, 95
Jimenez, Manuel H., **VI**, 252
John, Joseph P., **V**, 747
Johnson, A. W., **V**, 785, 1022
Johnson, B. A., **V**, 215
Johnson, Carl R., **V**, 791
Johnson, David K., **VI**, 791
Johnson, F., **V**, 614
Johnson, F. R., **II**, 449
Johnson, G. Dana, **IV**, 900
Johnson, H. B., **IV**, 804
Johnson, Herbert E., **IV**, 780; **V**, 654, 703
Johnson, Herbert L., **III**, 60
Johnson, John R., **I**, 111, 123, 133, 351, 440, 521; **II**, 8, 106; **III**, 641, 794, 806
Johnson, M. Ross, **VI**, 312
Johnson, Robert E., **VI**, 618
Johnson, William S., **IV**, 132, 162, 261, 390, 259; **V**, 245
Johnson, W. W., **II**, 343
Johnston, H. W., **III**, 167
Johnston, J. Derland, **IV**, 576
Johnstone, H. F., **I**, 158, 183
Jolad, S. D., **V**, 139
Jonczyk, A., **VI**, 897
Jones, E., **VI**, 558, 979
Jones, E. R. H., **IV**, 404, 755, 792
Jones, G. E., **VIII**, 63
Jones, G., **II**, **VII**, 102

Jones, J. B., **VII**, 406
Jones, J. E., **III**, 98, 590
Jones, L. D., **VII**, 326
Jones, M., Jr., **VII**, 470
Jones, Reuben G., **IV**, 824
Jones, T. K., **VII**, 524
Jones, Willard J., **V**, 845
Joos, Renato, **VI**, 56
Jordon, E. F., Jr., **IV**, 977
Jorgenson, M. J., **V**, 509
Josey, A. D., **V**, 716
Judge, Joseph M., **V**, 255
Julia, Marc, **VI**, 586
Julian, P. L., **II**, 391, 487
Jung, Michael E., **VI**, 353, 445
Juntunen, S. K., **VIII**, 540
Juve, H. D., Jr., **VII**, 282

Kabalka, George W., **VI**, 293
Kagan, H. B., **VIII**, 464
Kaiser, Carl, **VI**, 552, 910, 913
Kaiser, Edwin M., **V**, 559; **VI**, 259, 852
Kaiser, E. W., **III**, 209
Kaiser, H., **V**, 1088, 1106
Kaji, E., **VI**, 503
Kalaritis, P., **VII**, 258
Kalinowski, H.-O., **VII**, 41
Kalir, A., **V**, 825; **VII**, 433
Kalish, Joseph, **V**, 471
Kalm, Max J., **IV**, 398, 608, 616
Kamath, P. M., **IV**, 178
Kaminski, F. E., **V**, 263
Kamiya, Takashi, **VI**, 199
Kamm, Oliver, **I**, 25, 42, 104, 128, 250,
   372, 391, 392, 445
Kamm, W. F., **I**, 104
Kan, R. O., **V**, 612, 1051
Kane, V. V., **VII**, 470
Kaplan, L. J., **VI**, 163
Karabinos, J. V., **IV**, 506
Karlsson, Sune M., **VI**, 203
Kaslow, C. E., **III**, 194, 519, 580;
   IV, 718
Katagiri, T., **VIII**, 188, 190
Kauer, J. C., **IV**, 411; **V**, 1043
Kaufman, Daniel, **III**, 693

Kaufmann, W. E., **I**, 258
Kawada, K., **VIII**, 286
Kawamoto, Fumio, **VI**, 327
Kawanisi, M., **V**, 582
Kehm, Barbara B., **IV**, 515
Keinan, Ehud, **VI**, 43
Keiser, Jeffrey E., **V**, 791
Keith, D. R., **VII**, 59
Keizer, V. G., **VI**, 48
Keller, O., **VII**, 70
Keller, W. E., **VII**, 70
Kellert, M. D., **IV**, 108
Kellogg, R. M., **VIII**, 592
Kendall, E. C., **I**, 21, 131, 246, 256
Kende, A. S., **VII**, 208, 221, 241
Kendrick, D. A., **VIII**, 63
Kennard, Kenneth C., **IV**, 950
Kennedy, E. R., **II**, 351; **III**, 355, 482,
   483, 485
Kent, R. E., **III**, 490, 493, 591
Kenyon, Joseph, **I**, 263, 418
Kenyon, R. L., **III**, 747
Kenyon, William G., **V**, 559
Kern, J. W., **I**, 101
Kershaw, John R., **VI**, 115
Kesten, S., **VIII**, 68
Kester, E. B., **I**, 75
Khan, N. A., **IV**, 851, 969
Kharasch, Norman, **V**, 474
Kibler, C. J., **III**, 28, 108, 610
Kidwell, R. L., **V**, 918
Kiedrowski, G. V., **VIII**, 353
Kiefer, E. J., **V**, 22
Kienzle, Frank, **VI**, 709
Kim, C. U., **VI**, 220
Kim, G., **VIII**, 512
Kimball, R. H., **II**, 284, 487, 498
King, A. O., **VII**, 63; **VIII**, 430
King, B. T., **VII**, 59
King, Harold, **I**, 286, 298
King, Harold, S., **II**, 399
King, Harriette, **II**, 612
King, Mary S., **IV**, 88
King, R. B., **VI**, 39
King, W. B., **I**, 252
Kipphardt, H., **VIII**, 26, 403

Kirby, R. H., **I**, 361
Kirchner, Fred K., **III**, 468
Kirner, W. R., **I**, 374; **II**, 136
Kise, Mearl, **II**, 422
Kishida, Yukichi, **VI**, 704
Kitahara, T., **VII**, 312
Kitahonoki, Keizo, **V**, 83
Kiuchi, M., **VI**, 462
Klair, A., **V**, 130
Kleiderer, E. C., **II**, 538
Klein, G. W., **VI**, 101
Klein, R. A., **V**, 1058
Kleinberg, J., **III**, 516
Kleinberg, Jacob, **V**, 1001
Kleinfelter, Donald C., **V**, 852
Kleinschroth, J., **VII**, 485
Klinck, Jane, **V**, 121
Klingenberg, J. J., **IV**, 110
Klingsberg, Erwin, **III**, 119
Klotzer, W., **VII**, 8
Klug, W., **VI**, 731, 862
Kluge, A. F., **VII**, 160
Kluiber, Rudolph W., **IV**, 261
Kneisley, J. Wayne, **III**, 209
Knight, H. B., **IV**, 895
Knight, J. W., **VII**, 126
Knoeber, Sr. M., **VI**, 442
Knowles, W. S., **V**, 208
Knox, L. H., **V**, 194, 196, 689
Knutson, P. L., **VII**, 323
Kobayashi, Y. , **VIII**, 507
Kobe, Kenneth A., **III**, 653
Koch, H., **V**, 20
Koch, Tad H., **VI**, 226
Kocienski, P. J., **VI**, 991
Koelsch, C. Frederick, **III**, 132, 791
Kofod, Helmer, **IV**, 491
Kofron, William G., **VI**, 611
Kogan, T. P., **VII**, 495
Kogon, Irving C., **IV**, 182, 336
Kogure, T., **VII**, 417
Kohler, E. P., **I**, 78
Kohn Earl J., **IV**, 283
Kohn, E. D., **II**, 381
Kon, G. A. R., **I**, 192
Konieczny, M., **VII**, 18

König, C., **V**, 431
Kono, Hiromichi, **VI**, 919
Koo, John, **IV**, 327, 408; **V**, 550
Koppenhoefer, B., **VIII**, 119, 434
Korack, M., **V**, 414
Kormendy, Minerva F., **V**, 162
Korn, A., **VII**, 271
Kornblum, Nathan, **III**, 295; **IV**, 454,
    724; **V**, 845
Kornfeld, E. C., **III**, 413
Korosec, Philip S., **V**, 736
Korpics, C. J., **IV**, 93
Kosak, Alvin I., **III**, 14
Kosower, Edward, **III**, 510
Koten, I. A., **V**, 145
Kotera, Katsumi, **V**, 83
Koto, S., **VI**, 797
Kovacic, Peter, **V**, 35
Koyama, M., **VI**, 797
Kozyrod, R. P., **VII**, 229
Krantz, Karl W., **III**, 254
Krapcho, A. Paul, **V**, 198
Kraus, G. A., **VIII**, 357
Krebs, Edwin, **III**, 649
Kremers, Edward, **I**, 511
Kresze, G., **VIII**, 427
Krimen, Lewis I., **VI**, 8
Krimmel, C. P., **IV**, 960
Krolls, U., **VIII**, 68
Krolski, M. E., **VIII**, 357
Kropp, P. J., **VII**, 304
Krueger, Paul A., **II**, 349, 353
Krumpolc, M., **VII**, 114
Krynitsky, John A., **III**, 10, 164, 508,
    818; **IV**, 436
Kuehne, M. E., **V**, 400
Kuhlmann, H., **VI**, 866; **VII**, 95;
    VIII, 620
Kuhn, Stephen J., **V**, 66, 480
Kuhn, W. E., **II**, 447
Kumada, Makoto, **VI**, 407; **VIII**, 315
Kumar, A., **VIII**, 104
Kume, F., **VIII**, 302
Kumobayashi, H., **VIII**, 183
Kundiger, D., **III**, 123, 506
Kurita, Keisuke, **VI**, 715

Kuryla, William C., **V**, 648
Kurzer, Frederick, **IV**, 49, 172, 180, 213, 361, 502, 645, 934, 937; **V**, 966, 1046
Kuwajima, I., **VII**, 512; **VIII**, 277, 486, 578
Kyrides, L. P., **III**, 422

Labadie, J. W., **VIII**, 268
Lacey, A. B., **IV**, 396
Lachman, Arthur, **II**, 70, 234
Lack, Ruth E., **V**, 339
Lal, G. S., **VIII**, 104
LaLonde, John, **IV**, 757
LaMattina, J. L., **VII**, 149
Lamb, S. E., **VII**, 397
Lambert, B. F., **V**, 400
Lampman, Gary M., **VI**, 133, 179, 271
Landini, D., **VI**, 833
Lane, C. F., **VI**, 719
Lang, R. W., **VII**, 232
Lane, Stanley C., **I**, 140
Lange, B., **VII**, 27
Lange, N. A., **II**, 79
Lange, Richard M., **V**, 598
Langer, W., **VII**, 41
Langford, Robert B., **V**, 474
Langkammerer, C. M., **V**, 445
Langley, Wilson D., **I**, 127, 355; **III**, 334
Langman, A. W., **VI**, 184
Lapworth, Arthur, **I**, 181, 451
Larsen, A. A., **VI**, 529
Larsen, E H., **V**, 379
Larsen, S., **VII**, 277
Larson, G. L., **VIII**, 474
Larson, S. D., **VIII**, 31
Larsson, Lars-Eric, **VI**, 203
Lassalle, G., **VIII**, 467
Laurenzano, A. J., **VII**, 287
Lauer, W. M., **I**, 54; **III**, 194, 580
Lavety, J., **V**, 545
Lawesson, Sven-Olov, **V**, 70, 155, 379, 642, 924; **VII**, 372
Lawler, H. J., **I**, 222
Lawrence R. M., **VIII**, 516
Lawson, Alexander, **IV**, 645; **V**, 1046

Lawson, J. Keith, **IV**, 892
Lazier, W. A., **I**, 99; **II**, 142, 325
Leal, Joseph R., **IV**, 597
LeBel, Norman A., **VI**, 670
Le Coq, A., **VI**, 954
Lednicer, Daniel, **V**, 434, 578, 621
Ledon, Henry J., **VI**, 414
Lee, D. D., **V**, 572
Lee, Ving, **VI**, 613
Lee, Virgil., **VIII**, 512
Leermakers, Peter A., **V**, 297
Leete, J. F., **III**, 428
Leffler, M. T., **II**, 24; **III**, 544, 547
Le Gal, J. Y., **VIII**, 371
Leger, Frank, **II**, 196
LeGras, J., **VI**, 142
Leicester, Henry M., **II**, 238, 240
Leicester, James, **IV**, 525
Leimgruber, W., **VI**, 826; **VII**, 34
Leipold, Hans A., **V**, 80
Leiserson, J. L., **III**, 183
Lemmerman, Leo **V**., **III**, 44
Leonis, M., **VIII**, 77
Leopold, E. J., **VII**, 258
Lespieau, R., **I**, 186, 191, 209
Lesslie, T. E., **IV**, 329
Leston, Gerd., **IV**, 368
LeSuer, William M., **IV**, 919
Letsinger, Robert L., **V**, 450
Leubner, Gerhard W., **IV**, 866
Leuck, G. J., **I**, 283
Levene, P. A., **II**, 5, 88, 288, 320, 545
Levens, Ernest, **IV**, 68
Levin, Nathan, **III**, 191
Levine, R., **III**, 291, 405
Levis, W. W., Jr., **IV**, 307
Levy, Edward F., **III**, 456
Lew, Henry Y., **IV**, 545
Lewis, W. Lee, **I**, 70
Ley, Steven V., **VI**, 104; **VIII**, 71
Lhommet, G., **VIII**, 263
Li, George S., **VI**, 747
Liang, Poe, **III**, 803
Lieber, Eugene, **IV**, 380
Lieberman, S. V., **II**, 441
Liedhegener, Annemarie, **V**, 179; **VI**, 389

Lienhard, U., **VII**, 30
Liese, T., **VIII**, 124, 373
Liewen, Margo Beth, **VI**, 210
Lin, H.-S., **VIII**, 38, 543
Linden, Gustave B., **III**, 456
Lindlar, H., **V**, 880
Lindsay, R. O., **III**, 710
Lindsay Smith, J. R., **VIII**, 167
Linn, W. J., **V**, 1007
Liotta, Charles L., **VI**, 301
Liotta, E. L., **VII**, 77
Lipinski, Martin, **V**, 300
Lipshutz, B. H., **VIII**, 33
Lipsky, Sharon D., **VI**, 537
Lipton, Michael F., **VI**, 492
Liska, John, **V**, 320
Litle, Robert L., **V**, 1055
Little, R. D., **VII**, 56
Liu, H. J., **VI**, 1024
Liu, J., **VIII**, 87
Livinghouse, T., **VII**, 517
Lloyd, Winston D., **V**, 699
Locke, J. Michael, **V**, 723
Lockhard, Luther B., **III**, 463
Lockhart, L. B., Jr., **IV**, 784
Loder, D. J., **II**, 282
Loder, J. W., **IV**, 667
Loeliger, P., **VI**, 776
Loev, Bernard, **V**, 162
Logan, Ted J., **V**, 969
Logullo, Francis M., **V**, 54
Lohaus, G., **VI**, 304, 465
Lombardo, L., **VIII**, 386
Long, F. A., **II**, 87
Longley, R. I., Jr., **IV**, 302, 311, 660, 677
Lorenz, G., **VI**, 866
Lorette, N. B., **V**, 5, 25, 292
Loudon, G. M., **VIII**, 132
Lucas, H. J., **II**, 351; **III**, 355, 482, 483, 485
Lucier, John J., **V**, 736
Ludwig, R., **VIII**, 87
Lufkin, James E., **IV**, 872
Luicke, E., **V**, 533, 808
Luijten, J. G. A., **IV**, 881

Lukaszewski, H., **V**, 656
Lund Hakon, **II**, 594
Lusch, M. J., **VII**, 346
Lutz, Eugene, **V**, 562
Lutz, R. E., **I**, 49; **III**, 248
Lycan, W. H., **II**, 318
Lynch, J. E., **VIII**, 302
Lyster, Mark A., **VI**, 353

Macdonald, J. .E., **VIII**, 192
Macey, William A. T., **IV**, 525
MacKay, D. D., **II**, 580
Mackie, R. K., **V**, 941
Macleod, L. C., **II**, 62
Madan, P., **VIII**, 516
Magerlein, B. J., **V**, 946
Magnani, Arthur, **III**, 251
Mahadevan, A. P., **III**, 341
Mahood, S. A., **II**, 160
Maitte, P., **VIII**, 263
Majerski, Zdenko, **VI**, 958
Makosza, M., **VI**, 897, 940
Mallamo, J. P., **VII**, 495
Malone, T. C., **VIII**, 358
Mallory, F. B., **IV**, 74; **V**, 952
Mancuso, R. W., **V**, 32
Mandolini, L., **VI**, 698
Mansfield, G. H., **IV**, 128
Mansfield, K. T., **V**, 96
Manske, R. H. F., **I**, 404; **II**, 83, 154, 389
Mao, Chung-Ling, **VI**, 245
Marey, R., **IV**, 350, 601
Margaretha, Paul, **VI**, 971
Marguerite, J., **VII**, 131
Mariella, Raymond P., **IV**, 210, 288
Mark, V., **V**, 358, 602, 893, 901
Markham, E., **V**, 785
Marolewski, T. A., **VI**, 974
Marquet, A., **VI**, 175, 193
Marshall, J. L., **V**, 91, 424
Martell, A. E., **V**, 617
Martens, Ted F., **III**, 159
Martin, D., **VII**, 435
Martin, D. G., **IV**, 162
Martin, E. L., **II**, 499, 501, 560, 569; **III**, 465, 633

Martin, J. C., **VI**, 163; **VII**, 506
Martin, R. J. L., **V**, 175
Martin, S. F., **VIII**, 451
Maruoka, K., **VII**, 321, 568
Marvel, C. S., **I**, 25, 42, 48, 84, 94, 95, 99, 117, 128, 170, 177, 224, 238, 248, 252, 289, 327, 347, 358, 366, 377, 411, 435, 504, 531, 533, 536; **II**, 74, 76, 93, 310, 318, 371, 372, 374, 416, 558, 563, 602; **III**, 366, 495, 523, 705, 822, 848
Maryanoff, Bruce E., **VI**, 223, 376
Masamune, S., **VII**, 81, 87
Mascitti, A., **IV**, 788
Mash, E. A., **VIII**, 155
Mashburn, T. Arthur, Jr., **V**, 621
Masuda, T., **VIII**, 251
Matar, A., **IV**, 681
Mathai, I. M., **VI**, 925
Matlack, George M., **IV**, 914
Matsumura, Y., **VII**, 307
Matsushita, H., **VII**, 245
Matthews, A. O., **I**, 392
Maumy, Michel, **VI**, 675
Mauthner, F., **I**, 537
Maxwell, Charles E., **II**, 485; **III**, 305
May, G. E., **II**, 8
Mayberry, Gerald L., **V**, 456
Mayo, Joe, **V**, 166
Mazur, Yehuda, **VI**, 43
McAdams, Louis V., III, **VI**, 403, 555
McBee, E. T., **V**, 663
McCaleb, Kirtland E., **IV**, 281
McCann, S. F., **VIII**, 609
McCarthy, W. C., **IV**, 918
McCloskey, Chester M., **III**, 221, 434, 712, 668
McColm, E. M., **I**, 536
McCombs, Charles A., **VI**, 445
McCormack, W. B., **V**, 787
McCracken, J. H., **III**, 258
McCutcheon, J. W., **III**, 526, 531
McDermott, F. A., **II**, 365
McDonald, Richard N., **V**, 499, 985; **VI**, 276, 461

McElvain, S. M., **I**, 170, 385, 473; **II**, 114, 214, 272, 279, 419; **III**, 123, 251, 379, 490, 493, 506, 591; **IV**, 463, 662, 664
McEwen, W. L., **I**, 521; **II**, 53, 133; **III**, 227
McEwen, William E., **V**, 1001
McFarland, J. W., **IV**, 940
McGee, John J., **V**, 801
McGuire, M. A., **VIII**, 216
McIntosh, A. V., **III**, 234, 237
McKearin, J. M., **VII**, 501
McKee, G. H. W., **II**, 15
McKee, R. A., **II**, 100
McKeer, L. C., **VIII**, 167
McKennis, J. S., **VI**, 1002
McKenzie, B. F., **I**, 21, 131, 246, 256, 335
McKeon, Thomas F., Jr., **IV**, 683
McKillip, W., **V**, 758
McKillop, Alexander, **VI**, 36, 488, 549, 709, 791
McKusick, B. C., **IV**, 438, 746, 953
McLachlin, D. M., **III**, 28
McLaughlin, Keith C., **IV**, 824
McLaughlin, M. L., **VIII**, 223
McLaughlin, Thomas G., **VI**, 1028
McLeod, Donald J., **IV**, 510, 555, 635
McLeod, Gerald L., **IV**, 345
McLoughlin, J. I., **VIII**, 391
McMahon, Robert E., **IV**, 457
McMurry, John E., **VI**, 592, 648, 781, 799; **VII**, 1
McOmie, J. F. W., **V**, 412
McPhee, Warren D., **III**, 119
McRae, J. A., **IV**, 393
McShane, H, F., **III**, 197
McWherter, P. W., **I**, 245
McWhirter, J., **III**, 714
Medwid, J. B., **VI**, 101
Meek, John S., **V**, 126
Meerwein, H., **V**, 1080, 1096
Meienhofer, Johannes, **VI**, 252
Meinert, R. N., **II**, 541
Meinhardt, K.-P., **VII**, 15

Meinwald, J., **IV**, 738; **V**, 863, 866; **VI**, 840
Mekler, A. B., **V**, 743
Melby, L. R., **IV**, 276, 953; **V**, 239
Melton, Jack, **VI**, 648
Mendenhall, G. David, **V**, 829
Merenyi, Ferenc, **VI**, 28
Merritt, Charles, Jr., **IV**, 8, 711
Meth-Cohn, O., **VIII**, 350
Meuwsen, A., **V**, 43
Meyer, G. M., **II**, 288
Meyer, J. D., **II**, 468
Meyer, Karl, **III**, 637
Meyer, Kurt H., **I**, 60, 390
Meyers, A. I., **VI**, 64, 371, 905; **VII**, 530; **VIII**, 204, 241, 573
Meyers, Cal Y., **IV**, 39
Meyr, Rudolf, **V**, 300, 1060
Meystre, Ch., **V**, 692
Michaels, R. J., Jr., **IV**, 915
Michelotti, Franeis W., **IV**, 29
Mickel, S. J., **VIII**, 3
Mickey, Sally, **III**, 343
Micovic, V. M., **II**, 264
Middleton, W. J., **IV**, 243; **V**, 1013; **VI**, 161, 440, 664, 835; **VII**, 528
Midland, M. M., **VII**, 402; **VIII**, 391
Mieles, L. R., **VIII**, 474
Mierzwa, Sigmund A., Jr., **V**, 874
Mijs, W. J., **VI**, 412
Mikol, Cerard J., **V**, 937
Mikulec, Richard A., **IV**, 13
Milas, Nicholas A., **II**, 302
Miles, Marion L., **V**, 718, 721
Milewich, L., **VI**, 690
Milewski, Cynthia A., **VI**, 376
Milkowski, John D., **VI**, 5
Millar, I. T., **V**, 1120
Millard, Alan A., **VI**, 943
Miller, Charles S., **III**, 646
Miller, Ellis, **II**, 543
Miller, G. E., **II**, 576
Miller, Leonard E., **IV**, 39
Miller, M. J., **VIII**, 3
Miller, S. A., **VIII**, 556

Miller, S. R., **VIII**, 152
Miller, Sidney E., **IV**, 141
Miller, Sidney I., **V**, 921; **VI**, 925
Millet, W. H., **II**, 65
Mimura, S., **VIII**, 251
Miner Laboratories, **I**, 285
Minin, Ronald, **IV**, 182, 336
Minnis, Wesley, **II**, 357, 520
Minns, Richard A., **VI**, 1037
Minoli, G., **VII**, 23
Mintz, M. J., **V**, 184
Mirrington, R. N., **VI**, 859
Misco, P. F., **VI**, 220
Missbach, M., **VII**, 332
Mitchell, M. J., **V**, 944
Miyashita, M., **VII**, 396, 414
Miyaura, N., **VIII**, 532
Moffatt, John G., **V**, 242
Moffett, E. W., **II**, 121
Moffett, Robert Bruce, **III**, 234, 237, 562; **IV**, 238, 291, 354, 357, 427, 466, 605, 652, 834
Mohacsi, E., **VI**, 826
Möllering, H., **I**, 499
Monagle, J. J., **V**, 501
Montanari, F., **VII**, 367
Montes, de Lopez-Cepero, **VIII**, 1, 474
Montgomery, S. H., **VII**, 190
Moodie, I. M., **VI**, 558, 979
Mook, R., Jr., **VII**, 381
Moore, A. T., **V**, 586
Moore, H. W., **VI**, 210; **VII**, 116, 179
Moore, James A., **V**, 351
Moore, L. L., **V**, 687
Moore, Maurice, L., **III**, 599, 617
Moreau, C., **I**, 166
Morera, E., **VIII**, 126
Moretti, R., **VIII**, 33
Mori, H., **VIII**, 312
Mori, K., **VIII**, 312
Mori, S., **VIII**, 612
Moriarty, R. M., **VII**, 263; **VIII**, 43
Moriconi, Emil J., **V**, 623
Morris, Don L., **VI**, 932
Morrison, W. H., **III**, **VI**, 751

Mortensen, Harley E., **V**, 562
Morton, A. A., **III**, 831
Morton, G. H., **VII**, 315
Mosher, Harry S., **III**, 568; **IV**, 828
Mowry, David T., **IV**, 486
Moyer, W. W., **II**, 602
Moyle, M., **IV**, 205, 493, 700; **V**, 339
Mozingo, Ralph, **III**, 181, 258, 387, 576, 685
Muchmore, D. C., **VI**, 762
Mukaiyama, T., **VIII**, 323
Müller, Adolf, **II**, 535
Müller, Robert K., **VI**, 56
Mullican, M. D., **VIII**, 444
Munch, J. H., **VIII**, 80
Munch-Petersen, Jon, **IV**, 715; **V**, 762
Munsterer, H., **VIII**, 427
Murahashi, S.-L., **VII**, 172
Murakami, M., **VI**, 307
Murphy, Donald M., **IV**, 872
Murphy, M., **V**, 918
Murphy, William S., **V**, 523
Murray, Joseph I., **IV**, 744
Murray, Robert W., **V**, 166
Murray, T. F., **I**, 85, 272, 523
Musliner, Walter J., **VI**, 150
Musmanni, S., **VIII**, 226
Musser, John H., **VI**, 799
Muth, K., **V**, 944
Myers, R. R., **II**, 462
Myles, D. C., **VIII**, 13
Myles, J. R., **III**, 831

Nagashima, H., **VII**, 137
Nagata, Wataru, **VI**, 14, 307, 358, 436, 448
Nagy, S. M., **III**, 690
Nakagawa, M., **VI**, 462
Nakai, S., **VIII**, 568
Nakamura, E., **VII**, 512; **VIII**, 277, 578
Nakamura, T., **VIII**, 274
Nakane, M., **VII**, 323
Nakatsuka, Masashi, **VI**, 327
Nanninga, T. N., **VII**, 266
Naota, T., **VII**, 172
Narasaka, K., **VIII**, 210, 323

Narayanan, B. S., **VIII**, 602
Naruse, Norio, **V**, 266
Nasipuri, D., **V**, 384
Näsman, J. H., **VIII**, 396
Nassim, B., **VII**, 276
Natale, Nicholas R., **VI**, 293
Natelson, Samuel, **III**, 381
Nathan, E. C., **VI**, 991
Naylor, C. A., Jr., **II**, 523, 526
Neely, T. A., **V**, 1116
Neeman, M., **V**, 245
Negishi, E.-I., **VII**, 63, 245, 427; **VIII**, 295, 430
Neher, H. T., **I**, 56
Neises, B., **VII**, 93
Nelke, Janice M., **VI**, 167
Nelson, G. B., **III**, 392
Nelson, K. A., **VIII**, 155
Nemoto, H., **VII**, 137
Nenitzescu, Costin D., **II**, 496; **V**, 1106
Nesmajanow, A. N., **II**, 381, 432
Neubert, M. E., **VII**, 420,
Neuenschwander, M., **VII**, 30
Neuman, Fred W., **III**, 73, 519
Neumeyer, John L., **VI**, 115
Neveu, M., **VIII**, 309
Neville; Roy G., **V**, 801
Newkirk, D. M., **III**, 538
Newkome, G. R., **VI**, 12
Newman, Melvin S., **II**, 428; **III**, 502, 631; **IV**, 484; **V**, 486, 743, 1024, 1029, 1031; **VI**, 564, 613, 824
Newton, L. W., **III**, 313
Nicholas, K. M., 460, **VIII**
Nicholas, R. D., **V**, 16
Nichols, Gust, **III**, 159, 712
Nicolet, B. H., **I**, 410
Nielsen, Arnold T., **V**, 288
Nielsen, D. R., **V**, 414
Niemann, Carl, **III**, 11
Nierenstein, M., **II**, 557
Nierth, A., **VI**, 324
Nieuwland, J. A., **I**, 229
Nightingale, Dorothy, **III**, 42
Nilsson, Martin, **VI**, 28
Nishimura, Tamio, **IV**, 713

Nissen, B. H., **I**, 1
Nix, M., **VIII**, 148
Niznik, G. E., **VI**, 751
Noble, Paul, Jr., **IV**, 924
Noland, Wayland E., **V**, 567, 833;
    VI, 976
Noller, C. R., **I**, 109; **II**, 109, 184, 258,
    358, 478, 586; **III**, 835; **IV**, 545,
    571, 603
Nomura, G. S., **VII**, 346
Nonhebel, D. C., **V**, 206
Normant, J. F., **VII**, 290,
Normington, J. B., **I**, 205
Norris, James F., **I**, 144, 430, 548
Norton, A. J., **I**, 490
Norton, D. G., **IV**, 348
Norton, Richard **V.**, **V**, 709
Nowak, Robert M., **IV**, 281
Noyes, W. A., **I**, 457; **II**, 108
Noyori, R., **VI**, 520; **VIII**, 57
Nubling, C., **VI**, 799
Nugent, W. A., **VIII**, 112
Nyström, J. E,. **VIII**, 9

Oakwood, T. S., **III**, 112, 114
Obi, M., **VI**, 462
Ochiai, H., **VIII**, 274
O'Connell, E. J., Jr., **VI**, 401
Odom, H. C., **VI**, 883
O'Donnell, D. J., **VII**, 210
Oehler, Rene, **II**, 276; **III**, 401
Oertle, K., **VII**, 470
Oettle, W. F., **VI**, 652
Ogata, Yoshiro, **VI**, 90
Ogliaruso, Michael, **V**, 111
Ohme, Roland, **V**, 897; **VI**, 78
Ohno, Masaji, **V**, 266, 281
Oikawa, Y., **VII**, 359
Ojima, I., **VII**, 417
Okamoto, Masaru, **V**, 281
Okano, K., **VIII**, 501
Okuda, Takuo, **IV**, 566
Olah, G. A., **V**, 480; **VI**, 628; **VII**,
    254, 451
Olander, C. P., **I**, 447
Oldenziel, O. H., **VI**, 41, 987

Oliver, James E., **VI**, 424
Oliver, John J., **II**, 391, 487
Oliveto, Eugene P., **IV**, 104
Olmstead, M. P., **VI**, 913
Olmsted, Alanson W., **I**, 144
Olofson, Roy A., **VI**, 263, 571
Olomucki, M., **VIII**, 371
Olsen, Robert E., **V**, 541
Olson, Cecil E., **IV**, 898
O'Neil, T. H., **VII**, 66
Opie, J. W., **III**, 56
Opitz, G., **V**, 877
Oppenauer, R. V., **III**, 207
Orchin, Milton, **III**, 837
Orfanopoulos, M., **VII**, 393
Ort, O., **VIII**, 522
Ortar, G., **VIII**, 126
Orten, James M., **I**, 300
Orwoll, E. F., **III**, 130, 566
Osterberg, A. E., **I**, 185, 271
Osuga, David T., **V**, 126
Otsuka, S., **VIII**, 183
Ott, Erwin, **II**, 528
Ottenheijm, H. C. J., **VII**, 467
Otterbacher, T. J., **I**, 290; **II**, 317
Overberger, C. G., **III**, 200, 204; **IV**, 29,
    66, 182, 273, 274, 336
Overman, Larry E., **VI**, 95, 507; **VIII**,
    358, 609
Owsley, D. C., **VI**, 916
Oxley, P. W., **VIII**, 16

Pacaud, R. A., **II**, 336
Pachter, I. J., **VI**, 590, 1014, 1016
Padwa, Albert, **VI**, 893, **VII**, 203;
    VIII, 231
Pagano, A. S., **V**, 191, 367
Page, G. A., **IV**, 136
Paleveda, W. J., **VII**, 75
Palmer, Albert, **II**, 57
Palmer, C. S., **I**, 73, 75, 245
Panek, J. S., **VIII**, 597
Panella, J. P., **V**, 614
Panetta, C. A., **VI**, 1004
Papa, Domenick, **III**, 742
Pappalardo, Joseph A., **IV**, 186, 558

Paquette, Leo A., **V**, 215, 231, 408, 467; **VI**, 482; **VII**, 200, 453; **VIII**, 38, 223, 281, 298, 467, 543
Parham, William E., **III**, 395; **IV**, 295; **V**, 874
Paris, G., **IV**, 496
Park, Chung Ho, **VI**, 382
Parris, Chester L., **V**, 73
Parrish, D. R., **VII**, 363
Parshall, G. W., **V**, 1016
Parsons, P. J., **VIII**, 458
Partridge, J. J., **VII**, 339
Partridge, M. W., **IV**, 769
Paskins, K. N., **VI**, 181
Patrick, Tracy M., Jr., **IV**, 430, 980
Patsch, M. R., **V**, 859
Patterson, John M., **V**, 585
Patterson, J. W., **VIII**, 415
Patterson, L. A., **III**, 576
Pattison, F. L. M., **V**, 136, 580
Paufler, Robert M., **V**, 982
Paul, Raymond, **VI**, 675
Pauley, D., **VIII**, 208, 309
Paust, J., **V**, 859
Payne, George B., **V**, 805
Pearce, P. J., **VI**, 240
Pearl, Irwin A., **III**, 745; **IV**, 972, 974
Pearlman, W. M., **V**, 670
Pearson, David L., **IV**, 221
Pearson, D. E., **III**, 154; **V**, 117, 332, 890
Pecoraro, J., **VI**, 361
Pedersen, Charles J., **VI**, 395
Pedersen, Knud, **V**, 70
Pelter, M. W., **VIII**, 298
Penton, Harold R., Jr., **VI**, 788
Pericas, M. A., **VIII**, 161
Perkins, Edward G., **IV**, 444
Perkins, Matilda, **VI**, 278
Perkins, M. J., **V**, 51
Perni, R. B., **VIII**, 192
Perozzi, E. F., **VI**, 163
Perri, S. T., **VIII**, 179
Perry, B. J., **IV**, 325, 955
Perst, H., **V**, 1130
Peter, Hugo H., **IV**, 887
Peters, R. H., **VII**, 386

Petersen, James M., **V**, 909
Petersen, R. J., **V**, 929
Peterson, C. J., **V**, 758
Peterson, P. E., **VI**, 273
Peterson, W. D., **III**, 106, 242, 660
Pettit, D. J., **V**, 1099
Pettit, R., **VI**, 310, 422, 1002
Petty, C. Bruce, **VI**, 95
Phillipi, E., **II**, 551
Phillips, Donald D., **IV**, 313
Phillips, G. W., **VIII**, 451
Phillips, Ross, **II**, 232, 464, 578; **III**, 337
Phillips, W. V., **VII**, 346,
Photis, James M., **VI**, 482
Piasek, Edmund J., **V**, 80
Pickard, P. L., **V**, 520
Pier, Stanley M., **IV**, 554
Pieter, R., **VI**, 542
Pigon, Foster, **VI**, 1010
Pike, Arthur B., **II**, 284, 487
Pilgrim, Frederick J., **IV**, 451
Pinder, A. R., **VI**, 883
Pine, S. H., **VIII**, 512
Pinhey, J. T., **VII**, 229
Pinkney, P. S., **II**, 116; **III**, 401
Pino, P., **VI**, 338
Pinschmidt, Robert K., Jr., **VI**, 145
Pirrung, M. C., **VII**, 190
Ploss, G., **V**, 431
Pluygers, C. W., **V**, 223
Poindexter, Graham S., **VI**, 845
Politzer, Ieva R., **VI**, 905
Pope, Barry M., **VI**, 418
Pope, H. W., **V**, 117
Porter, H. D., **III**, 660, 708
Porter, P. K., **I**, 377, 411, 457
Posner, Gary H., **VI**, 248; **VII**, 495; **VIII**, 562
Posner, J., **V**, 514
Potts, K. T., **V**, 769; **VII**, 476
Poulter, C. D., **VIII**, 616
Poupart, M.-A., **VIII**, 467
Powell, J. Roy, **IV**, 213, 934
Powell Garfield, **II**, 449
Powell, S. G., **I**, 168
Prakash, I., **VII**, 263

Prakash, O., **VIII**, 43
Preuschhof, Helmut, **VI**, 78
Price, Charles C., **III**, 86, 174, 272, 649, 664, 667; **IV**, 186, 558, 566, 683; **V**, 255
Price, John A., **IV**, 285
Price, R., **V**, 785, 1022
Prichard, W. W., **III**, 302, 452
Prill, E. A., **II**, 419
Proctor, G. R., **V**, 545
Prosser, Thomas, **IV**, 169
Prout, Franklin S., **III**, 601
Prout, F. S., **IV**, 93; **V**, 572
Pulwer, M. J., **VII**, 203
Puntambeker, S. V., **I**, 524; **II**, 318
Purchase, Earl R., **III**, 430
Puterbaugh, Milton, **I**, 388
Putman, R. E., **V**, 235

Quast, H., **V**, 1018
Quayle, O. R., **I**, 211, 292, 294
Quici, S., **VIII**, 367
Quillinan, A. J., **VI**, 595

Raaen, Vernon, **IV**, 130
Raasch, M. S., **V**, 393
Raber, Douglas J., **VI**, 576
Rabjohn, Norman, **III**, 375, 557; **IV**, 441, 513
Rachlin, A. I., **VI**, 1007
Raecke, Bernhard, **V**, 813
Ragnarsson, Ulf, **VI**, 203
Raha, Chittaranjan, **IV**, 263
Rahrs, E. J., **I**, 91; **III**, 650, 652
Rainey, James L., **III**, 17
Rajagopalan, S., **III**, 425; **V**, 139, 504
Ralli, P., **VII**, 476
Ralls, J. O., **II**, 191
Ramachrandran, S., **V**, 486, 743
Ramalingam, K., **VII**, 210
Raman, P. V. A., **III**, 425
Ramarajan, K., **VII**, 210
Rambaud, M., **VIII**, 265
Rand, L., **V**, 663
Raneburger, J., **VII**, 8
Ranu, B. C., **VIII**, 309

Rao, C. N. Ramachandra, **IV**, 380
Rapala, Richard J., **III**, 820
Rappe, C., **VI**, 711
Rasmussen, J. K., **VII**, 521
Rasmussen, J. R., **VII**, 139
Rasmusson, Gary H., **V**, 324
Rassweiler, C. F., **I**, 10
Ratchford, W. P., **III**, 30
Ratcliffe, R. W., **VI**, 373
Rathke, Michael W., **VI**, 598, 943
Ratts, K. W., **V**, 361
Raube, Richard, **IV**, 288
Rauhut, M. M., **IV**, 114
Ray, F. E., **III**, 23, 420
Ray, Stephen J., **VI**, 36
Raymond, P., **VI**, 392
Raznikiewicz, T., **V**, 149
Read, R. R., **I**, 321, 514; **III**, 157, 444
Redemann, Carl T., **III**, 69, 656
Redemann, C. Ernst, **III**, 11, 59, 69, 244, 644, 656, 63
Redmond, W., **V**, 133
Reed, Donald E., **V**, 351
Reed, L. J., **III**, 395
Reetz, M. T., **VII**, 424
Reeve, Wilkins, **III**, 693
Reeves, Perry C., **VI**, 625
Regenye, R. W., **VII**, 461; **VIII**, 258
Regitz, Manfred, **V**, 179; **VI**, 389
Rehberg, Chessie E., **III**, 33, 46, 146
Reich, Hans J., **VI**, 23, 533
Reichardt, C., **V**, 1112, 1123, 1130
Reichert, J. S., **I**, 229
Reichle, Walter R., **V**, 1024, 1055
Reid, E. Emmet, **II**, 246, 338, 468, 474; **IV**, 47
Reif, Donald J., **IV**, 375, 860
Reifschneider, Walter, **V**, 107
Rein, T., **VIII**, 9, 137
Reineke, Charles E., **VI**, 461
Reisner, D. B., **IV**, 144
Renaldo, A. F., **VIII**, 268
Renfrew, W. B., Jr., **II**, 67, 268, 607
Renga, James M., **VI**, 23
Renger, B., **VI**, 542
Renoll, Mary, **III**, 502

Rens, M., **VI**, 282
Reverdin, Frederic, **I**, 219
Reynolds, Delbert, **III**, 432
Reynolds, George A., **III**, 374, 593; **IV**, 15, 633, 708
Reynolds, H. H., **I**, 476
Rhinesmith, Herbert S., **II**, 177
Rice, F. O., **III**, 244
Rice, P., **VIII**, 179
Rice, Rip C., **IV**, 283
Richards, D. H., **VI**, 240
Richman, J. E., **VI**, 652
Richter, Henry J., **IV**, 482
Rickborn, Bruce, **VI**, 312
Rideout, W. H., **V**, 414
Riebsomer, J. L., **III**, 326
Rieche, A., **V**, 49, 365
Riegel, Byron, **III**, 234, 237
Riegel, E. Raymond, **II**, 126
Rieke, Reuben D., **VI**, 845
Riener, Edward E., **IV**, 451
Riera, A., **VIII**, 161
Rietz, Edward, **III**, 851
Rieveschl, George, Jr., **III**, 23, 420
Rifi, M. R., **VI**, 153
Rigby, H. L., **VIII**, 309
Riley, H. A., **II**, 509
Rinderknecht, Heinrich, **V**, 822
Rinehart, Kenneth L., Jr., **IV**, 120, 444
Ringold, Howard J., **IV**, 221
Rinkes, I. J., **II**, 393
Riobe, Olivier, **VI**, 675
Ritchie, Bruce, **III**, 53
Ritter, E. J., **IV**, 307
Ritter, John J., **V**, 471
Robb, Charles M., **V**, 635
Robb, C. M., **III**, 347
Robbins, Frederick, **V**, 341
Robert-Lopes, M.-T., **VII**, 276
Roberts, B. W., **VII**, 142
Roberts, John D., **IV**, 457
Roberts, K. C., **I**, 235
Roberts, R., **III**, 244
Roberts, Royston M., **III**, 272; **IV**, 420, 464
Robertson, G. Ross, **I**, 52, 138, 396, 406

Robey, Roger L., **VI**, 791
Robinson, D. B., **III**, 103
Robinson, J., **II**, 47
Robinson, John C. Jr., **III**, 717, 720
Robinson, J. M., **VII**, 319
Robison, Bonnie L., **IV**, 947
Robison, Michael M., **IV**, 947
Robson, Jean H., **III**, 456
Rocek, J., **VII**, 114
Rodewald, P. G., **VII**, 400,
Rodgers, W. B., **VIII**, 586
Roebuck, Alan, **III**, 217
Rogers, Crosby U., **IV**, 884
Rogers, Richard J., **III**, 174
Rojas, A. C., **VI**, 786
Roll, L. J., **II**, 140, 266, 368, 382, 418, 460; **III**, 20
Rolla, F., **VI**, 833
Romanelli, Michael G., **V**, 552
Rondestvedt, Christian S., Jr., **IV**, 766, 846
Ronzio, Anthony R., **III**, 71, 438
Roos, A. T., **I**, 145
Roos, Jan, **VI**, 95
Ropp, Gus A., **IV**, 130, 727
Rorig, Kurt, **IV**, 576
Rosen, Milton J., **IV**, 665
Rosen, T., **VII**, 226
Rosenblum, M., **VIII**, 479
Rosendahl, Friedrich, **V**, 300
Rothenberger, S. D., **VIII**, 87
Rothrock, H. S., **II**, 408
Rouessac, F. P., **VII**, 99
Rousseau, Joseph E., **IV**, 711
Rowland, S. P., **III**, 549
Rubottom, G. M., **VI**, 106; **VII**, 282
Ruby, Phillip R., **IV**, 786, 798
Ruddy, Wayne A., **III**, 665
Ruffner, Robert J., **VI**, 737
Ruhoff, John R., **II**, 102, 246, 292, 315, 338, 468, 474
Russell, Alfred, **III**, 157, 282, 293, 463, 747
Russell, Glen A., **V**, 937
Rutenberg, M. W., **IV**, 620
Ruter, Jorn, **VI**, 389

Rutherford, K. G., **V**, 133
Rutledge, P. S., **VI**, 348
Rydon, H. N., **V**, 586; **VI**, 830
Ryerson, George D., **IV**, 957
Ryskiewicz, Edward E., **IV**, 831

Sabadie, Jean, **VI**, 856
Saegusa, J., **VI**, 462
Saegusa, Takeo, **VI**, 327
Sager, W. F., **V**, 595, 1011
Saha, M., **VIII**, 460
Sahyun, Melville R. V., **V**, 926
Sakane, S., **VIII**, 321
Sakurai, H., **VII**, 443
Salaun, J. R., **VI**, 320; **VII**, 117, 131
Saliani, N. A., **II**, 69
Salomon, R. G., **VII**, 177
Saltman, H., **V**, 658, 660, 665
Salzberg, Paul L., **I**, 46, 119
Sample, Thomas E., Jr., **VI**, 454
Samuel, O., **VIII**, 464
Sanchez, R. A., **V**, 32, 344
Sand, P., **VIII**, 493
Sandberg, Bengt E., **VI**, 203
Sandberg, M., **II**, 49
Sandberg, Rune, **V**, 155
Sandborn, L. T., **I**, 111, 133, 340, 526
Sanderson, James J., **III**, 254, 256
Sanderson, T. F., **IV**, 1
Sandin, R. B., **II**, 100, 196, 604
Sandler, Stanley R., **VI**, 187
Sandri, Joseph M., **IV**, 278
Sands, Lela, **I**, 67
Sankaran, D. K., **III**, 623
Santilli, Arthur A., **V**, 166
Santurri, Pasco, **V**, 341
Sapuppo, N., **VI**, 334
Sarda, P., **VI**, 769
Sasse, M. J., **VIII**, 16
Sasse, W. H. F., **V**, 102
Sato, F., **VIII**, 507
Sato, T., **VIII**, 441, 498
Sauer, J. C., **III**, 605; **IV**, 268, 560, 801, 813
Sauer, Robert F., **V**, 145
Saunders, J. H., **III**, 22, 200, 204, 416

Sauter, Frederick J., **V**, 258
Sauter, H., **VI**, 427
Savariar, S., **VIII**, 82
Sawyer, R. L., **III**, 276
Saxton, J. E., **V**, 769
Saxton, M. R., **IV**, 78
Sayre, J. L., **II**, 596
Scanlan, John T., **IV**, 317
Scarrow, J. A., **II**, 387
Schaefer, G. F., **IV**, 31
Schaefer, J. P., **V**, 142, 249, 285
Schaefer, W., **I**, 147
Schaeffer, J. R., **IV**, 19
Schafer, W., **I**, 364
Schaffner, P. V. L., **II**, 160
Scheibye, S. **VII**, 372,
Scheifele, H. J., Jr., **IV**, 34
Scheinbaum, Monte L., **VI**, 590, 1014, 1016
Scheinmann, F., **VI**, 595
Schenk, Erwin, **III**, 742
Schiemann, G, **II**, 188, 299,
Schiemenz, G. P., **V**, 496
Schiess, M., **VIII**, 201
Schiessler, R. W., **V**, 647
Schildneck, P. R., **II**, 236, 411
Schipp, Hubert, **V**, 813
Schlatter, M. J., **III**, 223
Schleyer, Paul V. R., **V**, 16, 852; **VI**, 378
Schloss, E. L., Jr., **II**, 457
Schmidt, A. C., **II**, 266
Schmit, C., **VIII**, 306
Schmitz, Ernst, **V**, 897
Schmitz, William R., **III**, 328
Schmutzler, R., **V**, 218, 1005
Schneider, H. J., **IV**, 98
Schneider, William P., **IV**, 132
Schoellkopf, U., **V**, 751, 859
Schoen, Kurt L., **IV**, 623
Scholz, K.-H., **VII**, 4
Schreiber, Jakob, **VI**, 56
Schreiber, S. L., **VII**, 168
Schroeder, E. F., **II**, 17, 137, 305, 307
Schroeder, Wesley, **II**, 574
Schuda, P. F., **VII**, 312
Schultz, Everett M., **III**, 343

Schultz, Harry P., **IV**, 364; **V**, 589

Schultze, L. M., **VIII**, 216

Schulze, F., **II**, 425

Schurig, V., **VIII**, 119, 434

Schurink, H. B. J., **I**, 425; **II**, 476

Schuster, R. E., **V**, 772

Schwartz, A., **VIII**, 516

Schwartz, George, **III**, 332

Schwartzman, Louis H., **IV**, 471

Schwarz, R., **IV**, 203

Schwarz, William W., **V**, 1116

Schweizer, Edward E., **V**, 874, 1145

Scilly, N. F., **VI**, 240,

Scott, L. T., **VIII**, 196

Scott, James E., **V**, 772

Scott, R. W., **I**, 179; **II**, 379

Scott, W. J., **II**, 128, 539

Scott, W. J., **VIII**, 97

Searle, N. E., **IV**, 424

Searles, Scott, **V**, 562

Sedergran, T. C., **VII**, 491

Sedlak, J., **IV**, 108

Seebach, D., **VI**, 316, 542, 556, 869; **VII**, 41, 153, 215, 447; **VIII**, 201, 332, 495

Segur, J. B., **I**, 372, 391

Seidl, Helmit, **V**, 957, 1124

Seidler, P. F., **VIII**, 505

Seifert, Wolfgang K., **VI**, 837

Seikel, Margaret K., **III**, 262

Seitz, Arnold H., **V**, 54

Sekera, V. C., **III**, 366

Sellas, J. T., **IV**, 702

Selman, L. H., **VI**, 744

Semmelhack, Martin F., **VI**, 722

Semon, W. L., **I**, 318; **II**, 204

Serratosa, F., **VIII**, 161

Seth, S. R., **II**, 522

Seven, Raymond P., **III**, 499

Seyed-Mahdavi, F., **VIII**, 373

Seyferth, Dietmar, **IV**, 258; **V**, 452; **VIII**, 343

Seymour, Dexter, **III**, 456

Shah, N. M., **III**, 280; **IV**, 836

Shama, S. A., **VII**, 256

Shapiro, Robert H., **VI**, 172

Shapiro, S. H., **II**, 531

Sharefkin, J. G., **V**, 658, 660, 665

Sharkey, W. H., **V**, 235, 445, 734

Sharpless, K. B., **VII**, 121, 223, 461

Shaw, A. C., **VIII**, 146

Shaw, B. L., **IV**, 404

Shechter, Harold, **IV**, 321, 323, 543; **V**, 328

Sheehan, John C., **V**, 555

Sheehan, J. T., **III**, 541

Sheibley, F. E., **II**, 79

Shemin, D., **II**, 1, 11, 330, 491, 519

Shen, Chah M., **VI**, 218, 951

Sheng, M. N., **VI**, 501

Shenoy, P. K., **V**, 142, 249

Shepard, Inella G., **V**, 839

Sheppard, A. C., **VIII**, 104

Sheppard, O. E., **II**, 440, 617; **III**, 37

Sheppard, William A., **V**, 396, 843, 959, 1085; **VI**, 875

Sher, P. M., **VIII**, 381

Sherman, Wm. R., **IV**, 247

Shibe, W. J., **III**, 260

Shields, J. E., **V**, 406, 1064

Shildneck, P. R., **II**, 345

Shioiri, T., **VII**, 206; **VIII**, 612

Shoffner, J. P., **V**, 572

Shono, T., **VII**, 307

Short, William A., **IV**, 706

Shriner, R. L., **I**, 101, 463; **II**, 140, 200, 266, 366, 368, 382, 453, 459, 538, 560; **III**, 73, 519, 562, 725, 737, 786; **IV**, 242, 786, 798, 910

Shrivers, J. C., **III**, 142

Shull, David W., **VI**, 737

Sidani, A. R., **VI**, 282

Sieburth, S. McN., **VII**, 456

Siegel, Elaine, **V**, 904

Sieloff, R. F., **VII**, 346

Siggins, J. E., **VI**, 529

Silloway, H. L., **III**, 82

Silverman, S. B., **VII**, 393

Silverstein, Robert M., **IV**, 831; **V**, 390

Simmons, Howard E., **V**, 855; **VI**, 382

Simms, N., **VIII**, 479

Simons, J. K., **IV**, 78

Simril, V. L., **III**, 157
Sims, Homer, **V**, 608
Sims, James J., **VI**, 744
Singer, Alvin, **II**, 214
Singh, Pritpal, **VI**, 82
Singh, R. K., **VII**, 411
Sisti, A. J., **V**, 46
Sjoberg, B., **VI**, 468
Sjostedt, George, **III**, 95
Skattebøl, Lars, **IV**, 792; **V**, 306
Skell, P. S., **V**, 929
Skelly, W. G., **IV**, 154
Skorcz, J. A., **V**, 263
Skulnick, H. I., **VII**, 213
Smart, B. W., **IV**, 55
Smiles, S., **I**, 7, 8
Smith, A. B., **VII**, 111, 271
Smith, Claibourne D., **VI**, 962
Smith, C. W., **IV**, 348
Smith, Douglas S., **IV**, 377
Smith, G. A., **VII**, 530
Smith, G. C., **VIII**, 1
Smith, Homer A., **VI**, 51
Smith, K., **VII**, 185
Smith, L. A., **II**, 242, 438
Smith, Lee Irvin, **II**, 95, 248, 254, 360; **III**, 56, 151, 302, 350, 673
Smith, L. H., **III**, 793
Smith, Lowell R., **V**, 204
Smith, Newton R., **IV**, 201, 337, 549, 731
Smith, P. A. S., **IV**, 75, 819
Smith, Paul V., **III**, 410, 735
Smith, Peter A. S., **V**, 612, 829, 1051
Smith, R. D., **V**, 855
Smith, Roger A., **VI**, 130
Smith, Ronald Dean, **IV**, 218
Smith, Walter T., Jr., **IV**, 345, 702
Smolin, Edwin M., **III**, 715; **IV**, 387
Snell, J. M.; **III**, 788
Snell, John M., **II**, 114
Snider, Theodore E., **VI**, 932
Snoddy, A. O., **IV**, 19, 529
Snyder, H. R., **II**, 171, 333, 489, 531; **III**, 471, 698, 717, 20, 798; **IV**, 539, 638
Soderquist, J. A., **VIII**, 19
Soffer, M. D., **IV**, 903

Soine, T. O., **IV**, 106
Sokol, Phillip E., **V**, 706
Sollazzo, R., **III**, 371
Solomon, S., **V**, 306
Solomonika, Eugen, **II**, 496
Somerville, L. F., **II**, 81
Sommer, Nolan B., **III**, 519
Sommers, Armiger H., **III**, 148, 446
Sondheimer, F., **VI**, 68
Sonneborn, H., **I**, 90
Sonnet, Philip E., **VI**, 424
Sorrell, Thomas N., **VI**, 807
Spangler, F. W., **III**, 203, 377, 621; **IV**, 433
Spartberg, M. S., **II**, 558
Spassow, A., **III**, 144, 390
Speer, John H., **III**, 39
Sperry, W. M., **I**, 95
Spessard, G. O., **VII**, 81, 87
Speziale, A. John, **IV**, 401; **V**, 204, 361, 387
Spillane, Leo, Jr., **III**, 302
Spitzner, D., **VIII**, 219
Sporikou, C. N., **VIII**, 606
Sprinzak, Yair, **IV**, 91
Srivastava, K. C., **VI**, 932
Sroog, C. E., **IV**, 271
Staab, Heinz A., **V**, 201
Stacy, Gardner W., **III**, 86, 667; **IV**, 13
Stadler, P. A., **VI**, 109
Stadlwieser, J., **VII**, 8
Stalick, W. M., **VI**, 564
Stamos, I. K., **VII**, 319
Stampfli, J. G., **II**, 175
Stang, Peter J., **VI**, 757; **VII**, 144
Starkey, E. B., **II**, 225; **III**, 665
Starr, Donald, **II**, 566, 571
Staskun, B., **VI**, 631
Staudinger, H., **II**, 573
Staunton, J., **V**, 8
Stearns, James A., **III**, 833
Steglich, W., **VII**, 93
Steiger, Robert E., **III**, 66, 84, 88, 91
Stein, Allan R., **V**, 1092
Stenberg, J. F., **IV**, 564
Stephens, Verlin C., **IV**, 333

Stephenson, Emily F. M., **III**, 475; **IV**, 984
Steppel, Richard N., **VI**, 276
Stetter, H., **VI**, 866; **VII**, 95; **VIII**, 620
Stevens, I. D. R., **VI**, 936,
Stevenson, H. B., **V**, 459, 734
Stewart, D., **VIII**, 77
Stewart, Jesse, **I**, 8
Stewart, Roberta A., **IV**, 903
Stier, M., **VIII**, 68
Stille, J. K., **VIII**, 97, 268
Stimmel, J. B., **VIII**, 132
Stirling, C. J. M., **VI**, 31
Stirton, A. J., **IV**, 862
St. John, Nina B., **II**, 425
Stöckel, K., **VI**, 68
Stone, Herman, **IV**, 321, 323, 543
Stork, Gilbert, **VI**, 242, 526, 638
Stormont, R. T., **II**, 7
Story, Paul R., **V**, 151
Stowell, J. C., **VII**, 59
Straley, J. M., **IV**, 415
Strating, J., **VI**, 981
Stringer, O. D., **VIII**, 546
Strube, R. E., **IV**, 6, 417, 967
Struck, H. C., **II**, 560
Stryker, J. M., **VIII**, 505
Stuart, Frank A., **III**, 668
Stubbings, Robert, **V**, 341
Stull, Arthur, **I**, 220
Stütz, P., **VI**, 109,
Suga, K., **VIII**, 188
Sugano, K., **VII**, 359
Sugasawa, Shigehiko, **IV**, 72
Sugimoto, Toshiyuki, **VI**, 90, 361
Sukornick, Bernard, **V**, 1074
Suleske, R. T., **VII**, 149
Sum, F. W., **VII**, 351
Sumitani, Koji, **VI**, 407
Summers, R. M., **IV**, 718
Sumpter, C. A., **VIII**, 196
Supniewski, J. V., **I**, 46, 119
Surrey, Alexander R., **III**, 535, 753, 759
Surrey, A. R., **V**, 668
Suter, C. M., **II**, 111, 112, 485; **III**, 288
Sutter, M. A., **VII**, 215
Suzuki, A., **VIII**, 532

Suzuki, H., **VI**, 700
Sveda, M., **III**, 536
Swaminathan, S., **V**, 743, 747
Swann, Sherlock, Jr., **II**, 276; **III**, 401,
    V, 463
Swern, Daniel, **IV**, 317, 895, 977; **V**, 904
Sy, J., **VIII**, 357
Synerholm, Martin E., **III**, 187
Szczepanski, S. W., **VIII**, 501

Taber, D. F., **VII**, 249
Takabe, K., **VIII**, 188, 190
Takahashi, T., **VIII**, 295, 430
Takaya, H., **VIII**, 57
Takayanagi, Hiroshi, **VI**, 662
Takeda, A., **VIII**, 251
Talaty, C. N., **V**, 504
Talbot, R. H., **II**, 258
Talbot, William F., **II**, 592
Talley, J. J., **VII**, 20
Tamao, Kohei, **VI**, 407; **VIII**, 315
Tamaru, Y., **VIII**, 274
Tanabe, M., **VII**, 386
Tanaka, J., **VIII**, 188, 190
Tani, K., **VIII**, 183
Tanigawa, Y., **VII**, 172,
Taniguchi, H., **VI**, 925
Tannenbaum, A. L., **I**, 435
Tarbell, D. Stanley, **III**, 267, 661, 809;
    IV, 136, 285, 807, 924; **VI**, 207, 418
Taschner, M. J., **VII**, 226
Taylor, E. Alan, **VI**, 788
Taylor, Edward C., **IV**, 166, 247, 654,
    704; **V**, 582; **VI**, 488, 549, 709, 791
Taylor, E. R., **I**, 115, 150, 153, 415; **II**,
    135, 588
Taylor, G. A., **IV**, 688
Taylor, G. N., **VI**, 101
Taylor, Harold M., **V**, 437
Taylor, J. M., **VIII**, 167
Taylor, R. J. K., **VIII**, 415
Taylor, R. T., **VII**, 200; **VIII**, 298
Taylor, S. J., **VIII**, 550
Taylor, S. L., **VII**, 506
Teague, Peyton C., **IV**, 706
Tebbins, W. G., **III**, 293

Teeter, H. M., **IV**, 125
Teeters, W. O., **IV**, 518
Teitel, S., **VI**, 1
Telinski, Thomas J., **IV**, 576
Teranishi, Roy, **IV**, 192
Terasawa, Isao, **V**, 266
Terry, Paul H., **V**, 166
Tertiuk, W., **V**, 966
Tetaz, J. R., **IV**, 690
Teuber, H.-J., **VI**, 480
Thacher, A. W., **V**, 509
Thal, A. F., **I**, 107, 270, 436
Thalmann, A., **VII**, 470
Thatte, S. D., **V**, 166
Thayer, F. K., **I**, 12, 117, 398
Thelen, R., **II**, 551
Theodoridis, G., **VII**, 476
Thielecke, Wilfried, **VI**, 378
Thielen, L. E., **IV**, 960
Thirtle, John R., **III**, 136
Thoman, Charles J., **V**, 962
Thomas, A. J., **VIII**, 192
Thompson, E. A., **VIII**, 132
Thompson, Ralph B., **II**, 10; **III**, 278
Thomsen, I., **VII**, 372
Thorpe, J. F., **I**, 192
Threlfall, T. L., **V**, 346
Threlkel, R. S., **VIII**, 505
Thurman, N., **I**, 231, 519
Thweatt, J. G., **V**, 277
Tibbetts, Fred, **V**, 166
Tichelaar, G. R., **IV**, 93
Tietze, L. F., **VIII**, 254, 353
Tijhuis, M. W., **VII**, 461
Tipson, R. Stuart, **IV**, 25
Tise, F. P., **VII**, 304
Tissot, Paul, **VI**, 971
Tobie, Walter C., **I**, 23
Toda, Masaaki, **VI**, 289
Todd, David, **V**, 617
Todd, H. R., **II**, 200
Tolbert, T. L., **V**, 520
Tolman, G. L., **VII**, 66
Tomita, K., **VIII**, 286
Tompkins, S. W., **II**, 106
Tonozuka, M., **VI**, 462

Totter, John R., **III**, 460
Tour, J. M., **VII**, 63
Towles, H. V., **IV**, 266
Townsend, L. B., **VII**, 334
Towson, J. C., **VIII**, 104, 110
Toye, J., **VI**, 282
Traylor, T. G., **VI**, 766
Tremper, Alan, **VI**, 893
Trevoy, L. W., **III**, 300
Trost, Barry M., **VI**, 364; **VII**, 266
Truesdale, L., **VIII**, 46
Trust, Ronald I., **VI**, 606
Tsai, Y.-M., **VIII**, 347, 471, 501
Tsang, S. M., **III**, 641
Tsubata, K., **VII**, 307
Tsuboi, S., **VIII**, 251
Tsuji, Jiro, **VI**, 662; **VII**, 137
Tucker, S. Horwood, **IV**, 160
Tuley, W. F., **I**, 289; **III**, 822
Tullar, B. F., **IV**, 671, 921
Turk, Amos, **III**, 121
Turner, A. B., **V**, 428
Turner, Leslie, **IV**, 828
Turro, Nicholas J., **V**, 297, 528
Tuttle, Neal, **I**, 199
Tuttle, Ned, **I**, 345
Tuzun, Celel, **V**, 111
Tyson, F. T., **III**, 479

Uff, Barrie C., **VI**, 115
Ugi, Ivar, **V**, 300, 1060
Ulbricht, T. L. V., **IV**, 566
Umbreit, M., **VII**, 121
Umemoto, T., **VIII**, 286
Underwood, H. W., Jr., **II**, 553
Ungnade, H. E., **III**, 130, 566; **IV**, 724
Urabe, H., **VIII**, 486
Urquhart, G. G., **III**, 363
Uskokovic, M. R., **VII**, 339

Vågberg, J., **VIII**, 5
Vaid, R. K., **VIII**, 43
Valenta, Z., **VI**, 1024
VanAllan, J. A., **III**, 1, 63, 76, 275, 353,
    394, 597, 726, 733, 765, 783, 824,
    827; **IV**, 15, 21, 245, 569, 633

van Bergen, T. J., **VI**, 601
Vandenberg, G. E., **V**, 946
van der Kerk, G. J. M., **V**, 223
Van Der Kerk, G. L. M., **IV**, 881
van der Plas, H. C., **V**, 977
Van Der Puy, M., **VII**, 251
Vanderwerf, Calvin A., **III**, 44; **IV**, 157
Van Deusen, S., **VIII**, 155
van Es, T., **VI**, 631
van Leusen, A. M., **VI**, 41, 981, 987
Van Lier, P. M., **VII**, 287
Vanlierde, H., **VIII**, 306
van Look, G., **VII**, 70
VanRheenen, V., **VI**, 342
van Tamelen, Eugene E., **IV**, 10, 232
Van Zyl, G., **IV**, 10
Varghese, V., **VIII**, 460
Varkony, Haim, **VI**, 43
Varland, Robert H., **III**, 829
Vassel, B., **IV**, 154
Vaughan, Herbert W, Jr., **V**, 1116
Vaughan, W. R., **III**, 329; **IV**, 594
Veber, Daniel F., **VI**, 5; **VII**, 75
Vedejs, E., **VII**, 277
Venegas, M. G., **VII**, 56
Venkataraman, K., **II**, 304, 522
Venkataramani, P. S., **V**, 747
Vesley, George F., **V**, 297
Vielt, E. B., **II**, 416
Villani, Frank J., **IV**, 88
Villieras, J., **VIII**, 265
Vishwakarma, L. C., **VIII**, 546
Vliet, E. B., **I**, 201, 203, 317, 360, 482
Voaden, Denys J., **V**, 962
Vofsi, D., **V**, 1076
Vogel, Arthur I., **IV**, 525
Vogel, D. E., **VIII**, 536
Vogel, E., **VI**, 731, 862
Vogel, K., **V**, 1135
Voss, E., **VIII**, 254, 353
Vogt, Paul J., **IV**, 420, 464
Voit, Axel, **II**, 594
von Braun, J., **I**, 428
Voong, Sing-Tuh, **III**, 664
Voorhees, V., **I**, 61, 280, 463
Vopel, K. H., **V**, 431

Vose, C. E., **I**, 179; **II**, 379
Voskuil, W., **V**, 211
Vrijland, M. S. A., **VI**, 727
Vyas, G. N., **IV**, 836

Waddey, Walter E., **III**, 560
Wade, Robert H., **IV**, 221
Wade, Ruth S., **VI**, 821
Wadsworth, Donald H., **VI**, 75
Wadsworth, W. S., Jr., **V**, 547
Waegell, B., **VI**, 142
Wagner, D. P., **VI**, 1007
Wagner, E. C., **II**, 65; **III**, 323, 488; **IV**, 383
Wagner, William S., **IV**, 892
Wakabayashi, Toshio, **VI**, 358, 448
Wakefield, B. J., **V**, 998
Wakeman, Nellie, **I**, 511
Walba, David M., **VI**, 567
Walborsky, H. M., **VI**, 751
Walker, G. N., **IV**, 327, 408
Walker, Joseph T., **II**, 169
Walker, K. E., **VI**, 965
Walling, C., **V**, 184
Wallingford, V. H., **II**, 349, 353
Walsh, W. L., **II**, 553
Walshe, N. D. A., **VIII**, 1
Walter, L. A., **III**, 757
Walti, A., **II**, 5, 545
Wamser, C. C., **VII**, 256
Wann, R. E., **V**, 893
Wanzlick, H. W., **V**, 115
Ward, A. M., **II**, 159
Ward, E. R., **III**, 341
Ward, H. P., **III**, 244
Ward, J. A., Jr., **IV**, 392
Ward, M. L., **III**, 549
Warnhoff, E. W., **IV**, 162
Waser, E. B. H., **I**, 499
Washburn, Robert M., **IV**, 68
Wasmuth, D., **VII**, 152
Wasson, Richard L., **IV**, 552, 957
Watanabe, S., **VIII**, 188
Watkins, Michael, **VI**, 628
Watts, C. T., **VI**, 936
Waugh, T. D., **III**, 438

Wawzonek, Stanley, **III**, 715; **IV**, 387, 681; **V**, 758
Waykole, L., **VIII**, 281
Wayne, Winston, **III**, 367
Webb, Carl N., **I**, 82, 217
Webb, K. S., **VIII**, 562
Weber, K. H., **III**, 379
Weber, R. H., **VII**, 215
Weber, William P., **VI**, 232
Webster, E. R., **IV**, 433
Wegner, P., **VI**, 1030
Wehrli, Pius A., **VI**, 615, 1010
Weichselbaum, T. E., **II**, 384
Weijlard, John, **IV**, 124
Weil, J. K., **IV**, 862
Weiler, L., **VII**, 351
Weiner, Michael A., **V**, 452
Weiner, Nathan, **II**, 279, 376
Weinreb, Steven M., **VI**, 492
Weinshenker, Ned M., **VI**, 218, 951
Weinstock, Joseph, **IV**, 641; **VI**, 552, 910, 913
Weinstock, L. M., **VI**, 620
Weis, C. D., **VI**, 427
Weisgerber, C. A., **III**, 112, 114
Weismiller, M. C., **VIII**, 104, 110
Weiss, Marvin, **II**, 590
Weiss, Richard, **II**, 61; **III**, 729
Weiss, Robert H., **VI**, 451
Weiss, U., **VII**, 50
Weissberger, A., **III**, 108, 183, 504, 610, 708, 788
Welch, E., **III**, 725
Wells, F. B., **II**, 221
Wendel, Kurt, **V**, 201
Wender, P. A., **VI**, 334; **VII**, 456
Wendland, Ray, **IV**, 757
Wenner, Wilhelm, **IV**, 23, 760
Werber, Frank X., **III**, 798
Werley, R. T., Jr., **VIII**, 391
Werner, Newton W., **V**, 273
Wersin, G., **VII**, 70
Wertheim, E., **II**, 471
Wester, R. T., **VIII**, 586
West, D. E., **V**, 412
West, Edward S., **III**, 800

West, Harold D., **III**, 774, 813
Weston, Arthur W., **III**, 288; **IV**, 915
Weyler, Walter, Jr., **VI**, 210
Whaley, A. M., **I**, 142
Wheeler, A. S., **I**, 450
Wheeler, Norris G., **IV**, 256, 857
Wheeler, Thomas N., **VI**, 840
Wheeler, T. S., **I**, 102, 296; **IV**, 478
White, D. A., **VII**, 181, 479, 482
White, Emil., **V**, 336, 797
White, R. A., **IV**, 700
Whitehead, Calvert W., **IV**, 515
Whitesell, J. K., **VIII**, 516
Whiting, M. C., **IV**, 128, 404, 755, 792; **VI**, 10, 392
Whitmore, Frank C., **I**, 56, 159, 161, 231, 325, 326, 401, 492, 519; **II**, 282, 317, 408; **III**, 568
Whitmore, Marion G., **I**, 401
Whitten, Charles E., **VI**, 248
Whyte, Donald Edward, **IV**, 234
Wiberg, Kenneth B., **III**, 197, 811
Widera, Ronald P., **VI**, 232
Wieland, P., **V**, 692
Wieleseck, R., **V**, 921
Wierenga, W., **VII**, 213
Wildeman, J., **VI**, 41
Wilds, Alfred L., **IV**, 281
Wiley, D. W., **V**, 1013
Wiley, Richard H., **III**, 560, 853; **IV**, 5, 201, 337, 351, 549, 731
Wilka, E.-M., **VII**, 41
Wilkinson, G., **IV**, 473, 476
Willard, Constance, **IV**, 831
Williams, A. L., **VII**, 400
Williams, Bill, **III**, 337
Williams, J. K., **V**, 459
Williams, Jonathan W., **III**, 10, 164, 508, 626, 818
Williams, W. Michael, **VI**, 7
Williamson, S. M., **V**, 12, 816
Willson, F. G., **I**, 102, 296
Wilson, C. V., **III**, 28, 358, 573, 575, 578; **IV**, 547, 564
Wilson, J. G., **IV**, 690
Wilson, J. W., **III**, 267

Wilson, W. C., **I**, 274, 276
Wilson, S. R., **VII**, 124; **VIII**, 235
Wilt, James W., **IV**, 254
Winberg, H. E., **V**, 883
Windus, Wallace, **II**, 136, 345, 411
Winkelmuller, W., **II**, 188, 299
Winslow, P., **VII**, 476
Wintner, Claude, **VI**, 56, 679
Wisansky, W. A., **III**, 138, 307
Wisegarver, Burnett B., **III**, 59, 644, 723
Wiselogle, F. Y., **I**, 90
Wittcoff, Harold, **IV**, 907
Witten, Benjamin, **IV**, 47
Witten, Charles H., **III**, 818
Wittenbrook, Lawrence S., **V**, 231
Wittig, G., **IV**, 964; **V**, 60, 751; **VI**, 901
Witzemann, E. J., **II**, 17, 137, 305, 307
Wohllebe, J., **VI**, 368
Wojcik, Bruno H., **II**, 260, 262
Wolf, Calvin N., **IV**, 45, 910
Wolf, Frank J., **III**, 737; **IV**, 124
Wollensak, J., **V**, 575
Womack, Ennis B., **III**, 392, 714
Wonchoba, Edward, **VI**, 875
Wong, Jack Y., **VI**, 218, 951
Wood, Clelia S., **V**, 952
Wood, Ernest H., **III**, 641
Wood, J. H., **III**, 98, 286
Wood, John, Jr., **III**, 444
Woodburn, H. M., **IV**, 271
Woodruff, Eugene H., **III**, 581
Woodruff, R. A., **VI**, 259
Woods, G. Forrest, **III**, 470; **IV**, 471
Woods, Sarah M., **VI**, 873
Woodside, A. B., **VIII**, 616
Woodward, F. E., **VIII**, 1
Woodward, Gladys, E., **I**, 159, 325, 408
Woodward, R. B., **III**, 50, 413, 453; **VI**, 263, 590, 1014, 1016
Woolfolk, E. O., **III**, 837
Woollett, G. H., **II**, 343
Worden, Leonard R., **V**, 251
Worrall, David E., **I**, 413
Wovkulich, P. M., **VII**, 271

Wynberg, Hans, **IV**, 295, 534
Wysong, Robert D., **V**, 332

Yamada, S.-L., **VII**, 206
Yamada, T., **VIII**, 190
Yamagata, T. , **VIII**, 183
Yamamoto, H. **VIII**, 141, 321, 568
Yamamoto, Yutaka, **VI**, 418
Yamamura, Shosuke, **VI**, 289
Yanami, T., **VII**, 396, 414
Yang, Dominic T. C., **VI**, 293
Yang, N. C., **VI**, 974
Yanko, William, **III**, 568
Yasuda, M., **VIII**, 547
Yeaw, J. S., **I**, 336
Yoda, Y. , **VII**, 417
Yokoyama, K., **VI**, 520
Yonemitsu, O., **VII**, 359
Yoshida, Z., **VIII**, 274
Yoshikoshi, A., **VII**, 396, 414
Yoshioka, M., **VI**, 14, 307, 436
Yoshioka, T., **VII**, 359
Young, D. M., **II**, 219
Young, R. J., **V**, 88
Young, S. D., **VII**, 381

Zajacek, J. G., **VI**, 501
Zambito, Arthur J., **V**, 373
Zartman, Walter, **II**, 606
Zaugg, Harold E., **III**, 820
Zaweski, Edward F., **V**, 324
Zeidman, Blossom, **III**, 328
Zen, S., **VI**, 503, 797
Zhao, S. H., **VIII**, 464
Ziegler, F. E., **VI**, 334; **VIII**, 586
Ziegler, Gene R., **V**, 921
Ziegler, K., **I**, 314
Zimmerman, F. J., **II**, 549
Zimmerman, Howard E., **V**, 892
Zingaro, Ralph, **IV**, 359
Zoellner, E. A., **I**, 524
Zubek, A., **V**, 39
Zuger, M. F., **VII**, 215
Zuidema, G. D., **IV**, 10
Zurquiyah, A., **V**, 993
Zwilgmeyer, F., **II**, 126